A
READER'S
GUIDE
TO THE
TWENTIETH-
CENTURY
NOVEL

A READER'S GUIDE

<hr>
TO THE
<hr>

TWENTIETH-CENTURY NOVEL

EDITOR

Peter Parker

CONSULTANT EDITOR

Frank Kermode

New York

OXFORD UNIVERSITY PRESS

1995

First published in Great Britain (as *The Reader's Companion to the Twentieth-Century Novel*) in 1994 jointly by Fourth Estate Limited, 289 Westbourne Grove, London W11 and Helicon Publishing Limited, 42 Hythe Bridge Street, Oxford OX1 2EP

Copyright © Fourth Estate Publishing Limited and Helicon Publishing Limited 1994
Preface copyright © 1994 by Frank Kermode

Editor's Introduction copyright © 1994 by Peter Parker

Published in the United States of America by Oxford University Press, Inc., 200 Madison Avenue, New York, New York 10016

Oxford is a registered trademark of Oxford University Press

Library of Congress Cataloging-in-Publication Data available upon request from the Publisher.

ISBN 0–19–521153–7

Printing (last digit): 9 8 7 6 5 4 3 2 1

Printed in Great Britain on acid-free paper

Page design by Roger Walker

Typeset by Mendip Communications Ltd
Frome, Somerset

Contents

Preface

The novel has been on the scene for two-and-a-half centuries, and although its demise has often been mourned or celebrated we continue to recognise it without difficulty, and still assume that of all the literary forms it is the one with which we remain on the easiest terms. The supply of new novels is apparently undiminished. Large numbers of people want to write them because the prestige attaching to the job description of writer, when associated with the possibility of making a fortune by being one, is greater, though not very great, if you are a novelist than if you choose poetry or criticism.

This continuing prestige is, on the face of it, rather surprising. What usually happens is that a medium supersedes an older one it borrows from, so that epic and romance had to be there before the novel could exist, and in its turn swallows them up. The novel had a marvellously free run in the nineteenth century, but at the beginning of the twentieth, perhaps as part of the extraordinary developments in technology generally, there appeared new ways of telling stories, making plots, examining characters. This development, commonly regarded as a marker of 'Modernism', tended to divide the audience by establishing new kinds of highbrow interest, and putting certain kinds of novel-writing outside the purview of the ordinary reader. And other more material forms of technology began to affect the novel, which became the natural food of films, just as television would in its turn absorb films and to some extent novels also. Both the cinema and, more overwhelmingly, television attracted vast audiences. Television is watched by some for several hours a day, ousting all other forms of domestic entertainment, reading included; people can't read books while they're watching the box, which, in any case, gives them in a new form the old fix of narrative.

So one might have supposed that the novel (and, in its turn, the film) would rapidly become obsolete, except as fodder or possibly as providing work for academics. That this has not happened is amply affirmed by *A Reader's Guide to the Twentieth-Century Novel*. It is true that, apart from heavily promoted best-sellers, novels sell less well than they did a century, or even seventy or eighty years ago, in the time of H. G. Wells and Arnold Bennett. I once heard a publisher telling Rebecca West that a living novelist, generally considered rather highbrow, had sold about 18,000 hardback copies of her latest book, whereupon Dame Rebecca expressed amazement that writers could sustain life on such small sales. She imagined that the publisher, far from boasting, was commenting on a catastrophic decline, since her own youth, in the sales of reasonably successful novelists. She had no idea that 18,000 is a figure most novelists can now only dream of. That may be a reason, though only one of several, why books have become more expensive, and why publishers feel a need to be careful about taking on new authors.

Yet the demand for new fiction, in this age when, as we are often told, the book itself is an obsolescent medium, continues to exist, with sufficient force to tempt cautious publishers; and a dauntingly large number of novels, more than even the most conscientious reviewer can hope to read, is published to meet it.

In his *Companion to Victorian Fiction* John Sutherland maps what he calls 'the lost continent' of the period 1837–1901, when 'somewhere around 60,000 works of adult and juvenile fiction' were published. He ventures 'an educated guess' that 'around 7,000 Victorians ... could legitimately title themselves "novelist"'. Sutherland heroically gives an account of 554 novels, apologising, though not very seriously, for his failure to read all the 197 novels of Annie S. Swan, and further provides biographies of 878 authors. The continent explored in *A Reader's Guide* is differently constituted. The sheer weight of print may be less, though the period covered is longer and the geographical scope is greater, since many of the novels considered are not by British writers but by writers in English, wherever they come from.

The number of contenders for a place in such a work as this is very large, and the criteria of selection include both quality and celebrity. Here is a point on which users of the *Guide* may instruct and amuse themselves by disagreeing with the editors. If they think of books that in their view ought to be in but aren't – and they should certainly do this as they critically browse – they can be seen as testifying to the truth that the twentieth-century novel in English represents one of the most extraordinary achievements in literary history; for long as it is, this book might, without dilution of quality, have been even longer.

I doubt if the work of the editors could be much better done, yet they remain calm in face of the certainty that they could not hope for total acceptance of their choices, and indeed regard it as part of their purpose to stimulate dissent and a desire to amend. One reader will regret that space has been found for *Lord Jim* but not for *Under Western Eyes*, in his view of the finest of Conrad's novels. Another may ask why Anthony Burgess's most ingenious achievement, *M/F*, is not included. Where, another may ask, are Graham Greene's *The End of the Affair*, Robert Graves's *Antigua, Penny, Puce*, John Updike's *Roger's Version*, Vladimir Nabokov's *The Real Life of Sebastian Knight*, Nicholas Mosley's *Impossible Object*, Muriel Spark's *The Driver's Seat*, David Lodge's *How Far Can You Go?*, Keith Waterhouse's fine first novel *There is a Happy Land*, Penelope Fitzgerald's *The Bookshop* – and so on and so on? Why are these not preferred to the choices actually made, from these authors themselves or from others? Making such lists is a way of adding indignant enjoyment to instruction; the book is intended to be sweet as well as useful – to make possible, to invite, such partisan protest, so enjoyable to the objector, so conducive to amiable argument. If all the novels held by the public at large to have been wrongly excluded could be added this would be an enormous book, the mere thought of which is an impressive testimony to the incredible richness of the store.

Reminded of that wealth, one can hardly resist a complaint against those who would, for their own convenience, establish a canon – of course all such selective lists are canons, but some are much narrower than others – attributing high value to a few great books and ignoring or disparaging the rest. For although, as this volume makes plain, there are certainly great books in the tradition, it is the sheer copiousness, variety and ingenuity of the twentieth-century novel that are so impressive.

And, for that reason, turning these pages, one ought to reflect on the patient labours of imagination and intellect that have gone to the making of the works they list and discuss. It is the point made in the celebrated fifth chapter of *Northanger Abbey*, where we are told that what a young lady has been taught to call 'only a novel' is 'only some work in which the greatest powers of the mind are displayed, in which the most thorough knowledge of human nature, the happiest delineation of its varieties, the liveliest effusions of wit and humour, are conveyed to the world in the best chosen language'. Here we see a just indignation, only a little, and understandably, overstated. Such high praise we tend to reserve for the masters, for Joyce and Proust and Henry James perhaps, maybe for Conrad and Musil, possibly for Lawrence, who was certainly thinking of other writers than these when he described the novel as 'the one bright book of life'. And we should be more generous. There remains that great mass of fiction, not a terra incognita but a land of which we nevertheless habitually fail to recognise the detail; or a crowd in which, because they are so many, we can sometimes fail to distinguish many talented individuals.

Evelyn Waugh, thinking of himself as well as of his double Mr Pinfold, once meditated thus: 'It may well happen in the next hundred years that the English novelists of the present day [1957] will come to be valued as we now value the artists and craftsmen of the late eighteenth century. The originators, the exuberant men, are extinct and in their place subsists and modestly flourishes a generation notable for elegance and variety of contrivance. It may well happen that there are lean years ahead in which our posterity will look back hungrily to this period, when there was so much will and ability to please'. But the lean years have not yet arrived, and among the anglophone novelists of the present day we may find as much to value as Waugh did among the English almost forty years ago. Some of the artists he had in mind are still with us (one, it may be said with safety, is Muriel Spark). Many more have joined their ranks. The names of these contrivers, together with some account of their craft, are on careful record in this volume.

FRANK KERMODE

Editor's Introduction

This book is primarily a work of enthusiasm. Contributors have been encouraged to write about books they admire. This is not to say that the entries are uncritical, but whatever reservations are admitted, each of the 750 or so novels is being recommended.

What this book is *not* is an academic guide to twentieth-century fiction in the English language: it is not a syllabus. To adopt for a moment an invidious distinction, all the 'major' authors – those whose standing is such that their books are studied at schools and universities and receive widespread coverage upon publication – are represented; but so are many 'minor' authors, whose works may be neglected but are nevertheless worthy of our attention. Examiners have to set papers and literary editors have to select a handful of titles to review from the torrent of new publications which flood their offices every week, but such necessities often mean that good books by 'minor' authors can lose out to indifferent books by 'major' authors. Even authors who now seem part of the fabric of twentieth-century fiction have often received less than their due. It now seems surprising, for example, that none of Olivia Manning's novels was ever accorded a separate review in the newspapers. Fashions change and perceptions shift. In years to come, no doubt, people will wonder why some contemporary writers whose books receive enormous attention in the press were ever thought worthy of consideration.

The distinction between 'major' and 'minor' persists, however, and this guide may go some way towards restoring the balance. Part of our purpose has been to provide a rather broader historical survey than is usual in literary encyclopedias. There should be room upon the shelves for the domestic, small-scale novel alongside the cosmically ambitious one: even when the arrangement is not alphabetical, Barbara Pym should be able to stand beside Thomas Pynchon. If nothing else, this book provides ample evidence of the range of fiction available to the reader between 1900 and 1993. The chronological arrangement emphasises this point: in 1916, for example, those who did not thrill to what might be described as the Buchaneering exploits of Richard Hannay and his chums in *Greenmantle* could plunge into the ever-rolling stream-of-consciousness provided by *Backwater*, the third novel of Dorothy Richardson's *Pilgrimage*; those who shied away from the brilliant artificiality and high camp intrigues of Ronald Firbank's *Inclinations*, could settle down with the solid provincial realism of Arnold Bennett's *These Twain*; Ada Leverson's splendidly Edwardian comedy of manners, *Love at Second Sight*, could be read alongside James Joyce's challengingly modernist *A Portrait of the Artist as a Young Man*; those who wanted to read a dissident account of the war which was then engulfing Europe could buy or borrow Rose Macaulay's *Non-Combatants and Others*.

Fiction reflects life, and to read this book is to see the century unfold, with its climacterics duly marked. The pre-war drudgery of the working classes in Britain and America, as recorded by Robert Tressell (*The Ragged Trousered Philanthropists*, 1904) and Upton Sinclair (*The Jungle*, 1906), stands beside portraits of a more leisured society by John Galsworthy (*A Man of Property,* 1906) and Edith Wharton (*The Custom of the Country,* 1913). The war in which Tressell's workers would fight alongside Galsworthy's officer-class is well represented by novels as diverse as Henry Williamson's *The Patriot's Progress* (1930), ee cummings's *The Enormous Room (*1922) and Siegfried Sassoon's *The Complete Memoirs of George Sherston* (1936). The 1920s are witnessed by F. Scott Fitzgerald, Edmund Wilson, Michael Arlen, Evelyn Waugh, Aldous Huxley and Anita Loos; the gathering crisis by Christopher Isherwood and (obliquely) T.H. White. The Second World War brings in a heavy crop of novels from all theatres, involving civilians and combatants – Eric Linklater, Martha Gellhorn, Henry Green, James Jones, Olivia Manning, Norman Mailer. . . . Alongside these well-known writers, others whose work deserves recognition and celebration provide a portrait of the world: J. Maclaren-Ross's sardonic account of *Love and Hunger* amongst vacuum-cleaner salesmen in the depressed 1930s; or Mollie Panter-Downes's record of *One Fine Day* in an English village in the first summer of the peace.

It was also important to represent other sources of English-language fiction than the predominating Anglo-American ones. While the majority of the books represented here come from the United Kingdom and the USA, we have also covered novels from and about Canada, Ireland, Australia, New Zealand, India, Africa and the Caribbean. The one stipulation was that all these novels had to be written in English, or have been subsequently translated into English by the author: this means, for example, that works by Isaac Bashevis Singer are included, but works by Jorge Luis Borges are not.

In spite of an attempt to represent as wide a selection as possible within these self-imposed limitations, as a list of titles was drawn up (many of them suggested by the contributors themselves) it soon became clear that in order to keep the book a manageable size certain ruthless editorial decisions would have to be made. The simplest course was to limit the book to novels which could be described as 'literary'. This meant that there would be no entries for genre fiction such as Romance or Historical. This is not, of course, to exclude literary novels with romantic themes or historical settings, such as Elizabeth Taylor's *Palladian* or Naomi Mitchison's *The Corn King and the Spring Queen.*

Crime and science-fiction novels were less easy to cast aside, largely because they have achieved a literary respectability not accorded to other categories of 'popular' fiction. A few representative entries are therefore included, the criteria being that these are novels that have been enjoyed by the general, as well as the specialist, reader. It often takes some decades before such novels can be described as 'classics', and it may be that in fifty years people will be reading the novels of Joseph Wambaugh with the same admiration as they now read those of Raymond Chandler. For our purposes, however, the latter is represented, but the former is not. A similar rule has been applied to sci-fi. Authors whose work is read by people who say 'I don't normally read science fiction, but . . .' are represented: there is an entry for Michael Moorcock's *The Dancers at the End of Time,* but not one for Brian Aldiss's *Helliconia Trilogy.*

Within these genres there were certain borderline cases, and those such as Daphne du Maurier's *Rebecca,* Agatha Christie's *Murder on the Orient Express* and Ian Fleming's *On Her Majesty's Secret Service,* books which some would regard as sub-literary, were included in a representative capacity and because no history of twentieth-century fiction would seem altogether complete without them.

Some books written for children have as much claim to literary merit as those written for adults, and we have admitted a small number of these. Most of us carry around memories of the books we read, or those read to us, as children, and novelists are no different, often acknowledging their debt to these early models. Several of the 1930s writers have saluted Beatrix Potter, and it may well be thought that some of the authors of novels for adults we have represented would have benefited from paying attention in their infancy to Potter's stylish economy. Occasionally, as in the case of Frances Hodgson Burnett's *The Secret Garden* or E. Nesbit's *The Railway Children*, adults enjoy the books as much as (perhaps, even more than) children, and these novels have found a place in this guide.

It will be noted that more novels are covered in the years after 1945 than in pre-war years. This is not because books have so improved that one is spoiled for choice; it is merely that the mixture has yet to settle and separate so that we can see clearly what is the cream and what is not. One has to make choices which in ten or twenty years' time may look perverse. That said, the chronological arrangement does show that, as with any other crop, there are good years and fallow ones. Even after more than thirty years, the late 1950s and early 1960s seem vintage; the early 1970s, on the other hand, seem not to have yielded a great deal. All this may look different in another thirty years, and no doubt people will disagree with this selection without waiting that long. All one can say is that we have tried to provide an interesting reading list, and that these are the books the contributors have liked. The editor's own prejudices (one of the few perks of the job) will be discovered by noting inclusions rather than exclusions.

Each entry provides a brief resume of the plot as well as some background information and critical analysis which may add to the reader's understanding or enjoyment of the novel described. It will no doubt be argued by some that there is little point in reading a book once you know the story. The plot of a good novel, however, is simply a framework which the author uses to support and give shape to ideas. It is the bare bones, which style fleshes over, which is given form by the author's own voice and viewpoint. If one imagines providing a plot for two writers and asking each of them to construct a book upon it, there is no doubt that the end results would be very different. It is not only in postmodernist times that the story has lost its pre-eminence. John Fowles may supply alternative endings for *The French Lieutenant's Woman* (1969), and other authors may play disruptive games with narrative, but then who ever read *Tristram Shandy* (1759–67) for its plot? Even the more conventional novelist, working in a narrative tradition, who has worked long and hard at his or her prose, may rightly object that plot is not everything. In Chapter XV of *Barchester Towers* (1857), for example, Anthony Trollope 'gives away' part of the story by informing the reader well in advance of the event that Eleanor Bold will marry neither Mr Slope nor Bertie Stanhope.

... And here, perhaps, it may be allowed to the novelist to explain his views on a very important point in the art of telling tales. He ventures to reprobate that system which goes so far to violate all proper confidence between the author and his readers, by maintaining nearly to the end of the third volume a mystery as to the fate of their favourite personage. .... And what can be the worth of that solicitude which a peep into the third volume can utterly dissipate? What the value of those literary charms which are absolutely destroyed by their enjoyment? ... And then, how grievous a thing it is to have the pleasure of your novel destroyed by the ill-considered triumph of a previous reader. 'Oh, you needn't be alarmed for Augusta, of course she accepts Gustavus in the end.' 'How very ill-natured you are, Susan,' says Kitty, with tears in her eyes; 'I don't care a bit for it now, Dear Kitty, if you will read my book, you may defy the ill-nature of your sister. There shall be no secret that she can tell you. Nay, take the last chapter if you please – learn from its pages all the results of our troubled story, and the story shall have lost none of its interest...'

That said, we have in a very few instances, when a novel turns upon a surprise or skilfully buried clue, withheld this information.

No doubt some people – students (and possibly even teachers) of English literature who are pressed for time – will use this book as a quick way of acquiring basic information without reading an entire novel. Indeed, one of its purposes is to be an *aide-memoire*: What was the name of the servant in *Death of the Heart*? Who did Nicholas Jenkins marry in *A Dance to the Music of Time*? What happens at the end of *1984*? None of the entries, however, is intended as, or can ever be, a substitute for the real thing.

Length of entry is no guide to the importance or merit of the book under discussion. Some books are easy to encapsulate in a couple of hundred words; others – perhaps with a large cast of characters or a large number of subplots – inevitably take more space. Novels which are straightforward, or about which there is a critical consensus, will need less background material than those over which opinions have differed.

Entries are often supplemented with additional information which few reference books include, for it has been an important goal to make this book as lively and entertaining as possible, something which can be read for pleasure as well as for information. The reader will learn here about the author who found his autobiographical novel in a bookshop upon a shelf labelled 'Of Interest to Students of Abnormal Psychology'; the author who said, on learning that her novel was selling well, 'how unexpected, how odd that people can read that difficult, grinding stuff'; the writer who committed suicide within two months of the publication of his only novel, a book which had been a critical and commercial success; the historical novel on a religious theme, described by its author as 'a very beautiful book about Penelope Betjeman's early sex life'; and the author who, late in life, said of the autobiographical novel which made her name that she thought the heroine 'a revolting character'. Adverse critical judgements have been quoted alongside complimentary ones.

Organisation and Arrangement

Within each year of the chronological arrangement, novels are ordered alphabetically by the author's surname. Where a single entry covers a trilogy, tetralogy or *roman fleuve*, the entry is placed in the year the final volume was published. The individual volumes are listed in their year of publication with a note directing the reader to the full entry. In a single case this guideline has been broken: it seemed important to place William Cooper's trilogy of novels about Joe Lunn in 1950, the year the first volume was published, since *Scenes from Provincial Life* heralded a new movement in British fiction, and was a precursor of the Angry Young Man movement. Furthermore, because of a libel action, the trilogy's volumes were published at irregular intervals and out of sequence, the second volume appearing twenty-one years after the third, and thirty-two years after the first.

Where novels have been gathered into one volume and given a new overall title – such as John Dos Passos's *USA*, published two years after the appearance of the trilogy's final volume – a cross-reference will direct the reader to the main entry. Authors have often taken the opportunity of changing the text – sometimes substantially – when arranging their novels for single-volume publication. R.H. Mottram and L.P. Hartley, for example, added linking passages when their *Spanish Farm* and *Eustace and Hilda* trilogies were republished in one volume, while Edna O'Brien added an epilogue to *The Country Girls Trilogy* twenty-three years after the publication of the final volume. Cross-references for such revised versions appear in their year of publication. Within the text, bold face italics, and a publication date in brackets, are used to refer the reader to any novel that has its own entry.

Since we all occasionally forget the exact title of a book, or the name of the person who wrote it, separate indexes of authors and titles are provided. There is also a section at the end of the book which provides a brief biographical note on each author listed. The second of these indexes includes alternative English or American titles: for example, *Mr Norris Changes Trains* also appears here under its American title, *The Last of Mr Norris*, and Jean Rhys's *Quartet* also appears under its original (but now discarded and forgotten) title, *Postures*.

PETER PARKER

Contributors

Phil Baker
Josie Barnard
Janet Barron
Caroline Bartlett
Stewart Bell
Peter Benedict
Kim Bishop
Daniella Cavallaro
Hazel Coleman
Geoffrey Cush
Geoffrey Elborn
Iain Finlayson
Mary Fogarty
Graham Grant
Prabhu S. Guptara
Georgina Hammick
Edward Harcourt
Alison Hennegan
Christie Hickman

C.A.R. Hills
John Jarrold
Michael Kerrigan
Shena Mackay
Martin Noble
Sandy Noble
Lawrence Norfolk
Peter Parker
Christopher Potter
Stephen L. Prasher
James Rogers
Hilary Spurling
Mary Stacey
Helen Szamuely
Martin Taylor
Antonia Till
Andrew Treip
Ian Wisniewski
James Wright

Titles of Books Listed by Year

1900

British Labour Party founded ● Hawaii is organised as a territory of US ● Boxer rising in China ● Max Planck elaborates quantum theory ● Sigmund Freud, *The Interpretation of Dreams* ● Paul Cézanne, *Still Life With Onions* ● Auguste Rodin exhibition establishes his reputation ● Gustav Mahler, Fourth Symphony ● Anton Chekhov, *Uncle Vanya* ● John Ruskin, Friedrich Nietzsche and Oscar Wilde die ● G.K. Chesterton, *The Wild Knight*

Joseph CONRAD   **1857–1926**

Lord Jim

Lord Jim is one of several books derived from Conrad's experiences as an officer on British merchant ships in the Far East in the 1880s. The story is based upon a real incident in which the *Jeddah*, a ship carrying pilgrims from Singapore, was abandoned by a Capt. A.P. Williams, whose subsequent career was rather less exciting than Jim's (although he made and lost a fortune and fathered sixteen children). Conrad began writing the book in 1898 and finished it in a twenty-one-hour burst of concentration in July 1900. The result did not altogether please him: 'I've been satanically ambitious,' he confessed to Edward Garnett, 'but there is nothing of a devil in me, worse luck. The *Outcast* is a heap of sand, the *Nigger* a splash of water, *Jim* a lump of clay.' One reviewer complained that the novel was in essence 'a short story which had got beyond the writer's control' (it is over 400 pages long).

Jim's story is related in the novel by Marlow, a character whom Conrad uses in a number of tales – including *Youth* (1902), **Heart of Darkness** (1902) and *Chance* (1913) – to explore the moral dilemmas in which people find themselves. Marlow appears to tell us the story in an attempt to understand himself more thoroughly; we see the story from Marlow's perspective, he comments on the events he narrates, and discusses the sources from which he has gathered his information. To a certain extent, therefore, Jim's story has to be pieced together by the reader, and the oblique manner of the telling raises the issue of how we can ever dis-cover the whole truth about anything.

Jim instinctively leaps to join other officers of the *Patna* who frantically take to a boat when the ship seems to be sinking, leaving the 800 or so Muslim pilgrims on board to their fate. If the ship had sunk, they would certainly have been drowned; however, the ship survives and is towed to port. Jim is an idealist and romantic, and his jump proves to be his 'fall': he is driven by his conscience to become a wandering do-gooder, attempting to compensate for that act of impulsive cowardice.

Eventually, by devoting himself to the service of the inhabitants of Patusan, he attains a degree of satisfaction with himself; and the natives come to trust him implicitly as *Tuan* (Lord). Jim now faces the second crisis of his life. A group of white men he befriends betray his trust and murder his best friend, Dain Waris. Jim's beloved, a local girl, pleads with him; but he cannot bring himself to run away in a crisis for the second time, and he wins back his self-respect by submitting with alacrity to native justice. His death is given additional poignancy because he has to be executed by Dain Waris's father, the old Chief Doramin.

Theodore DREISER   **1871–1945**

Sister Carrie

Dreiser's first novel, which remains his best known, is set in America in the 1890s, a dubious golden age of industrial expansion and *laissez-faire* economics in which 'the millionaire' first emerged as an icon for all those who would never themselves be wealthy. Dreiser knew this world of factories, sweat-shops, small hotels and smart department stores from his year as a journalist in Chicago and New York, and he saw very clearly the ways in which it raised and dashed the expectations of its inhabitants. With characteristic pessimism Dreiser's philosophy of life drew to these observations, seeing life as dictated by malignant twists of chance, and the individual powerless in the grip of forces much greater than himself: 'a leaf in a storm' is a metaphor reiterated again and again when Dreiser describes man's relations with this world.

Sister Carrie illustrates these themes through the stories of three individuals: Charles Drouet, George Hurstwood and Carrie Meeber. Of the three, only Drouet remains unchanged. Carrie meets him on the train to Chicago and later becomes his mistress rather than work in a shoe factory. He introduces her to Hurstwood, a wealthy saloon-manager, whose higher social standing lures Carrie away. Hurstwood's fall parallels Carrie's rise, as their affair scandalises his friends and breaks up his marriage. In some desperation, Hurstwood steals from the saloon and flees with Carrie to New York, where she works as a chorus-girl to support them both. As she rises towards stardom, Carrie sees Hurstwood only as a burden; when she leaves him, he deteriorates rapidly, sinking to begging in the streets, while Carrie stays in the best hotels and is courted by the wealthy families of New York. Drouet visits her, having heard of her success, but he no longer has anything she needs and is dismissed. At the novel's close, Hurstwood commits suicide and is buried anonymously as a pauper, while Carrie, who has achieved all she could wish for in terms of wealth and standing, remains dissatisfied, haunted by the feeling that there must be more to life.

Sister Carrie is the archetypal rags-to-riches story, but written realistically rather than as a fable. Dreiser's own travails in Chicago inform much of Carrie's relentless search for 'success', while her thoughtless, instinctual qualities are drawn from his sisters, who similarly prized new clothes and men above all else. Critical accusations of 'amorality' in the novel centred on Dreiser's frank depictions of Carrie's sexual manipulations, but behind that charge lay the point that Carrie's quest somehow slanders success itself – a particularly hallowed orthodoxy at the time. 'Wealth' and 'standing' leave Carrie dissatisfied, but the novel offers no satisfactory alternative, only a disconcerting void where a value-judgement is expected. In this omission, Dreiser is true to the method of the novelist he most admired, and against whom he is often measured, Emile Zola.

Sister Carrie was half-heartedly published by Doubleday Books, who released only 300 copies in an effort to suppress it. An abridged version (by Dreiser) sold well in England, but America effectively ignored the book until its reissue in 1907, when it was widely condemned for its amorality and crudeness of style. His writing has been criticised for being clumsy – even his supporters called it rough – and poorly balanced with its convoluted sentences containing long strings of subordinate clauses, jarring mixtures of poetic and colloquial diction and repetitions. Dreiser's fiction nevertheless opened the way for a whole school of realistic writers, notably Sinclair Lewis, who said of Dreiser in his Nobel Prize acceptance speech in 1930: 'Without his pioneering, I doubt if any of us could, unless we liked to be sent to jail, seek to express life, beauty and terror.'

Henry HARLAND   **1861–1905**

The Cardinal's Snuff-Box

This novel is usually considered to be the finest by its Anglophile American author whom contemporaries frequently compared to Walter Pater and Henry James. Although published in the first year of the twentieth century, the book bears many of the characteristics made fashionable by the aesthetes and decadents of the 1890s. Its setting – an Italy of snow-capped mountains, cascading rivers and limpid lakes, with which it is flatteringly assumed the reader is thoroughly familiar – is one they would have cherished, as is its attenuated tale of delicate sensibility and intricate misunderstandings stemming from confusions between Art and Life.

Peter Marchdale, an obscure English novelist who writes under the name of Felix Wildmay, is spending the summer in a rented villa which forms part of the estate of the young, widowed beauty, Beatrice, Duchesa di Santangiolo. On meeting her, Peter is astounded to recognise the woman whom he first glimpsed at the theatre in Paris some four years ago, and has seen – but never met – on various occasions thereafter. Lovestruck from the beginning, he has cherished her as her ideal and has enshrined her in a novel, *A Man of Words*. The woman herself, now so unexpectedly encountered, proves to be all that he has dreamed of, and she is clearly not indifferent to him. Moreover she is enraptured by the novel, whose authorship Peter does not reveal.

With each day, Peter becomes more entranced, but a declaration of love is, it seems, impossible. She is an extremely wealthy member of the Roman aristocracy (with an uncle who is both a prince and a cardinal); he is a modestly affluent English commoner. Wealth, rank and nationality need not, however, be insuperable barriers: indeed the Duchesa, it transpires, was born an Englishwoman and freely expresses her opinion that the hero of *A Man of Words* should have confessed his love. But the real difficulty lies deeper. She, the child of an old Roman Catholic family, is passionately devoted to her faith and unable to conceive of marriage to a Protestant. Peter, though suitably reticent on these matters, feels that his Anglicanism is something more than a garment to be discarded as convenience dictates.

Effectively silenced by self-doubt, over-scrupulous anxieties and religious conflict, Peter becomes steadily more morose and desperate; the Duchesa is first bewildered, then affronted, as she waits in vain for him to speak. Only the adroit intervention of her uncle-in-law, the Cardinal, breaks the impasse. As an added bonus, his saintly embodiment of Roman Catholicism in action brings about Peter's conversion.

Much praised on its first appearance, *The Cardinal's Snuff-Box* now reads unevenly. Delicacy, wit and humour were the characteristics which Harland's contemporaries,

including Henry James, found and appreciated in his work, but the wit is too often Wilde-and-water, and the delicacy overstrained. The language of the novel is often curiously disjunct, with its slightly obsolete and antiquarian usages, beloved by the aesthetes, crashing up against Peter's strangely juvenile diction – the cosmopolitan *homme du monde* suddenly topples into schoolboy hobbledehoy. There is, it is true, a good deal of genuine warmth and humour: in, for instance, Harland's depiction of Italian peasant farmers or child wayfarers; in Peter's impassioned monologues, often delivered to his bewildered but devoted servant, Marietta; and in the entirely sympathetic and delightful portrait of the Cardinal. But elsewhere the novel topples into sentimentality, racial stereotype and a casual but entrenched anti-Semitism; and although an exuberant Harland informed his publisher, John Lane, that the book was written 'in my most engaging and distinguished manner', the omniscient narrator's tones are all too often merely arch.

As an author Harland was something of a quick-change artist (a gentile, he made his name as 'Sidney Luska', an ersatz 'Jewish' writer), and *The Cardinal's Snuff-Box* sometimes seems to be a clever impersonation of a Jamesian novel of sensibility.

H.G. WELLS   **1866–1946**

Love and Mr Lewisham

Love and Mr Lewisham is the first of Wells's Edwardian social comedies, in which he turns away from his science-fiction stories to draw upon his own past, in particular his experiences at Midhurst Grammar School and the College of Science in South Kensington. Although the novel is carefully patterned, and in the conflict between the heart and the mind, the body and the intellect, presents a genuine drama, contemporary critics were unimpressed by the change of direction it represented, and it was poorly received.

The theme of youth and experience is suggested by the novel's subtitle, 'The Story of a Very Young Couple'. George Lewisham is a promising young scientist whose hopes of academic attainment are dashed by an unsuitable marriage. He is a pupil-teacher at Whortley Proprietary School, but becomes distracted from his 'Schema' – a plan of study that will ultimately lead him to academic glory and the writing of 'pamphlets in the Liberal interest' – by the beautiful but unintellectual Ethel Henderson (based on Wells's first wife, his cousin Isabel). He falls hopelessly in love, and is sacked after he has been on a 'scandalous' ramble with Ethel. He wins a scholarship to the Normal School of Science in London, where he meets Ethel's complete opposite in Alice Heydinger, a physically unattractive fellow student totally dedicated to self-improvement. This seems to be a marriage of minds, and Miss Heydinger and Lewisham dream of conquering the world together. When Ethel turns up again, however, Lewisham finds the old pull of sensuality as irresistible as ever.

Ethel is discovered in Chelsea assisting at a seance organised by her stepfather, James Chaffery, a plausible but fraudulent medium whom Lewisham exposes. He eventually marries Ethel to save her from being exploited by the swindling Chaffery. He also has to support Ethel's mother, Maggie, when Chaffery hypnotises Mr Lagune, a wealthy and credulous spiritualist, into signing a blank cheque, and flees the country in the company of his mistress. Lewisham keeps his marriage secret from his fellow students, but his work inevitably suffers, as does his relationship with Ethel.

Lewisham's most attractive quality is precisely the one that ruins his career; his capacity for pity. Ethel's most attractive quality is her vulnerability. She is acutely aware of her inability to satisfy her husband's intellectual needs, but Lewisham is reduced to passivity by her abject misery when he threatens to leave her. Miss Heydinger attempts to persuade him to review their partnership, but without success. She has a breakdown and fails her exams. Lewisham realises that he is trapped in marriage but finally accepts it, pinning his hopes on his unborn child. His youthful idealism and vague notions of a glorious future have been replaced by something more 'real' and an adult sense of responsibility. It is suggested in the final chapter that Mr Lewisham might, like Wells, become a writer.

1901

Death of Queen Victoria • President William McKinley is assassinated and succeeded by Theodore Roosevelt • The Social Revolutionary Party is organised in Russia • J.P. Morgan founds United States Steel Corporation • G. Marconi transmits messages by wireless telegraphy from Cornwall to Newfoundland • Max Weber, *The Protestant Ethic and the Birth of Capitalism* • Edvard Munch, *Girls on the Bridge* • A. Dvořák, *Russalka* (opera) • A. Strindberg, *Dance of Death* • André Malraux and Walt Disney born • Giuseppe Verdi and Henri Toulouse-Lautrec die

---

Miles FRANKLIN   1879–1954

My Brilliant Career

see **My Career Goes Bung** (1946)

Rudyard KIPLING   1865–1936

Kim

'There is a good deal of beauty in it, and not a little wisdom,' Kipling wrote of *Kim*; 'the best in both sorts being owed to my Father.' His 'vague notion of an Irish boy, born in India and mixed up with native life' was developed in long discussions with his father far away from India in Tisbury: 'I do not know what proportion of an iceberg is below water-line, but *Kim* as it finally appeared was about one-tenth of what the first lavish specification called for.' When the novel was complete, Kipling's father asked: 'Did *it* stop, or you?', and when told that it was *it*, said: 'Then it oughtn't to be too bad', a judgement that the book's enduring popularity has amply confirmed.

Impish Kimball O'Hara, the orphan of a soldier serving in an Irish regiment, leads the life of a street-arab in Lahore and is known as 'Little Friend of all the World'. He attaches himself as *chela* to a Tibetan lama who is travelling about north India in search of the mystic River of the Arrow. In the course of their numerous adventures, located in country scenes lovingly depicted by Kipling, they come across some soldiers pitching camp. Kim recognises a flag with a bull on it as having personal significance, and he

attempts to spy upon the regimental mess. In the dark, he is trodden upon by the Anglican chaplain, Mr Bennett, who mistakes him for an Indian thief, although he is surprised that the dark-skinned boy can speak English, 'the tinny, saw-cut English of the native-bred'. He then notices the amulet-case Kim has round his neck: inside it are papers, including the boy's baptismal certificate on which his father had repeatedly scrawled 'Look after the boy. Please look after the boy', signing this appeal with his name and

---

They followed the rutted and worn country road that wound across the flat between the great dark-green mango-groves, the line of the snow-capped Himalayas faint to the eastward. All India was at work in the fields, to the creaking of well-wheels, the shouting of ploughmen behind their cattle, and the clamour of the crows. Even the pony felt the good influence and almost broke into a trot as Kim laid a hand on the stirrup-leather.

KIM
BY RUDYARD KIPLING

---

regimental number. The regiment adopt the unwilling boy and send him to school, but in the holidays he continues to travel with his beloved lama. Col. Creighton of the Ethnological Survey comes to recognise that Kim might be of use to British interests and apprentices him to the British Secret Service ('The Game'), under the tutelage of the Indian agent, Hurree Babu. Kim's intimate knowledge of India makes him a valuable asset to the Service, and he wins renown while still a boy by capturing the papers of two Russian spies in the Himalayas.

The book is distinguished by its unparalleled picture of north Indian life under the open skies, in the bazaar and on the road. It has been described as the best picture of India by an English writer. Its picture of Indian religions is less reliable, however, and the narrative begins to lose its hold on the reader after Kim is sent to school; also the book has a poor trick ending. It remains, none the less, the most widely admired of Kipling's works and his only successful novel – a somewhat surprising fact in view of his prolific lifelong production of literary work, poetic, autobiographical and fictional.

Frank NORRIS   1870–1902

The Octopus

Following the success of his novel *McTeague* (1899), Norris planned a trilogy provisionally entitled 'The Epic of the Wheat'. The first part, *The Octopus*, describes the production of wheat in California, and the second, *The Pit* (1903), its distribution in Chicago. The third part, 'The Wolf', was to have been set in Europe, final destination of the wheat, but was unwritten at the time of Norris's early death in 1902.

The major plot in *The Octopus* concerns the dispute between a group of ranchers and the Pacific & Southwestern Railroad Company. The railroad, represented by the fat banker and real estate agent S. Behrman, exercises a monopoly over the transportation of the ranchers' wheat and fixes its prices accordingly. Against the advice of their senior member, Magnus Derrick, the ranchers use bribery to fix the elections to the Board of Railroad Commission, convinced that this is the only way they can 'buck the railroad'. Derrick's son Lyman, a San Francisco lawyer, is elected, but he betrays the ranchers. The railroad reneges on an earlier agreement concerning the sale of leased land, and the ranchers are threatened with eviction. Behrman seizes one of the ranches and installs his own man in it, and many of the ranchers are killed in the ensuing gun-

fight. They are left poor and homeless, impotent in the face of economic forces. Presley, a poet who had worked on Derrick's ranch, confronts the railroad boss, Shelgrim, who tells him: 'The Wheat is one force, the Railroad another, and there is the law that governs them – supply and demand. Men have only little to do in the whole business.' In a grim twist at the end of the novel, Behrman too is literally overpowered by the force of the wheat when he is crushed and suffocated in the grain-hold of his own ship.

Interwoven with this story are a number of sub-plots chiefly concerned with love and its absence. The misogynist rancher Annixter awkwardly courts his dairy-maid, Hilma Tree, who flees to San Francisco when she realises that his intentions are less than honourable. Transformed by a genuine love, he pursues her and they are married not long before he is killed in the gunfight. His friend, the poet Presley, is equally changed, finding in the plight of the ranchers a real subject and sense of engagement for his long-gestated 'Song of the West'.

Norris had come under the influence of Zola while an art student in Paris and became the pioneer and the most gifted practitioner of American naturalism. The two completed volumes of 'The Epic of the Wheat' suggest that the finished work would have been the great masterpiece of this school. His death at the age of thirty-two left his exact contemporary, Theodore Dreiser – an inferior writer – to become the chief exponent of naturalism. Norris's achievement was overshadowed by the controversy surrounding the publication of *Sister Carrie* (1900) and by Dreiser's comparative longevity, but his importance is now well established.

H.G. WELLS   1866–1946

The First Men in the Moon

The First Men in the Moon is an example of the 'scientific romances' with which Wells had made his name in the last decade of the nineteenth century. Although he continued to write futuristic books, he was by now drawing upon his own life in order to produce his comedies of lower-middle-class life, the first of which, *Love and Mr Lewisham* (1900), had been published to disappointed reviews. In *The First Men in the Moon*, Wells returned to the theme of people dramatically changing their lives through scientific inventions (rather than by education, marriage or good fortune, as in the Edwardian comedies).

Bedford, the central character, is a clerk who goes to Lympne believing that it will be the most uneventful place in the world, thus allowing him to write a play he hopes will make his fortune. However, he is disturbed at work on several occasions by Cavor, a local scientist, who stands in a field overlooked by Bedford's bungalow, hooting and gesticulating. Bedford goes out to complain about the noise, to be told by Cavor that his eccentric behaviour is an integral part of his scientific thought-processes, which have now been ruined by Bedford's complaint. Trying to make amends, Bedford offers to act as Cavor's partner.

At Cavor's laboratory, Bedford is intrigued by one of his inventions, a thin metal, also named Cavor, which makes the air pressure above it redundant and weightless. Bedford immediately considers the commercial possibilities of a substance which could be used to lift heavy objects quite effortlessly, but Cavor imagines using a sphere made of the metal to travel in space. He proposed a trip to the moon and after some hesitation Bedford agrees to join him.

Once there, they are fascinated by the extraordinary fauna and flora, and see an inhabitant of the moon, a strange insect-like being called selenite, from which they hide. Feeling hungry, they feast on what look like mushrooms and become intoxicated. In this comatose state they are taken prisoner by a group of selenites and imprisoned in a cavern beneath the moon.

Provoked by a selenite, Bedford lashes out and kills it, and a battle ensues before the two men eventually get back to the surface of the moon. They then split up to look for sphere. Bedford finds it, but in the meantime Cavor has been recaptured by the selenites.

Bedford returns to earth alone, landing just off the English coast. He brings back some gold he discovered on the moon, which enables him to live the high life, retire to Amalfi, and write his play. News of Cavor reaches him at the end of the novel. A scientist called Julius Wendigee has picked up messages sent by Cavor from the moon, which describe his present, rather civilised, life amongst the selenites.

1902

Boer War ends ● Theodore Roosevelt buys rights of the French Panama Company ● William James, *The Varieties of Religious Experience* ● Claude Monet, *Waterloo Bridge* ● Claude Debussy, *Pelléas et Mélisande* (opera) ● *The Times Literary Supplement* issued ● John Steinbeck born ● Cecil Rhodes and Émile Zola die ● Walter de la Mare, *Songs of Childhood*

Arnold BENNETT   1876–1931

Anna of the Five Towns

In a letter to his friend George Sturt, Bennett called this novel 'a sermon against parental tyranny'. As Bennett's biographer, Margaret Drabble, has pointed out, its theme of a young girl wooed for her money and dominated by a miserly father, against whom she finally rebels, has clear echoes of Balzac's *Eugénie Grandet* (1833). But it was George Moore, another realist, whose work encouraged Bennett to set his tales in his native Staffordshire, against the background of the stern Wesleyan faith in which he had been brought up. The early chapters of Moore's second novel, *A Mummer's Wife* (1885), are set in the Potteries; Bennett read it as a young man, reread it later, and admired it tremendously.

Wesleyan Methodism and the uncharitable face of materialism permeate *Anna of the Five Towns*. It is stuffed with characteristic detail – prayer meetings, rent collecting, Sunday-school activities – sardonically described. Anna Tellwright, a Sunday-school teacher, has 'a face for the cloister, austere in contour, fervent in expression, the severity of it mollified by that resigned and spiritual melancholy peculiar to women who through the error of destiny have been born into a wrong environment'. She is dominated by her father, Ephraim, whose riches 'made him notorious'. On coming of age, she discovers that she has inherited wealth from her mother, but this does not bring 'escape from the parental servitude', since Ephraim continues to control the money. When Henry Mynors, 'that symbol of correctness and of success', starts to take an interest in her she cannot help wondering whether 'he guessed that she was worth fifty thousand pounds, and her father perhaps more'.

The root of Anna's moral dilemma is that she has the power to destroy, but has no control of her destiny. Her father uses her to drive Titus Price, the old Sunday-school superintendent, to bankruptcy and suicide. 'Her conscience smote her for conniving at what she saw to be a persecution', and her eyes are opened at last by the predicament of Price's son Willie, 'so frank, simple, innocent, and big', before whom she 'almost trembled with the urgency of her desire to protect'. It is Willie's deplorable situation that causes her at last to defy her father, and helps her to define the limits of Mynors's understanding. When she asks Mynors what Willie will do now, he replies ('to soothe her'): 'surely you're not still worrying about that misfortune'. She also realises, too late, that her feelings for Willie are stronger than she had bargained for. Bennett concludes: 'Some may argue that Anna, knowing she loved another man, ought not to have married Mynors. But she did not reason thus ... She had promised to marry Mynors, and she married him. Nothing else was possible. She

who had never failed in duty did not fail then.' In fact, she has merely exchanged close confinement for an open prison.

In her biography of Bennett, Margaret Drabble writes, '*Anna of the Five Towns* was a very important step in [his] career. If he hadn't managed to write it ... he would probably never have tried another serious novel again.' Thus it is the curtain-raiser to *The Old Wives' Tale* (1908) and *Clayhanger* (1910).

Joseph CONRAD   **1857–1926**

Heart of Darkness

Heart of Darkness was written in ten weeks during the winter of 1898–9, and originally published in three parts in *Blackwood's Magazine* in 1899. Less than 40,000 words long, it is a tale told one evening on board a yawl off Gravesend. The teller is Marlow, an experienced sailor, and his audience consists of four men who share 'the bond of the sea'.

Marlow describes how he obtained the captaincy of a steamboat in the service of a trading company. The company's business is not precisely described, nor is the location of its operations. We may infer, however, that the trade is ivory and the place the Belgian Congo. Arriving in Africa, Marlow is delayed before embarking on his river-passage. During this delay he hears mention of a station manager, 'a very remarkable person' – Kurtz – whose name begins to resonate through the narrative. Marlow is told that Kurtz is 'an emissary of pity, and science, and progress ... a special being'. Although the exterior purpose of Marlow's journey is commercial, its interior motive is now 'a talk with Kurtz'.

Marlow finds Kurtz in 'the heart of darkness'. The phrase refers directly to the Congo yet, indirectly, its application is to Kurtz himself; for, in the wilderness, Kurtz has been accepted as the god-king of a tribe, has murdered and looted, has engaged in human sacrifice – even cannibalism. Taken to Marlow's steamer, he dies of a fever. His final words are, 'The horror! The horror!'

On one level, the novella is a critique of

colonialism. The white traders are greedy, squalid and vicious; their enterprise is robbery with cant. (If the natives are savages, they at least possess a dignity that is lacking in the Europeans.) On a second level, the tale presents a series of images suggestive of the futility of human action in the face of a geographical immensity – and that immensity is, in turn, 'some ghastly Nowhere', which stands as a symbol for the vacuity at the heart of Kurtz. When Marlow is called upon to persuade Kurtz to return to civilisation, he realises that he has no argument that can prevail against that nothingness. Ethical and moral contentions are as futile then as the efforts of the colonialists to impose themselves on the jungle. 'The horror! The horror!' is an epigraph for a consciousness without metaphysical or theological consolation.

The narrative derives its force in part from its method. Marlow relates not only the events but also their impact on him and, thus, Conrad conveys the extent to which Kurtz's amorality tempts and disturbs his narrator, and eloquent everyman. Hence Conrad can pose the question: How can we live when there is no philosophy or creed to sustain us? That this question is profound and pervasive is witnessed by the influence that *Heart of Darkness* has had on figures as diverse as E.M. Forster and Francis Ford Coppola. T.S. Eliot chose a line from the novel ('Mistah Kurtz – he dead') as an epigraph for 'The Hollow Men'.

In 1890 Conrad captained a steamer in the Congo. He said of *Heart of Darkness* that it was 'experience pushed a little (and only very little) beyond the actual facts of the case'.

Joseph CONRAD   1857–1926

Typhoon

This richly and consistently ironic tale is generally thought to inaugurate Conrad's period of artistic maturity. Chief mate Jukes tactfully tries to suggest to his captain, MacWhirr, that they should take a detour so that their steamer *Nan-Shan* can avoid the path of a typhoon that is thought to be impending. MacWhirr, an unimaginative

and imperturbable representative of British ordinariness, does not understand Juke's indirection. The typhoon turns out to be so extraordinarily fierce that even MacWhirr begins to doubt whether they will survive. The crisis brings out all of Jukes's racism and sense of superiority, while MacWhirr reveals an unexpected humanity and sense of fair play.

There are wonderfully comic scenes in which 200 Chinese coolies below deck roll to and fro and fight between themselves for their dollars while the typhoon is doing its worst. To control the fighting, MacWhirr orders that all their money be confiscated. Eventually, he has the money redistributed in equal shares to every Chinaman, because 'you couldn't tell one man's dollars from another's . . . and if you asked each man how much money he brought on board he was afraid they would lie'. MacWhirr masters the typhoon essentially by ignoring it; and, since the ship is now safely in Fu-Chau harbour, Jukes has to concede that MacWhirr 'got out of it very well for a stupid man'.

A. Conan DOYLE   1859–1930

The Hound of the Baskervilles

Sherlock Holmes is arguably the most popular, and enduring, of all fictional detectives, with Dr Watson (late of the Army Medical Department) as a perfect foil and narrator. Conan Doyle's ideas for the methods of deduction used by Holmes were originally inspired by Dr Joseph Bell's intensely logical methods of diagnosis, which Conan Doyle observed while studying medicine in Edinburgh. His first Sherlock Holmes story was 'A Study in Scarlet', published in 1887 in *Beeton's Christmas Annual*; it did not attract much attention. However, fifty-six short stories were subsequently published in the *Strand Magazine* between 1892 and 1927. The first few were enough to capture the imagination and affection of the public.

In 1892 the first twelve stories that had appeared in the *Strand* were published in book form. With the publication of *The Memoirs of Sherlock Holmes* in 1894, Conan

Doyle determined to end the Sherlock Holmes stories. Having grown tired of his creation, he had Holmes fall over the Reichenbach Falls, locked in a deadly conflict with his arch-enemy, Prof. Moriarty. Holmes's readers were greatly upset by this, and of the many protests sent to the author one began: 'You brute.' However, when Conan Doyle heard a friend's account of some of the legends associated with Dartmoor, he came up with the idea for *The Hound of the Baskervilles*, which appeared in the *Strand* and was subsequently published in book form in 1902.

'You interest me very much, Mr Holmes. I had hardly expected so dolichocephalic a skull or such well-marked supra-orbital development. Would you have any objection to my running my finger along your parietal fissure? A cast of your skull, sir, until the original is available, would be an ornament to any anthropological museum. It is not my intention to be fulsome, but I confess that I covet your skull.'

Sherlock Holmes waved our strange visitor into a chair.

'You are an enthusiast in your line of thought, I perceive, sir, as I am in mine,' said he. 'I observe from your forefinger that you make your own cigarettes. Have no hesitation in lighting one.'

THE HOUND OF THE BASKERVILLES
BY ARTHUR CONAN DOYLE

In the opening section of the story, Dr Mortimer consults Holmes about the mysterious death of his friend, Sir Charles Baskerville, who was found in his garden overlooking the moors, with the prints of a gigantic hound nearby. Dr Mortimer also narrates a Baskerville family legend in which an ancestor, Hugo Baskerville, carried off a young maiden and brought her to Baskerville Hall for his pleasure. While he was carousing with friends, she escaped across the moors. They gave chase and Hugo was separated from his friends, only to be found lying by the dead girl, with a giant hound tearing out his throat. Holmes initially wonders whether

Sir Charles's death was a supernatural occurrence.

Sir Henry Baskerville, the heir to the estate, arrives from Canada and shows Holmes a note he has received, warning him to keep away from the moors. Holmes insists that Watson accompany Sir Henry to Baskerville Hall, and on arrival in Devonshire they learn that a convict has escaped from a nearby prison. When they catch Barrymore, the family retainer, signalling across the moors, he confesses that the escaped convict is his wife's brother.

Watson meets Stapleton, a local naturalist and botanist, whose sister, thinking Watson is Sir Henry, warns him to leave the moors. When she meets the real Sir Henry a relationship develops between them, which, Watson observes, Stapleton is eager to curtail. It later transpires that she is really Stapleton's wife, though he has forced her to pose as his sister, to conceal their murky past.

Barrymore further confesses that Sir Charles received a letter from Laura Lyons, asking for a rendezvous at the time and place of his death. Questioned by Holmes, Laura Lyons denies having met Sir Henry at that time, and it emerges that Stapleton tricked her into making an appointment which he kept in her place.

A scream on the moors is heard, and Holmes and Watson set off; they discover a body which they initially think is that of Sir Henry, but which turns out to be the escaped convict. Stapleton arrives at the scene and is astonished to learn that the body is not Sir Henry's. Holmes suspicions of Stapleton are given further credence when it transpires that he is a distant relative, and thus an heir, of Sir Henry. Holmes then sets a trap, and as Sir Henry walks across the moors from Stapeleton's house after dinner, Holmes, Watson and Inspector Lestrade of Scotland Yard (who appears in many of the stories) wait in hiding. A giant hound, which has been secretly kept by Stapleton is released to chase Sir Henry. Holmes shoots the beast and sets off with his posse after Stapleton, but the villain perishes in the marshes on Grimpen Moor.

Henry JAMES   1843–1916

The Wings of the Dove

From an impoverished, though genteel, background, Kate Croy is housed by her aunt, the vulgar, affluent and pretentious Mrs Lowder. Secretly, and against her aunt's wishes, Kate engages herself to Merton Densher, a clever journalist, who is nevertheless without prospects. On a trip to the USA, Densher meets Milly Theale, the last surviving member of a plutocratic family, and the dove of the title. With her companion, Susan Stringham, Milly visits London, where the two Americans are 'taken up' by Mrs Lowder's set. Kate and Milly begin an intimate friendship. Meanwhile, it grows apparent that Milly is fatally if indefinably ill. That she is soon to die increases her attraction for such as the fortune-hunting Lord Mark. Naïve and generous, Milly is vulnerable to exploitation. Apart from Susan, her sole disinterested confidant is Sir Luke Strett, a consultant physician.

Sir Luke advises Milly that, to delay the progress of her disease, she must be happy: she must learn how to live. Accordingly, she and Susan travel to Venice and rent a palazzo. At Kate's subtle prompting, Densher follows the pair. Kate's scheme is that Densher should woo – and possibly marry – Milly. On Milly's decease, he will become a wealthy widower, and he and Kate can marry in comfort – and with Mrs Lowder's approval. The scheme seems to work. Lord Mark, however, also goes to Venice. He proposes, cynically, to Milly and – when she rejects him – reveals the secret betrothal of Kate and Densher. For Milly, this deception is to be fatal: she loved Densher genuinely.

After the exposure of his motives, and as Milly's health worsens, Densher returns to London. Once he was disapproved of by Mrs Lowder; now she values his acquaintanceship. She expects that in spite of Densher's duplicity, Milly, in her magnanimity, will leave him some of her riches. Densher learns, on Christmas Day, of Milly's demise – and Milly has indeed left him an ample sum. Nevertheless, he rejects it and offers Kate a choice: either she and he marry and do not touch Milly's bequest or they stay separate while Densher makes over the money to her. Kate ends her engagement to Densher. Whether she accepts Milly's gift is not fully clear.

Like the two novels it precedes – *The Ambassadors* (1903) and *The Golden Bowl* (1904) – *The Wings of the Dove* is notoriously a difficult book. Its characters' dialogue and cerebration are marked by evasion – evasion, always, of the concrete point. Unable to speak with candour, perhaps because of their own delicacy, perhaps because of their author's, they are the better able to mislead not only each other but even themselves; and, coupled with their doubtful honesty, their circumlocutions render their existences almost intangible. Hence, Milly's problem is that the life Sir Luke recommends her to lead is, in a double sense, evasive. Her wealth notwithstanding, she cannot grasp what surrounds her, either the nature of her situation or the material things offered by historic Venice and London. Her ambivalent lesson is that in order to possess she has to give. Her final gesture so sharpens Densher's moral dilemma that he is compelled to break with Kate and, posthumously, is 'acquired' by his benefactor.

A.E.W. MASON   1865–1948

The Four Feathers

Mason's best-known novel is a characteristically stirring, old-fashioned tale, in which the recovery of lost honour is paramount. The story has two heroes: Harry Feversham, a young officer who has fallen into disgrace and lost his fiancée, Ethne Eustace; and Jack Durrance, his devoted friend, who, in spite of his own feelings for Ethne, does much to restore Harry to his rightful place. Afraid that he cannot match his father's military courage, Harry resigns his commission, ostensibly because of his engagement to Ethne, but in fact because he has received news that his regiment is shortly to be sent on active service to Egypt. When his brother-officers Trench, Willoughby and Castleton discover his motive, they send him three white feathers to symbolise his cowardice. To these Ethne adds a fourth and severs all relations.

For Harry, a broken man, there is only one solution: he must persuade the officers to take back their feathers. Confiding only in the elderly Lt. Sutch, he sets off for Egypt. Castleton is killed in the Campaign, but Harry performs two acts of supreme bravery, feats that will put an end to Trench and Willoughby's suspicions that he is lily-livered. He restores some lost letters to Willoughby under very dangerous circumstances and rescues Trench from the grisly Omdurman prison.

Meanwhile, Durrance, unaware of Harry's disgrace, woos Ethne and only wins her consent after he is blinded by sunstroke in Egypt. With blindness comes greater perspicacity; piecing together various facts overheard or reported by others, and sure that he has seen Harry in Egypt, Durrance realises what has happened and resolves to help his friend. Ethne insists that she loves Durrance, but her violin cannot lie, and the sadness with which she plays a certain overture convinces him that she still loves Harry. Feathers are taken back, Durrance withdraws honourably and goes back to Egypt, and Harry, now worthy of his father's name, can marry Ethne.

Such tales of Empire have been perennially popular with both adults and children, and Mason's novel has been filmed three times.

Beatrix POTTER　　　　　　**1866–1943**

The Tale of Peter Rabbit

The first and most famous of Potter's classic tales of animal life was originally conceived as an illustrated letter written to amuse a sick child. It tells the story of Peter Rabbit, sent out blackberrying with his sisters, Flopsy, Mopsy and Cotton-tail. Peter ignores his mother's strict instructions to stay away from Mr McGregor's garden ('Father had an accident there; he was put in a pie by Mrs

McGregor'). While looking for parsley to counteract the effects of over-eating, Peter bumps into Mr McGregor and is chased around the garden. Badly frightened, he manages to escape, but leaves behind his shoes and his blue jacket, which are used by Mr McGregor for a scarecrow. Peter is put to bed with some camomile tea, whilst his obedient sisters are rewarded with bread and milk and blackberries. (Peter's clothes are retrieved with the aid of his cousin in *The Tale of Benjamin Bunny*, 1904.)

Once upon a time there were four little Rabbits, and their names were – Flopsy, Mopsy, Cotton-tail, and Peter. They lived with their Mother in a sand-bank, underneath the root of a very big fir-tree.

'Now, my dears,' said old Mrs. Rabbit one morning, 'you may go into the fields or down the lane, but don't go into Mr McGregor's garden: your Father had an accident there; he was put in a pie by Mrs McGregor.'

THE TALE OF PETER RABBIT
BY BEATRIX POTTER

Like its successors, the book is characterised by precision, economy, a dry sense of humour and a complete absence of whimsy. Unafraid of long words ('soporific') or complex social concepts ('infinitely superior company'), Potter never speaks down to her audience. If her characters sometimes wear clothes and talk, they also behave true to their animal natures. The author's beautiful watercolour illustrations, which are an integral part of the books, are further evidence of the unsentimental eye of a naturalist. Potter's skilful blend of an exact observation of animal behaviour and an acute eye for social foibles has made her one of the most admired of children's writers. Indeed, Graham Greene has described her as the author of some of this century's best novels, and both Auden and Isherwood have recorded their devotion to her work.

1903

Anti-trust laws in US reinforced and US regulation of child labour is introduced ●
Regulation of motor-cars in Britain, with 20 mph speed limit ● At London Congress the
Russian Social Democratic Party splits into Menshevists, led by G.V. Plecharoff, and
Bolshevists, led by V.I. Lenin and Leon Trotsky ● Mrs. Emmeline Pankhurst founds
Women's Social and Political Union ● G.E. Moore, *Principia Ethica* ● Frederick Delius,
Sea Drift ● Oscar Hammerstein builds Drury Lane Theatre, New York ● George Bernard
Shaw, *Man and Superman* ● Louis Leakey and Evelyn Waugh born ● Paul Gauguin and
Camille Pissarro die ● W.B. Yeats, *In the Seven Woods*

Samuel BUTLER   **1835–1902**

The Way of All Flesh

Butler's magisterial indictment of the social and religious hypocrisies of the Victorian era was written between 1872 and 1884, but not published until after the author's death in 1902. The story of Ernest Pontifex is narrated by his godfather, Mr Overton, who is supposed to have written the book in 1867, adding the final chapter as a postscript in 1882. What at first sight appears to be a novel very much in the Victorian 'family saga' mould, with a few eighteenth-century traditions thrown in, turns out, on closer reading, to be a vigorous satire, which embraced such 'modern' notions as the importance of the subconscious as a motivating force. It is in many ways very much a book of the twentieth century, exposing a revered age rather in the manner of Lytton Strachey's *Eminent Victorians* (1918). Butler's iconoclasm made him an influential figure upon later generations; E.M. Forster modelled Mr Emerson in ***A Room With a View*** (1908) on Butler and planned to write a book about him, while War Poets such as Sassoon and Graves saw him as a 'great spiritual father'. Although he did not revise the manuscript, Butler was dissatisfied with it; he none the less gave instructions on his death-bed that it should be published.

From a long line of Pontifexes springs Ernest, first son of the self-righteous and hypocritical clergyman Theobald, and his silly, pretentious wife, Christina. Ernest's totally repressed childhood, in which any spark of natural intelligence is nipped sharply in the bud, is followed by an equally oppressive school life, redeemed only by the attentions of his spinster aunt Alethea. Such an upbringing leaves Ernest ill-equipped to contest the assumption that after Cambridge he will follow his father into the church.

This was deplorable. The only way out of it that Ernest could see was that he should get married at once. But then he did not know any one whom he wanted to marry. He did not know any woman, in fact, whom he would not rather die than marry. It had been one of Theobald's and Christina's main objects to keep him out of the way of women, and they had so far succeeded that women had become to him mysterious, inscrutable objects to be tolerated when it was impossible to avoid them, but never to be sought out or encouraged. As for any man loving, or even being at all fond of any woman, he supposed it was so, but he believed the greater number of those who professed such sentiments were liars.

THE WAY OF ALL FLESH
BY SAMUEL BUTLER

Easily influenced, to the point of stupidity, Ernest falls under the power of Pryer, who helps him to lose his small inheritance from Grandfather Pontifex. He launches himself on an impoverished London household, presided over by the wonderfully vulgar Mrs Jupp, with the intention of saving their souls. This ends in disaster, however, when, confused by lust, he importunes an innocent lady lodger whom he takes for a prostitute and finds himself in prison. It is from this point that Overton, custodian of a large fortune left to Ernest by Aunt Alethea to be bestowed on him at the age of twenty-eight, begins to take an interest in his godson. In prison Ernest decides to turn his back on his family and his life in the church and make his way in the world as a tailor.

A chance meeting with Ellen, the Pontifexes' dismissed chambermaid, to whom Ernest has shown a past, boyish kindness, leads to a brief period of happiness and relative prosperity. Married, much against Overton's better judgement, Ernest is the last to realise that Ellen has returned to her drunken ways. Since honour forbids desertion, he resigns himself to ruin. Fortune once more intervenes in the form of John, a former coachman, who reveals that Ellen is his lawful wife. Ernest is free, Ellen is happy to leave him and their two children are farmed out to an honest bargeman and his wife. Ernest recovers his strength and dignity and is being unwittingly prepared for the revelation of his inheritance which, judiciously invested, has been maturing nicely. Duly enlightened, Ernest can now bring about a gracious, if reserved, reconciliation with his family, without the need for abject apology. Having, to date, 'failed in everything he has undertaken', he is now transformed into an urbane, handsome, self-sufficient fellow (mirror-image of the Towneley he admires so much) with a talent for satirical literature.

Based to some extent on personal experience (Butler's dislike of his father is well documented, as is his quarrel with received religion) this is, nevertheless, fiction. The plot may at times seem a little far-fetched, but it is used as a framework on which to examine the prevailing preoccupations and questions of the day, and the story itself is told with wit, insight, and at times with genuine compassion.

Erskine CHILDERS   **1870–1922**

The Riddle of the Sands

The invasion of an unprepared England by sinister foreign powers was a common theme in popular fiction in the early years of the century, notably in adventure stories written for boys. The Harmsworth Press in particular used juvenile fiction as a vehicle for alerting people to the danger posed by 'antagonistic foreign nations' – France, Russia and, increasingly, Germany – and the need for military training. John Buchan, Saki, C.M. Doughty, J.M. Barrie, William Le Quex and even (mockingly) P.G. Wodehouse all contributed to the genre. *The Riddle of the Sands* is an early example, sometimes regarded as the first modern espionage story.

The novel's heroes are two English gentlemen: the fastidious Foreign Office civil servant Carruthers, and the gauche, impulsive but principled Davies. The story begins languidly enough when Carruthers receives an invitation to join Davies aboard his yacht, the *Dulcibella*, for some duck shooting in the Baltic. Davies is evasive about the route they are to take, and it gradually becomes clear to Carruthers that there is more to this expedition than potting a few birds. Davies reveals that he is planning to retrace an earlier sailing to the seven Frisian Islands, when he was guided through a short-cut by Dollmann, an English salvage operator he had met on shore. The *Dulcibella* had nearly been wrecked on a bank, and Davies is convinced that Dollmann had been trying to lure him to his death. Dollmann appeared to be friendly with Von Bruning, the commander of a German gunboat; Davies believes that he is a spy and that the Germans are concealing warships in the Islands (the 'Sands' of the title). Neither English gentleman would consider spying on German defences, but Davies wants to shake up the Admiralty, and Carruthers, recruited because he speaks fluent German, is sufficiently shocked that a 'civilised government' appears to have made an attempt on his friend's life to fall in with the plan.

Davies has a secondary motive: he is in love with Dollmann's daughter, Clara, but concerned that she might be implicated in her father's treachery, a treachery confirmed when Carruthers overhears Dollmann discussing plots against the British with Von Bruning. Carruthers also witnesses a night-time rehearsal for the invasion of Britain, attended by the Kaiser himself. The Englishmen challenge Dollmann, and attempt to take him and his daughter back to England, but he jumps ship and is lost at sea.

Desert as I call it, it was not entirely featureless. Its color varied from light fawn, where the highest levels had dried in the wind, to brown or deep violet where it was still wet, and slate-grey where patches of mud soiled its clean bosom. Here and there were pools of water, smitten into ripples by the impotent wind; here and there it was speckled by shells and seaweed. And close to us, beginning to bend away towards that hissing knot in the north-west, wound our poor little channel, mercilessly exposed as a stagnant muddy ditch with scarcely a foot of water, not deep enough to hide our small kedge anchor which perked up one fluke in impudent mockery. The dull hard sky, the wind moaning in the rigging as though crying in despair for a prey that had escaped it, made the scene inexpressibly forlorn.

THE RIDDLE OF THE SANDS
BY ERSKINE CHILDERS

It is never stated whether or not Clara was aware of her father's activities. Childers created the character only at the insistence of his publisher and found her 'a horrible nuisance'. She is sketchily drawn, and her association with Davies is awkwardly presented. Childers pays far more attention to naval details, and accompanying charts help authenticate the story, which so alarmed the Admiralty that a naval base was built in Scotland to shore up defences in the North Sea. A tragically ironic postscript to the story is that Childers himself was later shot as a traitor because of his involvement with the IRA.

Henry JAMES   **1843–1916**

The Ambassadors

The Ambassadors is one of James's three late, great novels which he regarded as 'quite the best "all round" of all my productions'. Lambert Strether, the middle-aged editor of a serious New England journal, has been sent to Paris by his patron, the stern Mrs Newsome, to bring home to Woollet, Massachusetts, her son Chad so that he may take his place at the head of the family business. En route to France, Strether meets in England Maria Gostrey, a resident of Paris, who becomes his ally and confidante. When he is introduced to Chad in Paris, Strether is enchanted by the charm, taste and grace that Chad has acquired. He attributes this to the influence of Mme de Vionnet, towards whose young daughter, Jeanne, Chad is supposed by Strether to be romantically inclined.

Chad's friends, whom Strether likes and trusts, do little to enlighten him about the precise nature of Mme de Vionnet's relationship to Chad, and so instead of pursuing his mission, Strether enjoys new feelings of life's possibilities. In a memorable scene, which James pointed to as the core of the book, Strether exhorts an expatriate American, John Little Bilham: 'Live all you can: it's a mistake not to. It doesn't matter so much what you do – but live. This place makes it all come over me. I see it now. I haven't done so – and now I'm old ... You have time. You are young. Live!'

As a consequence of Strether's unwillingness to assert to Chad the claims of duty and materialism, Mrs Newsome dispatches her second embassy, which consists of her daughter, the formidably scornful Mrs Pocock, and her cynical, vulgar husband, Jim, both of whom have come to see in Paris the decadence and licence they despise – Jim with more relish than his wife.

When by chance Strether meets Chad and Mme de Vionnet on a boating trip outside Paris, it becomes clear to him that they are lovers; his innocence is dissolved but he clings to his belief that the relationship is one of beauty and exhorts Chad not to betray his lover by leaving her. Chad, although grateful for his education in manners, prepares to return to America, having succeeded in arranging a marriage for Jeanne. In spite of the knowledge that his future both as a possible husband of Mrs Newsome and as a benefactor are jeopardised, Strether returns to Woollet, rejecting the offer of a shared life with Maria Gostrey.

The Ambassadors, by virtue of its themes of American innocence meeting European sophistication, and of the dilemma between the idea of engaging with life rather than merely contemplating it, dramatises interests which abound in James's work. In its formal intricacy and in the technical brilliance by which the narrative unfolds through Strether's burgeoning perceptions of events it is an example of James's art at its most beautiful and accomplished.

Jack LONDON   1876–1916

The Call of the Wild

This fable about a dog called Buck owes something to Anna Sewell's *Black Beauty* (1877) and Rudyard Kipling's *The Jungle Book* (1894). London explores the interaction of animals with humans as a metaphor for the initiation of children into an adult world. It was also one of the first American novels to examine the quest of the pioneering individual who breaks away from the sheltered environment of civilisation and is romantically compelled to find freedom in nature. In the early part of the century this was considered the American dream.

The half-Alsatian Buck has enjoyed the first four years of his life in domestic comfort in San Francisco. Stolen from his caring master, Judge Miller, and taken to the Arctic, Buck is brutally forced to submit to the dangerous life of a working dog in a sledge-pulling pack in the Yukon. The security and friendship with humans the animal once knew, symbolised by the description of his home in terms of warmth, contrasts with the Arctic conditions, accentuated by images of dark, winter and cold.

Part of Buck's learning experience is to recognise the different types of human he encounters, and how he is at their mercy. After rebelling against his new master, Buck is savagely beaten, and has an inkling of one aspect of 'primitive law'. For the time being, the dog compromises, and co-operates with his master exactly to the degree that ensures safety and food. In relating this, London does not evoke sympathy for Buck, insisting instead how the dog must learn survival through adverse circumstances. Only through this process can the animal, like a child, understand the puzzling complexities of life.

As the novel progresses, Buck has several different owners, and learns to adapt his behaviour to each of their ways. Some are merely incompetent: Hal and Charles make no effort to understand their dog pack, while the sentimental Mercedes ineffectually tries to stop the whipping of 'the poor dears', while insisting the dogs carry her as well as their heavy load. The dogs have a certain pride in their work for humans, which gives order to chaos, but when their efforts are abused and exploited, respect gives way to disgust and hatred. Buck is freed from this treachery by John Thornton, with whom he builds a trusting relationship. They save each other's lives, and from Thornton Buck experiences 'passionate love ... for the first time' – even with Judge Miller he had only known 'a stately dignified friendship'. However, Thornton's benevolence places the dog in a dilemma.

Buck's long exposure to the open life gradually awakens in him 'the strain of the primitive', and the need to live in the wild as his ancestors once had. As long as Thornton is alive, Buck will not desert him, but when his master dies, the animal owes nothing to any human: 'The last tie was broken ... the claims of man no longer bound him.' Buck reverts to the wild, as leader of a pack of wolves, finally inheriting his ancestral birthright.

Both protection by humans and self-reliance have prepared Buck for his call. His experience is two-fold, for as he develops in civilisation, he regresses to nature. As children must, the animal has broken away from the protective but suffocating mother-influence, represented by Buck's contact with the human world, to take his place with his 'wild fathers'.

1904

Russo-Japanese War ● Theodore Roosevelt wins US presidential election ● Photo-electric cell, safety razor blades and ultra-violet lamp made ● Charles Rennie Mackintosh, The Willow Tea Rooms, Glasgow ● Giacomo Puccini, *Madama Butterfly* (opera) ● J.M. Barrie, *Peter Pan* ● Abbey Theatre, Dublin, founded ● Christopher Isherwood, Graham Greene and Marlene Dietrich born ● Samuel Smiles, Anton Dvořák and Anton Chekhov die ● George Russell, *The Divine Vision*

G.K. CHESTERTON   **1874–1936**

The Napoleon of Notting Hill

Like Orwell's dystopian fantasy, this, Chesterton's first novel, is set in 1984. The world has witnessed no wars for five years and politics have ceased to be contentious. England is truly a nation of shopkeepers. An efficient bureaucracy presides over affairs while the head of state is a monarch chosen by means of an alphabetical system.

Auberon Quin is an eccentric clerk – an antiquary with a penchant for the absurd. He is explaining apparently inexplicable anecdotes to his friends when he is proclaimed king. As he wanders through Notting Hill, a small boy, Adam Wayne, prods him with a wooden sword and announces: 'I'm the King of the Castle.' This gives Auberon an idea. He decrees a 'revival of the arrogance of the old medieval cities applied to our glorious suburbs'. He invents mythologies for Bayswater, Kensington and Knightsbridge, devises ceremonies, designs liveries and holds mock-feudal colloquies with his provosts. It is all a game, although an annoying one for most of his subjects. But ten years later Adam Wayne is provost of Notting Hill – and he plays the game in deadly earnest.

A business consortium wants to build a highway through the heart of Wayne's beloved district. Wayne will not permit this; he vows to defend his domain and raises a tiny army. His chief strategist is Turnbull, the owner of the local toyshop. The businessmen raise opposing armies, and three fierce battles are fought using spears and swords. King Auberon, in the guise of a war correspondent, reports the conflict. Wayne emerges victorious and a period of peace ensues. His values, however, have now infected London. Fierce local 'patriotisms' emerge and – after a dispute about a statue – battle is joined again. The army of Notting Hill is massacred. Only Wayne survives and, amidst the corpses, he and King Auberon discuss their actions. Auberon explains that he was playing 'a vulgar practical joke'; Wayne responds that, none the less, he has 'lifted the modern cities into ... poetry'. Auberon accepts the truth of this and the two march off together 'into the unknown world'.

Chesterton's *oeuvre* is preoccupied with a struggle against the everyday. Continually it strives to detect the peculiarity of the mundane and the mundanity of the peculiar. (Paradox is a favourite ploy.) *The Napoleon of Notting Hill* delights in inversion: the clerk becomes the king; the (apparently) stolid figures of the consortium are inflamed by Wayne's example, whereas the (apparently) lunatic monarch gradually grasps the gravity of the position; adults behave like children, and thereby attain romantic grandeur. Indeed, if the novel is a polemic against the banal, it is simultaneously a hymn to the heroism latent in 'the common man, whom mere geniuses ... can only worship'; if it finds farce in Wayne's fanaticism, it also finds tragedy in Quin's humour.

The book is dedicated to Hilaire Belloc, some of whose novels Chesterton illustrated.

Joseph CONRAD　　　**1857–1926**

Nostromo

Like **Lord Jim** (1900), *Nostromo* caused Conrad a great deal of anxiety in the writing, and friends were frequently regaled with letters reporting on his progress, or lack of it, throughout 1903–4. It was completed in the course of a literally agonising fortnight, during which a dentist had to be summoned to deal with the author's raging toothache. 'I have survived extremely well,' Conrad reported. 'I feel no elation. The strain has been too great for that ... my friends may congratulate me as upon a recovery from a dangerous illness.'

Nostromo's subject is nothing less than the material and intellectual forces that shape individual and national fates. Its narrative, accordingly, is complex – a canvas both immense and detailed. Switching back and forth in time, it often shows outcomes in advance of actions, thus to propose that means are mocked by ends. Subtitled 'A Tale of the Seaboard', the story centres on Sulaco, the chief port of Costaguana, Conrad's South American republic. Principal among its cast are Charles Gould, owner and administrator of the San Tomé silver mine; Martin Decoud, cynic, *boulevardier* and journalist; and Nostromo himself – Giovanni Battista Fidanza, immigrant Genoese sailor, idol of the common people, his indispensability to his masters acknowledged in his sobriquet: 'our man'.

San Tomé has brought wealth to the region. Furthermore, through astute and systematic bribery, Gould has propped up civilian rule. But stability is interrupted by a demagogic general, Montero. He rails against the influence exerted by Gould, his English and his US backers. Troops occupy Sulaco. Nostromo is entrusted with six months' production of silver, which must be saved from the insurgents. Accompanied only by Decoud (fleeing because, in his newspaper columns, he has roundly abused the Monterists), he takes the cargo to sea and, after a perilous escape, hides it on a barren island. Decoud is left in charge. In Sulaco, Nostromo learns that the risks he has run are irrelevant to Gould's broad strate-gies. Seeing that his trust has been abused – and not before an avaricious man – he decides at last to profit from his adventures, letting it be believed that the silver was lost. Meanwhile, Decoud commits suicide. He cannot bear his island solitude.

Peace returns to Costaguana. The Monterists are defeated and the economic interests represented by Gould establish a pliant government. Nostromo now is free to 'grow rich slowly', drawing on the treasure of whose hiding place he alone knows. He has become a slave to his secretive greed. His character is corroded. In a moment of cowardice, he pledges marriage to a girl, although he is enamoured of her sister. Finally, his duplicities lead to his death: the girls' father shoots him, mistaking him for a common thief or seducer.

But afterwards? he asked himself. Later, when a keeper came to live in the cottage that was being built some hundred and fifty yards back from the low light-tower, and four hundred or so from the dark, shaded, jungly ravine, containing the secret of his safety, of his influence, of his magnificence, of his power over the future, of his defiance of ill-luck, of every possible betrayal from rich and poor alike – what then? He could never shake off the treasure. His audacity, greater than that of other men, had welded that vein of silver into his life.

NOSTROMO
BY JOSEPH CONRAD

Decoud is destroyed by 'disillusioned weariness': Nostromo by 'disenchanted vanity'. Each suffers from 'want of faith' and the lack is contrasted with Gould's fanatical attachment to his mine. The owner of San Tomé perceives in its wealth a moral power: it will attract foreign investment which, enforcing the conditions of success, will make an end of coups and revolutions. Sure in this vision, Gould is sustained during the crises that beset him. None the less, Conrad suggests, his materialism is simultaneously harsh and sentimental for, while it puts people at the service of an enterprise, it fails to foresee the consequence. To Emilia Gould, already a patient victim of Charles's obsession, Conrad gives a resonant prediction: 'She saw the San Tomé [mine] hanging over ... the whole land, feared, hated, wealthy; more soulless than any tyrant, more pitiless and autocratic than the worst Government; ready to crush innumerable lives in the expansion of its greatness.'

F.R. Leavis described *Nostromo* as 'Conrad's supreme triumph'. It is a judgement with which most critics would concur.

W.H. HUDSON   1841–1922

Green Mansions

Hudson made his reputation as a naturalist, but he also wrote several novels, the best known of which is *Green Mansions*, subtitled 'A Romance of the Tropical Forest'. Its large sales went some way towards alleviating the poverty in which Hudson spent much of his life after coming to London from his native South America in 1874.

Set in Guyana, South America, in the mid-nineteenth century, the novel is the story of Abel Guevez de Argensola, a Venezuelan who, as a young man, fled his native country after his inopportune participation in an attempt to overthrow the government. Tired of searching for fabled gold, and giving up his idea of writing a book about his travels, Abel settles for a while with a tribe of Indians in a small village of some eighteen souls, about a league from the western extremity of the Parahuari mountains.

The natives avoid a nearby forest, which they suppose to be haunted by a malevolent spirit, but Abel takes to regular exploration on the fringes of the forest, increasingly aware of being shadowed by a presence, never visible, with a voice like a trilling bird. At length, after Abel has established his good standing in the forest by being peaceable, unlike the violent natives, he is granted a glimpse of the creature: 'a girl form, reclining on the moss among the ferns and herbage, near the roots of a small tree ... she was small, not above four feet six or seven inches in height, in figure slim, with delicately shaped little hands and feet'. The girl is Rima, a young woman brought up in the wood by an old man called Nuflo after the death of her mother.

Rima is Hudson's 'natural savage' – at one with her forest environment, living in harmony with the flora and fauna, feeling no fear and having cause for none. Abel falls in love with the girl, who, though most expressive in her whistling language, speaks his own tongue. Together, they discover from the old man that her mother came from a district known as Riolama, and Rima conceives an urgent and unassailable desire to go there to discover people of her own kind. She will not be dissuaded from the hazards of such a journey, and the three of them – Rima, Abel and Nuflo – set off. But there is nothing to be·found. Her mother's people do not exist, and in her despair Rima faints in Abel's arms, 'all that bright life seemed gone out of her'.

When she wakes, Rima appears to accept that her dream of finding her past is hopeless, and seems content to return to the forest and to wait there for Abel and Nuflo to catch up with her. She remembers, in every detail, the paths they have taken and can retrace them quickly by herself. When Nuflo and Abel reach the wood, days later, they find the natives hunting there, no longer afraid of the spirit that haunted it. They have exorcised the wood by trapping Rima in a tree and burning it. In revenge for her death, Abel persuades a neighbouring tribe, enemies of the Indians he has lived with, to attack and massacre the entire village. Nuflo is killed, and Abel intends to live reclusively in the forest with his grief.

'If I could only have faded gradually, painlessly, growing feebler in body and dimmer in my senses each day, to sink at last into sleep! But it could not be ... unless I quitted the forest before long, death would come to me in some terrible shape.' Abel, ill and sick at heart, quits the forest and returns to the world he has left, heavy with the memory of Rima and, 'self-forgiven and self-absolved' from his sins, at peace with his love for her.

Henry JAMES   **1843–1916**

The Golden Bowl

There is a broad critical consensus that James's last three novels are 'difficult' and that, of the three, *The Golden Bowl* is the most difficult. Often seen as the summation of all James's previous work, opinion is divided as to whether it repays the scrupulous attention it demands. It is long, densely written and tells only the most slender of stories. James himself summarised it in a line: 'the Father and Daughter, with the husband of one and the wife of the other entangled in a mutual passion, an intrigue'. Originally inspired by a newspaper story and conceived as a novella, the project grew out of these flimsy beginnings into a monumental examination of the lives, manners, thoughts and feelings of four characters.

Adam and Maggie Verver, widowed father and daughter, are wealthy American expatriates living in London. Charlotte Stant and Prince Amerigo are former lovers who, unable to marry each other for financial reasons, have parted and who later marry Adam and Maggie respectively. As the novel opens, Prince Amerigo, whose point of view dominates the first three sections, is preparing to marry Maggie, when Charlotte arrives for the wedding. She has become a friend to Maggie, who is ignorant of the former liaison. Later, Charlotte marries Adam, but her old romance with the Prince is rekindled. Through two incidents and an improbable coincidence involving the bowl of the title, Maggie learns of the affair. The final two sections, told from her point of view, chronicle her delicate manoeuvres to oust Charlotte from her husband's affections without alerting her father to the affair. The novel ends with Charlotte and Adam Verver departing from America and Maggie triumphant in her management of the situation.

Criticism of *The Golden Bowl* has focused on the character of Maggie, over whom opinion is divided. She is either a subtly brutal dictator or a woman in triumphant command of her situation, a manipulator or an agent of eventual harmony. The moral ambivalence is deep-rooted and half-acknowledged when she calls herself 'a mistress of shades'. Mrs Assingham, a garrulous choral figure, remarks of her that, 'She wasn't born to know evil', but James clearly knows that innocence might be quite as complex as experience. Maggie's father is a remote, uninvolving character and the relationship between the two is incestuously close. Charlotte and the Prince are far more engaging, but they are adulterers, perhaps even gold-diggers. The sympathies of James and his readers are given no easy resting-place in this novel.

Despite James's almost manic technical control, each character is allowed his or her full measure. The truth of the matter lies somewhere between the four of them, James is finally saying, and absolute judgement is misplaced. Historically, *The Golden Bowl* straddles the realistic and modernist novels. Its enormous attempt at pinning down exactly the thoughts of the characters is countered by the refusal of these thoughts to be pinned down. The by-product of this 'failure', a style developed precisely to minutely annotate the characters' thoughts and feelings, was later to re-emerge at the heart of the modern novel in the very different works of Joyce, Lawrence and other writers.

Fr ROLFE   **1860–1913**

Hadrian the Seventh

Perhaps best known through Peter Luke's 1967 stage adaptation, *Hadrian the Seventh* is a bizarre, autobiographical fantasy, in which Rolfe imagines himself catapulted from obscure poverty into the papal throne. 'Saint or madman?' wonders one of the cardinals of this unlikely pontiff; Rolfe leaves us in no doubt about his own view. The novel's extraordinary blend of passion and style, however, raises the novel far above mere apologia, and is particularly fascinating for its descriptions of Vatican life and papal elections.

George Arthur Rose, a failed priest and

unsuccessful writer, is visited in his lodgings by his friend Dr Talcryn, Bishop of Caerleon, who brings with him Dr Courtleigh, Archbishop of Pimlico (a caricature of H.E. Vaughan, Archbishop of Westminster, a former patron whom Rolfe believed had defrauded him). They have reviewed Rose's case and decided that a terrible injustice had been done in refusing him his ordination. Astonished that, in spite of numerous humiliations at the hands of his fellow Catholics, he has persisted in his vocation, they immediately grant him the priesthood and ask him to accompany them to Rome, where the papal election is about to take place.

After several failures to elect a successor to Leo XIII, nine cardinals, including Courtleigh, are appointed Compromissaries. Courtleigh feels moved to relate the story of Rose's devotion to the Faith, with the result that the newly ordained priest is instantly elected to Peter's Chair. As Hadrian VII, Rose horrifies the traditionalists by abolishing the *sedia gestatoria*, going on informal walkabouts and remaining an ascetic: he sells off the Vatican treasures, redecorates the papal apartment with brown wrapping paper, and installs a cold shower. He extends his patronage to such people as Mrs Dixon, his former landlady (whose gifts of pickles he buries in the Vatican gardens), and William Jameson, an *alter ego* whose financial security he arranges, and to his Gentlemen-in-Waiting, John Devine and Julio Carrino: the former is given the means to set up a farm with a muscular student called Hamish, while the later is entrusted with Hadrian's secret formula for colour photography and sufficient funds to marry. The Pontiff's enemies seem blind to his admiration for athletic young men, and are determined to discredit him by proving him a womaniser.

When a world war seems imminent, Hadrian acts as Supreme Arbitrator and blithely carves up mainland Europe between Italy and Germany, granting the latter control of Russia, while authorising Japan to annexe Siberia. Further acts of geographical generosity are thwarted by an old enemy, the mad, self-aggrandising 'haberdasher's bagman' and socialist, Jeremiah Sant (a portrait of Keir Hardie). Having failed to force the Pope to support his cause through blackmail

and libel, Sant assassinates him, and is subsequently torn to pieces by the enraged crowd. As he dies, Hadrian forgives both Sant and his accomplice, the grotesque Mrs Crowe, whose enmity is that of a spurned woman (Hadrian had found her advances 'as obvious, as abrupt and as shameless as a dog's'). The novel ends with the Pope delivering Apostolic Benediction, while the bewigged Mrs Crowe howls her love like 'a dog which breaks the cadence of Handel's *Largo* on arch-lutes'. 'Pray for the repose of His soul,' Rolfe commands the reader. 'He was so tired.'

H.O. STURGIS   1855–1920

Belchamber

Sturgis wrote only three novels, the first of which, *Tim* (1891), is a classic public-school romance set at Eton. The relationship between the sickly Tim and the athletic Carol of the earlier novel resurfaces in familial form in Sainty's physical subservience to his brother Arthur in *Belchamber*, which is distinctly Victorian in tone. The novel, which pits avarice and cynicism against generosity and honour, is an indictment of 'fast' society and the careless, philistine people who thrive in it. Although the book is now highly regarded, any success it might have had was spoiled for the author by his friend Henry James. 'It is a mere passage, a mere antechamber and leads to *nothing*', the Master complained to A.C. Benson, a criticism which so pained Sturgis that he attempted to withdraw the novel and never wrote another.

'Sainty' Belchamber is an unhappy heir, an aristocrat with radical sympathies, too idealistic for the corrupt society in which he finds himself. Picked upon as a prig at Eton, he is outclassed in all the gentlemanly pursuits by his younger brother, Arthur, whom everyone would have preferred to succeed as lord of the manor. When Sainty is crippled in a riding accident, he regards it as a blessed release from the duties of hunting and fishing, and his only desire is to die, so that Arthur might inherit his true role.

After an idyllic interval at Cambridge, Sainty's twenty-first birthday sweeps him back into the round of balls and parties, where he is shocked by the sex scandals of the County set. An effort to save Arthur from a disastrous affair with a music-hall actress fails spectacularly; Arthur is trapped into marrying her and is ostracised by their puritanical mother. Sainty's main obligation now is to prevent the vulgar actress from becoming the Duchess; but with his scrupulous morality and ugly appearance, he does not expect to find a wife for himself. Cissy Eccleston seems to see through to his real qualities, however, and Sainty begins to dream that the young woman could really love him for himself and not for his social status. Encouraged by their families, Sainty and Cissy announce their engagement; but in the weeks before the wedding, Sainty begins to suspect that she finds him repugnant, and that her affections already lie elsewhere.

On their wedding night, Cissy announces that she will never consummate the union, and rather than endure the scandal of a divorce, Sainty agrees to lead a separate life and keep the situation a secret. Cissy spends his fortune extravagantly, furnishing their London house *à la mode*, and her smart parties and loose behaviour begin to attract the notice of gossips. An affair develops between Cissy and Sainty's cousin, Claude Morland, and Sainty is duped into passing as the father of their baby. Ironically Sainty finds his only happiness in his love for the child that Cissy neglects – the only human being who shows pleasure in his company, unprompted by mercenary designs. The dilemma which develops is whether to let the fraud pass on to future generations, with the possible trauma of a chance discovery which will kill the boy's love for him; or to face the scandal and let Arthur's philistine family succeed to the title.

The situation is tragically resolved by the child's death in infancy. As Sainty mourns over the baby's coffin, Cissy announces her intention of leaving Sainty and going to 'the only man she has ever loved'. The ambitious Claude has already escaped her, however, through his engagement to a wealthy heiress, and Sainty and Cissy must live on together, locked into mutual hatred.

1905

Chinese boycott US goods • Workers in St Petersburg form first Soviet; mutiny on battleship *Potemkin* • Sinn Fein party is founded in Dublin • Sun Yat-sen organises a union of secret societies to expel the Manchus from China • V.I. Lenin, *Two Tactics* • Pablo Picasso, *Boy With Pipe* • Antoni Gaudi, Casa Milà, Barcelona • Richard Strauss, *Salome* (opera) • A five-cent cinema in Pittsburgh shows *The Great Train Robbery* • Arthur Koestler, Greta Garbo and Dag Hammarskjöld born • W.H. Davies, *The Soul's Destroyer and Other Poems*

E.M. FORSTER   **1897–1970**

Where Angels Fear to Tread

Although *Where Angels Fear to Tread* was the first novel Forster published, it was conceived after *A Room with a View* (1908). Both novels grew out of Forster's experience as a tourist in Italy during 1901 and 1902, a sojourn which was to have a profound effect upon him. The plot, inspired by some gossip Forster overheard about an Anglo-Italian *mésalliance*, centres upon the disastrous attempts of a middle-class English family, the Herritons, to 'rescue' a baby, the child of a hasty and brief marriage between Lilia (the widow of a Herriton son) and Gino Carella, the son of a small-town dentist of uncertain social status. The desire to break free from the constrictions of respectable Sawston had made Lilia susceptible to the charms of Italy in general and to the handsome Gino in particular. From this relationship the novel develops its central theme: the contrast between the passionate life of Monteriano and the claustrophobic existence of Sawston.

Forster, however, is too subtle an artist to construct a novel solely on this obvious contrast. Monteriano has its own snobberies and provincial limitations, as Lilia discovers when she tries to escape from domesticity and enter the life of the town; and Gino acquires as a husband qualities he lacked as a lover, notably a decidedly conventional view of his new role. But Italy has an instinctive sense of the importance of human emotions which England either lacks or stifles. When Caroline Abbott, who had accompanied Lilia to Italy, returns to Sawston, she realises that everyone 'spent their lives in making little sacrifices for objects they didn't care for, to please people they didn't love'. To complete her spiritual salvation, she must go back to Italy.

Lilia dies in childbirth and the Herritons are convinced that Gino, who was much younger than his wife, is incapable of bringing up the child correctly. Philip Herriton, who had already attempted to stop the marriage, is once more despatched by his mother from Sawston to Monteriano in order to persuade Gino to hand over the baby. He is accompanied by his humourless, xenophobic sister, Harriet, who is the embodiment of Sawston values. In Monteriano the repressions of Sawston are confronted by something overwhelmingly more vital. This conflict comes most obviously into focus when Philip, Harriet and Caroline visit the opera. During a performance of Donizetti's *Lucia di Lammermoor* the inhabitants of Monteriano display wholehearted, unembarrassed enjoyment, which shocks Harriet, for whom opera is 'culture', demanding a reverent silence. For Philip, however, it brings a new sensation, 'an access of joy' that 'promised to be permanent', and which leads to his transformation from being a spectator of life into becoming a full participant.

The following morning Caroline visits Gino, imagining that he is more likely to respond to her powers of persuasion than to

Philip's geniality or Harriet's rudeness. When Gino makes it clear that he loves his son, however, her resolve wavers. 'The horrible truth, that wicked people are capable of love, stood before her and her moral being was abashed', and she helps Gino bath the baby. When Philip arrives, he too makes little headway, incurring the fury of Harriet, who takes matters into her own hands by kidnapping the child. Philip assumes that she has persuaded Gino to part with the child, but when, on the way to the station, the carriage they are travelling in overturns and the baby is killed, Harriet cries out the truth. The horrified Philip feels responsible and determines to be the person to break the news to Gino, who takes out his anguish on Philip, brutally twisting the Englishman's broken arm. A reconciliation is overseen by Caroline, who persuades them to share a bowl of warm milk originally intended for the baby.

The novel begins and ends with a train journey, the first an unthinking departure, the second a sadder but wiser return. Despite the tragedies that occur, there is a conviction that action, even if it involves mistakes, is preferable to observation; and that mistakes can themselves lead to right action. Both Philip and Caroline, through their involvement in the life of Monteriano, have transfiguring experiences which enable them to rise above their personal disappointments. Philip vainly hopes that Caroline has fallen in love with him, but she has in fact fallen in love with Gino, whom she will never see again. The novel ends on a note of 'great friendliness', even towards those less able to respond to life's opportunities for spiritual development, as Philip and Caroline hurry 'back to the carriage to close the windows lest the smuts should get into Harriet's eyes'.

The novel was made into a distinguished film (1991), directed by Charles Sturridge from a screenplay he wrote with Tim Sullivan and Derek Granger.

H.G. WELLS 1866–1946

Kipps

The second of three comic novels which, with *The Wheels of Chance* (1896) and *The History of Mr Polly* (1910), sprang from

Wells's early experiences in the haberdashery trade, *Kipps* is remarkable for its outrageous plot and its consummate artistry. Even Henry James, forgiving its superficial faults, succumbed to its gusto: 'I am lost in admiration at the diversity of your genius,' he told Wells. The novel's great triumph is the character of Artie Kipps, battling in both poverty and wealth against a society in which only he is sane. The novel was twice filmed: the 1921 version was superseded in 1940 by a classic version directed by Carol Reed from a screenplay by Sidney Gilliat.

The Drapery bazaar at Folkestone, where the impoverished and ill-educated Kipps is indentured to the tyrannical Mr Shalford, corresponds in fine detail to the one at Southsea in which Wells worked for two years. Kipps's release is achieved most improbably through the agency of James Chitterlow, a flamboyant thespian and *deus ex machina*, who is as much a social outsider as Kipps himself. Chitterlow meets Kipps when he knocks him down with his bicycle, and revolutionises his life by producing an advertisement in an old newspaper which enables Kipps to claim a large legacy. Chitterlow has no compunction in persuading Kipps to invest in a play he is writing. Kipps's new wealth catapults him into Folkestone society where, guided by Chester Coote, a house agent and a 'conscious gentleman', he embarks upon a course of self-improvement. He becomes engaged to Helen Walsingham, an indigent but snobbish young woman with a taste for the arts, who sees Kipps as a way out of her financial embarrassments, even though his table manners cause her some concern. He is obliged to accompany his fiancée on the social round, but Kipps is ill-suited to being a 'gentleman', and it is only when he starts to follow his own instincts that things begin to come right.

When he discovers his childhood sweetheart, Ann Pornick, in service with Mrs Bindon Botting, he decides to marry her. Their attempt to adjust to middle-class life is cut short by Helen's brother, a feckless playboy to whom Kipps has unwisely entrusted his financial affairs. Having 'speckylated' with and lost the greater part of Kipps's fortune, Young Walsingham is obliged to flee the country. One consolation of this 'smash' is that Helen decides not to sue her former fiancé for breach of promise. Kipps buys a bookshop with the remains of his money, which will scarcely make a living, but once more Chitterlow comes to the rescue. The play Kipps invested in has been an unexpected success, and the royalties will keep Kipps, Ann and their adored infant son in the comfortable, modest seclusion they have always craved.

'I was thinking jest what a Rum Go everything is,' Kipps says to Ann. 'I don't suppose there ever was a chap like me.' Bewildered, vulnerable and honest, Kipps makes his way through the treacherous world with courage and dignity, and is undoubtedly the most touching of Wells's heroes. Although the plot, partly cannibalised from an early unfinished novel, 'The Wealth of Mr Waddy', requires considerable suspension of disbelief, *Kipps* provides an anatomy of the pre-1914 class structure which has never been surpassed.

Edith WHARTON   1862–1937

The House of Mirth

The House of Mirth takes its title from Ecclesiastes 7:4 – 'The heart of the wise is in the house of mourning; but the heart of fools is in the house of mirth.' It was Wharton's first big success, selling 100,000 copies within ten days of publication, owing largely to its 'scandalous' portrayal of society. Sex permeates the novel, and the story turns on the tolerance society shows towards the misbehaviour of those with wealth and power, and its intolerance towards the indiscretions of those, like Lily Bart, who have neither.

The novel opens with Lily, characteristically breaking the rules, being spotted leaving her friend Lawrence Selden's bachelor apartment. 'Why must a girl pay so dearly for her least escape from routine?' she asks herself. It is the question on which everything depends. Selden has everything she is looking for but money, and Lily must marry money to escape a dreary, precarious existence, dependent on rich friends and her mean old aunt, Julia Peniston. Being clever as well as beautiful, Lily is good at making opportunities, but always manages to spoil them either because she is too honest or because her sensibilities get in the way. First, at the Trenors' house-party she might have captured the milksop Percy Gryce, who has nothing to recommend him but his wealth. Having let him escape, she allows her hostess's lecherous husband, Gus, to 'invest' for her, and is shattered when he tries to claim his 'reward'. After this humiliating experience, she cries to her friend Gerty Farish, 'I am bad – a bad girl – all my thoughts are bad. I have always bad people about me.' Her anguish subsides but the shame of having taken money from a man stays with her literally to her dying day.

The Wetheralls always went to church. They belonged to the vast group of human automata who go through life without neglecting to perform a single one of the gestures executed by the surrounding puppets. It is true that the Bellomont puppets did not go to church; but others equally important did – and Mr. and Mrs. Wetherall's circle was so large that God was included in their visiting-list. They appeared, therefore, punctual and resigned, with the air of people bound for a dull "At home,"

THE HOUSE OF MIRTH
BY EDITH WHARTON

She might after this have married Simon Rosedale, the coming man, at the moment he needed her to advance his social ambitions, but she goes off instead with the spoilt, treacherous Bertha Dorset on a Mediterranean cruise. This is the decision that leads headlong to ruin. Bertha only wants Lily to occupy her husband, George, while she carries on an affair of her own. When Bertha's liaison is discovered, she ditches Lily. Worse still, rumours of scandal have already reached Lily's aunt, so that when she dies suddenly the handsome provision Lily might have expected does not materialise. Ostracised, with expectations of only $10,000 under her aunt's will – just enough to discharge her 'intolerable' obligation to Trenor – Lily dies at the moment when Selden is ready to come to her. Whether she takes the overdose deliberately is immaterial. Selden sees at last 'that all the conditions of life had conspired to keep them apart'. Lily Bart had too much heart, or too many scruples, and society has killed her.

The House of Mirth may not be quite a rounded work of art. ***The Custom of the Country*** (1913) is a more devastating satire on society as a whole, ***The Age of Innocence*** (1920) a more thorough exploration of the attitudes of old New York, but Wharton was never again to create a character of the magnitude of Lily Bart. She has the flaws and terrifying honesty of the tragic heroine. Her situation was not irredeemable; with one-tenth of the unscrupulousness of her enemy Bertha Dorset she might have won through in worldly terms, and indeed there are moments when she appears to be doing so. She is the victim of the missed opportunities that she herself has created within a corrupt society in which, with one side of herself, she is never quite at home.

1906

Alaska allowed to elect a delegate to US Congress • Alfred Dreyfus is rehabilitated • Trades Disputes Act legitimises peaceful picketing in Britain • Aga Khan founds All India Moslem League • US Pure Food and Drugs Act passed, following revelations in Upton Sinclair's *Jungle* of conditions in Chicago stockyards • HMS *Dreadnought* is launched • Georges Rouault, *At the Mirror* • Institute for Hampstead Garden Suburb, founded by Henrietta Barnett • A Mozart Festival is held in Salzburg • 'Everyman's Library' begun • Arthur Wing Pinero, *His House in Order* • Dimitry Shostakovich and Samuel Beckett born • Pierre Curie, Henrik Ibsen and Paul Cézanne die

---

William de MORGAN　　　1839–1917

Joseph Vance

William de Morgan was the leading designer of pottery in England in the second half of the nineteenth century, working with William Morris and others, and revolutionising the design of tiles and other decorative ceramics. After he retired in his mid-sixties, he embarked upon a new career as a novelist, 'this scribbling that keeps me quiet and prevents my being sulky', as he put it. His energy was undimmed, and between 1906 and 1914 he produced seven novels, few of which are now read.

Joseph Vance is a genial, somewhat complacent, Edwardian adventure-story, about a man who rises from the gutter to respectability after a life spend in virtuous struggle. Joe's father is a drunken builder, a shrewd businessman, but excluded from the future social successes of his son by his lack of education. Mr Vance wins a building contract from the altruistic Dr Thorpe, a scientist reminiscent of de Morgan's own father (who was a mathematician). Thorpe and his daughter Lossie take a liking to Joe, and set about educating him.

They send him to Oxford, where he excels in mechanics, but meanwhile Lossie breaks Joe's heart by marrying an Anglo-Indian general. Joe consoles himself with heroic feats of engineering, setting up a partnership with an old friend and designing a new brand of repeating rifle. His interest in women revives, and he proposes to the plain but intelligent Janey. She accepts, but trouble looms when she detects a lurking ulterior motive: Joe hopes that her influence will keep his widowed father away from the bottle. She breaks the engagement, but, fortunately, they meet again three months later, and this time get married.

Bliss ensues, briefly, until Janey is drowned at sea. Heartbroken again, Joe travels to Italy with Lossie's younger brother, Beppino, a dissolute aesthete. When they return, Beppino marries an heiress, but is quickly revealed to Joe as an evil monster: he has secretly married and impregnated another woman on the Italian tour, and is thus a bigamist. Poetic justice intervenes when Beppino dies on his honeymoon. The unfortunate Italian also dies, and Joe adopts her newborn baby, Cristoforo. He moves to Brazil, at a safe distance from the Thorpes and the widowed heiress, who are ignorant of the dreadful facts.

Lossie is widowed as well, and moves back from India to Italy, momentarily offering a new prospect of happiness to Joe. But fate is still unkind: Lossie hears rumours in Florence that Cristoforo is Joe's own illegitimate child, and she writes to him begging for a denial. Joe can now neither admit the truth nor hide it any longer without hurting the woman he loves, and ceases all further communications.

Years later, however, fate finally relents, since Lossie learns the truth from an old letter of Beppino's. She tracks Joe down in London, and they are united at last. The working-class lad has finally become worthy of society, after a lifetime's solid virtue and industry.

The novel is subtitled 'An Ill-Written Autobiography', and is tricked out with two appendices. A 'Note by the Editor' explains how 'the bulky MS. of which the foregoing forms part came into the possession of Mr F—— of Kensington', and how the publisher attempted to trace the people mentioned in it: 'The narrative is published now in the belief, on our part, that if it is, after all, a genuine one, the alteration of names is such that identification is impossible, and will remain so.' A 'Postscript by the Publisher', explains that 'a most embarrassing letter has come into the Editor's possession', establishing the true identity of Lossie. The publisher claims to have 'communicated with the writer and undertaken to suppress the work if she for her part will undertake to cover expenses up to date'. Clearly, no reply was forthcoming, and the letter – in which Lossie tells Joe that she regrets 'all these dreary years of darkness and misunderstanding' – is printed below. All this is of a part with the jocose chapter headings, of which the following is a brief, but otherwise characteristic, example: 'This chapter is really devoted to Dr Thorpe's opinions, although it pretends not at the beginning. Better skip them. A quotation from Tennyson. Janey and Joe make each a promise to the other.'

John GALSWORTHY   1867–1933

A Man of Property

see **The Forsyte Chronicles** (1929)

E. NESBIT   1858–1924

The Railway Children

Nesbit's best-loved novel for children originally appeared in serial form in the *London Magazine* in 1904, and is episodic in structure, charting the adventures of three children who move from the suburbs to a small country cottage when their father is obliged to 'go away'. It was dedicated to Nesbit's son,

Paul, 'behind whose knowledge of railways my ignorance confidently shelters'.

Never before had any one of them been at a station, except for the purpose of catching trains – or perhaps waiting for them – and always with grown-ups in attendance, grown-ups who were not themselves interested in stations, except as places from which they wished to get away.

Never before had they passed close enough to a signal-box to be able to notice the wires, and to hear the mysterious 'ping, ping,' followed by the strong, firm clicking of machinery.

The very sleepers on which the rails lay were a delightful path to travel by – just far enough apart to serve as the stepping-stones in a game of foaming torrents hastily organized by Bobbie.

Then to arrive at the station, not through the booking office, but in a freebooting sort of way by the sloping end of the platform. This in itself was joy.

THE RAILWAY CHILDREN
BY E. NESBIT

At Three Chimneys Cottage, the family live in reduced circumstances, largely supported (as Nesbit's own family was) by the income Mother gets for writing stories for magazines. The children, Roberta ('Bobbie'), Peter and Phyllis, spend much of their time watching trains and waving to them, hoping that the London-bound expresses will somehow take their love to their father, whose absence remains a mystery. One old gentleman on the train waves back, and the children, worried that their mother will be unable to pay medical bills, send a message to ask for the gentleman's assistance. He sends a large hamper, but when Mother discovers this, she is very angry. Charitable intentions are once again scrutinised when the children collect gifts from the villagers to mark the birthday of Mr Perks, the railway porter. At first Perks is angry, but he is eventually persuaded that the presents are a result of 'loving-kindness' rather than patronage.

The children avert a disaster by signalling a train to stop before it runs into a landslide, and are given gold watches as a reward by the old gentleman himself, who turns out to be a director of the railway company. Other adventures concern a Russian dissident writer in search of his family, a cantankerous bargee whose baby the children rescue from a fire, and a participant in a hare-and-hounds chase who breaks his leg in the railway tunnel. This boy, Jim, turns out to be the grandson of the old gentleman. Meanwhile, Bobbie has discovered from an old sheet of newspaper that Father is in gaol, convicted of treason. She enlists the help of the old gentleman to prove her parent innocent. One day, while watching the train, the children notice that all the passengers are waving to them, and the old gentleman is gesticulating with a newspaper. Bobbie alone realises that something significant has happened, goes to the station, and finds her father, whose conviction has been quashed.

Some (adult) readers have found the novel sentimental, and there is a certain wistfulness in this depiction of a perfect family and a cruelly wronged father. (Nesbit herself was married to an incorrigible philanderer and was obliged to rear two of his illegitimate offspring as well as her own children.) None the less, the book is both witty and self-aware and has proved perennially popular.

Upton SINCLAIR   **1878–1968**

The Jungle

The Jungle is a product of the school of socialist literature that emerged in the USA at the turn of the century, and which was characterised by a combination of nineteenth-century protest literature and twentieth-century revolutionary journalism. Of the novels of this school, *The Jungle* was the most effective, and its publication in 1906 achieved an effect comparable only to Harriet Beecher Stowe's *Uncle Tom's Cabin* (1852), which Sinclair had taken as a model. The book is a devastating and inflammatory critique of the American capitalist dream, and its appearance at a time of great social unrest in the USA caused Sinclair to be

stigmatised as a 'professional muckraker'. Nevertheless, his descriptions of the slaughterhouses of Chicago resulted in an immediate change in the food laws.

The Jungle tells the story of an archetypal immigrant family from Eastern Europe who come to live and work in the stockyards at Chicago. Their faith in the New World is quickly undermined as they are cheated and exploited at every turn by a system from which there is no respite and no escape: 'There would be no one to hear them if they cried out; there would be no help, no mercy. And so on until morning – when they would go out to another day of toil, a little weaker, a little nearer to the time when it would be their time to be shaken from the tree.' One character, who has become a prostitute to survive, remarks: 'We were too ignorant – that was the trouble. We didn't stand any chance. If I'd known what I know now we'd have won out.' While Sinclair accepts the truth of this admission on an individual and local level, he also sees it as the final victory of the system over the individual, and the whole drift of the novel is to prove that the only real answer is international socialism.

The Jungle is a researched book and as such lacks the painful intimacy of such novels as Robert Tressell's ***The Ragged Trousered Philanthropists*** (1914), which arose from the author's own experiences rather than from his observations. What *The Jungle* lacks in authenticity, however, it makes up for in passion. Indeed, Sinclair can become too rhetorical, and his feelings for the cornered immigrant family occasionally overwhelm his style. But his perorations are always supported by hard journalistic evidence in his detailed descriptions of the workings of the slaughterhouses and of the conditions of the operatives. Although the novel later degenerates into socialist polemic and the narrative at times resorts too often to the clichés of pamphlet literature, Sinclair's grasp of his subject is at once profound and moving. In his later novels Sinclair variously turned his attention to other abuses and corruptions in American society, but *The Jungle* remains his most focused and persuasive work.

1907

Triple alliance between Germany, Austria and Italy is renewed for six years ● Oklahoma is admitted as a US state ● Lenin leaves Russia ● Bakelite invented ● Boy Scouts founded ● US restricts immigration ● Henry Adams, *The Education of Henry Adams: a study of XXth Century Multiplicity* (privately printed) ● Exhibition of Cubist paintings, Paris ● Edward Elgar, March *'Pomp and Circumstance'* no. 4 in G (op. 39) ● J.M. Synge, *The Playboy of the Western World* ● W.H. Auden, Louis Macneice and Alberto Moravia born ● Edvard Grieg dies ● Hilaire Belloc, *Cautionary Tales* ● James Joyce, *Chamber Music*

Elizabeth von ARNIM   1866–1941

Fräulein Schmidt and Mr Anstruther

Fräulein Schmidt and Mr Anstruther ('being the letters of an Independent Woman by the Author of "Elizabeth and her German Garden" ', 1898) was von Arnim's seventh book, but it was the first to take her beyond the confines of the manufactured personality of her most famous work. Perhaps realising that the ersatz pantheism of her earlier works might become a mannerism, in March 1905 von Arnim literally closed the garden gate behind herself and became Miss Armstrong, a governess who was hoping to improve her German by spending her vacation as a servant to a German family. She experienced enough of German provincial life to lend verisimilitude to her novel; but more importantly she came to understand how the pressures of such a life impinged especially on the women of the community.

Von Arnim escaped her employment when the son of the house fell in love with her; there was no such escape for Rose-Marie, the Frl. Schmidt of the novel's title, who is trapped in Jena, a small provincial town. Her father writes unpublishable tomes on Goethe, and takes in English students such as Mr Anstruther to make money. Her stepmother relentlessly cooks and cleans, an example for Rose-Marie of what the future holds. That is until Mr Anstruther proposes and is secretly accepted an hour before his departure for England. But Mr Anstruther is not the same young man in England; and

Jena is not Clinches, the country house Mr Anstruther frequently visits, 'from which letters do not easily seem to depart'. Barely a month after his return, 'in the course of the longest letter you have written', he breaks the engagement. The remainder of the novel recounts Mr Anstruther's subsequent volte-face, his attempts to re-ingratiate himself into Rose-Marie's favours, and of Rose-Marie's struggle between the poles of emotional compromise and moral independence: 'Love is not a thing you can pick up and throw into the gutter and pick up again as the fancy takes you.'

Fräulein Schmidt and Mr Anstruther is von Arnim's most poised work. After the uneasiness of *The Benefactress* (1902), her first serious novel, she achieved here a wholly successful fusion of her favourite themes of pastoralism and feminism. The passion for the countryside displayed by Rose-Marie Schmidt is more persuasive from a girl in limited geographical and emotional circumstances than from the châtelaine of a *Schloss*; and the letter form, deliberately invoking the epistolary novels of the eighteenth century as well as the moral romances of Jane Austen, Maria Edgeworth and Fanny Burney (referred to in the novel by Rose-Marie), imposes a discipline absent from the meanderings of von Arnim's volumes of pseudo-autobiography. That one reviewer claimed the ending of *Fräulein Schmidt and Mr Anstruther* would send hordes of young men travelling as fast as they could towards Jena, is a testament not only to the wit and charm of Rose-Marie but also to those very qualities that distinguished the writing of her creator.

Joseph CONRAD   1857–1926

The Secret Agent

Conrad's 'Simple Tale' is a masterpiece of grim irony, concerned with anarchism in a densely evoked late nineteenth-century London. The germ of the story lay in a chance remark made to the novelist about a man who had been blown up by his own bomb in Greenwich Park: 'Oh, that fellow was half an idiot. His sister committed suicide afterwards.' Conrad's ironic detachment led to contemporary objections about the 'sordid surroundings and the moral squalor of the tale', objections he rebuffed in a note to the 1920 edition.

The agent of the title is Mr Verloc, over-weight, indolent, inefficient and in the pay of a foreign embassy. To the outside world Verloc is merely 'a seller of shady wares' from dingy premises in Brett St, Soho. However, his shop is also a meeting-place for the FP (Future of the Proletariat), a group of half-hearted anarchists, whose schemes and plottings Verloc reports to Chief Inspector Heat. Verloc is outraged when summoned to the embassy at an early hour to be sneered at and threatened by the suave and supercilious new First Secretary, Vladimir, who suggests that he does not earn his keep. In order to shock England out of its liberal complacence and bring it into line at an international conference for the 'suppression of political crime', Verloc must plant a bomb at the Greenwich Observatory. He obtains a bomb from the Professor, a sinister nihilist and explosives expert, and arranges for his trusting, half-witted brother-in-law, Stevie, to plant it. Stevie stumbles during his mission and is blown to fragments, one of which is the collar of his coat to which his protective sister Winnie has attached an address-tag in case the simpleton got lost.

The ambitious Chief Inspector Heat values Verloc as an informant and hopes to pin the crime upon Michaelis, a slug-like but harmless visionary anarchist, once imprisoned and now on ticket-of-leave. This brings him into conflict with the socially well-connected Assistant Commissioner, who knows Michaelis's 'influential and distinguished' patroness. Verloc realises that he will be arrested, but that his sentence will be comparatively light. His concern for the future outweighs the remorse he feels over Stevie's death, and, stricken by grief and outraged by her husband's callousness, Winnie murders him. In fear of the gallows, she rushes into the streets intending to drown herself, but meets the handsome and susceptible Comrade Ossipon, whose purely scientific interest in Stevie as a degenerate type she mistook for compassion. She persuades him to help her but, alarmed by her actions and her state, Ossipon fears for his safety. He accompanies her to the Channel train, but escapes before it moves off, taking Verloc's savings with him. He is last seen conversing with the Professor, haunted by the newspaper item recounting the mysterious suicide of a woman on a Channel steamer.

Conrad takes a melodramatic story but treats it with great subtlety, psychological insight and dry humour, gradually investigating the Verlocs' curious marriage as he pursues Winnie's story 'to its anarchistic end of utter desolation, madness, and despair'.

E.M. FORSTER   1897–1970

The Longest Journey

The Longest Journey is Forster's most auto-biographical novel. The story of Rickie Elliot, who struggles through life to maintain his human and artistic integrity, had more personal significance for Forster than any of his other novels, and he described it as the book 'I am most glad to have written. For in it I have managed to get nearer than elsewhere to what was on my mind.'

The novel is divided into three sections – Cambridge, Sawston and Wiltshire – which represent the three approaches to life offered to Rickie: of intellect, convention and instinct. Each section of the novel is also epitomised by a principal character – Stewart Ansell, Agnes Pembroke and Stephen Wonham – and the novel relates in personal terms the conflict between their worlds (in none of which Rickie feels fully comfortable) and his own preferred world of the imagination. Forster suggests that no one of these approaches is sufficient for life, but that all are necessary, although the attempt to achieve integration can, as in Rickie's case, lead to tragedy. Vulnerable from childhood as an orphan and a cripple, Rickie has a romantic nature which leads him to invest ordinary people with qualities they do not possess. He becomes finally a victim of his own faith in human nature.

The novel opens in Cambridge with a discussion of the nature of reality: Is the cow really there? Cambridge is portrayed as a place of ideal human fellowship, and specifically of male friendship, which by implication excludes heterosexual love. Into this charmed world breaks Agnes Pembroke, whom Rickie marries, despite Ansell's prognostications of doom. In the following section, Rickie's character deteriorates under the influence of Agnes and her brother, a housemaster at Sawston school where Rickie has become a teacher. The ideals of human fellowship and the pursuit of reality at Cambridge are replaced by the worthless ideals of Sawston and a life of compromise and defeat. Rickie is saved from all this when he learns that the coarse, virile young man, Stephen Wonham, who lives on his aunt's farm, is his illegitimate brother. Furthermore, Stephen is the son of Rickie's adored mother and not of his detested father. Shocked out of his complacency by this news, Rickie comes to accept Stephen as a symbol of his mother's love. This gives him the courage to leave his false life with Agnes and, in an attempt to recover all that he has lost, he goes to Wiltshire with Stephen.

In the last section of the novel Rickie dies saving Stephen from being run over by a train. Although disappointed by Stephen, who has broken his promise to remain sober, Rickie realises that he is the true 'inheritor' of England and that in him the continuance of life is assured. It is for this that Rickie sacrifices himself. Forster indicates that his pathetic gesture will not be in vain, and that for Stephen 'it had bequeathed salvation'.

Edmund GOSSE　　　　**1849–1928**

Father and Son

'The record of a struggle between two temperaments, two consciences and almost two epochs.' In *Father and Son* Gosse records his tormented mid-nineteenth-century childhood in a home dominated by the rigid and iconoclastic literalness of his parents' fanatical religion. In social exclusion, the boy's imagination runs on curious tracks as he begins to acquire 'a hard nut of individuality'. The mother's early illness clouds life further, and at her death he is possessed by unfocused, blind anger. The intense relationship between father and son is compacted with their removal to an insular Devonshire village and the society of the Plymouth Brethren, where Edmund lives and studies in the silent shadow of his father. The claustrophobia begins to lift as a cultivated Quaker stepmother modifies the father's strict educational plan, but fiction of any form remains banned. At his father's instigation Edmund is precociously elected to the saints of the sect, but the burden of guilt borne by the chosen few produces 'a faint physical nausea, a kind of secret headache'.

On the subject of all feasts of the Church he held views of an almost grotesque peculiarity. He looked upon each of them as nugatory and worthless, but the keeping of Christmas appeared to him by far the most hateful, and nothing less than an act of idolatry. 'The very word is Popish', he used to exclaim, 'Christ's Mass!' pursing up his lips with the gesture of one who tastes assafoetida by accident. Then he would adduce the antiquity of the so-called feast, adapted from horrible heathen rites, and itself a soiled relic of the abominable Yule-Tide. He would denounce the horrors of Christmas until it almost made me blush to look at a hollyberry.

FATHER AND SON
BY EDMUND GOSSE

The father's faith is challenged by the conflicting claims of Darwinism, and in assisting him in his work as a naturalist on the coast, Edmund finds an imaginative outlet in the worship of nature. Mirrored by the calm water, father and son peer into a submarine fairyland, but as they 'delicately lifted the weed-curtains of a windless pool ... all that panoply would melt away ... if we so much as dropped a pebble in to disturb the magic dream'. Echoes of *The Tempest* in the language here reflect Edmund's first fascination with literature. Looking back on his ado-

lescent self, he sees his distorted growth: 'Such a Caliban I had been.'

Out of memories of everyday trivia Gosse creates a parable of human individuality, in which the comedy was superficial and the tragedy essential: 'there gushes through my veins like a wine the determination to rebel'.

Dennis Potter's acclaimed television play, *Where Adam Stood* (1976) is based upon *Father and Son*, and in 1988 Peter Carey borrowed Gosse's childhood circumstances for the hero of his novel **Oscar and Lucinda**.

1908

South Africa constitutional convention agrees on a Union of South Africa ● *Daily Telegraph* publishes interview with Kaiser Wilhelm II in which he states the German people are hostile to Britain while he is a friend ● William Howard Taft defeats William Jennings Bryan ● Berlin Copyright Convention ● Jacob Epstein, 'Figures', for the British Medical Association, The Strand, London, causes indignation ● Peter Behrens, A.E.G. Turbine Factory, Berlin (first building of steel and glass) ● Béla Bartók, First String Quartet ● Jack Johnson becomes the first black world boxing champion ● Simone de Beauvoir, Ian Fleming and Lyndon Baines Johnson born ● Nicolai Rimsky-Korsakov and Joel Chandler Harris ('Uncle Remus') die ● Ezra Pound, *A Lume Spento*

| Arnold BENNETT | 1876–1931 |

The Old Wives' Tale

Bennett began writing *The Old Wives' Tale* on 8 October 1907 and finished it on 30 August 1908, a considerable feat considering its length and structural complexity. H.G. Wells wrote to him: 'it at least doubles your size in my estimation ... I am certain it will secure you the respect of all the distinguished critics'. Indeed, Frank Harris, Edward Garnett, Ford Madox Ford and Eden Phillpotts all praised it.

It is the story of two sisters, Constance and Sophia Baines, whose father keeps a draper's shop in St Luke's Square, Bursley, one of the 'Five Towns'. Since Baines is incapacitated by a stroke, the shop is actually run by Samuel Povey, 'a person universally esteemed ... absolutely faithful, absolutely efficient in his sphere'. Constance, the elder and more demure sister, marries Samual Povey, and on the death of her father inherits the shop. Their son Cyril, on whom Constance dotes, suffers from the removal of his father's firm hand when Samuel dies broken-hearted after his cousin Daniel is hanged for murder. Constance is left a widow – with a son who somewhat neglects his filial duties. She is desperately lonely and suffers cruelly from sciatica.

The restless and beautiful Sophia could never have been a stay-at-home. She runs off to Paris with the plausible commercial traveller Gerald Scales. He leaves her when the money runs out, but 'having no silly, delicate notions about stealing' from a man like Gerald she has previously provided herself with a small nest-egg. During the siege of Paris in 1870, she puts this to constructive use, enabling her, after the Commune, to buy Pension Frensham in the rue Lord Byron; she makes 'the name of Frensham worth something'. Pride has prevented her from telling her family of her whereabouts, but a native of the Five Towns turns up at the pension and betrays her. Worn out by relentless hard work, Sophia returns home and we have the spectacle of the two ageing sisters, with their ageing dogs, ending their lives together in their childhood home. The neighbourhood, and the house itself, is crumbling around them, but it is the timid, hypochondriacal Constance who has finally the stronger will, the greater endurance: 'Sophia had sinned. It was therefore inevitable that she should suffer.' She dies first, leaving Constance, who 'never pitied herself', to struggle on a while longer with all that she had cherished in ruins.

There is much incident – including the description of a public execution at Auxerre, though Bennett later admitted to Frank Harris that he had never seen one – and a large cast of supporting characters. Embracing the developing Industrial Revolution at home and political upheaval in Europe, the scope of *The Old Wives' Tale* is considerable. It placed Bennett in the first rank of his contemporaries.

G.K. CHESTERTON   **1874–1936**

The Man Who Was Thursday

Though it is Chesterton's poems and the Father Brown stories which retain their popularity, *The Man Who Was Thursday* is widely regarded as his best novel. This phantasmagoric farce is subtitled 'A Nightmare', and it was C.S. Lewis who first pointed out its Kafkaesque quality. Lewis thought, however, that it was more balanced than Kafka's work, because it admitted 'the exhilaration as well as the terror' of man's '(apparently) single-handed struggle with the universe'.

Gabriel Syme, poet and Scotland Yard detective, is led on by the anarchist poet Lucian Gregory into infiltrating the Central Anarchist Council, which is plotting to destroy the world. The seven members of the Council are named after the days of the week, and Syme thwarts Gregory's ambition to become Thursday by being elected in his place. He descends into increasingly convoluted hells of will and intellect as he discovers in turn that every other ordinary member of the Council is also a detective sent to unmask the plot under the orders of the Police Chief. Sunday, the President of the Council who has so far, eluded the joint efforts of the other councillors to uncover his identity, now reveals himself as the Police Chief who had given them their orders in the first place. Making his escape, Sunday uses, in the non-sequential manner of dreams, first a cab, then a fire-engine, then an elephant, and finally a balloon. The chase ends up at Sunday's estate, where Gregory, the only genuine anarchist, re-appears. He asserts that he would destroy the world if he could, charges the seven with having found happiness, and accuses them of never having suffered. However, the great face of the Police Chief now grows to an enormous size, filling the whole sky before everything goes black. 'Only in the blackness before it entirely destroyed his brain he seemed to hear a distant voice saying a commonplace text that he had heard somewhere, "Can ye drink of the cup that I drink of?" '

Though it has Christian overtones, the story is not a Christian allegory of any sort: no possible system of Christian beliefs can be uncovered under it. In later years, Chesterton found himself having to explain that 'the old ogre [Sunday] who appears brutal but is also cryptically benevolent is not ... God, in the sense of religion or irreligion, but rather Nature'. He explained that, at the time he wrote the story, he had just begun to change from a fatalistic-pessimistic pantheist into an optimistic pantheist. It was several years before he became a Christian, and he did not become a Roman Catholic until 1922.

E.M. FORSTER   **1897–1970**

A Room with a View

Forster's most popular novel was inspired by a long visit to Italy the author made with his mother beginning in October 1901. He began working on it before he had the idea for **Where Angels Fear to Tread** (1905), and during its long gestation referred to it as 'the "Lucy" novel'. (In 1904, in addition to these two novels, he was also working on an edition of the *Aeneid* and sketching out the plot of **The Longest Journey** [1907].) The book is also Forster's happiest novel. A social comedy in the spirit of Jane Austen and Meredith, it is none the less as committed as Forster's other novels to the idea that life is a battleground of conflicting impulses between freedom and truth, and convention and dishonesty.

The conflict in *A Room with a View* is summed up by its title. From the opening scene in the Pension Bertolini in Florence, where the unconventional Mr Emerson and his son, George, insist upon Lucy Honeychurch and her chaperone, Charlotte Bartlett, taking their rooms with a view, the title takes on a symbolic significance. Rooms stand for restrictive social conventions; views for naturalness, freedom and human potential. The confrontation between these two sets of values is variously repeated throughout the novel.

Like *Where Angels Fear to Tread*, *A Room with a View* is about the education of its

protagonist, and it too uses Italy as the place where this education can be begun. The contrast between Sawston and Monteriano in the earlier novel, however, is rather more simple than the situation described in *A Room with a View*. Lucy comes to Italy not from the restrictions of Sawston, but from the promisingly named Windy Corner, which has a view of its own over the Weald of Sussex; and Lucy has already shown her potential for greatness to one sympathetic character, the clergyman Mr Beebe, in her playing of Beethoven: 'If Miss Honeychurch ever takes to live as she plays, it will be very exciting.'

Lucy's subconscious encouragement of George Emerson comes to a climax when, at Fiesole, on a violet-covered hillside with an extensive view of Florence and the Arno Valley, George impulsively kisses her. But the spell is broken before Lucy has a chance to respond, when Charlotte appears on the scene, 'brown against the view'. Charlotte's sense of propriety and Lucy's own unwillingness to face her true feelings create 'a shame-faced world of precautions and barriers', and the two women escape to Rome.

Of course, it must be a wonderful building. But how like a barn! And how very cold! Of course, it contained frescoes by Giotto, in the presence of whose tactile values she was capable of feeling what was proper. But who was to tell her which they were? She walked about disdainfully, unwilling to be enthusiastic over monuments of uncertain authorship or date.

A ROOM WITH A VIEW
BY E.M. FORSTER

After her return to England and Windy Corner, Lucy begins to understand what Italy and George Emerson had offered her. Her path to love, freedom and truth, however, is made more difficult since she must first perceive the imperfections of Cecil Vyse, the supercilious aesthete to whom, in her retreat from George, she has become engaged. Caught in a web of petty lies, social embarrassment and sexual inhibition, Lucy refuses to acknowledge her real feelings, and it is only when the uncompromising honesty of Mr Emerson finally penetrates her conventional façade that she acquires the necessary 'courage and love' to save her from the fate of such as Charlotte Bartlett. The novel ends with Lucy and George married and back in Florence, looking out of their own room with a view and listening to 'the river, bearing down the snows of winter into the Mediterranean'.

The novel has been adapted for television and was made into a very successful film (1985) directed by James Ivory from a screenplay by Ruth Prawer Jhabvala.

Kenneth GRAHAME   1859–1932

The Wind in the Willows

The English countryside of the Edwardian era, with the river never very far away, is the idyllic setting for the adventures of the delinquent yet lovable Mr Toad, and his despairing companions, Badger, Mole and Rat. Unlike his friends, Toad moves easily between the animal world and that of the humans, and Grahame's skill is to persuade us without question of the co-existence of these two worlds.

Rich landowner Mr Toad becomes obsessed with driving motor-cars – the latest in a series of fads. After several smashes, his friends attempt to confine him to his bedroom in Toad Hall, his country seat, until 'the poison has worked itself out of his system'. Feigning illness, Toad escapes, and before long 'the old passion' leads him to steal a motor-car. He is arrested and sentenced to twenty years' imprisonment. While he sulks in his dungeon, Rat, Mole and Badger continue to have their own quiet adventures, centring on the peaceful domesticity of their homes.

It is significant that, with the exception of Toad, the animals live in underground seclusion, away from outside threats – 'There's no security, or peace and tranquillity, except underground.' Written during Grahame's unhappy marriage, *The Wind in the Willows* contains no female animals; the character of 'Mrs Mole' was deleted, and in his minutely

detailed account of Mole End he presents a bachelor's paradise that he perhaps longed to inhabit. The language powerfully evokes, in a peculiarly English manner, Grahame's deep love and knowledge of nature. His perfect world is confirmed by a vision of Pan, witnessed by Mole and Rat, to remind the reader of nature's inviolable and sacred aspects.

Meanwhile, in the profane world of humans, the artful Toad escapes from prison disguised as a washerwoman, and learns that Toad Hall has been occupied in his absence by the villains of the Wild Wood, the stoats and weasels. Grahame's dislike of socialism is evident when he allows the deposed Toad to be restored to his home after a furious battle; the vanquished upstarts are seen gratefully retreating after performing various servile tasks at Mole's command. The boastful Mr Toad is tempted to display his trait of *folie de grandeur* to excess during a victory banquet, but for once accepts the advice of his friends and, for that evening at least, he is a reformed character.

'Shove that under your feet,' he observed to the Mole, as he passed it down into the boat. Then he untied the painter and took the sculls again.

'What's inside it?' asked the Mole, wriggling with curiosity.

'There's cold chicken inside it,' replied the Rat briefly; 'coldtonguecoldhamcold beefpickledgherkinssaladfrenchrollscress sandwidgespottedmeatgingerbeerlemon adesodawater – '

'O stop, stop,' cried the Mole in ecstacies: 'This is too much!'

'Do you really think so?' inquired the Rat seriously. 'It's only what I always take on these little excursions; and the other animals are always telling me that I'm a mean beast and cut it *very* fine!'

THE WIND IN THE WILLOWS
BY KENNETH GRAHAME

The perceived flaws in *The Wind in the Willows* – its snobbery and chauvinism – hardly matter, for Grahame's genius lies in his presentation of a fantasy unmarred by whimsy. Equally enjoyed by adults – perhaps the test of an enduring children's classic – the book attracted several illustrators, including Arthur Rackham; but it is the drawings of E.H. Shepard that interpret most successfully the energy and spirit of the writing.

In 1929 A.A. Milne adapted the book for the stage as *Toad of Toad Hall*. This musical version of the story, with music by H. Fraser Simson, simplified the story by removing the mystical elements and concentrating upon Toad's adventures. It proved hugely popular and, like Barrie's *Peter Pan* (1904), became an immovable feature of the British theatre's Christmas season. In 1990 Alan Bennett, who had already recorded highly praised readings of the novel, wrote a very successful new adaptation for the National Theatre. His version retained Grahame's title and was closer in spirit to the original story than Milne, although it incorporated sly references to late twentieth-century politics (notably green issues: the weasels – one named after a former Chairman of the Conservative Party – depicted as property speculators; a debate between the merits of the railway and the car, with a car salesman named after particularly inept and unpopular Minister of Transport), and offended some purists with the faintly homoerotic colouring to the relationships between Mole, Rat and Badger.

Ada LEVERSON	1862–1933

Love's Shadow

see **The Little Ottleys** (1916)

Edith SOMERVILLE	1858–1949
and Martin ROSS	1862–1915

Further Experiences of an Irish R.M.

see **The Irish R.M. Trilogy** (1915)

1909

British alarm at growth of German navy ● Japanese dictatorship in Korea ● Women are admitted to German universities ● Girl Guides founded in Britain ● Sigmund Freud lectures in US on psychoanalysis ● Henri Matisse, *The Dance* ● Frank Lloyd Wright, Robie House, Chicago ● Sergei Diaghilev produces his Russian ballet in Paris including Michel Fokine's *Les Sylphides* (to Chopin's music) ● D.W. Griffiths transforms child actress Gladys Smith into Mary Pickford ● Barry Goldwater, Dean Rusk, Stephen Spender and Kwame Nkrumah born ● Algernon Charles Swinburne and George Meredith die ● William Carlos Williams, *Poems*

Gertrude STEIN	1902–1983

Three Lives

Stein's first book, written in Paris, was published at her own expense by the Grafton Press, a publishing house which found the manuscript rather disconcerting. The Press sent an American to see her in order to discuss 'corrections'. 'They think you are uneducated,' he explained, 'and that it is necessary to go over it with you.' Stein replied: 'No, you are mistaken about that, I am not uneducated. I have had more education and experience than they or you', and the book was published as written. Alice B. Toklas relates that H.G. Wells, to whom Stein had sent a copy of the book, 'thanked her, much later, with a short but appreciative letter which touched her'.

Stein's trilogy of narratives takes us into the minds of three women living in America, conveying their unhappy lives through the speech patterns of their idiolects. The first is 'The Good Anna', a German maid. Anna is unerringly loyal to her bourgeois mistress, Miss Mathilda, within whose household she constructs her own little realm of stray dogs and scolded underlings. As the years pass and she grows old, Miss Mathilda capriciously moves away, leaving Anna's life empty of its purpose.

The second story is that of Melanctha Herbert, a black girl whose sexual adventures have earned her a bad reputation. When Melanctha meets Jeff Herbert, a philanthropic doctor, she is unsure what sort of relationship she wants; the romance builds through their tortuous uncertainties, finally floundering as Jeff learns of her other affairs.

In the final narrative, 'The Gentle Lena', a young German girl is taken from her conventional home by a well-to-do aunt. In her new life in America, Lena complies with all demands through her gentle and passive obedience. Aunt Mathilda arranges for her to marry the son of a good German family, but Herman is reluctant. When the wedding at last takes place Lena moves into the house of her in-laws, taking on their slovenly habits and losing all interest in her own life. Three children are born, but with the fourth confinement Lena grows more and more unwell, picked on by her domineering mother-in-law. 'When it was all over Lena had died, too, and nobody knew just how it had happened to her.'

H.G. WELLS   1866–1946

Tono-Bungay

By 1909, Wells had already achieved considerable success with a series of science-based books. While *Tono-Bungay* contains a scientific element, the central theme of the novel is that people are 'Very much alike at bottom and curiously different in their surfaces.' He conceived it as a book 'on Dickens–Thackeray lines'.

Young George Ponderevo lives with his mother, a domestic in a vast country house called Bladesover, which gives him the opportunity to learn about life 'upstairs' as well as 'downstairs'; this drives his social aspirations. He quickly falls in love with Beatrice, a young girl who lives 'upstairs', but when George hits Beatrice's half-brother and then refuses to apologise, he is banished from Bladesover. At first he is sent to his Uncle Nicodemus Frapp, but he returns to Bladesover after a contretemps over his religious ideas, which are condemned as blasphemous. He is then despatched to his Uncle Edward Ponderevo, a chemist. Shortly afterwards, his mother dies, and only then does George realise that he loved her.

Uncle Edward, whom Wells portrays as a representative of the new entrepreneurial class, is dissatisfied with his life, and desperate to find a way of improving his professional and social standing. Meanwhile, George enjoys his new life, taking up Latin en route to becoming a chemist himself. Suddenly, Uncle Edward announces his bankruptcy, having lost everything gambling on the Stock Exchange. Forced to sell his business, he moves to London hoping to make his fortune there. George is initially left behind with the new owner of the business, before moving to London to study.

It is in London that George falls in love with Marian, who agrees to marry him, on the condition that he has a certain annual income. This prompts George to accept the offer of a job from his Uncle Edward, who has now invented Tono-Bungay, which is marketed as a wonder tonic but is in fact nothing more than coloured water. However, Tono-Bungay takes off and Uncle Edward starts building his diversified business empire. George grows less interested in the business and begins experimenting with aeroplanes and airships, while his marriage ends in divorce as he and Marian realise they have nothing in common.

Uncle Edward keeps buying ever larger homes, until he moves to a mansion. George meets his uncle's new neighbour, Beatrice, to whom he proposes. She agrees, but adds that she must explain something, at 'a later stage'. Uncle Edward, who has expanded his business too quickly, goes bankrupt, and George goes to Africa, hoping to plunder the mysterious but valuable quarp substance. This, however, fails. With an added charge of forgery against him, Uncle Edward flees with George in one of George's airships to France, where he soon afterwards dies. On his return to England, Beatrice refuses to marry George; his fallen social position and her being a 'spoilt' woman prevent it.

1910

Theodore Roosevelt propounds his concept of 'The New Nationalism' ● Marie Curie, *Treatise on Radiography* ● Manhattan Bridge, New York, opened ● Arthur Evans excavates Knossos ● Bertrand Russell and Alfred North Whitehead, *Principia Mathematica* ● 'Futurist Manifesto' published ● Igor Stravinsky, *The Firebird* (ballet) ● Mark Twain (Samuel Langhorne Clemens), Florence Nightingale, William James, Leo Tolstoy and Mary Baker Eddy die ● John Masefield, *Ballads and Poems*

Arnold BENNETT   **1876–1931**

Clayhanger

see **The Clayhanger Novels** (1916)

E.M. FORSTER   **1897–1970**

Howards End

Like Forster's earlier works, *Howards End* is built on a contrast between two different ways of life. The contrast, however, is no longer, as in the Italian novels, simply between spontaneity and convention, but between the weightier issues of culture and materialism, both in the lives of individuals and in the life of society. Whereas the earlier novels concentrated on the 'inner life' of the individual, *Howards End* displays a greater awareness of the outside world and its social and political implications for the individual, and considers whether or not the life of imagination and emotion can exist in tandem with the life of business and action. The mood is still one of social comedy, but the comedy is darker and embraces a broader spectrum of English society, so that the novel becomes a 'condition of England' novel, as well as an Edwardian comedy of manners.

Howards End describes two contrasting middle-class families, the intellectual and artistic Schlegels and the commercial and conventional Wilcoxes, and how they inad-

vertently combine to bring about the ruin of a third, working-class family, the Basts. The Schlegel sisters, Margaret and Helen, are initially attracted to the Wilcox family because they admire in them the practical qualities they themselves do not possess. Margaret in particular realises 'that there is a great outer world that you and I have never touched – a life in which telegrams and anger count. Personal relations, that we think supreme are not supreme there.'

The sisters' attempts to connect the 'inner world' of personal relations and the 'outer world' of telegrams and anger, to build a 'rainbow bridge that should connect the prose in us with the passion', has increasingly disastrous results. Helen's sight of the frightened Paul Wilcox at breakfast after they have become engaged reveals the 'panic and emptiness' that lies behind the Wilcox façade of power. Margaret's further contact with Paul's father, Henry, forces her to modify her notion of him as the embodiment of reality. His confident hold 'on the ropes of life' collapses when the past, in the shape of Jacky Bast, his former mistress, catches up with him. At the climax of the novel Leonard Bast dies, crushed under a falling bookcase, destroyed by the very culture he so ardently pursued.

The economic and moral ruin of the Basts expresses the failure of the Wilcox and Schlegel worlds of capitalism and liberalism to connect in any positive fashion. But the novel ends on a more optimistic note. Like

Wiltshire in *The Longest Journey* (1907), the Wilcoxes' house, Howards End (based on Rooksnest, Forster's childhood home), is Forster's symbol of England. It possesses spiritual qualities that transcend the materialism of the Wilcoxes and the intellectualism of the Schlegels, and it becomes the means of finally bringing together their disparate worlds. At the insistence of a changed Henry Wilcox, Howards End is to be inherited by the illegitimate child of Helen Schlegel and Leonard Bast, the result of the 'connection' of middle-class culture and working-class aspiration, and a plausible symbol of an achieved social harmony between culture, commerce and the common man.

Henry Handel RICHARDSON 1870–1946

The Getting of Wisdom

Henry Handel Richardson was the male pseudonym used by the Australian Ethel Florence Lindsay Richardson. Although considered by its author as 'a little book', *The Getting of Wisdom* is seen by critics as her masterpiece. A semi-autobiographical novel, it is set at the turn of the century, and follows the passage through the select Melbourne Ladies College of Laura Tweedle Rambotham. Through its portrayal of the intense emotional life of schoolgirls, it charts the relentless destruction of the child's innocence in her struggle to achieve academic success and popularity.

Laura is the daughter of a widowed gentlewoman living in straitened circumstances in a rural Australian township. For Laura, a wilful, proud and rather awkward girl, the transition from her position as the eldest of four children in a close-knit family, to the bottom class of the institutionalised and formal Ladies College, is painful and fraught with social disasters. She must first cope with her unwelcoming reception by Mrs Gurley, the lady superintendent of the school, and the gibes of the other girls and the teachers, who are instantly aware of Laura's naïvety and ignorance of school etiquette. Laura becomes acutely self-conscious of her home-made dresses in inappropriate colours and of her impecunious background. She lives in dread of the others finding out that her mother does paid embroidery work, after she rashly confides this fact to a roommate. However, Laura quickly learns that 'the unpardonable sin is to vary from the common mould' and she embarks with vigour on a programme of conformity. She masters the art of learning by rote, which ensures her success in all subjects other than mathematics, and assiduously courts the friendship of her classmates. But her freshness and spontaneity are not so easily suppressed and her 'luckless knack at putting her foot in it' by unwittingly breaching unwritten rules means that she is never as popular as she would like.

While other girls begin to mature and become interested in boys, Laura remains undeveloped and considers boys 'silly'. When her friend Tilly arranges an outing with her cousin, Bob, who is reportedly 'gone' on Laura, the encounter is an awkward failure because of her extreme self-consciousness and inability to think of anything to say. She marvels at the ability of her

---

You might regulate your outward habit to the last button of what you were expected to wear; you might conceal the tiny flaws and shuffle over the big improprieties in your home life, which were likely to damage your value in the eyes of your companions; you might, in brief, march in the strictest order along the narrow road laid down for you by these young lawgivers, keeping perfect step and time with them: yet of what use were all your pains, if you could not marshal your thoughts and feelings – the very realest part of you – in rank and file as well?

THE GETTING OF WISDOM
BY HENRY HANDEL RICHARDSON

---

peers to flirt and appear womanly, and tactically decides to nurture a crush on the suitably inaccessible married school curate, who is a popular object of infatuation amongst the schoolgirls. When Laura is invited to stay the weekend with his family (as he is the nephew of an old friend of her mother's), her standing in school is considerably raised and the fiasco with Bob forgotten. Although the curate proves very ill-natured and Laura is soon cured of her imaginings, in order to court popularity and to fulfil the others' fantasies, she concocts stories of a romance between them which she develops in the manner of a soap opera. Her deception is discovered when another girl, Mary Pidwall, a rigid moralist, stays with the curate. Laura is ostracised by her classmates for her dishonesty. To make matters worse, Chinky, a girl devoted to Laura, is expelled for stealing in order to buy a ring for her, the suspicion being that Laura encouraged her to do so. Laura's misery is compounded when she realises on holiday with her brothers and sister that she no longer holds them in thrall and that the family is not the safe haven it once was.

With the transiency of adolescent phases and her own efforts at ingratiation, however, she is gradually accepted at school once more, even by Mary Pidwall, and is invited to join the Literary Society. She becomes devoted to an older girl, Evelyn, but obsession and jealousy undermine her work, and only after a last-minute effort and the aid of a crib text, does she pass her final exams. She leaves school uncertain of what life holds in store and painfully aware of not having quite fitted in; but once out of the school gate, she is exhilarated by the freedom of the world.

The novel was made into a successful film (1977), directed by Bruce Beresford from a screenplay by Eleanor Witcombe.

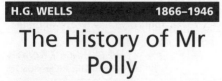

H.G. WELLS   **1866–1946**

The History of Mr Polly

Alfred Polly's quest for peace, associated as it is with the English countryside, exemplifies the Arcadian impulse which was so prevalent before the First World War. It opens with him cursing the town of Fishbourne ('*Beastly Silly Wheeze of a hole!*'), where he keeps a poky shop and is unhappily married, and ends with his sitting outside the Potwell Inn in the cheerful company of the 'fat woman', on a 'serenely luminous' evening, 'amply and atmospherically still, when the river bend was at its best'. Wells notes that: 'It was a plumper, browner, and healthier Mr Polly altogether than the miserable bankrupt with whose dyspeptic portrait our novel opened.' Mr Polly's aspirations are not, however, romanticised; indeed, they scarcely amount to aspirations, based as they largely are upon laziness, escapism, a vague wish for 'fun in companionship', and a desire to return to childhood, to a landscape 'as safe and enclosed and fearless as a child that has still to be born'. This is a very different journey from that taken by the eponymous hero of *Love and Mr Lewisham* (1900), whose youthful idealism yields to an adult sense of responsibility.

Mr Polly occasionally finds 'Joy de Vive' in his youth when, in his time off from the Port Burdock Bazaar, he roams the countryside with his friends Parsons and Platt, visiting inns and discussing literature. Another glimpse of Arcadia comes when, having inherited a little money from his father, he takes a holiday and meets an upper-class schoolgirl sitting on a wall, swinging her legs. Christabel is socially as well as physically out of reach and, after a ten-day romance, Mr Polly comes down to earth with a bump and marries his cousin, Miriam Larkin. For fifteen years he is a 'respectable shopkeeper' in Fishbourne, his digestion ruined by Miriam's cooking and his mind poisoned by her silent indignation. Spiritually and financially bankrupt, he determines to make 'an End to Things'. In a scene of high farce, he sets fire to his shop, but fails in his resolve to cut his throat. He is acclaimed as a hero when he rescues a deaf neighbour from the fire, which destroys not only his own shop but most of Fishbourne High St, but he has had a glimpse of freedom: 'When a man has once broken through the paper walls of everyday circumstance, those unsubstantial walls that hold so many of us securely prisoned from the cradle to the grave, he has made a discovery. If the world does not please you, *you can change it*.' Leaving Miriam the insurance money, he sets off into the countryside. At the Potwell Inn, he discovers contentment with the plump landlady, doing odd jobs, but there is a 'Drorback': the landlady's monstrous nephew Jim, a 'Reformatory Reformed Character' who takes exception to any man he finds around the place. After several skirmishes, Jim disappears, stealing Mr Polly's clothes. When, some time later, he is discovered drowned, he is mistaken for Mr Polly. Miriam is paid the life insurance and sets up tearooms.

'The bright side of Mr Wells's drama is the challenge to originality offered by the very stupidity of life,' wrote *The Times*'s critic; 'the dark side is the huge and pitiless waste which it involves.' Although Mr Polly finds peace in the end, the sense of 'pitiless waste' remains, a shadow cast across this genial, sunny and affectionate novel, which was filmed in 1949 by Anthony Pelissier from his own screenplay.

P.G. WODEHOUSE · 1881–1975

Psmith in the City

Psmith is the first of Wodehouse's great comic characters. He erupts into the middle of a school story, bringing with him a new sophistication. Halfway through *Mike* (1909, but originally serialised in *The Captain* in 1908), the eponymous hero is removed from Wrykyn, his beloved public school, and sent to Sedleigh, where it is hoped he will concentrate upon bookwork rather than cricket. Also newly arrived is another youth from Shropshire, an Old Etonian called Psmith. (The P is assumed and silent, and the subject of a nice typographical joke whereby only Wodehouse and Psmith himself acknowledge it in the text.) Languid, monocled, dandiacal, professedly socialist (addressing people as 'Comrade'), and exuding his old school's famously effortless superiority, Psmith is the antithesis of Mike Jackson, the charming, modest, chivalrous and loyal sporting hero of countless public-school stories of the period. As Evelyn Waugh noted in 1961, the entrance of Psmith marks the point in Woodhouse's fiction where 'the light is kindled which has

burned with growing brilliance for half a century'. (*Enter Psmith* was the title given to the second half of *Mike* when the original novel was reissued in two volumes in 1935.)

Psmith in the City opens with a cricket-match at which Mike is prevented from scoring a century by the inopportune appearance behind the bowler's arm of Mr Bickersdyke of the New Asiatic Bank. Bickersdyke has been summoned by Mr Smith in order to discuss openings for Psmith in 'the World of Commerce'. A similar fate awaits Mike, who is shortly informed that the Jackson fortunes have reached a low ebb, and that, instead of spending the next three years agreeably at King's College, Cambridge, he will have to earn a living in the City. He takes dismal lodging in Dulwich (so as to be near a sporting public school – Wodehouse's own), and starts work as a clerk at the New Asiatic (based upon the Hong Kong and Shanghai Bank, where Wodehouse had worked in a similar capacity). Psmith also joins the bank and instantly removes Mike to a flat in

It was one of Psmith's theories of Life, which he was accustomed to propound to Mike in the small hours of the morning with his feet on the mantelpiece, that the secret of success lay in taking advantage of one's occasional slices of luck, in seizing, as it were, the happy moment. When Mike, who had had the passage to write out ten times at Wrykyn on one occasion as an imposition, reminded him that Shakespeare had once said something about there being a tide in the affairs of men, which, taken at the flood, &c., Psmith had acknowledged with an easy grace that possibly Shakespeare *had* got on to it first, and that it was but one more proof of how often great minds thought alike.

PSMITH IN THE CITY
BY P.G. WODEHOUSE

Clement's Inn: 'I offer you the post of confidential secretary and adviser to me in exchange for a comfortable home.' Psmith is excited by the prospect of his new life and determined to ingratiate himself with his superiors, an ambition which wholly eludes him in the case of Mr Bickersdyke, who finds him insolent. Unfortunately for Bickersdyke, Psmith proves efficient at his not very onerous duties and gives no real cause for dismissal. Outside the office, Psmith makes the employer's life a misery, haunting him at their club and causing an affray at a political rally Bickersdyke is addressing in south London. When Bickersdyke sacks Mike – who in true public-school spirit has owned up to cashing a dud cheque, thus preventing the dismissal of his elderly, impoverished and widowed immediate superior – Psmith threatens to reveal his employer's past association with the Tulse Hill Parliament, a radical political group.

Mike finally escapes the City when offered a chance to play for his county at Lord's. Psmith slopes off work early in order to spectate, and manages to hand in his resignation before the flabbergasted Bickersdyke has the chance to sack him. 'I propose to enter the University of Cambridge, and there to study the intricacies of the Law, with a view to having a subsequent dash at becoming Lord Chancellor,' Psmith announces. He further persuades his father to pay for Mike to go as well so that the young cricketer can learn 'divers facts concerning spuds, turmuts, and the like' in preparation for a career as Mr Smith's estate manager.

In many ways, the novel is really a school story removed to another sphere: that of commerce. Its true glory, however, is Psmith himself. Wodehouse claimed that the character was 'the only thing in my literary career which was handed me on a plate with watercress round it'. He is based upon Rupert D'Oyly Carte (son of the founder of the Savoy Opera), who was at Winchester with one of Wodehouse's cousins. Young D'Oyly Carte 'was long, slender, always beautifully dressed and very dignified. His speech was what is known as orotund, and he wore a monocle. He habitually addressed his fellow Wykehamists as "Comrade." ' The ubiquitous lavender gloves may (like the name Rupert) have been borrowed for Psmith from this model, but the inventively elaborate mode of speech, which was to recur in various forms – notably in the fumblingly allusive Bertie Wooster and the gravely circumlocutory Jeeves – is Wodehouse's own brilliant creation. Psmith's subsequent career may be followed in *Psmith Journalist* (1915), where he goes to America, and *Leave it to Psmith* (1923), where he erupts into another Wodehouse world: Blandings Castle.

1911

Jean-Jaurès announces scheme for socialist organisation of France ● David Lloyd George introduces National Health Insurance bill in Britain ● US Supreme Court orders dissolution of Standard Oil Co. ● Wilhelm II speaks on Germany's 'place in the sun' which her navy will secure for her ● Suffragette riots in Whitehall, London ● Sun Yat-sen elected president of United Provinces of China ● Amundsen reaches South Pole ● Andrew Carnegie endows international peace foundation ● Georges Braque, *Man with a Guitar* ● Irving Berlin's *Alexander's Ragtime Band* ● Gustav Mahler and W.S. Gilbert die ● Edward Marsh (ed.), *Georgian Poetry*

J.M. BARRIE　　　　　　　**1860–1937**

Peter and Wendy

The character of Peter Pan first appeared in Barrie's novel *The Little White Bird* (1902), but the origin of 'the boy who wouldn't grow up' was buried deep in Barrie's childhood with the death of his thirteen-year-old brother: his mother found consolation in the idea that her dead son would remain a boy forever. Then in 1897 Barrie met and befriended two small boys, sons of Arthur and Sylvia Llewelyn Davies. He was eventually to adopt all five of the Davies brothers after their parents' early deaths, and George, Jack, Peter, Michael and Nico were to be a profound influence upon his life and work, in particular his masterpiece, *Peter Pan*, first performed at Christmas in 1904 and part of the theatrical calendar thereafter. In his 'Dedication' to the printed text, Barrie wrote that the play was 'streaky' with these boys, and every year he revised it for performance, incorporating remarks they had made and references to their lives and characters. The novel version, *Peter and Wendy*, allowed Barrie to fix the story and to comment upon the motivation and actions of his characters in his role as narrator, a creature at once whimsical and ruthless.

The plot concerns the three children of the Darling family, Wendy, John and Michael, who are taught to fly and taken to the Neverland by Peter Pan. It is here that Peter lives with 'the lost boys', a group of children who live in the home under the ground and conduct skirmishes with a motley crew of pirates, led by Captain Hook. Peter's jealous fairy, Tinker Bell, persuades the lost boys to shoot Wendy as she flies over the island. Fortunately, Wendy survives and becomes a surrogate mother to the boys. Peter nearly drowns after rescuing Tiger Lily, leader of the redskin Piccaninny tribe, from the pirates. He is marooned on a rock in the mermaids' lagoon, and faces his doom with the famous line, 'To die will be an awfully big adventure', but is saved by the Never Bird.

Hook's subsequent attempt to poison Peter is foiled by Tinker Bell, who drinks the potion instead. (In order to save Tinker Bell's life in the play, Peter makes his notorious appeal to the audience to clap if they believe in fairies. In the novel, he makes the same appeal to all dreaming children, and Barrie notes: 'Many clapped. Some didn't. A few little beasts hissed.') The children are kidnapped by the pirates, but saved from walking the plank by Peter who boards the ship, slaughters all but two of the pirates, and sends Hook into the jaws of a crocodile. The lost boys accompany the Darling children back to London, where they are all adopted by the happy parents. Only Peter escapes, recognising that adoption will lead to school, a career and adulthood. Mrs Darling agrees to allow her daughter to return to the Neverland once a year for spring cleaning, but Wendy grows up and loses the power to fly. Her daughter, Jane, takes her place, followed by *her* daughter, Margaret. 'When Margaret grows up she will have a daughter, who is to be Peter's mother in turn,' Barrie writes in the concluding lines of the novel; 'and thus it will go on, so long as children are gay and innocent and heartless'.

Although the story was a deeply personal one for Barrie, it also reflected a widespread Edwardian obsession with boyhood, as seen in the plethora of school stories, the works of Saki, and the revival of interest in Housman's 'lads who will die in their glory and never be old' in *A Shropshire Lad* (1896), a book which Barrie read every year. Dismissed by many as a piece of childish whimsy, and deprecated by others as the reflection of the author's 'warped' personality, the story makes uncomfortable reading, and has been issued in bowdlerised editions 'retold' for children, often accompanied by wholly inappropriate illustrations (most notoriously those of Mabel Lucie Attwell). Psychologically acute, and intimately concerned with the death of innocence, it is in fact a disturbing mirror held up to the Edwardian age, an age which worshipped youth but sent it to die in the trenches.

The book was subsequently reissued as *Peter Pan*.

Max BEERBOHM	1872–1956

Zuleika Dobson

Although purportedly set in the Edwardian era, Beerbohm's only novel owes much to the author's memories of Oxford in the 1890s, where he was an undergraudate at Merton College. Indeed, he began to draft this 'Oxford Love Story' shortly after he had gone down.

The eponymous heroine, the daughter of the Warden of Judas College, is a glamorous orphan who, after a brief and disagreeable spell as a governess, has pursued the career of a conjuror. The lack of originality and skill in her prestidigitation is more than compensated for by her great beauty, and when she arrives in Oxford for Eights Week, laden with the spoils of her career, she is 'the toast of two hemispheres'. She instantly catches the attention of two Judas undergraduates: dim and unprepossessing Mr Noaks and the 'orgulous and splendent' John Albert Edward Claude Orde Angus Tankerton Tanville-Tankerton, 14th Duke of Dorset.

Zuleika thrives upon being the cynosure of all eyes but can only love the man who does not cast himself down before her. The Duke, similarly afflicted, pretends to ignore her and thus, briefly, wins her heart. However, by apologising for his former boorish behaviour, he forfeits her admiration and consequently announces that he will kill himself for love. Upon sober reflection, the Duke becomes aware of Zuleika's all-consuming vanity and determines not to kill himself after all. Unfortunately a telegram arrives from the ducal home announcing that two black owls, harbingers of death, have been seen on the battlements, and thus the young man's fate is sealed. He bestows Zuleika's black and pink pearl earrings upon Katie Batch, his landlady's smitten daughter, and, magnificent in his Garter robes, makes his way to the river, where he plunges in and drowns – not before briefly resurfacing, putting the Magdalen crew off their stroke and thus posthumously ensuring his college's place as Head of the River. All the other undergraduates follow suit, except the cowardly Noaks, who is discovered by Katie behind the curtains. Zuleika conceives a passion for this one man who did not die for her, but recognising his cravenness rejects him, whereupon he hurls himself from an upper window. Zuleika threatens to enter a nunnery, but recants and is last seen preparing to depart for Cambridge.

Mainly architectural, the beauties of Oxford. True, the place is no longer one-sexed. There are the virgincules of Somerville and Lady Margaret Hall; but beauty and the lust for learning have yet to be allied. There are the innumerable wives and daughters around the Parks, running in and out of their little red-brick villas; but the indignant shade of celibacy seems to have called down on the dons a Nemesis which precludes them from either marrying beauty or begetting it.

ZULEIKA DOBSON
BY MAX BEERBOHM

In a note for the 1946 reissue of the book the elderly Beerbohm denied that it was 'intended as a satire on such things as the herd instinct, as feminine coquetry, as snobbishness, even as legerdemain'. It was, he claimed 'just a fantasy', and its dandified style, its air of elaborate, frothy artifice, and its apostrophes to classical gods and muses reinforce the author's definition.

Arnold BENNETT 1876–1931

The Card

Bennett grossly underestimated his achievement in *The Card*. He wrote in his journal: 'Stodgy, no real distinction ... but well invented and done to the knocker, technically, right through.' He need not have been surprised by the good reviews and its continued popularity as the excellent comic novel it truly is.

It is in effect a series of connected incidents in the rise of 'Denry' (short for Edward Henry) Machin, the aimiable 'card', who launches his career by adjusting his mark for geography in the scholarship exam, and then 'amending' the invitation list for the Municipal Ball to include himself; he even has the cheek to dance with the Countess of Chell, wife of a local magnate. Denry starts out as a rent collector, employing ingenious methods for making money for himself. He avoids marriage (twice) with Ruth Earp, 'that plain, but piquant' girl whose extravagance might so easily have proved his ruin after he had assisted her in 'a moonlight flit' in a pantechnicon. 'Three months of affianced bliss with Ruth Earp' in Llandudno would have exhausted his resources, but without her he recoups everything and more by buying 'the identical (guaranteed) lifeboat' that rescued the crew of the Norwegian ship *Hjalmar*, wrecked in a recent storm within sight of the town, and running trips out to her: 'Return Fare, with use of Cork Belt and Lifelines if desired, 2s.6d.' Back in Bursley, 'all the town knew that the unique Denry had thought of the idea of returning home to his mother with a hat-box crammed full of sovereigns'.

'The Five Towns Universal Thrift Club' is his next venture but, however wealthy, a card is no true card without the best patronage. Denry manoeuvres himself into Sneyd Hall, the Chells' seat, but his timing is awry: the Countess is just leaving and he finds himself locked in – an incident that gives rise to reports of an attempted burglary in the *Signal*. He is more successful at the ceremony at the Police Institute where from the platform he 'hears himself saying: "I have great pleasure in seconding the vote of thanks to the Countess of Chell" '; soon afterwards she agrees to be patroness of his 'Five Towns Universal Thrift Club'. In the 'Great Newspaper War', the upstart *Five Towns Daily* under his guidance deftly outwits the old-established *Staffordshire Signal*, and when he signs the best centre forward in England to revive the ailing fortunes of the town's football club, it becomes a foregone conclusion that he will be elected mayor.

Bennett drew on childhood experiences for some of the incidents in *The Card*: from his days as rent collector for his father, for instance, and the newspaper war, which derives from the battle between his father's paper and the *Sentinel*. He was accused of portraying the local Duchess of Sutherland as the Countess of Chell, while the identity of the card himself was claimed by an old schoolfriend H.K. Hales. Whatever his origins, the card is one of the most likeable rogues in twentieth-century fiction. His popularity led Bennett into the mistake of writing a sequel, *The Regent* (1913), which deals with Denry's adventures as an impresario in the London theatre.

Arnold BENNETT 1876–1931

Hilda Lessways

see **The Clayhanger Novels** (1916)

Frances Hodgson BURNETT 1849–1924

The Secret Garden

Like E. Nesbit and Richmal Crompton, Burnett had little success with her novels for adults, but her children's books have become classics. *Little Lord Fauntleroy* (1886), with its goodie-goodie, lace-and-velvet-clad hero has worn rather less well than *The Secret Garden*, which features two children who are initially unattractive, uncheerful and unpleasant. Burnett, who was herself a keen gardener, began writing the book in America, where she spent much of her adult life, but may have been motivated by nostalgic memories of gardens she had known in England.

She could see the tops of trees above the wall, and when she stood still she saw a bird with a bright red breast sitting on the topmost branch of one of them, and suddenly he burst into his winter song – almost as if he had caught sight of her and was calling to her.

She stopped and listened to him, and somehow his cheerful, friendly little whistle gave her a pleased feeling – even a disagreeable little girl may be lonely, and the big closed house and big bare moor and big bare gardens had made this one feel as if there was no one left in the world but herself.

THE SECRET GARDEN
BY FRANCES HODGSON BURNETT

'When Mary Lennox was sent to Misselthwaite Manor to live with her uncle everybody said she was the most disagreeable-looking child ever seen.' A sour and sickly little girl of ten, Mary is orphaned in an outbreak of cholera in India, where she was born. Neglected by her father, who held a post with the English government, and by her young mother, a beauty who cared only to go to parties, Mary is lonely, and tyrannical to the servants who brought her up. Misselthwaite Manor, 'the strangest house anybody ever lived in', standing on the edge of a Yorkshire moor, and with most of its rooms shut up, is as gloomy as its master, Archibald Craven. He has been a recluse living mostly abroad since the death of his young wife in a fall from a tree in the garden which she loved. The garden has been locked for ten years, and its key buried.

Mary is put in the charge of Mrs Medlock, the dour housekeeper, and Martha Sowerby, a little housemaid who is shocked at her helplessness and imperious ways. Despite herself, Mary is intrigued by Martha's tales of her mother, Susan Sowerby, a wise and kind woman, and her twelve children, especially Dickon, a twelve-year-old who has an affinity with plants and animals and who can charm wild creatures. On one of her lonely walks in the wintry gardens, Mary encounters Ben Weatherstaff, a crusty old gardener, and the robin who accompanies him while he works. The bird becomes Mary's first friend, and shows her the way into the locked garden, which she has longed to find. She lets Dickon into the secret and, as they tend the beautiful and neglected wilderness, Mary begins to bloom with the garden.

Mary is troubled by the noise of crying in one of the house's many long corridors but her questions are met with denials. One night, woken by the sound above the wuthering wind, she discovers her cousin Colin, an invalid who has been brought up to think he will not live, and who cannot bear to be looked at because he fears that he has a lump on his back. His father can hardly bear to see him because he reminds him too much of his dead wife. Colin, like Mary, is lonely, despotic and indulged by frightened servants. The two children become friends; only she can soothe the tantrums brought on by his fears, her hot temper is a match for his and it is she who screams at him that there is nothing wrong with his back and that all that ails him is 'hysterics and temper'. Mary and Dickon take Colin to the secret garden in his invalid chair, with elaborate subterfuge, and as its magic works on him he grows stronger with the progress of the spring, learning to walk and declaring that he will 'live for ever and ever'. Nobody must know of the change in him until his father's return, and the children have enormous fun deceiving the bemused servants, nurse and doctor. Ben Weatherstaff and Susan Sowerby, who provides food for their growing appetites, enter into the conspiracy.

Meanwhile, in Italy, Archibald Craven has a strange dream in which he hears his dead wife calling him back to the garden. A letter from Susan Sowerby consolidates his resolve to return to England. As he hurries to the secret garden, a boy bursts through the gate at full speed into his arms – his son Colin, 'as strong and steady as any boy in Yorkshire'. This story of two lonely children coming alive with a garden, redeemed by love, growing things and a sort of patheism, has lost none of its enchantment with the years, and is probably its author's masterpiece.

Mr Perrin and Mr Traill

Walpole's third novel, arguably his best, is an unusual contribution to the genre of the public-school novel in that it centres chiefly on masters rather than boys. Moffatt's is a minor and pretentious public school in Cornwall, staffed by men whose lives have been soured by teaching there. Vincent Perrin is a lonely, awkward bachelor who has been at Moffatt's for twenty years. Archie Traill, who is appointed in the autumn term, is a fresh young master, with conventional charm.

Mr Perrin hopes to improve his barren life during the term, and his plans centre on Isobel Desart, a young woman who stays at the school with her friend, the wife of one of the senior masters. Although Perrin hopes to marry Isobel, she naturally falls in love with Traill and becomes engaged to him. Tension has been growing between Perrin and Traill about this and other matters, and just before the engagement is announced Perrin physically attacks Traill in the Masters' Common Room; the incident that releases Perrin's hatred is the mistaken appropriation of his umbrella by Traill.

The school immediately splits into rival camps supporting one or other protagonist, and tension builds throughout the term. Perrin suffers something approaching a nervous breakdown, and believes that he is being shadowed by a double of himself and a ghostly Traill. In his confusion and unhappiness, he succumbs to his worst instincts and plots to kill Traill, although it seems clear that he is hardly capable of such dramatic action.

The book, which at this point seems about to peter out, in fact comes to a thrilling and moving conclusion. On the last morning of term Perrin shadows Traill to a clifftop by the sea, meaning to kill him. Traill attempts to reason with Perrin, the latter draws a knife, and Traill, startled, slips over the edge of the cliff. He lies injured and unconscious on the sand, threatened by an encroaching tide. Perrin is immediately filled with remorse, descends the cliff, and saves Traill's life at the expense of his own, thus in a moment of nobility redeeming his failure and pettiness.

Minor characters – masters and their wives, the malevolent headmaster, the boy Garden Minimus who switches his allegiance from Perrin to Traill – are brilliantly drawn, although Miss Desart herself is a cipher, and the romance with Traill is not fully 'done', as Walpole's mentor, Henry James, would have said.

Mr Perrin and Mr Traill is a powerful novel, written with controlled and often savage irony at the expense of the public-school system in which Walpole himself had briefly worked. It was the first of several indications that Walpole, who developed into a frankly popular and anodyne novelist, could have been something more.

The New Machiavelli

When *The New Machiavelli* was first published, its view on free love and the impermanence of marriage – a subject about which the author was eminently qualified to write – immediately outraged public opinion.

In the opening section of the novel young Dick listens to his father ask, 'What was I created for? Whatever you do boy . . . Make a good plan and stick to it. Find out what life is about . . . and set yourself to do whatever you ought to do.' And so, as his life unfolds, Dick follows a course of calculated ambition and self-centred instinct. This attitude is already apparent in his schooldays, when he says to his friends: 'Everything is for the likes of us. If we see fit, that is.'

After the death of his parents, Dick becomes closer to an uncle in Staffordshire, who is a prosperous, bigoted and insular manufacturer. They fall out when Dick decides to go to Cambridge University, rather than entering the family business as his uncle urges. At Cambridge Dick's political interests develop, and he embraces socialist principles. It is at this stage also that he starts to think seriously about what he is to do with his life. After a spell as a Cambridge don, Dick moves to London and begins to fall in with political and social sets. Through Altiora MacVitie, a political hostess, Dick meets Margaret Remington, also a Cambridge undergraduate, and after a short courtship they marry.

Realising his political ambitions, Dick enters Parliament as a Young Liberal. However, on arrival at the House his feeling is not one of triumph, but fear that his life may yet pass in vain; a meaningless existence. He soon begins to perceive his political associates as narrow and priggish, and feels that by working with them he is missing the very thing he is seeking. However, he reaches some resolution with the conviction that 'the real work of mankind is the enlargement of human experience, the release and intensification of human thought, the wider utilisation of experience and the invigoration of research'.

Disappointed by Liberalism, he resigns from the party and drifts towards Toryism, and to a scrutiny of the 'big people' – the wealthy and the influential. He begins writing for a Tory paper, *Blue Weekly*, and the success of his articles helps him to be elected as a Tory MP. Meanwhile, Dick and Margaret begin to drift apart and experience an estrangement within their marriage, partly because Margaret continues in her devotion to Liberalism.

Quite suddenly, Dick realises that he has fallen in love with Isabel Rivers, an assistant who worked for him when he was a campaigning Liberal. The impossibility of their relationship, with its political and domestic aspirations, dawns on them – as, shortly afterwards, does the public scandal of their affair. After initially agreeing to part and carry on with their own separate lives, they eventually run away together to the Continent, and leave everything behind them.

Wells drew upon his own sexual entanglements in depicting Dick's amatory adventures: Margaret is a portrait of his second wife, Jane, with whom he had eloped (while still married to his first wife) in 1894, while Isabel is Amber Pember Reeves, 'a girl of brilliant and precocious promise' twenty years Wells's junior, who bore his child in 1909. This last relationship, like that between Dick and Isabel, became a public scandal, largely through Amber's indiscretion, but although they escaped to Le Touquet, where they saw themselves 'wandering about the Continent, a pair of ambiguous outcasts', Amber soon returned to London, where her irate father married her off, while Wells took up with his wife and children once more.

1912

New Mexico and Arizona become US states ● Manchu dynasty abdicates in China and a provisional republic is established ● British House of Commons reject women's franchise bill ● Woodrow Wilson, Democrat, wins US presidential election, over Theodore Roosevelt, Progressive, W.H. Taft, Republican ● Arizona, Kansas and Wisconsin adopt women's suffrage ● SS *Titanic* lost on maiden voyage ● Marcel Duchamp, *Nude Descending A Staircase* ● Grand Central Railway Station, New York ● Frederick Delius, *On Hearing the First Cuckoo in Spring* ● Five million Americans visit cinemas daily ● London has 400 cinemas (90 in 1909) ● Charles Pathé produces first news film ● Tennessee Williams born ● August Strindberg, Octavia Hill and Jules Massenet die ● Harriet Monroe (ed.), *Poetry: A Magazine of Verse* (magazine) founded

E.C. BENTLEY　　　　1875–1956

Trent's Last Case

This influential novel – regarded by many as amongst the finest of all detective stories – opens with a taut, swift-moving first chapter sketching the history of Sigsbee Manderson, an American multimillionaire. The writing is charged with the author's passionate hatred of irresponsible capital, and of the arrogance and cruelty of great wealth. Manderson is a corrupt Colossus: 'Forcible, cold, and unerring, in all he did he ministered to the national lust for magnitude.' Wall Street trembles, the stock markets of the world are thrown into confusion and 'a waft of ruin, a plague of suicide' follows hard upon rumours and counter-rumours that 'the life had departed from one cold heart vowed to the service of greed'.

Manderson, we learn in the next chapter, is indeed dead, found murdered in the grounds of his English country house. Inspector Murch of the CID is detailed to find his killer. So, too, but more informally, is Philip Trent, the novel's true detective. Trent is an artist with a remarkable gift for criminal investigation. Like Holmes before him and Poirot after him, he is able to solve murders without leaving his own fireside, using only his powers of logic and deduction. Retained by Sir James Molloy, editor of the *Record* (a daily national newspaper clearly modelled on *The Times*), he has a number of

successful investigations to his credit. Once again Molloy (unlike Manderson, a benign version of the Great Man) calls upon Trent's services. Accordingly Trent departs for the Home Counties and Marlstone, the scene of the crime. Already awaiting him is Mr Nathaniel Burton Cupples, an endearingly benevolent old gentleman: Mabel, his beloved niece, is Manderson's widow, and she too would welcome Trent's expert, but decently unpaid, services.

'You could have the Buster, if you like' – Sir James referred to a very fast motor car of his – 'but you wouldn't get down in time to do anything to-night.'

'And I'd miss my sleep. No, thanks. The train for me. I am quite fond of railway travelling, you know; I have a gift for it. I am the stoker and the stoked. I am the song the porter sings.'

'What's that you say?'

'It doesn't matter,' said the voice sadly.

TRENT'S LAST CASE
BY E.C. BENTLEY

In the investigation that follows, Trent's powers of deduction lead him to the only conceivable solution, which, however, he despairs of revealing: he has fallen hopelessly in love with Mabel yet fears that she is implicated in the murder. Unable to confide in Murch, Cupples or Molloy, he takes himself to his Paris studio and throws himself into a frenzy of work. Meeting accidentally with Mabel (who, we quickly realise – although Trent is much more obtuse – is also in love with *him*), he is forced to admit his suspicions to her.

In a dazzlingly deft sequence of climax, anti-climax and true denouement, Bentley finally reveals the altogether unexpected truth. For Trent, so certain and so wrong, the real solution is, as the title of the final chapter states, 'The Last Straw'. Not too unhappily he relinquishes detection for more domestic pleasures.

Trent exemplifies many characteristics which were later to become essentially for the English detective in crime fiction of the Golden Age. He is a gentleman with a relish for friends in low places. His services are given free, unlike more stolid CID officers, who rely upon 'experience and method', Trent boasts 'the quicker brain and livelier imagination'. He is welcomed by the police as a fellow worker, allowed freedom of access to witnesses and information, and enjoys confidences vouchsafed by those who trust a gentleman but suspect a copper. Much given to quotation, whimsical soliloquy and constant, self-mocking commentary upon his own efforts, he foreshadows Sayer's Lord Peter Wimsey and Allingham's Albert Campion. Though physically courageous and impeccably heterosexual, his sensitivity is 'feminine' and, like Wimsey, he is capable of expressing his emotions in language of passionate and self-exposing intensity. Also like Wimsey, he does not believe that *all* murderers should be caught, or that colonial brutality is justified, provided it is British, or that capital punishment is a deterrent. Such maverick opinions voiced in a book published before the First World War are typical of the moral complexities often found in a genre frequently written off as reactionary.

Ada LEVERSON   1862–1933

Tenterhooks

see **The Little Ottleys** (1916)

SAKI   1870–1916

The Unbearable Bassington

H.H. Munro's talent was undoubtedly seen at its best in his short stories, and he only wrote two novels. Of these, *The Unbearable Bassington* is far superior to the stridently patriotic and uncharacteristically sentimental *When William Came* (1914), and its style and tone are far closer to those of the stories. It is interesting to note that Evelyn Waugh wrote an introduction for the 1947 reissue of this novel, for the character of Basil Seal in **Black Mischief** (1932) and **Put out More Flags** (1942) bears a marked similarity to Saki's young protagonist.

The Unbearable Bassington is an episodic, aphoristic, yet melancholy short novel depicting the fate of the eponymous hero who, like his near-contemporary, Peter Pan, 'wouldn't grow up'. His mother, the worldly Francesca Bassington, if asked 'to describe her soul, would probably have described her drawing-room', a sanctuary within Mayfair dominated by a large painting by Van der Meulen. In order to keep her house, it is necessary for her to marry off her wayward, faunlike son, Comus, one of those Edwardian youths 'who are Nature's highly finished product when they are in the schoolboy stage'. He thwarts every attempt she makes, fails through momentary greed to win the hand of an heiress, and pursues a life of extravagant self-indulgence in London society.

'You have all the charm and advantages that a boy could want to help him on in the world,' Francesca observes, 'and behind it all there is the fatal damning gift of utter hopelessness.' In despair, she arranges for Comus to become part of 'the black-sheep export trade', and sends him to the '*oubliette*' of West Africa, where he contracts fever and dies. The telegram announcing his death arrives on the same day that an expert declares the Van der Meulen to be a copy. (Saki originally intended to call his novel 'The Vandermeulen'.)

In spite of Saki's printed assurance that: 'This story has no moral', both Francesca's material values, to which she sacrifices her son, and the bored society in which she moves, enlivened only by barbs of vicious wit, are mercilessly exposed. The book is an acute Edwardian period-piece and a prophetic anthem for doomed youth.

| James STEPHENS | 1882–1950 |

The Charwoman's Daughter

The plot – by far the least important part – of this first novel is swiftly told. Sixteen-year-old Mary Makebelieve and her widowed mother occupy a wretched room in a Dublin tenement slum. Mrs Makebelieve was once a highly skilled dressmaker until her barely governable rage at the flaunting arrogance of the rich took her to the edge of sanity. Now she goes out charring, while her daughter, kept unnaturally young for her age and protected from the unremitting physical hardship which makes up her mother's life, spends her days walking in the streets and parks of Dublin. Among the city notables who populate her walks is the Policeman, a daunting Gargantua of a man, who controls traffic at the Grafton St crossing. One day, off duty and strolling grandly like a swell, he hails her, and they enter a state which any girl less naïve and immature than Mary would recognise as courtship.

When the Policeman discovers by chance that Mary is merely a charwoman's daughter, his initial reaction is one of anger and disgust, but eventually (and much impressed by his own magnanimity) he presents himself to Mrs Makebelieve as Mary's suitor. To his own disbelief and violent fury, he is rejected by Mary, who prefers to continue her developing relationship with the neighbour's lodger, a poverty-stricken young man of meagre stature but great spirit. Poverty and squalor are eventually put to flight when Mrs Makebelieve inherits a vast fortune from her brother, one of Ireland's more successful immigrants to America.

Such a prodigally happy ending is Makebelieve indeed: rightly so, for part of the novel's purpose is to maintain that everyday life is often magical and that the magical occurs so often as to be positively everyday. The book is a rich amalgam of fairy-tale, love story, psychological exploration, social realism, economic theory and political allegory, and its language is as varied as its modes. Stephens himself was an ardent Irish nationalist and in his hands this 'simple love story' becomes a vehicle for an examination of the parts power and powerlessness play in both intimate personal relations and in wider social and political ones.

All the main characters contain symbolic elements (two of them, the Policeman and the young man, remain nameless throughout). The Makebelieves' personal struggle against poverty (and that of their neighbours, the McCaffertys, and their six children) provides a microcosmic example of the brutal workings of unbridled capital. Mrs Makebelieve herself has a manic, elemental grandeur that makes her part-woman, part-fury, part avenging deity: 'her God was Freedom . . . Freedom! . . . She would oppose an encroachment on that with her nails and her teeth'. The young man, whose 'courage exceeded his strength, as it always should', is a passionate Irish patriot whose mettle might well be tried four years later in the Easter Rising of 1916.

Most fascinating of all is the relationship between Mary and the Policeman. He is, both literally and metaphorically, a member of the Force. The personal power of the man resides in his sex, his class, his job and his bulk. He is huge, powerful and menacing – a fact which sends 'a queer, frightened exaltation lightening through Mary's blood'. 'Everything desirable in manhood' seems to be 'concentrated in his tremendous body', and Mary speculates pleasurably on the 'immense, shattering blow that mighty fist could give!'

But pleasurable though the inexperienced Mary finds her masochistic fantasies, they are not endorsed by Stephens. For, as he goes on explicitly to show, male dominance exercised over women finds its echo in wealth's tyranny over poverty; and both are rehearsals for the larger violence enacted by an occupying colonial power (in this case, England) against a subjugated nation (Ireland). The book's light-hearted surface play of wit and whimsy does not conceal

Stephens's recognition of the often distressing link between male desire and violence (whether expressed by individuals or countries), and of many women's (and nations') equally distressing collusion with it.

1913

Mrs E. Pankhurst sentenced for inciting persons to place explosives outside D. Lloyd George's house • Reichstag passes bill to increase German army • First vessel passes through Panama Canal • Niels Bohr's model of the atom • Edmund Husserl, *Phenomenology* • Walter Sickert, *Ennui* • Cass Gilbert, Woolworth Building, New York • Alexander Scriabin, *Prometheus* • Sidney and Beatrice Webb found *The New Statesman* • The foxtrot sweeps to popularity • Richard Nixon, Benjamin Britten and Albert Camus born • J. Pierpont Morgan dies • James Elroy Flecker, *The Golden Journey to Samarkand*

Willa CATHER   1876–1947

O Pioneers!

Cather came from a family of farmers and pioneers. When she was nine, her family moved from the gently pastoral landscape of Virginia to Webster County, Nebraska, 'a country as bare as a piece of sheet iron'. Cather said that this forbidding landscape resulted in 'a kind of erasure of personality', but she was later to celebrate its harsh beauty in some of her best novels.

O Pioneers!, which takes its title from Walt Whitman's poem 'Birds of Passage' (1865), is Cather's second novel, coming after the false start of *Alexander's Bridge* (1912), a Jamesian story with a metropolitan setting. This second novel is set in the small town of Hanover, among Scandinavian farmers and French Catholics.

When John Bergson dies, worn out by the harsh Nebraskan land he has come from Sweden to farm, it is natural that his daughter, Alexandra, with her informed interest and growing respect for the landscape, should assume responsibility over his sons. The struggle is great and often disheartening, but it is a tribute to Alexandra's strength and determination that she persists while others around her are defeated.

Sixteen years on, the land, with its pros perous farms and rich fields, is barely recognisable. Alexandra's sense of self, however, is like an 'underground river' that come 'to the surface only here and there', her mind like a 'white book with clear writing about weather and beasts and growing things'. Her life has not been 'of the kind to sharpen her vision', and she has no idea of the strength of feeling that exists between her youngest, adored, brother Emil and her lively young Bohemian friend Marie, who has made an impetuous and unfortunate marriage with the jealous and irascible Frank Shabata. In spite of genuine efforts, such feeling cannot be suppressed, and when Frank finds Emil and Marie together, he shoots them in a moment of blind fury.

Life following the tragedy holds little meaning for Alexandra until, with characteristic compassion, she realises she must help Frank, who has been imprisoned for ten years. She is finally reunited with Carl Linstrum, a friend from her childhood who, frustrated in the rural community, had gone to the city to become an artist, and then spent some time prospecting for gold. Together they will settle in the 'rich and strong and glorious' land to which she belongs: 'Fortunate country, that is one day to receive hearts like Alexandra's into its bosom, to give them out again in the yellow wheat, in the rustling corn, in the shining eyes of youth!'

D.H. LAWRENCE   **1885–1930**

Sons and Lovers

Lawrence's third novel, completed in Italy in 1912, draws heavily on his own youth and early manhood in Eastwood. It is a rich evocation of the Nottinghamshire countryside, and of the small mining community known as Bestwood. Paul Morel is the third child of Walter, a miner, and Gertrude, a former schoolmistress's assistant from 'a good old burgher family' now fallen on hard times. After the initial passion, their marriage has deteriorated; he drinks heavily and is violent, and she devotes herself to the children, who turn against their father. The eldest son, William, gets a good job in London and is engaged to flippant, spendthrift Lily, but his death from pneumonia plunges Gertrude into a depression from which she recovers only after nursing Paul through a similar illness. Her hopes are now more than ever bound up with his life.

Paul leaves boarding school and takes a clerical post in Nottingham with Jordan's, manufacturer of surgical appliances, though he continues to dedicate his free time to painting and design. He meets a farmer's daughter, Miriam Leivers (based on Lawrence's first love, Jessie Chambers), and they take long walks engaged in passionate discussions of nature, poetry and the spirit: 'The fact that he might want her as a man wants a woman had in him been suppressed into a shame.' Frustrated, and knowing of his mother's animosity to Miriam, he ends their liaison. 'Why can't a man have a *young* mother?' he cries, and declares that he will never marry.

Through Miriam he has met Clara Dawes, an advocate of Women's Rights who is separated from her husband, Baxter. She persuades him to think of Miriam in less spiritual terms; he returns to Miriam, but the physical consummation of their love is predictably disastrous. He embarks on a passionate affair with Clara, and Miriam feels that he will return to her after this necessary 'baptism of fire in passion'. Baxter, a metal-worker at Jordan's, learns of the affair, and is dismissed after attempting to assault Paul. Later he gives Paul a good

thrashing, and Paul is strangely drawn to him. Paul is already drifting apart from Clara, unable once again to commit himself in love, and Gertrude is dying slowly from cancer. He hastens her death by administering morphine, but does not fulfil his intention of travelling abroad after her death. He stops painting and seems to have sunk into an aimless lethargy. He tries to persuade himself to marry Miriam, but knows it will not work, and is still, at the end, haunted by the powerful memory of his mother.

Lawrence's first major novel, *Sons and Lovers* is also arguably his best, and certainly one of the earliest and finest novels of British proletarian life. In particular, Lawrence's descriptive powers are combined with a devotion to his mother's memory in a moving and vivid recreation of the physical and emotional landscape of his childhood. Although frankly Oedipal, the story is not as misogynistic as later books; the portrait of Mrs Morel as an emasculator is considerably mitigated by Lawrence's deeply felt presentation of Paul's attachment to her. It is Mrs Morel, after all, who guides Paul towards being an artist by insisting that he is properly educated and does not follow the family tradition of mining, and by fostering his ambition. Something of Lawrence's ambivalence – the artist made and unmade by his mother – and an interesting insight into the novel can be found in his essay *Fantasia of the Unconscious* (1922).

Compton MACKENZIE   **1883–1972**

Sinister Street

see 1914

F.M. MAYOR   **1872–1932**

The Third Miss Symons

'I am not wanted,' cries Henrietta, the third Miss Symons, halfway through this novel, and she speaks no more than the truth. Born in the mid-nineteenth century to parents whose 'enthusiasm for babies had declined', she is nevertheless permitted the attentions traditionally paid to infants and small children. She reaches her zenith at five, has passed it by eight. Thereafter she will develop into a girl and woman unloved by any, save, intermittently, her younger sister, Evelyn.

If she is unloved, say some, it is because she is unlovable: a charge which Henrietta herself, with that fundamental humility which is one of her graces, often accepts. But, as Mayor points out, she is also unlovable because unloved; she cannot give what she has never received.

Henrietta is one of that growing army of 'superfluous' (that is, unmarried) middle-class women – upwards of a million of them by the end of the nineteenth century – upon whom Victorian novelists, philanthropists, divines and social theorists expended so much energy. Baulked of her almost-husband by the casually mischievous intervention of an older sister, Henrietta fails to fulfil the only destiny her parents ever envisaged for her. For the next four decades she tries, with little success, to find occupations, interests and friends. Her failure springs partly from temperamental flaws (her bad temper, never properly tamed, a certain stubborn self-righteousness), partly from family factors (her mother's dislike of daughters and contempt for female education), partly from larger social causes (middle-class horror at the idea of daughters supporting themselves by their own paid labours).

Such solutions as remain, Henrietta essays: settlement work in an East End slum, where her fussiness maddens fellow workers and her indiscriminate 'charity' undermines their efforts; foreign travel, where every sight is seen and nothing really understood; learning, imbibed via the lecture courses springing up everywhere to meet the hungry need of women seeking to make good educational defects. But for poor Henrietta, powers of concentration long ago eroded, pride too great to permit her to ask for the help she needs, lectures bring only defeat and self-contempt. Even the careers of aunt and godmother fail her, since two of her married sisters accept her very considerable material generosity but reject her uncertain, ill-expressed pleas for emotional involvement. Only with Evelyn, to whom she was once so close, does she achieve anything resembling genuine exchange, yet even Evelyn, who adored Henrietta as a child and turns to her once more on the death of a beloved infant daughter, never thereafter matches her older sister's feelings.

Mayor handles her indubitably sombre subject-matter with a skilful blend of satire and compassion. Her prose is rigorously controlled. Readers lulled by an almost deadpan smoothness are suddenly jolted by throwaway asides and cadences of shocking savagery. Equally shocking is Mayor's calmly matter-of-fact acknowledgement of things usually left unspoken at this period: the dislike many women feel for their daughters, the desire they feel for their sons; marriages conducted as mutual exploitation; the bitter resentment, masked as pity, which many married women feel for more fortunate sisters who have escaped the dubious blessings of matrimony.

Henrietta herself is dissected with unnerving thoroughness, but she is never mocked or scorned. Her fate was replicated many times over in her generation, for she is, as Mayor constantly reiterates, a very *ordinary* woman. Single women could and did escape her fate – many of Mayor's own aunts and great-aunts amongst them. But those who escaped were blessed with unusual gifts of intellect (such as Miss Arundel, Henrietta's former teacher), temperament or spirit (such as her credibly saintly old aunt): gifts which Henrietta, in company with many others, disastrously lacks.

The Third Miss Symons predates by nine years May Sinclair's **Life and Death of Harriett Frean** (1922) with which it inevitably invites comparison. Each is a work of rare quality, and Mayor's novel was shortlisted for the Polignac Prize. Five judges chose it. The sixth, John Masefield, who had provided an enthusiastic and perceptive introduction to the novel, felt it would be wrong for him to give his vote to a book with which he was so publicly associated. The prize went instead to Ralph Hodgson.

Edith WHARTON   1862–1937

The Custom of the Country

From 1910 to her death in 1937 Wharton made her home in France where *The Custom of the Country* was written. It was published in the year that she obtained her divorce. She became a prominent figure in American expatriate circles and Henry James, whom she had known from childhood, regarded her as an artistic equal. (There is an old joke that her novels are the kind James would have written had he been a man.) Indeed, their preoccupations are similar, rooted in their upbringing. In *The Custom of the Country*, the story of Undine Spragg's pyrrhic victory over society, Wharton displays a masterly understanding of the gradations of that society in New York, expatriate Paris and the French aristocracy.

When the story opens Undine Spragg has caused her newly rich parents to abandon Apex City for New York, but after two years in the city they are still 'stranded in lonely splendour in a sumptuous West Side hotel', depending on garrulous Mrs Heeney, the fashionable masseuse, to people 'the solitude of the long ghostly days with lively anecdotes' of the Van Degens, the Driscolls, the Chauncey Ellings, and other prominent families. But Undine will not accept defeat. Having discovered, as she thinks, the *crème de la crème*, she sets out to captivate one of them, Ralph Marvell, and not even the reappearance of Elmer Moffatt from her Apex past is allowed to spoil her triumph. Moffatt is persuaded to keep quiet about an old scandal, but on her wedding journey in Europe Undine discovers her mistake. 'To Ralph the Sienese air was not only breathable but intoxicating', while 'her mind was as destitute of beauty and mystery as the prairie school-house in which she had been educated'. Only in Switzerland and in Paris, among the raffish American expatriate set whom Ralph despises, does she come truly

alive. One thing is clear: Paris is where she has to be. For Paris she deserts Ralph Marvell, who is slaving away for her in his uncongenial job in New York; she is as indifferent to his tragic end (he gradually sinks into despair and commits suicide) as she is to the future of their son, Paul.

It is a setback when the womanising Peter Van Degen deserts her, and when her next victim, Raymond de Chelles, a proudly traditional member of the impoverished French nobility, imprisons her in his château of Saint Désert; but she is rescued by her old lover, Elmer Moffatt, who, after many ups and downs, is now one of the richest men in America and a serious art collector. The new and the vulgar have triumphed over the old and effete.

Undine and Elmer had, so it seems, been made for each other all along, and her happiness seems assured. After her whirlwind marriage to Elmer, we leave her dressing for dinner in her 'private hôtel overlooking one of the new quarters of Paris'. The house is stuffed with treasures, including the tapestries that Raymond has been forced to sell to pay his brother's gambling debts. She has just learned, however, that a dowdy friend has been made ambassador to the Court of St James. Elmer remarks that, married to a twice-divorced woman, he could never be an ambassador. Undine is furious. 'She had learned that there was something she could never get, something that neither beauty nor influence nor millions could ever buy for her. She could never be an Ambassador's wife; and as she advanced to welcome her first guests she said to herself that it was the one part she was really made for.'

This ending has all the brilliance of inevitability. The complete materialist, Undine wants things only until she has them. The irony of Wharton's beautifully constructed novel is that, married to one of the richest men in the world, having left a trail of destruction in her path, Undine can never be content.

1914

Archduke Franz Ferdinand of Austria and his wife assassinated at Sarajevo • Outbreak of First World War • US declares her neutrality • Ernest Shackleton leads Antarctic expedition • Oskar Kokoschka, *The Vortex* • Ralph Vaughan Williams, *A London Symphony* (no. 2) and *Lark Ascending* • Charlie Chaplin in *Making a Living* (film) • L. Baylis first produces Shakespeare at the Old Vic • Jonas Edward Salk born • George Westinghouse dies • Robert Frost, *North of Boston*

James Elroy FLECKER   1884–1915

The King of Alsander

Flecker's only novel was written, revised, partially lost and rewritten between 1906 and 1913. 'Here is a tale all romance – a tale such as only a Poet can write for you,' he wrote in the preface, and Flecker's difficulties with it stem from the uncongeniality of the novel form to a poet. His attitude to it vacillated: 'Nothing in God's earth can infuse any reality into the tale,' he wrote in 1912, yet a year later it was 'a very jolly and fantastic work'. Mundane jollity and a more exalted fantasy sit uneasily and interestingly together in this 'tale of madmen, kings, scholars, grocers, consuls and Jews'.

Norman Price is a grocer's son who becomes a king. A quiet life in the peaceful village of Blaindon is redirected by a forceful and mysterious old man who exhorts him to travel. Norman, and Flecker's imagination, are both set free from the constraints of rural English life to adventure in Alsander. This city has fallen into a decay symbolised by its half-witted king, Andrea, last of an inbred line, whose power has been usurped by his regent, Duke Vorza, who is determined to maintain the status quo. Norman pursues an abortive love-affair with his landlady's beautiful daughter, Peronella, while about him the Association for the Advancement of Alsander (the AAA) devises a plot to install him on the throne. This cabbalistic and shadowy organisation both threatens and fascinates Norman, who shrinks from his candidacy until its compelling androgynous leader, Arnolfo, reveals himself to be the Princess Ianthe, cousin to Andrea and in love with Norman. By combining seduction with ju-jitsu, she manipulates the unresisting Norman, and a bloodless coup takes place. Spurned by the Impostor-king at his coronation, Peronella subsequently leads a mob on the palace. In the ensuing battle she disappears, Duke Vorza, who has conspired with her, is killed, and Norman is left to govern Alsander with Ianthe as his queen.

In a bizarre coda, the old man (now revealed as the poet) revisits Blaindon to offer Norman's friend, the impoverished young aristocrat John Gaffekin, a hallucinatory vision of the artist's immersion in the protean sea of imagination: 'Then he thought he saw him rise naked among the flames and run towards the sea with a silver disc shining on his breast: and he began to swim out along the track of the moon.'

The novel as a whole marks a point, on the eve of the First World War, at which this sort of symbolist vision can no longer hold off the historical realities it shrinks from. Norman's desire to improve roads and drainage, to establish an aeroplane service and, most significantly, not to restore the baroque splendour of Andrea's throne-room, jars against the fairy-tale idea of kingship which the fantastic plot evokes. Against Vorza's determination to keep Alsander 'untouched by the world', Norman observes: 'Well, there's something to be said for being awake and something to be said for modernity.' The sensibility of the novel is tensed anxiously between realism and fantasy, but Flecker prefers to find pleasure rather than neurosis in this contradiction. Modernity was about to throw up events more unlikely even than Flecker could devise, a fact registered presciently and optimistically in 'a tale as joyously improbable as life itself'.

Sinister Street

Mackenzie's massive slab of Georgian prose charting the spiritual, intellectual and sentimental education of a naïve and idealistic young man at the turn of the century made his reputation instantly. Originally published in two volumes, it became a *succés de scandale* after the first part had been banned by the influential circulating libraries. The novel was highly praised at the time by Henry James and others, but Cyril Connolly's verdict, 'an important [because influential] bad book', is perhaps nearer the mark.

Sinister Street is divided into four books. 'The Prison House' deals with Michael Fane's 'deserted childhood', spent in gloomy Kingston with his younger sister Stella, and a succession of servants and nannies. Michael's father is assumed to be dead and his mother is frequently, and mysteriously, absent. His life is lightened by the arrival of Maud Carthew, an affectionate governess from a large and friendly family in Hampshire whom Michael visits. Stella, who is a musical prodigy, is sent abroad to study the piano and Michael begins at St James's Preparatory School as a day boy.

In the second book, 'Classical Education', Michael transfers to the senior school, where he meets Alan Merivale, an attractive athlete who is to be his best friend. The boys' sentimental education begins during a holiday at Eastbourne, where they meet two girls called Dora and Winnie. Rather more fruitful is the meeting between Miss Carthew and Alan's uncle, a soldier called Ross. Michael goes through the characteristic phases of intelligent *fin-de-siècle* adolescence, including Anglo-Catholicism and decadence. These two elements are combined in 'Brother Aloysius', otherwise Mr Henry Meats, a lost soul whom Michael encounters during a retreat at Clere Abbey. Michael's spiritual education continues under the guidance of the wordly Fr Viner, who runs a mission at Notting Hill, whilst a chance encounter with Arthur Wilmot, an 1890s aesthete, introduces Michael to the Yellow Age. Michael also experiences first love when he meets the beautiful young Lily Haden. Her mother, a vulgarian fearful of her reputation, objects to their relationship, and Michael renounces Lily, only to overhear her flirting with somebody else. Further sorrows are caused by the Boer War: Capt. Ross, who has married Maud Carthew and fathered her son, is killed in action, causing Michael to feel a revulsion for his contemporaries' jingoism. Also killed is Lord Saxby, to whom Michael had been introduced as a child. His mother now explains that Saxby was his father.

The third book, 'Dreaming Spires', charts Michael's Oxford experience and shows him acquiring a new circle of friends. He is at St Mary's College; Alan, whom he plans to marry off to Stella, is at Christ Church. Michael is aware of 'a boundless sense of communal life, a conception of boundless freedom that seemed to be illimitable for ever'. He does very little work. His circle includes Maurice Avery, the founder of the magazine *The Oxford Looking Glass*; a Balliol poet called Guy Hazlewood (who reappears in Mackenzie's *Guy and Pauline*, 1915); and a covey of Old Etonians: Lonsdale, the languid and unambitious son of the Duke of Cleveden; a bulky aesthete called Wedderburn; and Tommy Grainger, a genial oarsman. The unconventional Stella causes her conservative brother some qualms when she mixes in Bohemian circles, and Michael is astonished to discover that their guardian, an Albany bachelor called Prescott, has proposed to her. Prescott later kills himself, leaving Stella a considerable fortune. She buys an estate in Huntingdon, which Alan manages after their marriage.

The fourth book, 'Romantic Education', finds Michael aimlessly wandering around London, having left Oxford with a first-class degree in history. An acquaintance tells him that Lily Haden has been seen on the Orient Promenade, a notorious parade-ground for prostitutes. The chivalrous Michael blames himself for what he imagines to be Lily's decline and sets about trying to find her in order to marry and thus save her.

His descent into the London underworld

in pursuit of this bland Proserpina takes him to several vividly rendered locations, including Kentish Town ('recommended' by Meats) and Pimlico. He takes rooms in squalid boarding-houses, mixes with tarts and crooks and eventually bumps into Lily at a masked ball. She now lives with a raffish protector called Sylvia Scarlett (the eponymous heroine of Mackenzie's novel of 1918) in cheerful lodgings in Fulham and seems not to have fallen as low as Michael had feared. He none the less determines to marry her, despite the opposition of Sylvia, Stella and Alan, his mother and almost all of his friends. Lily is clearly just a pretty face, but Michael refuses to acknowledge this as he persists in pursuing his dream of first love and noble reclamation.

He agrees to postpone the wedding for a while, and sets Lily up in a large house, away from the influence of Sylvia. While he is visiting Guy Hazlewood, who has withdrawn to an idyllic Cotswold cottage in order to pursue his literary career, Lily takes a lover. Michael realises: 'She's only eaten seeds of pomegranate, but they were enough to keep her behind.' He leaves Lily to Sylvia and transfers his energies to reclaiming Meats (who now calls himself Barnes), without success, for the police catch up with this damned soul and arrest him for the murder of a prostitute. Although the case against Meats is cast-iron, in a final quixotic gesture Michael arranges to pay for a lawyer. The novel ends with Michael leaving the world behind and travelling to Rome in order to decide what to do with his life.

Frank NORRIS   **1870–1902**

Vandover and the Brute

Probably the first novel Norris completed, and written largely while he was a student at Harvard in 1895, *Vandover and the Brute* was published posthumously with a foreword by the author's brother, the novelist Charles Norris.

Vandover is a young man living with his widowed father, a property developer, in San Francisco. Having returned from Harvard, he resolves to be an artist and is funded by his father. With his friends Geary and Haight he frequents late-night restaurants and brothels, and 'the animal in him, the perverse evil brute, awoke and stirred'. Despite being informally engaged to Turner Ravis, he seduces another rich society girl, Ida Wade, who gets pregnant and commits suicide. Vandover is plunged into guilty remorse and, following his father's death, resolves to 'give over the vicious life into which he had been drifting' and dedicate himself to his art. Rumours of his involvement with Ida have spread, however, and he is socially ostracised. He plunges once again into dissipation: 'Drunkenness, sensuality, gambling, debauchery, he knew them all.' Ida's father brings a lawsuit against him, and Geary, as one of Wade's lawyers, swindles Vandover out of much of his fortune. With what is left he gambles and drinks ferociously, subject now to fits of insanity during which 'the brute' literally takes over his being: naked, he scampers round on all fours howling like a wolf. The years pass and he has gambled away all his money. Briefly he finds work as a commercial artist but is unable to keep it. Finally, penniless, he goes to Geary, who now owns many of the buildings Vandover's father once owned, to plead for work; Geary employs him as a cleaner, a humiliation he seems to accept without complaint.

For some time *Vandover and the Brute* was believed lost in the San Francisco earthquake of 1906; its rediscovery added a major work to Norris's comparatively small canon. Although Norris is generally regarded as the principal exponent of American naturalism, many would regard this more symbolic book – which might be seen as a broadly naturalistic reworking of Robert Louis Stevenson's *The Strange Case of Dr Jekyll and Mr Hyde* (1886) – as his best novel.

Robert TRESSELL   1868–1911

The Ragged Trousered Philanthropists

The only authentic novel of Edwardian working-class life, this book reflects the author's own experience of appalling social conditions in Hastings, where he worked as a house painter. First published in an abridged edition, the full text did not appear until 1955. Little is known of the author, whose real name was Robert Noonan, except that he was educated, and was a tireless exponent of socialism. He died in Liverpool in 1911 and was buried in an unmarked pauper's grave.

Tressell's intention was 'to write a readable story full of human interest and based on the happenings of everyday life, the subject of Socialism being treated incidentally'. The novel is unusual in that there is no plot, no 'love theme', and the characters do not 'develop'; Tressell simply relates a year in the lives of a group of men who work for an unscrupulous decorating firm, Rushton & Co. However, the varying natures of the workmen and their families contribute towards the drive of the novel, as does a meticulously described series of events designed to show how the cumulative burden of poverty and destitution affected the working classes at all stages of their lives.

Relentless competition in Rushton & Co., to offer the lowest estimates to gain contracts, results in work which has to be skimped to keep down costs. Skilled craftsmen are forced to take short cuts and have no pride in their workmanship. The fear of dismissal for trifling offences filters through from the highest ranks to the youngest apprentice. Hunger, poverty and ignorance create attitudes of mutual suspicion amongst the workmen in the interests of self-protection and survival.

The main character, Frank Owen, attempts to explain to his colleagues how *they* are the philanthropists who make their employers rich. Despite his explanation of socialism as a remedy for their abject poverty, Owen is largely sneered at, and the workmen are condemned to perpetuate their misery through their own unwillingness to listen. Owen himself exemplifies possibilities of triumph in adverse circumstances. Because of his artistic skill, he is asked to design and paint a complex 'Moorish' decoration in a room. The work has to be done well, and Owen's demands for certain conditions and materials are reluctantly met. Although he is only paid the standard rate, Owen achieves a rare feeling of satisfaction in his work which makes him forget his hardship and poor health.

Work, boys, work and be contented
So long as you've enough to buy a meal
For if you will but try, you'll be wealthy
  – bye and bye
If you'll only put your shoulder to the
  wheel.

As they sang the words of this noble chorus the Tories seemed to become inspired with lofty enthusiasm. It is of course impossible to say for certain, but probably as they sang there arose before their exalted imaginations, a vision of the Past, and looking down the long vista of the years that were gone, they saw that from their childhood they had been years of poverty and joyless toil. They saw their fathers and mothers, wearied and broken with privation and excessive labour, sinking unhonoured into the welcome oblivion of the grave.

And then, as a change came over the spirit of their dream, they saw the Future, with their own children travelling along the same weary road to the same kind of goal.

THE RAGGED TROUSERED PHILANTHROPISTS
BY ROBERT TRESSELL

The novel centres on the decorators, but Tressell introduces the rest of Mugsborough society – including ministers of religion, the local council and politicians – to expose corruption and vested interests. He is particularly vehement about the hypocrisy of the church and its version of Christianity, although he does not attack sincere religion. Despite the portrayal of an endless cycle of squalor in which wrecks of humanity attempt to survive, there is a balance of humour and an ending of positive optimism.

1915

Germans first use poison gas on Western Front ● First Zeppelin attack on London ● US loans $500 mill. to Britain and France ● Execution of Edith Cavell in Brussels ● Chinese princes vote for establishment of a monarchy ● Douglas Haig becomes British Commander-in-Chief in France and Flanders ● Einstein's General Theory of Relativity ● Wegener's theory of continental drift ● Marcel Duchamp, *The Bride Stripped Bare by Her Bachelors Even* (the first Dada-painting) ● Gustav Holst, *The Planets* (symphonic suite) ● D.W. Griffiths, *Birth of a Nation* ● Saul Bellow and Arthur Miller born ● Alexander Scriabin and James Keir Hardie die ● Rupert Brooke, *1914 and Other Poems*

John BUCHAN   1875–1940

The Thirty-Nine Steps

In a dedicatory letter, Buchan placed this novel in the genre of the 'shocker', a 'romance where the incidents defy the probabilities, and march just inside the borders of the possible'. Certainly the plot is resolved through remarkable coincidence and good fortune, but the novel's pace and excitement are maintained until the last page. This accounts for the book's continuing popularity, although Buchan's serious admirers prefer other works. It has survived merciless parody, several cavalier film adaptations, and serious objections to its apparent endorsement of unacceptable social attitudes.

When Richard Hannay returns to 'the Old Country' after making his 'pile' mining in Bulawayo, he is restless and bored, finding London 'as flat as soda-water that has been standing in the sun'. Almost immediately a terrified American called Scudder seeks refuge in his Portland Place flat, gibbering about the Balkans, 'Jew-Anarchists' and the planned assassination of Karolides, the Greek Premier. Shortly afterwards Scudder is found 'skewered ... to the floor'. Armed with Scudder's coded notebook, a tooth-brush and fifty gold sovereigns ('I'm a Colonial and travel light'), Hannay heads for Galloway to escape from suspicious police and Scudder's killers. He adopts a number of disguises and uses 'veldcraft' to evade his hunters. When he deciphers Scudder's notebook he realises that he has been spun a 'yarn' and discovers that there is an even more sinister conspiracy by German agents (the Black Stone) to infiltrate a secret Franco-British summit and steal military secrets, in particular details of the dispositions of Britain's fleet.

After numerous escapades, which include a bout of malaria, being stalked by a monoplane, and blowing up a house in which he has been held captive, Hannay makes his way back to England to warn the urbane head of the Secret Service, Sir Walter Bullivant. An enemy agent attends the summit in the guise of the First Sea Lord but is recognised by Hannay as he leaves. Recalling a reference in Scudder's book to a high-tide time and 'thirty-nine steps', Hannay pursues the agent to the east coast. The eponymous steps lead from a clifftop villa, where the agents pose as suburban tennis-players, to a private beach, off which a large yacht has dropped anchor. Hannay penetrates the disguise of the agents and of the apparently English captain of the *Ariadne* ('His close-cropped head and the cut of his collar and tie never came out of England'). The men are arrested and Britain goes to war with her naval secrets intact.

Victory

Generally regarded as Conrad's last major work, *Victory* reaffirms its author's abiding beliefs and concomitant pessimisms. It tells the story of Axel Heyst, a Swedish baron and son of a philosopher. Influenced by his father's absolute scepticism, Heyst has spent fifteen years wandering the Dutch East Indies. If he has a purpose, it is to remain 'the most detached of creatures in this earthy captivity'. He is, however, unable to disconnect himself utterly from the claims of humanity and gives financial assistance to a trader, Morrison, whose brig is about to be confiscated for debt. Overwhelmed by gratitude, Morrison procures for his benefactor the post of 'manager in the tropics' of the Tropical Belt Coal Company. Heyst sites the company's operations on the island of Samburan. Morrison returns to England, where he dies. The enterprise goes into liquidation and Heyst lives in its decaying headquarters with only a Chinese manservant, Wang. He is visited infrequently by Davidson, the captain of a steamer. From Heyst's point of view, such solitude is Utopian; but Schomberg, a hotelier in Java, propagates the rumour that Heyst has bled the mining company of its profits and as good as murdered Morrison.

Unaware of these calumnies, Heyst stays at Schomberg's hotel while visiting Java. There, he rescues Lena, a violinist, from a bullying orchestra leader and from the hotelier's odious attentions. Schomberg's dislike of 'that Swede' is now aggravated by a grievance. Heyst and Lena return to Samburan, where their precipitate alliance deepens into trust and – for Lena at least – love.

Mr Jones, Martin Ricardo and Pedro – respectively, 'evil intelligence, instinctive savagery ... [and] brute force' – next arrive in Java. The three so tyrannise Schomberg that he determines to be rid of them and persuades Ricardo that Heyst has hoarded the mine's profits on Samburan. The trio goes to the island to rob Heyst. Unarmed,

deserted by Wang, and disliking action, Heyst has no resources to counter this threat. He attempts to save Lena – who has become the object of Ricardo's rapacious infatuation – but she is shot by the misogynistic Mr Jones. In the ensuing catharsis of revenge and despair, the villains – and Heyst himself – perish.

Two images brood over the closing scenes: one, a portrait of Heyst's father; the other, the 'dull red glow' of a volcano. Destructive intellect and implacable nature – incarnated in the would-be robbers – are the witnesses and agents of Heyst's tragic end. Yet, while this symbolism adds a metaphysical dimension to the tale, the protagonist's fate originates directly in his own unheroic kindness and the malicious interpretation laid upon it. Companionship, instead of providing a refuge from an inimical universe, becomes the very fault which activates its violence.

Although early sections of the novel are narrated by an unidentified 'I', and some events reported by Davidson, the narrative is for the main part in the third person. Conrad was enjoying a measure of popular success by the time *Victory* was published, and obtained an advance of £850 for the book. Eight months before his death, the manuscript was sold for $8,100.

Worried that during the First World War a book called *Victory* might be mistaken for a work of patriotic propaganda, Conrad nearly changed the novel's title. Instead, he added the descriptive subtitle 'An Island Tale'. As Lena dies, Conrad writes that: 'The spirit of the girl ... clung to her triumph convinced of the reality of her victory over death', but this victory (like much in Conrad's fiction) is illusory – hollow, even. The real victory of the title is probably that of Heyst over his own tiresome mortality, or that of the malign implacability of the cosmos over poor mortals.

The Good Soldier

Ford was persuaded by his publisher that his original title, 'The Saddest Story', would prove unsaleable in the dark days of 1915. In a moment of irony – he was serving in the Coldstream Guards at the time – he suggested *The Good Soldier* as an alternative, and the date of the outbreak of the First World War, 4 August, is of recurring significance in the novel. If not perhaps the saddest, it is nevertheless a sad story, a formally perfect account, with autobiographical elements, of a fatal friendship which, as it unravels, mercilessly exposes the psychology, hidden motives and weaknesses of all its characters. Ford considered it his finest novel and contemporaries hailed it as great.

You see, Leonora and Edward had to talk about something during all these years. You cannot be absolutely dumb when you live with a person unless you are an inhabitant of the North of England or the State of Maine.

THE GOOD SOLDIER
BY FORD MADOX FORD

'Someone has said that the death of a mouse from cancer is the whole sack of Rome by the Goths, and I swear to you that the breaking up of our little four-square coterie was such another unthinkable event.' Florence and John Dowell, 'leisured Americans', are forced into permanent exile in Europe because to cross the ocean would place a fatal strain on Florence's heart, weakened during a storm at sea on their honeymoon. At Nauheim, a spa frequented by wealthy sufferers from heart complaints, they meet Capt. Edward Ashburnham (the good soldier of the title) and his wife, Leonora. The four become inseparable, but as John Dowell's narrative progresses, it becomes apparent that far from being 'dancers in a minuet', as he had imagined, they are locked together 'in a prison full of screaming hysterics'.

Edward, at first sight the perfect English gentleman, is a weak and sentimental philanderer. Leonora, 'so tall, so splendid in the saddle, so fair!', is a stern Roman Catholic. She orders her husband's affairs, amatory and monetary, settles his debts and pays off blackmailers, curbs his generosity to his tenants, and consequently inspires in him a humiliated resentment. She is also morally responsible for the death of an innocent young married woman, Maisie Maidan, whom she has attempted to procure for her husband.

John Dowell realises that he has been the perfect dupe when he discovers that Florence, to whose care he has devoted himself, has been conducting an affair with Edward (with Leonora'a knowledge), and that there is nothing wrong with her heart. After witnessing an unpleasant incident in which he struck his black servant, she manufactured her first heart-attack out of fear of what he might do if he discovered that she was not 'a pure woman' and the marriage has remained unconsummated. When an old acquaintance exposes the fact of her earlier affair, Florence kills herself with the prussic acid she has kept in readiness.

Leonora has a ward, Nancy Rufford, an intense and oddly beautiful girl, with whom John falls in love. Edward, on whom Nancy looks as a father, her own being unsatisfactory, is also in love with her, but too honorable to do anything about it, despite Leonora's tormented efforts to act as procuress. The bewildered girl is batted between them 'like a shuttlecock' until, half-demented, for she has realised that she loves Edward, and has received an upsetting communication from her degenerate mother whom she believed to be dead, she is sent to India to join her father.

Of the little coterie only Leonora emerges unscathed, to remarry and start a family. Edward, who has been dying slowly of love for Nancy, kills himself on the receipt of a would-be insouciant telegram from her. Nancy, on hearing the news, loses her mind and becomes a docile, almost speechless child. By a savage irony, John Dowell who spent so many years nursing a false invalid, is left as a 'horribly lonely' recluse, devoting his life to the beautiful but feeble-minded girl he had wanted to marry.

D.H. LAWRENCE   1885–1930

The Rainbow

Originally drafted as one novel, *The Rainbow* and *Women in Love* (1920) are connected by the characters of Ursula and Gudrun Brangwen, 'The Sisters', as Lawrence termed his work-in-progress. *The Rainbow* unfolds the lives and loves of three generations of the Brangwen family. Tom Brangwen courts and marries Lydia, a Polish emancipee, and the first section portrays their erotic self-absorption, establishing 'the arch' of the parental relationship in which Lydia's child, Anna, grows up.

With the birth of a second child, Anna and her stepfather become close companions in a relationship tinged with the Oedipal overtones of Lawrence's *Sons and Lovers* (1913). As Anna develops into a young woman, Tom has gradually to relinquish his role towards her and painfully acknowledge that she will be the partner of another man. Will Grangwen, Anna's cousin, comes to court her, and their marriage is symbolically foreshadowed by a ritualised scene in which they gather sheaves of wheat to make a shelter for themselves. With Anna's first pregnancy, Will feels jealous and excluded, and carves a 'Creation panel' for the village church, its biblical imagery infusing the prose along with the myth of the covenant and paradise lost.

Will and Anna's two eldest daughters, Ursula and Gudrun, are disturbed by their mother's continuing fertility, and form a partnership which maintains them as a separate presence in the house full of noisy children. While the sisters are young, their grandfather, Tom Brangwen, dies in a flood on the farm, and the memory haunts Ursula, forming a symbolic progression from the Edenic landscape of the novel's opening through the biblical image of Noah's ark and the rainbow.

The final section of the novel focuses on Ursula and her affair with Skrebensky, a family friend of her Polish grandmother, Lydia. Skrebensky is mostly away with his regiment, and Ursula's first experience of romance alternates with the erotic attraction of Winifred Inger, a fellow teacher at the school. Ursula looks back on this phase of her life in a chapter called 'Shame', reflecting Lawrence's fascination with and repulsion from feelings of homoeroticism. When Skrebensky returns she decides to marry him and their engagement is announced; but during a weekend on the east coast she begins to realise that she has made the wrong decision. She rejects him, but then finds she is pregnant. When she writes to him, full of remorse, Skrebensky replies that he has married the daughter of his colonel. Ursula suffers a miscarriage. The loss of this child means that when she reappears in *Women in Love*, she is once again a free agent.

The sexual explicitness of *The Rainbow* led to its prosecution for obscenity, and 1,000 copies were symbolically burned by the censor when the publishers agreed to withdraw the novel. Philip Morrell, the Liberal MP, raised the matter in the House of Commons, but Lawrence continued to experience censorship and rejection by the English establishment. The acquittal, after a celebrated trial, of *Lady Chatterley's Lover* (1928) at the start of the 1960s finally made his works available in unexpurgated editions.

W. Somerset MAUGHAM   1874–1965

Of Human Bondage

Maugham's longest and most ambitious novel, in which 'fact and fiction are inextricably mixed', draws heavily upon the author's own youth, with circumstances and names scarcely altered. An earlier version of the novel, 'The Artistic Temperament of Stephen Carey', was completed in 1898, but put on one side as 'a little strong'.

In the final version the hero has become Philip Carey, orphaned at nine and put in the care of his Uncle William, Vicar of Blackstable, and his Aunt Louisa, much as Maugham had been sent after his mother's death to live with his Uncle Henry, the Vicar of Whitstable. Uncle William's selfishness and lack of affection make Philip introverted, a state intensified by his experiences at King's School, Tercanbury (i.e. Canterbury). Philip has a club foot, which symbolises his sense of being an outcast (and is thought by some commentators to symbolise Maugham's own homosexuality). His religious faith wavers when he observes his uncle, who signally fails to practise what he preaches. A visit to Heidelberg, where he encounters a pretentious aesthete, Hayward, and a philosopher, Weeks, finally destroys all his beliefs, and he returns to England with a sense of liberation.

Philip is sexually initiated by an older woman, Emily Wilkinson, but it is an experience which leads to disillusionment, and he is glad when she departs. Miss Wilkinson epitomises the failure Philip will experience in his relationships with women, for he has made the mistake of replacing religion with idealism. It is this outlook which encourages Philip to abandon accountancy after a few months in order to study art in Paris. At first he is stimulated by the Bohemian company he keeps, but the death of a fellow student, Fanny Price, causes him to reappraise his own position. Fanny was completely without talent, yet never doubted she had genius. Unable to find money to pay for her lessons or to keep herself alive, she hanged herself rather than admit defeat. Philip realises that he, too, has no talent; he returns to England to be with his recently widowed uncle.

Always manipulated by stronger characters, Philip becomes infatuated with Mildred, a waitress he meets in London, where he has gone to study medicine. He cannot resist her anaemic complexion or her flat-chested appearance. (Her boyishness has led to suggestions that she was modelled on a youth.) She is selfish and grasping and does not love Philip; facts of which he is aware but powerless to escape. They live together chastely, and Philip nearly ruins himself attempting to support her. He is befriended by Thorpe Altheny, a copywriter whose extravagant personality combines both idealism and practicality. Altheny teaches Philip to achieve peace of mind by control of will.

Mildred becomes a prostitute, contracts a veneral disease and goes out of Philip's life. Freed from debt by his uncle's death, he successfully completes the medical studies he was forced to abandon and, realising that it is better to follow what life offers than to pursue an often elusive idealism, he marries Altheny's daughter and becomes a country doctor.

Of Human Bondage has no developing plot, only a series of incidents in England, France and Germany, connected by Philip Carey. An authentic sense of place gives the novel much of its drive and ensures its continuing popularity.

Dorothy RICHARDSON   1873–1957

Pointed Roofs

see **Pilgrimage** (1938)

Edith SOMERVILLE   1858–1949
and Martin ROSS   1862–1915

The Irish R.M. Trilogy

Some Experiences of an Irish R.M.
(1899)
Further Experiences of an Irish R.M.
(1908)
In Mr Knox's Country (1915)

E.A.Œ. Somerville and Violet Martin were Anglo-Irish cousins from Cork and Galway respectively. Under their joint pseudonym they collaborated on a number of novels, short stories and travel journals. Their most substantial novel was *The Real Charlotte* (1894), but they found enduring popularity with these thirty-three comic tales of Irish life. The idea for the series came while the two women were on a painting holiday in Etaples in 1898 (Somerville had studied art and later had a secondary career as a painter), and the first dozen stories were written while Ross was recovering from a serious hunting accident. They first appeared between October 1898 and September 1899 in the *Badminton Magazine*, and were collected in *Some Experiences*, published two months later to great acclaim. Another eleven stories (one of which, 'A Horse! A Horse!', was in two parts) were collected for a volume of *Further Experiences*, while the third and final volume of ten stories, published in the year of Ross's death, was in fact the work of Somerville alone. Although she continued to publish books under the joint pseudonym,

Somerville silently omitted these last stories from her own omnibus edition, *Experiences of an Irish R.M.* (1944). (All the stories, however, were gathered in the 1928 volume, *The Irish R.M.*)

An Irishman himself – although much Anglicised – Major Sinclair Yeates obtains the post of Resident Magistrate for a county in the rural west of Ireland. His post, in fact, is peripatetic: he travels the Petty Sessions Courts attempting to administer justice in trivial, but still baffling, disputes. As he hears his cases, he learns that things here are not as they are elsewhere. Familial relationships are indecipherably intricate; language is slippery; the clarities of English law are inadequate to the nuances of indigenous behaviour. Frequently he encounters 'the personal element', a feature of local life and character which sets routine at nought and undermines the best bureaucracy.

The bulk of the Major's memoirs (which are given in the first person) is concerned with hunting, shooting and fishing. Immediately upon his arrival at Shreelane, his dilapidated residence, he is sold a horse by Flurry Knox. It is made plain to the Major that his position in the community requires that he join the local hunt and, although at first reluctant, he is before long Deputy Master of Hounds. Thus he meets his most indeterminate acquaintances: there is Flurry himself, astute and affable, 'a gentleman among stableboys and a stableboy among gentlemen'; Flurry's formidable grandmother, 'a rag-bag held together by diamond brooches'; Miss Bobbie Bennett, whose observations are tart and whose motives obscure; Dr Jerome Hickey, who spins unlikely yarns only to discover that 'every [one] ... is true in spite of me'. Often the Major finds himself led astray – sometimes literally. Sometimes (and not always unwittingly) he wanders into illegalities, and sometimes into flirtations which his wife, Philippa, treats with icy forbearance.

The basic situation is familiar: an amiable confrontation between law and education on the one hand and native wit and custom on the other. The authors' success with this formula, however, is derived from their careful distribution of its constituents. The Major is by no means a gull or buffoon – nor is he invariably quite sharp enough. The contest is balanced and the outcome of each skirmish unclear until the closing paragraphs of the episode. Indeed, the victories recorded are rarely complete and the antagonists never so wounded that they cannot take up arms afresh.

If it is no longer possible to read *The Irish R.M.* without a trace of unease, that is because the genial struggle it portrays is the comic counterpart of a tragic reality. The Major was shortly to be recast as a defender of the Protestant ascendancy: the miscreants of his county would shortly be Republican guerrillas. Thereafter the subject could not be treated so lightly and the pleasure of these stories is in part the pleasure of nostalgia.

1916

Roger Casement lands in Ireland, is arrested and executed ● Sinn Fein Easter Rebellion in Dublin ● Italy declares war on Germany ● Woodrow Wilson re-elected US president ● The first military tanks ● Treatment of war casualties leads to development of plastic surgery ● Vilfredo Pareto, *Treatise of General Sociology* ● Claude Monet, *Water Lilies* ● Leoš Janáček, *Jenufa* (opera) ● Jazz sweeps US ● Harold Wilson and Yehudi Menuhin born ● Henry James and Horatio, Lord Kitchener die ● Robert Graves, *Over the Brazier*

Arnold BENNETT   1876–1931

The Clayhanger Novels

Clayhanger (1910)
Hilda Lessways (1911)
These Twain (1916)

On 19 November 1909, Bennett wrote in his Journal, 'Yesterday I finished making a list of all social, political and artistic events which I thought possibly useful for my novel between 1872 and 1882.' He began writing it at the Royal York Hotel, Brighton, on 5 January 1910 with a General Election in full swing – an election campaign was to be one of the features of his novel. By the middle of March he was in a panic: 'I was frightened by a lot of extraordinary praise of *The Old Wives' Tale* that I have recently had.' He finished in Paris in June – 'one week in advance of time'.

Bennett need not have worried. *Clayhanger*, with **The Old Wives' Tale** (1908), is the book on which his reputation as a novelist of social realism chiefly rests. He conceived it as the first part of a trilogy, but neither of its successors, *Hilda Lessways* and *These Twain*, recapture the same sustained intensity. The book is dominated by the relationship between Edwin Clayhanger and his father, Darius, the master printer who as a child was rescued from the workhouse by Mr Shushions. Now 'very old', Darius 'has survived into another and more fortunate age than his own', but his early experiences (which Bennett took directly from William Shaw's *When I Was a Child: Recollections of an Old Potter*, 1901) have left him permanently

scarred. He is a grim, overbearing father; there is no way that Edwin with his intellectual pretensions can escape from the tyranny of the printing works. 'What hypnotism attracted him towards the artists' materials cabinet which stood magnificent, complicated, and complete in the middle of the shop, like a monument?' A monument is what it remains; Edwin never gets away to train as an architect, his earliest aspiration,

---

And now his education was finished. It had cost his father twenty-eight shillings a term, or four guineas a year, and no trouble. In younger days his father had spent more money and far more personal attention on the upbringing of a dog. His father had enjoyed success with dogs through treating them as individuals. But it had not happened to him, nor to anybody in authority, to treat Edwin as an individual. Nevertheless it must not be assumed that Edwin's father was a callous and conscience-less brute, and Edwin a martyr of neglect. Old Clayhanger was, on the contrary, an average upright and respectable parent who had given his son a thoroughly sound education, and Edwin had had the good fortune to receive that thoroughly sound education, as a preliminary to entering the world.

CLAYHANGER
BY ARNOLD BENNETT

---

and while Darius has his health and strength he does not have much say in running the business either. It is only after Mr Shushions's death, when Darius goes into decline, that their roles are reversed and Edwin finds himself attending the annual dinner of the Society for the Prosecution of Felons, and acknowledged as one of Bursley's great tradesmen. Briefly, as Bennett did during the 1910 election, he even discovers in himself 'the symptoms of crass Toryism'.

He is cured of that soon enough, but as a prosperous bachelor with a decent library and his sister Maggie to look after him, he could hardly be described as living more than a comfortable, selfish existence. He might, when he was younger, have found salvation through the Orgreaves: the architect, Mr Orgreave, as a liberal father was the reverse of his own, and was most welcoming, as were his wife and children. However, Edwin has always been shy of them, particularly of Janet Orgreave who still loves him and has never married. It was Janet's friend, Hilda Lessways, 'fundamentally different from other women', who when they kissed had briefly 'inspired' him 'with a full pride of manhood'. But where was Hilda now? She had written to him from Brighton, where she had gone on some mysterious business, 'Every bit of me is absolutely yours.' After that there was silence; and so matters would have stood if, years later, with his father dead, he had not gone to Brighton to seek her out in her desolate boarding-house. Even then, she was hardly open with him, and they would never have come together had it not been for the nearly fatal illness of her son, George.

'The portrait of a civilisation' is how Frank Swinnerton described *Clayhanger*. Edwin, 'overawed by his tremendous father, struggling for independence', is a prototype of the middle-class young man in the Victorian industrial provinces. The novel's effect is cumulative: Bennett's list of 'social, political and artistic events', which he found so tedious to compile, and above all the heightened recollection of his own upbringing, combine to create a world which is compellingly authentic.

Hilda Lessways shows the woman's side of things. Bennett wrote in his Journal that he had conceived it as 'portraying the droves of the whole sex instead of whole masculine droves ... the multitudinous activities of the whole sex, the point of view of the whole sex,

against a mere background of masculinity. I had a sudden vision of it. It has never been done.' Hilda is a 'modern woman', inhibited by Victorian conventions but hungry for experience. When she learns that she is bigamously married to the newspaper man, Mr Cannon, she is 'dizzied by the capacity of her own body and soul for experience'. Margaret Drabble makes a plausible case for the novel (based on the proposition that Bennett, like Wells in *Ann Veronica*, 1909, 'dared to say things that women hadn't got round to saying for themselves') but *Hilda Lessways*, unlike *Clayhanger*, has not convincingly withstood the test of time. Perhaps because it is 'observed' rather than written out of an impassioned recollection of Bennett's own early experience, it lacks the dramatic impact of the earlier novel. Nevertheless Bennett does add a dimension to both characters by showing Edwin through Hilda's eyes – 'wistful, romantic, full of sad subtleties, of the unknown and the seductive, and of latent benevolence' – rather differently in fact from the way in which Edwin sees himself.

These Twain is the story of Hilda and Edwin's turbulent marriage; in it Bennett returns to experience – his own agonised marriage to Marguerite Soulié – and as a consequence the emotional tension rises again. Lying beside his wife, Edwin 'perceived that the conflict between his individuality and hers could never cease ... Hilda had divined nothing. She never did divine the tortures which she inflicted in his heart.' Nevertheless, to quote Bennett's modern champion, Margaret Drabble, their relationship is 'intensely real. With hardly an explicit word about sex, Bennett manages to convey [Hilda's] sensuality and their intimacy.' We see her involved critically with Edwin's family, the ageing figures of *Clayhanger*, now set in their ways: 'The book is full of the ordinary efforts of fairly ordinary people to amuse themselves, to look after one another, to achieve joy in a limited environment.' Written during the First World War, the novel has also to be seen in terms of a wider, more cataclysmic conflict; Bennett himself had moved on from the comparative certainties of the world of *Clayhanger* and, significantly, this was the last of his novels to be set in the Five Towns.

Tenuously connected to the trilogy by character, though not by setting, is *The Roll Call* (1919), which features Edwin's stepson, George Cannon, who fulfils the ambitions Edwin had for himself by becoming an architect. (His career is modelled on that of Bennett's friend, A.E. Rickard.) The novel is set among struggling artists in London, and it seems that Bennett paid the penalty of the provincial novelist who turns his back on his origins. Although it has good individual scenes, the book as a whole is wooden, and was poorly received by contemporary reviewers. It remains an inferior pendant to a trilogy rather than contributing a fourth volume to a potential quartet.

| John BUCHAN | 1875–1940 |

Greenmantle

It is November 1915 and Sir Walter Bullivant, head of the British Secret Service, is worried by rumours of a *jihad* being organised against the Allies in the East. All he has is a scrap of paper found on the body of his agent son on which is written: *Kasredin*, *cancer* and *v. I*. Fortunately, Major Richard Hannay is convalescing with a leg-wound incurred at Loos and is perfectly game to 'blow in on the Bosporus'. With his aristocratic chum, the Near East adventurer Sandy Arbuthnot, and a dyspeptic, portly but brave American called Blenkiron, Hannay sets off for Constantinople, gathering up his old veldcraft tutor, Peter Pienaar, en route.

More ambitious than *The Thirty-Nine Steps* (1915), this novel follows its middle-aged heroes through occupied Europe to the Turkish frontline city of Erzerum. The reader may share with Hannay the odd 'spasm of incredulity' as the tale unfolds assisted by the sort of coincidences which separate the world of Buchan from real life, but before the spasm registers the plot has leapt ahead. Amongst the villains encountered are Col. von Stumm ('hideous as a hippopotamus', but having 'a passion for frippery . . . a perverted taste for soft things'); the orientally effete 'jackal' of Enver Pasha, Rasta Bey ('an infernal little haberdasher . . .

a bandbox bravo'); and, deadlier than any of these males, the alluring Hilda von Einem whose 'inscrutable smile and devouring eyes' causes havoc among the bachelor heroes. It is this last (*v. I*) who is behind the attempt to consolidate Turkish support of the Germans

---

I must spare a moment to introduce Sandy to the reader, for he cannot be allowed to slip into this tale by a side-door. If you will consult the Peerage you will find that to Edward Cospatrick, fifteenth Baron Clanroyden, there was born in the year 1882, as his second son, Ludovick Gustavus Arbuthnot, commonly called the Honourable, etc. The said son was educated at Eton and New College, Oxford, was a captain in the Tweeddale Yeomanry, and served for some years as honorary attaché at various embassies. The Peerage will stop short at this point, but that is by no means the end of the story. For the rest you must consult very different authorities. Lean brown men from the ends of the earth may be seen on the London pavements now and then in creased clothes, walking with the light outland step, slinking in to clubs as if they could not remember whether or not they belonged to them. From them you may get news of Sandy. Better still, you will hear of him at little forgotten fishing ports where the Albanian mountains dip to the Adriatic. If you struck a Mecca pilgrimage the odds are you would meet a dozen of Sandy's friends in it. In shepherds' huts in the Caucasus you will find bits of his cast-off clothing, for he has a knack of shedding garments as he goes. In the caravanserais of Bokhara and Samarkand he is known, and there are shikaris in the Pamirs who still speak of him round their fires.

GREENMANTLE
BY JOHN BUCHAN

---

by using the seer Zimrud ('the Emerald', codenamed Greenmantle after an old Turkish miracle play called *Kasredin*) to fulfil a prophecy of deliverance. Zimrud is dying of *cancer*, so time is short. Sandy has penetrated the inner circle disguised as a leader of the Companions of the Rosy Hours, a group of whirling dervishes, but has been captivated by von Einem, agreeing to don the dead prophet's green cloak.

Captured by Stumm, now commander of Erzerum, Hannay, Pienaar and Blenkiron escape over the snowy roofs of the city, stealing a map of the Turkish defences as they go. Pienaar volunteers to take this vital information through the Turkish lines, across no man's land and into the Russian trenches, a task he accomplishes aided by a British spy who sends morse-coded messages to the Russians using a church bell. The novel ends after the heroes have managed to hold out against Stumm behind the natural fortifications of a hilltop *castrol*, during the siege of which von Einem is killed by stray shrapnel. The Russians break through the Turkish defences and their cavalry charges upon Erzerum, joined by Hannay (apparently unhampered by a gash in his thigh and a broken arm) and Arbuthnot, who, still in his green robes, thus fulfils the prophecy of deliverance.

The further adventures of Hannay and his friends are recounted in *Mr Standfast* (1919), *The Three Hostages* (1924), *The Island of Sheep* (1936), etc.

Ronald FIRBANK   **1886–1926**

Inclinations

Firbank's ambitious and elliptical Sapphic tragi-comedy is set in Greece and Yorkshire and was written in London and Oxford. Part I describes a disastrous holiday undertaken by the distinguished biographer Gerald(ine) O'Brookomore ('the authoress of *Six Strange Sisters, Those Gonzagas*, etc.') in the footsteps of her latest subject, Kitty Kettler. The new object of her inclinations is a country girl 'not yet fifteen' called Mabel Collins, who trifles with her companion's affections, but whose own interests are focused upon Count Pastorelli, whom she meets on the boat to Marseilles. The shady Count ('not so pastoral as he sounds') follows the women to Athens, where they mix with a number of Firbankian tourists: an unhappily honey-mooning couple called the Arbanels; Miss Mary Arne, a celebrated actress who is visiting Greece in preparation for her latest role as Lysistrata ('I mean to treat her as a character-part'); and a dipsomaniac Australian called Miss Ola Dawkins, who is searching for her lost parents. Among the social occasions described are an expedition to the Acropolis by a group of lady's maids and a catastrophic all-female duck-shooting party at Salamis, during which Miss Arne is accidentally (?) shot dead by Mrs Arbanel, who was aiming at a flying fish. Although Mabel feigns indifference when speaking of the Count to Gerald, she readily gives him her hand when he proposes. She elopes with him, leaving her benefactress a callous and illiterate note of farewell, thus provoking Firbank's most famous chapter (XX), in which Miss O'Brookomore's despairing exclamation 'Mabel!' is repeated eight times.

Part II is set at Bovon, the Collins family home, which they have been unable to let. Mabel (now the Countess Pastorelli) and her baby, Bianca, have returned, without the Count, a circumstance which gives rise to gossip amongst the neighbours. The Collinses give a face-saving dinner-party, at which Bianca circulates 'as might a fruit, from guest to guest along the table', Miss Dawkins gets drunk, and Mabel's bumptious 'kiddy sister', Daisy, captures the interest of the Farquahar of Farquahar. (In 1925 Firbank rewrote this chapter (IV) in a less telegraphic style; most editions print both versions.) The novel ends with news that the Collinses have finally let their house, and the Count sends word that he is arriving, thus provoking a discussion of his eating habits and the planning of a menu: ' "Then," Mrs Collins' voice rose as if inspired, "then Côtelettes – à la Milanaise ..." '

Less sprawling than its predecessor, *Vainglory* (1915), the novel is constructed almost entirely in dialogue, but suffered at the hands of its original publisher. 'I wanted so few "commas" more capital-letters & dots instead of dashes', Firbank complained.

James JOYCE — 1882–1941

A Portrait of the Artist as a Young Man

In the process of revising the rough manuscript of *Stephen Hero* to create *A Portrait of the Artist as a Young Man*, Joyce was not only revising his own biography, but his artistic practice as well, transforming himself from an Irish aesthete in the tradition of Wilde and Yeats into a fully fledged, experimental modernist writer. Like all of Joyce's later fiction, *A Portrait* (which was first published serially in *The Egoist* magazine from 1914 onwards) tells the story of a young Irishman's upbringing and eventual rejection of Catholicism for the secular religion of art.

The style of the narrative grows and develops, foetus-like, throughout the book, to suggest the development of Stephen Dedalus's consciousness from infancy to young adulthood. But at the same time, the narrative is riddled with holes, as if a Romantic *Bildungsroman* had been chopped up and re-edited, montage-style, into a series of microscope-sections through the development of a mind.

A recurrent pattern is recognisable from section to section, however: a movement from telling to experiencing, as Stephen keeps observing disjunctions between what he is taught and what he sees around him. Thus the first chapter begins with a fairy-tale narrated by Stephen's father Simon, an emblem of the adult manipulation of childish gullibility, and ends with Stephen's discovery of the unreliability of adults, as he is unfairly beaten by a teacher at Clongowes Wood College.

The second chapter similarly begins with the adolescent Stephen reading Victorian romantic fiction, and ends with his first sexual encounter, with a Dublin prostitute. In the third chapter, a hell-fire sermon terrorises Stephen into repentance, and then an oddly tangible, almost sensuous act of confession.

Confession is followed by penance and piety, but this period ends again in disillusionment and loss of faith. By the time Stephen's Jesuit teachers at Belvedere College invite him to join their religious order, it is too late, since he has already decided on a career as a writer. The chapter ends with a secular vision, of a girl Stephen sees wading on a beach, to complement the failed religious vision at the start of the chapter.

In the final chapter, Stephen has become a student at University College, Dublin, and is variously seen resisting atavistic Irish nationalism and religiosity, along with English cultural tyranny, then formulating aesthetic theories, writing poetry and, finally, jotting notes in a diary as he prepares to leave the confines of Ireland for the more liberated Continent. The last entry in the diary is an apostrophe to Stephen's namesake, the mythical Dedalus, who designed himself the wings with which to escape his imprisonment in a labyrinth. The novel thereby ends with myth, narrated by son to father, as it had opened with fairy-tale, spoken by father to son, suggesting Stephen's inheritance of both the paternal role of creator–artist, and the power to reconstruct, as Joyce himself had done, the narrative of his own existence.

Ada LEVERSON — 1862–1933

The Little Ottleys

Love's Shadow (1908)
Tenterhooks (1912)
Love at Second Sight (1916)

Leverson is remembered largely as Oscar Wilde's 'Sphinx', but she was also an accomplished writer, the author of six novels written after the death of Wilde and the collapse of her marriage. Her three best books feature Bruce and Edith Ottley, and were gathered together under the title *The Little Ottleys* in 1962 (with an introduction by Colin MacInnes). They display the charm and wit which earned Leverson the friendship and admiration of Wilde, and incidentally give an insight into the author's own marriage to the profligate Ernest Leverson.

Love's Shadow is ostensibly about the courtship and marriage of a young couple,

Hyacinth Verney and Cecil Reeve. Every-body is in love with Hyacinth except Cecil, who is infatuated with an older woman, the widowed Eugenia Raymond. Mrs Ray-mond, who is 'no longer young, and had never been beautiful', patiently rejects Cecil's advances, which only makes him more determined to win her. He none the less pays court rather dutifully to Hyacinth, encouraged by his uncle, Lord Selsey, who announces that he is to marry Mrs Raymond himself. Cecil and Hyacinth eventually marry, much to the regret of the latter's plain and plain-speaking companion, Anne Yeo, who is more than a little in love with Hya-cinth herself, and regards Cecil as no more than a pretty face. At first, Hyacinth is happy, but she is aware of her husband's obsession with the new Lady Selsey, and when by chance she catches sight of them together in a hansom, she assumes, incor-rectly, that they have been meeting each other in secret. She threatens to leave Cecil, but Anne Yeo intercedes, persuading Lady Selsey to take an extended yachting holiday to the Ionian Islands, thus putting herself out of harm's way. The novel ends with Hya-cinth begging Cecil to forgive her: ' "Well, I'll try," said Cecil.' The novel's title suggests both the shadowy, insubstantial quality of the Reeves' marriage and the shadow that Cecil's continuing love for Lady Selsey casts over it.

This rather commonplace drama takes centre-stage, but surrounding the young couple are some splendidly observed charac-ters, notably Hyacinth's guardian, the amia-ble Sir Charles Cannon, and his appalling, upholstered wife, whom he married 'through a slight misunderstanding in a country house'. The real protagonists of the book, however, are the little Ottleys themselves. Edith is Hyacinth's confidante, and much of the book's humour arises from scenes within the Ottleys' 'very new, very small, very white flat' in Knightsbridge. Bruce Ottley, a pomp-ous, overbearing, humourless, snobbish val-etudinarian, is Leverson's most brilliant creation. His wife regards him with amused exasperation as he huffs and puffs about writing a play (we know he never will) and about Mr Mitchell, a senior colleague at the Foreign Office. Bruce's attempts to involve himself in some amateur theatricals become one of the novel's running jokes; we only hear about them from Bruce himself, but Leverson manages to convey both his mass-ive self-deception and his pathetically trans-parent attempts to save face.

In *Tenterhooks*, the Ottleys' marriage is further eroded. At a party given by the Mitchells, they meet a handsome widowed barrister, Aylmer Ross, and it soon becomes clear that he and Edith have fallen deeply in love. Although Bruce accepts him as an *ami de la maison*, Aylmer is too honourable to act upon his passion for Edith, even though he knows it is reciprocated. He cannot accept mere friendship, and goes abroad, but returns when Edith writes to say that she misses him. Another chance sighting in the street nearly ruins their friendship, when Edith sees Aylmer in the company of her friend Vincy's protégée, a 'horrid little art-student' called Mavis Argles, whom he is in fact quite innocently escorting to the British Museum. Meanwhile, it transpires that Bruce has been enjoying a flirtation with the children's governess. Edith dismisses the woman and refuses to discuss the matter with her erring husband. She sorts out her misunderstandings with Aylmer, while Bruce elopes to America with Miss Argles, leaving a preposterous letter of explanation. Edith cannot bring herself to divorce Bruce, as he suggests, and as Aylmer urges, for she knows that her husband's relationship with Miss Argles will not last and that he will be very unhappy. The novel ends with Aylmer vowing never to see Edith again, and Bruce returning home alone, unchastened by his experience.

Love at Second Sight takes place during the First World War. Bruce's uncertain health (largely imagined) has made him unfit for service, but he takes a gloomy interest in the war. The English widow of a French wine merchant, Eglantine Frabelle, has been foisted upon the Ottleys as a house guest by Lady Conroy, the absurdly forgetful and disorganised Irish wife of a cabinet minister. Mme Frabelle credits herself with remark-able insights into the lives of her acquaint-ances, which invariably prove to be wrong, not that she ever admits as much, even in the face of incontrovertible evidence. She is convinced, for example, that Mr Mitchell is in love with Edith, and that the Ottleys' tone-deaf son, Archie, is a musical prodigy. Edith learns that Aylmer Ross, from whom she has not heard since he walked out of her life, is back in London recovering from a leg-wound received in the trenches. She visits him and, as the novel's title indicates, falls in love with him all over again. He is being tended by a young nurse, Dulcie Clay, who is also in love with him, a cause Edith attempts to promote. Eventually Edith realises that she is being given a second chance, and that she should leave Bruce for Aylmer. She is advised against this by her friend, the composer Sir Tito Landi, who feels that she would find the attendant scandal more than she could bear. Edith agrees that she will leave Bruce when Aylmer, now recovered, returns from the Front. The trilogy concludes triumphantly with Bruce running away to America with Mme Frabelle, thus releasing Edith to join Aylmer, who has been rejected for active service by the medical board, and who takes a desk job for the remainder of the war.

The trilogy's plot may occasionally seem novelettish, but Leverson uses it to create an irresistibly funny and acute comedy of manners. She allows her heroine what she never had in her own life, where (according to her daughter) she 'made the discovery that she was never to know the happiness of living with someone with whom she was in love; that the rest of her life would have to be a compromise'. If Edith herself is something of a self-portrait in pastels, the less admirable characters are painted in bold and vivid strokes, and in Bruce Ottley, Leverson created one of the funniest bores to be found in literature.

Rose MACAULAY   1881–1958

Non-Combatants and Others

Set in 1915 and published the following year, Macaulay's eighth novel was among the earliest to question the conduct and continuation of the First World War.

Alix Sandomir, a semi-crippled art student in her mid-twenties, is the child of politically ardent parents: her Polish father died in the Warsaw prison where his struggle for Poland's freedom ended; Daphne, her English mother, has for more than twenty years been associated in the public mind with her forthright and uncompromising championship of just but unpopular causes both at home and abroad. Alix and her older brother, Nicholas, are wary of external allegiances, living lives flavoured 'with the touch of irony which one often notes in the families of one or more very zealous parents'.

Alix has done her best to pretend the war does not exist, but her aunt Eleanor's house, where she is living when the book opens, is dominated by war work. Eleanor herself is secretary of the local Belgian Committee, with responsibility for refugees. Of her cousins, Dorothy nurses, Margot belongs to the Women's Volunteer Reserve and Betty is driving an ambulance in Flanders. Cousin John is home, briefly, recovering from wounds received in the trenches. Mlle Verstigel, a Belgian refugee, shares the house; Alix, feverishly intent on insulating herself from the war, 'hated and feared her whole nation; they had been through altogether too much'. Although John makes light of his experiences, Alix cannot fail to see that his hands never cease to shake, and his eyes are 'nervous and watchful'. Moreover, very much against her will, she discovers that at night he walks in his sleep, 'crying, sobbing, moaning, like a little child, like a man on the rack', and the things he says then bear no relation to the cheery anecdotes he tells his family.

Desperate to escape, Alix arranges to go instead to London, to stay with the middle-aged, widowed Emily Frampton (a relation only by virtue of her marriage to Daphne's cousin Laurence, who, in the eyes of his family, married hopelessly beneath him). 'Violette', Emily's Clapton villa, is an enclave of lower-middle-class respectability. It is also an entirely female household whose heart is basically sound but whose head rarely moves beyond the clichés and prejudices of jingoism and tabloid journalism. Yet, in effect, the war leaves them untouched, as it does the bulk of their bluffly complacent male friends and visitors, drawn, in these pre-conscription days, from the class 'which, for several good reasons, produces fewer soldiers than any other'. Alix's intellect may rebel but, emotionally, 'Violette' suits her well.

Only the shattering news of her younger, soldier brother's death finally jolts her into a recognition that the war is not to be evaded or ignored: Paul, a mere boy, commissioned straight from school, died, she eventually learns by chance, a death even more horrible than this war commonly provides. His dying forces Alix to rethink her attitude to the war. The book grows from its examination of the two utterly contradictory meanings of 'fighting the war': fighting in it, and fighting against it. Gradually, encouraged by her mother, Alix joins the war against war, fought by pacifists who include disillusioned young officers, international feminists and radical clergymen such as West, Nicholas's friend, who unwittingly becomes the means by which Alix makes her first tentative approaches to the Christianity her mother loathes.

Like most of Macaulay's novels, this one provides a vehicle for her characteristically sane ideas, pungently expressed in shrewd observations, lively dialogue, deftly economical characterisations and the occasional, utterly unabashed, mini-lecture from the author. As ever, humour, wit and satire are the allies Macaulay enlists in this passionate and early anti-war work.

Dorothy RICHARDSON   1873–1957

Backwater

see **Pilgrimage** (1938)

1917

February and Bolshevist Revolutions in Russia • US declares war on Central Powers • Germans intensify submarine warfare • Arthur Balfour's declaration on Palestine that Britain favoured the establishment of a national home for the Jewish people • 'Bobbed hair' sweeps Britain and US • Carl Jung, *The Unconscious* • 'Surrealist' first use of Picasso's sets and costumes for Diaghilev's ballet *Parade* • Piet Mondrian launches *De Stijl* magazine in Holland • Erik Satie's music for the ballet *Parade* • Pulitzer Prizes are first awarded in US • John Fitzgerald Kennedy born • Edgar Degas, Auguste Rodin, Elizabeth Garrett Anderson and W.F. Cody ('Buffalo Bill') die • T.S. Eliot, *Prufrock and Other Observations* • W.B. Yeats, *The Wild Swans at Coole*

Norman DOUGLAS   1868–1952

South Wind

Nepenthe, the setting for this novel, is Capri in 1906. Douglas, who lived there in his latter years until his death in 1952, musters his considerable scholarly knowledge of the history of the island and its extraordinary expatriate colony to create a novel with no plot to speak of, but rich in character, humour and the ideas that Douglas himself had evolved out of his own forceful, fantastical, hedonistic and complex character.

Mr Thomas Heard, the Bishop of Bampopo, has left his African see to visit his cousin, Mrs Meadows, who lives on Nepenthe. He is immediately caught up in the complicated public and private lives of the expatriate colony. Principal ornaments of this select group are the *soi-disant* Duchess of San Martino; Mr Eames the bibliographer; a pretty adolescent called Denis Phipps; the elderly Scot, Mr Keith; the impoverished Count Caloveglia; an American multimillionaire, Mr Van Koppen; disgraceful Miss Wilberforce; and representatives of the local clergy. Mr Heard, broadened by his

African experiences, is ripe for this cosmopolitan coterie and is not dismayed by their views and activities which, in sterner societies, might be described as bordering on the pagan.

He survives, with equanimity, a minor earthquake, a disturbing fall of volcanic ash, the whisky of the Alpha and Omega Club, several sophisticated philosophies agreeably presented over delicious dinners and luncheons, and his inadvertent observation of Mrs Meadows in the act of pushing her first husband to his death over a convenient precipice. Nepenthe, Mr Heard discovers, is not unlike Africa in certain respects: 'the same steamy heat, the same blaring noises, dazzling light, and glowing colours; the same spirit of unconquerable playfulness in grave concerns'.

In the end Mr Heard, 'drenched in volatile beauty', is seduced by the island and its inhabitants: 'How shall that come out of a man that was never in him? How shall he generate a harmonious atmosphere if he be disharmonious himself? ... All life is a concession to the improbable.' 'This is a good island,' remarks one of its lesser characters, 'We discourse like sages and drink like swine. Peace with Honour!'

| Ronald FIRBANK | 1886–1926 |

Caprice

Firbank's novel of a country innocent abroad in London is a theatrical tragi-comedy. The caprice of the title is that of Miss Sarah Sinquier, only daughter of a Westmorland canon, who leaves the cathedral close of Applethorp for a career on the stage. Taking some family silver and a valuable string of pearls, she travels to London for an interview with Mrs Albert Bromley, a dramatic teacher and agent, dismissed by another character as 'nothing but an old procuress'. Unfortunately, by the time Miss Sinquier arrives in Shaftesbury Avenue, Betty Bromley has died. Drifting into the Café Royal (which, in her rural naïvety she imagines to be 'some nice tea-shop, some cool creamery'), Miss Sinquier becomes embroiled in Bohemian society and is taken up by the sinister Mrs Sixsmith, a rapacious creature who offers to enlist the help of the shady banker, Sir Oliver Dawtry, in the disposal of the Sinquier silver.

Miss Sinquier undergoes an audition with Mrs Mary (a redoubtable actress famous for her 'positively roguish' performance as 'the "wife" in *Macbeth*', and who now runs a theatrical company), where she is asked to pronounce 'Abyssinia' and 'Joan' to determine whether she will suit tragic or comic roles. She is offered a few small parts, the 'leavings' of an actress on maternity leave, but none of these seems quite suitable. Meanwhile, her pearls and heirlooms have been sold and, under the entrepreneurial hand of Mrs Sixsmith, she leases the Source Theatre and puts on a production of *Romeo and Juliet*, casting herself as Juliet. This opens to great critical acclaim, despite some tricky casting and the drunken confusion of several roles (not all of them Shakespearean) by Mr Smee, a music-hall actor cast as Friar Lawrence. Miss Sinquier's own performance ('A decadent Juliet') and one of the love scenes ('... The Romeo kiss – you take your broadest fan') draw particular comment. Unfortunately this success is short-lived, because Miss Sinquier's habit of breakfasting in the theatre has attracted mice and on the morning after her triumph, she gets caught in one of the traps that have been laid and tumbles into the well beneath the stage. The novel ends with Mrs Sixsmith ingratiating herself with Canon Sinquier at his daughter's funeral and speculating upon the pickings to be had in the close at Applethorp.

The massy Dickens might seem an unlikely influence upon the svelte Firbank, but in character, setting and plot (the most straightforwardly linear of all his novels), *Caprice* looks back beyond Wilde to the great Victorian novelist. Indeed Miss Sinquier's theatrical experiences bear marked, and surely intentional, similarities to those of that other innocent abroad, Nicholas Nickleby.

| Dorothy RICHARDSON | 1873–1957 |

Honeycomb

see **Pilgrimage** (1938)

Alec WAUGH   1898–1961

The Loom of Youth

Waugh's precocious and libellously accurate novel of public-school life was written when the author was seventeen and had just left Sherborne. A *succés de scandale*, it broke the mould of celebratory accounts of school life which had been extremely popular since the 1850s, and provoked reams of comment in the press.

Gordon Carruthers (Waugh's *alter ego*) arrives at Fernhurst from his prep school eager to excel, but finds that his fellow pupils swear, cheat in class, foul in sports and tacitly approve of homosexuality, conforming to a code of expediency rather than to the pieties of *Tom Brown's Schooldays* (1857). The stranglehold of the Classics, the inefficiency of the Officers' Training Corps and the insincerity of public-school religion are all duly criticised, as is the worship of athletics in the person of 'The Bull', a fanatical Old Fernhurstian games coach.

When war is declared, Fernhurst experiences a short-lived keenness, but this rapidly dissipates; the boys resent being told that 'slackers will not be wanted in the trenches'. Carruthers pursues a liaison with a younger boy, and Tester, a doomed poet, returns to the school and inveighs against the casual sacrifice of a generation. Carruthers leaves Fernhurst with the feeling that he has sorted out his values and is prepared to face the world.

The novel's attack upon a system of education that was producing officers for the Front was considered to be in very poor taste at the time, and the book's controversy was compounded by a provocative Preface by the Sandhurst historian Thomas Seccombe, in which the public-school system was accused of being partly responsible for the First World War.

P.G. WODEHOUSE   1881–1975

Uneasy Money

William FitzWilliam Delamere Chalmers (Bill), Lord Dawlish, has a title and no money to support it. Golf is his passion and pursuing it one summer on the course at Marvis Bay he good-naturedly teaches a newly encountered fellow golfer how to stop slicing his approach shots. His pupil is far from effusive and, indeed, barely acknowledges him in the street a few weeks later.

But Bill's lessons are neither forgotten nor unappreciated. Unlike Bill, Ira Nutcombe, the misanthropic American millionaire who profited from his instruction, has *two* passions: golf, and changing his will. When he dies unexpectedly, his solicitors (Nichols, Nichols, Nichols and Nichols of Lincoln's Inn Fields) disclose that he has left Bill five million dollars. Which is marvellous for Bill, but tough on Elizabeth Boyd and Nutty, her younger brother, who had every reason to believe that they would be Uncle Ira's chief beneficiaries. It is particularly hard on Nutty, who not only has been saddled with a ridiculous name but also has had to act as guinea-pig for every one of his uncle's numerous crazes, ranging from an enforced and joyless vegetarianism to compulsory readings of Marcus Aurelius at first light on freezing February mornings. Not surprisingly, the still young Nutty took to drink years ago, although Elizabeth struggles hard to keep him sober and industrious at Flack's Farm, Brookport, Long Island, where she is gamely raising bees.

Bill, an unfailingly chivalrous man, is horrified to learn that he is the unwitting agent of Elizabeth's loss. He instructs his solicitors to offer her half the money and, when she refuses angrily, goes to America himself in hopes of persuading her to accept it. Heeding advice to conceal his title, he travels as plain Bill Chalmers.

In New York, Bill encounters an errant Nutty, without realising his true identity, and goes, as his guest, to Flack's Farm. From this point on, the novel, already generously supplied with sub-plots, takes off into areas of bewildering and labyrinthine complexity, where unfortunate misunderstandings, disastrously backfiring good intentions, mistaken identities and ill-timed revelations abound to an extent remarkable even for Wodehouse.

Predictably Bill falls in love with Elizabeth, and she with him, little realising that he is Lord Dawlish, in her eyes a wickedly defrauding schemer. The path of true love is strewn with obstacles which include Claire, Bill's fair-weather fiancée (sometimes ex-, sometimes not); a snake called Clarence, who has a lousy sense of timing; and Eustace, a monkey of uncertain temper, with a deep dislike of scullery-maids, a marvellous throwing action and a vicious way with eggs ('A yellow desolation brooded over the kitchen. It was not so much a kitchen as an omelette').

Eventually all snarls and tangles are triumphantly unravelled and the only remaining knot is the one which Bill and Elizabeth agree to tie on the penultimate page. A typically Wodehousian resolution of love's perplexities, but one reached only after an unusually highly charged and serious exchange between the lovers. Its four pages are entirely unrelieved by comedy, and pitched at a level of painfully intense self-examination rarely encountered in Wodehouse's work.

1918

Armistice signed between Allies and Germany ● Women over 30 gain the vote in Britain ● Three-colour traffic lights installed in New York ● Influenza epidemic ● Béla Bartók, *Bluebeard's Castle* (opera) ● Luigi Pirandello, *Six Characters in Search of an Author* ● European tour of the 'Original Dixieland Jazz Band' ● Leonard Bernstein and Billy Graham born ● Claude Debussy dies ● Gerard Manley Hopkins, *Poems* ● Edward Thomas, *Last Poems*

| Willa CATHER | 1876–1947 |

My Ántonia

While working on *The Song of the Lark* (1915), a long, semi-autobiographical novel about a Wagnerian opera singer, Cather returned to Red Cloud, Nebraska, to recuperate from an illness. This return journey to the landscape of her youth and the setting of *O Pioneers!* (1913), determined the subject-matter of her fourth novel.

'Ántonia had always been one to leave images in the mind that did not fade – that grew stronger with time.' Now middle-aged, Jim Burden looks back on his friendship with Ántonia Shimerda, who comes to live near Black Hawk, Nebraska, with her Bohemian family at the same time as Jim, aged ten and orphaned, arrives to stay with his grandparents. Plunged into difficult and poverty-stricken circumstances, the Shimerdas gladly receive help from Jim's generous and God-fearing grandparents and a bond soon develops between Jim and the slightly older Ántonia, in spite of her not always reasonable ways and her not always likeable family.

When her father, deeply unhappy and homesick, commits suicide, Ántonia must take even greater responsibility on the family farm, but her fortunes rise when the Burdens, who have moved from the country into the town, encourage her to take rather less gruelling domestic work with their neighbours, the Harlings. While Jim leaves Black Hawk to train as a lawyer, Ántonia, who is 'so sort of innocent', falls in love, is deceived and abandoned with a child, returning to her family with dignity and a refusal to be bowed. When she and Jim meet twenty years later, she has married a fellow Bohemian and produced a large and handsome family. Though physically no longer the 'lovely girl' she was, she has retained her 'inner glow', her ability to inspire and to 'stop one's breath for a moment by a look or a gesture that somehow revealed the meaning in common things'.

On one level the simple story of an enduring friendship, the novel also gives a delightful record of domestic life in small-town Nebraska. It is full of affectionate and vivid portraits, from Jim's family to the hired, mostly immigrant, country girls whose effect on the hypocritical social order of Black Hawk is 'considered as dangerous as high explosives'. The novel also possesses an evocative sense of childhood and a powerful feel for landscape and the seasons, so much a part of the fabric of Nebraskan life.

Wyndham LEWIS 1882–1957

Tarr

Lewis's Nietzschean first novel was inspired by the period he had spent painting and philandering in Paris from 1903 to 1905. Several of the characters are recognisable portraits of his associates, and the eponymous Tarr shares some of his creator's appearance and circumstances. Lewis completed the book in December 1915, but had difficulty in finding a publisher. Eventually, his friend Ezra Pound arranged for it to be serialised in *The Egoist* between April 1916 and November 1917, and it was subsequently published in book form in both Britain and America. Lewis designed the book's typography, as he was to do with subsequent novels. In spite of good reviews from (amongst others) T.S. Eliot, Rebecca West and Pound (who described it as 'the most vigorous and volcanic English novel of our time'), it sold only 600 copies in Britain.

The novel is set in Paris and opens with Tarr, an artist, asking a friend's advice about whether he should become engaged to Berthe, a German also living in Paris. Tarr visits Berthe, having already decided to separate from her. She senses a change in him, and he suddenly rounds on her, saying that he has only been acting the part of a fiancé, 'philandering with the idea of marriage'.

Kreisler, another artist, arrives from Rome seeking the help of his friend, Ernst Volcker, who has been a patron to him in the past. However, Kreisler finds that his position as a beneficiary of Volcker has been usurped by Soltyk, a Polish-Russian artist, and he is soon dropped by Volcker. Left to his own devices, Kreisler meets the attractive young Anastasya in a restaurant, though they part without making any arrangement; all Kreisler remembers is that she will be attending the Bonnington Club ball. Kreisler joins Fräulein Lipmann's party for the ball, lying about his ticket, which he cannot afford. En route to the ball he lingers with Berthe, who is one of the party, and kisses her in the street, much to Berthe's surprise and to the horror of Fräulein Lipmann's party.

Kreisler's drunken behaviour shocks everyone at the ball, and he is asked to leave before he has been able to make contact with Anastasya, who arrives at the ball with Soltyk.

Meanwhile, Tarr writes to Berthe saying he is going to England for a ten-month separation from her, though he does not in fact leave Paris. Berthe's friends tell her that Kreisler's offensive behaviour at the ball was due to the fact that he felt snubbed by Anastasya. Consequently, Berthe tries to test him at their next meeting, when an affair ensues between them. For his part, Tarr then engineers his first meeting with Anastasya.

Kreisler attacks Soltyk in a cafe, because of the rumours he has spread about Kreisler and Anastasya. This leads to a duel, in which Kreisler accidentally kills Soltyk. He then goes on the run before giving himself up to the police, and hanging himself in his cell.

Tarr and Anastasya begin an affair, although he continues to see Berthe. Despite the fact that Berthe is pregnant by Kreisler and that she knows Tarr loves Anastasya, they marry. However, they divorce shortly afterwards, though Tarr does not subsequently marry Anastasya.

Rebecca WEST 1892–1983

The Return of the Soldier

'Like most Englishwomen of my time I was waiting for the return of a soldier,' Jenny, the narrator of Rebecca West's first novel, reflects, for *The Return of the Soldier* tells of the devastation of the war from a woman's perspective. Christopher, a young officer, returns home from the Front suffering from amnesia induced by shell-shock. His former sweetheart, Margaret, comes to meet him. They had parted after a quarrel in which he falsely accused her of infidelity. Their romance is resumed, for they have continued to love one another, and they pass the days contentedly in Christopher's peaceful gardens. But Margaret is now married to a subsequent suitor, the shadowy Mr Grey – and Christopher's psychological damage has caused him to forget his wife, Kitty, and their dead son, Oliver.

Christopher's cousin, Jenny, has remained a spinster, and watches the renewed passion with the envy of an unloved woman. For the sake of Christopher's happiness she wishes that his romance were the truth; but with Kitty brooding in the empty nursery, pathetically begging for the return of his affection, Jenny knows that she must act. A psychologist is consulted, whose business is to 'bring people from various outlying districts of the mind to the normal. There seems to be a general feeling it's the place where they ought to be.' Margaret, responding to the doctor's plea, reveals to Jenny that she also has lost a son, and goes to meet Christopher in the garden to break the truth to him. The soldier returns, shocked, but having resumed a military manner which will make him meet his responsibilities in the future. The influence of the women has healed the damage of the war: but only to return him to the trenches and the tragedy.

Wealdstone is not, in its way, a bad place; it lies in the lap of open country and at the end of every street rise the green hill of Harrow and the spires of Harrow School. But all the streets are long and red and freely articulated with railway arches, and factories spoil the skyline with red angular chimneys, and in front of the shops stand little women with backs ridged by cheap stays, who tapped their upper lips with their forefingers and made other feeble, doubtful gestures as though they wanted to buy something and knew that if they did they would have to starve some other appetite. When we asked them the way they turned to us faces sour with thrift. It was a town of people who could not do as they liked.

THE RETURN OF THE SOLDIER
BY REBECCA WEST

1919

The Versailles Peace Conference ● Mussolini founds Italian Fascist Party ● Lady Astor elected first British woman MP ● John Maynard Keynes, *The Economic Consequences of the Peace* ● First successful helicopter flight; first motor scooter ● H.L. Mencken, *The American Language* ● Degas' figures of dancers and horses are cast in bronze ● The Bauhaus founded by Walter Gropius in Weimar ● Manuel de Falla, *Three-cornered Hat* (ballet) ● George Bernard Shaw, *Heartbreak House* ● Margot Fonteyn born ● Theodore Roosevelt, Andrew Carnegie and Pierre Auguste Renoir die ● Siegfried Sassoon, *War Poems*

Sherwood ANDERSON　　　**1876–1941**

Winesburg, Ohio

The originality and success of *Winesburg, Ohio* is as much the result of Anderson's deficiencies as his strengths. His later novels proved that he had little gift for sustained narrative, and the interlinked stories of *Winesburg, Ohio* can be seen as a formal device foisted on him by happy necessity. The book seeks to portray a small town in the American Midwest, the people who live in it and the growth of a young man, George Willard, who will later write about it. Some of the characters inhabit more than one of the twenty-three stories, but the central presence of the book is George. It is he who listens to the tales of lives broken down by convention and small-mindedness, and he who, as an artist, makes visible the brief, redeeming moments or epiphanies which illuminate those lives.

Significantly, these moments tend *not* to come at the end of the story; they are respites rather than triumphs. Anderson admired the rather formless realism of Dreiser and despised the contrivance of climaxes and trick endings in O. Henry's stories. In 'Hands', the first story to be written and one of the best, Wing Biddlebaum's hands are the object of the town's curiosity. Their quick, nervous movements seem to be linked to some undisclosed quirk of their owner's personality – they have a story. From a conversation with George, we learn that Biddlebaum was falsely accused of molesting a boy when he worked as a teacher, and was run out of town. Now he lives in

Winesburg under an assumed name. He is neither vindicated nor vilified in the story. At its close, he merely returns to the secretive, slightly shamefaced life he led before. The point of the story, indeed of all the stories, lies in the peeling away of other people's attitudes to reveal the complexity and potential of the man beneath. Anderson uses the word 'grotesque' to describe Biddlebaum and many other characters in the book, and by this he means a human spirit which labours under and is deformed by other people's (false) perceptions of it. Anderson decries the crudity of these views, but the idea that an individual is created by the perspectives which surround him or her finds an echo in the collection's structure.

Winesburg and George Willard both emerge from a large number of such perspectives, each tale revealing both in a slightly fuller light. The analogy with contemporary developments in Post-Impressionist and Cubist painting is strong, and Anderson is known to have visited Chicago's Armory Show (which exhibited such work) several times in 1914. The emphasis in revealing the object in space rather than charting its progress through time is endorsed by the design of *Winesburg, Ohio*, although it is also a *Bildungsroman* in the sense that George develops, rather than reveals aspects of himself that were already there.

In its contemporary context, *Winesburg, Ohio* was seen as a work of formal innovation and Anderson's influence on Hemingway, Faulkner and Wolfe in the 1920s was predicated upon this. The craft of fiction has moved on considerably since then, but because Anderson's experiment was a successful one the value of *Winesburg, Ohio* has remained intrinsic rather than being dependent upon a place in the history of fiction. It is now admired on its own terms for its unsentimental but passionate depiction of a way of small-town life which has almost disappeared.

Daisy ASHFORD   1881–1972

The Young Visiters

This novel of Victorian society life was in fact written in 1890 when the author was nine, but not discovered for twenty-nine years. Ashford, by now a married woman, found it amongst the papers of her mother, who had recently died. Amused to read this forgotten product of a literary youth, she passed the manuscript on to a friend, who showed it to Frank Swinnerton. The book was published with a distinctly whimsical preface by J.M. Barrie, vouching for the manuscript's authenticity (to little avail; many people believed that the book was by Barrie himself). It was an immediate and enormous success, running through eleven impressions in less than two months. Its popularity owes much to the fact that the publishers retained the author's spelling and punctuation (or lack of it) throughout, only supplying paragraphing. (In fact the title on the manuscript is 'The Young Viseters', and it is not clear why this spelling was changed.) Ashford was also an acute observer of the social scene and a skilled mimic of adult attitudes.

The novel is subtitled 'Mr Salteenas Plan', a reference to the attempts of 'an elderly man of 42', the son of a butcher, to get a position in 'Socierty'. 'I am not quite a gentleman but you would hardly notice it but cant be helped anyhow,' he confesses to his friend, Bernard Clark, with whom he goes to stay at Rickamere Hall. He is accompanied by his friend, Miss Ethel Monticue, an attractive, plainspoken seventeen-year-old. From Rickamere Hall, Mr Salteena is dispatched to London with a letter of introduction to the Earl of Clincham: 'Could you rub him up a bit in Socierty ways,' Bernard asks. The Earl agrees to introduce Mr Salteena into London Society for the sum of £42. Mr Salteena is instructed in 'clothes and etiquett to menials' by Edward Procurio, a supercilious 'half italian' manservant, and is taken to 'a levie' at Buckingham Palace. As 'Lord Hyssops' he is introduced to a sharply drawn Prince of Wales ('all I want is peace and quiut and a little fun and here I am tied down to this life he said taking off his crown being royal has many painfull drawbacks'). Meanwhile, Bernard and Ethel travel to London, book into the 'Gaierty Hotel' and attend one of Clincham's 'sumpshous' parties, at which Mr Salteena proposes to, and is rejected by, Ethel ('Be a man said Ethel in a gentle whisper and I shall always think of you in a warm manner'). During a picnic at Windsor Castle, Bernard's proposal meets with rather more success and he and Ethel get married in Westminster Abbey. Mr Salteena get a job 'galloping madly after the Royal Carrage in a smart suit of green velvit with knickerbockers compleat', marries Bessie Topp, a maid from the Palace, has ten children, grows morose, but finds 'relief in prayer'.

Arnold BENNETT   1876–1931

The Roll Call

see **The Clayhanger Novels** (1916)

A.P. HERBERT   1890–1971

The Secret Battle

'I am going to write down some of the history of Harry Penrose, because I do not think full justice has been done to him.' A.P. Herbert's passion for justice was the result of a degree in jurisprudence, gained from Oxford just before he sailed to Gallipoli with the Royal Navy Division in 1915. His experience of legal matters led to his participation in several courts martial when his batallion was transferred to France. He was disturbed by the conduct of these courts, in particular by the bias towards the prosecution, and he felt that justice was sometimes sacrificed to expediency. Although one of the three officers executed during the First World War was a member of the RND (a notorious case which led to questions in Parliament), Herbert's account of the unjust trial and execution of Harry Penrose is entirely fictional.

The title refers to 'the battle inside' of a man who acknowledges but must overcome his fears. Benson, the narrator, befriends Penrose during the Gallipoli campaign, attracted by his romantic outlook and his 'extraordinary vitality and zest'. The appalling conditions, the dust, the flies and the intense heat which takes a mere two hours to blacken a corpse, are vividly but unsensationally conveyed. The men fall into a 'dull coma of endurance', tempers fray and Penrose makes a fatal enemy of Burnett, a fellow officer whose pose as a 'man of action' masks laziness and cowardice. Penrose himself is a conscientious and popular officer who constantly drives himself, partly because of his 'terror of being a failure'. He blames himself when one of his men is killed on a scouting party and six more are killed by a shell. Although severely weakened by dysentery he refuses to report sick (unlike Burnett) until literally carried off the peninsula.

He declines the offer of a desk job, and when he rejoins his battalion in France he incurs the displeasure of Col. Philpott, one of the 'Old Duds', 'lazy, stupid, ignorant men, with many years of service – retired, reserve, or what not – but no discoverable distinction either in intellect, or character, or action'. Repeatedly picked on to lead dangerous operations, Penrose reaches breaking point, and his platoon's disarray under enemy fire is observed from the safety of a dug-out by Burnett, who reports the matter. Penrose is court-martialled and pronounced guilty; the recommendation for mercy is ignored, and he is shot. 'That is the gist of it,' the book ends; 'that my friend Harry was shot for cowardice – and he was one of the bravest men I ever knew.'

The book is unusual for its level tone, its description of Gallipoli and its early date, some nine or so years before the rash of war books at the end of the 1920s.

W. Somerset **MAUGHAM**   1874–1965

The Moon and Sixpence

Maugham's analysis of creative genius drew upon the life of Paul Gauguin, whose work he first encountered in Paris in 1905, and upon his own disastrous marriage. Indeed, Maugham had visited Tahiti in 1916 at the very moment Syrie Wellcome's divorce freed her to marry him. The novel was written in a burst of creative activity following a spell in a sanatorium, recuperating from tuberculosis, which he had contracted while working as a secret agent in Russia.

The novel's hero is Charles Strickland, a London stockbroker married to a literary hostess, Amy, who regards him as the epitome of dreariness. Amy likes to encourage the talents of the young and in all artistic matters her taste is regarded as impeccable. She dismisses her husband's holiday attempts at painting as mere 'daubs' and is therefore astonished when, after seventeen years of marriage, Strickland abandons her and their children in order to become an artist in Paris. In pursuit of his vocation, Strickland shows utter ruthlessness (he never sees his family again), and endures poverty and near-starvation in France. He is rescued by Dirk Stroeve, a mediocre artist, who insists that Strickland is a genius, and therefore accepts Strickland's constant rudeness and complete ingratitude. Stroeve is devastated, however, when his wife, Blanche, becomes Strickland's lover. Fearful of Blanche's possessiveness, Strickland dismisses her as he did his own wife, and shows no remorse at all when she commits suicide and Stroeve's life is ruined.

Strickland subsequently goes to Tahiti, where he marries a native girl, Ata, whom he treats like a servant. He continues to paint with demonic determination, uncaring that his genius is unrecognised, and eventually dies of leprosy. After his death, his work becomes famous and he is idolised by the art world. Unaware of the existence of Strickland's Tahitian wife, Amy slips easily into the role of the sorrowing widow of an acclaimed genius, entertaining professors who visit her to learn details of her husband's life. Her drawing-room is filled with reproductions of his paintings, since she cannot afford the originals; she now finds his work delightfully 'decorative'.

Maugham uses an unnamed first-person narrator to tell Strickland's story, which allows him to distance himself from his central character. It is clear, however, that Maugham believes that genius justifies Strickland's behaviour, and that those he destroys matter less than the creation of great art. Women, in particular, are seen as enemies of art, restricting their victims by sexuality or domesticity. Maugham is able in this novel to display a contempt for the conventional life of both the bourgeoisie and the fashionable haunters of *salons*, and to show a disregard for conventional morality that in his life he denied himself.

Dorothy RICHARDSON	1873–1957

The Tunnel

see **Pilgrimage** (1938)

Dorothy RICHARDSON	1873–1957

Interim

see **Pilgrimage** (1938)

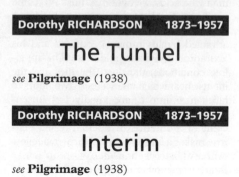

1920

Prohibition in USA ● Birth of League of Nations ● Women given the vote in US ● Northern and Southern Ireland each to have own parliament ● Oswald Spengler, *Prussianism and Socialism* ● John Taliaferro Thompson invents sub-machine gun ● Discovery of the skeletons of Peking Man ('*Sinanthropus*') ● Spectators at Cologne Exhibition of Dadaist art allowed to smash paintings ● First public broadcasting station in Britain opened by Guglielmo Marconi at Writtle ● First broadcasting station in US opened at East Pittsburgh by Westinghouse Company ● Paul Whiteman's band visits Europe and the rage for jazz becomes universal ● Mrs Humphry Ward and William Dean Howells die ● Edna St Vincent Millay, *A Few Figs from Thistles*

John GALSWORTHY	1867–1933

In Chancery

see **The Forsyte Chronicles** (1929)

D.H. LAWRENCE	1885–1930

Women in Love

Begun as a 'potboiler' intended for a young female readership, Lawrence's tale of two sisters and their lovers was originally published in the USA, and was condemned by the gutter press as obscene when published in Britain the following year. Ursula and Gudrun Brangwen, introduced in *The Rainbow* (1915) when Lawrence conceived of the two novels as one work, are now involved in contrasting relationships with Rupert Birkin, a school inspector, and Gerald Crich, son of the local mineowner.

' "Ursula," said Gudrun, "don't you *really want* to get married? ... You don't think one needs the *experience* of having been married?" "Do you think it need *be* an experience?" replied Ursula. "More likely to be the end of experience." ' Ursula, disillusioned at the end of *The Rainbow* by her unfortunate affair with Skrebensky, now falls in love with Birkin, who is trying to extricate himself from a tangled affair with Hermione Roddice, a wealthy and intimidating woman (whose character is partly drawn from Lady Ottoline Morrell). Gudrun, recently returned from London, where she has been training as an artist, is vague in her plans for the future, but as she watches Gerald mistreating his horse with cruel mastery she is both fascinated and repelled by him. Birkin, achieving with Ursula a union of 'stellar equilibrium', seeks a bond of 'blood-brotherhood' with Gerald, but is rejected.

The two couples go to Switzerland together on a holiday, when the antagonistic elements in the affair between Gerald and Gudrun erupt into a fateful quarrel. Gerald has turned to Gudrun to release him from himself, finding relief from the tragedies which have burdened him with guilt: the drowning of his sister and her sweetheart at a boating party and the subsequent death from grief of his father. But Gudrun now meets the sensual and sinister Loerke, a German artist whose independence strikes a chord in her own soul. Unable to bear the pressures of his own personality and incapable of giving himself in love, Gerald goes out into the mountains to submit to dissolution of the self, dying alone in the snow. As Gudrun prepares to leave Switzerland with Loerke, Ursula is content in the marriage she has achieved, but Birkin broods about the failed attempt at friendship and laments the impossibility of a true relationship between men. Ursula protests that ' "You can't have two kinds of love. Why should you!" "It seems as if I can't," he said. "Yet I wanted it." "You can't have it, because it's false, impossible," she said. "I don't believe that," he answered.'

The homoerotic element of *The Rainbow* recurs in male form in *Women in Love* in the relationship between Gerald and Birkin, notably in the celebrated chapter, 'Gladiatorial', in which they wrestle naked. The novel was to have opened with a chapter which gave further emphasis to this aspect of the two men's relationship, 'the transcendant intimacy which had roused them beyond the everyday life'. In particular, Birkin's barely repressed sexual attraction towards other men is dealt with in some detail in this 'Prologue', running in counterpoint to his unhappy affair with Hermione. Lawrence eventually decided to discard this chapter, beginning his novel instead with Ursula and Gudrun discussing marriage.

The novel was made into an award-winning film (1969), directed with uncharacteristic restraint by Ken Russell from a screenplay he wrote with Larry Kramer.

Sinclair LEWIS · 1885–1951

Main Street

Orphaned Carol Milford's liberal education has inspired her with a restless desire to do something socially useful with her life. After a year as a librarian in St Paul she meets and marries Dr Will Kennicott of Gopher Prairie, Minnesota, and begins life as a doctor's wife in this small Midwestern town. 'Life's so free here and best people on earth,' Will tells her, but her first impressions are of a dreary, ugly and soulless town. At a welcoming party she is introduced to 'the entire aristocracy of Gopher Prairie: all persons engaged in a profession, or earning more than twenty-five hundred dollars a year, or possessed of grandparents born in America'. Puritan and deeply conservative, the denizens of the town express their views of socialism ('hang every one of these agitators') and immorality ('I started a novel by this fellow Balzac you read about, and it told how a lady wasn't living with her husband, I mean she wasn't his wife. It went into details, disgustingly! And the English was real poor.').

Carol resolves to bring culture and sophistication to the town, but is humiliated when she learns that her 'Chinese' house-warming party was considered affected. Reluctantly, she joins the Jolly Seventeen bridge club, and goes to the weekly meetings of the Thanatopsis Club where the local matrons give dreary talks on the lives of the poets. She forms the Gopher Prairie Dramatic Association with the fulsome schoolteacher Vida Sherwin and the lawyer Guy Pollock, but her plan to direct Shaw goes awry when the committee votes for *The Girl from Kankanee*.

With the birth of her son, Hugh, Carol appears briefly to be drawn into the contented world of the gossiping matrons, but she is reading widely and quietly rebelling. When America enters the First World War, she remains unenthused by the bellicose patriotism which seizes the town. Her disillusionment grows when her friend, Bea, and her child, Olaf – Swedish immigrants – die of typhoid, ignored until the last by the local 'aristocracy'. The town's snobbery and hypocrisy are further exposed when the young schoolteacher Fern Mullins is sacked for purportedly corrupting a pupil, and when Carol's own romantic friendship with the tailor Erik Valborg makes her the target of wagging tongues. Through all these trials Will has remained doggedly in love with his wife, but he is incapable of understanding her discontent. Carol leaves him and goes to work in Washington, but returns after a year. She seems at least to be reconciled to Gopher Prairie, warts and all.

The novel was a huge critical and commercial success, establishing both Lewis's reputation and his fictional milieu. As E.M. Forster put it, Lewis's achievement was to 'lodge a piece of a continent in our imaginations'. Although the novel is subtitled 'The Story of Carol Kennicott', Lewis is less concerned with her as an individual than as an emblem of the modern American woman, just as Gopher Prairie (based upon Lewis's own Minnesotan birthplace, Sauk Center) is the archetypal Midwestern town: Main Street, USA. Carol's emergence into the post-war era is similar to that of Gopher Prairie (and by extension America), and the novel looks forward to **Babbitt** (1922), where such changes have resulted in the creation of the significantly named Zenith, a truly twentieth-century commercial city. For all his acquired eastern metropolitan sophistication, Lewis remained equivocal about his roots, and for all its satirical scorn, *Main Street* is in many ways an affectionate portrait. Carol's vacillating feelings about small-town America reflect Lewis's own.

| Edith WHARTON | 1862–1937 |

The Age of Innocence

The Age of Innocence, unlike the majority of Wharton's novels, is set in the past. The story begins at the turn of the twentieth century with Newland Archer in his study thinking back on the events of thirty years before, and the bulk of the novel takes place

in 1872. Newland is happily engaged to May Welland when her cousin Ellen, the Countess Olenska, arrives in New York in flight from her overbearing husband. The story charts the to-ings and fro-ings of the resulting love-triangle until Newland makes, or is forced to make, his choice. The complications arising from this arrangement are products of the characters' own complexities, and these are the novel's main subject.

Newland himself is recognisable as a type encountered in earlier novels. His passions and needs exist under a cerebral veneer which diffuses them into speculation and lackadaisical indecision. Wharton's skill is to root him firmly in his environment, the wealthy families of New York, while allowing him to look over horizons of which he knows nothing. The world beyond his own holds a strong attraction for him, not despite, but precisely because of, his ignorance. Untouched by fact, he can daydream it as he wishes, and this freedom is what he craves most.

The world of New York's upper class stands in stolid, if muted, contrast to Newland's flights of fancy. It was a world Wharton knew well and which she loathed. However, by the time she came to write *The Age of Innocence* she had largely escaped its banal routines and petty repressions. Her treatment of its institutions and characters is even-handed enough to make the point of Newland's self-deception. His attitude to May mirrors his attitude to the society itself, and he begins to see their marriage as an unbearable constraint on his 'freedom'.

When Ellen enters his life he invests her with all the vague aspirations that May and her world seem to deny. Ellen herself is suitably shadowy. She has an exotic but undefined past, and represents not Europe so much as American perceptions of it: she has undiscovered depths, and is inscrutable and fascinating because of this. Newland begins to think of life with Ellen in idyllic terms, never recognising that the possibility of freedom he sees in her might be illusory, the result of his inability to perceive the different (but similarly rigid) rules that govern her world. Trapped by his obligation to May, it seems that all hope of 'escape' is gone when she tells him that she is carrying their child.

The final chapter returns the action to the novel's opening situation. As Newland thinks back to their child's birth and upbringing, of his subsequent career and his relationship with his wife, it becomes clear that he has found neither the idyll of his daydreams nor the banal existence he had thought awaited him. His life has been full and whole. He has gained happiness with the years, and satisfaction. The novel is finally revealed not as a tale of love, but of maturity; in fact, a *Bildungsroman*.

The Age of Innocence was highly acclaimed on publication and won the Pulitzer Prize. Although the comparison with Henry James has misaligned criticism and damaged Wharton's reputation, it is a fair one in the case of this book. Her themes in it are the conflicts between aspiration and duty, responsibility and freedom, and finally between old Europe and new America. Wharton's great strength in *The Age of Innocence* is her articulation of these themes from within the narrowly compassed world of New York in the 1870s, and her uncovering of the potential within its seemingly narrower characters.

The novel was made into a sumptuous and highly successful film in 1993, directed by Martin Scorsese.

1921

President Harding declares US could play no part in the League of Nations ● State of emergency proclaimed in Germany in the face of economic crisis ● Britain signs peace with Ireland ● British Broadcasting Company founded ● Lytton Strachey, *Queen Victoria* ● Paul Klee, *The Fish* ● Eugene O'Neill, *The Emperor Jones* ● Rapid development of night clubs ● Engelbert Humperdinck and Camille Saint-Säens die ● Marianne Moore, *Poems*

Walter DE LA MARE   **1873–1956**

Memoirs of a Midget

De la Mare was one of English fiction's true originals, and *Memoirs of a Midget*, which was awarded the James Tait Black Memorial Prize, is characteristically *sui generis*. As with much of this author's work, it is set in a world at once recognisable yet dream-like, the odd perspective in this case provided by the tiny, alienated narrator, Miss M. Her exact size is never given, but seems physically improbable – certainly shorter than the world's record of 23.2 inches. 'Though your view may be delicate as gossamer and clear as a glass marble,' she is told, 'it can't be full-size. Boil a thing down, it isn't the *same*.'

The novel takes the form of a partial autobiography found after Miss M's unexplained disappearance ('I have been called away', her note states), and edited by her friend and executor Sir Walter Pollacke. Set in the Victorian era, it tells how Miss M was orphaned and thrust unprepared into the world at the age of nineteen, entrusted by her unsympathetic godmother to the care of Mrs Bowater, an old servant and the wife of an absent sailor. Miss M has her own room, her own door and her own front steps, all scaled down, but feels somewhat circumscribed, and so makes secret night-time sorties into the wild garden of a nearby deserted house. Mrs Bowater's beautiful but heartless stepdaughter Fanny, who teaches at a distant school, arrives for Christmas and Miss M falls instantly and hopelessly in love with her. Fanny addresses her as 'Midgetina', an unkind sobriquet the masochistic Miss M accepts. Miss M herself is loved by Mr Anon, the mysterious, all-seeing dwarf she has encountered in the ruined garden, but she is unable to return his devotion. Similarly thwarted is the local curate Mr Crimble; he has fallen in love with Fanny, but she remains indifferent to his protestations. So importunate is he that Fanny bullies Miss M into becoming a go-between. 'We mustn't care what she sees, what she thinks,' Miss M tells him, 'if only we can go on loving her.' Mr Crimble does care, and cuts his throat.

Not that in an existence so passive riddles never came my way. As one morning I brushed past a bush of lads' love (or maidens' ruin, as some call it), its fragrance sweeping me from top to toe, I stumbled on the carcass of a young mole. Curiosity vanquished the first gulp of horror. Holding my breath, with a stick I slowly edged it up in the dust and surveyed the white heaving nest of maggots in its belly with a peculiar and absorbed recognition. 'Ah ha!' a voice cried within me, 'so this is what is in wait; this is how things are';

MEMOIRS OF A MIDGET
BY WALTER DE LA MARE

Meanwhile, Miss M is taken up by Mrs Monnerie, a society hostess and connoisseur of minutiae, who takes her to London to add to her 'collections of the world's smaller rarities'. Miss M is lionised, and persuades her hostess to find employment for Fanny, who subsequently does her best to usurp Miss M's position as Mrs Monnerie's favourite. This London episode culminates in Miss M's disastrous twenty-first birthday party, featuring a grotesque cake decorated with statuettes of famous female midgets. Made horribly drunk by green chartreuse foisted upon her by Mrs Monnerie's loathsome nephew, Percy Maudlen, Miss M staggers down the centre of the dining-table towards Fanny, whom she hopes to 'save'. Removed to Mrs Monnerie's country house and running short of money (squandered upon metropolitan fripperies, and upon Fanny, who finally and brutally spurns her), Miss M briefly joins a circus, where she holds an audience and rides a pony. Mr Anon reappears and is horrified that Miss M should demean herself in this way, but she argues that it is no different to being shown off in London society. Unable to dissuade her from appearing again, he insists upon

taking her place in the circus ring, where he falls from the pony and subsequently dies. Realising that the world will always judge by appearances, Miss M returns to the wild garden, intent on committing suicide. She fails in this resolve, and eventually comes into sufficient money to live a quiet life where, looked after by Mrs Bowater, she writes her memoirs.

Like *Peacock Pie* (1913), de la Mare's haunting collection of poetry for children, this oblique novel is at once entrancing and disturbing. Miss M's world is minutely and exactly evoked, and the atmosphere of the book is fey in the old 'otherworldly' sense of the word, resembling nothing so much as Richard Dadd's asylum painting 'The Fairy Feller's Master-Stroke' (1855–64). The Oxford Paperbacks reprint of the novel (1982) contains an excellent preface by Angela Carter, whose own **Nights at the Circus** (1984) is clearly influenced by this dark and troubling tale.

John DOS PASSOS   1896–1970

Three Soldiers

Set during the last year of the First World War and in the aftermath of the armistice, *Three Soldiers* focuses on the experiences of three privates who represent a cross-section of American society. John Andrews is a New York based musician from an old Virginia family; 'Chris' Chrisfield an uneducated farm boy from Indiana; and Dan Fuselli a store clerk from San Francisco. The three first meet in training camp in the USA and are then shipped to Europe. Fuselli is consigned to a medical replacement unit at Cosne while Andrews and Chrisfield go into an infantry unit.

Dos Passos is at pains to examine the emotions and expectations of each. Fuselli is ambitious for promotion and inspired by army propaganda films with visions of 'jolly soldiers in khaki marching into towns, pursuing terrified Huns across potato fields, saving Belgian milk-maids against picturesque backgrounds'. Poor benighted Chrisfield is keen to see action and expresses his view of the Kaiser thus: 'They ought to torture him to death, like they do niggers when they lynch 'em down south.' Only the intellectual Andrews is untouched by this spirit; for him the war is an escape from the futility of his own life, a chance 'to humble himself into the mud of common slavery'.

Fuselli's disillusionment grows as he is passed over for promotion to corporal, and he is humiliated by the discovery that Yvonne, the French girl to whom he has proposed, is sleeping with half the company. Chrisfield's violent hostility towards his sergeant grows until, in combat, he avails himself of an opportunity to murder him with a grenade. The second half of the novel, following the armistice, concentrates exclusively on Andrews. Recovered from injuries received during fighting, he rebels against the machine-like regimentation of the army. Overcoming bureaucratic obstacles, he succeeds in getting on a detachment scheme to study music at the Sorbonne. This idyllic spell – during which he carouses round the bars and restaurants of Paris and rediscovers his musical ambitions – is cut short when he is arrested by the Military Police while in Chartres without a pass. He deserts from the labour battalion and returns to Paris, where he meets up with Chrisfield who has also gone AWOL. Having borrowed money, he disappears into the country and, claiming to have been demobilised, rents a room, where he works on a composition called, with grim significance, 'The Body and Soul of John Brown'. When he runs out of money his landlady reports her suspicions to the army. He is arrested and faces the prospect of a twenty-year prison sentence.

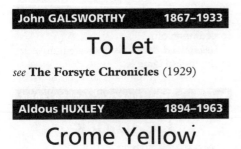

John GALSWORTHY   **1867–1933**

To Let

see **The Forsyte Chronicles** (1929)

Aldous HUXLEY   **1894–1963**

Crome Yellow

Huxley's first novel is brief and light-hearted, a Peacockian satire on the literary *salon* kept by Lady Ottoline Morrell at Garsington, and contains instantly recognisable caricatures of writers and artists of the period. A house-party of apparently indefinite duration is in progress at the mansion of Crome, where representative figures of the 1920s fill their days with brilliant talk. The arrival and departure of a young poet, Denis Stone ('he had been making an unsuccessful effort to write something about nothing in particular'), mark the beginning and the end of this almost plotless novel.

Denis's hosts at Crome are the ancestral owner, Henry Wimbush, with his impassive bowler-hat-like expression, and his wife, Priscilla, who predicts the football pools on the basis of astrology (Philip and Lady Ottoline Morrell). Their other guests include Henry's niece, Anne, a self-possessed young woman to whom Denis is hopelessly attached; Mary Bracegirdle, a young girl earnestly worried about the effects of sexual repression (Dora Carrington); Jenny Mullion, imprisoned in a world of deafness from which she escapes by making caricatures of her fellow guests (Dorothy Brett); Gombauld, an artist of Provençal descent (Mark Gertler); and Mr Scogan, a loquacious middle-aged man ('I am just *Vox et praeterea nihil*') of saurian appearance and sinister opinions (Bertrand Russell).

A visit to the estate's Home Farm to see the animals initiates the series of diversions with which successive chapters are filled. Gombauld paints Anne, while she attempts to fight off his advances; Henry Wimbush reads stories about his ancestors from his *History of Crome*, including the strange tale of the dwarf, Sir Hercules; one chapter is occupied by an eschatological sermon given by the local vicar, Mr Bodiham; another consists of an extended joke about the word 'carminative', which Denis has misused in a love poem. Other guests visit briefly: the spiritual journalist, Mr Barbecue-Smith (author of *Humble Heroisms* and *Pipe-Lines to the Infinite*), who recommends automatic writing to Denis, and the extravagantly romantic young dilettante, Ivor Newton, singer, composer, pianist, medium, telepathist, poet, actor, cook and 'symbolical' painter.

A country fair, in which the house-party rather condescendingly participates, provides a decorous climax to the final chapters, and in the last chapter, Denis, in despair over Gombauld and Anne's relationship, makes his escape by having a telegram ('Return at once. Urgent family business') sent to him by Mary. He leaves Crome, misappropriating Landor's 'I Strove with None': 'It sinks, and I am ready to depart.'

George MOORE   **1852–1933**

Héloïse and Abélard

By the time he wrote *Héloïse and Abélard*, Moore had discarded the Zolaesque realism of his nineteenth-century novels for a meticulous epic style, both mannered and lucid. His later fiction endorses the aesthetic values of the Pre-Raphaelites and (like their work) is concerned with the pastoral, the romantic and the remote. In the history of Héloïse and Abélard he found a subject which welcomed such a presentation and allowed him to meditate on the vulnerability of human will and desire.

Héloïse is born in 1100, the daughter of a knight killed in the crusades. She is educated in a convent at Argenteuil until, aged seventeen, she is brought to Paris by her uncle, Fulbert, the canon of Notre-Dame. One day she attends the school of the great and controversial philosopher, Pierre Abélard, and falls passionately in love with him. He reciprocates, attracted as much by her scholarship as by her beauty. Unaware of the nature of the couple's relationship, Fulbert invites Abélard to live under his roof as Héloïse's tutor. Abélard duly moves in. Soon Héloïse is pregnant. She and Abélard flee to Brittany.

Abélard shortly returns to Paris, believing that he must make peace with the influential canon. Fulbert insists that the lovers marry. Abélard accepts the justice of this. Héloïse, however, argues that bachelorhood is the only fit condition for a philosopher: marriage will jeopardise Abélard's reputation and render impossible ecclesiastical advancement. None the less, the two are duly married.

Almost immediately, they dissolve the tie. After the birth of Astrolabe, their son, Héloïse returns to Argenteuil: Abélard, it is intended, will be ordained and enter on a career within the church. The two, meanwhile, will meet illicitly.

Nine years pass in which Héloïse sees nothing of Abélard. She is puzzled to hear that he has become the abbot of a distant monastery. Astrolabe, who has been permitted to live with her, disappears from the convent, enticed to join one of several doomed children's crusades. The convent itself is closed after rumours of witchcraft, and Héloïse is briefly reduced to beggary. When Abélard finds her, he explains that he could not enter the priesthood because, after Héloïse had returned to the convent, he was castrated by Fulbert's hirelings. Héloïse exclaims, 'Is there no end to our misfortune?' She contemplates suicide but is dissuaded by Abélard, and they resolve to spend their remaining years apart in the service of God and in the hope of a perfect union in heaven. The world cannot accommodate their love; it must therefore be consummated afresh in a deferred Utopia.

The lovers' struggle is primarily against the church – here portrayed as a cruel and powerful institution. Rather than adhere to dogma, they follow the impulse of the heart and the guidance of the intellect. In that respect, their tragedy is a historical one: they are in advance of the society in which they exist. The church's medieval omnipotence, though, is merely a manifestation of the temporal and spiritual forces which allot to humanity an unkind fate. Escape is impossible, Moore suggests, except by a renunciation of earthly attachments. Abélard's error is not to perceive this law before asceticism is forced upon him.

Héloïse and Abélard is dedicated, mysteriously, to 'Madame X'.

Dorothy RICHARDSON   **1873–1957**

Deadlock

see **Pilgrimage** (1938)

1922

Irish Rebellion ● Mussolini's march on Rome ● Permanent Court of International Justice holds first sessions at The Hague ● British mandate proclaimed in Palestine while Arabs declare a day of mourning ● First insulin injection ● Discovery of Tutankhamūn's tomb ● Revival of Ku-Klux-Klan in US ● Marie Stopes holds meetings in London, advocating birth control ● Ludwig Wittgenstein writes *Tractatus Logico-Philosophicus* ● Marc Chagall, *Dead Souls* ● The debut of the 'cocktail' ● Kingsley Amis born ● Alexander Graham Bell and Marcel Proust die ● T.S. Eliot, *The Waste Land* ● A.E. Housman, *Last Poems*

Richmal CROMPTON   1890–1969

Just – William

'There is a theory that, on our way from the cradle to the grave, we pass through all the stages of evolution,' Crompton wrote in 1962, 'and the boy of eleven is at the stage of the savage – loyal to his tribe, ruthless to his foes, governed by mysterious taboos, an enemy of civilisation and all its meaningless conventions.' William Brown is the quintessential eleven-year-old, an anarchic middle-class schoolboy who lives in a Kent village with his long-suffering parents and his lovelorn older siblings, Robert and Ethel. He made his first appearance in a story entitled 'Rice-Mould' in the February 1919 issue of *Home Magazine*, which was an adult publication. In 1922 the stories began to appear in a sister publication, *Happy Mag.*, which was aimed at the entire family, and they ran there until 1940. Crompton's ironic wit and crisp style appealed then as now quite as much to adults as to children, and thirty-eight volumes of the William stories appeared between 1922 and 1970, the year after the author's death. Widely translated, they have sold millions of copies all over the world, and have been made into several films and television series.

Just – William, the first volume of the series, introduces William, his family, and his friends Henry, Ginger and Douglas, all members of a gang called the Outlaws. William disrupts one of Robert's romances, makes a disastrous New Year's Resolution 'to be p'lite', and acquires his faithful mongrel, Jumble. The book is vigorously illus-trated by Thomas Henry, who was as responsible as Crompton for fixing William in the public's imagination. Henry continued to illustrate the books until his death in 1962 at the age of eighty-three; the last five books were illustrated with rather less panache by Henry Ford. While the books' titles – *William – the Dictator*, *William and the Evacuees* and *William Does His Bit* (1938–41); *William and the Brains Trust*, *William's Television Show* and *William and the Pop Singers* (1945–65) – suggest the passing years, William himself does not change.

There were some quite interesting things to do outside. In the road there were puddles, and the sensation of walking through a puddle, as every boy knows, is a very pleasant one. The hedges, when shaken, sent quite a shower bath upon the shaker, which also is a pleasant sensation. The ditch was full and there was the thrill of seeing how often one could jump across it without going in. One went in more often than not. It is also fascinating to walk in mud, scraping it along with one's boots. William's spirits rose . . .

JUST – WILLIAM
BY RICHMAL CROMPTON

Plans to send him away to school in *William – The Outlaw* (1927) are satisfactorily foiled: 'Mrs Brown had a vague idea that some mysterious change of spirit came over a boy on entering the portals of a boarding-school transforming him from young savage to a perfect little gentleman, and she would have liked to see this change take place in William.' Crompton disagreed. She had studied classics and rejoiced in Plato's assertion that: 'A boy is, of all beasts, the most difficult to tame.' She always takes William's side against the ridiculous adults, and this explains the books' perennial popularity with children.

Several examples of the sort of 'little gentleman' both Crompton and William disliked are provided in the books, most entertainingly by Anthony Martin in *William – the Pirate* (1932). This horribly sleek child, the eponymous hero of books which bear an actionable resemblance to the works of A.A. Milne ('Anthony Martin is milking a cow'), is priggish and conceited and inclined to tantrums. The most famous of Crompton's monstrous children, however, is Violet Elizabeth Bott, daughter of a *nouveau riche* sauce magnate who comes to live at the Hall. First introduced as 'The Sweet Little Girl in White' in *Still – William* (1925), the lisping, frilly, soppy, tearful, blackmailing Violet Elizabeth ('I'll thcream and thcream and thcream till I'm thick. I can') represents everything that Crompton and (the occasionally gallant) William most disliked about small girls.

In William, Crompton created one of the twentieth century's most enduring and endearing fictional characters, one who has entered the national consciousness, frequently referred to in contexts far removed from the world of children's literature. She achieved here something which eluded her in the forty-one entirely forgotten books she wrote specifically for adults.

e e cummings   1894–1962

The Enormous Room

cummings's only novel, *The Enormous Room* is an experimental literary account of his imprisonment during the First World War. The narrator has been working as an ambulance driver, hating the pretensions of his fellow Americans and associating with the French. He and a friend run into trouble with the authorities for their pacifist views, and are incarcerated on a trumped-up charge.

Paradoxically, imprisonment becomes a release into personal freedom. As the whang-clang door closes on his solitary cell, the narrator is alone with *Ça Pue*, the stinking latrine in which the previous prisoner's turd sits as a reminder of human life. But the harshness of life in jail is more bearable than the military rule of life outside, and the prison, for all its filth and degradation, is free of the atrocities of war.

As the prisoner is transferred to a communal gaol, the enormous room of the title, his experiences take on the resonance of myth through allusions to Bunyan's *The Pilgrim's Progress* (1678–84). From the Slough of Despond of the ambulance service, through the stages of personal purgation, he moves towards the Delectable Mountains, which take the human form of the prison's bizarre and eccentric inmates.

Although the novel contains some vividly realised characters – the despised, degraded, excrement-encrusted Pole, Surplice, who is reclaimed by the narrator; the Jazz Age 'noble savage', Jean le Nègre; and the martyr-figure, Lena, who undergoes appalling punishment for a minor transgression – and paints a horrifying picture of the brutalising effects of imprisonment, it is chiefly remarkable for its use of language. Describing actual events, cummings abandons conventional chronology. His narrative is impressionistic and fragmented, often dispensing with normal syntax, employing inventive slang and improvisational effects, and juxtaposing radically different styles.

John GALSWORTHY   1867–1933

The Forsyte Saga

see **The Forsyte Chronicles** (1929)

David GARNETT — 1892–1981

Lady Into Fox

Garnett's 'first serious book', beautifully illustrated with woodcuts by his wife, Ray, remains his most highly regarded and popular novel. In 1923 it won both the Hawthornden and the James Tait Black Memorial Prizes. The story of Mrs Silvia Tebrick's sudden and unexplained metamorphosis is soberly recounted in a mock eighteenth-century style, likened by an admiring Virginia Woolf to that of Defoe. In theme the book lies somewhere between Gulliver's experience of the Land of the Houyhnhnms and J.R. Ackerley's account of the transference of love from human to dog in **We Think the World of You** (1960).

When his young wife becomes a fox, Mr Tebrick continues to love her for her remaining human qualities and characteristics. Although denied speech, she communicates her feelings with her eyes, dresses in a bed-jacket, plays piquet and enjoys piano music. Gradually, however, her feral instincts begin to predominate: she gazes hungrily at a pet dove, crunches up the bones of a proffered chicken wing, and messily slaughters a rabbit in the drawing-room. Worst of all, she displays the characteristics traditionally ascribed to her kind, becoming, indeed, a cunning little vixen. Mr Tebrick, who continues to expect her to be 'as candid and honest with him in all things as the country girl he had married', is driven to despair.

He takes her to the cottage where her old nanny lives. This has a walled garden 'where his wife could run in freedom and yet be in safety'. However, she attempts to tunnel under the wall and feigns death when she falls from a tree, in order to run away when her husband's back is turned. Eventually she escapes into the wild and Mr Tebrick realises that it is no longer his wife whom he misses – for he will never recover her in human form – but the beautiful creature she has become. When he tracks her down in a copse, he finds that she has had a litter and feels a pang of absurd jealousy over his vulpine rival for her love. Gradually, whilst playing with her cubs, he recognises that he has found true happiness, which 'is to be found in bestowing love'. He achieves Whitman's dream in 'Song of Myself': 'I think I could turn and live with animals, ... / they show their relations to me and I accept them, / They bring me tokens of myself, they evince them plainly in their possession.' The novel ends with Mr Tebrick vainly attempting to protect his vixen from the hunt.

Variously described as a parable or fable, the book defies interpretation, but was described by the author as a '*reductio ad absurdum* of the problem of fidelity in love'.

James JOYCE — 1882–1941

Ulysses

Ulysses established Joyce's reputation as perhaps the greatest writer of English prose in the twentieth century – arguably in any century. Joyce began initial work on the book as early as 1914, and during the composition process developed his mature, essentially comic, artistic vision, as well as his exceptionally inventive mature literary technique. Publication was fraught with difficulties, however, owing to the explicit language and to the sexual nature of some passages (notably Molly Bloom's erotic reverie, with which the novel concludes). Harriet Shaw Weaver of *The Egoist*, who had serially published **A Portrait of the Artist as a Young Man** (1916), was unable to find a printer willing to handle the new book. She eventually published four chapters, with some excisions, in 1919. By this time, the novel had begun to appear serially in America in the *Little Review*, starting in March 1918. Ezra Pound, the magazine's foreign editor, incurred Joyce's displeasure by removing lines from a passage about Mr Bloom's visit to the jakes without first asking the author's permission. Such caution did not prevent the US postal authorities seizing and burning three editions between 1919 and 1920. A formal complaint against one issue was launched by the secretary of the New York Society for the Suppression of Vice, and the resulting prosecution in February 1921 resulted in fines and an undertaking to publish no further episodes. A complete version of *Ulysses* was eventually published by Shakespeare and Company in Paris in February 1922 in an edition of 1,000 copies. Joyce worked on the book right up until the final moment, adding and altering (a circumstance which goes some way towards explaining why attempts to publish a 'definitive' text have been attended by controversy). The first English edition of 2,000 copies, published in Paris by the Egoist Press, followed in October, but 500 of these were burned by the New York postal authorities. Of a replacement edition of 500, which the Egoist Press hoped to send to America via England, 499 copies were destroyed by customs officials at Folkestone. The first unlimited edition was published in Paris in 1924, and although translations meanwhile appeared in Germany, France, Czechoslovakia and Japan, it was not until 1934 that the first American edition appeared. The government attempted to suppress it, were defeated in the courts, entered an appeal, and were defeated again. No unlimited edition was published in Britain until 1937.

The priest was rinsing out the chalice: then he tossed off the dregs smartly. Wine. Makes it more aristocratic than for example if he drank what they are used to Guinness's porter or some temperance beverage Wheatley's Dublin hop bitters or Cantrell and Cochrane's ginger ale (aromatic). Doesn't give them any of it: shew wine: only the other. Cold comfort. Pious fraud but quite right: otherwise they'd have one old booser worse than another coming along, cadging for a drink. Queer the whole atmosphere of the. Quite right. Perfectly right that is.

ULYSSES
BY JAMES JOYCE

The title of *Ulysses* signals a relationship with the whole epic tradition stemming from Homer's *Odyssey*. Whereas Homeric epic aimed to summarise the cultural life of a single nation, however, Joyce described his modern *Odyssey* as an epic of two races: of his native Irish, and also of the Jews. The hero of *Ulysses*, Leopold Bloom, is thus both

an Irishman and a Jew, unsettling the potentially atavistic nationalism of the epic genre. The novel's action spills fractionally over the confines of a single day, beginning early in the morning of 16 June 1904, and continuing into the small hours of the next day. But the narrative depends less on conventional action than on stylistic experimentation, beginning with Joyce's celebrated invention of the interior monologue, a style which presents to the reader the perceptions, thoughts, memories and desires of the various characters apparently unmediated by a controlling narrative voice. This and later stylistic devices helped Joyce to create not only an epic account of an individual hero named Leopold Bloom, but an epic of the human body as well, in which the conscious individual is seen in continuous relationship with the anonymous impulses of the unconscious.

The Homeric analogy is maintained throughout the novel: like his original, Mr Bloom is a wandering adventurer who is eventually reunited with a questing son and a patient wife who waits for him at home; but Mr Bloom navigates the hazardous cross-currents and exotic distractions of the modern city of Dublin, rather than the ancient Mediterranean Sea. Moreover, his 'son', Stephen Dedalus, is unrelated to him by blood, rather by temperament and chance encounter, while his wife, Molly, is faithful to him less in body than in spirit.

A brief summary cannot do justice either to the wealth of minor characters in the book, or to its extraordinary stylistic virtuosity, which is the result of a high-spirited reworking of all the avant-garde literary styles available to Joyce in the Paris and Zurich of the early twentieth century. Interior monologue is merely the initial style of the book, which becomes progressively more complex towards its centre, culminating in the phantasmagoric dream-theatre of the 'Circe' episode (chapter 15). In the final episode, 'Penelope' (chapter 18), the virtually unpunctuated flow of Molly Bloom's monologue fuses the realism of the initial episodes with the surrealism and fantasy of the central episodes to form a new whole, in which memory and imagination become indistinguishable. The ultimate achievement of

Ulysses is to project its psychological adventure through the lens of a parallel adventure of writing, an odyssey of literary technique.

D.H. LAWRENCE   1885–1930

Aaron's Rod

Aaron Sisson, a check-weighman in a Midlands colliery, brings home the Christmas tree on a snowy December night, to the delight of his two small girls but the disparagement of his wife, Lottie. After twelve years of marriage he feels trapped by the domestic scene, and determines to break free to form his own future. Leaving his money for his family, he sets out into the Bohemian world of upper-class life, meeting the enigmatic Rawdon Lilly and his liberated wife, Tanny.

Lilly and Tanny have learned to be apart, and Lilly now leads Aaron on his quest for individuality. But Aaron's identity is bound up in the struggle with Lottie: 'he considered himself as first and almost single in any relationship. First and single he felt, and as such he bore himself. It had taken him years to realise that Lottie also felt herself first and single: ... she, woman, was the centre of creation, the man was the adjunct.' One solution to this problem is the quasi-homosexual relationship which develops between the two men, who enter a state of cosy domesticity. Lilly takes on the 'female' role, tending Aaron when he is ill (a distinctly homoerotic massage works wonders: 'The spark had come back into the sick eyes, and the faint trace of a smile, faintly luminous, into the face. Aaron was regaining himself'), doing the cooking, even darning his friend's socks. They discuss the worrying ascendancy of women and deplore the fact that men's lives are circumscribed by their spouses and children. Even within this relationship, however, there is a struggle for supremacy, which does not resolve itself until the end of the novel.

The two men go to Italy, where Aaron earns his living busking with his flute. London had seemed to him stifling with its artificial lights, and in the aristocratic houses where he now finds he is welcomed as a guest, his creativity is given a generous home. Lawrence weaves a travelogue of Italy and the seeds of fascism into the biblical symbolism of *Aaron's Rod*: the biblical Aaron is the brother of Moses, and his blossoming rod is a miraculous symbol of authority. At the climax of the novel, during a riot in Florence, Aaron's flute is broken, and he is tempted by Lilly to submit to a new form of relationship, the passionate brotherhood of men: 'And then a man naturally loves his own wife, too, even if it is not bearable to love her.' The novel ends with a characteristic rebirth of hope through destruction, as Aaron enters the freedom of his uncertain future.

Sinclair LEWIS   1885–1951

Babbitt

George Follansbee Babbitt is a real-estate agent and 'a God-fearing, hustling, successful, two-fisted Regular Guy'. He lives in prosperous Zenith with his wife and three children. His politics are Republican; his religion is Presbyterian; his opinions are taken from the editorials of the *Advocate-Times*, and his aspirations from its advertisements. He is a member of three Rotarian-style organisations and his friends are by and large much like himself. But Babbitt, in middle age, is discontented. He dislikes his job, his family and himself. At night he dreams of a 'fairy child ... in the darkness beyond mysterious groves'.

The novel's first seven chapters portray a Babbittian day. From shaving to sleep, nothing extraordinary happens. Instead, the narrative concentrates on the ordinary activities of Zenith – the average dishonesties, the snobberies, the pleasures and strains, the boisterous society of the clubbable male. Babbitt appears to be the paragon of his tribe. However, his closest friend, Paul Riesling, is a misfit, a failed violinist married to nagging Zilla; to him, Babbitt is able to confess his secret and imprecise unease.

How Babbitt copes with that unease is described in the book's remaining twenty-seven chapters. At first, he frantically increases his prestige in the eyes of his peers: he speaks out on behalf of his profession, reorganises his local Sunday School and campaigns against Seneca Doane, the radical lawyer. He is freshly disturbed, though, when Riesling shoots Zilla. Riesling is gaoled and – his problems compounded by loneliness – Babbitt conducts an affair with Tanis Judique. Tanis and her 'set' represent Zenith's best offer of Bohemianism: they drink prohibition liquor and dance to gramophone records. Nevertheless, Babbitt's liaison rapidly turns into another version of the domesticity he is seeking to escape. There is a strike and Babbitt bumblingly supports the strikers – '[They're] not such bad people. Just foolish.' For this subversive crankiness he is ostracised; his business suffers and he is bullied by the Good Citizens' League. He acknowledges the errors of his thinking and resumes his previous ways. All the same, he tells his son, 'I've never done a single thing I've wanted!'

Around his hero's mild rebellion, Lewis assembles a critique of North American enterprise culture. His irony is heavy and he sacrifices character to theme. Moving from topic to topic – work, leisure, family, religion, class, politics – his manner is at once documentary and comic. Unable to perceive their own banality, his actors cloak their behaviour in Wild West metaphors and pioneer imaginings. If a part of the author admires their go-ahead vitality, his indictment is unmistakably made: Zenith's energy and affluence cannot conceal its lack of moral aim and spiritual value. No ready consolation is proposed: Babbitt's marital infidelity is depressingly conventional and he is ill-equipped (both by experience and education) for a romance with Seneca Doane's radicalism. He has no choice, save to participate in a dynamic individualism which sells its tasteless fruits at the price of conformity. His fairy child is denied to him.

May SINCLAIR　　　　　**1863–1946**

Life and Death of Harriett Frean

Sinclair's novels, which were highly regarded by her contemporaries, fell into neglect until rediscovered by feminists in the 1970s. Her work is very much of the twentieth century, employing modernist techniques – Imagism, symbolism, stream of consciousness (a term she coined) – and influenced by psychoanalysis. *Life and Death of Harriett Frean*, a masterpiece of economy which encompasses the whole life of a woman in fifteen terse chapters, is perhaps her best novel. Sinclair takes an essentially Victorian story and examines it from a distinctly modern viewpoint, and as such it is quite as much a critique of an entire age and society as it is of a particular individual.

Harriett Frean is the devoted daughter of a well-educated, intellectual apparently progressive parents. Her father, a stockbroker, reads 'dangerous books: Darwin, and Huxley, and Herbert Spencer', writes a study entitled *The Social Order* and contributes to the *Spectator*. Her mother insists upon behaving 'beautifully' at all times, regarding disobedience, selfishness and ill manners as aesthetic lapses. The compliant Harriett is determined to follow this code, all the more so when her one transgression – straying into a country lane where the sexually symbolic dark red campion grows – causes her parents pain: 'All very well to say that there would be no punishment; *their* unhappiness was the punishment.' At school she befriends the thin, shy, unhappy Priscilla Heaven, who swears that she will never marry because she loves only Harriett.

Priscilla does, however, become engaged, and Harriett falls in love with the fiancé, Robin Lethbridge. Although Robin returns her feelings, Harriett refuses to marry him, determined to sacrifice her happiness to that of her friend, a course naturally approved of by her parents. Her father loses his money (as did Sinclair's own father), but insists upon behaving with the utmost consideration to his creditors, even though he is dying. Harriett's mother also behaves beautifully as she succumbs to cancer.

As the novel progresses, Sinclair questions the whole notion of the sort of self-sacrifice followed by the heroines of countless Victorian romances. Harriett's altruism is seen to be less than pure: 'When she thought of Robin and how she had given him up she felt a thrill of pleasure in her beautiful behaviour and a thrill of pride in remembering that he had loved her more than Priscilla.' Priscilla realises that Robin is really in love with Harriett and develops a psychosomatic paralysis in order to gain her husband's attention. Robin's own health is undermined by the strain of looking after his increasingly invalid wife, while Harriett becomes a desiccated old maid, abandoned when her parents die, and clinging to the steadily diminishing reputation of her father. The truth about the past is gradually revealed, and we see how little of it Harriett has understood, and how misguided she has been. Robin's niece (the voice of another generation, and the author's own) denounces Harriett's actions as those of 'a selfish fool' thinking only of 'her own moral beauty'.

In her sixties, Harriett recalls the one occasion she disobeyed her parents by going into the lane: 'She had a fondness and admiration for this child and her audacity.' Like her mother, she develops cancer, a symbol of repression and thwarted desire, and drifts towards death, back through childhood, still behaving beautifully.

1923

The Union of Soviet Socialist Republics established ● British Matrimonial Causes Act gives women equality in divorce suits ● Adolf Hitler's *coup d'état* in Munich fails ● A birth control clinic is opened in New York ● Stanley Spencer, *The Resurrection* ● Maria Callas born ● Katherine Mansfield and Sarah Bernhardt die ● Edith Sitwell, *Façade*

Arnold BENNETT　　　　　**1876–1931**

Riceyman Steps

Henry Earlforward owns and lives above a secondhand bookshop in Riceyman Square, Clerkenwell. He is a miser (hoarding treasury notes and sovereigns) and a bachelor, but, in his forties, he sets out to woo Mrs Violet Arb, a widow and the proprietress of a confectioner's shop. Although considerable sums have been left her by her husband and by a great aunt, Violet fears penury. She sees in Henry a man who will be able to manage her wealth, and he discerns in her someone almost as stingy as he is. The progress of the couple's economical romance is facilitated by Elsie Sprickett, the young woman who chars for them both.

Herself a widow, Elsie spent a mere two nights with her husband before he was killed fighting in the First World War. Now she is being courted by Joe, a well-intentioned lad deranged by shell-shock. Joe works for Dr Raste, the local physician, until, in a fit of dementia, Joe waves a carving knife at Mrs Arb. Shortly after, he flees Riceyman Square, promising Elsie that he will return when he is cured.

Henry and Violet marry. Their honeymoon consists of a visit to Madame Tussaud's. Violet sells her business and moves into the bookshop; Elsie becomes the resident maid. Soon, however, even Violet is alarmed by Henry's meanness: he forbids fires, refuses to replace lightbulbs and declines all save the cheapest and most basic foods. (Plagued by hunger, Elsie is reduced to eating bacon raw and secretly.) Violet pleads and argues with Henry, yet always his 'soft obstinacy' prevails. In due course he and Violet fall ill. He refuses a doctor on the grounds of expense, but Elsie summons Raste. He diagnoses cancer in Violet and a serious, unspecified complaint in Henry. Violet is transferred to hospital where, under surgery, she dies. Had it not been for her malnutrition, she would have survived. Henry insists that Elsie nurse him at home. This she faithfully does. Her burdens are increased when Joe, suffering from malaria, reappears. Risking her reputation, Elsie hides him in her room while he recovers.

Henry, too, dies. His and Violet's fortunes

are inherited by the Revd Augustus Earlforward. Joe and Elsie marry and enter service in Raste's household. Joe is joyful: Elsie is 'apprehensive about future dangers ... but she was always apprehensive'. The novel ends by asserting that the newly-weds are characteristic of Clerkenwell.

Studied in its banality, *Riceyman Steps* is a comedy of social naturalism. With its small cast and restricted locality, it sketches the demise of a small corner of England. Shabby, respectable and insecure, the shopkeeper despises the working class (who swarm in the decay of Clerkenwell) at the same time as being dependent on their labour. If Elsie is too stupid to take the measure of her exploitation, she is none the less elevated by her stoical practicality. She and the masses like her (Bennett implies) possess a moral wealth, active and beneficial and sharply distinct from Henry's material accretions. Raste's employment of her is the nearest the novel can contrive to a 'happy ending', representing as it does an alliance between the wisdom of the educated bourgeoisie and the instinctive goodness of its social inferior.

The grotesque aspect of the story – two adults starved to death by an excess of thrift – is oddly reminiscent of the *Cautionary Tales* (1907) of Bennett's acquaintance and contemporary, Hilaire Belloc.

Ronald FIRBANK   1886–1926

The Flower Beneath the Foot

Firbank's 'Record of the Early Life of St Laura De Nazianzi and The Times in Which She Lived' is the logical culmination of his obsession with saints and their relics. It was conceived whilst the author was in Algeria, where he was much taken by the langourousness of the inhabitants. Although the setting is 'some imaginary Vienna', Pisuerga owes much to North Africa.

As a young 'lay boarder' at the Convent of the Flaming Hood, Laura had often gazed at the royal palace 'wondering half-shrinkingly what "life in the world" was like'; the novel shows that it is treacherous and ruthless and that she will be cast aside, the eponymous flower beneath the foot of society. Realising that her impulse to take the veil owed more to her idolisation of one of the nuns than to 'any more immediate vocation', Laura leaves the

---

The Countess became reminiscent.

'In Venice,' she said, 'the indecent movements of the gondolieri quite affected my health, and, in consequence, I fell a prey to a sharp nervous fever. My temperature rose and it rose, ah, yes ... until I became quite ill. At last I said to my maid (she was an English girl from Wales, and almost equally as sensitive as me): 'Pack ... Away!' And we left in haste for Florence. Ah, and Florence, too, I regret to say I found very far from what it ought to have been!!! I had a window giving on the Arno, and so I could *observe* ... I used to see some curious sights! I would not care to scathe your ears, my Innocent, by an inventory of one half of the wantonness that went on; enough to say the tone of the place forced me to fly to Rome, where beneath the shadow of dear St. Peter's I grew gradually less distressed.'

THE FLOWER BENEATH THE FOOT
BY RONALD FIRBANK

---

convent and enters the palace as a lady-in-waiting. She catches the eye of the philandering heir to the Pisuergan throne, his Weariness Prince Yousef. The Prince's mother, her Dreaminess the Queen (Lois), is appalled at the prospect of a morganatic marriage, but Yousef insists that Laura 'saves us from *cliché*'. When Laura learns from court gossip that not only was she 'not the first in the Prince's life, and that most of the Queen's maids, indeed, had had identical experiences with her own', but that Yousef had also had 'light dealings with the dancer April Flowers, a negress', she is distraught. Rumours of the Prince's betrothal to the 'none too engaging' Princess Elsie of England are rife, and Laura decides to re-enter the convent. This unorthodox hagiography ends with Laura watching the royal wedding and calling out her former lover's name in despair as she beats her hands against the broken glass which crowns the Flaming Hood's walls.

Much of this populous novel is taken up with Pisuergan court life, notably the state visit of King Jotifa and Queen Thleeanouhee of the Land of Dates. The exotic Queen shares the inclinations of the majority of the ladies at court and makes approaches to Lady Something, the unyielding wife of the British ambassador. Lady Something herself becomes involved in legal proceedings after inadvertently defaming the Kairoulla Ritz. Mme Wetme, wife of the proprietor of the Café Cleopatra and victim of 'the God *Chic*', attempts to bribe the indigent Duchess of Varna to present her at court. Her Dreaminess mounts an archaeological expedition to excavate the ruins of Chedorlahomor, 'a *faubourg* of Sodom', and attends the deathbed of the elderly Archduchess Elizabeth, who succumbs to a chill brought on by her passion for paddling in the palace grottoes. The Countess of Tolga's philanthropic visit to St Helena, where the 'fallen statesman' Count Cabinet lives in exile with a thieving choirboy, is foiled when her boat is becalmed, but she finds solace with her companion in distress, the delicious Mlle Olga Blumenghast. The English governess, Mrs Montgomery, is seduced by Dr Cunliffe Babcock, the royal physician, and there are glimpses of Mrs Bedley's circulating library and the Duchess of Varna's florist's shop, *Haboubet of Egypt*, which is staffed by exotic Eastern boys.

The novel is also partly a *roman à clef* in which diplomats, writers and the British royal family are portrayed, notably in the characters of Harold Chilleywater and Victoria Gellybore-Frinton (the Harold Nicolsons) and the Hon. 'Eddy' Monteith (Evan Morgan). Firbank takes his stylish literary revenge upon Morgan – who had churlishly refused the honour of having the play, *The Princess Zoubaroff* (1920), dedicated to him – by dispatching Eddy Monteith in a footnote which records that he succumbed to 'the shock of meeting a jackal while composing a sonnet . . . Alas, for the *triste* obscurity of his end!'

| Aldous HUXLEY | 1894–1963 |

Antic Hay

Huxley's second novel, which consolidated the reputation he had made with *Crome Yellow* (1921), is full not only of the social and intellectual verve of the 1920s but also its emotional unrest. The central character is Theodore Gumbril Junior, BA Oxon, a man in search of a role. We first meet him as a needy and discontented schoolmaster, disturbed by the hardness of the pew in the school chapel, dreaming of a pair of pneumatic trousers to soften the pain, and persecuted by a class of schoolboys' repetitive essays on Pio Nono.

He gives up his job and goes to London, meaning to patent the pneumatic trousers and make his fortune, but spends more hours in the pleasure-seeking social and intellectual world, the description of which includes portraits of several recognisable celebrities. It is an intoxicating, springtime London to which he comes ('Henry James's London possessed by carnival', as Evelyn Waugh described it), yet one haunted by emotional emptiness. Chief among the characters he meets is Myra Viveash, a fascinating society woman (based on Nancy Cunard), forever mourning the lover killed in the First World War, and suffering from a 'deathbed within'. She attracts almost all the male characters, including Gumbril, but can satisfy none of them.

Other denizens include: the pretentious Mr Mercaptan, who writes delicious 'middles' for the literary papers ('We needn't *all* be Russians, I hope. Those revolting Dostoievskys'); the clownish artist Casimir Lypiatt (Wyndham Lewis), tormented by his love for Myra Viveash and his lack of talent; the self-consciously diabolic Coleman (Philip Heseltine); the ponderous physiologue Shearwater (J.B.S. Haldane); and his bouncy little wife, Rosie, delighted with her pink underwear and passing happily from man to man. Rosie has an affair with Gumbril, who has grown a beard to prove his virility and calls himself Toto. The greatest intellectual seriousness is provided by Gumbril's father, who is obsessed by his model of an ideal London and the flight of birds in the garden of his square.

A different sort of life opens up for Gumbril through his brief affair with an innocent, sensitive girl, Emily, but he is prevented from keeping a rendezvous at her country cottage by meeting Myra, desperate for his company because of her boredom. The novel ends with Gumbril and Myra travelling around London at night by taxi, visiting all their friends, and ending up at the hospital where Shearwater is endlessly pedalling a stationary bicycle in a heated chamber to find out how much sweat the human body can produce. 'Tomorrow will be as awful as today,' comments Myra Viveash, yet the brilliance of the novel belies this dark message.

D.H. LAWRENCE   1885–1930

Kangaroo

Arriving in inter-war Australia from the exhausted intellectualism of the old world, Richard and Harriet Somers are disappointed to find, in place of the anticipated open bushland, a neat little town of dog-kennel bungalows. Their temporary home is quaintly called 'To rest in', but within its flimsy walls their struggle for supremacy in marriage, based on the Lawrences' own experiences, continues. Drawn into the increasingly violent internal politics of Australia, in a fictionalised version of Italian fascism, Somers becomes involved with the enigmatic Kangaroo, the Mussolini of the Diggers' movement. Harriet finds herself excluded from the rough male politics: 'I am nowhere. I just don't exist.'

Playing the intellect against the emotions, political involvement against aesthetic isolation, *Kangaroo* explores through the third-person narrative the themes of Lawrence's major novels. An episodic flash-back to wartime Cornwall sees Somers exposed to the indignities of patriotic militarism as he is declared unfit for a war he despises. 'They are *canaille*,' he reflects, echoing the sensitive and elitist Gudrun in **Women in Love** (1920). Lawrence drew upon his observations of the rise of fascism in Italy to create the major characters and events of *Kangaroo*, transferring them to Australia to set against the shimmering heat of the outback.

Instead of blood-brotherhood, Somers finds misfired democracy, which reduces everything to the lowest common denominator. Small-scale protest produces not revolution but sordid deaths, and, as Kangaroo lies dying, Somers refuses to confirm his love. 'Human love as a ritual offering to the God who is out of the light, all well and good. But human love as an all in all ... the strain and the unreality of it were too great.' Lawrence described this amalgam of autobiography and fiction as a 'funny sort of novel where nothing happens and such a lot of things *should* happen', and the lack of structural form reflects the final isolation of the two Europeans in 'God's Own Country' in an unfulfilled search for Utopia. Somers sadly concludes from his experiences: 'I am afraid, that like Nietzsche, I no longer believe in great events.'

Dorothy RICHARDSON   1873–1957

Revolving Lights

see **Pilgrimage** (1938)

1924

V.I. Lenin dies ● Britain, France and China recognise USSR ● Calvin Coolidge signs bill limiting immigration into US and entirely excluding Japanese ● Rioting between Hindus and Muslims in Delhi ● The first insecticide ● Gwen John, *The Convalescent* ● George Gershwin, *Rhapsody in Blue* ● British Empire Exhibition, Wembley ● Woodrow Wilson, Joseph Conrad, Anatole France and Giacomo Puccini die ● John Crowe Ransom, *Chills and Fevers*

Michael ARLEN   1895–1956

The Green Hat

Shocking in its time, *The Green Hat* achieved a reputation as the quintessentially brittle, cynical, glittering novel of the 1920s. Its status as a bestseller made its author (born Dikram Kouyoumdjiam in Bulgaria in 1895, but naturalised as a British citizen in 1922) a byword for cosmopolitanism, sophisticated European world-weariness, and negligent, aristocratic, elegant mockery of the traditional values of the English in the aftermath of the First World War.

The narrator's flat in Shepherd Market, Mayfair, is invaded late one night by Iris Storm, who has come looking for her alcoholic brother who lives in the upstairs apartment. Iris (who is based upon Arlen's lover, the poet and editor Nancy Cunard) is 'tall, not very tall, but as tall as becomes a woman. Her hair, in the shadow of her hat, may have been any colour, but I dared to swear that there was a tawny whisper to it. And it seemed to dance, from beneath her hat, a very formal dance on her cheeks.' She talks to this stranger, the author of her tragic history, and discovers a sympathetic intelligence to which she responds throughout her brief, brilliant career as a 'shameless, shameful lady', the personification of all the bright, beautiful – and doomed – young things of her generation.

Hints of her waywardness emerge from other sources as the narrator dines, drinks and dances across town: among them is the rumour that Iris has been 'seen night after night in a Russian cabaret in Vienna with an Italian Jew who is said to have made a fortune by exporting medicated champagne to America'. This is perhaps the least of her eccentricities. It is not, of course, the behaviour to be expected of an English aristocrat, daughter of an English peer. Her reputation is thoroughly compromised, and speculation inevitably mounts as to why her husband, Boy Fenwick, committed suicide by defenestration on their wedding night. Iris, it is natural to suppose, was the (literally) precipitating factor. She makes no attempt to refute this notion: her husband 'died for purity' – presumably, she lets it be inferred, after some intolerable confession by his new bride.

In the aftermath of a supposed affair with Napier Harpenden, who has always loved Iris but has married the beautiful Venice Pollen, Iris's secret is revealed (by Napier) to a gathering of their friends: Boy Fenwick, when he married Iris, had syphilis 'and went mad when he realised what he had done'. Iris, stripped of her secret, her holy possession, her defence of her husband and her whole *raison d'être*, merely whispers in her husky voice. 'You've taken from me the only gracious thing I ever did in my life.' She roars off in her yellow Hispano-Suiza into the night, closely followed by her alarmed friends. They are in time to watch the great car crash into a giant tree. On the grass beside the road, tossed from the car, lies Iris's green hat.

Ronald FIRBANK   1886–1926

Prancing Nigger

Firbank's penultimate novel was the first to be published commercially – that is, at the publisher's rather than the author's expense. This took place in America, after which it was issued in Britain as *Sorrow in Sunlight*, Firbank's preferred but now discarded title. Another of his studies of the destruction of innocents abroad, *Prancing Nigger* is the culmination of Firbank's negrophilia, and its comparative success was largely due to the fact that it reflected a period, Jazz Age *culte*. Indeed, there were plans (which came to nothing) for a 'Jazz Fantasy' based upon the book, with music by Gershwin. Much of the novel is written in a brilliantly sustained patois of Firbank's own devising.

Firbank's assertion that he had gone to the West Indies to collect material for a novel about Mayfair is borne out of this exotic study of social pretensions, set on the Cuba-like island of Tacarigua. Mrs Ahmadou Mouth is discontended with life in the provincial village of Mediavilla and is determined to persuade her pious husband (the 'prancing nigger' of the title) to move to the city of Cuna-Cuna thus 'allowing his offspring such educational advantages and worldly polish that only a city can give'. Whilst the thirteen-year-old Edna is excited by the prospect, and Charlie looks forward to exploring a new habitat for butterflies, their older sister Miami does not want to leave her boyfriend, a local fisherman called Bamboo. Regardless of the wishes of her family, Mrs Mouth takes a ninety-nine-year lease on the Villa Vista Hermosa, and the family dons clothes and trundles to the city in their cart.

Mrs Mouth attempts to get into 'S'ciety' by sending Edna to pay a call on Madame Ruiz, 'arbitress absolute of Cunan society'. Edna is immediately seduced by Vittorio, Mme Ruiz's equivocal son, who sets her up as a *poussin de luxe* in an apartment in the Avenue Messalina. Charlie fares no better and is last glimpsed 'fast going to pieces, having joined the Promenade of a notorious Bar'. Mrs Mouth manages to gain an invitation to a charity gala in aid of 'the famous convent of Sasabonsam', devastated by an earthquake, but encounters racial prejudice when she is worsted in an argument over a chair with the Duchess of Wellclose, and looked down upon by Mme Ruiz's 'bepowdered footman'. News comes that Bamboo had set out to sea in order to avoid the earthquake, 'but the boat had overturned, and the evil sharks ...' Mrs Mouth, who had seen death in her cards, observes that 'as de shark is a rapid feeder it all ober sooner dan wid de crocodile, which is some consolation for dose dat remain to mourn'; but Miami is not consoled and turns to religion. In the final scene Edna watches from a balcony as her sister takes part in a penitential procession, then goes off with Vittorio to the theatre.

Ford Madox FORD 1873–1939

Some Do Not

see **Parade's End** (1928)

E.M. FORSTER 1879–1970

A Passage to India

Forster's last and most highly regarded novel was published after a long silence. Apart from some journalism and a collection of short stories, he had published nothing since *Howards End* (1910). His visit to India in 1912–13 had inspired him to begin a novel, but he was soon overcome by a sense of failure. In 1921 he returned to India as secretary to the Maharajah of Dewas Senior, but this experience did little to renew his confidence. Encouraged by Leonard Woolf, however, he took up the abandoned novel once more. Further encouragement came from his new friend and protégé, J.R. Ackerley, who, at Forster's suggestion, had gone to India in 1923 to become secretary to the Maharajah of Chhatarpur. Ackerley wrote frequent letters to his mentor about the goings-on at the court of the eccentric maharajah, letters which Forster described as 'a godsend to my etiolated novel. I copied out passages and it became ripe for publication.'

The themes that characterise Forster's fiction receive their fullest and most profound expression in this novel. He found in the friendships and conflicts of the rulers and the ruled in British India a unique example both of human relationships and their inherent difficulties. In India the rich variety of races, religions, classes and types, and the sheer size of the country, came to represent, more than Italy or the English countryside, a metaphor for the universe itself, against which man's values and activities could be judged and tested. Out of the confrontation between the British and India, Forster made his most comprehensive and impressive statement of human possibilities and limitations.

Adela Quested comes to Chandrapore to marry Ronnie Heaslop, the city magistrate, chaperoned by her fiancé's mother, Mrs Moore. Against the advice of the British community, she pursues social contacts with the Indian population, a course which culminates in the incident that lies at the heart of the novel: the visit by Adela and Mrs Moore to the Marabar Caves. No other members of the British community at Chandrapore have visited the caves; but few others, in any case, would be as susceptible as Mrs Moore and Adela, who are both open to the experience of the 'real India'.

The 'extraordinary caves' represent the essence of India, unknowable and supernatural, and they have a malign effect on the two women. In their claustrophobic atmosphere Adela comes to believe that she has been assaulted by Dr Aziz. This accusation and the subsequent farcical trial (at which Aziz is acquitted) provoke all the latent hostility between the British and the Indians, even disturbing the friendship of Aziz and Fielding, Principal of Chandrapore's small college. Fielding believes that the world 'is a globe of men who are trying to reach one another and can best do so by the help of good will plus culture and intelligence – a creed ill-suited to Chandrapore' – and his friendship with Aziz is the one relationship that seemed to rise above the divisions of religion and race. Mrs Moore's experience generates a feeling of spiritual despair rather than an occasion of social panic: in the caves' dull, monotonous echo she hears with horror the nihilistic message: 'Pathos, piety, courage – may exist, but are identical, and so is filth. Everything exists, nothing has value.'

The pessimism that lurks at the edge of all human endeavour in *A Passage to India* is scarcely held at bay by the reconciliation of Aziz and Fielding, the *rapprochement* of Aziz and Adela, and the symbolic marriage of Fielding and Mrs Moore's daughter, Stella, which all occur at the end of the novel.

Although Fielding and Aziz resolve their differences, significantly not in British India but in the native state of Mau, it is with the recognition that 'socially they had no meeting place' and that 'the divisions of daily life' have made their friendship impossible to sustain. Some hope, however, can be derived from the individual salvation of the principal characters and, by implication, from the survival of the spirit of liberalism, tested and strained as it may have been by the threat of chaos and nothingness, exemplified by the extraordinary Marabar Caves.

John GALSWORTHY   1867–1933

The White Monkey

see **The Forsyte Chronicles** (1929)

Margaret KENNEDY   1896–1967

The Constant Nymph

The Constant Nymph was the outstanding bestseller of its time, selling more copies than any other novel in the decade of its publication. It tells the story of the Sanger family, and particularly of Tessa, the constant nymph of the title. Tessa is the daughter of a brilliant Bohemian composer, Albert Sanger, and his deceased second wife, a well-born Englishwoman. With his 'circus' of precocious children, his slovenly mistress and assorted hangers-on, Sanger lives for most of the year in a rambling chalet high in the Austrian Alps, where his admirers flock to see him. Chief amongst these is Lewis Dod, also a gifted composer and the object of the fourteen-year-old Tessa's devotion. Tessa is 'unbalanced, untaught and fatally warm-hearted', an unstable combination of innocence and determination: 'She would give herself to pain with a passionate readiness, seeing only its beauty, with the singleness of vision which is the glory and the curse of such natures.'

Tessa's idyllic world is shattered by the sudden death of her father, and the revelation of Lewis's love for her beautiful cousin

Florence, who has come to Austria to settle the family's affairs. Marriage to Florence in London, however, is too restrictive for the uncompromising Lewis, and he belatedly realises that he returns Tessa's love. They elope to Brussels where, in a scene reminiscent of high opera, Tessa dies of heart-disease: both emotional and cardiovascular.

He had always thought her the pick of the bunch. She was an admirable, graceless little baggage, entirely to his taste. She amused him, invariably. And, queerly enough, she was innocent. That was an odd thing to say of one of Sanger's daughters, but it was the truth. Innocence was the only name he could find for the wild, imaginative solitude of her spirit.

THE CONSTANT NYMPH
BY MARGARET KENNEDY

The principal theme of *The Constant Nymph* is the struggle in art and life for freedom from confining moral, social and religious attitudes. Both Tessa and Lewis are intended to appear almost as 'noble savages', innocent of convention. Tessa, in fact, is unconventional only in the mildest sense, except in the central matter of her obsession for Lewis. His unconventionality takes for the most part the form of a rather adolescent rudeness to and about people whose musical taste he despises. Florence, the figure of debilitating convention, is a more plausible character than either, and finally more sympathetic in her efforts to balance the demands of reason and licence.

It is a curiously naïve story, especially to a modern reader familiar with Nabokov's *Lolita* (1955); and it is perhaps the last quality novel of its kind before the English translations of Freud so diffused into proper culture as to make such a tale unbelievable. A later novel, *The Fool of the Family* (1930), about the subsequent fortunes of the Sanger family not surprisingly proved unsuccessful.

Like many of her readers, Kennedy did not want to look too steadily at the sexual aspects of the emotional situation she describes. The relationship between Tessa and Lewis significantly is not consummated; and Florence's belated accusation of impropriety is indicative of her jealousy rather than an acknowledgement of the sexual factor in the relationship. This central ambiguity, however, is integral to the novel's success, and in its abrasive sentimentality achieves an undeniable power and charm. It remains one of the most popular romantic novels ever written, and was adapted by the author and Basil Dean for an equally successful play (1926) and film (1933).

R.H. MOTTRAM   1883–1971

The Spanish Farm

see **The Spanish Farm Trilogy, 1914–1918** (1926)

Mary WEBB   1881–1927

Precious Bane

This novel is the story of Prudence Sarn, a young girl living in rural Shropshire at the time of the Napoleonic Wars, narrated by herself as an old woman and written in an approximation to the Shropshire dialect of the period. In reality it is a prose diction with much of the heightened quality of poetry, allowing many exalted descriptions of nature as well as the effective telling of a romantic story.

Prue lives with her mother and brother Gideon in the tiny village of Sarn. They are poor, but Prue is imaginative and intelligent. Unfortunately she has a hare-lip and this disfigurement makes the ignorant rural people among whom she lives believe she is a witch. Gideon, who accidentally killed their father after being threatened with a beating, is a hard, ambitious man, determined to improve their farm by growing more corn, the price of which has increased during the Napoleonic Wars. He enlists Prue to work beside him, although 'there's a discouragement about the place' and the land is reputed to be cursed.

Prue is taught to read and write by their neighbour, the atheistic and twisted Wizard Beguildy, with whose beautiful daughter, Jancis, Gideon is in love. At a love-spinning (a gathering of women to spin and play cards) Prue meets the weaver Kester Woodseaves, a kind and attractive man, with whom she falls in love, although she believes he cannot love her because of her hare-lip.

At a hiring fair, Jancis is apprenticed as a maid for three years, and Kester stops a bull-baiting by offering to chain the dogs single-handed. Prue kills a dog that is savaging him, and saves his life. When Gideon and Jancis are separated, Prue and Kester write letters for them, which are in reality a furtherance of their own developing love. Jancis runs away from her employment but Wizard Beguildy blames Gideon for this and sets fire to his corn-ricks. Angry and bitter, Gideon renounces Jancis; he also feeds his

mother poison because she is ill and threatening to become a burden. Jancis brings their baby to Gideon and pleads to be taken back, but he rejects her again, and she drowns herself and the child. In despair, Gideon also drowns himself. Prue goes to market, but is blamed by her neighbours for so many deaths, and is about to be put on a ducking-stool as a witch, when Kester, who has been in London learning coloured weaving, appears suddenly on a horse, fights and beats her persecutors, and rides away with her.

Ironically, *Precious Bane*, although a best-seller at its first appearance (partly thanks to the Prime Minister, Stanley Baldwin, who praised it in public), eventually became famous as the most celebrated of a school of rural novels satirised by Stella Gibbons in the novel **Cold Comfort Farm** (1932). This has led to a long-standing undervaluation of *Precious Bane*, the intense poetry of which merits wider readership and acclaim.

1925

Mrs Ross of Wyoming becomes first woman governor in US ● Leon Trotsky is dismissed from chairmanship of Russian Revolutionary Military Council ● Italy completes occupation of Italian Somaliland ● Adolf Hitler, *Mein Kampf* ● State of Tennessee forbids teaching of human evolution in schools ● Birdseye extends deep-freezing process to pre-cooked foods ● Constantin Brancusi, *Bird in Space* ● Aaron Copland, Symphony no. 1 ● *The New Yorker* is issued ● Sean O'Casey, *Juno and the Paycock* ● Charles Chaplin, *The Gold Rush* ● Peter Sellers born ● Sun Yat-sen, John Singer Sargent and H. Rider Haggard die ● e e cummings, *XLI Poems*

Willa CATHER	1876–1947

The Professor's House

With *The Professor's House*, Cather turns from the Nebraskan background of her early novels and looks forward to the New Mexican setting of **Death Comes for the Archbishop** (1927). Part of the novel records the discovery of an ancient Indian city in New

Mexico by a young man, Tom Outland, whose diaries the eponymous professor is editing.

Tom had been a student, and subsequently a friend of Godfrey St Peter, a professor of history. St Peter is reluctant to move into the new house he has built for himself and his wife Lillian, preferring instead to work in the attic-study he has shared for years with the sewing-woman, Augusta, and a leaking gas stove. Tolerating his family, though often disappointed by them, he observes the tensions which have arisen from the death of Tom, who had been engaged to his daughter Rosamond. She inherited Tom's substantial wealth (derived from the invention of a new type of engine) and has subsequently married the sincere but ingratiating Louie Marsellus. Her new husband has developed Tom's invention and speaks of him as an old friend, though he never in fact knew him. This deception infuriates Scott McGregor, one of Tom's fellow students, who is now married to the professor's younger daughter, Kathleen. It appears that Kathleen herself cared more for Tom than her family had realised.

Although Tom's story of the discovery and subsequent loss of the Indian city is absorbing and moving, it is the professor's fragile state of mind which occupies our attention. His desire to remain in the empty house seems to emphasise his sense of isolation and his attachment to the past. Editing Tom's New Mexico diaries, he contemplates past events, reflects on his childhood, and attempts to come to terms with the many painful compromises life has forced him to make. A brush with death, when he is almost suffocated by the fumes from his stove, marks a turning point. The professor accepts his situation – 'Theoretically he knew that life is possible, may even be pleasant without joy, without passionate griefs' – and resolves to face it 'with fortitude'.

| I. COMPTON-BURNETT | 1884–1969 |

Pastors and Masters

Pastors and Masters is a very early example of the witty, irreverent, high-spirited campus novel which did not become a recognised literary genre for another quarter of a century. It concerns a group of highly competitive Cambridge dons and their circle, a sardonic and godless crew, almost all unmarried and far more frankly interested in the opposite sex than was normal in conventional fiction of the early 1920s. All are gossips, much given to speculation about one another's private lives when not amusing themselves with topical questions such as equality for women, and the changing nature of God ('You can have him childless in these

days,' says Bumpus. 'But if you have him, I like him really. I like him not childless, and grasping, and fond of praise. I like the human and family interest.')

Nicholas Herrick, a distinguished scholar in his seventies, is profoundly embittered on account of his failure to produce a single original work in a lifetime of academic study. His younger colleague, Richard Bumpus, is a disappointed man for the same reason. Their friends are startled, and in some cases incredulous, when both men announce, one after the other, that each has at last fulfilled his lifetime ambition to produce a novel ('Real books coming out of our own heads,' says Bumpus. 'And not just printed unkindness to other people's').

Herrick meanwhile supplements his income by running a boys' preparatory school in the back parts of his house, acting as its nominal head while retaining a singularly underqualified and ill-paid staff to do the actual work under the wonderfully ingratiating Mr Merry. Goings-on at this sinister school – gruesome mealtimes, lessons, sports day, equivocal interviews with parents – are at once absurd, hilarious and cruelly accurate. Glimpses of the calamitous home life of one of the boys provide a foretaste of the domestic tyranny and terrorism on which this novelist was to concentrate for the rest of her career.

Herrick lives with his much younger, shrewder and more perceptive sister, Emily: 'Life is just labour and sorrow for us. Labour for Nicholas, and sorrow for me.' She is a philosophical humorist, first in the long line of astute, emancipated and contented spinsters, the characterisation of whom was to prove another of Compton-Burnett's specialities. Emily, having successfully concealed suspicions about her brother's book, is the one who saves his face in the humiliating denouement, in which it turns out that neither Herrick nor Bumpus has in fact written a new book: each is trying to pass off the same stolen or revamped manuscript as his own latest effort.

Theodore DREISER 1871–1945

An American Tragedy

The central event in *An American Tragedy*, Clyde Griffiths's murder of his pregnant girlfriend, Roberta Alden, is based on the much-reported case of Chester E. Gillette, who drowned one Grace Brown in a lake in 1906 and was executed in 1908. In accordance with Dreiser's belief in documentary naturalism, much else in this enormous novel is closely based on real people and events.

The novel opens in Kansas City, where Clyde's parents run an evangelical mission and live in ascetic poverty. Uneducated and beguiled by the wealth he sees around him, sixteen-year-old Clyde gets a job as a bell-hop in the prestigious Green-Davidson Hotel and is soon buying smart clothes and carousing in restaurants and brothels. He falls devotedly in love with a vain and avaricious shop-girl, Hortense Briggs, and spends his money on her even when he knows that his sister has returned to Kansas penniless and pregnant, abandoned by the actor she ran off with. He flees Kansas, having been a passenger in a car which knocked down and killed a young girl, and turns up three years later in Chicago. Working as a bell-hop again, he meets by chance his rich uncle, Samuel, who invites him to come east and try his hand in the family business, the Griffiths Collar and Shirt Company of Lycurgus, New York.

Clyde is running a department in the factory when, despite the warnings of his cousin Gilbert (who hates Clyde for looking like him – only somewhat better, according to the girls), he courts a factory girl, Roberta Alden. He has established her in a room so that he can visit her at night, but then he meets the beautiful and wealthy socialite Sondra Finchley. Through Sondra he mixes with the smart social set, and their romance is blossoming when Roberta informs him that she is pregnant. Their efforts at securing an abortion fail and Clyde is growing desperate; he reads a newspaper account of a drowning in which the victim's body was never recovered, and conceives his plan for disposing of Roberta. On a lake in the north of the state his resolve falters and he bodges the plan, but Roberta drowns all the same, and he proceeds to join Sondra's nearby party. He has left behind a trail of incriminating evidence and is soon arrested by District Attorney Orville Mason. The Griffiths

---

Dusk – of a summer night.

And the tall walls of the commercial heart of an American city of perhaps 400,000 inhabitants – such walls as in time may linger as a mere fable.

AN AMERICAN TRAGEDY
BY THEODORE DREISER

---

family hire a lawyer, who concocts a devious defence for the exhaustively documented trial, but the jury is unconvinced, the press has a field day ('MOTHER OF DEAD GIRL FAINTS AT CLOSE OF DRAMATIC READING OF HER LETTERS') and Clyde is sentenced to death. His mother fails to have the sentence commuted, and he goes to the electric chair.

F. Scott FITZGERALD 1896–1940

The Great Gatsby

Fitzgerald's brief, poetic second novel is his greatest achievement. It established his reputation both as a major novelist and as chief chronicler of the Jazz Age, and has some claim to being *the* Great American Novel. Some of Fitzgerald's own characteristics may be seen in the Great Gatsby himself (the title is, of course, ironic), but the author was also, when he wrote the book, like Nick Carraway, an outsider from the Midwest gazing at the wealth and sophistication of East Coast life. It is all the more remarkable, therefore, that Fitzgerald was not wholly dazzled by the spectacle. In Gatsby, he saw both the possibilities and the flaws in the American dream: Gatsby could rise from obscurity to become a figure of legendary riches and social glamour, but this worldly success, like his aristocratic manner (addressing everyone as 'old sport'), is no more than a veneer, beneath which is a void.

Crudely put, money and prestige do not bring happiness. Furthermore, his fortune is based upon shady business deals. It is ironic that much of Gatsby's bogus lustre has rubbed off upon Fitzgerald's dark fable, notably in film adaptations. There have been three, of which the first (1926) was silent, the second (1949) unmemorable, and the third (1974), directed by Jack Clayton from a screenplay by Francis Ford Coppola, a lavishly mounted, over-extended slab of Art Deco revivalism.

There was music from my neighbour's house through the summer nights. In his blue gardens men and girls came and went like moths among the whisperings and the champagne and the stars. At high tide in the afternoon I watched his guests diving from the tower of his raft, or taking the sun on the hot sand of his beach while his two motor-boats slit the waters of the Sound, drawing aquaplanes over cataracts of foam.

THE GREAT GATSBY
BY F. SCOTT FITZGERALD

The novel is narrated by the Midwesterner Nick Carraway, who has come East 'to learn the bond business'. He rents a house in the village of West Egg next door to a colossal mansion, the scene of dazzling weekend parties. Directly across the bay live Nick's brittle, spoilt cousin, Daisy, and her wealthy husband, Tom Buchanan. Tom is having an affair with Myrtle, the wife of George Wilson, a local garage proprietor. While visiting the Buchanans, Nick meets Daisy's long-standing friend Jordan Baker, who tells him that the vast mansion belongs to the mysterious Jay Gatsby, about whose origins and source of wealth there is much speculation.

Nick eventually meets Gatsby at one of the famous parties, and becomes intrigued by him. He is not altogether taken in by Gatsby's account of his life, particularly after he meets his new friend's shady business acquaintance, Meyer Wolfsheim. Meanwhile, Nick has developed a tepid romance based on mutual self-deception with Jordan, who fills in some of Gatsby's background. She also tells him that five years previously Daisy and the then impoverished Gatsby had been lovers. Gatsby persuades Nick to reintroduce him to Daisy and the old lovers resume their affair. Tom, who dislikes Gatsby for his ostentation and dubious claims to 'old money', finds out about the affair during a disastrous trip to New York, and there is a violent argument in the Plaza Hotel in which Tom accuses Gatsby of bootlegging and other nefarious dealings. He insists that Daisy and Gatsby's affair is over and sends them off home in the latter's yellow car.

During the journey they accidentally run over and kill Myrtle Wilson outside her husband's garage. They do not stop and Gatsby later admits to Nick that Daisy was driving. Nick tries to persuade Gatsby to go away, but he refuses to leave Daisy. George Wilson, aware of his wife's adultery and stunned by her death, asks Tom who owns the yellow car. Later, Gatsby's body is found floating in his swimming pool, with the body of George lying nearby. The Buchanans have vanished. 'They were careless people, Tom and Daisy,' Nick muses, 'they smashed up things and creatures and then retreated back into their money or their vast carelessness, or whatever it was that kept them together, and let other people clear up the mess they had made.'

Apart from Nick, the only people who attend the funeral are Gatsby's father, a humble farmer called Henry C. Gatz, who has arrived from Minnesota, the servants, the postman, and a sole representative of the dead man's many hangers-on. His affair with Jordan Baker over, and disillusioned by the East, Nick returns to the Midwest.

Ford Madox FORD   1873–1939

No More Parades

see **Parade's End** (1928)

William GERHARDIE   1895–1977

The Polyglots

Gerhardie was born in St Petersburg and educated in Russia before coming up to Oxford, where he wrote his first novel, *Futility* (1922), and a study of Chekhov. The influence of his Russian upbringing, his travels during the First World War as a military attaché, and a deep immersion in the works of the great Russian playwright can be seen in his second novel, *The Polyglots*. The novel was written at Innsbruck, where his father lay dying, and passages from the manuscript were read to the old man on his death-bed. It was hoped that publication would restore the family's fortune, lost during the Revolution, but the book's *succés* was entirely *d'estime*.

The novel is a parody of the linguistic limitations which embody human confusion. Georges Hamlet Alexander Diabologh, the narrator, has an ancestry as muddled as his name. Born in Japan of English parents, he was brought up in Russia. Now in his early twenties at the end of the First World War, he returns to his birthplace to join his extended family under the matriarchal rule of the *malade imaginaire*, Aunt Teresa.

Georges's Belgian cousin Sylvia Ninon Thérèse Anastathia Vanderflint is the natural object of his affections. In love with her name and enamoured of her person, he soon finds her extravagant tastes a drain on his solicitor's pay. But Aunt Teresa surprises them in an unchaperoned embrace and their engagement is announced.

Marriage seems a bad idea for Georges, who imagines himself to be a serious young man, 'a writer, an intellectual'. The wedding is in any case postponed by Uncle Lucy's suicide, a tragedy made farcical by his dressing in Aunt Teresa's underwear. While George procrastinates, Sylvia, impatient to be married, favours the unlikely but reliable Gustave, a choice endorsed by Aunt Teresa.

The night before Georges's intended departure, however, Aunt Teresa persuades her new son-in-law out of the way so that Georges and Sylvia can spend the night together. The affair is hypocritically tolerated by the family, who now depart for England, leaving Gustave behind to pay the bills. On the voyage Natasha, the daughter of a Russian exile, dies; her burial at sea is a further confirmation of life's ephemeral nature. As the boat arrives in port, Georges contemplates his circumstances: Sylvia is talking of divorcing Gustave, and Aunt Teresa has saddled Georges with responsibility for her host of dependants. If it were not all so pointless, he concludes, life would be a tragedy.

Anita LOOS   1888–1981

Gentlemen Prefer Blondes

At the suggestion of a 'gentlemen friend', Lorelei Lee decides to keep a diary. *Gentlemen Prefer Blondes* ('The Illuminating Diary of a Professional Lady') consists of four months' breathless, ungrammatical entries and alludes to the critical points of its heroine's past.

Lorelei was brought up in Little Rock, Arkansas. Her first romantic entanglement ends when she shoots her admirer; subsequently, her career blossoms. Judge Hibbard, who presides at her trial, arranges for her to go to Hollywood, where she meets Gus Eisman, the Button King. Concerned for her education, he sends her to New York. Lorelei, though, is a gregarious creature – happiest when her mind is being broadened by men who understand the joy of shopping – and she and her friend Dorothy vigorously pursue their education with the aid of the city's foremost intellectuals, politicians and bootleggers.

Lorelei meets an English novelist and consents to marry him. Learning of this, Eisman determines she ought to go on a cruise. Lorelei and Dorothy depart for Europe. They secure introductions to London society, but find it to be depressingly made up of chiselling aristocrats. (Lorelei nevertheless manages to extract a tiara from Lord 'Piggie' Beekman.) Paris, by contrast, 'is devine ... because shopping really seems to be what Paris is principaly for'. The girls next travel on the Orient Express through 'the Central of Europe'. Germans, they discover, are beer-swilling, sausage-munching boors; Austrians are idle cake-consumers. Lorelei is therefore cheered by the company of an American, the Presbyterian plutocrat Henry Spoffard.

Well yesterday Henrys letter came and it says in black and white that he and his mother have never met such a girl as I and he wants me to marry him. So I took Henrys letter to the photographers and I had quite a lot of photographs taken of it because a girl might lose Henrys letter and she would not have anything left to remember him by. But Dorothy says to hang on to Henry's letter, because she really does not think the photographs do it justice.

GENTLEMEN PREFER BLONDES
BY ANITA LOOS

Instantly reforming her behaviour – and with less ease reforming Dorothy's – Lorelei charms an offer of marriage from Spoffard. Eisman again intervenes. His button business can no longer stand the strain of Lorelei's transatlantic education, so she is shipped back to New York. Spoffard follows and Lorelei finally accepts his proposal. His parents may be demented, his sister alarmingly masculine, and he himself dull: his wealth, however, enables Lorelei to become a film star and to conclude that 'everything always turns out for the best'.

If Lorelei is a light-hearted version of Wedekind's Lulu, her journal conceals no parallel serious purpose. It portrays an artful *ingénue* whose diaphanous deceptions are matched by the semi-naked hypocrisy of her sugar daddies. Prevailing morality is as lax as Lorelei's spelling – and equally comic. That 'presence' is transmuted into 'presents' reflects the material bias in Lorelei's world.

The novel grew from a short story Loos sent to H.L. Mencken, who was at that time editor of *The American Mercury*. 'Little girl,' he replied, 'you're making fun of sex and that's never been done before in the USA. I suggest you send it to *Harper's Bazaar*,

where it'll be lost among the ads and won't offend anybody.' The editor at *Harper's* encouraged Loos to develop the story, and the novel first appeared as a serial in that magazine and was a great success. In book form, the first edition of the novel sold out on the day of publication. It was followed by a second instalment of Lorelei's diary, *But Gentlemen Marry Brunettes* (1927), which described Dorothy's exotic early career and demonstrated how little marriage had altered the diarist. More cynical and explicit than its predecessor, it never achieved the same commercial success. *Gentlemen Prefer Blondes* was filmed twice: first by Mal St Clair (1928), then as a musical (1953) directed by Howard Hawks and featuring Marilyn Monroe and Jane Russell.

Loos later revealed that Lorelei's given name was Mabel Minnow. In the course of a rather solemn interview on British television in the early 1960s Loos was asked: 'If you were to write such a book today, what would be your theme?' 'Gentlemen Prefer Gentlemen,' she replied.

R.H. MOTTRAM   1883–1971

Sixty-Four, Ninety-Four!

see **The Spanish Farm Trilogy, 1914–1918** (1926)

Liam O'FLAHERTY   1896–1984

The Informer

Set in Dublin, *The Informer* typifies a genre of Irish novel which emerged after the civil war of 1922. Ostensibly a political thriller, it expresses a deep dissatisfaction with what was alleged to be a 'new' Ireland, independent to a certain degree of British rule. Like much of O'Flaherty's work, this novel focuses on working-class society, and centres on a slum area of the city. This background is meticulously described in all its misery, and complements the grotesque features of the main character, Gypo Nolan. The theme is betrayal and forgiveness, while an unimportant sub-plot provides romantic interest.

A former policeman, Nolan is a muscular brute, alcoholic, generally penniless, and completely stupid. He once belonged to a secret Communist group known as the 'organisation', and is found one evening in the city doss-house by Frankie McPhillips, an organisation member, who is on the run for murder. Nolan betrays his friend to the police for £20, and during a shootout, McPhillips accidentally kills himself with his own gun. At first overwhelmed by his unaccustomed wealth, Nolan soon realises that he is unable to spend it without attracting attention; distraught with guilt he hears the word 'informer' in every conversation. He walks the streets, which assume a dark and sinister aspect, feebly trying to formulate a plan to save his own skin. He is under suspicion by the organisation, and its ruthless, intellectual leader, Gallagher, who has him watched. In desperation, Nolan accuses a man called Mulligan of 'informing', and fritters away the reward on himself and others.

Until Nolan commits his crime, he is described in animalistic terms: 'his jaws set like the teeth of a bear trap'. Afterwards, because Nolan has placed himself outside society, the narrator sympathises with him, just as Dostoevsky evokes the reader's pity for Raskolnikov in *Crime and Punishment* (1866). Pity is to Gallagher 'a ridiculous sensation', and he is despised by the narrator for this, and for his lofty ideals, never fully worked out and always in the 'theoretical stage'. Gallagher cares only about himself, despite loving McPhillips's sister Mary, who is educated and aspires to the middle class.

Pity is also felt for Nolan by a prostitute, who treats him with innocent affection in a squalid brothel, where he is captured. He is ordered at gun-point to attend a court of inquiry held by Gallagher, where Mulligan is questioned and freed after proving an alibi. Nolan, drunk and incoherent, is found guilty and sentenced to death. His pitiable state evokes the sympathy of the court, but not of Gallagher, whose lack of humanity is revealed when he ignores Mary's plea for clemency. (She does not want her brother's death avenged.)

Nolan uses his strength to escape from a makeshift cell, frightening Gallagher into speculating that he himself may be informed on. Like McPhillips, Nolan wanders Dublin, hunted like an animal, trying to find a way out of the city to hide in the mountains. Instead, he constantly returns to his starting-point. Exhausted, he shelters in a room belonging to his occasional mistress, Katie, and falls asleep. Representing the view of her own uneducated society, she betrays him to the organisation, who fire at Nolan as he tries to escape outside. He totters into a nearby church, where McPhillips's mother is hearing Mass. He confesses his crime to her, and dies in exultation, after she grants him the forgiveness he has been desperately and despairingly seeking.

Dorothy RICHARDSON　1873–1957

The Trap

see **Pilgrimage** (1938)

P.G. WODEHOUSE　1881–1975

Carry On, Jeeves

(Reginald) Jeeves, the immortal gentleman's gentleman, probably owes his wisdom and orotund delivery to Wodehouse's butler of the 1920s, Eugene Robinson. J.M. Barrie's butler, Frank Thurston, was said to shimmer into a room just as Jeeves went on to do. Characteristics of the butlers of W.S. Gilbert and Edgar Wallace may also be discerned in the Jeeves make-up. On the whole, though, Wodehouse did not draw from real life. The name Jeeves comes from the Warwickshire fast bowler Percy Jeeves.

Jeeves made his first entrance in a short story entitled 'Extricating Young Gussie' (1917), in which he had two lines. Wodehouse then realised that Jeeves could also be used to extricate young Wooster, and the character became more fully formed as the familiar indispensable valet in the story 'The Artistic Career of Corky' (1919). This story and four others which appeared in the collection *My Man Jeeves* (1919) were reworked for a collection made up entirely of stories about Jeeves and Wooster, *Carry On, Jeeves.*

The first unadulterated collection, however, had appeared in 1923 as *The Inimitable Jeeves*. The first Jeeves novel was *Thank You, Jeeves* (1934).

Jeeves's gentleman, Bertie (Bertram) Wooster, was educated at Eton and Magdalen but is, nevertheless, a bit short in the old brains department. In fact, though a jolly decent chap, he is a chump and much in need of his 'omniscient valet'. Bertie is a member of the Drones Club, as are fifty-two other characters who appear in the Wodehouse stories. They are outnumbered, however, by butlers, of which there are sixty-three, not including Jeeves.

Wooster and Jeeves, the greatest and most famous Wodehouse creations, do not always see eye to eye on the matter of clothes. Jeeves has infallible taste, which often gives him cause to doubt the appropriateness of Bertie's choice of attire. At various times they fall out over purple socks, Old Etonian spats, bright plus-fours, and a blue Alpine hat with pink feather. As nothing ever turns out quite the way Bertie plans it, Jeeves, with his knack of knowing everything, always keeps the upper hand, and each victory for Jeeves demands a sartorial concession from his master.

Bertie lives in fear of his Aunt Agatha, 5 foot 9 inches tall and mean. Aunts cause nephews to quiver 'like a jelly in a high wind', though his Aunt Dahlia is kind and hearty. Aunt Agatha, later Lady Agatha Worplesdon, wants Bertie to marry Honoria Glossop, a Girton girl, large, brainy and energetic. Honoria wants to take Bertie in hand. Her laugh is 'like cavalry clattering over a tin bridge'.

However, in the first of the stories in this collection, 'Jeeves Takes Charge', it is Lady Florence Craye who wants to take Bertie in hand, and improve his mind. The unfortunate engagement is, of course, eventually shaken off with the aid of Jeeves. The story contains a line as typical as any of the Wodehouse style: 'It was one of those evenings you get in Summer, when you can hear a snail clear its throat a mile away.'

The second Jeeves novel, *Right Ho, Jeeves* (1934), contains one of the greatest scenes in all comic writing: Gussie Fink-Nottle's prize-giving speech at Market Snodbury Grammar School.

| Virginia WOOLF | 1882–1941 |

Mrs Dalloway

In *Mrs Dalloway* Woolf breaks decisively with the fictional conventions of the realistic novel. Stream of consciousness now replaces the prose of *Jacob's Room* (1922), which, though impressionistic, still relied on an omniscient, controlling narrative voice to co-ordinate the characters. External events are now filtered almost entirely through the feelings, perceptions and memories of the individual characters, and transitions between scenes are replaced by glidings and leaps between radiant centres of consciousness.

The technique is almost orchestral, introducing and then interweaving the strains of the different characters' thoughts, and finally engineering, through a subtle sequence of readjustments and realignments, a new and delicate harmony between them at the close of the book. The story is nominally that of an individual woman, Clarissa Dalloway, the successful wife of a respectable Conservative politician, the book's focus is actually a network of shared sympathies, emotions and experiences which embraces not only Clarissa's present persona, but also her youthful past, her friends, and even the occasional complete stranger who wanders into her imaginative terrain.

The fluidity of the narration intimates that Clarissa cannot simply be pinned down to the role of mother and wife, and that she is still deeply entwined with other roles and affections. Her memories of her old friends – or perhaps lovers – Peter and Sally, render them as significant in her emotional make-up as is her husband. She has still not given up the more romantic and unconventional persona which for her these two embody, despite the fact that her outward actions in the course of the single day portrayed in the novel consist of preparations for her role as the 'perfect hostess' at a party to be given that evening, in the matrimonial home. Indeed, such is the force of Clarissa's past-directed imagination that it seems to conjure up Peter and Sally in the flesh, since they turn up unexpectedly at her party in the last section of the novel.

Other presences weave in and out of this spider's web of a text: Clarissa's independent daughter, Elizabeth, and her embittered, religiose governess, Miss Kilman; a latter-day Boadicea, Lady Millicent Bruton; and, more disturbingly, the psychotic Septimus Warren Smith, a shell-shocked victim of the First World War, who drifts helplessly through Clarissa's Westminister in a dark counterpoint to the empathetic radiance of her own thoughts. Learning of Smith's suicide from the eminent psychiatrist Sir William Bradshaw, Clarissa reveals an unexpected mood of complicity with Smith's despairing act, seeing it as a rejection, albeit, tragically extreme, of the social norms and expectations by which she too feels oppressed.

Mrs Dalloway thus initiated Woolf's sequence of radical experiments with literary form, embodying a striking combination of fluid sympathy and secret resistance in the mode of its narration as well as in the subject of the narrative. Through the novel's rapid transitions between apparently disconnected, but secretly related stories, Woolf was able to suggest the hazards of neatly pigeonholing human character according to social situation or gender.

1926

General Strike in Britain ● Ibn Saud becomes King of Hejaz and changes name of Kingdom to Saudi Arabia ● Expulsion of Leon Trotsky and Grigori Zinoviev from the Politbureau, following Josef Stalin's victory ● British legation in Peking declares Britain's sympathy with Chinese Nationalist movement (Kuomintang) ● John Logie Baird demonstrates television in London ● Henry Moore, *Draped Reclining Figure* ● Le Corbusier, *The Coming Architecture* published ● Alban Berg, *Wozzeck* (opera) ● Queen Elizabeth II born ● Claude Monet and Rainer Maria Rilke die ● Hart Crane, *White Buildings*

Ronald FIRBANK　　　　　**1886–1926**

Concerning the Eccentricities of Cardinal Pirelli

Firbank's last completed novel, which is suffused with intimations of mortality, was published posthumously. The eccentricities of the eponymous Spanish primate were such that Firbank's American publishers turned down the book 'on religious and moral grounds'.

The novel opens at St Eufraxia's Eve in the Mauro-Hispanic cathedral of Clemenza, where Andalucian society has gathered for the christening of a dog belonging to the Duquesa DunEden. This ceremony is being conducted under the disapproving eye of Monsignor Silex, a papal spy. The behaviour of the handsome Cardinal ('as wooed and run after by the ladies as any *matador*') is also being monitored by the 'venerable Superintendent-of-the-palace', Mme Poco. Feeling 'immured', the Cardinal has taken to making nocturnal sorties into the city, 'disguised as a caballero from the provinces or as a matron' (with some unfortunate results).

What come to be known as the 'scandals of Clemenza' form the principal gossip of the city and cause concern at the Vatican ('Why can't they all behave?' the Pope asks plaintively). Meanwhile, Don Moscosco, the overworked secretary of the chapter, is being petitioned by fashionable women for further canine baptisms. Gradually, the city falls into factions and Pirelli is summoned to Rome. He retreats to the mountain-top monastery of Desierto with Mme Poco and a provocative choirboy known as 'the Chicklet' ('an oncoming-looking child, with caressing liquid eyes, and a bright little tongue the colour of raspberry-cream – *so bright*') in order to prepare his defence. Before leaving for Rome, he conducts a 'service of "Departure"', during which the chorister pursues mice around the Basilica. Pirelli punishes the boy by locking him inside the building for the night, but then relents and, 'the heart in painful riot', chases the elusive and mercenary acolyte around the monuments until 'dispossessed of everything but his fabulous mitre', he drops dead in front of an El Greco. The novel ends with Mme Poco studying the 'serene, unclouded face' of her master, who has clearly obtained salvation.

Form and content are faultlessly combined in this brief and compressed novel, which is constructed upon a sequence of impressionistic scenes set in the cloisters, colleges and salons of Clemenza. The author's method of elision and suggestion perfectly evokes the atmosphere of intrigue. Comic, tragic and finally redemptive, this exuberant *danse macabre* is Firbank's most perfect work.

Ford Madox FORD   1873–1939

A Man Could Stand Up

see **Parade's End** (1928)

John GALSWORTHY   1867–1933

The Silver Spoon

see **The Forsyte Chronicles** (1929)

Ernest HEMINGWAY   1899–1961

The Sun Also Rises

A bestseller of the 1920s (subsequently published in Britain as *Fiesta*, 1927), *The Sun Also Rises* established both Hemingway's international reputation and the deceptively simple prose style for which he is famous. Opening in the world of American expatriates in the Montparnasse *quartier* of Paris, and moving thence to the world of Spanish bullfighting that Hemingway loved, the novel is written largely in passages of clipped dialogue and of unadorned but often poetical descriptive sentences: 'Standing on the bridge the island looked dark, the houses were high against the sky, and the trees were shadows.'

Jake Barnes is a newspaper correspondent in Paris: although all-American, he is war-wounded and impotent. He loves Lady Brett Ashley, an aristocratic Englishwoman, who responds to him, but also to many other men, being unable to control her impulses and desire for pleasure. She is engaged to marry a rich Scottish landowner, Mike Campbell. They also know a Jewish-American aspirant writer, Robert Cohn, talentless, unsure of himself, and unloved: he, too, loves Brett, and she tolerates him.

This group, accompanied by Jake's friend, the American writer and sportsman Bill Gorton, moves from the world of late-night bars in Paris to Spain, where they attend the seven-day fiesta in the town of Pamplona. Tension builds because Mike Campbell feels that Cohn is pestering Brett, but she steals a march on all three *beaux* by beginning an affair with a handsome nineteen-year-old Spanish bullfighter, Pedro Romero, who is making a great name for himself at the fiesta.

The novel seems to be building up to an impressive climax of passion and jealousy, but the plot rather peters out towards the end. While Brett is with Romero, Cohn, in a mad fit of jealousy, picks fights with Jake, Mike and the bullfighter then remorsefully leaves town. Brett goes off with Romero to Madrid, but later, having asked him to leave, summons Jake to her hotel. The novel ends with her planning to return to Mike, but saying to Jake that they could have 'such a damned good time together'. Although structurally flawed, Hemingway's first international success is rich in beautiful writing and atmosphere.

D.H. LAWRENCE 1885–1930

The Plumed Serpent

Lawrence's nightmarish novel created a half-infernal, half-paradisal vision of Mexico, which was to haunt the imaginations of later writers as diverse as Malcolm Lowry, Graham Greene and Aldous Huxley. In *The Plumed Serpent* Lawrence sees heaven and hell as eternal antagonists, and translates these elements into a battle between the sexes in which the urge to dominate is essentially cruel.

Kate, an Irish woman, has divorced her first husband and watched her second, a political agitator, die fighting for a revolution. The old world seems to her to be stale and played-out and, now beyond an age when she expects to remarry or bear children, she comes to Mexico in search of some new ideal to which she can commit herself. In the company of two American youths she witnesses a bullfight, where the power and impotence of the bull against the fat-hipped effeminate matadors revolts her; but the disgust she feels for the native traditions is intermingled with fascination and also respect.

The newspapers are reporting apparitions of the ancient gods, as the pagan deity Quetzalcoatl is returning to earth to expel the spirit of Christianity. Kate meets Don Ramon and Cipriano, the two men who are promoting the cult as a means of revivifying native Mexican power. Cipriano persuades her to marry him and draw her into the role of a goddess of the pagan pantheon, but the submission to phallic supremacy which Cipriano demands conflicts with Kate's independence as a western woman. She determines to leave Mexico and return to the safety of Europe, but on the eve of her departure she decides instead to stay, for Don Ramon's new wife Teresa, in the strength of her dedication to the 'ancient phallic mystery', seems to represent an affirmation of some deeper aspect of feminine nature which Kate has ignored in her own search for individual identity. 'I shall make my submission,' Kate concludes finally, 'but as far as I need, and no further.'

The Plumed Serpent might be seen as the culmination of Lawrence's preoccupation with 'the old supreme phallic mystery', in which the penis is finally elevated to a religious totem. Constance Chatterley kneeling before 'the primeval root of all full beauty', as proudly displayed by Mellors, is essentially indulging in a private form of worship; but the cult promoted by Ramon and Cipriano will clearly become the established church of their new political order. 'I feel one still has to fight for the phallic reality,' Lawrence wrote, but 'phallic fantasy' might be a more apposite description of the novel's underlying (and, some would assert, essentially fascist) message.

A.A. MILNE 1882–1956

Winnie-the-Pooh
1926

The House at Pooh Corner
1928

Milne's timeless children's classics, vigorously illustrated by E.H. Shepard, trace the adventures of the eponymous teddy bear and his friends. Pooh is a Bear of Very Little Brain, but a considerable poet with a 'Hum' for every occasion. He belongs to a small boy called Christopher Robin, based upon Milne's own son, and both had already appeared in Milne's book of children's verse *When We Were Very Young* (1924). The Pooh books are set in woodland based upon Ashdown Forest in Sussex, where the Milne family had a house.

The first volume introduces the major characters – some of whom are real animals, while others are soft toys – and recounts attempts to discover the North Pole and to trap a Heffalump. Eeyore, the lugubrious and fatalistic toy donkey, loses his tail and has a birthday, whilst the timid Piglet is rescued from a flood, and Pooh has to be pulled free from Rabbit's burrow after consuming too much honey. Motherly Kanga and her excitable son, Baby Roo, arrive in the forest, and Eeyore responds characteristically when told that a party is being held to honour Pooh: ' "Very interesting," said Eeyore. "I suppose they will be sending me down the odd bits which got trodden on. Kind and Thoughtful. Not at all, don't mention it."'

The title story of the second volume concerns the well-meaning attempts of Pooh and Piglet to rehouse Eeyore during a particularly cold spell. The irrepressible Tigger, a bumptious toy tiger, comes to stay with Kanga and Roo, and Eeyore inadvertently takes part in a game of 'Poohsticks' after being bounced into the river by him. The self-important Rabbit organises a hunt for one of his 'friends-and-relations', a beetle named Small who has gone missing. Owl's tree is blown down in a gale, an incident which leads to an act of bravery by Piglet, celebrated in a particularly long Hum by Pooh.

'I *think*,' said Piglet, when he had licked the tip of his nose too, and found that it brought very little comfort, 'I *think* that I have just remembered something. I have just remembered something that I forgot to do yesterday and sha'n't be able to do to-morror. So I suppose I really ought to go back and do it now.'

'We'll do it this afternoon, and I'll come with you,' said Pooh.

'It isn't the sort of thing you can do in the afternoon,' said Piglet quickly. 'It's a very particular morning thing, that has to be done in the morning, and, if possible, between the hours of – What would you say the time was?'

'About twelve,' said Winnie-the-Pooh, looking at the sun.

'Between, as I was saying, the hours of twelve and twelve five. So, really, dear old Pooh, if you'll excuse me."

WINNIE-THE-POOH
BY A.A. MILNE

The final chapter marks the end of Christopher Robin's carefree days of childhood and the beginning of his education. Intimations of this occur earlier, when the animals discover what the boy does in the mornings, and in the marked improvement in his spelling. Appropriately, it is Eeyore who takes on the mantle of laureate in order to compose a poem of farewell.

While some readers have been repelled by what they see as Milne's whimsy, there is no denying that the books have enough wit and energy to satisfy adults as well as children. And in the disconsolate Eeyore (apparently based upon Milne's boss at *Punch*, Owen Seaman) and the pompous Owl, the wise elder of the Hundred Acre Wood who never admits to his considerable fallibility, Milne created two splendidly comic characters.

R.H. MOTTRAM   **1883–1971**

The Spanish Farm Trilogy, 1914–1918

The Spanish Farm (1924)
Sixty-Four, Ninety-Four! (1925)
The Crime at Vanderlynden's (1926)

In the preface to the second volume of his trilogy about the First World War, Mottram writes that this part-fictional, part-autobiographical record was produced to warn future generations against the futility of armed conflict in which men's lives are wasted 'for the sole purpose of wholesale slaughter by machinery'. Although the three novels were conceived separately, in 1927 Mottram prepared them for a single volume and provided new interlinking sections.

Part one, *The Spanish Farm*, tells the story of Madeleine Vanderlynden, the daughter of a farmer, struggling to preserve Flemish traditions against the invasions of the Allied Armies. Her lover, the aristocratic Georges D'Archville, is called upon to defend the fatherland, and Madeleine's life becomes a vain search for love. When her widowed sister takes over the farmstead as her father's housekeeper, Madeleine becomes the patronne of the Lion of Flanders, an inn frequented by the troops. Still longing for news of Georges, she moves on to Amiens and finally to Paris, where she hears he has been killed in an aeroplane over the battlefields.

'D'Archeville: A Portrait' was added for the 1927 edition and describes Madeleine's lover from the point-of-view of an English officer: 'I am English, he was French. He *meant* his war.'

Part two, *Sixty-Four, Ninety-Four!*, takes its title from an army song: 'Sixty-four, ninety-four – / He'll never go sick no more, / The poor beggar's dead.' It develops the story of Madeleine's new lover, Lt. Skene, whom she has enlisted to help her in the search. Skene is a well-educated Englishman from the upper middle classes, an architect who reads *Tristram Shandy* in the trenches, and the poignancy of their wartime affair is the more apparent to him as he knows that under normal circumstances they would have nothing in common. On leave back in England Skene returns to his uncle's home in Cathedral Close, feeling 'as if he had just stepped out of a novel by Anthony Trollope or Bulwer Lytton', shocked to find that the war 'had not touched the people at home'. When he tries to describe his experience at the Front, his uncle responds by apologising that there is no cream to have with pudding: 'War-time, Geoffrey. We all feel it.' The narrative recounts in close detail the advances and retreats of the British Expeditionary Forces around the Somme, where Skene is wounded. His infatuation with Madeleine expresses the emotional void left by the war, as the soldiers desire only that someone should care about their face and look out for their letters. 'The War has outlasted all patience, and all interest,' he reflects as, after four long years, the demobilisation begins: it had 'become a cosmic stupidity like bad weather or high rates'. In his new feeling of weary cynicism, he reacts with contempt when Madeleine asks him to send her money, but posts off a hundred francs. After a final meeting with her at the Spanish Farm he realises that Madeleine has retreated into a calm but determined mood of self-defence, her only concern to keep the farm going. Back in England, the public are so tired of the war that there is no one to cheer the troops as Skene disembarks.

The second and third volumes are linked by another portrait, 'The Winner', in which the elderly, undistinguished Capt. Dakers is described. Clearance Officer and Skene's immediate superior, Dakers is an Englishman who has lived for forty years in Canada. He survives the war, but dies in the influenza epidemic: 'Of all the old and shaky and the new and gimcrack forms of civilization that may not outlast a decade, the England of which Uncle was a specimen will survive.

Who could hang on with so little fuss, and pass out so quietly?'

The final volume, *The Crime at Vanderlynden's*, tells of another soldier's unwilling involvement with the Vanderlynden family, as Stephen Doughty Dormer, an unimaginative bank clerk, is assigned to bring to justice a British soldier who has desecrated the shrine on the Spanish Farm. At first this offence against 'La Vierge' is mistaken as a case of rape of Madeleine, but she clears up the misunderstanding and explains that a muleteer with the BEF broke down the front of the Virgin's shrine to provide shelter for his wounded animals. The dossier of evidence haunts Dormer: 'The Crime at Vanderlynden's was the War, nothing more nor less', and his duties in the battalion are constantly interrupted by the demand that he prosecute a man he does not want to convict. 'The War was not going to be won in the trenches,' he concludes as, on a visit to England, he finds that profiteering is more likely to be the decisive factor than bayonets or bombing raids. The announcement of cease-fire peters out in a postponed football match; across Europe and Russia the surge for revolution is contrasted in Mottram's view with the docility of the English, who only want to return to work and oust non-union labour from their jobs. 'Anticlimax was the rule of the War.'

The book ends with a final portrait, 'The Stranger', in which an Englishman revisits the scenes of his war at Easter. Gazing at the neat cemetery, 'he understands why he is A Stranger. These have not relapsed into Peace and England, as he has. The War has survived them.'

William PLOMER 1903–1973

Turbott Wolfe

Plomer's first novel, dispatched to the Hogarth Press from Zululand, opens provocatively: 'I think Turbott Wolfe may have been a man of genius.' This was a very different view of Plomer's hero to that taken by the majority of his white fellow-countrymen. The novel caused a furore in South Africa, denounced not only in reviews but also in newspaper editorials. The *Natal*

Advertiser announced it as 'A Nasty Book on a Nasty Subject'; that subject was miscegenation.

Turbott Wolfe, lying upon his death-bed, recalls his experiences as a trader and artist in Lembuland. His white neighbours, horrified that he uses black Africans as models, write him threatening letters and report him to the Department for Aboriginal Protection, from whom he receives his trading licence. Wolfe falls in love with one of his customers, a statuesque girl 'fit to be the wife of an ambassador', called Nhliziyombi. She is already spoken for, but this unrequited passion leads Wolfe to collaborate with an English priest, Rupert Friston, in the formation of the Young Africa movement. 'I believe that the white man's day is over,' Friston declares. 'Anybody can see plainly that the world is quickly and inevitably becoming a coloured world. I do not assert yet that miscegenation should be actually encouraged, but I believe that it is the missionary's work now, and the work of any white man in Africa worth his salt, to prepare the way for the ultimate end.' This, in a community in which a doctor can boast of castrating a black man found making love to a white woman, is a radical point of view.

Friston's dreams are given a jolt when Mabel van der Horst, a forthright governess with whom he is in love, decides to marry Zachary Msomi, a handsome ordained native. Like Wolfe, he realises that 'It was one thing to talk glibly about miscegenation, ... another to find oneself face to face with the actual happening'. He takes to drugs, then vanishes and is later reported murdered. Wolfe sells up to return to England just before his trading licence is revoked.

Written in the robust idiom of its eponymous hero, the novel is a savage portrait of a country and its inhabitants. An indictment of hypocrisy and prejudice, the novel depicts a wide range of South African characters, from gently bemused missionaries to barbarous and degenerate farmers.

Carl VAN VECHTEN   1880–1964

Nigger Heaven

Nigger Heaven was a pioneering attempt by a white author to enter the consciousness of black Harlem during the renaissance of the 1920s. Van Vechten, besides being a novelist, was also a composer, critic and photographer, who did a great deal to widen public interest in black culture from the 1920s onwards. The novel, however, contains a peculiar blend of cerebral enthusiasm for and intuitive distrust of its characters, and was both an instant success and the subject of an immediate public controversy.

Theoretical discussion of the race problem jostles throughout with evocations of jazz and blues music: on balance, it is the music which takes centre-stage. There are evocations of street performances, rowdy house-parties, and night-clubs of all kinds, ranging from the urbane to the sleazy. Snatches of blues lyrics recur throughout, originally taken from recorded songs, but then hastily reworked by Van Vechten's friend, Langston Hughes, when a copyright suit was threatened.

The novel's protagonists, Mary Love and Byron Kasson, are both 'New Negroes', or members of the aspiring new black middle class of the period. Both are activists of a kind, but their strategies differ. Mary, who is a librarian, believes in passing on her understanding of white culture to the less educated, whereas Byron is a writer, and believes in cultivating consciousness of black difference. They quarrel over aesthetics, Mary arguing for detachment and irony, Byron for energy and anger. Byron drafts a story of a black man who, falling in love with a white woman, is then murdered by her brother, but the story is rejected by all the prominent magazines, evidently proving Mary's aesthetics to be superior.

Angered and frustrated, Byron deserts Mary for another woman, Lasca Ventoris, a wealthy and notorious widow. She quickly tires of Byron, and when she throws him out he irrationally blames another man, Randolph Pettijohn, a gambling tycoon known as the 'Bolito King'. Byron stalks Pettijohn to a night-club called the Black Venus, bent on killing him, but in fact another enemy of Pettijohn gets there ahead of him. Anatole Longfellow, 'alias the Scarlet Creeper', is a famous Harlem gigolo, and his grievance is genuine, since the Bolito King has rashly muscled in on the Creeper's latest woman. He shoots Pettijohn dead, but ironically, it is Byron who takes the rap, since it is Byron who is found standing over the corpse, gun in hand, when the police finally arrive.

Byron's fall seems to embody Van Vechten's view of the dangerous instability of black temperament, seen as at once spontaneous, energetic, creative, but also impulsive, rash, over-emotive. This dubious construction somewhat vitiates Van Vechten's loving evocation of black culture, achieved through a careful cataloguing of street talk, song, dance, fashion, physical characteristics and differing social attitudes.

Sylvia Townsend WARNER
1893–1978

Lolly Willowes

Warner's bestselling first novel is a feminist fable which takes the form of a witty and enchanting story of rural witchcraft. It triumphantly champions women – spinsters in particular – and unsentimentally celebrates the English landscape in rich, clear but allusive prose. She was prompted to write it after reading about the history and persecution of witches, and being convinced that for some women witchcraft 'was the romance of their hard lives, their release from dull futures'.

After the death of her father, twenty-eight-year-old Laura Willowes leaves Somerset for London. She is to live with her elder brother, Henry, and his wife, Caroline, representatives of the Willowes tradition which, once lively and eccentric, has atrophied to good conservative common sense. Laura's hosts make feeble attempts to marry her off, but she seems likely to settle into a comfortable spinsterhood, plying her needle and entertaining her nephew and nieces. Her future looks dull indeed.

Time passes, and although Laura accepts her new name and position at Apsley Terrace ('Aunt Lolly, a middle-aging lady, light-footed upon stairs, and indispensable for Christmas Eve and birthday preparations'), she is aware of some other self, a darker, more troubling spirit which responds to the gloomier quarters of the metropolis. She begins searching for this other self, almost discovering it at Bunhill Fields and at the railway goods-yard at Paddington. An extravagant purchase at a florist excites her curiosity about the Chilterns. When she announces that she intends to move to the village of Great Mop, her family – nephew Titus apart – are horrified, but she will not be dissuaded, even when she learns that Henry's financial incompetence has left her with little money.

'Is it true that you can poke the fire with a stick of dynamite in perfect safety? I used to take my nieces to scientific lectures, and I believe I heard it then. Anyhow, even if it isn't true of dynamite, it's true of women.

LOLLY WILLOWES
BY SYLVIA TOWNSEND WARNER

During her first winter in the unprepossessing village, Laura becomes aware of an unseen force in the surrounding landscape. When a kitten she finds in her rooms scratches and bites her, she recognises it as a familiar: at the age of forty-eight she has entered into a pact with the Devil. It transpires that almost the entire village is in league with Satan, and her landlady, Mrs Leak, takes her to a well-attended sabbat. She feels no more comfortable here than she has done at other social engagements, but takes heart when the following morning she meets the Devil himself in the form of a kindly gamekeeper ('The Loving Huntsman' of the novel's subtitle). Titus, who has annoyingly followed her to Great Mop in order to write a book about Fuseli, is finally driven back to London by witchcraft: his milk curdles, he is attacked by wasps and he falls in love. After seeing Titus and his fiancée on to a train, Laura once again meets the Devil, disguised this time as a gardener. Their conversation about the nature of witchcraft underlines the novel's feminist subtext. Women are like sticks of dynamite that can be thrust into a fire without exploding:

> But they know they are dynamite, and long for the concussion that may justify them. Some may get religion, then they're all right, I expect. But for the others, for so many, what can there be but witchcraft? That strikes them real. Even if other people still find them quite safe and usual,

and go on poking with them, they know in their hearts how dangerous, how incalculable, how extraordinary they are ... One doesn't become a witch to run round being harmful, or to run round being helpful either, a district visitor on a broomstick. It's to escape all that – to have a life of one's own, not an existence doled out to you by others.

1927

Economic conference at Geneva, attended by 52 nations, including USSR ● 'Black Friday' with collapse of Germany's economic system ● Josef Stalin's faction is victorious at All-Union Congress ● Kemal Atatürk, *The New Turkey* ● Adolf Hitler, *Mein Kampf*, Vol. II ● Heisenberg propounds 'the uncertainty principle' in quantum-physics ● Charles A. Lindbergh flies from New York to Paris in 37 hours ● Eric Gill, *Mankind*; also designs the 'Gill Sans' alphabet ● Sound films, popularised by *The Jazz Singer*, with Al Jolson ● Johnny Weismuller swims 100 yards in 51 seconds ● Babe Ruth hits 60 home runs ● Jerome K. Jerome dies ● W.H. Auden and Cecil Day Lewis (eds), *Oxford Poetry* ● James Joyce, *Pomes Penyeach*

| Elizabeth BOWEN | 1899–1973 |

The Hotel

Bowen's first novel had been preceded by two volumes of enthusiastically reviewed short stories. Her publisher encouraged her to attempt a full-length fiction, and the idea for *The Hotel* came to her 'in a flash', as she recalled her own childhood experiences of Riviera hotel life in the winter of 1921.

Recovering from a near-breakdown caused by over-studying, Sydney Warren, a 'very disappointable' twenty-two-year-old, has come to Italy with her sickly but kind married cousin, Tessa Bellamy. Staying in the hotel of the title, Sydney becomes the protégé of a powerful older woman, Mrs Kerr. However, when Mrs Kerr's self-consciously remote but surprisingly sensitive son, Ronald, arrives, Sydney feels neglected. Her enigmatic and abrupt manner has attracted the attention of James Milton, a naïve clergyman some twenty years her senior who, much to his own surprise, asks her to marry him after only a short acquaintance. Having refused initially, Sydney –

perhaps hurt by Mrs Kerr's desertion, and unsure of what she wants in life – accepts. Happiness is short-lived and, realising her mistake, Sydney breaks off the engagement, much to the embarrassment of everyone around her.

'I have often thought,' remarks Sydney, 'it would be interesting if the front of any house, but of an hotel especially, could be swung open on a hinge like a doll's house.' This in effect is what Bowen has done, since she examines in meticulous detail the lives of the hotel's guests: the inseparable spinster friends Miss Pym and Miss Fitzgerald; the anxious-to-please Lee-Mittisons; Victor Ammering, 'one of those young men of thirty (Public School and University education, active, keen sportsman, good general capacities) who advertise their willingness to try anything in the Personal column of *The Times*', and at dances 'talked to his partners most beautifully about the War'; and the gallant, tennis-playing Col. Duperrier, who has a querulous invalid wife and longs for the company of pretty young women like the three Lawrence sisters. Conveying their hopes and disappointments, with a recognition of the comic and a sympathy with the painful aspects of experience, Bowen probes at the undercurrents of her characters' lives with what can later be recognised as characteristic sensitivity and clarity.

Willa CATHER 1876–1947

Death Comes for the Archbishop

Cather made frequent trips to New Mexico, and became interested in the work of the French missionary priests of the nineteenth century, in particular the Jesuit Archbishop Lamy, who had built the Cathedral of St Francis in Santa Fé. In the summer of 1925, she returned to New Mexico, where she read a biography of the Archbishop's assistant, Joseph Machebeuf. By the end of her trip, she had gathered sufficient material to embark upon the writing of *Death Comes for the Archbishop*, a task she found extremely congenial. 'I had all my life wanted to do something in the style of a legend, which is absolutely the reverse of dramatic treatment. Since I first saw the Puvis de Chavannes frescoes of the life of St Geneviève [in the Pantheon in Paris] in my student days, I have wished I could try something a little like that in prose; something without accent, with none of the artificial elements of composition.' She was also influenced by the accounts of the lives of saints in *The Golden Legend*, a medieval miscellany published by Caxton in 1483, in which: 'the martyrdoms of the saints are no more dwelt upon than are the trivial incidents of their lives; it is as though all human experiences, measured against one supreme spiritual experience, were of about the same importance. The essence of such writing is not to hold the note, not to use an incident for all there is in it – but to touch and pass on.'

In 1851 Bishop Jean Marie Latour and his friend Fr Joseph Vaillant (based upon Lamy and Machebeuf respectively) are despatched to New Mexico to sustain and propagate the Catholic faith. They find themselves immersed in an alien land, remote in terms of distance and culture, which inspires respect for its traditions and beliefs. The two priests

possess very different but complementary natures; Latour, with his scholarly and distant demeanour, is 'gracious to everyone but known to very few', while Vaillant, stunningly ugly and utterly benign, is always ready to roll up his sleeves and dive into any situation, endearing himself to everyone. He is never ashamed to beg for anything for the furtherance of his faith, but has few personal possessions and is eager for 'very direct and spectacular' miracles. The two men are separated in later years when Vaillant is called to minister to the 'gold-crazed' population of Colorado, there meeting his death. Latour, now Archbishop, achieves his dream of building a cathedral and then retires for a period of calm contemplation before meeting a peaceful death.

The ride back to Santa Fé was something under four hundred miles. The weather alternated between blinding sand-storms and brilliant sunlight. The sky was as full of motion and change as the desert beneath it was monotonous and still, – and there was so much sky, more than at sea, more than anywhere else in the world. The plain was there, under one's feet, but what one saw when one looked about was that brilliant blue world of stinging air and moving cloud. Even the mountains were mere ant-hills under it. Elsewhere the sky is the roof of the world; but here the earth was the floor of the sky. The landscape one longed for when one was far away, the thing all about one, the world one actually lived in, was the sky, the sky!

DEATH COMES FOR THE ARCHBISHOP
BY WILLA CATHER

Cather presents Latour's death not so much as the climax of the novel as the title may suggest, but rather as the fitting close to a narrative crowded with stories and characters. It is a narrative in which the histories of dissolute priests, legendary or contemporary, or converts whose faith is supported by stories of miracles and visions of the Holy

Family, are juxtaposed with a strong sense of landscape and a belief in the simple nobility of an ancient culture.

Ernest HEMINGWAY   1899–1961

Fiesta

see **The Sun Also Rises** (1926)

Rosamond LEHMANN   1901–1990

Dusty Answer

Lehmann's first novel, published when she was twenty-six, instantly established her as a leading novelist, although the first reviews, she recalled, 'gave the impression that I had gravely offended against standards of womanly decorum'. A notice by Alfred Noyes in the *Sunday Times* which referred to it as 'quite the most striking first novel of this generation' (and bizarrely suggested that it was 'the kind of novel that might have been written by Keats'), attracted attention and the book became a bestseller, not only in England but also in France and America.

The title is taken from Meredith's *Modern Love* (1862): 'Ah, what a dusty answer gets the soul / When hot for certainties in this our life!' The romantic young heroine, Judith Earle, child of wealthy but largely absentee parents, becomes fascinated with her country neighbours, a family of cousins: the brilliant Etonian brothers, Julian and Charlie; kind but clumsy Martin; beautiful Mariella; and the enigmatic, artistic Roddy. She feels that she has found somewhere to belong, and is bereft when they move away.

Some years later, shortly before Judith goes up to Cambridge, she hears that the cousins have returned. Charlie has been killed in the war, but not before marrying Mariella and fathering a son, Peter. Julian has also been a combatant and bears permanent mental scars. Although outwardly friendly, there is something rather reserved about them all. Even so, Judith finds herself drawn further into their circle, because her childhood attachment to Roddy has now deepened into hopeless love. Roddy seems non-commital and teasing, perhaps more

involved with a Cambridge undergraduate, Tony Baring. Martin, however, who has grown into a huge, bluff young farmer, makes it clear that he is in love with Judith and there are also signs that the sophisticated, embittered Julian is interested in her. Mariella, bored with her child, has always been in love with Julian, but his interest centres far more upon his nephew, whom he will eventually adopt.

At Cambridge Judith reads English and forms a strong attachment to a fellow student, Jennifer Baird. In her turn, she is worshipped by the earnest, bespectacled and unappealing Mabel Fuller. The idyllic relationship between Judith and Jennifer breaks up when the latter is taken up by an older woman, Geraldine Manners, and leaves Cambridge suddenly. Further romantic disappointments await Judith at home, where in the course of a midnight boat ride she persuades Roddy to declare his love for her. She imagines that a true affair will be inaugurated, but the following day she finds Roddy totally withdrawn and realises that he will never commit himself to her.

On the rebound from this rejection she accepts a proposal from Martin, then changes her mind. Meanwhile, her father has died and she spends some time abroad with her elegant, socialite mother. At a French spa she meets Julian, who is touring Europe as a freelance musician. He almost persuades her to become his mistress, but then the shocking news reaches them that Martin has been killed in a boating accident. Back in England, Judith arranges to meet Jennifer in a Cambridge tea-shop, but her former friend fails to turn up. While waiting, Judith sees Tony and Roddy passing in the street: Roddy is as unobtainable as ever. Judith leaves Cambridge knowing that all the relationships which have taken up her youth are over: this is the 'dusty answer' to which the novel has constantly been moving.

Although the book has remained popular, praised for its sensitivity and the quality of its prose, Lehmann's opinion of it declined over the years. Looking at it again in 1953, she felt in spite of 'occasional boredom and irritation with the goings-on of a group of self-absorbed, emotionally immature and over-glamourised young people', she would read it to the end. Of Judith she commented: 'this girl has a great deal to learn, but she has shown promise'. By the 1980s, however, she had become completely disenchanted with her 'soppy' heroine, describing her as a 'revolting character'.

Elmer Gantry

Born in Sauk Center, Minnesota, Lewis owed everything to the American Middle West in much the same way as Arnold Bennett owed everything to the Potteries. In the five novels he published in the 1920s Lewis said all he had to say, and the twenty years until his death in 1951 represented a steady decline of his creative powers. *Elmer Gantry* was an instant bestseller. On publication Harcourt Brace took five billboards in New York; sales were further stimulated when Boston banned it and the evangelist Billy Sunday dubbed Lewis 'Satan's Cohort'.

'Elmer Gantry was drunk. He was eloquently drunk, lovingly and pugnaciously drunk.' Thus we meet him – a handsome hulk of football-playing college boy nicknamed 'Hell-cat' – in the company of his friend and mentor Jim Lefferts. Jim, an unbeliever and essentially sane, is 'Elmer's only friend; the only authentic friend he had ever had'. When Elmer is converted at a mass meeting of the Baptists ('Oh, come, come with us – don't stand there making Jesus beg and beg'), Jim warns him: 'They hypnotize themselves ... don't let them hypnotize you.' But Elmer, browbeaten by his mother (the only other influential person in his life), makes up his mind to betray Jim, the first of his many victims. 'Why did he feel so fine? Of course! The Call had come!' Yet conversion does nothing for Elmer's problem with drink and women. While attending the Mizpah Theological Seminary he is tempted by and falls for Lulu Baines, whom he gets pregnant. He saves himself by sidestepping neatly, but nothing can save him when he is discovered drunk and debauched in the city of Monarch, the sermon he is due to give undelivered. The Baptists fire him and for two years he is returned to the world

as a travelling salesman.

Nothing more might have been heard of the great preacher had he not encountered an evangelist of genius, Sharon Falconer. In a memorable scene, she seduces him before her altar, invoking the goddesses of fertility: 'O mystical rose, O lily most admirable, O wonderous union; O St Anna, Mother Immaculate, Demeter, Mother Beneficent, Lakshmi, Mother Most Shining; behold, I am his and he is yours and ye are mine!' He becomes her acolyte, her business manager, her lover, and might have remained in her shadow had she not burned to death in a fire at her Tabernacle.

The church, the Sunday School, the evangelistic orgy, choir-practise, raising the mortgage, the delights of funerals, the sniggers in back pews or in the other room at weddings – they were as natural, as inescapable a mould of manners to Elmer as Catholic processionals to a street gamin in Naples.

ELMER GANTRY
BY SINCLAIR LEWIS

Lewis had to kill off Sharon – she would otherwise have taken over the novel. However, with her gone, much of the novel's power is lost. Having failed to make it alone, Elmer joins the Methodists and, with the backing of the worldly-wise T.J. Rigg and others, sets out on his career of conquest, leaving a further trail of victims in his wake – in particular his pathetic wife, Cleo. The honest old Methodist minister Andrew Pengilly might ask, 'Mr Gantry, why don't you believe in God?' but for Elmer this is beside the point. He sees the righteous suffer – Frank Shallard, his old room mate at Mizpah, is driven out of the church, beaten half to death and blinded by Fundamentalist thugs for daring to express doubts – while he, the 'National Association of Purification of Art and the Press' practically in his pocket, has a vision of himself as 'emperor of America'. He is nearly conned out of it by a scandal engineered by the seductive Hetty, a most efficient secretary, but T.J. Rigg comes to his rescue and, in tears, he can declare from his pulpit: 'Oh Lord, thou hast stooped from the mighty throne and rescued thy servant from the assault of the mercenaries of Satan!'

A satire of American religious life, *Elmer Gantry* may be seen as a cautionary tale for 'born again' America, but it suffers from Lewis's lack of a positive belief. Old Pengilly's question is not answered; Frank Shallard's doubts are not developed to a point where they amount to a philosophical system; Jim Lefferts winds up a down-at-heel lawyer. Lewis is a journalist rather than an artist, a caricaturist rather than a developer of character; his novel has just one theme. Yet, as his biographer Mark Schorer puts it, Elmer's 'barbarous rise from country boob to influential preacher' was written in a state of 'almost pure revulsion', and some of the energy so generated is communicated to the reader.

R.H. MOTTRAM   1883–1971

The Spanish Farm Trilogy, 1914–1918

see 1926

T.F. POWYS   1875–1953

Mr Weston's Good Wine

Powys, who set the majority of his books in Dorset, was very aware of local associations with Thomas Hardy, having occupied the desk at Dorchester Grammar School where Hardy had sat many years before. Like Hardy's, Powys's books are coloured by a sense of *genius loci*, but whereas Hardy's characters are the playthings of an indifferent Fate, those of Powys receive divine visitations. This notion had been explored at the lyrical climax of his second book, *Soliloquies of a Hermit* (1916), in which he

imagines God descending upon East Chaldon (where Powys pursued a somewhat eremitical life) to lie among the buttercups and observe village life:

> I cannot see the least willingness on His part to leave the scent of the May clover, in order to go and look at old churches; but I do notice that He turns a little on one side to watch a young man and maid take the path that leads to the tavern; and He looks at them as though they really were His children.

In Powys's masterpiece, a distinctly personal Christian allegory, God (in the person of Mr Weston) comes to Folly Down in order to see what state the world is in. He arrives in his Ford motor car, accompanied by his beautiful assistant, Michael. It becomes clear that Mr Weston is more than a mere wine-salesman when a small girl whom he has accidentally run over gets up when bidden and sets off for home, laughing, 'no worse for the mishap'. He blazons an advertisement for 'Mr Weston's Good Wine' on the night sky and discusses the local inhabitants with Michael as he prepares to mete out his wines as he deems appropriate. Time stands still.

Mr. Weston noticed one child – a girl – whom, in coming round a sharp corner, he unexpectedly ran over. He looked round to where she lay and bade her pick herself up and run home, which she did, laughing and appeared no worse for the mishap. This slight incident, however, set Michael a-talking.

'A human girl-child,' he said, 'is a creature set in a dish for time to feed upon.'

MR WESTON'S GOOD WINE
BY T.F. POWYS

The Revd Nicholas Grobe, who lost his faith after the death of his sexy wife, Alice, tastes Mr Weston's wine, recovers his faith and is finally allowed to drink a very special draught which ensures eternal relief from worldly cares. The teetotal Luke Bird, a Francis of Assisi figure whose preaching to the birds and beasts has recently effected the conversion of a bull, determines to fill his well with wine in order to win the hand of the pub landlord's daughter, Jenny Bunce. Mr Weston saves him the trouble by delivering the sleeping Jenny to his door, marrying them on the spot, and sending them off to bed together. Grobe's lovely daughter, Tamar, who believes she is to marry an angel, finds one in Michael, and is carried off into the heavens by the stars after their mossy bed among the roots of the village green's oak tree is struck by lightning.

Not all the villagers receive their wish, however. The fornicating Mumby brothers, sons of the local squire, who were deaf to earlier warnings, are chased by a lion and reduced to gibbering wrecks; and the voyeuristic procuress Mrs Vosper, who lures young girls to their ruin, dies horribly and unexpectedly. His business concluded, Mr Weston leaves the village, and 'time moves again'. On Folly Down hill, he instructs Michael to drop a match into the Ford's petrol tank. As it explodes into smoke and flames, they disappear up to heaven rather in the manner of the Prophet Elijah.

Mr Weston is in part a God who failed, for it seems that those he created in his image are far from perfect. But he is also a forgiving God, who is prepared to make the best of a bad job, offering unconditional consolation to incorrigible mankind.

Dorothy RICHARDSON   1873–1957

Oberland

see **Pilgrimage** (1938)

Sylvia Townsend WARNER
1893–1978

Mr Fortune's Maggot

In 1925 Warner had a vivid dream in which she saw a middle-aged missionary who 'stood alone on an ocean beach, wringing his hands in despair'. This man became Mr Timothy Fortune, whose maggot ('a whimsical or perverse fantasy') is to convert the population of Fanua, a small Polynesian island. His Archdeacon looks upon this plan as 'at best, a sort of pious escapade', for the Fanuans are a childlike race whose 'language has no words for chastity or for gratitude'.

Undaunted, Mr Fortune travels to the island, where his attempts to instil the Fanuans with the Holy Spirit are greeted with friendly indifference by all apart from the young boy, Lueli. Rechristened 'Theodore', Lueli comes to live with Mr Fortune, who plans to use this convert as a 'holy decoy'. Mr Fortune's favourite psalm begins 'My soul truly waiteth still upon God', but after three years and no further conversions, his superiors become impatient. So patient is Mr Fortune that he decides to persuade Lueli to marry an equally 'biddable' girl and found a Christian dynasty, following God's example of Abraham and Sarah. However, Lueli loves only Mr Fortune. Worse still, it transpires that he has not abandoned his old gods, for he maintains a shrine in the forest. Angry at allowing himself to be deceived, Mr Fortune threatens to burn Lueli's idol, but is thwarted when an earthquake strikes. Lueli saves Mr Fortune's life, but loses his idol and consequently his will to live. Mr Fortune realises that he too has lost his God.

After Lueli attempts to drown himself, Mr Fortune reflects that Lueli's love for him was generous and accepting, whilst his love for Lueli was selfish and interfering: 'Because I loved him so for what he was I could not spend a day without trying to alter him. How dreadful it is that because of our wills we can never love anything without messing it about!' It is worth recalling that Mr Fortune is an Old Rugbeian (the author's father was a master at Harrow; her great-grandfather had served at the school under Dr Vaughan, one of Thomas Arnold's most illustrious protégés). The unspoken epigraph of this novel is Dr Arnold's admission that: 'My love for any place, or person, or institution, is exactly the measure of my desire to reform them.' Warner's dry account of Mr Fortune's maggot is a moving and funny rejection of such an ideal. Mr Fortune carves Lueli a new idol and leaves Fanua, loaded with gifts and the author's affectionate envoi, to face a world entering the First World War.

Thornton WILDER　　　**1897–1975**

The Bridge of San Luis Rey

Wilder had the idea for his philosophical novel, which became an unexpected and runaway bestseller, while a student at Princeton; it came to him as he was crossing a bridge on the campus there. He was studying French literature, and Prosper Merimée's play *La Carosse du Saint Sacrement*, which concerns the affair between the Viceroy of Peru and an actress, supplied Wilder with a setting for his book and two of its characters. The novel was widely reviewed, won a Pulitzer Prize, and sold over a quarter of a million copies during its first year. Wilder spent some of the proceeds on building himself a house, which became known as 'the house *The Bridge* built'. His sister Isabel recalled that the novel brought Wilder, at the time an unknown schoolteacher, 'worldwide fame weighted with a cumbersome baggage of perquisites: honors, privileges, dazzling opportunities balanced by loss of privacy and hazards to body, mind, and spirit'.

'On Friday noon, July the twentieth, 1714, the finest bridge in all Peru broke and precipitated five travellers into the gulf below.' The bridge, a construction of osier, thin slats and dried vine, had seemed eternal; it was unthinkable it should break. The catastrophe was witnessed by Br Juniper, a Franciscan friar, who determined 'to inquire into the secret lives of those five persons that moment falling through the air, and to surprise the reason of their taking off'. Given that the disaster had been an act of God, it should be possible to divine His purpose. For six years Br Juniper laboured, only to have his book and himself publicly burned as heretical.

Among those to die were the Marquesa de Montemayor and her young companion, Pepita, journeying back to Lima after praying for the safe delivery of the Marquesa's grandchild. Through letters of extraordinary brilliance to her daughter, the Marquesa had hoped to obtain the love and duty that – having read a letter written by Pepita to the nun who had brought her up in an orphanage – she recognises it is only possible to obtain through courage. She resolves to begin a new life, and perhaps, in her death immediately following this resolve, she does so. Pepita has attained wisdom and acceptance by deciding not to send her letter after all.

The third victim is Estaban. He and his twin brother Manuel had been inseparable from birth, but the actress La Perichole had captured a part of Manuel's heart – that part not wholly devoted to his brother – and this led indirectly to his death. Estaban, however, had no place in his heart for anyone but Manuel; assuming his brother's name, he haunted the streets of Lima until Pepita's nun, Madre Maria del Pilar, suggested he join a ship as a deck-hand. Estaban proved inconsolable, and tried to hang himself. Attempting to comfort him, the ship's captain suggested: 'we do what we can ... It isn't for long, you know. Time keeps going by. You'll be surprised at the way the time passes.' Indeed, Estaban's time is past: soon afterwards he falls with the bridge.

The final victims are Don Jaime and Uncle Pio. Gravely beautiful and liable to convulsions, Don Jaime is the son of La Perichole by the viceroy. His companion is the man who discovered, groomed and sustained La Perichole, acting as master of cremonies. When she retired at the age of thirty, Uncle Pio took care of her. One day, however, he found her applying make-up and was banished. Since La Perichole was unable to pay him back the money she owed, he asked for her son for a year. His wish was granted and the old man and the young boy were on their way out into the world when they tumbled to their deaths.

Br Juniper 'thought he saw in the same accident the wicked visited by destruction and the good called early to Heaven. He thought he saw pride and wealth confounded as an object lesson to the world, and he thought he saw humility crowned and rewarded for the edification of the city.' But Madre Maria del Pilar reflects:

> Soon we shall die and all memory of those five will have left the earth, and we ourselves shall be loved for a while and

forgotten. But the love will have been enough; all those impulses of love return to the love that made them. Even memory is not necessary for love. There is a land of the living and a land of the dead, and the bridge is love, the only survival, the only meaning.

Virginia WOOLF   **1882–1941**

To the Lighthouse

To the Lighthouse continued Woolf's sequence of experiments with narrative form begun earlier in the 1920s, through which she sought to express the feminine experience she saw as distorted or suppressed by traditional realist fiction. Once again, the novel is built up through a subtle network of perceptions, using a stream-of-consciousness technique which allows events and characters to interact and interpenetrate with a remarkable poetic fluidity.

Two main characters with little apparent connection are woven together through bonds of secret sympathy. In fact, their lives are not merely parallel to one another, but virtually sequels. Mrs Ramsay, based on Woolf's own mother, is an outwardly conventional Edwardian wife, married to an irascible and mildly tyrannical Oxbridge don. Lily Briscoe, in contrast, is a fictional self-portrait of the author, and is very much of the next generation – independent, single and an aspiring artist. Despite Lily's conscious rejection of most of what Mrs Ramsay stands for, the two women admire each other.

The book is in fact a kind of elegy, a lament by the younger woman for the elder, and for the loss of the conventional certainties possessed by Mrs Ramsay. It unites the two women across the barriers of social space and historical time, thereby symbolically enacting the sea-journey, a romantic journey across dangerous waters, to the lighthouse of the novel's title.

The form of the novel is painterly – just as that of *Mrs Dalloway* (1925) was musical – being arranged into three panels like those of a medieval triptych. Mrs Ramsay occupies the first panel, in which she presides over a holiday for the family and friends on a Scottish island. The narrative circles restlessly in this section around the viewpoints of the various characters. Lily occupies the third panel, in which she returns to the cottage with the Ramsays years later, after Mrs Ramsay has died. At the end of this final section, dropped threads from the first section are picked up once more: Lily goes back to work on a painting of Mrs Ramsay which she was unable to finish on the first visit, while Mr Ramsay takes the children across the bay to the lighthouse, thereby fulfilling a promise to his son James made years before by Mrs Ramsay herself. Both Lily and Mr Ramsay thus make their peace with the ghost of Mrs Ramsay, and in doing so seem to come to a silent understanding of each other as well. The narrative of this section swings back and forth, like a pendulum coming to rest, between the viewpoints of the Ramsay family in the boat at sea, and Lily at her easel on the shore.

The central panel of the book is dominated by absence, as a ghost-like narrator simply watches time pass, in the decaying, abandoned holiday cottage, after the death of Mrs Ramsay. The mood here is one of haunted, ethereal beauty, as if the image of a Madonna and Child had been erased from the central panel of a real medieval painting, leaving just a mysterious outline inside a gilded frame.

To the Lighthouse is perhaps the best appreciated of Woolf's challenging mature novels, its delicate balance between formal elaboration and narrative continuity attracting a wider readership than the more stylistically complex novels still to come.

1928

Women's suffrage in Britain reduced from age of 30 to 21 ● Chiang Kai-shek is elected President of China ● Herbert Hoover, elected US President against Alfred E. Smith ● Alexander Fleming discovers penicillin ● The 'Geiger counter' ● USSR first Five-year Plan ● Completion of *New English Dictionary* (begun 1884) ● Henri Matisse, *Seated Odalisque* ● Maurice Ravel, *Bolero* ● Kurt Weil and Bertolt Brecht, *The Threepenny Opera* ● Walt Disney makes first 'Mickey Mouse' film in colour ● Thomas Hardy, Douglas, Earl Haig, H.H. Asquith, Ellen Terry and Roald Amundsen die ● Edmund Blunden, *Undertones of War*

Ford Madox FORD　　　　　**1873–1939**

Parade's End

Some Do Not (1924)
No More Parades (1925)
A Man Could Stand Up (1926)
Last Post (1928)

Parade's End was first published in one volume under this title in America in 1950. Whether or not the book is a trilogy (the first three volumes are sometimes referred to as *The Tietjens Trilogy*) or a tetralogy depends upon whether one accepts the final volume as an integral part of the total design. *Last Post* was apparently written to satisfy a friend's curiosity about the fate of the characters, although financial considerations were probably a more pressing encouragement. Ford later came to regret the volume, which is, indisputably, very different in tone and technique from those it followed. Any consideration of *Parade's End*, however, must take *Last Post* into account, although it is probably best considered as an appendix or a coda to the tightly structured and interdependent preceding three volumes.

In different ways Ford's novels are all devoted to the 'condition of England' question central to the fiction of the day, and to its predominant theme of the movement from an older, caring world to a new world of mechanistic indifference. Like E.M. Forster, Ford was preoccupied with the idea of the connected society, wherein reason would be unified with emotion, commerce with art. *Parade's End* was to be Ford's *Howards*

End (1910), and he consciously set out with this novel to become the 'historian of his own time'. The subjects of *Parade's End* would be 'the public events of the decade' and 'the world as it culminated in war'; the method would be to filter these large social tendencies and forces through a central character who is located both in peace and war. To give his hero the right kind of connection between history and contemporary life, Ford not only made him an officer and a gentleman, but also a member of the political establishment and a country squire. Christopher Tietjens is a man torn in both his public and private lives by the pressures of the times. *Parade's End* describes his suffering under the forces of change that manifest themselves when the opulent world of Edwardian England becomes corrupt in the atmosphere of the First World War. His story is told against a panorama of a nation in decline with its debased ruling class: 'the swine in the corridors' of Whitehall and the incompetent General Staff on the Western Front. Yet these issues are never simply the backdrop to Tietjens' progress. Neither is *Parade's End* merely a diatribe against the iniquities of the system in peace and at war. All these issues have equal weight as part of the larger experience of the individual in society.

Tietjens suffers chiefly at the hands of his unscrupulous wife, Sylvia, both in peace and at war, for she stands behind his antagonistic military superiors at the Front and the unsympathetic civilians at home. Country squire and Christian gentleman, Tietjens is a man whose roots go deep in the traditions of

English society and life, the embodiment of the values the Edwardian world claimed but had long lost. At the opening of *Parade's End* he is already an anachronism: his chivalry, his code of honour, his monogamy and his chastity have become redundant, and he is lost among the new dispositions of power, the new sexual mores, the new habits and standards of Georgian England, all of which are embodied in the figure of his wife. It is Tietjens' uncorrupted and awkward integrity that antagonises his contemporaries, who are more at ease with the new social hypocrisy. Sylvia despises and torments him, but is unable to leave him completely because she recognises that other men, though easier to live with, are lesser creatures. Although made to bear an unusually wide range of significance, both Tietjens and Sylvia are intensely and individually realised. Tietjens convinces as that most difficult of all heroes, the good man; and Sylvia, hard, cruel and vicious as she is, persuades as a woman trapped and wasted in a cycle of destructive behaviour.

From the start, a certain absurdity attaches to Tietjens; he is 'a sort of lonely buffalo, outside the herd'. He is at once a romantic figure caught at the point of extinction, a man of parades in a new world which, socially as well as militarily, has no more need of them, and a comic character. *Parade's End* is indeed a sort of war comedy, a species that was to reappear in the parallel English trilogy for the Second World War, Evelyn Waugh's **Sword of Honour** (1952–61). In specific details and general themes Waugh's trilogy owes a great deal to Ford's masterpiece: it is difficult to believe that it could have been written without the example of *Parade's End*. Like Tietjens, Guy Crouchback is a man torn in public and private life by the pressures of his time. He is also a comic figure, absurd and anachronistic in his chivalry and integrity, yet moving and dignified in the fortitude with which he bears his increasing disillusionment. Both novels also share an unacknowledged debt to Howard Sturgis's **Belchamber** (1904), whose central character is a cripple and a cuckold, a good man victimised by the unscrupulous forces at large in society.

Tietjens, whose presence provides the necessary unity for the first three volumes of *Parade's End*, is largely absent from *Last Post*. The last scene of *A Man Could Stand Up* shows Tietjens and Valentine Wannop dancing amidst the confusion of armistice night, poised between the unresolved past and the unknown future. It is clear from the first page that *Last Post* is different. The narrative is related largely from the point of view of Tietjens' brother, Mark, who took to his bed and refused to speak when he learnt of the peace terms. He is the new representative figure of the generation that never recovered from the war. Our final glimpse of Tietjens shows him to have retired into quiet country life with Valentine Wannop, the former suffragette, and as such, one of the few good new forces in society. The chaste, emotional relationship that has paralleled Tietjens' unhappy marriage throughout the novels is finally consummated. Tietjens is now a purely private man, and his new career as smallholder and antique dealer is an ironic comment on his former representative status. Sylvia has at least lessened her hostility, and this perhaps represents a triumph for Tietjens, of survival if not of happiness. But at no point could *Parade's End* be described as a simple fable of moral growth, with Tietjens winning through at the end: his disappearance from centre-stage in *Last Post* prevents this. That the focus has moved from Tietjens suggests also that life moves on, indifferent to our interest in particular individuals, and that Tietjens' story is simply one of many.

The great strength of *Parade's End* is its detached but committed tone. It can be read as a tragedy of Tietjens' accumulated suffering, a comedy of his emergence into a new life, a conservative attack on the decline of English society, or a radical exposé of the ruling class and its incompetent handling of both peace and war. *Parade's End* is none of these novels and all of these novels. It is the *Iliad* of the First World War, and Ford's greatest achievement.

John GALSWORTHY   1867–1933

Swan Song

see **The Forsyte Chronicles** (1929)

Radclyffe HALL   **1880–1943**

The Well of Loneliness

'I would rather put a phial of prussic acid in the hands of a healthy girl or boy than the book in question.' Condemned thus by the *Sunday Express*, withdrawn at the order of the Home Secretary (the notorious Joynson Hicks), defended by Arnold Bennett and Havelock Ellis, *The Well of Loneliness* remains the classic novel of lesbian love. The book may no longer shock but, despite its often overwrought prose, it retains its power to move. Hall waited to write it until she had established a reputation with four earlier books, the better to plead overtly the case of the 'invert'.

The heroine is Stephen, daughter of Sir Philip and Lady Anna Gordon of Morton Hall, who grows up scorning dolls and pretty clothes, preferring to fence and hunt. Lady Anna feels an increasingly guilty antagonism towards her unconventional daughter, but Sir Philip becomes Stephen's companion and advocate. He has read Krafft-Ebing and suspects the truth about his daughter. Stephen develops a childhood passion for Collins, a pretty young housemaid, but her heart is broken when she discovers the maid in the arms of a footman.

In Stephen's adolescence, her indulgent French governess, Mlle Duphot, is replaced by the redoubtable 'Puddle', whose own nature gives her an insight into Stephen's. Without disparaging Stephen's athletic interests (much frowned upon by the County), Puddle nurtures her pupil's intellectual side and helps her discover a talent for writing. Stephen finds her first real friend in Martin Hallam, a young farmer and arborist; but when Martin declares his love, Stephen flees in confusion and horror to her father, who shrinks from enlightening her as to why she should have this reaction. Sir Philip is later killed by a falling cedar-branch, and the grief-stricken Stephen rides once more to hounds on her beloved hunter, Raftery; but now she identifies with the pursued, and vows to hunt no more.

Soon afterwards, she falls in love with Angela Crossby, beautiful but unhappily-married, who accepts Stephen's lavish gifts and initially responds to her devotion. Eventually, however, Angela makes it brutally clear that they have no future together. In a hideous replay of the scene with Collins and the footman, Stephen finds Angela in the arms of an old enemy, Roger Antrim. Angela panics and in order to save her reputation tells her husband about Stephen's infatuation. He sends one of the letters that Angela had received from Stephen to Lady Anna, who disowns her daughter. Before leaving Morton, accompanied by her horse and her governess, Stephen reads the hidden books in her father's library and discovers the truth about herself.

She held up her hand, commanding silence; commanding that slow, quiet voice to cease speaking, and she said: 'As my father loved you, I loved. As a man loves a woman, that was how I loved – protectively, like my father. I wanted to give all I had in me to give. It made me feel terribly strong ... and gentle. It was good, good, *good* – I'd have laid down my life a thousand times over for Angela Crossby. If I could have I'd have married her and brought her home –'

THE WELL OF LONELINESS
BY RADCLYFFE HALL

In London she writes an acclaimed novel. Raftery grows old and lame and, in the novel's most poignant passage, Stephen has to shoot him. She moves with Puddle to Paris and mixes in the homosexual *demi-monde*, which Hall populates with recognisable portraits of her contemporaries. During the First World War, Stephen joins an ambulance unit, where she encounters many women like herself. Her face is scarred by a piece of shrapnel and she is decorated for heroism. Most importantly, she meets Mary Llewellyn, whom she brings back to Paris to share her home. Stephen is determined to behave honourably, and much agonising precedes the historic words; '... and that night they were not divided'.

They are idyllically happy for a time, but

as Stephen (by now a famous author) grows ever more absorbed in her work, Mary becomes bored with her domestic role and, when Martin Hallam reappears, Mary unwillingly falls in love with him. Aware that Martin can give Mary the family life and social acceptance she craves, Stephen pretends that she has been unfaithful in order to release her lover. The novel ends with Stephen's resolution to speak for those legions who have no voice, and her plea: 'We have not denied You, rise up then and defend us, oh God, before the whole world. Give us also the right to our existence!'

Aldous HUXLEY　　　　**1894–1963**

Point Counter Point

This huge *roman à clef* has an enormous cast but no tightly knit plot: instead, influenced by André Gide, Huxley attempts to make fiction approximate to contrapuntal music, counterpointing the affairs of his characters in a complex scheme. Much of the novel consists of brilliant conversation, from which, eventually, the elements of an exciting storyline take shape.

The novel's chief characters are mainly from three families: the upper-middle-class Bidlakes and Quarleses, and the aristocratic Tantamounts. The first 160 or so pages occupy a single evening, in which a dazzling society party is being held at Tantamount House in London. Walter Bidlake, a sensitive young literary journalist, is in love with Lucy Tantamount, a heartless siren (based upon Nancy Cunard), who causes him to neglect his pregnant mistress, Marjorie Carling. After the party, Walter and Lucy go on to the fashionable restaurant, Sbisa's. At these two venues we see a large proportion of the characters in action, a legion of self-obsessed emotionally unsatisfied cranks, poseurs and fanatics.

As the novel progresses, the emptiness and baseness at the heart of most of the characters' lives is shown up. Walter earns little but torment from his relationship with Lucy, who briefly becomes his mistress but then goes off to pursue sensation and casual sex in Paris. Philip Quarles (a partial self-portrait of the author) is a novelist, and

selections from his notebooks are reproduced thoughout the novel. His inability to relate to others in anything other than intellectual terms almost drives his loving wife Elinor into an affair with the leader of the proto-fascist British Freemen, Everard Webley. Other characters include the diabolic Spandrell (inspired by the character of Baudelaire), who gains satisfaction from corrupting little girls; the ingratiating literary editor Denis Burlap (John Middleton Murry), who indulges in childish sex games with his frigid editorial assistant Beatrice Gilray; and the comically self-deceived *rentier* Sidney Quarles, Philip's father. To such figures is counterpointed the Lawrentian writer and artist Mark Rampion, who preaches a gospel of healthy acceptance of the instincts.

Towards its close, the novel inaugurates the taste for incidents of violence that marks Huxley's later work. Philip and Elinor's young son Philip ('little Phil') falls ill with meningitis at a moment when Elinor has arranged an assignation with Webley. She rushes to her child's bedside and leaves the keys of her London house with Spandrell, who, in turn, gives Frank Illidge, a socialist who has expressed a wish to murder Webley, the chance to make his wish a reality. When Webley arrives at the house to meet Elinor, the pair kill him. Little Phil dies agonisingly of his disease, and a third death occurs when Spandrell achieves a personal apotheosis by deliberately arranging to be shot dead by members of the British Freemen.

Point Counter Point provides a sharply satirical portrait of London society in the 1920s and, although partly a novel of ideas, it shows more human understanding and sympathy than some of Huxley's other books. Mark Rampion may indeed, as Huxley confessed, be 'just some of [D.H.] Lawrence's notions on legs', but other people in the book are more fully characterised. And in the portrait of the over-intellectual Philip Quarles, and the extracts reproduced from his notebooks, Huxley offers a penetrating self-criticism of the sort of fiction he wrote himself.

D.H. LAWRENCE   1885–1930

Lady Chatterley's Lover

Printed privately in Florence in 1928, Lawrence's last novel has achieved notoriety as a result of being prosecuted for obscenity, not appearing in unexpurgated form in the UK until the Penguin edition of 1960. Lawrence intended his explicitly erotic writing to celebrate rather than degrade human sexuality, his provisional title being 'Tenderness', and he wrote the novel three times before achieving the balance of personal and public themes of *Lady Chatterley's Lover*. (The earlier versions were subsequently published as *The First Lady Chatterley*, 1944, and *John Thomas and Lady Jane*, published in an Italian translation in 1954, and in English in 1972.)

Clifford and Constance Chatterley spent one month on honeymoon before Clifford had to return to the army; six months later, he was shipped home to England, paralysed from the waist down. After two years, he is beginning to accept his condition, although confinement to a wheelchair produces spasms of impotent rage, the legacy of the First World War scarring his spirit as well as his body. Sexually unfulfilled, Connie experiences a similar imprisonment in the bleak environment of Wragby Hall. When Clifford's Cambridge contemporaries visit, she sits silently as they talk of the life of the mind, an intellectual creed of sexual bohemianism which she herself had followed before the war. A brief affair with Michaelis, 'the eternal outsider', provides temporary and empty satisfaction. Meanwhile Clifford becomes 'the most modern of modern voices', a successful writer of clever stories devoid of humanity, and a progressive industrialist dragging his coal-mines into a profitable future.

With 'the sudden rush of a threat out of nowhere', Connie becomes aware of the presence of Mellors, Clifford's gamekeeper, whose broad Derbyshire speech symbolises the deliberate adoption of an alternative approach to life. He looks like a free soldier rather than a servant, Connie muses; he might almost be a gentleman. A secret and tender affair results in Connie's pregnancy, and she contemplates bringing up the child as the heir to Wragby, a principle which Clifford had theoretically encouraged. Whilst on holiday in Italy, however, where she had intended Clifford to believe she had found a lover, she learns that scandal is developing at home. Rather than resume her cold life with Clifford, Lady Chatterley chooses divorce and a future with Mellors, the satisfaction of true sexuality.

Penguin's decision to make the book available to a wide audience became a test case for the recently introduced Obscene Publications Act of 1959. The trial opened at the Old Bailey on 20 October 1960, with the case for the Prosecution presented by Mervyn Griffith-Jones, who reminded the jury that not only did the novel contain thirteen acts of sexual intercourse 'described in the greatest detail', but that Lawrence's text was strewn with words 'not voiced normally in this Court ... The word "fuck" or "fucking" occurs no less than thirty times. I have added them up, but I do not guarantee that I have added them all up. "Cunt" fourteen times; "balls" thirteen times; "shit" and "arse" six times apiece; "cock" four times; "piss" three times, and so on.' He was much ridiculed for being out of touch with the real world in his addresses to the jury: 'For those of you who have forgotten your Greek, "phallus" means the image of a man's penis'; 'Is this a book you would wish your wife or even your servant to read?' Thirty-five 'distinguished men and women of letters, moral theologians, teachers, publishers, editors, and critics' – including E.M. Forster, Richard Hoggart, Rebecca West, Walter Allen, Helen Gardner and the Bishop of Woolwich – appeared for the Defence, testifying to the book's high moral purpose and great literary merit. ('By the way,' Forster wondered, 'did D.H. Lawrence ever do anything for anybody? Now that we have been sweating ourselves to help him, the idea occurs.') Penguin were acquitted and the following year issued a transcript, *The Trial of Lady Chatterley*, edited by C.H. Rolph.

The House at Pooh Corner

see **Winnie-the-Pooh** (1926)

Quartet

First published as *Postures* – and reissued as *Quartet* in 1969 – this was Rhys's first novel, though in the previous year she had published a collection of short stories entitled *The Left Bank*, with an enthusiastic introduction by Ford Madox Ford. Neither this nor the books that followed were successful on first publication, but all are now firmly established in the repertoire of modern classics. *Quartet* announces Rhys's central theme of a girl impoverished and on her own, the victim of those who offer her 'protection' in a world that is 'cruel and horrible to unprotected people'; indeed Marya Zelli might almost be a study for the young Sasha Jenson in **Good Morning, Midnight** (1939).

Marya (her pet name is Mado) had met Stephan Zelli in London while 'resting' as a chorus-girl. He is 'a short, supple young man of thirty-three or four, with very quick, brown eyes and eager but secretive expression', who tells her very little about himself except that 'he was of Polish nationality, that he lived in Paris, that he considered her beautiful and wished to marry her'. Also that he was a *commissionaire d'objets d'art* – that is, he sold 'pictures and other things'. It was to Paris that he took her and for a time she felt that her marriage, 'though risky, had been a success ... Sometimes they had a good deal of money and immediately spent it. Sometimes they had none at all and then they would skip a meal and drink iced white wine on their balcony instead.' But he leaves her alone a good deal while pursuing his 'business deals', mainly in cafés, which is why she takes up with the shiftless English-speaking community in the first place, and catches the eye of the corrupt, predatory Heidlers.

Stephan is arrested and sentenced to a year's imprisonment in Fresnes to be followed by exile from France. Alone and destitute Marya can only comply when Lois Heidler and her husband 'H.J.' offer her 'the spare room in the studio', and proceed to master her. From the start Marya is rarely fooled by Lois, whom she comes to loathe. 'Lois was extremely intelligent ... And gave a definite impression of being insensitive to the point of cruelty.' She plots with her husband to get 'Mado' into bed with him – indeed, that is the whole point of taking her up. They are 'inscrutable people, invulnerable people, and she simply hadn't a chance against them, naïve sinner that she was'. 'You have made an arrangement!' she accuses Lois. 'Not in so many words, perhaps, a tacit arrangement. If he wants the woman let him have her.' 'A profound sense of the unreality of everything possessed her. She thought: "I wonder if taking opium is like this." ' Worse still, despite his not being a good lover – 'He didn't really like women. She had known that as soon as he touched her' – she falls in love with H.J., and her disgust with Lois and herself reaches desperation. Her position becomes intolerable; she is in effect their slave. 'You've smashed me up, you two,' she tells them.

All this time she has been faithfully visiting Stephan in Fresnes. These visits are dreadful but she has reached the point where she 'thought of her husband with a passion of tenderness and protection' ... which 'she extended ... to all the inmates of the prison, to all the women who waited with her under the eye of the fat warder, to all unsuccessful and humbled prostitutes, to everyone who wasn't plump, sleek, satisfied, smiling and hard-eyed'. But Stephan is not to be relied on either. On being released from prison his situation is desperate, all his plans come to nothing and he has been brutalised. When she tells him about her affair with H.J. he knocks her down and goes off with another woman. The Heidlers have won, as they were bound to. Mado is destroyed.

In *Quartet* Jean Rhys may not have gained the utter conviction of *Good Morning, Midnight* or the mastery of **Wide Sargasso Sea** (1966) – at points the story lacks the relentless logic that she was to achieve – but the novel contains scenes of chilling conviction. Above all she has established a world and a voice that is uniquely hers.

Siegfried SASSOON   **1886–1967**

Memoirs of a Fox-Hunting Man

see **The Complete Memoirs of George Sherston** (1936)

Edith SOMERVILLE   **1858–1949**
& Martin ROSS   **1862–1915**

The Irish R.M.

see 1915

Evelyn WAUGH   **1903–1966**

Decline and Fall

Waugh's first novel, largely written in a Dorset pub, was based upon his experiences as a schoolmaster in Wales and as a socialite in London. It was turned down by Duckworth (the publisher in 1928 of his biography of Rossetti) on the grounds of obscenity, and was eventually published by his father's firm, Chapman and Hall, which subsequently published all his novels. Chapman and Hall demanded a number of cuts, many of which were restored when the book was reissued in the uniform edition of Waugh's work (1962). Waugh provided illustrations and designed the wrapper of the first edition, dedicating the book to his Oxford contemporary, Harold Acton.

Paul Pennyfeather is the prototype of all Waugh's innocents, a passive young man who becomes caught up in the world of those more sophisticated than himself. His first misfortune is to be sent down from Scone College, Oxford, where he has been studying theology, after being debagged by some fellow undergraduates. He finds employment at Llanabba Castle, an eccentric preparatory school in North Wales run by Dr Augustus Fagan. Amongst the staff there are Capt. Grimes, an Old Harrovian pederast with a wooden leg (based upon one of Waugh's colleagues at Arnold House); Mr Prendergast, a bewigged former clergyman; and Solomon Philbrick, a criminal now employed as butler. One highlight of the novel is the description of the chaotic school sports, during which Philbrick shoots little Lord Tangent in the foot. The young aristocrat's subsequent demise is reported in laconic asides, a technique Waugh borrowed from Firbank. In spite of his devotion to young Percy Clutterbuck, Grimes finds himself engaged to Fagan's unprepossessing elder daughter, Flossie, an entanglement he escapes by faking his own suicide.

One of Paul's pupils is Peter Pastmaster, whose glamorous mother, Margot Beste-Chetwynde, electrifies the sports day by arriving on the arm of a black musician. She invites Paul to spend the holidays at her modernist country house, King's Thursday, ostensibly to coach her son. Paul becomes engaged to Margot and is offered a job with her Latin-American Entertainment Co Ltd., a white-slave racket masquerading as an employment agency. The police investigate this business, and on the morning of his wedding, Paul is arrested. He is sent to Blackstone Gaol, where Prendergast has secured the post of chaplain, and where Philbrick is serving a sentence for arson, having burned down Llanabba Castle after an unsuccessful attempt to buy it. Prendergast has his head sawn off by a lunatic prisoner, and Paul is transferred to Egdon Heath Penal Settlement, where he meets Grimes, who has been convicted of bigamy. Margot and her new lover, Alastair Digby-Vane-Trumpington (who was responsible for Paul's Oxford disgrace), arrange for Paul to be transferred to Cliff Place, a private sanatorium at Worthing run by Dr Fagan, in order to have an appendectomy. Papers are forged to the effect that Paul has died under anaesthetic and, posing as his own distant cousin, Paul returns to Oxford. He is visited by Peter Pastmaster, now an undergraduate and a rowdy member of the Bollinger Club, who informs him that Margot has married the Minister of Transport, becoming Viscountess Metroland. Upon hearing that

Paul is to enter the church, Peter remarks: 'Damn funny that. You know you ought never to have got mixed up with me and Metroland.'

A remarkably assured fictional debut, *Decline and Fall* established Waugh as a vigorous new comic talent and remains one of the funniest novels of the twentieth century. Several of the characters, including Alastair, Peter and Lady Metroland, reappear in Waugh's subsequent novels.

Virginia WOOLF   **1882–1941**

Orlando

Along with *A Room of One's Own* (1929), *Orlando* was written in an interlude between two major creative outbursts which produced **To the Lighthouse** (1927) and **The Waves** (1931). It is a spoof biography, and also a homage to Vita Sackville-West, the woman with whom Woolf had fallen passionately in love in the early 1920s. (Angela Carter provocatively described the novel as 'a slobbering valentine' to the upper classes.)

Orlando is a creature of dubious gender, who lives for an equally improbable 400 years, thus situating his (or her) origins in a mythical past, before the constrictions of Victorian femininity had imposed their corset-like hold on the female population of England. Orlando's career begins as that of a strangely beautiful young nobleman in Elizabethan England, a kind of boy-woman from a Shakespeare play. Shakespeare himself makes a cameo appearance near the start of the novel, and becomes a surreptitious mentor for Orlando's later career as a poet. Orlando undergoes numerous adventures, including encounters with a beautiful but faithless Russian princess, a scurrilous Elizabethan poet, Good Queen Bess (who is as virile as Orlando is effeminate), large sections of London's low-life, an eccentric (but amorous) Romanian Archduchess, the Sultan of Turkey, and a tribe of Oriental gypsies. While in Turkey, he also changes into a woman, and simultaneously changes centuries.

Returning to England as a woman, Orlando again steers agilely between literary salons (now those of Johnson, Addison and Pope) and the seedier parts of London. The archduchess reappears, to reveal that she has been a man disguised as a woman all along. Orlando is unimpressed, however, and trades in the eighteenth for the nineteenth century.

The Victorian age assails Orlando with a tide of pompous architecture, crinolines, bustles and wedding rings, to which she at first appears to succumb. However, the gender of her newly acquired husband is almost as dubious as her own, and since he is a sailor the marriage functions mainly as a convenient cover for Orlando's literary activities. The scurrilous poet, now a respectable Victorian critic, reappears just in time to publish Orlando's now finished masterpiece, a poem called 'The Oak Tree', which has taken her several centuries to revise and complete.

Finally, the twentieth century rolls around, and the last glimpse of Orlando is of her lying beneath her favourite oak tree at the ancestral home (a portrait of Vita Sackville-West's beloved Knole), arbitrating between the multiple selves she has acquired over the centuries, and awaiting the return of her sea-capain.

The mood of *Orlando* is always comic, but the comedy displays a wide range of tonalities: sometimes it is charming, sometimes infantile, sometimes bitingly satirical, and sometimes almost private (as with the photographs of 'Orlando' included in the first edition of the book, which were actually photographs of Vita Sackville-West dressed in masculine clothing). *Orlando*, in short, is a popular historical novel imbued with a serious message: that everyone is secretly androgynous, and that gender is ultimately indeterminable.

The novel was made into a remarkable and acclaimed film (1993), adapted and directed by Sally Potter.

1929

Wall Street Crash • Leon Trotsky is expelled from USSR • Arab attacks on Jews in Palestine, following disputes over Jewish use of the Wailing Wall, Jerusalem • Basle Bank for International Settlements is founded to deal with Germany's reparation payments • The term *Apartheid* is first used • Kodaks develop a 16 mm colour film • *Graf Zeppelin* airship flies round the world • Heidegger, *What is Philosophy?* • Piet Mondrian, *Composition with Yellow and Blue* • Audrey Hepburn, John Osborne and Jacqueline Bouvier Kennedy born • Millicent Garrett Fawcett and Serge Diaghilev die • D.H. Lawrence, *Pansies* • Louis MacNeice, *Blind Fireworks*

Richard ALDINGTON   1892–1962

Death of a Hero

Aldington's intemperate account of the death of Edwardian England 'galloped off the typewriter', he claimed, and was written with the 'mysterious sense of somebody dictating'. It was not published in its unexpurgated version until 1965, three years after Aldington's death. Until then each edition of the novel was prefaced with a Note '*en attendant mieux*' explaining why the text was littered with asterisks: 'It is better for a book to appear mutilated than for me to say what I don't believe.' The novel is a grotesque puppet-show in which Aldington's delicacy, as seen in his Imagist poetry, is sacrificed to coarse sarcasm as he dissects the life and times of his eponymous hero, George Winterbourne.

The novel opens with a prologue (*allegretto*) in which Winterbourne's death in action during the last week of the First World War is announced. The narrator describes the reactions of the people who knew him: his father, a Catholic convert and 'an inadequate sentimentalist'; his mother, an adulterous nymphomaniac; his wife, Elizabeth, who conned him into marriage by means of a phantom pregnancy; and his mistress, Fanny, Elizabeth's worldly best friend. Their reactions betray their selfishness and hypocrisy, and it becomes clear that George was driven to join up and to commit 'suicide' (by unnecessarily exposing himself to enemy fire) largely because his home life had become intolerable.

Part one (*vivace*) is an attack upon 'Cant' in the form of a history of Winterbourne's parents ('eponyms of sexual infelicity'), and a description of George's public-school education, where he learnt to dissimulate and 'played the fine, healthy, barbarian schoolboy'. Part two (*andante cantabile*) follows George's '*naïf* peregrinations' as painter and journalist in artistic London, and his tangled relations with Elizabeth, Fanny and a Cambridge intellectual called Reggie, who becomes Elizabeth's lover. The attack upon Bohemian London prefigures *The Apes of God* (1930), and, like his friend Wyndham Lewis, Aldington pillories his contemporaries in recognisable caricatures. (Aldington's old friend Erza Pound appears as the fraudulent Upjohn, while Ford Madox Ford, D.H. Lawrence and T.S. Eliot are lampooned as Shobbe, Bobbe and Tubbe.) In Part three (*adagio*) Aldington confronts a subject worthy of his rancour and his descriptive powers: the war itself. George has joined up as a private in the Pioneers and sails to France feeling 'almost happy'. Looking at his fellow soldiers he thinks: 'I swear you're better than the women and the half-men, and by God! I swear I'll die with you rather than live in a world without you.' Fanny and Elizabeth fade to become 'memories and names at the foot of sympathetic but rather remote letters.' George becomes a runner and is persuaded by Evans, a young public-school officer who admires him, to take a commission. When he returns to England for training he finds Elizabeth and Fanny occupied with other men, and his old acquaintances bored by the war. Newly commissioned, he returns to the Front, 'a wrecked man, swept along in the swirling cataracts of the War', and is killed in an attack.

The novel's title is, of course, ironic. Aldington explains: 'George's death is a symbol to me of the whole sickening bloody waste of it ... Somehow or other we have to make these dead acceptable, we have to atone for them, we have to appease them ... somehow we must free ourselves from the curse – the blood-guiltiness.' Aldington's impassioned but bilious expiation is perhaps the most angry book to emerge from the First World War.

William FAULKNER   1897–1962

The Sound and the Fury

Faulkner's fourth novel is a masterpiece of construction. Divided into four, its first, third and final sections cover three consecutive days, while the second goes back eighteen years. The first, 'April Seventh, 1928', is written from the point of view of Benjy Compson and suggests the reason for the novel's title – Macbeth's summation of life: 'It is a tale/Told by an idiot, full of sound and fury,/Signifying nothing.' Benjy is indeed a congenital idiot. His world is built upon associations of sound, sight and smell. (He haunts a golf course so that he can hear the golfers call the name of his beloved sister, Caddy.) Since he has no conception of linear time, his narrative constantly slips from present to past and, though he discloses fragments of the Compsons' history, he does so with an extreme and baffling subjectivity lent coherence only by the subsequent parts of the saga.

'June Second, 1910' is set at Harvard and describes the deeds and thoughts of Benjy's older brother, Quentin, on the day of his suicide. Like Benjy, Quentin is obsessed with Caddy – but Quentin's obsession is sexual, frustrated and jealous. Before going up to Harvard he has told their father that he and Caddy have committed incest. Quentin is correctly disbelieved; his falsehood is the product of ill-defined yearning and unattributable guilt. Nevertheless, when Caddy – already pregnant – marries a bluff, deceitful businessman, Quentin crumbles into desolation. A sophisticated parody of Benjy's, his mind is filled with aimlessness and grief.

'April Sixth, 1928' is narrated by Jason. The third of the Compson brothers, Jason is a cold, self-righteous tyrant. In his pragmatism (his mother claims) he shows the lineage of her family, and not that of the fatalistic Compsons. Through his embittered recollections it is revealed that Caddy has been thrown out by her husband, who suspects that Caddy's child is not his. The child – a daughter, also named Quentin – is brought up in the Compson household. Caddy's visits to see her daughter exacerbate Benjy's obsession: impervious to the logic of time, he looks for Caddy among the local schoolgirls, and when he runs after them it is assumed that his purpose is sexual assault. He is castrated. Jason, for all that he has a regular whore in Memphis, persecutes Quentin for her promiscuity. He steals the money Caddy sends to her, losing some on the stock exchange and hoarding the rest. Quentin eventually steals her money back and runs away with a musician. Jason's attempts to recapture her fail miserably.

The three-quarters began. The first note sounded measured and tranquil, serenely peremptory, emptying the unhurried silence for the next one and that's it if people could only change one another for ever that way merge like a flame swirling up for an instant then blown cleanly out along the cool eternal dark instead of lying there trying not to think of the swing until all cedars came to have that vivid dead smell of perfume that Benjy hated so.

THE SOUND AND THE FURY
BY WILLIAM FAULKNER

In the fourth part, 'April Eighth, 1928', which is narrated in the third person, the Compson remnants are seen from without. Jason, imbecile Benjy and their permanently invalided mother are the sole inheritors of a diminished wealth and standing. The rest of the Compson dynasty has perished or gone and, if its decline symbolises the decline of the Confederate South, it conveys with equal power Faulkner's vision of moral entropy. Doomed by its own desires, doomed in its struggle to escape from them, humankind 'must just stay awake and see evil done for a little while its not always and it doesnt have to be even that long'.

John GALSWORTHY   1867–1933

The Forsyte Chronicles

The Man of Property (1906)
In Chancery (1920)
To Let (1921)
The White Monkey (1924)
The Silver Spoon (1926)
Swan Song (1928)

Galsworthy's sequence of novels, which became a phenomenal popular success when adapted for television in the 1960s, has claim to be regarded as the first literary soap opera. It spans forty years in the lives of the Forsytes, an English upper-middle-class family who first made their appearance in *The Man of Property*. Other novels followed, then as now largely unread, until in 1918 Galsworthy conceived the idea that he would 'make *The Forsyte Saga* a volume of half a million words nearly; and the most sustained and considerable piece of fiction of our generation at least'. The saga comprises six novels, which were eventually arranged, with linking 'interludes', into two trilogies: *The Forsyte Saga* (1922) and *A Modern Comedy* (1929). D.H. Lawrence did not share Galsworthy's high opinion of this enterprise and declared: 'The story is feeble, the characters have no blood and bones, the emotions are faked, faked, faked. It is one great fake.'

The saga runs from 1886 to 1926, but the historical events of this period play little part in the fiction itself. Nostalgia for a recent past is suggested by the gradual change in attitude to what once seemed of permanent value to the middle classes. For the Forsytes this is marked by their declining interest in, and decreasing attendance at, family funerals as the decades pass. Lengthy and idyllic descriptions of the English countryside stress the importance to Galsworthy of a beauty unaffected by social or political change.

The saga opens in 1886 with the first novel of the first trilogy, *The Man of Property*. Several generations of Forsytes gather to celebrate the engagement of June Forsyte to an impecunious architect, Philip Bosinney. The family are not especially bonded by love, but are united in their concern for money. The older members, such as Swithin, James, Timothy and 'old Jolyon' Forsyte represent the solid Victorian values of the lawyer, stockbroker and publisher, living in houses near Hyde Park with spinster sisters. Their investments have to yield ten per cent, and great faith is placed in Consols. Second-generation Forsytes include Winifred, wife of the dissolute Montague Dartie, who, with George Soames, a horse-fancier and bachelor, provides the novel with a comic element.

Fleur was still gracefully concealing most of what Michael called 'the eleventh baronet,' now due in about two months' time. She seemed to be adapting herself, in mind and body, to the quiet and persistent collection of the heir. Michael knew that, from the first, following the instructions of her mother, she had been influencing his sex, repeating to herself, every evening before falling asleep, and every morning on waking the words: 'Day by day, in every way, he is getting more and more male,' to infect the subconscious which, everybody now said, controlled the course of events; and that she was abstaining from the words: 'I *will* have a boy,' for this, setting up a reaction, everybody said, was liable to produce a girl.

A MODERN COMEDY
BY JOHN GALSWORTHY

Galsworthy jibes at the foibles of the elderly Forsytes but also notes a breach in middle-class respectability caused by young

Jolyon, a second-generation Forsyte, who has eloped with a 'foreign woman' when already married, and whose moral conduct has exiled him from his daughter June's engagement celebration.

Most disgruntled of the party is the central character, Soames Forsyte, brother of Winifred. A lawyer and art collector, and named 'man of property' by old Jolyon, he has married the beautiful Irene, and although he loves her passionately and considers her one of his acquisitions, she is not only cold and remote towards her husband but despises him, refusing him his conjugal 'rights'. Bosinney is commissioned to design a large house for Soames, and subsequently enjoys an adulterous affair with Irene. When Soames is driven by despair to reassert his claim on his wife, and rapes her, she escapes to Bosinney's house for refuge. Bosinney is upset by Irene's degradation and is killed when he blunders into a cab in thick fog.

The theme of silent suffering in broken marriage in an age in which divorce was socially unacceptable, is questioned, but there is no happy resolution at this point. Irene lives alone, Soames moves to Brighton, never occupying his new house, hardly aware that he is at least morally responsible for the architect's death.

In Chancery opens twelve years later. Soames is determined to divorce Irene in order to marry a French woman, Annette Lamotte, whom he hopes will provide him with an heir. Although capable of bargaining astutely in legal matters, Soames can not understand or cope with the emotions of others. This inability results in Irene uniting with Jolyon, now a widower. Galsworthy becomes less concerned with the past, which he evoked with frequent interior monologue in the earlier volume, and introduces several integrated sub-plots against the background of the Boer War. The divorced Soames finally marries Annette a few weeks before the death of Queen Victoria. Irene is happily married to Jolyon and, through an irony of fate, living in the house built by Soames and designed by her former lover. The Edwardian age is further confirmed with the birth of a daughter to Soames, and the death of his father, James, on the same day.

By opening in 1920, *To Let* omits the First World War and its immediate aftermath. This last section of the first trilogy is dominated by events of the past which continue to dictate the course of the present. The Labour Government's increase in the level of income tax causes Soames to feel concern for the future of his capital and his considerable art collection which he wishes to conserve for his daughter, Fleur. Now an adult, Fleur accidentally meets and falls in love with Jon, the son of Jolyon and Irene. They want to marry but, shortly before dying, Jolyon leaves his son a letter detailing the disastrous relationship of Soames and Irene, and explaining how ruinuous the effect of a union between Jon and Fleur would be for his wife. For the sake of his daughter only, Soames approaches Irene for her views on the proposed marriage of their respective children. Although the idea is as repugnant to her as to Soames, Irene leaves the decision to Jon, who respects his father's wishes and turns down Fleur.

The first trilogy concludes with Soames's troubled reflections on what life 'might have been' with Irene. With Annette, at least there is no pretence of love on either side. Soames is concerned only with Fleur, who hastily marries Michael Mont, a publisher and the son of a baronet. With the death of old Timothy Forsyte at 101, the remaining second and third generation of the family are securely in the Jazz Age, with the scene prepared for *The White Monkey*, the first volume of the second trilogy, *A Modern Comedy*.

Bored with her marriage, Fleur attempts to establish a *salon* in which she presides as hostess. Galsworthy felt distaste for contemporary literature, art and fashion, and satirises the 'moderns', as he calls those who visit the Monts. Writers and artists are portrayed as sterile, and both the books discussed and their authors have ridiculous names, such as *Nesta Gorse* or Walter Nazing, whose work sometimes resembled Shelley's poetry 'and sometimes . . . the prose of Marcel Proust'. D.H. Lawrence appears in conversation as a fashionably shocking writer under the initials 'L.S.D.'

The Monts' house, sparsely decorated 'in not more than three styles' is the antithesis of the over-filled dwellings inhabited by first-generation Forsytes. Galsworthy's dislike of the 'modern' is particularly defined through Fleur, in her indiscriminate acceptance of the modish. His own sympathy is for Soames, with his love for Gauguin, Turner and Monet, but whose taste according to his daughter has 'no sense of the future'. A portrait of a monkey with vacant staring eyes which Soames buys for Fleur, symbolises the confused idealogy of the 'modern' mind.

Another side of society is represented in the difficulties of the unemployed Tony Bicket and his wife Victorine in a lengthy sub-plot. Fleur falls in love with Wilfrid Desert, a poet, but is abandoned. She steadies her shaky marriage by producing a son.

The Silver Spoon takes Britain's economic decline and weakening influence as a world power to parallel a shift in society's view of acceptable morality. Michael Mont becomes an MP and tries to popularise a revolutionary economic and social plan. Fleur is libelled in the gossip column of an evening paper and later slandered in her own drawing-room by Marjorie Ferrars, a clever aristocrat noted for her promiscuity. Soames wins an action for the libel on a question that hinges on the interpretation of morality, but it is a pyrrhic victory. As a supposed champion of modernity, Fleur's stance on strict morals makes her a laughing-stock in the society she alleged to represent.

Swan Song concludes the second trilogy, and the story of the Forsyte family. The General Strike is a convenient means of bringing the Forsytes together, mirroring the family gathering with which the *Chronicles* opened forty years earlier. The cyclic pattern is confirmed when the crisis of Irene and Soames's relationship is re-enacted in events forced by the unresolvable love of Fleur and Jon. Inheriting the same passion which drove her father to rape, Fleur seduces Jon, hoping as Soames did, to possess what she loves. As Irene left Soames, so Fleur loses Jon. Aware of what he has allowed to happen, Jon informs Fleur that he will never see her again. Distraught and suicidal, Fleur accidentally starts a fire, and Soames dies saving his daughter's life.

The repercussions of Soames's behaviour towards Irene which recur throughout the complete saga finally cease. Soames belat-

edly pays retribution to Irene, with a death caused by the delayed consequences of his violence to her, and the Forsyte drama ends.

An inferior third trilogy, *The End of the Chapter* (1935), is loosely connected to the Chronicles by the reappearance of Wilfrid Desert, who falls in love with the central figure, Dinny Cherrell. In a foreword, Galsworthy's widow suggests that the trilogy, which comprises *Maid in Waiting* (1931), *Flowering Wilderness* (1932) and *Over the River* (1933), should be considered part of the story, since the characters are 'all links in the long chain that binds the Forsytes and their collaterals and descendants'. This suggestion is generally ignored.

Henry GREEN   1905–1973

Living

This boldly impressionistic account of factory life in Birmingham was written out of Green's own experience with the family firm of H. Pontifex & Sons, where he worked on the shop-floor after leaving Oxford. He wrote it in the evenings in his working-class lodgings, and it was published when he was twenty-four. Astonishingly mature, it contained the elements which were to become familiar in his other novels, notably the idiosyncratic style in which elision and compression produce sentences from which articles, nouns and even verbs are often omitted, but sense and sensation are forcibly conveyed.

The novel is set in the offices and workshops of the foundry of Dupret & Son, general engineers, and in the contrasting houses of the owners and employees. The cast is large, but two families provide the focus: the upper-class Duprets and the working-class Gateses. Young Dick Dupret, his mind futilely preoccupied with the glamorous Hannah Glossop, is in the process of taking over the company from his ailing father and is determined to make his own decisions, in spite of countermands from the paternal sickbed. While the senior employees regret and fear Dick's inexperience, young malcontents like Mr Tarver, who dislike the old order presided over by the works manager, Mr Bridges (or 'Tis'im', as he is known to the workforce), attempt to influence the new boss. Meanwhile, Lily Gates, who lives with her father Joe in the house of old Mr Craigan, pursues an affair with Bert Jones. Craigan, who regards Lily as his daughter, wants her to marry the lodger, a steady but dull young man called Jim Dale, but Lily's head is full of cinema romance and she runs away with Bert.

Old Dupret dies and Dick lays off the older men, including Craigan and Gates. The household falls apart: Craigan is ill in bed, Gates is briefly gaoled for abusive language, and the disappointed Jim Dale moves out. Lily's dream crumbles when Bert is unable to find his parents in a grimly realised Liverpool, and she returns to Birmingham, like one of the homing pigeons which provide one of the book's most striking images.

Much of the action takes place on the factory floor amidst the 'wild incidental beauty' of functional machinery and the vividly colloquial conversation of the workforce as they pursue their grudges, air their discontents and hatch their plots. The resonant title refers not only to the work by which the characters earn their livelihood but also to the densely rendered texture of day-to-day existence which was to preoccupy Green throughout his career.

Ernest HEMINGWAY   1899–1961

A Farewell to Arms

Like the hero of this novel, Hemingway volunteered for an ambulance unit in Italy during the First World War; like him, he was badly wounded. Being brought fact to face with death had a profound effect on Hemingway: in his own words, 'he ceased to be hard-boiled', and discovered the nature of his own vulnerability. He became obsessed with the need to confront death, to exorcise fear. All his best work is concerned in one way or another with the nature of mortality.

The plot of *A Farewell to Arms* is very simple. Frederic Henry, the narrator, is an American volunteer in charge of a small ambulance unit of the Italian army. Before going into action near the Austrian border he meets and is attracted to an English VAD called Catherine Barkley. He is badly wounded in the leg and transferred to a hospital in Milan where Catherine joins him. They fall in love and before he has to return to the Front, in time for the retreat from Caporetto, she becomes pregnant. He escapes being shot by the 'battle police' during the retreat and, disgusted by the war, deserts. He meets up with Catherine again on Lago Maggiore and they escape together to Switzerland where, after an idyllic interlude, she dies in childbirth.

At the beginning of their relationship Catherine says to Henry, 'Oh, darling ... You will be good to me, won't you? ... Because we're going to have a strange life.' This anticipates the principal theme of the novel: the greyness, the periodic boredom, the fear, the pain, the disillusionment of war. Passini, one of the ambulance drivers, says, 'There is nothing worse than war.' To which Henry replies, 'Defeat is worse.' But Passini will not have it. 'I do not believe it ... What is defeat? You go home.' This is before Henry is wounded, before the retreat from Caporetto, before his disillusionment is turned to disgust by the random shooting of the ragged exhausted men by their own 'battle police'. In civilian clothes, after his desertion, he decides 'to forget the war; he has made a separate peace'.

The central love-affair may strike readers who have never experienced war as sentimental; certainly Arnold Bennett's view that the novel is 'utterly free of sentimentality' is nonsense, and Catherine's death in childbirth in neutral Switzerland is perhaps just too much. Hemingway does succeed, however, in conveying the desperate need of the lovers to achieve permanence in an impermanent world, and there is much else to admire in *A Farewell to Arms*. The retreat from Caporetto is one of the great war sequences in literature, and Hemingway's symbolic use of landscape and the weather is masterly. When they reach Switzerland Catherine remarks: 'Isn't the rain fine? They never had rain like this in Italy. It's cheerful rain', and one understands what they have left behind and that, for a season at least, they have escaped the inescapable. Perhaps after all Catherine does have to die, since happiness is seen to be only on a shortlease.

Richard HUGHES   **1900–1976**

A High Wind in Jamaica

Hughes's working life is divided into two phases: a precocious and prolific early one and a painfully slow later one. *A High Wind in Jamaica* belongs to the author's productive youth. It was published when he was twenty-nine, by which time he had already published two volumes of poetry and had three plays produced. The novel which made his name is remarkable for its entirely (and, for the time, shockingly) unsentimental depiction of children. Instructive comparisons might be made with popular boys' adventure stories of the period and with J.M. Barrie's celebrated account of children and pirates in *Peter and Wendy* (1911).

The Bas-Thorntons are an English family living in reduced circumstances in nineteenth-century Jamaica. The five children virtually run wild in their tropical environment, although they are instilled with a very British sense of what is proper – bare feet shock them, as do unseemly displays of feeling in adults. A violent thunderstorm brings to an end the children's Jamaican life. When their house is ruined by gale-force winds, the Bas-Thorntons decide that their children would be safer in England. They place them on board the *Clorinda*, along with two other friends, but the ship does not get far before it is ambushed by pirates. The children are not alarmed by the change in their circumstances, and settle down quickly to the new routine of the pirate ship. The disappearance of John, the eldest boy, is a mystery that they pass over in silence: they do not allow anything they cannot understand to enter their untrammelled lives.

Emily is the most clearly defined character in the book, and it is she who is at the centre of its eventual tragedy. She is a strangely self-possessed little girl, whose emotional life is concealed behind a calm and composed façade. She deals impulsively and often violently with difficult situations (for instance, biting the captain's finger when he drunkenly attempts to caress her), and subsequently experiences agonies over her behaviour. After receiving a ghastly leg injury, Emily is confined to bed in the captain's cabin. She is instructed to guard the captain of a Dutch ship, which the pirates have captured and are now using for an animal-baiting show. Emily becomes increasingly terrified of the bound captain, and when he attempts to reach for a knife, she grabs it and stabs him to death. The pirates blame one of the other children, Margaret, and Emily says nothing.

The atmosphere on board changes and the pirate captain, Jonsen, decides to rid himself of the children. He manages to transfer them to a steamer bound for England. He is subsequently arrested for the murder of the Dutch captain, and in the ensuing court case in London, Emily is called as a vital witness, although she seems to remember nothing. In a dramatic scene her composure finally breaks down and she hysterically recalls seeing the victim's bloody corpse. Although she is unaware of the fact, this is enough to convict Jonsen. The book ends with Emily's new headmistress hoping that her pupil will forget the 'terrible things she has been through'; Emily, in the midst of new friends, has already wiped everything from her memory.

A film adaptation of the novel was made in 1965, directed by Alexander Mackendrick from a screenplay by Stanley Mann, Ronald Harwood and Denis Cannan, and notable for the appearance of the young Martin Amis in the role of the doomed John.

Frederic MANNING   **1882–1935**

The Middle Parts of Fortune

This dispassionate ranker's-eye-view of the First World War was originally produced in a limited edition of 520 copies for subscribers only, and its reputation derived from the 1930 bowdlerised trade edition issued under the title *Her Privates We*, 'by Private 19022'. The author's identity was not acknowledged until 1943, eight years after his death, and the unexpurgated version was not republished until 1977.

Manning was an Australian *littérateur*

based in England who served as a private in the King's Shropshire Light Infantry Regiment, and the novel is based upon his experiences on the 'Somme & Ancre, 1916'. It opens and closes with attacks and in between these two events describes the day-to-day life of a group of soldiers in and out of the line. Bourne is a middle-class ranker and something of an existentialist who repeatedly refuses the commission urged upon him by his seniors, notably Capt. Malet. He has thrown in his lot with the men, in particular with his friends Shem and 'little Martlow': 'It had been just a chance encounter. They had been three people without a thing in common; and yet there was no bond stronger than the necessity which had bound them together.' The attitude of the men is one of philosophical resignation rather than heroic stoicism; the voice of dissent is provided by a splendidly realised malcontent called Weeper Smart. The punning title is taken from *Hamlet* and allies the infantrymen with those interchangeable nobodies Rosencrantz and Guildenstern, who realise that Fortune is 'a strumpet'. The soldiers' world is self-contained and self-referring, a hierarchical society so cut off from the home front that we learn little about the backgrounds the men have left behind them. For better or worse (usually the latter) they are in the army and must abide by its absurdities and injustices whilst retaining their individuality.

Although several of the characters are killed, Martlow and Bourne amongst them, the novel is without special pleading, accepting war as a fact of life. 'War is waged by men; not by beasts, or by gods', Manning wrote in his preface. 'It is a peculiarly human activity. To call it a crime against mankind is to miss at least half its significance; it is also the punishment of a crime.' The Shakespearean epigraphs to each chapter point up the continuity of war as a human activity, as well as signalling the author's consciousness of the novel's literary heritage.

J.B. PRIESTLEY   **1894–1984**

The Good Companions

Priestley's first popular success is a picaresque tale of vaudeville low-life in which three unlikely companions find their paths crossing as they unexpectedly depart from the everyday world. Jess Oakroyd clatters along the Great North Road on a midnight lorry; Elizabeth Tranter bowls away from her Cotswold home in an Italian two-seater; and Inigo Jollifant strides away from the prunes and pianos of an east coast boarding-school with his knapsack on his back. Life on the road proves every bit as exciting as Elizabeth's historical romances, and their separate stories are interwoven as they tumble into the wonderland of the Dinky Doos, a bankrupt pierrot troupe who sound, as Elizabeth reflects, like an infectious disease.

In this glamorous fairy-tale, the adventurers are ensured of a happy ending as Priestley controls the puppet-strings like a latter-day Fielding. Transformed into 'The Good Companions' the troupers rocket to success with the launch of Elizabeth's money, boosted by Inigo's unsuspected - talent for song-writing and serviced by Jess's down-to-earth carpentry skills. Yet, as Jess remarks, there is 'summat i' t'air', and the cold Midlands wind seems increasingly unlikely to soften into a balmy Bournemouth summer. When Elsie Longstaff triumphantly exchanges seedy boarding-house lodgings to become queen of the Black Horse saloon as Mrs Bert Dulver, she is replaced by the insufferable Mamie Potter, a platinum blonde and 'a born putter-up of backs'. Dreams beyond the end-of-the-pier show, long submerged, now bob to the surface. Elizabeth glimpses her long-lost suitor; Jess thinks wistfully of reunion with his daughter in Canada; but the biggest waves are made when Inigo stops behaving like a lame duck and begins to swim after success.

Spotted by a talent scout, Inigo finds the London agents clamouring for his songs; but he is only interested in the Big Time if it includes Susie Dean, the dark and vivacious comedienne. Tying his career to hers, Inigo drags Monte Mortimer, king of the Charing Cross Road, to the Gatford Hippodrome to watch Susie perform. The magical circle is about to be shattered, 'the sands are running out, so that every grain has some significance', and for the last time the curtain rises on the Good Companions.

The day was just crossing the little magical bridge between afternoon and evening. The early autumn sunlight was bent on working a miracle. A moment of transformation had arrived. It hushed and gilded the moors above, and then, just when Mr. Oakroyd's tram reached the centre of the town, passing on one side the Central Free Library and on the other the Universal Sixpenny Bazaar, it touched Bruddersford. All the spaces of the town were filled with smoky gold. Holmes and Hadley's emporium, the Midland Railway Station, the Wool Exchange, Barclays Bank, the Imperial Music Hall, all shone like palaces.

THE GOOD COMPANIONS
BY J.B. PRIESTLEY

Concluding this fable of phantasmagoria, Priestley addresses an epilogue to 'those who insist upon having all the latest news'. The young stars now glitter in the firmament far above the concert-party orbit. Inigo's love for Susie will soon be rewarded with marriage; Jess is known in Canada as a grand old man of the theatre; and Elizabeth is happily settled in Scotland with her sweetheart. The old troupers have dispersed to their natural habitat of seaside resorts well satisfied by having rubbed shoulders with fame. Yet the play has been enacted against the backdrop of the 1930s depression, its unreality reinforced by the constant direction of the narrator, and beneath the high-flown fantasy we are warned that perfection is not to be found in 'these stumbling chronicles of a dream of life'.

Edmund WILSON   1895–1972

I Thought of Daisy

Like Cyril Connolly – who was to some extent his English counterpart – Wilson was not by temperament a writer of fiction. Although *I Thought of Daisy* provides a fascinating portrait of life among the artists in 1920s Greenwich Village, it is as much a critical meditation upon intellect versus materialism (as embodied in its two female protagonists) as a novel of character and incident. In retrospect, Wilson himself thought the book too schematic, and later revised it.

'It was Hugo who had taken me around and who told me what to think of what I saw; and I had seen through Hugo's eyes.' Robert, the narrator, is an impressionable young man with vague ambitions to be a poet. In the meantime he has taken a safe job in a publishing house, in sharp contrast to his old college friend, Hugo Bamman (based on John Dos Passos), whose social realist novels have made him the hero of the liberal establishment in Greenwich Village after the First World War.

Robert is attracted to the eccentric poet Rita Kavanagh, but when he becomes her lover he realises she will always give more attention to her work. Daisy Coleman, a chorus girl, is a very different prospect: where Rita is strangely beautiful, with her *jolie-laide* face, Daisy is pretty and petite, and dances to Robert's desires like a doll. When he thinks of her he can escape into fantasy, though her habits of using alcohol and opiates take him into disturbing company.

'Beyond publishing a few satiric verses in a radical magazine,' Robert reflects, 'I had never myself struck any blow in the war for humanity.' Hugo, now rich from the royalties of his novels, decides to depart to experience the political unrest in the Middle East, and Daisy begs him to take her with him. Robert begins to realise that in seeing the world through Hugo's eyes he is imposing limitations on his own vision: both Rita and Daisy have appeared to him as symbols rather than as people. Rita retires to her aunt's home upstate to gain the peace she needs to work on her new collection; Daisy, in search of another form of retreat, moves in with an idealistic man who believes he can survive through small-scale agriculture as part of an older American tradition. Now neglected by both of his mistresses, Robert is able to see himself and others more clearly. The narrative is a record of growth from adolescence to maturity, in which Robert learns to interpret experience for himself.

| Thomas WOLFE | 1900–1938 |

Look Homeward, Angel

see **Of Time and the River** (1935)

1930

Mahatma Gandhi opens civil disobedience campaign in India ● Name of Constantinople changed to Istanbul and of Angora to Ankara ● Ras Tafari becomes Emperor Haile Selassie of Abyssinia ● In German elections National Socialists (Nazis), denouncing Versailles Treaty, gain 107 seats from Moderates ● Acrylic plastics are invented (UK perspex; US lucite) ● Amy Johnson's solo flight, London to Australia, in 19½ days ● Sigmund Freud, *Civilisation and its Discontents* ● Shreve, Lamb and Harmon, Empire State Building, New York ● William Randolph Hearst owns 33 newspapers with total circulation of 11 mill. ● Noël Coward, *Private Lives* ● Josef von Sternberg, *The Blue Angel* with Marlene Dietrich ● D.H. Lawrence, Robert Bridges and A. Conan Doyle die ● W.H. Auden, *Poems*

| Stella BENSON | 1892–1933 |

Tobit Transplanted

Living in the Kanto region of Manchuria in the 1920s, Benson observed the White Russians exiled from the Soviet Union in the wake of the Revolution, and saw in their plight a parallel with the story of the exiled Jew Tobit in the *Apocrypha*. The plot of *Tobit Transplanted* closely follows that of the Book of Tobit, but the perspective is wider, compensating for what Benson described in her introduction as the 'patriarchal bias' of Tobit's narrative.

Old Sergei Malinin (Tobit) is mysteriously blinded after fighting off a group of Chinese soldiers who were digging up a Russian grave. His shop and its stock are partly destroyed when the soldiers retaliate, and he wallows in self-pity, speaking only of his wish for death. To alleviate his financial problems he wants to send his son Seryozha (Tobias) to Seoul to collect the money he lent a friend some years back, but his wife

Anna is fearful of the dangers her only son will encounter on the journey. Seryozha persuades her to allow him to travel with Mr Wilfred Chew, an English-educated, Christian Chinese barrister (the angel Raphael, disguised as Azarias in Tobit). They set off on foot with Seryozha's dog, and stay near Seoul with distant kinsmen of Anna, Pavel and Varvara Ostapenko. The Ostapenkos' beautiful daughter Tanya has already rejected a succession of suitors who – to the vast distress of her father – have subsequently committed suicide or (nearly as bad) joined the Chinese Army. Promising her freedom within marriage, Seryozha wins her heart immediately and they are betrothed that night. Chew is sent on to Seoul where, with some difficulty, he collects Sergei's money. The party returns home rich with a dowry and Sergei's money, and Seryozha cures his father's blindness by applying the rotting gall of a fish, an ancient Chinese remedy given him by Chew. (In the Book of Tobit Raphael says to Tobias: 'Therefore anoint thou his eyes will be the gall, and being pricked therewith, he shall rub, and the whiteness shall fall away, and he shall see thee.') The family fortunes are restored and there is great feasting and celebration.

China, where Benson spent the latter part of her life (her husband was in the Chinese customs service), also provides the setting for a number of her short stories, which are not much read today. *Tobit Transplanted* (first published in America as *The Faraway Bride*) won the Femina Vie Heureuse Prize and remains Benson's best-known novel. Within three years of its publication she died from pneumonia, aged only forty.

Kay BOYLE　　　　　　　1903–1992

Plagued by the Nightingale

Boyle's lyrical first novel tells the Jamesian story of a young American woman encountering the jaded culture of the Old World for the first time. Its corruptions are incarnated in the form of a hereditary bone disease, which runs in the family of Bridget's French husband, Nicolas.

The young couple marry in California, but return to France because of the progress of Nicolas's disease. Bridget and the family are initially delighted with each other, and Bridget becomes especially close to Charlotte, Nicolas's sister, who is married to her cousin Jean, another sufferer of the disease. Bridget's other favourite is Luc, who is not an actual member of the family, but has practically been adopted by them, and visits every summer with Nicolas's brother Pierre, whose medical practice Luc shares. The unspoken assumption within the family is that Luc will eventually propose to one of the three unmarried daughters, and Marthe in particular suffers agonies in her efforts to secure Luc's affections for herself alone.

The initial pastoral serenity of life in the Breton village begins to disintegrate, however, as it becomes evident that Nicolas hates his family for reminding him of the source of his affliction. He cannot get away, however, because he has no income, and his father will only provide him with one if he agrees to have a child. Nicolas does not want to inflict the disease on yet another generation, and refuses. His cousin Jean and uncle Robert also have money, but refuse to help him. Only Charlotte understands his need for independence, and she promises to arrange for Nicolas to manage one of Jean's farms in North Africa.

Before she can persuade her husband to agree, however, she dies, during a difficult pregnancy, and there is no question of Jean honouring her word. Meanwhile, Luc, instead of proposing to one of the three daughters, has begun to fall in love with Bridget. She is also attracted to him, increasingly so as Nicolas retreats into a private world of hatred and bitterness. Luc announces that he is going away to Indo-China, and asks Bridget to go with him. Her final choice, however, is to stay with Nicolas, and to have a child, so that she can remove him from the influence of the family.

The nightingale of the novel's title plagues by its silence: it appears as a caged pet, silent because captive, reflecting the stifling atmosphere of a family that is closed in on itself to the point of self-extinction. The atmosphere is effectively created in the novel through the treatment of speech, the general cloying sweetness of which is progressively invaded by cruelty and rage, and also through landscape, the lyricism of which is punctuated by stray images of muted horror.

John COLLIER　　　　　　　1901–1980

His Monkey Wife

Against a background as colourful as a Rousseau painting, Collier's most famous novel tells the story of Emily, a monkey who educates herself, falls in love with her owner, prevents his marriage to an unsuitable girl and wins him for herself. It is also a cynical portrait of a society in which the only truly good and civilised character is a chimpanzee.

Alfred Fatigay, a schoolmaster dismissed to the Congo by his flighty fiancée Amy Flint, has no inkling of his pet's devotion, nor her intelligence. She has learned to read, and only the ability to speak eludes her. Emily's secret dreams of marriage are shattered when she discovers Alfred's correspondence with Amy, although she soon establishes that her master's blind adoration is not reciprocated. With characteristic selflessness she resolves to serve Alfred when they return to England. That he intends her to become Amy's maid is a blow indeed, particularly since Amy, perceiving the chimp's depth of feeling, proceeds to make her life a misery. Emily cheers herself up with the occasional somersault and secret forays to the reading room at the British Museum, but her hopes are finally dashed when Amy agrees on a date for the wedding. A foggy day and an ill-lit church enable the desperate chimp, in an act of inspired deviousness, to swap places with Amy and marry Alfred. When Alfred realises what has happened, he banishes his pet. Amy, furious to have been taken for an ape and anxious to save face, claims it was her idea and leaves Alfred for ever.

Months later, reduced to abject poverty, Alfred comes face to face with Emily as she steps from a limousine. She whisks him off to her 'little home' in South Audley Street, and nurses him back to health. With the help of her typewriter, Emily relates the circumstances by which she has become a famous dancer, and is in the process of revealing the truth about the wedding when Amy, annoyed to hear of her former fiancé's change of fortune, arrives to tell him herself. She is dismissed by a now-enlightened Alfred, and the Fatigays return, triumphant, to Africa. As Emily snuffs out the bedroom candle with her 'prehensile foot' we can be in no doubt that Mr Fatigay and his monkey wife will live happily ever after.

Collier's deliberately named characters – the exhausted, passive Mr Fatigay and the tough, impervious Miss Flint – have led some commentators to discern serious intent behind his story (cf. David Garnett's **Lady Into Fox**, 1922). Man in harmony with nature? Man facing his Darwinian origins? The noble savage taken to its logical conclusion? It may, of course, be no more than its subtitle suggests, a whimsical love story, in which the hero is 'Married to a Chimp'.

E.M. DELAFIELD   **1890–1943**

Diary of a Provincial Lady

Delafield had already written eighteen books when she was asked by Lady Rhondda to contribute to the weekly *Time and Tide*. Two collections resulted: *Diary of a Provincial Lady* and *The Provincial Lady Goes Further* (1932). Two more collections of diaries – *The Provincial Lady in America* (1934) and *The Provincial Lady in Wartime* (1940) – followed after the column moved to *Punch*, and all four books were collected into one volume in 1947.

The author herself was very much a provincial lady, with many of the same preoccupations as her fictional counterpart. She admitted that the Provincial Lady's children, Robin and Vicky, were 'mild likenesses' of her own two children. Her fictional husband, Robert, monosyllabic, unobservant, conservative and usually hidden behind a copy of *The Times*, is a cartoon of the typical middle-class husband of the interwar years (cf. Laura's husband, Fred, in the 1945 Noël Coward / David Lean film *Brief Encounter*).

These diaries are a record of middle-class peculiarities minutely examined. The Provincial Lady herself is never described, but she is the archetype of a certain sort of woman, as familiar now as then: she is acutely aware of not underdressing (common) or overdressing (vulgar, or possibly continental, which amounts to the same). Nevertheless, this is not to say that she does not worry a great deal about the way she looks. In fact, she worries a great deal about everything: minor worries like planting the bulbs in time, whether or not the cook has taken umbrage, or whether the children are getting out of control, and the ever-present worry over money, particularly in the forms of the overdraft and the bank manager: 'Financial situation very low indeed', 'old friend the pawnbroker'. If life does get her down, it never does so more than can be lifted up again by the purchase of a new hat. Life throws up no greater concerns than how to put the snobbish Lady Boxe in her place (perfect retorts are invariably thought up too late), or how to avoid the visits of 'Our Vicar's Wife'. Small scenes of social embarrassment are commonplace: '[She says she] has heard all about how very, very amusing I am. Become completely paralysed and can think of nothing to say except that it has been very stormy lately.' She is self-deprecating and self-deflating: everyone else she thinks of as wittier, wealthier, more composed, better dressed and with better-behaved children. Her pride is pricked, her dignity affronted, but these occasions are recorded and recalled, a peculiarly British trait (cf. Mr Pooter's attempting to leave a room with quiet dignity and catching his foot in the carpet; G. and W. Grossmith, *The Diary of a Nobody*, 1892). The humour of the diaries derives from perception rather than profundity, and from ironic understatement.

June 27th. – Cook says that unless I am willing to let her have the Sweep, she cannot possibly be responsible for the stove. I say that of course she can have the Sweep. If not, Cook returns, totally disregarding this, she really can't say what won't happen. I reiterate my complete readiness to send the Sweep a summons on the instant, and Cook continues to look away from me and to repeat that unless I *will* agree to having the Sweep in, there's no knowing.

This dialogue – cannot say why – upsets me for the remainder of the day.

DIARY OF A PROVINCIAL LADY
BY E.M. DELAFIELD

The diaries are full of rhetorical memos and queries: '(Query: Is not a common hate one of the strongest links in human nature? Answer, most regrettably, in the affirmative.)'; '(*Mem*: A mother's influence, if any, almost always entirely disastrous. Children invariably far worse under maternal supervision than any other.)'

As Nicola Beauman has pointed out in the introduction to a reissue of the one-volume edition, the four books chart the progress from narrow-focused Provincial Lady to classless British housewife: 'In the first vol-ume the Provincial Lady ventures beyond Devon only on special occasions; in the second she spends part of her time in London; in the third she goes to America; and in the final volume she works, in trousers, outside the home.'

John DOS PASSOS · 1896–1970

The 42nd Parallel

see **USA** (1936)

William FAULKNER · 1897–1962

As I Lay Dying

This is one of the series of novels and stories that Faulkner set in 'Yoknapatawpha County', north Mississippi. In it, he explores the values that he spoke of when he was awarded the Nobel Prize for Literature in 1949: 'courage and honour and hope and pride and compassion and pity and sacrifice'.

Addie Bundren has laid down and set to die having made her husband Anse promise to bury her with her own people in Jefferson – a good day's journey from their home, which takes even longer when the storms break. The story of her family, their desires and fears and rivalries, is revealed as husband, children and neighbours each recount episodes of the death and burial in the vernacular speech of the South.

Addie and Anse had married in loneliness and gone to live on an isolated mountain, where they had five children. When Addie had Cash (who is now building her coffin, right outside the window), she 'knew that living was terrible and that this was the answer to it'. After Darl, who loved her, and Jewel, whom she loved, she knew that her father was right: 'that the reason for living was to get ready to stay dead a long time'. And then came a girl, Dewey Dell, and last of all, young Vardaman, still small enough to hide behind his father's leg.

To earn just three dollars extra, Anse sends Darl and Jewel on an abortive wagon trip that causes Jewel to miss his mother's death and delays the start for Jefferson.

Meanwhile, the waters have risen and washed away two bridges. They must go miles out of their way, until they decide to swim the ford, then find the road washed out and have to go clean round by Mottson. Even though the corpse is days old and the buzzards are gathering, turning back is never an option, because of Anse's promise.

During the journey there are a number of disasters. Cash's leg is broken and, after it is set in cement bought from a store, it goes bad; Darl (who folks have always thought queer, not so much for anything he said or did, but for the way he looked at people) goes crazy and is sent to Jackson on a train; Jewel is nearly drowned and then badly burned while twice saving the coffin, and must sell his own horse to buy a team of mules to replace the team lost in the ford; it becomes clear that Dewey Dell, only seventeen and unmarried, is pregnant and unable to get rid of the child.

Only Anse remains relatively unscarred, physically and emotionally, but it is at the expense of those around him. After Addie is finally buried, he takes the ten dollars Dewey Dell had set aside for an abortion, and buys a set of false teeth, something he has been waiting to do for fifteen years, and returns to the wagon with a woman he introduces as the new Mrs Bundren.

Dashiell HAMMETT   1894–1961

The Maltese Falcon

The experience Hammett gained as a detective for the famous Pinkerton agency had already been put to good use in a number of short stories written for pulp magazines. In *The Maltese Falcon*, however, Hammett produced one of the first novels of the 'hard-boiled' school of American detective fiction. Characterised by a realistic treatment of violent crime, and witty, laconic dialogue, novels such as this adapted very well for, and were further popularised by, the cinema. (John Huston's 1941 film, in which Humphrey Bogart played Sam Spade, is as much a classic of the genre as Hammett's original novel.)

The plot is complex, not to say convoluted, and occasionally perplexed even the author, who leaves some minor loose ends hanging in the San Francisco air. Enter, on the first page, the troubled and attractive Miss Wonderly, who claims to be looking for a wayward sister. Miles Archer, Sam Spade's associate in the private detective agency, is killed while trailing Floyd Thursby, the alleged seducer of Miss Wonderly's sister. Miss Wonderly, who also uses the name Leblanc, is pressed to confess that her story, and both her names, are false: she is in fact a Miss Brigid O'Shaughnessy.

Enter next one Joel Cairo, attempting to recover a valuable statuette of a falcon which he suspects Miss O'Shaughnessy illicitly possesses. A meeting between them is arranged, interrupted by the police investigating the death of Thursby, whom Spade now knows to have been Brigid's associate. Later, Spade is telephoned by a mysterious Mr Gutman, who makes a financial offer for the elusive falcon and explains its provenance and immense value. He attempts to make a deal with Spade, eliminating Brigid and Cairo.

The fat man was flabbily fat with bulbous pink cheeks and lips and chins and neck, with a great soft egg of a belly that was all his torso, and pendant cones for arms and legs. As he advanced to meet Spade all his bulbs rose and shook and fell separately with each step, in the manner of clustered soap-bubbles not yet released from the pipe through which they had been blown.

THE MALTESE FALCON
BY DASHIELL HAMMETT

Investigation apparently reveals that Thursby killed Miles Archer and was in turn murdered by a 'punk' employed by Gutman. By dint of some intelligent thinking, Sam Spade acquires the falcon and in due course becomes involved in a meeting between Gutman, Brigid, Cairo and the punk gunman. Sam's principal aim is to disembarrass himself of the police's suspicion that he has murdered Miles and, possibly, Thursby. Conditions are set for a four- or five-way trade-off, which degenerates into double-

and triple-crossing, switching of allegiances, and an eventual confession from Brigid that she herself killed Miles, hoping to frame Thursby. The punk gunman shoots Gutman offstage, Cairo survives, and Sam, handing over the beautiful but deadly Brigid to the police, congratulates himself on not having been played for a sap.

Sam Spade is a modern folk-hero who walks up to trouble and punches it right in the mouth. His hard, fast, moral, straight-shooting and straight-talking remains distinctive despite many subsequent imitations.

Wyndham LEWIS   1882–1957

The Apes of God

Lewis conceived *The Apes of God* as a decisive broadside in the war of words he waged against the London-based literary coteries who, he believed, blocked the recognition and advancement of greater talents such as himself. The novel is a massive *roman-à-clef* in which many of Lewis's contemporaries are portrayed as vain pseudo-artists or self-obsessed grotesques. At the centre of this satire are the Sitwells, against whom Lewis kept up a series of attacks throughout the 1930s and 1940s, and by whom he was attacked in turn. Several of the Bloomsbury group are also vilified, as well as many people who believed themselves (with some justice) to be Lewis's friends.

Reading *The Apes of God* outside its original context reveals a very odd novel. The plot is slight (notwithstanding the books' length) and takes the form of an innocent's progress through the 'Ape-world' of London. Dan Boleyn, under the guidance of Horace Zagreus, is sent through the studios and salons of the 'Apes', 'prosperous mountebanks who alternately imitate and mock at and traduce those figures they at once admire and hate'. Dan is a young man of nineteen, an aspiring writer, impossibly delicate and subject to nose-bleeds whom Zagreus believes to be a genius, although no evidence is offered to support this view. Dan stumbles, faints, is humiliated or embarrassingly exalted as he progresses through the Ape-world in a state of near-perpetual terror. At 'Lord Osmund's Lenten Party' (which

makes up the second half of the novel) he is forced to dress in women's clothes, is propositioned, threatened by a real ape and finally collapses in a drunken heap (thus forfeiting the dubious benefits of Zagreus's sponsorship). The Ape-world proves itself resilient to these incursions and remains unchanged at the novel's close.

Lewis's satirical method is based loosely on his theory of painting. He rejects the Proustian and Joycean emphases on inner states of thought and feeling in favour of objective description. Everything is conveyed in terms of surface; as Lewis put it: 'The ossature is my favourite part of the living animal organism, not its intestines.' In practice this results in long, detailed descriptions of each character's appearance in which every feature, mannerism and action is subjected to intense scrutiny. This programme, coupled with Lewis's extraordinary, baroque prose, makes grotesques of them all and contributes to the near-unreadability of several sections. Sentences such as: 'But now the last milkman of this auroral ballet had long yodelled himself out of the square' are not untypical. The novel becomes a monumental and pitiless still-life of its victims: Lewis's targets are engulfed by his prose rather than dissected by it, and now that those targets are not readily identifiable the novel can seem beside the point. The central flaw is the discrepancy between Lewis's talents as a writer and the meagreness of his victims, who do not seem to justify the trouble taken over them. F.R. Leavis, who took an interest in the controversy aroused by the novel, wrote of the Sitwells that they 'belong to the history of publicity rather than of poetry'. This damning epigram summarises the central message of *The Apes of God* and suggests the root cause of its failure. Lewis's case against the pseudo-artists he lambasts hardly needs proving, and to prove it at such length and with such ferocious talent leaves this daunting, experimental novel strangely empty. When Lewis has demonstrated the worthlessness of his subjects, his own novel (which is built about them) cannot help but seem devalued in turn.

Amongst those Lewis lampoons are Stephen Spender, in the character of Dan

Boleyn; Clive and Vanessa Bell, who appear as the pretentious pseudo-Bohemians Mr and Mrs Jonathan Bell; and Osbert, Edith and Sacheverell Sitwell, who appear, respectively, as Lord Osmund, a puffed-up relic of the 1890s Aesthetic movement, Harriet Finnian Shaw, a 'flying harpy', and Osmund's cousin, Lord Phoebus. The unconventional relationship of Lytton Strachey and Carrington is travestied in the mismatched pair, Matthew ('The Virgin') Plunkett and the 'doll-woman' Betty Bligh, while Edwin Muir is discernible in the henpecked Eddie Keith of Ravelstone. The smooth, clever American Mr Horty is an irreverent (though accurate) character-sketch of T.S. Eliot, while the amoral and prolifically successful seducer, Zulu Blades, is a reasonably flattering portrait of Lewis's friend and defender, Roy Campbell. Horace Zagreus is an amalgam of George Borrow and Horace de Vere Cole, while the publisher and writer Jimjulius Ratner appears to combine the characteristics of Leopold Bloom and his creator, James Joyce. The least provoked, and least forgivable, portrait is of Lewis's long-suffering and generous patrons, Sydney and Violet Schiff: their wealth, their love of Proust, and Sydney's lack of artistic talent, are brutally guyed in the characters of Lionel and Isabel Kein.

W. Somerset MAUGHAM   1874–1965

Cakes and Ale

Cakes and Ale offers a view of literary life by presenting us with three novelists: Willie Ashenden, Edward Driffield and Alroy Kear. It caused a minor scandal on publication because Edward Driffield was perceived as a composite portrait of Thomas Hardy and H.G. Wells. Equally scandalous, and far crueller, was Maugham's instantly recognisable caricature of Hugh Walpole in the character of the snobbish and talentless Kear. Walpole recorded in his diary that he sat up at night reading an advance proof of the novel 'with increasing horror. Unmistakable portrait of myself. Never slept.'

The story is narrated by Ashenden, who shares not only Maugham's first name, but also something of his background, notably an unsympathetic uncle who is vicar of Blackstable, and an early career in medicine (see *Of Human Bondage*, 1915). Driffield's widow, Amy, invites Kear to write a biography of her late husband, who is revered as a Grand Old Man of Letters. Early in his career, Driffield had been scorned for writing novels reflecting his knowledge of the working classes, amongst whom he lived in Blackstable. At that time he was married to the cheerfully promiscuous Rosie, 'The Skeleton in the Cupboard' of the novel's subtitle. Ashenden is one of the few people who knows the truth about Driffield's past, since he was befriended by the novelist in Blackstable and later became one of Rosie's lovers. Driffield's first marriage ended in divorce, and his second wife, Amy, had been industrious in reshaping her husband's career and tailoring his manners so that he became eminently respectable and consequently respected. Kear is well chosen as a biographer, for he would also like to suppress the facts about Driffield's early life, which he finds too 'vulgar' and 'common' to find a place in the bland book he intends to write. Maugham attacks Kear (and Walpole) as a talentless social climber, more concerned that writers should be gentlemen than to tell the truth.

Driffield wrote the best of his fiction while living with Rosie, because he was in touch with real life. Amy has had the opposite effect: by creating a myth for his later career and for posterity, she drained his life of imagination. Neither she nor Kear can accept that a woman like Rosie could be charming or likeable. They assume she is dead, but Ashenden knows otherwise and tracks Rosie down in Yonkers, where, fat, painted and still surrounded by admirers, she lives with fond memories not of Driffield, but of a loudly dressed coal merchant with whom she lived. She loved him, she declares, because he 'was always such a perfect gentleman'.

In Rosie, Maugham created the only convincing and fully rounded woman to be found in his novels. Her good humour, lack of snobbery and unashamed willingness to sleep with men simply to give them pleasure makes her an endearing figure, far more plausible   than   most   of   Maugham's cardboard females. As in *The Moon and Sixpence* (1919), Maugham satirises the insincerities and pretensions of contemporary literary life, contrasting the deceptions involved in respectability with the truth of heterodoxy, a theme of particular resonance for a writer who was unable to deal openly with what he perceived as the skeleton in his own cupboard: his homosexuality.

Seigfried SASSOON   1886–1967

Memoirs of an Infantry Officer

see **The Complete Memoirs of George Sherston** (1936)

Dorothy L. SAYERS   1893–1957

Strong Poison

Sayers believed that writing detective fiction was a good way to embark upon a career as a professional writer, and although she later became bored with the genre, it is for her novels featuring the aristocratic amateur detective, Lord Peter Wimsey, that she will be remembered rather than her numerous books of Christian apologia. Wimsey made his first appearance in Sayers's first novel, *Whose Body?* (1923). He is the younger brother of the Duke of Denver, and is a man of independent means, with a taste for first editions and incunabula, and a tailor in Savile Row. He affects a monocle, and cultivates a bantering, sub-Wodehouse manner of speech that belies a quick intelligence and steely determination. He is a polymath, liable to know a great deal more about any given subject than his effete, affable, silly-ass demeanour might suggest. He is irresistibly attractive to bluestocking women (including his creator who, it has been claimed, was 'in love' with him), who admire his mind as much as his refined good looks, superior breeding, blond hair, height and impeccable manners.

One such woman is Miss Harriet Vane, who makes her first appearance in *Strong Poison*. In many particulars, she is an echo of her creator: highly intelligent, Oxford-educated, principled, plain rather than beautiful (although her features, in themselves, are beautiful), and a writer of detective novels. She is on trial for her life, accused of the murder of her lover, Philip Boyes. The jury is unable to agree a verdict, and when a retrial is ordered, Wimsey offers his help to the defence counsel. He is convinced of Miss Vane's innocence, but he has also fallen in love with her.

Wimsey enlists the aid of Miss Climpson whose secretarial bureau (which Wimsey funds) is in fact an effective investigation agency. He is also helped, as usual, by his impeccable butler, valet and man-of-all work, Bunter. Between them, they establish a motive of financial gain for Norman Urquhart, a solicitor and cousin of Boyes, with whom Boyes had dined on the night of his death. Urquhart has been embezzling clients' funds and has forged the will of an elderly relative, who wished to leave a large inheritance to Boyes, in his own favour. Boyes, had he lived, might have suspected the forgery and claimed the legacy. Wimsey has reason to suspect that Urquhart administered arsenic to Boyes in an omelette and that Urquhart, to immunise himself from fatal effects, had taken regular doses of arsenic over a long period of time. But in order to exonerate Harriet Vane beyond a shadow of a doubt, Wimsey resorts to a ruse that induces Urquhart to confess to the murder.

The murderer is perfectly obvious at an early stage of the novel, and apart from the techniques of sleuthing, the interest for the modern reader lies in the very precise period detail (the novel is set in 1929) and an examination of contemporary morals among a rather Bohemian set of artists and writers in what is now known as Bloomsbury. Here, too, Sayers gives the first exchanges between Vane and Wimsey that, several novels later, lead to their marriage.

Evelyn WAUGH   1903–1966

Vile Bodies

Waugh's second novel, written as his first marriage failed, is at once a glittering social comedy and a sour dissection of contemporary mores. It begins with an emetic Channel crossing, ends on 'the biggest battlefield in the history of the world', and charts the giddy capers of a number of significantly named members of the Younger Set upon the edge of an abyss. 'I am sure you will disapprove of it,' Waugh told Harold Acton (dedicatee of *Decline and Fall*, 1928). 'It is a welter of sex and snobbery written simply in the hope of selling some copies.'

The nominal hero, Adam Fenwick-Symes, is a young writer attempting to scrape together enough money to marry Nina Blount. Like *Decline and Fall*'s Paul Pennyfeather, he is a fool of fortune: the manuscript of his autobiography is destroyed by passport officials; his publishers cheat him; he wins money only to lose it to a mysterious 'drunk Major'; he is given a dud cheque by his eccentric prospective father-in-law; and he eventually loses Nina to his rival, 'Ginger' Littlejohn. The characters stagger from party to party ('Masked parties, Savage parties, Victorian parties, Greek parties, Wild West parties, Russian parties, Circus parties ... all that succession and repetition of massed humanity ... Those vile bodies'), their antics watched by baffled members of the older generation and reported at length in the columns of the *Daily Excess*.

There is talk of 'radical instability' and this is demonstrated by a number of thematic threads which run through the story: the government changes every week; the aristocracy is reduced to writing (frequently fictitious) gossip columns; parents fail to recognise their own children; and religion is represented by the grotesque and acquisitive American evangelist, Mrs Melrose Ape, and a ludicrous film purportedly representing the life of Wesley.

Agatha Runcible, brightest of the Bright Young Things, dreams upon her death-bed that 'we were all driving round and round in a motor race and none of us could stop, and there was an audience composed entirely of gossip writers and gate crashers ... and people like that, all shouting at us to go faster, and car after car kept crashing'. The frenetic pace and superficial emotions which dominate the lives of the characters are reflected both in the headlong, elliptical narrative (which owes much to Firbank), and the artificial dialogue, with its promiscuous use of intensifiers ('Too, too shaming ... too, too sick-making').

Henry WILLIAMSON   1897–1977

The Patriot's Progress

Although generally credited to Williamson, this book is a genuine collaboration between writer and artist, both veterans of the First World War. J.C. Squire had been impressed by a series of lino-cuts made by the Tasmanian artist William Kermode and suggested that Williamson could provide a text to accompany them. Williamson likened the result to 'a line of German mebus, or "pillboxes" ... [Kermode's] lino-cuts would be shuttering to my verbal concrete'. The image is exact and the book achieves an expressionism rare in English art and fiction.

It is subtitled 'Being the Vicissitudes of Pte John Bullock' and portrays the war of a City clerk, an emblematic patriot whom Arnold Bennett dubbed 'Everysoldier'. Indeed, Bullock has no discernible character and is less a man of action than one of Wilfred Owen's passive victims, 'those who die as cattle' (his name suggests a diminutive John Bull bred for slaughter). He joins up when war is declared and his suffering begins almost immediately as he is reduced to 19023 Pte J. Bullock and undergoes a harsh course of training. He serves on the Somme, where his chum Ginger is shot through the head, then takes part in an attack at Ypres in 1917, in which he loses a leg. There are vivid descriptions of life in billets and in the trenches, of annihilating route marches, a dismal visit to a prostitute, and the notorious mutiny at Etaples. Williamson achieves his effects by mixing catchphrases and clichés with jagged interior monologues and onomatopoeic effects. The result is mimetic of the warfare it describes – the confusion and violence conveyed by shellbursts of language – and the book is one of the most authentically horrible accounts of that most squalid of wars. Kermode's lino-cuts are equally bold, suggesting a dark underworld populated by faceless automata pitted against infernal machinery. The book ends with Bullock confronting an armchair patriot during the Armistice celebrations. 'We'll see that England doesn't forget you fellows,' says the 'toff'. 'We are England,' Bullock replies.

The allusion to Bunyan (like that of e e cummings in his novel of the First World War *The Enormous Room*, 1922) is of course ironic. Like Christian, Bullock sets out on his journey with 'a great burden upon his back' (the ranker's 70lb of equipment), but he does not reach the Celestial City. At the end of his pilgrimage he returns to the City of London and his job as a clerk, as Kermode's final lino-cut, the book's wordless 'Epilogue', shows.

E.H. YOUNG   **1880–1949**

Miss Mole

Like the elusive creature whose name she bears, Miss Mole works underground to surprising effect and can, if need be, disappear with remarkable speed. Fortyish when we meet her, she already has behind her two decades of paid domestic servitude which would have broken the spirit, dulled the intellect and quenched the hope of a lesser woman. But Hannah Mole is not quenchable. Somehow, despite poverty, loss and betrayal, she has discovered and maintained a spirited independence which outfaces the bullying, exploitation and disdain attempted by a long line of defeated employers. Living a life which is 'not hampered by man's conventional morality', in which unkindness is the greatest sin, and literal truth, 'as human beings knew it, . . . a limiting and embarrassing convention', Hannah is an uncomfortable companion to the hypocritical, malicious and self-loving. For the reader she is a source of unnerving delight: underlying her surface wit and gaiety are other, more disquieting traits, with hints of earlier secrets and catastrophes which Miss Mole – and her author – keep hidden for as long as possible.

About to be sacked again when the book opens, Hannah is taken firmly in hand by Mrs Spenser-Smith, wife of one of the most prominent citizens, pillar and patron of the most select Chapel of Radstowe (the setting of most of Young's novels, and based on Bristol, where she lived for many years). Less elevatedly, Mrs Spenser-Smith is also Hannah's cousin Lillas, though she will dissemble as desperately as her Non-Conformist conscience allows to keep the fact a secret. Best for everyone if Hannah is installed as soon as possible as housekeeper to Robert Corder, the widowed minister of Lillas's Chapel, and his two daughters, Ethel and Ruth.

Corder, 'a man of great energy but no intellect', is adored by his flock, less easily worshipped at home. His capacity for coercion and constraint are prefigured in his name. So, too, more subtly, is his possession of a not altogether unworthy heart. Early attempts to patronise and intimidate Hannah (who commits the double sin of being more intelligent than he and unattractive to him physically) fail utterly. But skirmishes with her employer, which bring victory to Hannah and enormous pleasure to the reader, are less important than doing what she can to help his motherless daughters. Ethel, the elder, weak but stubborn, affectionate but distressingly capable of easy treacheries, will never be a rewarding task; but Ruth, still stunned by grief, subject to night-terrors and daytime wretchedness, can clearly profit from the new spirit which Hannah brings to the house. Gradually, harsh evangelical certainties are tempered by the values of Hannah's 'vague and tender God'.

Vagueness and tenderness are not, however, the qualities most apparent in the Non-Conformist community and Corder's own household when rumours begin to circulate about Hannah's long-guarded secret. (The destructive power of rumour was something Young knew well. For twenty years she lived, as his lover, with the headmaster of Alleyn's School and his wife.)

Trying to decide how best to deal with the threat and fact of scandal forces Hannah to examine more closely certain tenets of her personal creed. Materially she has always travelled light, collecting few possessions, ready to move on at a moment's notice. Emotionally she has been heavily burdened. Perhaps, she comes to think and Young most certainly suggests, her vagabondage has been at least as much a flight from further hurt as a spiritual rejection of worldly snares. Mr Blenkinsop, who enters the novel as a mildly peevish figure of fun and leaves it very differently, certainly seems to think so and, Hannah finally concedes, he is probably right.

Miss Mole offers us one of fiction's most remarkable and difficult heroines, disquieting in her blend of 'frankness and slyness', always exhilarating, often discomforting, tremendously attractive but even less knowable than most human beings. The novel, its author's seventh, was a worthy winner of the 1930 James Tait Black Memorial Prize.

1931

Financial collapse of Central Europe begins ● Britain and Japan abandon gold standard ● British government's economy measures provoke riots in London and Glasgow and mutiny at Invergordon ● Treaty of friendship between Egypt and Iraq, the first pact between Egypt and an Arab state ● Noël Coward, *Cavalcade* ● Boris Karloff, *Frankenstein* ● Anna Pavlova, Nellie Melba, Arnold Bennett and Thomas Edison die ● Wilfred Owen, *Collected Poems*

John HAMPSON　　　　**1901–1955**

Saturday Night at the Greyhound

Published to widespread critical praise, Hampson's first and best-known novel became an unexpected bestseller for the Hogarth Press. Within ten days of publication it had been reprinted twice, and it started a brief vogue for novels of regional working-class life. The novels of Hampson and his associates in 'the Birmingham Group', anticipated by some twenty years the more celebrated work of the Angry Young Men.

Based upon Hampson's own experiences in the licensing trade, the book is a stark and spare account of one evening's events at a failing pub on the Derbyshire moors. The publican is Fred Flack, a vain and idle charmer whose irresponsibility dooms his prospect amongst the dour, wary and dishonest villagers of Grovelace. Most of the work is done by his wife, Ivy, and her doting young brother, Tom, who is unable to undermine his sister's hopeless infatuation with Fred, despite the fact that Fred is ruining their lives. Tom and Ivy have been brought up in the trade and know the business backwards, whereas Fred sees the pub as an opportunity to drink freely, gamble recklessly and seduce the barmaid, Clara. Like Fred, Clara cares only for herself. She is the illegitimate daughter of the late Squire and the dreadful Mrs Tapin, who works as a cleaner at the pub. Mrs Tapin spends most of her time stealing from her employers,

---

Women were the very deuce; he could not resist them. Fortunately he knew how to treat them. Time given to the study of individual women always repaid him. The thought made him chuckle. The way in which old Aunt Susan responded to his wheedling tricks had saved his backside from many a tanning. A bit of thought, and any woman could be got round.

If only men could be got round in the same fashion, life would be very pleasant, but men were different, worse luck.

SATURDAY NIGHT AT THE GREYHOUND
BY JOHN HAMPSON

---

whom she despises, reserving particular scorn for Tom, 'an oily-headed, fancy-socked little snot' condemned for being immune to Clara's charms.

As the pub prepares to open for business, the characters muse upon their present circumstances and the events of the past that have brought them there. An altogether different perspective is supplied by Ruth Dorme, the bored and frivolous girlfriend of the present Squire, who stays the night and, having 'stumbled accidentally on a patch of human misery', becomes 'anxious to unravel the history' of her hosts. The climax of the novel takes place after Clara has announced that she is pregnant by Fred and, seeking revenge for his indifference, enlists the support of her mother. Mrs Tapin, who has already killed the Flacks' pet greyhound, Pertinax, with casual brutality, has no qualms about informing the police about Fred's after-hours drinking and gambling sessions. A visit by PC Gaunter puts an end to the Flacks' life at The Greyhound.

The atmosphere of the grim pub and its surly drinkers is vividly evoked, as is the despair of the principal characters. Ivy and Fred, blinded respectively by pathetic optimism and bleary self-interest, fail to recognise the gradual slide towards nemesis that is obvious to Tom and Ruth. Their story has all the doomed inevitability of a classical tragedy.

James HANLEY   1901–1985

Boy

At the age of thirteen Hanley ran away to sea, sailing round the world until he jumped ship at New Brunswick to join up and fight in the trenches with the Canadian army. The sea was to be a frequent feature of his fiction, most notably in his second novel, *Boy*, which was originally published in an edition of 145 copies for subscribers only. An expurgated trade edition followed, but when in 1934 it was issued in a cheap edition, copies were seized by the police from a public library and the book was successfully prosecuted for obscenity. The publishers were fined £400 and copies of the book were burned. Although Hugh Walpole denounced the novel and destroyed a copy in a bookshop, *Boy* was defended by T.E. Lawrence and E.M. Forster. At the International Congress of Writers in Paris in 1935, which was devoted to the 'Defence of Culture', Forster (who was accompanied by Hanley) made a speech entitled 'Liberty in England', which criticised the workings of the Obscene Publications laws with reference to the trial. Hanley forbade republication of the novel during his lifetime – one reason, perhaps, that it remains his most famous book – and it was not reissued until 1990.

Arthur Fearon, nearly always impersonally referred to as 'the boy', or by his surname, is forced to leave school at thirteen by his bullying father and ineffectual mother because the family needs him to start bringing in a wage. Intelligent, physically small, and extremely sensitive and shy, he takes a job at the local docks for which he is completely unsuited. After one day's labour he resolves never to return to this degrading work, a situation his mother understands more than he is able to realise. Caught between the hell of his violent father and a job which will kill him, he stows away aboard a ship bound for Alexandria. To Arthur going to sea is a romantic adventure, fuelled by his father's stories of his own experiences, and he imagines a fulfilled and happy future. He is unprepared for the rough hostility of the sailors who, when they discover him, laugh at his weakness and abuse him sexually. Ignorant and lacking direction from the captain, who is constantly drunk, the seamen treat Arthur as they feel they have been treated themselves. At several points in the novel, Arthur tries to assert himself, but no one supports him, and he realises that he might have been better off at home despite the misery he knew there. Objectively, however, at no point in his life has he ever known happiness, so that can only be conjecture. Despite the abuse, Arthur tries to please the crew, but always fails.

Endless sexual talk drives Arthur to sleep with a young prostitute in Cairo, in his anxiety to prove his manhood. His awakening feelings of tenderness, happiness and desire are abruptly destroyed when, after sex, she throws him downstairs. In despair, he romantically idealises his home life, becoming delirious as undiagnosed syphilis takes hold. He recalls the advice of the crew that he should 'jump overboard', for he can have no happy future on board ship.

By the time the syphilis is finally diagnosed, it is incurable. Arthur feverishly repeats the word 'boy', the word that taunts him, and represents the state of innocence that has destroyed him. He is quietly suffocated by the captain, who wishes to avoid a scandal, and who wires to Arthur's parents that their son has been washed overboard.

Naomi MITCHISON   1897–

The Corn King and the Spring Queen

Usually a rapid and prolific writer, Mitchison took five years over this long, ambitious, wide-ranging novel set in the third century BC, and generally agreed to be her *chef d'oeuvre*. At first lively and sensuous – reflecting her interest in tribal communities and her sense that women are the repositories of instinct and intuition, symbolised by the magic in which she herself to some extent believed – the novel's tone deepens into a more sombre quest, for an understanding of the individual's relationship to community and society, and for the 'just society', a concept which echoes the author's increasing commitment to socialism.

The heroine is Erif Der, a young witch in the imaginary Scythian kingdom of Marob on the Black Sea, the outermost fringe of Hellenic culture. Its ruler, Tarrik, is both chieftain and the Corn King on whose powers the land's fertility depends, and who will be put to death if they fail. Erif, Tarrik's Spring Queen and partner in the seasonal rituals, finds her loyalty torn between her father, who wants to depose Tarrik, and her young husband, and her magic fails at a crucial moment. Knowing he must find the best form for inevitable change, and influenced by the theories of a Stoic philosopher, Tarrik travels to Sparta, where the idealistic new king, Kleomenes, plans to free the helots and establish community of property. Erif, pregnant, remains in Marob.

The austerities of Sparta provide a strong contrast to the barbaric richness of Marob, but Erif has a counterpart in Philylla, a young patrician committed to Kleomenes' revolution. Freeing herself with difficulty for her reactionary family, Philylla is eventually married to Anteus, the king's companion and lover. Tarrik and his retinue are made welcome, and join the fight with hostile neighbouring states. Tarrik is captured, to be found and freed by Erif, who, when her baby was murdered by her father, joined her husband. On their return to Marob, Tarrik reclaims his kingship. But, during the harvest festivities, when both men enact the Corn King's death, Erif as Spring Queen kills her father. Because Erif has brought 'death into the cycle of seasons', she must leave Marob to be cleansed. She makes first for Sparta, when she and Philylla renew their friendship, but this rational society with its reforms and battles has no help for her. She goes to consult the oracle at Delphi. Characteristically it is riddling, and even the young Epicurean philosopher who has befriended her cannot help. He does, however, take messages to Marob and, among savage hardships, learns respect for Tarrik, who is consciously forming himself into a just ruler. Meanwhile, the Spartans are retreating before an alliance of their enemies. Kleomenes and his followers, including Philylla's husband, Anteus, flee to Alexandria, seeking help from the new Egyptian king, Ptolemy. Returning to Sparta, Erif finds Philylla and escapes with her to Alexandria.

The Alexandrian court is frivolous, decadent and corrupt, with arcane and vicious politics which baffle the upright Spartans. But Erif finds help for her quest in some aspects of Egyptian life, especially the cult of the mother goddess, Isis. When, finally, the Spartan companions, betrayed, fall on their swords, and Philylla is slaughtered with Kleomenes' family, Erif is granted an enlightening vision from her dead mother. The Delphic oracle is fulfilled. She falls into a deep trance while her *ka* (spirit) takes the form of a snake guarding the body of Kleomenes lashed to a stake, and wakes to find that Tarrik and her young son have come for her. An epilogue shows the continuity of the seasons in Marob, while the images of Kleomenes' passion and death have spread throughout the Mediterranean world.

The quest and adventure strands of *The Corn King and the Spring Queen* make for a thoroughly enjoyable historical novel, but it is Mitchison's inquiring and generous temperament which lift it to the status it now enjoys. While she had read extensively (Plutarch is an important source, and she had devoured Frazer's *The Golden Bough* while still in her teens), Mitchison's characteristics quality is a vital empathy with the worlds about which she writes. This if found in her many historical novels as well as in an early example of women's space fiction, **Memoirs of a Spacewoman** (1962). The texture of life is evoked through vividly sensuous descriptions of fabrics, buildings, objects and animals, while the customs and ceremonies of the three contrasting societies are described with documentary immediacy. As can be seen, Mitchison's religious views are eclectic, with these two kings 'who die for the people' being identified with such deities as Adonis and Christ. Even today, Mitchison's celebration of sexuality in many forms and manifestations seems hearteningly liberal. Above all, it is the spirited warmth and energy, embodied in the character of Erif Der, which not only make this an attractive feminist text but a leader in the admittedly slender first rank of twentieth-century historical fiction.

Anthony POWELL   1905–

Afternoon Men

The first of the five novels published by Anthony Powell before the Second World War, *Afternoon Men* immediately establishes his characteristic social world: a shabby London artistic Bohemia given a limited *éclat* by the presence of some people with money or birth. The novel begins and ends in an early-evening drinking-club in Soho, and between these two points it describes a perfect circle: a busy but meaningless round of parties, drinking bouts, gossip and seductions. Narrative is minimal, and the novel proceeds largely through long passages of spare dialogue.

The initial scene in the drinking-club introduces us to the two principal characters, Raymond Pringle, an inadequate red-haired minor artist, and William Atwater, who, having failed to get into the diplomatic service, has obtained a post in a museum through influence. The scene moves to a bottle-party, where Atwater meets a gauche girl called Lola, who is interested in Bertrand Russell, and glimpses the beautiful and enigmatic Susan Nunnery ('She had rather big eyes that made her seem as if she were all at once amused and surprised and at the same time disappointed').

The scene next moves to the museum where Atwater works lackadaisically, and is persecuted by the eccentric Dr Crutch, who wants to buy the museum's exhibits. Atwater starts to take Susan out, but she makes only limited response to his growing attachment, and he is obliged to find mechanical sex with Lola and others. Pringle takes a house in the country to which he invites a number of friends for Spartan relaxation ('We none of us wash much here'). He discovers that his mistress, the man-hungry Harriet Twining, is deceiving him with a fellow artist, Barlow. He attempts to commit suicide by drowning, but cannot carry it through; in the meantime, the shabby house-party has reacted callously, destroying his suicide-note and failing to inform the police. Atwater returns to London and finds that Susan has gone to America with another man.

A large number of minor characters flit across the scene: the good-time American publisher Scheigan; Naomi Race, an elderly Bohemian who knew Rossetti; and Fotheringham, a paste-up-artist eternally looking for another job ('The aura of journalism's lower slopes hung round him like a vapour'). The sense of a time and place exactly evoked, the elegant contempt, and the occasional spare lyricism of the writing, all add to this powerful picture of an atomised and decaying society.

Dorothy RICHARDSON   1873–1957

Dawn's Left Hand

see **Pilgrimage** (1938)

E. Arnot ROBERTSON   1903–1961

Four Frightened People

Robertson made her reputation with this, her third novel, which became one of the very first Penguin Books (No. 15). It is dedicated 'To, and in reproof of, Henry Ernest Turner, my sailing partner, who said: "The trouble with you is that you're a prematurely mouldy intellectual, and at the touch of your pen romance goes rancid."' The first half of Turner's remark also opens the novel itself. Robertson had met Turner during a sailing expedition and married him in 1927. The unpleasantness of the sea voyage which opens the novel may be traced to Robertson's claim that: 'Family pressure, always exercised towards more and tougher sailing, resulting in this one sea-sick member developing an abiding dislike of the sea.'

Recovering from a broken love-affair, Judy Corder, a newly qualified doctor, is sailing down the east coast of the Malay Peninsula with her cousin, Stewart, on a dilapidated steamship. She is one of the few English people willing to admit that she endures the intense heat and substandard conditions not through choice, but for financial reasons. In the company of the cerebral Arnold Ainger, she witnesses three corpses being thrown overboard, thus confirming Arnold's suspicion that bubonic plague has broken out amongst the coolies. Judy, Stewart and Arnold decide to leave the ship secretly but, discovered mid-escape, are forced to take a fourth person with them, the indefatigably optimistic and jolly Mrs Mardick.

Arnold's plan to walk thirty miles to the next port is scuppered by the hostile terrain, and what should be a journey of a few days becomes an endurance test of some weeks. Their guide, Deotlan, does not excel in orienteering and – despite some sustaining if incongruous conversations about the state of art and youth, the unmerciful burlesquing of Mrs Mardick and her 'unquenchable vitality', Judy's practical outlook, Arnold's ability to interest himself in obscure dialects even in a crisis, and Stewart's notions of adventure and heroism, nurtured by a boyhood diet of *Young England* – nothing can prepare them for the physical hardship and danger, or the mental anguish of finding themselves in the alien jungle, with its overwhelming sense of invisible menace. Deotlan and his companion Wan Nau are eventually killed and, perhaps influenced by the will to survive, and with a 'ruthlessness nearly equal to that of the land itself', the group abandons Mrs Mardick near a village while she is sleeping. In the midst of truly catastrophic events, the forces of sexual attraction assert themselves and Judy, who has few illusions about herself, and a refreshingly frank appreciation of sex and the male physique, embarks upon an affair with Arnold. Judy knows that the affair will be short-lived for Arnold is married, but she consoles herself with observations once made over the dissecting table ('Curiously tough thing, a human heart') and prepares for disappointment. Back in England, and more or less in one piece, Judy meets Arnold at Simpson's for a farewell lunch (how often the thought had sustained them as they chewed on a jungle root). She finds him 'behaving very funny' (a waiter tries to divert her from the giggling man to her usual table) and discovers that his wife is divorcing him.

Virginia WOOLF   1882–1941

The Waves

Avoiding a direct presentation of events and place, *The Waves* begins with an elaborate, italicised description of a sunrise over the sea. The modulation of the radiance, its effects on the water and the shoreline and – moving inland – on a garden, a house, a window, are exactly depicted. Then (in roman type) Woolf records the self-addressed thoughts of her characters: Bernard, Susan, Rhoda, Neville, Jinny and Louis. This pattern is repeated eight times: a description of the sun, as it travels through the sky to nightfall, followed by reflections which, although reported as speech, are supposed to be unspoken. A single, italicised sentence concludes the novel.

The transition from dawn to dusk provides a kind of stage lighting, indicating the emotional colour of the scene to come and marking the phases of the characters' lives. At the outset, the six are children in the country. They experience intensely the external world, and the language in which they strive to capture their sensations is highly wrought and compressed. It is also a shared language: while the characters nurse different attitudes and ambitions, each 'speaks' very much like the rest. Phrases recur from voice to voice. An innocent kiss between Jinny and Louis – traumatically merged with the violent kiss of two servants – at once disturbs the circle and compels it to define its inner relations. In the subsequent scenes, through school days to death, these founding relations never substantially shift.

A few focal events are obliquely reported. Percival – Neville's lover, loved by the whole group – is killed in India. His extinction is the subject of a collective and ideal grief. Susan marries a farmer. Jinny becomes a society beauty, Louis a financier and Neville a successful man of letters. Bernard, having imagined himself a Byronic figure, appears to finish nothing that he writes, but to him is given the novel's long, final soliloquy. The friends meet in middle age at Hampton Court. Some time after that, miserable and isolated, Rhoda commits suicide. Nevertheless, when Bernard at last greets death, he does so on behalf of the complete circle, saying: '[We] are divided; we are not here. Yet I cannot find any obstacle separating us.' Thus he touches on the quick of the novel's concern – namely, the paradox of identity. If identity is the unique possession of the individual – the product of his or her circumstance, knowledge and choice – it is also contrived and maintained for the benefit of others. Simultaneously public and private, it engenders in the protagonists a need for both solitude and society and, accordingly, the six are bound together by the very force which sunders them.

Formally, *The Waves* alludes to Greek drama. Stylistically, with its dense, overlapping images and its refusal of narrative conventions, it is a sustained prose poem. At first fractured and disorientating, it gains in ease and lucidity as it proceeds, closing with a melancholy wisdom.

On learning that the book was selling quite well, Woolf wrote in her diary: 'how unexpected, how odd, that people can read that difficult grinding stuff!'

1932

Japanese occupy Shanghai ● Adolf Hitler rejects German chancellorship ● Franklin Delano Roosevelt wins US presidential election in landslide over Herbert Hoover ● Catalonia is granted autonomy, with its own flag, language and Parliament ● Famine in USSR ● Chadwick's discovery of the neutron ● Karl Jansky pioneers radio-astronomy ● Alexander Calder exhibits 'stabiles' and 'mobiles' ● Bertolt Brecht, *The Mother* ● First 'Tarzan' film and Shirley Temple's début ● Lytton Strachey and Kenneth Grahame die ● W.H. Auden *et al.*, *New Signatures* (anthology)

| Erskine CALDWELL | 1903–1987 |

Tobacco Road

Set near Augusta, Georgia, during the dust-bowl Depression, *Tobacco Road* begins by describing the theft of a sack of turnips. The thief is Jeeter Lester, the victim is Lov Bensey, and the motive is imminent starvation.

For years, Jeeter and his wife, Ada, have lived in worsening poverty, unable to borrow money to plant a cotton crop which, in any case, would never prosper. Infrequently, when he stirs from his profound laziness, Jeeter tries to sell scrub oak for a few cents.

Of his seventeen children, only adolescent Dude and hare-lipped Ellie May are still at home – as is Jeeter's harmless, much-loathed mother. Twelve-year-old Pearl has married Lov (she refuses, however, even to look at her spouse); Tom is reputed to manage a trading camp; the rest have either died or gone away, and Ada and Jeeter no longer recall all their names.

Jeeter returns from the woods, where he has been eating Lov's turnips, to find Sister Bessie on his porch. Bessie is a middle-aged widow and she now wants Dude for a husband. Jeeter lets her have him. Intending to become travelling preachers, the newlyweds buy a car with Bessie's savings. A black is killed in one of several accidents that follow. Nobody seems concerned. Jeeter insists on a trip to Augusta, during which the car runs out of oil, and its engine is irreparably damaged. After a quarrel, Jeeter throws Dude and Bessie off his land. Dude drives over his grandmother, who dies slowly and wholly ignored, and Lov arrives to complain that Pearl has run off. He is given Ellie May in compensation.

The end of the ploughing season passes. Yet again, Jeeter has planted nothing. Around him, the annual brush fires have started and one night his house is set alight. He and Ada are incinerated. Their remains are buried by neighbours the next day. As Dude inanely dreams of farming cotton, Bessie refuses to say a few words at the grave.

Jeeter Lester's plight is not unique. Indeed, the novel's subject is the general eradication of the small-time farmer, ruined by his own inefficiency and the flow of capital and labour to the cities. Illiterate, and weakened by hunger, he is as ripe for exploitation as he is unfit for employment. His creed is evangelical fatalism, accepting the misery of the present in the sure and certain hope of future reward. The characters' behaviour, though, is close to bestial, violence is endemic, and sex is indiscriminate and instinctive. Deformities and simple minds abound, while the repetitious cadences of speech serve to emphasise the speakers' limitation. None the less, *Tobacco Road* is comic, its characters grotesques, so blunted in their sensibilities that they have no true capacity for pain. They represent a dreadful innocence, free of

the virtues and vices of civilisation; but their adherence to their grisly Eden is an anachronism, reminding the world beyond of its brutish origins and replaying a past against which progress is measured.

E.M. DELAFIELD   1890–1943
The Provincial Lady Goes Further
see **Diary of a Provincial Lady** (1930)

John DOS PASSOS   1896–1970
1919
see **USA** (1936)

James T. FARRELL   1904–1979
Young Lonigan
see **Studs Lonigan** (1935)

Lewis Grassic GIBBON   1901–1935
Sunset Song
see **A Scots Quair** (1934)

Stella GIBBONS   1902–1989
Cold Comfort Farm

Shortly before Mary Webb's death in 1927, Stanley Baldwin publicly praised *Precious Bane* (1924) and made it a bestseller. A few years later Gibbons responded to all the fuss with her riotously funny tale of a sensible young woman who sorts out the bucolically deranged Starkadders of Cold Comfort Farm. She satirises Webb's melodramatic manner, her use of rural dialect, her preoccupation with superstition and unhealthy rites, and her lack of concern for probability. In a cod introductory epistle to her friend Anthony Pookworthy, Esq., ABS, LRR, the author explains that she has 'adopted the method perfected by the late Herr Baedeker, and firmly marked what I consider the finer passages with one, two or three stars'. The most egregiously Webb-like (or indeed Lawrentian) passages thus come ready-graded. Gibbons was paid £32 for the novel, which was her first and remains far and away her best known. It was deservedly an enormous popular and critical success, winning the Femina Vie Heureuse Prize for 1933. A late sequel, *Conference at Cold Comfort Farm* (1949), made less of an impression.

'My little mop!' He stood staring at it in a dream.

'Yes. It's to cletter the dishes with,' said Flora, firmly, suddenly foreseeing a new danger on the horizon.

'Nay ... nay,' protested Adam. ''Tes too pretty to cletter those great old dishes wi'. I mun do that with the thorn twigs; they'll serve. I'll keep my liddle mop in the shed, along wi' our Pointless and our Feckless.'

'They might eat it,' suggested Flora.

'Aye, aye, so they might, Robert Poste's child. Ah, well, I mun hang it up by its liddle red string above the dish-washin' bowl. Niver put my liddle pretty in that gurt old greasy washin'-up water. Aye, 'tes prettier nor apple-blooth, my liddle mop.'

COLD COMFORT FARM
BY STELLA GIBBONS

When Flora Poste is orphaned, she inherits 'a strong will' from her father and 'a slender ankle' from her mother, but no property and an income of only £100 a year. When asked what she will live on, Flora replies: 'I am only nineteen, but I have already observed that whereas there still lingers some absurd prejudice against living on one's friends, no limits are set, either by society or by one's own conscience, to the amount one may impose upon one's relatives.' The most promising reply to the letters she writes to her relatives is from Judith Starkadder of Cold Comfort Farm, Howling, Sussex: 'Child, my man did your father great wrong. If you come to us I will do my best to atone, but you must never ask what for. My lips are sealed. We are not like other folk ...'

Indeed they are not. Flora is met at Howling Station by Adam Lambsbreath, ancient guardian of the ancient cows Graceless, Pointless, Feckless and Aimless. He tells Flora of the curse upon the farm: 'The seeds wither as they fall into the ground ... The cows are barren and the sows are farren and the King's Evil and the Queen's Bane and the Prince's Heritage ravages our crops.' There is no question of selling up – as Flora sensibly suggests – because Aunt Ada Doom, shut away upstairs, would never allow it: 'There have always been Starkadders at Cold Comfort.' Flora sees that there is much tidying up to do and, with the help of her favourite book, the Abbé Fausse-Maigre's *The Higher Common Sense*, she sets about it. 'Do what you please, Robert Poste's child, if so be you don't break in on my loneliness', she is told by Judith, who mopes in a room decorated with 200 photographs of her favourite son, Seth.

Flora introduces the philoprogenitive hired help, Meriam, to contraception, and encourages Judith's husband, Amos, to find a wider audience for his sermons by going round the country in his Ford van, preaching on market days. Her greatest triumph, however, is to transform Adam Lambsbreath's beloved Elfine, who is 'as wild and shy as a pharisee of the woods'. Elfine grows accustomed to houses, gives up poetry and, in a fifty-guinea dress bought for the occasion by Flora, captures the heart of Dick Hawk-Monitor at his family's ball. This development brings Ada Doom downstairs to hold 'the Counting', a yearly headcount of the family. Recalling the occasion when, 'no bigger than a titty-wren', she 'saw something nasty in the woodshed', she insists the whole family must remain with her, an edict ignored by Amos, who goes out into the night crying: 'I mun go where th' Lord's work calls me and spread th' Lord's word abroad in strange places.' (He goes to Chicago.) The others swiftly follow suit: Seth is taken to Hollywood by a talent scout; the distraught Judith is entrusted to the care of a Viennese psychoanalyst; and 'poor daft' Rennet is whisked off to Charlotte Street by the Lawrentian writer Mr Mybug. Cheated of Elfine, Urk finds consolation in Meriam and his water-voles; turned down by Flora, Reuben inherits the farm. Flora's final act is to spend nine hours closetted with Aunt Ada, armed with the latest copy of *Vogue* and a prospectus for the Hotel Miramar in Paris. The sensation of Elfine's wedding breakfast is the spectacle of Aunt Ada, 'dressed from head to foot in the smartest flying kit of black leather', taking off in an aeroplane. Her task accomplished, Flora herself departs in an aeroplane piloted by her second cousin and future husband, Charles Fairford.

Ellen GLASGOW   1873–1945

The Sheltered Life

In her autobiography *The Woman Within* (1954) Glasgow records her view that the best of her work consisted in the five novels, starting with *Barren Ground* (1925), and ending with *Vein of Iron* (1935), which she wrote during her fifties. Set in her native Virginia, *The Sheltered Life* is probably the best expression of the stultifying, peculiarly Southern atmosphere of her work.

Two families, the Archbalds and the Birdsongs, remain in the decaying splendour of Washington St, Queensborough, resisting or simply ignoring the 'change and adversity and progress' epitomised by the smell wafting from the new chemical factories. In the first part of the novel nine-year-old Jenny Blair Archbald observes the world around her: her grandfather, the General; her widowed mother Cora; and her plain Aunt Etta, languishing in loveless misery. Through her eyes too we see Eva Birdsong, once considered the finest beauty of her age, and her philandering husband, George. One day Jenny strays into a poor area, falls from her roller-skates and is rescued by a black laundress. Inside the house she sees George Birdsong, minus his coat, who suggests that they both keep secret their visit here. Eva rigorously keeps up appearances and does not acknowledge George's infidelities, but after a ball at which he has misbehaved, Jenny experiences 'the feeling of moral nakedness that came to her whenever the veil slipped away from life and even grown-up people stopped pretending'.

The second and third parts are set eight years on in 1914. Jenny's desire to escape to New York is echoed in the General's memories of his own youth, when he had wanted to be a poet and had been branded a milksop by the ferocious grandfather. Like everyone else, he is besotted with Eva's beauty and feels that the age of glory has passed and everything, even beauty, is declining into mediocrity. Eva goes into hospital for an operation and George is plunged into a guilty remorse.

Jenny, who dislikes boys of her own age, develops a painful infatuation with George. Upon her return from recuperation, Eva sees George embracing the distraught Jenny. Jenny flees, and when she returns finds Eva sitting with a fixed smile on her face and George dead on the floor, apparently – so everyone will have it – as a result of an accident with his gun. Appearances are maintained even in death.

Mrs Birdsong was one of those celebrated beauties who, if they still exist, have ceased to be celebrated. Tall, slender, royal in carriage, hers was that perfect loveliness which made the hearts of old men flutter and miss a beat when she approached them. Everything about her was flowing, and everything flowed divinely. Her figure curved and melted and curved again in the queenly style of the period; her bronze hair rippled over a head so faultless that its proper setting was allegory; her eyes were so radiant in colour that they had been compared by a Victorian poet to bluebirds flying.

THE SHELTERED LIFE
BY ELLEN GLASGOW

Aldous HUXLEY   1894–1963

Brave New World

After the success of his first three novels, Huxley abandoned the fictional *milieu* of literary London and directed his satire towards an imagined future. He admitted that the orignal idea of *Brave New World* was to challenge H.G. Wells's Utopian vision, in particular the desirability of eugenics. The novel also marks Huxley's increasing disenchantment with the world, which was to result in his leaving England for California in 1937 in search of a more spiritual life. The book was immensely successful, selling 23,000 copies in the first two years of publication, and has remained popular ever since.

The hands of all the four thousand electric clocks in all the Bloomsbury Centre's four thousand rooms marked twenty-seven minutes past two. 'This hive of industry,' as the Director was fond of calling it, was in the full buzz of work. Every one was busy, everything in ordered motion. Under the microscopes, their long tails furiously lashing, spermatozoa were burrowing head first into eggs; and, fertilized, the eggs were expanding, dividing, or if bokanovskified, budding and breaking up into whole populations of separate embryos.

BRAVE NEW WORLD
BY ALDOUS HUXLEY

'Everybody's happy now,' the citizens of the Utopian world state assert; genetic engineering reinforced by lifelong conditioning makes people accept their inescapable social density. Physical pain and old age have been entirely eradicated, and the Freudian dangers of family life are avoided by the abolition of all emotional attachment, love reduced to sex and passion to 'soma', the perfect drug of eternal oblivion. Babies are bred in bottles, processed into standard men and women in uniform batches, predestined for their part in the industrial structure through gradations from alpha pluses to epsilon minuses.

The only threat to this artificial stability lies in the necessary intelligence of the upper caste. Encircled by faces flushed with the inner light of universal benevolence, Bernard Marx does his best to share in the pleasure of the social orgy; but the pneumatic beauty Lenina Crowne is horrified to discover that Bernard longs to be more than just a cell in the social body. On holiday at the Savage Reservation in New Mexico, Bernard and Lenina encounter Linda, the long-lost lover of the Director of Hatcheries and Conditioning, accidentally marooned on a similar holiday twenty-five years ago. Despite scrupulous adherence to her Malthusian drill, Linda had found herself pregnant and far from the comfort of the Abortion Centre. Bernard now rescues her and her son John, returning them to the sterilised civilisation in which they can never be other than peep-show freaks.

'Oh brave new world,' rejoices John in eager anticipation, echoing Miranda in *The Tempest*; but the people in it prove to be identical obscenities. His passion stimulated by his early reading of Shakespeare – now a redundant and forbidden author – John finds Lenina's easy sexual availability symbolic of the social madness. As Linda dies in a painless soma coma, John's anger at this perversion of the human spirit erupts into an inevitably successful rebellion, which results in his alpha-plus associates being expelled from the dubious Eden. Attempting to live as a self-chastising hermit, John becomes a savage relic of long-forgotten folklore, hounded by reporters and voyeurs; alone in self-determination, his suicide completes this science-fiction tragedy.

Christopher ISHERWOOD 1904–1986

The Memorial

'If it isn't at once recognised as a masterpiece,' W.H. Auden wrote of Isherwood's second novel, 'I give up hope of any taste in this country.' Although something of a transitional work between the oblique *All the Conspirators* (1928) and the Berlin stories in which the authentic Isherwood voice is sounded, *The Memorial* is a crucial 1930s text, in which the antagonism between the pre- and post-war generations is explored from both sides of the divide. It is a memorial both to those like Isherwood's father, who was killed at Ypres and is the novel's dedicatee, and the author's own 1920s generation, deeply affected by a conflict in which they were too young to take part.

The novel is divided into four non-sequential 'books' and the events are seen from the viewpoints of a number of inter-related characters. Principal amongst these are Lily Vernon, a sentimental war-widow (heartlessly drawn from the author's mother); her awkward, idealistic son, Eric; her down-to-earth Bohemian sister-in-law, Mary Scriven, and her two children, Maurice and Anne; Edward Blake, a decorated air ace who was the late Richard Vernon's best friend and has now gone to pieces; and Margaret Lanwin, a painter who is in love with Edward, although she knows that he is homosexual. The lives of Mary and Lily are contrasted in the opening section (1928), where the former is observed organising a recital in Chelsea, whilst the latter makes genteel archaeological expeditions in London with her elderly admirer, Major Charlesworth. In Berlin Edward is undergoing analysis and fails in a messy attempt to commit suicide.

The second section (1920) is set at the Hall, the Vernons' Cheshire home, on the day the war memorial of the title is to be unveiled, and once again contrasts the characters of Lily and Mary. Lily is lost in swathes of sentiment and snobbery (objecting that officers and men are undifferentiated on the cross), whilst Mary observes the proceedings sardonically and recalls her own marriage to a philandering Irishman who deserted her and was also killed in the war. Eric suffers pangs of guilt because he would prefer to spend this sacred day with the Scrivens rather than with his mother.

The third section (1925) is largely set at Cambridge where the repressed Eric is studying mathematics, whilst the popular and reckless Maurice passes the time in acts of undergraduate hooliganism and scrounging. Eric attempts to persuade Edward to keep away from Maurice but is aware that he is prompted more by feelings of jealousy than by moral concerns.

The fourth book (1929) is a final glimpse of the characters at the end of the decade. The vulgarian, mill-owning Mr Ramsbotham ('Rams B') has bought the Hall from the Vernons and Anne is to marry one of his sons. Maurice has become a garage mechanic, and his mother, realising that 'the Past couldn't hurt her now', attempts a reconciliation with Lily. Eric has left Cambridge

without a degree to do social work and becomes a Catholic. Major Charlesworth recognises that his relationship with the saintly Lily must remain a friendship, whilst Edward and Margaret struggle with their asexual partnership, interrupted by assorted young men, one of whom provides the novel's final words: 'That War ... it ought never to have happened.'

Julia STRACHEY   1901–1979

Cheerful Weather for the Wedding

Nineteen years separate the two novels on which Strachey's reputation rests, a period during which she published a number of short stories but nothing of full length. Indeed, *Cheerful Weather* is of novella length, very much a Bloomsbury product, published by the Woolfs with a wrapper by Duncan Grant. Virginia Woolf, with whose own work the novel has an affinity, described it as 'complete and sharp and individual'.

The wedding of the title is that between Dolly Thatchman and Owen Bigham and takes place in far from cheerful weather on a windy clifftop. Cheerfulness is generated by Dolly's mother, a widow determined that all should be well in spite of numerous intimations of disaster. The sun shines, but there is an icy March wind, and these conditions are reflected within the Thatcham household. Mrs Thatcham, relentlessly sunny, muddles the servants and everyone else, assigning several guests to the same bedroom, changing her mind about where food should be served, and exclaiming the while about the extraordinary behaviour of everyone else. The bride is in her room, fussing over her clothes and vaguely hoping that a former suitor, Joseph, might rescue her; she has consumed half a bottle of rum. Joseph, a saturnine student of anthropology, mopes about downstairs, feeling that the wedding should be stopped, but lacking the confidence to take any action. Dolly's cousins, Tom and Robert, are arguing about the latter's emerald socks, which his brother feels will disgrace them in front of their fellow Rugbeians in the congregation. Dolly's younger sister, Kitty, looks forward to meeting some naval lieutenants who, unlike her sister's circle, will be clean-minded English gentlemen. Mad Nellie from the village is laying tea in the wrong place, railing against marriage and children. The groom does not have a ring.

Joseph fails to halt the marriage, despite the fact that Dolly's arrival at the church is delayed when she drunkenly spills ink down her dress. He does, however, deliver a few home truths to the infuriating Mrs Thatcham. Strachey observes the crises with an amused but affectionate eye, deftly manipulating a large cast of characters. The tensions of family and friends, servants and lovers, are evoked in less than seventy pages.

Evelyn WAUGH 1903–1966

Black Mischief

While attending the coronation of the Emperor Haile Selassie of Abyssinia in 1930, Waugh informed his parents that he had 'the plot of a first rate novel'. The result was this energetic and splendidly tasteless black comedy which, as Waugh was forced to explain when it was denounced in the *Tablet*, 'deals with the conflict of civilisation, with all its attendant and deplorable ills, and barbarism'.

Basil Seal, bored with London, travels to Azania, an island off the east coast of Africa, where a fellow Oxonian, Seth, has become Emperor. Determined to drag his country into the twentieth century ('I am the New Age. I am the Future'), Seth appoints Basil Minister of Modernisation and instigates a programme of increasingly preposterous reforms, culminating in a Birth Control Pageant. This is interrupted by a coup, organised by dissident factions, including his former ally, Gen. Connolly, the scheming French ambassador, M. Ballon, and the Patriarch of the Nestorian church. The rightful heir to the throne is released from long imprisonment and, senile and bewildered, is crowned during a ceremony of inordinate length, only to die at the moment of coronation.

British residents and visitors have fled to the English Legation, where the indolent and ineffectual ambassador, Sir Samson Courteney, is an unwilling host. They are finally airlifted to Aden, but the plane carrying Prudence, the ambassador's young daughter who Basil has taken as a mistress, goes missing. Basil pursues Seth, who has fled into the jungle, but arrives to find that the deposed emperor has died in mysterious circumstances. At the funeral feast, in which Basil partakes with relish, the baked meats include Prudence, 'stewed to a pulp amongst peppers and aromatic roots'. Azania becomes a joint protectorate of English and France, entirely losing its identity and run by faceless colonial servants. Basil returns to London, where none of his friends wishes to hear of his travels.

As with Waugh's two previous novels, *Black Mischief*'s style is clearly influenced by Firbank, particularly in the use of compression and omission. Long passages are relayed in telegraphic sentences and overheard fragments of conversation; the coup is seen entirely through the eyes of two distressed and inebriated English gentlewomen; important incidents are reduced to asides. 'Civilisation' is mocked as Seth's subjects continually misunderstand his innovations and remain barbarous. Where the law of the jungle quite literally applies, only ruthless entrepreneurs like Basil and the Armenian café proprietor, Mr Youkoumian, survive.

1933

Hitler becomes German Chancellor • Japanese advances in China • 'New Deal' in US • Persecution of Jews begins in Germany • Unrest in Palestine grows • US repeals prohibition • Leon Trotsky, *History of the Russian Revolution* • Anderson and Robert Millikin discover positrons • The first commercially-produced synthetic detergent is made • Richard Strauss, *Arabella* (opera) • *King Kong* is made • Calvin Coolidge and John Galsworthy die • Geoffrey Grigson (ed.), *New Verse* (magazine) founded

---

Lewis Grassic GIBBON   **1901–1935**

Cloud Howe

see **A Scots Quair** (1934)

Walter GREENWOOD   **1903–1974**

Love on the Dole

Greenwood's first novel has become a classic text of social realism, chronicling the human tragedy behind economic depression. Written in 1932 and published the following year, *Love on the Dole* portrays a grim panorama of working-class life in Salford.

Harry Hardcastle is determined that 'they'll not send me out clerking'. When he leaves school he chooses instead the glamour of the engineering works mythologised by his street-corner peers. But the seven-year apprenticeship leads not to a man's wage but to the dole queue, a capitalist con to ensure a supply of exploited labour. In this position of poverty, plans for marriage look impossible: but when Harry's girlfriend becomes pregnant they are forced into a life of rented rooms and Poor Relief.

This is the fate which Larry Meath, a self-educated socialist, is anxious to avoid. Putting his energy into political agitation, he tries to break the pessimistic apathy of the Salford men. Harry's sister, Sally, finds Larry an inspiring alternative to her customary boorish lovers, and Larry is torn between his intellectualism and his passion for Sally. Eventually she persuades him to chance marriage despite the hardship: but just before the wedding Larry is also made redundant.

The people's protest against the injustice of the system explodes into political action as the men form a march for jobs. But as they approach the town hall the police intervene violently, and Larry is severely beaten and arrested. His treatment precipitates fatal pneumonia, and Sally sits by his bedside as he dies.

Dully, insistently, crushing came the realization that there was no escape, save in dreams. All was a tangle; reality was too hideous to look upon: it could not be shrouded or titivated for long by the reading of cheap novelettes or the spectacle of films of spacious lives. They were only opiates and left a keener edge on hunger, made more loathsome reality's sores.

When you went to the public baths and stood, after a day's work, in a queue waiting your turn until the attendant beckoned you to a cubicle where was a bath half-filled with dirty water left by a girl or woman just quitted the place – you couldn't by any stretch of imagination, even when the attendant had drained off the water and washed the muck away, see it as anything other than what it was, Hanky Park, the small corner of the wide, wide world where you lived.

LOVE ON THE DOLE
BY WALTER GREENWOOD

Sally, now supporting the entire family, struggles on with her job at the mill. Her youth is passing in perpetual drudgery and her future looks bleak. In a desperate effort to break free of her circumstances she finally turns to Sam Grundy, the local bookie anxious to make her his mistress. Local gossip condemns her as a whore yet acknowledges that only cynical immorality offers a way out of the poverty trap. Grundy's wealth offers a comfortable life, and his contacts provide jobs for her father and brother. Greenwood's novel ends on this note of bitter worldly wisdom, as the patterns of life in the Salford slums settle back into their cycle.

Greenwood was brought up in Salford and the novel attempts a realistic impression of Northern working-class life, notably in its use of dialect. Unfamiliar vocabulary and expressions are glossed in parentheses, which somewhat hinders the narrative flow, but the novel proved popular, and was later adapted for both the stage and the screen. Greenwood himself wrote the screenplay for the 1941 film, directed by John Baxter, in which Deborah Kerr played Sally.

John Cowper POWYS   1872–1963

A Glastonbury Romance

At over 1,100 pages, *A Glastonbury Romance* is a long and rambling novel, which moves at an extremely slow pace dramatically to establish the progression of time. Powys aimed to promote 'an acceptance of . . . human life in the spirit of absolute undogmatic ignorance . . . to convey a jumbled-up and squeezed-together epitome of life's various dimensions'. This included his own pantheistic belief in a dual-natured power he calls the 'First Cause' or the 'Unknown Ultimate'. Essentially evil, it may be overcome by good. Nature is used to represent human emotion, where trees 'think', the dead 'speak', and within a personalised cosmological structure the sun, moon and stars also have a mystical influence. Powys explains this directly by 'author' statement, then demonstrates how his characters are affected by such spiritual force.

The purpose of the plot, which ostensibly follows John Geard's efforts to establish a mystical community in Glastonbury, is to let the main characters examine their different attitudes to the Holy Grail. The Grail is deliberately vaguely defined, but each individual is 'attracted [to it] by a magnetism too powerful . . . to resist'.

The action develops by means of a series of episodes, generally complete within each chapter. The first section introduces the Crows: Elizabeth, aunt to John, Philip, Mary and Persephone, all cousins to each other. They have returned to their ancestral county of Norfolk for the funeral of Canon William Crow. The principal beneficiary of the large estate is John Geard of Glastonbury, the canon's secretary. Norfolk has instilled in the Crows an awareness of the heathen, or profane, and constant reference is made to this as a point of secure reality against the supernatural that dominates Glastonbury.

To show different reactions and possible responses to circumstance, each main character has an 'opposite', or counterpart. Philip, the local Glastonbury factory owner, mines for tin in the Wookey Hole caves, which he owns, and intends to build new roads and a bridge. He is disliked by his workers, whom he wishes to rule, and his authority is challenged by a Communist who leads a strike. Philip's counterpart is Geard, who becomes the Mayor of Glastonbury, and is known as 'Bloody Johnny' because of his obsession with the Blood of Christ. Geard also hopes to dominate the town, but through love and by demonstrating visible proof of his mystical beliefs. He forms an economic community partly financed by an arts-and-crafts factory he establishes, and with his inherited Crow money plans a pageant that includes details of the Arthurian legend and the Crucifixion.

Geard employs as his secretary John Crow, a man who is cynical and raffish and has no religious beliefs. Crow dislikes the Glastonbury cult, but has a psychic awareness, ultimately rewarded by a vision of the casting aside of Excalibur. He is in love with his cousin Mary, whom he later marries. John Crow's counterpart is the vicar's son, Sam Dekker. For him, God is the evil 'First Cause': instead of looking for any goodness in God, Dekker ignores Him, and supports the suffering Christ – unlike Geard, who exploits Christ as an equal. Dekker's identification with Christ proves a barrier to his own deeply felt love for Nell Zoyland, who is married to Will, the illegitimate son of a local marquis (known only as Marquis P). Will is aware of his wife's infidelity with Sam Dekker, and asks her to choose between her lover and himself.

The development of religious awareness corresponds with dilemmas in the main personal relationships. Mary Crow resists the influence of the Glastonbury cult, but is in any case naturally generous and good, tolerating and loving John Crow, who is unpopular because of his outspoken personality. She has a perfect understanding of Crow's unrealised homosexual love for a mutual Norfolk childhood friend, Tom Barter, and is herself loved by her elderly lesbian employer, Miss Drew.

The pageant forms the conclusion of the first volume, in which the themes and characters are fully established. The entertainment reaches an unexpected crisis with its effect on Owen Evans, an antiquarian, who takes the part of Christ. A repressed sado-masochist, he is tormented by the desire to kill, and longs for oblivion, to absolve his guilt. Emotionally charged with the theme of suffering, Evans collapses during his 'death' on the cross and is taken to hospital.

Geard's ascendancy over Philip, which began when he relieved the pain of a woman dying of cancer, is materially confirmed by the pageant. It continues in the second part of the novel, when he cures the woman, and apparently raises a boy from the dead. His religious faith is practical, and he does not seek the Vision of the Grail. Evans, who is romantic, tries and fails, as does his parallel figure, Persephone. She discovers from sexual encounters with Philip and Will Zoyland that she has an inability to love; realising she must search for some other reality, she hopes she will find this politically in Russia. Of those who desire the Grail, only Sam Dekker succeeds in seeing it clearly, because he searches for Christ. Like Mary, he has an absolute determination of purpose. She 'sees' the Grail as a mystical display of light, while Dekker witnesses an appearance of the Chalice itself. Unable to choose between

Nell, who has a son by him, and his desperate need for proof of Christ's existence, he looks after the sick. Although the Grail supplies this proof, he repents his neglect of Nell too late, and loses her to Will.

The main events, concerned with the sacred and the profane, are interspersed with reference to nearly forty other inhabitants of Glastonbury. The local procuress, the innkeeper, a gang of children and their respective dwellings, are all described in detail, and the complete evocation of community life conveys a strong sense of earthbound reality.

The final section brings the various crises to a resolution. Philip's material ambitions are thwarted by the decline of his tin-mine and Communist opposition arranged by an anarchist lawyer. Evans achieves partial fulfilment of his sadism when he witnesses the brutal killing of Barter, who is brought to the scene of his death by the equivalent of a Grail Messenger. Evans is saved, because he does not commit the crime himself. A flood, in which many drown, destroys Philip's new bridge, and with it his final hopes of becoming a rich magnate. He is rescued by his enemy, Geard, who is at his peak of success, but who willingly dies saving Philip's life. As strangers to Glastonbury, the Crows are defeated by it, and must leave. Elizabeth will die of a heart-attack, and John has lost his opportunity of achieving spiritual reality. Nevertheless, he must ultimately gain something from Mary, with whom he returns to Norfolk. The new Glastonbury of its builders, Philip Crow and Geard, is destroyed, but like Stonehenge the old ruins outlive them, with the lasting reality of eternal Grail.

Helen WADDELL　　　　**1889–1965**

Peter Abelard

'The grief that is to come will be no less than the love that went before it.' The great love between the twelfth-century scholar, Peter Abelard (d. 1142) and his young pupil, Héloïse, resulted in a famous collection of letters. Their story also formed the basis of Pope's poem 'Eloisa to Abelard' (1717), George Moore's **Héloïse and Abélard** (1921), and the medieval scholar Helen Waddell's only novel, an extraordinarily powerful work which, by a combination of simple prose and vivid detail, gives a convincing picture of life in medieval France.

At the age of seventeen, Héloïse is sent by her doting uncle, Fulbert, to have lessons with Abelard. To Gilles de Vannes, Canon of Notre-Dame and Abelard's mentor, the love between teacher and pupil is inevitable and forgivable. The rumours and bawdy songs which begin to circulate about them pose no threat, but when the unsuspecting Fulbert finds the couple together, he suffers a stroke, and the guilt-ridden situation reaches crisis point. Héloïse goes to Abelard's sister in Brittany, where she gives birth to a son, while Abelard attempts to pacify the broken Fulbert by promising to marry his pupil. This is against the wishes of Héloïse and the advice of Gilles, since it will ruin Abelard's reputation. The marriage takes place in secret, but does not appease Fulbert. After Héloïse is forced to deny the marriage in public, Fulbert carries out his horrible and long-planned revenge: he has Abelard castrated.

The convent to which Héloïse has fled to escape her uncle's deranged and violent behaviour becomes her permanent home, while Abelard, who cannot face the world in his impaired state, enters a monastery. With a charge of heresy close on his heels, he eventually settles at a humble retreat with his devotee, Thibault, and there finds peace of a kind. For Héloïse, who in her despair admits that she has 'taken the veil not for God's sake, but for Abelard's', this news of her lover, who sends no word to her, is of small comfort. Only Gilles understands that Abelard cannot bear to communicate with his former lover.

Nathanael WEST　　　　　**1903–1940**

Miss Lonelyhearts

'Miss Lonelyhearts' is a male agony columnist, whose non-professional name is never disclosed. He works for the New York *Post-Dispatch*, where he receives about thirty letters a day. Each one cries out for help or begs the solution to an insoluble problem: each, in its honest, semi-illiterate way, is a tale of cruelty and deprivation. Miss Lonelyhearts answers through the newspaper, offering optimistic platitudes. His editor, Shrike, tells him that agony columnists are 'the priests of twentieth-century America', and Miss Lonelyhearts, the son of a Baptist minister, feels his task is – or ought to be – a sanctified one. He knows, however, that he is not adequate to it. He is a prophet without a message, a saint without miracles. He drinks too much, and relationships are failures, disturbed by sadism and self-disgust.

West's tale describes, in economic episodes, how Miss Lonelyhearts lurches from crisis to crisis. He finds an old man in a public lavatory, bullies and assaults him and, the same night, gets attacked in a bar. He almost rapes Shrike's wife. Later, he has cursory sex with Fay Doyle, a woman who has sought advice from him. Repeatedly, in a short period, he falls ill. After the second illness, Betty, his sometime fiancée, takes him to a Connecticut farm to recuperate, but, on his return to New York, he imagines the people on the streets have 'broken hands and torn mouths'. He meets Fay Doyle's lame, pathetic husband, visits Fay and, when she tries to seduce him, beats her unconscious. Doyle threatens revenge. At a party, Miss Lonelyhearts sees Betty again. She tells him she is pregnant. He agrees to marry her and comforts himself with images of domestic bliss and stability. Then, on the steps outside his apartment, Doyle shoots him. Whether the shooting is fatal is not made clear.

Physically and mentally sick, the hero is a sacrificial victim. His violence constitutes a search for punishment, a desire symbolically to bear the guilt of big-city evils. Really, he can do no more than despair. Thus, if West's story is most immediately a psychological portrait, it is also a sketch of a society that offers no hope to its literal and metaphorical cripples. As Shrike tells Miss Lonelyhearts, 'neither the soil, nor the South Seas, nor Hedonism, nor art, nor suicide, nor drugs, can mean anything to us'. What remains is religion, but for Miss Lonelyhearts at least, religion is not a matter of belief: it comprises dark obsessions, paranoid visions and futile gestures. All it suggests is the form of redemption, and this is transmuted into brutal behaviour. Degraded, atavistic ritual becomes the replacement for the American dream.

Miss Lonelyhearts was West's second novel. Its brevity (it is about 25,000 words long) gives it a sharpness of focus lacking in his other works.

Antonia WHITE 1899–1980

Frost in May

White's greatest achievement as a writer was a quartet of autobiographical novels, of which *Frost in May* was the first. Her father converted to the Catholic faith when she was seven years old, and she was educated at a convent. She began writing *Frost in May* when she was only sixteen, but did not return to the manuscript until a year after her father's death in 1929. She was encouraged to finish the novel by her first husband, the journalist Tom Hopkinson. Although the book was very well received, White was prevented by a recurrence of mental illness from writing any further novels until *The Lost Traveller* (1950). This second novel and its sequels – *The Sugar House* (1952) and *Beyond the Glass* (1954) – continued the story of the first, following the heroine through adolescence, a stage career, love, marriage and mental breakdown and recovery. White was determined that these later novels should be less transparently autobiographical than her first, and to that end altered or invented many episodes and changed her heroine's name to Clara Batchelor.

Frost in May follows the fortunes of Fernanda Grey, daughter of a Catholic convert, who is sent to the Convent of the Five Wounds at Lippington, near London, at the age of nine. The uncluttered, evocative prose draws us into Nanda's world of beeswax and mysticism, invites us to view it through her eyes, to smart at her humiliations and to feel outrage at her final downfall.

Converted only a year previously, Nanda finds this a mysterious world with an alien language, a world she is desperate to inhabit but from which she feels ultimately barred. Her status as convert is continuously held up by nuns and girls alike as the root of her failure. The absence of the 'Catholic breed-ing' which her fellow pupils, daughters of aristocratic, European Catholic families, so effortlessly possess, must always make her the outsider.

That she has managed to maintain her faith at all seems miraculous in the light of a thousand subjugations – from the physical stringency of a cold bath to the ban on individual friendships. That she develops associations with the sultry Spanish Rosario, the boyish, aristocratic Leonie de Wesseldorf, and the boorish Clare Rockingham (who, being a Protestant with an ardent desire to convert against the wishes of her family, has all the 'glamour of a secret sorrow'), is perhaps an indication of the 'spiritual pride' and obstinacy which the spiteful, mocking Mother Frances and the calculated disciplinarian Mother Radcliffe (who is ready to sacrifice justice for the sake of principle) conspire to conquer. Nanda is hardly a natural rebel, and her good behaviour seems as culpable as her bad. While her religious faith becomes the bedrock of her existence, her feelings for Lippington vacillate until a suggestion is made that it is time to continue her studies elsewhere – then she realises how 'thoroughly' Lippington has 'done its work'.

A bout of measles is the idyll before the fall. From the relative comfort and geniality of the infirmary, Nanda works on her 'novel', the story of a giddy, pleasure-seeking heroine and a thoroughly wicked hero who will eventually be transformed by the blinding light of faith. No one in authority, however, is interested in the intended outcome when this 'vulgar filth' is discovered in Leonie's desk, nor is anyone prepared to see it as merely the immature fruit of an adolescent mind. On her fourteenth birthday Nanda is not only removed from the convent under the blackest of clouds, but is also expected to believe that it is 'good' for her. There is no real pity to soften the blows – she is a 'germ carrier', a taint on those who come into her circle. Clare's long-awaited permission to convert (an event for which Nanda has long wished and prayed) takes on an unbearable irony – is this the price that Nanda has to pay? Wounded, blighted, Nanda knows that nothing for her will 'ever be the same again'.

1934

Adolf Hitler becomes Führer ● Federal Farm Mortgage Corporation set up in US ● Disarmament Conference ends in failure ● Germany suspends all cash transfers on debts abroad ● Fascist and anti-Fascist demonstrations in Hyde Park, London ● Jean Frédéric Joliot and Irène Curie-Joliot discover induced radioactivity ● Reinhold Niebuhr, *Moral Man and Immoral Society* ● Sophia Loren born ● Edward Elgar, Gustav Holst, Frederick Delius and Marie Curie die ● Karl Shapiro, *Poems*

James M. CAIN　　　　　**1892–1977**

The Postman Always Rings Twice

Frank Chambers is a drifter. One morning he drifts into Nick Papadakis's roadside diner, intending to bum a meal and be on his way. He stays though – because Nick offers him a job and because he sees Cora, Nick's wife. He does not want the job, but he does want Cora. After scant and brutal preliminaries, he and she have sex. Shortly they decide to murder Nick. The attempt is a failure. Nick's skull is fractured (he thinks he slipped in his bath) and, while he is in hospital, Frank and Cora decide to pack up and leave. But when the moment arrives, Cora can't do it. She stays: Frank hitches off alone. He wins and loses money playing pool; then, by chance, he meets Nick again. Nick invites him to return to work at the diner. Frank accepts.

He and Cora resume their affair. One day Nick takes them both to a fiesta in Santa Barbara. On the way back, Frank kills Nick with a wrench and fakes a motor accident to cover the crime. He beats Cora up – so that she seems authentically hurt in the crash – has sex with her beside the wreckage and next inflicts some injuries on himself.

The Los Angeles District Attorney, Sackett, is suspicious. He pressures Frank into signing a statement saying that Cora planned and executed Nick's murder – so Cora makes a statement accusing Frank. However, Katz, a tricksy defender, ensures that no charges are brought against Frank and

gets Cora a suspended sentence on the reduced charge of manslaughter.

I began slipping off her blouse. 'Rip me, Frank. Rip me like you did that night.'

I ripped all her clothes off. She twisted and turned, slow, so they would slip out from under her. Then she closed her eyes and lay back on the pillow. Her hair was falling over her shoulders in snaky curls. Her eye was all black, and her breasts weren't drawn up and pointing up at me, but soft, and spread out in two big pink splotches. She looked like the great grandmother of every whore in the world. The devil got his money's worth that night.

THE POSTMAN ALWAYS RINGS TWICE
BY JAMES M. CAIN

Now Frank and Cora are free, and rich from Nick's insurance policies. Their relationship festers, however, since they distrust each other. Frank wants to leave California; Cora wants to stay. Soon they are entangled in blackmail, further violence and infidelities. As Cora says: 'God kissed us on the brow ... We had all that love, and we just cracked up under it.' She tells Frank that she is pregnant and offers him the chance to murder her too. He turns it down. Instead, they marry and spend an idyllic afternoon at a nearby beach. On the drive home to the diner, there is a genuine collision. Cora is killed and Frank is convicted of her murder. As he waits on death row he wonders: 'Do you think she knows I didn't do it? ... Maybe it went through her head, when the car hit, that I did it anyhow.'

Cain uses Frank as first-person narrator. The result is powerfully bleak: description is sparse, dialogue stark, motivation simple. The characters enact their destinies against a barely populated landscape – one with few geographical and no moral reference points. It is a tale, simply, of obsession and an urge to self-destruction. The intertwining of the two is suggested by Cora's repeated pleas to Frank of 'Rip me! Rip me!'

The Postman Always Rings Twice was Cain's first novel. One publisher rejected it on the grounds that at 30,000 words it was 'too short'. Nevertheless, it became an immediate bestseller, and has been filmed three times: by Luchino Visconti (as *Ossessione*, 1942), Tay Garnett (1946) and Bob Rafelson (1981). If Visconti's cerebral and romanticised version is the most acclaimed (and Rafelson's marred by noisy over-acting), Garnett's (from a screenplay by Harry Ruskin and Niven Busch) is truest to the *noir* spirit of the original. John Garfield and Lana Turner play the lovers, conveying – in spite of censorship restrictions – a dangerous passion and intensity.

Agatha CHRISTIE   1890–1976

Murder on the Orient Express

Classic Christie, *Murder on the Orient Express* is one of the most famous of 'Golden Age' detective novels. It is set in the closed environment of a train snowbound in the Balkans, and features the fastidious Belgian detective Hercule Poirot. Retired from the Belgian police force, Poirot is rather vain and proud of his elaborate moustaches. Logical and tidy-minded, he has an unpredictable command of English idioms. He claims to use 'the little grey cells' as his sole aid to investigation, but nevertheless he occasionally descends to a Holmes-like investigation of material clues when method and order are insufficient. His passion for justice, his professional expertise, his meticulous deductions, and his complete self-confidence have made him arguably the most famous fictional detective since Sherlock Holmes.

An odiously rich American tycoon, Ratchett, is murdered in his sleeping compartment on the Orient Express. He has been stabbed to death, and each of the occupants of the other compartments on the coach is found to have a motive for his murder. It is established by Poirot that Ratchett had been connected with a famous case of kidnapping in the United States; it dawns on him that:

'For so many people connected with the Armstrong case to be travelling by the same train by a coincidence was not only unlikely, it was *impossible*. It must be not chance, but *design* ... A jury is composed of twelve people – there were twelve passengers – Ratchett was stabbed twelve times. And the thing that had worried me all along – the extraordinary crowd travelling in the Stamboul–Calais coach at a slack time of year – was explained.'

Before working this out, Poirot has been led a devious dance by Christie's improbable cast of twelve, which comprises several nationalities: an English valet, an English colonel, a Swedish nanny, a Russian princess, a Hungarian count and his countess, several Americans, and various European supernumeraries. The characters are pretentious and cosmopolitan almost to the point of caricature, but the plot is deft, complex and genuinely baffling. Raymond Chandler, exasperated by the solution, commented: 'only a half-wit could guess it'. It is in the class of surprise denouements that occur in other famous teasers – *The Murder of Roger Ackroyd* (1926), *Ten Little Niggers* (1939), *Endless Night* (1967), and, of course, *Curtain* (1975). *Murder on the Orient Express* is unusual in that the murderers are not brought to judicial reckoning or death; but the moral principle remains that they are found out, obliged to recognise their actions, and are brought to account by Poirot. Like many of Christie's books, *Murder on the Orient Express* was made into a highly successful film (1974), directed by Sidney Lumet from a screenplay by Paul Dehn.

1934

E.M. DELAFIELD — 1890–1943

The Provincial Lady in America

see **Diary of a Provincial Lady** (1930)

James T. FARRELL — 1904–1979

The Young Manhood of Studs Lonigan

see **Studs Lonigan** (1935)

F. Scott FITZGERALD — 1896–1940

Tender is the Night

When *Tender is the Night* first appeared, its tone was ill-suited to a reading public in the throes of the Depression. The novel captures the spirit of the decadent Jazz Age, the wealth and dissipation of an American elite. The narrative was revised by Fitzgerald several times; he experimented with scenes on luxury cruise liners and in high-class hotels, before deciding on the structure as it finally appears in the edition of 1948.

Dick Diver, a brilliant young psychoanalyst, is 'already too valuable to be shot off in a gun' at the time of the First World War. Sent to Zurich to complete his studies, he becomes involved with the schizophrenic Nicole Warren, whose adolescence has been destroyed by her father's sexual abuse of her. Baby Warren, Nicole's sophisticated elder sister, is looking for a reliable husband for the unstable girl, an arrangement which she expects to be able to purchase through the enormous family wealth she now controls.

Dick's motives are emotional not mercenary: he wants to love and to be loved, and Nicole's vulnerability appeals to his protective instincts. Their marriage, however, traps them in the roles of doctor and patient, and, as Nicole's mood swings between elation and despair, he is unable to treat her as an adult. One summer on the French Riviera he meets Rosemary Stoyt, a beautiful but immature

eighteen-year-old film-star; as an affair develops, Nicole looks on disconsolately.

Dick's career, meanwhile, is being jeopardised by his drinking habits, and over the next few years his mental and physical prowess fails. Nicole's father becomes his patient temporarily, but then refuses treatment and disappears. Dick's partner in the Swiss clinic forces him to sell his share, and the Divers drift through Europe, mingling with high society, although their wealth can no longer buy them respect. Meeting Rosemary again, Dick finds that he is now pitied by both his wife and his mistress, and determines to bring the situation to a climax.

Nicole, gaining a sense of her own identity as Dick sinks into depression, takes her admirer Tommy Barban as a lover. Tommy forces their tangled relationships into a denouement, claiming Nicole from Dick, who decides to depart quietly. 'I'm going to him,' Nicole pleads with Tommy, seeing Dick about to leave: '"No, you're not," said Tommy, pulling her down firmly. "Let well enough alone."' After the divorce, Dick slides into obscurity, unable to sustain his once brilliant reputation, while Nicole, now happy in her new marriage, tries to help him with rejected offers of money. 'When she said, as she often did, "I loved Dick and I'll never forget him," Tommy answered, "Of course not – why should you?"'

Lewis Grassic GIBBON — 1901–1935

A Scots Quair

Sunset Song (1932)
Cloud Howe (1933)
Grey Granite (1934)

Gibbons' trilogy (collected under this title in 1946) follows the personal quest of Chris Guthrie, a Scotswoman, to resolve the conflict of either remaining in the country and accepting its narrow and limited philosophy, but also its powerful allure, or betraying his inheritance of her ancestors, for a wider but alien, hostile urban existence. The novel as a whole is narrated in a colloquial style, reflecting local speech patterns, and alternates between the second and third person to distance or involve the reader by turns.

Sunset Song opens in Kinraddie, with a 'Prelude', 'The Unfurrowed Field', which relates the history of the area (the Mearns, of Kincardineshire, a former county of eastern Scotland) and describes past and present inhabitants, now mainly crofters. Seen by all the generations is an ancient circle of standing stones, representing the changeless and enduring nature of the land. The novel itself begins in 1911 at Blawearie, a particularly desolate smallholding, only recently occupied by John Guthrie, a newcomer from Barmekin. Embittered by years of struggling to make a living from the soil, he is gripped by a Calvinistic faith, and a selfish passion for his wife, Jean. She has borne him four children in succession and, against her wishes, conceives twins after an interval of seven years. Guthrie is despised by Will, his sixteen-year-old son to whom he is violent, and barely tolerated by Chris, his fourteen-year-old daughter. She in turn is ridiculed by her father, who detests her education, and her prospect of a university career. Broken in spirit, and once again pregnant, Jean kills the twins and commits suicide.

And then a queer thought came to her there in the drooked fields, that nothing endured at all, nothing but the land she passed across, tossed and turned and perpetually changed below the hands of the crofter folk since the oldest of them had set the Standing Stones by the loch of Blawearie and climbed there on their holy days and saw their terraced crops ride brave in the wind and sun. Sea and sky and the folk who wrote and fought and were learnèd, teaching and saying and praying, they lasted but as a breath, a mist of fog in the hills, but the land was forever, it moved and changed below you, but was forever, you were close to it and it to you, not at a bleak remove it held you and hurted you.

A SCOTS QUAIR
BY LEWIS GRASSIC GIBBON

Chris is forced to abandon her education and remain with her father. Her sexuality is partly awakened by her response to nature, but she is tormented by the knowledge that the crofting women are destroyed by constant childbearing and by despotic husbands. Gossip centres on private lives, and is encouraged by the local minister, who falsely believes Will has made a tinker girl pregnant. Eventually, unable to live with his father and unwilling to inherit the harsh responsibilities of the croft, Will elopes abroad with another woman. Education has made Chris aware of what she could achieve: a refined world which includes English values, but is irrelevant to her crofting life. Kinraddie as a community survives by expressing contradictory values. Religious hypocrisy and troublemaking co-exist with loyalty to and support of those in crisis. Only through the death of her father is Chris able to determine her own future. By marrying Ewan, a Highlander and outsider, she takes the first step towards separating herself from her neighbours. While each asserts their own wary independence, the continuance of a rooted lifestyle in the croft seems assured after they have a son, also called Ewan.

The outbreak of the First World War abruptly propels Kinraddie into recognising a world it has hitherto wished to ignore. War prompts discussion of socialism, of those who exploit circumstances to become rich, and it returns men from the Front, disillusioned with its purpose. Despite the Conscription Act, Ewan as a farmer is exempt but, tired of taunts of cowardice, he departs for the Front, without telling Chris. As the war progresses, many of the inhabitants of Kinraddie repudiate the English patriotic values they were told to accept, and disbelieve the anti-German ranting of their minister. This instinctive understanding of the uneducated, which also repudiates the war propaganda of an English-dominated government, is symbolised by Ewan, who had enlisted unwillingly, from a sense of duty. Duped by indoctrination, he returns from France on leave fired with enthusiasm for the war, drunk and unpleasant, a hostile stranger to his wife. He is later shot as a deserter, realising too late his mistake in fighting for a king and country that had no part in his life and was alien to Kinraddie, where he truly belonged. When Chris learns what has happened, she understands that he deserted because of her, and their love is reaffirmed in death.

Sunset Song ends with an 'Epilude' also called 'The Unfurrowed Field', which prepares us for *Cloud Howe*. War has destroyed the old life of Kinraddie and removed familiar characters, including the minister, who vacates the pulpit. He is replaced by Robert Colquohoun, a socialist, who has fought in the war and is considered Bolshevik by his parishioners. After a short romance, he marries Chris, and she is again alienated by the community of Kinraddie, who regard her as an opportunist. Colquohoun rejects the weeping angel that is proposed as a war memorial, and instead has commemorative lettering cut on a standing stone to honour the Kinraddie dead. His speech at the ceremony explains the title of the first book. Those who died were the last of a particular kind of Scotsman, who 'in the sunset of an age and epoch ... died for a world that is past'. Thinking of a new and, it is implied, socialist future, Colquohoun suggests that those who died would have 'from the places of the sunset' wanted a 'greater hope and a newer world'.

This is the hope for *Cloud Howe*, which is prefaced by a 'Proem' delineating the people and history of Segget, a small mill village where Colquohoun comes as minister. Chris's search for fulfilment remains unrequited, as it did in Kinraddie. The grim unity of those who farmed and understood the seasons gives way to disjointed small-town life, corrupted by greedy property owners who maintain pitiable slums. Colquohoun's religion is not shared by his wife, who thinks such beliefs are simply pillars of cloud followed by men who are doomed. Ideologically separated from Robert, she senses that his efforts to improve material conditions in Segget will fail. As in *Sunset Song*, descriptions of the hopelessness of the millworkers' circumstances are relieved by sketches, often humorous, of local life. The premature birth and death of a son on the day of the strike, which Robert supports, epitomises the gulf between Chris and Robert, whom she never deeply loves. As the country falls into economic decline during the MacDonald Coalition Government, Robert is powerless to help the poor; the visions of Christ he 'sees' are a compensatory effort to believe in the goodness of mankind, but is shocked into reality when he hears of an evicted family whose baby was killed by rats. He neglects his own health, which is declining owing to lung damage caused by gas in the war. *Cloud Howe* ends with Robert dying in the pulpit, while attempting to preach an angry sermon about needless death in an alleged Christian world. Always independent, Chris is undisturbed by the satisfaction given to some by Robert's death, and her loss of social position and income.

Unlike the first two book of *A Scots Quair*, *Grey Granite* has no prelude or postlude. Chris and her son move to the city of Duncairn, whose inhabitants have no common background and therefore no communal history. Ewan, once obsessed with the past and determined to become an archaeologist, gives up his education to become an apprentice in a steel mill. Chris helps to run a boarding-house where they live, but Duncairn is a disaster, and forces mother and son apart. Moved by the poverty and unemployment of the 1930s, Ewan becomes a devout Communist and an agitator at his work. Chris reminds him of the endless search of men to find a faith, and for her Ewan's principles, like Robert's religion, are 'just another dark cloud'. All the men close to Chris fail to understand that as a woman she has a need for the primitive, and accepts life without any compulsion to find an explanation for it. Isolated in the dreary city, Chris marries desperately and briefly. Ake Ogilvie, an acquaintance from Segget, gives her capital and leaves for Canada after deciding that the marriage was a mistake. Chris has no reason to remain in Duncairn. She agrees to lead a life separate from Ewan and completely withdraws from social activity. Her acceptance of her fate of isolation is confirmed by a cyclic return to the croft in Barmekin where she was born, and where she will die, understanding at last not to fear 'Change ... whose right hand was Death and whose left hand Life'.

Robert GRAVES   1895–1985

I, Claudius

Claudius the God

The hairy fifth to enslave the State,
To enslave the State, though against his
    will,
shall be that idiot whom all despised …

Thus, according to Robert Graves, ran part of a curious Sybilline prophecy about the emperors of Ancient Rome, which came into the possession of the Emperor Claudius (AD 41–54). He was himself 'the hairy fifth', the lame, stammering member of the imperial family who, despised by his relations as an idiot, eventually succeeded to the vacant throne when they had all been killed off in decades of fratricidal warfare. In *I, Claudius* and its successor, *Claudius the God*, Robert Graves wrote two of the most resonant historical novels of the century, based on the material provided by the Roman historians Tacitus and Suetonius and several other ancient authors. Claudius, himself an accomplished historian, narrates his own story in a secret autobiography which he intends to leave around for remote posterity to read.

In impersonating Claudius, Graves displays a scholarly and imaginative understanding of the mores and ethos of the Roman world, and sensationalises his material only as much as must be expected from a novelist with a living to earn. Thus, in *I, Claudius* he builds on the dark hints of Tacitus to portray Livia, wife of the Emperor Augustus and Claudius's grandmother, as a systematic poisoner, who kills off the imperial family wholesale so that her son, Tiberius, can succeed to the Purple, an interpretation by no means accepted by modern historians. Graves even makes Livia poison Claudius's childhood sweetheart, whom he is about to marry. Yet Graves's people are real Romans, and Claudius's autobiography is credible as the product of a Roman mind. Claudius is not credited with modern ideas of human psychology: he divides his contemporaries into noble and depraved characters, with some who are mixtures of good and evil in various proportions (his own family, for instance, are divided into the 'good' and 'bad' Claudians). The drama of the novel lies in watching how the evil members of the imperial family contrive the disgrace and death of the noble and statesmanlike, only to fall victim inevitably in their turn. Claudius, sympathetic to the forces of good but disqualified from the power-game because he is agreed to be an idiot, contents himself with writing histories and has a quiet sympathy for a return of the Republic. He tells us his story (which he is writing in Greek) largely in long passages of dignified historical narrative, although diversified by some telling and often witty stretches of *oratio recta*. ('"Men such as Antony, real men, prefer the strange to the wholesome," Livia finished sententiously. "They find maggoty green cheese more tasty than freshly pressed curds." "Keep your maggots to yourself," Octavia flared at her.')

I, Tiberius Claudius Drusus Nero Germanicus This-that-and-the-other (for I shall not trouble you yet with all my titles), who was once, and not so long ago either, known to my friends and relatives and associates as 'Claudius the Idiot', or 'That Claudius', or 'Claudius the Stammerer', or 'Clau-Clau-Claudius', or at best as 'Poor Uncle Claudius', am now about to write this strange history of my life.

I, CLAUDIUS
BY ROBERT GRAVES

Claudius's humiliations reach their height during the brief reign of his demented nephew Caligula, who throws him into the river Rhone at Lyons, whereupon Claudius clambers out dripping and quotes Homer to him, because he must flatter Caligula's belief that he (Caligula) is a God. But Caligula is murdered, and the Praetorian Guard, finding Claudius hiding behind a curtain, proclaim him Emperor ('"You'll be all right, sir, once you get accustomed to it," Gratus said grinning. "It's not such a bad life, an Emperor isn't."'). Graves leaves Claudius riding on their shoulders and reflecting that at last he will be able to get people to read his history books.

Claudius the God begins where the earlier novel ended, but it quickly breaks off from this exciting point for Claudius to tell the story (in four chapters of around sixty pages) of his friend, the Jewish king Herod Agrippa, who is to play an important role in the book. Perhaps this deliberately un-novelistic technique is intended as an imitation of the historical Claudius's own writing, which, if fragments of surviving speeches are anything to go by, was marked by a certain honest awkwardness. The material of this second novel is less intrinsically dramatic than that of *I, Claudius*, because it deals with the successful reign of a just emperor. But Graves once again holds the reader's attention with his mastery of classical detail and significant incident.

After telling of Herod, Claudius goes back to how attempts to contest his new-found authority were beaten off, and how he stifled his own sincere Republicanism, always intending to restore the Republic at some suitable opportunity. He then goes on to deal with his methods of rule and reform. For the Roman historians Tacitus and Suetonius, Claudius was a fool, and for Suetonius he was naturally cruel: Graves does not falsify the historical record – he shows Claudius as ready to execute senators by the handful and to order gladiators' throats to be cut if he believes they are shamming serious injury – but he also allows Claudius to demonstrate that he is a capable and humane emperor. We hear of Claudius's public works, such as the building of the harbour of Ostia, of his successful conquest of Britain, of his liberal attitude to provincials, of his allowing himself to be mocked in the Law Courts. Graves gives him disquisitions on many topics of interest, such as the origin of the Celts and Roman poetry, where he shares Graves's own distaste for Virgil, preferring Ennius.

The main emotional interest of the book lies in Claudius's relations with Herod Agrippa and with his third wife, Messalina, in both of whom he puts total trust, only to find himself betrayed. Claudius confirms Herod in several Jewish kingdoms, but Herod develops the belief that he is the Messiah (the beginnings of Christianity are also mentioned in the novel), and plots to take the East from Rome; however, he dies as he is about to proclaim his Godhead. Claudius is deeply in love with the much younger Messalina, but, unknown to him, she commits adultery on a large scale, and eventually stages a feeble *coup d'état* with an aristocratic lover, which leads to her unmasking and execution. This takes place in AD 48, near the end of the novel. After Messalina's death, Claudius, although now worshipped as a god by the Britons, loses hope that the empire can be ruled justly, and believes that he must sit it out like 'Old King Log', allowing matters to get much worse before they can become better and the Republic be restored. He marries his vicious niece, Agrippinilla, and prefers her equally vicious son, Nero, to his own son, Britannicus. At the end of the novel, she is preparing to murder Claudius, and Claudius's plan to save Britannicus (later murdered by Nero) by hiding him in Britain is foiled. The novel ends on a tragic note as Claudius, shortly to die, lays down his pen with the words: 'Write no more now, Tiberius Claudius, God of the Britons, write no more.'

The novels were to have been filmed by Alexander Korda as 'Claudius', with Charles Laughton in the title role and Josef von Sternberg as director. Almost from the start, the project ran into difficulties. Graves was appalled by Carl Zuckmayer's screenplay – 'absolute cheap nonsense strung on historical absurdities' – and was hired to rewrite it himself. Von Sternberg attempted to produce a shooting script using both versions, and then, just as the film went into production in early 1937, the female star, Merle Oberon (who was also Korda's wife), was involved in a serious car crash. The film was eventually abandoned, and is considered by some film historians as the greatest picture never made. The books were subsequently adapted by Jack Pulman for a hugely successful television series transmitted by the BBC in 1976.

James HILTON   1900–1954

Good-bye, Mr Chips

Hilton's brief, telegraphic novella, which first appeared in *The British Weekly* magazine, is a sentimental paean to the schoolmastering profession and to the place of the public schools in English history. It consists of the fireside reminiscences of the eponymous schoolmaster, now retired and living in a cottage at the school gates with his housekeeper, a former domestic employee of the school called Mrs Wickett. Mr Chipping's career at Brookfield ('a good school of the second rank') spans forty-three years (1870–1913) and is recounted against the tickertape of history: 'Strikes and lock-outs, champagne suppers and unemployed marchers, Chinese labour, tariff reform, H.M.S. *Dreadnought*, Marconi, Home Rule for Ireland, Doctor Crippen, suffragettes, the lines of Chatalja ...' Those who visit the crusty old man, who blends his own tea and serves walnut cake with pink icing, imagine that he is a bachelor. However, at the age of forty, just as he had 'begun to sink into that creeping dry-rot of pedagogy that is the worst and ultimate pitfall of the profession', he had met a young radical called Katherine during a Lake District holiday. Their marriage was brief, but happy, for she tempered his conservatism and won the hearts of both colleagues and pupils, before dying in childbirth. When the First World War breaks out, the ancient Chips is asked to help out and responds 'with a holy joy in his heart', eventually becoming acting headmaster. On his death-bed he overhears a master saying that it is a pity Chips never had children, to which his celebrated response is: 'Yes – umph – I have ... Thousands of 'em ... thousands of 'em ... and all boys ...'

The book was written at a time when the Romilly brothers were launching their violent assault upon the public schools in the magazine *Out of Bounds*. Hilton's defence of the traditional values of decency, conformity, loyalty and patriotism harked back to the complacency of Ian Hay's 1914 bestseller *The Lighter Side of School Life* ('If this be mediocrity, who would soar?'); it was curiously old-fashioned but perhaps reassuring in the face of growing fears about another war. If Chips seems to exemplify what E.M. Forster diagnosed as 'the undeveloped heart', there is no doubt that Hilton succeeded in creating an endearing and enduring archetype, humble, affectionate but firm, admired for what he stands for rather than for his achievements.

Good-bye, Mr Chips has been adapted for the cinema twice by two writers eminently suited to the task. The 1939 film, written by R.C. Sheriff and directed by Sam Wood, is far superior to the one directed by Herbert Ross in 1969, which was written by Terence Rattigan, but tricked out with forgettable songs in an ill-advised attempt to turn the story into a musical.

Storm JAMESON   1891–1986

Company Parade

see **Love in Winter** (1935)

F. Tennyson JESSE   1888–1958

A Pin to see the Peepshow

By the time Jesse came to write her celebrated novel, she was already an expert criminologist, having studied law under the Attorney-General, Sir Rufus Isaacs, and edited two volumes of the *Notable British Trials*. *A Pin to see the Peepshow* is a substantial novel (over 400 pages) closely based on actual events which culminated in the trial and execution in 1923 of the twenty-nine-year-old Edith Thompson and Frederick Bywaters, her lover, and eight years her junior. They were both found guilty of murdering Mrs Thompson's husband (stabbed in the open street by an overwrought Bywaters), even though Mrs Thompson almost certainly knew nothing of her lover's intentions. It is now generally agreed that she was in effect hanged for having committed adultery with a man younger than herself.

Despite its conventional claims to purely fictional status, *A Pin to see the Peepshow* offers in the person of its protagonist, Julia

1934

Starling (*née* Almond), an imaginative reconstruction of the influences and impulses which bring Julia/Edith to the gallows. It also provides a vividly detailed, lovingly recreated portrait of a period of flux in English social (and especially sexual) mores.

Julia picked the glass up and began to drink. As the fluid became less she found she was staring at Herbert through the bottom of the tumbler. Like all short-sighted and astigmatic people, she could see better even when looking through the bottom of a tumbler, or between two finger-tips held close together against her eyes, or between the chinks in the brim of a coarsely-woven straw-hat, than she could without anything to narrow down and focus the field of her vision. Now she stared at Herbert and saw him distorted through the curved glass at the bottom of the tumbler, but much more clearly and sharply than she could have if she had been looking at him in the ordinary manner. There was a little flaw in the bottom of the glass, and she sipped very slowly so as to be able to see his face as it changed, as the flaw caught it now at one place, now at another. Quite suddenly the whole of his lower jaw would swell out till his face looked like that of a hippopotamus, a tiny tilt of the tumbler, and he would have practically no jaw at all, only a thin little slit of a mouth, and a bulbous forehead.

A PIN TO SEE THE PEEPSHOW
BY F. TENNYSON JESSE

Julia Almond is the only child of two lack-lustre, lower-middle-class Londoners whose reverence for respectability is strong but who have little genuine moral sense and therefore instil none in Julia. Patchily educated but a voracious reader, the young Julia rapidly learns to live mainly in a dream-world of escapist fantasy fuelled by third-rate novelists for whom love – always noble, often adulterous – is woman's whole existence.

In her imagination Julia is always centre-stage in a perpetually unfolding drama of her own creation: for her the boundaries between fact and fantasy are often dangerously blurred, and the ease with which she can summon her dream-worlds is matched – or caused? – by a frequent failure of imagination in her emotional dealings with other people. Hence her disastrous marriage to the widowed Herbert Starling, whose unwanted suit she finally accepts because it solves family problems caused by her father's death. The mistake is more easily made because, though well read in novelettish notions of love, she is ignorant of genuine passion and of her body's capacity for sexual pleasure. (The novels have been strong on sentiment, weak on physiology.)

When Leonard Carr, a young sailor and distant relation, enters her family circle, she is powerfully drawn to him. Her junior in age, he is years older in experience, and they become lovers. Of necessity much of the affair is conducted by correspondence, and in her letters Julia gives full rein to every extravagant and melodramatic sentiment, including fantasies about murdering the increasingly enraged and jealous Herbert. She destroys all Leonard's letters. He, with meanly calculating motives and subsequently devastating results, keeps hers and they eventually provide damning evidence against her.

Herbert's death is the deed that ends the dream Julia has lived, and the last five chapters of the book describe, in often harrowing detail, Julia's bewildered, and eventually sedated, struggle to disentangle fact and fantasy, shadow and substance, during her interrogation, remand, trial, imprisonment, and final hours of life. Other people's fantasies – those of prosecuting counsel, jury, judge, newspapermen – make her struggle harder, and Jesse is at pains throughout the book to identify socially sanctioned nonsenses: that marital rape is 'love'; that an eighteen-year-old youth who has fornicated his way around the Mediterranean must be less experienced than a woman whose first lover he is; that adultery in a woman is equivalent to murder.

Class, also, has its own double standard. Julia's destiny, Jesse insists time and again, would almost certainly have been very different had she been born (as were her employers at the fashionable dress-shop whose business she manages so astutely) into an upper class where condoned adulteries, female freedom of movement, combined with financial independence, and ever-easier divorce were becoming commonplace.

Edith Thompson's execution, and well substantiated rumours of appalling events surrounding her death, contributed greatly to growing revulsion for capital punishment. *A Pin to see the Peepshow* provided additional impetus for the campaign to abolish it.

Henry MILLER   **1891–1980**

Tropic of Cancer
1934

Tropic of Capricorn
1939

'What strikes me most, rereading *Cancer* after thirty years', wrote Colin MacInnes, 'is that it is a prophetic book: a warning of what deadens life, an affirmation that it can yet be lived, though with extreme difficulty, in an age whose sterile non-cultures seek to thwart all mainsprings of fertility ... For twenty years [Miller] lived in appalling material conditions to discover his own truth and he endured obloquy and denigration from the ignorant and condescending. His reputation is now well into harbour because life has caught up with art, and a new generation see the value of his prophecies.'

Tropic of Cancer was first published in Paris by the Obelisk Press in 1934, but banned in Britain for nearly thirty years, finally appearing in 1963. It is set in an inter-war Paris of Russian emigrés, American expatriates and whores, all outlaws. 'It is no accident that propels people like us to Paris. Paris is simply an artificial stage, a revolving stage that permits the spectator to glimpse all phases of the conflict.' Though classed as a novel it is really a fictionalised autobiography, episodic and plotless, the incidents selected and arranged not haphazardly, as in life, but so as to develop a coherent narrative and support Miller's argument: 'For a hundred years or more ... *our* world has been dying. And not one man

... has been crazy enough to put a bomb up the asshole of creation and set it off.'

It is a surreal world, spilling over with disgust. Roaming the street alone, his belly empty, Miller observes: 'The city sprouts out like a human organism diseased in every part, the beautiful thoroughfares only a little less repulsive because they are drained of their pus.' People swim in and out of view: Boris, who discovers he is 'lousy'; Tania, for whose sake Miller 'would become a Jew'; the 'cunt struck' Van Norden, with whom he shares a room in a seedy hotel where Maupassant once lived. 'When he opens the door of 57 I have for a fleeting moment the sensation of going mad.' Van Norden's sex is joylessly obsessive; watching him 'tackle' a whore, Miller observes: 'As long as that spark of passion is missing there is no human significance in the performance ... these two are like a machine which has slipped its cogs. It needs the touch of a human hand to set it right. It needs a mechanic.' There are others, including the weak-willed Fillmore, scion of a wealthy family back in America, who has to be saved from an unfortunate marriage. But when, after a winter's purgatorial exile in Dijon, Miller returns in the spring to Paris and by a lucky chance finds himself in possession of 'exactly 2875 francs and 35 centimes', for a moment the city looks different. 'So quietly flows the Seine that one hardly notices its presence. It is always there, quiet and unobtrusive, like a great artery running through the human body.' He concludes: 'Human beings make a strange fauna and flora. From a distance they appear negligible; close up they are apt to appear ugly and malicious. More than anything they need to be surrounded by sufficient space – space even more than time.'

Tropic of Cancer is in the tradition of European bawdry – its liberal use of obscenities was what most worried the censors of the 1930s – but it is also an essentially American book and thoroughly of its time, conveying sharply the desperation of the inter-war years. It is sprawling, repetitious, often infuriating; but one cannot deny Miller's stamina and courage in writing it.

The companion volume, *Tropic of Capricorn*, explains something of the narrator's American background, and his preparations for becoming an artist. It details his work with the Cosmodemonic Telegraph Company of North America, and memories of childhood, including the day he and a friend murdered a member of a rival gang: 'It had a slight taste of terror in it which has been lacking every since.' Miller writes: 'Cancer is separated from Capricorn only by an imaginery line ... You live like a rock in the midst of the ocean; you are fixed while everything about you is in turbulent motion ... To radiate goodness is marvellous, because it is endless and requires no demonstration.' The critical reputation of the books lies somewhere between feminist dismissal of Miller as a misogynistic, sex-obsessed fraud and Lawrence Durrell's absurdly hyperbolic claim that 'American literature begins and ends with the meaning of what Miller has done.'

| George ORWELL | 1903–1950 |

Burmese Days

Burmese Days was Orwell's first novel. Set in Kyauktada, a trading outpost of the British empire, it tells the story of Flory, a prematurely middle-aged timber merchant, mild socialist and reader of serious literature. Flory hates 'the atmosphere of imperialism', and the Anglo-Indian bores who generate it. Nevertheless, he has remained in Burma for ten years, despising himself for his participation in the heavy drinking, casual whoring and brutal racism which characterise colonial life. A facial birthmark functions as the symbol of his moral disfigurement while reinforcing his alienation from his compatriots. His closest friend is an Indian, Dr Veraswami, and in the all-white expatriate club Flory is referred to as the 'nigger's Nancy Boy'.

At the start of the novel, U Po Kyin, the villainous local magistrate, is plotting to destroy Veraswami's reputation. This he attempts to do by circulating anonymous letters alleging that the doctor is corrupt. Seeking to protect his friend, Flory nominates him for membership of the club, which high officialdom has instructed to admit a token 'native'. If Veraswami is accepted, his augmented prestige will safeguard him from U Po Kyin's libels. U Po Kyin therefore determines to destroy Flory as well.

In the middle of these machinations, beautiful, shallow Elizabeth Lackersteen arrives. She is looking for a husband and Flory is the only prospect. A romance of misunderstanding blossoms. Flory believes that Elizabeth can offer him intellectual companionship; Elizabeth loathes Flory's ideas and conversation but admires his courage – particularly, his hunting skills. Flory banishes his Burmese mistress, and his engagement to Elizabeth seems imminent. Then Lt. Verrall, a minor aristocrat, comes to Kyauktada. Encouraged by her snobbish aunt, Elizabeth jilts Flory and transfers her attentions to the lieutenant. He, however, is obsessed with polo ponies, physical fitness and little else, and duly makes his escape. Flory now plays what Elizabeth perceives to be a heroic part in quelling an indigenous rebellion (covertly organised by U Po Kyin). Once again, it appears inevitable that Flory will propose to Elizabeth and that she will accept. The malignant magistrate, however, bribes Flory's former mistress publicly to denounce Flory. As much in self-disgust as desolation, Flory commits suicide. U Po Kyin briefly prospers, Veraswami is ruined, and Elizabeth marries a dull Scottish administrator.

Flory's undoing is his ambivalence. Too thoughtful to accept the arrogant banalities of the empire, he is simultaneously too weak to reject them. *Burmese Days* shares his confusion. On the one hand, the colonialists are cruel and vulgar; on the other, the Burmese are either devious and avaricious or fatalistic and incompetent. Orwell's indictment of the rulers is undercut by his less explicit contempt for the ruled, and Flory's final reaction of despair enacts the author's violent uncertainties. Indeed, the novel's least troubled passages occur when Orwell abandons his blurred introspections for the clearer vision of reportage and journalistic aside.

Orwell served in the Indian Imperial Police in Burma from 1922 to 1927. Reflecting on that period, he wrote: 'With one part of my mind I thought of the British Raj as an unbreakable tyranny ... with another part I thought that the greatest joy in the world would be to drive a bayonet into a Buddhist priest's guts.'

Henry ROTH   1906–

Call It Sleep

The Schearls – Albert, Genya and their son, David – are Austrian Jews. Albert has emigrated to New York, where he has found work as a printer, and in May 1907 he is joined by his wife and twenty-one-month-old child. The family's reunion, on the deck of the *Peter Stuyvesant*, is tense with suppressed recrimination. In a fit of anger, Albert hurls his son's hat into the sea.

The story moves forward five years. Through David's shrewd but bewildered observations, it describes the protagonists' relationship to each other: David's clinging dependence on his mother; his terror of his unpredictable, dangerous father; and his mother's nearly limitless patience towards her son and her husband. David has a few friends: some from the neighbourhood, some from the Hebrew classes he attends. He finds these other boys charmless, however. Genya withdraws from her neighbours; Albert is acutely sensitive to real and imaginery slights. Indeed, although living in Brownsville and, later, the even more bustling lower East Side, the Schearls remain isolated and introspective, their present overshadowed by events that occurred in their pre-emigré existence.

Genya's sister, Bertha, arrives from Austria. Loud and vulgar – and loathed by Albert – she pesters Genya to confide the story of a scandalous romance in Genya's past. This Genya does, and David half hears and half understands. He gleans, too, that his father is guilty of a violent and unspecified betrayal. Bertha marries a Russian widower. David, meanwhile, grows increasingly preoccupied with the partial enlightenment he has received and associates it with the story of the Prophet Isaiah, whose lips an angel touched with burning coal. Wandering by the docks, David learns that he can create a spark by inserting a piece of metal into the tramlines. The act assumes a mystical significance, suggesting to the boy a glimpse of all the truths that are denied to him. After a cathartic family quarrel, David returns to the rails and once more creates a spark – this time electrocuting himself. Close to death,

he has a vision of the vitality and chaos around him. He is at last receptive to the energies beyond his home, while his parents, shocked by his almost fatal mishap, realise afresh his preciousness to them.

Oblique in method, *Call It Sleep* presents the history of a consciousness imperilled by its own fixations: for, as long as David idealises his mother and dreads his father, his emotions are precariously stable. It is only when he begins to understand that Albert and Genya are mortally damaged that his balance collapses into confusion. This 'rite of passage' is both traumatic and liberating, and the intensity of the novel's concentration upon it is allied to a highly charged prose style. Immediate sights and sounds are magnified: background disappears. The resulting narrative has the quality of dream and, if it describes a world of vivid fragments, it also evokes with relentless accuracy the gigantic fears, extreme attachments and dreadful misconceptions of childhood.

Call It Sleep, which has been described as one of the 'greatest achievements of American writing this century', received little attention until it appeared in a paperback edition in 1964. Roth was rediscovered, but did not publish another novel until 1994, when a *Mercy of a Rude Stream*, the first volume of a projected sequence of six novels, appeared.

Evelyn WAUGH   1903–1966

A Handful of Dust

During a gruelling and miserable trip to Brazil in 1932, undertaken with 'a heart of lead' after the collapse of his first marriage, Waugh met a religious maniac and had the idea for a story, in which a man is held captive by a Dickens-loving lunatic. This story eventually served as the denouement ('a "conceit" in the Webster manner') of his fourth novel: 'I wanted to discover how the prisoner got there and eventually the thing grew into a study of other sorts of savages at home and the civilized man's helpless plight amongst them.'

The novel opens amidst these savages at the house of Mrs Beaver, a voracious interior designer involved in vandalising London property, carving up elegant houses to make brash *pieds-à-terre* for the adulterous. Her son, John, a dull, parasitic man-about-town, visits Hetton Abbey, a vast 1860s Gothic pile where 'the civilized man', Tony Last, lives in uncomfortable seclusion with his wife, Brenda, and their young son, John Andrew. The Lasts dread the prospect of this guest, but for Brenda he brings a whiff of the London social life she misses. On her next visit to London, Brenda meets Beaver once again and before long they have embarked upon an affair. Brenda acquires one of Mrs Beaver's flatlets and tells Tony that she is starting a course in economics. He remains in the country, missing Brenda and only seeing her when she blows in with her shrieking coterie of women friends led by Lady Polly Cockpurse. After the death of John Andrew in a hunting accident, Brenda feels free to dissolve her last links with Hetton and Tony. Gentlemanly beyond the call of duty, Tony agrees to allow Brenda to divorce him and spends a farcical weekend in Brighton with a prostitute and her appalling, gap-toothed, eight-year-old daughter in order to provide grounds for the petition.

However, when Tony learns that Brenda has been advised to seek a settlement so large that he would have to sell Hetton to pay her, he refuses to co-operate and goes on an expedition to South America with an incompetent explorer called Dr Messinger, who intends to charm the natives with clockwork mice. While Tony penetrates deeper and deeper into the Amazon jungle in search of a legendary city, Brenda and Beaver drift apart. Messinger is drowned and Tony, racked with fever, stumbles upon the settlment of Mr Todd, a sinister, philoprogenitive old half-caste with a passion for Dickens, who nurses him to health and then refuses to help him find his way back to civilisation. Tony is last seen embarking upon yet another reading of *Little Dorrit*. Back in England Tony is presumed dead, Brenda marries her old flame, Jock Grant-Menzies, and Hetton passes to another branch of the family and becomes a silver-fox farm.

The novel is complex and ambiguous, clearly influenced by the failure of the author's own marriage. Although Brenda is condemned out of hand for her triviality and selfishness, part of the blame is put upon the 'madly feudal' Tony. Hetton may be preferable to Mrs Beaver's flatlets but, far from being a genuine architectural treasure, it is a Victorian fantasy, described in the county guide book as 'devoid of interest'. It serves as a reminder of Tony's immature and naïve preoccupation with a romanticised past which so disastrously blinds him to the realities of the present.

Thornton WILDER   1897–1975

Heaven's My Destination

Wilder's picaresque satire (subsequently published in the USA in 1935) is the quixotic tale of George Marvin Brush, traveller in school textbooks, follower of the pacifist teachings of Gandhi and Tolstoy, and self-styled evangelist. On his travels across the Midwest he follows the advice of the course he took at college entitled 'How to approach strangers on the subject of Salvation', berates women for the ungodly practice of smoking cigarettes, and gets into arguments when he inscribes biblical messages on hotel blotting pads.

On the day of his twenty-third birthday George resolves to follow Gandhi in taking a vow of voluntary poverty and goes to the bank to withdraw his money so that he can use it for worthy purposes. The bank manager calls the police when George refuses on principle to take the interest which has accrued on his $500; these are the days of the Depression with banks frequently closing down, and the policeman who runs George out of town immediately withdraws his own money from the bank believing George knows something he doesn't. At every turn George's ideals spark fear and suspicion in others. Later he is arrested, having given money to a burglar who was holding up a store, and an uncomprehending courtroom looks on while George explains Gandhi's principle of *adhimsa*, which advocates non-violence to all.

Underpinning these peripatetic adventures is George's quest for Roberta, the girl with whom he committed his single lapse from chastity one night in a Kansas barn. He is intent upon marrying her and establishing an 'American Home' filled with children. He finally tracks her down working in a restaurant, but she does not want to see him, sharing everyone else's view that he is crazy. In order to be reconciled to her father she agrees to marry George and look after the orphaned child he had adopted. She soon leaves him, however, and he loses his faith and plunges into a near-fatal illness (having never before suffered a day's ill-health in his life). However, he regains both faith and health and returns to his evangelical mission.

By the end of the novel, considered by many to be Wilder's best, George is a strangely sympathetic character who seems no more than the other people in the world he inhabits.

1935

Germany repudiates disarmament clauses of Versailles Treaty ● Anti-Roman Catholic riots in Belfast ● Franklin Delano Roosevelt signs Social Security Act ● Italy invades Abyssinia ● Persia changes its name to Iran ● The 35 mm 'Kodachrome' film devised ● Margaret Mead, *Growing up in New Guinea* ● Salvador Dali, *Giraffe on Fire* ● George Gershwin, *Porgy and Bess* (folk opera) ● T.S. Eliot, *Murder in the Cathedral* ● Françoise Sagan born ● T.E. Lawrence dies ● William Empson, *Poems*

Mulk Raj ANAND　　　　　1905–

Untouchable

Untouchable is the most compact and structurally satisfying of Anand's many novels. It covers the events of a single day in the life of a low-caste boy, Bakha, who starts each morning by cleaning out lavatories. This is a task he accomplishes with vigour and grace: 'Each muscle of his body, hard as rock when it came into play, seemed to shine forth like glass ... "What a dexterous workman!", the onlooker would have said.' However, his assiduousness must remain unrewarded in a caste-dominated society, whose realities Bakha begins to discover on this particular day: he is abused for 'polluting' a high-caste child by physical contact while trying to help when the child is injured during a game; the temple priest attempts to rape his sister, and then hypocritically shouts 'Polluted' when Bakha steps across the temple threshold to rescue her.

Three alternative solutions to Bakha's predicament are explored at the end of the novel: Christianity and Gandhism are rejected, more by Anand than by Bakha – and neither of them for any really sound reasons – while modern sanitation is presented as the solution to the social predicament in which Bakha's people find themselves.

Bakha typifies the victim–protagonist in Anand's work, which divides the world neatly into oppressors (who oppose change and progress) and good men (social workers, labour leaders, all those who believe in progress, particularly of the technological variety).

Untouchable is representative of Anand's deeply felt compassion for 'the dignity of ... [the] weak'. Along with *Coolie* (1936) – a more varied work exploring the lot of another class of underprivileged people – *Untouchable* is Anand's best work.

Enid BAGNOLD　　　　　1889–1981

National Velvet

This vigorous novel, full of surprises, is the story of fourteen-year-old Velvet Brown and one eventful year by the end of which 'she had become an heiress, got a horse for a shilling, and won the Grand National'.

Her inheritance is five ponies, bequeathed to her by their elderly, ailing owner seconds before he shoots himself. The horse, won with Ticket 119 in a raffle, is The Piebald, a garish, ugly beast, with a great and loving heart and the power to jump a fence higher than the National's highest and still have something in reserve. The raffle brings him to the Brown household on the South Downs: four daughters, Edwina ('Dwina), Malvolia (Mally), Meredith (Merry) and Velvet; Donald, a formidable four-year-old; Mr Brown, a slaughterer-butcher, and Mrs Brown, now an enormous, silent woman of some twenty stone – fat which was once the muscle that enabled her, at nineteen, as Araminty Potter, to swim the Channel, breast-stroke, against the tide, in a storm. Permanently attached to the family is Mi Taylor, son of Araminty's coach, a rolling stone who rolled one day to the Browns' door, recognised Araminty and stayed.

Horses dominate Velvet's life: 'I don't like people,' she says, 'except us and mother and Mi. I like only horses.' Before her inheritance, she already cherished her own stable – of paper horses, pasted on cardboard, harnessed in cotton. Her only mount had been the cantankerous Miss Ada, who, in company with all the other animals in this book, receives from Bagnold the same intelligent observation and sympathy as the human characters. Mi, equally passionate about horseflesh, has encyclopaedic knowledge of the history of the Turf – over two centuries of horses, owners, jockeys, prizes, pedigrees and stable lore. Together he and Velvet dare to plan and perform the unthinkable: to enter a piebald, of no known pedigree, in the world's greatest race, and, most shocking of all, to let Velvet – female and therefore ineligible – ride him.

In utterly convincing detail Bagnold records the planning, conniving, setbacks and stratagems which bring them to Aintree. The £100 entrance fee, a seemingly insuperable obstacle, is provided by Mrs Brown: the golden sovereigns of Araminty Potter's prize money. ('"Kept it," said Mrs Brown, "Thought I might. Thought I would ... Queer thing. I had a feeling."')

Apparelled, barbered, coached and valeted by the ever-resourceful Mi, who has also provided her with the official papers rightfully belonging to an Anglo-Russian jockey already on his way back to the Soviet Union, Velvet rides – and wins – as 'James Tasky'. But Mi forgot one crucial warning: a rider who dismounts before weighing-in will be disqualified. Velvet slides, in exhaustion and relief, from her saddle and loses the race: officially, that is. In the popular mind she is established as National Velvet.

Velvet's dreams were blowing about the bed. They were made of cloud but had the shapes of horses. Sometimes she dreamt of bits as women dream of jewellery. Snaffles and straights and pelhams and twisted pelhams were hanging, jointed and still in the shadows of a stable, and above them went up the straight damp oiled lines of leathers and cheek straps. The weight of a shining bit and the delicacy of the leathery above it was what she adored.

NATIONAL VELVET
BY ENID BAGNOLD

But the novel is more than just a rattling good yarn. Characterisation, especially of Araminty, Velvet and Donald, is outstanding. The boundaries between animals and humans blur constantly in the writing, without sentimentality or condescension, and physical facts of animal life – mating, gelding, dogs hunting bitches in heat – are included with an easy frankness unusual in a 'children's story'. Running through the book is a strong egalitarian impulse: a piebald ('circus horse'), of unknown origin, owned by a child from a class which does not usually provide the winners of classics, is ridden to (moral) victory by a female jockey trained by a man who has never ridden in his life.

How has it been achieved? 'It's the mare that done it,' says Mi, to the National Hunt Committee which has summoned him and Velvet to account for themselves. Velvet's courage and endurance come directly from Araminty. 'The father, he's added nothing.'

And it is Araminty, immense, broodingly powerful but largely inarticulate, who dominates the book.

E.F. BENSON — 1867–1940

Mapp and Lucia

Elizabeth Mapp of Mallards, a Queen Anne house in Tilling (modelled on Benson's own house and town, Lamb House in Rye) is bossy and cunning. When she is first introduced in *Miss Mapp* (1922) she 'might have been forty, and she had taken advantage of this fact by being just a year or two older. Her face was of high vivid colour and corrugated by chronic rage and curiosity ... [However,] anger and the gravest suspicions about everybody had kept her young and on the boil.' In *Mapp and Lucia* she meets someone worthy of her Machiavellian scheming – and her match.

Emmeline Lucas, 'universally known to her friends as Lucia', of The Hurst, Riseholme, made her first appearance in *Queen Lucia* (1920) and subsequently in *Lucia in London* (1927). She is forceful and domineering but she is also the lifeblood of Riseholme, and without her life would be very dull indeed.

Mapp and Lucia were made for, or at least deserve, each other. The success of their first meeting resulted in two sequels: *Lucia's Progress* (1935) and *Trouble for Lucia* (1939). *Lucia in Wartime* (1985) and *Lucia Triumphant* (1986), by Tom Holt, are further sequels, written in the Benson style.

Mapp and Lucia begins in Riseholme a year after the death of Lucia's husband, the Italian poet Peppino, author of the slim volume *Pensieri Persi*. During her year of mourning, her position as undisputed 'queen' of Riseholme has been challenged by hopeless Daisy Quantock, who is trying to organise an Elizabethan pageant but making a hash of it. Lucia seizes this opportunity to re-enter society. Usurping Daisy is not difficult and Lucia soon knocks the pageant into shape – casting herself as Elizabeth I, of course – and scores triumphantly.

A holiday in Tilling with her companion Georgie Pillson convinces her that Riseholme has become boring and that Tilling offers a new challenge. Georgie Pillson is the only other substantial character in these books. His proclaimed ambition is 'to live quietly and do my sewing and sketching', but he recognises in Lucia a galvanising source of inspiration. He is a bit of a dandy: 'He had just time to change into his new mustard coloured suit with its orange tie and its topaz tie-pin.' 'Tepid water and fluff on my

---

Then forth she went for the usual shoppings and chats in the High Street and put in some further fine work. The morning tide was already on the ebb, but by swift flittings this way and that she managed to have a word with most of those who were coming to her po-di-mu to-morrow, and interlarded all she said to them with brilliant scraps of Italian. She just caught the Wyses as they were getting back into the Royce and said how molto amabile it was of them to give her the gran' piacere of seeing the Contessa next evening: indeed she would be a welcome guest, and it would be another gran' piacere to talk la bella lingua again. Georgie, alas, would not be there for he was un po' ammalato, and was going to spend a settimana by the mare per stabilirsi.

MAPP AND LUCIA
BY E.F. BENSON

---

clothes' is as miserable a life as he can imagine. He is rather despised for his unmanliness by his counterpart, the fraudulent Major Benjy Flint, consort to Mapp. Major Benjy has supposedly served many years in India and calls 'Qui-hi' when he wants breakfast. Georgie passes for Lucia's lover but physical intimacy is the last thing either of them wants. The shadow Freud cast over the twentieth century has not fallen on Tilling: Lucia speaks for them all when she talks of 'that nasty thing Freud calls sex'. As Georgie says, it's 'horrible to think of', and the subject is usually avoided altogether.

Certainly no one remarks on the painter Irene Coles (based on Benson's Rye friends Radclyffe Hall and Una Troubridge, whom he referred to as 'the girls') and her predilection for dressing in men's clothes; though Mapp does address her as 'Quaint' Irene. She paints vaguely modernist works with titles like 'Women Wrestlers'.

Nancy Mitford wrote of the characters that: 'None of them could be described as estimable and they are certainly not very interesting, yet they are fascinated by each other and we are fascinated by them.' The characters also possess a certain brutalism which when manifested is really rather shocking. Lucia hardly seems to care when it seems likely that Georgie will have to remain in Riseholme when she moves to Tilling, and later Lucia's 'death' is swiftly accommodated by Georgie. Even when Miss Mapp is at her most vanquished, we are urged not to feel sympathy toward her since 'defiance and hatred warmed her blood most pleasantly'. Spitefulness and guile seem only to be expected. Perhaps it is this 'honesty' which makes the novels so absorbing.

Certainly nothing very much happens. Lucia scores more victories over Mapp, but sails a tight ship. Mapp anyway gains her prize, the recipe for Lobster à la Riseholme, which Lucia had tried to keep secret. There are often-repeated jokes: the 'Italian' and 'baby-talk' of 'Lulu' and 'Giorgino mio'; the cod Scots of the padre, Kenneth Bartlett, who is only Scottish by way of Birmingham; the endless performances by Lucia of the slow movement of the Moonlight Sonata; the duets of Lucia and Georgie, which both practise madly but affect not to have practised at all; the 'Royce' of the Wyses; and so on.

The Wyses are crashing snobs. Mrs Wyse is always wrapped in sable even in summer, and is usually wearing her MBE decoration, when it is not out on display ('so like the servants to leave that about').

The novels were very popular in their day – rather to the annoyance of the author, who would have preferred fame for his biographies (whose subjects were as varied as Charlotte Brontë and Alcibiades) – but they seemed old-fashioned and too flippant for the austere world of the post-war years.

From the 1970s, however, Benson's 'Lucia' books have once again achieved a considerable following, as have his novels written in a similar vein, such as *Paying Guests* (1929) and *Secret Lives* (1932).

Elizabeth BOWEN   1899–1973

The House in Paris

In 1933 Bowen embarked upon an affair with a young academic, Humphrey House, who was eight years her junior and about to get married. The relationship lasted for some three years, and some of its circumstances are reflected in this novel, not only in the triangle between Karen, Naomi and Max, but also in the relationship between Max and the much older Mme Fisher. Bowen disliked attempts to relate her work to her life ('Why can't they just read my books – if they care to – and leave it at that?'), but admitted that writing *The House in Paris* was almost like transcribing a dream.

The novel tells the powerful story of nine-year-old Leopold, who is waiting at the Fishers' Paris house to meet his mother for the first time in his life. Precocious and highly sensitive, he has a perplexed awareness of the 'awkward tangle of motives' operating in the adult world. In a penetrating study of the 'terror strange children feel of each other', he is shown to use the presence of another visitor, eleven-year-old Henrietta, to test his own feelings. It is in the central section of the novel that the circumstances of his birth are revealed, forcing us to reassess our initial judgements and illuminating the present behaviour of some of the characters.

Although engaged to Naomi Fisher, Max Ebhart (greatly influenced by Naomi's malevolent and possibly sexually jealous mother) realises that he is in love with his fiancée's friend, Karen Michaelis. After two secret meetings, Karen decides she cannot marry her own fiancé, Ray Forrestier, and Max resolves to end his engagement. A confrontation with a delighted Mme Fisher exacerbates his despair and he commits suicide in front of her, unaware that Karen is expecting his child. Perhaps realising that her power has 'overreached itself', Mme Fisher becomes an invalid shortly afterwards. When Leopold is born, the ever-dedicated Naomi supervises his welfare and, with Karen's consent, arranges his adoption by an American couple living in Italy. Karen has married the all-forgiving Ray who over the years becomes obsessed with the idea of adopting Leopold, their marriage being childless. When Karen fails to meet Leopold, Ray does so instead and resolves, without his wife's knowledge, and with Leopold's complicity, to prevent him from returning to Italy.

James T. FARRELL **1904–1979**

Studs Lonigan

Young Lonigan (1932)
The Young Manhood of Studs
  Lonigan (1934)
Judgement Day (1935)

The *Studs Lonigan* sequence is set in Chicago during the period 1916–31. Its first volume describes William 'Studs' Lonigan's fifteenth summer. The eldest of four children and the son of a prosperous Irish Catholic, Studs is glad to graduate from St Patrick's Grammar: 'He ... wanted to become a big guy ... He told himself that he'd have to go out now in the battle of life and start socking away.' He hopes to begin working, but his father is keen that he pursue his education and his mother that he train to be a priest. Unable to decide his future, Studs mooches around the neighbourhood, fighting and dreaming of sex. He beats up Weary Reilly, participates in a 'gang-shag' and earns his notoriety as 'a scrapper'. All the same, part of him is attracted to the respectable and pretty Lucy Scanlan, and he and Lucy spend a chaste, idyllic afternoon sitting in a tree in Washington Park. The experience is fixed in Studs' recollection – his moment of purest happiness, never to be recaptured.

Young Lonigan concludes in November 1916, with its hero playing truant from the high school his father has compelled him to attend. The second volume continues the story in April 1917. The USA enters the war in Europe. Studs and two of his friends try to enlist. Under age, they are rejected, and the conflict finishes before Studs can join up. Rarely at school, he spends more and more time at the local pool hall. One night, following a family row, he attempts a robbery. This strike for independence fails laughably and old man Lonigan employs his son as a painter in his decorating business. Believing he is finally an adult, Studs brawls, whores, drinks heavily and sleeps little. He worries over the deterioration of his physique; he is reproached by his parents, his church and his conscience; several of his contemporaries fall prey to consumption, syphilis or both. Nevertheless, Studs does not alter his ways. Striving to resume his relationship with Lucy, he escorts her to a dance, then gropes her brutally in a taxicab. It is the last he sees of her. His existence slides on a downward and futile spiral, reaching its lowest point on New Year's Eve 1919. He collapses in the street after a party – vomiting, battered and insensibly drunk.

Judgement Day takes up Studs's career twelve years later. He is still living at home; his health is ruined; his sisters are successfully married; his brother is drinking and whoring (as Studs used to do), while Studs is tamely engaged to a stenographer. Luck and lack of foresight conspire against him: he suffers a mild heart-attack, loses most of his savings on the stock market and impregnates his fiancée. A hasty wedding is planned. The Depression, however, has forced old man Lonigan to the brink of bankruptcy, so Studs is shortly broke and out of a job. Searching for work, he gets pneumonia. The illness proves fatal and, as Studs lies dying, his mother blames his fiancée for his sad state. His father witnesses a left-wing demonstration and struggles to forget his grief in booze.

By the close of the trilogy, Studs is a pitiable figure. He has not 'become a big guy'. If anything, he has diminished in stature; even his roughhouse reputation has gone. What renders his decline significant, though, is the extent to which it represents the fate of the people to whom he belongs – for Farrell's Irish are not saints and scholars. Rather, they are a stupid and violent clan:

'beer guzzlers, flat-feets, red mugs and bone-heads'. Catholicism keeps them intellectu-ally backward, instilling an unthinking hostility to secular ideas and radical change. Virulently racist, they envy the Jews, despise the Poles and fear the blacks. Their pride, all the same, is the pride of the doomed. The Jews prove more commercially resourceful, the Poles assume control of City Hall and the blacks drive out the Irish from their enclaves. When old man Lonigan watches the hetero-geneous and (he thinks) Bolshevik pro-testors, he is reduced to terror and confusion – a dinosaur, discarded not by revolution but by evolution.

The monotony of Studs' behaviour often impels the tale towards the repetitious. Such is the price of protracted naturalism and, to avoid its full exaction, Farrell inserts other kinds of discourse into his narrative: ser-mons; semi-illiterate letters; fragments of surreal conversation; pastiche film plots; advertisements; newspaper stories. In this (as in much else) he echoes the technique of Sinclair Lewis and – again, like Lewis – manages simultaneously to suggest the paucity and promiscuity of North American culture. To say that *Studs Lonigan*, in its scope and detail, accurately reflects this displacement of quality by quantity is to bestow on it at best ambiguous praise.

John GALSWORTHY   1867–1933

The End of the Chapter

see **The Forsyte Chronicles** (1929)

Graham GREENE   1904–1991

England Made Me

Walt Disney's claim that 'All the world owes me a living' is used as an ambiguous epi-graph to this dense and complex novel. Many people do indeed owe their living to Krogh, the friendless Swedish financier; but the epigraph applies equally to Anthony Farrant, the sponging twin-brother of Krogh's mistress, who does not believe in having to work in order to survive, and scrounges his way around the world, unable to keep a job.

Kate Farrant is not only Krogh's mistress but also his closest adviser, a position she has attained largely in order to help Anthony, whom she loves with a consuming passion. 'We are all thieves,' she says. 'Stealing a livelihood here and there and everywhere, giving nothing back.' She even connives in Krogh's crooked deals and endorses the zestless grandeur of his headquarters and flagrant tastelessness of his clothes. She gets her brother a job as Krogh's bodyguard, and his reckless charm briefly captivates and liberates the dour millionaire. But Anthony has met an English tourist called Lucia, whom he imagines he loves. 'Incorrigibly conventional' after a public-school edu-cation, Anthony takes up Lucia's suggestion that his job is not that of a gentleman. He also learns of Krogh's crooked dealings and when the great man refuses to see one of his workers and announces that he is to marry Kate, Anthony is outraged. He decides to resign and follow Lucia back to England. Hall, an old associate and now an employee of Krogh whose devotion to his master is without scruples, realises that Anthony knows too much and so murders him, arranging the death so that it seems acciden-tal. Kate leaves Sweden and Krogh: 'I'm simply moving on. Like Anthony.'

Much of the action is observed by Minty, a down-at-heel Old Harrovian journalist whose meagre life is sustained by feeding his editor with scraps of information about Krogh, and by religion, which he practises with 'the dry-mouthed excitement of a secret debauchee'. Despite Anthony's fake OH tie, Minty feels a kinship for him. Both Minty and Anthony have in their different ways been made (or destroyed) by England, in particular by their education, and this is reflected in the notions of loyalty and honour (particularly that amongst 'thieves') which run through the novel. At its centre is the doomed relationship of Anthony and Kate, one of the most moving and disturbing explorations of sibling love in fiction.

In 1961 there were plans, which came to nothing, for Greene to collaborate with his cousin Christopher Isherwood on a screen-play based on the novel. By a curious irony, the much underrated film of the novel which eventually appeared in 1972, directed by Peter Duffell from a screenplay he wrote with Desmond Cory, suffered from being released the same year another film set in the 1930s with Michael York in the lead role: *Cabaret*, adapted from Isherwood's Berlin novels.

Christopher ISHERWOOD 1904–1986

Mr Norris Changes Trains

The first of Isherwood's Berlin stories, this novel (subsequently published in the USA as *The Last of Mr Norris*) once formed part of an ambitious novel, *The Lost*, which was never completed. It was inspired partly by the Irish writer, traveller and eccentric Gerald Hamilton, who was the delighted model for Mr Norris, and partly by visits to Dr Magnus Hirschfeld's splendidly named and splendidly appointed Institut für Sexualwissenschaft, the museum of which was crammed with fetishistic paraphernalia. Isherwood described the novel as 'a sort of glorified shocker; not unlike the productions of my cousin Graham Greene', and later came to dislike it. 'What repels me now,' he wrote in 1956, 'is its heartlessness. It is a heartless fairy-story about a real city in which human beings were suffering the miseries of political violence and near-starvation.' It is this (perhaps heartless) objectivity which is in fact the novel's great strength, and much of the book's comedy arises from the phlegmatic reaction of its protagonist to the bizarre atmosphere of Weimar Berlin.

Returning by train to Berlin, William Bradshaw meets Arthur Norris, a man of uncertain years whose wig and suspicious behaviour arouse the young teacher's interest. They strike up a friendship which flourishes against the backdrop of political unrest of 1930s' Berlin. Norris claims to be in imports and exports, but is evasive when pressed about his activities. Bradshaw, who sees himself as a 'connoisseur of human nature', is warned against Norris by various friends, whom he ignores. Norris is always surprising Bradshaw: at one moment dis-covered kneeling at the booted feet of the prostitute Anni, the next addressing a meeting of the Communist Party upon 'the crimes of British Imperialism in the Far East'. Through Norris Bradshaw meets Kuno von Pregnitz, a monocled homosexual aristocrat, who works for the government and divides his leisure time between boys' adventure stories and a gymnasium. Bradshaw also meets Ludwig Bayer, the inscrutable leader of the Berlin Communist Party, and Helen Pratt, a tough young left-wing journalist.

Norris's domestic and financial circumstances are as mysterious as his business activities. His 'secretary' is a hideous and sinister young man called Schmidt, who appears to control the money. In fact Norris is hopelessly in debt, living in continual fear of the bailiffs and of Schmidt, to whom he owes nine months' wages. Occasionally Norris disappears, usually abroad, in order to flee his creditors or to pursue business contacts. He persuades Bradshaw to accompany Baron von Pregnitz to Switzerland on a skiing trip in order to bring him into contact with a man called 'Margot', who Norris claims wants to take over a glass factory in which von Pregnitz has an interest. In fact 'Margot' is a member of the French secret service who wants to buy government secrets. Bradshaw is told of Norris's plot when recalled to Berlin by Bayer, who has been using Norris to spread false stories to Party enemies. Bradshaw is furious that he has been used, but finds that he cannot long remain angry with the absurd, conscienceless Norris.

Norris's activities have come to the attention of the authorities and his is forced to leave Berlin for South America. He cannot, however, escape Schmidt, who pursues him like one of the Furies. Von Pregnitz had been selling state secrets in order to pay off Schmidt, who was blackmailing him, and when his treason is discovered, he shoots himself. Bradshaw returns to England, where he receives a succession of postcards from Norris, asking what he has done to deserve his fate.

Storm JAMESON   **1891–1986**

Love in Winter

In the second volume of Jameson's autobiography, *Journey from the North* (1969), she explains that her trilogy, *The Mirror in Darkness*, is a *roman-à-clef*, centred upon her marriage to the historian Guy Chapman, 'a happy difficult second marriage, part of the nervous system of my mind'. *Love in Winter* is the central novel of the trilogy. The first, *Company Parade* (1934), chronicles Hervey Russell's arrival in London from Yorkshire in the month after the Armistice, her disastrous marriage to the feckless Penn Vane, and the tentative establishment of her career as a novelist.

In *Love in Winter* we have reached 1924. Hervey has met and fallen in love with her cousin Nicholas Roxby while Penn is living it up in Oxford. Nicholas's grandmother had wanted him to run the family shipping company of Garton but, his energy sapped by the war, he has lost all ambition and settles instead for dealing in antique furniture. He is separated from his silly wife, Jenny, but will he have enough energy to withstand the horrors of divorce? He makes excuses to Hervey: 'I am no good to you, or to any woman. I don't want to be troubled to make love to anyone, or . . . to live with anyone so unspoiled, so *good*, as you are.' But nor can he live alone.

Hervey's other problem, equally poignant, is guilt at leaving her son Richard in the North with her mother while she earns a living in London. She says that Richard is 'the only person to whom I am bound and no pretence about it', but after she has spent Christmas with him, she returns to find Nicholas distraught at being left alone. He eventually recognises that he must divorce Jenny and commit himself to Hervey. Two badly wounded people, deeply in love, Nicholas and Hervey begin living together; the 'happy difficult' marriage has begun. The progress of this marriage against the background of the General Strike is described in the trilogy's third volume, *None Turn Back* (1936).

Love in Winter also deals with Nicholas and Hervey's relationships with other characters. Hervey is employed as secretary to the untrustworthy and egocentric literary editor Evelyn Lamb, who hands over all the work, and sacks her when she refuses to take responsibility for a libel. It is Evelyn who tells Penn about Hervey's involvement with Nicholas. Another thread of the novel concerns Nicholas's erstwhile friend, William Gary, a mine owner, whose war wound has left him impotent and destroyed his relationship with Nicholas's sister, Georgina. Gary's sex drive is diverted into the business of becoming a man of power and influence. 'A curious intimacy – without warmth' springs up between him and the ship owner, Thomas Harben, and they take up the Labour MP Louis Earlham. 'It is even possible that you and I want the same sort of world,' Gary tells Earlham. 'Without brass hats, and with plenty to eat for the men.' Under their influence, Earlham begins to forget the people who elected him, a development his friend David Renn watches with sardonic detachment. It seems that Renn, who is 'so thin that you could think that he has happened to leave his body at home and come out in his bones', might find salvation with Hannah Markham, a woman of great warmth who knows her own mind. Other characters include Marcel Cohen, the Jewish financier and newspaper proprietor, and the sinister, limping Julian Swan, who dreams of becoming Britain's Mussolini, and employs Timothy Hunt,

---

A fog had extinguished London. People moved about in it half choked, looking to each other like maggots. It was as thick as pitch in north London and Hervey groped her way up the unlit staircase of Nicholas's mother's flat. She felt exasperated and dirty – London is perhaps fit for human beings ten days in the year, not more.

LOVE IN WINTER
BY STORM JAMESON

---

wife-beater and former Black-and-Tan, to break up Labour meetings. Swan is sorted out by the novelist Frank Ridley (drawn from J.B. Priestley) at one of Evelyn's literary parties.

The novel perhaps contains rather too many characters, and the connections between the political parts of the novel and the central love affair are often somewhat tenuous. Yet they are joined in a thematic sense: everyone in some way or another is a victim of the war. It is true that, writing the novel at a period when the effects of Hitler's rise to power were already evident, Jameson has the benefit of hindsight; nevertheless, the sense of exhaustion on the one hand and manic energy on the other is valid. The characters are for the most part, as Auden wrote on the eve of the Second World War, 'Children afraid of the night/Who have never been happy or good'.

John MASEFIELD   **1878–1967**

The Box of Delights

Masefield's novels for adults are not much read now, but his two fantasy stories for children have become classics, frequently reprinted and dramatised for radio and television. They have also been much admired by adults, just as his 'adult' books have been enjoyed by children; indeed, several of the same characters feature in both genres. The narrative skills of such long poems as *The Everlasting Mercy* (1912) and *Reynard the Fox* (1919) are equally apparent in *The Midnight Folk* (1927) and its sequel, *The Box of Delights*. Like his creator, Masefield's young hero, Kay Harker, is an orphan and the author drew upon memories of his own governess to create the odious Sylvia Daisy Pouncer, who was unmasked as a witch in the first book.

The eponymous box is owned by Cole Hawlings, an old Punch-and-Judy man Kay meets on a platform when he is travelling back home from school for the Christmas holidays. Hawlings asks Kay to carry a message to a woman he will find at the local baker's: 'The Wolves are Running.' This is a warning concerning the activities of Kay's old adversary, Abner Brown, an American magician and jewel thief who is determined to steal the box. Brown (now married to Sylvia Daisy Pouncer) is masquerading as the Revd Dr Boddledale, head of a theological training college, and his agents have adopted clerical garb.

Kay's glamorous young guardian, Caroline Louisa, has invited the Jones children to spend Christmas at Seekings, and Peter, Jemima, Susan, and the splendidly naughty Maria (expelled from three schools) are enlisted to help defeat Brown and his gang. In their search for the box, Brown's men 'scrobble' (kidnap) first Hawlings, then anyone to whom he might conceivably have passed the precious object. Peter, Maria, Caroline Louisa, and the Bishop of Tatchester, along with every single cathedral servant and chorister, disappear, and are eventually traced to a network of limestone caves beneath Chester Camp. It is Kay who has the box which, at the flick of a switch, can make him 'go small' or 'go swift'. It also allows him to enter the past and witness such events as the sacking of Troy. On one journey he meets the creator of the box, a philosopher of the Middle Ages called Arnold of Todi, whom he brings back to the present, assisted by two of Hawlings's associates, Herne the Hunter and the Lady of the Oak Tree. Hawlings is in fact Ramon of Lully, a contemporary of Arnold who had been searching for the Elixir of Life.

She fetched the little barrels and they divided them up among the three ships, and they put raisins and currants and bits of biscuits in each barrel.

'I vote,' Jemima said, 'that the other barrels shall be filled with ham, which we will pretend is salt pork.'

'They don't take salt pork any more,' Peter said. 'They take pemmican, which is beef chopped up with fat and raisins and chocolate and beer and almonds and ginger and stuff. It must be a sickening mess, but it's very nourishing. It's supposed to be what the ancient Britons had. They could take a piece as big as a currant and live on it for a week.'

THE BOX OF DELIGHTS
BY JOHN MASEFIELD

After numerous adventures, Kay rescues the prisoners, helped by the fact that Brown's gang have mutinied because of the bad press their 'scrobbling' is earning them. The clergy are restored to Tatchester cathedral in time for the Millenial Christmas Service, in which Kay takes part – only to wake up at the station and discover that the entire story was a dream.

This feeble ending apart, *The Box of Delights* lives up to its title, crammed with character and incident and with plenty of wit to keep adult readers amused.

George ORWELL   **1903–1950**

A Clergyman's Daughter

The first quarter of this, Orwell's second novel, describes the daily grind of Dorothy Hare. Aged twenty-seven, she lives with her father, the rector of St Athelstan's, in the Home Counties. Her activities are dictated by his demands, the needs of his parishioners and her own devout faith. She rises at half-past five for a cold and penitential bath; then she attends early Communion, prepares breakfast, visits the sick, measures costumes for a children's drama, and ashamedly avoids the many tradesmen with whom the rectory has run up debts. Her father's doctrinal squabbles – he is High Anglican – have isolated her from most local church-goers. Indeed, her only friend, albeit an unlikely one, is Mr Warburton, an ageing and affable roué. Warburton has a Spanish mistress and three children whom he calls 'the bastards'. He sexually assaulted Dorothy when they first met. Nevertheless, while his behaviour and beliefs are uncongenial to her, his unrepressed and irrepressible nature appeal. She visits him in the evening of her day, and once again he tries to seduce her. Sexually repressed, she repulses him.

About a week later, she is in London. She has no memory of her past existence – no idea, even, of her name. She has lost her faith and, in her dazed state, joins some itinerant labourers walking to Kent. For a while she picks hops. Discovering her true identity, she writes to her father. He does not reply: he believes she is engaged in some sinful adventure. When the hop-picking season ends, she

travels to Lambeth, renting a room there in a slum boarding-house. Her money is soon exhausted, and she begins to sleep rough, has to beg – and is arrested. Her plight comes to the notice of a cousin, who rescues her and helps her find a job.

She teaches for a couple of terms at Ringwood, a fraudulent and squalid private school. Its owner, Mrs Creevy, is concerned far more with fees than with education; still, though half-starved and miserably paid, Dorothy tries to stimulate her pupils. Mrs Creevy intervenes to stop her. Shortly after, Dorothy is sacked. She is homeless again, and facing poverty, when Warburton appears as her saviour.

By now, Dorothy knows that she can go back to the rectory. However, on the train to Knype Hill, Warburton describes to her what her future will be if she does so: she will become a pitiable spinster, an unpaid drudge in perpetuity. He proposes marriage instead – and Dorothy is on the brink of accepting when a 'wave of disgust and deadly fear went through her ... The harsh odour of maleness forced itself into her nostrils.' Tamely she returns to her former routines.

Dorothy has visited the Orwellian underworlds of destitution and wage slavery (and thus allowed her author to grind a few axes), but her crisis has not changed her much. Her piety has vanished, but her practicality prevails. Doing replaces believing. Faith can be (quite literally) forgotten, and hedonism – embodied by Warburton – decisively rejected. Orwell suggests that what is needed for survival in mean-minded, middle-class suburbia is a wilful narrowing of horizons. Thus Dorothy's mental wellbeing depends ironically on a kind of spiritual amnesia – a blotting-out of anxieties and desires beyond the small, the immediate and the mundane.

Dorothy RICHARDSON   **1873–1957**

Clear Horizon

see **Pilgrimage** (1938)

Look Homeward, Angel
1929

Of Time and the River
1935

Look Homeward, Angel and *Of Time and the River* were the only full-length novels published by Wolfe during his short lifetime. Taken together they tell the story of the Gant family from the battle of Gettysburg to the departure of the youngest member, Eugene, from Europe in the 1920s. In *Look Homeward, Angel*, the troubled relationship between Eugene's drunken father, a stonecutter, and his shrewd, penny-pinching mother gradually become the background to his story as he grows up in Altamount and later attends the State University. Eugene's adolescence is full of grand day-dreams, hopes and ambitions which contrast starkly with his home life. The buried conflict becomes explicit near the end of the book when he bitterly rejects his parents together with all they stand for and asserts his individuality. Characteristically though, Wolfe has it both ways and the book's climax is reached in a meeting between Eugene and the ghost of his dead brother, Ben. The ambiguous wish to escape the ties of birth or upbringing and yet retain some sustenance from those ties, a sense of belonging, is one of two enduring and related themes in Wolfe's work.

The other is nothing less than America itself. 'They were all there – without coherence, scheme or reason – flung down upon paper like figures blasted by the spirit's lightning stroke, and in them was the huge chronicle of the billion forms, the million names, the huge, single and incomparable substance of America.' These lines from towards the end of *Of Time and the River* describe the literary outpouring of Eugene, a figure closely identified with Wolfe himself. He has left Altamount for Harvard, worked in New York and travelled from there to England and France. If his journeying has

had any object, it has been to find a place from which America itself can be comprehended, named and written down. Here again is a contradiction, for Eugene's character, which is based on his omnivorous appetite for experience, depends ultimately on America's never being exhausted, never wholly 'named'. This is a genuine tension; Wolfe truly believes that it might be possible to complete this 'naming' process. His faith in this hyperbolic enterprise is what sets him uniquely apart from his contemporaries. It is also the root-cause of his many structural and stylistic excesses.

Everything that Wolfe wrote was too long, and the extravagant thanks to his editor in *Of Time and the River* bear oblique witness to Maxwell Perkins's lack of editorial ruthlessness. Too many passages take the form of rhapsodies that leave the original scene far behind and Wolfe's efforts to prune his manuscripts often ended in his adding thousands of words. His prose is often massively self-indulgent, untrammelled by notions of internal proportion. At the same time, and despite his admiration for Joyce's *A Portrait of the Artist as a Young Man* (1916), there is not a trace of irony in his protagonist's flights of fancy. Wolfe takes himself, his hero and his enterprise seriously. Although it is not difficult to find passages of surpassing skill in both novels – the death of Ben in *Look Homeward, Angel*; the race between the locomotives in *Of Time and the River* – these cannot account for his novels' massive and enduring popularity. Those who read Wolfe's books and like them, like the whole of each one, faults and all. Not only inclusive themselves, they urge the reader to be inclusive in turn, and in this fact lies Wolfe's particular genius. The effects that he sought in his novels are not achieved despite those novels' faults but, paradoxically, because of them. William Faulkner understood this very well when he wrote:

Man has but one short life to write in, and of course he wants to say it all before he dies. My admiration for Wolfe is that he tried his best to get it all said; he was willing to throw away style, coherence, all the rules of preciseness, to try to put all the experience of the human heart on the head of a pin.

1936

German troops enter Rhineland ● Spanish Civil War begins ● Abdication of Edward VIII ● Chiang Kai-shek is forced to declare war on Japan ● John Maynard Keynes, *General Theory of Employment, Interest and Money* ● Frank Lloyd Wright, Kaufmann House, 'Falling Water' in Bear Run, Pennsylvania ● Allen Lane founds Penguin Books, starting the paperback revolution ● Olympic Games in Berlin ● Rudyard Kipling, A.E. Housman, G.K. Chesterton, Maxim Gorky and Luigi Pirandello die ● Michael Roberts (ed.), *Faber Book of Modern Verse* ● Robert Frost, *A Further Range*

Djuna BARNES　　　　　**1892–1982**

Nightwood

At the time of its publication, *Nightwood* was widely held to be the perfect expression of its literary milieu – the circle of artists and writers who gathered in Paris between the wars. Admired by James Joyce and championed by T.S. Eliot, the novel remains famous but is little read today. A brief resurgence of interest prompted by feminist critics in the 1970s failed to restore the novel to the modernist canon, and many readers' knowledge of it now derives from its use as an example in Joseph Frank's essay 'Spatial Form in Modern Literature'. Its neglect can be traced first to Barnes's retirement from the literary coteries of Paris shortly after the book's publication, secondly to the lack of an easily definable progression in her oeuvre, and thirdly to its classification as a 'poetic novel'. Eliot's laudatory but pompous preface, which first applied the term 'poetic' to it, is unenticing: 'To say that *Nightwood* will appeal primarily to readers of poetry does not mean that it is not a novel, but that it is so good a novel that only sensibilities trained on poetry can wholly appreciate it.'

Readers of *Nightwood* will find themselves cast adrift in an uncharted, urban landscape. They will re-arrive at points they visited only briefly many pages before and they will have to pick the right details (from a vast number of primarily decorative ones) by which to orientate themselves. That said, the novel has a fairly extensive and recognisable plot. Broadly, it concerns the reactions of four people – Felix Volkbein, Nora Flood, Jenny Petherbridge and Dr Matthew O'Connor –

to Robin Vote, a young woman whose character hovers uncertainly between humanity and bestiality. Divided into eight sections, the novel opens with 'Bow Down', in which Felix Volkbein, fake Baron and archetypal Wandering Jew, is born, grows up and travels to the Paris of the 1920s, where he meets the three characters named above. 'La Somnambule' introduces Robin when she is revived from a faint by Dr O'Connor and Felix. O'Connor falls in love with Robin, marries her but fails to keep her, even though Robin has a child. 'Night Watch' reveals that Robin's wanderings have taken her to America and into the arms of Nora, but this relationship also deteriorates and finally breaks down in 'The Squatter', in which Nora is replaced by Jenny in Robin's involuntary affections.

From here on, the novel loses any real sense of continuing time, events seeming to occur in the consciousnesses rather than the world of the characters. 'Watchman, What of the Night?' tells how the distraught Nora seeks solace from Dr O'Connor, only to be treated to a long rhapsody on night, day and the nature of time. 'Where the Tree Falls' propels Felix and his retarded son, Guido, into the narrative. Guido's soul prompts a long discussion between Felix and Dr O'Connor, which in turn prompts a flashback to a meeting between Felix and Jenny. 'Go Down Matthew' returns to the meeting between O'Connor and Nora in 'Watchman ...' Nora's grief and understanding of Robin's nature is again flooded by O'Connor's verbal onslaught on the nature of 'the night', although this does not prevent him from being cowed by her despair. In 'The Possessed', the last and shortest section,

Robin has abandoned Jenny in America and now roams like an animal until arriving at Nora's estate, where Nora finds her in a bestial confrontation with her dog. At the close, Robin finally sinks to a sub-human level, none of her lovers having been able to control or retain her. The novel's incidents consistently point up the 'between-ness' of modern experience. The characters' sexual inversions confound the polarities of male and female, active and passive, while their thoughts seem often to be suspended between real and dream worlds. Felix Volkheim remarks that 'an image is a stop the mind makes between uncertainties', and Barnes carries this idea through, arresting the action with images that, for brief moments, crystallise the sensations and ideas of the characters. *Nightwood* is a fantasia underpinned by the rigour of a modernist aesthetic – derived largely from Imagism – and even its most overladen, 'symbolist' passages are subject to Barnes's unobtrusive but rigid control. The prose is certainly poetic and the effect dream-like, but the dream is sustained and controlled, while the poetry is subsumed within the stylistic demands of the novel.

Cyril CONNOLLY   1903–1974

The Rock Pool

Connolly's only completed novel, an account of Bohemian life in the Mediterranean, was turned down by Faber on the grounds of obscenity. Connolly was then approached by Jack Kahane of the Obelisk Press in Paris, which was renowned for publishing such books. Kahane later described the novel as a 'disgrace to his list', because it was so unsalacious. Others thought it disgraceful for different reasons: 'Even to want to write about so-called artists who spend on sodomy what they have gained by sponging betrays a kind of spiritual inadequacy,' George Orwell proclaimed. Connolly ascribed the novel's 'moral tone' (as he ascribed much else) to his classical education, in particular to the discovery that 'we had been learning by heart the mature, ironical, sensual and irreligious opinions of a middle-aged Roman [i.e.

Horace], one whose chief counsel to youth was to drink and make love to the best of its ability, as these were activities unsuitable to a middle-age given over to worldly-wise meditation and good talk'. The novel was eventually published in Britain in 1947, and was the first part of a projected trilogy, to be completed by 'The English Malady' and 'The Humane Killer' and published under the anagrammatical pseudonym, Lincoln Croyle. A nine-page fragment of the latter volume was published in *London Magazine* (August/September 1973).

The title is taken from 'a little monograph' the protagonist is intending to write about Trou-sur-Mer, a decaying French resort which seems to him 'a microcosm cut off from the ocean by the retreating economic tide'. Naylor is a pretentious and conceited Wykehamist, who is working on a biography of the nineteenth-century 'banker bard' Samuel Rogers, and living on a tiny private income. He is staying at Juan-les-Pins, but, during a day in Trou, he meets the painter Rascasse and, through him, Toni van Schaan, a young girl of impoverished Baltic baron stock to whom Naylor is attracted. Naylor is in revolt against his conventional English background ('elbowing like tadpoles in a jar'), and determines to stay and study the community of Trou as a naturalist might study a rock pool.

The novel has little plot, but is carried along by unfailingly funny conversation and narration ('He escaped three hundred francs to the bad, the equivalent of two women and an indecent cinema'), through scenes of parties, drinking, casual sex and no less casual violence. Among the denizens of the rock pool are the German girl Sonia, who has hitch-hiked across Europe to be with her friend Toni; the peroxide-haired homosexual and 'old-fashioned sadist' Jimmy; the lesbian couple Duff and Varna, who keep the Bastion Bar; the model Lola, attached to a fierce young Corsican; and the rich, kilt-wearing Mr Foster, with whom Naylor engages in snobbish sparring matches. Naylor sits for his portrait to Rascasse, with whom he stays briefly, has vain designs on Toni, and one-night stands with Lola and the American divorcee Ruby, as well as with a half-caste woman, Tahiti, with whom he believes himself in love, although she deserts him on their second meeting. He experiences money difficulties and moves to cheaper rooms in Trou. As winter approaches, and the inhabitants of the rock pool begin to disperse until spring, Naylor increasingly spends his days in a gloomy Pernod haze. In Nice he has a dispute with a taxi-driver, is hit by him, but is still forced to pay out almost all his remaining money to avoid court and prison.

At the novel's conclusion, Naylor is living in a drunken, quarrelsome liaison with Ruby, who spills blood and Pernod over his manuscript, and he is dismissed by some visiting English people as 'just another bum'. Witty, allusive, and tinged by melancholy, *The Rock Pool* is an astringently memorable indication of what Connolly's fictional talent might have been.

John DOS PASSOS　　　　**1896–1970**

USA

The 42nd Parallel (1930)
1919 (1932)
The Big Money (1936)

USA (which was first published under this title in one volume in 1938) features the author's characteristic style, a mixture of factual narrative (including newsreel sections and accounts of historical figures such as the Hearst family) and fictional characters. This technique produces a constantly changing perspective, charting a comprehensive view of American society over the first twenty-five years of the century. It also reveals a disintegration of the ideals and potential of the American dream.

The opening section of *The 42nd Parallel* features Fainy O'Hara, born into a poor Irish family, who moves from Middletown, Connecticut, where there are no prospects for him, to Chicago, in order to work for his uncle, who apprentices him as a jobbing printer. After his uncle's bankruptcy, he goes on the road, taking various jobs including, rather unwittingly, selling pornographic novels. He also teams up with various drifters, from whom he parts again just as casually.

In San Francisco he meets Maisie, and after a short courtship, they decide to marry. However, his deep commitment to the trade union movement drives him to leave Maisie and work for a trade union paper in Goldfield, Nevada; he has a strong sense that he belongs there, and that he is working for a cause in which he believes.

Maisie writes informing him of her pregnancy, however, and after some thought he finally returns to marry her. Their marriage soon becomes a life of drudgery for Fainy, as he works hard to support Maisie's growing material ambitions; these include Maisie pushing him into accepting an offer of help from her brother to house them in Los Angeles, and effect some social mobility. An argument over money leads to Fainy taking off for Mexico to join the revolutionaries, attempting to fulfil his belief that he should rise with the ranks, and not from them.

A theme maintained through *1919* and *The Big Money* is that of the protagonist being forced to choose between fulfilling ideals and sacrificing them, and settling for material benefits. In *The Big Money*, which covers the crucial post-war period, a researcher, Mary French, meets George Barrow, a politician who enlists her help in tracking unemployment. But after his overtures of marriage she suddenly leaves town, determined to experience the lives of industrial workers and to help their lot.

Mary soon becomes a publicist for a trade union office, but she finds the all-embracing poverty and oppression too harrowing an experience. When she meets Barrow again, she accepts his offer to become his personal secretary. After a further offer of marriage from him, however, she flees once more, this time to New York, to continue her activities in the trade union movement, as well as Communism. Her decline of a third proposal of marriage from Barrow represents her final rejection of his material world, and intensifies her commitment to her cause.

William FAULKNER　　　　**1897–1962**

Absalom, Absalom!

On completing this, his ninth novel, Faulkner exclaimed, 'I think it's the best novel yet written by an American', a view which over 300 essays and books have gone some way towards confirming. The novel's title echoes the cry of grief and loss uttered by King David at the death of the son who rebelled against him. Faulkner's book tells the story of Thomas Sutpen, a man with a 'design', who rides into Yoknapatawpha County – the fictional environment of almost all Faulkner's work – in order to realise it. Without regard for his own past or the surroundings which are his present, he founds his future by force. Wresting a wife, Ellen Coldfield, from the town, land from the Indians, and a mansion, Sutpen's Hundred, from the ground itself, he establishes himself by brute strength of will. He fathers two daughters and the son who will carry on the line, but a son from his unacknowledged past reappears to disrupt his private schema, while the civil war begins to fracture the country. The relationships between his children are made destructive by the divisive, prohibitory contexts of miscegenation, illegitimacy and incest. The historical realities of the South cut across the desires of the novel's individuals, turning the potentials for filial and sororal love into a reality of fratricide. Sutpen's design, rotting from the inside out, comes ultimately to nothing.

Faulkner sets this tale within a complex structure of narrators, all with varying degrees of competence, all with more or less vested interests to protect. The principal narrator (who also features in *The Sound and the Fury*, 1929) is Quentin Compson, who is told parts of the story by Rosa Coldfield forty-three years after her involvement in the action (as Sutpen's sister-in-law and, later, putative fiancée) has come to an end. Other parts are told to him by his father, whose father told it to him in turn. Quentin retells it to Shreve McCannon, his roommate at Harvard, and between them they imaginatively reconstruct Sutpen's story. Minor factual discrepancies and major judgmental ones emerge between these differing versions, and no one version reigns supreme. Sutpen's battle with the problem of lineage is re-fought in the novel's structure as the tale is handed down problematically from the narrator of one generation to that of another.

The story resists rational explanation by its narrators; its loose ends, ambiguities of motive in particular, return as obsessive and malignant forces through Faulkner's coercive prose to haunt its tellers. Quentin's 'very body was an empty hall echoing with sonorous defeated names', while Rosa speaks of the South as a whole, 'peopled with garrulous outraged baffled ghosts'. The traffic between past and present is both fractured and fracturing; the single point of view which might hold the truth about the South and its people is dispersed amongst competing narrators. The sympathy necessary to understand the issue precludes the objectivity necessary to judge them, thus Shreve's plea to Quentin to explain 'the South' is answered: 'You can't understand it. You would have to be born there.' Faulkner's major achievement in this novel is his offering the reader a chance to synthesise the multiple versions, mythologies even, of the South, to redeem an area of value, in short, to answer Shreve's questioning: 'I just want to understand it if I can and I don't know how to say it better. Because it's something my people haven't got.'

Winifred HOLTBY   1898–1935

South Riding

Holtby's last and most substantial novel was completed only one month before her death in 1935 and won the James Tait Black Prize in 1937. Drawing upon her own background in the East Riding of Yorkshire, where her mother was an alderman on the county council, the novel uses the fictitious South Riding to portray a wide social panorama spanning the classes from landed gentry to impoverished slum-dwellers. It was intended to promote social reform, but is perhaps more widely read as a popular romantic novel.

Sarah Burton, a native of the area, returns from London to take up the post of headmistress of Kiplington Girls' High School. She is appalled at the condition of the school, reflective of 'a world besieged by poverty, ugliness, squalor and misfortune', and campaigns for money from the council. Most resistant of all the councillors is the traditionalist Tory land-owner Robert Carne, whose financial difficulties have forced him to send his own hysterical daughter to the school. His mentally ill wife is in an expensive nursing home and his farm is losing money. Other members of the council include the Liberal and Methodist minister Huggins, who is being blackmailed by a girl he has made pregnant. To pay her off he borrows money from Councillor Snaith, who advises him to speculate on land which will go up in value when the housing committee, headed by Snaith, approves a new housing scheme. Carne hears of this and resolves to expose the corruption, and Snaith changes the location of the buildings to deflect his attack, thereby ruining Huggins.

Despite their ideological hostility, Sarah finds herself falling in love with the dark and mysterious Carne. After a number of encounters, they run into each another one night in a Manchester hotel and have retired to her room when Carne is struck down with a severe attack of angina. Knowing he has not long to live, he has taken out life insurance policies to provide for his wife and daughter, an act which provokes rumours of his suicide or staged disappearance when he dies shortly thereafter, his horse having slipped from a subsiding cliff. Knowing the truth, Sarah denounces the rumours, but she believes that Carne died despising her as a loose woman. She learns finally of his high regard for her from his old friend Alderman Mrs Beddows.

Aldous HUXLEY　　　　　**1894–1963**

Eyeless in Gaza

While *Eyeless in Gaza* is not autobiographical, some of its protagonists' circumstances and dilemmas correspond with Huxley's. He began writing the novel at Sanary in 1934, drawing upon his travels in Mexico the previous year. The theme of the novel, he wrote, 'is liberty. What happens to someone who becomes really very free ... The rather awful vacuum that such freedom turns out to be.' It is also prophetic, pointing towards the pacifism Huxley himself was to adopt in the company of Gerald Heard (some of whose characteristics Dr Miller shares). As well as marking a development in Huxley's own life and beliefs, the novel is very much of its period, pursuing the theme of 1930s man in search of a set of values by which to conduct his life in uncertain times.

In the glittering world of the inter-war avant-garde Anthony Beavis is a cynical success. His work as a sociologist provides a psychological excuse for his fear of involvement with people; but in personal relationships his stance of detached observer proves cruel and manipulative. Through a narrative structure of constantly changing viewpoints Huxley explores the processes by which personality is formed. The novel's title is taken from Milton's *Samson Agonistes*, and suggests that to live in a world without value or meaning is to be imprisoned in a state of blind helplessness. As Milton's Samson achieves an inner calm at his death, after a prolonged struggle to come to terms with divine providence, so Beavis's search for a mystical wholeness is rewarded at the end of the novel when he achieves peace of mind. He escapes from the prison of his own creating by having the courage to reconstruct his self.

An unhappy childhood has left Beavis emotionally scarred. He has never expressed grief over his mother's early death, and an unbridgeable distance has developed between him and his father. At Oxford he takes refuge in empty intellectualism, which commands superficial respect, but the split between mind and body is continually widening. An affair with Mary Amberley, an unstable hedonist, ends in tragedy as she dares him to tempt away the fiancée of his closest friend, Brian Foxe (a portrait of Huxley's brother, Trevenen). When Brian believes the girl no longer loves him he commits suicide. As Mary's drug addiction begins to destroy her, Beavis becomes involved with her daughter, Helen. But he is still incapable of love beyond sex and inevitably loses her.

The tortuous drama of the rich Bohemians is played out against the backcloth of social poverty. Even the Communist sympathies of Beavis's former schoolfriend, Mark Staithes, are the mental masturbation of an intellectual, dangerous because superficial. Despairing over the purpose of his life, Beavis goes to Mexico with Staithes to join the revolutionaries.

In this very un-Lawrentian version of the new world, Beavis finds an unexpected peace. A riding accident prevents them from joining the rebels, and as Beavis seeks help for his wounded friend he encounters an English doctor, Miller. Almost Quaker-like in his calm simplicity, Miller has a commitment to the human good which transcends any form of religious faith. With the guidance of this modern *salvator mundi*, Beavis begins to take control of his own destiny, reuniting mind and body through courage and trust.

Storm JAMESON   1891–1986

None Turn Back

see **Love in Winter** (1935)

Sean O'FAOLÁIN   1900–1991

Bird Alone

The struggle for Irish autonomy – particularly, the disgrace and death, in 1891, of Charles Stewart Parnell – provides the political context for *Bird Alone*, O'Faoláin's second novel. It is narrated by Cornelius 'Corney' Crone from his old age. Born in 1873 into a family of building contractors, Corney describes his childhood in the town and countryside of Cork – his home, his ever-anxious father, his bibulous mother, his friends and his haphazard education. Two figures overshadow his memories: one is his grandfather; the other is Elsie Sherlock, the daughter of a water bailiff and Corney's first and only love.

For 'old Phil', the grandfather, the church has betrayed the nationalist cause. *Inter alia*, it has pronounced the Fenian oath contrary to Catholic teaching. By his unwavering allegiance to Fenianism, therefore, old Phil has put himself outside devout society. He reads secular literature and often tells Corney the story of Faust. Moreover, he offers his grandson an example of proud isolation, and this leaves a more permanent mark on Corney than the political rhetoric. Indeed, following Parnell's death, Corney is overwhelmed by a defeatist malaise. Looking out over Cork, he meditates: 'it was right it should all be dead, when he was [dead] who might have brought it all to life.'

The Crone building business goes into decline. Old Phil, before he dies, confesses to a Parnellite priest. (However, he will not be buried on Catholic ground.) Meanwhile, after a protracted, tentative flirtation, Corney and Elsie have started to sleep together. Elsie knows this is sinful: she is a pious daily worshipper at mass. Still, in time she becomes pregnant. She conceals her pregnancy, frightened of the distress it would bring to her father. Corney takes her to a seaside cottage. There, one night, during a storm, she tries to drown herself. She fails, but goes into premature labour. Corney fetches a doctor and a priest. Both arrive too late to be of service. Elsie dies unabsolved: her baby is stillborn. Back in Cork, Corney is of course shunned. He gathers some carpenter's tools and takes to the road, resigned to being a journeyman. The rest of his days, it seems, are spent wandering alone – although, by the conclusion of the book, he has returned again to the family home and to Cork.

Like his grandfather, Corney enters a kind of internal exile. Like Parnell, he is the victim of conventional morality (and yet, also like Parnell, he has fetched disaster on himself). If he has been faithful to his grandfather – to that fierce mixture of personal and political independence – he has simultaneously betrayed Elsie, with her dependence on the church and its rites. Through the dilemma, O'Faoláin diagnoses a contradiction in Irish consciousness: rebelling against England's embrace, it clings to Rome's bosom. It is at once dissenting and conformist. Unable to exist with this opposition, Corney has to set out to find his 'new faith'. He never finds it and, weighing his deeds, ambivalently ends, 'I have denied life, by denying life, and life has denied me. I have kept my barren freedom.'

Siegfried SASSOON **1886–1967**

The Complete Memoirs of George Sherston

Memoirs of a Fox-Hunting Man
(1928)
Memoirs of an Infantry Officer (1930)
Sherston's Progress (1936)

Sassoon's skilfully fictionalised autobiography traces the spiritual development of a sporting young man, whose life of barely ruffled indolence is rudely interrupted by the First World War. Intimations of impending disaster are clear from the outset, since the first volume bears the Shakespearean epigraph: 'This happy breed of men, this little world.' This ironic appropriation of Richard II's eulogy of England (which continues 'This other Eden, demi-paradise, / This fortress built by Nature for herself / Against infection and the hand of war') should be kept in mind whilst reading the lengthy account of the fox-hunting, cricketing and steeple-chasing life of George Sherston in the Weald of Kent. Indeed, the narrative is carefully threaded with images which achieve their full ironic resonance as Sherston's prelapsarian world is overwhelmed by the war. 'Those who expect a universalization of the Great War must look elsewhere,' Sassoon writes. 'Here they will only find an attempt to show its effect upon a somewhat solitary-minded young man.' However, Sassoon later described Sherston as a 'simplified version of my "outdoor self"', and he thus becomes an avatar of the scores of young men, unremarkable in themselves, who led volunteers and conscripts to a costly victory.

Sherston's formal education is dismissed in two asides concerned with 'Ballboro'' and Cambridge. The trilogy's concern is with his education in the ways of the world and his 'emancipation from the egotism of youth'. In peaceful Kent, 'the only link with Europe' is a record of Dvořák's 'New World' Symphony, and the Coal Strike means little more than a disruption of trains which results in

Sherston missing 'some of the best hunts of the seasons'. The war soon cures him of such insularity as he enters the 'new world' created by high-explosive shells. Encountering the smashed body of a private, he begins to realise that 'life, for the majority of the population, is an unlovely struggle against unfair odds, culminating in a cheap funeral'. Sassoon is a sardonic biographer of his former self, deprecating the many acts of heroism, undermining brave gestures with accusation of feeble prevarication, and mercilessly analysing Sherston's attitude to the war: 'I wanted the War to be an impressive experience – terrible, but not horrible enough to interfere with my heroic emotions.' However, he is even less merciful to the Home Front, where the rural idyll, dreamed of in the trenches, is transformed into an alien world in which the soldier has no place.

Sherston is a subaltern in the Flintshire (i.e. Royal Welch) Fusiliers, and takes part in a number of major engagements, including the Somme and Arras. He wins the Military Cross, suffers from trench fever and German measles, and is wounded. Whilst convalescing in England he broods upon the war and upon the large number of his friends who have been killed. A leader in the *Unconservative Weekly* (i.e. *The Nation*) acts as 'a sort of divine revelation' and he determines to contribute an article intended to destroy the complacency of those at home. The editor, Markington (i.e. Hugh Massingham) advises him to speak to the pacifist intellectual Thornton Tyrrel (i.e. Bertrand Russell), after which he issues a statement outlining his reasons for refusing to fight any further. His regiment treats him with embarrassing (and embarrassed) decency, but his grand gesture is defused by the intervention of his fellow officer, David Cromlech (i.e. Robert Graves), who tricks him into attending a medical board. It is decided that Sherston is suffering from shell-shock, and he is sent to Slateford War Hospital for treatment.

The final volume reproduces large sections of Sassoon's diaries virtually unaltered, unlike its predecessors in which the material has been carefully selected and ordered. For Sassoon, this period was immensely important and productive, for it was while he was at the hospital that he met Wilfred Owen. Sherston, 'denied the complex advantage of being a soldier poet', merely recuperates under the care of the psychologist W.H.R. Rivers, and rejoins his regiment in Limerick. He gets in a bit of hunting before being posted to Palestine, where he delights in the bird-life but sees no action. He goes once again to France. Returning from a patrol, he is accidentally shot in the head by one of his sergeants and invalided home, where he feels disorientated and defeated by the war. The trilogy ends with a visit from Rivers, whose 'smile was benediction enough for all I'd been through'.

Although deprecated by Robert Graves, whose *Goodbye to All That* (1929) provides a fiercer (and more fanciful) account of the war he shared with Sassoon, *Memoirs of a Fox-Hunting Man* was one of the earliest and most artful of all the books written about the trenches. Sassoon's deft handling of his material may be gauged by comparing the trilogy with his *Diaries 1915–1918* (1983), and with *Siegfried's Journey* (1946).

Stevie SMITH   1902–1971

Novel on Yellow Paper

When Smith submitted a sheaf of poems to a publisher in 1935 she was told to go away and write a novel. Using the yellow paper reserved by her employers for carbon copies, she wrote an extraordinary account of the life and opinions of her *alter ego* Pompey Casmilus. Often compared with Sterne's *Tristram Shandy* (1759–67), it shares that novel's discursive manner and a narrative that is guided by association rather than chronology.

Pompey is the private secretary of Sir Phoebus Ullwater Bt, a magazine publisher and 'the only man with whom I have consistently ... behaved myself, as an efficient worker, as a willing donkey, as a happy, equable creature'. Pompey goes back and forth over her life, and offers opinions on love, marriage, death, the church, and the superiority of Racine over Shakespeare as a tragedian. 'For this book is the talking voice that runs on, and the thoughts come, the way I said, and the people come too, and come and go, to illustrate the thoughts, to point the moral, to adorn the tale.' Amongst the people who come and go are William, convalescing in a large house during the First World War and proving 'a very great nuisance, and a stumbling block to the VAD nurses' because of his 'arrangement dementia'; blue-eyed Karl, a student of philosophy whom Pompey visits in Germany as Nazism is gaining power; Miss Hogmanimy, the teetotal sex educationalist; Pompey's absent sailor father; her mother, dying of heart trouble; the 'Lion Aunt' with whom she lives in the north London suburb of Bottle Green; and 'sweet boy Freddy', to whom she is engaged.

The reader is put in the position of Coleridge's wedding guest. Pompey darts from topic to topic, mock-apologising for her garrulity: 'And you Reader, whom I have held by the wrist and forced to listen, I am full of regret for you ...' She retells Greek myths and Russian plays and slips in a few of her own poems. The book is lively and loquacious, but shot through with melancholy, 'richly compostly loamishly sad', never more so than when being stoic in the face of disaster, as when she breaks off her engagement: 'And I take the ring and put it back on the insignificant finger. *Adieu, éclat de fiancée.*'

Smith later disparaged the novel, which had been written under the influence of Dorothy Parker, whose work she had been reading. She felt that it had 'this sort of pseudo (for me) American accent'. It now seems, like her poetry, *sui generis*, the voice not Parker's but unmistakably Smith's own.

Over the Frontier (1938) is a sequel, in which Pompey, escaping from the misery brought about by the end of her affair with Freddy, travels to Europe and becomes embroiled in espionage.

1937

Japanese take Peking, Shanghai and Nanking ● Rebel victories in Spain ● Franklin Delano Roosevelt signs US Neutrality act ● Purge of USSR generals ● Italy joins German–Japanese Anti-Comintern Pact ● Martin Buber, *I and Thou* ● Nazi exhibition of 'Degenerate Art' in Munich ● Paris World Fair ● Jean Renoir, *The Great Illusion* ● Vanessa Redgrave born ● J.M. Barrie, Guglielmo Marconi, Edith Wharton, Lord Rutherford and Maurice Ravel die ● Wallace Stevens, *The Man with the Blue Guitar and Other Poems* ● Isaac Rosenberg, *Collected Works*

Morley CALLAGHAN	1903–1990

More Joy in Heaven

The last of a cluster of novels produced in the 1930s before a long period of silence, *More Joy in Heaven* deals with the criminal world, a subject which recurs in Callaghan's fiction. The story draws upon the life of the infamous bank robber Red Ryan, who was released from an Ontario penitentiary in 1935. Ryan vowed that he was a reformed man and was supported by a priest, a senator and even the Canadian Prime Minister, R.B. Bennett. A year later he was shot dead by police while robbing a liquor store.

Unlike Ryan, Callaghan's hero, the erstwhile bank robber Kip Caley, is genuinely reformed when he returns from prison to Toronto. He wants a job and a simple life, but his picture is in all the papers and people treat him as a celebrity. Sen. MacLean, who helped secure his release, urges him to take a job with hotel-keeper Harvey Jenkins, who wants him as a front-of-house man to draw the punters into his restaurant. Kip is the star attraction at a New Year's Eve party for the city's dignitaries. Overcoming his unease, he attempts to use his position to be 'a link between the two worlds, the outcasts and the right-thinking people', and seeks a position on the parole board. Only his priest, Fr Butler, questions his zealous hope: 'Maybe the prodigal son had a job going from feast to feast,' he comments. 'I wonder what happened to him when the feasting was over.'

Kip's attempt to get on to the parole board is blocked by the reactionary Judge Ford, and he is further disillusioned when Jenkins tries to involve him in a series of rigged wrestling bouts. Two of his old criminal cronies, Foley and Kerrmann, want him to take part in a bank job and he refuses; but his girlfriend Julie overhears their conversation and, fearing the worst, goes to Fr Butler. When Kip learns that the priest has tipped off the police he goes to the bank to warn Foley and Kerrmann and is shot at in the ensuing gun fight. He shoots a policeman and escapes wounded. He returns to Julie and she is shot by the police as she rushes to him. Now vilified as passionately as he was once lionised, Kip dies in hospital without giving the authorities the satisfaction of hanging him.

Callaghan's devout Catholicism and his experiences as a journalist and lawyer form the background to the novel, as they do to similar, early, moralistic books like *Such is My Beloved* (1934). His later work has divided critics, but there is no doubt that his novels deserve wider recognition than they have so far received outside his native Canada.

I. COMPTON-BURNETT	1884–1969

Daughters and Sons

'Well, gapy-face,' said Mrs. Ponsonby. A girl of eleven responded to this morning greeting.

'I wasn't yawning, Grandma.'

'Do not lie to me, child; do not lie; do not begin the day by lying,' said Sabine Ponsonby, entering her dining-room and glancing rapidly round it. 'And it would be better to shut your mouth and open your eyes. You will have us confusing one part of your face with the other.'

Domestic tyranny is immediately established in the opening lines of this, Compton-Burnett's seventh novel. Sabine goes on to assert her malign authority over her grand-children in the course of the first chapter, but she is an old woman, and her grip upon the household will gradually slacken throughout the novel.

Sabine's son, John, is a widowed novelist, whose sales have begun to decline. He has five children, whose ages range from twenty-five to eleven; as the author explains: 'The spaces in age were caused by the deaths of infant children, and Muriel had reversed the family rule and survived at her mother's expense.' The late Mrs Ponsonby's duties have been taken on by John's possessive and overbearing sister, Henrietta (Hetta). Like most Compton-Burnett juveniles, the children have formed a protective alliance, commenting upon the excesses of their elders in witty, penetrating, but usually murmured asides. Chilton – 'as much like Sabine as a boy of eighteen could be like a woman of eighty' – finds relief in the verbal bullying of his handsome younger brother, Victor. Of the two older daughters, France takes refuge in her writing, but Clare finds little to alleviate the oppressive atmosphere of the house. The gapy-faced Muriel is too young to understand much of what is going on, but is learning fast. Her latest governess, Miss Bunyan, is driven from the house by Sabine's rudeness, just as her predecessors had been, and is replaced by Edith Hallam. France's qualities as a writer, publicly aired when scenes from her novel are acted at a village function, rouse the professional jeal-ousy of her father. When the book is com-plete, therefore, she submits it to a publisher using Edith's name. It is accepted and earns her £2,000, which she sends to her unde-serving father anonymously, as if from an appreciative reader. Sabine, who regularly steams open other people's mail, is led to believe that Edith is her son's mysterious benefactor and suggests that he should pro-pose to this prosperous governess. He does so and is accepted, to the open disgust of Hetta, who feels herself displaced.

In a bid for attention, Hetta disappears, leaving a suicide note. Sabine alone is truly distressed by this event; the remainder of the family soon discover that life goes on as usual. Hetta returns, hoping to have taught her family a lesson, and is bluntly told by her brother: 'You have taught us that your service to us is not worth the price we have paid.' Hetta – who shares her mother's belief that knowledge is power, and has been doing her own detective-work – makes a final bid to reclaim her lost authority. During a dinner-party, in an extraordinary loss of control, she reveals that John had married Edith for money she does not have and that France is the true benefactor. This outburst proves fatal to Sabine, who quietly expires at table, an event which both family and guests take in their stride. Less easy to accept is the news that Sabine has left everything to Alfred Marcon, the young man who had been engaged as tutor to Chilton and Victor and who had become devoted to their grand-mother. Alfred announces that he intends to keep half the money, thus enabling him to propose to Clare as a man of means. How-ever, it transpires that he is merely an executor, and has been left a legacy rather than the entire estate, and is thus in the unhappy position of having lost both his money and his good name. Clare accepts the hand of an elderly neighbour, Sir Rowland Seymour, whose nephew Evelyn had partly engineered Alfred's disgrace, and Hetta marries the vicar, Dr Chaucer, whose niece, Miss Bunyan, is reinstated as Muriel's governess.

This ferocious (and ferociously funny) novel is given an added dimension by Comp-ton-Burnett's clear identification with France, whose writing is dismissed as a waste of time by her relations, rather as her own was dismissed by her companion, Margaret Jourdain, a prolific and successful author of books on furniture. 'Of course I never read Ivy's trash,' Jourdain purportedly remarked, and although she could be more generous in her response (and did indeed read the novels), she was clearly disturbed by her friend's literary endeavours. The depiction of Alfred's aunt, Charity Marcon, who researches and writes non-fiction books at the British Museum, is an affectionate joke at Jourdain's expense.

M.J. FARRELL 1905–

The Rising Tide

In this novel, set in Ireland in the first four decades of the twentieth century, places and Time are at least as important as any human character and, in most instances, far more powerful. Two houses in particular dominate the book, providing both cause and symbol of events. They are adversaries, and those who find their spiritual home in one will attempt to inhabit the other at their peril.

Garonlea, the first, is the seat of the French-McGraths, a house whose very fabric is steeped in an all-pervasive misery which even exorcists have failed to dispel. Lady Charlotte, its châtelaine when the book opens, is worthily matched with her marital home. A despot whose cruelty has sexual roots she will never understand, Lady Charlotte tyrannises her husband and four daughters – Muriel, Enid, Violet and Diana. Desmond, her son and Garonlea's heir, she adores.

Now and again Garonlea throws up an heir who pits himself against the house's malevolent powers and briefly stems the traditional tide of wretchedness. In Desmond, perhaps, Garonlea will meet its master, especially since, on his succession, he will bring a very different mistress: Cynthia, whom we first meet as a newly engaged woman, and who will go on to dominate people and events as surely as Lady Charlotte does now.

Cynthia's challenge to Lady Charlotte and Garonlea begins from the moment of her first visit, although only the most percipient discern it. With powers of manipulation better suited to the highest levels of international diplomacy, she woos, wins, forms alliances, creates rifts, and eventually, makes for herself a second court, at Rathglass, a family house on the other side of the valley where even the air is clearer, fresher, freer. (That Rathglass must first be emptied of an impoverished maiden aunt who, even in the process of eviction, becomes one of Cynthia's most ardent admirers, is testimony to both Cynthia's ruthlessness and irresistible charm.)

Beautiful, and able, it seems, to annexe the love and loyalty of most members of either sex, she is generous in her concern for those who have consented to be enslaved, sunnily skilful in her management of those who remain unwon. In ten years Lady Charlotte's overthrow has been virtually encompassed. Cynthia has produced two children, and she and Desmond remain as passionately in love as first they were, united particularly in their almost mystical love of horses and hunting.

But with Desmond's death in action in 1915 Cynthia's character begins a decline so gradual as to be at first imperceptible, and made all the more confusing because so many of her vices mimic virtues. Her own equestrian skill and splendid physical courage in the hunting field deflect attention from the cruelty with which she schools her less proficient children, who come to hate her. Her spirited rejection of stifling conventions, small-minded snobberies and meaningless taboos goes hand in hand with an increasingly frenetic pursuit of hopelessly unsuitable sexual adventures, and the alcohol she turns to never adequately numbs the pain of each new affair's inevitably disastrous outcome. Most chillingly of all, Cynthia, who had from the first sworn to conquer Garonlea and its ghosts, brings to her task a single-mindedness and strength of will hideously reminiscent of Lady Charlotte – and of Simon, Cynthia's son, who watches unforgivingly the lovers, the drink, the cruelty and egotism.

Hard, absurd, vain, stupid, sordid, silly, cheap and cruel – all these adjectives, and worse, are used by Farrell of Cynthia. To others, especially Diana, who is more than half in love with her and long ago deserted Garonlea for Rathglass, she is goddess, queen and Venus (to Farrell, too, when in softer mood). The book's final pages are devoted to Simon's elaborate plan to end his mother's monstrous sway; in the event he unleashes Garonlea's sleeping powers, dormant under Cynthia's rule. Only Cynthia – the most troubling and morally ambiguous of all Farrell's characters – averts disaster, whilst finally acknowledging her own defeat.

V.S. PRITCHETT   1900–

Dead Men Leading

Harry Johnson is an explorer whose missionary father, Alexander, disappeared in the Amazon jungle seventeen years ago. Restless in England, Harry has an affair with Lucy Mommbrekke, persuading her, on one occasion, to tie him up and, on another, to sail with him during a violent storm. Soon and without regret (for he is a misogynist), he leaves on an expedition to the region of his father's disappearance.

The purpose of the expedition at first is unclear; but Lucy's stepfather, Charles Wright, is to be its leader. Its third member is Gilbert Phillips, Lucy's lover before Johnson intervened. (Phillips harbours no resentment: Lucy was a tribute to his charismatic friend.) Phillips and Johnson journey upriver to meet Wright. Johnson tortures himself by supposing that Lucy is pregnant, a supposition that becomes an unsubstantiated certainty. Reluctant to face Wright, he tries to persuade Phillips that they should skip the rendezvous. Phillips is dismayed. Johnson falls ill, and is taken, delirious, to the town where Wright awaits them – to the house from which his father last departed.

When he recovers, he escapes his companions and, with a few natives and a guide, begins to retrace his father's route. Wright and Phillips follow and finally catch up with him. Wright reveals that he is searching for gold. Then, on a hunting sortie, he is shot – possibly by Johnson (Pritchett is careful to blur this critical scene). The guide, Silva, and the natives desert for the town and Silva spreads the rumour that Wright's wife was expecting Johnson's child. Meanwhile, Johnson bullies Phillips into travelling with him overland. As ever, the explorer's motives are opaque. He may believe that he can find his father; he may desire his own and Phillips's destruction. Phillips fears the latter; Johnson's contempt for him grows. Nevertheless, the two stagger on, suffering from boils, hunger and thirst. Phillips attempts – and fails – to kill Johnson. They both imagine that they hear laughter. Close to death, Johnson sets off alone to look for water and – like Alexander – disappears. Phillips is rescued

by the seasonal rain, and a party of Germans returns him to civilisation. Back in London, he offers Lucy a memento – Johnson's gun. She declines it. She is married and now genuinely pregnant.

Her gesture is a rejection of the vanished man. Johnson has been drawn into the vacuum of Alexander's indeterminate demise. His previous explorations have been merely rehearsals for a suicidal and mysterious quest. Cruel and masochistic, he has been attempting to recapture the masculinity which – literally, in the person of his father – he lost as a youth. The obsession has endangered those around him. Furthermore, in so far as he has emulated Alexander, Johnson has sought to be a similar temptation – to pose a kind of riddle for the living and thus, in an apparent act of abnegation, vainly to extend his hold on them. Hence, Phillips keeps the gun uneasily: it is an emblem of Johnson's paradoxes, the eerie remnant of his influence and simultaneously the proof of its end.

Pritchett told Paul Theroux (who provided an introduction to the 1984 reissue of the novel) that in *Dead Men Leading* he had 'attempted a psychology of exploration'. He had read several books about missionaries and explorers and decided that what they shared was a certain masochism. The novel was partly inspired by Peter Fleming's classic travel book about the search for a missing explorer, *Brazilian Adventure* (1933), a useful source since Pritchett had not himself visited Brazil. Echoes of both books may be found in Justin Cartwright's excellent novel *Interior* (1988), in which a man sets off into the heart of Africa in search of his missing father.

Forrest REID   1876–1947

Peter Waring

This work, strongly autobiographical in mood and feeling if less so in actual events, is Reid's revision of his fourth novel, *Following Darkness*, published twenty-five years earlier in 1912, and one of the first books in English to take as its main subject the particular emotional and intellectual crises of male adolescence.

Set in Belfast and rural County Down, the novel tells of one year in the life of Peter Waring, a physically ungainly, intellectually gifted boy who lives with his schoolteacher father, David, in the mean, gimcrack little schoolhouse of Newcastle. His mother abandoned husband and son years ago, and the gloomy, shoddy house is an appropriate setting for the narrowly joyless brand of Protestantism which David Waring espouses. His undoubted love for his son is barely perceptible beneath doom-laden prophecies for his future and constant disapproval of his present. The straitened ugliness of the house contrasts strongly with the beauty, space and freedom of Derryaghty, the home of Mrs Carroll, a recently widowed neighbour of means and influence. Sympathising discreetly with Peter's unhappiness, whilst recognising the intellectual abilities which she is determined to foster and support, she plans to send him eventually to Oxford or Cambridge, at her own expense. (Reid himself would in later life fulfil just such a tutelary role for a number of promising young men.) Peter, made free of Derryaghty's light and airy rooms (two of them set aside for his especial use) revels in the attics filled with entrancing lumber, and the walls covered with haunting portraits of long-dead Carrolls. He develops a relationship with the house and its 'ghosts' which complements his mystical response to the mountains, rivers, shores and sea of the surrounding countryside. An unusually severe attack of scarlatina, with its attendant fevers and hallucinations, heralds a new period in Peter's life, when his already acute perceptions reach a pitch of sometimes intolerable intensity.

Into this mass of heightened emotions come Kathleen Dale and her twin-brother Gerald, young relations of Mrs Carroll, visiting Derryaghty for the summer. Kathleen is friendly, natural and straightforward; Gerald is difficult, confusing and intimidating. It is Kathleen with whom Peter falls in love, although a never fully articulated undercurrent of tension in the book links Peter to Gerald, almost, it seems, against the author's will. The summer is an idyll, darkened only by the knowledge that the coming academic year must be spent in Belfast, where Peter will live, as a lodger, with his father's unappetising relations: Uncle George, kindly but ineffectual; Aunt Margaret, vicious and unjust; and Cousin George, superficially amiable, but fundamentally meretricious and wearyingly dirty-minded.

Only the friendship of Owen Gill, a fellow pupil, makes life bearable and Peter invites him to spend part of the next summer in Newcastle where, once again, the Dales will be. But where the previous summer was paradise, this one is hell. Owen gets on alarmingly well with Peter's father. Even more disturbingly, Owen is clearly attracted and attractive to Kathleen. Peter rages, and by outbursts of bitter jealousy and humiliating importunity sabotages his friendship with the hopelessly uncomprehending Kathleen. Gerald watches all with the quick and subtle compassion which underlies his carefully affected cynicisms.

With the total despair of adolescence, Peter makes an unsuccessful suicide attempt and is nursed back to health by Mrs Carroll. Her feeling for him, he realises, constitutes 'a kind of surety', bringing with it 'if not positive happiness, at any rate a sense of healing and of peace'. It is a sombrely measured conclusion, after the turbulent extremes of exaltation and abasement which have gone before.

E.M. Forster, to whom both *Following Darkness* and *Peter Waring* were dedicated, advised against Reid's revision, saying he 'didn't see how you can possibly bring *Following Darkness* up to date'. In the event, *Peter Waring* has proved to be one of Reid's most popular works, achieving numerous reprints.

John STEINBECK   1902–1968

Of Mice and Men

Steinbeck's tale of landless farm-labourers focuses on the character of Lennie Small, a simple-minded giant who relies on his friend George Milton to keep him out of trouble. After an incident upstate, Lennie has been falsely accused of rape, and the two friends are now on the run from the lynch-mob.

George and Lennie dream of the time when they will have a small plot of land together, and live a simple life of self-sufficiency, where Lennie, with his child's delight in petting little animals, will get to tend the rabbits. 'He's jes' like a kid, ain't he?' a ranch-hand reflects. 'Sure he's jes' like a kid,' George assures him. 'There ain't no more harm in him than a kid, neither, except he's so strong.' In pursuit of this fantasy, George finds them work on another ranch, where they will live quietly until they have got the money together. But as the boss's son, Curley, starts to pick a fight with Lennie, the atmosphere turns sour; and Curley's flirtatious young wife is showing a dangerous interest in the ranch-hands.

In a scene foreshadowing the novel's ending, Candy, an old ranch domestic, agrees to have his sick dog shot. Lennie is given a puppy from a new litter, but his constant attentions to it threaten to destroy it. Curley goads Lennie until, in self-defence, Lennie crushes the smaller man's hand in his grip; shortly afterwards, his clumsy strength kills the puppy too. Curley's wife finds him alone in the barn, mourning the death of the dog and fearing that George will now forbid him to look after the rabbits. She taunts the simpleton with her sexuality, letting him stroke her hair, but his fascination frightens her and as she begins to scream Lennie smothers her.

Hiding out in the brush, Lennie waits for George to join him so that they can escape together. But George, watching Curley and his men prepare to kill the dangerous man, realises that Lennie can no longer be allowed to live. Soothing his friend with the fantasy of their little farm, he puts a pistol to the back of Lennie's head. 'Le's do it now,' Lennie begs, 'Le's get that place now.' 'Sure,' George agrees. 'Right now. I gotta. We gotta,' he says as he pulls the trigger.

The novella's title refers not only to the mice Lennie kept as a boy, but also to Robert Burns's poem 'To a Mouse': 'The best laid schemes o' mice an' men / Gang aft a-gley.' Not only is George and Lennie's scheme to become self-sufficient farmers doomed to failure, but in a more general sense, humankind is at the mercy of an indifferent fate, the same one that has given a simple-minded man enormous physical strength. The more specific references to the Arthurian legend in

Tortilla Flat (1933) may also be discerned in *Of Mice and Men*, most particularly in the character of George, who moves among the sex-obsessed ranch-hands like an unspotted Sir Galahad. His protection of Lennie is loosely chivalrous, while their journey towards the grail of a smallholding has something of the knight's quest. The book made Steinbeck's reputation, was made into a highly successful play the same year, a film in 1939, directed by Lewis Milestone from a screenplay by Eugene Solow, and in 1970 was the basis of an opera.

J.R.R. TOLKIEN 1892–1973

The Hobbit

Originally written for children, Tolkien's fantastical adventure story became a cult book among adolescents and adults in the 1960s and 1970s. It can be read as a prelude to *The Lord of the Rings* (1954–5). The hobbit in question is Mr Bilbo Baggins of Bag End in the Shire, a district of mythical Middle-Earth. Into Bilbo's blameless, middle-aged, bachelor life suddenly comes Gandalf, an old, itinerant magician who needs his help in an adventure. Bilbo's reluctance to leave the comforts of his burrow is overcome when Gandalf sends thirteen dwarves to stir his imagination with wild songs describing a wider, more dangerous world beyond the Shire.

He agrees to help the dwarves in their attempt to regain their ancestral halls and mines from the great dragon Smaug, who has driven them out from the Lonely Mountain and appropriated their treasure. Bilbo's journey with Gandalf and the thirteen dwarves becomes a terrifying quest involving encounters with elves, goblins, huge and malevolent spiders, and a peculiar creature who calls himself Gollum.

Separated from his companions, Bilbo finds himself in deep caverns under the Misty Mountains. Groping in the darkness, his hand finds a ring which, as it turns out, belongs to Gollum, who treasures it as his sole possession of value. In a rather unfair contest, Bilbo wins the right to keep the ring and escapes from Gollum, who ever after bears a grudge against Bilbo. Gandalf has some inkling of the ring's power, but decides to say nothing and allow Bilbo to keep it.

The remainder of the story is devoted to the journey towards the Lonely Mountain, Bilbo's encounter with Smaug in his treasure cave, the death of the dragon, and the restoration of the dwarves to their mountain kingdom.

Virginia WOOLF   1882–1941

The Years

A great popular success, *The Years* was at the same time critically the worst received of Woolf's mature novels, owing to its superficially simpler and less lyrical style. It was innovative in other ways, however, since it offered a perceptive portrait of the English middle classes and their shifting attitudes and experiences from 1880 to 1937. Woolf referred to the first draft as an 'essay novel', in which sociological essays were intercut with fictional illustrations, but the essays were eventually, after a painful and laborious process of revision, merged into a larger fictional narrative.

The novel describes three generations of a family, the Pargiters, and especially the middle generation, a generation of idealistic belief in social change. The women of this generation tend towards feminism. Rose and Eleanor remain unmarried: Rose in order to pursue a career as a Suffragette campaigner; Eleanor involuntarily, because the early death of her mother saddles her with the responsibility of raising the family and housekeeping for her father. Delia's sense of injustice leads her to Parnellism and marriage to an Irishman, while Milly, the more conventional sister, settles down to domesticity with a north-country squire.

The men of the generation are outwardly more complacent: Edward is a classics scholar at Oxford, Morris a barrister, and Martin a soldier in India. But they too are individualistic beneath the surface. Edward and Martin are also unmarried, Edward possibly due to unstated homosexual preferences, Martin because of a broad dislike of English family life. Morris is the exception, being a married man with children. His children are another unconventional generation, however. Peggy, as a doctor, achieves the social status denied her aunts by the conventions of the previous epoch. North reaches adulthood in the trenches of the First World War, and afterwards rejects England for life on a remote farm in South Africa.

The novel has three broad sections, one for each generation. In the first, the family assembles for the funeral of Mrs Rose Pargiter, the wife of Col. Abel Pargiter, an event symbolising the gulf between the Victorian parents and their restless children. The next section is broken up into a string of brief encounters between the seven brothers and sisters and their various cousins, and is further dislocated by the intrusions of the First World War. The last section, echoing the first, crystallises around a family gathering, this time a generally good-natured family party in 1937. Unlike the silence dividing the Victorian parents from their children, the Edwardians survive into the modern world with the ability to communicate with the next generation still intact. Peggy and North are sceptical, anti-Utopian, but they still admire the conviction and courage of their elders.

The panoramic sweep of *The Years* allows Woolf little time to develop the interiority of her characters, but her writing is still highly innovative. It is illuminated by flashes of Proustian reminiscence, which recurrently enliven the characters' sense of wonder at their own lives, and understanding of each other's.

1938

Germany annexes Austria ● The 'Munich' crisis ● British national register for war service ● Ball-point pen invented ● Benny Goodman's band dominates Broadway ● Edward Hulton starts *Picture Post* ● Thornton Wilder, *Our Town* ● New York World Fair ● Constantin Stanislavsky and Kemal Atatürk die ● Laura Riding, *Collected Poems*

Samuel BECKETT   **1906–1989**

Murphy

Beckett's skills as a dramatist are clear in this black comedy, in which each scene is described with remarkable attention to detail. The dialogue is unnaturalistic, all the characters speaking with the author's personality and considerable vocabulary, and this produces its own peculiar humour. The scene in which Murphy successfully defrauds a tea-room of less than tuppence is an exquisite comic vignette at the centre of this very amusing first novel.

Murphy is an expatriate Irishman, living in a London bedsit, and pursuing an affair with a prostitute called Celia. His determination to avoid work is matched by the ability to embezzle the tiny sums of money that sustain him: 'The last time I saw him, he was saving up for an artificial respiration machine to get into when he was fed up breathing.' Naked and tied to his rocking chair when we first meet him, Murphy is trying desperately to answer the telephone, so that his landlady, Miss Carridge, will not enter the room to do so and find him in this exposed, if self-imposed, condition. He succeeds, but subsequently tips over the chair, knocking himself unconscious.

Back in Dublin, Murphy's old teacher, Neary, is in despair over his love for a Miss Counihan, who is herself smitten with the absent Murphy. He enlists the help of his one-eyed, alcoholic manservant, Cooper, and another former pupil, the treacherous Wylie, to find Murphy, so that it can be established that he is either dead or no longer cares for Miss Counihan.

Meanwhile, Murphy deserts Celia, who has been badgering him to get a job, when he agrees to take over the nursing duties of an old acquaintance, the homosexual Austin Ticklepenny. He applies to the Magdalen Mental Mercyseat, an insane asylum, working under the head-nurse, Bim, who has a fancy for Ticklepenny 'not far short of love'. Murphy strikes up a friendship with one of the patients, a schizophrenic called Mr Endon, who, it is feared, might attempt suicide by apnoea (cessation of breathing). Nurse and patient play protracted games of chess. Murphy installs himself in an attic room at the hospital, along with his rocking chair, to which he has tied himself when he is killed by a gas explosion.

In Dublin a week later, that would be September 19th, Neary minus his whiskers was recognized by a former pupil called Wylie, in the General Post Office, contemplating from behind the statue of Cuchulain. Neary had bared his head, as though the holy ground meant something to him. Suddenly he flung aside his hat, sprang forward, seized the dying hero by the thighs and began to dash his head against his buttocks, such as they are.

MURPHY
BY SAMUEL BECKETT

Cooper is entrusted with Murphy's ashes, which, in accordance with the deceased's wishes, are to be flushed down the lavatory of the Abbey Theatre in Dublin. On the way to perform this duty, Cooper stops at a pub and, as the result of a fight, Murphy's ashes are there scattered amongst the spit and sawdust on the floor.

Elizabeth BOWEN   1899–1973

The Death of the Heart

Although *The Death of the Heart* was a critical and popular success, Bowen liked it least of all her novels, dismissing it as 'an inflated short story'. Essentially a moral tale about the corruption – and corrupting power – of innocence, it is a tragi-comedy which displays to best advantage the author's distinctive style and dry wit. The sharply drawn character of Eddie is based upon Goronwy Rees, with whom Bowen had fallen in love only to lose him to her fellow novelist and friend, Rosamond Lehmann. Rees recognised himself and determined to sue for libel, a course of action from which he was dissuaded by E.M. Forster.

After the death of her mother, sixteen-year-old Portia Quayne is landed upon her reticent half-brother, Thomas, and his cool, sophisticated wife, Anna. 'Conceived among lost hairpins and snapshots of doggies in a Notting Hill Gate flatlet' (as Anna puts it), Portia is the result of a liaison between Thomas's father and a socially inferior woman whom he was obliged to marry when his wife – 'all heroic reserve' – arranged a divorce. Portia has been living a peripatetic life in hotels on the Continent, and cannot get used to the quiet grandeur of the Quayne's Regent's Park house, or the ways of the world in which she finds herself. For her part, Anna feels uncomfortable and exposed with Portia in the house, a feeling exacerbated when she reads Portia's diary. 'It's given me a rather more disagreeable feeling about being alive – or, at least, about being me', she complains to Thomas and her friend, the novelist St Quentin Miller.

Portia falls in love with Eddie, an unpredictable, vain, neurotic (but curiously innocent) twenty-three-year-old who has enjoyed a flirtatious relationship with Anna and is somewhat unwillingly employed by Thomas in his advertising agency. When Thomas and Anna go to Capri, Portia stays at the coast with Anna's former governess, Mrs Heccombe, and her two stepchildren, Dickie and Daphne. Although the comically

rumbustious household of 'Waikiki' seems less restrained than Windsor Terrace, it turns out to be governed by a stricter code of decorum, as Portia discovers when she tells Daphne that she has seen her holding Eddie's hand in the cinema. 'People creeping and spying and then talking vulgarly are two things that I simply cannot stick,' Daphne says angrily.

Portia is convinced that everyone is discussing her and laughing at her, and St Quentin's revelation that Anna has been reading her diary confirms her worst fears. Believing that Eddie, who seems incapable of returning her intense love, is in conspiracy with the Quaynes, she seeks refuge at the Karachi Hotel in Kensington with the unfortunate, unemployable Major Brutt ('makes of men date like makes of cars'), an acquaintance of Anna. With unintentional cruelty, she tells him that he too is an object of mockery, and thus a kindred spirit, and wildly suggests they marry, a proposal Brutt declines with dignity. Portia's refusal to return home throws the Quaynes into confusion, and after an amusing scene in which various resentments and preoccupations rise to the surface, it becomes clear that only the inexorable Matchett, a servant inherited with the furniture, who takes 'suggestions only' from her employers but has a close relationship with Portia, is equal to the task of negotiating the girl's return. The novel ends with Matchett travelling by taxi to the Karachi.

Lettice COOPER   1897–

National Provincial

Cooper's most substantial novel is set in Leeds (which she calls Aire) in the middle 1930s, and draws upon her own experience a few years earlier as a young woman returning from Oxford to her native city. It is a highly political novel, intended, in the words of Cooper's preface to the 1987 reissue, as 'a microcosm of the world in general'.

In 1935 journalist Mary Welburn, an Oxford scholarship girl of humble origins, reluctantly returns from London to Aire to nurse her ailing mother now that her sister is about to be married. Mary gets a job on the conservative *Yorkshire Guardian*, and her first assignment is to cover a charity performance of *A Midsummer Night's Dream* in the grounds of Greenoak Hall, home of land-owner William Marsden, a Liberal 'more Conservative than any Tory'. Many of the multitudinous cast of characters are here introduced, among them Marsden's unhappily married nephew, Stephen Harding, who works for the self-made clothing manufacturer Alfred Ward. Mary's uncle, John Allworthy, a Labour councillor and shop steward, works in Ward's factory where he enjoys the grudging respect of his old sparring partner. Mary's relationship to her uncle places her in a socially ambiguous position when she is invited to dinner with the Wards. Stephen and Mary are drawn towards one another, and Stephen is led into left-wing politics much to the consternation of his family.

But it was not surprising, he thought, that strikes arose and bred hatred and bitterness and were difficult to settle! Not surprising that nations persuaded themselves that they went to war for an ideal, or that governments acting against the interests of two thirds of the population honestly believed that they were doing their best for the whole nation. It was not surprising that in a world of plenty large numbers of men, women and children went hungry. He dealt every day with people who had only to walk out of their home or put out their hand to enjoy a free, full and active life, but who did not walk out or put out their hand.

NATIONAL PROVINCIAL
BY LETTICE COOPER

Allworthy has a young protégé in the factory, Tom Sutton, who leads an unofficial strike against the wishes of the union. The traditionalist Labour view of these militants is reflected in Grace Allworthy's opinion of the Left Discussion Group: 'Not a solid Labour man among them! Just these young whipper-snappers of I.L.P. boys and Communists.' The strike is broken with All-

worthy's help and Sutton stands against him in the 1936 council elections, thereby splitting the Labour vote and allowing the Tory candidate, Stephen's brother, to get in. With the war in Abyssinia raging, and fascism ascendant across Europe, families are torn apart by political disagreement. Ward's flippant son becomes a Communist while at university and has his allowance withdrawn: 'If you'd any real principles, you wouldn't touch my capitalist money,' his father tells him. Everywhere, Cooper suggests, the spectre of English 'respectability' prevents real political change or personal happiness, as witnessed in Mary and Stephen's doomed love. At the end of the novel she returns to London, unable to live in the same place as him.

John DOS PASSOS   1896–1970

USA

see 1936

Daphne DU MAURIER   1907–1989

Rebecca

The famous opening line, 'Last night I dreamt I went to Manderley again', is as well known from Alfred Hitchcock's film (1940), starring Joan Fontaine as the nameless heroine, as from the novel. The story subtly subverts the classic rags-to-riches or Cinderella theme to provide an eerie, romantic, almost Gothic narrative. The atmospheric dream described in the opening chapter gives a sonorous clue to the grim reality of the last: 'The house was a sepulchre, our fear and suffering lay buried in the ruins. There would be no resurrection.'

The heroine is employed as a companion by the odious Mrs van Hopper, a rich, gossipy, snobbish and inconsiderate American. In Monte Carlo she is courted by the glamorous widower Maxim de Winter, and swept off as his second wife to the exquisite house Manderley in Cornwall. She feels herself to be dull and mousey in comparison to Maxim's first wife, Rebecca, by all accounts a creature of impulse and beauty, who was drowned while sailing at night near the estate. The spirit of Rebecca haunts the heroine, the house and her widower. It also obsesses the housekeeper, Mrs Danvers, who devotes herself to the memory of her late mistress.

The grandeur of Manderley, the introversion of her husband, the animosity of the housekeeper, and her own self-doubt when she compares herself unfavourably with Rebecca, conspire to intimidate the heroine until, to her astonishment, Rebecca is shown to have been faithless, capricious and malicious. Further, it is revealed that she was hated by her husband.

It transpires that Maxim has murdered Rebecca and, when her body is found in the cabin of her sunken boat, things begin to look bad for the de Winters. To the relief of all, it is discovered that Rebecca had consulted a doctor just before her death, and that he had given her only a few months to live. This appears to confirm the Coroner's verdict of suicide. While investigations are being pursued in London, Mrs Danvers, alone at Manderley and thirsting for revenge on the

---

He did not look at me, he went on reading his paper, contented, comfortable, having assumed his way of living, the master of his house. And as I sat there, brooding, my chin in my hands, fondling the soft ears of one of the spaniels, it came to me that I was not the first one to lounge there in possession of the chair; someone had been before me, had surely left an imprint of her person on the cushions, and on the arm where her hand had rested. Another one had poured the coffee from that same silver coffee pot, had placed the cup to her lips, had bent down to the dog, even as I was doing.

REBECCA
BY DAPHNE DU MAURIER

---

de Winters, has set fire to the house, destroying Manderley. As the de Winters drive towards the house on their return from London, 'the ashes blew towards us with the salt wind from the sea'. If the dream is destroyed, so too is the spirit of Rebecca, which had driven a wedge between de Winter and his wife. They go into a rootless exile: 'We can never go back again, that much is certain. The past is still too close to us. The things we have tried to forget and put behind us would stir again, and that sense of fear, of furtive unrest, struggling at length to blind unreasoning panic – now mercifully stilled, thank God – might in some manner unforeseen become a living companion, as it had been before.'

Graham GREENE   **1904–1991**

Brighton Rock

Greene's first explicitly Catholic novel uses the genre of the thriller to explore the reality of good and evil in relation to an *âme damnée*, the young gangster Pinkie Brown. Greene himself described the novel as an examination of 'the effect of faith on action'.

It is set in a richly realised pre-war Brighton amongst gangsters: the upmarket Colleoni gang, and a mob led since the death of its former head, Kite, by the seventeen-year-old Catholic, Pinkie Brown. The novel opens with a newspaper employee known as 'Kolley Kibber' laying a trail of cards around Brighton for tourists to find and cash in as part of a publicity drive. This man is really Fred Hale, who is in fear of his life because of the death of Kite, whose gang are in pursuit of him. He picks up blowsy Ida Arnold, hoping that the gang will not strike if he is with a woman, and telling her that he is a dying man. While she visits the lavatory, he disappears, and is later found dead from a heart attack. Ida is suspicious. One of the gang, an incompetent called Spicer, becomes frightened, and Pinkie arranges for him to be killed by Colleoni's gang at the races. In the event, Pinkie is attacked and slashed with a razor, while Spicer escapes. Pinkie is obliged to murder Spicer, arranging the death to look like an accident. Ida pursues Pinkie, determined to uncover the real facts of Hale's death.

Meanwhile Pinkie has been forced into a relationship with a dim little waitress called

Rose who has evidence which might lead to the discovery that Hale was murdered. Pinkie is repelled by sex, having witnessed as a child his parents' weekly exertions, but he arranges to marry Rose to prevent her being able to give evidence against him in court. Even then he cannot trust her and he attempts to set up a suicide pact in which Rose will kill herself first, after which he will be free. Rose is persuaded because she, like Pinkie, is Catholic and believes herself to be in mortal sin (their marriage is not consecrated by the church). She would prefer eternal damnation with him to salvation without him. Ida arrives in time to prevent Rose's suicide, whilst Pinkie, burning already with the effects of the acid he had attempted to throw in another gangster's face, topples over the cliff. The novel ends with the pregnant Rose at confessional being told by the priest that Pinkie's love for her might prove his redemption, this being part of 'the ... appalling ... strangeness of the mercy of God'. But she returns home 'towards the worst horror of all': a recorded message Pinkie made in which she expects him to declare his love, but which will reveal his hatred of her.

'Of course it's true,' the Boy said. 'What else could there be?' he went scornfully on. 'Why,' he said, 'it's the only thing that fits. These atheists, they don't know nothing. Of course there's Hell. Flames and damnation,' he said with his eyes on the dark shifting water and the lightning and the lamps going out above the black struts of the Palace Pier, 'torments.'

'And Heaven too,' Rose said with anxiety, while the rain fell interminably on.

'Oh, maybe,' the Boy said, 'maybe.'

BRIGHTON ROCK
BY GRAHAM GREENE

As a thriller the book is effective, cynical, squalid and brutal, but its ultimate success (like that of **Brideshead Revisited**, 1945, by Greene's friend and co-religionist Evelyn Waugh) depends upon the reader's acceptance of the theology behind it. Jansenist in

outlook, it suggests that Pinkie's grasp of the reality of eternal good and evil – the acceptance of the latter presupposing the existence of the former – is, paradoxically, preferable to Ida's cheerful, *worldly* (and therefore ultimately worthless) sense of right and wrong.

R.C. HUTCHINSON   1907–1975

Testament

A panoramic novel set against the background of the First World War and the Russian Revolution, *Testament* is the narrative of Capt. Alexei Alexeivitch Otraveskov. Following the trials of 1909, he has been in political exile in Siberia but is granted a conditional pardon in order to serve in the war. Wounded in the retreat from Lensie, he is taken prisoner by the Austrians and in Krozkohl gains the friendship of another prisoner, Count Anton Antonovitch Scheffler, a barrister renowned for his advocacy of political defendants. As a result of an exchange of wounded prisoners, the two are repatriated to the clearing station at Mariki-Matesk. On leave Otraveskov travels to Petrograd and his home town of Voepensk searching for his wife, Natalia, and their young son, Vava, who have disappeared. Petrograd is a comic chaos of bureaucracy run wild and frivolous high society frantically pretending that the war is not happening. With the help of Scheffler's wife, Otraveskov finds Natalia in hospital, where she has lost her memory and believes him to be dead, and Vava, who is crippled by a spinal defect, living in a squalid slum.

On his return to Mariki Otraveskov becomes involved in the court martial of Scheffler, who has refused an order to send unfit soldiers back into action. Despite face-saving offers, Scheffler refuses to compromise his moral position. Discharged from the army, Otraveskov returns to Petrograd where he attempts to wield influence to save his friend from execution. The situation changes with the February Revolution, a newspaper declaring: 'The Cadets hold in a Czarist Dungeon a famous Advocate of the Workers' Rights.' Otraveskov and the partially recovered Natalia travel to Moscow –

where Scheffler is now imprisoned – to seek medical help for Vava. Sheffler is released after the October Revolution and appears at first to be in cahoots with the Bolsheviks, though once again his conscience makes him speak out against injustice. Following his unsuccessful defence of Natalia's father, he is again imprisoned and charged with being a Czarist collaborator. After a show trial – during which falsified evidence purportedly given by Otraveskov is used against him – he is executed. Otraveskov manages to flee to Paris with his family.

To me, it seemed to make no difference. The bitter wind that drove in through the broken window would grow no milder, the pain of my leg would not be slighter or the carriage wall less hard if all the thrones in Europe were overturned. I wanted only to lie at full length in some warm place and go to sleep. But an odd phrase of the drunken boy's, 'Wait till you see how it is in Petrograd!' was repeated in my mind's ear like the dreary call of a cuckoo, reminding me that Petrograd was my destination; and as soon as my restless thoughts were shown that track they dashed full speed along it.

TESTAMENT
BY R.C. HUTCHINSON

Testament is a large, ambitious novel, rich in its descriptions of the Russian landscape and the chaos and hardship of the times, which seeks to examine the agonising moral dilemmas people confront in the face of political upheaval and corruption. It won Hutchinson public acclaim and the *Sunday Times* Gold Medal for fiction, and remains, along with the unfinished *Rising* (published posthumously in 1976), the best known of his seventeen novels.

Raja RAO 1909–

Kanthapura

Rao's first novel, like his later short-story collection *The Cow of the Barricades* (1947), draws upon his own involvement in the Indian independence movement of the 1930s. It is set in the southern Indian village of Kanthapura, in a poignantly evoked landscape 'with chillies and coconut, rice and ragi, cloth, tamarind, butter and oil, bangles and kumkum, little pictures of Rama and Krishna and Sankara and the Mahatma'. Narrated by an elderly Brahmin woman, Achakka, the novel is steeped in Hindu legend and myth, and was warmly praised by E.M. Forster.

The young men of the village, particularly the university-educated Dorè and Moorthy, join the Congress party and espouse the doctrines of Gandhi. The merchant and money-lender Bhatta is implacably opposed to these modern ideas, especially as they regard the caste system and the position of the pariahs or untouchables: 'The Mahatma is a good man and a simple man,' he says. 'But he is making too much of these carcase-eating pariahs.'

Moorthy is excommunicated by the local Swami (who, it transpires, is in the pay of the British) for attempting to educate the pariahs who live on the outskirts of the village. His mother is so mortified that she takes her own life. Despite police harassment he continues to organise resistance against the authorities, and even elderly and conservative Brahmins like Achakka are drawn into the nationalist struggle. The villagers refuse to pay their taxes and the young men boycott the 'toddy shops' from which the 'Red-man's' government makes so much money. The government responds by sending in the police to confiscate the land of those refusing to pay their taxes. The novel culminates with a protest march during which the police and the army beat and arrest the protestors, finally driving many from the village.

Dorothy RICHARDSON 1873–1957

Pilgrimage

Pointed Roofs (1915)
Backwater (1916)
Honeycomb (1917)
The Tunnel (1919)
Interim (1919)
Deadlock (1921)
Revolving Lights (1923)

The Trap (1925)
Oberland (1927)
Dawn's Left Hand (1931)
Clear Horizon (1935)
Dimple Hill (1938)

'There is no drama, no situation, no set scene,' a reviewer wrote when the first part of *Pilgrimage* appeared, remarking on the 'moments tense with vibration, moments drawn out fine, almost to snapping point ... It is just life going on and on.' Spanning the early decades of the century, from the First World War to the eve of the Second World War (when *Pilgrimage* was revised and printed as a complete sequence), Richardson's autobiographical, stream-of-consciousness narrative extends to over 2,000 pages and contains twelve novels within the four volumes.

Volume one opens with *Pointed Roofs*. Miriam Henderson, the central character, begins her young adult life by leaving the comfort of her upper-middle-class family to work as a teacher in Germany. Her stay at Waldstrasse introduces her to the new horizons of a different world, as she ceases to regard music as an embarrassing performance art and comes to understand its 'expression, style and phrasing'. Throughout Richardson's work, music provides both a social experience through the life of artists and a conscious metaphor for creativity.

Miriam is popular amongst her pupils, who are girls of almost her own age, and her acceptance is consolidated as she is moved from her initial lodging in an attic and into the central dormitory of the exclusive boarding-school. But after the first term her contract is not renewed; the headmistress, Frl. Pfaff, suspects her liberal behaviour of infectious moral laxity and resents the romantic affection which Pastor Lehmann begins to display.

In *Backwater* Miriam, accompanied by her mother, attends an interview for a teaching post in suburban North London. Before she takes up the job at Banbury Park her parents throw a sumptuous summer party at which Miriam unintentionally damages her relationship with her first love, Ted, by appearing to flirt with another young man.

At Banbury Park Miriam's dislike of sub-urban values manifests itself as snobbery towards the girls, whom she sees as 'knowing and hard'. Her rejection of bourgeois conformity develops into the main theme of *Pilgrimage*, and in *Backwater* she begins to challenge organised religion, her views verging on atheism. She becomes friendly with a devout teacher, Miss Haddie, but quickly tires of religious texts and joins a lending library, in a passion for romantic fiction.

In *Honeycomb* Miriam moves to the gentrified country environment of Newlands as she takes up a post as governess to the Corrie family. The children are difficult to teach, and she soon finds 'it was impossible and always would be impossible to make their two hours of application anything but an irrelevant interval in their lives'. Mrs Corrie includes her in dinner parties and social events, but Miriam has begun to realise that her intelligence and independence tend to offend both men and women. She admires Felix Corrie, the husband who is generally away on business trips in London, but despairs of Mrs Corrie, who is an entirely conventional woman.

On a shopping trip to London with Mrs Corrie, Miriam is left on her own to wander around and explore the city, and the experience gives her a great sense of adventure. During her time at Newlands Miriam's home life is fragmenting; her father goes bankrupt and has to sell the house, and two of her sisters get married, 'thus rescued from poverty and fear'. Miriam herself is courted by Bob Greville, a family acquaintance, and by the more impudent and insistent Tremayne, for whom she feels little attraction. But marriage seems less attractive to her than the freedom of remaining single and their offers are decisively rejected.

However, her mother is declining into mental illness, and after the joint wedding of her sisters Miriam does not return to Newlands. Furious that the family's changed fortunes mean they cannot afford adequate treatment, Miriam stays to nurse her mother, borrowing money from her new brothers-in-law to take her for a holiday on the south coast. Richardson's own mother committed suicide, cutting her throat with a kitchen knife in a similar situation, and though in the novel the event is not described, the closing passage of grief makes it clear that this is what has happened.

Volume two opens with *The Tunnel*. Miriam has spent some months moving between a succession of cheap London lodgings before moving into an attic room in Mrs Bailey's boarding-house just off Euston Rd. *The Tunnel* describes in minute detail her job as an assistant and clerk in a Wimpole St. dental practice, a post she holds for the next ten years, until the final section of *Pilgrimage*.

One of the partners, Mr Hancock, begins to introduce her to London society, taking her to public lectures and to musical and cultural gatherings. Her admiration for him turns to love, but she still treasures her independence and enjoys the company of two women friends, Jan and Mags. The three of them spend their evenings talking and smoking, and they encourage Miriam to learn to bicycle to increase her freedom.

After an office outing to a performance of *Hamlet*, Miriam starts going to the theatre, determinedly not minding sitting alone in the cheap seats. She inadvertently renews her acquaintance with an old schoolfriend, Alma, now a wealthy married woman. Alma invites her to stay for the weekend at a house-party, where Miriam meets Alma's husband, Hypo Wilson, a writer and critic who makes a profound impression on her, especially as he suggests that she should try to write. (Hypo is based upon H.G. Wells, with whom Richardson had an affair.)

Mr Hancock abruptly changes his behaviour towards Miriam after his aunts advise him it is not decent to associate with a 'shabby clerk'. With this harsh return to 'official' manners, Miriam's enjoyment of her job declines, and increasingly her satisfactions are found outside work. An old friend, Miss Dear, who is sick with consumption, drags her into an embarrassing situation; Miriam begins to nurse her, and finds that her patient has mistaken the benevolence of a vicar for sexual interest. The strain ends as Miss Dear is sent on a convalescent holiday to the coast, and Miriam is relieved to learn that, although Mrs Bailey has plans to make the boarding-house more upmarket, she will be welcome to stay on in her beloved attic room.

In *Interim* Mrs Bailey's house is now full of wealthy foreign boarders, who mistakenly believe it to be a typical middle-class household; Miriam starts to feel like a poor relation. Where *The Tunnel* dealt with Miriam's life outside her lodgings, *Interim* explores the social tensions of the awkward evenings, and Miriam becomes friendly with a group of Canadian doctors lodging there.

On a cycling holiday with the doctors, Miriam's affection for Dr von Herber begins to take on a romantic note. But she finds that she has become the object of scandal through an association with an irascible Frenchman, Mr Mendizabel, in whose company she had enjoyed risqué evenings in Bohemian cafes. Dr von Herber, shocked by this reputation, disappears back to Canada without saying goodbye, but an explanatory letter from Miriam brings a swift reply: 'He was coming, carrying his suitcase out of the hospital, no need for the smart educated Canadian nurses to think about him. Taking ship ... coming back.'

In *Deadlock*, the first book of volume three, Miriam's romance has again come to nothing, and she is still in Mrs Bailey's lodgings surrounded by foreign guests. 'There is no need to go out into the world,' she thinks: 'Everything is there without anything; the world is added.' In the space of time between *Interim* and *Deadlock* she has met the intriguing Russian Jew Michael Shatov.

Shatov has introduced her to Russian writing, a great new adventure, and *Deadlock* focuses on the fresh impressions of life she is forming as a result. But his Judaic commitment makes her uneasy, and she feels that the status of Jewish women makes it impossible for her to become his wife: 'She pulled up sharply. If she thought of him, the fact that she was only passing the time would become visible.'

Miriam becomes involved with the group of 'Lycurgans', Richardson's fictional version of the Fabians. In *Revolving Lights* their ideas are debated and challenged as Miriam begins to realise their limitations and the narrow morality offered by their beliefs. 'Oh, the hopeless eternal inventions and ignorance of men; their utter cleverness and ignorance. *Why* had they been made so clever and yet so fundamentally stupid?'

Michael Shatov takes Miriam to meet a Russian Jewish friend in exile, living in penury with his young wife. But though she enjoys being introduced as Shatov's companion, the sight of the Jewish bride reinforces her sense of the impossibility of marrying Michael and becoming passive and subordinate. The attraction Miriam had felt for Alma's husband, Hypo, now develops into a potentially adulterous liaison: 'complications are enlivening'.

Miriam moves into a new flat in Flaxman's Court which she shares with the wealthier Miss Selina Holland. *The Trap* explores the uneasy relationship between the two women, with Miriam taking on a 'man's role' by going out to work and dealing with the difficult landlord over the payment of rent, while Miss Holland, leisured and artistic, does the shopping and prepares the meals. They talk intimately but awkwardly, wary of intruding too far into each other's privacy, the fragility of their arrangements emphasised by the dividing curtain which Miriam hangs between their beds. Richardson often reflects on the lack of communication between men and women, but here she examines the difficulty of really knowing any other individual, even when free, as here, from erotic overtones. 'Suddenly she found herself wanting to say outrageous things,' Miriam feels, as Selina Holland's tactful formality is oppressive: 'The decorous voices sounding all about her seemed to call for violence.'

In *Oberland*, the first book of the final volume, Miriam flees from *The Trap* for a holiday in Switzerland, which Richardson describes through sensual impressions of the stillness and the snowscape. The sense of freedom that Miriam first felt when she learned to cycle is now extended as she ventures nervously into the sports of skiing and tobogganing. As she arrives at the station in Oberland after the long trek across Europe, she finds herself being appropriated by a bourgeois family from Croydon, the epitome of the standards she despises in the suburban English.

At the *Schloss* where she is staying Miriam meets Vereker, a young Cambridge man who intrigues her yet makes her feel intellectually inferior because of her lack of education. Her resentment at her own impoverished life mounts, as she looks back on the Lycurgan meetings and the dreary London societies in which talking and lecturing has become a chore instead of an excitement. The Alpenstock offers her sophistication and potential for growth, and when her holiday is over she re-encounters the Croydon family, but now feels not threatened by their moral judgements but rather superior and contemptuous of their narrow lives.

In *Dawn's Left Hand* Miriam is plunged back into the daily routine of work in the Wimpole St. practice, but her vision of freedom in Oberland has refreshed her. She learns of a minor tragedy: her consumptive friend Miss Dear has married but died soon afterwards. In Miriam's own emotional world the affair with Hypo is becoming more intense and they finally have sex; Alma, aware of what is happening, is silent, and Miriam realises that it means nothing to Hypo, who regards it as a form of love free from any commitment.

Selina Holland is determined that the two of them will not re-enter *The Trap* which had developed before Miriam's holiday, and asks her to move out, saying she needs more space for herself. The eighteen months they had spent together in Flaxman's Court now seems like an interval with few memories to Miriam, who moves back into Mrs Bailey's boarding-house with all its old associations.

Mrs Bailey has a new lodger, a young half-French, half-Irish girl called Amabel, who is living in London on a moderate allowance from her family. Before Miriam moved out of Flaxman's Court she had already met Amabel, who sneaked into her room and wrote the impassioned message 'I love you' with soap on her mirror. Miriam tries to put her feelings into words but the process seems to her false: 'The torment of *all* novels is what is left out.'

In *Clear Horizon* Michael Shatov returns, and Miriam arranges a meeting between him and Amabel, hoping that her two friends will share the love she feels for them both. To Miriam's confusion Amabel insists that she should have married Michael. Her state of health, however, is deteriorating, and she goes to see Dr Densley, who is taking care of her sister, Sarah, during a period of illness.

Densley advises Miriam to rest and to go on leave from the Wimpole Street surgery, offering her the excuse that she has experienced a nervous breakdown.

Miriam goes to visit the Babingtons, friends of the family in her childhood, and sees again the privileged lives lived by the wealthy. She feels 'a faint nostalgia for shelter and security and, with it, pity for all the confined lives around her'. Grasping at Dr Densley's advice she draws on a sum of money she had almost forgotten about, which had been invested in an insurance fund at the behest of Jan and Mags, and sets out on the 'clear horizons' of an indefinite holiday.

'Never, since childhood, had she known freedom in July.' In *Dimple Hill* Miriam goes to the coast to rest, and moves in with a family suggested by Michael Shatov; they live simply and honestly as part of the Quaker brethren, but Miriam is amused to find that Michael did not mention the sons, with one of whom, Richard Roscorla, she now falls in love.

Letters from Amabel keep her in touch with London life, and with some satisfaction but also shock she learns that Amabel is now deeply involved in the Suffragette movement and her family have cut off her allowance as a punishment. After a demonstration, Amabel is arrested and force-fed on hunger strike in prison, and Miriam goes to see her, sensing defeat and martyrdom. However, the friendship between Amabel and Michael has developed, and they marry, Amabel finding emotional and financial security in being the wife of a wealthy Jewish intellectual.

When Amabel and Michael visit Miriam at the farm of the Quaker family, it appears that she is also about to marry Richard Roscorla. But Miriam feels she has falsely represented herself to the good family: 'What ought I to do?' she wonders: 'Tell these folks that I am not what I seem, am, from their point of view, a wolf in sheep's clothing?' While Richard is away on a business trip, a young American, Luke Mayne, comes to stay with the family, and Miriam finds herself asserting her individuality by partaking in intellectual debates with him which shock the Quakers. The younger brother, Alfred, warns her that this is not

permissible, and hints that it is being misinterpreted as a sexual overture. Miriam's first relationship with Ted was destroyed by just such insinuations; men are ready to believe that intellectual independence signals sexual immorality. The rumours of Miriam's behaviour reach the ears of the Quaker mother who tells Richard to break off the engagement. Yet Miriam's disappointment in him is transformed into the triumphant assertion of freedom, as she departs for Switzerland once more at the end of *Pilgrimage* to renew her acquaintance with the Cambridge intellectual, Vereker.

Dimple Hill was never published separately, but appeared as the final section of the four-volume omnibus edition of *Pilgrimage*. Richardson's publisher had assumed that it would be the final novel of the sequence, and had planned the new edition accordingly. When they discovered that the author intended to continue the story, they suggested delaying publication of the omnibus until it was complete. Richardson had struggled to finish *Dimple Hill* while suffering from 'acute neurasthenia', and felt that she could not guarantee a delivery date for a final volume of the story. She was also concerned that unless the omnibus edition appeared soon, her work would simply slip out of print and be forgotten. The approach of war made her keener still to see the new edition published, which it duly was, without the concluding volume. Consequently, the story seemed to finish unsatisfactorily, and this was commented upon by the reviewers. Richardson was sufficiently discouraged to wonder whether there was any point in completing her original design. The new edition had sold badly, and her publisher was unprepared to publish a further volume unless it contained at least three novel-length 'chapters'. Although three sections of what eventually became *March Moonlight* appeared in *Life and Letters* in 1946 as 'Work in Progress', Richardson lost heart, and the rest of the book remained unpublished until 1967 when a new omnibus edition of *Pilgrimage* appeared. By this time, Richardson was dead; she had revised only three of *March Moonlight's* seventeen chapters, declaring that she would 'leave it to take its posthumous chance'. It describes Miriam's attempts to become a writer, completing the circle at the point when she embarks upon writing her first novel, which the reader will recognise as the beginning of *Pilgrimage* itself. Richardson's biographer, Gloria G. Fromm, has commented: 'What she cared about most was the survival of *Pilgrimage* in the future, thereby ensuring her own survival. *March Moonlight* had in her mind a problematical relation to the novel that posterity would judge as *Pilgrimage*, and it suffered accordingly. After 1946 she could no longer take the fragment seriously.'

Sylvia Townsend WARNER   1893–1978

After the Death of Don Juan

As the author explained in a letter to Nancy Cunard, this work is 'a parable, if you like the word, or an allegory or what you will, of the political chemistry of the Spanish [Civil] War, with the Don Juan – more of Molière than of Mozart – developing as the Fascist of the piece'. Warner, a member of the Communist Party in the 1930s, loved Spain. She made and kept a vow never to return there while Franco ruled.

The novel grows from three familiar characters: Don Juan, the heartless libertine, a monster of coldly lustful egotism; Leporello, his servant and corrupted sidekick; and Doña Ana, the most recent in a long line of lecherous, treacherous and foolish women besotted and abused by the Don. Don Juan, as all the world knows, got his just deserts, spectacularly hustled off to hell by a covey of demons wreathed in clouds of smoke and sulphur. Or so Leporello, safely ensconced in the adjoining room, always maintained. But Leporello, like his master, is an incorrigible liar . . .

The novel opens with a most unusual honeymoon. Doña Ana has married Ottavio, an army officer, although she remains devoted to Juan's memory. She insists on making the long and arduous journey to Juan's native village, where they will bring news of the Don's death to his father. Don Saturno bears the news with a degree of composure not altogether explained by aristocratic sang-froid. His son has long been an expensive and dangerous liability, never bothering to conceal his arrogant contempt for the peasants, his utter indifference to their well-being and that of the thankless, arid land on which they toil so hard for such little reward.

Saturno himself is no model landlord but his feeling for the villagers is real, and grudgingly accepted as such by them, even as they curse him for his fickle, whirligig enthusiasms which buffet him from one aborted scheme to another and which leave the village strewn with the wreckage – often human – of his abandoned projects. Don Francisco is one of his disasters, the schoolmaster imported some thirty years ago with such high hopes of universal literacy, who festers now in increasingly insurgent bitterness.

Irrigation, so essential for the region's parched earth, is of all Saturno's ill-fated projects the one dearest to the peasants' hearts and the one which Saturno most nearly effects. (Throughout the book, irrigation holds both literal and allegorical significance, representing the flow of knowledge, the sharing of resources and the recognition of mutual dependence.) With Juan dead, no longer leeching on the estate and checking his father's more generous impulses, there is hope. But Juan, it proves, is altogether too alive. Unexpectedly he too is won by the virtues of irrigation – but only because he can see a way of appropriating the peasants' land, using their forced labour, and reaping large profits from the newly enriched soil.

Events culminate in a bloodily ill-matched confrontation between primitively armed peasants and fully equipped soldiers. The book ends with two villagers, one dying, the other doomed, looking 'at each other long and intently, as though they were pledged to meet again and would ensure a recognition'. By implication, that meeting comes some two centuries later in the Spanish Civil War.

Written in an austerely timeless prose, constantly irradiated by wit and irony, and typical of all this author's fine recreations of earlier historical periods, the novel is filled with characters who carry complete conviction as individuals, whilst also symbolising the complex relations between those groups and interests in Spanish society destined eventually to confront each other in a particularly bitter and bloody civil war.

Evelyn WAUGH   1903–1966

Scoop

Subtitled 'A Novel about Journalists', *Scoop* was Waugh's fourth book, based upon his experiences as a *Daily Mail* correspondent covering the civil war in Abyssinia. Waugh gleefully depicts the international press corps as unscrupulous, cynical and dishonest, more concerned with scoops and sensationalism than with such moral values as truth and good style. His aim was 'to expose the pretensions of foreign correspondents' and to this end he allows an inexperienced but honest journalist to get the scoop of the title.

William Boot, rural correspondent of the *Daily Beast*, is mistakenly sent to cover a civil war in the African Republic of Ishmaelia. The paper's previous correspondent, Sir Jocelyn Hitchcock, had left the *Beast* for its rival, the *Daily Brute*, after a disagreement with the megalomaniac proprietor, Lord Copper, about the date of the Battle of Hastings. On the advice of Mrs Stitch, the socialite wife of a cabinet minister (whose influence is as unreliable as her driving, which is largely confined to pavements and public conveniences), Lord Copper decides to hire John Boot, a successful novelist who has become a foreign correspondent in order to escape from an unhappy romance. The two Boots are confused by the grovelling, unambitious foreign editor, Mr Salter. Fearful of openly disagreeing with his employer, Mr Salter responds to the magnate's most egregious misapprehensions with the phrase 'Up to a point, Lord Copper'; this becomes one of the novel's running jokes.

Hung about with all manner of safari gear, including a collapsible canoe, William sets off for Africa. En route, he is introduced to 'the craft of journalism' by the morally and sartorially squalid Corker, who works for the Universal News press agency. In Ishmaelia, the journalists await developments and keep an eye on Sir Jocelyn, who is reputed always to be the first to know of any news. William gets to know Corker's associates, Shumble, Whelper and Pigge; the American Wenlock Jakes, who is much admired for 'reporting' non-existent events; and Eric Olafsen, Swedish Vice-Consul, journalist, head surgeon at the Swedish Mission Hospital and proprietor of 'the combined Tea, Bible and Chemist shop' in Ishmaelia's capital, Jacksonville. William also falls hopelessly in love with Kätchen, the German 'wife' of a bigamous geologist and prospector.

Dr Benito, Minister of Foreign Affairs and Propaganda, leads the Marxist 'Young Ishmaelites' in a coup to overthrow the government headed by Mr Rathbone Jackson. The revolution is brief, however, and the Jacksons are restored to power by 'Mr Baldwin', a mysterious, pseudonymous *eminence grise* who, despite his polyphonic accent and exotic appearance, claims to be English. A fortuitous combination of coincidence, naïvety and prep-school connections leads to William being the first journalist to report the defeat of the Marxist faction and the securing of valuable mineral rights for Britain.

He returns to England a hero in the eyes of the credulous, metropolitan public, but, in spite of the entreaties of Mr Salter, refuses to attend Lord Copper's celebratory banquet. A knighthood is offered in error to John Boot, who accepts it. William is happy to retire to his dilapidated country seat, Boot Magna, where, surrounded by his eccentric relatives, he continues to write his column, 'Lush Places'.

1939

Dismemberment of Czechoslovakia ● End of Spanish Civil War ● German invasion of Poland begins World War II ● Franklin Delano Roosevelt signs bill enabling Britain and France to purchase arms in US ● DDT invented and Polythene produced ● First regular commercial transatlantic flights ● George S. Kaufman and Moss Hart, *The Man Who Came to Dinner* ● *Gone With the Wind* ● W.B. Yeats, Ford Madox Ford, Havelock Ellis and Sigmund Freud die ● Louis MacNeice, *Autumn Journal*

Rumer GODDEN   1907–

Black Narcissus

Godden's third novel is set in India between the wars. The plot is simple and known to many through the classic Powell and Pressburger film of the same title (1947), an adaptation Godden herself loathed. Born in India, the author was sent to a convent school in England, which she hated, vowing that one day she would have her revenge. When she came to write *Black Narcissus*, however, she 'mysteriously' found this impossible.

An Anglo-Catholic order sends a young Sister Superior and five nuns to establish a convent in an isolated hill community facing the Himalayas. The combination of climate, cultural differences, human fallibility and barely supressed sexuality defeats their purpose, and within a year they recognise their failure and leave. From the outset Godden builds layers of obstacles the sisters must face and overcome, not least of which is the location of the mission. Known as the General's Palace at Mopu, it is little more than a ramshackle house, always creaking and straining in the winds as though it had a power of its own. In its heyday, the Palace was filled with women of a different sort, the mistresses of Gen. Toda's licentious father. This history seems to haunt every corner of the Palace and lives on in the grumpy old Ayah who remains there as housekeeper.

Sister Clodagh and her nuns arrive knowing that a previous order of Brothers survived at Mopu for only five months. They intend to start a hospital and a school, but, distrustful of western education and fearful of western medicine, the natives want neither. The nun's only contact with the outside world, and their only support, is Mr Dean, the General's agent. Dean has a reputation as an objectionable fellow, blasphemous and alcoholic, who has 'gone native'. His presence stirs violent emotions, further exacerbated by the arrival of the General's exotic young nephew, Black Narcissus, who has come to the nuns to be educated.

She stood on the horse-block ringing the Angelus. She must look, she thought, like a fly that had fallen into a green and blue bowl; the midday light ringed the tea bushes with myrtle green; she had not seen tea growing before; it looked a little silly on that wild hill, she thought, like rows and rows of neat green buttons. Who needed buttons on a naked hill?

BLACK NARCISSUS
BY RUMER GODDEN

Dean's presence stirs unhappy memories in the beautiful Sister Clodagh, who had joined the order after Con, the young man everyone thought she would marry, left rural Ireland without her. The sharp and uncomfortable Sister Ruth, who resents Clodagh's authority, becomes obsessed by Dean and imagines that the other nuns, in particular the Sister Superior, are trying to steal him away from her. She refuses to eat or sleep, oblivious of the fact that Dean is clearly more fascinated by Clodagh.

It is Sister Briony, the oldest of their

group, who sees the tensions mounting, but she is poorly equipped to deal with it. Sister Laura loses her early enthusiasm and becomes ill and irritable; the steadfast gardener, Sister Philippa, realises that she has lost sight of God and asks to be transferred; and pretty, sentimental Sister Honey, against all advice, ministers to a sick child, who dies. As a result the nuns are ostracised by the community. With her simple narrative style and careful attention to detail, Godden has prepared the reader for the novel's inevitable violent conclusion.

Sister Ruth goes to Dean to declare her love and is rejected. As the nuns go in search of her, she lies in wait by the bell which had rung out over the mountains so regularly to announce the nuns' fruitless mission. Despairing in her search, Sister Clodagh goes to ring the Angelus. The insane Sister Ruth attempts to push her over the railings but in the struggle slips and falls to her death.

Rayner HEPPENSTALL   1911–1981

The Blaze of Noon

This, Heppenstall's first novel, was written during the summer of 1938, 'much of it in Newport (Mon.), about whose streets in the evening I would get my wife to lead me with my eyes shut'. On publication day the *Evening Standard* carried the headline: 'Frankest Novel is Challenge to the Censor'; with subheadings, 'An Affront to Decency'; 'Story of Poultry-Yard Morals'. As a result the novel sold out on publication. Other reviewers, however, 'said that they could not understand what the *Evening Standard* had been talking about. The book was serious and distinguished ... So I had a *succès d'estime* as well as a *succès de scandale*.'

The novel is a celebration of 'profane' love experienced through the tactile senses of a blind man, the narrator. (The title is taken from Milton's *Samson Agonistes*.) Louis Dunkel went blind at the age of twenty-three – 'they said I should be a good surgeon, possibly a great one, for I was musical too and had a musician's hands'. After initial despair he learned detachment and how to apply his firm, sensitive touch as a masseur. Now a fashionable practitioner in London,

he is summoned to a remote part of Cornwall to work on Mrs Nance, an old friend: 'she had a magnificent pair of legs. It was a pleasure to thump them!' The land here is very old: 'I knelt for a while to feel the surface of the granite with my fingers ... the weird, druidical forms around me were rather felt presences than seen'. Living with Mrs Nance are her nephew, John Maldon, and her niece, Sophie. Life terrifies John and he is liable to fall victim to any experience he does not understand. At the moment he is under the thumb of 'a brassy upper-class tart' called Betty de Voeux, who sleeps with him 'when she feels like it'. Sophie has more substance, but is also intensely vulnerable. She tells Louis that at fourteen, 'I thought I was made for love. I had awful fantasies. I wanted to be a whore. I thought it was my vocation, like taking the veil.' She is currently experimenting with a local quarryman. She fascinates Louis, whose whole life depends on being in control of things. Her sexual education to the point where she is 'one and indivisible, a whole' becomes his entire purpose and they ultimately reach a plateau of sensuality on which they can lose themselves, while disturbing the others, especially John. But the idyll is shattered by the arrival of Amity Nance, who is not only blind but deaf and dumb. Louis finds her deeply disturbing – 'I could not help finding a partial image of myself in whatever I might feel about Amity Nance' – and as a result loses for a time his essential control. The effect on his relationship with Sophie is catastrophic; it takes a removal from the pagan landscape of Cornwall to London, 'a place which is entirely man's creation', to repair it.

The solution may be trite, and Louis's central philosophy, summed up by 'In man's decision is woman's peace', may not strike a perfectly happy note today; nevertheless *The Blaze of Noon* does pass the test of time if only because of the extent to which Heppenstall, a sighted person, gets into the mind of a blind man determined to be in full control of his life. The description of Louis gradually 'getting to know' Sophie through his other senses and his dangerous 'journeys' through the wild Cornish countryside are magnificent.

Geoffrey HOUSEHOLD   1900–1988

Rogue Male

Household always thought his short stories represented his best work, but with his second novel, *Rogue Male*, he hit upon a highly successful formula for narratives of pursuit and revenge. The novel was serialised in *Atlantic Monthly* and remains his best-known work, although he wrote twenty further novels in this style, including a late sequel, *Rogue Justice* (1983). By the time the book was published, Household himself was involved in international sabotage, serving with a commando unit in Romania planning to blow up oil wells.

The plot concerns the attempts of an unnamed English gentleman abroad to assassinate a dictator at a country house. The dictator similarly lacks a name, but is clearly intended as Hitler. At the beginning of the novel, which takes the form of a diary, the Englishman's motives appear to be simply those of the sportsman (as the title suggests): 'I am not an obvious anarchist or fanatic, and I don't look as if I took any interest in politics.' It is later revealed, however, that he had been briefly in love with, and planned to marry, a foreign woman murdered by the dictator's agents. The secret police capture the intruder when he has his quarry in his gunsights. They torture him and attempt to murder him by pushing him over a cliff. Unknown to them, he manages to break his fall and, although badly wounded, survives. He manages to escape back to England, smuggled aboard a British ship by its first officer, Mr Vaner, and put ashore near Chelsea.

His troubles are not over, however: the dictator's secret police are still on his trail. With the intention of going to ground, he visits Saul, his solicitor in Lincoln's Inn, where he becomes aware that he is under the surveillance of an enemy agent calling himself Major Quive-Smith, and his accomplice. He attempts to avoid them by using the underground system, but is forced to kill the accomplice at Aldwych underground station, thus putting the British police as well as the enemy agents on his trail.

He goes to ground in a remote part of rural Dorset, and builds himself an underground bunker in a country lane, where he finds company in a cat whom he names Asmodeus (after the devil in the Book of Tobit from the Protestant *Apocrypha*). Major Quive-Smith and another agent track him to this lair and are able to hole him up in it, keeping watch on the entrance until they can make him surrender. Quive-Smith kills Asmodeus and throws his body into the now stinking and confined lair, an action which angers the prisoner into determining to kill him. He uses Asmodeus's skin to make a primitive ballista and shoots Quive-Smith through an entrance at the top of the lair. He then wins over the other agent, an obedient Swiss named Müller, who drives him to Liverpool. Both men board a ship bound for Tangier, the hero puts Müller ashore secretly near Lisbon, and changes identity and goes to ground on arrival in Morocco. He sends his diaries to his London solicitor, from which we gather that he is living in a Latin country but still intends to return to finish the dictator in his own time.

Household, who was himself a country gentleman fond of shooting, perfectly captures the laconic, stoic, very English voice of his diarist who, stalking the dictator, considers that he is 'taking quite a conventional course: to go out and kill something in rough country in order to forget my troubles'. The narrative is replete with images of country sportsmanship. Although the title makes the reader think of the dictator, 'rogue male' in fact refers to the diarist, as an epigraph makes clear: 'The behaviour of a rogue may fairly be described as individual, separation from its fellows appearing to increase both cunning and ferocity. These solitary beasts, exasperated by chronic pain or widowerhood, are occasionally found among all the large carnivores and gramnivores.'

Christopher ISHERWOOD   1904–1986

Goodbye to Berlin

Like *Mr Norris Changes Trains* (1935), the six sections of this book originally formed part of an ambitious, never completed novel of late Weimar Republic life, *The Lost*. Four of the sections were first published in John

Lehmann's *Penguin New Writing*, while the most celebrated section, 'Sally Bowles' (turned down by Lehmann because it contained an abortion) was first published as a novella in 1937.

The book opens and closes with two extracts from 'A Berlin Diary', dated 'Autumn 1930' and 'Winter 1932–3'. These impressionistic accounts of what was happening in the city as Hitler rose to power contain Isherwood's famous statement: 'I am a camera with its shutter open, quite passive, recording, not thinking.' Isherwood later qualified this, but it gives some sense of both the cinematic quality of the writing and the narrator's mood of disengagement. We are introduced to his sentimental but pragmatic landlady, Frl. Schroeder, and his fellow lodgers: the wan barman Bobby; the genial tart Frl. Kost; and the anti-Semitic music-hall artiste Frl. Mayr.

Everything in the room is like that: unnecessarily solid, abnormally heavy and dangerously sharp. Here, at the writing-table, I am confronted by a phalanx of metal objects – a pair of candlesticks shaped like entwined serpents, an ashtray from which emerges the head of a crocodile, a paper knife copied from a Florentine dagger, a brass dolphin holding on the end of its tail a small broken clock. What becomes of such things? How could they ever be destroyed? They will probably remain intact for thousands of years: people will treasure them in museums. Or perhaps they will merely be melted down for munitions in a war.

GOODBYE TO BERLIN
BY CHRISTOPHER ISHERWOOD

In 'Sally Bowles', the narrator meets a preposterous and promiscuous English cabaret artiste, whose evident lack of talent does not prevent her from believing in a future glittering career in the movies. A great believer in her own wickedness, Sally attempts to involve the narrator in her incompetent attempts at gold-digging, notably with the wealthy drunkard Clive, who abandons them both.

'On Ruegen Island' is set in one of the city's holday resorts and charts the disintegrating relationship between Peter Wilkinson, a neurotic upper-class Englishman, and Otto Nowak, a working-class teenager, who shamelessly exploits his friend whilst relentlessly pursuing women.

In the next section, 'The Nowaks', the narrator has moved back to Berlin and, temporarily embarrassed, moves in with Otto's family, who live in a cramped tenement. Fr. Nowak, an exhausted consumptive, rails at Otto for his laziness, comparing him with his industrious older brother, Lothar, a solemn Nazi. Isherwood can no longer work or sleep in the atmosphere of their flat, and moves out when Fr. Nowak is sent to a sanatorium. The section ends with Isherwood and Otto visiting her in the bleached hopelessness of the hospital.

In 'The Landauers' the narrator becomes friendly with a wealthy Jewish family who own a chain of department stores. He finds the young daughter, Natalia, bossy and prim, like an elder sister, and the older cousin, Bernhard, oblique, apathetic and removed from life. Bernhard invites him twice to his 'country cottage', a sumptuous villa on the Wannsee. During the first visit Bernhard confesses something of his past, attempting to achieve some form of intimacy; the second occasion turns out to be a party held on the eve of the Brüning referendum. Isherwood sees the largely Jewish company as doomed, the occasion as 'the dress-rehearsal of a disaster . . . the last night of an epoch'. Some time later he overhears two businessmen discussing Bernhard's death from 'heart failure'. The implication is that he is an early victim of the Nazis.

The characters reappear within the interlinking stories, which are related with wit and insouciance. The book is Isherwood's most celebrated, partly because it fixed Weimar Berlin in the public imagination, but also because of several highly successful (and cavalier) adaptations of the stories for the theatre and cinema: *I am a Camera* by John Van Druten (1951) was filmed in 1955 by Henry Cornelius from a screenplay by John Collier; *Cabaret* (1966) is a musical version with songs by John Kander and Fred Ebb and a book by Joe Masteroff, which became an award-winning film (1972) directed by Bob Fosse from a screenplay by Jay Presson Allen.

James JOYCE   **1882–1941**

Finnegans Wake

Where *Ulysses* (1922) is an epic account of a day, Joyce's subsequent novel, *Finnegans Wake*, is a book about the night. Joyce spent the years between 1923 and 1939 elaborating the punning night-language, or not-language, of his uncompromisingly difficult and experimental last book. A rough draft of the entire text had already been completed by 1926, and the following thirteen years were spent simply developing and expanding this early structure. Working like a collage artist rather than a conventional novelist, Joyce travelled back and forth through his manuscript, adding layer upon layer of new verbal decoration, in the process nearly obliterating the early framework.

In a sense it is the language itself that is the book's real hero, rather than any particular character: the multiple puns embedded in each word frustrate any attempt to pin down a stable set of characters, or a single plot-line. Instead, Joyce was attempting to reconstruct verbally what he called the 'great myth of everyday life,' or the shared unconscious experience of the human race. To achieve this, he collected, pieced together, and superimposed verbal fragments of every conceivable kind, including words from all the world's languages, snatches of idiomatic conversation and slang, political and advertising slogans, abstruse theological terms, quotations from Shakespeare, popular song titles, hieroglyphs, rebuses and even telephone numbers.

Not that there is any lack of plot or characters – on the contrary, there are actually too many of them. At one moment, the book may seem to be the story of a night's sleep and dreams, but at another a history of the world, a murder mystery, a Freudian case-study, an ersatz Bible or cabbalistic scripture, a law report, a picture-postcard, an Irish bar-room ballad (such as 'Finnegan's Wake', from which the novel is named), or even several of these simultaneously. By this perpetual reference to different forms of narrative, the book turns itself into a kind of story-machine, a Heath-Robinson-style factory, which endlessly reprocesses and recycles old narratives into impossible new combinations.

A handful of recognisably differentiated characters does occasionally float to the surface of this Irish stew of verbiage. Thus Humphrey Chimpden Earwicker, Anna Livia Plurabelle, Shem, Shaun and Issy do sometimes seem to constitute a fictional family, comprising father, mother, sons and daughter respectively, as might be expected in a conventional novel. However, the members of this family are in constant danger of melting into one another, of exchanging roles, or of splitting up into multiple new identities. The two sons often fuse to take the place of the father, for example, while the daughter tends to fragment schizophrenically to become a classful of twenty-nine identical schoolgirls. The family are equally liable to exchange roles with characters from history, or from popular or religious myth. Earwicker, for example, is repeatedly confused with Charles Stewart Parnell, the Irish nationalist politician; with Finn MacCool, the mythical Irish warrior; and with Tin Finnegan of 'Finnegan's Wake', the bricklayer who falls from a ladder to his death, only to be revived again at his wake by a splash of whiskey.

No summary of *Finnegans Wake* can do justice to its seemingly endless diversity. Every reader's experience will be different, depending on the particular elements he or she retrieves from Joyce's bag of literary tricks. The main organising principle of the book is perhaps its use of cyclical repetition: it ends and begins in the middle of the same, vast, circular sentence, and the cycles of sleep and wakefulness, death and resurrection, and the rises and falls of history itself play endlessly through the book. But this wealth of repetition may simply have the effect of mocking the reader's desire for familiar landmarks, around which to stabilise Joyce's disturbingly fluid panorama, or 'langscape,' of half-familiar words.

| Henry MILLER | 1891–1980 |

Tropic of Capricorn

see **Tropic of Cancer** (1934)

| Flann O'BRIEN | 1911–1966 |

At Swim-Two-Birds

At Swim-Two-Birds was the first novel of an Irish civil servant called Brian O'Nolan. He had written most of the book while at University College, Dublin, where he was rewriting an M.A. thesis on 'Nature in Irish Poetry', which his tutor had originally rejected. The novel was also rejected (by Collins), but subsequently published by Longman on the recommendation of Graham Greene, then a publisher's reader for the firm. Although praised by such reviewers as Dylan Thomas, *At Swim-Two-Birds* did not sell at all well, on top of which most of the first edition was destroyed when the publisher's warehouse was bombed. The author received an advance of £30, which, as he pointed out, was £20 less than he was then receiving for the detective stories he wrote for pulp magazines. The pseudonym he was obliged as a civil servant to adopt for this book became the one by which he was generally known.

Substance of reminiscence by Mr Shanahan, the comments of his hearers being embodied parenthetically in the text; with relevant excerpts from the public Press: Do you know what I am going to tell you, there was a rare life in Dublin in the old days. (There was certainly.) That was the day of the great O'Callaghan, the day of Baskin, the day of Tracy that brought cowboys to Ringsend. I knew them all, man.

AT SWIM-TWO-BIRDS
BY FLANN O'BRIEN

At Swim-Two-Birds is a novel about writing novels which write themselves. This *reductio ad absurdum* of the principle of self-reference is made possible by the novel-within-a-novel structure. The narrator's own experience, both as a writer and a member of a fast-drinking cast of students, is intertwined with the stories of the fictional characters he creates: a restless throng of cow-punchers, clerics, fairies and Irish legendary heroes. These are not his direct creations, but the product of the imagination of one of his characters, the eccentric and despotic Dermot Trellis, who is said to 'hire' his characters and keep them locked up, lest they should creep out of their appointed roles. Omniscient though this author may be, he cannot avoid his characters' revolt against him when his own weapon is made to boomerang. Relying on his son Orlick's talent for literary composition, 'they suggest that they utilize his gift to . . . compose a story on the subject of Trellis, a fitting punishment indeed for the usage he had given others'.

A literary juggler, O'Brien rejects the sense of illusion inherent in the mimetic conception of art: we are delighted and amused, but there is never any question of our suspending our disbelief. Our attention is repeatedly drawn towards purely technical aspects of the narrative, such as the stylistic devices employed: 'I (for my part) was no Rockefeller, thus utilizing a figure of speech to convey the poverty of my circumstances. *Name of the figure of speech*: Synecdoche (or Antonomasia).' On the other hand, accepted conventions can be reinvested with a new, jocular authority, as in the case of the theory of 'aestho-autogamy', whereby a human being can be produced 'from an operation involving neither conception or fertilization'. O'Brien is elevating to mock-scientific status one of the most obvious truths of artistic creation, since *any* character can be brought into the world 'already matured, teethed, reared, educated'.

| Jean RHYS | 1894–1979 |

Good Morning, Midnight

After the publication of this, her fifth book, Rhys disappeared for almost twenty years. Her novels gradually went out of print and she was almost forgotten, presumed dead, until traced to Cornwall by Selma Vaz Dias, who had adapted this novel for radio in 1957.

These people all fling themselves at me. Because I am uneasy and sad they all fling themselves at me larger than life. But I can put my arm up to avoid the impact and they slide gently to the ground. Individualists, completely wrapped up in themselves, thank God. It's the extrovert, prancing around, dying for a bit of fun – that's the person you've got to be wary of.

GOOD MORNING, MIDNIGHT
BY JEAN RHYS

By 1939 the autobiographical figure who came to be known as 'the Jean Rhys woman' was well established. Given a different name in each of the preceding novels, she appears in *Good Morning, Midnight* as Sasha Jensen, who, rescued from drinking herself to death in Bloomsbury, is back in Paris for 'a quiet, sane fortnight'. In her hotel room on the Left Bank 'there are two beds, a big one for madame and a smaller one for monsieur', but the room is 'a place where you hide from the wolves outside': there is no monsieur, only memories. Memories of her husband, Enno, who 'knew so exactly when to be cruel, so exactly how to be kind'; of the rooms they had shared ('The musty smell, the bugs, the loneliness ... this is all I want from life'); of the people who pass through and the money that changes hands; of her baby: 'He has a ticket tied round his wrist because he died. Lying so cold and still with a ticket round his wrist because he died.' As she wanders through the streets from bar to bar, places remind her of the past, of humiliating jobs, of men who had spurned her when she was starving, of where she had cried. She still has her old fur coat, she has had her hair dyed *blond cendré*, she buys a new hat, and a painting of 'an old Jew with a red nose ... standing in the gutter, playing a banjo'. It is the fur coat that gives the gigolo René the idea that Sasha has money. She tells him: 'I'm afraid of men. And I'm even more afraid of women ... Who wouldn't be afraid

of a pack of damned hyenas?'; but he finds his way into her room – 'this damned room ... all the rooms I ever slept in'. She tells him that she can just about afford to give him a 1,000 franc note which he will find in her dressing case, but implores: 'For God's sake leave me the others, or I'll be in an awful jam.' This is how things repeat themselves – so it begins all over again.

The theme of the novel – a woman's struggle against isolation, poverty and exploitation – is central to Rhys's work. She brilliantly evokes the Paris of *entre-deux-guerres*, with its Bohemians preying on the rich and each other, its supercilious waiters and cynical owners of the seedier sorts of hotel. Sasha is at once contemptuous and very much afraid of everyone, including the cold-blooded English 'just across the Channel' – everyone, that it, except the old Jew painted on canvas, who is 'gentle, humble, resigned, mocking and a little mad'. It is a world of pure feeling, of memory in which the past and present mingle, of sights and smells and distant longings, of profound sadness minutely dissected. In short Rhys's voice is unique; she preserves a world that is of one time and place and gives it universality with her precise analysis of the nature of loneliness and its companion, fear.

John STEINBECK · 1902–1968

The Grapes of Wrath

Steinbeck's famous novel about agricultural migration from the dustbowls of Oklahoma towards the promised land of California (made into an equally celebrated film, directed by John Ford from a screenplay by Nunnally Johnson, 1940) takes its title from the 'Battle Hymn of the Republic': 'Mine eyes have seen the glory of the coming of the Lord: / He is trampling out the vintage where the grapes of wrath are stored.' An angry, compassionate novel, it transforms the Joad family into representatives of 'the people', with whom the future of America lies.

Paroled from gaol, where he was serving a sentence for killing a man in a drunken fight, Tom Joad finds his family home deserted and in ruins. He joins up with Jim Casy, a once fervent preacher who has lost his faith in a God who allows injustice and suffering, and together they set out in search of the family. At Uncle John's farm the Joads are reunited; but times have changed, and the homesteaders are being forced off their land.

'Okie use' ta mean you was from Oklahoma. Now it means you're a dirty son-of-a-bitch. Okie means you're scum.' Through the story of a single family, Steinbeck traces the suffering of the dispossessed in the Depression, and the capitalist exploitation responsible for their starvation. The Joads sell up their goods and depart for the fertile land of California, lured by the promise of employment; but as their old jalopy labours through the desert, they find a quarter of a million others on the move, hungry for work at any price. The squabbling grandparents die on the road and are buried in paupers' graves; the children grow sick with malnutrition. Ma tries to hold the family together, but her belief in the individualist spirit of the pioneers wavers. Tom is slow to realise the political reality, though Casy, with his solitary meditations, begins to identify the cruelty which underlies their treatment. Shanty towns called 'Hoovervilles' provide a makeshift life for the migrants, gypsy camps hated by the locals and burned out by the corrupt police.

The Joads obtain a camping place on a government reservation, a utopian syndicate of self-regulation and co-operation. But as the authorities seek to smash the collective power by singling out the leaders, Casy is arrested and Tom involved in the fight. His sister is pregnant and ill, and Ma decides they have to move on; survival must take precedence over politics.

The need for food overrides all else, and the Joads move into a labour settlement, unwittingly breaking a wage strike. Casy, now released from prison, is leading the protesters, and as Tom sees him murdered by the authorities his anger finally breaks out and he kills an officer. On the run from the law, Tom becomes committed to the ideals which Casy had preached. A brief period of cotton-picking provides relative prosperity, but once the harvest is over, a winter of unemployment looms, with disease and death. Tom leaves the family, determined to fight against the system, and his sister's child is stillborn. But in the gathering anger, Steinbeck sees a future strength. 'We ain't gonna die out,' Ma declares, 'People is goin' on – changin' a little, but goin' right on.' The suffering breeds a positive vision of human dignity and compassion. In a shattered barn with the flood waters rising, the bereaved young mother gives her milk to a dying man; the will to survive is triumphant.

Dalton TRUMBO　　　　　**1905–1976**

Johnny Got His Gun

Trumbo's famous anti-war novel was, by chance, published in the USA on 3 September 1939 – the day on which England and France declared war on Germany. Trumbo, who would later be imprisoned as one of the communist Hollywood Ten, found himself championed by the American Civil Liberties Union and other left-wing groups opposed to American involvement in the war. He regarded this war as entirely different to the 'romantic' First World War depicted in the novel, and later responded to the rumour that the novel had been suppressed (it had merely gone out of print) by saying: 'If . . . it had been banned and I had known about it, I doubt that I should have protested very loudly.' The novel was turned into a film, directed by the author from his own screenplay, in 1971.

The novel begins as twenty-year-old recruit Joe Banham emerges slowly into consciousness in a hospital bed, his mind filled with half-hallucinatory memories of his earlier life. He gradually realises the extent of his injuries: he is deaf and blind and has no arms, legs or mouth. His initial terror is replaced by an ironic detachment: 'There weren't many guys the doctors could point to and say here is the last word here is our triumph here is the greatest thing we ever did and we did plenty.' Joe learns to distinguish his different nurses by the vibration of their feet, and to calculate the passage of the days from their visits: 'When a new nurse came in he always knew what she would do first. She would pull the covers off him and then she would make no movement for a minute or two and he would know she was looking at him and probably getting a little sick.' In the fourth year of his 'new time world' he is aware of a group of men pinning a medal to his chest. It occurs to him that he can communicate by tapping out Morse code with his head, and finally one of the nurses understands his intention. A man is brought and Joe frantically explains that he wants to go free, that he will support himself as a living exhibit testifying to the effects of war. The message comes back: 'WHAT YOU ASK IS AGAINST REGULATIONS WHO ARE YOU.' Joe is sedated and returns into his interior universe, all hope of communication with the outside world seemingly quashed.

Nathanael WEST   1903–1940

The Day of the Locust

Given the large number of novelists who have worked in Hollywood, it is surprising how few of them have used the experience to produce good novels. Although some critics regard Scott Fitzgerald's unfinished *The Last Tycoon* (1941) as a masterpiece, West's apocalyptic novella is perhaps the only great book to come out of America's film capital. West was able to get beyond the self-pity which assails most writers who worked for the studios and so produce a genuinely horrifying vision of the place where the American dream is manufactured. 'In America violence is idiomatic,' West stated; 'in America violence is daily.' His novel builds through images of appalling but small-scale physical and psychological violence to a final scene of mob hysteria, the logical outcome of the dream factory's manipulation of the public.

Tod Hackett is a painter who has been brought to Hollywood by a talent scout for National Pictures to work as a designer. He falls in love with Faye Greener, a platinum blonde who works as an extra but aims to be a big star: 'But she wouldn't have him. She didn't love him and he couldn't further her career. She wasn't sentimental and she had no need for tenderness, even if he were

capable of it.' In fact, Faye is the ultimate bitch goddess of American fiction, a woman whose power and influence are wholly destructive. Also in love with her is Homer Simpson, a shy, ungainly Midwesterner who has retired early and come to California for his health. After her father dies, Faye goes to live with Homer, who agrees to keep her until she becomes a star. This is strictly a 'business arrangement', although Homer clearly hopes for more. Faye soon becomes restless and takes out her frustration upon her servile benefactor, treating him with terrible and humiliating cruelty, which he appears to accept. She eventually walks out on him after a drunken party which ends in a brawl as Faye's grotesque admirers fight over her.

The novel reaches its bloody climax at a movie première. The scenes there resemble 'The Burning of Los Angeles', an ambitious picture Tod is in the process of painting, which depicts a surging, torch-bearing mob of 'all those poor devils who can only be stirred by the promise of miracles and then only to violence'. Unhinged by Faye's desertion, Homer drifts to the première half-dressed, carrying suitcases and claiming that he is returning to his home town. While sitting in a daze on a bench, he is tormented by his neighbour, the repellent child star Adore Loomis. When Adore throws a stone at him, Homer loses control: attempting to escape, the boy trips over and Homer stamps him to death. Tod attempts to intervene, but is swept away by the crowd. His last sight of Homer is of his being set upon by the murderous fans, disappearing beneath them as they fall upon him, a vision which is the apotheosis of the book's title. Tod himself is rescued by the police, and the novel ends with his being driven home, loudly imitating the siren's scream.

Perhaps appropriately, *The Day of the Locust* was West's last book; he was killed in a car accident the year after it was published. The movie world provided him with the perfect setting in which to exercise his taste for the grotesque and macabre, which was partly derived from his association with the Surrealists in Paris during the 1920s. West's vision may be extreme, but his story has a perfectly plausible progression, with intimations of its resolution carefully placed throughout the text. In 1975 it was made into an overblown film, directed by John Schlesinger from a screenplay by Waldo Salt.

The Sword in the Stone

see **The Once and Future King** (1958)

1940

Germany invades Norway and Denmark ● Winston Churchill becomes Prime Minister ● Dunkirk ● Fall of France ● Battle of Britain ● The Blitz ● US sends destroyers to Britain ● Howard Florey develops penicillin as an antibiotic ● Vasily Kandinsky, *Sky Blue* ● Michael Tippett, *A Child of Our Time* (oratorio) ● Charlie Chaplin, *The Great Dictator*, Alfred Hitchcock, *Rebecca* and Walt Disney, *Fantasia* ● Leon Trotsky, Neville Chamberlain and Eric Gill die ● C. Day Lewis, *Poems in Wartime*

Raymond CHANDLER   1888–1959

Farewell, My Lovely

Chandler's second novel is usually regarded as his best, a judgement with which the author himself concurred in 1949. 'I think *Farewell, My Lovely* is the top,' he wrote to someone who had written to give that accolade to *The Little Sister* (1949), 'and that I shall never again achieve quite the same combination of ingredients. The bony structure was much more solid, the invention less forced and more fluent, and so on.' It is certainly, along with *The Big Sleep* (1939) and *The Long Goodbye* (1954), a masterpiece of the 'hard-boiled' school of American detective fiction. Based upon two previously published short stories, 'Try the Girl' and 'Mandarin's Jade' (both 1937), it was originally to be called 'The Second Murderer', a title (as Chandler explained to his publisher) derived from *Richard III*. The setting is Bay City, a thinly disguised Santa Monica, which Chandler disliked because of the collusion he saw between its wealthy inhabitants and its corrupt police force.

Six foot five and 'not wider than a beer truck', Moose Malloy has spent eight years in gaol. Now he is out and looking for his Velma. He visits a pool hall where she used to work. The owner pulls a gun on Moose. Moose kills him, and Philip Marlowe – Chandler's narrator and his most famous creation – happens to be a witness to most of the scene. The police assign the case a low priority. It is passed to an idle no-hoper, Lt. Nulty, who asks for Marlowe's help. Marlowe agrees to give it – he badly needs a friend in the department – and traces Mrs Jessie Florian, the widow of Velma's last employer. She tells him that Velma died five years ago, but Marlow steals Velma's photograph in any case.

I got up on my feet and went over to the bowl in the corner and threw cold water on my face. After a little while I felt a little better, but very little. I needed a drink, I needed a lot of life insurance, I needed a vacation, I needed a home in the country. What I had was a coat, a hat and a gun. I put them on and went out of the room.

FAREWELL, MY LOVELY
BY RAYMOND CHANDLER

Soon afterwards, he is telephoned by Lindsay Marriott, a slick playboy who wants

Marlowe to go with him to a rendezvous. There he will turn over a ransom in exchange for the return of a jade necklace recently stolen from one of Marriott's friends. Although Marlowe dislikes and mistrusts Marriott, he takes on the job. The two attend the rendezvous and, as they wait, Marlowe is blackjacked unconscious. When he comes round, he discovers Marriott dead, with his skull smashed in. Prompted partly by curiosity and partly by his sense of having failed, Marlowe sets about finding the killers.

Through Anne Riordan, a freelance journalist, he is introduced to the owner of the necklace, Mrs Lewin Lockridge Grayle, the young and beautiful wife of an elderly millionaire. She is Marlowe's next client and, nearly, his lover. Marlowe continues his investigations: he prefers to work alone; he usually carries a gun; he keeps a bottle of Scotch in his desk; his judgement often errs; his guesses are wrong. He hounds a charlatanish spiritualist, gets badly beaten up and locked in a private asylum and – escaping from it – glimpses Moose. He forges a wary and intermittent alliance with Randall, who is an altogether sharper lieutenant than Nulty. Mrs Florian is murdered. Rapidly it becomes clear that there is a connection, however convoluted, between Velma and Moose on the one hand and Marriott and Mrs Grayle on the other. 'What I like about this job,' Marlowe says, 'everybody knows everybody.' Pursuing his leads and hunches – from City Hall to offshore gambling ships – he uncovers the explanation of these entwinements, and justice is imperfectly dispensed.

Although he has 'solved the mystery', Marlowe cannot extirpate the institutional corruption underlying it. The primary villains – the holders of office and the possessors of wealth – are beyond his reach; and it is his realisation of his inability to touch them that makes him a distinctively modern detective. His fictional predecessors righted wrongs: he believes that what he does is largely futile. If this provokes habitual cynicism (and the hard-boiled wisecracks that go with it) it also leaves him struggling to identify his motives. Thus, why he acts – placing himself in danger for little reward – is the mystery he can never explicitly solve.

The reader must deduce from behavioural clues that, beneath the cynicism, he remains a sentimentalist.

Farewell, My Lovely has been well served by Hollywood. First filmed in 1944 by Edward Dmytryk, from a screenplay by John Paxton, and remade in 1975 by Dick Richards, from a screenplay by David Zelag Goodman, its Marlowes were Dick Powell – perfectly cast – and an elderly but robust Robert Mitchum.

Walter van Tilberg CLARK 1909–1971

The Ox-Bow Incident

Although Clark went on to write two more interesting novels, and published some verse and a number of short stories, he is largely remembered for *The Ox-Bow Incident*, his first book, which was subsequently made into a highly original Western film (1942, released in Britain as *Strange Incident*), directed by William Wellman from a screenplay by Lamar Trotti. The story is unusual in that it attacks the basic assumptions of the genre about frontier law and natural justice. This is all the more remarkable in a book published while a world war was raging in Europe, and it raises the novel far above the level of most Westerns.

The novel is set in 1885 in Nevada, where Clark was brought up, and the plot is as simple as its message. Two roving cattlemen, the narrator, Art Croft (who faithfully records all he observes without always understanding its significance), and his buddy, Gil, stop off at Bridger's Wells, a small town in cattle country. In Canby's saloon the talk is of rustling. A boy brings news that Kinkaid, whom everyone likes ('there was something about him … a gentle permanent reality that was in him like his bones and his heart') has been murdered by a party of rustlers. Risley, the sheriff, is out of town and in any case the men have no faith in the process of law; it operates too slowly and uncertainly. They decide to take the law into their own hands. Old Davies argues strenuously against this – as do Osgood, the Baptist preacher, and, when he gets to hear of it,

Judge Tyler; but Davies is the one who cares most deeply. It becomes a battle between him and Tetley, 'the biggest man in the valley', who wants a lynching. It is unclear at first which way the rustlers have gone, but further intelligence works in Tetley's favour and the party sets off.

Encamped in the mountains at the Ox-Bow, they find a young man called Martin and his two hands, a feeble-minded old man and a Mexican who at first pretends not to understand English. They have cattle with them, and, though Martin claims to have bought them from a man called Drew, he has no bill of sale. Tetley persuades himself that these are the rustlers that murdered Kinkaid. Davies argues that they must be brought back for trial, but succeeds only in persuading Tetley, who wants to hang them immediately, to wait until morning. During the night Martin writes a letter to his wife which he entrusts to Davies, who is much moved by it and more than ever convinced that the young man could not possibly be guilty. He cannot, however, prevail against Tetley, and the men are duly hanged.

On the ride back to Bridger's Wells, the lynch-party run into the sheriff, who is accompanied by Kinkaid, very much alive, and Drew, who confirms that he sold Martin the cattle. In spite of the judge's urging, the sheriff proposes to take no action, and advises everyone to keep quiet about the lynching. However, Tetley's sensitive son, Gerald, who was bullied not only into accompanying his father but into actually hanging Martin, a job he does inefficiently, cannot live with the memory. Sickened by the incident, he afterwards, at the second attempt, kills himself. Tetley, who appears to be as unmoved by the news of his son's death as he is by the fact that the other men have turned against him, also commits suicide.

Clark's economy, his masterly handling of a necessarily large cast of characters, his ear for speech rhythms, the atmosphere he creates – particularly in the climactic scene with the three men hanging in the snow – and above all the strong moral values he espouses, all go to make this an almost unbearably moving novel.

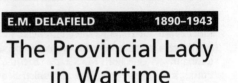

E.M. DELAFIELD   1890–1943

The Provincial Lady in Wartime

see **Diary of a Provincial Lady** (1930)

Martha GELLHORN   1908–

A Stricken Field

As a reporter working for *Collier's* magazine, Gellhorn visited Prague in October 1938, one week after the signing of the Munich Pact. In a letter to her mother she wrote: 'The League High Commissioner for Refugees, one Sir Neill Malcolm, showed up for two days in Prague and didn't see a single refugee. But I had seen them, all of them, all over the place and I was sick at heart and very angry.' She persuaded Malcolm and Gen. Faucher, the former head of the French Military Mission, to see Prime Minister Sirovy of Czechoslovakia in an effort to stay the expulsion order applying to German and Austrian refugees and to those from the annexed area of the Sudetenland, many of whom faced the prospect of torture and the concentration camps. The effort was unsuccessful. This incident is at the centre of *A Stricken Field*, written in Cuba in 1939.

Mary Douglas, an American foreign correspondent, arrives in Prague in October 1938, one of the 'camp followers of catastrophe' ensconced in a luxurious hotel amidst the squalor and decay of the city. She runs into Rita Salus, a Communist activist she had known earlier in Berlin, and witnesses the appalling plight of the refugees, now living underground to avoid the Gestapo. She resolves to help Rita and her lover, Peter, and applies to Lord Balham (a barely disguised version of Sir Neill Malcolm), an effort laughed at by her facetious colleagues: 'Are you going to make a plea for suffering humanity?' one asks.

The expulsion order goes ahead and Peter is arrested by the Gestapo. Mary intends helping Rita to escape to Paris but she too has disappeared. Suffering from a fever, Rita searches for Peter and finds the house in which the Gestapo are interrogating him.

Concealed in the basement she listens as they torture him. He remains silent and dies from the injuries inflicted on him. Mary is forced to abandon her search for Rita, and flies to Paris, smuggling with her a package of documents testifying to the Nazi atrocities. Unknown to her, Rita has been arrested as she wandered the streets, having fled the house where Peter died.

No propaganda, they would say. We want the inside story. Make it clear, make it colorful, make it lively. If I knew how, I would write a lament. I would tell how I heard the children sing once in Spain, in Barcelona, that cold and blowing March when the bombers came over faster than wind, so that it would all happen in three unending minutes, but if you saw them they were hanging in the sky not moving, slow and easy, taking their time, you'd think, not worried about anything.

A STRICKEN FIELD
BY MARTHA GELLHORN

The novel's title is taken from 'a Medieval Chronicle' describing a battle scene. Some young knights 'who had never been present at a stricken field' turn away from a massacre of prisoners, and are castigated by a Burgundian veteran: 'Are ye maidens with your downcast eyes? Look well upon it. See all of it. Close your eyes to nothing. For a battle is fought to be won. And it is this that happens when you lose.' Gellhorn's career is a testament to looking well upon the battlefields of the twentieth century and closing her eyes to nothing. Her original intention upon leaving university had been to work as a reporter and recreate what she had seen in fictional form. Rereading *A Stricken Field* some forty-five years after its first publication, she commented: 'I am proud of it. I am glad I wrote it. Novels can't "accomplish" anything. Novels don't decide the course of history or change it but they can show what history is like for people who have no choice except to live through it or die from it.'

Graham GREENE   1904–1991

The Power and the Glory

Set in a remote southern state of Mexico, *The Power and the Glory* was written after Greene had visited Mexico in 1938 to report on the religious persecution described in his travel book *The Lawless Roads* (1939). The unnamed protagonist – a drunkard and father of a daughter by a peasant woman – is the last Catholic priest on the run from the authorities. He moves from village to village granting confession and communion to the faithful peasants, but when he discovers that his chief pursuer, the lieutenant, has started taking hostages in an effort to find him, he is faced with this dilemma: 'If he left them, they would be safe, and they would be free from his example ... But it was from him that they took God – in their mouths ... Wasn't it his duty to stay, even if they despised him, even if they were murdered for his sake?'

Fleeing to the capital city he picks up and gets rid of a mulatto who has recognised him and wants to collect the reward for his capture. There is a prohibition on alcohol and the priest is arrested and imprisoned for possessing a bottle of brandy, but remains unrecognised despite revealing his identity to his impoverished fellow prisoners. As he leaves the prison, the unsuspecting lieutenant takes pity on him and gives him some money. The lieutenant is a compassionate man, sincere in his zealous humanism: 'He was a mystic, too, and what he had experienced was vacancy – a complete certainty in the existence of a dying, cooling world, of human beings who had evolved from animals for no purpose at all.'

Heading for the border, the priest returns to the banana plantation where he has earlier found sanctuary with the strangely precocious girl Coral Fellows, but finds it now abandoned. Continuing, he encounters a peasant woman whose child has been shot by a gringo fleeing from the police. He crosses the border and recovers from his arduous journey, but just as it seems he has escaped his pursuers, the mulatto tracks him down and tells him that the dying fugitive is a

Catholic who wishes to confess before death. Despite his suspicion that he is being led into a trap, the priest cannot risk allowing the man to die in a state of mortal sin. The fugitive *is* a Catholic, but he refuses absolution, and after his death the priest is arrested by the lieutenant and returned to the capital for his trial and execution.

The Power and the Glory won the Hawthornden Prize in 1940. As in **Brighton Rock** (1938), Greene is concerned in this novel with the co-existence of good and evil. The origins of his 'whisky priest' may be found in references in *The Lawless Roads* to an alcoholic priest who lived on the run. Greene also borrowed some features from a far more respectable model, Padre Miguel Pro, a young Jesuit who had conducted services and heard confessions in spite of the brutal persecution of the Catholic church by Mexico's President Calles. Pro dressed as a civilian and managed to evade the state police for some time, but was eventually caught and executed. Further back still, Greene was drawing upon childhood memories of relatives who returned from holidays in Spain shocked by the fact that village priests kept mistresses. 'I saw no reason why a man should not be different from his function,' Greene reflected, 'that he could be an excellent priest while remaining a sinner.'

Ernest HEMINGWAY	1899–1961

For Whom the Bell Tolls

Hemingway's most ambitious novel – the title of which is taken from Donne's famous sermon – was written as a result of his experiences as a correspondent in the Spanish Civil War. The war had a personal as well as a political significance for him, since he was accompanied by the journalist and novelist Martha Gellhorn, the dedicatee of *For Whom the Bell Tolls*, whom he married shortly after the book was published. Hemingway wrote the novel in Havana in a burst of creativity: 'the words poured out of him at a rate of six to seven thousand a week', according to his biographer, Kenneth S. Lynn, 'and new characters and new episodes

kept mutiplying in astonishing fashion'. When he had finished, he told his publisher: 'I was as limp and dead as though it had all happened to me. Anyhow it is a hell of a book.' This vainglory was wholly justified by the sales. 'Book selling like frozen Daiquiris in hell,' Hemingway informed his first wife, and by 1943 its combined American and British sales had reached 885,000 copies. Hemingway sold the film rights for $100,000; the resulting movie, directed by Sam Woods from a screenplay by Dudley Nichols (1943), was a box-office success, but is not highly regarded. The action of the novel covers a period of just four days, the central characters are a young American guerilla leader and a band of peasants in a mountain cave, and yet the story seems to contain the whole agony of Spain in the first year of the war. In spite of its often absurd dialogue and a psychologically improbable love-affair, it is an impressive performance.

Robert Jordan's mission to blow up a bridge is vital to the success of the coming Republican offensive. An old guide, Anselmo, leads him to the cave of Pablo, whose band have recently joined forces with that of El Sordo to blow up a train. Pablo is uneasy about the mission: he has lost his nerve and is intelligent enough to see that 'this of the bridge' will bring disaster to his band, for whom he feels more than for the Republican cause, unlike his large, fifty-year-old '*mujer*', Pilar: for her 'the Republic is the bridge'. Pilar has gypsy blood, can read palms and seems harsh, but she speaks of her first husband with deep compassion, and nurses back to health the nineteen-year-old Maria, who has been raped by the Falangists. During the four days they have together, Maria and Jordan have an affair that must stand for their whole lives.

Pablo is an ever-present threat to the success of the operation and when El Sordo's band is annihilated by a detachment of Nationalist cavalry, he deserts, throwing Jordan's detonator in the river. His return just before Jordan and the men set out for the bridge is prompted not so much by a change of heart, but because he has worked out a plan of escape. He has recruited five men to replace El Sordo's band, men he secretly plans to kill once the mission is accomplished in order to leave enough horses for his own men to ride away to safety. Jordan will have to use a grenade to set off the dynamite.

The action at the bridge is an exciting sequence in Hemingway's best manner, and the final scene of Jordan lying in the pine forest, his leg broken, after the others have made their escape, has a haunting quality. The story is given depth by its many flashbacks: to Pilar's life before the war with a tubercular bullfighter who lived in a perpetual state of fear; the massacre of the 'fascists' at Avila at the outbreak of war, organised by Pablo, which ended in a drunken orgy of senseless brutality; Maria's horrifying account of having her head shaved before being raped; Jordan's memories of Hotel Gaylord's in Madrid where he was 'indoctrinated' by Karkov, the cynical Soviet journalist; the young peasant Andrés' surreal journey with Jordan's important message for the general, which takes him via the anarchists who might easily have shot him, and briefly into the hands of a murderous political commissar of the International Brigade. It is from such material that the book develops richness – though at times Hemingway piles detail upon detail at the expense of the main narrative – and allows the reader to feel that this is indeed the whole of Spain.

Michael INNES　　　　　　　**1906–**

The Secret Vanguard

Innes's fifth novel, and the first of his spy stories, *The Secret Vanguard* takes the material of such novels as John Buchan's **The Thirty-Nine Steps** (1915) and irradiates it with Innes's characteristic ready fantasy and intellectual wit. It is 1939, and an organisation of German spies, 'the secret vanguard', wants to obtain a formula possessed by a scientist, Orchard, who has hidden himself in a remote Highland cottage. On a train, members of the group pass information about Orchard's whereabouts in some lines of poetry they claim to be by a minor poet, Philip Ploss. He happens to be present at the time, and is subsequently murdered while sitting in the gazebo of his country home. Innes's detective hero, intellectual policeman Inspector John Appleby, is put on the case.

Meanwhile, Sheila Grant, a young English gentlewoman with all the intrepidity of her breed, is travelling in another train over the Forth Bridge when other members of the vanguard are passing information about Orchard's whereabouts, this time in bogus lines added to Swinburne's *Forsaken Garden*. Realising that something is wrong (perhaps only in Innes could the plot turn on a character's ability to distinguish genuine from pastiche Swinburne), she is kidnapped, and wakes up a prisoner in a cottage on a lonely Scottish moor with Dick Evans, a young American studying Caravaggio. Evans had also been on the train and had pluckily intervened when he saw her seized. He helps Sheila escape from the cottage, and they separate on the moor to avoid joint capture. Sheila eludes her pursuers by hijacking a railway truck and, on arrival at a branch-line station, is shielded by a blind, half-mad poet and fiddler, Harry McQueen ('Poems,' he cries 'The poems of a Moray loon.'). She is lured to the vanguard's headquarters, the sinister Castle Troy, by a member of the gang pretending to be a friend; on meeting the vanguard's leader, Belamy Mannering, she realises her mistake and makes her escape by motor launch. Passing a solitary beach, she sees a man there reciting Homer in Greek ('A foreign language, but one which was more reassuring than any English could have been'), who turns out to be the scholar Ambrose Hetherton, a friend of Appleby.

They meet a person claiming to be Orchard, but Appleby, arriving on the scene, unmasks him as an impostor ('I'm not a traitor ... My pedigree's Mitropa out of Wagon-Lits'). Arriving at Orchard's cottage, the group finds it empty, but Dick Evans had sketched a collection of 'Caravaggio' drawings there which contained a coded version of the formula; this is now, with Orchard, in the possession of the vanguard. The forces of justice (now also including Evans and the debonair young Scot Alaster Mackintosh) prepare for an assault on Castle Troy, where a garden-party is in progress. Mackintosh disguises himself as a vendor of shoelaces and lets down the vanguard's tyres; when they try to escape in a powerful boat, Appleby, disguised in female clothing, passes a signal from a boat to an aeroplane, and the spies are eventually captured. Order has been restored by Britain's gentlemen and intellectuals, and Appleby himself now has nothing more to detain him than 'an overdue appointment with a burglar in Putney'.

Arthur KOESTLER   1905–1983

Darkness at Noon

Written originally in German, but first published in English translation, *Darkness at Noon* tracks the agony of a member of the Bolshevik old guard, N.S. Rubashov, whom Party loyalists arrest and interrogate during the Stalinist purge. Ostensibly, Rubashov is arrested for disloyalty, but in reality he is arrested because his idealistic socialism cannot any longer be accommodated to the political pragmatism of Stalin's autocratic rule. A member of the first generation of Russian revolutionaries, Rubashov has built his life on the principle that the end justifies the means. Now he finds himself a victim of the amoral, arbitrary political machine he had helped to establish. Without compunction he had sent other people to their deaths, simply because it was expedient. He is now without any grounds on which to object to his own expedient death.

The power of the book lies in Koestler's careful depiction of the uneasy, self-deceptive workings of Rubashov's mind, as his conscience slowly awakens to the human cost of what he has done in the past. The full irony of his situation comes home to him with the realisation that not only does the Party intend to consign him to the rubbish-dump of history, but is demanding his consent to this. Step by step, Rubashov's mind and body are worn down until he succumbs and signs a false confession which ensures his execution.

Based on Koestler's experiences as a member of the Communist Party in Hungary, *Darkness at Noon* influenced George Orwell's two best-known books, **Animal Farm** (1945) and **Nineteen Eighty-Four**

(1949). It is also reminiscent of Kafka's vision of isolated individuals menaced by an incomprehensible society, and of Dostoevsky's treatment of doubt, anguish and guilt.

Carson McCULLERS   1917–1967

The Heart is a Lonely Hunter

The pattern for all McCullers's major fiction was established with this, her first published novel. Her principal themes of loneliness and alienation, communication and love, are related through the story of five isolated people in their search for social progress and spiritual faith. McCullers declared that the book was about 'man's revolt against his inner isolation and his urge to express himself as fully as is possible'.

The story revolves around John Singer, a deaf-mute living in a small town in the American South. Singer unwittingly becomes the focus for a group of disparate individuals who share one common characteristic: loneliness. These individuals include: Mick Kelly, a fourteen-year-old girl; Benedict Mady Copeland, a black doctor; Biff Brannon, a widowed café owner; and Jake Blount, a Communist drunk. They represent, respectively, the problems attendant on adolescence, race, sexuality and politics in an unsympathetic society. Because of Singer's disability his outward character is vague and undefined. His friends are able therefore to impute to him all the qualities they wish him to have. The four people create their understanding of Singer from the reflection of their own desires. Singer's disability allows them to articulate their feelings without constraint and he becomes the repository for their most personal thoughts. They come to believe that Singer possesses the inner serenity and completeness lacking in their own lives.

The central irony of the novel is that although the major characters believe Singer to be a figure of encompassing wisdom, he is as lonely as they. His only friend, Spiros Antonapoulos, another deaf-mute, has been sent by his cousin to an insane asylum. Singer imputes to Antonapoulos qualities similar to those the other characters see in him. In reality Antonapoulos had been as uncomprehending of Singer's concerns as Singer is of the concerns of the characters that are drawn to him.

Singer commits suicide after learning of the death of Antonapoulos, unable to bear the burden of his own loneliness. Only the café owner comes to suspect that Singer was largely a creation of their own individual imaginations, a reflection of their hopes and desires. After managing Singer's funeral, Biff finds that 'There was something not natural about it all – something like an ugly joke.' Although all the major characters are defeated in their struggle – Mick Kelly is forced to abandon her education, Benedict Mary Copeland is dying, and Jake Blount leaves town after a riot – the overwhelming impression is not one of futility. The novel ends on a note of wary affirmation with Biff opening his café and waiting for the morning sun:

> For in a swift radiance of illumination he saw a glimpse of human struggle and of valour. Of the endless fluid passage of humanity through endless time. And of those who labour and of those who – one word – love. His soul expanded. But for a moment only. For in him he felt a warning, a shaft of terror. Between the two worlds he was suspended.

| C.P. SNOW | 1905–1980 |

George Passant

see **Strangers and Brothers** (1970)

| Christina STEAD | 1902–1983 |

The Man Who Loved Children

Unlike Stead's previous novel, *House of All Nations* (1938), which had been highly praised and became a bestseller, *The Man Who Loved Children* was a critical and commercial failure when it was first published. Randall Jarrell's 1965 essay on the novel was entitled 'An Unread Book', and it was only in 1970, when it was reissued in paperback, that it reached a new, more appreciative audience. Although set in America, the novel draws upon Stead's own childhood circumstances in Australia, when her widowed father remarried and she became 'the eldest of a large family'.

'The man who loved children' is Sam Pollit, unhappily married to Henny, and father of a large brood of children, including Louie, the ugly-duckling daughter from his first marriage. The other children are: Ernie, who is 'everyone's darling' and the most scarred by poverty; Sam's favourite '*Little-Womey*', Evie; the twins, Saul and Little-Sam; and Tommy, with his 'great round black eyes', who is too young to understand what is going on most of the time. Unable to communicate with his wife, Sam addresses the children in baby-talk: 'Bring up tea, Looloo-girl: I'm sick, hot head ... nedache ... dot pagans in my stumjack ... want my little fambly around me this morning ...' Sam believes himself to be 'a public character and a moralist of a very saintly type', but Henny, who has come down in the world by marrying him, finds him intolerable, and spends much of her time playing patience alone in her room, only speaking to him through the children. What Henny describes as Sam's 'mawkery' is rooted in his infinite capacity for self-deception: 'Mother Earth ... I love you, I love men and women, I love little children and all innocent things ... I feel I am love itself – how could I pick out a woman who would hate me so much.' When Sam is offered a longed-for appointment with the 'Anthropological Mission to the Pacific', Henny wonders what she is going to live on in his absence.

When Sam returns from Malaya in disgrace and is sacked, he feels it is beneath him to answer the charges against him. In any case, he has achieved his heart's desire to spend all his time with his children, including the newly born Charles-Franklin ('Chappy'). Henny does not receive an expected legacy when her indulgent father dies a bankrupt, and the family moves to the ramshackle Spa House: 'Henny awoke from a sort of sullen absence and knew what was happening; her heart was breaking. That moment, it broke for good and all.' They sink into abject poverty and Louie, unable to bear 'the daily misery', retreats into her own world, making up stories, writing plays, even inventing her own language, only coming alive at school with her teacher Miss Aiden, with whom she falls in love, and her friend Clare Meredith. In despair, Louie decides to murder her parents – 'If I killed them both we would be free' – but succeeds only in giving cyanide to her depressed stepmother, who is a willing accomplice in her own death. It is assumed that Henny has committed suicide, and when Louie tells Sam that she had intended to poison them both, he angrily dismisses her confession as an 'incredible, insane, neurotic story'. He decides to remove her from school so that she can 'recover' at home in his company, but Louie leaves home to go to her mother's family.

As Henny sat before her teacup and the steam rose from it and the treacherous foam gathered, uncollectible round its edge, the thousand storms of her confined life would rise up before her, thinner illusions on the steam. She did not laugh at the words 'a storm in a teacup.' Some raucous, cruel words about five cents misspent were as serious in a woman's life as a debate on war appropriations in Congress: all the civil war of ten years roared into their smoky words when they shrieked, maddened, at each other; all the snakes of hate hissed.

THE MAN WHO LOVED CHILDREN
BY CHRISTINA STEAD

Stead's novel is a savage portrait of a family in crisis, destroyed by the mutual hatred of parents. It is all the more powerful because, as Jarrell notes, it makes the reader 'a part of one family's immediate existence as no other book quite does'.

T.H. WHITE   1906–1964

The Witch in the Wood

see **The Once and Future King** (1958)

Richard WRIGHT   1908–1960

Native Son

Native Son tells the story of Bigger Thomas, a Mississippi-born black who has migrated with his family to the slums of Chicago, where he grows up. The novel is divided into three sections, 'Fear', 'Flight' and 'Fate', which chart Bigger's progress towards his own destruction. 'Fear' tells of his hard home life and how he becomes involved with a street gang. When he takes a job as a chauffeur with a wealthy white family, the Daltons, his fellow gang-members fall away from him. Wright builds up the psychological tensions in this seemingly mundane history until Bigger becomes literally crazed by fear. The culminating incident of this section takes place after Bigger has escorted home the drunk daughter of his employer and is almost discovered in the bedroom by her blind mother. Although innocent of any malicious intent, Bigger becomes hysterical in this situation and accidentally smothers the girl while trying to keep her silent. Realising that he has killed her, Bigger cuts off her head and burns her body in a furnace.

'Flight' takes up the story of Bigger's crime as he tries to evade suspicion by faking a blackmail note. His black mistress urges him to collect the money, but before he can do so the charred remains of Mary Dalton are discovered, and he flees with his mistress in tow. Realising that she will be a burden to him, Bigger murders her and becomes a more monstrous figure than before. Inevitably, his flight ends in capture by the police.

The first two sections present the action from within Bigger's consciousness alone, giving the story a fierce, almost coercive single-mindedness. The third, 'Fate', links Bigger's actions to a much wider social and political context. The black community, the church, judiciary, former friends and the Klu Klux Klan all have opinions of Bigger Thomas and give vent to them during his trial. Bigger is sentenced to death and at the last is shown to have conquered the fear which had led him to commit his atrocities, accepting in his execution a Kierkegaardian *sein zum Tode* which raises him above simple criminality.

Wright's greatest achievements in the novel are its structure and the creation of the main character. Organising the book into three sections, Wright welds his headings into a gruesome and wholly convincing cycle. There is a powerful sense that Bigger could not have done other than he did within this framework. The character of Bigger is the linchpin of the book though, and Wright's minute psychological notation as he allows seemingly random incidents to accumulate in the killer's psyche generates huge tensions. The understanding that he brings to Bigger is constantly pitted against the enormity of Bigger's crimes; this battle is finally won by Bigger when he gains a measure of self-knowledge shortly before his death.

Native Son proved a huge success on publication, selling over 200,000 copies in only a few weeks. Early reviewers saw it as an outstanding contribution to the novel of social protest. This view of the book became the orthodoxy when the fact that it was based on a real case (that of Robert Nixon in 1938) came to light. Critics have also noted similarities between Wright's own life and the background which produces Bigger Thomas in the novel. However, the most acute criticism came from James Baldwin when he stated that 'no American Negro exists who does not have his private Bigger Thomas living in his skull'. This fearsome judgement of the position of blacks in America attests not only to the range and accuracy of Wright's creation, but also (and more disturbingly) to its continuing relevance within that society.

He was tired, sleepy, and feverish; but he did not want to lie down with this war raging in him. Blind impulses welled up in his body, and his intelligence sought to make them plain to his understanding by supplying images that would explain them. Why was all this hate and fear? Standing trembling in his cell, he saw a dark vast fluid image rise and float; he saw a black sprawling prison full of tiny black cells in which people lived; each cell had its stone jar of water and a crust of bread and no one could go from cell to cell and there were screams and curses and yells of suffering and nobody heard them, for the walls were thick and darkness was everywhere. Why were there so many cells in the world?

NATIVE SON
BY RICHARD WRIGHT

1941

Germany invades Russia ● Japanese bomb Pearl Harbor and US enters the war ● Beginning of 'Manhattan Project' for atomic research ● 'Utility' clothing and furniture in Britain ● Étienne Gilson, *God and Philosophy* ● Benjamin Britten, Violin Concerto ● Noël Coward, *Blithe Spirit* ● Orson Welles, *Citizen Kane* ● Lord Baden-Powell, Amy Johnson, James Joyce, Sherwood Anderson, Virginia Woolf, Ignaz Jan Paderewski and Rabindranath Tagore die ● W.H. Auden, *New Year Letter* ● Theodore Roethke, *Open House*

Joyce CARY　　　　　　**1888–1957**

Herself Surprised

see **First Trilogy** (1944)

Patrick HAMILTON　　　　**1904–1962**

Hangover Square

J.B. Priestley classed *Hangover Square* and *The Slaves of Solitude* (1947) 'among the minor masterpieces of English fiction', and it is true that while Hamilton's reputation has fluctuated over the years, his work has always had distinguished champions. He is particularly adept at evoking the lives of those at the margins of society, people who shuttle between the bogus conviviality of noisy pubs and the despairing solitude of dismal boarding-houses and cheerless hotels. His fictional locales are carefully chosen: Brighton, Maidenhead, Reading and the bedsitter areas of London. *Hangover Square*, a tragi-comic tale of schizophrenia and sexual obsession, is subtitled 'A Story of Darkest Earl's Court'. It is set in 1939; the country is on the brink of war, but this fact hardly impinges upon the lives of Hamilton's shiftless characters.

George Harvey Bone suffers from schizophrenia, but all he knows is that sometimes something 'snaps' in his head and he goes into what his friends call one of his 'dumb moods'. In his dumb moods George has one overarching purpose in life: to murder Netta Longdon, the girl he loves, and then to go to Maidenhead. When he returns to normal he can never remember what he felt or did in these lost stretches of time.

He has lived in Earl's Court for eighteen months, an endless succession of pubs and restaurants gradually eroding his money, and is unable to escape because of his obsession with Netta. Netta's friends are the thuggish 'blond fascist', Peter, and the spectacular drunkard Mickey: 'Your hangover was never so stupendous as Mickey's, nor your deeds the night before so preposterous.' George hates them both and hates himself for mixing with them: 'Drunken, lazy, impecunious, neurotic, arrogant, pub-crawling cheap lot of swine – that was what they all were.'

The wheels and track clicked out the familiar and unmistakable rhythm – the sly, gentle, suggestive rhythm, unlike any of its others, of a train entering a major London terminus, and he was filled with unease and foreboding as he always was by this sound. Thought and warmth must give place to action in cold streets – reality, buses, tubes, booking-offices, life again, electric-lit London, endless terrors.

HANGOVER SQUARE
BY PATRICK HAMILTON

Netta is an aspiring actress, and when she meets George's old schoolfriend John Littlejohn and discovers that he works for a famous theatrical agent, she finds a new interest in George. She agrees to a quiet assignation in Brighton, but turns up drunk with Peter and a pick-up, with whom she noisily makes love in the hotel room next to George's. In a dumb mood, George returns to London and goes to Netta's flat to kill her and Peter. He returns to normal at the last moment and thereafter resolves to break with Netta and the whole drunken crowd. But Netta lures him back and then cancels another assignation. In Brighton he confirms his suspicion that she is really pursuing the theatrical agent Carstairs and this time the dumb mood lasts long enough for him to dispatch Peter with a golf club and drown Netta in the bath. He sets off on foot for Maidenhead, gradually realising what he has done; in a hotel room he seals the windows and turns on the gas. War has broken out and the story merits a mention in only one newspaper under the headlines:

SLAYS TWO
FOUND GASSED
THINKS OF CAT.

| Kate O'BRIEN | 1897–1974 |

The Land of Spices

This, commonly regarded as one of the finest of O'Brien's nine novels, was the second (of two) of her works to be banned for 'immorality' by the Irish Censorship Board. (The other was *Mary Lavelle*, in 1936.) One sentence incurred their wrath and it holds the key to the novel's central character. Helen Archer, Revd Mother of an Irish house of the Order of Saints Famille. In 1906, when the book opens, she is forty-three. She has been in exile for four years, an Englishwoman (albeit raised in Brussels), sent from Belgium by the head of her Order, to live amongst people to whom her nationality, voice and manner are a perpetual affront, and to work in proximity with a male church hierarchy resentful and suspicious of her Order's freedom from their jurisdiction.

Not only is Revd Mother English, she is also European, and that, too, in a period of rapidly intensifying Irish nationalism renders her suspect. Steadfastly refusing to allow herself or her convent to be co-opted by narrowly sectarian notions of a National Religion, she is an obstacle and an irritant to the Bishop, and a succession of chaplains, seminarians and parish priests.

Revd Mother is 'a cold fish', say many of her pupils and her nuns (although the humbler and more timid members of the Community find themselves unexpectedly confident in her presence). Cold she has indeed been, frozen at the age of eighteen, by a discovery which caused her to hate the father she had hitherto loved above all others. Unable to tell him what she has learned and knowing no human comforter to whom to turn, she enters the Order within weeks.

The vows of poverty and chastity hold no terrors for her. Obedience is very much harder, and the injunction to love hardest of all. 'She is afraid of love, even of the love of God,' says one of her superiors. And, indeed, despite her abundant sense of irony and humour, her swift response to beauty of thought and language, her compassion for the vulnerable, Revd Mother's public virtues remain the rather chilly ones of diligence, efficiency and justice. But over the years she has come gradually to suspect that her frenzied retreat, at the age of eighteen, was not only cruel to her earthly father but also discourteous to her heavenly one: 'a soul should not take upon itself the impertinence of being frightened for another soul; ... God is alone with each creature.'

A thaw is signalled and the chief agent of it is introduced on the second page. Through the six-year-old Anna Murphy, the convent's newest and youngest pupil, Revd Mother will learn to risk loving once more. She will resolve the torment of her adolescent trauma and be fully reconciled with her father. And, because Anna is there, she will even elect to stay and wrestle for eight more years with her recalcitrant Irish Community, thereby incidentally fitting herself for the highest office of all. On the very eve of the First World War, when the book ends, she is due to return, as Mother General, to Belgium. (O'Brien does not emphasise the timing but, for the reader, a sense of foreboding must accompany the seemingly triumphant ending.)

The often sombre but ultimately satisfying account of Revd Mother's spiritual odyssey unfolds against a vivid backcloth of exuberantly characterised girls, nuns, parents and clerics, and gleefully satirised provincial snobs and priestly politicians. Beneath the energy and humour, however, lies a complex examination of the relations between secular and divine love, and the meaning, for women, of freedom and liberty – qualities which both Revd Mother, in her enclosed Order, and Anna, destined for the world, desire and seek in very different ways.

The Aerodrome

The third of Warner's novels exploring the conflicts between political theory and the individual, *The Aerodrome* shows a concern for balance. The idyllic village life, with its inefficiencies and inequities, is rejected in favour of the ordered and visionary world of the aerodrome which gradually engulfs it; but this act of usurpation has in it the seeds of totalitarianism and the novel ends with an affirmation of life unconstrained.

The conflict is explored in the character of Roy, the young narrator, brought up in the village, but attracted to the aerodrome. The ethos of the new order, as expounded by the Air Vice-Marshal, is: 'To be freed from time, ... From the past and from the future. From shapelessness.' To this end, family ties and sexual bonds are discouraged and pro-creation is banned. The impressionable, hero-worshipping Roy, who has recently learned that he was adopted and that Bess, the woman he married secretly, is not only in love with someone else but is also his sister, is ripe for conversion to this creed. He is taken up by the Air Vice-Marshal, becoming his private secretary, and enjoys an unfettered affair with Eustasia, the glamorous wife of the aerodrome's Chief Mathematician. As his star rises, so that of the Flight Lieutenant, whose dashing recklessness first caught Roy's imagination, falls, and the two young men exchange roles. The Flight Lieutenant accidentally shoots Roy's adoptive father, and is made secular rector of the village in his place. He gradually wins over his congregation, siding with them against the depredations of the aerodrome.

Human relations are not so easily put aside, however, and when Roy visits Bess, who has become catatonic with incestuous guilt, his affection is rekindled and he realises that: 'We in the Air Force had escaped from but not solved the mystery ... We had abolished inefficiency, hypocrisy, and the fortunes of the irresolute or the remorseful mind; but we had destroyed also the spirit of adventure, inquiry, the sweet and terrifying sympathy of love that can acknowledge mystery, danger, and dependence.' Further irony colours the complex web of familial relationships which underlies the story: it transpires that both the favoured Roy and the disgraced Flight Lieutenant (who is killed when he deserts with Eustasia) are the sons of the Air Vice-Marshal. After declaring their opposition to the aerodrome, Roy and his mother are saved from punishment by the death of the Vice-Marshal in a plane-crash. Roy remarries Bess, who turns out not to be his sister after all, and the novel ends with the affirmation of the world: 'Clean indeed it was and most intricate, fiercer than tigers, wonderful and infinitely forgiving.'

The Ill-Made Knight

see **The Once and Future King** (1958)

Between the Acts

Woolf's last novel was published posthumously, and although she repeatedly revised it, she was prevented from making a final draft by her last illness. The technique shows a shift from the earlier stream-of-consciousness novels, now balancing introspection against a new interest in dialogue, though this is dialogue of an oddly fractured, abortive quality.

The theme of dislocated speech is embodied in the village pageant which unfolds through the course of the narrative. Staged in the grounds of a country house, family home of the Olivers, the pageant is continually being disrupted by over-long intervals, bad weather, mishaps with costumes and other dramaturgical setbacks. It is also broken up by other, equally abortive, dramas, which press for the reader's attention between (and during) the officially staged scenes. The chief of these is the tense silence between Isa and Giles Oliver, for whom the entire day is an interval, an uneasy truce within their disintegrating marriage.

The only communication between them during the day is accidental and oblique, conducted through intermediaries. Thus the unfaithful Giles flirts with the fulsome Mrs Manresa, while Isa discovers an unspoken empathy between herself and Mrs Manresa's companion, William Dodge. This second couple is as ill-assorted as the first, since Mrs Manresa is in determined pursuit of admiration, while Dodge is homosexual. Hence the four characters are woven into a strange pattern of crossed purposes and vain bids for attention.

The pageant itself, a condensed history of England, penned by another misfit character, Miss La Trobe, is equally diverted, derailed and misconstrued. At first, the local gentry who make up the audience are pleasantly lulled by Miss La Trobe's seemingly innocent snapshots of quainte olde Englande, but they become progressively less easy as the action continues. They are dismayed when the Victorian virtues of piety, patriotism and purity, from which they are not yet conscious of having dissociated themselves, are impishly satirised. Then, to make matters worse, they themselves are projected on to the stage, when the actors hold up mirrors to the audience. The comfortable pageant threatens at this point to come dangerously to life. Attempts are made to patch up the social occasion: the gramophone plays, the local vicar moralises, but the latter is interrupted in turn by the sound of aeroplanes overhead, a reminder that the Second World War is about to disrupt *all* scripts.

Afterwards, Miss La Trobe retreats to the village pub to nurse her damaged pride. She feels that, against her artistic intentions, the pageant has turned into a mouthpiece for obtuse village solidarity. Yet, despite all this, another play, another scene, is already beginning to form in her imagination. Finally, when everyone else back at the house is asleep, Isa and Giles return to the marital battlefield, and start their first conversation of the day.

The focus of *Between the Acts* is typically Woolfian: a social event is treated as a kind of tribal festival, and also imbued with a vital aesthetic radiance. At the same time, an unusually simple plot-line is joined to a newly dark and fractured emotional rhythm, which ominously foreshadows Woolf's own impending personal tragedy.

1942

Fall of Singapore ● Germans reach Stalingrad ● Battle of El Alamein ● US task force at Guadalcanal beats off Japanese ● US war factories achieve maximum production ● The atomic age begins ● The V-2 rocket launched ● William Beveridge, *Report on Social Security* ● C.S. Lewis, *The Screwtape Letters* ● Dmitry Shostakovich, 7th Symphony ('Leningrad') ● Jean Anouilh, *Antigone* ● Walter Sickert and Michel Fokine die ● John Berryman, *Poems* ● T.S. Eliot, *Four Quartets* (1935–42)

Joyce CARY   **1888–1957**

To Be a Pilgrim

see **First Trilogy** (1944)

Maura LAVERTY   **1907–1976**

Never No More

This very successful first novel, set in Ireland in the small town of Ballyderrig, County Kildare, spans the years from 1920 to 1924. It begins and ends with a death. The first, that of the narrator's father, a small and unsuccessful draper, heralds momentous changes for his family. His widow moves back to Kilkenny, taking all her children, bar Delia, the narrator. Delia's relationship with her mother has always been uneasy ('deep love, but no liking'); it is her maternal grandmother whom she loves best, and her feelings are fully reciprocated, though never articulated by either until the book's close.

Gran, sixty-eight when the book opens, 'unloved by the slovenly and malicious and foolish', is all compact of energy, wisdom, common sense and kindliness. The book is Delia's loving and grateful celebration of the way in which Gran's offer to keep her granddaughter with her 'changed the whole world for me, bringing me from despair to wild happiness'. The teenage girl and the old woman find in each other the perfect mother and daughter which Delia's own mother could never be for either of them. The four years with Gran bring Delia close to an Ireland already threatened. Subtitled 'The Story of a Lost Village', the book is in part a long elegy for old ways waning, and a passionate declaration of love for the actual

soil and landscape of a physically beautiful country.

Derrymore, Gran's house, is run with a superb efficiency rooted in the incessant and all-encompassing labour of its womenfolk, whether in the henyard, dairy, garden, still-room, laundry or kitchen. Gran is as skilful with a homemade prescription, recorded in her own *Materia Medica*, as she is with a recipe: 'Her head was a card-index file of amazing old herbal remedies. Many of them, I now suspect, were more amazing than remedial.' Of gran's prowess as a cook, however, there could be no two opinions. She is magnificent, and the book is liberally sprinkled with detailed recipes which cry out to be tried immediately – for hot seedy cake, stewed eels, washday coddle, potato thump, ash cakes, mushroom ketchup, flummery and apple dumplings. Food, its cultivation, preparation, consumption and exchange, becomes a metaphor for human life and love: 'Human nature is like bread, I think. Soda-bread calls for buttermilk, and baking powder bread for new milk. Use the wrong kind of milk and the bread is sodden. Gran was the right kind of milk for me.' (The book is dedicated 'To Thrush, my brother, who has always been the buttermilk for my soda-bread'. Laverty herself published a number of cookery books.)

A string of marvellously vivid, swiftly told stories thread their way through the book, some of them tiny episodes in Delia's daily life, others part of the town's rich store of communal tales, stretching back over many decades and losing nothing in the telling. The scale of each story may be tiny – a tale of brutal seduction, or of a quietly devoted courtship lasting thirty years, of 'go-boys' meting out their own rough justice to faithless lovers and social miscreants, or of IRA supporters denied burial in their local churchyard – but cumulatively they uncover the intricate web of loyalties, secrets, ancient feuds, shared suffering and communal memories which together make up the township's history and present.

Gran's own death brings the book to a close and, by implication, her passing marks the end of an era whose life has been miraculously prolonged but whose eventual dissolution was inevitable. Sean O'Faoláin, in his enthusiastic preface to the first edition, recognises the book's mythic function – one shared by a long and distinguished line of twentieth-century Irish autobiographical novels of which *Never No More* is an outstanding example.

Evelyn WAUGH 1903–1966

Put Out More Flags

The outbreak of the Second World War finds the Bright Young Things of Waugh's earlier novels somewhat tarnished and older. Basil Seal alone retains an unsettling juvenility. While others search for honourable occupations to suit their age and station, Basil determines to become 'one of those people one heard about in 1919; the hard-faced men who did well out of the war'. To this end he elects himself billeting officer whilst staying in the country with his doting sister, Barbara Sothill. His secret weapon is a trio of delinquent evacuees, the Connollys, whom he forces upon genteel families and then removes (at a price) after the children have driven their hosts to breaking point.

Meanwhile Ambrose Silk, 'a cosmopolitan, Jewish pansy ... all that the Nazis mean when they talk about "degenerates"', relinquishes his tiresome young associates and founds a *Horizon*-like arts magazine, *Ivory Tower*; Peter Pastmaster is in a crack regiment and in search of a wife; and Alastair Digby-Vane-Trumpington has enlisted in the ranks and is about to become a father. Surprised by a genuine billeting officer, Basil returns to London, where in his absence his long-serving mistress, the chic and aloof Angela Lyne, has begun to drink heavily. He

joins the Ministry of Information, bluffing his way into the Department of Internal Security. Partly in order to fulfil his Richard Hannay fantasies, but chiefly to seduce a secretary, he tricks Ambrose into altering an anti-Nazi essay about a former German boyfriend in *Ivory Tower*, and then shops the unfortunate writer to his superiors as a fascist propagandist. He assists Ambrose to flee to Ireland, then takes over his flat and his monogrammed underclothes. Angela's ineffectual architect husband, Cedric, is killed in action, leaving her free to marry Basil. He agrees, but immediately volunteers for special service. The novel ends with the optimistic assertion that: 'The war has entered into a new and more glorious phase.'

In *Put Out More Flags* the 'phoney war' is seen largely through the eyes of Basil and Ambrose, and scenes of high comedy are juxtaposed with passages of reflected, malancholic soliloquy. In tune with the times, Waugh's sympathies are with the men of action, men such as Alastair (and, of course, Waugh himself) who adapt at some cost to the new conditions. However, if aesthetes such as Ambrose and Cedric Lyne (a specialist in reconstructing grottoes), who 'belong, hopelessly, to the age of the ivory tower', are ruthlessly eliminated, they are also portrayed with remarkable sympathy.

Eudora WELTY 1909–

The Robber Bridegroom

Based on the Brothers Grimm folk-tale of the same name, Welty's first novel uses the archetypal figures of a fairy-story – the trusting father, the wicked stepmother, the beautiful daughter and the good-hearted bandit – but explores the background and personality of the characters in a way a fairy-story does not. Likewise, despite all the plot contrivances of disguise and mistaken identity, and the use of phantasmagoria, the story is firmly rooted in the Natchez Trace area of Mississippi at the end of the eighteenth century, with real historical figures appearing albeit in a legendary guise. As

Welty commented in an address to the Mississippi Historical Society, this is not 'a *historical* historical novel', but one which seeks to describe a time and place through the use of legend and fable.

The innocent planter Clement Musgrove lost his first wife when they were set upon by Indians, and he and his beautiful daughter, Rosamond, escaped only with the assistance of the hideously ugly Salome, whom he married in gratitude. Unbeknown to trusting Clement, Salome detests Rosamond and endlessly seeks her destruction by sending her out on errands into the wild woods, where she is protected from the beasts by her mother's locket. One day Rosamond is ravished by a bandit and Clement seeks the assistance of big, burly Jamie Lockhart in tracking down her assailant, unaware that it was Jamie who did the terrible deed. Rosamond goes to live with Jamie and his band of thieves, though both are entirely unaware of their previous encounter.

There ensues a fabulous story of crossed paths and mistaken identities which culminates in all the characters being kidnapped by murderous Indians. Rosamond is freed by Salome's malicious side-kick, Goat, on the understanding that she will become his wife, and Jamie breaks free; each thinks the other is dead. Clement is freed by the petrified Indians after Salome, naked, has danced herself to death in an attempt to stop the sun in its tracks. Some time later, he comes across the happy couple, now living in New Orleans, Jamie a prosperous merchant with one hundred slaves, and Rosamond the happy mother of twins.

1943

Russian victory at Stalingrad • Germans surrender in North Africa • Allies invade Italy • Mussolini falls • Churchill, Roosevelt and Stalin meet at Teheran for planning overthrow of Germany • Streptomycin discovered • Henry Moore, *Madonna and Child* • Aram Khachaturian, *Ode to Stalin* • Rodgers and Hammerstein, *Oklahoma!* (musical) • *Colonel Blimp* • Frank Sinatra becomes first pop idol of the teenager • Sergei Rachmaninov and Beatrice Webb die • Keith Douglas, *Selected Poems* • Kathleen Raine, *Stone and Flower*

Nigel BALCHIN　　　　　　　**1908–1970**

The Small Back Room

Of Balchin's seventeen novels, *The Small Back Room* and *Mine Own Executioner* (1945) are perhaps the best. Most of his books were written out of his own wide experience ('there is practically nothing in which I am not or cannot be intensely interested,' he commented), and *The Small Back Room* draws upon his war service as a psychologist in the personnel department of the War Office and as Deputy Scientific Advisor to the Army Council. The novel was made into a successful film (1949) directed from their own screenplay by Michael Powell and Emeric Pressburger, who were no doubt attracted to the project partly because, like their 'war films', the book is questioning and ambivalent rather than straightforwardly, unthinkingly 'patriotic'.

Sammy Rice is a scientist working for a research unit based in Whitehall during the Second World War. Following an amputation, Rice has an aluminium foot, which is frequently painful, and is struggling to contain a drinking problem: he stares at a bottle of whisky 'like a rabbit looking at a snake'. He lives with his girlfriend, Susan, a secretary at the unit, but will not marry her because, with characteristic self-pity, he cannot believe she could long love someone as contemptible and weak as himself.

The novel, which is narrated by Rice, deals primarily with the petty personal and inter-departmental rivalries which hamper any efforts at achieving useful work. Those who are actively self-seeking tend towards a tone of weary cynicism, and Rice reflects: 'It isn't easy to do anything to win this war, is it? There are such a hell of a lot of people who'd rather lose it than let you help.' His own development of a low-temperature lubricant is threatened by departmental infighting, and Waring, an advertising man in civilian life, tells him: 'You can have ideas that'd win the war four times over and it still wouldn't do anybody any good if you can't sell them.' One of the more absurd parts of the unit's work is a weekly meeting to discuss a file they call the 'Keystone Komics', a list of suggestions – for things like poisoned barbed wire and bomb-carrying trained dogs – sent in by members of the public.

Rice has been advising Stuart, a young officer involved in bomb disposal, on a series of mysterious small bombs which have killed a number of people. At the end of the novel two unexploded bombs are found and Stuart is killed trying to defuse the first of them. Rice sees the challenge of defusing the second as a means of conquering his weakness. 'I am scared, but it doesn't really matter fundamentally, as I do not mind much if I am killed.' He overcomes his fear and defuses the bomb, but chastises himself for lack of guts because he needed help at a crucial moment. The novel ends on a curious note of resignation mingled with self-disgust as he returns to Susan.

Jane BOWLES   1917–1973

Two Serious Ladies

Two Serious Ladies is the only novel that Bowles, a short-story writer, completed before being incapacitated by illness. It is a humorous account of the lives of two women of very different characters, both sliding into alcoholism and a Bohemian existence. Christina Goering is rich and unmarried, obsessed by the spiritual significance of accidents in her childhood, but her search for sainthood leads her to flee from her privileged background and buy a rundown house

on Staten Island. Two chance acquaintances, Miss Gamelon and Arnold, develop intense but sceptical affection for her and move into the house, Arnold exhilarated by the contrast with his job at the bank, Miss Gamelon more begrudging about their unnecessarily sacrificed comfort. Christina's friend Frieda Copperfield leads an equally bizarre existence, after separating from her stingy husband on a holiday in the red light district of Panama.

The interwoven narratives of these two serious ladies embody the theme of women's desire, not just for sexual fulfilment but for sexual autonomy. The religious Miss Goering becomes a high-class whore, spurred on by the passion to explore experience to the full, while Mrs Copperfield moves into a brothel after a pleasant cup of tea with the madam. Embracing this atmosphere of liberation, Frieda Copperfield finds a satisfactory outlet for her capacity to love in the figure of the teenage prostitute Pacifica, witnessing with bewilderment the girl's infatuation with her brutal and unreliable lover. 'I can't live without her, not for a minute,' she says of this socially impossible companion. 'I have gone to pieces, which is a thing I've wanted to do for years . . . but I have my happiness, which I guard like a wolf.' Bowles's chaotic, comic novel is a celebration of the defiant mores of her own life, proclaiming a right to female freedom which shocked her contemporaries.

Henry GREEN   1905–1973

Caught

Elizabeth Bowen and Henry Green are the two novelists who most perfectly evoked London during the blitz. Green was one of several writers who joined the Auxiliary Fire Service, which was formed to supplement London's firemen, whose resources were thought inadequate to deal with the air raids war with Germany would inevitably bring. Green was to begin, but never complete, an official history of the AFS, and recorded his experiences in several short stories (some of which were published in *Penguin New Writing*) and in *Caught*, a novel of the 'phoney war', written in his characteristically elliptical, highly wrought prose. He was obliged by his publisher, John Lehmann, to rewrite the novel in order to excise the protagonist's adulterous relationship: in the original version, Dy was not Roe's sister-in-law, but his wife.

The book opens with Richard Roe, an upper-class widower, spending his first leave from the AFS visiting his five-year-old son, Christopher, whom he has sent out of London to the country to be looked after by his dead wife's sister, Dy. Christopher's safety had previously been threatened when he was abducted from a London toyshop by a disturbed woman who was subsequently confined to an asylum. By an awkward coincidence, Pye, the voluble, socialist Regular fireman who commands the substation at which Roe is serving, is the kidnapper's brother.

Much of the book describes the long hours of inaction and boredom at the station, and Green takes the same delight in detailing the mechanics of the job as he did in portraying factory life in **Living** (1929). Unused to authority, and uneasy in particular with those under his command who are better educated than he is, Pye is unsuited to his position, and his competence is further undermined by personal anxieties. In an attempt to overcome the embarrassment of serving under Pye, Roe attempts to do him a favour by introducing him to two women, 'a couple of enthusiastic amateurs' who live near the station. Pye becomes involved with one of them, whose company costs him more than he can really afford. He is already worried that he will have to pay large amounts of money for his sister's upkeep, and is further disturbed when a doctor's question about her past prompts him to imagine that (supposing her to have been someone else) he might unwittingly have had sex with her in a dark lane one night.

The District Officer, Trant, is down on Pye, who always appears to be absent when random checks are made. Pye's position is further eroded by Piper, an ancient Auxiliary and bore who is forever attempting to curry favour with his superiors. Piper persuades Pye to take on a female crony of his in the kitchens, even though the woman cannot cook. She subsequently goes AWOL on a mission to remonstrate with her feckless son-in-law, an absence Pye does not immediately report. Piper betrays Pye's failings to Trant, whose room he is redecorating. Meanwhile, Roe embarks upon an affair with Hilly, the sub-station's young mess manager. In delineating this equivocal, patrician, night-club romance, Green wittily and touchingly suggests the war's sense of *carpe diem*.

The novel concludes with Roe on leave suffering shell-shock after being knocked unconscious by a bomb during a raid. At last the firemen have seen action, and Roe describes to an uncomprehending Dy their chaotic and vain attempts to control a vast conflagration at some dockyard timber warehouses as bombs fell around them. Amongst those killed in the raid are Piper – dying as he had lived, 'making up to the Regulars' – and Shiner Wright, a heroic former sailor. We also learn, almost incidentally, that Pye has committed suicide.

Green's concern to reproduce the speech patterns of people from widely differing backgrounds and to engage the reader by elaborate description are both given full rein in the novel. The easy, matey atmosphere of the sub-station is contrasted with the reticence between Roe and his family, in particular the mutual shyness between father and son (the novel is dedicated to Green's own son). Given the novel's setting of London in the blackout, it is appropriate that Green's binding image is the light in which the characters are 'caught' at various moments: the windows of Tewkesbury Abbey, where the young Roe experienced vertigo; the light passing through the stained-glass windows of the toyshop, colouring its wares and adding to Christopher's sense of wonder; the firelight in which his abductor's eyes glitter; the moonlight by which Pye glimpses his sister returning from a night-time rendezvous; the torches of pedestrians in the dark streets; the spotlights and table-lamps of the nightclub; and above all, the beautiful but hellish flames of the fires the AFS attempt to quench.

1944

Normandy landings ● Germans driven from Russia ● V-bombs on England ● US monetary and financial conference at Bretton Woods, New Hampshire ● Franklin Delano Roosevelt, Democrat, wins US Presidential election, for a fourth term ● Jean-Paul Sartre, *Huis Clos* ● Tennessee Williams, *The Glass Menagerie* ● Laurence Olivier, *Henry V* ● Edwin Lutyens, Edvard Munch and Ethel Smyth die ● W.H. Auden, *For the Time Being: A Christmas Oratorio*

H.E. BATES   **1905–1974**

Fair Stood the Wind for France

In 1941 Bates joined the RAF as a flight lieutenant and later rose to the rank of squadron leader. His short stories of service life, collected in *The Greatest People in the World* (1942), *How Sleep the Brave* (1943) and *The Face of England* (1953), were published under the pseudonym of 'Flying Officer X'. He did not, however, retain the pseudonym for *Fair Stood the Wind for France*, a novel which had to be passed by the censor before publication.

Returning from a bombing mission Capt. John Franklin is forced by an engine failure to crash-land his Wellington bomber in occupied France. Aboard the plane are four sergeants, Godwin, Sandy, O'Connor and Taylor. Franklin's arm is badly injured and he is helped by his crew to a remote farm where a young girl, Françoise, and her family offer them sanctuary. Mindful of German atrocities against the French, Franklin is struck by Françoise's cool assurance and bravery as she takes him to the doctor in the town. Her father arranges false papers through the local Resistance and the sergeants leave for Marseilles, but Franklin has a fever from his infected arm and cannot travel. When his arm has to be amputated he is compelled to remain recuperating even longer.

Against the background of an idyllic summer in the country, Franklin and Françoise fall in love, and he is reluctant to face the prospect of leaving this peaceful place to return to the business of war. Driven to despair by the execution of friends, Françoise's father commits suicide, using Franklin's revolver. Françoise resolves to escape France with Franklin, and the two set off on bicycles for Marseilles. Once there, Franklin runs into the bedraggled and penniless O'Connor and all three take the train to Madrid. Françoise and Franklin are separated, and she is picked up by the *gendarmerie*. O'Connor runs from the train so as to distract their attention and is shot, and Franklin and Françoise are safely reunited on board. 'She was not crying for O'Connor, shooting and being shot at, doing a stupid and wonderful thing for them ... what she was really crying for was the agony of all that was happening in the world.'

Joyce CARY   **1888–1957**

First Trilogy

Herself Surprised (1941)
To Be a Pilgrim (1942)
The Horse's Mouth (1944)

Cary's three novels, published in one volume under this title in 1958, were conceived as a trilogy from the outset. 'What I set out to do,' he wrote in the preface to the 1958 edition, 'was to show three people, living each in his own world by his own ideas, and relating his life and struggles, his triumphs and miseries in that world. They were to know each other and have some connection in the plot, but they would see completely different aspects of each other's character.' Cary described the novels as a 'Triptych' (a title by which the trilogy is sometimes known), in which the central volume is flanked by the other two, 'each a narrow side panel', although the arrangement of two men of very different character linked by love for the same woman, places Sara Monday, the protagonist of the first volume, at the emotional centre of the work.

The trilogy is based upon the thesis that 'we are alone in our worlds'. The title of *Herself Surprised* refers to Sara's intuitive response to the world. She is a representative of 'sensuous domesticity', and it is this sensuousness that gets her into difficulties. When, as the novel opens, the judge who sends her to prison describes her as 'another unhappy example of that laxity and contempt for all religious principle and social obligation which threatens to undermine the whole fabric of our civilization', she is rightly upset. 'I have never been against religion,' she complains; 'far from it.' As a simple country girl in service, she discovered the works of the High-Anglican novelist Charlotte M. Yonge, and over the years these have given her 'strength in adversity'. There is no doubt, however, that sex plays an important part in her life. The put-upon son of the household, Matt Monday, is so attracted to her that he marries her. She also appeals to Mr Hickson, a millionaire patron of the arts, through whom she meets the painter Gulley Jimson. When Matt dies she is left with very little money and Jimson, after much pestering and in spite of already having a wife, marries her. Her attempts to make him a bankable artist fail and in the ensuing argument he punches her. They part company and Sara returns to domestic employment.

Her employer is the lawyer Tom Wilcher, 'a little man with a bald head and round black spectacles', who, in spite of being religious, has an unfortunate tendency to molest young women. He has difficulties with his family on that account, particularly with 'Mrs Loftus', the wife of his 'sleepy' nephew, Mr Loftus Wilcher. He finds consolation in Sara, attracted by both her religious susceptibilities and her flesh. In her own interests, Mrs Loftus keeps her uncle out of gaol, but shows no such mercy to Sara, who has been pilfering for years in order to support Jimson. Although his heart is broken at this turn of events, the conservative Tom cannot bring himself to intervene: he 'had always a great opinion of the law and did not like to interfere with it because it didn't suit him'. In

prison, Sara heeds the chaplain's words and is optimistic about her return to society: 'A good cook will always find work, even without a character, and can get a new character in twelve months, and better herself, which, God helping me, I shall do, and keep a watchful eye, next time, on my flesh, now that I know it better.'

A different perspective is provided in *To Be a Pilgrim*, the title of which is taken from Bunyan's famous hymn in *The Pilgrim's Progress* (1678–84), and refers to Tom Wilcher, driven by a sense of duty into the law and to take charge of the family fortune. This volume covers a period of more than half a century, from the age of Gladstone to the rise of Hitler. Now an old man with a weak heart, in the custody of his niece Ann, a qualified doctor, Wilcher sits in the crumbling family home, Tolbrook Manor, awaiting Sara's release from gaol, and musing on the past. Ann is unhappily married to his nephew Robert, who works all hours in an attempt to make the estate a paying proposition, and is pursuing an affair with one of his female helpers.

Wilcher is haunted by memories of his father, a retired soldier who did not believe in sparing the rod; of his tempestuous sister, Lucy, who went off with a wandering religious sect, the Benjamites, whose 'master' preached hellfire and abused her; of his charming, extravagant brother, Edward, whose radical politics were ahead of his time and who died defeated; and of his equally unworldly older brother, Bill, who chose Amy, the plainest and least guileful of local girls for his wife. All are now dead. Even Wilcher's relationship with Edward's former mistress, Julie, which he kept going for thirty years out of habit, has come to an end. Only Sara can save him now, he believes. But when he does briefly escape to her, she hands him back to Ann and to death, which is the pilgrim's only release.

If the style of Sara is warmly domestic, and the style of Wilcher repressively biblical, the style of Gulley Jimson in *The Horse's Mouth* is vividly impressionistic. The title, Cary explained, refers to 'the voice that commands Gulley to be an artist, and makes him struggle to realise his imagination in spite of all discouragement' (a phrase which relates him to the pilgrim Wilcher: 'There's no discouragement / Shall make him once relent'). Cary was inspired by two visionary artists, William Blake, whom Jimson reveres, and Stanley Spencer, some of whose characteristics Jimson shares. (Spencer employed a cook, whom he often painted, called Elsie Munday.) When the novel opens, Jimson has emerged from prison, where he has been incarcerated for making threatening phone calls to Mr Hickson, who had acquired by stealth a considerable collection of early canvases. Jimson is preoccupied with his present masterpiece, 'The Fall': 'Crimson apples. Eve terracotta with scarlet reflection.' Distractions include the boy Nosy, who wants to be an artist, and – more seriously – Coker, the 'long-bodied and short-tempered' barmaid at the Eagle to whom he owes fourteen guineas. Pregnant and deserted by her boyfriend, Coker is desperate for money and drags Jimson off to see Sara Monday in a forlorn attempt to recover his pictures. 'She's got thief written all over her'. Coker comments when 'the old boa constrictor' refuses to yield the paintings, but then Coker is not blessed with imagination. She next takes him to see Hickson, where Jimson finds 'Sara in the bath', hanging 'Middle of the wall between Goya and Tiepolo. Best light in the room.' In spite of Jimson's doubts, the meeting with Hickson goes quite well until the police arrive and discover Jimson's pockets crammed with netsukes and snuff-boxes.

When he next emerges from prison, Jimson finds Coker's mother occupying his boathouse studio. She has used 'The Fall' to patch up the roof. Fortunately, his patrons, the Beeders, and the 'art-cricket', Prof. Alabaster, who wants to write Jimson's biography, are on hand to assist his work on 'The Raising of Lazarus'. When the Beeders return to the flat they have lent him, they discover it in turmoil. Worse still, only the central figure's feet have been finished. Jimson's last masterpiece, 'The Creation', in which only the whale is finished, is accompanied by the trumpets of fame, but interrupted by the artist's death, which appears to be imminent as the novel closes.

His demise is brought about by Sara, 'the everlasting trumpets, a challenge to battle and death'. While attempting to snatch back a study for the picture of her in a bath, Jimson breaks her back, and only escapes an accusation of murder because she has already given the police a description of her assailant that does not fit the artist. He suffers a fatal fever: 'You are the only one I ever loved, Gulley,' he hears Sara say to him in his delirium. 'I certainly kept you busy, didn't I? I was a full-time occupation.'

Although *The Horse's Mouth* (which was made into a film, 1959, directed by Ronald Neame from a screenplay by Alec Guinness, who also played Jimson) remains the most popular 'panel' of the triptych – and indeed of Cary's *oeuvre* – many critics prefer the first two volumes. Cary was himself an accomplished painter, who had once intended to make this his career, and he drew upon his experiences at art school and in Bohemian Paris to create the central character. Some commentators have suggested, however, that while Gulley Jimson is vividly realised as an artist, Cary fails to convince the reader of the painter's *genius*. That said, the entire trilogy, with its tension between conservatism and innovation, its moral and philosophical debates about art and individuality, and its rumbustious comedy and characterisation, is a remarkably ambitious and largely successful enterprise, which has secured Cary his place in literature.

I. COMPTON-BURNETT   1884–1969

Elders and Betters

'Wars come and go, nations fall, but Miss Compton-Burnett goes imperturbably on her way,' wrote Edwin Muir, reviewing *Elders and Betters* in *The Listener*. 'We should be grateful for it. There is not another contemporary novelist who is so close to life and so remote from events.' If contemporary events impinged at all upon this grimly funny novel (which, like all the author's books, is set in the Victorian era), it is in the unrelenting darkness of its view of human relations. Families rarely behave well in Compton-Burnett's novels but, as Violet Powell nicely put it: 'In no other work does [she] allow such a consistently low standard of behaviour to prevail among the principal characters.'

The Donne family have moved to an incommodious new house in the country to be near their cousins, the Calderons. Benjamin Donne is a widower, and his children are the complacent thirty-two-year-old Bernard: the implacable, plain-speaking, thirty-year-old Anna; Esmond, who is twenty-six but childishly spiteful; and Reuben, a lame thirteen-year-old who is morbidly and self-indulgently aware of his slight disability ('I like as many apron-strings as I can get'). Since the death of their mother, the children have been looked after by their father's first cousin, Claribel. The housekeeping is overseen by Maria Jennings ('Jenney'), assisted by the servants, Cook and Ethel, women of uncertain age and unusual intimacy ('If it was hinted that their devotion bordered upon excess, Ethel would reply with quiet finality that they were first cousins'). The head of the Calderon household is Thomas, a writer, who is married to Benjamin's unstable older sister, Jessica. Their elder children – both in their twenties – are Tullia, for whom Thomas feels more than he does for his wife, and the idle, unmanly Terence, to whom Jessica is devoted. Julius, eleven, and Dora, ten, are – like most Compton-Burnett children – sophisticated beyond their years. Overseen by an amateur governess of means called Miss Lacey, they are often left to their own devices, which largely revolve around the cult of an imaginary deity, whose altar is a rock in the garden. The other member of the household is Jessica's younger sister, Sukey, a former beauty who is fatally ill and making the most of the opportunities her situation affords, frequently rewriting her will when vexed. (She and Reuben, who would be objects of pathos in a genuine Victorian novel, exemplify their creator's rigorous and modern eschewal of sentimentality.)

Anna befriends Sukey, who in a fit of pique makes a new will in her niece's favour. She subsequently relents and instructs Anna to destroy the new will, upon which she conveniently expires. Anna destroys the old will made in favour of Jessica and thus becomes Sukey's beneficiary. Jessica rightly suspects foul play and unwisely attempts to make Anna give up her inheritance, insisting that Sukey must have been confused. Anna launches a devastating attack upon her aunt's character, which leaves Jessica so debilitated that she commits suicide. This clears the way for Anna to marry Terence, falsely telling him that this was his mother's wish. Before Anna's intervention, Terence looked set to marry Miss Lacey's niece, Florence; his father now announces that *he* is to marry the young woman, a decision the spoiled, possessive Tullia treats with ridicule. Bernard takes advantage of the fact that Tullia's pre-eminence in the affections of her father appears to have been usurped, proposes to her, and is accepted. In fact, Florence has quickly come to recognise that she has no hope of competing against Tullia for Thomas's love and breaks the engagement, whereupon she is snapped up by Esmond: ' "One announcement more or less cannot make much difference," he said, in an almost cold tone.' Thomas is thrown back upon his two youngest children who, shocked by the unseemly behaviour of their elders (who are very far from being their betters), resign themselves to acting their allotted roles as doting, innocent children, praying to their god to assist them through the 'long courses of hypocrisy' ahead of them.

L.P. HARTLEY   **1895–1972**

The Shrimp and the Anemone

see **Eustace and Hilda: A Trilogy** (1947)

W. Somerset MAUGHAM   **1874–1965**

The Razor's Edge

The title of Maugham's most ambitious novel is taken from the *Katha Upanishad* in which it is suggested that the road to enlightenment is like the edge of a razor, narrow and painful. Maugham consulted Christopher Isherwood (whom some critics incorrectly cited as a model for the protagonist) and Aldous Huxley about Indian mysticism, and read an enormous amount on the subject before embarking upon his novel. When asked how long it had taken him to write, he replied: 'Sixty years.'

Larry Darrell is a young American, whose search for spiritual truth cause him to reject social conventions, a decision that places him outside the sympathy and understanding of his friends. Profoundly disturbed by the death of a friend in the First World War, who was killed saving his life, Darrell declines the offer of a job from his millionaire friend, Henry Maturin. He prefers to study philosophy, and leaves Chicago for Paris to continue his inner quest.

He is later joined there by his fiancée, Isabel Bradley, who hopes she can persuade him to return to America with her, but Darrell feels that his self-exploration has scarcely begun, and he breaks off the engagement. His behaviour infuriates Isabel's uncle, Elliott Templeton, who is Darrell's worldly opposite. Based on the famous socialite 'Chips' Channon, Templeton is absurdly snobbish, obsessed by the glittering life of Parisian *salons*, where he delights in spreading malicious gossip.

Maugham himself narrates the novel; as a writer and traveller he can plausibly follow the fortunes of his characters. Darrell is absent from much of the action of the novel, elusively travelling through Europe to Asia, and living in penury, and it is only in the final chapter that we learn fully about him and his lessons from a guru in India. Instead, Maugham details Isabel's marriage to Maturin's son, Gray, and their survival of the crash of 1929; her successful attempt at preventing Darrell from marrying an alcoholic; and the happy story of an artist's model. These events suggest modes of existence that Darrell might have followed had he not been so determined upon the spiritual life.

It is Templeton, however, who remains the most memorable character of the novel. He is greatly upset when the declining aristocracy he loves is replaced by a society of the *nouveaux riches*, who laugh at him and exclude him from their gatherings. As Templeton declines, so Darrell nears his goal of understanding. Unlike Darrell, whose search for spiritual salvation is a struggle, Templeton believes that he can buy his. He is buried in the church he has built, attired in the splendour of a fake antique Spanish costume, while Darrell, having found inner peace, is content to imagine a future in overalls, working in a garage or as a taxi driver in New York.

1945

Germany surrenders • Clement Attlee forms Labour Government • The atom bomb • Japan surrenders • Franklin Delano Roosevelt dies and is succeeded by Harry S. Truman • Independent Vietnam Republic formed with Ho Chi Minh President • France enfranchises women • Karl Popper, *The Open Society and Its Enemies* • Marc Chagall, sets and costumes for *The Firebird* ballet • Benjamin Britten, *Peter Grimes* (opera) • 'Bebop' dancing in US • David Lloyd George, Franklin Delano Roosevelt, Benito Mussolini, Adolf Hitler, Paul Valéry, Béla Bartók and Theodore Dreiser die • Vernon Watkins, *The Lamp and the Veil*

Eleanor DARK　　　　　　　1901–

The Little Company

The Little Company draws upon Dark's own upbringing in an intellectual and political family (her father was a Labour politician) and is set in Sydney and the Blue Mountains, where she lived for much of her married life. She had spent much of the 1930s opposing the rise of fascism in Europe and this novel, perhaps her most political, was written during, and in direct response to, the war. To work on it, she interrupted her famous trilogy, *The Timeless Land*, having only completed the first of its volumes.

It is 1941 and Australia stands on the brink of entering the war. Gilbert Massey is a middle-aged novelist suffering from writer's block: he has produced nothing since *Thunder Brewing*, a state of affairs he attributes to the world political situation. As a young man he married the dull and neurotic Phyllis, whom he has now grown to 'heartily dislike' for her prurience and conservative, 'genteel' attitudes. She, in turn, dislikes the radical politics of Gilbert and his brother and sister and feels 'assailed by ugly, tiresome words – unemployment, slums, exploitation, markets, undernourishment, venereal disease, war'. When she suggests that they move out of Sydney to the weekend house in the mountains, Gilbert agrees and they let the Sydney house. After Pearl Harbor, petrol rationing provides him with an excuse for staying increasingly often in Sydney (where he runs the family bookselling business) with his daughter or brother. He starts an affair with Elsa Kay, whose mother was once married to Scott Laughlin, a writer and political radical who was the formative influence on the Massey children despite the objections of their overbearing, evangelical father.

Phyllis discovers Gilbert's affair and becomes hysterical, then retreats into her habitual martyrdom with 'the terrible pride of the self-consciously inferior'. Their wayward daughter Virginia goes into hospital for what is apparently an appendectomy, and dies from what, it transpires, is a ruptured ectopic pregnancy. Phyllis blames the pregnancy on Gilbert's liberal upbringing of the children and cracks up completely. Gilbert learns of her disappearance from the house and a police search finds her in the mountains where she has tried to commit suicide. These experiences have convinced Gilbert that the political analysis to which he has always subjected everything may not be an adequate account of life, and he finds fresh inspiration for his writing in the life of Scott Laughlin.

F.L. GREEN   **1902–1953**

Odd Man Out

The raid on the mill has gone badly wrong. A man has been shot dead by Johnny, 'Chief of the militant Revolutionary Organisation', who is in turn severely wounded and on the run; the 'polis' are out in force all over the city of Belfast. Green's novel chronicles the last eight hours of Johnny's life, and the actions of the people who have an interest in him. The efforts of members of the 'organisation' (never actually called the IRA) to find and rescue their chief are in vain – hopelessly outnumbered, they are either shot down in gun battles or captured by the police. Of those who accidentally or purposefully make contact with the dying man, one of the most purposeful is 'wee' Shell. 'Wispish altogether, he was suggestive of all forms of life that are minute, fearful, and yet possessed proportionately by the same compelling desire for existence and self-expression as are the mighty beasts of the field, and the forceful personalities of city life.' He is at Fr Tom's house when Johnny's girl, Agnes, goes to visit the old priest, carrying a bird cage with a sick bird in it. He knows where Johnny is and is after the £2,000 reward for him. But all Fr Tom – so old and wise that he is no longer concerned with conventional morality – has to offer him is 'a precious particle of faith'. In Agnes Fr Tom discovers a purer purpose, and tells Shell to bring him to them at midnight. 'You and I are both looking for Johnny,' he later tells the Inspector of Police, who has been watching these comings and goings. 'You want his body. I want his soul.' So in her way does Agnes.

Other characters whom Johnny encounters include Rosie, the Protestant wife who briefly succours him, and sees, when she finds out who he is, the suffering human being; Fencie, the shady publican, who sees himself caught between the police and the organisation when Johnny takes refuge in a crib in his pub; Lukey, the artist, whose sole interest is in painting the dying man; and Tober, the embittered medical student who, after patching Johnny up, wants to get him to hospital and let the law take its course. At moments we enter Johnny's head as he drags himself about the city, or rests in the shadows fighting his battle with pain. In the end, Shell leads him to the square near Fr Tom's house. Agnes gets to him minutes before the police, and by having the charity to kill Johnny and herself, according to the old priest, she redeems him.

'God help you,' she cried, 'what am I to do with you? What did you come till here for?'

She sighed heavily, sniffed, and dried her eyes quickly. He was nothing more than a figure of tragedy and pain. The whole tale of his life with its furious challenge to authority was inscribed on his body where it was ending in a thin trickle of blood from a wound.

ODD MAN OUT
BY F.L. GREEN

The atmosphere of the Belfast streets, the bitter weather, the quirky, sinister people who pass through the city, all make this a novel of great intensity, partially captured by Carol Reed's film (1947), scripted by the author and R.C. Sheriff. Its power and pathos, and the universal truths it airs, make it as relevant today as when it was written.

Henry GREEN   **1905–1973**

Loving

Green's fourth and best-known novel (inaccurately castigated by his friend Evelyn Waugh as 'an obscene book ... about domestic servants') is apparently less opaque than its predecessors, opening and closing with two direct statements borrowed from the fairy-tale: 'Once upon a day an old butler called Eldon lay dying in his room attended by the head housemaid, Miss Agatha Burch' and 'The next day Raunce and Edith left without a word of warning. Over in England they were married and lived happily ever after.' Bracketed between these not unrelated statements is an evocation of 'loving' in its many and various forms.

The setting is Kinalty Castle, a large ornate and anachronistic country house in neutral Ireland during the Second World War. It is owned and staffed by English exiles and, although apparently distant, the war is like a circling storm which gradually affects most of the characters. Above stairs is old Mrs Tennant, worried, in her languid way, about the servants and about her son Jack, who has enlisted. Her daughter-in-law, Violet, conducts a passionate and secret affair with a local widower, Capt. Davenport. Green perfectly captures the bored, upper-class drawl of the gentry, and contrasts it with the rich patois of below stairs, the atmosphere of which hums with jealousies and intrigues, amorous and otherwise. When Eldon dies, Charley Raunce, the middle-aged footman, steps into his shoes, inherits his books and continues the old tradition of fiddling the accounts. The two lively and attractive under-housemaids, Kate and Edith, fantasise about sex and romance, and go about their duties in a state of suppressed hysterics, constantly bursting into giggles. 'Mr Raunce's Albert', a gormless adolescent employed as pantry boy, moons after the beautiful Edith, but she is in love with Raunce himself, as is Kate. The girls' friendship is damaged when Raunce woos Edith, and Kate takes refuge in a passion for one of the house's few Irish servants, the Neanderthal-like Paddy, who is employed as lampman and devoted keeper of the peacocks. The kitchen is presided over by Mrs Welch, who is obsessed by a fear that the tradesmen are all members of the IRA and by the disappearance of some water-glass.

Edith surprises Violet in bed with her Captain; the ancient nanny, Miss Swift, goes into a decline; old Mrs Tennant's sapphire ring disappears, is found, and vanishes once more; a sinister, lisping insurance investigator pays a call; Mrs Welch's nephew, a foul-mouthed little horror from London, comes to stay and strangles one of the peacocks, but wins the heart of Violet's daughter, Moira. And swirling in a great white mass around these events is Kinalty's vast flock of doves, emblem of all these amours. As the war encroaches, Mrs Welch takes to drink and Miss Burch to her bed; Mrs Tennant begins to distrust her entire staff; the pantry boy gives notice to enlist and

Raunce elopes with his Edith. Green charts the manoeuvrings of his characters with wit and affection in his inimitable and luminous prose.

Nancy MITFORD   1904–1973

The Pursuit of Love

Mitford's fifth novel, published in the wake of the Second World War, combines autobiography and wish-fulfilment in an engagingly frothy and funny mixture which exactly suited the mood of the times.

The story of Linda Radlett's incurable romanticism is narrated by her favourite cousin, Fanny. Deserted by her mother, 'the Bolter', Fanny is brought up by a spinster aunt but spends much of her childhood at Alconleigh with her Radlett cousins, an extensive aristocratic family presided over by the lovably ferocious 'cardboard ogre', Uncle Matthew, and his blithely vague wife, Aunt Sadie (based upon Mitford's own parents). Reality rarely intrudes into the hermetically sealed Cotswold world of the Radletts, and Linda and Fanny are determined to fall in love at the earliest opportunity. Through an eccentric neighbour, Lord Merlin (based on Lord Berners), Linda meets Tony Kroesig, a rich young man who wins her heart by rescuing a drowning hare. Linda marries him, but discovers almost immediately that he has inherited the dull solemnity of his banking father. A near-fatal pregnancy is rewarded by an intolerably stolid child, whom Linda vengefully christens Moira, then deserts, running away with a handsome young Communist, Christian Talbot. The couple marry, then go to Perpignan to help the Spanish refugees. Christian is more devoted to causes than individuals and spends most of his time with a numbingly sensible childhood neighbour of the Radletts called Lavender Davis. Linda goes to Paris alone and is discovered weeping amidst her luggage on the Gare du Nord by 'a short, stocky, very dark Frenchman in a black Homburg hat'. In this improbable figure, Fabrice, Duc de Sauveterre (a flattering portrait of Mitford's horrible lover, Gaston Palewski), Linda finds true love. He sets her up in a chic *apparte-*

ment, then sends her back to England when the German army marches upon Paris. Alone, miserable and pregnant, Linda stays in Cheyne Walk, waiting for the phone to ring. Fabrice arrives to declare his love, then disappears, after which the house is bombed. Dug out of the rubble, Linda joins Fanny, also pregnant, at Alconleigh. Doctors had warned Linda that she should be unable to have another child, and she dies whilst giving birth to a son. Fabrice becomes a Resistance martyr and Fanny adopts their baby.

The novel contains a gallery of comic characters, largely drawn from life, notably Fanny's hypochondriacal Uncle Davey (Eddie Sackville-West), her mother 'the Bolter' and the Radlet family, whose closeness to their models may be judged from Jessica Mitford's equally funny but rather less rosy family memoirs, *Hons and Rebels* (1960).

George ORWELL   1903–1950

Animal Farm

Orwell's bitter political allegory is written in the form of an extended fable. Several characters – revolutionary animals who take over their farm – are caricatures of contemporary figures. Orwell draws a pessimistic picture of dictatorship sustained by a population that is apathetic, ignorant or just too frightened to rebel.

Old Major (Marx), an elderly boar, knowing his death to be imminent, calls a meeting of all the animals on Manor Farm to pass on the required wisdom of twelve years. His message is one of revolution, and it does not take much to persuade his audience that they have been abused and exploited for too long. A few days later, Old Major dies, but the seeds of rebellion have been planted, so that when one day the drunken farmer (Tsar Nicholas II) neglects to feed the livestock, they break into the store-shed. When the farm-hands attack them with whips, the beasts chase the humans from the farm. The pigs (the Communist Party), who have, since Old Major's speech, been making preparations for this eventuality, immediately assume their natural roles as leaders and set about organising the animals into a co-operative workforce so that the renamed

'Animal Farm' can become self-sufficient. The lesser creatures are quite amenable to this and even accept that their leaders should be allowed special rations to help their 'brain-work'. The humans attempt to re-invade but are repulsed in a bloody battle, during which one of the pigs, Snowball (Trotsky), displays extraordinary heroism.

Afterwards Squealer was sent round the farm to explain the new arrangements to the others.

'Comrades,' he said, 'I trust that every animal here appreciates the sacrifice that Comrade Napoleon has made in taking this extra labour upon himself. Do not imagine, comrades, that leadership is a pleasure! On the contrary, it is a deep and heavy responsibility. No one believes more firmly than Comrade Napoleon that all animals are equal. He would be only too happy to let you make your decisions for yourself. But sometimes you might make the wrong decisions, comrades, and then where should we be?

ANIMAL FARM
BY GEORGE ORWELL

Later, Napoleon (Stalin), another pig, who is ambitious for supreme power, stages a coup and Snowball, eloquent and innovative, and thus a threat, is set upon by a pack of dogs that Napoleon has secretly trained. The hero flies the farm and is subsequently denounced by Squealer – pig in charge of propaganda (representing *Pravda*) – as a traitor. His plans for building a windmill, initially derided by Napoleon, are quickly taken up; however, the enormous effort involved in building it results in shortages, and Napoleon opens trading negotiations with neighbouring farms. It is at this point that the pigs move into the farmhouse and start sleeping in beds. The working animals recall these things having been forbidden after the revolution, but Squealer 'corrects' them.

The windmill is destroyed in a storm, but Napoleon blames Snowball for the catastrophe. There follows a horrifying purge, in which numerous suspected supporters of Snowball are slaughtered. A second windmill is built, but Frederick, a local farmer who has successfully defrauded Napoleon in a business deal (and who represents Hitler's Germany), has it blown up. Several animals are killed in the subsequent fighting, but the men are badly beaten and make no further attacks. Napoleon orders another windmill to be built and work commences immediately; Boxer (the proletariat), a horse whose industry and dedication have been an example to the whole workforce, collapses as a result of injuries sustained during the recent battle, combined with an excessive workload. The pigs sell him to a horse butcher whilst Squealer reassures the workers that he had in fact been taken to the village hospital, where he made a peaceful death.

Slowly, the tenets of the revolution are eroded to the point where pigs are walking on their hind legs and using whips. The humans are invited to inspect the farm and join the pigs in the farmhouse for a drink and a game of cards. As they look through the window, the animals outside realise that they can no longer distinguish between the faces of the pigs and those of the men.

Elizabeth SMART **1913–1986**

By Grand Central Station I Sat Down and Wept

Although a Canadian, Smart spent much of her teens in Europe and, during one of her stays in London, she came across a volume of poetry by George Barker. After corresponding with the author she arranged – in 1940 – for he and his wife to fly to California, where she was then living. Less a novel than an extended prose poem, *By Grand Central Station* is an oblique and passionate recollection of the affair that ensued.

Told in the first person, it provides only filtered glimpses of the geometry of the relationship which, in itself, is not extraordinary. The mistress vacillates between contempt and compassion for the wife, who, in turn, is both self-pitying and a genuine martyr; the husband's loyalties and desires are torn; the world is censorious of such unconventional love (and pruriently fascinated by it). There is an episode of the grotesque, when the lovers are arrested and briefly imprisoned for crossing state lines for an immoral purpose, and an episode of the macabre when the husband attempts suicide. The narrator visits him in a Stygian place that is neither quite a hospital nor an asylum. Later, she is pregnant, and the unborn child is equally a compensation for and a reminder of the man from whom she is now separated. The story ends with what seems to be an omen of a final separation: a rendezvous is missed and the author sheds her tears at Grand Central Station.

The book derives its power from its exposure of the beatings of the heart. With remarkable and passionate candour it reveals the contradictory mechanisms of love: how love is selfish and selfless, callous and tender, creative and destructive; how it demands the totality of its object and yet – simultaneously – yearns to consume it. The lover's identity thus comprises opposing extremes. The author is alternately Penelope (her lover's destination) and Circe (his temptation and diversion). Reflecting on the Second World War, she becomes sometimes the redeemer of its agonies and sometimes too the bearer of them all. If the image is deliberately Christian, it is, once again, ambivalent – either revelation or blasphemy.

This flayedness is shown through a densely woven garment of allusion. Smart takes her title from the Book of Psalms (137:1). After her arrest, her unspoken answers to her interrogator are from the Song of Solomon. Comparisons with Greek myth and Shakespearean tragedy are implied – often to comic effect, for Smart has a lively sense of the absurdity of love. Her writing also draws strength from Walt Whitman, Robert Frost and (perhaps surprisingly) Wallace Stevens. It is a further paradox that, in order to express a uniquely private experience, it is necessary to invoke so many precedents from public culture.

Largely ignored on its initial appearance, the novel gained a wider audience and a greater acclaim on being reissued in 1966. Brigid Brophy has described it as 'One of the half a dozen masterpieces of poetic prose in the world' – although Alice Van Wart, the editor of Smart's journals, has called it 'a slim work with a long title'.

Elizabeth Smart had four children by George Barker.

Evelyn WAUGH   1903–1966

Brideshead Revisited

Waugh wrote his most ambitious novel in 1944 while he was on leave from the army. 'I feel full of literary power,' he confided to his diary, and in letters he referred to the evolving book as 'M.O.' (Magnum Opus) and 'G.E.C.' (Great English Classic). However, although the book proved immensely popular, Waugh ruefully claimed that it 'lost me such esteem as I once enjoyed amongst my contemporaries', and he spent much time revising it, publishing a new edition in 1960. The first edition was issued with a 'Warning' that the novel is 'an attempt to trace the workings of the divine purpose in a pagan world, in the lives of an English Catholic family, half-paganized themselves, in the world of 1923–1939'. Many critics have baulked at both the theme and Waugh's treatment of it, deploring its purple patches and its apparent genuflections before the aristocracy and the Catholic faith. If it is in places overwritten, even after Waugh's revisions, the novel is also more considered, more complicated and more ambiguous than its detractors sometimes claim.

Subtitled 'The Sacred and Profane Memories of Captain Charles Ryder', the novel is nostalgically conceived and (in the 1960 version) divided into three books, framed by a prologue and an epilogue. It opens during the war when Charles's company is billeted at Brideshead Castle. In book one, 'Et in Arcadia Ego', Charles recalls his time at Oxford where he was romantically involved with Sebastian Flyte, the charming but dissolute younger son of the Marchmain family. The two friends visit Brideshead and gradually Charles meets the rest of the family.

Lord and Lady Marchmain are separated; his lordship lives 'in the conventional Edwardian style' in Venice with his Italian mistress, Cara, whilst her ladyship remains at Brideshead with her four children: the dull heir, Lord Brideshead ('Bridey'); the beautiful and wordly Julia; Sebastian himself; and the ebullient, horsey and pious Cordelia. Sebastian resents the way Charles is absorbed into the family, seeing it as a betrayal, but Charles swears to honour 'Sebastian *contra mundum*'. In book two, 'Brideshead Deserted', Sebastian has been taken in hand by the preposterous, snobbish young don Mr Samgrass, but has also taken to drink. He eventually decamps to Morocco where he lives with a

---

'It is typical of Oxford,' I said, 'to start the new year in autumn.'

Everywhere, on cobble and gravel and lawn, the leaves were falling and in the college gardens the smoke of the bonfires joined the wet river mist, drifting across the grey walls; the flags were oily underfoot and as, one by one, the lamps were lit in the windows round the quad, the golden lights were diffuse and remote, like those of a foreign village seen from the slopes outside; new figures in new gowns wandered through the twilight under the arches and the familiar bells now spoke of a year's memories.

BRIDESHEAD REVISITED
BY EVELYN WAUGH

---

degenerate German. Lady Marchmain dies and Cordelia talks to Charles about vocations. In book three, 'A Twitch Upon the Thread', Charles and Julia meet again aboard a transatlantic liner. Both are unhappily married: Charles, who is now a successful if limited architectural painter, to Celia, a silly and adulterous society woman; Julia to an ambitious Canadian vulgarian called Rex Mottram. They begin a doomed affair, arrange their divorces, and plan to marry. Julia is aware of the very real nature of her 'sin', however, and when her father returns from his exile in Italy, she is determined that he should die in the faith he had abandoned.

He does so, and Charles and Julia separate. The epilogue depicts Charles, now a Catholic, visiting the family chapel to find that something good had come out of 'the fierce little human tragedy' of his involvement with the family: the 'small red flame' of the faith burns there once more.

Waugh investigates parallels between bad faith and bad art and shows through his principal characters 'the operation of divine grace' and the way people eventually conform to God's will ('a twitch upon the thread'). In spite of an almost overwhelming sense of loss (the loss of love, friendship, youth and faith), the novel is in parts high comedy and is studded with splendid secondary characters, notably Charles's eccentric father and the outrageous Oxford aesthete, Anthony Blanche, who acts as chorus and warns Charles against 'charm'. It was adapted for a lavish and highly acclaimed television series (1981), scripted by John Mortimer.

Denton WELCH   1915–1948

In Youth is Pleasure

Although narrated in the third person, Welch's only novel is as clearly autobiographical as his other full-length works. Indeed, it was originally written in the first person, but altered on the advice of Welch's nervous publishers. The reason for this, and some idea of the book's contents, may be gauged from the fact that Welch later discovered the book for sale under the label: 'Of Interest to Students of Abnormal Psychology.' In fact, what the book does is to reproduce with startling and unflinching accuracy the inner life of a lonely and imaginative adolescent boy: a life dominated by the erotic and the macabre, described with an acute and fresh eye.

Fifteen-year-old Orvil Pym is spending the school holidays with his widowed father and two older brothers at a Surrey hotel. Left largely to his own devices, Orvil explores the local countryside and encounters several people whose behaviour is obscurely sinister or threatening; amongst these are an ancient dowager who mistakes him for a girl, a schoolmaster who ties him up in a hut, and

Lord Alfred Douglas, depicted (but not named) as a mad old man on a train. Still mourning his mother, Orvil is drawn to glamorous older women, including Aphra, a member of his hated elder brother's circle, and Constance, the sister of a prep-school friend whom he visits in Hastings. He lives almost entirely in his imagination, filtering the mundane through the acute sensibility of adolescence and thus transforming it. His life is ruled by fantasy in which pleasure and pain (both physical and psychological) jostle for ascendancy. He bargains with God for the return of his mother; is discovered by his brother Ben chained to a garden roller

pretending to be a slave; locks himself in the bathroom and, daubed with lipstick, performs naked dances; and imagines a regiment debauching its mascot, a 'delicate-stepping, wicked little goat', during a military tattoo. Alienated from the real world, he confesses: 'I don't understand how to live, what to do.'

Welch's total identification with his fifteen-year-old self avoids the collusion between knowing author and sophisticated reader which is the pitfall of books about adolescence, whilst his sureness of touch and mastery of imagery ensures that the book is never artistically naïve.

1946

First meeting of UN General Assembly ● The Nuremberg Trials ● Juan Perón elected President of Argentina ● USSR adopts fourth Five-year Plan ● British National Health Act in force ● General Douglas MacArthur purges extreme nationalists in Japan ● Xerography invented ● US Supreme Court rules the segregation of blacks on interstate buses unconstitutional ● Bertrand Russell, *History of Western Philosophy* ● Frederick Ashton, *Symphonic Variations* (ballet) ● Eugene O'Neill, *The Iceman Cometh* ● Frank Capra, *It's a Wonderful World* ● John Maynard Keynes, Gertrude Stein, Herbert George Wells and Damon Runyon die ● Dylan Thomas, *Deaths and Entrances*

Miles FRANKLIN　　　　　　**1879–1954**

My Brilliant Career
1901

My Career Goes Bung
1946

Franklin's first novel, *My Brilliant Career*, and its sequel, *My Career Goes Bung*, form a teasing sequence of fact-and-fiction masquerades and games-playing. Both purport to be semi-autobiographical and are told through the narrator, Sybylla Melvyn, who at the start of *My Brilliant Career* is an ugly, adventurous, discontented writer living in the Australian bush in the 1890s.

Chafing at the restrictions placed on her

by poverty and feminine propriety, Sybylla presents a frank portrait of adolescent frustration in a colloquial and consciously naïve outpouring of desire. Her father is a drunkard, her mother a prude, and Sybylla, the eldest, tomboyish daughter, feels unwanted and unappreciated. Longing for a refined environment where music, literature and art are valued, she is sent to stay with her grandmother and her Aunt Helen at their home at Caddagat, where she meets Harold Beecham, owner of Five-Bob Downs.

Sybylla is wary of marriage, fearing that a husband will object to her literary ambitions and enslave her unintentionally through the burdens of poverty and childbearing. However, when Harold loses his fortune, she consoles him by agreeing to a secret engagement, promising to marry him when she is twenty-one. In the meantime, her father's drinking has brought the family into debt, and Sybylla's blissful interlude of the good life at Caddagat gives way to her new life as governess to the M'Swats, a vulgar squatter family to whom her father now owes money. Sybylla's disgust at the filth in which they live brings her close to a mental and physical breakdown, and she returns to her own family at Possum Gully, where her mother's cruel pragmatism fails to reconcile her to her fate. Her pretty sister, Gertie, has taken her place at Caddagat – and, Sybylla hears, in the affections of Harold, who has regained his fortune through an unexpected inheritance. Harold rides to Possum Gully and asks Sybylla to marry him, but though she is astonished, her own resistance to social convention still dissuades her from marriage. Harold promises that she could continue to write and that he will support her, but she fears he will change his mind after the wedding and try to force her into the role of wife and mother. At the end of *My Brilliant Career* Sybylla remains poor but optimistic, determined through her own efforts to make her way in the world and never sink to being an 'appendage' to a man.

If the souls of lives were voiced in music, there are some that none but a great organ could express, others the clash of a full orchestra, a few to which nought but the refined and exquisite sadness of a violin could do justice. Many might be likened unto common pianos, jangling and out of tune, and some to the feeble piping of a penny whistle, and mine could be told with a couple of nails in a rusty tin-pot.

MY BRILLIANT CAREER
BY MILES FRANKLIN

Franklin completed the sequel to *My Brilliant Career* in 1946. It was 'planned as a corrective' to her first novel, because she had been deluged by letters from young women who had taken it for autobiography and had written to share their confidences. Franklin gave the sequel the title 'The End of My Career', and it might well have been that, for it was turned down by her publisher, who thought it 'delicious', but 'too audacious for publication'. It was not only audacious, but also libellous, since the character of Goring Hardy was based upon 'Banjo' Patterson, with whom Franklin had been entangled, and the book contained recognisable portraits of other figures in Sydney society. It was also, for its time, a challengingly feminist book, which asked frank questions about the relationship between the sexes. A determined extrovert, Franklin was not put off her brilliant career by this rejection and went on to write other books. The 'pother' raised by the first volume, however, had been unnerving, and for a time Franklin believed the manuscript of the sequel had been destroyed. When she finally decided to publish *My Career Goes Bung*, she added an engagingly bumptious preface, 'To all young Australian writers Greetings!', in which she explained that the book was 'now an irrefutable period piece, and, in the light of EXPERIENCE, it is to be discerned that while intentionally quite as little, unintentionally it was equally as autobiographical as my first printed romance; no more, no less'. She added: 'I have kept faith with that girl who once was I. I have not meddled by corrections which would have resulted, probably, in no better than the substitution of one set of solecisms and clichés by another, for such abound in even the greatest English novels.'

My Career Goes Bung represents an ironic shift in perspective, as Franklin explodes the romances she had concocted in *My Brilliant Career* and exposes Sybylla as a mistress of artifice who had fabricated her own story. Now the only child instead of one among many, Sybylla revises her opinion of her parents and presents them as a liberal, gentle couple, though convention still makes her mother narrow-minded. The owner of the 'real-life' Five-Bob Downs, Henry Beauchamp, is intrigued by the young woman who had so flatteringly portrayed him in her now-famous novel which had catapulted her to literary success. Instead of the ugly duckling of *My Brilliant Career*, Sybylla is revealed here as a fascinating original, though she compares herself unfavourably with the society beauty, Edmee, whom she meets at the house of Mrs Crasterton in Sydney. Sybylla's stay in Sydney as a literary celebrity sees her fêted as a great Australian writer, but her notoriety brings problems as well as fame. 'Big Ears', one of Edmee's admirers, believes he is in love with her and threatens suicide; 'Gaddy', Mrs Crasterton's brother, also proposes marriage; but Sybylla hopes to meet a cultivated and enlightened partner who will not be threatened by her creative gifts. Goring Hardy, a novelist, takes her into his patronage in an erotic but unconsummated liaison, trying to persuade her to cramp her style to meet the tastes of the English establishment. Sybylla's month in Sydney ends in disappointment, as the people from whom she expected so much turn out to be merely bourgeois; but a brief meeting with an elderly poet and journalist, Renfrew Haddington, holds out the promise that if she sticks to her own style she will one day be rewarded with a man who is worthy of her.

'Bung went my Sydney career. I was back at Possum Gully,' Sybylla writes at the end of her story, wooed by Henry Beauchamp who remains convinced that her literary ambitions are a passing aberration and will end with the birth of her first child. She is now twenty, and the need to make a decision about marriage is weighing heavily upon her; but her determination to succeed on her own leaves her convinced that the way forward is through escape to literary London.

| Lewis Grassic GIBBON | 1901–1935 |

A Scots Quair

see 1934

| L.P. HARTLEY | 1895–1972 |

The Sixth Heaven

see **Eustace and Hilda: A Trilogy** (1947)

| Eric LINKLATER | 1899–1974 |

Private Angelo

Like his hero, Linklater served in Italy in the Second World War, though in only one army, and his satirical, picaresque, anti-war

novel, begun in Rome in August 1944 and finished in Orkney a year later, was written when memory was green, and before the ravages of battle could be repaired. The book was filmed in 1949, directed by Peter Ustinov and Michael Anderson from their own screenplay.

Though, as he is the first to admit, he lacks the '*dono di coraggio*', the engaging Angelo is most definitely a man of conscience. As such he is put upon by everyone and lives in a state of perpetual trepidation. We first find him on the mat before his colonel, Count Agisilas Piccologrando of Pontefiore in Tuscany, having 'run all the way from Reggio to Rome'. Fortunately he is the natural son of the Count, who remembers his dead mother with affection, and has little to boast of himself, having commanded his regiment from Rome; besides, before the end of the interview, news reaches them that Italy has signed an armistice. If only the blinkered Tedeschi were not determined to go on fighting they might both have gone back to Pontefiore in peace. As it is, Angelo has to choose between slave labour in Germany and joining the German army; reluctantly he chooses the latter.

He deserts, assisted by a fair dosage of Spanish Fundador, only to find himself with the British in the company of acting Major Simon Telfer and other people who had actually 'enjoyed the war in Libya'. He does get back to Pontefiore to marry his sweetheart, Lucrezia Donati, but it is in the full knowledge that she has made love to Cpl Tom Trivet, 'purely out of charity', and produced little Tommaso. She has also been raped by the *Morocchini*; she does not tell him that the resulting baby, whom he insists on calling Otello, is black. Before making this last discovery Angelo loses a hand in the service of the newly constituted Italian army, and saves from starvation the beautiful Annunziata, who 'felt sorry' for a lonely Pole and has produced a baby called Stanislas. Before making love to her, Angelo insists that it is because *he* is sorry for *her* and not the other way round. Reflecting on the condition in which he embarks on 'peace', he tells himself, 'I have served in three armies ... without wishing to ... and now have three children, none of whom I desired.'

Nevertheless he settles down to farm at Pontefiore with the divine head of the Madonna from Piero della Francesca's *Adoration of the Shepherds* and two women who simultaneously produce more children at regular intervals; he is persuaded that he has the '*dono di corragio*' after all.

The numerous other characters are caricatures in the picaresque manner, designed in an exaggerated way to represent their national characteristics. The pragmatic Count, whose philosophy is sufficiently broad to justify any action which seems to his advantage, is really an older version of Angelo, possessing a heart but little conscience. The scene in which he is taken hostage by the Germans and thinks he is going to be shot is one of the best in the book. His English countess, formerly a schoolteacher in Bradford, is hard working with a strong sense of responsibility but takes to drink after the retreating Germans have destroyed her beloved edition of the novels of Ouida. The German general by whom Angelo is pressed into giving Italian lessons tells him: 'I have a remarkable character, and you will be deeply interested to hear about it. I am, for instance, equally capable of tragic perception and Homeric laughter.' Schlemmer, another German officer, 'a lover of the arts', refuses to believe that the *Adoration of the Shepherds* is a genuine Piero because the Countess tells him it is, and shoots it full of holes. The Englishman, Simon Telfer, knows 'at least three hundred' influential people by their Christian names. Among these eccentrics Angelo, always engaging, always himself, moves in perpetual bewilderment.

Carson McCULLERS   **1917–1967**

The Member of the Wedding

McCullers's third novel continues her exploration of spiritual isolation. Like its predecessors, it is set in the Deep South, where she was born.

The story takes place over a weekend in the dog days of a long, troubled summer and follows the fantasies of twelve-year-old Frankie Addams, a tomboy who is experiencing growing pains. Much of the action takes place in the Addams kitchen, where the motherless Frankie passes her time in the company of her six-year-old cousin, John Henry West, and Berenice, the black cook. Frankie has shot up several inches during the year, and sees herself at 'five feet and three-quarter inches tall, a great big greedy loafer who was too mean to live'. She feels herself to be an outsider who can identify only with the fairground freaks (a typically Southern Gothic touch) because of her height, or with prisoners because she once stole a penknife. She feels that she 'belonged to no club and was a member of nothing in the world'. The only glimmer of hope is the approaching wedding of her older brother, Jarvis, a soldier serving in the war in Alaska. She sees Jarvis and Janice, his fiancée, as '*the we of me*', saving her from an 'I' existence, and she intends to run away with the newlyweds immediately after the service.

It was the year when Frankie thought about the world. And she did not see it as a round school globe, with the countries neat and different-coloured. She thought of the world as huge and cracked and loose and turning a thousand miles an hour.

THE MEMBER OF THE WEDDING
BY CARSON McCULLERS

In part two her identification with the couple is consolidated by her adopting a name which also begins with *Ja*, and she becomes 'F. Jasmine Addams'. She wanders around the small town, buys an unsuitable dress for the wedding, and is picked up by a drunken soldier whom she agrees to meet later in the evening. She then goes to see Big Mama, Berenice's elderly bedridden mother who is a fortune teller, but she is disappointed in the old woman's forecasts. She decides to keep her promise to the soldier, but when he attempts to molest her, she bites his tongue and knocks him unconscious with a glass pitcher.

In part three she has metamorphosed into 'Frances Addams' and returns from the wedding, her plan to escape with the bride and groom humiliatingly thwarted. She attempts to run away but gets no further than the town's Blue Moon Café. The novel ends some months later when Frankie and her father are to leave their home and move to a new house with John Henry's parents. John Henry has meanwhile died hideously of meningitis, and Berenice is leaving to marry her elderly fiancé. However, Frankie has found a new friend, a slightly older girl from her school, with whom she intends to go round the world – a fantasy rather more grounded in reality than the one to be a member of the wedding.

The novel is generally considered to be McCullers's finest achievement and was much admired by fellow writers such as Tennessee Williams and Edith Sitwell. In 1950 she adapted it for the stage, winning the New York Critics' Award and a new, wider audience, and in 1952 it was made into a film by Fred Zinnemann from a screenplay by Edna and Edward Anhalt.

| Mervyn PEAKE | 1911–1968 |

Titus Groan

see **The Gormenghast Trilogy** (1959)

| Elizabeth TAYLOR | 1912–1975 |

Palladian

The second of Taylor's twelve novels, a short but formally intricate book, *Palladian* takes the material of the Gothic and romantic novel and, in the manner of Jane Austen in *Northanger Abbey*, bends it to the author's ironical purposes. (Taylor, like Austen, had been a pupil at the Abbey School, Reading.) In Taylor's case, she places her anachronistic material in opposition to the modern world. Cassandra Dashwood, a literary and melancholy young girl, recently orphaned, travels across the sodden English countryside to a crumbling Palladian mansion, to be governess to Sophy, daughter of the widowed landowner, Marion Vanbrugh. 'She was setting out with nothing to commend her to such a profession, beyond ... a very proper willingness to fall in love, the more despairingly the better, with her employer.'

Marion Vanbrugh turns out in line with Cassandra's aspirations: he is handsome in the tradition of the English upper classes, and his main activity is to read the Greek classics in the original, while eternally mourning his beautiful dead wife, Violet. His first move is to offer Cassandra Greek lessons. A variety of strange relations are also living in the house: his half-senile old aunt, Tinty, eternally singing Victorian songs; Tinty's daughter, Margaret, an acerbic doctor locked into a seemingly interminable pregnancy; and Margaret's brother Tom, a drunken but not unintelligent idler engaged in a punishing affair with Mrs Veal, the vulgar wife of the local publican. Cassandra's pupil is a precocious child who is writing a novel, while a vengeful and snobbish old nanny and the cleaning lady form a sort of Greek chorus. Rain drips down the windows of the collapsing house 'like gin'.

As the weeks pass, Cassandra's love for her bookish employer hesitantly blossoms, although she realises that all the inhabitants of the house are haunted by the memory of its beautiful former mistress. The melancholy atmosphere of the house is set off against the raucous vulgarities of the pub where Tom goes daily, and the life of the nearby town, where the MGM film of *Pride and Prejudice* is showing. The nanny attempts to accuse Cassandra of stealing. Marion alone with Cassandra in the library, gives her a chaste kiss while all the lights in the house have fused.

A *denouement* is reached when Sophy falls from collapsing statuary on the house's terrace and is killed. After her death, Tom confesses to Marion that Sophy was really his daughter, and the ghost of Violet is laid. Cassandra has fled to the house of her former headmistress, but Marion pursues her and wins her in marriage. All seems in line with the conventional pattern of the romantic novel, but a strong element of farce enters the book in its last pages, giving a new twist to the atmosphere of this strange novel. Artificiality and artifice are of the essence of *Palladian*: Taylor shows how such qualities can do as much as stark realism to illuminate the world in which we live.

Robert Penn WARREN   1905–1989

All the King's Men

The genesis of Warren's third and most successful novel lies in a group of poems on the nature of democracy which he wrote in 1937 while he was teaching at Louisiana State University. These ideas resurfaced in a play called *Proud Flesh* (1947), which was abandoned, and his second novel *At Heaven's Gate* (1943). In the 1953 preface to *All the King's Men* Warren wrote that it was 'a continuation of the experience of writing *At Heaven's Gate*'.

The novel tells the story of Willie Stark, Governor of Louisiana, and his sycophantic factotum, Jack Burden. Burden's task is to do Stark's dirty work, uncovering truths and manipulating those around him into doing his master's bidding. Burden gradually uncovers the truth about himself during the story. Running parallel with this narrative is that of Stark's rise, fall and violent death at the hands of his mistress's brother. One of Warren's major successes is to present Burden through his narration of Stark's political chicanery. Burden collects all the facts that his master and the novel's reader alike require. His deficiency in assembling those facts lies at the heart of his character and functions as an invitation to the reader to make judgements for him. Burden is slow in identifying the causal links which unite the facts he finds out, and does not grasp the consequences of his own actions until after they have happened. When they do become clear, his continuing incomprehension compromises him in the eyes of the reader. The use of a morally deficient narrator, and the subtle manipulation of the reader to compensate for this, are crucial technical achievements, for together they create an ethical perspective from which the politicking can be judged.

The sophistication of Warren's political analysis does not match that of its presentation. The essential debate – whether or not the end justifies the means – is far less complex than the dilemmas it throws up for the characters. This, however, is the main point of the novel. Warren's plea is for the responsible involvement of human beings in their own affairs (that is, history). Despite Warren's own political commitments at the time (he was a prominent figure in the Southern Agrarian movement), *All the King's Men* features no special pleading. The story he tells concerns Burden and Stark, the 'king' and his man. External political battles (such as left versus right or North versus South) are seen only as a backdrop to internal power struggles. The implication of this is that ideologies and issues are only counters in the game of politics and that the business of securing power takes precedence. The popular appeal of *All the King's Men* probably depends on this fact: Stark is a figure who is possible in any political arena.

When the book first appeared, many reviewers seized on the parallels between Stark and the case of Huey Long, who had been Governor of Louisiana in the 1930s. Like Stark, Long had been a ruthless and charismatic operator who had been shot by someone with a personal grudge. Warren used Long's story to construct that of Stark, but to the charge that he had written an apologia for Long, his response was unequivocal: 'Willie Stark was not Huey Long. Willie was only himself, whatever that self turned out to be, a shadowy wraith or a blundering human being.' He later identified William James and Machiavelli as closer analogues to his protagonist. These seem far-fetched, but clarify his intent to write a novel about the nature of political manners rather than the specific political situation of Louisiana in· the 1930s. Machiavelli and James, like Stark, are enduring, almost universal figures and this forms the basis of *All the King's Men*'s success.

Eudora WELTY   1909–

Delta Wedding

Welty has lived in the Mississippi area most of her life, written stories and novels about it, and taken photographs of its people. Unlike other Southern writers she is uninterested in the civil war and uncertain about the supremacy of Southern culture. She is, none the less, fascinated by its originality, self-absorption and anachronism, qualities displayed by the Fairchild family at the centre of *Delta Wedding*.

The novel is an account of the last few days before the marriage of Dabney, the prettiest of the Fairchild daughters, to Troy Flavin, the overseer from the hills, a man well below her socially. Her marriage causes consternation, for the family does not like its individual members to break conventions. At the same time, since they all love one another – or, at least, each of them loves the rest of the family – they cannot refuse. Dabney is married with parental blessing, grudgingly given. In her account of the last days before the wedding of the wedding itself, Welty builds up a mosaic picture of the life of one Southern family. There are family discussions, meals, small rituals, childish games. Though the action takes place in the 1920s, the plantation, the blacks, the social structure, and, above all, the land around them are timeless. Emancipation, reconstruction, the First World War have come and gone, but nothing can change the life of the Fairchild family or the myths it lives by. Only one thing threatens them: love for an individual – Dabney's for Troy.

At various times members of the family had asserted their individuality by loving and marrying the 'wrong' person. One was Denis Fairchild, whose wife dropped their daughter on her head, making the girl into an idiot. But Denis was killed in the First World War and could, therefore, be elevated into a family saint, 'the sweetest man in the Delta'. His brother George is a more difficult problem. He came back from the war and has been the focus of the family's love and attention since then. But there has always been one thing wrong with George, as Dabney knows: 'all the Fairchild in her had screamed at his interfering – at his taking part – *caring* about anything in the world but them ... George loved the *world*, something told her suddenly. Not them! Not them in particular.' George, too, marries beneath him – a shop girl, Robbie Reid – but their love lives despite family disapproval, Robbie's occasional vulgarity and her exasperation at the Fairchild in George. To her they are 'a spoiled, stuck-up family that thinks nobody else is really in the world!' But she recognises that their strength is more than snobbery: 'You're just loving yourselves over and over again.' In the same three days Robbie leaves George, frustrated by his overwhelming allegiance to his family, and comes back, unable to overcome her love.

The book is written in a slow rhythm that conjures up the hot September of the Mississippi, with detailed description of the colours and fragrances of the delta. Several points of view are voiced: Robbie – hard, miserable, full of hate but also love; Shelley, the eldest Fairchild girl – bewildered and rebellious (she, too, understands the problem and begs for a chance to love everyone individually); Ellen, her mother, who has long ago accepted the Fairchild code but is often clear-sighted and compassionate towards those who do and those who do not; finally ten-year-old Laura, who loves her cousins, their home and their family, but resents their corporate view of the world which destroys her own importance as a recently orphaned child.

1947

The Truman Doctrine ● Marshall Aid ● Independence of India and Pakistan ● US abandons efforts at mediation in China ● Britain refers Palestine question to UN ● First supersonic air flight ● Discovery of main series of Dead Sea Scrolls ● Alberto Giacometti, *Man Pointing* ● Tennessee Williams, *A Streetcar Named Desire* ● Henry Ford, Fiorello La Guardia, Max Planck, Sidney Webb and Alfred Whitehead die ● Stephen Spender, *Poems of Dedication*

John Horne BURNS   1916–1953

The Gallery

The Gallery is not a novel in the usual sense, but a collection of portraits unified by its setting – the Galleria Umberto Primo in Naples in 1944 – and in its themes. The Galleria is an arcade through which the various characters pass, or where they meet.

On 3 October 1943, Naples fell to the US Fifth Army. The supposed 'liberation' of the city may mark the end of war for the Neapolitans, but it is also the beginning of a new struggle: against the prejudice and mere presence of Americans. The GIs' comparative wealth throws the city into economic chaos. Life is a 'minute to minute struggle, in which any problem five minutes hence seemed a lifetime removed'. It is not possible to live honestly.

The irreconcilability of the army and the civilian population is one of the novel's most powerful themes. The vitality of the Italians – 'the vitality of the damned' – is contrasted with the shallowness of the lives of certain sorts of Americans who lack self-knowledge. For most Americans the Italians are just gooks. Louella, for example, in the portrait of that title, is a volunteer who runs the worst American Red Cross Club in Italy. She thinks she is 'so old and mellow and wise' when really she is suffering from 'that sacred and awful thing, American loneliness'. She sees herself as one of the great pioneer women, or as a modern Joan of Arc, gracious and loved by all. She is entirely deluded, and almost an alcoholic, like her pitiful flatmate, Ginny, who gives herself to any passing GI because she feels sorry for him. Real feelings and thinking pass them both by. There are a few, however, who come to learn the lesson that E.M. Forster also taught: that the English-speaking puritan ethic is sterile but can be countered by absorption of the deep-stirring sensuousness and passion of the Italians. In Brigid Brophy's estimation, Burns's imagination 'made him, by sympathetic adoption, an Italian'.

In three linking passages called '(Naples)', the 'I' narrator becomes at the end finally identified as 'John', the novelist himself, who has learnt not just this lesson but also 'in the twenty-eighth year of my life I learned that I must die'. Further, he comes to appreciate the unrepeatable lesson of war, that life is precious and more than a physical reality.

Institutionalised religion gets short shrift: 'I'd like to shake hands, the chaplain said, withdrawing. But God has made certain diseases highly infectious.' The disease is VD: 'like a toothless old woman dragging her skirts in a black corridor'. The novel is redolent with such striking similes: 'a voice as rich and persuasive as a pie bursting with suet'; 'He was as gay as one coming out of an anaesthetic.'

Sex, too, unifies the stories: bisexual in the excellent story set in a VD clinic in which a US sergeant offers himself to the new patient infected by an Italian girl as the best way of preventing any recurrence of the disease; homosexual in the remarkable story set in a gay bar. Momma, aged forty-six and whose 'only selfish desire was to be renowned as a great hostess', is presented as either an innocent or splendidly indifferent to the sexual inclinations of the customers at her bar, 'the most celebrated in Naples'. In fact we only gradually appreciate the bar's specialised nature.

The rhythms of war resonate in everyday

Neapolitan life. Living consists of an ever-shifting present heightened by the ubiquity of death. Indeed, the novel prefigures the author's own early death in 1953, at the age of thirty-seven.

I. COMPTON-BURNETT   1884–1969

Manservant and Maidservant

This novel, the author's favourite, won Compton-Burnett widespread acclaim both in England and America, where it appeared under the title *Bullivant and the Lambs*. The Lambs are an extended family of father, mother, five children and the father's dependent cousin and aunt. Bullivant is the dignified butler, who presides over a small staff of cook, footman, skivvy and a couple of uncharacterised maids. 'I think it's a story with a good straightforward plot, don't you?' the author commented to Cicely Grieg, who had typed the manuscript for her. By Compton-Burnett standards the plot *is* reasonably straightforward; it is also characteristically perfunctory and unlikely, since the author is less concerned with the mechanics of story-telling than with the power-structure of a late-Victorian household.

'We must consider dumb animals,' said Horace, in a tone of rejoinder, 'and these horses are valuable beasts.'

A twitch crossed Bullivant's lips at this statement of the case.

'I think too much compensation is made to animals for being dumb,' said Mortimer. 'After all, it is not our fault. Neither is it that we are articulate; we can be too much penalised.'

MANSERVANT AND MAIDSERVANT
BY I. COMPTON-BURNETT

The marriage of Horace and Charlotte Lamb had been based upon her love of him, and his love of her money: 'The love had gone and the money remained, so that the advantage lay with Horace, if he could have taken so hopeful a view of his life.' Horace's

view of life is that it is expensive and difficult (a reflection of the post-war period during which the novel was written). Like several Compton-Burnett tyrants, Horace believes himself to be an aggrieved party, whose family spend their time attempting to thwart him, principally by 'squandering' his (in fact, Charlotte's) money. Charlotte is called away to attend her ageing father, who conveniently lives 'on the other side of the earth', and during her absence a tutor, Gideon Doubleday, enters the household. Gideon's sister, Magdalene, is introduced to Horace's cousin, Mortimer, and falls in love with him. She learns that Mortimer is in love with the absent Charlotte and manages to intercept a letter, which she leaves for Horace to find.

The discovery briefly reforms Horace's character, and Charlotte returns to a house no longer darkened by the shadow of her husband's tyranny. None the less, she realises that her place is with Horace and the children, rather than with Mortimer, whom Horace banishes from the house. Mortimer considers marrying Magdalene but discovers her betrayal. Horace's two eldest sons, twelve-year-old Jasper and eleven-year-old Marcus, fearful of a return of the bad times, fail to warn their father when he goes for a walk which will take him over an unsafe bridge spanning a ravine. Horace returns safely, but is appalled to find that his children could contemplate, or even collude in, his death with such equanimity. The young footman, George, is discovered stealing food and, under the threat of an interview with Horace, contemplates suicide. Eventually he decides not to throw himself down the ravine, but to remove the warning notice so that Horace might fall to his death. A knife he has stolen from Marcus is discovered by Horace, who believes that his sons have made another attempt upon his life, but this is disproved by the intervention of Gideon's mother, who had seen George with the knife. Horace succumbs to an illness and makes a deathbed repentance, asserting that were recovery possible he would treat his family better. He does in fact recover, but whether or not he keeps his promise is left to the reader's imagination.

Horace is one of Compton-Burnett's most ferocious tyrants, a compendium of pedantry, parsimoniousness, self-deception and self-pity. His bullying, hectoring manner alienates the affections of everyone but Mortimer and Bullivant, whose conversations can be regarded as a chorus upon the action. For Horace, the perfect conclusion to any argument is 'when his opponent wept, and as he always pursued one to this point, [he] had no experience of defeat in words'. The children, too, are splendidly realised, particularly the bookish, plain-spoken, ten-year-old Tamasin ('Oh, not a real book?' she says when her tutor announces that he is writing a collection of essays), and seven-year-old Avery, whose favourite reading is the Book of Job. (He is 'concerned with the things that had value for him, and his father's opinion was not among them'.) In its combination of stark horror and high comedy, the scene in which Marcus makes a wax effigy of his father – ' "Now where are the pins?" he said, with a twitch of his lips. "They will not be wasted. It is an honourable purpose for them" ' – is characteristic of this dark dissection of family life.

Patrick HAMILTON　　　　**1904–1962**

The Slaves of Solitude

During his last years, which were spent in an alcoholic haze, Hamilton wrote three novels featuring his most famous creation, the con-man Ernest Ralph Gorse. *The West Pier* (1951), *Mr Stimpson and Mr Gorse* (1953) and *Unknown Assailant* (1955) are not, however, in the same class as **Hangover Square** (1941), which is set in 'Darkest Earls Court in the Year 1939', or *The Slaves of Solitude*, which is set in Thames Lockenden (an undisguised Henley-on-Thames) in 1944.

The slaves of the title are the inhabitants of the Rosamund Tea Rooms, a depressing boarding-house run by a widow called Mrs Payne, famous for her notes, which have long since stifled any attempts at innovation on the part of the residents. The main characters are Miss Roach, a thirty-nine-year-old former schoolmistress, now a secretary at a London publishing house, who is

'too amiable and [tries] too hard in company and conversation', and Mr Thwaites, a large, narrow-minded man in his sixties, who has been a bully all his life, a 'lifelong trampler through the emotion of others'. Miss Roach is the main target of Mr Thwaites's scorn: 'Sitting at the same table with Mr Thwaites, and having to talk to him directly, was very much like being called out in front of the class at school.' Mr Thwaites is also famous for his turns of phrase:

> 'I Hay ma Doots, that's all ...' said Mr Thwaites. 'I Hay ma Doots ...'
>
> (He is *not*, thought Miss Roach, going to add 'as the Scotchman said', is he? *Surely* he is not going to add 'as the Scotchman said'?)
>
> 'As the Scotchman said,' said Mr Thwaites. 'Yes ... I Hay ma Doots, as the Scotchman said – of Yore ...'
>
> Only Mr Thwaites, Miss Roach realised, could, as it were, have out-Thwaited Thwaites and brought 'of Yore' from the bag like that.

Among the other boarders are Miss Steele, sixtyish, who has a predilection for 'fun' and the company of young people, neither of which ever come her way (she is reading a *Life of Katherine Parr*), and Miss Barratt, sixty-fiveish, and hardly distinguished from Miss Steele, except that her preoccupation is with minor ailments.

The pattern of daily life is disturbed when two American servicemen arrive at the Tea Rooms. One of them, Lt. Dayton Pike, befriends Miss Roach. For him, nothing is of any consequence; for Miss Roach, everything is. He turns out to be fickle and promiscuous. When Miss Roach's German friend, Vicki Kugelman, arrives at the Tea Rooms, she insinuates herself into the lieutenant's company and reveals loathsome verbal mannerisms to match those of Mr Thwaites, one of which is to emphasise the indefinite article 'in the most repulsive way'. She is painfully arch, and Miss Roach grows to hate her. She is also, perhaps, a Nazi sympathiser, as is (if inconsistently) Mr Thwaites.

Goaded beyond endurance, Miss Roach pushes Mr Thwaites, who falls over. He is unharmed, but immediately (and coincidentally) develops peritonitis and dies slowly and in agony in hospital. Miss Roach, who is too weak to wholly engage the reader's sympathy, comes into an inheritance, and leaves the Tea Rooms to stay, if only temporarily, at Claridge's.

An oppressive atmosphere of hopelessness and privation makes *The Slaves of Solitude* one of the best fictional evocations of wartime Britain, with its petty restrictions and large frustrations. The malevolence of Mr Thwaites is in tune with the malevolence of the times, and Hamilton unsparingly demonstrates how easily the weak can be isolated and put upon.

L.P. HARTLEY   1895–1972

Eustace and Hilda: A Trilogy

The Shrimp and the Anemone (1944)
The Sixth Heaven (1946)
Eustace and Hilda (1947)

Hartley's trilogy, rather Jamesian in both its background and perceptive observations of middle-class Edwardian manners, describes the progressively destructive relationship of a brother and sister, Eustace and Hilda Cherrington. Although this theme develops in the second book, and is concluded in the last, all the elements of the novel are present in the opening volume of the trilogy, *The Shrimp and the Anemone*.

Eustace is a weak and sickly child of nine when the novel begins. His closest companion is his sister, Hilda. Four years older, she is allowed to dominate her brother by their charming but weak father, Alfred, who is a widower. Hilda determines to shape her brother and meets little resistance, because: 'The effort to qualify for his sister's approval was the ruling force in Eustace's interior life.' Their puritanical Aunt Sarah, who manages their poky seaside house in Anchorstone, Norfolk, repels Eustace, who is 'chilled by her shadow'. A baby sister, Barbara, is nursed by an old retainer, Minney, who also mollycoddles Eustace.

When in a rock-pool the children find a shrimp being devoured by an anemone,

Hilda tugs them apart. In the process, she kills the shrimp and maims the anemone, symbolically indicating the unavoidable course of her relationship with Eustace. In justification Hilda states: 'We had to do something . . . we couldn't let them go on like that.' Eustace limply replies: 'They didn't mean to hurt each other.'

The use of symbolism, especially the waters of the rock-pool and, later, Venice, underlines the incestuous attraction Eustace feels for Hilda, but this desire is only recognised by him subconsciously, emerging in daydreams and interior monologues. Because Eustace is unable to interpret these, he is destined to remain infantile and unable to confront reality. An explanation of his interior world is hinted at in epigraphs from Emily Brontë: 'I've known a hundred kinds of love, / *All* made the loved one rue.' Eustace's childhood efforts to escape from his sister are marked by various rebellious acts which end in failure, followed by guilt which he interprets as punishment for his sins. A forbidden paper-chase with Nancy Steptoe, a local girl whom he loves, but who is hated by Hilda as a rival, results in a heart-attack and a lengthy illness.

Sliding back into his former self was a sensation as grateful as putting on an old suit of clothes; he suddenly realised what a strain his new deportment had put on all his moral muscles.

EUSTACE AND HILDA
BY L.P. HARTLEY

Hilda punishes her brother's disobedience by forcing him to befriend an old woman, Janet Fothergill, whose twisted appearance has given her a reputation amongst children as a witch. Eustace is happy to be indulged in her comfortable servanted house, the significantly named Laburnum Lodge, where he plays cards and amuses the invalid. Nevertheless, Miss Fothergill knows that she is, with Hilda, Minney and Aunt Sarah, one of the 'designing women' in Eustace's life, and tells him: 'You mustn't let yourself be sucked in by us.' When Miss Fothergill dies, she leaves Eustace a large legacy, and with it a posthumous curse, for financial independence further shields him from accepting responsibility.

In *The Sixth Heaven*, Eustace is a contented undergraduate at Oxford, where he delivers a paper on nineteenth-century mystics. For a time, Hilda's nun-like dedication to a private home she runs satisfies her, but she then agrees to accompany Eustace for a weekend at Anchorstone Hall, the home of the aristocratic Staveleys. For Eustace the visit fulfils a childhood obsession about the place, for he was rescued after his collapse during the paper-chase by the handsome, virile eldest son, Dick, whom he has since regarded as a hero in the chivalric mould. When Dick takes Hilda for a flight in an aeroplane, Eustace feels 'a strange exultation ... a sense of fulfilment ... the ecstasy of achievement', responding to these circumstances as though he, not Dick, were sexually involved. Thus Eustace is in his 'Sixth Heaven'.

Eustace and Hilda opens in Venice where Eustace is spending a holiday with Lady Nelly, an attractive, middle-aged widow related to the Staveleys. Another mother-substitute, she dominates the willing Eustace, who is now in his late twenties. Venice offers sexual and other temptations (including a symbolic ritual bathe in the Lido during the Feast of the Redeemer), to which the prudish Eustace refuses to succumb. When he encounters Nancy, the only threat in his life to Hilda's love, he bravely cancels dinner with Nelly, and dines with her instead. Nancy has married and left her husband, and indicates that she still feels something for Eustace. Eustace can project in his imagination that they are married, but misunderstands an invitation to her bedroom. This is the testing crisis of the novel, for in rejecting the opportunity for sex, Eustace remains possessed by Hilda. Meanwhile, Hilda has been seduced and then abandoned by Dick after she attempted to possess him as she had possessed her brother. The shock of his desertion has left her paralysed. Eustace blames himself for neglect and returns to Anchorstone, where Hilda is temporarily living with Barbara and her husband and Minney.

Barbara has challenged the Victorian values of Aunt Sarah, rebelled and married a social inferior, the garage owner, James Crankshaw. Her newborn son, also called Eustace, will redeem the sterile waste of his namesake's past, and fulfil what he might have been. For Eustace, the only meaningful expiation of his guilt is his death. He pretends to tip Hilda from her bath chair over the cliffs, providing the shock needed to cure her psychosomatic illness. Eustace resolves that Hilda will no longer exercise 'her habit of lordship' over him, but it is too late. The events have been too much for Eustace's weak heart and, dreaming of the rock-pool, he dies in his sleep. Eustace feeds the anemone with his finger rather than the shrimp, and his sacrifice to Hilda, passing beyond the symbolic, is now complete.

Like the anemone, Hilda is partly destroyed, and having rebuffed the gentle advances of her lawyer, a perceptive Oxford friend of Eustace called Stephen Hilliard, she returns to work in the sheltered nunnery of the home she founded with the help of her brother's money.

When the trilogy was published in a single volume in 1958, Hartley added a chapter linking the first two books. 'Hilda's Letter' outlines her vain attempts to keep Eustace from being sent away to preparatory school, frightening him with such public-school stories as F.W. Farrar's *Eric: or, Little by Little* (1858). As a final resort, she writes to Dick Staveley, whom she hates, in an attempt to enlist his support, but is subsequently told by Eustace that she too is to go away to school. Relieved that she will not have to 'sacrifice' herself to Dick, she manages to retrieve the letter before it reaches its destination.

Philip LARKIN　　　　　　**1922–1985**

A Girl in Winter

John Bayley described this, the second of Larkin's two novels, as 'one of the finest and most sustained prose poems in the language'. Larkin was just twenty-five when he published it. After reading his first novel, *Jill* (1946), his friend Barbara Pym wrote to him: 'remembering *A Girl in Winter* one wonders why you didn't go on writing fiction, and regrets it. I suppose you were *too good* and didn't perhaps sell enough, and then you preferred writing poetry? But couldn't you possibly give us a novel now and again ... surely you are being rather *selfish?*' One suspects that Pym may have put her finger on the reason for his not going on.

A Girl in Winter is set during the war in an unnamed provincial town. Katherine Lind, in her early twenties, is a refugee and entirely alone. We are not told which country she came from, what she suffered, how she got out, nor what happened to her family; but at one point, when they are punting on the river near his home, her friend Robin Fennel remarks, 'If we'd lived in prehistoric times, before England was an island, I could nearly have taken you home. The Thames used to flow into the Rhine.' One has to assume then, that she is German (and probably Jewish) – or possibly Dutch.

She has a dreary job in a lending library, ruled over by the tyrannical Mr Anstey, who resembles 'a clerk at a railway station who had suffered from shell-shock'. While he is lecturing her interminably about a minor misdemeanour, Katherine looks at him 'as if he were an insect she would relish treading on'. We are given to understand that he owes his position to the fact that all the young men have been called up. The forward action encompasses a single Saturday in winter (a full working day at the library). Sandwiched between the forenoon and the afternoon/evening is an extended flashback to the visit that Katherine, aged sixteen, paid to the Fennels at their country house in the Thames Valley one summer before the war.

The Fennels had been remote and very English. Having met them all, Katherine 'was left with the absurd feeling that the most important person, her real friend, had not yet appeared'. As for Robin, the pen-friend who had invited her here, 'it couldn't be natural for anyone of sixteen to behave like a Prince Regent and foreign ambassador combined'. The Fennels seemed to care as little for each other as they did for her, but Jane, the older sister, aged twenty-five, 'sallow and irritable', had accompanied them everywhere. Had she been acting as a chaperone? Not a bit of it; it was simply that she was tremendously bored, and though she did not much enjoy being with them, it was better than being alone. It was Jane who had 'made' Robin invite Katherine in the first place. When he had got her he had used her for free language lessons, which he hoped would be useful in his future career.

It is small wonder then, that, despite chronic loneliness, Katherine has been eighteen months in England as a refugee before she writes to the Fennels; and when Robin, now in the army, replies suggesting a meeting, she takes no steps to see that it is effected. But there he is, waiting outside the house where she has an attic flat, when she returns after a particularly taxing day. He has changed. 'He had lost the self-possession he had moved with when a boy, and was given over to the restlessness of his body.' The war had 'mucked everything up ... Broken the sequence, so to speak ... my career and so forth ...' It soon appears that he has come all the way from wherever it is he is stationed for the sole purpose of sleeping with her. In the end Katherine – in the words of one of Larkin's early poems 'the less deceived' – gives in, knowing that 'he could not touch her. It would be no more than doing him an unimportant kindness, that would be overtaken by oblivion in a few days.'

A Girl in Winter is about loneliness. None of the Fennel family connects with one another, the people who work at the library only combine in their hatred and fear of the ultimately pathetic Anstey; even the bitter winter weather is a factor in separating people. But it is Katherine's desolation, the phases of which Larkin describes in minute detail, that we care about. He makes us feel what it is like to be friendless in a foreign country without any hope of escape. With the quality that made him the most distinctive of the 'Movement' poets, he invests all with his own brand of lyricism.

Malcolm LOWRY　　　　　**1908–1957**

Under the Volcano

Lowry's best novel was conceived in Mexico, where the author had gone in November 1936, after a brief spell spent in New York's Bellevue Hospital in a vain attempt to cure his alcoholism. In a cottage with a view of Popocatepetl, he set about writing a short story entitled 'Under the Volcano', from which the novel gradually evolved through a fog of alcohol. It went through four complete drafts, the third of which was offered to and rejected by four publishers. Jonathan Cape rejected a fourth draft in 1945, which spurred Lowry to write a long and detailed defence of the novel (see Lowry's *Selected Letters*, 1965), during the composition of which he made a drunken and half-hearted attempt to kill himself. He pointed out that the book's apparent incoherence was part of its poetic quality, and painstakingly explained his method and the novel's symbolism. The passages of drug-induced narration by the Consul were not simply an overlong elaboration, as Cape's reader had suggested, but were intended 'to symbolise the universal drunkenness of mankind during the war, and during the period immediately preceding it'. Impressed by this spirited apologia, Cape agreed to publish the novel as it stood. It proved a great critical success, hailed as a truly visionary novel, but this did not much cheer the author: 'Success is like some horrible disaster / Worse than your house burning', he wrote in a poem about the book's publication.

On the Day of the Dead in 1938, Geoffrey Firmin sits in a bar in Quauhnahuac in a haze of mescalin. He has resigned his post as British Consul and pursues his vice, awaiting dissolution. Propped between the bottles are the unanswered letters of his estranged wife, Yvonne, pleading for a reconciliation.

Yvonne returns to the shadow of the volcano, determined to save her husband. She enlists the help of his half-brother, Hugh, her ex-lover. Lowry weaves the three streams of consciousness into a narrative of the carnival day, as Yvonne, 'the Consul' and Hugh all seek to escape the inevitable. In the macabre celebrations, children beg for pennies to buy chocolate skulls. The Consul flees from their importuning hands on to a fairground ride, a mechanical roundabout which plunges him between the states of heaven and hell in his mind. Hugh and Yvonne pursue him on his drunken tour of the Mexican dives, aware that their efforts will prove futile. For the Consul, 'nothing is altered and in spite of God's mercy I am still alone. Though my suffering seems senseless I am still in agony. There is no explanation of my life.' A recurrent motif is the appearance of a man on a horse apocalyptically branded with the number seven. On an afternoon excursion the man is found dead by the roadside, his pockets robbed by the vigilante police. Yvonne and Hugh, shocked by a landscape of corruption, find themselves joining the Consul in his search for oblivion.

The evening brings a heightened sense of hopelessness as the Consul breaks the bonds of affection and plunges blindly into the night. As Yvonne and Hugh go after him, the riderless horse reappears, and the Day of the Dead claims Yvonne as its victim. Unaware of his wife's death, the Consul continues to drink, believing salvation has come too late. In a final twist of fate, the lives of the half-brothers change places; the vigilantes, suspecting Hugh of communism, execute the Consul by mistake. The last two pages follow his thoughts as he lies there, feeling 'his life slivering out of him like liver, ebbing into the tenderness of the grass', and are juxtaposed with a brutal final sentence: 'Somebody threw a dead dog after him down the ravine.' Then, after a blank page, the sign on the park railings is reproduced, which translates as 'You like this garden? Why is it yours? We evict those who destroy!' Seeing these earlier, the Consul had thought: 'Simple words, simple and terrible words, words which one took to the very bottom of one's being, words which, perhaps a final judgement on one, were nevertheless unproductive of any emotion whatsoever.'

Compton MACKENZIE   1883–1972

Whisky Galore

In an 'Author's Note', Mackenzie writes: 'By a strange coincidence the SS *Cabinet Minister* was wrecked off Little Todday two years after the SS *Politician* with a similar cargo was wrecked off Eriskay; but the coincidence stops there, for the rest is pure fiction.' It is possible that he had a dram or two in him, as well as tongue in cheek, when he wrote this apparently conventional disclaimer. It would be more appropriate to call his story fact made credible with a thin layer of fiction. This is not to belittle Mackenzie's achievement; never has a journalist been more puzzled by an embarrassment of riches. Nobody will ever know for certain precisely what happened; each surviving *bodach* (old man) and *cailleach* (old woman) would have been ashamed not to improve even a good story – it is doubtful whether they even remember the literal truth themselves. Mackenzie's responsibility was to make the 'historical' incident believable to his readers. That apart, living as he did on Barra, among the people he is writing about, he felt obliged to leave a great deal out: for instance the skill and daring with which the islanders lifted the cargo, knowing that at any moment the ship might slip off the ledge where it was lodged and plunge into deep water, and the equal skill with which they hoodwinked the excisemen, are necessarily avoided.

Many romantic pages have been written about the sunken Spanish galleon in the bay of Tobermory. That 4000-ton steamship on the rocks off Little Todday provided more practical romance in three and a half hours than the Tobermory galleon has provided in three and a half centuries. Doubloons, ducats, and ducatoons, moidores, pieces of eight, sequins, guineas, rose and angel nobles, what are these to vaunt above the liquid gold carried by the *Cabinet Minister*?

WHISKY GALORE
BY COMPTON MACKENZIE

In the novel are two islands: Great Todday, which is Presbyterian, and Little Todday, which is Roman Catholic. It is February 1943 and the topic of conversation on both Toddays is the whisky drought: 'this is a terrible war, Iosaiph, right enough,' Roderick MacRurie, owner of the Snorvig Hotel, complains. 'My brother Simon said we would have to pay for it by going to war on the Sabbath.' Everybody is depressed. Capt. Waggett, commander of the Home Guard, retired Sassenach stockbroker, is concerned

about 'defeatism' in G Company. But Dr Maclaren's problems are real. He has already lost one patient, the 'Ancient Mariner' Capt. MacPhee, who dies of shock when Roderick has to tell him that he cannot even have his customary third pint of beer. 'And I'm not surprised,' says the doctor, completing his diagnosis; 'he drank his three drams and three pints ... every night of his life and on such a tonic he might have lived to be a hundred.' And the white-haired crofter Hector MacRurie, who had fought in the Boer War, is likely to go the same way. (The scene in which the doctor consoles old Hector is one of the tenderest in the book.) Even the sport to be had making fun of the Intelligence officer (posing as Anthony Brown, tweed merchant) who has been sent at Capt. Waggett's instigation to investigate 'careless talk', is small consolation. Before divine intervention, in the shape of fog which causes the *Cabinet Minister* to lose her way, can rescue the bereaved islanders, they must first suffer a disappointment. The metal urn which is washed ashore is found to contain neither the hoped-for gold nor whisky but the 'ashes of H.J. Smith, cremated at Toronto'. Even when the wreck is sighted, the cargo known, there is still the agonising wait through the longest Sabbath anyone can remember.

But with the whisky (or St Minnie, as it comes to be known on Little Todday) safely ashore – even Capt. Waggett can see that there is no point in setting G Company to guard the wreck – miracles start to occur. George Campbell, well fortified, at last defies his tyrannical old mother – who had once, sergeant and headmaster though he is, locked him in his room rather than let him go on a Home Guard exercise on the Sabbath – and names the date of his marriage to Catriona Macleod for Easter Week. It takes a tragedy, however, to ensure the other miracle. Joseph Macroon, the postmaster of Little Todday, has so far resisted all Sgt-Major Odd's entreaties to name the date of the wedding to his daughter Peggy, even though Odd is received into the Catholic church and has Father Macalister to intercede for him. However, when Peggy and her sister, seeing as they think the excisemen approach the post office, 'pour away two hundred and sixteen bottles, and twelve of them Stalker's Joy, the best of all', he might have decided that he could manage without such a daughter after all. Finally Duncan Ban, crofter-poet, is given the strength to roam at night again, communing with *uisge beatha* (the water of life), and the fairies. One likes to think that Gaelic literature is the gainer.

Whisky Galore is not just a splendid romp; it captures, without exaggeration, a precious culture that is fast disappearing – if it has not already gone.

J. MACLAREN-ROSS   1912–1964

Of Love and Hunger

The title of Maclaren-Ross's first and best novel is taken from W.H. Auden and Louis MacNeice's *Letters from Iceland* (1937), a book to which the narrator is introduced by his lover:

Adventurers, though, must take things as they find them,
And look for pickings where the pickings are.

The drives of love and hunger are behind them,
They can't afford to be particular ...

Unlike Auden and other 1930s writers, Maclaren-Ross's picture of the era is based upon experience rather than imaginative sympathy. Funny, unsentimental, and narrated in a laconic, demotic prose, *Of Love and Hunger* is one of the most evocative novels of the Depression. When Graham Greene heard that Maclaren-Ross was working as a vacuum-cleaner salesman, he asked: 'Are you doing it to get material?' 'No,' Maclaren-Ross replied, 'I'm doing it because I wouldn't have any money otherwise.' In spite of a disclaimer that 'all the characters, vacuum-cleaner firms, etc., in this novel are completely fictitious', the novel is an accurate recollection of the author's experiences in Bognor Regis.

The story is set in 1939 against the growing threat of war. The narrator is Richard Fanshawe, who has returned from Madras in disgrace, sacked from his job as a

journalist, and with malaria, a broken engagement, and a nasty dose from a native woman (painfully cured) to his credit.

There were a lot of young men in the world like me. You see 'em in the vacuum-cleaner schools, selling second-hand cars, Great Portland Street, silk stockings from door to door: *Young man, public school education, can drive car, go anywhere, do anything.* Living on hope. Something'll turn up. Luck's bound to change. And nothing'd turn up except the war.

Behind with his rent, owing money everywhere, he ekes out a grim existence giving 'dems' to unenthusiastic 'prospects' in an attempt to make them buy the machines. When a Conrad-reading colleague, Derek Roper, gets the sack and joins a liner as a steward, he entrusts his well-read, left-wing wife Sukie to the unwilling Fanshawe's care. Bored and depressed, Fanshawe and Sukie gradually embark upon an affair. Sukie confesses that she has never loved her husband, but this does not entirely absolve the lovers of feelings of guilt.

Fanshawe himself is sacked in due course, after which he goes to work for a rival firm, Sucko, where he is rather more successful, made Leading Salesman for the Worthing Area. This does not, however, lead to much in the way of salary and he is thrown out of his digs. He lodges with his untrustworthy group leader, Smiler Barnes, who runs an illegal sideline in selling secondhand cleaners. Sukie encourages Fanshawe to get on with the novel he is writing, but the prospect of Roper's return from abroad unsettles her. Eventually, she breaks off her relationship with Fanshawe, who is subsequently sacked when Barnes's racket is uncovered. Fanshawe has meanwhile attracted the attention of one of his clients, the glamorous Jackie Mowbray, a widow from 'out east' who owns a flower shop. They go to a dance attended by other 'prospects' who are scandalised that Jackie could be consorting with a salesman. On the way home, Fanshawe makes a pass at Jackie, which is eagerly reciprocated: 'She knew her stuff, no doubt about that. All smooth, scented, and warm and a kiss like a mouthful of rice pudding. Her teeth didn't get in the way and she knew what to do with her tongue.' Jackie's expertise, however, does not drive out the ghost of Sukie. On his return, Roper discovers Fanshawe's letters to Sukie, who insists that she had been pursued unsuccessfully, a version of events Fanshawe gallantly confirms when Roper confronts him. Fanshawe is saved from the dole-queue by war.

An epilogue, set three years later, finds Fanshawe a captain, having risen through the ranks; he plans to marry Jackie after the war. On the last day of his embarkation leave, he bumps into Sukie, who has finally fallen in love with her husband, now a sub-lieutenant in the navy. They discuss the past and Fanshawe discovers that Sukie no longer obsesses him: 'I'd carried it with me so long I expected it to come alive immediately, but it didn't. Because it was dead now, at last. Died a natural death, too.' 'The past's over and done with,' he tells her sardonically. 'A happy ending is being had by all.'

Vladimir NABOKOV 1899–1977

Bend Sinister

Prof. Adam Krug is an internationally renowned philosopher and the most – perhaps, the only – famous citizen of his fictional country. At the opening of *Bend Sinister*, he contemplates a puddle outside the hospital where his wife has just died and then, half-drunk and grieving, walks home. Attempting to cross his customary bridge, he is hampered by revolutionary guards of the Ekwilist party.

Led by Paduk, a childhood enemy of Krug, the Ekwilists make a virtue of mediocrity. All the same, they are eager to attach the professor's prestige to their cause and try to bully him into co-operating with their regime. His university colleagues beg him to sign a manifesto proclaiming loyalty to Paduk's doctrines. His closest friends are arrested. He is betrayed by an antique dealer – whom he believes will help him to escape abroad – yet still declines to compromise his intellectual independence. But, when his beloved eight-year-old son, David, is seized by the police, Krug agrees to 'speak, sign, swear – anything the Government wants'. Krug is promised his son's return. However,

in a mix-up over names, the wrong boy is released and David is gruesomely killed. Viewing his own fate with detachment now, Krug refuses further compromise, preferring execution for himself and bringing it, incidentally, upon his friends.

Although the structure of the tale is tragic, its tone is none the less perversely comic. Laced with anagrams, spoof translations and allusions – notably to *Hamlet*, **Finnegans Wake** (1939) and *L'Après-midi d'un faune*, – the novel mocks with playful brilliance the grim quotidian life of the state. As in *Invitation to a Beheading* (1935) and 'Tyrants Destroyed' (1938), aloof and private laughter is the weapon against which totalitarianism has no defence. Resistance assumes the form of subtle lampoons. By contrast, cliché and mixed metaphor bedevil Padukist propaganda, and this verbal bungling does justice to the incompetence of the bureaucracy. Indeed, the narrative's ruling irony is that, while officialdom strives to enforce normality, its gaffes become increasingly grotesque. It is appropriate, therefore, that the book's final scenes are impelled by the twin logics of nightmare and farce.

Here, as elsewhere, Nabokov eschews 'the literature of social comment', declaring in his 1964 preface that the 'main theme of *Bend Sinister* ... is the beating of Krug's tender heart'. He cannot, however, avoid excoriating the lumpen culture upon which repression thrives; and, while scorning didacticism, he allows the reader to infer that bad aesthetics and brutality are partners.

Bend Sinister was the first novel Nabokov wrote in the USA. Nymphetologists will be delighted by the presence in it of Mariette – Krug's faithless servant-girl – an older, plainer sister of Lolita.

Mollie PANTER-DOWNES 1906–

One Fine Day

As its title suggests, the action of this enchantingly perfect book unfolds within twenty-four hours, 'one fine day' in one of the first summers of the peace that follows the Second World War. Set in and around Wealding, a village near the Sussex coast, the novel offers us one 'ordinary' middle-class

Englishwoman's response to the irreversible changes the war has brought.

Laura Marshall, wife of Stephen, and mother of the ten-year-old Victoria, is one of a host of middle-class woman facing life in large houses which will never again hold the small army of servants which polished, dusted, cooked and gardened before the war. Borders, shrubs and lawns riot, scarcely registering the efforts of jobbing gardeners, and the house itself is 'slowly giving up, loosing its hold, gently accepting shabbiness and defeat'. But this is not a nostalgic threnody for the Good Old Days, though loss – often grievous – there has most indubitably been.

The day begins on Barrow Down, the neighbourhood's highest point, populated only by rabbits and larks who, like the land itself, were here before humans and may well survive them: 'Up here, on the empty hilltop, something said I am England, I will remain.' A few hours later, down in the village, Laura begins a day which will take her, in reality, recollection and imagination, through a substantial chunk of one microcosmic community's peacetime present and pre-war past. It will include time with Mrs Prout, who comes in 'to do', three mornings a week, where once two maids, Ethel and Violet, spent all of every day and Mrs Abbey, the cook (killed by a flying bomb whilst visiting her niece in Putney), ruled the kitchen. It will take her to Bridbury and the queues for rationed cakes and fish, and home on the bus where a young, male hiker, just demobilised, studies his map, 'reading England with rapture'.

Her quest for a new part-time gardener will bring her to Mrs Porter's externally picturesque, internally squalid cottage where toddlers begotten by departed Allied soldiers tumble merrily, and older sons plan the exodus to towns, cities and other lands which will leave the English countryside denuded of its young. Her journey home includes a chance meeting with Mr Vyner, the vicar, a genuinely good man whose beautiful parish church grows steadily emptier.

A telephone call from her mother, Mrs Herriot, prompts recollections of Laura's own girlhood and a social training so unhelpfully predicated on endless leisure and private means. As the blistering mid-day heat begins to wane, Laura sets off in search of Stuffy, her unregenerate bitch who, as is her custom on coming into season, has made a dash for the dogs kept by the gypsy who lives at the foot of Barrow Down. Passing Cranmer, the village's Big House, she sees its owners preparing to move out: they have given it to the National Trust, in exchange for a flat in the converted stables, yet old Mrs Cranmer remains curiously unmoved, still secure in her own strong, unsentimental sense of self and her place in the new order.

Reunited with a sated and only mock-repentant Stuffy, Laura, prompted by the gypsy, climbs to the top of Barrow Down and finds herself exalted by its overwhelming beauty. Glimpsing the sea, she follows in her mind's eye the coast of Britain and feels a surge of gratitude for its survival and for the Marshalls' 'wonderful, stupendous luck': she, Stephen and Victoria, still living, still together. Blessed with such bounty, who could be defeated by the difficulties or terrified by the challenge of the post-war world? Not Laura, certainly, who welcomes the future her mother fears.

One Fine Day is an outstanding piece of work, distinguished not only by its flawless construction, but also by its magnificent descriptions of places and people, observed with a poet's precision and couched in richly sensuous prose.

C.P. SNOW　　　　　　　　**1905–1980**

The Light and the Dark

see **Strangers and Brothers** (1970)

1948

Brussels Treaty ● End of British Mandate in Palestine ● Berlin blockade ● Mahatma Gandhi is assassinated ● Harry S. Truman wins US presidential election against Thomas E. Dewey ● Invention of long-playing record ● Alfred Kinsey, *Sexual Behaviour in the Human Male* ● Jackson Pollock, *Composition No. 1* (tachisma) ● Christopher Fry, *The Lady's Not for Burning* ● Prince Charles born ● Ali Jinnah and Orville Wright die ● W.H. Auden, *The Age of Anxiety: A Baroque Eclogue*

Jocelyn BROOKE   1908–1966

The Military Orchid

see **The Orchid Trilogy** (1950)

Truman CAPOTE   1924–1986

Other Voices, Other Rooms

Capote's first novel made his reputation and marks the culmination of what he called the first phase of his writing career. A shimmering, ambiguous and dreamlike exercise in Southern Gothic, it traces the journey towards adulthood and identity made by a delicate thirteen-year-old boy, Joel Knox.

When Joel's mother dies, he is sent from New Orleans to the house of his estranged father in a remote Southern town, Noon City. Skully's Landing is an isolated and decaying mansion choked by luxuriant vegetation and gradually subsiding into the landscape. Its inhabitants are similarly decrepit and exotic: his stepmother, Miss Amy, is frail and distrait; his cousin Randolph is a faded Southern Belle, prone to asthma and alcohol; the servants consist of a pygmy-sized tottering ruin called Jesus Fever, and his young granddaughter, Zoo, who lives in constant fear of the man who once cut her throat.

Joel's enquiries about his father elicit evasive replies from members of the household, until the man makes his presence known by red tennis balls, which he rolls out of his bed in order to attract attention, for he is entirely paralysed. Randolph later reveals that he shot Joel's father, who had been the manager of a boxer whom Randolph worshipped. Joel's first attachment is to Zoo, but she leaves when her grandfather dies. An attachment to Idabel Thompkins, a rampaging, foul-mouthed local tomboy (partly based on Harper Lee) with whom he hopes to run away, is thwarted by Miss Wisteria, a fairground midget from Baltimore. His father remains indifferent to whether Joel reads him romances or recipes. Joel's final and most significant attachment is to Randolph and the novel ends with Joel accepting his circumstances and pausing upon re-entering the house: 'He stopped and looked back at the bloomless, descending blue, at the boy he had left behind.'

Oblique or not, this extraordinary account of a young boy coming to acknowledge himself as homosexual caused considerable controversy, carefully fuelled by the author who had himself photographed for the jacket in the guise of a debauched child. Not everyone was beguiled: *Time* magazine's reviewer complained of the book that 'the distasteful trappings of its homosexual theme overhang it like Spanish moss', while George Davis, fiction editor of *Mademoiselle* and a one-time mentor of Capote, merely commented: 'Well, I suppose someone had to write the fairy *Huckleberry Finn*.' Others were more enthusiastic, however, and the book sold extremely well. No one was left in any doubt that Capote had arrived on the literary scene and intended to stay there.

Graham GREENE   1904–1991

The Heart of the Matter

The Heart of the Matter is set in a port town of colonial West Africa during the Second World War. Its atmosphere and landscape – of malarial torpor, and rain-beaten Nissen huts – are a vivid expression of familiar Greene terrain. Major Henry Scobie, Deputy Commissioner of Police, has been here for fifteen years, fourteen of them married to Louise, and their one child is recently dead. Louise is Catholic, and Scobie converted to Catholicism prior to their marriage. Louise is unhappy, friendless and derided at the club, and her misery is increased when Scobie is passed over for promotion. He no longer loves her but pities her, 'bound by the pathos of her unattractiveness'. When she wants to travel to South Africa, he is forced to borrow the money for her fare from the disreputable Syrian merchant Yusef, who is suspected of involvement in the diamond-smuggling operations which Scobie is attempting, unsuccessfully, to prevent.

After her departure, Scobie is involved helping the survivors of a sunken ship, and is drawn towards a helpless young widow, Helen Rolt, whose loss seems to echo that of his own daughter. They become lovers and, despite their precautions against detection, are watched by Wilson, a government spy whose interest in Scobie's malfeasance is increased by his own boyish infatuation with Louise. Yusef acquires a love-letter and blackmails Scobie into giving a package (almost certainly containing diamonds) to the captain of a ship. Unbeknown to Scobie, Louise learns of the affair; she returns, and is insistent that they go to mass together, presumably thinking that this will force her husband to take communion and thereby repent his sin. Scobie knows that he cannot genuinely repent: 'I can regret the lies, the mess, the unhappiness, but if I were dying now I wouldn't know how to repent this love.' Eventually he takes communion and believes himself to be damned. After his servant Ali is murdered because of his involvement with Yusef, Scobie wants

nothing more but to escape this world where he seems capable only of bringing suffering. Feigning illness, he gathers a collection of sleeping pills and stages his suicide so that it will look as though he has died of angina. Louise discovers what has happened and is convinced he is damned; the priest, Fr Rank, expressing a heretical view clearly closer to Greene's, is less sure, doubtful that the church can ever comprehend the extent of God's mercy.

In the course of a long review in *Commonweal*, Evelyn Waugh suggested that '*The Heart of the Matter* should be read as the complement of *Brighton Rock*' (1938). Comparing Scobie with Pinkie, Waugh commented:

> Both believe in damnation and believe themselves damned. Both die in mortal sin as defined by moral theologians. The conclusion of the [later] book is that no one knows the secrets of the human heart or the nature of God's mercy. It is improper to speculate on another's damnation. Nevertheless the reader is haunted by the question: Is Scobie damned?

This would seem to be the heart of the matter. Waugh went on to suggest that Greene believed Scobie to be a saint, a notion prompted by the novel's epigraph, taken from Charles Péguy's *Nouveau Théologien*: 'Le pécheur est au coeur même de chrétienté . . . Nul n'est aussi compétent que le pécheur en matière de chrétienté. Nul, si ce n'est le saint.' Greene subsequently wrote to Waugh: 'I did not regard Scobie as a saint, and his offering his damnation up was intended to show how muddled a man full of goodwill could become once "off the rails".'

Ross LOCKRIDGE   1914–1948

Raintree County

Lockridge's only novel is a compendious and hugely ambitious work which he intended would be 'the first real representation of American culture in fiction'. Despite its critical and commercial success (it was a Book-of-the-Month Club selection in 1948), Lockridge committed suicide only two months after publication. Set on a single day

– the auspicious date of Independence Day, 4 July 1892 – in a small Indiana county, the novel is a collage of dream sequences, excerpts from imagined works of literature and newspapers, and flashbacks. It is laden with allusions to Greek and biblical mythology.

For the Independence Day celebrations the county's famous sons – Sen. Garwood Jones, the financial Cassius Carney, and the Civil War hero Gen. Jacob Jackson – have each returned to the county seat of Freehaven. The principal character, however, is John Wickliff Shawnessy, a schoolteacher who rises early and goes to the station to meet his old schoolfellow Sen. Jones. He also meets their old teacher, Prof. Stiles, and they reminisce about old times. In the afternoon there is the Grand Patriotic Program, and in the evening a fireworks display and a meeting of the Literary Society. The day's events are simple and serve primarily as a structuring device for the series of flashbacks which recount John Shawnessy's life.

His was a romantic, dream-filled youth, reading Shakespeare and Virgil on the banks of the river and dreaming of being a great poet himself. One day he glimpsed Nell Gaither bathing naked in the river and was transfixed by her beauty. On Graduation Day they bathed together and declared their love for one another, but a few weeks later John got drunk and went out to Paradise Lake with the newly arrived sultry Southern belle Susanna Drake and there – under what he believed to be the mythic Raintree – he lost his innocence. Susanna told him she was pregnant (though she was not) and so trapped him in a loveless but passionate marriage. Haunted by the dark secrets of her family and the fear that she might have Negro blood, she went mad after the birth of her child and fled south, having burnt down their house.

John went off to fight for the Union in the Civil War. He was wounded, and planned to have his marriage annulled upon his return home so that he could marry Nell, but she had heard a report that he had been killed in action, and had married Garwood Jones. When John returned to Freehaven she was a week dead, having died in childbirth. He remained in Raintree County, became a schoolteacher, and married in later life. Jones cynically pursued his political career; Carney went east and prospered in business; only John Shawnessy was left behind, his dreams of being a great poet now long over.

The novel was filmed in 1958 by Edward Dmytryk from a screenplay by Millard Kaufman.

Norman MAILER   **1923–**

The Naked and the Dead

This, Mailer's first novel, was published to general critical acclaim. It seemed that a 'normal', competent novelist of some substance had appeared on the American literary scene and Mailer's public persona played along with this: 'I've got all the average middle-class fears', he told an interviewer. The reception of the book was important for it is one of only three of Mailer's works to be unaccompanied by the Mailer ego. Contemporary interviews record him declining to attack the few voices hostile to the book and withholding comment on contentious issues. If his readers doubt the 'wild man of literature' tag attached to him in the 1950s and beyond, then the industrious, earnest Mailer of *The Naked and the Dead* is, with hindsight, no less incredible.

The novel itself nevertheless outlines many of the major themes which will preoccupy Mailer in later works: notably the conflict between man and his environment, and the angst arising from his own sense of mortality. *The Naked and the Dead* is set amongst the men of a fourteen-strong platoon who form part of an invasion force on the island of Anopopei during the assault on the Philippines in the closing stages of the Second World War. The novel opens with a card game, and Mailer's narrative viewpoint follows the action impartially, taking brief glances at what each man is holding, assessing their chances and recording their successes or failures. The members of the platoon represent American society writ small: a bible-bashing farmer and pleasure-loving wastrel from the Deep South, a Brooklyn Jew (like Mailer himself), an ill-treated travelling labourer, a Mexican–American, a Chicago wide-boy, a bigoted Irishman and a salesman from Kansas. Mailer does not orchestrate their interactions schematically, but their differences, quarrels and fights, a sense is given of the tensions which exist within American society. The most important characters are the officers, though, for it is through Gen. Cummings, Sgt Croft and Lt. Hearn that Mailer expresses the struggle between the human and natural orders that is the novel's controlling theme. The attempt by Croft to climb the island's central feature, Mount Anaka, is presented as the expression both of the animal will to conquer and of the spiritual aspirations of the men.

Mailer makes no judgement on the actions of his characters, and one of the faults of the novel was said to be the rapidity of the cuts between the men's viewpoints: the reader's sympathy is never given time to settle. In this, Mailer was simply following the technical experiments made by Dos Passos (an acknowledged influence) in his *USA* (1936), but the effect, coupled with the extreme violence and brutality of some of the action, was to make the book seem nihilistic. Mailer insisted that the novel

> offers a good deal of hope. I intended it to be a parable about the movement of man through history. I tried to explore the outrageous proportions of cause and effect, of effort and recompense in a sick society. The book finds man corrupted, confused to the point of helplessness, but it also finds that there are limits beyond which he cannot be pushed, and it finds that even in his corruption and sickness there are yearnings for a better world.

This manifesto was to set the agenda for his later, more ambitious works.

Alan PATON   **1903–1988**

Cry, the Beloved Country

Paton's novel about the state of South Africa was acclaimed on its publication as an outstanding work of libertarian sympathy for the difficulties which beset his beloved homeland. The story is told with compassion rather than condemnation, and the book does not supply any easy answers. The only solution, Paton proclaims, is for blacks and whites to begin to love one another; however, the likelihood is that by the time the white South Africans begin to exercise toleration, the blacks will have succumbed to hate.

The great red hills stand desolate, and the earth has torn away like flesh. The lightning flashes over them, the clouds pour down upon them, the dead streams come to life, full of the red blood of the earth. Down in the valleys women scratch the soil that is left, and the maize hardly reaches the height of a man. They are valleys of old men and old women, of mothers and children. The men are away, the young men and the girls are away. The soil cannot keep them any more.

CRY, THE BELOVED COUNTRY
BY ALAN PATON

Stephen Kumalo, the *umfundisi*, or parson, of Ndotsheni, receives an unexpected letter from Johannesburg, telling him that his only son Absalom has gone missing there. Absalom had left the village to search for Kumalo's sister, Gertrude, whom he suspected may have fallen into a life of prostitution. The message is from a missionary, Msimangu, who tells him that Gertrude is very ill and asks him to come to the city at once. Arriving in Johannesburg, Kumalo is confused by the unfamiliar environment but thrilled by his sense of self-importance. He finds his sister and offers to bring her home to live with her young son under his protection; but first he must discover the whereabouts of Absalom, who is rumoured to be keeping bad company. Eventually tracking his son down with the help of a white prison officer, Kumalo learns that he is living with a girl who is expecting his child; however, any shame that these discoveries might be expected to result in Kumalo's community is outweighed by the dreadful news he now receives. Absalom has been involved in a burglary and, frightened by being accidentally caught, he has shot and killed a white man, a liberal of reformist views who was working for the rights of the black people.

The dead man's father, Jarvis, is Kumalo's neighbour in the district of Ixopo, and Kumalo anticipates that he will be forced to leave his home. At the trial, however, despite the anger and grief on both sides, no recriminations are forthcoming. Kumalo arranges for his son to marry the girl while an appeal against the death sentence is awaited; Kumalo's own nephew has got off the charge, causing a bitter rift with his brother, but Msimangu finds a lawyer who is willing to take the case 'For God', or free of charge. No mercy, however, can be obtained, and Absalom is executed. Gertrude disappears, preferring her luxurious life in the brothel in Johannesburg to the boredom of respectability in her home village. Taking Absalom's young wife and his sister's son, Kumalo returns to his village, to be told by the biship of his church that it would be better for the community if he were to move on. Yet out of this series of tragedies comes evidence of a form of grace, though its origin is human and not divine. Jarvis's son rides out from his grandfather's house and seeks the friendship of the killer's father; gifts follow, and positive help for the village, as Jarvis senior pays for a farming instructor to teach them to make the most of the land.

In this affirmation of the possibility of reconciliation between the races, Paton suggests that there is hope for the future; but Kumalo has gained this at the cost of being a white man's lackey, even though he is allowed to remain in his native land. 'God Save Africa' is the prayer which runs through the novel, undercut by the perpetual fear that the help will arrive too late.

Paton himself adapted the novel for a film (1951; US title *African Fury*), directed by Zoltan Korda. The book is also the basis of the 'Musical Tragedy' *Lost in the Stars*, first produced on Broadway in 1949 and often revived thereafter. Although Paton was moved by Kurt Weill's music, he had distinct reservations about Maxwell Anderson's book and lyrics, which sidestepped the novel's explicitly Christian viewpoint.

Evelyn WAUGH   1903–1966

The Loved One

The Forest Lawn cemetery in Los Angeles had been recommended to Waugh as an outstanding example of 'religion and art brought to their highest possible association'. Unsurprisingly, Waugh disagreed, but was delighted to find 'a deep mine of literary gold' there went he visited it. He described the book as 'a study of the Anglo–American cultural impasse with the mortuary as a jolly setting'. It was first published in *Horizon*, and for many readers it marked a return to savage satiric form after what had been perceived as the lapse of ***Brideshead Revisited*** (1945).

As in several of Waugh's novels, the setting is one of the 'barbarous regions' of the world, and indeed references to 'native huts' on the first page tease the reader into believing that he is back in Azania, rather than in California. Dennis Barlow is a former war poet who had come to Hollywood to write a biopic of Shelley, but is now employed at the Happier Hunting Ground, a pets' cemetery, and is therefore considered an embarrassment by the cricket-obsessed English colony. When his protector, Sir Francis Hinsley, is driven to suicide by Megalopolitan Studios, Dennis goes to Whispering Glades Memorial Park to arrange the funeral ('Normal disposal is by inhumement, entombment, inurnment, or immurement,' he is told). During his tour of the Administrative Building he meets Aimée Thanatogenos, a cosmetician who is being courted by the mother-dominated Senior Mortician, Mr Joyboy. Intrigued by Aimée, Dennis woos her with classics of English poetry, which he cynically passes off as his own compositions. Aimée is torn between Dennis and Joyboy and confides in Guru Brahmin's agony column. Dennis talks of setting himself up as 'non-sectarian' parson, but Aimée feels that he does not take religion sufficiently seriously. She is further appalled to discover, whilst attending the obsequies for Mrs Joyboy's bald parrot, that Dennis works for the Happier Hunting Ground. In despair she telephones 'Guru Brahmin' (in fact an alcoholic journalist called Slump) and is told 'to go take a high jump'. She takes a fatal dose of cyanide in Mr Joyboy's workroom and the Senior Mortician, fearful of scandal, pays Dennis to dispose of the body in the pets' crematorium. Further funds are provided by the Cricket Club, who are anxious to repatriate Dennis, and he returns to England.

The appalling Whispering Glades, where death is sanitised and travestied by the mortician's 'art' is a perfect setting for Waugh's portrayal of Americans as 'exiles uprooted, transplanted and doomed to sterility'. Art, religion, language ('the Loved One' is a euphemism for a corpse), nature and humanity are systematically debased in the nightmare world of the cemetery, created by a man ironically called 'The Dreamer'. Whilst Waugh's narrative is laconic and heartless, he clearly revels in his subject-matter: his civilised distaste is palpable on every page. The novel was made into a truly dreadful film (1965), directed by Tony Richardson from a screenplay by Terry Southern and Christopher Isherwood.

Patrick WHITE   1912–1990

The Aunt's Story

'But old Mrs Goodman did die at last.' The opening sentence of White's first major novel sets up claustrophobic horrors: the bourgeois family life that stifles individuality and eccentricity. The malicious old invalid hangs on in an upstairs bedroom, while her unmarried daughter withers among the sideboards. Theodora Goodman is one of the many lonely, solipsistic people in White's fiction searching for significance in their lives.

An awkward girl with a long, yellow face, she grows up in Meroë, an Australian farmhouse named after the ancient, dead city in the Abyssinian desert. Her bookish father has 'no sense of responsibility to his own land', according to his neighbour, Parrott; he prefers to read Homer and Herodotus. When a gold-prospecting friend from his past visits and he offers him food, Mrs Goodman exclaims: 'You are a romantic, George ... I refuse to sit down to table with every tramp who comes along.' Theodora suggests that the man be given his dinner on the verandah. When he leaves, he promises

to return, but Theodora knows that he will not. She dances with Parrott's son, Frank, and is 'released from her own body and imprisoned in [his] molten gold', but when they go shooting she is unable to hide her superior skill, and Frank marries instead her dainty sister, Fanny.

When her father dies, Theodora is taken to live in Sydney by her mother. Her remoteness appeals to the cultivated taste of the family solicitor, Huntly Clarkson, but no romance flourishes. 'Theodora is a fool,' Mrs Goodman declares. 'She is a stick with men.' When told about Greece by a visiting cellist, Theodora retorts: 'I have not seen it. I have seen nothing.' In the 1930s, however, Mrs Goodman finally dies and the forty-five-year-old Theodora travels to Europe, settling in the *Hôtel du Midi* at a French resort. The lives of her disturbed fellow residents invade her own. Gen. Sokolnikov, a refugee from Russia, imagines that she is his sister Ludmilla, executed by the Bolsheviks, and informs her that she is dead. The American Mrs Rapallo awaits the imminent arrival of her daughter, supposedly an Italian princess by marriage, but in fact a figment of her drug-crazed imagination. The Desmoiselles Bloch, elderly Jewish sisters, worry about the disappearance of pens and toothbrushes and that: 'Walls are no longer walls ... are at most curtains. The least wind and they will blow and blow.' A crisis is reached after the Englishman Wetherby, spurned by the mad painter Lieselotte, coldly seduces Katina Pavlou, a young girl who writes secret poems and longs to fall in love. Theodora, who loves Katina, is revolted, but Lieselotte kills Wetherby and sets the hotel ablaze. All the guests escape, expect for Mrs Rapallo, who appears in the 'jewelled splendour' of a blazing window, 'as if she were not watched, but watching something that was taking place'.

Time passes, and Theodora is gradually slipping into madness. 'I have seen and done,' she writes to Fanny, 'and the time has come at last to return to Abyssinia.' Travelling through the American mid-west, intending to take a ship from California, she is frightened by the motion of the train, alights and walks into the mountains, tearing up her steamer ticket. She is given a meal by a poor farming family called Johnson, amongst whom is a 'rich dark child' called Zack, the last of Theodora's *alter egos*. She then goes on and at the top of the mountain road finds a large empty house, which she sets about tidying. Holstius, a man who has 'the familiar texture of childhood', and smelled of horses and leather and guns', appears. She wants to trust him, but fears the inevitability of loss: her father had died, and the man he invited to dinner never returned. 'You cannot reconcile joy and sorrow,' Holstius tells her. 'Or flesh and marble, or illusion and reality, or

life and death. For this reason, Theodora Goodman, you must accept.' She falls asleep with her head against his knee. In the morning he is gone, but she knows he will return, as he does, bringing her the message: 'They will come for you soon, with every sign of the greatest kindness ... They will give you warm drinks, simply nourishing food, and encourage you to relax in a white room and tell your life. Of course you will not be taken in by any of this, do you hear? But you will submit.'

1949

North Atlantic Treaty • Apartheid • Establishment of Communist Republic of China under Mao Tse-tung • Britain recognises independence of Eire, but re-affirms position of Northern Ireland within the UK • Transjordan is renamed the Hashemite Kingdom of Jordan • Communist purge in Hungary • Nationalisation of British iron and steel industries • Cortisone discovered • Arthur Miller, *Death of a Salesman* • Carol Reed, *The Third Man*, with Orson Welles • Maurice Maeterlinck and Richard Strauss die • Langston Hughes (ed.), *The Poetry of the Negro*

Elizabeth BOWEN   1899–1973

The Heat of the Day

Although Bowen had published several volumes of stories and non-fiction since *The Death of the Heart* (1938), *The Heat of the Day* was her first novel to be published for eleven years. It was dedicated to Charles Ritchie, a Canadian diplomat with whom Bowen had an intense friendship, and whose knowledge of the workings of MI6 contributed towards the novel's plot.

Begun in 1944, and set in 1942, the novel reflects the sombre mood of wartime London and is one of the best evocations of that place and period in fiction. The main plot is a sinister variation upon the eternal triangle. Stella Rodney, a divorcee who works at the War Office, refuses to believe that her colleague and lover, Robert Kelway, is a traitor. Her informant is the enigmatic and shabby Harrison, who seems to be

known to, or encounter, several of the novel's other characters. He suggests that if she has an affair with him, he will guarantee Robert's safety. She acquiesces, but too late, for Harrison has been removed from the job. When Stella confronts Robert, he attempts to explain his behaviour, ascribing it partly to injuries he received at Dunkirk, and partly to his upbringing. Believing he has been followed, Robert falls to his death while trying to escape across the rooftops.

Meanwhile, Stella's twenty-year-old son, Roderick, who has inherited a house in Ireland from his father's cousin, uncovers the real circumstances of his parents' divorce, which refute the generally held belief that his late father was the innocent party.

A largely comic parallel plot concerns the feckless Louie Lewis – 'all over herself she gave the impression of twisted stockings' – a country girl who feels displaced in London, but is looked after by her friend, Connie, an ARP warden. Missing her husband, Tom, who is in the army, Louie has affairs with other men, and ends up pregnant. She tries to befriend Harrison, with little success, but is able to pass off her pregnancy as legitimate when Tom is killed.

Alone, she could hope to keep cool and wear a face. As the train slowed down for the station, she hastily glanced at herself in her handbag mirror, trying on an expression of imperviousness – even, should it be necessary, of effrontery. The London black suit she wore, though severe and matt, somehow failed altogether to look like mourning.

THE HEAT OF THE DAY
BY ELIZABETH BOWEN

While the novel ends in apparent harmony – Stella is to remarry, Louie is to return to Kent to become an 'orderly mother' – its great strength lies in the author's attention to shifting nuance and the subtleties of experience. In her own words, she is dealing with a 'subsidence of the under soil' which affects and alters her characters' lives, 'without the surface having been visibly broken'. Amongst the most stylistically difficult of Bowen's books, *The Heat of the Day* was none the less very popular, its large sales enabling the author to install bathrooms at her Irish house, Bowen's Court.

Paul BOWLES   1910–

The Sheltering Sky

Bowles first had the idea of writing a novel when watching his wife, Jane Bowles, at work on *Two Serious Ladies* (1943). Although a very private person, he has acknowledged the crucial importance of her roles as inspiration and critic in his writing career. *The Sheltering Sky* was first conceived in 1947 in New York. An advance on the project funded a trip to Tangier where the book was written. The story (Bowles rejected the word 'novel' as too daunting) proceeded in Bowles's head and was written down with only negligible revision. In outline it is very simple. Port Moresby, together with his wife, Kit, and friend, Tunner, have come to

Morocco to keep as far away as possible from war, 'one facet of the mechanised age he wanted to forget'. Port is a traveller rather than a tourist, a man who compares his civilisation with others 'and rejects those elements he finds not to his liking'. The trio journey with two ill-met companions, Eric Lyle and his travel-writer mother, to the south, where the appalling conditions bring the tensions between Port and Kit to a head. Kit is less open to new conditions than her husband: 'It'd be abnormal if I were able to adapt myself too quickly to all this. After all, I'm still an American, you know. And I'm not even trying to be anything else.' Complications arise when they lose their passports and Port falls ill. Far into the desert, remote from help, Port succumbs to an infection – perhaps meningitis – and dies. Tunner, following them, discovers the distraught Kit, who, in the 'first moments of a new existence, a strange one in which she already glimpsed the element of timelessness that would surround her', evades Tunner and runs away to the desert where she joins a merchant caravan. She is first raped by one of the Arab merchants, Belquassim, to whom she becomes emotionally attached. More or less imprisoned by him as a concubine, she contrives – owing to the jealousy of Bequassim's harem – to escape. By now quite mad, Kit wanders until discovered and restored to European care.

Behind, around, in fact all about this simple story, lies the desert. Bowles called it 'the main protagonist'. A minor character remarks: 'Nothing ever happens the way one imagines it is going to. One realises that most clearly here; all your philosophic systems crumble.' The desert represents the forces of chaos – unimaginable forces which, in being faced, give meaning to human existence. The sheltering sky of the title is only a partial shelter and at root only an illusion. Kit's madness is caused by a glimpse of the world beyond that shelter. Her experience of life is more authentic for this insight, but the insight itself is unbearable, and she cannot withstand it. Unsurprisingly, Bowles has been dubbed an existentialist writer – there are marked parallels between his novels and those of Camus in terms of landscape and theme.

The strength of *The Sheltering Sky* can almost be summed up in the word 'description'. The intimate annotations of Kit's and Port's psyches combine subtlety with intensity, and Bowles's prose is always accurate. The evenness of tone (despite some lurid subject-matter) never lets up and finally becomes a version of the desert itself, dispassionate in the face of human frailty, omnipresent and inescapable. *The Sheltering Sky* can not be called an optimistic novel; rather, it is an overwhelmingly honest one, in which the human condition is examined in the unsentimental light of the desert sun and found wanting. It was made into an admired film directed by Bernardo Bertolucci (1990).

Jocelyn BROOKE   **1908–1966**

A Mine of Serpents

see **The Orchid Trilogy** (1950)

Nancy MITFORD   **1904–1973**

Love in a Cold Climate

Fearing that she had exhausted her material in *The Pursuit of Love* (1945), Mitford was wary of begining another novel. However, the mixture as before proved enormously popular and, in spite of the strictures of Evelyn Waugh (who discerned a Jamesian subject-matter which deserved greater application than Mitford was prepared to invest), the novel was highly acclaimed. Once again, the narrator is Fanny Wincham, and the younger Radletts, Jassy and Victoria, make a notable contribution.

The action runs parallel to that of *Pursuit*, but largely concerns the Montdores, neighbours of the Radletts, whose only child, a chilly beauty called Polly, seems uninterested in love until her uncle-by-marriage, 'Boy' Dougdale, is unexpectedly widowed. Known by the Radletts as 'the Lecherous Lecturer', the snobbish and unsavoury Dougdale is notorious for his more-than-avuncular interest in small girls and is rumoured to have once been Lady Mont-

dore's lover. From every aspect the match is deplorable, but Polly is determined to marry Boy even when she is cut off without a penny by her grieving parents. Boy's country house is let and the couple go abroad to live on the proceeds.

The food at dinner, served by the slut in a gaunt dining-room, was so terrible that I felt deeply sorry for Mrs Cozens, thinking that something must have gone wrong. I have had so many such meals since then that I do not remember exactly what it was; I guess, however, that it began with tinned soup and ended with dry sardines on dry toast, and that we drank a few drops of white wine. I do remember that the conversation was far from brilliant, a fact which, at the time, I attributed to the horrible stuff we were trying to swallow, but which I now know was more likely to have been due to the presence of females; dons are quite used to bad food but become paralysed in mixed company. As soon as the last sardine tail had been got down, Mrs Cozens rose to her feet, and we went into the drawing-room, leaving the men to enjoy the one good item of the whole menu, excellent vintage port. They only reappeared just before it was time to go home.

LOVE IN A COLD CLIMATE
BY NANCY MITFORD

The bereft Montdores decide to send to Nova Scotia for the distant cousin who will now inherit what Lady Montdore refers to expansively as 'all this'. Instead of the backwoods lumberjack they expect, Cedric Hampton is a cultivated and flamboyant young man whose education has taken place at the hands (as it were) of a number of European aristocrats, and whose beauty (carefully maintained by the liberal application of creams and lotions) is almost a match for Polly's. (His character and habits are based upon the aristocratic wastrel, the Hon. Stephen Tennant.) He shares his uncle's taste in French furniture and his aunt's habit of self-regard, and he revitalises their lives. The furnishings at Hampton are cleaned and restored by a young lorry-driver whom Cedric has picked up, and a similar transformation is effected upon both the appearance and the temperament of Lady Montdore. Even the stuffy country neighbours are won over by Cedric's easy charm. Polly is left money by an aunt and returns from Sicily. Heavily pregnant and clearly bored by Dougdale, she forms an attachment to the young Duke of Paddington, whose family history Dougdale is researching. Her child is stillborn. Cedric takes Dougdale off her hands ('the kind of looks I adore, stocky and with deep attractive furrows all over his face'), returning him to Hampton, where he is happily reunited with Lady Montdore. Polly goes to live with her duke.

John O'HARA   **1905–1970**

A Rage to Live

O'Hara's fourth novel is set in the small Pennsylvania town of Fort Penn, which was clearly drawn from Harrisburg where O'Hara lived at the time. *A Rage to Live* opens on the Fourth of July 1917 at the Caldwell Farm, home to Grace Caldwell Tate and her husband, Sidney. They are amongst the richest and most prominent of Fort Penn's citizens and are playing host to the whole town to celebrate Independence Day. O'Hara inserts a huge flashback into this scene which describes the history of the Caldwell (now Tate) family from 1883. Grace's birth, adolescence and marriage are chronicled, intercut with scenes, passages of dialogue and events which amount to a history of the town. Sidney Tate's inherited wealth and New York background are contrasted with the more established lineage and even greater wealth of the Caldwells. The crucial event of this long, retrospective section is Grace's affair with Roger Banon, a building contractor, which has led to the Tates' estrangement. Their marriage is now one of convenience only and when the novel's action finally returns to the Independence Day festivities, the two are revealed to be playing a marital role for the benefit of the townspeople. The remainder of the novel tells how the marriage deteriorates further

before both Sidney and the couple's son die of polio. Grace pursues other affairs with her social inferiors despite the disapproval of Fort Penn's citizens. New money moving in to usurp the position of the old parallels Grace's fall in the town's esteem and the major part of *A Rage to Live* ends with her all but ostracised. A further, brief, section finds Grace thirty years on in 1949 living in New York. Her family is split by racial and financial tensions and she herself is involved in an unsatisfactory affair with a married man. The Caldwell estate has dwindled to a Manhattan apartment.

At nearly 600 pages, *A Rage to Live* is almost certainly too long. We learn far more than we need about the historical minutiae of life in Fort Penn, and some reviewers suggested that the book should have ended with Sidney's death. O'Hara's aim is to identify Grace's fortunes with those of her nation and this explains in part the huge breadth of his canvas. Many passages are slackly written and much of the characterisation (the vivid, amoral Grace excepted) is flat. The novel's style is uniformly realistic and its values undisguised: social forms should be maintained, not altered, material wealth is a valid index of human worth and the break-up of 'the Family' is synecdochal of failings within American society as a whole. O'Hara himself said of his books: 'Everything I have written since 1948 has had a secondary purpose; I have deliberately attempted to record the first half of the century in fictional forms but with the quasi-historical effect that, say, Dickens achieved.'

A Rage to Live marked O'Hara's return to the novel form eleven years after *Hope of Heaven*. The book marked a watershed in his career by far exceeding his earlier works in both scope and scale, and those that followed were to be similarly expansive. On publication, the novel was savaged by the critics, who derided its socio-historical pretensions; but it sold over 100,000 copies in its first two months. In retrospect it becomes clear that *A Rage to Live* provided O'Hara, his critics and his readers with something like a model. Almost without exception, O'Hara's subsequent novels were to meet with critical failure – and huge commercial success.

Olivia

This autobiographical, first-person narrative (whose author was later revealed as Dorothy Bussy, Lytton Strachey's sister) conveys, in just one hundred pages, the rapture, pain, confusion and terror experienced by the sixteen-year-old Olivia when she falls in love for the first time. Events which in reality occurred in southern England are fictionally transposed to Les Avons, a girls' school just outside Paris, in the last years of the nineteenth century.

Mlle Julie, whom Olivia loves, is one of two principals; Mlle Cara the other. Mlle Julie's capital, intellectual brilliance and magnificent teaching, combined with Mlle Cara's entrancing social skills, have made their school celebrated. Moreover, for more than a dozen years the deep, palpable happiness radiating from the two women's personal relationship has proved irresistible to teachers, pupils and parents alike – until the advent, some three years before Olivia's narrative begins, of Fr. Riesener, the German teacher. Deftly she engineers dissension, playing on Mlle Cara's jealousy, encouraging invalidism and hysteria.

Olivia enters a school where pupils and staff are either Julie-ites or Cara-ites, and where little-understood emotions run dangerously high. It is Olivia herself who, unwittingly, helps bring matters to a tragic climax. Her love for Mlle Julie grows rapidly. No mere 'schoolgirl crush', its power is felt and answered by Mlle Julie who battles with herself to keep the relationship safely contained with professional, pupil–teacher bounds.

Against a background of Mlle Cara's steadily mounting hysteria and denunciation, skilfully exacerbated by the ever-helpful Fr. Riesener, Mlle Julie struggles to decide what course of action is kindest, fairest and most honourable for herself, for Mlle Cara and for Olivia. Unable to resume the old relationship with Mlle Cara, yet denying herself the right to act upon her feelings for Olivia, Mlle Julie determines to leave her school, emigrate to Canada and start afresh, accompanied by only one teacher, the little Signorina Baietto, an Italian-Jewess eternally and selflessly devoted to Mlle Julie who once saved her and her family from destitution. But just before their departure, Mlle Cara dies from an overdose of chloral, in circumstances never satisfactorily explained: the formal verdict is misadventure, the assumption is suicide, the possibility is murder. Once settled in Canada Mlle Julie refuses to receive letters from Olivia; the only source of news is Signorina Baietto, who writes, some four years later, to tell Olivia of Mlle Julie's death from pneumonia.

Olivia's brevity belies its remarkable power and complexity. Its turbulent events are chronicled in a prose whose classical simplicity and restraint echo that of the great French tragedians whom Mlle Julie takes as her primary teaching texts. Books and their power – to shape, usurp and pre-empt readers' own emotions and experience – are an ever-present force in *Olivia*. So too is the relationship between love and intellectual energy: Olivia's mind stirs fully into life only when Mlle Julie has stirred her heart, an experience shared, to a lesser extent, by all her most successful pupils. 'Love has always been the chief business of my life,' says the narrator, and *Olivia* explores the many faces love may wear: the Signorina's self-obliterating devotion; the transcendental, deep but passionless love of Laura, whom Mlle Julie also loves and who becomes, in adulthood, Olivia's closest friend; Fr. Riesener's coldly monstrous self-love, and the young American Cécile's consuming narcissism; Mlle Cara's love, vitiated by doubt and egotism; Mlle Julie's own anguished battle against the physical desires which she fears corrupt love; and the sexually ignorant, inexperienced Olivia's own perpetual efforts to comprehend needs and emotions for which she has no language, and which even the poets do nothing to explain.

Fittingly, *Olivia* was dedicated 'to the beloved memory of V[irginia]. W[oolf].', and published by her Hogarth Press.

George ORWELL   1903–1950

Nineteen Eighty-Four

By 1984, Orwell's fictional year, the nations have coagulated into three superpowers: Eastasia, Eurasia and Oceania. Each of these blocs is totalitarian, through each adheres to a different doctrine. Oceania's is Ingsoc, a corruption of English Socialism. Overseen by Big Brother and cowed by the Thought Police, Oceanian society comprises the Inner Party, the Outer Party and the proles.

Winston Smith belongs to the Outer Party. He inhabits a dimly recognisable London in a defaced Great Britain (now called Airstrip One). Sharing in neither the privileges of the Inner Party nor the degraded pleasures of the proles, his life is regimented and joyless. His work at the Ministry of Truth has taught him that Ingsoc's version of history is a self-serving falsehood. But Winston is under constant surveillance. He knows that if he betrays the least sign of dissent he will be annihilated. He confines himself, therefore, to recording his hatred of Big Brother in an illicitly purchased album.

Sexual activity for pleasure is a 'thought-crime', but Winston snatches interludes of passion with Julia – outwardly an eager member of the Anti-Sex League – until the two of them are arrested. Winston is taken to the Ministry of Love. There he is tortured. He renounces the truths he had perceived, betrays Julia and accepts the omniscience of the party. Released, he spends his days drinking gin in a seedy cafe. 'He [has] won the victory over himself. He [loves] Big Brother.'

If 6079 Smith W. is an everyman, the spirit of everyman is destructible. Love, loyalty, integrity – a short prescription of pain soon remedies these. Ingsoc is invulnerable. The proles are kept quiescent on beer and porn. The wars in which Oceania is always engaged, far from being a threat, are used to stimulate patriotism. Similarly, the heresy of Goldsteinism provides a focus for collective hatred and justifies the ceaseless persecutions. Language is manipulated; propaganda overwhelms fact; technology eliminates privacy; and 'doublethink' – the simultaneous acceptance of contradictory notions – ensures the stupefaction of the mind.

The techniques of control Orwell describes reflect the methods of Nazism and Stalinism, and the novel's fame rests in part on the precision of its chilling caricature. Its style is stark; at times it approaches the simplicity of fable. Many of its coinages have entered into common currency. It expresses a democratic nightmare and warns of the worst direction of the political will. None the less, as a piece of literature it is flawed. On the one hand, the humanity of its central characters is uncomfortably at odds with the rigorous dehumanisation Ingsoc is supposed to have wrought; on the other, the masses are too co-operative in their own unfreedom to elicit the sympathy which their function in the narrative demands. Thus, enveloped in its pessimism, *Nineteen Eighty-Four* pays a small price for the seamless oppression it so harshly contrives to depict.

William SANSOM   1912–1976

The Body

Sansom made his reputation during the Second World War with stories about his experiences in the London Fire Service. His novels are less well known, apart from *The Body*, his first and best.

Henry Bishop, obsessive botanist, proprietor of two hairdressing shops and narrator of the novel, has been happily married to his wife, Madge, and living with her in a London suburb for twenty years. However, while studying a fly in his garden, he sees a stranger from the next-door lodging-house appearing to spy on Madge in her bath. Immediately, his suspicions of both his wife and the stranger are aroused. Matters are made worse when the man, a garage-owner called Charley Diver, comes round to borrow a screwdriver and succeeds in inveigling the Bishops into having dinner with him that Saturday. The dinner with Diver is torture for Henry: he is humiliated by Diver's practical jokes, is introduced with reluctance to the other lodgers in the house, and ends the evening standing miserably in a lavatory imagining what Madge and Diver may be doing in a nearby bedroom. Over the next few weeks he takes various steps to confirm his belief that they are having an affair: these include arranging to be away from home and arriving back early, taking out one of the next-door lodgers, a young girl called Norma, to pump her for information, and snooping around Diver's flat.

The novel builds to its climax when Henry invites the next-door lodgers, including Diver, to accompany him and his wife on a day's boating on the Thames near London. While the party is in a pub in the evening, Norma, who has not been with them, bursts in with the news that she has won £443 on the football pools (a substantial sum in 1949). A little later, Henry, walking near the pub, sees Diver and Madge exchanging a kiss. This seems the final confirmation of his suspicions. He spends the next day wandering about London in a daze, getting drunk and meditating on his revenge. In the evening he arrives at Diver's flat and smashes a tank of goldfish. He is surprised by Madge

and the landlady, Mrs Lawlor, with the news that Diver and Norma had gone off with the money from the pools win but have suffered a car-crash: Norman is dead and Diver is critically injured. The kiss, Madge reveals, was only a momentary impulse: there never has been any affair between her and Diver. The Bishops' marriage is restored, although it cannot be on the old terms.

The Body is a distinguished novel of obsession and physical disgust: if its plot never finally delivers what it seems to promise, the book's beautiful writing, its effortless evocation of atmosphere and its understanding of the motivation of ordinary people, make it unfailingly compelling and memorable.

C.P. SNOW   **1905–1980**

Time of Hope

see **Strangers and Brothers** (1970)

1950

Korean war begins • Riots in Johannesburg provoked by racial policy • Marshal Voroshilov states USSR possesses the atomic bomb • Britain recognises Israel • Legal Aid comes into force in Britain • Thor Heyerdahl, *The Kon-Tiki Expedition* • Bernard Berenson, *Aesthetics and History* • UN Building, New York, completed • Akira Kurosawa, *Rashomon* • George Orwell, Harold J. Laski, Léon Blum, Jan Christian Smuts and George Bernard Shaw die • Ezra Pound, *Seventy Cantos*

Jocelyn BROOKE   **1908–1966**

The Orchid Trilogy

The Military Orchid (1948)
A Mine of Serpents (1949)
The Goose Cathedral (1950)

Brooke's trilogy made his name and enabled him to buy himself out of the army and devote his life to writing. Combining

elements of autobiography and fiction, the books form an overlapping, non-sequential account of the author's childhood, schooldays and army service, 'sets of variations upon the same or similar thematic material'.

Tracking back and forth through his life, Brooke portrays himself as a precocious and solitary child, pathologically shy of strangers, loathing school and living only for the holidays, when he could ramble in the Kentish countryside in pursuit of rare plants. He endures an appalling prep school and runs away twice from his public school before settling down at the progressive Bedales, where botany is taken seriously. He goes to Oxford and cultivates a literary persona, later exposed as bogus, sees service as a 'pox-wallah' in the RAMC during the Second World War, then re-enlists in the ranks after demobilisation, trusting the army to provide some sort of framework for his life.

The narrative proceeds by association, with memories triggered by some sensual experience – a sight, sound or smell – a series of Proustian madeleines that evoke some vivid fragment of the past. He creates his own world, dominated by obscure objects of desire which the child's imagination transforms into totems, notably the elusive plant, spectacular firework and gothic folly which lend their names to each volume. The effect is self-consciously nostalgic, a lament for unobtainable, because evanescent, happiness. Along with Gerard's *Herball* and other botanical texts, the works of Firbank, Beatrix Potter, Proust and Housman are a constant source of reference. The bloom that withers as it is picked and the firework which achieves its glory at its moment of destruction are potent metaphors for transience, and, more particularly, the *paradis perdu* of early childhood. The 'Goose Cathedral', a lifeboat-house which looks like a chapel and later becomes tea-rooms, is a reproachful symbol of Brooke's own 'irretrievably bogus' persona of a young *littérateur* who will never write his projected Proustian masterpiece.

The curious sense of other worlds, populated by sinister or absurd people, is emphasised by the eccentrics whose lives impinge upon that of the narrator. The irresistible rise of Bert, picked up whilst a young soldier and employed as a valet by prim, elderly 'Pussy' Wilkinson, punctuates the narrative. Whisked off and wed by Pussy's sister, Moira Wemyss, who leaves him her considerable fortune, Bert keeps appearing in Brooke's life, fatter, fruitier and more prosperous, as he rises through the ranks of the airforce, ending up as an unlikely squadron leader. There is also Hew Dallas, an Oxford aesthete and 'pyrotechnophil', who turns up in a number of guises, and Basil Medlicott, the object of the child Brooke's hero-worship, who is gradually revealed as a preposterous fantasist.

Entirely *sui generis*, *The Orchid Trilogy* combines *Bildungsroman*, travel writing, literary criticism and botany to form a vivid account of one man's life and obsessions.

Jocelyn BROOKE   **1908–1966**

The Image of a Drawn Sword

Brooke's best-known book after **The Orchid Trilogy** (1948–50) is a mysterious and atmospheric novel which depicts the gradual surrender of a repressed and disaffected young bank-clerk to the confusing world of military authority. It is, perhaps, psychologically (rather than factually) autobiographical, going some way to explaining Brooke's own unlikely career as a soldier. In the unexpected encounter with Roy Archer, a friendly young officer, Reynard Langrish sees a chance of escape from the dreary predictability of his life. Almost unwittingly, he agrees to undertake a course of military training in preparation for some unspecified but imminent 'state of emergency'. He is prevented by influenza from reporting for enlistment, and his life returns to normal until he is arrested by soldiers while out walking. Accused of evading military service, he finds himself unable to communicate with those, like Archer, who are in authority, and he enters a terrifying spiral of disorientation, violence and death.

Although Reynard has the image of the title tattooed upon his forearm as a mark of his acceptance of a new life, he none the less determines to break bounds and visit his mother. He is arrested by military police and sentenced by someone who may or may not be Archer to corporal punishment and detention. He escapes and eventually rediscovers his old house, neglected, decayed and containing the remains of his mother. He is discovered there by Archer, whom he shoots dead, but whose last orders (to return to the army camp with a message that the war has started) he obeys.

Brooke's native Kent is transformed into a shifting and insubstantial world, suspended in time and space, a place of sudden mists, eerie bugle calls and the drone of unseen aircraft. Whether viewed as a Kafkaesque nightmare or as a disturbing allegory of sexual guilt, the novel is both haunted and haunting.

William COOPER 1910–

Scenes from Provincial Life
1950
Scenes from Metropolitan Life
1982
Scenes from Married Life
1961

H.S. Hoff's wordly and funny trilogy concerning Joe Lunn was written under the pseudonym by which the author subsequently became known. *Provincial Life* paved the way for Amis's **Lucky Jim** (1954) and a whole new wave of post-war British fiction. The novel, as its assertive title suggests, is set in an unnamed provincial town where the streets are empty at 6 p.m. because 'everyone was having an evening meal'. Set in 1939, it is narrated in the first person by its anti-hero, Joe Lunn, an irreverent and amoral physics teacher in a run-down grammar school. The trilogy follows Joe's enjoyable, but somewhat caddish, involvements with a series of women, beginning with Myrtle, a successful commercial artist. She wants to get married, but Joe is also an egocentric novelist who refuses to be 'tamed', particularly by someone who does not sufficiently appreciate his work. They spend blissful, but ultimately fruitless, weekends in a cottage which Joe rents for this purpose, sharing the rent with his friend Tom, who uses it as a venue for his own affair with a vain seventeen-year-old called Steve.

Joe and Tom, advised by an older friend called Robert, who is Dean of a small Oxford college, are planning to become 'prospective political refugees', fleeing to America. Tom, an irascible, Jewish chartered accountant who has also written a novel, claims to understand the human heart in a way Joe does not, and is forever lecturing his friend upon how he should conduct his professional and personal life, whilst his own remains unstable. He even contemplates marrying Myrtle, during a period when Steve is becoming difficult and discovering girls. The novel ends with a look into the future in which all the characters, except Joe and Robert, are married.

Although set in 1946, and written in the 1950s, *Metropolitan Life* was not published until 1982, for fear of a libel action. Joe has moved to London, where he is working in 'one of the slums of the Civil Service', a department concerned with scientific research. Robert is a senior colleague and both men spend most of their time commiserating with each other about their romantic misadventures and looking forward to 'the day of liberation through the art of letters'. Joe is still involved with Myrtle, although she has since married his provincial rival, a journalist called Haxby who has yet to be demobilised. The tables have turned, since it is Joe who now wants to marry, but is unable to tell Myrtle the melodramatic untruth: 'I can't live without you.' Haxby, when informed of his wife's request for a divorce, has no such qualms. 'The old battle of wills was on again,' Joe reflects, 'as fiercely and as lastingly as ever.' Meanwhile, Robert is considering marriage to Julia, a highly sexed, hard-drinking, but beautiful young woman who may or may not be married to a Pole, whose surname she has legally adopted. These prevarications are set against a background of the absurd bureaucracy and Machiavellian manoeuvres of the Civil Service. The novel ends with the two men once again womanless, but determined to leave their jobs and start new novels.

At the beginning of *Married Life*, set in 1949, Joe is involved in a desultory and long-standing affair with a beautiful but myopic librarian from his home town called Sybil. Although fond of her, he is beginning to embark upon a search for a wife. Meanwhile, Robert marries Annette, a much younger woman who is the daughter of a senior civil servant. At a party given by an old schoolfriend, the meddlesome Harry, Joe meets a young teacher called Elspeth. His reaction is commonplace but immediate: 'This one's the right one for me.' He gently disentangles himself from Sybil and embarks upon a relationship with Elspeth, steeling himself to propose ('Marriage, I thought ... the very word was like a knell'). His proposal is accepted, and he soon realises that in all his other relationships there has been 'a clash of wills', but that 'MARRIED LIFE IS WONDERFUL', pacific and rewarding. Unfortunately, the same cannot be said of his professional life: his job in the Civil Service is endangered by two disapproving superiors, Spinks and Murray-Hamilton, and his latest novel has to be rewritten to avoid a prosecution for obscene libel. The original reply to his application for the Service had been headed 'MISC/INEL', standing for Miscellaneous/Ineligible. 'That', Joe reflects at the end of the trilogy, 'was the theme of my life. MISC/INEL, trying to be EL. In this simple statement were embodied the poetry, the dynamism, the suffering of one man's existence.'

A companion volume to the trilogy depicts the characters in *Scenes from Later Life* (1983).

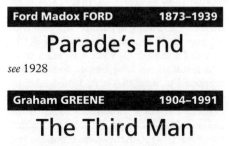

Ford Madox FORD 1873–1939

Parade's End

see 1928

Graham GREENE 1904–1991

The Third Man

'*The Third Man* was never written to be read but only to be seen,' Graham Greene explains in his preface to the novel: it was a means of developing the characters and plot which formed his screenplay for Carol Reed's classic film (1949). Told from the point of view of Calloway, a Scotland Yard officer, *The Third Man* is set in post-war Vienna, where the 'peacekeeping' pact between the Western Allies and the Russians is dissolving into suspicion and rivalry.

'I have reconstructed the affair as best I can,' Calloway begins; 'an ugly story if you leave out the girl: grim and sad and unrelieved, if it were not for that absurd episode of the British Council lecturer.' Rollo Martins, a writer of Wild West yarns, has arrived in Vienna to stay with Harry Lime; but Harry has been killed in a street accident, and as Martins dashes across the city just in time to see the coffin being buried, he realises that he has stumbled into a plot.

All of the coincidences seem to point to murder. Harry's own chauffeur was driving the car which hit him, his doctor appeared on the scene, apparently strolling by, and the caretaker of the block of flats, Herr Koch, reveals that a third man was present when the body was moved – a statement which rapidly leads to Koch's own death. Martins begins to enquire further, and becomes involved with Anna, Harry's girl. Waiting outside her flat one night, he catches a glimpse of what he thinks is Harry's face, and pursues the fleeing figure until it vanishes into thin air by the side of an advertisement kiosk in a market-place. Telling Calloway of these events, Martins believes he has seen a ghost, but the policeman reveals that his old friend was a criminal, and the place where the figure vanished was the entry to the city's sewers, an underground network used by army deserters and gangsters.

Martins, meanwhile, is becoming caught up in his own sub-plot, a sinister intrigue which turns into a farce of mistaken identity. A black car whisks him away from his hotel and delivers him to the British Council, where a lecturer has confused Martins with a famous English experimental writer, and makes him sign copies of a novel he has not written.

Martins goes to meet Harry by the ferris wheel of a funfair, and as they revolve high above the city he confronts Harry with the truth. Harry had been selling watered-down penicillin for profit, and the sick children who were treated with the drug have been left insane and incurable. Harry shows not remorse, and Martins determines to bring him to justice, finally having to accept the fact that his old friend is evil. Calloway sets up a further meeting and lures Harry down into the sewers where, amid the rushing water in the mysterious half-light, Martins

Martins kills him. At the second funeral, in which Harry is now genuinely buried, Martins and Anna are at last free of their delusions about his character, and walk off arm in arm, into the start of another story.

| C.S. LEWIS | 1898–1963 |

The Lion, the Witch and the Wardrobe

see **The Chronicles of Narnia** (1956)

| Mervyn PEAKE | 1911–1968 |

Gormenghast

see **The Gormenghast Trilogy** (1959)

| Barbara PYM | 1913–1980 |

Some Tame Gazelle

'The new curate seemed quite a nice young man, but what a pity it was that his combinations showed, tucked carelessly into his socks, when he sat down.' In the first sentence of Pym's first published novel, the richly comic tone of her whole *œuvre* is set. Begun soon after she went down from Oxford in 1934, but only published, in a revised draft, in 1950, *Some Tame Gazelle* stands slightly apart from the rest of Pym's work. It is set in the type of timeless English village familiar to readers since the days of Oliver Goldsmith and Mary Russell Mitford, and is specifically a *roman-à-clef*: the two chief characters, Harriet and Belinda Bede, are based on Pym's sister and herself, and the sentimental, literary people surrounding them are their undergraduate friends imagined, with characteristic originality and poignancy, as they would be when middle-aged.

Harriet and Belinda are spinsters and gentlewomen in their early fifties. The plump, exuberant Harriet ('radiant in flowered voile') dotes on young curates. The more retiring and sensitive Belinda ('Belinda took out her blue marocain, a rather dim dress of the kind known as "semi-evening"') gains emotional satisfaction from a devotion

to the 'greater English poets' and her long unrequited love of thirty years for the local archdeacon, Henry Hoccleve, who is married to Agatha, the dominating daughter of a bishop.

Agatha goes to Karlsbad for a cure ('She says she is getting on quite well speaking Anglo-Saxon and Old High German in the shops'), which offers Belinda the chance to enjoy some chaste time in the Archdeacon's company, as he recites to her from Edward Young's *Night Thoughts*. An old friend of Belinda, a university librarian called Dr Nicholas Parnell, and his deputy, Nathaniel Mold, visit the village. Mold, a ladies' man but not quite a gentleman, proposes to Harriet, who, despite her initial enthusiasm, rejects him. Agatha returns from Karlsbad, bringing with her Theodore Grote 'Mbawama', now bishop of a large African diocese, but once cherished by Harriet when he was a handsome young curate. In unattractive middle-age he looks and sounds like a sheep, but has a high opinion of his own charms, and proposes to Belinda, who, like her sister, declines to exchange the familiar and sentimental satisfactions of her present life for unknown trials.

Marriage is in the offing, however, for the curate, Edgar Donne, who is carried off by Agatha's niece, Olivia Berridge. The book ends with the wedding ceremony and the curate's impending departure, but also with the arrival of a new curate for Harriet to cherish ('although his voice was rather weak as a result of his long illness, Belinda was overjoyed to hear that it had the authentic ring'). Harriet and Belinda need 'something to love', but yet not too much: such is the emotional character of this novel, in which the characters retain the warm emotions and literary enthusiasms of undergraduates, yet have the knowledge of personal limitation that comes to those who have lived.

| Evelyn WAUGH | 1903–1966 |

Helena

Waugh's own favourite, but least popular, novel started out as a straight hagiography of St Helena, but was soon recast in the form of historical fiction. His researches included extensive reading, numerous letters to Robert Henriques on Jewish matters, and 'a very interesting correspondence with Mrs Betjeman about horses & sex'. Waugh takes the bare outline of what is known of Helena's quest for the True Cross and manages to create a short but dense and vividly imagined account of her life.

Helena, the daughter of 'Old' King Coel, is first encountered as a spirited, boyish and well-educated adolescent. She identifies with her Grecian namesake and is mad about horses (Waugh once described the novel as 'a very beautiful book about Penelope Betjeman's early sex life'). She meets Constantius, who marries her and takes her away from Colchester to Ratisbon, on the Danube. He leaves her there whilst he goes campaigning, telling her that he expects daily to be called to Rome and made Emperor. History unfolds, brought to Helena by messengers and rumour, and the years pass without her becoming an empress. When Constantius is finally given power, he divorces Helena and remarries, leaving her to create and manage an estate.

She only becomes (Dowager) Empress when her son, Constantine, comes to power. From the sidelines she watches developments in politics and religion – battles, assassinations and martyrdoms – horrified at the prospect of 'a whole world possessed of Power without Grace'. She is baptised as a Christian and, despite her considerable age, sets out on a quest to find the True Cross: 'Just at this moment when everyone is forgetting it and chattering about the hypostatic union, there's a solid chunk of wood waiting for them to have their silly heads knocked against. I'm going off to find it.' She does so, aided by a dream in which the Wandering Jew takes her to the cistern where the wooden pieces have lain hidden.

Waugh rationalises the numerous legends (to the saint's advantage: for example, no unfortunate elderly rabbi is thrust down a well) and creates believable characters from these ancient figures, notably the vain and vulgar Constantine ('a shit', Waugh asserted), his appalling second wife, Fausta ('Empress of the world; like a doll floating on the water where a ship had foundered'), and the indomitable Helena herself. He regarded

it as 'the best written' of his novels, and certain scenes (the death of Fausta and Helena's prayer to the Magi) are superb, but imperial purple occasionally strays from the plot into the prose.

Antonia WHITE   **1899–1980**

The Lost Traveller

see **Frost in May** (1933)

Tennessee WILLIAMS   **1911–1983**

The Roman Spring of Mrs Stone

Williams, who began by writing stories, had always been interested in the medium of prose fiction; this short novel, is his masterpiece in the form. It is set in post-war Europe, not the American Deep South of the plays; the theme of an innocent American woman brought to confusion by corrupt Europeans has echoes of Henry James, but Williams takes it to limits that would have made 'the Old Pretender' shudder. 'Rome is three thousand years old, and how old are you? Fifty?' The gigolo Paolo throws this at Mrs Stone at a terrible moment towards the end of the book. The cruelty of old Europe proves altogether too much for her.

Mrs Stone, once a famous stage actress, has lost her looks, abandoned her profession and been recently widowed. She has come to Rome to escape the penalties of her past glory, taking an apartment in an ancient palazzo that flanked 'the immense cascade of stone that descended from Trinità di Monte to the Piazza di Spagna'. On the whole she is good at escaping the American tourists who recognise her, but she cannot get away from the persistent Miss Bishop, a 'woman-journalist' who prides herself on her clarity of mind and whose clothes give her 'a rather shocking transvestite appearance'. As girls they had sometimes 'shared a bed in the dormitory of an eastern college'. The encounter is brief but telling. Having been exposed to Mrs Stone's Roman friends, Miss Bishop proclaims herself 'shocked and revolted at what you seem to be doing with yourself'.

If Miss Bishop had been less awful, Mrs Stone might have listened to her – though the seeds of corruption are within her, as seen from two episodes in her past. However, she has already fallen under the spell of the 'Contessa', who procures beautiful young men for wealthy American women of a certain age. Through her, Mrs Stone meets Paolo, who dubs her 'the wealthiest of the lot and the only one whose interest in him appears to be rooted in something deeper than concupiscence'. Nevertheless her great white bed does yawn. She gives herself to him and though he never gets from her the ten million lire he is seeking, he is still able to destroy her, as he has previously destroyed a certain 'Signora' Coogan on Capri. 'Paolo is by way of being a little *marchetta*!' the Contessa tells her when, for a time, the two bloodsuckers fall out. 'That is one word for a boy who has no work and no money but lives very well without them.'

Mrs Stone is followed everywhere by an uncouth youth with 'the sort of beauty that is celebrated by the heroic male sculptors in the fountains of Rome'. He is also distinguished by the 'dreadful poverty of his clothes and his stealth of manner'. He appears at the beginning of the story, and he is there at the end, but he is never named. He is observed by Miss Bishop 'about to make water against the wall' of the palazzo. Later Mrs Stone discovers him doing the same thing against the window of the shop where she is buying suits for Paolo. On occasions he exposes himself and makes rude signs to her. After the last terrible scene in which Paolo abuses her in front of his Roman friends before going off with a film star, she is '*drifting, drifting*'; her bed is 'a landscape of snow, a stretch of pure desolation', as she signals to the youth waiting outside. 'In a few minutes now the nothingness would be interrupted, the awful vacancy would be entered by something.' Awaiting the final degradation, Mrs Stone, for whom 'plays have been written', may well be left pondering, as she had earlier, on how 'she had arrived where she was'. She might also have remembered a remark that Paolo had once thrown at her: 'I don't suppose it has occurred to you that women of your kind are often found murdered in bed?'

All this is seen against the background of Mrs Stone's marriage and her stage career; the restlessness of post-war Rome, with its rich tourists crying out to be preyed upon, is brilliantly conveyed. The verdict of Williams's friend Carson McCullers is just: 'It stands as a work of art with *Daisy Miller* and *Death in Venice*. There is in this book the hallmark of the masterpiece.'

1951

Harry S. Truman relieves General Douglas MacArthur of command in Far East • Éamon de Valéra returns to power in Eire • Guy Burgess and Donald Maclean flee to USSR • US–Philippines mutual defence pact • Electric power produced from atomic energy • David Riesman, *The Lonely Crowd* • Igor Stravinsky, *The Rake's Progress* (libretto by W.H. Auden and Chester Kallman • Rodgers and Hammerstein, *South Pacific* (musical) • John Huston, *The African Queen* • André Gide and Arnold Schönberg die • Robert Frost, *Complete Poems* • Marianne Moore, *Collected Poems*

James JONES   1921–1977

From Here to Eternity

Jones's first novel shocked the American public, for it provides a brutal and realistic account of life in the United States Army. Set in the fictitious Schofield Barracks in Hawaii, it draws on many of Jones's own experiences at the time of the bombing of Pearl Harbor.

Robert E. Lee Prewitt ('Prew') has resigned his post as a regimental bugler and determines to return to the ranks. But the peacetime bravado of wrestling championships and false honours is frustrating to the soldiers, and bitter feuds develop. Prew's Top Sergeant, the tough-minded but humane Milton Warden, sees Prew's rejection of promotion as an arrogant display of independence; but as the two men clash in their professional rivalry, they come to develop mutual respect.

Warden begins an affair with his Captain's wife, the lonely Karen Holmes, whose contacts with the army's top brass offer him the chance of taking a commission. Prew's prickly personality, meanwhile, lands him in the Stockade, a cruel army prison, where he meets up again with Maggio, an Italian youth whose gentle nature has been transformed into tough masculinity by the harsh treatment he has received. On his brief release from prison, Prew gets into a fight with a belligerent fellow soldier and takes his revenge against injustice by murdering him. On the run from further punishment, Prew takes refuge with his mistress, Lorene, a prostitute in Mrs Kipfer's establishment.

With the Japanese attack, however, Prew determines to rejoin his regiment and play his part in the war, a hopeless gesture since his papers reveal his guilt. As the Military Police close in on him, he forces them to shoot him, maintaining his own integrity rather than surrender and go back to the Stockade. In his pocket are the lyrics of a song he has composed with his mates, 'The

---

It was perhaps the stringing of the barbed wire, more than anything else, that ate into the patriotism of the troops in the next few days. The men who had acquired the new unknown disease of aching veins in their arm joints from the building of these positions now found it coming back on them doubly powerfully from putting up barbed wire to protect these positions. So that even when they were not pulling guard at night, they couldn't sleep anyway. The stringing of the barbed wire, after the first day, was an even more powerful astringent to the patriotism 'than their getting crummy with no prospect of a shower, or their getting itchy with beard and no prospect of a shave, or their having to sleep on the rocks with nothing but a single shelterhalf and two blankets over them when it rained.

FROM HERE TO ETERNITY
BY JAMES JONES

Re-Enlistment Blues'. Warden finds the lyrics and arranges for Prew to be buried honourably, and in a testament to GI solidarity decides to turn down his own commission.

The year after the novel was first published, it was made into a hugely successful film by Fred Zinnemann from a screenplay by Daniel Taradash. Chiefly remembered for a scene in which Burt Lancaster and Deborah Kerr roll around in the surf, it won a large number of Academy Awards.

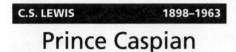

C.S. LEWIS   **1898–1963**

Prince Caspian

see **The Chronicles of Narnia** (1956)

Carson McCULLERS   **1919–1967**

The Ballad of the Sad Café

The Ballad of the Sad Café is the most representative of McCullers's major works. As with many of her novels, it is set in a small, bleak southern town which survives rather than thrives, and which is inhabited by a population of freaks and grotesques locked in physical and spiritual isolation. To this barren world invariably comes love, and McCuller's novels detail love's inevitable thwarting and failure, while acknowledging its power and grace. Without love, the individual and the community disintegrate before the pressures of social, economic, racial and emotional injustice. Yet love also carries with its flowering, the seeds of its own destruction.

The Ballad of the Sad Café depicts a simple emotional triangle: Miss Amelia, the gaunt and masculine owner of a small-town store, becomes obsessed with Cousin Lymon, a vain and malicious hunchback, who transforms her store into a café. Cousin Lymon, in turn, becomes obsessed with Miss Amelia's formerly adoring, and now vengefully returning, husband, the 'bold and fearless and cruel' Marvin Macy. The novel is a meditation on the contradictory nature of love, and on the tragedy of the fact that 'in a deep and secret way, the state of being loved is intolerable to many'.

The eponymous café, born of the love of Miss Amelia for Cousin Lymon, represents the transforming power of love. Its existence gives the town a sense of pride and self-esteem; it becomes a place where 'for a few hours at least, the deep bitter knowing that you are not worth much in this world could be laid low'. The café is both created by love and destroyed by love, literally, when it is vandalised by Cousin Lymon and Marvin Macy. With the destruction of the café, Miss Amelia becomes a recluse, now physically as well as spiritually isolated, and the town 'rots with boredom' until it dwindles into 'a place that is far off and estranged from all other places in the world'.

McCullers once defined the essential matter of her fiction: 'Love, and especially love of a person who is incapable of returning it, is at the heart of my selection of grotesque figures – people whose physical incapacity is a symbol of their spiritual incapacity to love or to receive love – their spiritual isolation.' In *The Ballad of the Sad Café* the dual themes of love and isolation emerge more vividly than in any of her other novels, partly because of its shortness, but also because of the manner of its telling. The tale is imbued with the archaic atmosphere of the popular ballad rather than with the political consciousness of the realistic novel. It is, of course, a prose rather than a verse ballad, but it shares many of the characteristics of the form. It is at once dramatic and impersonal, melancholy and heroic, folkloric and literary; and it culminates in an epic battle worthy of the genre. This poetic quality binds the novel together with hypnotic effect, and the envoi to the piece, 'The Twelve Mortal Men', succinctly defines the story's peculiar success: 'It is music that causes the heart to broaden and the listener to grow cold with ecstasy and fright.'

In 1963 the novel was adapted for the stage by Edward Albee, and in 1990 it was made into a successful film, directed by Simon Callow from a screenplay by Michael Hirst.

Olivia MANNING   **1908–1980**

School for Love

A subtle and touching novel, *School for Love* is set in Jerusalem during the earlier months of 1945. Felix Latimer, a young English boy on the verge of adolescence, comes here from Iraq, where both his parents have died; he is particularly mourning his mother, to whom he has been extremely attached. He is to stay at the house of an English spinster, Miss Bohun, who was once an adopted child of his grandparents. Felix is one of the most brilliantly rendered characters in modern English fiction: the feelings of a sensitive and warmhearted boy, as well as his occasional sillinesses, are captured by Olivia Manning with great sympathy and without sentimentality.

Miss Bohun is another complex character: a loveless woman who dominates a small religious community called the 'Ever-Rea-dies', she wants to do well by Felix but is handicapped by her eccentricity and extreme meanness. Felix transfers much of his affection to her small Siamese cat Faro, and fascinatedly observes the other lodgers in the house: the Lesznos, a difficult Jewish mother and son from Poland, and an old man, Mr Jewel.

Miss Bohun takes in another lodger, Mrs Ellis, a young woman stranded in wartime Palestine, whose husband has been killed, and who turns out to be expecting a baby. Miss Bohun is initially enthusiastic about letting to Mrs Ellis, but the relationship between the two women soon sours. Felix, however, develops a strong attachment to Mrs Ellis as soon as he meets her and they become friends, although she sometimes treats him only with toleration and his alien-ated feelings then turn back to the cat Faro. The year goes on, the war in Europe ends, and there is tension in Palestine as the country prepares for its own war between Arabs and Jews.

Towards the end of the novel, Felix is about to be given a passage back to England. Miss Bohun has turned out Mr Jewel and now wishes to evict Mrs Ellis. She has also given Faro away: she is unable to prevent herself alienating those she has initially tried to help. Miss Bohun and Mrs Ellis have a violent argument, during which the latter falls down the stairs of the house, and loses her baby. Felix abducts Faro and gains Miss Bohun's reluctant permission to take the cat home to England with him. The novel ends with Mr Jewel (for whom Felix has won a substantial inheritance, by writing to England on his behalf) being invited back to the house by Miss Bohun: it seems that these two castaways are to find consolation in an eventual marriage.

Anthony POWELL   1905–

A Question of Upbringing

see **A Dance to the Music of Time** (1975)

J.D. SALINGER   1919–

The Catcher in the Rye

Salinger's first novel is *the* archetypal tale of American adolescent angst and disaffection. The sixteen-year-old narrator, Holden Caulfield, has been 'flunked out' of yet another expensive private school and is depressed by the prospect of returning to New York to confront his parents (about whom we are told none of 'that David Copperfield kind of crap'). On the Saturday evening before the start of the Christmas vacation, Holden takes the train to New York and checks into a hotel. Returning from a club, he agrees to the elevator boy sending a hooker up to his room, then finds he is too depressed to take the opportunity to lose his virginity. On Sunday he meets a former date, Sally Hayes, decides he is in love with her even though she is stupid, suggests they elope together, then calls her 'a right royal pain in the ass'. That evening he gets drunk, runs out of money, walks round the frozen park until he thinks he is going to die of pneumonia, sneaks into his parents' apart-ment to talk to his kid sister, Phoebe, and goes to stay with an old teacher, Mr Antolini, who makes a pass at him. More confused than ever, he flees in the early hours of the morning and resolves to hitch-hike out west, pretend to be a deaf-mute so he never has to talk to anyone again, get a job at a gas station, and build a log cabin in the woods. He goes to say goodbye to Phoebe at her school only to find her hauling a suitcase along having decided to come with him. She stubbornly refuses to desist from this plan, and he is forced to take her home. His parents send him off for psychiatric treatment in a clinic, which is where he is now writing.

The show wasn't as bad as some I've seen. It was on the crappy side, though. It was about five hundred thousand years in the life of this one old couple. It starts out when they're young and all, and the girl's parents don't want her to marry the boy, but she married him anyway. Then they keep getting older and older. The husband goes to war, and the wife has this brother that's a drunkard. I couldn't get very interested.

THE CATCHER IN THE RYE
BY J.D. SALINGER

Holden's reflections on life dwell on all that is 'phoney', 'corny' and 'old bull': his brother D.B., a writer who is prostituting himself in Hollywood; 'all those Ivy League bastards' who sip cocktails and have pretentious conversations in loud voices; and above all, the movies. In his naïve and desperate way he is searching for anything which is innocent and sincere. He is possessed by the memory of his younger brother Allie, who died of leukemia, and never fails to be 'killed' by the innocent antics of children. The title of the novel derives from an image (inspired by the Robert Burns poem 'Comin' through the Rye') of himself standing on the edge of a cliff catching the playing children who fall from a field of rye: 'That's all I'd do all day. I'd just be the catcher in the rye and all. I know it's crazy, but that's the only thing I'd really like to be.'

The Catcher in the Rye is undoubtedly a twentieth-century classic. It struck a popular note, particularly with young readers, who strongly identified with Holden Caulfield and his yearning for lost innocence. Four years later, the youth of America was to identify similarly with the character played by James Dean in the film *Rebel Without a Cause*. Salinger's novel was, and continues to be, a phenomenal success, the author's reclusiveness contributing significantly to the cult aura which surrounds his work. Some adult readers have found the book sentimental, and while there is no denying Salinger's skill as a ventriloquist of adolescence, the pervasive narcissism of the novel (and, indeed, of much of Salinger's fiction) is not to everyone's taste. Whatever its detractors may say, however, *The Catcher in the Rye* has secured a place in the hearts of successive generations, and no amount of critical reservations seem likely to dislodge it.

C.P. SNOW   **1905–1980**

The Masters

see **Strangers and Brothers** (1970)

1952

Eisenhower elected US President ● Accession of Queen Elizabeth II ● Chinese delegates arrive in Moscow ● Arab League Security Pact comes into force ● State of emergency in Kenya because of Mau Mau disturbances ● Contraceptive tablets made ● The first H-bomb ● 'Smog' in London ● Agatha Christie, *The Mousetrap* ● Eva Perón and Chaim Weizmann die ● David Jones, *The Anathemata* ● Dylan Thomas, *Collected Poems 1934–52*

Ralph ELLISON   **1914–1994**

Invisible Man

Set against a background of racial prejudice, *Invisible Man* revolves around the themes of 'seeing' and 'not seeing', of light and darkness. It is the story of a black man whose experiences lead him to the conclusion that he is 'invisible': white people do not see *him*, they see a black man. He is also invisible as an individual because he has let other people see him merely as a vehicle by which to express their ideologies.

The 'invisible' man tells the story from his home, a dark New York basement into which he was hounded during a race-riot, and which he has wired with 1,369 stolen light bulbs, illuminating both the room and his understanding of himself. He cites three main instances of what he now sees as his allowing other people to force their versions of reality upon him. Expelled from a state college for Negroes for unconsciously tampering with its spurious racial Utopia, he begins to understand racial prejudice after working for Liberty Paints, a company which bears the slogan 'If it's Optic White, it's the Right White'. White being right, and black being unimportant, he does not appreciate the extent of his 'invisibility' until he has been working for some time for the 'Brotherhood', an underground political organisation supposedly supporting the poor and oppressed of all races. He believes that he is being accepted as an individual, when he is merely a mouthpiece for the Brotherhood ideology.

His confusion over individuality is illustrated in a scene in which he is repeatedly mistaken for someone else; each mistaken recognition being of a different facet of the man's personality. Betrayed by the Brotherhood, the 'invisible man' realises that he has to distance himself from other people in order to see and understand himself. This he

---

Then the door banged behind me and I was crushed against a huge woman in black who shook her head and smiled while I stared with horror at a large mole that arose out of the oily whiteness of her skin like a black mountain sweeping out of a rainwet plain. And all the while I could feel the rubbery softness of her flesh against the length of my body. I could neither turn sideways nor back away, nor set down my bags. I was trapped, so close that simply by nodding my head, I might have brushed her lips with mine. I wanted desperately to raise my hands to show her that it was against my will. I kept expecting her to scream, until finally the car lurched and I was able to free my left arm. I closed my eyes, holding desperately to my lapel. The car roared and swayed, pressing me hard against her, but when I took a furtive glance around no one was paying me the slightest attention. And even she seemed lost in her own thoughts.

INVISIBLE MAN
BY RALPH ELLISON

---

achieves in the symbolic light of his stolen bulbs, deep in the solitary hideaway from which he decides to emerge and face the world again.

Ambitious in its scope, and impressive in its achievements, *Invisible Man* is considered by many to be the most important novel to emerge from black America. It is the only novel Ellison completed, although he was for many years working upon another one, segments of which have been published.

G.F. GREEN   **1910–1977**

In the Making

'It is the eye of a child that informs the artist,' Green wrote, 'and it is his first view of the world which gives the work of art its newness, its power to compel us into a unique, original and personal planet.' His best novel (much admired by E.M. Forster and others) turns away from his customary landscapes of the industrial cities and the exotic East to that of his own childhood.

'The magic securities of childhood' end for Randal Thane when his older sister, Kit, returns from school and he finds their relationship changed. He embarks upon 'his sentence of perpetual dependence on another, to whom his renunciation would mean nothing, and he enter[s] gladly the prison of another's life'. Sent to prep school, he falls unwillingly under the influence of Felton, a charismatic older boy whose conceit and bravura make him something of a school hero. When Felton befriends him, Randal experiences the lover's sense of exclusivity and apartness, the almost metaphysical creation of a new world, which Green suggests in a series of magical scenes: a snow-muffled half-holiday; drives in the car of Little Willie, a sentimental master who dubs Randal 'Greco' because he resembles 'that cold self-conscious page from sombre Escorial'; and a dazzling Hallowe'en party at which Randal and Felton are dressed as Pierrot and Harlequin. As with Kit, Randal 'never reached the fact that this new world which he shared with Felton did not exist for Felton ... He never discovered that this happiness, keener perhaps than any he would experience again, was derived from his own imagination.' The boys quarrel and Randal gets into trouble at school. The novel ends with Randal travelling to his public school, 'filled with an extraordinary knowledge of himself and the permanence of his life', having sorted out its patterns and Felton's place in them.

This intense, evocative and exquisitely written 'Story of a Childhood' is partly autobiographical, representing Green's own search into his childhood and schooldays (the Quantocks substituting for the Malverns). Unlike Randal, who reads Tennyson's 'Ulysses' to Little Willie in the garden, 'only vaguely touched' by the sense of the line 'I am part of all that I have met', Green recognises what 'makes' his *alter ego*.

Henry GREEN   **1905–1973**

Doting

Green's last two novels, *Nothing* (1950) and *Doting*, disappointed many of his admirers. Comedies of manners, they seemed less rich and challenging than his early work, but they were in fact equally experimental. In a BBC radio broadcast, 'A Novelist to His Readers' (published in *The Listener*, 9 November 1950), Green explained his change of approach:

> Reading is a kind of unspoken communion with print, a silent communion with the symbols which are printed to make up the words; if that is so, then how is the reader's imagination to be fired? For a long time I thought this was best lit by very carefully arranged passages of description. But if I have come to hold, as I do now, that we learn almost everything in life from what is done after a great deal of talk, then it follows that I am beginning to have my doubts about the uses of description. No; communication between the novelist and his reader will tend to be more and more by dialogue ...

Doting is divided into short scenes consisting almost entirely of dialogue, often rounded off with a terse narrative 'curtain': 'She then changed the conversation adroitly and they talked of musical comedy until the time came for her to go back, late, to work.' The dialogue appears to be straightforward – a laconic, slangy, upper-class patois, very much of the 1920s – but is in fact dense with (usually deliberate) obliquities and evasions. 'It's simply this ...' one character begins to explain; but it rarely is, and it is never entirely clear how much each speaker is aware of the real motive of the other.

Set in London in 1949, the novel opens in a night-club, where an attractive middle-aged couple, Arthur and Diana Middleton, have taken their sixteen-year-old son, Peter, on the first evening of the school holidays. Accompanying them is Annabel Paynton, the late-teenage 'favoured daughter of a now disliked old friend', whose evident charms distract Arthur Middleton from the floor-show. Annabel has a penchant for older men and agrees to lunch with Middleton the following week. The Middletons have an agreement whereby each partner is allowed to see friends of the opposite sex unchaperoned, but this sophisticated arrangement is put to the test when Diana returns unexpectedly to their flat to find her husband kneeling at the feet of Annabel, who has taken off her skirt. Middleton's (true) explanation that Annabel had spilt coffee on herself and that they were trying to remove the stain is disbelieved. (He does not admit that the coffee was spilt whilst he was kissing the young woman on the sofa.)

Middleton asks a widowed family friend, Charles Addinsell, to take Diana out and reassure her about her husband's devotion. Charles and Diana so much enjoy their meeting, which ends with some passionate but not very serious embraces, that they begin to see each other regularly. Middleton then introduces Annabel to Addinsell, ostensibly so that his friend can tell the young woman that her admirer is being obliged by his wife to see less of her. The real motive, however, is to use Annabel to lure Addinsell away from Diana. This works all too well, so much so that Diana, whose motives are equally mixed, persuades her husband to warn Annabel off Addinsell. A further complication arises when Diana invites Addinsell and Annabel's plump confidante, Claire Belaine, to join the party for Peter's final evening out before the new term. Determined to scupper whatever schemes Diana is pursuing, Annabel asks Claire to pretend to make a pass at Addinsell. Claire does so in earnest and she and Addinsell embark upon an affair, the only one in the novel which is consummated. Both the Middletons and Annabel have been thwarted in their machinations, and the novel ends with very little resolved: 'The next day they all went on very much the same.'

This deliberately inconsequential last line is characteristic of the novel's approach to its subject. When Annabel tells Middleton that a friend is compiling an anthology of love poetry entitled 'Doting', he replies: 'Well, you know, doting, to me, is not loving ... love must include adoration of course, but if you just dote on a girl you don't necessarily go so far as to love her. Loving goes deeper.' Feelings rarely go deeper in *Doting*, but it is not exactly a satire, for Green scrupulously avoids providing any moral standard. Rather, it is an elaborate game of feint and counterfeint, which the author uses to examine the whole notion of communication between friends, lovers and spouses. The result is a novel more complex than its witty, insouciant manner at first suggests.

Ernest HEMINGWAY   1899–1961

The Old Man and the Sea

This story of an old Cuban fisherman's struggle with a great marlin is perhaps Hemingway's finest achievement in terms of taut narrative, and it won him the Nobel Prize. He lived in Cuba for many years, and his intimate knowledge of deep-sea fishing and the waters of the Gulf Stream give the work absolute authority. This is Hemingway at his most involved.

He was an old man who fished alone in a skiff in the Gulf Stream and had gone eighty-four days now without taking a fish. In the first forty days a boy had been with him. But after forty days without a fish the boy's parents had told him that the old man was now definitely and finally, *salao*, which is the worst form of unlucky, and the boy had gone at their orders into another boat which caught three good fish the first week.

With this opening Hemingway immediately establishes circumstances and theme: the old man's heroic struggle against ill luck. On this, the eighty-fifth day, he hooks a giant marlin ('eighteen feet from nose to tail', as we eventually discover) and 'four hours later the fish was still swimming steadily out to sea, towing the skiff, and the old man was still braced solidly with the line across his back'. He hangs on despite cramp in his left hand, the endurance of both man and fish reaching epic proportion so that a bond is forged between them. 'He is my brother. But I must kill him and keep strong doing it.' When at last he does bring the fish to the surface – 'the speed of the line' cutting his hands – and harpoons him, the fish is so big that the old man cannot bring him on board. The struggle has taken its toll. As he prepares the nooses and rope to lash the fish alongside he says: 'I am a tired old man. But I have killed this fish which is my brother and now I must do the slave work.' He misses the boy's company and assistance now as he has missed him all along.

His moment of triumph is short-lived. He is far out at sea and on the way back the fish is attacked by a succession of sharks. The old man fights them off as best he can but they keep coming until they have eaten all the flesh. The old man is at last forced to accept defeat. 'It is easy when you are beaten, he thought. I never knew how easy it was.' The boy finds him lying in his shack, his hands lacerated; in the night he felt 'something in my chest was broken'. Although the boy tells him, 'you must get well fast for there is much that I can learn and you can teach me everything', one wonders whether the old man will ever fish again.

The Old Man and the Sea is the apotheosis of Hemingway's preoccupation with the primitive. In it he raises to heroic heights the battle of a poor man for survival. The boy tells him, '*He* didn't beat you. Not the fish.' 'No,' the old man replies. 'Truly. It was afterwards.' The sharks deprive him of his livelihood, but not his dignity.

Doris LESSING 1919–

Martha Quest

see **Children of Violence** (1969)

C.S. LEWIS 1898–1963

The Voyage of the *Dawn Treader*

see **The Chronicles of Narnia** (1956)

Flannery O'CONNOR 1925–1964

Wise Blood

O'Connor came from an old Catholic family; but although she is concerned in her work with religion, she is not a Catholic writer in any obvious sense. Rather, she is a Gothic writer, who deals with religious extremism in the Bible-Belt of the Southern states, where the dividing line between the genuinely visionary and the frighteningly insane often seems wafer-thin. She described *Wise Blood*, her first and best known novel (which reached a new audience when successfully filmed by John Huston in 1979), as 'a comic novel about a Christian *malgré lui*'.

Hazel Motes is the grandson of an itinerant preacher who 'had Jesus in him like a sting', and has been brought up to be aware of Christ as a constant presence at the back of his mind. He returns from the war consumed by nihilism and takes a train to Taulkinham, Tennessee, where he baptises himself in sin with Mrs Leonora Watts in 'the friendliest bed in town'. The following evening he encounters a blind preacher, Asa Hawks, and, determined to found his own ministry, inaugurates the Church Without Christ. 'I'm member and preacher to that church where the blind don't see and the lame don't walk and what's dead stays that way,' he tells an indifferent crowd in front of a movie house. 'I'm going to preach that there was no Fall because there was nothing to fall from and no Redemption because there was no Fall and no Judgement because there wasn't the first two. Nothing matters but that Jesus was a liar.' The only person who takes any interest in Haze is Enoch Emery, a disturbed eighteen-year-old who lives in 'two minds' and claims to have 'wise blood like his daddy'.

Haze buys an old car, 'a high rat-coloured machine with large thin wheels and bulging headlights', which he drives round until he discovers where Hawks and his fifteen-year-old daughter, Sabbath, are lodging. He gets a room of his own in the same house, intending to seduce Sabbath: 'He thought that when the blind preacher saw his daughter ruined, he would realize he was in earnest when he said he preached the Church Without Christ.' The degenerate Sabbath proves only too willing, since she wants someone to look after her when her father moves on. Haze discovers that Hawks is a fraud: the preacher had intended to burn his eyes out with quicklime to prove his faith, but his nerve had failed: behind his dark glasses he can see perfectly well. His deceit exposed, Hawks leaves town.

Meanwhile, in a scene of high comedy, Enoch Emery inexpertly disguises himself in order to steal a mummified corpse from the local museum, which, somewhat the worse for wear, is delivered to Haze as 'the new jesus'. Another unwelcome disciple is Hoover Shoats, a man who, as 'Onnie Jay Holy', hopes to make some money from the crowds by claiming that he has been saved by Haze. When Haze refuses to co-operate, the man finds a new 'Prophet', dressed to resemble Haze, with whom he sets up a rival *Holy* Church of Christ Without Christ. Haze follows the 'Prophet' – a pathetic consumptive hoping to earn some money to support his family – and runs him over. He then sets off in search of a new city in which to preach, but is stopped by a patrolman who, when he discovers Haze has no licence, tips the car over an embankment.

Haze makes his way back to town, buys some quicklime, and blinds himself. Sabbath, who 'hadn't counted on no honest-to-Jesus blind man', goes off in search of her father, and Haze is left to the care of his rapacious landlady, who hopes to marry him

in order to get her hands on his army disability pension. When the landlady proposes, Haze wanders out into the night and is found two days later, dying in a ditch. The novels ends as it had opened, with a woman staring into the depths of Haze's unfathomable eyes. Mere sockets now, they seem to 'lead into a dark tunnel where he had disappeared'.

O'Connor makes no judgements and provides no explanations: 'Free will does not mean one will, but many wills conflicting in one man,' she wrote. 'It is a mystery and one which a novel, even a comic novel, can only be asked to deepen.' However, her acute ear for the cadences of speech, her ability to light upon the exact, striking detail, and above all her imaginative empathy with even the most grotesque of her characters, convince the reader of the fictional 'truth' of this essentially Gothic tale.

Anthony POWELL   1905–

A Buyer's Market

see **A Dance to the Music of Time** (1975)

John STEINBECK   1902–1968

East of Eden

'Even as in Biblical times, there were miracles on the earth in those days.' Steinbeck's saga of two families, the Hamiltons and the Trasks, spans the lives of settlers in the Salinas ·Valley through the course of three generations. The author himself appears as the narrator, reconstructing his childhood memories and the traditions of his family from the years of hardship at the end of the nineteenth century into the era of the automobile and Henry Ford's production line.

Samuel Hamilton, Steinbeck's grandfather, is an impoverished but imaginative inventor and water-diviner. Bringing up his many children on a poor homestead, he bequeaths to his offspring qualities of personality which can be beneficial or disastrous. Will has a shrewd head for money-making and becomes a wealthy businessman in the town; Tom, a dreamer like his father, carries on the farm; and Gussie, after an attempt to run a dress-making shop, returns to the family home to die. Sam, meanwhile, has to acknowledge that his life is over, and gives up his struggle to act as the patriarch for a retirement which he finds humiliating. Before he dies, however, his friendship enriches the life of Adam Trask, whose strange career provides the central theme of the novel.

Returning from years of vagrancy and penal servitude in road gangs, Adam finds that his father, a corrupt government employee, has left a fortune to him and his brother Charles. The two men had always been rivals for their father's affection, and they live together now uneasily on the farm, until the arrival of Kathy, the victim of a violent attack. Adam falls in love with her and she agrees to marry him, but he knows nothing of her past.

Kathy is a prostitute and murderess, who has killed her own parents: the narrator's initial impression is that she is 'a monster' devoid of normal human affection. On her wedding night Kathy drugs her husband with sleeping tablets and slips into bed with his brother, recognising in him some demonic streak of nature to which she can respond. When Kathy becomes pregnant, Adam sells his share of the farm to Charles and buys a new homestead in the Salinas Valley where he plans to establish his dynasty: but the family are tainted by the myth of Cain and Abel which is played out through the next generation.

As soon as twin sons are born, Kathy deserts Adam and runs away to a brothel, where her sadistic sexuality gives her a lucrative profession. Adam, destroyed by her cruelty, leaves the care of the twins Aron and Caleb to his Chinese servant Lee, who has struck up a close friendship with Sam Hamilton. The shame of Kathy's actions leads Adam to lie to his sons, but as they grow up they learn the truth, to devastating effect. Kathy worms her way into the affections of the madam, feeding her poison while adopting the role of daughter and being rewarded by inheriting the business. Over the years, Adam gradually regains an urge to live,

encouraged by Sam Hamilton's friendship, and he moves his household to the town of Salinas, where Kathy is now running the brothel.

Just as Adam idolised Kathy, his son Aron idealises Abra, the girl he plans to marry. Caleb, jealous of his brother's good fortune and unfairly excluded from his father's love, tries to take the girl away from Aron. Going into partnership with the businessman Will Hamilton, Caleb makes a fortune from war-time profiteering and tries to buy the love of his family, giving Adam a gift of money which is rejected, as Adam's own present to his father was unjustly preferred. Caleb offers to pay for a college education for his brother, but as Aron becomes obsessed with puritan spirituality, he also refuses Caleb's generosity. To wound him, the more worldly Caleb takes his innocent brother to the brothel, confronting him with the truth, which he cannot accept. Abra, stifled by the role which Aron has assigned to her, and feeling unworthy of his high idealisation, tries to find a way to break off the relationship; her crisis is cruelly solved as Aron, fleeing his own confusion, runs away to join the army and is killed. At the deathbed of Adam Trask, Abra declares her love for Caleb, whose sense of guilt over his past seems to her more human than his brother's spiritual evasion. As Adam dies, he is finally able to absolve his living son from the irrational burden of guilt which has been passed on through the family. The sadistic, murderous Kathy is, Steinbeck concludes, not an aberration of nature, but an extreme example of the darker side of the soul which is present in us all.

The novel was made into a long film (1955), directed by Elia Kazan from a screenplay by Paul Osborn, chiefly memorable for the first starring appearance of James Dean.

Evelyn WAUGH   1903–1966

Men at Arms

see **The Sword of Honour Trilogy** (1961)

Antonia WHITE   1899–1980

The Sugar House

see **Frost in May** (1933)

Angus WILSON   1913–1991

Hemlock and After

Wilson's first novel, written in the wake of two highly praised volumes of short stories, addresses the problems of liberal humanism, a theme which was to preoccupy the author in subsequent books. The novel displays Wilson's gifts for mimicry and high comedy – the opening ceremony at Vardon Hall is an early example of the brilliantly rendered social catastrophes which litter his novels – and his ability to create believable grotesques, notably the monstrous procuress, Mrs Curry. As an embodiment of peculiarly English evil, Mrs Curry is one of the great characters of twentieth-century fiction. The fact that Wilson himself had been blackmailed over a homosexual relationship contributed to his portrait of her criminal activities.

Vardon Hall, a manor house in the Home Counties, has been acquired by Bernard Sands, a successful novelist in his sixties, as a retreat for young writers. For Sands, this is the crowning achievement of his career, and a personal triumph against local opposition and governmental bureaucracy, a victory for the liberal humanism which has directed his life.

Bernard's personal life is less well ordered. His wife, Ella, is suffering from the after-effects of a serious breakdown, and his adult children, the ambitious Tory barrister James, and the hard, cynical journalist Elizabeth, are suffering from the after-effects of a haphazard upbringing in which liberal ideals substituted for strong family affection. Bernard is homosexual, currently involved with a beautiful young bookseller, Eric Craddock, whom he is attempting to set up in London. Bernard's proclivities become known to Mrs Curry, a country neighbour whose 'many interests' include procuring and blackmail. While waiting to meet a former boyfriend in

London, Bernard witnesses the arrest of a man for importuning and is shocked to feel a thrill of sadistic pleasure. He suffers a mild heart-attack.

Bernard looked down from the window of the flat on to Bloomsbury lying embalmed in its Sunday death. The rows of scarlet and lemon dahlias, the heliotrope carefully tended and the little green chairs carefully painted for Festival visitors decorated its stillness with a strange air of smartness – embalmed in best Sunday clothes, no doubt, to accord with the conventions of American visitors. From the ninth floor the sordid remnants of Saturday lost the squalor of greasy paper, refuse ends, dirt and spit which they showed to the pedestrian and merged into a general impression of dust and litter. The district was dead but recently, and these seemed the tag ends of life's encumbrance. But already the corpse was stirring into maggoty life – a paperman distributing the sheets that, crumpled and tired-looking, would add to the stuffy Sunday muddle of bed-sitting rooms and lounges; knots of earnest foreign tourists collected for an 'early start' on hotel steps; disconsolate provincial families already straggling and scratchy at the day before them; a respectable nineteenth-century couple, who Bernard hoped might be Irvingites bent on worship in Gordon Square.

HEMLOCK AND AFTER
BY ANGUS WILSON

The opening ceremony at Vardon Hall is a grand set piece in which the two halves of Bernard's life – the professional and the personal – come together in an atmosphere of mistrust and recrimination. Too many people, insufficient catering and a faulty public address system add to the stresses of the day. Shocked by his recent experience, Bernard delivers a speech which is considered defeatist and inappropriate. Tempers ignite, VIPs are insulted, and assorted skeletons come tumbling out of cupboards.

In the aftermath, Bernard retreats into himself, taking refuge in country walks, and ignoring his London circle. Ella emerges from her neurotic state and attempts to sort out Bernard's personal affairs – in particular the problems of Eric, whose selfish, sentimental mother wants him to stay at home – and urges Bernard to expose Mrs Curry. When Bernard has a fatal heart-attack, Ella pursues her plans, which include the arrest of Mrs Curry, whose schemes for procuring a schoolgirl for a local architect have been overheard by Bill Pendlebury, Ella's drunken, sponging brother. Mrs Curry is given a severe gaol sentence and passes the time recruiting a set of willing girls; the architect hangs himself; and Eric is set up in rooms in London. Bernard's place on the board of Vardon Hall is taken by a friendly civil servant, but its Utopian vision fails as cuts are implemented. Ella reflects that 'doing' does not last, but that Bernard's influence upon people will.

1953

Death of Joseph Stalin ● Coronation of Queen Elizabeth II ● Korean armistice ● Egypt becomes a republic ● Rise of Nikita Khrushchev ● Dwight D. Eisenhower proposes an international control of atomic energy ● Edmund Hillary and Sherpa Tenzing climb Everest ● USSR explodes a hydrogen bomb ● B.F. Skinner, *Science and Human Behaviour* ● Barbara Hepworth, *Monolith Empyrean* ● Henry Koster, *The Robe* (first film in 'Cinemascope') ● Hilaire Belloc and Arnold Bax die ● D.J. Enright, *The Laughing Hyena and Other Poems*

Ray BRADBURY   1920–

Fahrenheit 451

Published five years after Orwell's *Nineteen Eighty-Four* (1949), Bradbury's allegorical novel presents a similar, if less savage, vision. In his America of the near future *panem et circenses* consist not of endless wars against Eurasia intermixed with Stalinist false production claims, but of gross material satiation, symbolised by (literally) wall-to-wall television entertainment. The state operates a Big Brotherly control over the media and people's thoughts, and has banned books altogether.

Guy Montag is a fireman whose job, now that all houses are fireproofed, is to burn books. (The novel's title is taken from the degrees of heat needed to burn paper.) Initially he enjoys his work, but he is intuitively aware that his life lacks meaning. A series of encounters with Clarisse McClellan, a poetic seventeen-year-old intellectual, acts as a catalyst upon his dormant rebelliousness. Assisted by Faber, a timorous former professor of English, he plots a literacy campaign. Their plans are thwarted by Capt. Beatty, the sinister and equivocal chief fireman, who appears to hate books, but is capable of quoting from a wide range of literary sources. Beatty has recognised Montag's backsliding and plays a cat-and-mouse game with him, alternately soothing and needling his victim. Montag begins to fear for his resolve – he is no match for Beatty in debate – and grows progressively more reckless. Eventually, enraged by his wife and her friends who are incapable of discussion and meet merely to plug into the television, he causes outrage by reading them 'Dover Beach'.

Back at the station, the alarm bell rings and the firemen set off, only to arrive at Montag's own house. Beatty orders Montag to burn his house and his small hoard of books, after which he will be placed under arrest. Montag acts upon Beatty's dictum, 'If there is a problem, burn it away', by setting fire to his captain, after which he escapes. After a harrowing chase, he makes contact with the literary underground, where former academics preserve literature by memorising whole books. The ever-lurking threat of war is realised and the city is destroyed in a nuclear strike. The after-blast purifies all, and makes real Montag's dim sense of his humanity and his capacity for intelligent thought. He rises from the ground, a latter-day Sir Percival, the unspoken new leader of the underground, and sets off to found a new society.

They hurried downstairs, Montag staggered after them in the kerosene fumes.

'Come on, woman!'

The woman knelt among the books, touching the drenched leather and cardboard, reading the gilt titles with her fingers while her eyes accused Montag.

'You can't ever have my books,' she said.

'You know the law,' said Beatty. 'Where's your common sense? None of those books agree with each other. You've been locked up here for years with a regular damned Tower of Babel. Snap out of it! The people in those books never lived. Come on now!'

FAHRENHEIT 451
BY RAY BRADBURY

Shorn of its apocalyptic ending, the novel was made into a memorable film, directed by François Truffaut (1966).

Brigid BROPHY   1929–

Hackenfeller's Ape

Brophy's slim first novel (which won the Cheltenham Literary Festival Prize in 1954) was inspired by living opposite the Regent's Park zoo, and began life as 'a narrative poem on the (alas, still visionary) theme of the liberation of animals'.

The novel concerns the attempts of a zoologist, Prof. Clement Darrelhyde, to prevent scientists sending the ape of the title into space. Hackenfeller's Ape (*Anthropopithecus hirsutus africanus*) is the size of a gorilla, but has the characteristics of a chimpanzee, and its closeness to humans makes it the target of experimentalists. Darrelhyde spends hours

before Percy and Edwina's cage, hoping to be the first man to observe the species mate, and meditating upon the genius of *The Marriage of Figaro*. He is approached by Kendrick, an ambitious and cocky young scientist, who tells him that Percy's destiny is a space research station in Northolt. In an attempt to prevent Percy being sent *per ardua ad astra*, Darrelhyde attempts to enlist the support of several people, each of whom refuses to help and criticises his idealism. His sister argues that people are more important than animals; Post, the Co-ordinator of Scientific Studies, talks of the mark of Cain and the inevitability of human cruelty; the ironically named Col. Hunter of the League for the Prevention of Unkind Practices to Animals is more concerned with public relations than animal rights. Eventually Darrelhyde is abetted by Gloria, a girl who has attempted to pick his pocket. Together they break into the zoo and liberate Percy, who bounds over the roofs of the enclosures, returns to mate with Edwina, and is finally felled by the rifle of his keeper.

Percy, bounded by his cage, had not seen and could not imagine the other species displayed in the Zoo. He could not recognise in them the footprints of his own ancestry. Time past meant nothing to him, least of all time recorded in compressed millennia, in fossils and rocks. Nevertheless, he came to know the word 'Evolution' from its recurrence in the Professor's musings.

HACKENFELLER'S APE
BY BRIGID BROPHY

Kendrick steals Percy's skin and, dressed in this, achieves his ambition by being launched from Northolt, never to return. Darrelhyde loses his job, but is called by the keeper to assist at Edwina's accouchement at Easter. The novel ends with an astonishing 'Soliloquy of an Embryo' as Edwina and Percy's son forces his way out from the womb into the prison of captivity, where: 'It woke all the other animals and set them gibbering as it let out its first roar of wrath.'

This witty and impassioned novel, narrated from the viewpoints of both *Homo sapiens* and *Anthropopithecus*, unsentimentally challenges assumptions of human superiority. It contains themes upon which the author was to enlarge, and which were to become her trademarks, in later works: notably animal rights and Mozart. The arguments against which Darrelhyde strives are posited in Brophy's essay upon 'The Rights of Animals' (1965, collected in *Don't Never Forget*, 1966).

William S. BURROUGHS   1914–

Junkie

Originally published under the pen-name of William Lee, Burroughs's first novel is an account of the experiences of a heroin addict. The story is related in the first person by 'Bill', who is born in 1914 into a normal bourgeois American family. Many of these incidental details correspond accurately with the details of Burroughs's own life, and later interviews have established that much of the novel is autobiographical. In fact, Burroughs shifts between a number of frameworks, uncertain as to which (if any) can best redeem the radically outcast mentality of the junk-user. The idioms of the novel – auto-biography, documentary and even medical case-history – are all used as vehicles for encapsulating and presenting the experience of addiction from the inside.

Burroughs offers a harrowing, picaresque tale populated by a rogues' gallery of hookers, addicts, pimps, police and gangsters; but they are all sideshows to the real action, which is the feeding of the addiction. The presence of junk permeates the story at all levels, just as it permeates the body of the junkie: the story and the body both proceed by 'injection'. As Burroughs explains: 'When you stop growing you start dying. An addict never stops growing. Most users periodically kick the habit, which involves shrinking of the organism and replacement of the junk-dependent cells. A user is in a continual state of shrinking and growing in his daily cycle of shot–need for shot completed.' The first edition saw the need to provide an admonitory footnote to this stating, 'The foregoing is not the view of recognised medical authority'. Burroughs's treatment of Bill as an organism is the basis of the novel's refusal to engage in any ethical debate. Addiction is nowhere advocated or condemned, it is merely described as a physical event, the action of a drug on a collection of cells.

The drug sub-culture in which the story takes place is radically separated from normal life in post-war America. Burroughs does not here equate sub-culture with any revolutionary breeding ground, as he does in later novels, nor does he draw comparisons between junkies and 'healthy' people. Junkies are *completely* changed, placed completely outside normality by their habits: 'Junkies run on junk-time and junk metabolism. They are subject to junk-climate. They are warmed and chilled by junk.' Junk even has its own language of coded slang in which the experiences of junk can be classified and described. Burroughs immerses himself completely in this environment and yet still manages to represent it starkly and unambiguously in painfully comprehensible language. Perhaps significantly, this most technically straightforward of Burroughs's works is the only one written under the influence of heroin. As both addict and chronicler of his addiction, the narrator presents an eerie double-vision which is the defining mark of this repellent and compulsive novel.

L.P. HARTLEY   1895–1972

The Go-Between

'The past is a foreign country: they do things differently there.' The celebrated opening of Hartley's most popular novel sets the scene for an old man's confrontation of the traumatic events in which he became involved as a child at the beginning of the century. Leafing through an old diary, Leo Colston stirs the ghosts from this distant past and begins to understand why his adult life has been emotionally sterile: 'You flew too near the sun, and you were scorched. This cindery creature is what you made me.'

In the blazing July of 1900 the twelve-year-old Leo was visiting Brandham Hall, the Norfolk home of his schoolfriend,

Marcus Maudsley. His hosts are impressed by his reputation as a magician, gained after putting a curse upon two bullying school-fellows who subsequently fell off a roof. He is befriended by Marcus's older sister, Marian, and unwittingly becomes caught up in a web of adult intrigue when he agrees to carry letters and messages between her and a local farmer, Ted Burgess. Although he has been told that his errands must remain a secret, he believes that he is merely conveying business communications between Marian and Ted. Gradually this innocence is destroyed, and with it the lives of those involved in the deception.

The indomitable Mrs Maudsley intends her daughter to marry the heir to the Brandham estate, Hugh, 9th Viscount Trimingham, a debonair soldier whose face has been badly scarred, but Marian is in love with Ted. Leo is enchanted by Marian (whom, ironically, he identifies with Virgo), admires Trimingham and is fond of Ted, and thus torn by conflicting loyalties. When both Marian and Ted display their anger, Leo feels betrayed and exploited and places a curse upon them in the hope of averting disaster. Leo's innocence and experience are those of twentieth-century man: 'In my eyes the actors in my drama had been immortals, inheritors of the summer and of the coming glory of the twentieth century.' In fact they prove all too mortal, for when the affair is discovered, Ted shoots himself. Trimingham marries Marian despite the fact that she is bearing Ted's child, a child that is to die in the Second World War. Trimingham dies in 1910 and both Marian's brothers are killed in the First World War.

After recalling these events, Leo revisits Brandham and finds Marian, old and alone, wrapped in delusions about the events of that summer. He agrees to perform one last 'errand of love' for Marian, reassuring her grandson that he is not under a curse and is free to marry.

The novel was awarded the Heinemann Foundation Prize, and was made into a very distinguished film (1970) directed by Joseph Losey from a screenplay by Harold Pinter.

Rosamond LEHMANN   1901–1990

The Echoing Grove

Lehmann wrote what is often regarded as her best, and certainly most ambitious, novel shortly after the end of her affair with C. Day Lewis, and something of the characters' circumstances are derived from that complicated entanglement, although the book is not in any real sense autobiographical. (Lehmann, who described the novel as 'dark and terrible', had originally wanted to call it 'Buried Day', a Freudian slip which had to be pointed out by her publisher.) Its relationship to *Dusty Answer* (1927) was noted by Lehmann in a radio broadcast in 1953:

> I began to feel that, more than any of the others, this novel had something to do with the first I ever wrote. Not the same one in a fresh guise; not even a development from it; but more as if somehow – I cannot exlain why – some cycle of experience that had opened when I was a girl was now coming to a close.

The complicated structure of the novel reflects this cycle of experience, divided as it is into five sections – 'Afternoon', 'Morning', 'Nightfall', 'Midnight' and 'The Early Hours' – the first and last of which, set in the present, frame the central three, in which events of the past (narrated unchronologically in both the first and third person) gradually unfold from the differing viewpoints of the three principal characters.

Dinah Sandhurst arrives at her sister Madeleine's country cottage to sort out the tensions between them. After a fifteen-year estrangement, the two women have recently met at the death-bed of their mother. Madeleine's husband, Rickie, had been Dinah's lover for years before his early death, and the incestuous overtones of the triangle make it difficult for the sisters to be honest about their feelings for each other. Rickie was 'a man of his word to both of them – it was farcical', and *The Echoing Grove* is a meditation on the reverberations of their interrelationships.

The sisters' mutual suspicions emerge through casual details as they discuss their different memories of the trivial events which threatened to expose the truth of the affair. Dinah returns, without explanation, a pair of Rickie's cufflinks, and the dual narratives reconstruct the past, sifting through half-forgotten nights and blatantly false excuses. Dinah had become pregnant by Rickie but their baby died at birth, while Madeleine has a grown-up family: a daughter, Clarissa, smothered by misdirected love, and a son missing in army service in Africa. After the affair with Rickie temporarily broke up, Dinah had married Jo, a political idealist who died within months of their marriage, fighting with the Republicans in the Spanish Civil War.

He had not come back. Hall, study, dining-room, drawing-room: she opened doors and saw the face of each in turn, sunk in lugubrious hostility, emanating greyness and decay, like animals behind bars, or manic-depressives in a private bin. Already midsummer dawn fainted prenatally in the high uncurtained windows. He was upstairs ... But already she knew along the vibrations of her nerves that there was nobody on the next floor. The house of a widow.

THE ECHOING GROVE
BY ROSAMOND LEHMANN

In this intricate pattern of jealousies and rivalries, the two women become aware of Rickie's involvements with other mistresses: 'he had an innate resistance to furthering his own interests'. Madeleine becomes involved with Jocelyn Penrose, a pretentious young man from the London literary world, and Dinah reveals her anguish over her love for her youthful flatmate, an unstable homosexual called Robert. The novel ends with nothing resolved, the sisters enmeshed in their painful, contradictory reflections on the past: 'There was never any knowing what went on between these sisters.'

C.S. LEWIS   **1898–1963**

The Silver Chair

see **The Chronicles of Narnia** (1956)

John WAIN   **1925–1994**

Hurry on Down

Wain's first novel (published in America as *Born in Captivity*) was hailed after the event as the opening shot in the Angry Young Man movement. In fact, as he points out in his introduction to the reissue in 1978, this 'didn't come into existence until after the success of John Osborne's *Look Back in Anger* in 1956, and even its nebulous predecessor, known simply as "The Movement", was not talked about until 1954'.

The novel is in the picaresque tradition. Its hero, Charles Lumley, an engaging rogue with an uncomfortable conscience in moments of moral crisis, turns his back on the middle class, engages in a series of bizarre adventures and gets the girl in the end. As freelance window-cleaner – an occupation with clear potential for freedom when one has 'come down from the University with a mediocre degree in History' – he is saved from a casual beating in the street by the 'cloth-capped' Ern Ollershaw, with whom he goes into partnership. For the sake of the beautiful Veronica, with whom he has a torrid affair under the nose of her possessive 'uncle', the factory manager Mr Roderick, Charles gets into drug-running, to make real money. This career comes to an abrupt end thanks to the obstinacy of the wilful young journalist Harry Dogson, who, seeking 'the road to Fleet Street', insists on snooping down at the docks when a 'drop' is being made, and is murdered.

Charles is badly hurt in the getaway; he ends up in hospital – physically and morally at rock bottom – and when he has recovered, hides from the world as an orderly: 'Anonymity, obscurity, a relief from the strain, the situation was exactly what he had prescribed for himself.' He becomes involved with Rosa, who works at the hospital and wants to marry him. His heartless treatment of her – how could his behaviour have been otherwise once he had met her depressingly petit-bourgeois family? – necessitates another bunk, and he finds himself employed as chauffeur by the rich but gloomy Mr Braceweight, of Braceweight's Chocolates. A quiet life in the country is

Why was this? Why had he failed? he asked himself as he dragged the heavy suitcase down the main street towards the station. The answer, like everything else, was fragmentary: partly because the University had, by its three years' random and shapeless cramming, unfitted his mind for serious thinking: partly because of the continued nagging of his circumstances ('Go out this morning or she'll know you haven't a job – come to a decision today before you waste any more time – look at the papers to see what sort of jobs are offered'), and partly for the blunt, simple season that his problems did not really admit of any solution.

HURRY ON DOWN
BY JOHN WAIN

disturbed when his arch-enemy, the intellectually snobbish George Hutchins, turns up as tutor to Braceweight's son Walter, who only wants to be a mechanic. Hutchins and his poisonous girlfriend, June Veeber, do the dirty on Charles, who again leaves in a hurry.

Down and out in London, he gets a job through an old acquaintance, the genial Mr Blearney, as chucker-out in the Golden Peach Club. He is found there by his former room-mate, the once lugubrious Froulish. His great novel has been abandoned and he now makes a fair living in the 'gag-making business'. Froulish gets him on to the team as 'seventh man' and, with a three-year contract under his belt and reunited with Veronica (who has finally left her 'uncle'), Charles is happy at last.

With its huge cast of characters appearing or disappearing and reappearing, *Hurry on Down* takes a swipe at everything in sight and, for all its faults, has an energy that is not always present in Wain's later work.

1954

Gamal Abdel Nasser gains power in Egypt ● US–Japanese mutual defence agreement ● Dien Bien Phu falls to Communist Vietnamese ● US Senate condemns Joseph McCarthy ● France sends 20,000 troops to Algeria ● Richard Wright, *Black Power* ● The connection between smoking and lung cancer is first seriously suggested ● Temple of Mithras is uncovered during excavations for rebuilding in City of London ● Paul Tillich, *Love, Power and Justice* ● Lennox Berkeley, *Nelson* (opera) ● Dylan Thomas's dramatic poem *Under Milk Wood* broadcast ● Elia Kazan, *On the Waterfront* ● Henri Matisse dies ● Wallace Stevens, *Collected Poems*

Kingsley AMIS 1922–

Lucky Jim

Amis's first and most famous novel was hailed as an innovation in British fiction: the central character a manifestation of what came to be known as the Angry Young Man. Its setting, although academic, was the ungilded provinces, and its protagonist was neither personable nor heroic. Indeed, Jim Dixon, with his repertory of grimaces to denote varying degrees of rage and disgust, is

the prototype of a now familiar Amis hero, a disgruntled and irascible philistine.

Dixon is a junior history lecturer on probation at an unnamed provincial university. His status, made even more precarious by his sloth, necessitates his keeping in with Prof. 'Neddy' Welch, the absurd head of department. Dixon's dislike of Welch is as nothing compared with his loathing of Welch's son, Bertrand, who conforms to the philistine's image of a painter: large, bearded, talentless and philandering. Bertrand comes to stay with his parents, bringing with him his current girlfriend, Christine Callaghan. Dixon is also staying with the Welches, enduring a grotesque musical weekend, from which he briefly escapes to the pub. On his return he drunkenly attempts to seduce Margaret Peel, a dowdy colleague recovering from a suicide attempt, with whom he has become unwittingly entangled. Dismissed to his own room, he falls asleep whilst smoking and awakes to find that he has burned holes in the bedclothes and a rug, and scarred the bedside table. He tries to cut the charred areas out of the blankets and rug, and Christine helps him hide the table in the attic.

At a college ball, Bertrand attempts to ingratiate himself with Christine's uncle, Mr Gore-Urquhart, 'a rich devotee of the arts', in order to secure a job. Dixon leaves with Christine, returning her to the Welches in a taxi hijacked from a senior colleague. To his surprise, Christine is attracted to him and they arrange to meet for tea at a hotel. Meanwhile Margaret throws a hysterical fit in Dixon's rooms when he informs her that their relationship is at an end. Horrified by her reaction, and unable to separate Christine from Bertrand, he feels obliged to relent. Christine has also decided that they should not see each other, but Bertrand finds out about their meeting and attempts to warn Dixon off. By his eponymous good fortune, Dixon manages to floor Bertrand in the course of a fight. However, the lecture he is to give, which will determine whether or not his contract will be renewed, is a fiasco. His nerves fortified by alcohol, Dixon delivers the speech in a drunken stupor, unintentionally imitating the styles of both Welch and the Principal. He is sacked, but is promptly offered the job Bertrand had been pursuing. One of Margaret's former boyfriends reveals that her suicide attempt was not serious and merely a neurotic gesture to shame both himself and Dixon. Christine announces that she is leaving Bertrand and she and Jim set off for London, leaving the Welches well routed.

The novel provoked a great deal of controversy, much of it prompted by writers of an older generation shaking their venerable heads over the follies of their juniors. The most memorable comment was published in the *Sunday Times* on Christmas Day 1954, in which Somerset Maugham expressed his fears for the future of a country populated by Jim Dixons, where people failed to absorb the civilised values traditionally inculcated by a university education: 'They do not go to the university to acquire culture, but to get a job ... They have no manners, and are woefully unable to deal with any social predicament ... They are mean, malicious, and envious ... They are scum.' Amis was vigorously defended by such writers as C.P. Snow, John Wain and Philip Larkin, and the novel became a classic.

William GOLDING   **1911–1993**

Lord of the Flies

'Absurd and uninteresting fantasy about the explosion of an atomic bomb on the colonies and a group of children who land in jungle country near New Guinea. Rubbish and dull. Pointless.' The story of how Golding's first novel, now widely regarded as a masterpiece, came to be rescued from a publisher's 'slush pile' is one of the industry's best known cautionary tales. Faber's reader was not alone in her opinion of a dog-eared manuscript entitled 'Strangers from Within', which had clearly been seen and rejected by other publishers. Fortunately, Charles Monteith, who had been at Faber less than a month, decided to take the manuscript home in order to read beyond the opening pages, which contained a long account of atomic war. 'As I read on I found that, reluctantly, I was becoming not merely interested but totally gripped,' Monteith recalled. He recognised that the book needed considerable revision, and persuaded his reluctant colleagues to allow him to discuss this with the author. Severe cuts were made, and the character of the charismatic chorister, Simon, was made less obviously Christ-like. The novel was well received in Britain, but in America it became an enormous popular success amongst students. It subsequently appeared on the syllabuses of English literature courses at schools and universities, was translated into some twenty-eight languages, and was twice filmed, most hauntingly by

Peter Brook in 1963.

The novel powerfully combines a children's adventure story with anthropology, demonstrating what Golding calls 'the end of innocence, the darkness of man's heart'. The narrative loosely follows the pattern of the biblical story of the Garden of Eden, which illuminates the innate human tendency to abuse freedom and to follow evil rather than good. Golding called his books fables and used them to express the savagery that lies only just below the surface of our civilised veneer. His pessimistic vision was formed during the Second World War, in which he served in the navy, and in the classroom, where he taught small boys. The title is a literal translation of 'Beelzebub'.

This toy of voting was almost as pleasing as the conch. Jack started to protest but the clamor changed from the general wish for a chief to an election by acclaim of Ralph himself. None of the boys could have found good reason for this; what intelligence had been shown was traceable to Piggy while the most obvious leader was Jack. But there was a stillness about Ralph as he sat that marked him out: there was his size, and attractive appearance; and most obscurely, yet most powerfully, there was the conch. The being that had blown that, had sat waiting for them on the platform with the delicate thing balanced on his knees, was set apart.

LORD OF THE FLIES
BY WILLIAM GOLDING

In the course of some future war, a plane-crash leaves a group of English boys stranded on a paradisal island. Ralph and the fat, bespectacled, asthmatic Piggy valiantly attempt to set up a civil and democratically run society but, in spite of their best efforts, most individuals disintegrate, regressing rapidly into barbarism, ritual murder and allegiance to 'The Lord of the Flies'. Jack, a head chorister, takes over as dictator, unleashing the equivalent of a reign of terror. The safe ordinariness of everyday objects takes on a new dimension of threat and

menace as the true nature of human beings is revealed: 'I'm afraid of *us*,' says Ralph.

When an unexpected ship arrives to rescue the boys, Piggy and Simon have been killed, Ralph is being hunted for his life and the island is in flames. The naval officer who comes ashore is shocked at what has transpired: 'I should have thought that a pack of British boys ... would have been able to put up a better show than that', especially where their situation was 'like the Coral Island' – a reference to R.M. Ballantyne's novel *The Coral Island* (1857), in which a group of castaways enterprisingly recreate British society in the wilderness, complete with religious ideals, cold baths and training in observation. Golding deliberately names several of his characters after those in Ballantyne's book, to make it quite clear that his own novel is intended as a powerful reply to Ballantyne's naïve Victorian optimism.

Randall JARRELL   1914–1965

Pictures from an Institution

The poet and critic Jarrell's only novel is a comic satire set on the campus of Benton College, a well-to-do progressive women's college near New York, and clearly draws upon his own experience as a professor of English. Narrated by an unnamed poet who teaches at Benton, the novel is almost entirely devoid of plot; it takes the form of a series of anecdotes about various members of staff (hardly any attention is paid to the students, though at one point we do glimpse a student handing in her Kafka-influenced story concerning 'a bug that turns into a man') and achieves most of its effect through witty one-liners and literary jokes.

The central character is Gertrude Johnson, the fairly famous novelist currently 'in between' novels and teaching creative writing for the year. Hers is a dim view of humanity ('People had always seemed to Gertrude rather like the beasts in *Animal Farm*: all equally detestable, but some more equally detestable than others') and she delights in shocking people. She soon realises, though, that Benton is Material and hosts a dinner-party to Collect for the Book: 'Gertrude was never polite to anything but material: when she patted something on the head you could be sure that the head was about to appear, smoked, in her next novel.'

When well-dressed women met Flo they looked at her as though they couldn't believe it. She looked as if she had waked up and found herself dressed – as if her clothes had come together by chance and involved her, an innocent onlooker, in the accident. If a dress had made her look better than she really did, she would have felt guilty; but she had never had such a dress.

PICTURES FROM AN INSTITUTION
BY RANDALL JARRELL

The staff at Gertrude's dinner ('There was very little of Gertrude's dinner, but what there was was awful') include the president, Dwight Robbins; the elderly Austrian composer Gottfried Rosenbaum ('an exponent of what the French call *la musique sérielle et dodécaphonique*'); and the sociologist Jerrold Whittaker. Benton, with its progressive ideals, we discover, rarely succeeds in teaching anyone anything; rather, 'the girls were encouraged to have problems (one famous student had so many that her adviser said to her at last, *If I were you I'd commit suicide*; but he was not one of the real faculty)'.

For all this, Benton is a place where, as the narrator advises Gertrude, 'nothing ever happens'. How, he wonders, can she make a novel out of such unpromising material? Gradually he discovers that she is concocting an elaborate and scandalous plot, and that Benton is for her 'a giant nursery of facts, facts that would cover, with their mild academic ivy of verisimilitude, the girders of a plot that could have supported the First National Bank'.

The Tortoise and the Hare

The best of Jenkins's many quietly distinguished novels, *The Tortoise and the Hare* is a story of traditional femininity and how it may fail to meet challenges presented on its own domestic ground. Imogen Gresham is an attractive, timid, middle-class woman devoted to creating a serene domestic background for her hard-driven barrister husband, Evelyn. 'She tried hard to foresee what should be done and to carry out his requests; but she never acquired that unthinking grasp of practical matters that creates confidence.' She is treated by her irascible husband and pre-teenage son Gavin, a small replica of his father, with a condescension bordering on contempt.

Imogen gradually becomes aware that a neighbour, Blanche Silcox, is playing an increasing role in her husband's life. Blanche is an inelegant woman, who wears tweeds and intimidating hats, shoots and fishes, drives a Rolls Royce, and seems to lack all the conventional feminine charms that Imogen (some fifteen years her junior) possesses. But the novel justifies its title by giving a detailed and convincing account of how Blanche, using a clear knowledge of what she wants and undoubted practical capacities, wins Evelyn from Imogen. Advised by her friends, the quietly adoring doctor Paul Nugent and the understanding young woman Cecil Stonor, Imogen can only watch in passive anguish as the gruff Blanche invades her comfortable Berkshire home.

Gradually the routine of comings and goings between the two houses, telephone calls, and clandestine evenings for the lovers in London, develops into a situation in which Evelyn is spending part of the week at Blanche's London address in Halkin St, and Imogen must decide whether to accept the situation or make a decisive break. Gavin has injured another boy at his prep school and, Evelyn being in court, Paul Nugent, goes to the school to sort matters out. Evelyn asks Imogen why she did not ask Blanche to go, which insult leads her to break down and demand a divorce.

In spite of her efforts to be active and useful, her nights became more and more disturbed by dreams. Wild, low-lit landscapes and ruined towers merged into realistic scenes. Once she saw herself late at night in a closed-off part of a street, lit by lanterns, where road-menders were digging fast in the bottom of a pit. She stood looking down at them, till they laid aside their picks. One of them said: 'Now it's ready,' and they held up their arms to help her down into it.

THE TORTOISE AND THE HARE
BY ELIZABETH JENKINS

Paul, who has been suffering from a terminal illness, dies; Cecil marries Evelyn's friend, Hunter Crankshawe; and Imogen goes to live at Cecil's former flat. Imogen has befriended Gavin's schoolfellow Tim Leeper, the neglected son of progressive parents, and the novel comes to a touching interim conclusion when he arrives at the flat where she is living, valuing her for herself and proposing to desert his parents and live with her. *The Tortoise and the Hare* is marked by beautiful writing and keen insight, and draws great emotional power from the familiar modern theme of marital breakdown.

Doris LESSING   1919–

A Proper Marriage

see **Children of Violence** (1969)

C.S. LEWIS   1898–1963

The Horse and His Boy

see **The Chronicles of Narnia** (1956)

C.P. SNOW   1905–1980

The New Men

see **Strangers and Brothers** (1970)

J.R.R. TOLKIEN   1892–1973

The Fellowship of the Ring
The Two Towers

see **The Lord of the Rings** (1955)

Antonia WHITE   1899–1980

Beyond the Glass

see **Frost in May** (1933)

1955

Treaty for European Union ratified ● West Germany enters NATO ● Jonas Salk's anti-poliomyelitis vaccine ● Phenomenon of 'flying saucers' attracts attention ● John Bratby, *Still Life with Chip-Fryer* ● Frederick Gibberd, London Airport Buildings ● Samuel Beckett, *Waiting for Godot* ● Nicholas Ray, *Rebel Without a Cause* ● Bill Haley, *Rock Around the Clock* ● Alexander Fleming, Albert Einstein and Thomas Mann die ● R.S. Thomas, *Song at the Year's Turning* ● Stephen Spender, *Collected Poems 1928–1953*

Samuel BECKETT   **1906–1989**

Molloy

see **The Beckett Trilogy** (1958)

Elizabeth BOWEN   **1899–1973**

A World of Love

Written in the wake of her husband's death, *A World of Love* is notable for its strong sense of atmosphere and a symbolic juxtaposition of internal and external events. Explaining that it was a 'deliberately' short book, Bowen commented: 'It's on the periphery of a passion – or, the intensified reflections of several passions in a darkened mirror.'

The action of the novel takes place in the 'mirage-like shimmer' of an unprecedented heatwave, which reflects the growing tensions within Montefort, an Irish country house. Twenty-year-old Jane Danby discovers a cache of love-letters in the attic, written by Montefort's previous owner, Guy, who was killed in the First World War. Guy had been engaged to Jane's mother, Lilia, but had also been pursuing a relationship with his cousin, Antonia, who inherited the house. Jane's father, Fred Danby, is an illegitimate cousin of Antonia and farms the estate. He assumes that the letters are addressed to Lilia, and they are an unwelcome reminder that he is an inadequate substitute in her affections for the dead Guy. Jane discovers that the letters are addressed neither to Lilia nor Antonia, but to an unidentified, possibly imaginary, woman. When Lilia finally admits that she knew of the relationship between Guy and Antonia,

the usually abrasive Antonia shields her from Jane's discovery concerning the letters' real recipient.

For Jane, whose unconscious sexual power is not unnoticed by others, the letters become a focus for her romantic imagination, so that Guy becomes an imaginary lover, a circumstance mirrored by her pugnacious twelve-year-old sister, Maud, who has an imaginary companion, 'Gay David'. Memory and fantasy together appear to evoke Guy's spirit in a tangible way, so that it hangs 'in the air, over scene and people, going on affecting them, working on them' in the oppressive heat. Jane is taken up by Lady Vesta Latterly, a manipulative, *nouveau-riche* neighbour, who dispatches her to meet a former lover, an American called Richard Priam, at Shannon airport. The novel's final line is: 'They no sooner looked but they loved.' Jane's romantic obsession with Guy has been a 'trial lesson in love', preparing her for this *coup de foudre*. Jane's real story begins only after the novel ends, and Bowen leaves the reader to decide whether it will be a happy one.

J.P. DONLEAVY   **1926–**

The Ginger Man

Like many books which timorous publishers considered obscene, *The Ginger Man* was originally published in Paris. While Donleavy's first novel was not quite as much of a coup for the Olympia Press as Vladimir Nabokov's **Lolita**, which they published the same year, it was nevertheless more literary than some of their publications, and added a certain lustre to their list. Unfortunately, it also led to the Press's downfall, when the author sued the proprietor, Maurice Girodias, for breach of contract. Girodias was obliged to sell his firm's title and had the further humiliation of seeing Donleavy buy it at auction. The novel was subsequently published in the USA in 1958, in a revised edition, and a fully unexpurgated edition followed in 1965. The attendant publicity (fuelled further when a stage adaptation had to be withdrawn from the Dublin stage in 1959) ensured that Donleavy's reputation as a writer of bawdy farces became firmly established in the reading public's mind, and he has remained a popular novelist thereafter.

Inspired by the author's memories of his undergraduate years at Trinity College, Dublin, *The Ginger Man* concerns the adventures of Sebastian Dangerfield, who discovers what he perceives to be the great tragedy of the Irish nation: 'sixty-seven per cent of the population have never been completely naked in their lives'. Divesting female members of that population of their inhibiting garments becomes his mission. An example of Irish prudery occurs when Sebastian accidentally exposes the Ginger Man of the title on public transport and he is hounded through the streets by enraged locals. His long-suffering wife, Marion, has already taken steps to distance herself from her philandering husband, and his abominable fellow student at Trinity, Kenneth O'Keefe. The final straw was when Sebastian disconnected the lavatory waste-pipe, pouring down further shame upon the heads of his wife and daughter.

Meanwhile, Sebastian pursues his sexual obsessions with characteristically Donlevian abandon. He finds consolation in the bed of Chris, an orphaned laundry girl whose unshaven legs drive him mad with desire. The neglected Marion writes to her wealthy father-in-law as her husband, still hot from Chris's embraces, falls into the broader arms of Mary, whom he cons out of her savings when disinheritance threatens. Deserted by his wife and cast off by Chris, Sebastian considers a new life with Mary. This nightmarish fantasy becomes a reality, when he

realises that Mary is in fact the only friend he has left, and he is obliged to join her in London. Wondering where next his errant flesh will land him, he takes his leave of the reader with a final plea:

> God's mercy
> On the wild
> Ginger Man.

William GADDIS   1922–

The Recognitions

The Recognitions is a hugely ambitious novel of labyrinthine complexity and vast scope, which paved the way for the fictional experimentation of John Barth, Thomas Pynchon and others in the decades following the 1950s. In many ways it surpasses the achievements of these later authors. The story begins with the Revd Gwyon and his wife, Camilla, sailing from America to Spain. Camilla contracts appendicitis and dies when the ship's surgeon bungles the operation: the surgeon then reveals himself to be a counterfeiter disguised as a doctor to throw his pursuers off the scent. This is the first instance of counterfeiting in a novel crammed with forgeries, fakes, disguises, aliases, *noms de guerre* and *de plume*, imitations and simulacra. Its hero is Gwyon's son, Wyatt, who takes up the counterfeiting of old masters as an almost spiritual activity, but whose talent is, of course, used more cynically by those around him. Wyatt sees in his counterfeit Van Eycks and Van der Goes an intuition of the original impulses and emotions which long ago produced them, but these recognitions of reality are quickly submerged in the opportunistic fakery of the post-war urban America that Gaddis describes.

A well-heeled rogues' gallery of unscrupulous art-dealers, brokers and pretend-connoisseurs surrounds Wyatt, while a substratum of penniless young artists, writers and musicians revolves about Otto, his 'surrogate-son' in the novel. Otto has written an atrocious play which everyone recognises is plagiarised (although from whom, no one is quite sure) while his rather uncaring 'friends' are the butt of Gaddis's most pitiless satire in their hyping of (generally worthless) artistic theories and practices. Around this 'artistic' community, American consumer culture belches out its complementary messages, summed up in a radio advertisement for '*Necrostyle*, the wafer-shaped sleeping pill. No chewing, no after-taste'. Religion has become an anodyne substitute for feeling: nothing is real, everything makes you 'feel good'.

The second half of the novel charts the gradual drift of the characters towards their different destinies and destinations. Otto, having accidentally been given 5,000 counterfeit dollars by a man he mistakes for his father (actually Camilla's murderer of twenty years ago) ends up injured and on the run in a revolution-torn Central American state. Wyatt, deranged and divorced, travels to his mother's grave in San Zwingli, where he is mixed up in a plot to fake an Egyptian mummy, and is last seen in the monastery his father visited twenty years ago, 'restoring' its pictures by scraping off the paint. Meanwhile, in Rome, Stanley, the one authentic artist of the novel, finally finishes his organ oratorio and prepares to play it; a monk explains the structural weaknesses of the church, the danger of the bass-pipes, but Stanley does not understand Italian. The piece literally brings the house down, kills Stanley and ends the novel.

The Recognitions is an astonishing first novel. Its length (almost 1,000 densely written pages) may seem off-putting and probably accounts for its neglect when first published. However, the subsequent course of post-war American fiction has confirmed Gaddis's achievement: his erudite and sophisticated satire on the mechanisation of human emotion and commercialisation of religious and artistic experience provides the model for many later (and better-known) writers. With its huge cast of characters and multiple areas of interest (everything from the hideousness of Sacré Coeur to chastity belts for poodles and Mithraism in New England), *The Recognitions* is a richly comic meditation on the vanity of human wishes in post-war Europe and America. Gaddis's foresight makes his first novel if anything *more* relevant today than when first published.

Graham GREENE   1904–1991

The Quiet American

Set in Vietnam, *The Quiet American* explores the unheroic world of post-colonial compromise through the relationship between Thomas Fowler, a seasoned war correspondent, and Alden Pyle, the quiet American of the title. In Fowler's narration of events we witness his growing awareness of the ambiguities involved in all actions, as he is forced from his creed of neutrality (in both politics and sex) to an uneasy partisanship. 'Sooner or later . . . one has to take sides. If one is to remain human.' The reporter is made to become the leader-writer.

Ostensibly employed to bring economic aid to Vietnam, Pyle is an idealistic graduate who attempts to apply academic theory to chaotic reality. His naïve enthusiasm involves him in political terrorism, with the sinister support of the American government. A *Boy's Own* character, 'like a prize pupil', he saves Fowler's life and dies a hero's death in the cause of democracy. In the gradual reconstruction of events, however, we sense a bitter underlying reality. Fowler's experience of the war has made him conscious of the sufferings of the innocent, and it is Pyle's continuing blindness to the pain he causes which leads Fowler to become implicated in his death: 'I never knew a man who had a better motive for all the trouble he caused.' Fowler finds he holds the fate of another man in his hands and, unable to believe in God, he cannot abdicate responsibility. His moral guilt is paralleled by the theme of American involvement in Vietnam. Trying to hide from himself the result of his actions, Fowler leads Pyle into the trap of a sordid political killing; stabbed with a rusty bayonet, Pyle meets his death in the mud of the river. His murder is investigated by Inspector Vigot, who uses psychology rather than clues in his pursuit of truth. A reader of Pascal, he does not fit the brutal police, and in another age and country would have been a priest.

Love and war are closely related, and the complexity of motivation is increased as Pyle falls in love with Fowler's mistress, Phuong. 'She'll just have to choose between us,' says

Pyle, but Fowler knows that there can be no black-and-white choice. Pyle's love for Phuong is as ill-conceived as Fowler's own image of her as an extension of the opiate. In his life in the East, Fowler is living out his desire to escape the transience of human relationships. It is through the death of Pyle that he is forced to realise his moral responsibility. Finally, Fowler is free to marry his mistress, but it is an empty happy ending: 'You spoke his name in your sleep,' he tells her, and she replies, 'I never remember my dreams.' In a Godless world Fowler can find no confessor.

C.S. LEWIS　　　　　　**1898–1963**

The Magician's Nephew

see **The Chronicles of Narnia** (1956)

Naomi MITCHISON　　　　**1897–**

To the Chapel Perilous

Mitchison's astonishingly versatile publishing history, which began in 1923, has included many fine historical novels. *To the Chapel Perilous* is one of the most surprising and high-spirited. Based on the innumerable myths surrounding King Arthur, the Round Table and the Quest for the Holy Grail, the novel explores the links and conflicts between legend and history, truth and fact, religion and magic, journalism and literature – all subjects which have long preoccupied the author.

Dutifully stationed outside the Chapel Perilous when the book begins are Dalyn, the man from the *Northern Pict*, and Lienors, the *Camelot Chronicle* girl. Surprising characters, perhaps, to be encountered in the Middle Ages but this, as the reader swiftly realises, is a novel which plays exuberant tricks on time, and anachronisms abound. Cameras exist, but cars do not, and the medieval world proves to be as abundantly supplied with warring newspaper empires, trial by media and strict 'news management'

as our own. (The book is dedicated 'To the *Manchester Guardian* and the *Daily Mirror* affectionately'.)

Over the months Dalyn and Lienors have watched many a gallant knight enter the Chapel and none has yet emerged with the Grail (although plenty 'had burst out of the door screaming or hideously dumb'). Dalyn's and Lienors's newspapers, The Big Two, are the only ones which can afford to keep journalists permanently on the story. Although 'the Grail Quest was still news', both writers have seen their copy suffer much at the hands of sub-editors back at the office 'who were likely to angle it so that it fitted in with whatever they thought up or had handed to them', usually by their formidable editors and proprietors. The *Chronicle* even has Merlin on its staff, and its editor and owner is Lord Horny, who comes disconcertingly equipped with 'a thick black tail'. His second-in-command is 'a competent elderly devil' who 'always had a pretty little toadstool in his button-hole and he shared Lord Horny's point of view' – which is devilish indeed.

Subs, editors and proprietors such as these begin the centuries-long process whereby leading figures in the Arthurian legends – Kay, Lancelot, Guinevere, Arthur himself – have received 'a good press', or a bad one, according to the prejudices and vested interests of the chronicler.

The two journalists' vigilance is suddenly rewarded when not one knight but several emerge unscathed from the Chapel, each, it seems, clutching the Grail. But which is the true Grail, and who is its true knight? Gawain, Lancelot, Perceval, Bors, Galahad – these are names to conjure with and each of them, it seems, has achieved the Grail. In pursuit of the *real* Grail story, Lienors, Dalyn and a bevy of other, provincial, journalists make their way to the castles, forests and courts which house the various grails and their knights. Each grail, it seems, brings blessings, or curses, tailored to the knight who found it. The saintly Galahad's brings peace to the uneasy dead whilst that of the just and equable Bors increases the fertility of his already abundant and harmonious demesne.

But grail is pitted against grail as champions and detractors clash and the 'news managers' go about their malevolent business. Lord Horny and the church form an unholy alliance, and help set in train the events which will destroy Arthur, the Round Table and Camelot itself.

This remarkably complex novel, which wears its considerable Arthurian learning so lightly, not only breathes new life into old legends: it also uses the tale of their constant transmogrifications to illuminate present-day media-manipulation of news and events.

Brian MOORE   1921–

The Lonely Passion of Judith Hearne

Judith Hearne is a spinster in her early forties. Convent educated, devoutly Catholic, plain and shabby-genteel, she has moved into a boarding-house on Belfast's Camden St. Among her possessions – beside a picture of the Sacred Heart – is a photograph of her Aunt D'Arcy, who brought Judith up, instilling in her a narrow snobbery and, through manipulative, chronic illness, keeping her housebound for her marriageable years. Since her aunt's death Judith has survived on a small annuity, which she augments by giving piano lessons. She is desperate for a man. She has, Moore intimates, a little weakness. Her only friend, Edie, is hospitalised.

James Madden has just returned from the USA. The brother of Mrs Rice, Judith's landlady, he too is lodging in Camden St. Judith is charmed by his boasts of New York prosperity and excited when she learns that he is a widower. He, for his part, fudges the fact that he was merely a doorman. Embittered by his failure in the land of opportunity, he suspects that this spinster may provide him with capital for a business venture. Meanwhile, he is flattered that she listens to his tales – although he will, of course, avoid marrying her. A bleak kind of comedy ensues: Judith's bathetic pursuit, Madden's evasions. The situation is exacerbated by Bernard, Mrs Rice's pampered and poisonous son: his dalliance with the servant-girl has been hampered by Madden's prurient, sadistic interference. He hopes to manoeuvre Madden out of the house.

A quarrel with Mrs Rice prompts Judith to indulge her 'little weakness' – a fanatical appetite for whiskey. This precipitates further disasters. Overcome by pious remorse, Judith is rejected by Madden and ordered to leave Camden St. She embarks on a self-destructive binge, squandering money she can ill afford. Her emotional crisis mutates into religious agony: God, she decides, has never heard her prayers. Drunkenly, she seeks guidance from her priest, companionship from Edie and sympathy from a tolerant long-time acquaintance, Mrs O'Neill. No one can truly help or solace her. In a final effort to provoke God to speech – and probably to elicit some mortal attention – she desecrates the altar of a church.

Outside the confessional, the children appeared to be engaged in a form of musical chairs. As each small penitent left the box, the remaining queue moved up one bench nearer. Thus, every few minutes the crocodile reared up, weaving in and out of the benches, and Miss Hearne, a tall vertebra in the crocodile's tail, was obliged to move in turn. It was distracting and she found little time for sustained prayer.

THE LONELY PASSION OF JUDITH HEARNE
BY BRIAN MOORE

Bernard gets his wish: Madden flees to Dublin. Judith recovers from her nervous and alcoholic exhaustion in a hospital bed paid for by Mrs O'Neill. Nuns dispense spiritual encouragement. The priest who was useless before visits to offer inapposite platitudes now. Yet Judith may regain her faith. Indeed, the novel ends much as it began – with the picture of the Sacred Heart and the photograph of Aunt D'Arcy regarding Judith from a dressing-table.

Christ and the aunt have both misled Judith. The former has aroused spiritual expectations, the latter social ones. Neither has helped Judith address a practical question: How is she to survive in 1950s Belfast? On the contrary, her loneliness has been buttressed by her absurd guilts and foolish perception of her own status. Hence, Moore's criticism thrusts neither at the illogicality of Catholicism nor at the pretentiousness of class. Its target is the irrelevance of the two systems.

The story is told from shifting third-person perspectives, predominantly those of Judith and James Madden. (There is, however, an interlude of Joycean monologues.) The novel, which was originally published simply as *Judith Hearne*, won the Authors' Club First Novel Award.

Vladimir NABOKOV   1899–1977

Lolita

Nabokov's dense, allusive and linguistically playful account of a middle-aged Englishman's obsession with a twelve-year-old American girl was responsible for his popular, international reputation. The novel started life as a thirty-page short story, set in France and written in Russian (1939–40). Rewritten at full-length and in English ('the language of my first governess'), it was offered to, and turned down by, four horrified American publishers, but snapped up by the enterprising French underground house, Olympia Press. Its publication in Britain was attended by a great deal of controversy and delayed until 1959.

Subtitled 'The Confessions of a White Widowed Male', it takes the form of the prison confessions ('Oh, my Lolita, I have only words to play with!') of Humbert Humbert, a European-educated scholar whose sexual desires were fixed at the age of thirteen when his young girlfriend died of typhus. Since then he has been attracted only to pubescent sirens: 'nymphets'. After a disastrous marriage, he goes to America to take advantage of an inheritance. There he finds his ideal in Dolores (Dolly, Lo, Lola, Lolita) Haze, the daughter of a widow with whom he lodges. With wit and lyricism the first half of the book charts his infatuation. Lolita recognises and encourages his interest in her, but her mother, Charlotte, is also drawn to him. In order to take advantage of the daughter, Humbert marries the mother and plans to murder her. His nerve fails, but a fortuitous accident ('hurrying housewife, slippery pavement, a pest of a dog, steep grade, big car, baboon at its wheel') removes Charlotte just as she has discovered Humbert's real interests by reading his journal. Humbert and Lolita take off on a car journey across America and consummate their relationship.

In part two Lolita is enrolled at the advanced Beardsley School, the curriculum of which concentrates upon 'the four D's: Dramatics, Dance, Debating and Dating'. Gradually she is growing up and away from Humbert, turning from delicious nymphet into 'that horror of horrors', a college girl. Bored with Humbert, whom she sarcastically addresses as 'Dad', she eventually takes off with the repulsive author of the school play, Clare Quilty, who is the nephew of her old dentist. Humbert goes in pursuit of them from motel to motel in 'a cryptogrammic paper chase' which the taunting Quilty wins. Over the next few years Humbert tracks down Lolita, now married to a deaf young man called Richard Schiller, and hugely pregnant. It transpires that Quilty had abandoned her when she refused to take part in a pornographic film. Humbert tracks Quilty to his sumptuous mansion and messily murders him. Humbert dies in prison while awaiting trial, and Lolita dies giving birth to a stillborn daughter.

In spite of Humbert's fate, Nabokov insisted that the story has no moral: 'For me a work of fiction only exists in so far as it affords me what I shall bluntly call aesthetic bliss, that is a sense of being somehow, somewhere, connected with other states of being where art (curiosity, tenderness, kindness, ecstasy) is the norm.' Authentically erotic, but never merely pornographic, its central relationship provides a paradigm of the author's love affair with the English language.

Anthony POWELL   1905–

The Acceptance World

see **A Dance to the Music of Time** (1975)

J.R.R. TOLKIEN   1892–1973

The Lord of the Rings

The Fellowship of the Ring (1954)
The Two Towers (1954)
The Return of the King (1955)

Tolkien's epic three-volume sequel to **The Hobbit** (1937), was begun at the request of his publishers and took some twelve years to write. Originally conceived as a short book for children, it gradually evolved into a much larger (and, some would argue, less satisfying) work. Daunted by the increasingly ambitious scope of his undertaking, Tolkien wrote the book in fits and starts, breaking off to write other stories for children. Episodes were read out at meetings of 'The Inklings', a donnish bachelors' society in Oxford attended by Charles Williams, C.S. Lewis and others. It was Lewis who gave Tolkien the greatest encouragement, although during one of Tolkien's readings of work-in-progress, he was heard to mutter: 'Not another fucking elf.'

Although the book was an overwhelming commercial success, and became a cult novel, particularly on campuses – undergraduates were seen wearing sweatshirts pronouncing 'Gandalf for President' during Union elections – critical opinion remains divided. Some have seen the book as a masterpiece in the traditions of the medieval romance; other readers, more resistant to the mythology of Middle-Earth, have echoed Lewis's groan.

Sixty years after the adventure described in The Hobbit, Bilbo Baggins vanishes mysteriously from his birthday party at Bag End. Before disappearing, he has made young Frodo Baggins his heir and entrusted the Ring to his care. Gandalf suspects it to be the One Ring of Power forged long ages ago for the Dark Lord, Sauron, who urgently wishes to possess it again in order to increase his strength and to dominate Middle-Earth. The Dark Lord, says Gandalf, must be thwarted, and Frodo, with three hobbit friends (Pippin, Merry and Sam Gamgee), sets out for the White Council at Rivendell. After some horrifying adventures, they reach the

---

Day was waning. In the last rays of the sun the Riders cast long pointed shadows that went on before them. Darkness had already crept beneath the murmuring fir-woods that clothed the steep mountain-sides. The king rode now slowly at the end of the day. Presently the path turned round a huge bare shoulder of rock and plunged into the gloom of soft sighing trees. Down, down they went in a long winding file. When at last they came to the bottom of the gorge they found that evening had fallen in the deep places. The sun was gone. Twilight lay upon the waterfalls.

[. . .] Merry looked out in wonder upon this strange country, of which he had heard many tales upon their long road. It was a skyless world, in which his eye, through dim gulfs of shadowy air, saw only ever-mounting slopes, great walls of stone behind great walls, and frowning precipices wreathed with mist. He sat for a moment half dreaming, listening to the noise of water, the whisper of dark trees, the crack of stone, and the vast waiting silence that brooded behind all sound. He loved mountains, or he had loved the thought of them marching on the edge of stories brought from far away; but now he was borne down by the insupportable weight of Middle-earth. He longed to shut out the immensity in a quiet room by a fire.

THE RETURN OF THE KING
BY J.R.R. TOLKIEN

Council of elves and magicians who decide that the Ring must be destroyed. But none of them wishes to possess the Ring even for a short time – it perverts and corrupts even the strongest, most altruistic mind. The Ring can be destroyed only by throwing it into the Crack of Doom, the volcano in which it was originally forged – and the Crack of Doom is in Mordor, the stronghold of the Dark Lord himself. Frodo volunteers to take the Ring, and a fellowship of nine is formed to go into Mordor: Frodo, Merry, Pippin, Sam, Legolas the elf, Gimli the dwarf, Gandalf, Aragorn the Ranger, and Boromir, man of Gondor.

The story takes the form of an epic quest, the fellowship beset by enemies and terrors as they make their way towards Mordor. A seemingly fatal blow to their plans is struck when, deep in the mines of Moria, Gandalf is attacked by a demon, a Balrog, and falls with it into a fathomless abyss. Aragorn takes over the leadership of the group, and they seek sanctuary in the magical Lothlórien where they are received by Queen Galadriel who gives them vials of water from her fountain, to be used only in times of greatest need. The baleful influence of the Ring affects Boromir, who becomes envious of Aragorn and quarrels with him. Then Boromir attempts to take the Ring by violence from Frodo who, with Sam, runs away to pursue his quest independently.

Immediately after the departure of Frodo and Sam, the fellowship is broken up – Pippin and Merry are captured by Orcs (tough and swarthy goblin troops of Sauron and his vassal, Saruman, a magician) and Boromir is killed trying to protect them. In *The Two Towers*, Aragorn, Legolas and Gimli follow the trail of the Orcs and come to the territory of the Rohirrim, warrior horsemen, who are suspicious of them until they convince Eomer, Lord of the Mark of Rohan, that they are friends and not in league with the Dark Lord. The two hobbits, meanwhile, have escaped from the Orcs and find themselves in the Forest of Fangorn where they encounter Treebeard, an ancient creature resembling a tree, of the race of Ents. They rouse him to anger against Saruman, and Treebeard leads a company of Ents against Isengard, Saruman's stronghold. Aragorn and his two companions also enter Fangorn where they see a mysterious figure who, to their astonishment, reveals himself as Gandalf. He tells of his return

from the battle with the Balrog, and hints at a profound transformation – he has become Gandalf the White, ousting Saruman as foremost magician of his Order, and more powerful than before.

This second volume in the trilogy is mostly concerned with the war and individual battles fought against Sauron by the races of Middle-Earth. Saruman's power is broken, and some advances are made against the power of Sauron. However, Frodo and Sam, at the border of Mordor, are pursued by Gollum who seeks his treasured ring ('my precious'), and as they attempt to enter Mordor through a dark tunnel in the border mountains they are attacked by the monstrous spider Shelob, who apparently mortally wounds Frodo and is driven back only by Sam wielding the blazing phial of Galadriel and by grievous blows of his sword.

Sam, convinced that Frodo is dead, determines to carry on the quest alone. He takes the One Ring from Frodo's finger, and reluctantly leaves the body of his friend. Some passing Orcs discover Frodo and carry him off – merely wounded, as Sam overhears them say, and not yet dead. Frodo, to Sam's horror, is taken into an Orc fortress.

After the destruction of Saruman's citadel by the Ents, the races of Middle-Earth gather to do battle with the forces of Sauron. In *The Return of the King*, Rangers, men of Gondor, men of Rohan, a ghostly host of the dead from the Haunted Mountain, Gandalf, Merry, Pippin and Aragorn all combine to beat off the forces of Sauron which, in addition to his Orcs, include the ghostly Nazgûl, terrible and faceless creatures of heartless evil. The Lord of the Nazgûl is wounded by Merry and slain by the Princess Eowyn on the field of battle. The battle won, the forces of Middle-Earth ride to Mordor to the very gates of Sauron's stronghold where they learn that the Ring-Bearer, Frodo, is Sauron's captive.

Sam, wearing the One Ring, has not been idle. He enters the Orc fortress and succeeds in escaping with Frodo who, bone weary and profoundly affected by the burden of the One Ring, nevertheless struggles with the last of his strength and courage to the lip of the Crack of Doom where, in the last moment of the quest, he is finally seduced from duty by the power of the Ring and says, 'I do not choose now to do what I came to do. I will not do this deed. *The Ring is mine!*' Sam, aghast, is suddenly struck down by a blow from behind – Gollum has tracked them to the end, and is determined to have his prize. He bites off Frodo's finger, and gloats over the Ring he now possesses – but, losing his footing, falls with the Ring over the edge into the Crack of Doom, where the Ring is finally destroyed.

With the Ring, Sauron's power is also destroyed. The forces of Middle-Earth triumph over the Dark Lord and Aragorn is revealed as the King of Gondor. Those most closely connected with the Ring leave Middle-Earth for the elven lands of the Grey Havens, and Sam returns to the Shire.

Evelyn WAUGH	1903–1966

Officers and Gentlemen

see **The Sword of Honour Trilogy** (1961)

1956

Suez crisis • Hungarian Revolution • Nikita Khrushchev denounces policies of Joseph Stalin • Pakistan becomes an Islamic Republic • US presidential election, Dwight D. Eisenhower, Republican, re-elected over Adlai Stevenson, Democrat • Transatlantic telephone service • Discovery of the neutrino and the anti-neutron • CND members march from Aldermaston in protest against nuclear arms • Colin Wilson, *The Outsider* • John Osborne, *Look Back in Anger* • Ingmar Bergman, *The Seventh Seal* • 'Rock 'n' Roll' dominates dance-floors • Max Beerbohm and Walter de la Mare die • Allen Ginsberg, 'Howl'

James BALDWIN	1924–1987

Giovanni's Room

Baldwin's major novels dramatise the plight of the individual for whom the question of personal identity is at odds with that of social survival. In his first novel, *Go Tell it on the Mountain* (1953), the question is one of race; in his second, *Giovanni's Room*, it is of sexuality. Later novels would discuss both these questions in tandem, but at this stage in his career Baldwin was concerned not to over-reach himself artistically, and his first two novels are, therefore, complementary tales of individuals who sacrifice personal fulfilment on the altar of social conformity.

The success of *Giovanni's Room* depends largely upon Baldwin's ironic handling of two favourite American themes: the innocent from the New World being put to the test of experience by the Old World; and the impossibility of creating a separate peace from the pressures of society. Both David and his girlfriend, Hella, are representative Americans who have come to Europe in search of themselves through sexual and social freedom; but when this search raises questions which are too disturbing to answer, they retreat into conformity, and reassume the American values they had thought to leave behind. David comes to believe that 'nothing is more unbearable, once one has it, than freedom'. Hella admits: 'I'm not really the emancipated girl I try to be at all. I guess I just want a man to come home to me every night.' Her fears of not being a 'real' woman are matched by David's of not being a 'real' man. He feels that his affair with Giovanni has caused him to slip through the 'web of safety' and he longs to be secure again, 'with my manhood unquestioned, watching my woman putting my children to bed'.

The real innocent is Giovanni, a fact David recognises upon their meeting in the homosexual underworld of Paris, where Giovanni works in a bar. Giovanni's honesty, integrity and dignity are thrown into relief by the sordidness of his surroundings, like a figure in a religious icon, with 'all the light of that glowing tunnel trapped about his head'. After David leaves him in order to become engaged to Hella, Giovanni returns to the streets. He attempts to get back his job in the bar, and is tricked into bed by the owner, Guillaume, who subsequently refuses to employ him. Giovanni strangles him, goes on the run, but is subsequently caught and condemned to death. The novel opens and closes with David's thoughts as he imagines Giovanni being led to the guillotine.

Another character, Jacques, voices Baldwin's own philosophy of homosexual love when he urges David to seize this opportunity for happiness; 'you can make your time together anything but dirty ... if you will not be ashamed, if you will not play it safe'. But this is precisely what David is incapable of doing: the influence of the Old World is ineffective against the power of the New and the separate peace, as represented by the room of the title, is invaded and destroyed.

At the outset Giovanni's room is a sanctuary from social pressures, but as David's sense of sexual shame at his defection from the rigid, puritan, masculine code returns, it becomes a trap. In his later novels, Baldwin's characters would rebel against the destiny marked out for them: *Giovanni's Room* reveals the tragedy that awaits those incapable of rebellion, the passive victims of their own limitations and the proscriptions of society.

Samuel BECKETT   **1906–1989**

Malone Dies

see **The Beckett Trilogy** (1958)

Sybille BEDFORD   **1911–**

A Legacy

Bedford, who was born in Charlottenburg in 1911, drew upon family memories in this, her best-known novel, a richly detailed recreation of German upper-class life before the First World War. A tragi-comedy, it is narrated by Francesca, but most of the story takes place before her birth and is concerned with the lives of her father and his first wife's family:

> What I learnt came to me ... at second and at third hand, in chunks and puzzles, degrees and flashes, by hearsay and tale-bearing and *being told*, by one or two descriptions that meant everything to those who gave them ... Which memories are theirs? Which are mine? I do not know a time when I was not imprinted with the experiences of others. In a sense this is my story.

Hence the book's title.

When Julius von Felden marries the much younger Mélanie Merz he is unexpectedly accepted by her family. The Merzes are members of the Jewish *haute bourgeoisie*, wealthy, metropolitan and philistine; the von Feldens are francophile (and francophone) aristocrats from Baden in the south, 'undilutedly Catholic' and highly cultured. The marriage is promoted by Sarah, an aniline heiress, who collects Impressionist paintings and is unhappily married to Edu, the Merzes' second son. Julius arrives in Berlin with his pet chimpanzees to discover that the naïve Mélanie imagines that she has sorted out their religious differences by being received into the Protestant Church. The marriage is disastrous and Mélanie swiftly succumbs to consumption, despite which Julius continues to be regarded as a valued member of the Merz family. Sarah next introduces him to an English woman called Caroline, who is unhappily embroiled in an affair with a married man. Caroline becomes Julius's second wife and the narrator's mother.

Meanwhile the death of Julius's gentle younger brother, Johannes, gives rise to 'the Felden Scandal'. When he ran away from his cadet school, he suffered a breakdown and refused to return. To avoid a scandal, he had been given a commission and quietly put in charge of a stud-farm. When the new colonel of the regiment, an officious Prussian, insists upon Johannes being brought to him, Johannes refuses to come and is shot in an ensuing brawl. Whether or not this is a case of murder becomes a national issue, and both families are vilified in the press. The government is expected to fall as a series of indiscretions and errors are revealed, ultimately involving the Kaiser.

This motor seemed to make itself a place. Whenever, later on, Caroline tried to think about that period of her life, she saw Sarah's motor. Sarah's motor (it was its sole identity) with herself in it going somewhere at great rate; Sarah's motor being sent for her, Sarah's motor waiting at all the doors. It solved and created many of their problems – being in two places at one time, or almost; then failing altogether to get them to a third, and it seemed to impose its pattern on their days. The motor, Sarah made known over the telephone, should be taken out, should not be taken out.

A LEGACY
BY SYBILLE BEDFORD

Like most family histories, this one is complicated, and there are numerous interrelated threads concerning other members of both families, notably Julius's selfish and unprincipled elder brother, Gustavus, who is married to the prim and devout Clara, and commits suicide when his treachery towards Johannes is exposed; Sarah's husband, Edu, bankrupted by gambling debts; Jeanne, Friederich Merz's French mistress, who is not admitted to the family home; and Grandpapa and Grandmama Merz, whose household and affairs are run with insolent efficiency by Gottlieb, the outspoken butler. As well as the division within families, the story also illustrates the division within the Kaiser's Germany between the gentle south and the Prussian north.

Patricia HIGHSMITH   1921–

The Talented Mr Ripley

'Was that the kind they sent on a job like this, maybe to start chatting with you in a bar, and then *bang*! – the hand on the shoulder, the other hand displaying a policeman's badge. *Tom Ripley, you're under arrest.*' In *The Talented Mr Ripley*, the first of Highsmith's disturbingly objective psychological thrillers about the eponymous hero, Tom Ripley is introduced as a young man with a psychopathic streak of criminality. (Highsmith's work, from the stories she wrote as a teenager onwards, has been influenced by Karl Menninger's *The Human Mind*, a study of psychological abberation she had discovered at the age of nine.) Living on the fringes of New York society, he engages in extortion as a tax-office clerk, unable to cash the cheques he collects, but doing it for the thrill of taking the risk. However, the father of an old school acquaintance tracks him down to send him on a mission which will change the course of his life.

Dickie Greenleaf has everything Tom Ripley lacks: a wealthy family and privileged background, and the economic freedom to lead a Bohemian life in Europe, where, despite his unpromising talent, he is trying to be an artist. Dickie's father disapproves of his son's activities, and offers Tom the bribe of a lavish expense account if he will go to Italy to persuade Dickie to come home and work in the family business. In the little town of Mongibello, Tom reacquaints himself with his old friend and wins his trust by telling him the truth about his father's plans. Dickie's girlfriend Marge is suspicious of Tom's motives, and warns him that she believes Tom to be homosexual.

Dickie feels oppressed by Marge's adoration of him, and agrees to accompany Tom on a sightseeing tour for a few days' break. The two young men take a motorboat out into a harbour, and Tom seizes the opportunity to act out his sinister fantasy. Murdering Dickie and anchoring the corpse in the deep water, he takes over his friend's identity, drawing Dickie's allowance and wearing his clothes.

In Rome, Tom passes himself off as the wealthy Dickie Greenleaf, until an old friend arrives and the deception is revealed. To evade arrest Tom commits a second murder, bringing 'Dickie' under suspicion. Marge, arriving unexpectedly, is sure that her boyfriend cannot be a killer, but Tom (*in propria persona*) plants doubts in her mind about Dickie's mental stability. However, the police are making enquiries about forged signatures in Dickie's cheque book, and Tom is frightened into giving up the game and reporting to the police in Venice as Tom Ripley himself, whom they had assumed to be a missing person.

Living on the remains of the money he has stashed away, Tom convinces both Marge and Dickie's father that Dickie has committed suicide. But the comfortable standard of living to which he has become accustomed makes him loath to return to his old status. Forging a will in which Dickie bequeaths everything to him, Tom hazards all on the remote possibility that his story will be believed. As Highsmith's narrative builds up to its tense climax, it looks as if Tom's deadly deception will be discovered; but the talented Mr Ripley has hoodwinked everyone, and he inherits the large estates of his victim.

C.S. LEWIS   **1898–1963**

The Chronicles of Narnia

The Lion, the Witch and the Wardrobe (1950)
Prince Caspian (1951)
The Voyage of the **Dawn Treader** (1952)
The Silver Chair (1953)
The Horse and His Boy (1954)
The Magician's Nephew (1955)
The Last Battle (1956)

At a debate at the Socratic Society in Oxford in 1948, C.S. Lewis's argument for the existence of God was roundly crushed by the then Professor of Philosophy at Cambridge, the formidable Elizabeth Anscombe. Lewis recognised the force of the professor's argument and felt utterly humiliated. The incident was just the motivation Lewis seems to have needed to resume the writing of *The Lion, the Witch and the Wardrobe*, which he had put to one side in 1939. In these subversive Christian allegories (a term Lewis himself thought inappropriate), he was able to wreak revenge not only upon those who undermine faith through logic (there are numerous occasions where logical intuition is proved both wrong and the work of the Devil), but upon Prof. Anscombe in particular and women in general. Lewis was of the opinion that women's minds were intrinsically inferior to those of men. At the same time he found women frightening.

Lewis claimed to have written the Narnia stories within a couple of years, but in reality *The Last Battle*, the final story in the sequence, was not finished until March 1953. They were probably written in imitation of J.R.R. Tolkien's **The Hobbit** (1937). Although a close friend, Tolkien disliked the Narnia stories intensely. He would have particularly objected to the coarseness of much of Lewis's writing and the inconsistencies between the books. The stories' continuing popularity with children, if not with adults, is probably due to the child's acute sensitivity to what is truly felt. As Tolkien observed, it is impossible to write 'for' children. That Narnia was created to satisfy Lewis's own needs as much as to entertain children is evident throughout these stories. Ovid, Malory and Spenser are just a few of the sources Lewis the scholar drew on, if in an unscholarly fashion. As a storyteller he modelled himself on E. Nesbit and George MacDonald, though the influence of *Gulliver's Travels* (1726), the Alice stories, and the Arabian Nights is also apparent. His debt to Nesbit is acknowledged in *The Magician's Nephew*, which explicitly refers to the Bastable children, heroes of *The Story of the Treasure-Seekers* (1899) and other books.

The series starts with the creation of Narnia and ends with its destruction, but the individual volumes were published out of chronological sequence. In *The Lion, the Witch and the Wardrobe*, often considered the first volume, Lucy Pevensie discovers Narnia at the back of a wardrobe in a large country house where she and her three elder siblings had been spending their holidays. However, *The Magician's Nephew* is a 'prequel' to the series, in which we discover that the old, unnamed professor who owns the house is Digory Kirke, who had as a boy been present when Narnia was created by Aslan the lion. Digory is living with his Aunt Letty and Uncle Andrew, who are brother and sister. His father is away in India, and his mother is dying in an upstairs bedroom. Digory becomes friendly with the girl next door, Polly Plummer. Uncle Andrew is a useless magician, whose godmother had fairy blood and had given him some magic rings. Too cowardly to wear them himself, he makes the children wear them. They find themselves at 'the Wood between the Worlds': a place of trees and pools. The pools are ways to other worlds, and through one of them they arrive at a world just as it is ending. They escape, but with the Amazonian Jadis, Queen of the House of Charm, in tow. She wreaks havoc in London before the children can drag her back to the between world, but this time they also have with them a cabman, his horse, Uncle Andrew and an iron bar which the Queen has wrenched off a lamp-post.

Plunging into another pool they find themselves at the beginning of Narnia, witnesses of the First Voice, Aslan singing. Aslan, who represents Christ in these stories and is said to have come to Lewis after a nightmare about lions, breathes life into the creatures of Narnia and separates out the talking animals from the brute beasts. (Talking, for Lewis, is akin to having a soul.) For a while, everything planted in the soil grows – including the iron bar, which grows into the lamp-post that marks the edge of Narnia and will guide the Pevensies back to the wardrobe. The cabman and his wife (whom Aslan calls to Narnia) become the first King and Queen of Narnia, and Jadis the first presence of evil. Digory is given a magic apple which will cure his mother, and the core is planted in the garden. Years later the tree blows down; the wood is used to make the wardrobe through which the Pevensies later find themselves in Narnia.

In *The Lion, the Witch and the Wardrobe*, Lucy learns something of the history of Narnia from a faun, Mr Tumnus, who is in thrall to the White Witch. The Witch is descended from the daughter of Adam's first wife, Lilith (a Jinn, one of the fiery mountain spirits in Muslim mythology), on one side, and from Giants on the other. (Lewis is obsessed with nobility and blood lines. He says of another character: 'Not very clever, perhaps ... but an old family. With traditions, you know.') She has ruled Narnia for 100 years, during which time it has been always winter but never Christmas, and her reign can only come to an end when the four thrones at the castle Cair Paravel are occupied by Sons and Daughters of Adam: Peter, Susan, Edmund and Lucy Pevensie. In order to prevent this, Mr Tumnus has instructions to capture any human children he finds and deliver them up to the Witch; he is too kind-hearted, however, and allows Lucy to return home.

At first the other children find Lucy's story too incredible, but when Edmund, the younger of the two brothers, discovers Narnia for himself, Lucy imagines she has an ally. Edmund, however, is an unpleasant child, who betrays her, just as he betrays Narnia by going over to the White Witch. The Witch exacts a penalty for this: Edmund belongs to her, and to be freed someone else must die in his stead. It is Aslan who lays down his life for the children of Adam in a scarcely veiled re-enactment of the Crucifixion. 'Deep Magic' (the 'Emperor's Magic'), stronger even than Aslan, and capable of reversing death, restores Aslan to life.

The Horse and His Boy is set during Narnia's Golden Age, the reign of High King Peter and his brother and sisters. It is the least successful story, probably because the Pevensies hardly appear and in any case they are now adults, who talk in a cod courtly manner. The story, which adds little to our knowledge of Narnia, concerns the adventures of a boy called Shasta, who is about to be sold into slavery, but escapes from the land of Calormen upon Bree, his talking Narnian horse. On their travels, Shasta and Bree meet a princess, Aravis Tarkheena, and her horse, Hwin. In order to escape an arranged marriage, she too is fleeing

Calormen. After various adventures, in which Bree is taught a lesson by Aslan for having doubted his 'lionness' (in other words, the corporeality of Christ), Shasta discovers that he is actually one of the twin sons of King Lune of Archenland. The Pevensies enter the story via a sub-plot involving an attempt by Prince Rabadash, the son of the Tisroc (the ruler of the city of Tashbaan in Calormen), to coerce Susan into marrying him.

In *Prince Caspian*, the Pevensie children discover that it is now many generations on from their reign. Caspian the Tenth is too young to rule and so the country is being run by his wicked uncle, King Miraz, and Queen Prunasprismia. Miraz has wiped Narnia from the history books, and the talking creatures have either gone underground or have been killed, but Caspian is secretly taught something of the country by his tutor, Dr Cornelius, who is really a Narnian dwarf in disguise. Caspian also learns of a plot to kill him and he escapes. The Pevensie children are on their way to school at the beginning of a new term when suddenly they are called to Narnia by Susan's horn, which she had been given by Father Christmas in *The Lion, the Witch and the Wardrobe* and which is now owned by Caspian. During their adventures, Lucy is again doubted (except by Edmund, who recalls his previous betrayal) when only she can see that Aslan is with them. (Lucy's sufferings relate to a story Lewis was told by his mother, who as a girl on holiday in Rome had seen a statue move. She had not been believed and the feeling of humiliation had stayed with her throughout her life.)

After various adventures there is the inevitable battle: after the battle of Aslan's How (a mound where the magic stone lies on which Aslan was killed in the earlier story), Narnia reawakens and Bacchus goes abroad with his Maenads, 'his fierce, madcap girls'. 'I say Lu,' says Susan, 'I wouldn't have felt safe with Bacchus and all his wild girls if we'd met them without Aslan.' 'I should think not,' says Lucy. Caspian is installed as king and Aslan reveals that Peter and Susan are now too old for further adventures in Narnia.

Edmund and Lucy, however, return to Narnia in the next volume through the painting of a ship, the *Dawn Treader*. The painting, which suddenly comes to life as they look at it, hangs in the house of their Uncle Harold and Aunt Alberta Scrubb. The children are spending the school holidays with them and their unpleasant son, Eustace, who is indistinguishable from the Edmund of *The Lion, the Witch and the Wardrobe*. Eustace's parents are 'modern', something to be despised: 'They were very up-to-date and advanced people. They were vegetarian, non-smokers and tee-totallers, and wore a special kind of underclothes.' Edmund and Lucy have been called to help Caspian find the seven lost lords, friends of his father, who were banished to the Eastern Islands by King Miraz. By the end of the story, in which Eustace has been turned into a dragon and back again, but as a better person, the seven Lords have been accounted for. The children meet a lamb, who turns into Aslan. This is Edmund and Lucy's last adventure in Narnia; when they meet Aslan again, it will be in his own country (Heaven), where, he says: 'I have another name.'

In *The Silver Chair*, Eustace is at a mixed school ('some said it was not nearly so mixed as the minds of the people who ran it'), where he befriends a girl who is being bullied, Jill Pole. Escaping from the bullies through a gate in the school wall which is usually locked, they find themselves in Narnia at the end of Caspian's reign. Caspian's son Rilian has disappeared during a journey to search for his mother's killer, a venomous green worm. Eventually, after avoiding being made into Man-pie and eaten by Giants at their autumn feast, Jill and Eustace find themselves in the Underland of the Lady of the Green Kirtle, who had directed them to the Giants in the first place. The passive knight by her side turns out to be Rilian, who has been bewitched. Every so often Rilian must be bound to a silver chair to keep him under her spell. He is freed by Eustace and Jill under the guidance of Aslan. They also release the Lady's subjects, the Earthmen, who belong to an even deeper world – the Land of Bism, to which they return. As the Lady's spell is broken, she turns into the worm and is beheaded by Rilian. At one point the Lady tries to make them believe that hers is the real world and Narnia only a dream. One of the characters, as he resists her charms, says revealingly: 'I'm going to live as like a Narnian as I can even if there isn't any Narnia.'

The Last Battle is the counterpart of *The Magician's Nephew* (published the previous year), and Lewis's illustration of his Manichaean beliefs. King Tirian, the last king of Narnia, is typical of Lewis's masculine ideal: 'between twenty and twenty-five years old; his shoulders were already broad and strong and his limbs full of hard muscle'. A false prophet is abroad. An ape, Shift, has persuaded a dim donkey called Puzzle to wear a lion's skin and pretend to be Aslan. Polly and Digory, now in old age, have called all the visitors to Narnia together, except Susan, who has stopped believing and is 'no longer a friend of Narnia': 'She's interested in nothing nowadays except nylons and lipstick and invitations.' They all die in a train crash: Polly and Digory, Peter, Edmund and Lucy go straight from the 'Shadowlands' of Earth to Aslan's kingdom, where they will meet Jill and Eustace, who have been called through death for one last adventure in Narnia. The Calormenes are felling the Lantern wastes where the Dryads live; the Narnian dwarves have for the large part deserted to the Calormene army. Tirian is struggling to raise an army in the last battle against the Calormene captain, Rishda Tarkaan. Rishda claims that a certain stable houses the great god Tash, whom he has supposedly summoned. Although he does not realise it, the stable actually contains the entrance into Aslan's kingdom. '"Yes," said Queen Lucy. "In our world too, a stable once had something inside it that was bigger than our whole world."'

When Tash really does arrive, he claims Rishda for his own kingdom of the dark. The last battle is won by Aslan, and he leads all who are with him into his kingdom as Narnia comes to an end. Aslan's kingdom is discovered to be the world of Platonic Ideals ('It's all in Plato'), which contains both our world and Narnia. As everyone travels higher up into the world, it gets bigger but more and more connected: 'The term is over: the holidays have begun.'

Rose MACAULAY 1881–1958

The Towers of Trebizond

Ostensibly the story of a group of eccentric people 'riding about Armenia hawking the C. of E. to infidel dogs who thought we were mad and were probably right', *The Towers of Trebizond* is a poetic, witty, tragicomic novel primarily concerned with the loss of faith. Macaulay herself had abandoned the Anglican church in similar circumstances to her narrator, although by the time she came to write the book she had returned to its fold. The novel achieves a perfect balance between high, often black, comedy and a deeply felt meditation upon love and religion, and was highly praised by Macaulay's fellow writers: Elizabeth Bowen, John Betjeman, Compton Mackenzie, and Anthony Burgess, who listed it 'among the twenty best novels of the century'. It was awarded the James Tait Black Memorial Prize, became a bestseller in the USA, and is Macaulay's masterpiece.

Its comic tone is established in the celebrated opening sentence: ' "Take my camel, dear," said my aunt Dot, as she climbed down from this animal on her return from High Mass.' The narrator is a young woman called Laurie, who joins her indomitable aunt, Dorothea ffoulkes-Corbett, a handsome but neurotic camel, and Revd the Hon. Fr Hugh Chantry-Pigg ('an ancient bigot who had run a London church several feet higher than St Mary's Bourne Street and some inches above even St Magnus the Martyr') upon an expedition to Turkey and the Black Sea. Funded by the Anglo-Catholic Missionary Society, they are hoping to convert the women of Turkey to Christianity and liberation. Aunt Dot is compiling a book about her experience to be entitled *Women of the Euxine today*, for which Laurie is providing illustrations, while Fr Chantry-Pigg is hoping to collect relics for the private oratory of his Dorset manor house.

As they approach Trebizond, Laurie is taken to task by the priest for 'shutting the door against God'. Her reason for doing so is an adulterous affair, for love (as

Macaulay knew) 'was the great force, and drove like a hurricane, shattering everything in its way, no one had a chance against it, the only thing to do was to go with it, because it always won'. 'The real point of the story,' Macaulay wrote, 'is a great nostalgia for the church, on the part of the central character, who is lapsed from it', and the legendary city of Trebizond becomes a symbol of Laurie's agnosticism, with its 'disused, wrecked Byzantine churches that brooded, forlorn, lovely, ravished and apostate ghosts, about the hills and shores of that lost empire'.

The trouble with countries is that, once people begin travelling in them, and people have always been travelling in Turkey, they are apt to get over-written, as Greece has, and all the better countries in Europe, such as Italy and France and Spain. England has not been over-written, at least not by foreigners, on account of its not being very attractive, what with the weather and the Atlantic Ocean and the English Channel and the North Sea and the industrial towns and not having many antique ruins, but above all the weather, for no one from abroad can stand this for long, and actually we can't stand it for long ourselves, but we have to.

THE TOWERS OF TREBIZOND
BY ROSE MACAULAY

The expedition is joined by Xenephon Paraclydes, a Greek student who has 'a strong hereditary objection' to the Moslem faith, and Dr Halide Tanipar, a patriotic Turkish feminist and Christian convert. Like Laurie, Halide is torn between *agape* and *eros*, for the man she loves is Moslem and wants a Moslem wife. (She eventually renounces Christianity to marry him.) Laurie also meets Charles Dagenham, a Cambridge contemporary who is travelling around the Black Sea with his friend David Langley in order to write a travel book. After a quarrel, the two men part company and Charles is subsequently killed by a shark.

When aunt Dot and Fr Chantry-Pigg disappear across the Russian border, Laurie is left to fend for herself. David has been passing off his late friend's work as his own in the newspapers, but Laurie has inherited Charles's notebook, and uses it to blackmail David into looking after her. She eventually makes her way back to England, accompanied by an ape, whom she instructs in chess, gardening, painting, driving and religion.

She rejoins her lover, but when he is killed in a car accident that seems almost willed, she is left without either love or faith: 'If the object of pleasure be totally lost, a passion arises in the mind which is called grief. Burke: and he did not overstate.' Aunt Dot, safely returned, suggests that Laurie should come back to the church now that the chief obstacle no longer exists; to do so, however, would be a denial of the intense happiness this relationship had given her. Laurie will stand for ever outside the gates:

> At the city's heart lie the pattern and the hard core, and these I can never make my own: they are too far outside my range. The pattern should perhaps be easier, the core less hard.
> This seems, indeed, the eternal dilemma.

Mary RENAULT   1905–1983

The Last of the Wine

One of Renault's many fictional reconstructions of classical Greece, *The Last of the Wine* is narrated by an Athenian citizen, Alexias, who tells of his youth and young manhood at the time of the Peloponnesian War with Sparta (431–404 BC). The novel centres on his relationship with his lover, Lysis (who is the young man in the dialogue by Plato called *Lysis*), but also brings Alexias (an imaginary character) into contact with many of the leading historical figures of fifth-century Athens: Socrates (mentor to the two young men), Xenophon, Alcibiades, Phaedo, Euripides, Critias, Plato.

Alexias tells of his upbringing by his father, the stern Myron, his early initiation into the circle of Socrates, his becoming a champion runner, and his progress through the stages of Athenian youth from school to military training to war against the Spartans. Meeting Lysis in the circle of Socrates, who brings the two of them together, he serves under the command of his rather older lover in battle.

The novel follows Alexias and Lysis through a period of about a dozen years. It shows them joining the Athenian navy and fighting under Alcibiades, only to live through starvation and family tragedy during the final siege of Athens by the Spartans. After the capitulation, during the ensuing regime of the Thirty Tyrants, Alexias' father, Myron, is murdered at the instigation of the tyrant Critias. Alexias and Lysis join the forces fighting to restore democracy, but at the battle of Munychia (403 BC), while Alexias kills Critias, Lysis is also killed. Alexias is left to mourn and eventually to marry Lysis' young wife.

If so far I have mentioned none of my suitors by name, you will understand why. Only their numbers had been pleasing to me in some degree, as a mark of success, as if so many trophies had been awarded me for my looks; and even so, the crowns I had won for running had pleased me more, being a thing in which my father had not excelled before me. Yet I was civil to them, even to the most foolish, out of regard for my good name; so that people said I was not spoiled by admiration, which was as I wished.

THE LAST OF THE WINE
BY MARY RENAULT

Some critics of the novel have found it carefully researched but rather stiff and lacking life, and certainly the unfailing rectitude of the two protagonists sometimes seems more redolent of the Edwardian public school than ancient Athens. But the intensity of the relationship described means that Alexias and Lysis do not forfeit our sympathy, and the writing, while formal and slightly old-fashioned, seems nevertheless to match the subject, and rises at times to impressive poetry. An example occurs after the final naval defeat of Athens: 'All night in the streets you could see lighted windows, where those who were sleeping had rekindled the lamps: but on the High City night only, and silence, and the slow turning of the stars.'

C.P. SNOW   **1905–1980**

Homecomings

see **Strangers and Brothers** (1970)

Angus WILSON   **1913–1991**

Anglo-Saxon Attitudes

Wilson identified the themes of *Anglo-Saxon Attitudes* as self-deception and class: 'I have tried in this book to show the degree to which people are limited and are able to deceive themselves, because they belong to a particular group which never associates with other groups; then they are brought into conflict with those other groups, often with explosive results.' The novel is, amongst other things, a richly detailed social comedy, with a labyrinthine plot and a very large cast of characters, most of whom have some connection with an archaeological fraud which occurred in 1912. Although the main action of the book takes place in a contemporary setting, many of the events take place much earlier and are revealed piecemeal, in flashback.

During an excavation on land belonging to Canon Portway at Melpham before the First World War, a pagan idol was found in the tomb of a Christian missionary, a circumstance which caused considerable controversy. The idol had in fact been placed there by Gilbert Stokesay, the destructive son of the archaeologist in charge of the dig. During the war in which he is to be killed, Gilbert drunkenly boasts of the hoax to his old schoolfriend, Gerald Middleton, and admits the deception to his father in a letter from the trenches. Portway also learns of the hoax through his chauffeur, who had assisted Gilbert, but both he and Lionel Stokesay decide not to disclose the facts, for fear of discrediting archaeology. Middleton, who was Stokesay's favourite pupil and has an affair with Gilbert's widow, Dollie, also remains silent about the hoax.

Middleton has meanwhile married a sentimental Dane called Inge and fathered three children. Inge accepts her husband's adultery and insists upon hiding nothing from the children. Middleton's affair with Dollie gradually disintegrates, and he fails to live up to his early academic promise, preferring to collect paintings than to devote himself to history. Although the Middletons are wealthy, Inge refuses to employ a nanny, one result of which is that through her carelessness their daughter Kay is maimed in a domestic accident. When the children are grown up, Gerald and Inge separate.

When the novel opens, the Middletons' eldest child, Robin, is director of the family firm, unhappily married to a French Catholic. He is having an affair with Elvira Portway, who is related to the Canon and is the secretary of Robin's brother, John. John is a crusading popular journalist who is investigating a case of Civil Service incompetence in which the land of a market gardener has been mistakenly expropriated by the government. The market gardener is married to one of Canon Portway's former housemaids, whose father (now incapacitated) was instrumental in the hoax. Robin intends to offer the civil servant his brother is persecuting a job with the family firm, but is foiled by the calculated indiscretion of Kay's husband, Donald, a sociologist, whom Robin has employed to lecture the workforce. John discovers that the gardener is the father-in-law of one of his former boyfriends.

The discovery of another pagan idol in the tomb of a missionary in Heligoland stirs memories of Melpham. Gerald, now the reluctant editor of *Medieval History*, is moved to attempt to investigate the circumstances of the earlier discovery. The search for the truth becomes a voyage of personal discovery as Gerald uncovers a complicated web of blackmail, deceit, coincidence and unsuspected relationships, including those of his sons. His belated attempt to be less aloof from his family and their problems is firmly rebuffed. He turns to Dollie, hoping to revive their former relationship, but realises that this would be impossible. The hoax is finally made public and Gerald accepts his nomination as the new President of the Historical Association of Medievalists.

Wilson acknowledges that the truth is rarely pure and never simple. Gerald's determination finally to acknowledge the hoax is admirable and necessary, but it is achieved at considerable cost. In particular, it destroys Rose Lorimer, a disorganised, eccentric lecturer in medieval history with a genuine love for her subject (unlike her arid colleague, Prof. Clun). Her hobby-horse is the pagan influence in medieval Christianity, and Lionel Stokesay had been her hero. Gerald's disclosures unhinge her, and at the end of the novel she is confined to an asylum near Whitby.

1957

Harold Macmillan succeeds Anthony Eden ● Rome Treaty for Common Market ●
Eisenhower and Macmillan re-establish special relationship between Britain and US ●
Wolfenden Report published ● Sputnik I launched into space ● Desegregation crisis in
Little Rock, Arkansas ● Richard Hoggart, *The Uses of Literacy* ● Francis Bacon, *Screaming
Nurse* ● Lerner and Loewe, *My Fair Lady* (musical) ● David Lean, *The Bridge on the River
Kwai* ● Arturo Toscanini, Jean Sibelius and Dorothy L. Sayers die ● Ted Hughes, *The
Hawk in the Rain* ● Stevie Smith, *Not Waving But Drowning*

James AGEE　　　　　　　　**1900–1955**

A Death in the Family

A Death in the Family was published two
years after Agee's death; left unfinished by its
author, the published text was collated from
his manuscripts and drafts. The variants and
alternative versions of certain key episodes,
coupled with barely decipherable hand-
writing of Agee's manuscripts, added to the
difficulty of producing a coherent text, but
most critics have found themselves in broad
agreement with the original editors' decisions. Agee's own intentions clearly
changed during its writing and several early
outlines project a much longer work than
that eventually published.

The central story is very simple. Jay Follett
is called to visit his father on his death-bed in
hospital. Returning home, he is involved in a
car accident and killed. The other members
of the Follett family wait for him with a
mounting sense of foreboding until the news
of his death arrives. These events are narrated by Rufus, Jay's son, in three sections
that describe their life together, the confrontation with Jay's death and the preparations
for his funeral. This last involves the whole of
the Follett family, so that Jay's interment acts
as a kind of catalyst to bring family tensions
and truths to the surface. Agee's analysis of
the family focuses primarily on the tensions
and differences between its members (particularly religious ones) and he broadens these
tensions into a wider-ranging inquiry. The
contrasts of the South, where the novel is set,
are everywhere apparent. Black and white,

rich and poor, town and country: Agee
makes the point that these contrasts are not
outside forces, a backdrop to family life, but
are actually at work deep within the structure
of the family itself.

The most important oppositions are even
wider-ranging than these. At the centre of
the novel lies the opposition between life and
death. Jay's death prompts an examination
of his life by the family, an evaluation which
is a type of reincarnation. In their memories,
this is what he will become. Their reactions,
ranging from religious faith to atheistic cynicism, lay out the range of responses available
to human beings in face of the fact of death.
Rufus himself is shattered by the death, but
the knowledge that he will go on to take a
hand in the making of the novel, that he will
come to terms with its events, provides a
muted guarantee that his own life will
continue.

Agee's manipulation of Rufus's role as
narrator is one of the novel's greatest
achievements. When the book opens, Rufus
seems only to be part of the prelude leading
up to Jay's death, but he subsequently comes
to seem the main focus of the book. Italicised
'interchapters' delve into the incidents and
feelings of his earlier boyhood and these
details resonate with those of the main
narrative. Rufus himself begins to grow into
manhood, and it gradually becomes clear
that the events are being narrated from a
much greater distance than was at first
apparent. In this way, Agee allows a third
temporal dimension to grow out of the first
two, and suggests through this growth a
continuity between all three.

On the novel's appearance, most
reviewers noted that its structure was poetic

rather than novelistic. Much of the material in the interchapters had little narrative function in the strict sense and the novel's overall form (as a *novel*) owed a great deal to the work of its editors. The roots of the novel lay in Agee's own experience of his father's death and the immediacy of this experience, even near his own death, informs this study of bereavement with power and conviction.

John BRAINE   1922–1986

Room at the Top

Braine's first novel was a critical and commercial success, instantly establishing him alongside the Angry Young Men amongst his contemporaries. It sold some 35,000 copies during the first year of publication, was serialised in the *Daily Express*, and in America was compared to Dreiser's *An American Tragedy* (1925). It was subsequently made into an award-winning film (1959), directed by Jack Clayton from a screenplay by Neil Paterson. Although subsequent novels – including a sequel, *Life at the Top* (1962) – proved popular, none made the same impact as *Room at the Top*, which most commentators regard as Braine's best book. Part of its commercial success may be attributed to what was considered at the time its daring depiction of sexuality, but this is merely one aspect of the realism which gives the novel its force. The book provides a sharp picture of post-war northern provincialism with its subtle indicators of class, and is precisely detailed in its cataloguing of evocative brand-names (something for which Braine's late books, with their slick arrays of up-market consumer goods, were criticised). By a nice irony, the book's success led to Braine leaving working-class Bingley for eminently middle-class Woking.

The novel is narrated by Joe Lampton, who is recalling events which took place ten years previously, shortly after the end of the war. Joe is an orphan, who has been brought up by his Aunt Emily in the poor, dismal Yorkshire town of Dufton. While a prisoner of war, Joe has studied for his accountancy exams, and he subsequently moves to the affluent part of Warley ('T' Top' in local parlance), and gets a job as a town hall accountant. His landlady introduces him to the Warley Thespians, and thereby to a previously unfamiliar world of affluence: big cars, posh girls and membership of the Conservative Club. It's all a long way from Aunt Emily and her tea so strong 't'spoon'll stand up in it'.

Through his involvement with the Thespians, Joe meets Susan Brown, the beautiful daughter of a rich industrialist. She remains remote and prim and seems destined to become engaged to rich Jack Wales, an officer and gentleman, who inspires hatred and envy in Joe. Joe becomes the lover of an older, married woman, Alice Aisgill, but starts to take Susan out once Jack has gone to Cambridge. 'I was the devil of a fellow,' he recalls, 'I was the lover of a married woman, I was taking out the daughter of one of the richest men in Warley, there wasn't a damn thing I couldn't do.'

I wanted an Aston-Martin, I wanted a three-guinea linen shirt, I wanted a girl with a Riviera suntan – these were my rights, I felt, a signed and sealed legacy.

ROOM AT THE TOP
BY JOHN BRAINE

He parts from Alice after an argument and sees more of Susan, but her father seems to be using his clout as a local councillor to apply pressure to end the affair. Joe is once more passionately engaged with Alice when he gets an unexpected summons to lunch with Councillor Brown, who offers to set him up in business if he will stop seeing Susan. 'If I were a younger man, I'd knock you down, by God I would!' Joe declares indignantly. This is precisely the kind of manly Yorkshire talk the councillor admires, and he now reveals that Susan is pregnant. It is arranged that Joe will marry Susan and go to work for his new father-in-law, on condition that he 'leave off Alice Aisgill'. Alice is distraught, gets drunk, and kills herself by driving her car into a wall. Joe is genuinely remorseful and goes on a drunken binge.

The novel is inconclusive in that although ten years have passed since the incidents described, we do not learn what has happened to Joe, except that he is still (in every sense) at the top and still unsatisfied. More of a Ruthless than an Angry Young Man, Joe is politically conservative and motivated by social ambition – rather like his creator, who became an intemperate right-winger and (according to his friend Kingsley Amis) half-jokingly dreamed of marching through Bradford in 'triumphal procession ... flanked by a pair of naked beauties draped with jewels'.

John CHEEVER   1912–1982

The Wapshot Chronicle

see **The Wapshot Scandal** (1964)

Lawrence DURRELL   1912–1990

Justine

see **The Alexandria Quartet** (1960)

Patricia HIGHSMITH   1921–

Deep Water

Victor Van Allen is thirty-six. Living in Little Wesley, New England, he breeds snails for a hobby and, with his substantial private income, runs the small, loss-making Greenspur Press. His domestic arrangements are curious: he sleeps in a bedroom annexe while his wife – boozy and beautiful Melinda – entertains her lovers in the house. (Vic and Melinda have not slept together for several years – not since the birth of Beatrice, their daughter.) Vic appears untroubled by the situation. His tolerance is remarked on by his friends.

Melinda's latest lover is Joel Nash. Annoyed by Nash's crassness, and wanting to frighten him, Vic claims to have murdered his predecessor, Malcolm McRae. This claim is credible – McRae was battered to death by an unknown assailant – and Nash flees. He is soon replaced by Charles de Lisle, a cocktail pianist. To Vic's embarrassment, Melinda brings de Lisle to a party and makes a show of her attachment to him. Vic

drowns de Lisle in a swimming pool. The action is not witnessed. Vic registers no outward sign of guilt (nor, to his faint surprise, does he feel any inwardly). On no stronger evidence than her suspicions, Melinda denounces Vic as de Lisle's murderer, but a verdict of accidental death is recorded. Vic's social circle deplores Melinda's behaviour – while Vic, forbearing as ever, excuses his wife.

Melinda and Don Wilson, a seedy writer of detective fiction, hire a private eye to investigate Vic. Vic is aware of the investigation; he finds it silly and contemptible and lets his opinion be known. The detective is called off. Frustrated by Vic's imperturbability, Melinda takes a further lover: Anthony Cameron, a boisterous, boorish property developer. Melinda says she wants to marry him. Vic, however, puts a stop to that: he drives Cameron to a disused quarry, kills him and dumps the body in the flooded workings. Vic is not sure, though, that his crime was perfect. He returns to the scene to check that no bloodstains remain and, as he is doing so, discovers he is being watched by Wilson. Panicked, Vic hurries back home. All self-possession gone, he strangles Melinda and smiles at the policemen as they arrive.

For Vic (and for many of Highsmith's protagonists), murder is easy – a mere matter of seizing the opportunity when it arises. The deed's commission is accompanied by no heightening of consciousness and no intensification of prose and, accordingly, it derives its horror not from its exceptionality but from its mundanity. Furthermore, in *Deep Water* the reader's sympathies are from the outset tilted in favour of Vic: he is a cultured and generous eccentric, while de Lisle and Cameron are vulgar opportunists. None the less, as the narrative progresses it subtly signals Vic's disturbedness. His teasing jokes, his fascination with minutiae, even his jovial, adult attitude to Beatrice, begin to seem less charming, more macabre. At last, the reader grasps that they are symptoms of a rigorous pathological repression. Yet, along with Vic's friends, the reader has been alarmingly misled: like the workings in which Cameron sinks, Vic's mind is truly deep and murky water.

Jack KEROUAC   **1922–1969**

On the Road

Between 1947 and 1950, Neal Cassady and Jack Kerouac took off on a freewheeling journey through the USA and Mexico in search of something outside their domestic experience. Ten years later their adventures were related in *On the Road*, with Kerouac appearing under the name Sal Paradise, Cassady as Dean Moriarty. The novel's composition has become a well-known anecdote in its own right. Returning home from his wanderings. Kerouac spent almost a year pondering how (specifically, in what form) he might convey the life he had been living. Several false starts were made, but in April 1951 he fed a 120-foot roll of teletype into his typewriter, typed for three weeks and the result, largely unrevised, was *On the Road*. Six years went by until in 1957 it was finally published by the Viking Press.

The style of the novel is undisciplined, raw and fragmentary – the very antithesis of the academic writing bequeathed by T.S. Eliot and the 'Fugitives' which was then the orthodoxy. Kerouac called his style 'spontaneous prose', and the writing process 'sketching'. Later editing was anathema, the point being to record the multiplicity of the writer's experience with the maximum of intensity, much as jazz (a favourite metaphor of Kerouac's) is 'composed'. Some aspects of this technique were developed further by Kerouac's friends Allen Ginsberg and William Burroughs, both of whom have acknowledged *On the Road* as an influence on their work. The genesis of spontaneous writing was probably a happy accident, although Kerouac's reading and rereading of Proust during the 1940s and 1950s made it a more likely one – an open capacity to absorb and transform their experiences characterise both these otherwise dissimilar writers. Two years after writing *On the Road*, Kerouac codified the process in *Essentials of Spontaneous Writing*, but this was a retrospective analysis, not a prescription for writing. When the novel appeared and the controversy over its merit began, it was the style which Kerouac's defenders chose as their battleground, although in a famous televised put-down Truman Capote described it as not writing 'but typing'.

> I woke up as the sun was reddening; and that was the one distinct time in my life, the strangest moment of all, when I didn't know who I was – I was far away from home, haunted and tired with travel, in a cheap hotel room I'd never seen, hearing the hiss of steam outside, and the creak of the old wood of the hotel, and footsteps upstairs, and all the sad sounds, and I looked at the cracked high ceiling and really didn't know who I was for about fifteen strange seconds.
>
> *ON THE ROAD*
> BY JACK KEROUAC

The book's content provided much cannon-fodder for Kerouac's critics. Although the 1960s' counter-culture, which grew out of the Beat Generation's experiences, was later to familiarise the lifestyle that Kerouac had described, in the 1950s it was still alien, disturbing and thus objectionable to contemporary reviewers. Undoubtedly, this was also part of the attraction for its readers (and it sold in huge numbers), but the literary merits of the book were buried beneath its function as a kind of alternative social document.

Given the nature of Kerouac's composition technique, it is unsurprising that the novel is uneven. Several passages fall flat and a certain 'cliquishness' is apparent in the familiar allusions to a lifestyle outside the experience of most readers. At times it is hard to avoid the feeling that Cassady, Ginsberg and Burroughs (the latter appear as Carlo Marx and Old Bull Lee in the book) would have enjoyed the novel more than anyone else. At other times the style triumphantly carries the reader along, almost literally in the case of Dean and Sal's cadillac ride from Denver to Chicago, and with a perfect match of style and subject in the virtuoso descriptions of live jazz. Passages such as these are not happy accidents: Kerouac empathises with what he describes without trying to analyse it; his openness to his material is a passive achievement, but an achievement none the less.

City of Spades

see **The London Trilogy** (1960)

The Assistant

Times are bad, and business in Morris Bober's grocery store worse than bad. His wife Ida 'still hoped to sell. Every day she hoped. The thought caused him grimly to smile . . . if he miraculously did, where would he go?' Like so many Jews before him, Morris is caught in a trap. He blames Karp, the liquor dealer, for his troubles: 'He promised me he wouldn't put in a grocery around the corner, but what did he put? – a grocer . . . Why didn't he keep out the German around the corner?' 'God bless Julius Karp, the grocer thought. Without him I would have my life too easy. God made Karp so a poor grocery man will not forget his life is hard.'

When two 'holdupniks', young Ward Minogue and his semi-reluctant partner, Frank Alpine, break into Morris's store to find only seven dollars in the till, Ward refuses to believe it. The victim of his father, a neighbourhood cop who beats the hell out of him, he takes out his anger and frustration on the Jew. But Frank feels only pity. While Morris is laid up as a result of the attack, Frank volunteers to help in the store. And so begin his first tentative steps towards redemption through love and toil. Business picks up, not, as it turns out, so much due to his efforts as to the German grocer falling sick, thus enabling Frank – while his thieving habits are still with him – to liberate a few dollars from the till; he returns them with interest when times are bad again, but by then it is too late: Morris knows. Ida never wanted Frank around, fearful for her daughter Helen with whom he falls deeply in love. She wants Nat Pearl, the educated Jewish boy, for her daughter, not this no-good Italian boy. Their love blossoms slowly but

Helen keeps Frank on a string and when he takes advantage of her after saving her from being raped by the drunken Ward Minogue, she shrieks 'uncircumcised dog' at him. Though at this point she was probably coming to him of her own accord, he has in her eyes committed the unforgivable crime.

The relationships between Frank and Morris, and Frank and Helen, are at the centre of the novel. After Morris dies of pneumonia when at last on the brink of a sale which, true to form, subsequently falls through, Ida and Helen are entirely dependent on Frank. He labours on without thanks from the girl who steadily ignores him until she is forced to see what he has become: 'groggy from overwork, thin, unhappy, a burden lay on her because it was no mystery whom he was working for'. Such is his love, that he has himself circumcised. 'For a couple of days he dragged himself around with a pain between his legs. The pain enraged and inspired him. After Passover he became a Jew.'

The Assistant, which won the Rosenthal Award and Daroff Memorial Award, has a biblical quality; there are clear echoes, for instance, of the *Genesis* story of Jacob, 'who served seven years for Rachel'. Malamud is a master of parable. He shows here, as in his finest short stories that only Saul Bellow, among the American urban Jewish writers, can rival his understanding of the nature of Jewishness, the plight of the impoverished immigrant. The characters are unforgettable: Morris Bober, incapable of dishonesty, imprisoned in his grocery store; Helen, wronged and slow to forgive, with the intelligence to yearn for a better life; Frank Alpine, the exile everywhere and intruder in this enclosed Jewish world, who by love and toil redeems his sins and ultimately becomes part of it; Ida, the harassed wife and mother whose nagging derives from fear and pain. All Malamud's finest talents – his acute moral sensibility, his compassion, his subtle humour – are poured into this, probably his finest novel.

Anthony POWELL   **1905–**

At Lady Molly's

see **A Dance to the Music of Time** (1975)

Elizabeth TAYLOR   **1912–1975**

Angel

When in a Book Council promotion in 1984 *Angel* was chosen by the panel as one of the best British novels published since the Second World War, critical eyebrows were raised. It was argued that the novel was too small in scope and that Taylor's talent was far too *English*, in the pejorative, restrictive sense. Good as *Angel* is, few of Taylor's admirers would choose it as her most characteristic book. It does, however, feature one truly memorable character, Angelica Deverell. Some of the sharpness found in Taylor's letters is apparent in this portrait of a romantic novelist, drawn from such egregious examples as Ethel M. Dell and Marie Corelli, but altogether her own creation. Taylor succeeds in exposing her protagonist as a monster, without entirely alienating the reader's sympathy, a considerable achievement.

The rise and fall of Angelica Deverell, a popular novelist who believes every word of the tawdry nonsense she writes, is mercilessly charted. Snobbish, arrogant, ruthless, and without the faintest glimmer of self-knowledge or a sense of humour, 'Angel' regards herself as a genius. Unlike her creator, she is no observer of the world, but lives entirely within her own preposterous imaginings: 'At sixteen, experience was an unnecessary and usually baffling obstacle to her imagination . . . She had removed herself, romantically, from the evidence of her senses.' This dislocation persists, ruining both her life and her work. In an attempt to escape her humble background, she transforms her extravagant fantasies into a novel, not surprisingly rejected by the Oxford University Press, but accepted by Gilbright & Brace, and published to derisory reviews but considerable sales.

When she meets Esmé Howe-Nevinson, 'an unknown minor painter of erratic talent and . . . the very slackest of habits', she imagines that she has fallen in love. His sister, Nora, a devotee of the Deverell *œuvre*, becomes Angel's companion, and lives with the couple when they marry and acquire Paradise House, the large mansion which was the subject of Angel's childhood fantasies, but has since fallen into decay. A gambling and adulterous wastrel, Esmé is none the less shrewd: he dubs his portrait of Angel 'Study in Solitary Confinement' and says of her writing: 'the secret of your power over people is that you communicate with yourself, not with your readers'.

Gilbright & Brace had been divided, as their readers' reports had been. Willie Brace had worn his guts thin with laughing, he said. *The Lady Irania* was his favourite party-piece and he mocked at his partner's defence of it in his own version of Angel's language.

'Kindly raise your coruscating beard from those iridescent pages of shimmering tosh and permit your mordant thoughts to dwell for one mordant moment on us perishing in the coruscating workhouse, which is where we shall without a doubt find ourselves, among the so-called denizens of deep-fraught penury. Ask yourself—nay, go so far as to enquire of yourself—how do we stand by such brilliant balderdash and *live*, nay, not only live, but exist too. . . .'

'You overdo these "nays"', said Theo Gilbright. '*She* does not.'

'There's a "nay" on every page. M'wife counted them. She took the even pages, I the odd. We were to pay a shilling to the other for each of our pages where there wasn't one, and not a piece of silver changed hands from first to last.'

ANGEL
BY ELIZABETH TAYLOR

This unconcern with her audience leads to a decline in her popularity, particularly after she adopts (for all the wrong reasons) such causes as pacifism and vegetarianism and begins to litter her novels with proselytism.

Deluded to the end, she sinks into decay with her house, devoted to the memory of Esmé (who has drowned), surrounded by a pack of verminous cats and cut off from human society.

The Ordeal of Gilbert Pinfold

Waugh's frank 'account of my late lunacy' is also his most penetrating self-portrait. 'Mr Pinfold's experiences were almost exactly my own,' he admitted; indeed the novel began as a therapeutic exercise, written at the suggestion of a psychiatrist.

The book opens with a 'Portrait of the Artist in Middle-age'. Gilbert Pinfold is a reclusive Catholic novelist suffering from acute accidie: 'His strongest tastes were negative … It was never later than Mr Pinfold thought.' In an attempt to keep rheumatism, insomnia and terminal boredom at bay he has been imbibing an unappetising cocktail of bromide, chloral and *crème de menthe*. In order to improve his health, complete a novel and escape a harsh English winter, he books a passage upon the SS *Caliban*.

The ship, whose unpropitious name is matched by that of its captain, Steerforth, is 'without pretence to luxury', so that Mr Pinfold is not particularly surprised when he is disturbed by music and conversation from other cabins, conveyed, he supposes, by some freakish accident involving the ship's wartime communications system. Gradually the disturbances increase and take on a more sinister character: he overhears an accident in which one of the Lascar crew is severely injured; another crew-member is apparently being tortured by the Captain and his mistress. Attempts to confront his fellow passengers and the crew with what he has heard are met with embarrassed bafflement. Voices then begin to accuse Pinfold of being Communist, homosexual, impotent, talentless, snobbish, fraudulent, arriviste and conscienceless. As the *Caliban* cruises slowly towards Rangoon, so Pinfold slips gradually (and unaware) into madness.

Three voices eventually emerge from the babel of persecution: a BBC interviewer called Angel; the captain's vicious 'doxy' whom Pinfold dubs 'Goneril'; and the one friendly character, 'Margaret', a young woman who claims to love him. Pinfold has a vague notion that the voices might be connected with 'The Box', a diagnostic contraption which one of his neighbours had insisted was capable of curing ailments by transmitting 'life-waves'. In an attempt to escape his tormentors, he disembarks at Port Said, but the voices follow him all the way back home, whither he has been summoned by his anxious wife. Finally convinced that he has been suffering from hallucinations, he recovers completely and settles down to write an account of his ordeal, which becomes the present novel.

The Fountain Overflows

'We could not advance in intelligence and wordly knowledge without becoming daily more conscious of how much less he was doing for us than other fathers did for their children.' West's novel about a family growing up in seedy eccentricity is told from the point of view of Rose, a young girl who childhood has been overshadowed by insecurity as a result of her father's disastrous gamblings on the stock market. Her earliest memories are of being forced to move from the family home in Scotland to a rented house in a London suburb, where her father, Piers, has been given an editorial job on the local newspaper by his benefactor, Mr Morpurgo.

Moving to London brings their mother, Clare, into contact with her old friend, Constance, who is now unhappily married to the violent Uncle Jock, and whose home is haunted by poltergeists. The two Scottish women share a power of parapsychology which they fear may have been handed down to their daughters, and at a children's birthday party Rose begins to demonstrate a disturbing talent for reading people's minds. Mrs Phillips, their hostess, tries to persuade

Rose to tell her fortune, and so begins the family's involvement in a notorious murder case. Beneath the apparently calm surface of the Phillipses' home life there are chaotic hatreds and frustrations. Mrs Phillips poisons her husband, Harry, and hides away in a seaside resort, while her sister, 'Aunt Lily', turns to Rose's family for help.

Piers, with his talent for rhetoric and impassioned speech-making, takes on the task of challenging the death sentence on Mrs Phillips which has been passed by an elderly and insane judge. Manipulating old connections, he takes Rose with him to the House of Commons, where he persuades Mr Pennington, the nephew of the Home Secretary, to insist on bringing the case to a Court of Appeal. Rose witnesses her father's performance with admiration, but is made aware of how little room there is in his idealistic commitments for the usual responsibilities of a father to his wife and children. Piers is willing to go to prison and see their home broken up, and Rose recognises how much strain their mother has been under in her marriage, after giving up her career as a musician to bring up the family.

Rose and her sister Mary have inherited their mother's muscial aptitude, and through these difficult years they practise constantly until they are finally rewarded by gaining scholarships to London music colleges. Their sister Cordelia, however, who is beautiful but untalented, is wrongly encouraged by her adoring teacher, Miss Beevor, and eventually has a nervous breakdown.

For Rose, growing up has been a process of realising her own limitations and learning to accept that, although her family make her feel unique, the future in the professional world is going to be demanding. Piers absconds from his family, and they are saved from financial ruin only by their mother's prudent action in lying to her husband about the value of the family paintings. Rose's pleasure in the inspiration of music provides her with the promise of happiness, for as she plays it seems to her that 'the fountain overflows'.

Patrick WHITE　　　　**1912–1990**

Voss

Voss is an historical novel set in nineteenth-century Australia. It was conceived during the blitz in a London bedsitting room, where White sat reading contemporary accounts of the explorer Leichardt's expeditions across the Australian continent. It was 'nourished by months spent traipsing back and forth across the Egyptian and Syrenaican deserts' and 'influenced by the arch megalomaniac of the day'. It expresses horror at the exaltation of the average, an affliction White considers to be particularly – but not entirely – Australian. *Voss* is about human will, and about suffering as a means of reaching a state of grace.

The structure of the novel is straightforward: first there is the preparation for Voss's expedition into Australia's interior; then there is the expedition itself in 1845, which ends in disaster; and finally, a return to urban, Victorian Australia with an attempt to assess the expedition. The style is poetic and elaborate, and has caused some critics to level at White the charge of intellectualism. Others compare him to Joyce for his experimental approach, to Wordsworth for his lyricism, and to Dickens for the depth and scope of his characterisation.

The main players in the drama are Johann Ulrich Voss, Laura Trevelyan and Australia itself. Voss, a German, is an outsider by birth and inclination. He is full of pride and, knowing that he must exercise his will, he identifies as his only fitting opponent the vast and fearful Australian continent. Mr Bonner, a wealthy draper, finds that it flatters his vanity to sponsor Voss's expedition, unaware that he and his family represent precisely the pompous mediocrity that Voss despises. Mr Bonner's niece, Laura Trevelyan, is isolated by her intelligence and spiritual purity. She and Voss are kindred spirits, and they communicate throughout the whole expedition, physically by letter but also through a direct link of minds. They become spiritual man and wife, and she is the key to his redemption.

Voss is accompanied on the expedition by six men, and later joined by two Aboriginals.

In this section of the novel Australia becomes a palpable presence, sensual and malicious by turn, beautiful and darkly ugly. The Aboriginals they come across are at one with the earth, often indistinguishable from it. The expedition has invaded the land, and is a brutal intrusion into the Aboriginals' culture. When disaster strikes, it is largely through the Aboriginals' hands. Three men mutiny, and the novel tells us, with relentless attention to detail, how two of the remaining party are murdered, one commits suicide, and Voss is beheaded. The expedition, in real terms, has failed, but Voss in his last moments learns humility: 'God has become man to become God again.' Nothing tangible can be learned afterwards about the fate of the expedition, but some years later, Laura learns from the one surviving mutineer that Voss has become a myth in Aborigine culture – that 'he is there, in the land, and always will be.'

All his days were wasting away in precise acts. His feet were heavy with dust as he tramped between shed, tent, and stockyard. Now his distaste for men returned, especially for those with whom he had surrounded himself, or, to be more accurate, with whom an ignorant jackass had surrounded him against his will. Blank faces, like so many paper kites, themselves earth-bound, or at most twitching in the warm shallows of atmosphere, dangling a vertebral tail, could prevent him soaring towards the apotheosis for which he was reserved. To what extent others had entangled him in the string of human limitation, he had grown desperate in wondering.

VOSS
BY PATRICK WHITE

The critical reaction to White and his writing has been mixed – generally aggressive in Australia ('a parodist's push-over'; 'difficult to start and difficult to finish') and adulatory in Britain ('genius'; 'in size, intention and achievement *Voss* is the work of a man for whom Tolstoy is the only fitting rival'). In 1973 he was awarded the Nobel Prize for literature 'for an epic, psychological narrative art which has introduced a new continent into literature'.

John WYNDHAM   1903–1969

The Midwich Cuckoos

Richard and Janet Gayford live in Midwich, a tiny village 'where things [do] not happen'. Yet, returning from London one morning, they find the roads are sealed by police. At 22.17 the previous night, Midwich's population fell asleep. It remains enclosed within an invisible bubble that has the properties of an anaesthetic. Impotent and baffled, the government imposes a reporting ban. Then, thirty-six hours after arriving, the bubble departs and the sleepers awaken. They seem to have come to no harm. The village reverts for a time to its normal condition.

Meanwhile, Richard, Wyndham's first-person narrator, has agreed to notify military intelligence of future events. He relays the news that almost all the women in the village are simultaneously pregnant. The women support each other in their predicament, while most of the men accept the situation, knowledge of which is kept from the rest of the world. Gordon Zellaby, an eminent philosopher and Midwich resident, calms the locals' superstitious fears.

In due course, thirty-one boys and thirty girls are born. They appear normal – except for their golden eyes and lucent skin – but they develop at twice the rate of normal children and can manipulate human behaviour. Zellaby discovers that each sex shares 'a common consciousness . . . what we have are *two* entities only – *a* boy and *a* girl'. The children, Zellaby argues, are of a higher order than mankind.

The Gayfords go to Canada, and Richard resumes the story when he visits Midwich eight years later. The children have detached themselves from their mothers and are living together at the Grange, where they are educated and observed. Their formidable wills are becoming malignant. One of their number is injured in a car accident and the driver is telekinetically murdered. The violence escalates. The villagers try to burn down the Grange; instead, they fight among themselves and four are killed. It is disclosed that similar attempts at colonisation have occurred across the globe – and have been dealt with ruthlessly. The British authorities, though, are barred by liberal tradition from such action and so Zellaby – to defend his species – blows up the children and himself with them.

Wyndham thought 'science fiction' an unfortunate term and, if *The Midwich Cuckoos* in an enduring example of the genre, it nevertheless plays on atavistic anxieties: the myth of the incubus lies behind the book. For the novel's protagonists, however, the children are a modern alien culture – harbingers of youth and foreignness intruding on the timeless rural landscape. They are alike immune to the philosopher's arguments, the vicar's theology and the chief constable's threats. Reason, religion and law – the triple pillars of society – are threatened with demolition. Zellaby's suicidal sacrifice, therefore, is warranted both by a paradoxical Darwinism and by the need to safeguard civilisation. A crude means secures a sophisticated end, and the preservation of Zellaby's values is secured by his finally stepping beyond them.

1958

Nikita Khrushchev in power in Russia ● Election of Pope John XXIII ● Charles De Gaulle elected President of France ● Alaska becomes 49th state of US ● Race riots in Nottingham and in Notting Hill, London ● J.K. Galbraith's *The Affluent Society* ● Stereo gramophone recordings ● Victorian Society founded in London ● Mies Van der Rohe and Philip Johnson, Seagram Building, New York ● The 'Beatnik' Movement spreads to Britain ● Leonard Bernstein, *West Side Story* (musical) ● Ralph Vaughan Williams and Marie Stopes die ● Gregory Corso, *Gasoline*

H.E. BATES   **1905–1974**

The Darling Buds of May

The first of five short novels recounting the comic adventures of the Larkin family, this marked a distinct change of direction from Bates's earlier work, and was inspired by a real family glimpsed by the author one summer afternoon near his home in Kent: father, enormous mother, and a multitude of children consuming a mountainous quantity of ice-cream and potato crisps. In similar manner we are introduced to the Larkins: Pop, Ma ('almost two yards wide') and their six children, Mariette, Zinnia, Petunia, Primrose, Victoria and Montgomery. Pop surveys his junk- and nettle-strewn farm in the May sunshine and conceives of it as paradise: everything is 'perfick'. The arrival of Mr Cedric Charlton, a nervous and underfed officer of the Inland Revenue, might seem to cast a dark shadow over Pop's paradise, but when Cedric casts his eyes on the beautiful Mariette even this turn of events proves to be perfick, for Mariette thinks she is pregnant (although she does not know by whom) and is in need of a father for her child.

Within a few weeks Cedric is a changed man, lusty and radiantly healthy. This transformation is wrought by a combination of Mariette's kisses, long balmy days of strawberry-picking, Ma's cooking, and Pop's lethal arsenal of cocktails – Rolls Royces, Chauffeurs and Red Bulls. The novel ends on the day of the annual gymkhana, which Pop is hosting because all the local gentry are now too impoverished by the burden of taxation and rationing, phenomena which seem never to impinge upon the Larkins. During the post-gymkhana cocktail party, Cedric (now dubbed 'Charley') and Mariette announce their engagement and Ma announces that she's having another baby. 'If this lark goes on much longer,' Pop comments, 'you and me'll have to get married as well.' Charley advises that this would be financially disadvantageous, and Pop, despite having no plans to pay any tax, concedes that there would be little point anyway, considering that everything is perfick as it stands.

The adventures of the Larkins continue in four further novels: *A Breath of French Air* (1959), *When the Green Woods Laugh* (1960), *Oh! To Be in England* (1963) and *A Little of What you Fancy* (1970). A phenomenally successful television adaptation of the stories in 1991 brought the word 'perfick' into the language via the medium of the tabloid press.

Samuel BECKETT   1906–1989

The Beckett Trilogy

Molloy (1955)
Malone Dies (1956)
The Unnamable (1958)

Beckett's blackly comic and surreal trilogy, which prefigures the themes of his plays, was originally published in French as *Molloy* (1951), *Malone meurt* (1951) and *L'Innommable* (1953). Beckett himself subsequently translated the novels into English, assisted with the first volume by Patrick Bowles. Essentially monologues, they become increasingly bizarre and incoherent, as dreams or stories – often inconclusive – interrupt or entirely take over the narrative.

Molloy is an intriguing puzzle in two sections, narrated by two different characters whose identities merge by the end of the book. The first narrator, we eventually learn, is called Molloy; the second Jaques Moran. At the beginning of part one, Molloy tells us that he is in his mother's room, that he cannot walk, and that he has to write out his story for a man who collects his pages once a week. He attempts to remember an encounter with two strangers and a dog, but can recall little beyond the fact that he had acquired a bicycle, on which – despite his dependence on crutches – he is travelling to visit his mother. When stopped by a policeman and asked to produce his papers, he is nonplussed and presents the officer with some bits of old newspaper he keeps to clean himself after defecation. Arrested, he is questioned at the police station, where he gives equally little satisfaction, and is passed on to a social worker, who tries to press unappetising refreshments on him. 'To him who has nothing it is forbidden not to relish filth', but he flings the mug of grey liquid against the wall in disgust, and is released.

Out on the road again, he promptly runs over a dog, 'an ineptness all the more unpardonable as the dog, duly leashed, was not out on the road, but in on the pavement, docile at its mistress's heels'. An angry mob is prevented from tearing Molloy to pieces by the dog's owner, who reveals that she was taking it to be put down on account of its age and infirmity: this lucky accident has relieved her of the unpleasant task. She asks Molloy to help her bury the dog, and he brings the corpse on his bicycle to her house. He stays on as a guest, but after a while tires of his hostess and slips away.

Minus his bicycle, he swings along on his crutches through a forest until he meets a charcoal-burner. He responds to the man's friendliness by cracking him on the head with his crutch. When Molloy has ascertained that the man is not dead, he proceeds to kick him with his otherwise useless legs. He goes on his way, abandoning the usual method of propulsion in favour of dragging himself along the ground, using his crutches as grapnels, until he emerges from the forest, tumbling into a ditch. It is from this ditch, evidently, that he is rescued and taken to his mother's room, not knowing what has become of her in the meantime.

The second narrator, Jaques Moran, is a very different character. He receives instructions from his chief, via a messenger called Gaber, to set out at once and locate an individual called Molloy. Accompanied by his young son, Moran sets off for Ballybaba, despite the fact that his leg has suddenly developed pains. Some days into the journey, the pains become more pronounced and Moran sends his son to buy a bicycle. While the boy is absent, Moran has two visitors to his makeshift camp. The first is an elderly man with a stick, who asks for food and goes away. The second, a man whom Moran recognises as having features reminiscent of his own, enquires after the first man. Moran is not aware of what follows this conversation, but finds the stranger lying dead before him, his head beaten to a pulp. 'He no longer resembled me', adds Moran with evident relief. He conceals the body in a copse, and when the son returns with a cycle, Moran argues with him and continues his journey alone. He is intercepted by Gaber, who orders him home. He travels slowly, his health and clothes deteriorating.

Shabby and crippled, Moran sits alone in his house, writing his report. After having spent the summer in his garden, he is now deciding to leave on another journey, to try to become free. The second part, as the first, ends where it began.

Not only is *Molloy* an interesting formal achievement, it is also a clear statement of leading Beckettian preoccupations: the problem of exile or alienation, and the impotence of language to convey anything of importance. In a world of the imprecision of the self and the uncertainty of knowledge, the only constant is the tyrant, the mysterious chief whose orders and rules must be obeyed.

Malone Dies is the final monologue of a dying man, in which the speaker – who is possibly a reincarnation of Molloy – pauses in his storytelling to reflect upon his own condition: 'It is in the tranquillity of decomposition that I remember the long confused emotion which was my life.' Malone lies in bed, calmly anticipating death. He does not know where the room in which he lies is located, and has only the vaguest recollections of how he came to be there. He attempts to tell several stories, wondering as he does so whether they are fiction or autobiography. In the first, the academic shortcomings of a boy named Saposcat are a great worry to his parents.

Instead of working for his examinations, he broods upon himself, wondering how he will live in a world full of strangers. In the summer holidays, his parents pay for him to have private tuition. After these lessons, he makes visits to the present family of Mr Lambert, 'a bleeder and disjointer of pigs'.

Malone abandons the story to describe a Jew who tried to teach his parrot to speak Latin, before returning to the subject of his present state: 'If I had the use of my body I would throw it out of the window. But perhaps it is the knowledge of my impotence that emboldens me to that thought.'

The next story concerns Macmann, an itinerant who is evidently Saposcat. He wakes up one day in an asylum, the House of St John of God, and is put under the personal supervision of Moll, 'a little old woman, immoderately ill-favoured of both face and body'. His clothes are taken away, but Moll breaks the regulations by allowing him to keep his hat. They embark upon a sexual relationship, which flourishes until Moll becomes ill and is replaced as keeper by Lemuel, a man who has taken to beating himself with a hammer, and inflicts brutalities of a similar nature upon his charges.

Malone interrupts the story, first to say that the mysterious old woman who has hitherto brought him his soup and pot has stopped doing so, and then to mention that he has been visited by a mysterious, funereal stranger, who struck him on the head with a rolled umbrella, stayed silently in the room for several hours and then left. Malone then returns to the story of Macmann, who now regularly strays from his room into the grounds. Lady Pedal, a patron of the asylum, sponsors an excursion for five inmates, to be supervised by Lemuel. The narrative suddenly accelerates. They travel by boat to an island, where Lemuel kills Lady Pedal's two manservants with a hatchet. She faints. The men leave the island in the boat.

Malone's words runs out, and he is left, like the men, drifting in the open sea.

The final volume, *The Unnamable*, is a plotless monologue that attempts to define the agony of a soul without identity. It is painful and uncomfortable to read, and its ideas have influenced the work of several major dramatists, including Pinter and

Ionesco; Beckett himself has recycled some of them in his work for the stage.

It is a difficult work to summarise satisfactorily. In the opening pages, the Unnamable is describing a place where he exists, and is telling us of his 'delegates', who have enlightened him about the world of human beings and their ways. One of these delegates is named Basil, although his name soon changes to Mahood. The Unnamable then narrates, but in the first person, 'one of Mahood's stories', in which he is seen as a man with one leg, who took so long to make a journey to his family that he arrived to find that they had all died of food poisoning. He then begins the return journey.

When Mahood is next described, it is as a limbless torso in a jar, which stands outside a squalid eating-house. Later on, Mahood's name is changed to Worm, a creature who not only lacks any semblance of a body, but even all understanding and faculty of perception. Worm, understanding nothing, can be used as a shield against the numbing facts: 'the inability to speak, the inability to be silent, and solitude', 'nothing but this voice and the silence all around'. Truly alone, with no reprieve possible, the Unnamable knows that he cannot stop, yet cannot go on:

> If only I knew if I've lived, if I live, if I'll live, that would be simply everything ... what's a door doing here? ... perhaps they have carried me to the threshold of my story, before the door that opens on my own story ... I don't know, I'll never know, in the silence you don't know, you must go on, I can't go on, I'll go on.

Truman CAPOTE   1924–1986

Breakfast at Tiffany's

This novella, which first appeared in *Esquire*, marked the end of 'the second cycle' of Capote's writing and introduced the world to Holly Golightly, an enchantingly insouciant and rootless girl-about-town, who lives on her wits and the bemused generosity of her countless admirers, and whose card reads: 'Miss Holiday Golightly, Travelling'. Although several of Capote's women friends imagined that they were the model for Holly, other commentators have suggested that the 'real' model was Isherwood's Sally Bowles in **Goodbye to Berlin** (1939). Capote was a friend of Isherwood and an admirer of his work.

> The instant she saw the letter she squinted her eyes and bent her lips in a tough tiny smile that advanced her age immeasurably. 'Darling,' she instructed me, 'would you reach in the drawer there and give me my purse. A girl doesn't read this sort of thing without her lipstick.'
>
> *BREAKFAST AT TIFFANY'S*
> BY TRUMAN CAPOTE

When shown a photograph of an African carving strongly resembling Holly, the narrator, Buster, recalls the early years of the Second World War when he lived in the same New York brownstone building as she did. At first their relationship simply involves his letting her into the block when she loses her key, but one night she takes refuge in his apartment from a drunken and violent lover. Drawn to the narrator because he resembles her adored brother, Fred, who is later killed in action, she tells him something of her life. She also listens without enthusiasm to one of his short stories, which she dismisses as 'Brats and niggers. Trembling leaves. *Description.*' Buster is introduced to her world of admiring men-friends, some of whom, Holly suggests, might help his career. Amongst these is her former Hollywood agent, the bald, dwarfish, Pekinese-eyed O.J. Berman, who lovers her because 'she's a *real* phoney', and 'Rusty' Trawler, a fat 'middle-aged child', celebrity millionaire – and latent homosexual, according to Holly, who none the less plans to marry him. Buster also meets an immensely tall *jolie laide* called Mag Wildwood, who moves in to share Holly's apartment.

Alternately infuriated and enchanted by Holly, Buster charts her raffish progress and begins to piece together her past. Doc Golightly, an old hick who adopted then married Holly when she was only fourteen and called Lulamae Barnes, solicits his help in reclaiming her, but she resists ('You can't give your love to a wild thing: the more you do, the stronger they get. Until they're strong enough to run into the woods'). She exchanges Trawler for Mag's former lover José Ybarra-Jaegar, a sauve Brazilian-German diplomat. Engaged to, and pregnant by, José, she plans to live with him in Brazil and have lots of children. This scheme is thwarted when her past catches up with her, for one of her 'jobs' was to visit a Sicilian racketeer called Sally Tomato in Sing-Sing, unwittingly assisting him to continue to run a drugs ring. She is arrested, but has a miscarriage. Berman arranges bail, which she skips, flying to South America and out of the lives of her friends.

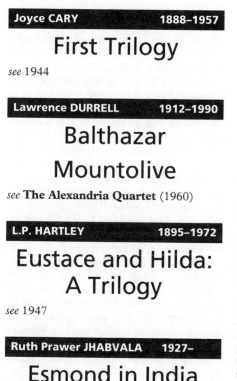

Joyce CARY   **1888–1957**

First Trilogy

see 1944

Lawrence DURRELL   **1912–1990**

Balthazar
Mountolive

see **The Alexandria Quartet** (1960)

L.P. HARTLEY   **1895–1972**

Eustace and Hilda:
A Trilogy

see 1947

Ruth Prawer JHABVALA   **1927–**

Esmond in India

'Myself in India', the title Jhabvala gave to an essay written to introduce a collection of short stories in 1971, seems a deliberate echo of the title she gave to her third novel. 'I have lived in India for most of my adult life,' she wrote. 'My husband is Indian and so are my children. I am not, and less so every year.' Esmond Stillwood, the English protagonist of this novel, lectures with great expertise on all aspects of Indian culture at the Western Women's Organisation, but is increasingly revolted by Indian life, with its alien smells, tastes and customs, all of which are embodied in, and all of which he attempts to eradicate from, his beautiful but indolent Indian wife. 'It is not always easy to be sensitive and receptive to India,' Jhabvala has written, and she never romanticises the country, preferring instead to highlight the absurdities of life there, and the clashes between Indian and European cultures.

This characteristically sardonic comedy of Indian manners depicts the lives of two families, once closely involved in the struggle for independence in 1947, but now separated by wealth and status. Ram Nath went to prison for his beliefs and now lives in reduced circumstances, having withdrawn from the world; his best friend from their student days at Cambridge was Har Dayal, who avoided arrest and now lives in some splendour, serving on committees and 'doing important work in upholding, as he put it to himself, standards of culture and refinement'. A chance to reunite the families was lost when Ram Nath's niece, Gulab, chose to marry Esmond rather than Har Dayal's son, Amrit. Although Amrit subsequently made a successful marriage to the biddable Indira, who spends most of her time planning shopping expeditions, his elegant mother, Madhuri, considers that her family has been slighted, and finds her husband's continued hero-worship of Ram Nath in poor taste.

A second chance to unite the families arises when Ram Nath's son, Narayan, a philanthropic doctor who is working in remote countryside among peasants, tells his parents that he is seeking a wife. His long-suffering mother, Lakshmi, decides that he should marry Har Dayal's daughter, Shakuntala, a romantic idealist whose dowry would restore the family's fortunes. Lakshmi assumes that Har Dayal will be unable to refuse this plan, and she is almost right. Har

Dayal has always felt that he would do anything for Ram Nath, but to his deep shame finds himself unable to grant the only favour his old friend has ever asked. Shakuntala, meanwhile, has met and fallen in love with Esmond, whose exasperation with his wife has reached breaking-point. Esmond carelessly sleeps with Shakuntala, who attaches much more importance to the affair than he does: 'Gulab behind him and Shakuntala before him: and all he wanted was to be free.' A genuine companion is the flippant Betty, who urges him to return home with her to England, even offering to lend him the fare.

Steeped in the belief that a husband is God and may do as he pleases, Gulab endures Esmond's constant goading and ill-temper, and refuses to return to her family as her mother, Uma, urges, enticing her with forbidden spicy food. It is only when her new servant molests her that she packs her bags; protection is the one thing a wife may expect, and Esmond has failed to provide it. The novel concludes with Esmond, unaware of her departure, and humouring Shakuntala's dreams of a life together, deciding to take up Betty's offer.

Both Ram Nath and Har Dayal come to see themselves as out of touch with the main stream of life, and Jhabvala's sympathy is evenly distributed throughout the novel. Esmond may be chilly and supercilious, but Gulab is staggeringly idle and stubborn, while Shakuntala may be well educated, but is naïve and absurdly rhapsodic. The dilemmas facing Esmond and Gulab are similar to those Jhabvala later confronted more squarely in 'Myself in India':

> To live in India and be at peace one must to a very considerable extent become Indian and adopt Indian attitudes, habits, beliefs, assume if possible an Indian personality. But how is this possible? And even if it were possible – without cheating oneself – would it be desirable? Should one want to try and become something other than what one is? I don't always say no to that question. Sometimes it seems to me how pleasant it would be to say yes and give in and wear a sari and be meek and accepting and see God in a cow. Other times it seems worthwhile to be defiant and European and – all right, crushed by one's environment, but all the same have made some attempt to remain standing.

Doris LESSING   1919–

A Ripple from the Storm

see **Children of Violence** (1969)

Iris MURDOCH   1919–

The Bell

Murdoch's fourth novel – rather more straightforward than some of her books – is set at Imber Court, a lay community attached to an Anglican convent. As in other Murdoch novels, a disruptive individual acts as a catalyst upon a closely knit group of people; in this case, the principal character is a young woman, Dora Greenfield, who (the novel's opening informs us) 'left her husband because she was afraid of him' and 'decided six months later to return to him for the same reason ... She decided at last that the persecution of his presence was to be preferred to the persecution of his absence.' Her art historian husband (thirteen years her senior) is Paul, attached to the Courtauld Institute, but now researching manuscripts at the Gloucestershire convent.

The community is preparing for the arrival of a new bell. Legend has it that in the fourteenth century a young man had died while attempting to scale the abbey walls in order to meet one of the nuns. No one would admit to being the lover of this importunate suitor and a bishop had cursed the abbey, whereupon the great bell 'flew like a bird out of the tower and fell into the lake'. The guilty nun subsequently drowned herself. The imaginative Dora secretly plans to effect a 'miracle' by dredging the old bell from the lake (assisted by another young guest, Toby

Gashe, who has discovered its whereabouts) and substituting it for the new bell at the baptismal ceremony. As her ill-conceived plan comes to fruition, tensions mount and the already precarious structure of the community begins to crumble.

The leader of the Imber Court community is Michael Meade, a former teacher whose involvement with a manipulative pupil caused him to abandon his plans to become a priest. The pupil in question, Nick Fawley, now a drunk and dissipated young man, has come to the community at the request of his twin-sister, Catherine, who is shortly to take her vows. With a terrible sense of history repeating itself, Michael finds himself drawn to Toby Gashe. Alarmed by Michael's advances, Toby attempts to fall in love with Dora. All this is observed by the destructive Nick, who bullies Toby into confessing about Michael's bungled pass to his sponsor, a community member called James Tayper Pace.

It seems that the abbey's grim legend is also being repeated during the inauguration of the new bell, a scene of high comedy where disgruntled morris dancers jostle some elderly ladies, and the bell, sabotaged by Nick, topples into the lake. Catherine, who has fallen in love with Michael, attempts to drown herself, but is rescued by Mother Clare, an 'intrepid and amphibious nun', and sent to London for drug therapy. Her brother, however, commits suicide, adding a further scandal to the goings-on at the abbey, which have been gleefully reported in the press by Noel Spens, a former lover of Dora who was covering the ceremony at her invitation. The Abbess decides to disband the community, leaving Michael once more attempting to rebuild his life. Paul accuses Dora of being the instrument of these catastrophic events and suggests she sees a psychiatrist. She decides to take a job in Bath.

Murdoch has been criticised for creating implausible plots peopled with mere puppets, but no such accusations could be levelled against *The Bell*. By turns deeply moving and extremely funny, it fruitfully explores the confrontations between religious faith and suppressed sexuality, and remains one of her best novels.

Barbara PYM   1913–1980

A Glass of Blessings

Pym's beautifully realised world of genteel high-church intrigue, where the arrival of a new assistant priest is a major event and giving blood cheers up a dreary November day, is meticulously observed in this novel. Described by Lord David Cecil in 1977 as one of 'the finest examples of high comedy to have appeared in England during the past seventy-five years', *A Glass of Blessings* takes its title from George Herbert's *The Pulley*:

> When God at first made man,
> Having a glasse of blessings standing by;
> Let us (said he) poure on him all we can:
> Let the world's riches, which dispersed lie,
> Contract into a span.

Unlike most of Pym's heroines, Wilmet Forsyth is young, smart, good-looking and married. Although her life revolves around the local Anglo-Catholic church, Wilmet has always been aware that she is not as excellent a woman as other members of the congregation. She imagines that this is because she is too sophisticated and cynical for this role, unlike her friend, the 'dim and mousy' mother-ridden Mary Beamish. To her astonishment, however, it is Mary who captures the handsome new curate, Fr Ransome. Romance in Wilmet's own life is restricted to shared memories with her friend Rowena Talbot of the Wrens in Italy and of Rocky Napier (a glamorous flagship lieutenant who features in *Excellent Women*, 1952), and the 'heavy Edwardian style' flirtation of Rowena's husband, Harry. Far more intriguing is Rowena's glamorous but 'vaguely unsatisfactory' brother, Piers Longridge, who Wilmet imagines will provide her with an opportunity to do good in an interesting way. She finds herself attracted to the gallant, charming, aquiline Piers, who works as a proof-reader and a teacher of modern languages at night-school, but whose personal life is shrouded in mystery. Interspersed with Wilmet's chaste 'romance' are glimpses of life at the priest-house, where Wilf Bason's appointment as housekeeper leads to scampi replacing boiled cod on the

Friday dinner menu, and to the disappearance of Fr Thames's Fabergé egg.

'Won't you at least have a drink before you go?' Sybil asked. 'I'm sure you'll need it.'

I refused, thinking that it might not mix very well with the refreshments I should get at the parish hall, and it occurred to me that one could perhaps classify different groups or circles of people according to drink. I myself seemed to belong to two very clearly defined circles – the Martini drinkers and the tea drinkers though I was only just beginning to be initiated into the latter. I imagined that both might offer different kinds of comfort, though there would surely be times when one might prefer the one that wasn't available.

A GLASS OF BLESSINGS
BY BARBARA PYM

It is eventually revealed that Piers lives with one of his former pupils, a beautiful but rather boring young man called Keith, who works in the Cenerentola coffee bar and models for books of knitting patterns. Attempting to hide her surprise and disappointment, Wilmet is tactless, and then criticised by Piers for the narrowness of her world. 'Some people are less capable of loving their fellow human beings than others,' he says, 'it isn't necessarily their fault.' Bruised by this encounter, Wilmet reflects: 'Perhaps I had never really known him, or – what was worse – myself.' Pym, who herself had complicated friendships with a number of homosexual men, described Wilmet as one of her two favourite heroines.

Alan SILLITOE   **1928–**

Saturday Night and Sunday Morning

The popular success of Sillitoe's first novel is an example of the power of the cinema. Although it was trumpeted by his publisher as 'the best first novel we have seen in years', and respectfully reviewed in the press, it achieved real fame when the paperback edition was published two years later to coincide with the première of Karel Reisz's film of the book, which Sillitoe himself had adapted for the screen. The film's immense popularity helped to boost sales to a record 5,000,000 copies. Sillitoe's classic, unflinching account of provincial working-class life in the new post-austerity era earned him a high public profile, and led to his being numbered among the Angry Young Men.

'There are no flies on me,' boasts the novel's protagonist, Arthur Seaton, a lathe-operator in a Nottingham bicycle factory. A sharp-suited teddy boy with a fat wage-packet, he swaggers in never-had-it-so-good prosperity, but he is no Joe Lampton on the up-and-up. Although he chooses as his quasi-political motto 'Don't let the bastards grind you down', work is hard, and his only escape is into the weekend haven of booze and sex, his mind becoming a magic lantern of mucky memories played back 'in glorious loony-colour'. No matter how drunk he gets, he will always land safely in one soft bed or other. 'I'm just too lucky for this world,' he says.

He walked towards Slab Square, his bones aching for the noise of a public house, wanting to lose himself in a waterfall of ale and laughter. The main road was lit by overhead lamps furtively shining, as if ready to fall into darkness at a moment's notice on sighting a reconnoitring member of the Lord's Day Observance Society. Sunday, he thought bitterly, even preferring Monday though it meant the first grinding day of the week.

SATURDAY NIGHT AND SUNDAY MORNING
BY ALAN SILLITOE

His luck runs out, however, when he gets the wife of a workmate pregnant. Brenda has to undergo a gin-induced abortion, while Arthur seeks solace with her married sister, Winnie. He guiltily extends the hand of false friendship to Brenda's husband, Jack, who works on a night-shift. This pattern of sleeping with other men's wives eventually brings down 'the forces of righteousness' on Arthur's head as Mrs Bull, the local gossip, gets busy. He had reckoned without Winnie's squaddie husband, Bill, who beats Arthur up in an alley. Arthur takes to his bed, reflecting: 'A dangerous life.'

A safer bet altogether is Doreen, who is unmarried and knows just how far to go to get her man. Arthur announces their engagement in appeasement of breached morality. Marriage, his only safe option, looms as a trap which will keep him forever in his own class and city.

C.P. SNOW   **1905–1980**

The Conscience of the Rich

see **Strangers and Brothers** (1970)

T.H. WHITE   **1906–1964**

The Once and Future King

The Sword in the Stone (1939)
The Witch in the Wood (1940)
The Ill-Made Knight (1941)
The Candle in the Wind (1958)

Written in the shadow of the Second World War, White's retelling of England's greatest patriotic epic was inspired by Malory's *Le Morte D'Arthur* (1485). In both accounts, Arthur is the illegitimate child of Igraine and King Uther Pendragon. Given into Merlyn's care at birth, he is entrusted to Sir Ector and brought up alongside Sir Ector's natural son, Kay. Whereas Malory says little of his hero's youth, White (himself a schoolmaster) devotes his first book to the subject.

Sir Ector is a gentle, bumbling soul and his friends – King Pellinore and Sir Grummore Grummursum – amiable parodies of chivalric convention. The world seems innocent and benevolent; Arthur himself is generous, modest and bold. One day, lost in a forest, he

meets Merlyn again, who guides him back to his guardian's castle and is engaged as his tutor. Knowing his pupil's destiny, the wizard educates him unconventionally – changing him into a fish, a hawk, a badger and even an ant. He means to extend the range of Arthur's sympathies and to teach him the virtues of peace as well as of war. Though Arthur learns, he does not understand what his learning is for: he expects to rise no higher than Kay's squire. Then, after Uther Pendragon's death, he accompanies Sir Ector and Kay to a tournament, where there is a sword, embedded in an anvil and a stone. Whoever frees it will be king of England, and only Arthur can perform the deed. Thus he is recognised as Uther's heir. *The Sword in the Stone* concludes with his coronation.

Tilting and horsemanship had two afternoons a week, because they were the most important branches of a gentleman's education in those days. Merlyn grumbled about athletics, saying that nowadays people seemed to think that you were an educated man if you could knock another man off a horse and that the craze for games was the ruin of scholarship – nobody got scholarships like they used to do when he was a boy, and all the public schools had been forced to lower their standards – but Sir Ector, who was an old tilting blue, said that the battle of Crécy had been won upon the playing fields of Camelot. This made Merlyn so furious that he gave Sir Ector rheumatism two nights running before he relented.˙

THE ONCE AND FUTURE KING
BY T.H. WHITE

The second book describes Arthur's conquest of the feudal barons, the inception of the Round Table and the comic adventures of Pellinore, Grummursum and a third paladin, Palomides, in Lothian. The trio encounters four brothers – Gawaine, Agravaine, Gaheris and Gareth – whose sinister mother is Arthur's half-sister, Morgause. For reasons national and personal, the Scottish family loathes its English kin. Nevertheless,

Morgause travels south, seduces Arthur and bears him a son, Mordred. Eventually, Mordred and his half-brothers are to join the Round Table and to be the ruin of the kingdom.

Lancelot's long affair with Guenever provides the substance of *The Ill-Made Knight*. Lancelot is an obsessive: his noble deeds, coldly accomplished, are a compensation for facial ugliness and spiritual lack. Although cruel by temperament, in combat he is always merciful, and his loyalty to the king is as fixed as his love for the queen. Arthur, for his part, chooses to ignore the adultery, wishing to quarrel with neither his friend nor his wife. Lancelot is tricked into the bed of Elaine – the daughter of King Pelles – and fathers a son, Galahad, by her. (Later he unhappily marries Elaine.) Galahad grows to be the purest knight: he it is who finds the Holy Grail (and vanishes in distant Palestine) while Lancelot is barred from this sacred success, unable to renounce his habitual sins.

In *The Candle in the Wind*, Mordred and Agravaine conspire to confront Arthur with Guenever's unfaithfulness. Mordred is now a bitter demagogue, whom White explicitly compares to Hitler. The king is forced to try his queen for treason. She is about to be executed when (as Arthur hoped would happen) Lancelot rescues her. In the rescue, however, Lancelot kills Gaheris and Gareth. Gawaine is sworn to avenge his brothers, and Arthur is compelled by honour to assist. The two lay seige to Lancelot's castle in France and, in their absence, Mordred seizes the throne. Arthur returns to England, and Gawaine dies of a wound inflicted by Lancelot. On the eve of his final battle, old and exhausted, Arthur reviews the course of his life with despair. His best knights are gone; his order is destroyed; he has failed to bring peace and law to his lands. He summons his page, narrates the saga to him, and commands him to remember the ideals and feats of the past. The page's name is Thomas Malory.

White's Arthurian England is fantastic: unicorns and griffins inhabit the woods, and Robin and his men are already in Sherwood. Merlyn, on his backward voyage through time, peppers the tale with anachronisms. It is a mock-medieval kingdom and, if *Le Morte D'Arthur* is a lament for lost values, *The Once and Future King* is a lament for values that never existed. Furthermore, White suggests, Arthur's Utopian strivings are doomed to defeat. Man is at best, like Lancelot, a compromised creature; at worst, like Mordred, corrupted utterly by lust and ambition. To be perfect, like Galahad, is to be removed from the realm of practical action – to seek fulfilment in a mystic embrace. In the constant crises evil presses on humanity such a pious solution has no appeal.

The fourth volume was never issued separately, only as 'Liber Quartus' of the one-volume edition of the tetralogy, in which *The Witch in the Wood* was retitled *The Queen of Air and Darkness*. The first book inspired Walt Disney's animated feature *The Sword in the Stone* (1963), and the whole saga formed the basis of a saccharine musical play, *Camelot* (1960) by Alan Jay Lerner and Frederick Loewe (filmed 1967). After White's death in 1964, a fifth instalment of the story was found amongst his papers. This was subsequently published as *The Book of Merlyn* (1977).

Angus WILSON　　　　　　**1913–1991**

The Middle Age of Mrs Eliot

Wilson's third novel is a muted study of a woman's painful journey towards self-knowledge. It is a dense and more narrowly focused novel than its predecessors, but once again the central character is an intelligent person who has been self-deceived, 'practising the final hypocrisies of the educated and worldly'. Wilson's portrait of Meg Eliot has been compared to *Madame Bovary* (1857), and he has echoed Flaubert in describing his heroine as a self-portrait.

Meg is a contented, competent and managing person whose world is brutally exposed by the death of her husband Bill, a wealthy barrister accidentally killed during an assassination attempt in the Far East. The first intimations of her inability to order the world come when she fails to prevent the execution of her husband's killers. She discovers that Bill's gambling, which she had indulged, has left her with heavy debts and she has to sell her large house and seek temporary accommodation in a Kensington hotel, becoming, like her 'lame duck' friends, a member of the 'new poor'. She imagines that her love and knowledge of china and her experience on the panel of a private charity might be turned to financial advantage, but learns that she has been little more than a dilettante, and so she enrols at the Garsington Secretarial College. Her lame ducks are astonished by this decision, but the gregarious Bohemian world of Poll Robson, a failed artist, is not for Meg, nor is the genteel poverty of Lady Viola Pirie or the embittered widowhood of the ostensibly 'plucky' Jill Stokes. Meg's notions of managing people come to be seen as crude and tactless interference. Her tenancy with Lady Pirie ends abruptly when Meg's attempt to befriend Lady Pirie's rancid and idle son, Tom, leads to his making a pass at her. Her well-meaning bid to breach the gulf between Jill Stokes and her smug, selfish son-in-law is equally disastrous. The resulting row leads to a breakdown and Meg is taken in by her younger brother.

David has also been recently bereaved, for his lover, Gordon, with whom he ran the Andredaswood Nursery on the South Downs, had died of cancer. 'The wires of communication' between brother and sister 'had simply rusted away', but now they are re-established. Each is initially wary of the other, for Meg's prosperous social round has been very different to David's quietist rural community (both, according to Wilson, hells of their own devising). Gradually they rediscover the shared past of their childhood: 'back', as Wilson punningly suggests 'to the nursery'. Meg encourages David to take up and complete his doctoral thesis on the eighteenth-century novel, but realises that they are both retreating into a destructively cosy idyll. The past is not to be recaptured (a scheme to restore a Jekyll garden for a neighbour also comes to nothing) and Meg decides to return to the real world, becoming a freelance secretary.

Wilson has complained that David has often been seen as the real hero of the book, that readers have failed to recognise the sterility of his passive stoicism (epitomised by the chastity of his relationship with Gordon), which should be seen as 'a final abnegation of life'. Bill's death releases Meg from 'the hell of total social commitment' so that she can rebuild her life upon new foundations.

1959

Fidel Castro becomes premier of Cuba ● Hawaii becomes 50th state of US ● Nikita Khrushchev addresses UN General Assembly ● UN decides not to intervene in Algeria ● USSR sends dogs in orbit ● First section of London–Birmingham motorway opened ● Iona and Peter Opie, *The Lore and Language of Schoolchildren* ● Pope John XXIII announces the first Vatican Council since 1870 ● Frank Lloyd Wright, Guggenheim Art Museum, New York ● Shelagh Delaney, *A Taste of Honey* ● Arnold Wesker, *Roots* ● Jean Cocteau, *Le Testament d'Orphée* ● Jacob Epstein and Stanley Spencer die ● Michael Horovitz (ed.), *New Departures* (magazine) founded ● Gary Snyder, *Riprap*

Malcolm BRADBURY   1932–

Eating People is Wrong

Bradbury began writing his first novel in the early 1950s, but it underwent numerous revisions throughout the decade. He has acknowleged his debt to F.R. Leavis, whose influence was at its peak in the 1950s, and has described *Eating People is Wrong* as an attempt at 'reading an entire decade in which critical energy had directed itself less through politics than morals, less through aesthetics than practical literary criticism, and when the role of the writer, the intellectual and the academic had greatly changed'.

Prof. Stuart Treece, an eighteenth-century specialist on the verge of middle age, has largely resigned himself to processing the accountants of the future in a converted padded cell (his university was formerly a lunatic asylum). His inter-war idealism is expressed by championing the culturally underprivileged, and he finds himself defending Louis Bates, an earnest, wet-lipped, working-class boy, whose egocentricity is seen by Treece's colleagues less as a manifestation of genius than an indication of insanity. Bates becomes obsessed by Emma Fielding, a postgraduate student of Virginia-Woolf-like beauty who is researching the imagery of fish in Shakespeare's tragedies. Treece is also attracted to Emma, and encourages her to shake off the skackles of middle-class morality. Unfortunately, Emma has spent the last ten years trying to

eliminate her inherited snobbery and now finds it impossible to be discriminating amongst people of a different race or class. As well as Bates, her admirers include Eborebelosa, the son of an African chief, who aims to make her his fifth wife. In order to escape Eborebelosa, Emma claims that Treece is her favoured suitor, whereupon the African makes an effigy of the professor and sticks pins into it. This is a sore trial of Treece's liberal conscience, but it soon becomes the least of his problems. As Bates and Treece vie for possession of her body and mind, Emma is encouraged to consider her own happiness by Dr Viola Masefield, a liberated but confused feminist, and she learns that causing a little harm is inevitable.

Treece's bafflement in the face of the modern world now manifests itself in guilt and a sense of isolation; on the verge of collapse, he is admitted to a (politically obligatory) National Health Service ward. Amongst the working-class staff and patients, his academic authority carries little weight: 'the ordinary laws of sound human contact were slipping'. The suicidal occupant of the next bed turns out to be his rival-in-defeat, Bates, who is now destined for the local asylum. When Emma visits Treece, her tears and his despair are testimony to the absurd ironies of well-meaning intellectual liberalism: 'I suppose all you can say for us is, at least we feel guilty.'

Bradbury, who later satirised 1970s' university politics in ***The History Man*** (1975), is often regarded as the founder and principal exponent of the 'campus novel', a term he dislikes. Both Mary McCarthy in *The*

Groves of Academe (1952) and Randall Jarrell in **Pictures from an Institution** (1954) had already used university settings to explore the nature and limits of post-war liberalism. Bradbury chose the campus (a location with which, as a teacher, he has been intimately involved) because it was 'a place where people did discuss ideas, theoretical and aesthetic, contemplated literary and cultural theory, and experienced and responded to the large intellectual and social changes that have shaped our late twentieth-century world'.

William S. BURROUGHS   1914–

The Naked Lunch

The Naked Lunch is an important book, not only for its style and content but also because it provoked intense literary discussion when, four years after publication by the Olympia Press, in Paris, it was issued in Britain by the avant-garde publisher John Calder. The *Times Literary Supplement* review was headlined 'Ugh', and proceeded to analyse the reasons for the reviewer's 'steady nausea' and 'finally boredom' with the 'texture of the grey porridge in which Mr Burroughs specialises'. The question arose: 'is there a moral message? And how about if the moral message is itself disgusting?' Burroughs, in defence, did not dispute that the novel might be 'brutal, obscene and disgusting'. His purpose was to expose '*the junk virus*' as '*public health problem number one of the world today* . . . Sickness has often repulsive details not for weak stomachs.' 'Ugh' might be the reasonable response of a faint-hearted reader unprepared to recognise the effects of drug addiction.

Burroughs knew all about the junk virus – he was suffering from addiction and his attempts to withdraw from dependence during a sojourn in Tangier where *The Naked Lunch* was written and pieced together by Allen Ginsberg and Jack Kerouac from the pages dropped at random on the floor. It is not a connected narrative, but it achieves a weird coherence as themes and characters submerge and surface again in improbable but characteristic situations of surrealist – almost Dadaist – fantasy. Sections of the book are homoerotically explicit, achieving a Bosch-like effect, and Burroughs is a master at creating not so much characters as states of degenerate being: 'A Near East Mugwump sits naked on a bar stool covered in pink silk. He licks warm honey from a crystal goblet with a long black tongue. His genitals are perfectly formed – circumcised cock, black shiny pubic hairs. His lips are thin and purple-blue like the lips of a penis, his eyes black with insect calm.' The Mugwump, in pursuit of sexual satisfaction, is merciless: 'The boy looks into Mugwump eyes blank as obsidian mirrors, pools of black blood, glory holes in a toilet wall closing on the Last Erection.'

> The Inspector opens his fly and begins looking for crabs, applying ointment from a little clay pot. Clearly the interview is at an end. 'You're not going?' he exclaims. 'Well, as one judge said to the other, "Be just and if you can't be just be arbitrary." Regret cannot observe customary obscenities.' He holds up his right hand covered with a foul-smelling yellow ointment.
>
> *THE NAKED LUNCH*
> BY WILLIAM S. BURROUGHS

Burroughs has a witty, baroque style, decadent and rich, scatological, sometimes ecstatic, sometimes clinically cold. The novel is episodic in order to create a dream/nightmare landscape of hallucinatory, shifting effects which are relentlessly tied to the reality of bodily functions and desires.

Barbara COMYNS   1909–

The Vet's Daughter

Comyns's work is not easily categorised; beneath its apparently simple surface it has the power to shock, amaze and horrify without obvious effort or contrivance. Speaking out with an unnerving naïvety, her characters – all victims in one way or another – are capable of profound perceptions. An innocuous event can become a nightmare in the space of a sentence. Horror becomes farce. The poetic is frequently counter-balanced by description which is at times so stark that the reader is startled into laughter. Her work is not, however, a belly-laugh at human existence, but rather the wry, ironical smile of comprehensive and compassion.

The Vet's Daughter, her fourth and perhaps best-known novel (it has been adapted for both television and the stage), is surreal in tone, with an ending that leaves the reader curiously discomfited. Alice Rowlands is the painfully innocent daughter of a violent vet who has no apparent fondness for animals, and a downtrodden, dying mother ('if she had been a dog my father would have destroyed her'). Temporary relief from her miserable life is provided by Lucy, her deaf and dumb friend who is soon to become a seamstress and be corrupted by the outside world, and the cloth-capped char, Mrs Churchill. When Alice's mother dies and her father goes away, Henry Peebles ('Blinkers') the earnest, blinking locum, appears on the scene and grows fond of her. Although she is grateful for his kindness, he is not the knight in shining armour she dreams of, reminding her rather of 'the bark of a tree, all covered in tweed'. When her father returns with Rosa, the 'strumpet from The Trumpet' pub, Alice's life as a lackey worsens and reaches crisis point when she is more or less raped by Rosa's friend, the creepy Cuthbert.

Alice takes refuge at Mrs Churchill's where her levitatory powers reveal themselves, allowing her literally to rise above her situation. Taken to the comparative safety of Blinkers's mother's half-burnt home as companion, Alice remains puzzled by her new-found skill. Is it normal? Certainly not, pronounces Mrs Peebles, as she crunches on her boiled egg shell. Blinkers finds the subject uncomfortable, and Nicholas, the white-jumpered sailor who suddenly appears in Alice's life and exits almost as quickly, is positively disgusted by it. Mrs Peebles, a chronic depressive, drowns herself, and Alice has to return to her father's house until Blinkers can claim her for his wife. Her mistake is to levitate in front of her father as an act of defiance, for he at once sees the financial possibilities of her 'peculiarity'. Now more prisoner than slave, Alice is treated with snivelling respect, while Rosa

and her father plot to expose her to the world as a circus curiosity.

On Clapham Common Alice is led like a 'reluctant bride' before the crowd which has gathered in response to an enigmatic newspaper advertisement. Her smuggled-out plea for help to Blinkers seems to have gone unheeded and she has no choice but to perform as bidden. The circus act distorts into nightmare. Seeing Blinkers and Mrs Churchill running towards her, too late, Alice plunges from her elevated position into the crowd where she, Rosa and a stranger are trampled to death in the panic. For the first time in her life Alice is unafraid. The inaccurate, detached newspaper account with which the novel ends, suggests that Alice's own account has come from some other, unearthly, dimension.

Pamela Hansford JOHNSON 1912–1981

The Unspeakable Skipton

The wittiest and also artistically the most perfect of Johnson's many novels, *The Unspeakable Skipton* is set in Bruges and centres on an English expatriate writer, Daniel Skipton, living in poverty in a boarding-house. Skipton is a rogue and con-man, but also utterly self-deluded in his conviction that the novel on which he is interminably working is a masterpiece which will end his financial worries for good. As Johnson herself acknowledged, Skipton is largely based on the writer Frederick Rolfe ('Baron Corvo'), and particularly on his impecunious and sponging last days in Venice, although certain of his characteristics are omitted (notably his promiscuous homosexuality), while others are added.

In the Grand' Place in Bruges, Skipton contrives an introduction to a party of four English tourists at a nearby table: Dorothy ('Dotty') Merlin, mother of seven sons and execrable poetess of motherhood; her much older husband, the shrewd bookseller Cosmo Hines; their friend Duncan Moss, a photographer and lover of girls and pleasure; and Matthew Pryar, a cultured socialite and habitué of country houses. (These characters recur in two other novels by Johnson, *Night and Silence, Who is Here?*, 1962, and *Cork Street, Next to the Hatter's*, 1965; the three novels are consequently sometimes known as the 'Dorothy Merlin trilogy', although their stories are not in fact closely linked.) The plot of the novel, which is intricate if small-scale, centres on Skipton's attempt to make money out of his association with these four, and with Conte Flavio Querini, an Italian nobleman and singer. Skipton's squalid and ultimately doomed manoeuvres are set against many exquisite cameos of the dignified and unchanging beauty of Bruges. He arranges to take the four English tourists to a sex-show of 'Leda and the Swan', for which he collects a commission, as he does for arranging Cosmo Hines's visit to a prostitute. However, his allowance from his English cousin, 'Flabby Anne', is cut off when he writes her a letter threatening to expose her to the income-tax authorities. To recoup, he determines to earn a commission by inducing Conte Querini to buy a fake Flemish-master painting, from a dealer. He lays out most of his remaining cash on a lunch for Querini, but the sudden death of the brothel-keeper from whom he has been collecting commissions makes his position even more desperate.

Matters are improved when a rich visiting Englishwoman, Mrs Jones, is persuaded to lend Skipton some money, and a loan arrives from his English publisher. Believing he is shortly to cap his triumph by disposing of the painting, Skipton is extremely rude to Dotty Merlin at a party. Querini turns out to be short of ready cash, although he promises he has it at home, and Cosmo Hines advises Skipton to lend him the deposit for the picture; when this is done, Querini hurriedly leaves town. A letter from Cosmo Hines informs Skipton that the English tourists have also gone, reveals that Hines knew that Querini was in fact penniless, and reminds the defeated and struggling author of his nickname in England, 'the unspeakable Skipton'. The comedy of the artistic life, desperate yet always amusing, has come full circle.

Laurie LEE 1914–

Cider with Rosie

Cider with Rosie originated in a series of periodical articles which Lee adapted into a loose chronology. At the age of three he was brought to the Gloucestershire village of Slad on a carrier's cart, rolled out of a Union Jack in a scene which marks the birth of consciousness, a 'fat young cuckoo' in a cottage which seemed like 'a still green pool flooding with honeyed tides of summer'. The village school instilled the simple education necessary for country life, an unthinking array of facts and figures 'absorbed ... as primal truths declared by some ultimate power'. It was a 'conveyor belt along which the short years drew us' towards the stirrings of adolescence and 'the first faint musks of sex'. The taste of cider with Rosie in the hay-waggon was a 'long secret drink of golden fire ... never to be forgotten or even tasted again'.

The expanding horizons of the growing boy coincided with the first intrusions of the modern world, as charabancs and motorbikes turn miles into minutes. Yet the nostalgia for the vanished world combines with a clear vision of his harsh realities: 'our village was clearly no pagan paradise', but the scene of murderous ambush, incest, suicide and superstition. There was 'an acceptance of violence as a kind of ritual which no one excused or pardoned', unpoliced by the state: 'it is not crime that has increased, but its definition'.

An impoverished childhood in the overcrowded cottage is mitigated by the sparkling wit of the mother, 'muddled and mischievous as a chimney-jackdaw'; she 'would sacrifice anybody for a rhyme', breeding in the young Laurie a poetic talent which survived the mechanical scansions of poems written 'a dozen an hour' in the schoolroom. Her intellect had been curbed in its ambition by early domestic duties, and marriage to a widower had left her in charge of seven children. But she could not sustain 'the protective order of an unimpeachable suburbia' in Morden, and, rejected by the ambitious husband whom she never ceased to love, she was packed off to her native county. Lee's affectionate tribute to 'this servant girl born to silk', an extravagant and romantic buffoon, encapsulates this autobiography 'distorted by time'.

Each morning was war without declaration; no one knew who would catch it next. We stood to attention, half-crippled in our desks, till Miss B walked in, whacked the walls with a ruler, and fixed us with her squinting eye. 'Good a-morning, children!' 'Good morning, Teacher!' The greeting was like a rattling of swords. Then she would scowl at the floor and begin to growl 'Ar Farther ...'; at which we said the Lord's Prayer, praised all good things, and thanked God for the health of our King. But scarcely had we bellowed the last Amen than Crabby coiled, uncoiled and sprang, and knocked some poor boy sideways.

CIDER WITH ROSIE
BY LAURIE LEE

Skilfully capturing a rural England that seemed forever lost to the car and to industry, *Cider with Rosie* was, and continues to be, a phenomenal success, frequently reissued in numerous illustrated and 'gift' editions, and used as a set text for schools. It earned Lee the W.H. Smith Award and has become firmly established as a modern classic. *As I Walked Out One Midsummer Morning* (1969) and *A Moment of War* (1991) are sequels.

Colin MacINNES   **1914–1976**

Absolute Beginners

see **The London Trilogy** (1960)

Mervyn PEAKE   **1911–1968**

The Gormenghast Trilogy

Titus Groan (1946)
Gormenghast (1950)
Titus Alone (1959)

Although now regarded as a trilogy, Peake's three Gormenghast novels would almost certainly have been part of a far larger scheme had not Parkinson's disease and premature death intervened. The author had made notes for a fourth volume, and his widow believes that he would have continued the saga, describing the hero's adventures into old age. As it was, *Titus Alone* went to press when Peake was already considerably incapacitated by illness, and a revised and expanded edition – nearer to his original intentions and containing one entirely new episode (chapter 24) – was published posthumously in 1970. The genesis of the series may be found in Peake's short story 'Boy in Darkness', published with stories by William Golding and John Wyndham in *Sometime Never: Three Tales of Imagination* (1956), and in the fragment of a novel, 'The House of Darkstones' (written *c.* 1938–40). The eponymous boy is recognisably Titus, while the fragment's Lord Groan is a sketch for Lord Sepulchrave. The first volume of the trilogy was very well received, but did not make the author much money; the second, along with Peake's poem *The Glassblowers*, was awarded the W.H. Heinemann Foundation Prize; by the time the third volume was published, Peake had a substantial cult following, one that was to increase as the years passed until the trilogy became widely acknowledged as a highly individual masterpiece.

The books create a world of Gothic fantasy on the fringes of twentieth-century life. Gormenghast, the home of the Groan family, is a rambling and ramshackle estate, and the lives of its inhabitants are enmeshed in a rigid hierarchy ordained by centuries of lore. In *Titus Groan*, the birth of the heir, Titus, seems to ensure the continuity of the old order: but rebellious elements are emerging. Titus is nursed at the breast of Keda, a commoner from the settlement of the Bright Carvers which clings to the walls of Gormenghast castle. With her milk he sucks in a spirit of freedom, which aligns him with the passion and violence of the lower class rather than the artificial rituals of the aristrocracy. The greatest threat to the traditional life, however, is Steerpike, a servant lad who has escaped from the boiling underworld of the kitchens presided over by the violent and drunken chief Swelter. Steerpike begins to exert a sinister influence over the castle's inhabitants as he worms his way into the dubious confidence of the higher orders.

'Isn't he sweet, oh isn't he the sweetest drop of sugar that ever was?' said Mrs Slagg.

'Who?' shouted the Countess so loudly that a string of tallow wavered in the shifting light.

The baby awoke at the sound and moaned, and Nannie Slagg retreated.

'His little lordship,' she whimpered weakly, 'his pretty little lordship.'

'Slagg,' said the Countess, 'go away! I would like to see the boy when he is six.'

TITUS GROAN
BY MERVYN PEAKE

Lord Sepulchrave, Titus's ageing, melancholy father, spends his gloomy days in the castle library surrounded by the mildewed tomes of sacred ritual, while his wife, the massive and majestic Countess Gertrude, attends to her birds and cats. Steerpike learns that Sepulchrave has two sisters, Cora and Clarice, identical twins who dwell in a lonely wing of Gormenghast nursing their grievance at being excluded from power. Steerpike persuades the sisters to reassert their claims to the family inheritance by staging an arson attack on Sepulchrave's library. Worked into a frenzy by his promises of power and glory, they become pawns in his Machiavellian mastergame. As fire engulfs the library, Steerpike appears to rescue the family, who are gathered there for a ceremony. Yet his triumphant coup of self-promotion sows the seeds of doubt in the mind of Prunesquallor, the devoted family doctor, who has foolishly befriended him. Steerpike's manipulation of the sisters, too, has created problems for the ambitious youth. As he fails to deliver on his promises of a golden life, the twins begin to blab of their part in the fire, and Steerpike's unmasking seems close.

With the destruction of his library, Sepulchrave's melancholia intensifies. Titus's adolescent sister, Fuchsia, watches with horror as her father slides into insanity. Sepulchrave loses any identification with the human race, believing himself to be the child of the death owls who haunt the high turrents. A growing sense of evil hangs over Gormenghast, erupting into a long night of terror and death. Sepulchrave's manservant, Flay, is embroiled in a feud with Swelter, who dogs his steps and plots murder as the faithful Flay sleeps on the threshold of his master's door. Steerpike's malevolence stirs up further havoc as, mocking Sepulchrave's insanity, he provokes the valet into a violent remonstrance which results in Flay's dismissal. Flay's loyalty remains untarnished despite this harsh treatment: improvising shelter in the nearby woods he returns at night to protect the family from the malignant youth.

In the nightmarish confusion of the final night of Sepulchrave's life, Flay's feud with Swelter reaches its climax. They fight in the weird cobwebs of the Hall of Spiders, but Flay evades his murderous enemy and Swelter drowns in the waters of a freak storm. Sepulchrave stalks through the castle's disasters on an impersonal mission to his own doom. Now entirely in the grip of insanity, he climbs the Tower of Flints to be torn apart by his kindred death owls. *Titus Groan* concludes with death and destruction inexplicable to the people of Gormenghast, and the child Titus flings unconscious rebellion in the face of tradition.

The second volume, *Gormenghast*, sees the growth to maturity of Titus, now the seventy-seventh Earl of Gormenghast. Steerpike has risen to a position of power during Titus's minority, but his hold over castle life is increasingly precarious.

Sepulchrave's sisters are secretly imprisoned in an isolated wing of the castle, believed by the family to be dead. As they mull over their experiences, their hostility to Steerpike grows, the sexual undertones of submission and perversion reversing into a terrifying sense of power over his vulnerability.

Suspicions of Steerpike are spreading elsewhere. From his exile in the woods Flay acts as the family's secret protector, stalking through Gormenghast, his cracking knees bandaged to muffle their noise. Prunesquallor, Fuchsia's favourite, also recognise Steerpike's sinister motives. Fuchsia has long been disturbed by Steerpike, and his sexual threat now becomes explicit as he plans to rape her. Flay's vigilance foils the youth's plot, and Steerpike is caught like a rat in a trap, hated and feared by everyone, as a second freak storm floods the castle. An attempt to silence the sisters misfires as they turn on their tormentor. Steerpike escapes their elaborate booby-trap and bricks them up to die in their cell, martyrs to their own alienation. As the rising waters of the flood flush evil into the open, Titus tracks Steerpike to his lair. Gormenghast mounts an extended campaign against the malicious outsider, but Titus emerges as the true hero, destroying his enemy in a final struggle. Again the future looks assured; but Titus leaves Gormenghast for ever to venture into the modern world.

The world that he discovers in the final volume of the trilogy, *Titus Alone*, is a science-fiction framework of twentieth-century horrors. The book was written after Peake had visited Belsen as a war-reporter, and in it the Gothic grotesques of Gormenghast are translated into contemporary forms. In a state controlled by the surveillance of futurist technology, the citizens are divided between the successful bourgeoisie and the rebels of the Under River, a secret refuge of slum habitations.

Titus is discovered drifting in the river by Muzzlehatch, an eccentric zoo-owner hated by the authorities. Leaving Muzzlehatch's protection, Titus comes across a party of glamorously inane socialites, whom he observes through a skylight; the skylight collapses, and he is arrested as an alien. Muzzlehatch's mistress, Juno, like him a covert rebel, takes a liking to the handsome youth and offers him her sexual favours together with the freedom of her house. Titus is soon bored by the older woman, however, and asks Muzzlehatch to help him leave. Muzzlehatch provides him with a pass to the hidden underworld.

In the strange regions of the Under River Titus finds himself faced by a madman, whose hapless female victim he had tried to protect. Muzzlehatch again comes to the rescue, but Titus now finds his friend greatly changed; the authorities have destroyed his animals. Titus journeys on to a new world of pleasure and privilege. His memories of Gormenghast now constitute a country of the mind, disbelieved by everyone, and his sense of identity is crumbling. Sinking into delirium, he is found by Cheeta, the capricious daughter of a wealthy factory magnate, and Titus becomes her plaything, entertaining her with ramblings about his fabulous past.

Cheeta knows that she cannot control Titus, and her sexual fascination conceals a savage malevolence. As Titus prepares to depart she stages a psychological masque mocking the scenes of his youth, a parody of the Gormenghast spectacle. Muzzlehatch and Juno reappear to rescue Titus from his impending insanity, flying him to a strange land where he begins to perceive a familiar landscape. On the slopes of Gormenghast mountain he turns away from the castle he knows lies beyond; Gormenghast has become an identity he carries within himself.

Peake illustrated his own books as well as those of other authors, and the manuscripts of these novels contain drawings of the characters, several of which are reproduced in some editions. His imagination was essentially visual and he sometimes made sketches of his characters in order to understand them before committing them to the written word. One result of this is that the novels teem with richly imagined people, who seem to have evolved organically from the text, like gargoyles thrusting out from the castle walls: Prunesquallor's vain sister, Irma, too myopic to see how ugly she is; Sourdust, the bad-tempered octogenarian Master of Rituals, who is killed in the library fire and succeeded by his septuagenarian nephew, the crippled

Barquentine; Mrs Slagg, the ageing nanny; Rottcodd, the solitary and forgotten curator of the Hall of Bright Carvings; and the Gormenghast Professors, elaborated from Peake's own teachers at the public school: Bellgrove, Opus Fluke, Perch-Prism, Shrivel, Shred, Spiregrain, Splint, Throd, Crust, Cutflower and Mulefire, all overseen by the ancient, wheelchair-bound head-master, Deadyawn. Such names suggest a fairy-tale quality, and Peake's story contains all the darkest and most violent images from this rich heritage, gathered to create a par-able of the contemporary world, where the polarities of good and evil inextricably merge.

Alan SILLITOE   1928–

The Loneliness of the Long-Distance Runner

The long-distance runner – we learn eventu-ally that his name is Smith – is at war with the governor of the Borstal to which he has been sent as a result of the 'bakery job'. 'I'm alive and he's dead,' says the boy; 'it's blokes like him as have the whip-hand over blokes like me'. The conflict has everything to do with 'honesty'. It is a question of whether his 'Out-law' brand of it is more valid than the governor's 'In-law' brand. The latter, who talks to him as if he were a 'prize race horse', is counting on him winning the long-dis-tance 'All England' running cup for his Borstal and the boy seems to be playing along – 'they can't take an X-ray of our guts to find out what we're telling ourselves' – even to be enjoying it. He goes out every morning 'frozen stiff, with nothing to get me warm except a couple of hours' long-dis-tance running before breakfast' and feels 'like the first bloke in the world . . . fifty times better than when I'm cooped up in the dormitory with three hundred others'. What is more, he has a plan. 'Cunning is what counts in this life,' he tells us at the outset, 'and even that you've got to use in the slyest way you can.'

The day of the big race arrives. He lets the 'pop-eyed potbellied governor', sitting in the grandstand next to 'a pop-eyed potbellied Member of Parliament', believe he is going to walk it, even become some kind of pro-fessional runner when he gets out. 'I'd hold my trump card until later.' His only possible rival is Gunthorpe, 'one of the Aylesham trusties', and he is not in the same league. 'You can always overtake on long-distance running without letting the others smell the hurry in you . . . I never race at all; I just run, and somehow I know that if I forget I'm racing and only jog-trot along . . . I always win the race.' Soon he is out by himself, thinking his thoughts, remembering 'the Out-law death my dad died, telling the doctors to scat from the house when they wanted him to finish up in hospital'. He arrives back in the sportsground miles ahead of anyone else and marks time in full view of everyone, and, knowing that for the last six months of his sentence he will get all the dirty jobs going, waits an age for Gunthorpe to catch up. 'I'll show him what honesty means if it's the last thing I do . . . By God I'll stick this out like my dad stuck out his pain and kicked them doctors down the stairs.'

This novella, published as the title story in a collection of shorter pieces all about work-ing-class life, is a *tour de force*. Told almost entirely from within the boy's head, it is one of the most profound studies of the rebel mind ever written. It was made into a celebrated film (1962) directed by Tony Richardson from a screenplay by Sillitoe himself.

Muriel SPARK   1918–

Memento Mori

Spark's third novel is an exuberant *danse macabre*, a ruthless black comedy of the kind at which she was to excel. Beneath the dark humour, however, is a serious meditation upon man's preparedness for the inevita-bility of death, an awareness of which is necessary to bring savour to life. The novel is prefaced with three quotations: Yeats on the 'absurdity' of age 'that has been tied to me / As to a dog's tail'; Traherne on those 'Vener-able and Reverend Creatures . . . the Aged'; and the *Penny Catechism*'s instruction to remember the Four Last Things.

The story concerns a group of people who, at the beginning of the century, had been a 'progressive set'. Now in their seventies and eighties, their only progress is a slow, arthritic shuffle towards death. The friends assemble at the funeral of one of their number, a patron of untalented practitioners in the arts called Lisa Brooke, whose will provokes a row between her family, her companion-housekeeper, Mabel Pettigrew, and a former husband, the ancient roué and critic Guy Leet. Done out of her expected inheritance, Mrs Pettigrew agrees to help Godfrey Colston, a wealthy brewing magnate, look after his wife Charmian, a former popular novelist now lost in the mists of senile confusion. It is Godfrey's sister, Dame Lettie, a former penal reformer and welfare worker, who is the first of the circle to receive the anonymous telephone call instructing her: 'Remember you must die.' Meanwhile Mrs Pettigrew has learned about Godfrey's ancient infidelities with Lisa Brooke and threatens to tell Charmian unless he gives her money and alters his will in her favour. Charmian rallies considerably when her books are reissued and enjoy cult status, but receives the phone call and uses this as an excuse to escape into a home, leaving Godfrey in the clutches of Mrs Pettigrew. One by one, all the characters receive the phone call, and Dame Lettie, getting little satisfaction from the police and becoming increasingly paranoid, calls in an old acquaintance, the retired Inspector Mortimer, to investigate the case. The Inspector concludes that since each victim of the caller hears a different voice, the message (which he regards as salutary: the contemplation of death intensifies life) must come from God.

Miss Taylor, Charmian's former paid companion, now confined by chronic arthritis to the Maud Long Medical Ward of a public hospital, would agree with him. She is a still centre amidst a group of old ladies who consult horoscopes, complain about the staff, and gradually die off. In order to save Godfrey, she sends details of Charmian's own infidelities with Guy Leet. With considerable relish the characters eye each other slipping into senescence, as old wounds are opened and old scores settled. Their activities are observed by Alec Warner, an old man himself, who is conducting an obsessive study in gerontology, using his friends as guinea-pigs. Dame Lettie's fretting about the phone calls leads to her brutal death at the hands of a burglar, and the novel ends with a brusque account of the subsequent deaths of each of the remaining characters.

Keith WATERHOUSE 1929–

Billy Liar

Keith Waterhouse's best-known novel is partly derived from James Thurber's celebrated short story, 'The Secret Life of Walter Mitty' (1947), but is transferred from America to Stradhoughton, a small town in Yorkshire, where the fantasy life of young Bill Fisher is given a contemporary social dimension. Like other novels and films of the period (in collaboration with Willis Hall, Waterhouse adapted the book for both the stage, 1960, and a classic film, directed by John Schlesinger, 1963), *Billy Liar* explores the aspirations and frustrations of northern working-class youth, but does so through high comedy. The potential for escape, heralding the new freedoms of the 1960s, is embodied by Liz (memorably portrayed in the film by an icon of the period, Julie Christie); but the downbeat ending reminds the reader that for many people like Billy, freedom must remain as illusory as the imaginary country of Ambrosia, where Billy is Prime Minister.

The novel opens with Billy lying in bed, telling himself that it is a day for big decisions. Stradhoughton is no place for a brilliant comedy scriptwriter, and the crumpled letter in his coat pocket confirms the offer of a big job in London. Shadrack and Duxbury's can stick their job. Besides, life in Yorkshire is becoming complicated; the Guilt Chest under his bed is overflowing. As if problems at the undertakers' firm, where he works as a clerk, are not enough to mortify the flesh, there is also 'the Witch', Barbara – a sexless girl fond of Jaffa oranges and sentimental gravestones – who is now his fiancée.

During the coffee-break at work, Billy finds himself engaged to Rita, waitress at the Kit Kat Cafe and 'a natural for every beauty contest where personality was not a factor'.

At lunchtime in the cemetery the passion pills he pushes into an orange cream appear to have no effect on Barbara. Only plump, scruffy Liz ('Woodbine Lizzy'), who regularly vanishes into an unknown sexual life in Doncaster or Welwyn Garden City, understands his boredom. That evening at the Roxy, his sins find him out as all the inhabitants of his self-created hell encircle him. Barbara and Rita engage in a battle of ownership, while *coitus* with Liz in Foley Bottoms becomes *interruptus* when his colleague Eric Stamp spies on him. Real life, or rather death, intrudes as Gran is admitted to hospital. Billy awaits the London train with eight pounds in his pocket, but his late-adolescent confusion will not allow him to make the final momentous decision. The job in London as scriptwriter for comedian Danny Boon is an uncertain illusion, while marriage to Liz is too adult. When the train arrives, Billy returns to the infirmary.

Much of the comedy lies in the characters Waterhouse creates, notably Mr Shadrack, the forward-looking undertaker, who is inspired by Evelyn Waugh's **The Loved One** (1948) to design glass-fibre coffins, and Billy's parents: his irascible father, owner of a haulage firm, whose favourite adjective (and adverb) is 'bloody', and his long-suffering mother, who spends much of her time writing letters to 'Housewives' Choice'.

My mother shouted up the stairs: 'Billy? *Billy*! *Are* you getting up?' the third call in a fairly well-established series of street cries that graduated from: 'Are you awake, Billy?' to 'It's a quarter past nine, and you can stay in bed all day for all I care,' meaning twenty to nine and time to get up. I waited until she called: 'If I come up there you'll *know* about it' (a variant of number five, usually 'If I come up there I shall *tip* you out') and then I got up.

BILLY LIAR
BY KEITH WATERHOUSE

1960

Sharpeville shootings ● Congo Crisis ● John F. Kennedy elected US President ● Sit-in campaign by blacks in US lunch-counters begins ● Russian Communist Party condemns Mao Tse-tung ● Twenty artificial satellites in orbit ● Surgeons develop a pacemaker for the heart ● Harold Pinter, *The Caretaker* ● Federico Fellini's *La Dolce Vita* ● Alfred Hitchcock, *Psycho* ● Boris Pasternak and Sylvia Pankhurst die ● Sylvia Plath, *The Colossus*

J.R. ACKERLEY 1896–1967

We Think the World of You

Ackerley's 'fairy story for adults' is his only novel, but like his other works it is firmly grounded in fact; so much so that it had to undergo considerable revision for libel. He described it as: 'Homosexuality and bestiality mixed, and largely recorded in dialogue: the figure of Freud gleefully suspended above.' It is based upon his relationship with a guardsman deserter he had met in 1942. The young man's wife was extremely jealous and detested Ackerley, even though he financially subsidised the family, taking on extra work to do so. Missed appointments and other upsets put an intolerable strain upon the relationship. The young man was eventually arrested for burglary, which is the point at which the novel opens.

We Think the World of You is a compact, intense and beautifully written book about possessiveness in which Ackerley portrays himself as Frank, a waspish middle-aged civil servant in love with a feckless young sailor called Johnny. Their relationship has been eroded by the latter's marriage to a jealous slattern called Megan, a portrait in which the author's fastidious distaste is apparent in every line. When Johnny is arrested for housebreaking, he asks Frank to look after his dog, Evie, since Megan already has three children and is expecting another. Frank refuses, and Evie is dispatched, along with Johnny's youngest child, Dickie (one of fiction's most repellent babies), to the care of Johnny's mother, Millie. Formerly Frank's char, Millie is now married to the loathsome Tom Winder, who sees Evie as an investment: she can guard the house and produce litters of profitable puppies. When visiting the Winders, Frank becomes enchanted by the young bitch and, appalled that she is kept in a yard, offers to exercise her. Gradually, Evie replaces Johnny in Frank's affections, just as Ackerley's Alsation bitch, Queenie, became the centre of his own emotional life. Frank offers to buy the dog, but Johnny (virtually incommunicado in prison, since Megan never passes on Frank's messages) refuses. 'I think the world of her,' he insists, employing a cliché that is frequently repeated and debased by himself and his relations. Johnny also thinks the world of Frank, but his behaviour makes it clear 'what the world amounted to', which is not a great deal.

Now I had the pleasure I had promised us both, the pleasure of setting her free upon grass. And her reward was mine. Across the open spaces of the park the rough wind blew with its full strength, and she became a part of the dancing day, leaping and flying among the torn trees, wild in her delight. And her gratitude was as boundless as her happiness.

WE THINK THE WORLD OF YOU
BY J.R. ACKERLEY

Frank eventually buys the dog and is obliged to allow his cousin Margaret (based upon the book's dedicatee, Ackerley's sister Nancy) to live with him to look after Evie

while he is at work. The two possessive females, dog and woman, fight for Frank's attention; Evie wins and Margaret moves out. More jealous than any wife – 'She is Eve, the prototype, Shaw's tigress,' Ackerley explained – Evie dominates Frank's life, preventing him from seeing his friends and even destroying his mail, much as Queenie herself did. Possessed and bewildered, Frank submits to the dog's will.

We Think the World of You is a meticulously crafted book, with Ackerley compared with 'an eighteenth-century cabinet, everything sliding nicely and full of secret drawers'. He managed to conceal his art so that the story lost none of the naïve, embarrassing quality which he thought necessary for the book's impact. By concealing the novel's pattern, Ackerley also ensured that the casual reader would, like Frank, plough blindly on until the final page, and only then realise that Frank's fate was predestined and clearly signalled. A second reading reveals verbal and imagistic echoes which increase the book's central irony that Megan and Evie are both (in Ackerley's terms) jealous, possessive bitches. Frank constantly bewails the fate of Johnny at the hands of his wife without realising that his own fate at the paws of the dog is exactly the same. The novel earned Ackerley the W.H. Smith Award in 1962, and was made into a film (1989), directed by Colin Gregg from a screenplay by Hugh Stoddart.

Lynne Reid BANKS   1929–

The L-Shaped Room

Jane Graham, a twenty-seven-year-old public relations officer, is pregnant and wants nothing to do with Terry, the father. Her own father, a widowed medium-ranking civil servant, banishes her from the family home. She moves into a semi-slum in Fulham, renting the L-shaped room of the title. Initially, she means to have little to do with the other tenants of the house, but they infiltrate her life.

Jane tries to conceal her pregnancy both from her new acquaintances and from her colleagues at work. Suffering from severe morning sickness, she considers (and rejects) the possibility of an abortion. She, forms a triangular friendship with John, a black musician, and Toby, a struggling Jewish writer. When Jane and Toby sleep together, however, John is appalled and distraught. He has guessed Jane's condition and tells Toby that she is a whore. Toby temporarily deserts her.

Dismissed from her job after an unfortunate vomiting episode, Jane survives on severance pay. She and Toby resume their relationship, but Toby finds it stymies his writing. Repenting of his interference, John redecorates Jane's room. Shortly before Christmas, Jane survives a near-miscarriage. While recuperating at her Aunt Addy's country cottage, she types the manuscript of Addy's novel. She returns briefly to Fulham then, reconciled with her father, moves back to his house and anticipates fondly the baby to whose arrival she was at first indifferent. Terry reappears and he and Toby have a fist-fight, prompting a premature labour. A healthy son, David, is born, and the story concludes with Jane's problems almost resolved: she is bequeathed Addy's cottage (plus any royalties which might accrue from the novel); her father (recovered from depressive alcoholism) dotes on David; and, while Toby declines to commit himself to her idyll, Jane feels 'so happy about him, it [doesn't] matter too terribly'.

Banks employs her heroine as first-person narrator. The resulting style is fluent, uncomplicated and intimate but, as in a personal letter, much of the content is embarrassing. Jane remarks repeatedly on John's 'powerful negro odour'; she is 'disgusted' by 'a queer called Malcolm'; if she harbours a giddy prejudice in favour of Jews, this is not of itself sufficient to engage the reader's sympathy. She remains a curiously compromised figure. Venturing into seediest west London, she turns her L-shaped room into a little nest of bourgeois treasures – just as, in a final reversal, she introduces her socially stigmatised child to suburbia. Neither the fallen woman of Victorian myth nor the independent one of feminist celebration, she registers a moment of transition. Indeed, the book derives what significance it retains from its historical acuity: its perception that an old morality would not be able to

sustain the habits of a coming generation.

The L-Shaped Room was Banks's first novel, and remains her best-known book. Although derided by the critics, it was a popular success, became a Book Society choice, and was made into a much-praised film (1962), directed by Bryan Forbes from his own screenplay. Jane's story is continued in two subsequent novels: *The Backward Shadow* (1970) and *Two is Lonely* (1974).

Stan BARSTOW   1928–

A Kind of Loving

Like *Billy Liar* (1959), Stan Barstow's tragi-comic slab of northern life, also narrated in the first person, was made into a celebrated film (1962), once again with a screenplay by Keith Waterhouse and Willis Hall and directed by John Schlesinger. The son of a Yorkshire miner, Barstow was one of the leading 'regional' novelists of the period. He published *A Kind of Loving*, his first and best-known novel, when he was thirty-three.

Immensely funny and extremely depressing, it is the story of Vic Brown's disastrous infatuation with Ingrid Rothwell. Vic's earthy humour ('I get up to dance and pick a bird who looks okay from a distance and pongs like beef gravy gone off close to') and amusing observations – his tale is full of marvellous cameos of huffy relatives, of people on a bus – make light of his situation. His tentative note of optimism at the end of the book, however, rings hollow: if somewhat wiser, he is certainly no happier.

Twenty years old, son of a miner from Cressley, Vic is a draughtsman in an engineering office and obsessed with Ingrid from the typing pool. Begun hesitantly, their relationship soon reveals a sad imbalance, with Ingrid talking of love and Vic, confused and unhappy, beginning to understand the difference between love and sexual attraction. Torn between his notion of true love, his strong – if contradictory – moral sense and his overwhelming desire for Ingrid, Vic is never quite able to leave her, and when she becomes pregnant, marriage is the only answer.

Since they are obliged to live with Ingrid's dreadful mother, their marriage, shaky to start with, is doomed to failure. Ingrid's miscarriage causes further misery and tension, while her willingness to side with her mother finally drives Vic away. Where he hopes to find sympathy, however, there is none; once married, he finds that he is expected to remain so. A reconciliation is agreed and Vic hopes that in a flat of their own, away from his monstrous mother-in-law, he and Ingrid will discover 'a kind of loving' to carry them through: 'What it boils down to is you've got to do your best and hope for the same. Do what you think's right and you'll be doing like millions of poor sods all over the world are doing.'

John BARTH   1930–

The Sot-Weed Factor

In 1708 Ebenezer Cooke wrote 'The Sot-Weed Factor', a long satirical poem chronicling his various mishaps and adventures in Maryland. One way to read Barth's novel is as an 800-page exegesis of the original, wherein every allusive aside or stray thought is tracked down, traced back or extrapolated forward until every part is (somehow) related to every other. The sot-weed factor (or tobacco merchant) of the title is Ebenezer Cooke, whose life Barth relates in three sections. The first, 'The Momentous Wager', establishes Ebenezer as an innocent hero of great intelligence and little common sense. A poetaster and virgin, he resolves that these two states are mysteriously linked. He sails for America, but only after having himself appointed Poet Laureate of Maryland with the avowed intention of writing an epic poem, 'The Marylandiad', extolling its virtues. Despite the cornucopia of sexual activities that Barth puts on display in the novel, Ebenezer manages to remain a virgin for a great part of the book, but his cherished 'Marylandiad' is eventually written as the chastening satire of 'The Sot-Weed Factor'.

The greater part of the novel ('Going to Malden' and 'Malden Earned') is set within, or on the voyage to, Maryland, where Ebenezer is party to several different plots (interrelated to varying degrees) involving (amongst other things) piracy, land rights, his inheritance, virginity (again) and a Catholic conspiracy against the Protestants. Behind several of these plots (and many of the sub-plots) lies the chameleon-like figure of Henry Burlingame – once Ebenezer's tutor, sometimes his adversary, on several occasions his saviour. A master of disguise, Burlingame is an equivocal figure who guides Ebenezer through his travails in the manner of someone teaching a novice the rules of an impossibly complex game. Burlingame is also one of only a very few loose ends in Barth's plot – he simply disappears towards the end of the book, although he will reappear, in various ancestral guises, in later novels.

The plot itself is horribly complicated but almost perfectly worked out. Real historical incidents from the founding of Maryland are seamlessly integrated into the fictional activities of the characters. It is this integration which sets *The Sot-Weed Factor* apart from one of its major antecedents: Fielding's *Tom Jones* (1749). Barth's plot echoes Fielding's in several ways: the essential innocence of the hero; his battle to attain his rightful inheritance; even the lurking possibility of incestuous love. Barth's postmodernity means, however, that he never simply gives free rein to the exuberance of his invention. For him, the art of storytelling is closely bound up with the creation of history, and history is never more than the most credible version of the story.

With its digressions, inserted tales, homiletic episodes and occasional excesses, *The Sot-Weed Factor* is constantly stretching its readers' patience. Apostrophising himself as 'The Author', Barth often professes to be in just as great difficulties and presents himself as a hapless narrator overwhelmed by the sheer complexity of the tale. Naturally this is a pose. Barth exerts a rigid, organising control over the hundreds of characters in the book, many of whom are little more than marionettes attached to this or that strand of the plot. Language, too, is rigorously controlled. The whole novel is written in a flawless pastiche of eighteenth-century English, its vocabulary replete with the argot, erudition and technical terms of its supposed time. The book's weaknesses are, unsurprisingly, those of the eighteenth-century novel in general: many characters lack depth and there is an over-reliance on plot to solve problems of motivation and psychological credibility. This apart, *The Sot-Weed Factor* is a brilliant piece of literary forgery which, amongst the rhetorical conjuring tricks, raises important questions about the relationship between history and fiction and the centrality of storytelling in the way men and women negotiate their lives.

H.D.　　　　　　　　**1886–1961**

Bid Me to Live

Though executed in the same post-Imagist style as Hilda Doolittle's better-known poetry, *Bid Me to Live* – like *End to Torment* (1979) – is almost straight autobiography. Drafted in 1927, 1939 and 1943, it charts H.D.'s relationships with her husband Richard Aldington, D.H. Lawrence and the composer Cecil Gray, in the last years of the First World War. The rest of H.D.'s circle of intimates are also included: Frieda Lawrence, Aldington's lover Dorothy Yorke, and the novelist Brigit Patmore. Other writers hover in the wings, such as Ezra Pound, her former colleague and fiancé, the Imagist poet John Cournos, and the poet and critic Amy Lowell.

The setting is Bloomsbury towards the end of the war: not Woolf's Bloomsbury, but the 'Other Bloomsbury', a second artistic circle, sharing the aesthetic and sexual progressiveness of the Bloomsbury Group proper, if not its affections and social activities. Complex relationships, *ménages à trois* (or *à quatre*) were the unstated norm, made socially less than unacceptable by the siege-like, improvised conditions of wartime.

The novel begins with a series of painful meetings between Julia and Rafe Ashton (H.D. and Aldington) during Rafe's home leaves from the front line, which are intercut with happier memories of times together in Italy and Dorset. Relations deteriorate with each encounter: Rafe's sexual attentions turn to Bella Carter (Dorothy Yorke), living upstairs from Julia. At first, Bella preoccupies only Rafe's body, while his soul remains loyal to Julia, but this convenient sexual geometry soon crumbles, and Julia's attentions turn elsewhere too.

Frederick and Elsa (the Lawrences) arrive in London, and invite Julia to join their own *ménage*: 'kick over your tiresome house of life', says Frederick–Lawrence. 'That will leave me free for Vane' (Cecil Gray), adds Elsa. But Frederick's daring spontaneity proves to be more theoretical than practical: when Julia makes a physical advance, he seems to shrink away, alarmed. In the end, Julia chooses Vane (Cecil Gray) over Frederick, leaving London to join him in Cornwall.

The new affair is short-lived, and despite his stereotyped view of femininity, it is actually Frederick who leaves a lasting impression on Julia. The novel closes with a long, unposted letter of Julia's to Frederick, a complex declaration of affection tinged with suspicion. She acknowledges his sensitivity, and his power of artistic intuition, identifying him with Van Gogh in his ability to project his emotions, in his writing, outward into other characters and landscapes. But she also asserts the female artist's right to stylistic autonomy, just as earlier she had needed to wrest from her husband the power to organise her own emotional life. Frederick–Lawrence may 'Bid [her] to Live', and to write, but she responds by living, and writing, on her own unique terms.

The palms of her hands cupped the crystal goblet. She held solid crystal in her hand. A shifting plane of gold that had been Rhineland grapes steadied her, concentrated her, held her to her centre. Cyclone-centre, she had thought. But to achieve the very centre, it had been necessary for a million young men to die. It had been necessary ... but don't think.

BID ME TO LIVE
BY H.D.

H.D.'s artistic independence is conveyed by the highly wrought style of the novel as well, which owes as much to Ezra Pound's telegraphic use of images as to Lawrence's diffuse emotionalism, and more again to neither.

1960

Lawrence DURRELL **1912–1990**

The Alexandria Quartet

Justine (1957)
Balthazar (1958)
Mountolive (1958)
Clea (1960)

In his preface to the one-volume edition of *The Alexandria Quartet* (1962), Durrell writes: 'This group of four novels is intended to be read as a single work ... a suitable subtitle might be "a word continuum".' The ancient, cosmopolitan, amoral city of Alexandria provides the all-important landscape, and the period is the 1930s and 1940s, but Durrell deals relativistically with time and shifts the viewpoint to explore his characters 'with a truly Proustian ferocity', as Philip Toynbee put it. 'The central topic of the book is an investigation of modern love', Durrell proclaimed, while a subsidiary theme is the nature of fiction itself.

The narrator of *Justine* is Darley, an aspiring young novelist who is purportedly writing this book in retreat on an Aegean island. He has been living with the vulnerable, sickly, night-club dancer Melissa, but loves Justine, a beautiful Jewish woman married to a Coptic banker, Nessim Hosnani. Darley feels that he has had possession of Justine's body, but not of her soul, and suffers increasing guilt about Melissa, and fear of the gentle, suave, but furiously jealous Nessim. His book is an exploration of Justine, who has already featured in the novel *Moeurs*, written by her first husband, Jacob Arnauti. Groping through the maze of stories and the paradoxes surrounding Justine, Darley wonders whether the conclusions he reaches are largely mistaken. Had she really been raped by one of her relatives? What is behind the mystery of her stolen child, whom Nessim is helping her to seek? Is she, as Arnauti suggested, a nymphomaniac? What is abundantly clear is that Justine is tortured and consequently destructive. 'Who invented the human heart ... ?' she cries. 'Tell me, and then show me the place where he was hanged.'

Around these central characters, acting, commentating, themselves waiting to be revealed, are other Alexandrians. Balthazar is a homosexual Jewish doctor specialising in veneral diseases. He is also a student of the cabbala, and Darley is recruited by Scobie, a 'nautical man' working for the Egyptian secret service, to report on his cabbalistic meetings. Clea is a 'gentle, lovable, unknowable' young artist, who has painted Justine's portrait. Was Justine her one and only lover? Pursewarden is a successful novelist who unexpectedly commits suicide and leaves Darley a small legacy. Capodistria (known as Da Capo), a wealthy collector of pornography – and possibly Justine's rapist – is apparently killed at Nessim's duck-shoot. (It later transpires that he has simply disappeared.) It is revealed that Melissa and Nessim have had an affair, which results in the birth of a child.

> When you are in love with one of its inhabitants a city can become a world. A whole new geography of Alexandria was born through Clea, reviving old meanings, renewing ambiences half forgotten, laying down like a rich wash of colour a new history, a new biography to replace the old one. Memory of old cafés along the seafront by bronze moonlight, their striped awnings a-flutter with the midnight sea-breeze.
>
> *CLEA*
> BY LAWRENCE DURRELL

Justine leaves many questions unanswered and much that looks solid enough proves to be illusion. As Pursewarden points out, Darley has much to learn, and in *Balthazar* the young novelist sees himself in a new context after reading the eponymous doctor's extensive and brutal criticism of his work to date. Much of what has been taken for granted in the first novel is stood on its head in the second. The relationship between Nessim and Justine is quite different and infinitely more complex than the reader had been led to suppose. They are drawn together by politics and are secretly plotting a coup on behalf of the Jewish and Coptic minorities. Nessim is tormented not by jealousy but by the fear that his activities will be discovered, while Justine, for whom intrigue acts as an aphrodisiac, has been using Darley as part of their cover. The only person she has really loved appears to have been Pursewarden, who did not care for her at all. 'Truth has no heart,' Pursewarden wrote. 'Truth is a woman. That is why it's enigmatic.'

Other members of Nessim's family are introduced: his mother, Leila, a former beauty disfigured by smallpox, who cannot understand why her son should choose a Jewish bride; and his younger brother, Narouz, who lives the life of a Coptic squire on the family estate at Karm Abu Girg, travelling to Alexandria once a year for the carnival so that, his hare-lip hidden by a mask, he can gaze upon Clea, whom he adores. The novel reaches its climax during the carnival, where 'the grim velvet domino which shrouds identity and sex, prevents one distinguishing between man and woman, wife and lover, friend and enemy'. It seems the ideal setting for Scobie, who admits that he 'slips on female duds and my Dolly Varden ... when the Influence comes over me'. Persuaded to borrow Justine's ring, and consequently mistaken for her, he is murdered by Narouz, who subsequently visits Clea to declare his love for her before giving himself up to his brother. Clea is disgusted. Recovering from the pain of what he has learned about Justine, Darley notes another of Pursewarden's observations:

At first we seek to supplement the emptiness of our individuality through love, and for a brief moment enjoy the illusion of completeness. But it is only an illusion. For this strange creature, which we thought would join us to the body of the world, succeeds at last in separating us most thoroughly from it. Love joins and then divides. How else would we be growing?

Of the four novels, only *Mountolive* is narrated in the third person, and in it Darley becomes an incidental character. Developing the theme of political intrigue, it centres on David Mountolive, the stiff British ambassador, and Pursewarden, now revealed as a diplomat as well as a writer. In London ('home of the eccentric and the sexually disabled') Mountolive had become a friend of Pursewarden and his blind sister, Liza. The true nature of the Hosnani conspiracy and the circumstances of Pursewarden's suicide are revealed. It transpires that as a young diplomat Mountolive had had an affair with Leila, which had certain parallels with the relationship between Darley and Justine. In the years of his absence from Egypt, Leila has written him 'long, wonderful' letters, but by the time he returns as ambassador, she has had smallpox, and she only agrees to see him when she attempts to plead for Nessim. Pursewarden has been defending Nessim against the vigilant War Office intelligence officer, Brig. Maskelyne, who is compiling a dossier on the Coptic conspiracy. It is clear that one or other man has to go, and Mountolive unwisely dismisses Maskelyne.

In Alexandria, Pursewarden picks up Melissa at her night-club, and during the night confesses to her that he and Liza have been lovers: 'We shall never be able to love other people.' In the morning, Melissa tells Pursewarden that her former lover, the Jewish furrier Cohen, has been running arms to Palestine on Nessim's behalf. Pursewarden's personal and professional failures prove too much, and he commits the 'gross solecism' of killing himself, leaving a note for Mountolive stating that he is 'not equal to facing the simpler moral implications raised by this discovery' of Nessim's activities.

The violence latent in Narouz erupts when he becomes an inspired preacher under the influence of Taor, a 'famous woman saint', and incites his followers to revolution. He is deaf to his brother's pleading, and is eventually shot by 'unknown men' at Karm Abu Girg, dying in agony as he waits for Clea to come to him. Although Nessim had told Justine that he was 'faced with the terrible possibility of having to do away with Narouz', the assassins might as easily have been sent by Maskelyne, or even by Memlik, the sinister Egyptian Minister of the Interior.

In *Clea*, Darley is once again the narrator, and the action progresses within a regular time-span. Durrell is less concerned here with the relativistic exploration of relationships than with their development into the future, as the novel describes the effects of the action already charted in earlier volumes. The novel opens with the war at its height. Alexandria is being bombed nightly, and Nessim, stripped of his wealth, is working as an ambulance driver at the docks. Justine, her relations with her husband soured, is virtually under house arrest at Karm Abu Girg. She now disgusts Darley, who has returned from his Aegean island accompanied by Melissa's child. 'I cheated you,' she tells him, 'you cheated yourself.' Only at the end of the novel, when she is discovered working on the detested Memlik, as she once worked on Darley, do the Hosnani fortunes revive. Other characters are also suffering. Leila, driven into exile, dies, while Balthazar is recovering from a breakdown which has aged him terribly. Mountolive is hopelessly in love with Liza, who oversees the burning of her brother's letters in order to thwart a projected biography instigated by Pursewarden's former wife.

The central issue of the novel is an affair between Darley and Clea, who descends from her ivory tower of chastity. She observes that he has changed: he no longer stoops nor wears spectacles. Darley replies that he broke the glasses some time ago and finds that he no longer needs them. Darley records that when they make love: 'She shivered at the first terrible howl of the sirens but did not move; and all around us the city stirred to life like an ants' nest . . . I knew that Clea would share everything with me, withhold nothing.' Their relationship appears serene, but he occasionally hears Clea weeping. 'We aren't quite ripe for each other yet,' Clea concludes. 'It will come.' She finds the small island of Narouz where, away from Alexandria, they are temporarily fortified, but their eventual parting, which 'after such a momentous relationship' seems oddly listless, is foreshadowed. They take Balthazar to their island, where he accidentally pins Clea's hand to an underwater wreck with a harpoon. Clea nearly drowns and, after being cut free, loses her hand. In hospital, it is revealed that she and Dr Amaril, who is treating her, were once lovers. She dismisses

Darley, but eventually writes to him, describing her slow recovery, and declares: 'I wait, quite serene and happy, a real human being, an artist at least.'

Wilson HARRIS   **1921–**

Palace of the Peacock

see **The Guyana Quartet** (1963)

Harper LEE   **1926–**

To Kill a Mockingbird

Lee's only novel had an immediate and continuing success, selling well over five million copies, winning the Pulitzer Prize for the best novel of the 1960s, and being made into an award-winning film (1962), directed by Robert Mulligan from a screenplay by Horton Foote. The book, which draws upon Lee's own Deep South background, is set in Maycombe, Alabama, and concerns the trial of a black man, Tom Robinson, accused of raping a white woman, Mayella Ewell. The events of the trial, and the various reactions to it of the people of Maycomb, are described by Jean Louise ('Scout'), the six-year-old daughter of Robinson's unsuccessful defence lawyer, Atticus Finch. Scout is pugnacious and quick-witted, seen by her father and his conventional sister, Aunt Alexandra, as worryingly precocious. Her ten-year-old brother, Jem, is thoughtful and gentle – very much in his father's mould – and Scout finds a more natural alliance with the seven-year-old Charles Baker Harris, otherwise known as Dill, who is a fantasist with a considerable imagination (based upon Lee's friend Truman Capote).

Although clearly innocent, Robinson is found guilty, and is subsequently shot dead by prison guards when attempting to escape during an exercise period. The townspeople of Maycomb are largely indifferent to this death ('Typical of a nigger to cut and run. Typical of a nigger's mentality to have no plan, no thought for the future, just run blind first chance he saw'), but a local newspaper editor, Mr B.B. Underwood, likens it to 'the senseless slaughter of songbirds by hunters and children'. Reading the editorial, Scout recognises Underwood's point that 'Atticus had used every tool available to free men to save Tom Robinson, but in the secret courts of men's hearts Atticus had no case. Tom was a dead man the minute Mayella Ewell opened her mouth and screamed.' She had in fact been raped by her father, Bob, a violent, racist drunk, forever raging with inchoate grievances, who browbeats his slatternly daughter into accusing Tom of the crime. He swears vengeance upon Atticus, but is killed when he attempts to attack the Finch children after a Hallowe'en party. The children are saved by the intervention of Boo Radley, a mysterious and reclusive neighbour, whom they have spent much time attempting to lure out into the open. The sheriff, Heck Tate, insists that Ewell fell upon his own knife, and so a form of rough justice appears to have been done, although Atticus is uneasy about this.

People moved slowly then. They ambled across the square, shuffled in and out of the stores around it, took their time about everything. A day was twenty-four hours long but seemed longer. There was no hurry, for there was nowhere to go, nothing to buy and no money to buy it with, nothing to see outside the boundaries of Maycomb County.

TO KILL A MOCKING BIRD
BY HARPER LEE

Lee prefaces the novel with Charles Lamb's observation that 'Lawyers, I suppose, were children once', and it is the child's perspective of Southern justice which makes the book so compelling. Children have a relentless logic and ask straightforward, unanswerable questions, unhampered by the qualifications and complexities which confront adults, such as the pragmatic, troubled Atticus.

Colin MacINNES   **1914–1976**

The London Trilogy

City of Spades (1957)
Absolute Beginners (1959)
Mr Love and Justice (1960)

Generally referred to as a trilogy, MacInnes's London novels are in fact connected by setting and mood rather than by character or theme. MacInnes was a prolific social journalist who immersed himself in the marginal worlds of blacks and teenagers, and these novels have been hailed as realistic documentaries of the youth culture of the late 1950s. They certainly evoke a period, and were immensely popular when first published; but they subsequently fell out of favour, remaining cult novels, until they were rediscovered, along with 1950s' fashion, in the 1980s. This revival reached its apotheosis (or, perhaps, its nadir) in 1986 when *Absolute Beginners* was made into a much-heralded but widely derided film.

MacInnes has been praised for having his finger on the pulse of 1950s' popular culture, for creating a literary voice for blacks and teenagers, and for instinctively sensing that music was the key to grasping the teenage revolution. This last is true enough, and quite astonishing when one considers that he was writing before the era of the Beatles. The received image of the period is greatly influenced by his essays and novels, and his best writing accurately reflects the era's innocence. Central to MacInnes's vision, however, was the notion that corruption could be innocent, or at least more admirable than toeing the establishment line; that semi-criminality was both daring and creative, a worthy escape from form-filling and tax-paying which ended when one became cannon-fodder. Perhaps it is this sense of 'revolution' which most dates his work, and makes his optimism about a new generation – which grew up much like the old one, settling into complacent middle age during the Thatcher years – seem rather naïve.

The title of *City of Spades* refers to the African and Caribbean community of London, and the novel is set in the real pubs, bars and clubs where blacks ('spades') met, although the names are changed. Montgomery Pew is a welfare officer with special responsibility for immigrants who gradually becomes involved in their social world, and the novel's strength is that it looks at events from the black insider's point of view as much as from that of the white outsider. MacInnes's negrophilia is everywhere apparent, but its sexual basis may be seen in the character of Montgomery's friend Theodora Pace, who is in love with the African Johnny Fortune (based on one of MacInnes's own lovers). As MacInnes was to do throughout his life, Theodora loves and loses her man, but ascribes this to the differences in outlook, which she sees as racial and insurmountable.

These differences erupt violently in *Absolute Beginners*, which reaches a climax amidst the Notting Hill race riots of 1958. MacInnes drew upon accounts of the riots in the *Manchester Guardian*, fictionalising incidents he found reported there. The 'absolute beginners' of the title are teenagers, and MacInnes's empathy with a much younger generation is remarkable. The novel is narrated by an unnamed youth who works as a pornographic photographer and represents the more or less cheerful anarchism which typified the youth of the period. The novel covers four months in his life, when his amiable lifestyle and cutting wit are thoroughly and violently subverted as the latent racism of the time bursts out into the open.

Mr Love and Justice is once again set in the criminal underworld but, although sociologically precise in its details, is an allegory about a ponce, Frankie Love, and a policeman, Edward Justice. The theme of the novel is that the label-names tell only half the story, that Love recognises justice and Justice is swayed by love. The novel ends with both men lying wounded in hospital, 'where each man finally becomes, as the result of his material fall and inner illumination, identical with the other', as MacInnes explained in a *Spectator* article.

Although MacInnes's picture of immigrant life was much praised at the time, some later commentators have been less impressed. 'With friends like MacInnes ...' was how one black British writer of the 1980s put it. At the time he was criticised for showing the immigrants in an unflattering light (a criticism which, years later, was made by Asians of Hanif Kureishi's film, *My Beautiful Laundrette*, 1986). Replying to an African woman talking about *City of Spades* on a radio programme in 1957, MacInnes objected that her remarks were: 'prompted by an instinct to suppose that because [Johnny Fortune] is set in a delinquent world, that the book implies that this is a natural thing to happen to any African coming to England – in other words, you're leaping to the defence of your people'. MacInnes thought this unnecessary, since Johnny is intended as a heroic figure. It is this hero-worship which colours, and perhaps compromises, MacInnes's vision, rather as the socialism of some middle-class writers in the 1930s was coloured, and perhaps compromised, by their sexual predilections for working-class men.

There is no doubt, however, that at his best MacInnes vividly conjures up a London which few other writers were exploring at the time, and that he fixed a place, a period and a generation in the public imagination.

Olivia MANNING   1908–1980

The Great Fortune

see **The Balkan Trilogy** (1965)

Edna O'BRIEN   1936–

The Country Girls

see **The Country Girls Trilogy** (1964)

Flannery O'CONNOR   1925–1964

The Violent Bear It Away

Although she knew she had not long to live, O'Connor spent seven years perfecting *The Violent Bear It Away*, an intense and oblique novel about the warring forces of good and evil. She was troubled by the character of Rayber, and worried that it was not sufficiently clear that he was supposed to represent the Devil, since she had used his viewpoint for the second part of her novel, and treated him with a true artist's sympathy.

Tarwater is a young orphan, who had been looked after by his uncle, Rayber, until abducted by his elderly great-uncle, Mason. Rayber had 'taken him in under the name of Charity, [and] had at the same time been creeping into his soul by the back door'. Appalled by this, Mason, who claimed to be a prophet, had brought Tarwater to 'the farthest part of the backwoods', and raised him 'to expect the Lord's call himself and to be prepared for the day he would hear it'. When Rayber – accompanied by a 'welfare-woman', to whom he was married just long enough to sire a half-witted boy, Bishop – had come to reclaim Tarwater, Mason had shot at him twice, taking a wedge out of his right ear which, it later transpires, has rendered him deaf. Mason had warned his great-nephew about Rayber: 'He don't believe in the Resurrection. He don't believe in the Last Day. He don't believe in the bread of life.' Before he dies, Mason lays two injunctions on the fourteen-year-old Tarwater: Mason must not be cremated by Rayber, but must be buried according to his instructions in a box he has made himself; and Tarwater must baptise Bishop.

The old man said that as soon as he died, he would hasten to the banks of the Lake of Galilee to eat the loaves and fishes that the Lord had multiplied.

'Forever?' the horrified boy asked.

'Forever,' the old man said.

The boy sensed that this was the heart of his great-uncle's madness, this hunger, and what he was secretly afraid of was that it might be passed down, might be hidden in the blood and might strike some day in him and then he would be torn by hunger like the old man, the bottom split out of his stomach so that nothing would heal or fill it but the bread of life.

THE VIOLENT BEAR IT AWAY
BY FLANNERY O'CONNOR

Tarwater attempts the first task, but fails. 'In the darkest, most private part of his soul,

hanging upside down like a sleeping bat, was the certain, undeniable knowledge that he was not hungry for the bread of life.' After a few hours' digging, he is seduced by Mason's still – 'he's the only prophet I ever heard of making liquor for a living' – and in a drunken frenzy sets fire to the house in which the old man's body is lying. The first part of the novel ends with Tarwater arriving at the house in the city where Rayber and his son live: 'The boy had a vision of the school-teacher and the child as inseparably joined. The schoolteacher's face was red and pained. The child might have been a deformed part of himself that had been accidentally revealed.'

The second part of the novel deals with Rayber's 'pain' and the accomplishment of Mason's prophecy that 'THE PROPHET I RAISE UP OUT OF THIS BOY WILL BURN YOUR EYES CLEAN.' 'It had taken [Rayber] barely half a day to find out that the old man had made a wreck of the boy and what was called for was a monumental job of reconstruction.' But Tarwater, ridiculously wedded to his tall hat, will not let Rayber get close: 'At every turn an almost uncontrollable fury would rise in Rayber at the brand of independence the old man had wrought – not a constructive independence but one that was irrational, backwoods and ignorant.' Rayber recognises Tarwater's detestation of Bishop, who, in his feeble-minded way, is obsessed by water and so ripe to be baptised. He is determined to 'cure' Tarwater of his great 'affliction', but gets nowhere and eventually has to acknowledge that 'the only feeling he had for this boy was hate'. His defeat is complete when Tarwater takes Bishop out on the lake, and Rayber knows 'with an instinct as sure as the dull mechanical beat of his heart that [Tarwater] had baptised the child even as he had drowned him, that he was headed for everything the old man had prepared him for, that he moved off now through the black forest toward a violent encounter with his fate'. Indeed, the last part of the novel describes Tarwater's purgatorial return to 'the farthest part of the backwoods' in preparation for that fate.

Rayber, or the Devil, has lost. Alternatively, reason has failed to triumph over inspired madness. O'Connor herself recognised this ambiguity: 'The modern reader will identify with the schoolteacher,' she wrote, 'but it is the old man who speaks for me.' Readers who do not share her religious convictions may find her vision hard to accept. However, as in **Wise Blood** (1952), her control over language, her acceptance of the absurd and her ability never to let it get out of hand, enables her to create a world uniquely her own and utterly compelling.

Anthony POWELL   1905–

Casanova's Chinese Restaurant

see **A Dance to the Music of Time** (1975)

Raja RAO   1909–

The Serpent and the Rope

The storyteller, Rama (or, properly, Rama-swamy), is a young Indian who goes to France to pursue historical research and falls in love with Madeleine, a lecturer in history. They marry, but begin to drift apart as Rama comes to appreciate the gulf between Indian and western conceptions of love, marriage and family – particularly after he meets Savithri, a Cambridge-educated girl who manages to combine Indianness with militantly modern views. Madeleine finally withdraws not only from Rama but also from the world, while Rama realises that his love for Savithri is not of a physical kind, or even platonic, but spiritual. The novel ends with Rama setting out to go to his guru, so that he can undertake his spiritual quest with guidance of which he feels confident.

Philosophical novels are not everyone's cup of tea; novels of Indian philosophy present additional challenges for readers. The first is Rao's linguistic style: he tries to graft Sanskrit rhythm on to English by means of lengthy and complex sentence-structures. This results in an effect that is sonorous, repetitive and swift, and that is made richer by the importation of Indian

expressions. The second hurdle is the structure of the novel as a whole; imitating some Indian classics, it blends story, philosophy and religion, interspersing the narrative with poetic verse and pithy dialogue on philosophical questions. Overall, the unsympathetic reader will find the book diffuse and garrulous, full of statements such as: 'Meaning is meaningful to meaning.' Rama can be an irritating storyteller, constantly parading his learning, and reducing our attentiveness by his intellectual arrogance and bouts of self-pity. For those inclined to admire philosophical novels, however, *The Serpent and the Rope* is an epic performance which combines profundity and symbolic richness, lyrical beauty and descriptive power. Its daring experiments with form and style then become advantages rather than obstacles.

Conrad RICHTER   1890–1968

The Waters of Kronos

The Waters of Kronos was an outsider in the competition for the 1961 National Book Award. John Updike's **Rabbit, Run** (1960) and Harper Lee's **To Kill a Mocking Bird** (1960) were also entered, and the decision to award the prize to Richter proved a controversial one. Part of the surprise can be attributed to Richter's low public profile. He never courted publicity in either his life or his work, which remains traditional in both form and subject.

Richter's *œuvre* is dominated by two groups of interrelated novels. The first is his historical trilogy about an immigrant family in Ohio; the second is made up of two closely linked novels, *The Waters of Kronos* and *A Simple Honorable Man* (1962). All five books draw heavily on the history of Richter's family, and his career can be viewed as a progressively more direct examination of his own immediate ancestry. *The Waters of Kronos* has its roots in a short story, 'Doctor Hanray's Second Chance', which appeared in 1950. It features a scientist who meets himself as a young boy and is spirited back to the time of his childhood where his parents (naturally enough) fail to recognise him.

This story provided the central conceit of *The Waters of Kronos*, in which John Donner, an old and ailing writer, is likewise transported back to the time of his childhood when he accepts a ride on an old mine-wagon after visiting his family's cemetery. His experiences amongst the family and friends of his past are all coloured by his gift of hindsight, which Richter manipulates to good ironic effect. Donner arrives in 1889 the day before his grandfather's funeral. He is not recognised by his family and is brushed aside by his own father, from whom, as a young man, he had become estranged. His wanderings through the town take the form of a double quest: to find out whether he is indeed his father's son and to identify the strange face which has appeared in his dreams. Taken ill in a neighbour's house, he is visited by a young boy (himself at an earlier age) who prompts him to a kind of spiritual acceptance of his father. When the boy leaves, Donner approaches death and realises that the face in his dreams was the spectre of his own mortality. In this way, the ambiguity of the title (Kronos as 'time' and the tyrannical father of Greek mythology) is neatly resolved at the book's close.

The Waters of Kronos often borders on sentimentality when its themes of forgiveness and acceptance are given direct expression by Richter, but the strong autobiographical sources of the story reinforce the real difficulties in the relationship between father and son. It is this realism which gives the novel's resolution its conviction and lends force to the reconciliations between father and son, and between Donner and his own mortality.

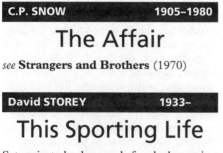

C.P. SNOW　　　　　**1905–1980**

The Affair

see **Strangers and Brothers** (1970)

David STOREY　　　　　**1933–**

This Sporting Life

Set against a background of rugby league in a bleak northern town, David Storey's first novel was classed with the work of Braine, Sillitoe, Barstow and Waterhouse, as a grittily realistic depiction of working-class life. It was a background Storey knew well: the son of a Yorkshire miner, he had himself been a professional footballer. The novel won the Macmillan Award for fiction, and in 1963 Storey adapted it for a celebrated film, directed by Lindsay Anderson.

For Arthur Machin, rugby represents an escape route from a miner's life dominated by routine and a lack of money. When he auditions successfully for the local rugby club, Primstone, he is quickly swept along by the adrenalin of his ambition. Talent is one means of entry into Primstone, while, for the committeemen, wealth is another. Arthur soon realises how dependent he is on the favouritism of his wealthy peers at Primstone, such as Mr Weaver and Mr Slomer. Indeed, Arthur declines the advances of Mrs Weaver in case it jeopardises his rugby career. He enjoys moving in the higher social circles to which success has brought him, but he is soon put in his place, and made to feel like a clumsy court jester amusing his superiors.

Arthur's parents are not as excited about his rugby success as he expects them to be. They are more concerned about the moral corruption which stems from his good fortune, and are suspicious of him mixing with people who are 'not his kind'. Unable to find a confidante, Arthur turns to his landlady, Mrs Hammond. However, she is a very withdrawn woman, and Arthur cannot understand her quiet, almost resigned approach to life, which contrasts with his own driving ambition. Using his increased income from rugby, Arthur keeps Mrs Hammond like a mistress, and the more he spends on her the more dictatorial he becomes. He desperately wants some evidence of her feelings for him, but she is afraid of investing her feelings in someone who, like her dead husband, may leave her.

The tide eventually turns against Arthur: he is no longer the golden boy at Primstone; his relationship with Mrs Hammond founders, and he moves out. He is also suspected of having made Judith, a local girl, pregnant. Again, Arthur immerses himself in his rugby career to avoid facing his predicament.

Mrs Hammond's unexpected death acts as a catalyst, giving Arthur a different perspective on life. In the raw honesty of the Primstone dressing-room, Arthur watches Arnie, the eager young hopeful, and Frank, who is terrified of becoming a has-been. Arthur is left feeling detached from his team-mates, reflecting on rugby merely as an aspect of his life, and no longer a driving force. 'Why be burnt up about it?' he asks himself.

John UPDIKE   **1932–**

Rabbit, Run

see **The Rabbit Tetralogy** (1990)

1961

S. Africa leaves the Commonwealth ● The Berlin Wall is erected ● John F. Kennedy establishes Peace Corps ● Yuri Gagarin is orbited in space ● Structure of DNA determined ● Eero Saarinen, TWA Building, Kennedy Airport ● *Beyond the Fringe* (revue) ● François Truffaut, *Jules et Jim* ● Carl Jung and Ernest Hemingway die ● Thom Gunn, *My Sad Captains*

William COOPER   **1910–**

Scenes from Married Life

see **Scenes from Provincial Life** (1950)

Wilson HARRIS   **1921–**

The Far Journey of Oudin

see **The Guyana Quartet** (1963)

Joseph HELLER   **1923–**

Catch-22

The year is 1944. Yossarian, a US Air Force bombardier stationed on Pianosa, has flown forty-eight bombing raids over Italy and France. Like most of 256th Squadron, he is shortly due to be taken off combat duty. Col. Cathcart, however, keeps raising the number of missions required of the crews. Accordingly, Yossarian decides to go crazy because, Doc Daneeka explains, 'I have to ground anyone who's crazy.' But there is a catch – Catch-22: Yossarian has to *ask* to be grounded and, if he asks, he must be sane. If he is sane, then he cannot be grounded – although he must be crazy if he flies.

Cutting back and forth in time and place, Heller's narrative proceeds by adeptly related anecdote. It portrays its characters drunk, brawling, psychotic, whoring, exuberant and desolate and alone. Predominantly comic, it is spiked with moments wherein it is horribly chill, moments which serve to suggest the roots of the proliferating madness. Milo Mindbender, the clear-eyed profiteer, puts Europe on a free-market footing and unites the warring armies under the banner of M & M Enterprises. Yossarian's rebellion remains central. His friends depart the story one by one. Haunted by two deaths for which he feels in part responsible, he finally, bluntly, refuses to fly any more. The authorities offer him a deal. All he has to do is promise to like them and they will have him posted home a hero. Yossarian turns them down. He decides to desert, thereby retaining his integrity.

The book is full of *trompe l'oeil* reasoning and neatly balanced absurdities: Major Major, who can be seen when he is not there yet, when he is there, cannot be seen; McWatt, whose virtuosity precedes disaster and Orr, whose failure precedes success; a living man who is (bureaucratically) dead and a dead man who (bureaucratically) never existed; Generals Dreedle and Peckem, their energies dedicated to subverting each other, and Pte First Class Wintergreen (demoted), a figure of greater influence than both. Indeed, *Catch-22*'s underlying proposition is perverse for, in the conflict it describes, the real threat to Yossarian comes from his own side. The US military – incompetent, inflexible, internecine and paranoid – compels him to endanger his life. The Germans, by contrast, appear merely as lines on a map or as bursts of flak in the sky: extremely adverse flying conditions, no more.

Major Major had been born too late and too mediocre. Some men are born mediocre, some men achieve mediocrity, and some men have mediocrity thrust upon them. With Major Major it had been all three. Even among men lacking all distinction he inevitably stood out as a man lacking more distinction than all the rest, and people who met him were always impressed by how unimpressive he was.

CATCH-22
BY JOSEPH HELLER

Thus, eliding the evils of Nazism, Heller is able to sidestep questions which might embarrass his protagonists into commitment. He neither entangles himself in moral ambiguities nor succumbs to the temptation of platitude. The result is appropriately paradoxical: a novel which suppresses its own subject. Its achievement is to create a closed system within which farcical persecutions and power struggles are substituted for the historical ones beyond; but it can maintain its sardonic poise only by excluding from its comedy the fascist threat which brings it into being. Heller is oddly caught in his own Catch-22: to contemplate the war he must evade it – although evasion makes it bearable.

The novel was made into a long and rather unsatisfactory film (1970) directed by Mike Nichols from a screenplay by Buck Henry.

Richard HUGHES   1900–1976

The Fox in the Attic

This novel was intended as the first volume in a projected trilogy or tetralogy, 'The Human Predicament', described by Hughes as an 'historical novel of my own times culminating in the Second World War'. Unfortunately, Hughes was a notoriously slow writer; born in 1900, he had published only two novels before this one. He said that he would have to live to be 120 in order to complete his design, and achieved neither ambition. A second volume, *The Wooden Shepherdess*, was published in 1973, and is generally considered inferior to the first, while fragments of a third novel were in existence when Hughes died in 1976.

The Fox in the Attic is set in England and Bavaria in 1923, and mixes mostly fictitious characters with a few real ones, such as Hitler and Ludendorff. The central character is Augustine Penry-Herbert, a young Welsh squire just down from Oxford, a prototype of the liberal idealist of the 1920s: 'He could not get beyond his starting-point that all previous generations had been objects, whereas *his* were *people*.' Augustine walks into the novel carrying a dead child on his shoulder: this is Rachel, who has been killed accidentally in Pembrokeshire marshes near the decaying property of Newton Llantony, which Augustine has inherited and where he is living as a recluse. He incurs the hostility of the people of the neighbouring township of Flemton, who suspect his involvement in the death of the child. To escape the incident, he goes to the Dorset country house of Mellton, where his sister Mary lives with her husband Gilbert, a Liberal MP. The attachment of Augustine to his little niece Polly, who calls him 'Gusting', is stressed, and the fate of the living and dead child are contrasted. It turns out that Rachel is the niece of the housekeeper at Mellton, Mrs Winter.

Seeking further escape, Augustine goes to stay with distant German cousins, the von Kessens, gentlemen-reactionaries who live in the castle of Lorienburg in Bavaria. He falls in love with Mitzi, the young daughter of the castle's owner, Walther von Kessen. She is in the final stages of going blind. Augustine has arrived at the time of Hitler's beerhall *Putsch* in November 1923, and the novel gives an account of this, largely through the eyes of a young Nazi, Lothar. Lothar's rather older brother, Wolff, a right-wing soldier and killer, is in hiding in the attic at Lorienburg (he is the 'fox in the attic' of the title), where he accidentally hangs himself. The novel ends with Augustine visiting Munich and returning to find that Mitzi's family, with her consent, intend her to become a nun.

Next Sunday he announced from the pulpit a momentous discovery: Johns the Baptist and Evangelist were one and the same person! He was stuttering with excitement, but Augustine heard no more because Uncle William, startled at the news, dropped his eyeglass in his ear-trumpet and began fishing for it with a bunch of keys. Uncle Arthur in his senior corner of the family box-pew kept commenting 'Damn' young fool!' (he was unaware of the loudness of his own voice, of course) 'Oh the silly damn' fool!' then snatched the ear-trumpet from his brother's hand and dislodged the eyeglass by putting the trumpet to his lips and blowing a blast like the horn of Roland.

THE FOX IN THE ATTIC
BY RICHARD HUGHES

As it stands, *The Fox in the Attic* perhaps lacks a clear emotional centre, but it showed immense promise as the initial volume of an extended *roman fleuve*. Among other things, it is magnificently written, in a prose style that moves easily from the seemingly casual to the significant, and allows subtle character analysis as well as political and philosophic discussion.

V.S. NAIPAUL 1932–

A House for Mr Biswas

Naipaul's best-known novel relates the life story of Mr Mohan Biswas, from his inauspicious birth in an obscure village to his death in his own house in the city forty-six years later. The novel charts his tragi-comic progress through assorted jobs, from cane fields to journalism, towards the longed-for goal of being a house owner.

Biswas is placed in the dense and changing background of the rural Tulsi household into which he is married. The Tulsi's old way of life has been forged over three generations in a colonial context. It crumbles owing to the demise of authority – represented by the Tulsi matriarch, who does not survive a move to a new home – and the subsequent outbreak of opportunism and individualism in the clan. Though living in a house in Port of Spain, the families of the various Tulsi daughters become separate economic units, with the children competing with each other and the rest of their generation for education – which will itself scramble the old values with new ones.

A historical novel, the book stands broader interpretation. Mr Biswas's strivings for a house of his own are Everyman's striving for material success. His resistance to being absorbed into anonymity by the Tulsi household can be viewed as everyman's resistance to the depersonalised modern state. However, Naipaul is also writing about the experience of acculturation – specifically, the adaptation of the culture of his own childhood to a broader West Indian culture. Finally *A House for Mr Biswas* is a gradual revelation of the hypocrisy and invalidity of colonial Hindu culture, and a sad indictment of his own East Indian community.

Muriel SPARK 1918–

The Prime of Miss Jean Brodie

When the intellectual Sister Helena of the Transfiguration is asked who most influenced her schooldays, she replies: 'There was a Miss Jean Brodie in her prime.' Spark's most popular novel (subsequently adapted for stage, cinema and television) describes the effect of an Edinburgh schoolteacher upon five of her pupils in the 1930s by tracking back and forth through their lives.

Jean Brodie has little time for the prevailing ethos of Marcia Blaine School, maintaining a staunch belief in the primacy of the individual ('Cleopatra knew nothing of the team spirit'). What the headmistress disparages as 'the Brodie set' is a hand-picked elite: 'I am putting old heads on your young shoulders, and all my pupils are the crème de la crème.' In place of the curriculum, Miss Brodie regales her pupils with descriptions of her European holidays and her love-affair with an idealised young man who 'fell on Flanders' Field'. Compromised by a 'defective sense of self-criticism', she insists that her principals of education are the correct ones: 'The word "education" comes from the root *e* from *ex*, out, and *duco*, I lead. It means a leading out. To me education is a leading out of what is already there in the pupil's soul.' Unfortunately her enthusiasm for Italy embraces Mussolini, and the fascism of *il Duce* gradually supplants the ideals of *e duco*.

While her pupils compose romantic stories about their teacher's life, she has attracted the devotion of Gordon Lowther, the music master, and glamorous, one-armed Teddy Lloyd, the art master, much to the disapproval of her colleagues, who resent her influence and vainly plot her downfall. When she is finally dismissed, the headmistress, Miss Mackay, vindictively informs her

---

Get out your history books and prop them up in your hands. I shall tell you a little more about Italy. I met a young poet by a fountain. Here is a picture of Dante meeting Beatrice – it is pronounced Beatri*chay* in Italian which makes the name very beautiful – on the Ponte Vecchio. He fell in love with her at that moment. Mary, sit up and don't slouch. It was a sublime moment in a sublime love.

THE PRIME OF MISS JEAN BRODIE
BY MURIEL SPARK

---

that it was one of her set who betrayed her. Miss Brodie spends her retirement wondering which of her six girls it could be. Poor, stupid Mary Macgregor, exacting revenge

for Miss Brodie's unremitting scorn before dying in a hotel fire? Or Rose Stanley, 'famous for sex', who posed naked for Teddy Lloyd? Or Eunice Gardiner, a natural athlete who was 'attracted to the team spirit'? Or Jenny Gray, who disappointed her teacher by training to be an actress? Or brainy Monica Douglas, for 'there is very little Soul behind the mathematical brain' and she was given to temper? The only one Miss Brodie does not suspect, until she is on her death-bed, is piggy-eyed, all-seeing Sandy Stranger, later to become Sister Helena, who sees through her teacher and tells the headmistress: 'You won't be able to pin her down on sex. Have you thought of politics?'

'Give me a girl at an impressionable age,' Miss Brodie boasted, 'and she is mine for life.' The final twist in this compact fable of spiritual pride is that in the case of Sandy, her least tractable pupil, Jean Brodie was right.

Evelyn WAUGH   **1903–1966**

The Sword of Honour Trilogy

Men at Arms (1952)
Officers and Gentlemen (1955)
Unconditional Surrender (1961)

'Here love had died between me and the army,' Charles Ryder reflects in the prologue to ***Brideshead Revisited*** (1945), the novel Waugh wrote while on leave during the Second World War. His later trilogy of novels expands this theme, tracing a civilian's romantic involvement and subsequent disillusionment with both the army and the war. This process is recorded in the novels' titles: *Men at Arms* suggests the yeoman tradition, stretching back over centuries; *Officers and Gentlemen* is bitterly ironic, since it soon becomes apparent that not only are officers no longer necessarily gentlemen (in any sense of the term), but gentlemen signally fail to behave like officers;

Unconditional Surrender to new inimical values is the only option. The book certainly reflects Waugh's own bruising experience of the modern army – where he was considered insubordinate and abrasive by his superiors and was heartily disliked by the men under his command – and in many details Guy Crouchback's war is identical with the one endured by his creator. The other guiding theme of the novels is the Catholic faith, treated with more restraint than in *Brideshead Revisited*, but no less important.

It seems likely that one literary (as opposed to auobiographical) source of the trilogy was Ford Madox Ford's tetralogy about an English gentleman and the First World War, **Parade's End** (1924–8). Like Ford, Waugh had difficulty in deciding a final form for his sequence, and in 1965 reissued the novels in a single volume, *Sword of Honour*. He had been particularly dissatisfied with *Men at Arms*, and took the opportunity, both in this novel and its companions, to remove minor characters and make numerous other excisions and alterations, few of which are considered improvements. Although, for convenience, the trilogy is being treated here as a single narrative, the texts followed are those of the original novels.

Guy Crouchback is the only surviving son of an old Catholic family, and has been living in lonely exile in Italy since the failure of his marriage. He responds with joy and relief to news of the outbreak of war, news which brings (in a telling echo of Rupert Brooke) 'deep peace to one English heart'. Although he is nearly thirty-six, he returns from abroad (where his chivalric notions about warfare are suggested in the prayers he offers at the tomb of a crusader) and contrives to enter the Royal Corps of Halberdiers. He receives little encouragement: 'All that sort of thing happened in 1914 – retired colonels dyeing their hair and enlisting in the ranks,' his brother-in-law warns. 'All very gallant, of course, but it won't happen this time.' Indeed, his First World War ideals about honour and heroism are consistently undermined by experience. He suffers a prolonged and dismal period of training amongst younger and uncongenial fellow officers. His only ally is an eccentric of similar age called Apthorpe, the proud owner of 'porpoise' boots and a 'thunder box' (portable latrine), which, in a farcical episode, is booby-trapped by Brig. Ben Ritchie-Hook, a ferocious veteran of the First World War.

The Halberdiers sail for Africa and, after many delays and disappointments, Guy becomes involved in an unofficial raid on Dakar during which Ritchie-Hook returns with the head of a native soldier. This incident invokes the displeasure of the authorities and when the fever-stricken Apthorpe dies after drinking Guy's illegal gift of whisky, Guy is sent back to England in disgrace.

In the second volume, Guy is awaiting instructions, while searching for 'Chatty' Corner, Apthorpe's chief legatee. He is sent to the Scottish island of Mugg to join a Commando force being formed under Tommy Blackhouse, an old friend and also a former husband of Guy's ex-wife, Virginia. Amongst Guy's fellow officers is Trimmer, a former hairdresser who had failed to get his commission in the Halberdiers, and is now masquerading under the name McTavish. Whilst on leave, Trimmer meets Virginia and they have a brief affair. Also at Mugg is Ivor Clare, a dandy whom Guy befriends. After many false alarms, the Commando force sails to Egypt and thence to Crete. Trimmer is left behind, taking part in a survival course. In Egypt, Major 'Fido' Hound joins the force as brigade-major and Blackhouse breaks a leg. The Commandos, without Blackhouse, arrive in Crete to find that they are to cover the evacuation of the island and then surrender to the advancing enemy. Clare deserts his troop and boards one of the evacuation craft. Hound also deserts and is never seen again. It is to be assumed that he was murdered by Ludovic, a sinister and godless corporal major. Guy joins Ludovic on a small boat after the evacuation has been completed and they sail to the African coast. During the voyage, suffering from hunger and exposure, Guy becomes delirious and does not witness the disappearance of a sapper from the boat, presumably murdered by Ludovic.

Meanwhile Trimmer has been sent with a tiny force to blow up a lighthouse on an uninhabited Channel Island as a publicity exercise. Ian Kilbannock, a press officer and a friend of Guy, accompanies the mission, which gets lost and lands in Occupied France. Kilbannock writes up the episode and Trimmer becomes a national hero who is sent on propaganda tours. He falls in love with Virginia, who has been divorced by her third husband, a rich American called Troy.

Recovered from his experiences, but disillusioned by the war, Guy rejoins his regiment in England. The Soviet alliance has put an end to Guy's belief in the justice of the war; all that is left now is personal honour.

The third volume opens two years later with Guy left behind in England, employed as a liaison officer at the Hazardous Offensive Operations HQ. By chance, his name is put forward for work in Italy. He is sent on a parachute course where he meets Frank de Souza, a young cynic who had trained with Guy at the beginning of the war. The course commandant is Ludovic, who believes that Guy suspects him of murdering Hound and the sapper. He keeps to his room and displays signs of severe mental disturbance. Guy injures himself during training and convalesces in the London flat of his Uncle Peregrine.

Meanwhile Virginia has become pregnant by Trimmer. She attempts unsuccessfully to procure an abortion, then asks Guy to remarry her. Guy sees this as an opportunity to perform an entirely unselfish act and so agrees. He is posted to Yugoslavia almost immediately as a liaison officer.

Virginia has a son, whom she sends to the country to be looked after by Guy's sister. A flying bomb hits the London flat killing both Virginia and Peregrine. In Yugoslavia, Guy's attempts to help a group of Jewish refugees are continually thwarted by the partisans. An exercise to demonstrate to the Allies the proficiency of the partisan forces is organised. Ritchie-Hook contrives to become actively involved and is killed whilst assaulting a blockhouse. Mme Kanyi, one of the Jews, tells Guy that the war is the result of a universal death-wish and Guy realises that the satisfaction of personal honour is as unjust a motive as any other in war. Although the majority of the Jews are liberated, Mme Kanyi and her husband are tried (and presumably executed) by the partisans for reasons to which Guy unwittingly contributed.

In an epilogue it is revealed that Guy has married the daughter of a Catholic woman who has been looking after Virginia's baby. By accepting Trimmer's child as his own, Guy has ensured the continuity of faith in the Crouchback family. Trimmer has vanished, possibly having jumped ship in South Africa. After the success of his wartime journal, Ludovic has written a very bad but very popular novel (which bears a marked similiarity to *Brideshead Revisited*) and has bought the Crouchbacks' Italian castle on the proceeds. Guy returns to a house on the family estate, his social and spiritual exile at an end.

Although Waugh is addressing large and solemn themes, his writing still has enormous comic energy. The saga of Apthorpe, about whom there is 'a sort of fundamental implausibility', and the equally explosive antics of the Laird of Mugg and his Scottish Nationalist niece, are broad comedy; elsewhere there are satirical swipes at such contemporaries as Cyril Connolly (in the character of Everard Spruce, editor of *Survival*) and Harold Nicolson (in the character of the influential, homosexual diplomat, Sir Ralph Brompton), as well as more affectionate portraits of such friends as Robert Laycock, the dedicatee of the second volume and recognisable as Tommy Blackhouse, and Lady Diana Cooper, appearing (as she did in *Scoop*, 1938) as the beautiful, scatty and madly indiscreet Mrs Algernon Stitch.

1962

Independence of Algeria • Cuban crisis • U.S. military council established in South Vietnam • de Gaulle escapes assassination • Kennedy and Macmillan agree that U.S. provide Britain with Polaris missiles • Telstar launched • Anthony Sampson, *The Anatomy of Britain* • T.H. White, *The Making of the President* • U.S. astronauts John Glenn and M. Scott Carpenter are put in orbit • Pan-American Airways Building, New York, provides world's largest office accommodation • *Private Eye* is issued • Edward Albee, *Who's Afraid of Virginia Woolf?* • R.H. Tawney, Eleanor Roosevelt and Niels Bohr died

---

Dorothy BAKER   **1907–1968**

Cassandra at the Wedding

Baker's fourth and last novel is acknowledged to be her best. Its themes – the quest for identity, human and artistic; the betrayal of trust; the choices we make between good and evil, between being pro- and anti-life – are to an extent the themes of her earlier novels: *Young Man with a Horn* (1938), inspired by the music, and based loosely on the self-destructive life, of Bix Beiderbecke; *Trio* (1943), the story of a young woman graduate's rescue (by an all-American male) from a sado-masochistic liaison with a French literature don (female); and *Our Gifted Son* (1948), in which a young classical musician's mother is driven to suicide by the cruelty of the boy's father.

In *Cassandra at the Wedding* Baker examines the twin relationship, what it means to the twins themselves and the effects their bond has on the people closest to them. Cassandra Edwards, an intelligent, rebellious, homosexual (so far) postgraduate literature student at Berkeley University, drives home (in her dead mother's old Riley) to her father's Californian ranch for the wedding of her identical twin, Judith, to a Connecticut doctor. Unable to face the idea of losing – as she sees it – her twin, Cassandra attempts suicide by swallowing a bottle of the Nembutal her Berkeley analyst, Vera Mercer, has prescribed for her depression and

insomia. (We learn later that Cassandra has, at some time, tried – and failed – to seduce Vera.) In the end, Cassandra is saved from death (by, irony of ironies, her twin-sister's husband-to-be), braves the wedding and decides to have another go at independence and at life. This sounds like the stuff of melodrama, but the novel avoids it by means of Baker's skilful characterisation, the self-mocking tone she allows her protagonist, and the irony that informs the book as a whole.

While *Cassandra at the Wedding* cannot be said to be autobiographical in any real sense, it is unquestionably a reinventing of autobiographical material. The ranch which is the setting for the novel is in every detail the one at Terra Bella where Baker and her husband, the poet and critic Howard Baker, lived. The models for the twins were their daughters who, not twins themselves, were unusually alike in physical and emotional ways as children and as adolescents. The brandy-and-soda-drinking philosopher father of the novel is, according to Howard Baker, a portrait of himself. The twins' dead mother, Jane, whose headlong ghost haunts the proceedings, is in all essentials Baker herself, who during the writing of the book was suffering from terminal cancer. Howard Baker saw the novel as a projection of the two sides of the author, whose usual zest for life was undermined by periodic bouts of depression and whose bisexuality made for a complex and at times stormy marriage.

Throughout her professional life, Baker lectured and taught at various American universities. The classical simplicity she admired in the writing of others is a hallmark of her own. In *Cassandra at the Wedding*, her punchy short sentences (often verbless and occasionally allowed a paragraph to themselves) and her use of conjunctions, rather than clauses, in long sentences; give her prose a 1990s immediacy, and make it hard to remember that the book was published in 1962.

James BALDWIN   1924–1987

Another Country

In the late 1940s Baldwin began work on a novel provisionally titled 'Ignorant Armies'. He later realised that he was in fact writing two different novels, one of which became **Giovanni's Room** (1956), while the other eventually became *Another Country*. Like its predecessor, *Another Country* was to have been set in Paris, where Baldwin had lived, but he eventually decided to set it in America, the title referring to the country as seen from the outside. A surprising influence upon the book was Henry James, and Baldwin prefaced his novel with a Jamesian quotation about the emotional inarticulacy of Americans, which he thought applied equally to the present day. Several of the characters are based upon people he knew, and the central scene of Rufus's death (which Baldwin described as 'the hardest thing I ever wrote') was inspired by the suicide of a young socialist friend, Eugene Worth, who, in similar circumstances, threw himself off the George Washington Bridge. The novel had a very mixed reception, disappointing some critics, and outraging others because of its inter-racial, bisexual couplings. Charges that the book was obscene, and attempts to ban it in New Orleans, ensured high sales.

The powerful driving force behind the novel is Baldwin's own profound knowledge of how it feels to be an outsider – black, homosexual, poor, creative – living and dealing with social pressures and prejudices. The catalyst that brings together a group of friends in New York in the 1950s is the suicide of Rufus Scott, a young black jazz musician. Desolated by the violent failure of an affair with Leona, a white Southern girl,

haunted by poverty, racial prejudice, and self-disgust, and alienated from the support of family and friends, Rufus has thrown himself off the Washington Bridge.

Rufus's friends – Vivaldo Moore, Cass and Richard Silenski – and his sister Ida are devastated. His death provokes a ruthless examination of their lives and values, their attitudes towards lovers, children, friends and the world at large. Vivaldo and Ida begin an affair tense with suspicion and memories: Vivaldo is white, Ida is black; Vivaldo is poor, a struggling writer living a Bohemian life, and Ida is about to make a successful career as a singer – though to achieve her ambitions she will have to sacrifice some hard-won and deeply held values. Richard and Cass have to come to terms with tensions within their marriage. Richard's successful novel is despised by Cass and Vivaldo as a second-rate piece of work, and Cass has an affair with a bisexual actor, Eric, who has returned to New York from France to play an important leading role in a new play. There is, too, an unresolved tension between Eric and Vivaldo which comes to a head when they get drunk and end up in bed together.

He had often thought of his loneliness, for example, as a condition which testified to his superiority. But people who were not superior were, nevertheless, extremely lonely—and unable to break out of their solitude precisely because they had no equipment with which to enter it. His own loneliness, magnified so many million times, made ·the night air colder. He remembered to what excesses, into what traps and nightmares, his loneliness had driven him; and he wondered where such a violent emptiness might drive an entire city.

ANOTHER COUNTRY
BY JAMES BALDWIN

The violence of these encounters forces Baldwin's characters towards rare honesty, towards dealing finally with the truth of their lives, towards integration of their needs and their courage and cowardice when faced with near-intolerable pressures and guilt.

They draw together in their grief and confusion, and are strengthened in their true commitments. A mood of fierce outrage informs the novel – it pulses with the bleak music of urban streets, with passion and melancholy, with ardent rhythms that underlie the darkest emotions and the most intense tenderness.

Brigid BROPHY   1929–

Flesh

When Nancy takes up with Marcus nobody can understand what she sees in him. Brophy's third novel is an elegant, funny, erotic and beautifully observed account of his transformation from agonised awkwardness into a voluptuous voluptuary. Marcus, the indulged son of a wealthy Jewish businessman and a hippopotamus-like mother, has been unable to find a career to suit his refined sensibilities. He is an unprepossessing young man whose nose hints at proboscidian circumcision. He is also a martyr to the Firbankian complaint, hyperaesthesia, and suffers 'a delight intense to the point of agony' in beautiful things, especially in the 'great blonde areas of Rubens flesh'. He fears that he will never meet anyone who gets the point of him.

However, the brisk and attractive Nancy, whom Marcus at a party where he is enduring his usual social torture, realises at once that his passive appreciation of beauty is the perfect complement to her own active talent, which is for sexual intercourse. Nancy is received with joy by Marcus's parents in their hideously comfortable house near Ken Wood, with its smell of central-heating oil and furniture polish and 'jazzy' orange 1930s décor. Marcus perceives that his sister, a compulsively knitting divorcee, would, like his father, fancy Nancy, if there were any provision for this in the Jewish Tradition. (He is proved correct later in the narrative.) Nancy in turn introduces Marcus to her parents' bleak Bohemia in Belsize Park. Both families attend a dance at which Nancy bullies Marcus into taking the floor. He loses his clumsiness in her arms, 'this publicly permitted parody of an experience he had never had, sexual intercourse, at last liber-

ated his physical response to Nancy – and he asks her to marry him.

His appearance undergoes subtle improvements during their engagement, and their honeymoon in Lucca, where he proves as apt a pupil as Nancy is a teacher, changes him utterly. Filial guilt, occasioned by news of his father's death, leaves Marcus temporarily impotent, but he is once again rescued by Nancy. They set up home in Chelsea and Marcus's sensuality now extends to an appreciation, to the point of gluttony, of exotic food. He finds congenial employment in the fantastic Wigmore St emporium of the mysterious Polydore, a restorer and faker of antiques, and grows ever more plump and contented, secretly supplementing the diet on which Nancy has placed him with stout and pickled onions. When Nancy becomes pregnant, it is Marcus and not she who swells. Bored by motherhood, Nancy returns to work and Marcus's new sensuousness expresses itself in love in the afternoon with the beautiful German *au pair*.

His metamorphosis is seen to be complete on the day that Nancy encounters him bulging horribly in his swimming trunks and accuses him, in mingled horror and desire, of looking like a woman, provoking Marcus to chuckle: 'It's a process of empathy. I've *become* a Rubens woman.'

Anthony BURGESS   1917–1993

A Clockwork Orange

Principally because of Stanley Kubrick's spectacular (and subsequently suppressed) film of the book (1971), this is one of the most popular of Burgess's many novels. Alex, who narrates the story, leads one of the gangs of delinquents who afflict an England of the future. Theft, rape, torture and murder are the order of the day. The violence of the gangs is an attempt at deliberate self-assertion. As Alex explains: 'What I do I do because I like to do.'

Their rebellion is partly justified by being aimed against a soulless, technological society of advanced urban decay, marked by the unrestrained sale of drugs and crass commercialism. Alex's vitality, charm and wit are more attractive then the inertia of his society, and his constant complicity with his readers, whom he addresses as 'my brothers' and 'only friends', also blunts the repulsiveness of his violence. Moreover, the novel's most interesting feature is the gang's language 'nadsat', which reflects Alex's enormous energy and style (he prefers classical music): the language functions as another device to distance readers from the results of the violence in which Alex is involved.

Finally, however, Alex is arrested and sentenced to the 'humane' and 'liberal' Reclamation Treatment, which transforms him – via electric shock therapy and films of Nazi-like horrors – into an emotionally neutered creature, sickened even by art, music and sex. He has become a piece of machinery, a 'clockwork orange'.

There are two versions of the story's end, in both of which Alex recovers his humanity. The first shows Alex contemplating marriage and a settled life. However, Kubrick's film, and all subsequent editions of the book, end with Alex returning to a full and joyful renewal of his choice of evil.

In 1990 the novel was also adapted for a much heralded but short-lived stage show, chiefly remarkable for its sets.

HAN Suyin   1917–

Winter Love

'Only when my mind goes back to that London winter do I feel alive, instead of merely knowing as a fact that I live', says Red, the narrator of this powerful, short novel set in wartime London, where its Chinese author did her own medical training.

Horsham College of Science, where Red studies zoology, is an all-female institution whose students are no strangers to passionate 'situations' – the college's code word for the 'quarrels or tight-lipped scenes' which indicate that yet another Special Friendship between two of the women has ended. The passions are ubiquitous and readily accepted, though few of the students are, as Red puts it with characteristically uneasy evasion, 'permanently like that'. Red herself

is 'like that' and the tragedy that unfolds in this novel is the bitter fruit of her inability to acknowledge and accept the fact. Badly damaged in childhood and adolescence by a succession of indifferent adults who provide material care but no genuine affection, Red, when at last offered the love she seeks, can only destroy it, impelled by cruelties and distrust rooted in her own self-contempt.

Mara, whom Red loves, is an unhappily married fellow-student. She is several years older than Red but, unlike her, has never before loved another woman sexually. She is gentle, vulnerable and too honest and straightforward to need the self-protective psychological games which Red, unable to believe that she is truly loved, plays constantly. Too honourable to remain in a marriage where sexual love is travestied by a quietly sadistic husband, and too badly hurt by Red's own brand of emotional sadism, Mara quite simply disappears.

The novel is, in effect, Red's despairing threnody for the love she laid waste and for the suffering she inflicted on Mara. It is also a demanding act of personal dissection in which Red painfully uncovers the reasons why love and cruelty long ago fused for her. Self-examination forces the recognition of her own lesbian sexuality. That, in its turn, brings a realisation that her marriage, treacherously engineered by her in a frenzy of social panic, cannot last: 'And though he keeps me safe, I know I'll leave him one day, walk out of this safety which is a mess.'

Winter Love offers a particularly convincing description of Red's moral and emotional perplexities, and sets them in their context of larger social confusions. Love between women, and the perennial possibility of its sexual expression, is everywhere – at Red's school, at Horsham College, amongst Red's cousins, even between the owners of the isolated Welsh guest-house where Red and Mara spend a disastrous fortnight's holiday. But refusing to name it neutralises it, or so Red has mistakenly believed: 'So many things could be done, and if one didn't talk about them, didn't think about them, one could live with them, they would be quite all right … So the important thing was not to call things by their name.'

But the safety of silence is a delusion, and a dangerously destructive one, as Mara realises early on. Red's cowardice freezes Mara out, freezes Red's own capacity to love. Coldness, bleakness and sterility permeate *Winter Love*. Its physical setting – London, gutted, bone-weary at the fag end of a relentlessly grinding war – is one of desolation and deprivation. Despite the resilience and warmth still displayed by individual Londoners, the city itself becomes the perfect symbolic landscape for Red's enforced journey as 'autumn deepens and darkness draws in to another constricting winter, towards a cold so cold, short days so short', that her only respite is to 'turn on the lights, and the radio loud, and draw the curtains against the night'. Yet those words, the novel's last, must be set against the many signs, scattered throughout the book, that although Red's pain is irreversible, her defeat is not. Her new self-knowledge, so hardly won, heralds the possibility of change.

Wilson HARRIS   **1921–**

The Whole Armour

see **The Guyana Quartet** (1963)

Ken KESEY   **1935–**

One Flew Over the Cuckoo's Nest

In common with several other American writers (such as Robert Pirsig and Erica Jong) Kesey has achieved phenomenal popularity with one book, while its successors have gained only modest readerships. *Zen and the Art of Motorcycle Maintenance* (1974), **Fear of Flying** (1973) and *One Flew Over the Cuckoo's Nest* have all sold millions of copies and run to many editions. The factor that these otherwise dissimilar works have in common seems to be their relevance to specific and previously unarticulated needs felt widely at the time of publication. The worth of all three is dependent, to a certain extent, upon historical accident. When *Cuckoo's Nest* was published, it found its first constituency of readers on the campuses of the American universities, and from there its popularity was quickly exported to a much wider (but still predominantly American) audience.

Set in a mental hospital (the 'cuckoo's nest' of the title, which is taken from the traditional nursery rhyme 'Hinty Minty'), the action begins with the admittance of Randle Patrick McMurphy, the novel's hero. A free spirit and determined upsetter of convention, McMurphy dares to laugh at the absurdities of the hospital regime, and sets about the task of 'liberating' his fellow inmates from the institutional constraints they have accepted, organising a number of vaguely therapeutic, hilarious escapades and generally cocking a snook at the authorities.

The hospital authorities are personified in 'Big Nurse', and McMurphy's struggle with her is increasingly presented as a struggle between good and evil. Big Nurse is offered as a perversion of femininity, her large breasts frigidly constrained within a starched, white uniform. An enemy to her own nature, she also opposes the natural spontaneity urged by McMurphy. Kesey has been accused of misogyny in his portrayal of women and this is probably just. Most of the women in the book are seen as agents of

---

Yes. This is what I know. The ward is a factory for the Combine. It's for fixing up mistakes made in the neighbourhoods and in the schools and in the churches, the hospital is. When a completed product goes back out into society, all fixed up good as new, *better* than new sometimes, it brings joy to the Big Nurse's heart; something that came in all twisted different is now a functioning, adjusted component, a credit to the whole outfit and a marvel to behold.

ONE FLEW OVER THE CUCKOO'S NEST
BY KEN KESEY

---

repression; only the two prostitutes who give their sexual favours freely to the inmates in one of McMurphy's escapades are presented sympathetically. The sexist underside of

1960s' radicalism is clearly on show, and of the values *Cuckoo's Nest* promotes, this is the one that dates it most. McMurphy finally wins a pyrrhic victory over Big Nurse. At the novel's close, many of the inmates have been revitalised, and have learnt to laugh at the system. McMurphy himself, though, is lobotomised – only the shell of him remains.

The novel's messages are that an institution is not a monolith, is not immutable and, above all, is not necessarily right; the individual can mobilise support against what is 'given', can change things. Its final point, that the agent of change might have to sacrifice himself to his cause, is potentially tragic and helps to anchor its other, more optimistic affirmations in harsh reality.

One Flew Over the Cuckoo's Nest can stand independently of its historical context, but much of its relevance does derive from the ideas put forward in it, which were shortly to gain much wider currency. It remains a striking anticipation of the beliefs which informed both the radicalism of America in the 1960s and the counter-culture within which Kesey himself was shortly to play a starring role.

Doris LESSING   1919–

The Golden Notebook

This is an immense book in every sense – not least in its intention which, in the author's own words, sets out to portray the 'ideological feel of our mid-Century'. As such, it could be said to have dated, as Lessing predicted. In the United Kingdom the sweep of change brought about by socialism in the 1950s and 1960s can hardly be imagined after the decade of Thatcherism; the frankly promiscuous lives of the characters seem somewhat reckless in an era overshadowed by the threat of Aids. Some things, however, have not changed and are universals: the struggle between male and female, truth and lies, reality and fiction, fidelity and betrayal. The way Lessing's book has been consistently misunderstood (her dismay over this prompted a later explanatory preface) is regrettable but not surprising since the work

(novel? psychological/sociological drama?) is almost overwhelming in its complexity and intellectualism.

Its 'skeleton' is a 'conventional short novel', 'Free Women', which is intercut with excerpts from the notebooks of Anna Wulf, one of its central characters. Anna, a novelist apparently suffering from a writer's block, and Molly, an actress, are both single parents, left-wing and essentially feminist. Molly, though, bullied by her ex-husband Richard and tyrannised by her son Tommy, whose suicide attempt has left him blind, is only really content when she is about to remarry at the end of the story. Anna, despite numerous affairs, is deeply – permanently – wounded by the break-up of her relationship with her lover, Michael. Only Marion, Richard's second, betrayed, alcoholic wife – and motif for all the wives that Anna and Molly have helped betray – achieves a kind of tenuous freedom, with Tommy's help, in the form of a growing, if naïve, political consciousness.

Anna's attempts to avoid inner chaos are embodied by four Notebooks, the colour of each representing a different aspect of her life. Black is for novelist Anna, author of *Frontiers of War*, an inter-racial love story set in Central Africa where she has spent some years of her life. This is a highly political work and not the 'Brief Encounter with Wings' (the English character is an RAF officer) that most television and film producers would like to turn it into. The block, for which she is receiving psychoanalysis, is more a deliberate decision to avoid the pernicious nostalgia which she feels is pervading her work. Red is for political Anna, who joins the Communist Party as a cultural adviser and leaves it, disillusioned, four years later. It is also for the Anna who is appalled, paralysed, at the horror and violence of world events – the McCarthy witch hunts, the threat of war, the H-bomb. Yellow is written by the Anna who makes stories out of her own experience, private stories which at once reflect her life, reveal truths about it and finally exceed it. Initially in the form of a novel – *The Shadow of the Third*, in which Ella is rejected by her lover Paul – it tails off into synopsis and story-lines which are linked directly to her Blue Notebook. This latter 'tries to be a diary' but, unable to record experience truthfully within the parameters of language, pained at the 'thinning of language against the density of our experience', Anna descends into the hell of madness. Out of the chaos is created the Golden Notebook, the 'central point', the 'essence' which says 'implicity and explictly that we must not divide things off, must not compartmentalise'.

Connections and interconnections are made. Events from the Notebooks coincide with, contradict, events in 'Free Women'. Anna's breakdown is reflected in the breakdown of boundaries between one Notebook and another, one fiction and another. Characters are interchangeable; mad becomes sane. Saul, fellow traveller in Anna's Blue-Notebook-Madness, becomes her saviour and suggests the first line of a new novel (the first line of 'Free Women') and thus the end is the beginning. The shape of the book has, as Lessing intended, made its own comment, expressed its own theme.

Ada LEVERSON   **1862–1933**

The Little Ottleys

see 1916

Olivia MANNING   **1908–1980**

The Spoilt City

see **The Balkan Trilogy** (1965)

Naomi MITCHISON   **1897–**

Memoirs of a Spacewoman

The imaginative empathy Mitchison employs to such effect in her historical novels is evident in this early example of women's science fiction, which explores the emotional and social implications of intergalactic travel rather than hard-edged techno-thrills. The first-person narrator, Mary, is a Terran and an explorer, trained specifically as a communicator to establish

contact with other life-forms, however alien. Though we are told little of other Terrans' lives, it can be inferred that explorers are in an aristocracy, remaining young much longer than their earthbound coevals because of the 'time blackout' involved in each space flight. Her first expedition demands intuition of the thought processes of 'Radiates' whose starfish-like five-sidedness undermines Mary's bi-lateral human thinking. Another expedition is undertaken with Martians, whose communication is exclusively tactile. Comfort from one of these after a disaster makes Mary pregnant: the ensuing child, Viola, is a haploid, formed from Mary's genetic material only. A grafting experiment with a primitive, worm-like life-form wakes unexpectedly basic maternal feelings, and extreme grief when the independent graft deliquesces. Mary undertakes other expeditions, and has other children by different fathers, with ensuing 'stabilising time'.

A long central section shows an all-woman biological team establishing great sympathy with some creative caterpillar-like creatures. Both caterpillars and women are harassed by brilliant giant butteflies, whose weapon is the telepathic imparting of agonising guilt and blame. When it appears that this anger is directed at the caterpillars' charming creativity because this is deemed to injure the ensuing butterfly, we realise that Mitchison is weaving an elaborate allegory about creativity, dualistic religion and the ethics of short-term repression for long-term benefits. One of the women commits the gravest crime in this avowedly non-colonialist culture by interfering with the life processes of another world. She is imprisoned and loses her status and longevity. Now comes the time for an extended experiment with the deliquescing grafts. Mary offers herself again, along with a Martian and various laboratory animals, of whom Daisy the labrador is a particular friend. All the hosts become oppressively maternal and irrational, and it emerges that this is a dangerously opportunistic parasite. The experiment ends in sorrow and shame for the surviving hosts. Mary retreats to Peder Pedersen, the Norwegian captain by whom she has already had one child.

Though its episodic nature and unvarnished narrative do not make this a sophisticated literary affair, its confident sense of genre and the many questions it raises assure it of cult status. There is plenty of laconic humour, often arising from Mary's matter-of-fact delivery, and the exploration of female feelings – culminating in the uncomfortably vivid pregnancy analogue – is handled deftly and lightly. Questions of ethics, society, gender and religion are elegantly adumbrated, and each episode is enlivened by scientific imaginativeness of a high order.

Penelope MORTIMER   1918–

The Pumpkin Eater

'I suppose you would like to be something useful,' the doctor said sadly. 'Like a tea cosy.' Mrs Armitage, the narrator of Penelope Mortimer's novel, has been persuaded to take her problems to an analyst. 'I thought I was meant to lie on a couch and talk about whatever came into my head,' she points out when he attempts to make her be more constructive. She has been through several marriages and given birth to a brood of children, but now, against all the considerations of common sense, she wants to have another baby. Her husband Jake, intent upon his career, is not convinced that this would be a good idea – he has nothing *against* children, he explains; it is just that there are so many of them. Mrs. Armitage, however, has no sense of identity apart from her babies. She and Jake were both only children, she tells the doctor, and she has a house with eight bedrooms and a queue for the bathroom. As her past is unravelled on the analyst's couch, the doctor thinks she will be able to make progress, but her obsession with reproducing is beyond cure. She understands the loneliness which drives her into it, the unloved childhood and the failed relationships, but still she is not convinced that she exists at all unless she has a baby to take care of.

Running away from her problems she retreats into the country, to a tower which Jake is building as a holiday home for the family, the symbol of her dreams which she has never occupied. Reality is lost in the consciousness she now inhabits, as she waits for Jake to come and find her: 'He had already incapacitated me, harried me, cut away most of my illusions and some of my ignorance; he had already so weakened me that I was falling back on myths, words, mysteries to replace what I had lost.' A window is smashed upstairs and Jake has come to the rescue, but her dreams and desires cannot be answered by his riddles:

> Peter, Peter, pumpkin eater,
> Had a wife and couldn't keep her.
> He put her in a pumpkin shell
> And there he kept her very well.

The Pumpkin Eater remains Mortimer's best-known novel, partly because the fact that she has six children has fuelled speculations about the novel's autobiographical content, but also because the book was made into a highly successful film (1964) directed by Jack Clayton from a screenplay by Harold Pinter – a film, Mortimer complained, which 'had no connection whatsoever with what I'd written'.

Vladimir NABOKOV   1899–1977

Pale Fire

Nabokov's *Pale Fire* comprises John Shade's 'Pale Fire' – a 999-line poem – together with foreword, notes and index, contributed by Charles Kinbote. In his reminiscent and inapposite commentary – through which the burden of narrative is obliquely conveyed – Kinbote discloses that he and 'his' poet are colleagues at Wordsmith College, New Wye. There Kinbote, who has rented a house belonging to the fearsome Judge Goldsworth, teaches Zemblan, the language of his native country, Zembla. His obsession, however, is Shade. Kinbote spies and intrudes upon the poet and tells him tales of the recent (1958) revolution in Zembla and of the flight and exile of the king, Charles Xavier Vseslav. It becomes obvious that Kinbote believes himself to be the deposed monarch and that he expects Shade's 'Pale Fire' to be based on his fantastic anecdotes.

All the while (according to the annotations), Gradus, a revolutionary, is approaching. Typical of Nabokov's politically fixated psychopaths, Gradus is a grubby mediocrity; but his incompetence is outweighed by his good fortune and, in due course, he arrives at his goal. He makes his way to Kinbote's house, finds Shade on the doorstep – and shoots him. Although Kinbote mourns the death of his friend (and contrives to filch the manuscript of 'Pale Fire'), he remains convinced that the assassin's bullet was meant for him. The reader may prefer to surmise that Gradus was an escaped criminal seeking revenge on Goldsworth and mistaking the poet for the judge.

Eluding Shade's literary executors, Kinbote departs to a mountain cabin, where he discovers that Shade's poem is meditative, autobiographical and scarcely alludes to Zemblan events. The spurious academic now comes to the fore, treating the verse as a prompt for his obsessions – even, perhaps, rewriting it to suit them; and, if the novel is *inter alia* a parody of scholastic publishing (and Shade a revenant of Robert Frost), then Kinbote is the lengthened shadow of those parasitic critics whom Nabokov so dreaded and despised.

Lines 939–940: Man's life, etc.

If I correctly understand the sense of this succinct observation, our poet suggests here that human life is but a series of footnotes to a vast obscure unfinished masterpiece.

PALE FIRE
BY VLADIMIR NABOKOV

The structure and subject of *Pale Fire* place it in the rank of 'modernist texts'. Flaunting its protagonist's monstrosity, revelling in its formal brilliance and referring to little beyond its own protocols, it is situated neither on the caricature campus of New Wye, nor in Kinbote's Zenda-resembling realm, but in a purely literary kingdom. Its approbation is allowed to *Timon of Athens* (source of the novel's title) and Pope's *Essay on Man* – works of Olympian misanthropy –

whereas Eliot's *The Four Quartets* and William Faulkner's *œuvre* are covertly derided. Paradoxically, the novel travesties the genre it inhabits, and this recalcitrance – coupled with its anti-academicist contempt – has rendered it something of an embarrassment to those who might otherwise have embraced it.

Nabokov was always swift to deny that any close relationship existed between himself and his characters. He evinces, however, a certain glee over Kinbote's talent for scandalising his fellow lecturers – just as the author himself, when teaching at Cornell, scandalised his peers with such dicta as: 'Style and structure are the essence of a book: great ideas are hogwash.'

Edna O'BRIEN　　　　　**1936–**

The Lonely Girl (Girl with Green Eyes)

see **The Country Girls Trilogy** (1964)

Katherine Anne PORTER　**1890–1980**

Ship of Fools

Porter's allegorical novel takes its title from Sebastian Brant's satire *Das Narrenschiff* (1494), freely translated by Alexander Barclay as 'The Shyp of Folys' (1509), in which a number of 'types' are depicted sailing to the Land of Fools. Its genesis, however, was a voyage the author took aboard an old German ship in 1931, sailing from Vera Cruz to Bremerhaven. She spent much of the journey observing her fellow passengers, writing about them in a letter to a friend. On her return to America, she retrieved the letter, but did not begin writing the novel until 1942. Progress was slow and erratic – she often left the manuscript for years at a time before returning to it – and it became clear that the short novel she had planned would have to expand in order to encompass the forty-five or so main characters.

The book portrays the 'simple almost universal image of the ship of the world on its voyage to eternity'. Aboard the North-German Lloyd SA *Vera* are crowded passengers from several countries, making the long ocean crossing from Mexico to Bremerhaven. Over them all hangs the threat of a world in transition: it is the summer of 1931, the eve of the rise of Nazism. In Porter's narrative we glimpse the lives of the passengers, both their present and their past experiences, with suggestions of brief encounters and jealousy, bourgeois snobbery and political revolution.

Father Carillo had talked to him, in abominable German, as if the fight were not merely a low brawl; he mentioned several times such terms as syndicalism, anarchism, republicanism, communism, besides atheism, above all atheism, the common root of such pernicious theories; and he seemed very instructed in the fine shades of opinion on all subjects existing among the rabble of the lower deck. According to Father Carillo, the lower classes were being led astray by a thousand evil influences from every direction, and a great many dangerous subversive elements were on the steerage deck. These should be watched carefully, not only for the sake of the first-class passengers, the crew, the ship herself, but for the protection of the innocent poor below, those good and harmless people who wished only to be allowed to obey the law and to practise their religion in peace.

SHIP OF FOOLS
BY KATHERINE ANNE PORTER

We are allowed to see some of the characters' lives in more detail. Jenny Brown and David Scott ('Jenny angel' and 'David darling') are two young artists and lovers on their first voyage to Europe. Jenny's allusions to past affairs, and her romantic flirtation with another passenger, Herr Wilhelm Freytag, fuel David's self-mistrust. Their temperaments clash, and their relationship swings between moods of happiness and hopelessness. Freytag is travelling to Germany to bring his wife and mother back to Mexico. His idealisation of his Jewish wife

and identification with her social disadvantages conflict with his feelings for Jenny, tempting him to abandon the diffcult life he has chosen. When Freytag shocks the captain's table by revealing his wife's religion, he is ostracised by his fellow passengers.

William Denny is David's American cabin-mate. A small-town youth from a God-fearing background, he lusts after women and drinks himself into a stupor. Jenny's cabin-mate is Elsa Lutz, the ungainly daughter of a failed Swiss hotelier. Her attempts to improve her appearance with Jenny's guidance on clothes and cosmetics fail to attract the attentions of a young student she admires. Her parents, meanwhile, plan a respectable marriage for her, encouraging the suit of Arne Hansen, a Swede whose affair with a Spanish dancer makes the bourgeois passengers resentful and suspicious. A feud develops between Hansen and Herr Siegfried Rieber, publisher of a ladies' garments trade magazine. His opportunistic mistress, Frl. Lizzi Spöckenkieker, shares a cabin with Mrs Treadwell, a divorcee of forty-five, who is returning to Paris.

Also on board are: Herr Wilibald Graf, a dying religious enthusiast with belief in his own powers of healing; his nephew Johann, a rebellious and frustrated youth awaiting his uncle's death; Herr Professor Hutten, a retired schoolmaster, and his wife, whose childlessness is solaced by their love for Bébé, a pampered white bulldog which dies during the voyage; Herr Julius Löwenthal, a Jewish manufacturer and salesman of Catholic church furnishings, despised for his religion and low social status; Herr Karl Baumgartner, a drunkard, and his wife and young son; Dr Schumann, the gentle ship's doctor, too ill to cure himself, who falls in love with La Condesa, a *déclassée* noblewoman and political exile from Cuba, who, to her admirer's distaste, has taken up with a group of six Cuban medical students; and a zarzuela company, singers and dancers from Spain. The men, Pepe, Tito, Manolo and Pancho, serve as pimps to their mistresses, Amparo, Lola, Concha and Pastora. Lola's obnoxious six-year-old twins, Ric and Rac, cause mischief and accidental death, and steal La Condesa's pearls.

In steerage are 'eight hundred and seventy-six souls', deportees from Cuba as a result of the failure of the sugar market. Their life below decks fascinates the privileged passengers, providing a hellish dimension to Porter's microcosm of the world.

Anthony POWELL   1905–

The Kindly Ones

see **A Dance to the Music of Time** (1975)

Isaac Bashevis SINGER   1904–1991

The Slave

'As is usual in the affairs of men, the relationships were complex, and all were based on deception.' Singer's tale of faith and transgression is set in seventeenth-century Poland, where the exiled Jew Jacob is living in slavery amongst the myths and superstitions of the peasants. His master's daughter Wanda, a beautiful young widow, falls in love with him and persuades him to sleep with her, and Jacob finds his religion difficult to sustain now that he is sinning in his 'lust for a forbidden woman'.

A chance encounter with a group of travelling musicians brings Jacob back into contact with his old community. His wife and children have been murdered in the Chmielnicki massacre, but the Jews now arrange for him to marry a wealthy businesswoman. On the eve of the wedding, however, Jacob sets out to rescue his former mistress from the poverty of her home, and brings her to live amongst the Jews, disguised as the deaf-mute Jewess Sarah. The couple gain in respectability, but are surrounded by the suspicious and hypocritical worthies of the town: when Sarah becomes pregnant, their discovery is inevitable, for she cannot maintain her silence through the labour.

Jacob confesses their secret and is sent into exile once more, and Wanda dies in childbirth. Bribing the family with whom the elders have placed his son, Jacob takes the boy and makes a new life in Israel, where his son becomes a powerful figure in the synagogue. In his old age, Jacob returns to the town, seeking the grave of his beloved Wanda. She has been buried outside the cemetery in disgrace, but as Jacob dies in a poorhouse the grave is accidentally discovered, and the lovers are reunited in death, in a miracle of divine grace which gives Jacob absolution from his sins.

Although he wrote in Yiddish, Singer translated his own work into English (often with assistance) and gained an enormous international reputation, winning the Nobel Prize for literature in 1978. Apparently a straightforward fable of the seventeenth century, *The Slave* (translated into English by the author and Cecil Hemley) is also a complex meditation upon Judaism itself. It looks back to the Bible and forward to the Holocaust, the two most significant periods in Jewish history. The Polish peasants recall the idolators of the Old Testament, whom Jews had to resist, and from whom in this story Jacob rescues Wanda, bringing her to the Jewish faith, while the Cossack massacres prefigure the fate of European Jewry at the hands of the Nazis.

Edward UPWARD   1903–

In the Thirties

see **The Spiral Ascent** (1977)

1963

Assassination of President John F. Kennedy ● Test Ban Treaty ● Britain refused entry to Common Market ● Winston Churchill becomes an honorary citizen of US ● Profumo affair ● Blacks killed by bomb in Birmingham, Alabama ● Beeching Report on *The Reshaping of British Railways* ● Robbins Report on higher education ● Nobel Committee inquires into moral impact of TV on the young ● Rachel Carson, *The Silent Spring* ● John Robinson, *Honest to God* ● Hilton Hotel, Park Lane, London ● Benjamin Britten, *War Requiem* ● Joseph L. Mankiewicz, *Cleopatra* ● Tony Richardson, *Tom Jones* ● The Beatles make the Liverpool sound international ● Robert Frost, Georges Braque and Aldous Huxley die ● Kenneth Rexroth, *Natural Numbers: New and Selected Poems*

Anthony BURGESS   1917–1993

Inside Mr Enderby

see **The Enderby Novels** (1974)

Hortense CALISHER   1911–

Textures of Life

Like Henry James, Calisher evokes emotional responses from her reader where the material, dispassionately considered, treats only the mundane. Sensibilities are minutely explored and modulated. Sentence by sentence, too, James is recalled in cascades of sub-clauses: 'He was like any traveller tracking what he had been led to believe was a subcontinent, whose daily pounding of its humps slowly hinted him the true enormity, on whom it was finally borne, as he climbed its scaly rocks and mapped the fixed hairs of its forests, that what he stood on was Leviathan.' The language of the novel is peculiar, deliberately illiterate sometimes (a technique used by Henry Green) as a way of heightening intensity: 'Far below them, the immense of the Pacific moves its colors or lay still thereunder.' She also uses an absorbed biblical language to reinforce the novel's power as a searcher after truth.

The narrative opens with the wedding of David and Elizabeth. They move into a flat and have a baby. Not much else happens. David's mother is dead and Elizabeth's father is dead. At the wedding, David's father, Jacques, and Elizabeth's mother, Margot, start a relationship which leads to their marriage – a marriage that shocks their children. Margot notices how weddings force consideration of the wedding bed, a disconcerting thought for any parent. It is for the same reason, acknowledged this time by Elizabeth, that the marriage of their parents is shocking to her and David.

The relationship between Margot and her daughter is uneasy, but the reader divines that their seemingly polarised attitudes spring from a common source. Margot loved too much, and the cycle is beginning to repeat itself in Elizabeth and her daughter May now that the child is becoming no longer just a 'thing'. 'For she was their thing, their greatest possession and expression.'

Textures of Life is about things: '"things" were not merely a taste or a boastfulness in [Margot's] life – or even a love – they were her gravitation'. Margot and Jacques, according to David, 'haven't a thing ... in common. Not a thing.' Jacques is 'sort of above things'. He is a moral centre from which the others, directly or indirectly, are changed. He is above things only to best realise the importance of the particularity of things and possessions. For their lack of possessions, Margot expresses a passing interest in being a Hindu. Jacques's proclamation, 'It's no sin to love the particular' is, however, the truth of the novel. That this love is of 'the surface frivol of existence' is denied; it is, instead, what keeps man 'from flying off the planet altogether, or from sinking into the saw-toothed fires at his own core'.

For David and Elizabeth the things they want are 'the kind of stuff they pretend isn't there'. They are artists: Elizabeth a sculptor and painter; David, the least substantial (and successful) character, a photographer and film-maker. It is he who wishes to make a film about the textures of life itself, whether of life now (New York at the time of writing) or life as it was in the fifteenth century, say: 'we want to do a picture about people, mainly . . . by means of the objects they live through'. The texture of life, however, 'is so tight that one could never see . . . where any one part had begun . . . All that happened was that now and then the serpentine shift of the ordinary . . . suddenly . . . added up.'

Ian FLEMING   **1908–1964**

On Her Majesty's Secret Service

This is one in a series of 'snobbery with violence' thrillers featuring the tall, dark, urbane British Secret Service agent 007, James Bond. In the 1960s, the Fleming novels were runaway bestsellers and have inspired numerous films loosely based on their plots and characters. The books themselves, once famous for casual sex, random violence and expensive living, now seems less exciting, more traditional and more perfunctory than they did on publication. They have been supplanted by more realistic spy novels, but Bond retains a powerful glamour as a modern folk-hero.

Contemplating resignation from the Service, Bond visits the Casino at Royale-les-Eaux where he saves a beautiful young woman, Therese ('Tracy') Draco, from dishonour when she cannot meet a gambling debt. Tracy offers him in return 'the most expensive piece of love of your life. It will have cost you forty million francs. I hope it will be worth it.' It is – Bond falls in love with Tracy, and he is offered her hand, together with a million pounds, by her father, Marc-Ange Draco, who happens to be *capo* of the Union Corse – the Corsican equivalent of the Mafia. Bond declines the offer, but enlists Draco's aid in tracing an old adversary, the villainous Ernst Blofeld, an agent of SPECTRE (Special Executive for Counterintelligence, Terrorism, Revenge and Extortion), the international organisation whose aim, if not to control the world, is at least to destroy democracy and exercise power over governments.

It was then, on a ten-mile straight cut through a forest, that it happened. Triple wind-horns screamed their banshee discord in his ear, and a low, white two-seater, a Lancia Flaminia Zagato Spyder with its hood down, tore past him, cut in cheekily across his bonnet and pulled away, the sexy boom of its twin exhausts echoing back from the border of trees. And it was a girl driving, a girl with a shocking pink scarf tied round her hair, leaving a brief pink tail that the wind blew horizontal behind her.

If there was one thing that set James Bond really moving in life, with the exception of gun-play, it was being passed at speed by a pretty girl.

ON HER MAJESTY'S SECRET SERVICE
BY IAN FLEMING

Blofeld has built himself a skiing club atop a mountain in Switzerland, and Bond insinuates himself into the closely guarded enclave by passing himself off as a herald from the College of Arms. (Blofeld has expressed interest in tracing his family tree.) Exposed as a spy, Bond is forced to flee on skis, narrowly missing an avalanche, and returns to London to tell M, head of the Secret Service, that Blofeld plans to wage biological warfare on the UK. By destroying British agriculture, Blofeld intends to bring the country to its economic knees. The agents of destruction are ten British girls who have been hypnotised by Blofeld and ordered to contaminate crops and animals. They are picked up when they attempt to enter the country, and Bond sets off again to Switzerland to arrest or kill Blofeld – in any event, to thwart his dastardly plans.

With the help of Draco and several Union Corse thugs, Bond lands a helicopter on Blofeld's alpine hideaway and takes it over. In pursuit of Blofeld, he has to take to a bobsleigh run and avoid a grenade that Blofeld pitches in his path. By the time Bond has recovered from the buffeting he has suffered as a result of the bobsleigh run and the explosion, Blofeld has escaped. Bond gives up the chase, and takes a few days' leave to marry Tracy, who has been recovering from a threatened nervous breakdown. Shortly after the marriage ceremony, while driving off on honeymoon, their car is overtaken by a red Maserati and a shot is fired at them. Bond, stunned by the impact of his head against the windscreen, recovers to find Tracy dead in the passenger seat, shot by the vengeful Blofeld.

Wilson HARRIS   **1921–**

The Guyana Quartet

Palace of the Peacock (1960)
The Far Journey of Oudin (1961)
The Whole Armour (1962)
The Secret Ladder (1963)

Each of the novels comprising *The Guyana Quartet* presents an independent sequence of events and a different constellation of characters. Accordingly, the unity of the work derives not from narrative but from recurrent images and common themes. The Guyanese rivers, 'arteries of God's spider', provide a metaphorical web for Harris's work, while his concern is with the strands that connect people to people, to place, history and myth.

Palace of the Peacock is told in the first person. The narrator describes a series of dreams in which he repeatedly dreams that he wakes, only to dream that he is still dreaming. Awake, he reveals that he is going blind in his left eye. His blindness, though, is a blindness merely to the physical world and, as he sees less of the real, his spiritual perceptions become more intense. In one of his dreams Donne, a ruthless planter, is shot from his horse.

Donne sets out by boat up a dangerous river. His purpose is twofold: to find Mariella – with whom he is obsessed – and to press-gang labour for his farm. Accompanied by his crew of seven, he kidnaps an old Arawak Indian to assist in his search. She tells Donne that he will catch up with Mariella and the labourers in seven days' time. Donne drives the crew on; its members die or desert. On the seventh day, the journey reaches a mystical conclusion. Donne has a vision, fusing Christian and Caribbean motifs. He sees the truth of his own cruel nature and sees his crew in 'the windows of the universe'. His colonial ideology merges with, and is transformed by, the native belief and practice of Guyana.

It is not clear who the 'I' of the story is. He may be Donne's brother, a member of the crew – or all of the characters may be ghosts. In Harris's Guyana, the line between living and dead is indistinct.

The Far Journey of Oudin begins with Oudin's ghost finishing the rum he had been drinking when he died. Oudin leaves a pregnant widow, Beti, and a son; he also leaves a contract which he signed with a moneylender, Ram. Fearful of any document, Beti eats the contract.

The narrative reverts to a time before Oudin's arrival in the region. Beti is living on a farm owned by Mohammed, her father's cousin. Mohammed is the last of four brothers. The first was murdered by the other three. The second died of a stroke and the third in a fire. When Oudin enigmatically appears, he bears a resemblance to the murdered man. Mohammed is heavily indebted to Ram, who employs Oudin to steal from the debtor. Finally, intent on producing an heir, Ram persuades Oudin to kidnap Beti. Oudin does so but, instead of delivering the girl to his employer, he flees with her into the jungle. The couple are taken prisoner by a woodman. Mohammed is killed by the woodman's bull while searching for them. The woodman risks losing his livelihood because Ram is about to foreclose on loans to him. Oudin negotiates with Ram on the woodman's behalf, and Ram realises how greatly he depends on Oudin. Oudin and Beti return to Oudin's shack.

The narrative returns to where it began, disclosing that Oudin agreed to give his unborn child to Ram in exchange for some land. Ram's economic power is the web in which generations are ensnared; it is a malign parody of the river system – promising prosperity, yet oppressive in its extremes of abundance and drought.

Oudin may be intended as an everyman (his name is a play on the Greek *outis* – 'no man'). The central character of *The Whole Armour*, Cristo, is plainly allegorical. He is accused of murder and takes refuge with Abram, an old man who may be his father. Abram is killed by a tiger and Cristo, with the connivance of Magda, his prostitute mother, dresses Abram's corpse in his own clothes. The corpse is discovered and assumed to be Cristo's. Cristo hides in the jungle, shoots the tiger and dresses in its skin. Magda holds a wake for her son. Among the guests are Sharon – whose fiancé Cristo is said to have murdered – her new fiancé, Mattias, and Peet, her father. Peet is humiliated by Magda and provokes a quarrel with Mattias. Mattias is killed. Cristo emerges from the jungle and enjoys a short idyll with Sharon. As the police encircle Cristo, Magda begs him to escape, but he gives Sharon his tiger skin and waits for arrest and probable execution. He is clothed in an armour 'superior to the elements of self-division and coercion'. His role is to expiate the uncertainties and rivalries of a heterogeneous society: to be a scapegoat – just as, in his animal disguise, he was a terrifying, predatory creature.

Counterbalancing the Christian resonance of *The Whole Armour*, *The Secret Ladder* displaces Greek myth to a Caribbean setting. Fenwick is a government surveyor commissioned to chart the Canje river. He has under his command seven men – a devious and discontented crew. Torn between authoritarianism and sympathy, Fenwick begins to lose control of them. On a surveying expedition he meets Poseidon, an old fisherman, 'the black king of history'. Fenwick finds Poseidon absurd and cannot understand what Poseidon tells him. Bryant – one of the crew – translates: Poseidon complains that Fenwick's storekeeper is paying an unfair price for fish; moreover, the fishermen are afraid that the government means to divert the Canje for an irrigation project which will benefit only the rich farmers. Poseidon becomes the figurehead of a growing local opposition. Fenwick struggles to persuade himself that he is acting in the cause of progress. Some of his surveying equipment is wrecked by the river people. The crew urges decisive action. Fenwick tries to reason with the fishermen, but fails. (Odysseus, in Homer's *Odyssey*, attempts to placate his enemy Poseidon.) When one of the crew steals some turtle meat, violence breaks out. A member of the crew is almost killed and Catalena Perez, the wife of another, is abducted.

Bryant sets out to rescue Catalena and accidentally kills Poseidon. Catalena – about to be raped by her abductors – is saved by mystical and mechanical intervention. Demoralised by Poseidon's death, the river people disperse into the jungle. Bryant and Catalena vanish to 'make swift love on every trail' and Fenwick dreams that 'an inquisition of dead gods and heroes had ended . . . [He] awoke. It was the dawn of the seventh day.'

Fenwick's dilemma – that of an educated liberal caught between alien progress and indigent tradition – is not fully resolved. Instead, Harris presents a general resolution – the triumph of 'the science of the invader' over the ancient economy of the Caribbean. It is, however, only a partial triumph. Bryant and Catalena represent a hedonism which cannot be accommodated in the world of material conquest. Fenwick perceives the Canje as 'one of the lowest rungs in the ladder of ascending purgatorial rivers'; for him, this episode is the beginning of a rationalisation of Guyana. *The Secret Ladder*, contrasting Donne and Fenwick and their crews, similarly returns to the quartet's starting-point. On the one hand, it registers the distinction between civilised Fenwick and rapacious Donne; on the other, it acknowledges that the pair are merely versions of each other. Harris's provisional conclusion is that Guyana – with its mixed bloods and incomplete traditions – has to remain a site of conflict. A final conclusion depends upon a further ascent of the secret ladder, the arteries of God's spider.

Like the jungle it frequently describes, the style of *The Guyana Quartet* is complex and dense, poetic in its compression and luxuriance. Harris's oblique narrative method does not guide the reader straightforwardly through this terrain; rather, it is a map charting the unsettled geography of the land, the people and the spirits which animate both.

B.S. JOHNSON   **1933–1973**

Travelling People

In the prelude to this, his first novel, which won the Gregory Award – the judges included T.S. Eliot – B.S. Johnson tells us, 'I decided that one style for one novel was a convention that I resented most strongly.' He therefore elected to make the 'style of each chapter ... spring naturally from its subject matter'. He concluded that it was 'not only permissible to expose the mechanism of a novel', but that by doing so the author might 'come nearer to reality and truth ... Pursuing this thought, I realised that it would be desirable to have interludes between chapters in which I could stand back, so to speak, from my novel, and talk with the reader.'

A product of the 1960s, Johnson was full of theories. His method, which harks back to the eighteenth century, has the advantage of flexibility. In addition to these 'interludes' the story is told in journal form, by letter, by interior monologue, by film script, as well as by conventional narrative. The method allows for changes of mood and viewpoint, but dramatic unity is destroyed and the reader is given the choice of taking the book in several ways. One such is as comedy and social satire.

Henry Henry is twenty-six years old, stands 5 foot 10⅞ inches tall, is overweight, has 'flintgrey' eyes, an 'unnaturally unhealthy' complexion, 'black, overlong, riotous' hair, a 'quietly desperate' disposition and a 'torpid' bearing (see interlude one). The examiners of the philosophy faculty of London University have recently placed him in the first class, though Henry has no great opinion of them, and it has not helped him to find a job – not that he has tried very hard, being taken up with the process of eliminating the jobs that for conscientious reasons he can not take. He is a travelling person. He travels in and out of relationships, and, more literally, he also travels by road, always hitchhiking, preferring lorries to private cars; they are slower but more interesting and tend to land him in less trouble. We first meet him hitching a particularly smelly lift in a lorry loaded with dead dogs; 'We boils 'em up. For glue,' the driver tells him. This causes Henry to fantasise about a possible career: '*Use Henry's Glue. See that Your Pet has a Happy Home in Henry's Glue.*' His next lift, in a private car, is smoother but, true to form, dangerous. The driver is Trevor Tuckerson, manager of the exclusive Stromboli Club in North Wales, and it is there that Henry ends up: a barman and general handyman for several weeks in the summer. At the Stromboli Henry finds peace for a while, but there are factions. On one side there is the ageing Maurie, owner of the club, and the girl, Kim, who sleeps with him; they are pro-Henry. On the other side are Trevor and Mira ('I can feel how evil she is ... How she longs for me to make one mistake, even a small one ... that would put me in her power') who are distinctly anti-Henry. Unhappily the 'fun, fireworks, and high jinks at North Wales' newest nitespot', to say nothing of sleeping with the nubile Kim, are too much for Maurie's tender heart. He suffers one attack (conveyed visually by a page and three-quarters of speckled grey) and, the following night, death (two-and-a-half pages of jet black). This occurs with a distraught Kim in his bed and she summons Henry to cover up the evidence. The event leaves them free to indulge in the odd bout of passion, but Mira now definitely has the upper hand. Henry finds himself once more on the road, and travelling in a different direction from Kim.

There is plenty of incident, as one might expect in a place where 'the blueblood sport

... is pushing people into the pool, fully clothed'. On the 'gala night' with fireworks a version of the Venus de Milo (with arms) is 'unveiled' and finishes up on its back, symbolising 'the everlasting struggle of Man to raise himself towards Heaven'. There is the inevitable orgy in the course of which Mira is unfaithful to Trevor. There is the night, after Maurie's death, when to revenge themselves on Trevor and Mira, Henry and Kim smuggle in their less than exclusive mates from the pub. Johnson was more interested in form than content – the novel is very much a period piece, and as such it dates. Nevertheless there are deft touches and *Travelling People* remains an enjoyable curiosity, provided one largely forgets about 'reality and truth'.

Mary McCARTHY   1912–1989

The Group

McCarthy began writing *The Group* in 1952, but abandoned it after a few chapters because, she said: 'I thought it was just so terrible, so sad and I felt, these poor girls ...' In 1959, her marriage crumbling, she took up the book once more, eventually finishing it in 1962. Critics were divided over the book, some seeing it as a sell-out to the popular form, others recognising it as a serious and witty work of art. The *Alumnae Magazine* of Vassar, alma mater of the novel's characters, received a great deal of mail condemning the book, and even suggesting that McCarthy's degree should be withdrawn. (One Vassar alumna hoped that 'the noble image of Jacqueline Kennedy' would do something to counterbalance this disgraceful slur upon the university.) Most people objected to the sexual details, the *Times Literary Supplement* reviewer noting that he learned 'much more about Dottie's vagina than about Dottie', but the book sold enormously well and made McCarthy a literary celebrity.

The Group of the title consists of eight women who graduate from Vassar in 1933, having roomed together and been the object of envy and dislike for their sophistication – mostly spurious as it turns out. Though the women themselves and several other charac-ters in the novel think of them as 'the Group' they appear as such only twice, at the wedding and funeral of Kay Petersen, the outsider from the Midwest, a pushy, deter-mined and tragic figure. Ironically, her wed-ding, at the beginning of the novel, is a disaster – awkward, clumsy, misarranged – while her funeral, at the end, is a triumph: the climax of Kay's social ambition and an emotional focus for her beloved Group. Between these two events we see the women, individually or in pairs, and follow their efforts to achieve happiness and fulfilment in Roosevelt's America.

The worst fate, they utterly agreed, would be to become like Mother and Dad, stuffy and frightened. Not one of them, if she could help it, was going to marry a broker or a banker or a cold-fish corporation lawyer, like so many of Mother's generation. They would rather be wildly poor and live on salmon wiggle than be forced to marry one of those dull purplish young men of their own set, with a seat on the Exchange and bloodshot eyes, interested only in squash and cockfighting and drinking at the Racquet Club with his cronies, Yale or Princeton '29. It would be better, yes, they were not afraid to say it, though Mother gently laughed, to marry a Jew if you loved him – some of them were awfully interesting and cultivated, though terribly ambitious and inclined to stick together, as you saw very well at Vassar: if you knew them you had to know their friends.

THE GROUP
BY MARY McCARTHY

They are largely defeated in their attempts by the men around them who use the women's strongest individual characteristics against them. Kay, for example, is destroyed by her egregious husband, Harald. Priss, the conscientious and ardent supporter of the New Deal, is forced into using her own maternity as a medical experiment by her insensitive, paternalistic, Republican doctor husband. Libby, the most successful in her career, is almost raped by a man who is temporarily fooled by her veneer of sophistication. Polly's strongly developed maternal feelings are taken advantage of by a vacillating left-wing publisher and by her mentally unstable father.

The book's structure resembles that of a daisy-chain. After several chapters about one woman, another member of the Group turns up, say, at a party, and the emphasis is shifted to concentrate on the second. The language changes with the emphasis; McCarthy uses an ironic version of undergraduate chatter, adapted to reveal the character of each of her heroines. Thus the chapters about Libby McAuleskey, the would-be litterateuse, speak of a 'spiffy apartment in the Village'; Libby is described as speaking 'a breathless Italian with a nifty Tuscan accent'. The rather dull New Dealer Priss can only think about her child's discomfort in the language of a political pamphlet: he 'was making a natural request in this day and age; he was asking for a bottle'. Only when they reassemble for Kay's funeral, overwhelmed by the tragedy of her death and the greater tragedy of the war in Europe (the year is 1940), and subdued by their own frustrations, do the women abandon their Vassar mannerisms.

The least successfully delineated character is Lakey, the 'Mona Lisa' of the smoking room, because her own language is never used. Although she appears only at the beginning and end of the book, she remains the pivotal character of the groups but is described entirely from the outside, without irony, the author joining the group in their adulation. She is always thought of as somehow different from the others, and we later learn that she is lesbian. At first disconcerted, her companions recognise that Lakey has more common sense than any of them. She takes charge of Kay's funeral, making it into a social and emotional success. She is also the only one who can face up to Harald, fight his aggressive, predatory masculinity, and defeat him.

Sylvia PLATH   **1932–1963**

The Bell Jar

Originally published under the name of 'Victoria Lucas', *The Bell Jar* is instantly recognisable as the work of Sylvia Plath. She apparently chose a pseudonym because she did not regard the novel as 'serious', as she did her poetry, and was nervous of its critical reception. The book is transparently autobiographical, with episodes from the author's own life vividly recreated. The title refers to the central image: the way in which a person suffering from mental illness is separated from the world, as if encased in a glass dome. In spite of several very funny episodes, the novel, like Plath's poetry, is suffused by death, opening with the words: 'It was a queer, sultry summer, the summer they executed the Rosenbergs.'

The novel is narrated by the nineteen-year-old Esther Greenwood, a bright but quiet Bostonian who has won a fashion magazine's writing competition. The prize is a month's work in New York, 'expenses paid, and piles and piles of free bonuses'. While everyone envies Esther, imagining that she must be 'steering New York like her own private car', she feels that she is not even steering herself: 'I felt very still and very empty, the way the eye of a tornado must feel, moving dully along in the middle of the surrounding hullabaloo.' Nor can she feel any real interest in the prospect of getting married to the extrovert, tubercular medical student Buddy Willard, whom she can only associate with dissected cadavers and pickled foetuses with piggy smiles (images familiar from Plath's poems). When he takes off his clothes, Esther is merely reminded of 'turkey neck and turkey gizzards'. At a party she is almost raped by a 'woman-hater' called Marco.

The hope that she may be accepted for a creative writing course keeps Esther going through the absurd pageant of the New York experience, but she returns home to find that she has been rejected, a blow that precipitates her sense of worthlessness and pushes her further and further towards the periphery of existence. Gradually, she loses her ability to sleep, eat, write or even read. She sees a psychiatrist and embarks upon a series of half-hearted suicide attempts, defeated by fear of discovery, physical cowardice or by her own body: 'I saw that my body had all sorts of little tricks, such as making my hands go limp at the crucial second.'

Her last attempt, an overdose of pills taken in the cellar which nearly succeeds, marks the beginning of a grim pilgrimage from one psychiatric ward to another, cared for by a succession of ogre-like, coldly tolerant or genuinely loving doctors and nurses. Under the care of Dr Nolan, Esther responds well to electrotherapy: 'The bell jar hung, suspended, a few feet above my head. I was open to the circulating air.' At Belsize Hospital, a stepping-stone to the outside world, she is befriended by the infuriating and clinging Joan Gilling, a former girlfriend of Buddy who has also attempted suicide and is now lesbian. Determined to lose her virginity, Esther picks up a young professor called Irwin on an outing to the town. Her defloration is messy, leading to severe bleeding. Joan later hangs herself, and Esther is visited by Buddy, who wants reassurance that he is not to blame for the fact that two of his girlfriends have ended up in asylums. The novel concludes with Esther appearing before the hospital board. She recognises that even if she returns to the world, there is no guarantee that 'the bell jar, with its stifling distortions, wouldn't descend again'.

Within a month of the book's publication, Plath committed suicide.

Thomas PYNCHON   1937–

V

Pynchon's first novel opens with Benny Profane, ex-seaman turned roadmender, making his way to New York via a string of unsuccessful seductions and dead-end jobs. With Profane's arrival there, Pynchon begins to extend the scope of the novel massively, exchanging the knockabout comedy of Benny's misadventures for a host of other idioms, each suited to its particular locale in the novel. The action ricochets from New York in the 1950s to Malta in the 1940s, from Paris at the turn of the century to south-west Africa in the 1920s, and from a German bierkeller in Egypt to the second Eden under the South Pole. Pynchon's manufacture of a history and ethos for each of these is a virtuoso piece of literary forgery.

The structuring principle of the novel is the quest for V pursued by Herbert Stencil through a succession of key sites and times. 'What is V?' is a question underlying much of the frantic plotting, obsessive deciphering and recoding indulged in by the characters. V is variously a woman (under different names and guises at different times); a place (Vheissu) which may or may not be real; a quasi-robotic construction of which only the teeth definitely exist; possibly a rat converted to Roman Catholicism in the sewers of New York; or even the Maltese town of Valletta. At other levels of interpretation, V is the convergence of two vectors in a point, a precarious happy medium between conflicting impulses: black jazz saxophonist McClintic Sphere's catchphrase, 'Keep Cool, But Care', is an example. It may be the primal symbol of 'Woman', or perhaps the stain on a plate in a Cario beer hall. Stencil's father notes in his journal: 'There is more behind and inside V than any of us had suspected. Not who, but what: what is she.' His son adds the coda: 'in this search the motive is part of the quarry'.

For Benny Profane the world is as it seems, a fact reinforced by his generally hostile encounters with its material objects. For Stencil, *nothing* is as it seems; everything is a symbol leading more or less circuitously to V. Between these two views (of the world as brute matter or sign) Pynchon treads an often delicate path. The antics of the Whole Sick Crew or Profane's old shipmates from the USS *Scaffold* are at once good, clean, dirty fun and *wasted* activity, a symptom of some more deep-seated malaise. Likewise, when Kurt Mondaugen monitors atmospheric radio waves against the background of a genocidal subjugation of an African rebellion, the possibility arises that the quest for V, truth or whatever is actually an avoidance of the real world.

Pynchon leaves most of these issues unresolved; Stencil is still looking for V at the end of the novel, Profane is no wiser as to the point of his own life. Of the remaining 200 or so named characters, the Europeans generally involve themselves in pointless and circuitous plots while the Americans move towards a symbiosis with the inanimate world of televisions, false teeth and artificial limbs. Fusion of the animate and inanimate is a large theme in the novel, which includes several semi-robotic characters and two actual 'robots'.

V offers two distinct visions of the twentieth century: the first, a product of an heightened, paranoid consciousness, is that its history is a conspiracy, secrets lurk in the least events, and the world harbours a wealth of threatening meanings which we must decipher. The second is brutally materialistic: there is no conspiracy and if there was it would not matter, the real world is inanimate, things not thoughts are what counts. Between these two, Pynchon steers his modern ship of fools through a macabre, licentious and violently funny course. Inconclusive, gripping, confusing in outline but convincing in detail, *V* is a blackly comic odyssey into the occult corners and bizarre anxieties of twentieth-century Europe and America.

Muriel SPARK 1918–

The Girls of Slender Means

The death of Nicholas Farringdon, a Catholic missionary murdered in Haiti, stirs the memories of a group of women, former inhabitants of the May of Teck Club. Most of the action of this characteristically compact and witty tragi-comedy is set in 1945, when 'all the nice people in England were poor'. The Club has been founded in Kensington 'for the Pecuniary Convenience and Social Protection of Ladies of Slender Means below the age of Thirty Years, who are obliged to reside apart from their Families in order to follow an occupation in London'. The novel depicts the lives of the members as they cope with rationing, discover sex, swap clothing coupons and boyfriends and share a single Schiaparelli dress. Jane Wright, a plump and eager initiate in 'the world of books', is employed by an inept Rumanian confidence-trickster called Rudi

Bittesch, for whom she writes letters to famous authors in order to secure saleable holograph replies. Marginally less dubious is her job with Guy (or George) Throvis-Mew, a shady publisher who employs her to discover potential authors' weak spots which can later be exploited during contract negotiations. Nicholas is one such author, and Jane introduces him to the Club when she is going through his manuscript, *The Sabbath Notebooks*. Although attracted to him, she loses him almost at once to the enviably slim Selina Redwood, who has achieved Perfect Poise by way of a correspondence course.

Another member of the Club who intrigues Nicholas is Joanna Childe; the daughter of a rural vicar, she is training to become an elocution teacher and gives lessons to the domestic staff and a flat-vowelled fellow member called Nancy Riddle. Her pupils' training consists of reciting Joanna's favourite poems, notably *The Wreck of the Deutschland*. Joanna's fate is similar to that of the nuns in Hopkins's poem, for an unexploded bomb wrecks the Club. While most of the members escape, eleven girls are trapped on the top floor, not slender enough to escape through the bathroom window. Joanna recites a psalm as firemen attempt to break in through a bricked-up skylight. Eventually all the girls but one are rescued: 'The house sank into its centre, a high heap of rubble, and Joanna went with it.'

During the rescue attempt Nicholas had involuntarily crossed himself, and Nancy, 'the daughter of a Midlands clergyman, now married to another Midlands clergyman', claims that is was Joanna's example, like that of Hopkins's tall nun, that converted him. However, he had crossed himself when Selina had returned to the building in order to save the Schiaparelli dress, and Jane points to 'a note in his manuscript that a vision of evil may be as effective to conversion as a vision of good'.

Kurt VONNEGUT 1922–

Cat's Cradle

'A cat's cradle is nothing but a bunch of Xs between somebody's hands, and little kids look and look and look at all those Xs . . .'

'And?'

'No damn cat, and no damn cradle.'

In this novel, the illusory quality of the familiar tangle of Xs becomes a metaphor for the principles of religion and science, the butts of Vonnegut's dark satire. If *Cat's Cradle* is a comedy, then it is a very black one, and its final vision that of a waste-land in the grip of '*ice-nine*'. Responsible for the discovery of this 'new way for the atoms of water to stack and lock, to freeze', is Dr Felix Hoenikker, one of the fathers of the atomic bomb. After his death, *ice-nine* is shared between his three children: Newt, Angela and Frank, the 'Minister of Science and Progress' in the Republic of San Lorenzo. The island is ruled by 'Papa' Monzano and the beautiful Mona, with whom the narrator falls in love at the sight of her picture in a Sunday newspaper supplement. Here the main characters come together and the narrator is acquainted with 'Bokononism', the nihilistic philosophy (both universally condemned and universally followed) which matches ideally the final vision of negation:

And I remembered *The Fourteenth Book of Bokonon* ... [It] is entitled, 'What Can a Thoughtful Man Hope for Mankind on Earth, Given the Experience of the Past Million Years?' It doesn't take long to read *The Fourteenth Book*. It consists of one word and a period.

This is it:

Nothing.

Frank Hoenikker has foolishly entrusted 'Papa' with his own crystal of *ice-nine* and when the old dictator, terminally ill, swallows it to put an end to his agony, he freezes solid, and his own corpse, thrown into the sea, causes the whole ocean to freeze. The narrator and Mona re-emerge from their bomb shelter into the lifeless world: 'It was winter now and forever'. Mona joins Bokonon's disciples, who have poisoned themselves with *ice-nine* in a common grave, while the narrator, rescued by Newt, decides to record his experience. His book, like Bokononist thought, will centre on a paradox: 'the heart-breaking necessity of lying about reality, and the heartbreaking impossibility of lying about it'.

1964

US Civil Rights Bill ● Fall of Nikita Khrushchev ● China explodes atom bomb ● Nelson Mandela sentenced to life imprisonment ● Martin Luther King is awarded Nobel peace prize ● Close-up photographs of the moon ● The Brain Drain alarms British government ● The Verrazano–Narrows Bridge (the world's longest) opens ● Roman Catholic hierarchy rule against use of contraceptive pill ● Peter Brook, *Lord of the Flies* ● Stanley Kubrick, *Dr Strangelove* ● The Beatles in *A Hard Day's Night* ● Cassius Clay defeats Sonny Liston in World Heavy-weight Championship ● Brendan Behan, Douglas MacArthur, Ian Fleming, Clive Bell, Sean O'Casey, Herbert Hoover, Cole Porter and Edith Sitwell die ● Philip Larkin, *The Whitsun Weddings* ● Elizabeth Jennings, *Recoveries*

Saul BELLOW   1915–

Herzog

Moses Herzog (Jewish, forty-eight) sits alone in a dilapidated house in Ludeyville, Massachusetts, engaged in the characteristic Bellovian task of trying to make sense of his life: 'If I am out of my mind, it's all right with me, thought Moses Herzog.' The novel's action takes place over a few weeks, the majority of it within a few days, and is initiated by the hero: 'Late in spring Herzog had been overcome by the need to explain, to have it out, to justify, to put in perspective, to clarify, to make amends.' This is broadly the position at the end of the novel too. In between, Herzog quits his job as a lecturer at an adult education class, travels to Martha's Vineyard to visit an old friend but leaves immediately to return to New York, where he visits his lover, Ramona, before travelling to Chicago to visit his ex-wife, Madeleine, and launch a half-baked plan to gain custody of his daughter. He fails in this, is arrested after a minor road accident, and leaves Chicago for his abandoned house in Massachusetts. There he is visited by his brother, Willie, and Ramona. The novel ends with Herzog waiting for her to come to dinner.

As this story unfolds, Herzog's mind wanders back and forth: to his first marriage which lost out to his studies; to his parents and stepmother; his lovers past and present; but most of all to Madeleine. Madeleine is a bitch; everyone in the novel agrees on this.

She has dismissed Herzog (in an episode he remembers as 'one of the very greatest moments of her life') and taken up with his best friend, the one-legged Valentine Gersbach. Herzog is reduced to writing letters (which he never sends) to friends, enemies, people in public office, even to 'the famous dead'. In them he reveals his disquiet and incomprehension of the world about him.

The author of *Romanticism and Christianity*, a book which looked at the past 'with an intense need for contemporary relevance', is ill-equipped to deal with his own present circumstances. Herzog's extensive research does, however, provide Bellow with a background of ideas against which to pursue one of his favourite theses: that the complexity of modern life has far exceeded the ability of philosophy to make sense of it. Herzog writes letters to Heidegger, Nietzsche and Eisenhower, whom he lectures on Hegel. The historical failure which Bellow diagnoses pervades the action without contributing to it, and the central focus of the novel is provided by Herzog's relationships with women. While Herzog vacillates, the women of the novel are remarkably constant. Madeleine is a capricious adulteress, but she is consistently this person and does not change. Ramona is kinder and more obviously seductive: 'Her theme was her power to make him happy.' Daisy, Herzog's first wife, was housewifely, and Sono, his Japanese lover before Madeleine, was all Oriental submissiveness. Inevitably, it is the beautiful, calculating, manipulative and unavailable

Madeleine that Herzog wants most. He can only play the victim in his relations with her and this he does, rhapsodising at one point about 'The knife and the wound aching for each other.' Similarly, Madeleine's lawyer, accountant and psychoanalyst all also gain the upper hand with Herzog in one way or another. His indignation always comes too late.

Because the vast majority of the novel is taken up by Herzog's own thoughts, Bellow has some difficulty distancing himself from his hero, and Herzog's ironic self-examinations become crucial. Herzog's prose (being Bellow's) is better than he knows: subtle reversals of word-order, non-sequiturs in the middle of paragraphs and sly juxtapositions of words, thoughts and deeds are all calculated to throw Herzog's complacency off-balance. Searching for his own defining boundaries in either his domestic situation or the ruins of European philosophy, Herzog remains a sympathetic character. Both these environments having failed his hero, Bellow's major theme and achievement in *Herzog* remains that difficult, even paradoxical, project, the definition and sustenance of the modern individual.

Elizabeth BERRIDGE   1921–

Across the Common

Although she began writing in the 1940s, Berridge made her reputation with her three novels of the 1960s: *Rose Under Glass* (1961), *Sing Me Who You Are* (1967) and *Across the Common*, which won the *Yorkshire Post* Novel of the Year Award. The Common of the title is based upon Wandsworth Common, near to which Berridge was brought up.

At the centre of the novel is a Victorian villa, The Hollies, a shelter from the world and simultaneously a visible embodiment of a curse that seems to have settled on the Braithwaite family who inhabit it. The youngest member is Louise, who is married to an artist. Unable to have children and jealous of her husband's students, she despairs of her marriage and returns to The Hollies, which she still considers her home. The house is now inhabited by her old aunts: Aunt Rosa,

who has only one eye; Aunt Seraphina, Louise's favourite and a keen gardener; and diminutive, malicious Aunt Cissie, confined to a wheelchair after breaking her hip 'running across the road at Eastbourne'. Obscurely Louise feels that if she could find in the history of the house, or the family, the reason for the spiritual blight on it, she could abandon the past and construct her own successful existence.

On one level, therefore, the novel is a mystery. Guided by her curiosity and a letter from her long-dead father, Louise unravels a terrifying tale of sexual neurosis, voyeurism, rape, murder and suicide. The process of discovery gives Louise, and, in some ways, her old aunts, the strength to break away from the stifling past.

At the same time, *Across the Common* is a complex pattern of time and memory, best expressed in Louise's favourite quotation (from T.S. Eliot's *Burnt Norton*, and used as the novel's epigraph): 'Time present and time past/Are both perhaps present in time future.' Louise and her aunts are in thrall to the past, but can escape by knowing the truth about it and speaking of it. The past then becomes a source of strength. In the same way, the outside world, represented by the Common, can be terrifying if its savagery and passion are not acknowledged. It is Louise's husband, Max, who urges her to escape from the past, and promises a future in which all the conflicting strands of past and future will be woven into one: 'I'm in the present, and I love you and I want you to get unstuck. You must do it yourself or it won't be any good … Oh, and we'll have a child, even if we have to conceive it on the common. I'm sure of it. And it won't be the first to be gotten that way.'

Elizabeth BOWEN   1899–1973

The Little Girls

After her mother died in 1912, the thirteen-year-old Bowen went to live with her Aunt Laura in Hertfordshire. While a pupil at Harpenden Hall, Bowen oversaw the burying of a box of objects, a ritual that is at the centre of the novel she wrote some fifty years later. She intended to write an objective novel, in which her characters' actions were described, but their thoughts left unrevealed. She found the novel's conclusion – 'like a comedy version of the mad scene in *The Duchess of Malfi*' – technically challenging, and altered it several times, even after the book had been set up in type by her publishers.

Like Bowen herself, eleven-year-old Diana ('Dicey') Piggott had been a 'born ringleader' at St Agatha's School, Southstone, in Kent. With her friends, Clare ('Mumbo') Burkin-Jones and Sheila ('Sheikie') Beaker, she buried a box of objects for posterity in the school grounds. Since 1914 she has lost touch with the two girls, but spurred on by present-day events, and against the advice of her friend, Major Frank Wilkins, Diana (now widowed and styling herself Dinah Delacroix) decides to trace them. She places notices in the personal columns of various newspapers, including the *Southstone Herald* where one appears 'in the "Miscellaneous", among the rabbit advertisements'. The undignified tone of these advertisements is the first of many things which irritate Clare and Sheila. The former is now the owner of a chain of 'Mopsie Pye' speciality gift shops, while Sheila is married to a local estate agent called Trevor Artworth (who, as a small boy, had been terrorised by Mumbo, Dicey and a friend called Muriel Borthwick) and still lives in the vicinity of the now demolished St Agatha's. Dinah's determination to retrieve the buried box seems at once an attempt to make sense of the past and to establish a present identity. Although the other two disapprove, they cannot resist participating in the nocturnal exhumation of the box from what is now someone's garden.

Finding the box empty proves too much for Dinah and shortly afterwards she appears to have a breakdown. She is watched over by the two other women, and the crisis passes. We are left with the feeling that some, at least, of Dinah's ghosts have been exorcised and that, for better or worse, there will always be a bond between the three women.

Half-remembered, half-explained events from the past give the novel a curious tone (Dinah is shown to operate 'as though from another planet'), but it is lightened by some splendid comedy, particularly in the form of Dinah's temperamental Maltese house-boy, a frustrated secret agent, and in the revelation that the secret object buried by the all-too-proper Sheila, who in 1914 had been a dancing prodigy, was her extra toe which had been surgically removed and pickled.

Brigid BROPHY 1929–

The Snow Ball

A rich, dazzling and erotic comedy of manners, 'deliberately constructed as a baroque monumental tomb', *The Snow Ball* is set at a New Year's Eve costume ball, given by the wealthy, seal-like and much-married Anne. The party's theme is the eighteenth century, to match the elegant house (presided over by a statue of battered, wormy Cupid, 'phallic to his wing tips') where Anne lives with her fourth husband, Tom-Tom. Although obsessed by white (from her quilted, snowy bedroom to her little baskets of peppermint creams), for this evening Anne is swathed in lamé ('a solid gold orb') to represent the queen whose name she shares. Her best friend (with whom she has a former husband in common) is Anna, a devotee of Mozart whose perfect pitch can set Anne's chandeliers ringing, and who comes as Donna Anna. Amongst the guests are a masked man dressed as Mozart's Don (empathising with the character as social outcast rather than seducer), and the Blumenbaum family: Rudy in a kilt ('Rabbi Burns'); his self-effacing wife, Myra, inconspicuous in twentieth-century grey; their teenage daughter, Ruth, dressed in white as Cherubino; and her ambitious boyfriend, Edward, her negative in black as an immature Casanova.

The novel is prefaced by a quotation from the author's own study of *Mozart the Dramatist* (published the same year) concerning the question of whether, when the curtain rises upon *Don Giovanni*, the hero 'has just seduced or just failed to seduce Donna Anna'. Dr Brompius, a musicologist who is sorting through Tom-Tom's collection of scores, merely restates the question to show more clearly that there is no way of knowing. His exegesis is interrupted by the arrival of 'Don Giovanni' to claim Anna (who shares her creator's preoccupations with 'Mozart, sex and death'). He eventually seduces her, but her plan to commit an act of 'perfect bad taste' by making love in her hostess's bedroom is foiled by the fact that Anne and Tom-Tom are already there, having escaped from the appalling French cabaret to which they have subjected their guests. After making love in the Don's flat, they return to the party, passing the Blumenbaum Bentley, where Ruth and Edward have just divested each other of their virginity and are now playing in the newly fallen snow. The Don wants his affair to continue, seeing himself as Anna's victim, rather than vice versa, because he loves her. Anna flees from the party and from her lover after a fellow guest drops dead, driving into a hideous dawn and letting herself into her flat, 'thinking about death'.

Except for the final phrase, Anna's description of her idealised self might stand for the novel: 'a decoration, something very contrived, very highly wrought, that wouldn't touch the heart at all'. For all its elaborate plotting, its concern with costume, disguise and witty conversation, the book, like a Mozart opera, is finally very moving.

John CHEEVER   **1912–1982**

The Wapshot Chronicle
1957

The Wapshot Scandal
1964

Cheever was known to his readers as a writer of short stories until the publication of *The Wapshot Chronicle*. Those of his critics who doubted his ability in the longer form were confounded when it achieved popular and critical acclaim, winning the National Book Award in 1958. The story begins and ends in the small New England town of St Botolphs, home of the Wapshots: Leander and Sarah; their cousin Honora; and Leander's sons, Moses and Coverly. Honora holds the family

purse-strings and settles a trust on both her nephews to be paid to them only when they produce male heirs. This they do, and the story forms the skeleton on which Cheever hangs a succession of interlinked short stories. After Moses and Coverly's respective departures from their small town to the big city (Washington and New York), *The Wapshot Chronicle* moves episodically between their separate stories and events in St Botolphs. Both sons earn a living, Moses in banking and Coverly in a missile-testing programme; both get married; and both eventually return, although too late for their father, Leander, who, true to his classical predecessor, drowns while swimming the river he used to work as a pleasure-boat captain.

This story allows Cheever to point up the ironies of life in St Botolphs, and his characters to reveal themselves, usually comically, in environments for which their small-town upbringing has imperfectly equipped them. Significantly however, it is the values of St Botolphs which are consistently affirmed over those of the city, or the missile site, or even the decaying old-world mansion in which Moses courts Melissa. Criticism of the novel centred on its structure. At times, the episodes of the Wapshots' adventures seem too self-contained and unrelated to one another. A girl whose car crashes near to the family home is simply spirited away by her parents after she has coupled with Moses (witnessed by Honora from within a closet), and is never seen again. The strength of Cheever's characterisation glosses over such discontinuities, but the structure remains loose. This slackness does allow the inclusion of material which one would not be without. One of the happiest devices is Leander's journal, written in his inimitable, adverbless style, and his advice to his sons, which is discovered after his funeral and which ends the book:

Never wear red necktie. Provide light snorts for ladies if entertaining. Effects of harder stuff on frail sex sometimes disastrous. Bathe in cold water every morning. Painful but exhilarating. Also reduces horniness. Have haircut once a week. Wear dark clothes after 6 p.m. Eat fresh fish for breakfast when available. Avoid kneeling in unheated stone churches. Ecclesiastical dampness causes prematurely grey hair. Fear tastes like a rusty knife and do not let her into your house. Courage tastes of blood. Stand up straight. Admire the world. Relish the love of a gentle woman. Trust in the Lord.

In a short story written for the *New Yorker* magazine in 1962 called 'The Embarkation for Cytherea' a rich, middle-aged woman escorts her grocery-boy to the family holiday-home for a night of illicit passion. When they reach the cottage, however, the erotic *frisson* between them has gone; the woman only feels old, the boy would rather have gone to a football game. This affair, slightly reworked, reappears as 'the scandal' in *The Wapshot Scandal*, a loose sequel to *The Wapshot Chronicle*, several of whose characters reappear. Cheever's interest focuses on Coverly, Honora and Melissa Wapshot, the latter being Moses' wife and the woman of the earlier story.

What a tender thing, then, is a man. How, for all his crotch-hitching and swagger, a whisper can turn his soul into a cinder. The taste of alum in the rind of a grape, the smell of the sea, the heat of the spring sun, berries bitter and sweet, a grain of sand in his teeth – all of that which he meant by life seemed taken away from him.

THE WAPSHOT CHRONICLE
BY JOHN CHEEVER

Set rather vaguely in the 1950s, the book opens with a collage of scenes from St Botolphs on Christmas Eve. The town seems idyllic, with its Christmas trees, carolsingers and falling snow, but there are hints of decay. The priest might be a drunk; one family keeps a secret locked in their parlour which nobody knows; and an old man, taking a sack of kittens to drown in the river, drowns himself instead. Cheever comments: 'no one heard him, and it would be weeks before he was missed'. Much of the novel takes place outside St Botolphs, but other locations are implicitly measured against its

image. Towards the end of the novel, St Botolphs itself is visibly falling short of that image, the touchstone crumbling.

The disintegration of this venerable, high-church community is set against the vigorous growth and challenge of others. Coverly lives in a town specially constructed for workers at the nearby missile site. He and his wife, Betsey, try to break the coldness of his community by giving a party. Nobody comes and their relationship suffers badly under the pressure exerted by their vacuous surroundings. Meanwhile, Melissa Wapshot has become involved with Emile, a nineteen-year-old delivery-boy. She flees to Rome when the affair is exposed and near the end of the book 'purchases' Emile in a beauty-contest auction. Honora too travels to Rome, but her flight is from the US Tax Inspectorate, which has threatened her with gaol. Eventually she is extradited (not before seeing the Pope) and starves herself to death back in St Botolphs. Neither Europe nor the Catholic faith offer sanctuary from the incursions of bureaucracy. Melissa and Emile find a kind of contentment there, but only at the expense of living in an environmental and moral vacuum.

The vision Cheever offers is of a certain way of living under siege from other, more aggressive, ways. He has been criticised for a social narrowness in that it is the upper-middle-class families, founded on tradition and 'old money', which usually carry the banners of gentility and decency in his work. Outside this world the superhighways, airport lounges and third-rate hotels cluster in menacing fashion, but the equation of monetary and moral poverty also makes 'the poor' part of this featureless and hostile landscape. In the relationship between Melissa and Emile it is implied that Melissa's moral fibre is being corrupted by her young (and penniless) lover, not the other way about.

Both *The Wapshot Chronicle* and *The Wapshot Scandal* interweave the strands of their plots in such a way as to suggest that a story is taking place, but in fact the method of both novels is closer to collage than to narrative. Cheever is adept at juxtaposing different environments so as to bring out the comparative moral values implicit within them, and he does this without recourse to overt symbolism or extravagant rhetoric. His classically disciplined and concise style has been developed through the numerous short stories he has written. *The Wapshot Chronicle* and *The Wapshot Scandal* deploy these resources subtly, using conventional novelistic devices in deceptively unconventional ways.

Agatha CHRISTIE   1890–1976

A Caribbean Mystery

Miss Jane Marple is an English spinster, devoted to gardening and domestic economy. Gossipy, shrewd, insatiably curious, and ruthless in a manner that belies her soft, genteel, middle-class, somewhat old-fashioned appearance, Miss Marple tends to rely on her intimate knowledge of human nature, her intuitive wisdom and her rather worldly view of human frailty to make connections that Christie's more rigorous detective, Hercule Poirot, would hesitate to consider as firm evidence. Miss Marple is an inspired amateur of crime, but always 'gets her man'.

The majority of Christie's books featuring Miss Marple are centred on the emblematic village of St Mary Mead, where the elderly sleuth lives. Her close scrutiny of village life provides parallels which assist her in solving the most brutal of crimes. In *A Caribbean Mystery*, however, Miss Marple is on holiday in the rather unlikely setting of the island of St Honore.

A fellow guest, Major Palgrave, is murdered in his bungalow at the Golden Palm Hotel run by Molly and Tim Kendal, an attractive young couple in their twenties. The other guests include an English tycoon, Mr Rafiel; his masseur, Jackson; his secretary, Mrs Walters; botanists Edward and Evelyn Hillingdon; two rather brash Americans, Greg and Lucky Dyson; Canon Prescott and his sister; and a South American woman who is generally surrounded by gigolos.

This characteristic Christie collection of suspects is observed by Miss Marple, who knows that one of them at least has committed a murder in the past – Major Palgrave has been just about to show her a photograph of the guilty party before his death. Two further deaths (of a chambermaid and Lucky Dyson) take place before Miss Marple arrives at a neatly worked-out solution. The murderer is Tim Kendal, who is exposed as a ruthless good-for-nothing who has married Molly merely for her money, and has been trying to murder her so that he can marry another woman.

The clue to the murder, for the sharp-witted, is given on page 2; from the beginning, however, Christie calmly points our gaze in quite another direction. Miss Marple, as Nemesis in fluffy pink wool, enjoys yet another quiet triumph.

Several of the Miss Marple novels have been adapted as highly successful films and television plays, with notable performances in the central role by Margaret Rutherford and, most recently and perhaps definitively, by Joan Hickson.

Christopher ISHERWOOD   1904–1986

A Single Man

Like all Isherwood's later novels, *A Single Man* grew out of several ideas. In February 1962, when he was teaching English Literature and Creative Writing at Los Angeles State College, he had an idea for a 'novelette', in which a middle-aged professor becomes sexually involved with one of his former students. By the end of March, he had had a new idea for a novel, 'The Englishwoman', about a woman who had come to America as a GI bride and now feels adrift in California; the professor would be tutor to her son. Progress on this new novel was slow, since Isherwood was undergoing domestic crises and was also working on his biography of Ramakrishna. In August, however, he reread Virginia Woolf's **Mrs Dalloway** (1925), and wondered whether he might attempt something similar. 'I want to write about middle-age, and being an alien. And about the Young,' he noted.

Like Woolf's novel, *A Single Man* records the life and thoughts of its protagonist in the course of one day. Although George shares many of his creator's circumstances, characteristics and preoccupations – he is a fifty-eight-year-old English professor living and teaching in California; he is physically vain, gleefully misogynistic, homosexually militant, and has a love–hate relationship with England – he is not the 'Christopher Isherwood' of earlier books. The novel is narrated in the third person, which makes it more genuinely objective than previous books derived from autobiographical material. It opens with George's ageing body stirring to life, and follows him to work, where he gives a lecture on Aldous Huxley, to the gym, where he works out, and to dinner with Charley, the GI bride of 'The Englishwoman', now relegated to (in every sense) a supporting role.

George's American lover, Jim, has been killed in a car-crash, and George, although still in mourning, is defiantly living in and for the present. Charley has been deserted by both her husband and son, and in the course of a drunken evening attempts to seduce Geo, as she calls him. George escapes and weaves his way to a bar, where he finds one of his students, Kenny Potter, who has come in search of him, clearly wanting to talk. Although Kenny has a girlfriend, and seems respectful towards George (insisting upon addressing him as 'Sir'), his manner is intellectually – even physically – flirtatious. Tutor and pupil skinny-dip together, then return to George's house, where George confesses that he is unable to give Kenny whatever it is he wants: the benefit of experience, perhaps, for there is a suggestion of a Platonic dialogue in their conversation. Increasingly drunk, George ends by haranguing the boy for his unserious, flirtatious approach to life and relationships, then passes out. He awakes to find that Kenny has put him to bed, left a friendly note, and departed. He lies in bed and masturbates, stimulating himself (when images of Kenny prove ineffective) with fantasies about two tennis-playing students he has observed earlier in the day. The novel ends with George's body relaxing into sleep, and follows what might happen to that body if it died during the night. The closing pages (which possibly derive from Huxley's description of the death of Everard Webley

in *Point Counter Point*, 1928) speculate on George's passage from the individual self to a greater collective consciousness, and here Isherwood's interest in mysticism, which has surfaced intermittently throughout the novel, emerges as a major theme.

A Single Man is certainly the best book Isherwood wrote in America, and is perhaps his masterpiece. Economical, moving and savagely funny, it is less 'about' homosexuality than such universal concerns as human survival, the importance of the present (as opposed to the past), and the often anomalous position of the individual in society: single people in a world of couples and families; thoughtful people in a country obsessed by instant gratification; Jews, blacks, nisei and homosexuals in a predominantly WASP America; the minority of the living vastly outnumbered by the dead.

B.S. JOHNSON   **1933–1973**

Albert Angelo

Composed of five sections – 'Prologue', 'Exposition', 'Development', 'Disintegration' and 'Coda' – *Albert Angelo* was Johnson's second novel. Formally audacious, its content is resolutely mundane. Twenty-eight years old, Albert Albert – so named 'to emphasise his Albertness' – lives alone in a flat in Islington. He is an architect by training but he works as a supply teacher in North London. The schools to which he is sent are shabby and demoralised; his pupils both frighten him and arouse his pity. In his free time he designs buildings, drinks in his local pub and visits clubs and late night cafés with his friend Terry. He has no sexual partner, preferring to be a spectator of human affairs. However, he constantly recalls his relationship with Jenny, particularly the six days he and she spent camping beside a loch in Ireland. It appears that Jenny separated from Albert to go back to a previous lover. Since then, Albert has tormented himself with memories of his brief and lost utopia.

Simple though the story is, Johnson employs a multiplicity of techniques to tell it. Thus, 'Prologue' consists of a dramatic dialogue – comic and naturalistic – followed by a rare instance of conventional descriptive writing. 'Exposition' contains second person narration as well as first and third. 'Development' includes children's essays, an excerpt from an eighteenth-century textbook on anatomy, a poem of Audenesque beauty and a short account of the killing of Christopher Marlowe. A lengthy passage is printed in two columns, the left hand giving the direct speech of a geology lesson, the right hand giving Albert's simultaneous thoughts; rectangles are cut through four pages, allowing the reader a metaphorical glimpse of the future – which proves, ironically, to be a glance at the past. The final words of 'Development' interrupt Albert's tale with a cry of frustration: 'OH, FUCK ALL THIS LYING!' Regardless of the strategies invoked, the fiction cannot sustain its fictionality and 'Disintegration' denounces the preceding sections for 'covering up covering over pretending pretending' – for seeking the truth through the invention of 'an arbitrary pattern'. All the same, 'Coda' returns to Albert Albert: he is thrown into a canal by five of his pupils. The novel concludes with a semi-literate paragraph on an old woman's funeral.

The juxtaposition invites the reader to infer that Albert has drowned. That he was only ever imagined diminishes the pathos of his probable death, without wholly erasing it. Indeed, it is a paradox of the novel that, in spite of its exposure of its own deceit, it achieves many of the effects normally associated with realism. The authorial voice admits this confusion: 'what im ... trying to write about is writing ... about the fragmentariness of life ... another of my aims is didactic'. The impossibility of unifying such disparate purposes may be the cause of *Albert Angelo*'s self-dissatisfaction; yet that same impossibility generates an extraordinary resourcefulness and innovation.

Like the rest of Johnson's work, *Albert Angelo* has received more critical than popular acclaim. On its publication, Edward Lucie-Smith commented that 'it [has] a kind of crispness and clarity we're not accustomed to in novelists'.

Norman MAILER　　　　**1923–**

An American Dream

When first published in the USA, *An American Dream* evoked a wide range of response, from condemnation to excitement. It was Mailer's sixth novel, and is now considered to be a landmark of modern American literature.

Stephen Richard Rojack, formerly a war-hero and congressman and now a psychology professor in New York, contemplates committing suicide by leaping from his balcony. Although he feels that his life is over, he holds back, convinced that there are still many things he must achieve. In reviewing his life at this point, he reveals the depth of resentment he feels against his estranged wife, Deborah Kelly. At the same time he cannot deny how helpful her social position has been for him professionally.

When Rojack visits his wife they soon argue, and this leads to a violent fight during which he murders her. In desperation he throws her body over the balcony on to the street below, to make it seem as though she had committed suicide. Still in shock, he promptly has a bestial sexual encounter with Deborah's maid, Ruta. Only then does he call the police.

A sequence of coincidences and encounters then unfolds. The police question passengers from the car pile-up which results from Deborah's fallen body, and this introduces two other characters: Cherry, a lush chanteuse, and Eddie Ganucci, a Mafia prince. The police try to force a confession of murder from Rojack, but eventually release him, due to, it later transpires, the intervention of an unknown but influential figure. On his release Rojack heads for the night-club where Cherry is performing, and an affair promptly ensues.

Cherry turns out to have many shady connections, including Shago Martin, a black singer and her ex-lover. Cherry was also the mistress of Barney Oswald Kelly, Deborah's tycoon father. When Shago jealously confronts Cherry and Rojack, the two men fight, with Rojack almost killing Shago.

The series of deaths in the novel concludes with the murder of Shago, and Cherry's death at the hands of a confused friend of Shago.

Rojack's life begins to crumble. He is dismissed from his television psychology show, and forced to take a leave of absence from his college. He also learns that his father-in-law was behind his release from custody, but he is none the less anxious about meeting up with him, when he is summoned to Kelly's apartment. Ruta is at the apartment, in the course of her affair with Kelly. Ganucci, who turns out be an old friend of Kelly, is also present. Acting on his suspicions, Kelly manipulates Rojack until he admits that he murdered Deborah. Following the admission, Kelly's preoccupation is with Rojack attending Deborah's funeral, which Rojack resists. As Kelly says: 'Public show is the language we use to tell our friends and enemies that we still have order enough to make a good display.' This is how Kelly has covered up his own suspicious liaisons.

I took a step toward him. I did not know what I was going to do, but it felt right to take that step. Maybe I had a thought to pick up the whiskey bottle, and break it on the table. The feeling of joy came up in me again the way the lyric of a song might remind a man on the edge of insanity that soon he will be insane again and there is a world there more interesting than his own.

AN AMERICAN DREAM
BY NORMAN MAILER

While some readers objected to the explicit nature of the novel's sexual episodes, a far more damaging criticism of Mailer's books has been made by feminist critics, notably Kate Millett, with whom Mailer has clashed publicly on several occasions. Millett's pioneering *Sexual Politics* (1970) opens with a lethal analysis of *An American Dream* which she describes as 'an exercise in how to kill your wife and live happily ever after', pointing out that it lacks the 'humanist convictions' which underlie *Crime and Pun-*

ishment (1866) and Dreiser's *An American Tragedy* (1925), where the murderers accept responsibility for their actions, and eventually atone for them. In *An American Dream*, 'the reader is given to understand that by murdering one woman and buggering another, Rojack has become a "man"'; worse still, 'Mailer transparently identifies with his hero'. Millett persuasively describes Mailer as 'a man whose powerful intellectual comprehension of what is most dangerous in the masculine sensibility is exceeded only by his attachment to the malaise'.

Edna O'BRIEN   **1936–**

The Country Girls Trilogy

The Country Girls (1960)
The Lonely Girl (Girl with Green Eyes) (1962)
Girls in their Married Bliss (1964)

Oh, God, who does not exist, you hate women, otherwise you'd have made them different . . .

The Country Girls was O'Brien's first novel, written during the first few weeks she spent in London after leaving her native Ireland in 1960. She has described it as 'my experience of Ireland and my farewell to it', and it caused something of a sensation in her family's village, where several copies of the book were burned in the churchyard by the parish priest. In England it established her reputation, and formed the first part of a trilogy. While O'Brien admits that the books draw upon her own background, she insists that they are in no other sense autobiographical.

The three novels trace the lives of Caithleen Brady and Baba Brennan from adolescence to their mid-twenties, and examine many values through the experience of girls who have been raised within a very specific social and religious framework, and who are often victims of the contradictions between their conditioning, expectations and true desires.

Her life clouded by the drinking of her father and the accidental drowning of her mother, shy, awkward Caithleen is dominated by her 'coy, pretty, malicious' friend, Baba, the daughter of a serious-minded vet and an embittered, somewhat loose mother, who later becomes a reformed character. Following a deliberately provocative act (instigated, of course, by Baba), which secures expulsion from their convent school, the girls go to Dublin and look forward to a life full of excitement and eligible men. While Baba quickly becomes embroiled in a rather sordid world of married men not keen on preliminaries, Caithleen pursues a romantic involvement with her 'Mr Gentleman', a lawyer from her home town, whose marriage eventually proves to be the greater commitment.

It is the only time that I am thankful for being a woman, that time of evening, when I draw the curtains, take off my old clothes and prepare to go out. Minute by minute the excitement grows. I brush my hair under the light and the colours are autumn leaves in the sun. I shadow my eyelids with black stuff and am astonished by the look of mystery it gives to my eyes. I hate being a woman. Vain and shallow and superficial.

THE COUNTRY GIRLS
BY EDNA O'BRIEN

In the second novel (reissued in 1964 as *Girl with Green Eyes*, the title of the 1963 film, directed by Desmond Davis from O'Brien's own screenplay), the quest for the perfect man continues. Caithleen meets Eugene Gaillard, a meticulous and quietly tyrannical man who is separated from his wife and child. He undertakes Caithleen's education, immediately changing her name to Kate, and protects her from the moral outrage of her father and his friends. Ever diffident, Kate none the less finds Eugene's demands and the competition from his friends too much, and eventually leaves him, secretly hoping that he will come to reclaim her. When he fails to do so, she agrees to go to London with Baba in search of a better life.

The title of *Girls in their Married Bliss* is savagely ironical. Baba, whose fierce and cynical manner appears to conceal a genuinely good heart, marries Frank Durack, a wealthy builder who is 'thick, but nice too', though totally inexperienced sexually, which leaves her bored and frustrated. Kate has been reunited with Eugene, become pregnant and married him, but, finding his 'little dictatorship' hard to tolerate, becomes romantically involved with another man. When it transpires that Eugene has been plotting and documenting her movements, she leaves him and is forced to surrender her son, Cash, whom Eugene finally removes beyond her reach to Fiji. While Kate struggles to survive and come to terms with her emotional needs, Baba, equally a victim, becomes pregnant by another man. Forgiven by Frank, she is 'cornered in the end by niceness, weakness, dependence'. When, at the age of twenty-five, Kate has herself sterilised, even Baba, who feigns hardness and nonchalance, recognises that she is 'looking at someone of whom too much has been cut away', and that the 'dead dopey' Kate she knew has somehow been eliminated in the process of gaining experience.

A sour epilogue added to the one-volume edition of the trilogy (1987) is narrated by Baba: 'I am at Waterloo again, the railway station where Kate gashed her wrists ... Nearly twenty years ago. Much weeping and gnashing between.' Kate survived that attempt, but has now drowned herself at a health farm, and Baba is waiting for the coffin to arrive by train. Baba's own history has not been much happier: her daughter left home before she was thirteen, and Frank has recently suffered a debilitating stroke. We discover that the two women had lost touch over the years – 'different lifestyles and so forth' – but that Kate, thin and nervous, had visited Baba a week before her death. 'I could see it coming,' Baba recalls. 'I knew there was some bloody man and that he was probably married and that she saw him once a fortnight or less, but of course saw him in street lamps, rain puddles, fire flames, and all that kind of Lord Byron lunacy.' Kate did indeed appear to have been hallucinating, but in a moment of lucidity provided her own epitaph: 'She put her hand to her heart and said she'd like to tear it out, stamp on it, squash it to death, her heart being her undoing.'

While, even without the addition of the epilogue, the trilogy's descent into disillusionment makes bleak reading, the novels contain some splendid comic characters: Joanna, the excitable East European Dublin landlady; Harvey, the drummer with a predilection for long boots and plastic rainhats who, given that he shows more promise than prowess, becomes the father of Baba's child more by luck than anything else; and Mrs Cooney, Baba's huffy char who considers herself a cut above her mistress. O'Brien's depiction of the pursuit of experience through half-innocent, half-shrewd eyes is the source of many comic moments and perceptions.

Anthony POWELL   1905–

The Valley of Bones

see **A Dance to the Music of Time** (1975)

Simon RAVEN   1927–

The Rich Pay Late

see **Alms for Oblivion** (1976)

Hubert SELBY   1928–

Last Exit to Brooklyn

Selby's first novel became a *cause célèbre* when it was tried for obscenity following its British publication in 1966. In March of that year Sir Cyril Black MP (aided by prosecution witnesses who included Sir Basil Blackwell and Robert Maxwell MP) brought a successful private prosecution which applied only in the area of the Marlborough St Magistrates' Court where the case was heard. The Director of Public Prosecutions then reversed his earlier decision and prosecuted under Section 2 of the 1959 Obscene Publications Act, and the publishers, despite an impressive array of scholarly and literary defence witnesses, were found guilty at the

Old Bailey in November 1967. With the assistance of the newly founded Defence of Literature and the Arts Society they successfully appealed the following year, with John Mortimer QC leading for the defence. Lord Justice Salmon's judgment was widely held to clarify many areas of ambiguity in the 1959 Act.

The novel, written in a vividly rendered vernacular, consists of a series of loosely linked episodes portraying the squalid lives of a group of characters living in Brooklyn. The Greeks all-night diner is the chief location, the favoured hang-out of 'hustlers, pushers, pimps, queens and would be thugs' including the 'hip queer' Georgette, and Vinnie, the small-time hoodlum whom 'she (he)' adores. It was here, back in the war years, that fifteen-year-old Tralala (who, in perhaps the most brutal scene of the novel, is beaten senseless and gang-raped) began her career of thieving and prostitution.

The longest and most sustained of the chapters, 'Strike', concerns a trade union shop steward, Harry, who, in the midst of organising a strike, grows increasingly disillusioned with his wife and nauseated by their sexual relations. He finds himself reluctantly fascinated by a 'pretty young fairy' in the Greeks, and one drunken night goes home with 'Alberta'. Once Harry has 'come out', the tone of the prose becomes unusually tender, with Harry 'hoping he might meet someone who would want to live with him and they could make love everynight or just sit and hold hands ...' (The prosecution in the obscenity trial had worried that this might lead an unhappily married man to conclude that his path to happiness lay in homosexuality.) The chapter ends with Harry being beaten up by the denizens of the Greeks after he has attempted to fellate a ten-year-old boy.

The final chapter describes life in a low-rent Brooklyn housing project, and is a harrowing catalogue of routine domestic violence, racial antagonism and all manner of brutality. At one point a crowd of women watch expectantly in the hope that a baby will drop from a ledge on to which he has climbed, and are disappointed when his mother rescues him.

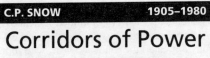

C.P. SNOW 1905–1980

Corridors of Power

see **Strangers and Brothers** (1970)

William TREVOR 1928–

The Old Boys

H.L. Dowse, a housemaster at an unnamed public school, once observed: 'The world is the School gone mad.' Trevor's immaculate black comedy (which received the Hawthornden Prize) explores the truth of this remark by charting an enfeebled power-struggle taking place between Dowse's pupils, now in their mid-seventies and serving on the committee of the Old Boys' Association. Nox has never forgotten the treatment he received when acting as Jaraby's fag in 1907: 'Jaraby was a ghost he had grown sick and tired of, which he could lay only by triumphing in some pettiness.' Over fifty years later Jaraby is standing for the post of President of the Old Boys' Association, and Nox's moment has arrived. He offers a private detective called Swingler lessons in Italian in exchange for information with which to discredit his old enemy.

Jaraby's own energies are engaged in having his wife certified. They live in cantankerous and elaborately expressed enmity in Crimea Rd, SW17, constantly duelling with their blunted wits. Jaraby plans his election campaign, finding consolation in Monmouth, his enormous, ravening cat. Mrs Jaraby knits and watches the television with the sound turned off, while she plots to reinstate their middle-aged son, Basil, in the house. Jaraby's contemporaries lead equally dispiriting lives: Mr Sole and Mr Cridley live in the Rimini, a residential hotel owned by a rapacious Miss Burdock, attempting to stem boredom by replying to mail order offers; Mr Turtle, lonely, confused and failing, lives in terror of his grasping char, a Mrs Strap.

Mrs Jaraby recognises that: 'As the future narrows one turns too much to the past,' a habit tediously exemplified by her husband. She takes the opportunity of his absence at Old Boys' Day (an occasion at which Mr Turtle dies) to sedate and drown Monmouth, so that Basil can return to the house, bringing with him his collection of diseased caged birds. Unfortunately, the police catch up with Basil, who is a confidence trickster and child molester. His arrest is observed by Swingler, who approaches Nox and offers to ensure maximum publicity, then approaches Jaraby, promising to keep the case out of the papers for a certain sum. Richer by £665, he departs for Italy. At an explosive meeting of the Old Boys' Association, Jaraby's bid for the presidency is defeated. He returns to Crimea Rd and his wife's chilling reminder that: 'We are bystanders now ... We cannot move events or change the course already set. We are at the receiving end now.'

Angus WILSON 1913–1991

Late Call

The influence of E.M. Forster is apparent in this acute, funny and moving novel about an old woman in a New Town. The clash between Harold Calvert, a progressive headmaster whose religion is the 'serious social experiment' of 'multi-status' Carshall New Town, and his raffish parents who come to live with him, show Wilson's recurrent preoccupation with liberalism and human values, and the novel might well borrow the famous epigraph of **Howards End** (1910): 'Only connect ...'

The story is told through Sylvia, Harold's mother, a hotel manageress forced to retire because of ill-health, whose life has been circumscribed by her husband. Arthur is a flatulent and fruity victim of the First World War's 'temporary officer' class, whose life revolves around tall stories and gambling. The recently widowed Harold's plan for his children's domestic self-sufficiency leaves little for Sylvia to do apart from watch 'the tele' and read historical romances. She is aware that her life is like a badly knitted jumper and that 'to weave all the threads together again, she needed to return to the country world of her childhood' (an unsentimental world of harsh rural poverty depicted in a prologue). During a walk in the countryside, she gets caught in a storm and saves a small child, Mandy, from lightning, earning the gratitude and friendship of the girl's family, Canadian farmers called Egan. Sylvia reaches back into her past for stories with which to entertain the child and gains strength and understanding with which to confront the disintegrating world.

Harold, more than any of Wilson's characters, exemplifies the failure of liberalism. His children treat him with a mixture of awe and tolerant amusement as he organises the household roster, lectures them about social responsibility and attempts to save from the developers a strip of meadowland in Carshall (leased by the Egans). Each child eventually exposes Harold's social vision as soulless and meretricious: Ray, the golden boy, lives dangerously in the homosexual underworld; spotty, scowling, rebellious Mark leaves home; and Judy becomes attached to the County set of her friend Caroline Ogilvie. Harold is a hypocritical bully whose obsessive concern with the issue of Goodchild's Meadow, sanctified by the supposed approval of his dead wife, overrides any real interest in his family. His jocular dismissal of Judy's concerns as 'snobbish' mask his genuine class hatred, and his professed pride that

Mark 'should have turned out such a rebel' is belied by his opposition to his son's plan to give up a career in engineering to work for famine relief. Perhaps most telling of all is his reaction to the news that Ray is homosexual. Carshall is not some prejudiced backwater, he claims; a psychiatrist will soon 'cure' Ray. This revelatory scene occurs at the climax of the novel, when Harold has lost his temper at a planners' meeting and Arthur has collapsed with a fatal stroke. Sylvia, who has 'never been clever, never been anything really', and would be condemned by Harold's ideals to be 'a superflous, fat old woman', takes control – but only temporarily, for the novel ends with her, newly widowed, determined to live on her own. In coming to terms with, and finally giving vent to, her feelings, Sylvia displays a genuine humanity which is in sharp contrast to the sanitised Utopia of Carshall.

Wilson's gift for social mimicry and for capturing scenes of toe-curling social unease contribute to his unerring evocation of the not-so-brave new world of the New Town.

1965

US offensive in Vietnam ● Malcolm X shot dead ● Race riots in Los Angeles ● Ho Chi Minh rejects peace talks offered by US ● MBE awarded to the Beatles ● Edward White walks for 20 minutes in space from US Gemini IV ● Soviet Antonov AN-22 makes flight with 720 passengers ● Piet Mondrian designs dominate fashion ● *Dr Who and the Daleks* ● The State Funeral of Sir Winston Churchill ● T.S. Eliot, Nat 'King' Cole and Albert Schweitzer die ● Sylvia Plath, *Ariel*

Jerzy KOSINSKI   **1933–1991**

The Painted Bird

This, Kosinski's first novel, takes its title from the practice of painting a bird and releasing it back into the wild. When the bird tries to rejoin its flock it is invariably torn to pieces for being different. The book is set in wartime Eastern Europe and tells the story of an unnamed six-year-old boy who is sent into the countryside by his parents to escape

the worst effects of the Nazi invasion. He quickly becomes separated from his companions and begins a nightmarish odyssey lasting six years (the length of the war), which forms the bulk of the narrative. Believing himself either a gypsy or a Jew, the boy avoids contact with the German forces at all cost. Even so, he is captured twice, but each time manages to escape. The boy is thrown back on the 'charity' of various peasant communities, who treat him with universal suspicion and frequently appalling cruelty.

On one occasion, villagers try to drown him in a pit of manure. On another, a carpenter attempts to kill him because he believes the boy's dark hair has attracted a lightning-strike on his barn. The boy, however, pushes him into a bunker filled with trapped rats and watches while he is eaten alive. Much of the time, he sleeps rough in the countryside. He loses the power of speech and becomes little better than an animal. Adopted as a mascot by passing Russian soldiers, he is eventually deposited in an orphanage where his parents discover him. The boy proves incapable of living a domestic life but after a serious accident he regains the power of speech and begins, slowly, to be reintegrated into society.

When *The Painted Bird* was published, critical reaction was at first appreciative, citing the psychological plausibility and sensitivity of the main character's portrayal. However, the undeniable brutality and bestiality of the story prompted a wave of revulsion, particularly from the ranks of American writers. The fact that much of the novel is informed by Kosinski's own experiences during the war seemed to provoke rather than pacify this distaste. In return, Kosinski pointed out that very few Americans had any experience of life within Eastern Europe during the war and that, in effect, they were objecting to the unpalatable nature of the world rather than the novel. However, Kosinski's account of the relationship between his own life and the book has been inconsistent; he has reacted angrily to suggestions that the two are one, but also tacitly promoted the idea. That most of the incidents in the novel happened to Kosinski is clear; the question of whether he wishes the book to be seen as an autobiographical testament or a work of fiction remains unresolved.

Although *The Painted Bird* was Kosinski's first novel, it was preceded by two very successful sociological works (as Joseph Novak) in which the values of the novel and its attitudes to society are explored with some theoretical rigour. Whether its power derives from historical fact or artistic fiction, *The Painted Bird* remains a compelling, often horrific and disturbing, work. The only untruth in it – the title – is its main source of optimism. Koskinski is, finally, not the painted bird, having survived the depredations of the flock and lived to tell the tale.

Doris LESSING   1919–

Landlocked

see **Children of Violence** (1969)

John McGAHERN   1934–

The Dark

'Bless me, father, for I have sinned. It's a month since my last Confession. I committed one hundred and forty impure actions with myself.' McGahern's second novel, set in rural West Ireland in the 1950s, is about growing up. The narrator – a bright scholarship boy, raised a Catholic – is torn between the priesthood and a life of learning. His decisions are complicated by adolescent lust, experienced within the confines of a puritanical and passionate religion, and by an intense love–hate relationship with his widower father, Mahoney.

The Mahoneys live in an isolated and claustrophobically small house. McGahern's terse narrative style vividly conjures up a childhood marked by violence and hardship, one that is at once repressive and secure. Short scenes dramatically convey his themes without recourse to dry commentary. The clash between the boy's academic ambition and the father's stubborn pride in his farming roots, for example, is manifested in a frenetic and rain-soaked nocturnal potato-stacking scene, labour imposed by father on son after a long day studying.

Fr Gerald, a cousin of Mahoney, encourages the boy to follow his calling. Young Mahoney's sexual appetite – for womankind in general and Mary Moran on her bike in particular – is an obstacle. It is a source of fear and darkness; he thinks constantly of sin, death, hellfire and damnation. But it is not the niggling worry that he might never appreciate the softness of women's flesh that settles his mind against the priesthood; it is the discovery that Fr Gerald has doubts, that priests are still fallible.

Similarly a life of studying proves flawed. At the school, Fr Benedict pushes young Mahoney through two scholarships and on to university in Galway. There, the other students' main concerns are base and worldly: picking up girls at dances, drinking and becoming wealthy. The boy realises that neither God nor books can provide an answer in life that is pure and indestructible.

The ending of the novel is not happy – but it *is* triumphant. The narrator decides to leave university and take a clerkship in Dublin with the ESB. This seems, on the surface of it, to be the most pedestrian option (a settled job with money and good prospects for promotion), but it represents a turning-point. The young Mahoney has found his own way (he can, after all, leave the job and change direction). In the process, he has conquered his fear of the dark, accepting the unknown that life holds in store for him before death. The job also ensures a practical independence from Mahoney that makes possible real forgiveness and love at last between father and son.

The book caused a furore when published in Ireland, and was banned by the State Censorship Board on the grounds of obscenity. McGahern, who worked as a teacher at a church-run primary school, lost his job and left Ireland in a blaze of publicity. The ban lasted for seven years, and McGahern did not return to his native country until the early 1970s. He disliked the notoriety all this brought to his novel, and found that the scandal and its ensuing disruption to his life meant that it would be some four or five years before he felt able to write again.

Olivia MANNING   1908–1980

The Balkan Trilogy

The Great Fortune (1960)
The Spoilt City (1962)
Friends and Heroes (1965)

The Balkan Trilogy is the first half of a sequence of six novels depicting the mar-

riage of a young English couple, Guy and Harriet Pringle, set against the backdrop of the Second World War in Europe and Africa. The entire sequence, completed by *The Levant Trilogy* (1980), is entitled *Fortunes of War*, and is based upon the author's own experiences. Just before the Second World War, Manning married R.D. Smith, a British Council lecturer based in Bucharest, and, like her heroine, spent the first years of her marriage moving from Rumania, through Greece to Egypt just ahead of the advancing German army. The novels are one of the most impressive accounts of the war, a brilliantly managed tragi-comedy seen almost entirely from the point of view of expatriate non-combatants. Apart from the Pringles, Manning deploys an enormous cast of engaging, eccentric, absurd and despicable characters, each one a memorable individual who emerges triumphantly from the vast anonymity of war.

The Great Fortune takes its title from a notion that Rumania is like a person who inherits a great fortune, but loses it through folly. Initially pro-British, the country is fatally fickle, changing allegiance when it imagines that Germany will be a strong ally against the threat from Soviet Russia. However, there is a secondary meaning, taken from Ruskin's assertion that: 'There is no wealth but life.' Life, indeed, is the only valuable possession left to the Pringles at the end of the trilogy: 'a depleted fortune, but a fortune'. The action is seen from the contrasting viewpoints of the acerbic Harriet Pringle and a sentimental White Russian, Prince Yakimov. Harriet has married Guy without really knowing him and is unprepared for his compulsive gregariousness as he immerses himself in Bucharest life. He is working as an 'Organisation' (i.e. British Council) lecturer and has time for everyone but his wife, forever deserting her for some new excitement or acquaintance. Blundering and myopic, he is a Marxist idealist who sees the best in everyone, even in those who have no good in them, naïvely imagining that generosity will be repaid in kind. Harriet knows better, and attempts to protect her husband from people like the opportunistic Sophie Oresanu, who had hoped that marriage to Guy would provide her with a passport, and the scrounging Yakimov, who will betray any friend or cause for the price of a dinner. Apart from a harebrained scheme to blow up oil wells, mounted by a Secret Service man called Sheppy, Guy's chief preoccupation is a prestige-boosting production of *Troilus and Cressida*, which (playing to a packed house as Paris falls) mirrors both the characters of the amateur actors and the real, unheroic war which is about to engulf them all.

The Spoilt City is Bucharest, which has lost its fortune and sold its soul to the Nazis. The once abundant food is now scarce, the vacillating king abdicates and the fascist Iron Guard grows in strength, murdering British Legation employees and smashing up Professor Inchcape's Bureau of Information. The Pringles have taken in Yakimov, but are also harbouring Sasha Drucker, the son of a disgraced Jewish financier. A former student of Guy, Sasha had been drafted into the army, but has subsequently deserted. Although disillusioned with Guy, Harriet resists the advances of his colleague Clarence Lawson, a self-loathing young man in charge of Polish refugees. Yakimov discovers a document concerning the abortive plan to blow up oil wells and shows it to a former friend, Freddi von Flügel, who is now a *Gauleiter* in Transylvania. Guy's name is put on a Gestapo hit-list, and when the Pringles return from a holiday, they find their flat raided and Sasha gone. In the wake of the German advance, Harriet flies to Athens leaving Guy behind her.

Friends and Heroes opens with Harriet anxiously waiting for Guy to join her. She has found Yakimov, who is now employed by the British Information Office distributing news-sheets, and has forgotten her quarrel with him. Officially Guy should proceed to Cairo and this puts him in a weak position when he arrives in Athens and applies for work at the Organisation. This is being run by the languid Colin Gracey, aided by two nonentities, a simple-lifer called Dubedat and an incompetent called Toby Lush, both of whom Guy had employed in Bucharest, but who, alarmed by the German advance, had left him in the lurch. Harriet is appalled that Guy should have to ask these men for work, and they are disinclined to help. However, due to the machinations of the widow of a former Director of the Organisation, Mrs Brett, Guy is eventually promoted to become Chief Instructor in the school, answerable to the preposterous Prof. Lord Pinkrose. Guy joins forces with another left-wing idealist, Ben Phipps, a misogynist who resents and despises Harriet. She befriends Alan Frewen, the Grecophile British Information Officer, who gives her a job in his office, from which she is eventually sacked by Pinkrose.

Athens is at first a pleasant contrast to Bucharest, with the victorious Greeks repelling Italian troops. However, as the Greeks, now joined by the British, are beaten back by the Germans, food shortages and air raids destroy the idyll. Once again abandoned by Guy, who has been busy mounting an entertainment for the RAF, Harriet teeters on the brink of an affair with a young British officer, Charles Warden. As the Germans advance, it is discovered that there are no boats to evacuate the British. Warden goes off with the British troops and Yakimov is accidentally shot dead in the blackout. Eventually Phipps discovers two boats secretly hired by the Organisation's effete old regime and commandeers them. The novel ends with the Pringles sailing through treacherous waters from Piraeus towards Egypt.

Although Manning's account of the war is in itself remarkable, the great strength of the trilogy lies in her depiction of the characters she sets against this richly detailed backdrop. Even the most insignificant of players upon her stage are brought vividly alive, while Yakimov and the Pringles have a substance rare in the twentieth-century novel. Significantly long-sighted, Harriet sees what her husband misses, in particular the greed, duplicity and selfishness of her fellow humans, few of whom are ennobled by their experience of war. Harriet does not believe that mankind is susceptible to improvement, recognising in cruelty to animals an index of ingrained brutality. (Manning herself was a staunch supporter of animal's rights.) Loyal to the point of martyrdom to Guy, Harriet rejects the advances of men who pay her her proper due. Unafraid to speak her mind, she alarms many of Guy's associates when, like the kitten she loves and adopts, she turns from an apparently vulnerable young creature into 'a mad little bundle of pins'. Contrasted to the idealistic (and significantly short-sighted) Guy, she seems cynical, but is so only in the definition of Ambrose Bierce: 'a blackguard whose faulty vision sees things as they are, not as they ought to be'. As she puts it to Guy: 'You're interested in ideas; I in people. If you were more interested in people, you might not like them so much.' While Guy appears to be driven by inexhaustible enthusiasm, it is left to Harriet to reveal 'the prosaic wiring that lay behind the star-bursting excitements of life'.

Guy is a refreshingly sharp study in masculine complacency. Although many people regard him as a saint, his clay feet are all too apparent to Harriet, who sees him rather as a holy fool in need of protection from those less decent and more wily than himself, 'a husband made unreliable only by his abysmal kindness' to others. Compulsively gregarious, he frequently abandons the young wife he has taken into war-harried Eastern Europe: 'the thing he loved most was the fatuous good-fellowship of crowds'. He is at once endearing and infuriating, and it says much for Manning's skill that however much we may share Harriet's exasperation with her husband, we equally understand why she puts up with him.

Nicholas MOSLEY　　　1923–

Accident

Mosley's novels, being experimental, intellectual and elliptical ('difficult'), are not as well known as they might be. *Accident* is perhaps the most celebrated because it was made into a particularly fine film (1967), directed by Joseph Losey from a screenplay by Harold Pinter. (The author himself may be glimpsed in a cameo role.)

A short novel of some 150 pages, it opens with the accident of the title, a car-crash involving two pupils of the narrator, Stephen Jervis, a philosophy fellow of St Mark's College, Oxford. Stephen is writing his account of the accident, and the circumstances which led up to it, three years after the event. He discusses the pupil–teacher relationship, in particular the element of flirtation and the burden of responsibility it involves; he wonders what philosophies the new generation will bring to displace those of his own; he regrets his own loss of youth and yearns for a new life. At the time of the crash, Stephen lives in an old farmhouse, Palling Manor, just outside Oxford, with his wife, Rosalind, and their two children. Although Rosalind is pregnant, the marriage is stale. Stephen's greatest friend is a writer called Charlie: they had been students together at Oxford, both scholarship boys. Charlie is married to Laura and has three children; their marriage has reached the stage where love exists alongside boredom and irritation. Stephen's two pupils are William Codrington, the younger son of a family whose house the public pay half-a-crown to visit in the summer, and Anna von Graz und Leoben, who may be Austrian or German, and who is the novel's catalyst. William is in love with Anna and wants Stephen to play Cupid, but Stephen finds himself drawn to the girl, as does Charlie, who intends to leave Laura.

While in London to take part in a television programme, Stephen sleeps with an old girlfriend, but immediately regrets this. He returns to Oxford to discover that his wife has almost miscarried and is with her mother, and that Charlie and Anna have taken advantage of Stephen's empty house to consolidate their affair. Stephen is left with the task of trying to restore order, travelling back and forth between Palling, Charlie and Laura's house, and his mother-in-law's. His exhaustion is mirrored in the novel's style: at the beginning it is mannered and dry, but by the end of the novel it has become broken and hysterical, reflecting Stephen's awareness of the human mess academic game-playing leaves in its wake.

Charlie decides to return to Laura, appreciating like Stephen the value of marriage, safety and the old order. To make Charlie jealous, Anna decides to marry William. It is on their way to announce this to Stephen that the accident occurs, and William is killed. Stephen and Charlie face a moral dilemma: they are responsible both for William's death and for Anna's future. The police do not know who was driving the car: it was in fact Anna, who was drunk and did not have a licence. Do Stephen and Charlie turn her in or protect her? What are their real motives? 'We can help by deceiving: help ourselves by helping others. I had an ear to the ground. No trouble; no publicity.' They choose abstract justice and allow the police to assume that William had been driving.

While Stephen has been dealing with the accident, Rosalind has given birth prematurely and it seems the child will die. For Stephen, the death of William and the life of his child are inextricably linked: if the baby lives, his guilt will be assuaged. The novel ends on a note of optimism. Although William is dead, Anna is safe (returned to Germany, secure in everyone's sympathy for the loss of her fiancé), Charlie and Laura are back together again, and Stephen's child will live. 'The earth has changed. I was sorry, but how could you have life without this ache, this terror. How could you have joy?'

NGUGI Wa Thiong'o   1938–

The River Between

Ngugi has written a variety of works – novels, diaries, plays, essays – mainly in English, though in recent years he has turned to writing in his own language, Kikuyu. His first novel, *Weep Not, Child* (1964), which tells the story of a schoolboy who loses his opportunity for further education when his family is torn apart during the Mau Mau rebellion, was the first novel in English to be published by an East African.

The River Between is Ngugi's second novel, written while he was an undergraduate. It deals with an unhappy love affair in a rural community divided between those who still follow their ancestral religion, and those who have become Christians. Makuyu, the Christian village, is led by Joshua, while Kameno is led by Chege, a descendant of a local prophet. Between these villages flows the river which should unite them because it is a common source of life, but which in fact divides them: the waters of circumcision for Kameno are the waters of baptism for Makuyu.

The hero, Waiyaki, is a young non-Christian schoolteacher, who tries to unite his people through western education. He is a member of the Kiama, a society dedicated to the preservation of Gikuyu religion, but he hopes that his love for Nyambura, one of Joshua's daughters, will help heal the division in the tribe. Waiyaki's rival for Nyambura's hand is Kamau, who ultimately betrays Waiyaki to the Kiama – partly out of spite and partly because his father, Kabonyi, is jealous of Waiyaki's success and influence. The very people who benefited most from Waiyaki's education turn on him because they construe his love for Nyambura as a betrayal.

The themes which preoccupy Ngugi in all his work are: the divisiveness of tribalism; the European contribution to Kenya's social, cultural and political disintegration; the passing of the colonial inheritance to Kenya's elite, with wealthy landowners and predatory political leaders capitalising on the miseries of others and thereby perpetuating economic inequality and social injustice; sympathy with the poor, the weak and the oppressed, which leads to an exposure of the cruelties they suffer; the land, which comes to acquire an almost religious significance in Ngugi's work; the role of education, which he regards as vital but not salvific; the necessity of tempering one's hopes and of shaping one's conduct by living vividly in the present; and, finally, disillusionment – all his major characters, types as much as individuals, have their expectations thwarted.

Simon RAVEN   1927–

Friends in Low Places

see **Alms for Oblivion** (1976)

Frank SARGESON   1903–1982

Memoirs of a Peon

Sargeson's considerable reputation was made in the 1930s with his short stories and novellas, but he had a late flowering in the 1960s, producing autobiographies, plays and novels, amongst which is this, a novel Sargeson himself held in high regard. 'My publishers think my longest book, *Memoirs of a Peon*, is one of the funniest books written anywhere this century. They may be right,' he once said of his comic romp. Narrator Michael Newhouse is a New Zealand farmer's son born at the start of the twentieth century, but his memoirs, written in an ornate and archaic language, clearly draw upon the eighteenth-century confessional memoirs of figures such as Casanova and William Hickey. The chief comic effect is one of deflation, for Michael's exploits rarely match those of his literary models.

After the death of his mother Michael is sent from the farm to the home of his well-to-do grandparents in Wellington, where he receives a classical education from his uncle. The first words we hear the young boy speak are: 'They seem to be three merry young ruffians, don't they, Granny?' After the death of his grandparents and a brief return to his father's home, where his reading is further encouraged by a teacher, he returns as a student teacher to Wellington, where 'I was expected to devote myself to over-large classes of very young children; and I must risk disapproval by honestly saying that without any qualification or exceptions I detested them heartily.' He sets about his career as a social adventurer by cultivating the family of beautiful fellow student Betty Gower-Johnson, and on a summer holiday with the family his amorous advances towards both mother and daughter are thwarted. He consoles himself with the Maori maid and contracts a painful 'unmentionable malady'. He lodges next with a poor laundress's family and seduces her daughter, only to find his next advance rebuffed. Having swindled money out of a thief he takes up with a girl called Moira and, in her absence, seduces a young island girl. 'I was suddenly conscious of myself as a creature of monstrous depravity ... endowed with a carnal appetite which it was virtually useless to satisfy,' he says, though his seductions to date number a fairly paltry four. Moira leaves, and he runs out of money and must take up a post at a remote country school. His inattention to his pupils soon sees him dismissed, and he begins a life on the road before returning on the novel's last page to Betty and a marriage of the 'shotgun variety'; her father conveniently dies having 'provided the wherewithal for the purchase of every amenity and convenience'.

Wole SOYINKA   **1934–**

The Interpreters

In 1986 Soyinka became the first black African to win the Nobel Prize for Literature. Though he regards himself primarily as a playwright, it is almost certainly his fiction rather than his other work (plays, poems, essays, translations, autobiographical works, journalistic pieces) that will be regarded as the test of his reputation by most readers, as well as by posterity. He has published a large body of work, but in all of it his inventiveness of symbol and language place him beyond the reach of a wide public.

The Interpreters at least has the advantage of being in the form of a novel. It is, however, his most complex narrative work. Intricate and seemingly chaotic in structure, it is peopled with tantalisingly emblematic characters. At the centre, like a set of refracting lenses, is a group of six young intellectuals: an engineer, a journalist, an artist, a teacher, a lawyer and an aristocrat. They meet and get drunk fortnightly in clubs in Lagos and Ibadan; between binges, their ordinarily disorganised lives naturally bring them into contact with people from different sections of urban Nigerian society. The book captures, largely through their dialogue, the idealism of the six and their hopes for the development of a new Africa.

The contrast between their apparently idealistic talk and their hedonistic lifestyle appears not to strike them, their author, or indeed most readers and commentators. Beyond those common ideals and lifestyle, little unites the six protagonists. Sometimes together and sometimes apart, in agreement or at odds, they interpret and reinterpret, to themselves and to us, their experiences as well as the world in which they live. We are left to gauge the value of their interpretations for ourselves.

The implications of Soyinka's work are essentially conservative, concerned as he is with personal survival rather than with wider issues. His pessimism seems to deny any possibility of a renewal of Nigerian society – by Muslims, Marxists or Christians. We are left with material sufficiency and personal satisfaction as the only goals. However, it makes for complex and challenging literature from this imaginative and lyrical writer.

Evelyn WAUGH   **1903–1966**

Sword of Honour

see **The Sword of Honour Trilogy** (1961)

P.G. WODEHOUSE   **1881–1975**

Galahad at Blandings

Clarence, 9th Earl of Emsworth, 'vague and dreamy peer . . . not one of England's keenest brains', lives at Blandings Castle, grey-stoned, mid-fifteenth-century with fifty-two bedrooms and staterooms, near Market Blandings in the Vale of Blandings, Shropshire. Richard Usborne, who has made a study of the topography of Blandings, has identified Market Blandings as Buildwas, ten miles from Shrewsbury. As a child, Wodehouse lived for a time in the village of Emsworth on the border of Hampshire and Sussex, in a house called Threepwood, Lord Emsworth's family name. Owen Dudley Edwards has suggested in *P.G. Wodehouse* (1977) that Blandings takes its name from Beatrix Potter's *The Tale of Pigling Bland* (1913). Novels and short stories set at Blandings Castle span sixty-two years of Wodehouse's career, beginning with *Something Fresh* (1915, published in the USA as *Something New*), a novel about a stolen scarab, and ending with the unfinished *Sunset at Blandings* (1977).

Lord Emsworth attended Eton in the 1860s, where he was nicknamed 'Fathead'. He is a member of the Senior Conservative Club in London, extremely wealthy, in excellent health, and he loves flowers. Even more, he loves his prize pumpkin and his prize sow, the Empress of Blandings, thrice winner of the Fat Pig Class at the annual Shropshire Agricultural Show. His favourite book is *On the Care of the Pig* by Augustus Whipple (though called Whiffle until this novel). *Pigs at a Glance* has become his favourite book by *A Pelican at Blandings* (1969).

His son, Freddie Threepwood, was expelled from Eton and sent down from Oxford but goes on to become Vice-President of Donaldson's Dog Joy, an American dog-biscuit manufacturer. Clarence has a younger brother, Galahad, known as Gally, a small, dapper man who wears a monocle. In his youth he was a great sportsman and was once a leading light of the now defunct Pelican Club. Gally and Clarence have four troublesome sisters: Julia, Constance, Dora and Hermione. Hermione is married to Col. Egbert Wedge and their only daughter is the beautiful but dim Veronica. (Sebastian) Beach, Lord Emsworth's imperious butler, is in evidence. 'Ice formed on the butler's upper slopes', was Wodehouse's famous description of Beach's response to having once been addressed with the words: 'Hoy, cocky.'

As is usual in Wodehouse, *Galahad at Blandings* (first published in the USA as *The Brinkmanship of Galahad Threepwood*) is an elaborate web of plot and sub-plot. Tipton Plimsoll, of Tipton's Stores, the largest chain of supermarkets in the USA, is engaged to Veronica. Owing to a misunderstanding on the part of Clarence, it is mistakenly supposed that Tipton has lost all his money. As the story opens, Tipton is sharing a cell overnight in a New York police station with Wilfred Simmons, who turns out to be a cousin of Veronica. Wilfred is in love with Lord Emsworth's pig-girl, Monica Simmons (whom Wilfred describes as 'majestic' and Tipton as looking like an 'all-in-wrestler').

In another sub-plot Clarence's unwanted secretary, Alexandra ('Sandy') Callender, has broken off her engagement to 'Sam' (Samuel Galahad) Bagshott, son of the now deceased 'Boko'. 'Boko' was once Galahad's best friend, so Gally is only too pleased to try and help Sam win Sandy back again.

Meanwhile, Hermione is trying to marry her friend Dame Daphne Littlewood Winkworth to Clarence. Daphne had once been a girlfriend of Clarence a long time ago. She is now the widow of a historian and running a school for large girls where Wilfred is due to become a teacher next year.

Needless to say, after various crises and much confusion everyone gets their just deserts.

1966

France withdraws from NATO command structure ● Cultural Revolution in China ● US bombs Hanoi and Haiphong ● Luna IX lands on the moon ● Theodor Adorno, *Negative Dialectics* ● Jacques Lacan, *Ecrits* ● Joe Orton, *Loot* ● Films: *A Man for All Seasons, A Man and a Woman, Alfie, Georgy Girl* ●Television: *Cathy Come Home*, first episodes of *Batman; Thunderbirds; The Monkees; Star Trek* ● Motown sells most singles ● England wins World Cup ● Walt Disney dies ● Seamus Heaney, *Death of a Naturalist*

John BARTH   **1930–**

Giles Goat-Boy

Barth's fourth novel takes the basic proposition of the campus novel – that the life of a university is a metaphor for life itself – and pushes it to its logical conclusion: the university, not the world, is everything. The story begins in the goat-barns a short distance from New Tammany College where George, the hero of the novel, has been raised as a goat by Max Spielman, a (wrongly) disgraced and exiled professor. *Giles Goat-Boy* traces George's progress from the simple, goatly pleasures of his 'kidship', through his journey to the West Campus of the College and his candidacy for the post of 'Grand Tutor', to his final stoic acceptance of the inherently 'flunked' nature of human experience.

On his way to this insight George becomes embroiled in the internecine world of campus politics. The university has been divided into East and West Campuses, each controlled by a computer (known respectively as EASCAC and WESCAC) whose motives and actions are shrouded in mystery. WESCAC may even be George's father, through a long-abandoned insemination programme known as The GILES. Meanwhile, false Grand Tutors assert their claims over George's, and the West Campus degenerates within while weakening in its opposition to the East Campus. Barth's evocation of a society, with its complex politicking and factional splits, based entirely on the model of a university is impressive and inventive. East and west, 'informationalism' and 'Student-Unionism', 'Enochism' and

'Moishianism' (and many others) are variants on the fundamental division of western liberalism and socialism. George's negotiations through these opposing viewpoints are the basis of his journey towards self-knowledge.

At the centre of the book is 'The Tragedy of Taliped Decanus', a play peformed on the West Campus, which is Barth's brilliant pastiche of *Oedipus Rex* as it might have occurred on campus rather than in Thebes. The tragic stoicism of Taliped is the philosophical core of the novel, towards which George moves and beside which the mundane disputes of the university pale into insignificance. However, the presentation of the play is far from serious and is made ironic by the generally farcical tone of the labyrinthine plot of which it forms a part.

The question of how seriously to take the issues raised by *Giles Goat-Boy* is a vexed one. The narrative is told exclusively from George's point of view and is conveyed in goatish prose which tends to divorce its subjects from context – everything becomes a curiosity. Furthermore, the tragic (or just serious) side of life is almost always presented comically; rapes, murders, blindings and lynchings are recounted as mock-heroic episodes. The problem of whose view is the legitimate one is complicated by a bewildering array of editor's and author's prefaces, a 'Posttape', 'Postscript to the Posttape' and even a 'Footnote to Postscript to the Posttape'. All these complicate rather than explicate the text of the novel and variously nominate George, Anastasia Stoker, Anastasia and George's son and even WESCAC as the 'real' authority behind the work.

Whether this determined confusion represents a crisis of confidence on Barth's part (a refusal to commit himself to any one view) or a frank acknowledgement that life, on or off campus, is at root irresolute, is the dilemma upon which the worth of the book depends. Barth, a university professor himself, has been accused of writing novels whose relevance is only to themselves; the relentless irony, pervasive comedy and refusal to look beyond the campus wall of *Giles Goat-Boy* seem to bear this out. Yet within its original context, America in the late 1960s, when students clashed with the national guard and used the campus as a political, even literal, battleground, Barth's novel becomes more central to the concerns of its time. Although it has dated surprisingly quickly in this respect, *Giles Goat-Boy* remains an inventive, erudite and frequently hilarious work, which, without a single contemporary reference, remains strangely evocative of its time.

Maureen DUFFY   1933–

The Microcosm

The Microcosm began life as a non-fiction project, documenting the London lesbian scene of the 1960s through taped interviews. Turned down by various publishers at the time as too controversial, the material was then developed into a novel as singular in style as in subject-matter. Consciously harking back to Joyce and Woolf, Duffy exploited stream-of-consciousness, allusion and pastiche, to create an 'energised realism', in which 'concrete images function like metaphor in poetry', and which would be imaginatively spacious enough not only for 'geographical spread but also historical reconstruction.'

In place of a single narrative thread, the novel loosely assembles the stories of a group of women, most of whom gravitate towards a gay bar in London known popularly as the 'House of Shades'. This human microcosm provides both a refuge from an alternative to the 'straight' world outside.

The central figure is Matt, a 'butch' who acts as spokesperson and agony aunt for the other clubgoers, and yet who is full of self-doubt. 'His' partner, Rae, in contrast, is fully content with her job and home life. Rae is a museum artist, whereas Matt, though a trained archaeologist, works as a petrol pump attendant.

Other clubgoers and characters include Steve, a school games teacher, who is attracted to the visiting French teacher, but is hesitant about revealing her feelings in their professional environment; Marie, an unhappy housewife, who is unable to come to terms with her feelings for women, and is driven to attempt suicide; Cathy, a younger woman who succeeds in discovering herself, and leaves the north of England for a new life in the capital; Sadie and Jonnie, a couple of blue-collar workers undergoing the age-old teething problems of 'married' life; and, finally, Judy, the night club's *femme fatale*.

Characters from other marginal worlds are also woven into the realistic tapestry: the historical Charlotte Charke, a strolling actress and writer who adventured her way across eighteenth-century England dressed in masculine clothing; an aboriginal tribe, struggling for survival in the face of colonialist expansion; Boadicea warring against the Romans; and, briefly, Florence Nightingale, in love with a younger female nurse.

The novel is framed by Matt's story, and, at the end, after a troubled visit to an old lover of Rae's, Matt decides to depart the microcosmic world of the House of Shades, and so takes an archaeological job abroad.

The stylistic originality of *The Microcosm* consists in a subtly controlled alternation between stream-of-consciousness and more linear types of narrative allowing the reader to see the characters simultaneously both from the outside and the inside. This double vision succeeds in evoking the dilemmas of a minority group torn between conflicting needs, between the need to assert a separate identity, and the need to adapt to the wider demands of society.

John FOWLES   1926–

The Magus

Fowles has described *The Magus* as a first novel in every respect except date of first publication, since he began to write it in the early 1950s, and assembled numerous drafts for the first edition of 1966. It was revised again for the 1977 edition. The story is a modernised, wordly-wise version of *The Tempest*, set on the imaginary Greek island of Phraxos, and bathed in an atmosphere of mystery, eroticism and paranoia.

Its hero, Nicholas Urfe, takes up a teaching job in the island's school (as Fowles himself did on the island of Spetsai), largely to escape the last throes of a messy love-affair. Phraxos's Prospero is Maurice Conchis, an eccentric millionaire who begins to entertain Nicholas in lavish style at his villa. Nothing Conchis says can be trusted, however, and the visits turn into a mental sparring-match, in which Nicholas punctures one layer of lies after another, to reveal yet more untruths about Conchis's fantasy-world hiding beneath.

The main incentive to continue the charade is the presence of the island's Miranda, a beautiful woman introduced as Lily Montgomery. Nicholas is first asked to believe that Lily is a ghost, a psychic trace from Conchis's past. Then she becomes a psychiatric patient whose real name is Julie Holmes, and Conchis her doctor. Finally, she mutates into an actress, who, along with a twin sister called June, has been hired to act in a Stanislavskian 'meta-theatre', a real-life drama without limits staged by Conchis as an experiment in human psychology. Nicholas is asked to act out a *ménage à trois* with the two sisters.

At this point, Nicholas learns that Alison, an old girlfriend in England, whom he has met again briefly in Athens, has committed suicide, after his refusal to renew their relationship. Despite the shock and guilt, he carried on with the masquerade on Phraxos. He finally persuades Julie to drop her staged persona, and coaxes her into bed. But this real Julie is no more reliable than the previous ones: Nicholas is kidnapped, put on trial by Conchis and the other actors, and declared psychologically, morally and sexually inadequate. As a 'cure' he is forced to watch Julie make love to another man, the novel's Caliban, who is a black bodyguard of Conchis, and has probably been Julie's lover all along.

Nicholas returns to England, humiliated, and depressed about Alison's death. Events on the island seem to have acted as a lesson, teaching him the value of the woman he has rejected and now lost for ever. Months later, however, he tracks down the mother of Julie and June. She reveals that Alison is not actually dead at all, but had been persuaded to fake her suicide as part of Conchis's charade. Finally, Nicholas meets Alison herself, and, in a reversal of roles, asks her to return to him.

In the novel's second edition, some of the original moments of dramatic incompletion – especially in the erotic scenes and the ending – were redrafted, to render Fowles's festival of ambiguity a fraction less ambiguous.

William GASS   1924–

Omensetter's Luck

The American Midwest of the 1880s is the setting for Gass's first novel. Brackett Omensetter drives into the town of Gilean, on the banks of the Ohio river, in his wagon with a pregnant wife and two daughters. There he sets up home and waits for the birth of his son. He rents a house from Henry Pimber, who later disappears and hangs himself. His son is born, almost dies of diphtheria, and Omensetter is nearly blamed for Pimber's death. This is the substance of the story from which grows the local legend of Omensetter's luck. Gass narrates this from three points of view, each of them acute psychological portraits of the teller, as well as different versions of the tale.

The first (and shortest) version is 'The Triumph of Irabestis Tott', which Tott relates years after the events at the auction of Pimber's deceased widow's belongings. Each item prompts a memory, a story that survives its selling-off. Tott's triumph is to remember: 'He walked through the whole of his storied past, greeting everyone.' The second section, 'The Love and Sorrow of Henry Pimber', tells how Omensetter first arrived, found a house and work, and how his wife bore their son, all through the alienated eyes of Henry Pimber. A wasting illness has drained Pimber of the capacity for feeling; he sees in Omensetter's natural ease and harmony with his surroundings an unbearable image of what he has lost. The third, final and by far the longest section is 'The Reverend Jethro Furber's Change of Heart'. Embroiled in an excruciating psychological stew of repressed desires, lusts and hates, Gilean's preacher covets and loathes Omensetter's easy acceptance of life. His attempts to poison popular opinion against Omensetter almost result in the latter's indictment for Pimber's 'murder', but at the last moment Furber desists, accepting rather than damning Omensetter, learning the secret of Omensetter's luck, which is simply this acceptance.

Between these three views lies Brackett Omensetter and beyond him the American Midwest of which he is something close to an embodiment. Omensetter becomes an almost mythical figure in his obsessive telling and retelling by the other characters, but also someone radically at odds with being reduced to a neat story (it remains possible that he *did* kill Pimber at the end of the novel). Gass's subsequent work has been more and more concerned with these theoretical questions, such as how 'tellable' a story really is, or how much of a person can actually be communicated in words, but in *Omensetter's Luck* these debates remain a shadowy background of ideas to the real action. To the fore of the story is the rude particularity of the region and its inhabitants. Gass exploits the trick, probably learnt from Faulkner, of turning provincial eccentricity into intense, credible psychology and in the Revd Furber creates a labyrinthine personality of Gothic proportions and complexity. *Omensetter's Luck* is successful as a meditation on the nature of fiction, a character-study and even as a whodunit, but its brilliance lies in its depiction of the American Midwest, which Gass anatomises as accurately as William Faulkner and Flannery O'Connor have done the Deep South.

Thomas PYNCHON   1937–

The Crying of Lot 49

Pynchon's second novel seems almost insubstantial sandwiched between the consider-

ably larger volumes of *V* (1963) and *Gravity's Rainbow* (1973), yet this interlude picks up many of the themes which are initiated in his first novel and developed in his third. Its relative brevity apart, a major difference between *The Crying of Lot 49* and Pynchon's other works is its location. The novel is set on the west coast of America, which Pynchon sees as in flight from reality and thus highly susceptible to the devices of fiction. The plot, which is impossibly complex, concerns Mrs Oedipa Maas, who is named as the executor of the estate of Pierce Inverarity, her one-time lover. Pierce was a major stockholder in Yoyodyne Inc. (a company which appeared as a malevolent presence in the background in *V*), an avid philatelist and a sponsor of innumerable businesses and properties around San Narciso, where Oedipa travels to unravel his affairs. In the course of her (largely unsuccessful) efforts she stumbles across an underground postal network called the Trystero, which exists in defiance of the government monopoly, issuing its own stamps and linking together various groups of social, political and sexual deviants.

At first it seems that the Trystero is an unlicensed extension of the Yoyodyne internal post, but later its traces are seen in a group of counterfeit Indians who challenged the Wells Fargo Monopoly in the 1850s, a faction who aided the Confederates in the Civil War (incidentally initiating the first Russo-American conflict in the process), and a shadowy organisation which ran a subversive postal network in seventeenth-century Europe. Oedipa goes to see an Elizabethan play called 'The Courier's Tragedy' (a brilliant pastiche by Pynchon) in which there seems to be a reference to the Trystero, but the book in which the play was printed has used a corrupt text and the shop in which the book was bought is burned to the ground, and then Oedipa discovers that it was leased from Pierce Inverarity . . . There is a possibility that the Trystero, intimately entwined with Pierce's affairs, is nothing more than a creation of his imagination, fabricated for his posthumous amusement. A major clue rests in Pierce's stamp collection which includes stamps issued by the Trystero. Oedipa attends the auction when they are sold in the belief that the identity of their buyer will reveal more of the Trystero's secrets. The novel ends just before the lot, lot 49, is auctioned.

Yet at least he had believed in the cars. Maybe to excess: how could he not, seeing people poorer than him come in, Negro, Mexican, cracker, a parade seven days a week, bringing the most godawful of trade-ins: motorized, metal extensions of themselves, of their families and what their whole lives must be like, out there so naked for anybody, a stranger like himself, to look at, frame cockeyed, rusty underneath, fender repainted in a shade just of enough to depress the value, if not Mucho himself, inside smelling hopelessly of children, supermarket booze, two, sometimes three generations of cigarette smokers, or only of dust – and when the cars were swept out you had to look at the actual residue of these lives.

THE CRYING OF LOT 49
BY THOMAS PYNCHON

This summary of the plot excludes many sub-plots and related incidents. The labyrinth of clues, hints and suppositions which Oedipa negotiates exceeds comprehension in its complexity. The wealth of invention is overwhelming. Bizarrely credible characters such as Oedipa's acid-peddling, ex-Nazi psychotherapist, Dr Hilarius, Genghis Cohen, the stamp expert, or Metzger, her paedophiliac lover and co-executor, all populate a society in which paranoia is the only justifiable state of mind. Pynchon's trick of siting the action on the edge of absurdity without letting it fall off is carefully performed. The immaculate pastiche of 'The Courier's Tragedy' and the worked-up 'history' of the Trystero together with 'substantial' scientific elements ranging from Maxwell's Demon (a kind of perpetual motion machine) to the manufacture of ink from human bones, all lend the outrageous interconnectedness of the plot an important credibility. In one sense *The Crying of Lot 49* is the last will and testament of Pierce Inverarity, and Oedipa's quest for its mean-

ing the disposal of its one real asset, information. The way in which she deals with the flood of different kinds of information is a metaphor for dealing with an increasingly incredible reality, and it is the insight which converts the novel from a *roman à clef* to a trenchant and exuberant account of the nature of life in America. As Oedipa herself puts it: 'She had dedicated herself, weeks ago, to making sense of what Inverarity had left behind, never suspecting that the legacy was America.'

Anthony POWELL   1905–

The Soldier's Art

see **A Dance to the Music of Time** (1975)

Simon RAVEN   1927–

The Sabre Squadron

see **Alms for Oblivion** (1976)

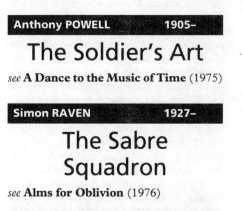

Jean RHYS   1894–1979

Wide Sargasso Sea

Wide Sargasso Sea is set in Jamaica and Dominica during the 1830s. Rhys, who was half Creole, was born in Dominica and spent her childhood there. She supplemented her personal knowledge of the West Indies with a reading of their history in order to explore in fiction the life of one of the mad Creole heiresses of the early nineteenth century. The novel's central character, Antoinette Cosway, is inspired by the figure of the first Mrs Rochester, the mad wife in Charlotte Brontë's *Jane Eyre* (1847), who had haunted Rhys for many years. The style is very different from that of Brontë's classic, however: Rhys's dream-like prose is a distinctive mixture of quivering immediacy and glassy objectivity. *Wide Sargasso Sea* is also very different from Rhys's other work, drawing as it does on memories of her youth and not on her gloomy experiences of disastrous affairs in later life.

The novel is divided into three parts. The first, told in the heroine's own voice, presents Antoinette's childhood. She and her family are the products of an inbred, decadent, expatriate community, resented by the recently freed slaves (whose superstitions they share), and despised as 'white niggers' by the English. They live in the heart of the lush Jamaican landscape at Coulibri. Antoinette's mother – a penniless widow who is beautiful, and terrifying, and entirely self-absorbed – is courted by the rich and generous-hearted Mr Mason, whom she marries, to the surprise and horror of white society. But her good fortune is short-lived. Coulibri is attacked and burned down by a gang of blacks whose resentment is fired by the family's new wealth. The blaze kills Antoinette's idiot brother; she herself is very ill; and her mother grows violently mad, then dies.

The middle section of the novel is told largely from the viewpoint of young Mr Rochester, who is considered hard, stupid, self-centred and greedy. Mr Mason has died, leaving half his wealth to Antoinette, whom he adored. His son, Richard, has sold Antoinette off in a marriage that gives her no legal claim on her dowry. The marriage had taken place only a month after Rochester's arrival in Jamaica, and for nearly three weeks of that he had been in bed with a fever – an inauspicious beginning. They are each uncomfortable about the other's country, and the clash between them is indicative of the wider conflicts inherent in nineteenth-century colonialism.

At first Rochester is sexually enchanted by Antoinette, but the happiness he gives her is fragile, for his mind is soon poisoned against her by tales of madness in the Cosway family. In desperation Antoinette begs her loyal servant, Christophine, to use magic to make Rochester love her. Christophine is reluctant, and her premonitions are borne out. As a result of her white powder, Rochester does become passionate but exercises it on one of the black maids, and when he discovers that his wife had turned to *obeah* – which is part of the culture he has come to find extremely threatening – he begins to hate her. Antoinette's sense of isolation is complete. She wonders 'who I am and where

is my country and where do I belong and why was I ever born at all'. She makes herself drunk with rum and hatred, and before leaving for England, Rochester colludes with Richard and the doctors to certify her and keep her money.

The last part, by far the shortest in the book, is told in Antoinette's voice. Her mind is now locked in a state of unreality, and she is prisoner in the attic of Thornfield Hall, with Grace Poole her gaoler. One night she dreams that by setting fire to Thornfield Hall she can return to Coulibri. The novel ends as she wakes and leaves the room with a candle, moving inevitably towards the final conflagration.

Rhys's literary career was renewed with the publication of *Wide Sargasso Sea* after twenty years' silence. The book won the Royal Society of Literature Award and the W.H. Smith Award – her only comment on the latter being: 'It has come too late.'

Paul SCOTT   **1920–1978**

The Jewel in the Crown

see **The Raj Quartet** (1975)

Christina STEAD   **1902–1983**

Cotter's England

First published in America as *Dark Places of the Heart*, a title Stead's publisher believed would appeal to that market, the book was published in Britain under Stead's preferred title in 1967. It is a naturalistic story of vanity, greed and betrayal, set within the post-war Labour movement in England. Stead appears to have used real individuals as research material: in particular, the heroine and her family are partly modelled on a Communist acquaintance named Anne Dooley (*née* Kelly) and her paternal family. In 1949 and 1950 Stead visited the Kellys, who lived near Newcastle, with the project of writing a novel about the English working class consciously in mind. *Cotter's England* was drafted shortly afterwards, in the early 1950s.

Nellie Cook (*née* Cotter) is a journalist on a left-wing London paper. An incurable talker, she shovels out the rhetoric in conversation as well as in print in order to fuel a personality cult. Her current project is breaking in a new disciple called Caroline, a young woman from the north of England who has moved to London in order to recover from the failure of her marriage. Nellie's younger brother, Tom, reappears after an interval of some years, during which he has been nursing a dying lover. Nellie classifies Tom as one of her lapsed followers, but he quickly asserts his independence and enrages his sister by encroaching upon her female friendships. This sibling rivalry reaches crisis-point when Tom speaks to Nellie of marrying Caroline. Nellie counter-attacks by attempting to seduce Caroline, both physically and spiritually, while Caroline is ill and feverish in bed. The success of the physical seduction is in doubt, but when Caroline commits suicide, Nellie considers this a sign that she had gained possession of Caroline's soul. The real reason for Caroline's death is more to do with the intolerable pressures she has suffered, caught in the crossfire between brother and sister.

An almost equally vampiric tale is unfolding in the Cotter family in the northern town of Bridgehead. When the alcoholic 'Pop' Cotter dies, he leaves behind his prematurely aged wife, Mary, drifting into senility; his disturbed younger daughter, Peggy, who may be the victim of sexual assaults by other members of the family (the suspicion of incest still pervades the entire household); and his embittered, spinsterish brother-in-law, Simon, who supports the entire family on his meagre pension. When 'Ma' Cotter dies, Peggy immediately steals Uncle Simon's savings and throws him out of the house. Tom and Nellie protest, but not forcefully enough to prevent Simon being sent to an old people's home, where he is left to die in penury. Tom is too busy getting married to help out, and Nellie is preparing to leave England. Her union activist husband, George, has moved to Geneva, to take up a comfortable job as a Euro-socialist bureaucrat, and despite her loud protestations that he has betrayed the English working class, Nellie goes after him when George finds her a job there as well.

However, the new life on the Continent is nipped in the bud by George's death in a skiing accident, and Nellie returns to England. Listlessly, she drifts towards a new circle interested in mysticism, in questions 'that the professors and scientists could not solve'.

Stead's vision of contradictory forces, of love turning into sadism, heroic vision into image-manipulation, stems partly from her Marxist politics, but her writing skilfully portrays these dialectics as they operate within individuals and individual families.

1967

Six-Day War ● Civil war in Nigeria ● Racial violence in US cities ● Che Guevara executed ● Demonstrations against Vietnam War ● First heart transplant operation ● Sexual Offences Act decriminalises homosexual acts between consenting adult males in Britain ● Introduction to mammography ● Desmond Morris, *The Naked Ape* ● Jacques Derrida, *Of Grammatology* ● Andy Warhol, *Marilyn Monroe* ● Anthony Caro, *Prairie* ● Karlheinz Stockhausen, *Hymnen* ● Harold Pinter, *The Homecoming* ● Films: *The Dirty Dozen*, *Bonnie and Clyde* ● Television: *The Forsyte Saga*, *The Prisoner* ● The Beatles, *Sergeant Pepper's Lonely Hearts Club Band* ● The Doors, *The Doors* ● The Jimi Hendrix Experience, *Are You Experienced?* ● John Masefield, Joe Orton, Clement Attlee die ● Adrian Henri, Roger McGough, Brian Patten, *The Mersey Sound*; *The Liverpool Scene*

Angela CARTER   1940–1992

The Magic Toyshop

Several of the themes which were to pre-occupy Carter are already apparent in this, her second novel, which won the John Llewellyn Rhys Prize, and was later adapted by her for a film (1986), directed by David Wheatley. The theatrical element was to resurface in both **Nights at the Circus** (1984) and *Wise Children* (1991), while fairy-tales were to be a constant source of reference, particularly in her post-Freudian collection, *The Bloody Chamber* (1979). 'Fairy stories talk about life and death as though they matter, as though they are important,' she said, and she often used fantasy, the Gothic and the grotesque, as she does here, as metaphors for more mundane human experience, in particular growing up.

Fifteen-year-old Melanie's parents are away for the summer and she and her younger siblings are being looked after by a nanny. One night she dons her mother's wedding dress and walks in the garden, only to discover she has locked herself out and must climb the apple tree to get through her bedroom window. The dress is ripped to shreds, and when a telegram arrives in the morning she is immediately aware of its contents: her parents have been killed in an air-crash. Up to this point, Melanie has led a day-to-day, linear kind of existence, engrossed in herself and her experience of change: 'Since she was thirteen, when her periods began, she had felt that she was pregnant with herself, with the slowly ripening embryo of Melanie-grown-up inside herself for a gestation time the length of which she was not precisely aware.' The radical change which follows the accident throws Melanie into a grotesque re-enactment of family life. The orphaned children are dispatched to South London to live with their mother's brother, Philip Flower, a toy-maker. Melanie's new family consists of the taciturn, Bluebeard-like Uncle Philip; his terrifying and terrifyingly attractive deaf-mute Irish wife, Margaret; and Margaret's brothers, Finn and Francie. Philip does not so much as acknowledge the existence of the children as he presides regally at the dinner table, and Melanie soon learns of the tyrannical regime he operates in the large house

---

She felt lonely and chilled, walking along the long, brown passages, past secret doors, shut tight. Bluebeard's castle. Melanie felt a shudder of dread as she went by every door, in case it opened and something, some clockwork horror rolling hugely on small wheels, some terrifying joke or hideous novelty, emerged to put her courage to the test. And now she was entirely alone [...]

THE MAGIC TOYSHOP
BY ANGELA CARTER

---

above the toyshop, and of his ruling obsession – the elaborate, life-sized puppet show solemnly performed in the basement to the captive audience of his family.

She is drawn into the 'charmed circle' of Margaret and her brothers, and even comes to welcome smelly Finn's romantic overtures (if not his habit of watching her undress through a peep-hole). Finn's hostility to the tyrant grows when Philip recruits Melanie to play the part of Leda in his new production of *Leda and the Swan* ('You're well built for fifteen,' says Philip. 'I wanted my Leda to be a little girl. Your tits are too big'). After the show Finn gets drunk and destroys the swan in Philip's absence. An anarchic carnival spirit breaks out, and Philip returns to discover his wife and Francie in incestuous union on the parlour floor; in a fit of rage he burns down the house and Melanie and Finn escape over the roof-tops, liberated from his tyranny.

The Magic Toyshop follows the stylised pattern peculiar to fairy stories, tracing the heroine's attempts to overcome a series of obstacles and eventually triumph. The villain is destroyed and hero and heroine are united. The trigger for the young protagonists' self-awareness, however, is a family drama involving the discovery of something which could not exist outside the family, and yet radically challenges its whole concept: incest.

Margaret DRABBLE   1939–

Jerusalem the Golden

Drabble has been seen, not always flatteringly, as a chronicler of late twentieth-century, middle-class Britain, using her fiction to investigate contemporary social issues. Her earlier novels were more autobiographical and dealt with the need to escape the limitations of provincial life. *Jerusalem the Golden*, her fourth novel, draws upon her mother's background as the first member of a Methodist, working-class family to go to university, and is also strongly influenced by the novels of Arnold Bennett, whose biography Drabble was to write. It is an account of the sentimental education received by Clara

Maugham in London, the Jerusalem of the title, and more particularly from the literary, artistic, close-knit Denham family. Clara comes from the North, the child of a loveless marriage. Disparaged and thwarted by her mother, she nevertheless manages to win a place at London University. More, she manages to retain the idealistic expectancy of a better, richer, more exciting, 'thicker' life. She finds this when she meets the artist Clelia Denham, daughter of a poet, and is drawn into the family. Clara finds the brew almost too rich, for the family has everything she has looked for: intelligence, talent, sophistication, ease with the world, above all, love. A momentary doubt that it is perhaps material well-being that is at the root of the Denham self-assurance is silenced: 'there had been love and at every stage'.

After a while Clara becomes the mistress of Clelia's brother, Gabriel, who is unhappily married to a girl clearly in need of psychiatric help. On a trip to Paris Clara attempts to become a liberated sophisticated personality, but this collapses into a sordid farce and brings her own affair with Gabriel to an end. For Clara is a mediocrity and her greatest attraction lies in the absence of any characteristics.

On returning from Paris she finds that her mother is ill with cancer. Her hasty, guilt-ridden, fearful trip back to Northam teaches her two things. She finds her mother's old notebooks, which reveal that long before her marriage Mrs Maugham had attempted to write, but her talents had withered and her nature become embittered through lack of encouragement. Clara also overhears two elderly ladies discussing their grandchildren with pride and affection. 'Clara, overhearing them, suddenly wondered if her whole vision of Northam might not after all have been a nightmare, and that the whole city might have been filled with warm preoccupations, a whole kind city shut to her alone, distorted in her eyes alone.'

The first few paragraphs of *Jerusalem the Golden* imply that Clara achieves a wondrous, enviable social position. At the end she asserts her determination to overcome her family, her home town, her mother's life and death, to use her relationship with the Denhams for the creation of 'a bright and peopled world, thick with starry inhabitants, where there was no ending, no parting, but an eternal vast incessant rearrangement'.

Alan GARNER   1934–

The Owl Service

Garner's best-known novel is a subtle and oblique book, purportedly written for children but complex and ambiguous enough to satisfy an adult audience; it won both the Carnegie Medal and the *Guardian* Award for Children's Fiction. As is Garner's other books, the dividing line between the past and the present, between myth and reality becomes blurred when three children are caught up in an ancient myth.

When Alison is left a large house in a Welsh valley by her father, she inherits more than a holiday home, as she discovers when staying there with her mother, her new stepfather, Clive, and his son, Roger. The first intimation of the valley's strange atmosphere is a sound of scratching in the loft above Alison's bedroom. She asks Gwyn, the housekeeper's clever grammar-school son, to investigate, and he finds a large dinner service. The stylised floral pattern of the plates reminds Alison of owls, and their discovery has a disturbing effect upon the household. Gwyn's cantankerous mother, Nancy, forbids the children to have anything to do with the service, but Alison has already traced the pattern to create paper owls, and released the powers of an old myth.

This is the old Welsh story from the *Mabinogion* of Lleu Llaw Gyffes, who was given a wife made of flowers by a wizard. She fell in love with Gronw Pebyr, who decided to kill Lleu by hurling a spear at him. Lleu appeared to die, but was transformed into an eagle, then back into a man by the wizard. He got his revenge by throwing a spear back at Gronw, who stood behind a rock, but was killed when the spear passed through it. The unfaithful wife was turned into an owl by the wizard in accordance with her name, Blodeuwedd. The pierced Stone of Gronw has already been discovered by the children and they also uncover a wall-painting of Blodeuwedd in the billiard-room. Huw

Halfbacon, the mysterious gardener whose lapidary mutterings suggest lunacy, is the only person who understands that the myth repeats itself down the centuries. He tells Gwyn that he is his heir, 'lord in blood' to the valley, and that the three children are fulfilling the triangular pattern of the myth, because Alison saw owls rather than flowers in the dinner-service decoration. Nancy attempts to escape, but is prevented from leaving the valley by the power. The novel ends in violence as the powers, flowers and feathers, battle for supremacy in Alison during a terrific storm. Gwyn thinks

---

There were no clouds, and the sky was drained white towards the sun. The air throbbed, flashed like blue lightning, sometimes dark, sometimes pale, and the pulse of the throbbing grew, and now the shades followed one another so quickly that Gwyn could see no more than a trembling which became a play of light on the sheen of a wing, but when he looked about him he felt that the trees and the rocks had never held such depth, and the line of the mountain made his heart shake.

THE OWL SERVICE
BY ALAN GARNER

---

that he has been mocked by Alison and cannot help her, but Roger succeeds in persuading her that it is flowers not owls she has seen, thus breaking the pattern.

Flann O'BRIEN   1911–1966

The Third Policeman

The novel many consider to be O'Brien's masterpiece was completed in 1940, but was turned down by his publishers, who suggested that he should write something rather more straightforward than the curious manuscript, three-quarters of which, a friend recalled, was footnotes. O'Brien himself thought the plot was the best thing about the book, but this tale of murder is encrusted with a critical apparatus relating to the theories of an imaginary philosopher called de Selby. Shocked by the novel's rejection, O'Brien consigned it to a drawer, pretending that it had been mislaid in a pub or on a country drive, and it was not discovered until after his death.

In the preliminary section of the novel, the narrator plans and commits the murder of a wealthy cattle dealer called Mathers. The motive is robbery, for the narrator needs money in order to buy all the available commentaries upon the works of de Selby, a philosopher he discovered at the age of sixteen, when he stole a copy of *Golden Hours* from the science master's study. The idea of murder is put into his head by his friend and accomplice John Divney. Having battered the old man to death with a spade, the narrator goes to Mathers's house, searching for a money-box containing some £10,000 in negotiable securities. He encounters his victim's ghost, who directs him to a police barracks. After spending the night in the house, the narrator sets off on his journey, conversing the while with an inner voice (perhaps his soul?) whom he dubs 'Joe'. He eventually comes upon an almost illusory police station. From this point on, the dreamlike plot develops in a perfect circle, gradually returning to the approach to the barracks. The reason for this becomes apparent at the end of the book: the narrator is dead, 'and all the queer ghastly things which have been happening to him are happening in a sort of hell which he earned for the killing', as O'Brien explained. 'Hell goes round and round. In shape it is circular and by nature it is interminable, repetitive and very nearly unbearable.'

'When you are writing about the world of the dead – and the damned – where none of the rules and laws (not even the law of gravity) holds good,' O'Brien remarked, 'there is any amount of scope for back-chat and funny cracks.' His surreal and comic novel satirises local police forces, contains a great deal about bicycles, and makes constant reference to the philosophy of de Selby, who seems to have an observation for every

occasion. The narrator eventually returns to the house where he used to live and meets John Divney, who dies of fright: although the narrator thinks he has been gone three days, he has in fact been dead for sixteen years, killed by a bomb Divney had planted in Mathers's house. He sets out on his journey once again, this time accompanied by Divney. The same macabre tale is set to begin all over again.

Simon RAVEN   1927–

Fielding Gray

see **Alms for Oblivion** (1976)

Dorothy RICHARDSON   1873–1957

March Moonlight

see **Pilgrimage** (1938)

Angus WILSON   1913–1991

No Laughing Matter

Considered by many people to be Wilson's great achievement, *No Laughing Matter* is at once a hilarious, incisive critique of English cultural life through the twentieth century and the narrative of a quest for authenticity. It loosely follows the format of a family saga through four generations, using the shifting perceptions of each to expose the illusions of its predecessors. Wilson described it as 'a kind of anti-*Constant Nymph* [1924] and anti-*Forsyte Saga* [1929]', and the parodies which Wilson had used (unnoticed by most critics) in earlier novels, such as *Late Call* (1964), were made more explicit in this one, with sections of the novel presented as charades in the form of plays by contemporary dramatists: Maugham, Shaw, Chekhov, Rattigan and Beckett. There are also references to Shakespearean plays and the cinema. These techniques were intended 'as a means of alienating the reader, of ensuring that he does not become over fond of characters, of preventing him from regarding the book as a "good read" and from being unable to see where he's going because he's too absorbed in it'. That acknowledged, the book, like all Wilson's novels, is immensely readable. A further link with Wilson's other novels is his preoccupation with evil, in this case embodied in political and social creeds (Nazism, Stalinism, anti-Semitism, homosexual persecution) rather than in individuals.

William ('Billy Pop') Matthews is a complacent, idle Edwardian man of letters, and his wife, Clara ('the Countess'), is an acerbic, witty woman who lives in a style to which her husband is unfortunately incapable of ever becoming accustomed. Their children determine to transcend the hypocritical bohemianism and liberalism of their parents, but eventually turn out to be new, if more successful, versions of them.

The eldest, Quentin, becomes a maverick left-wing journalist, who falls out with most of the British left of the 1930s over his early warnings about the totalitarianism of the Soviet regime. Rupert becomes a theatrical star, after a first success playing a Chekhov character on the model of his own father's genteel ineffectuality. Margaret writes satirical stories about a loveless bourgeois English family called the Carmichaels. Suky simply aims to provide her own family with the love and stability which was absent from her own childhood, while Gladys founds a pioneering employment agency for women. Marcus, the youngest, stranded at home, becomes a spare time rent-boy, before moving in with a male lover, and then making a fortune selling contemporary art.

The idealism of the new generation is severely tested as time wears on. Quentin gradually becomes a television personality, adored by the public for his savaging of politicians of every ideological colouring. Suky's reserves of maternal love are dangerously over-stretched by the advent of the Second World War, and she invests all her affections in her youngest son, who is not old enough to be removed from her by the fighting. Ironically, the two elder boys survive the war, while the youngest is killed shortly afterwards during military service in Palestine. Rupert trades on his initial success in serious theatre to launch a lucrative career making trashy movies. Gladys, who had suffered both sexual abuse and financial leeching from Billy Pop, falls for a man who is almost equally exploitative, and who lands her in jail for a fraud from which he benefits. Margaret's novels gradually mutate from direct, acid satire, towards inaccessible 'writers' writing'.

Marcus follows an opposite trajectory to that of his brothers and sisters, from apolitical dilettantism, towards increasing radicalism. He opens a perfume factory in Morocco, which he runs on co-operative lines. Suky's grandson Adam visits him with friends, after visiting Suky at her villa in Portugal, and the youngsters decide that Marcus is easily their favourite amongst that generation. What they like about him, strangely, is both his sincerity and the bitchy sense of humour he has inherited from the Countess.

No Laughing Matter is in fact full of humour, ranging from literary parody and delicate irony to satire and crude farce, yet paradoxically its quest for personal, artistic and political authenticity is a theme of deadly seriousness.

1968

Assassination of Martin Luther King ● 'May Events' in Paris ● Soviet invasion of Czechoslovakia ● Manned mission to the moon ● Pope Paul VI issues encyclical *Humanae Vitae* ● Abortion becomes lawful in Britain ● Michel Foucault, *The Archaeology of Knowledge* ● Jürgen Habermas, *Knowledge and Human Interests* ● Richard Hamilton, *Swinging London* ● John Hancock Building, Chicago ● John Tavener, *The Whale* ● End of theatre censorship in Britain ● *Cabaret, Hair* ● Films: *If . . ., 2001, A Space Odyssey, The Good, The Bad and The Ugly, The Graduate* ● Robert Kennedy, Tony Hancock, Upton Sinclair, Enid Blyton, John Steinbeck die ● Allen Ginsberg, *Planet News; Airplane Dreams; Ankor Wat*

Anthony BURGESS　　　　　**1917–1993**

Enderby Outside

see **The Enderby Novels** (1974)

Frederick EXLEY　　　　　**1929–**

A Fan's Notes

Subtitled 'A Fictional Memoir', Exley's first novel was hailed on its appearance as a minor American classic; it won the William Faulkner Award and the National Institute of Arts and Letters Rosenthal Award, and was nominated for a National Book Award. It is the story of a man almost destroyed by his craving for fame and by his inability to accept what he portrays as the overwhelming crassness of American society. Exley has since built on it as the first book in an autobiographical trilogy.

The opening shows Exley as an alcoholic English lecturer with an almost religious devotion to the New York Giants football team and particularly to running-back Frank Gifford, whose career he follows vicariously. He is drinking in a bar and waiting for the football game on television when he suffers a painful seizure which he thinks is a heart-attack. In the following chapters he recalls the life that has led to this condition. He remembers his football-star father and his unintentional legacy: 'Other men might inherit from their fathers a head for figures, a gold pocket watch . . . or an eternally astonished expression; from mine I acquired

this need to have my name whispered in reverential tones.' Exley is unable to take the mundane world of employment seriously, which leads him first to a sofa at his mother's house for six months and then to repeated hospitalisations, during which he is subjected to insulin coma therapy and electro-shock therapy. He also recalls his love for the beautiful but mindless Bunny Sue, with whom he is impotent, and he digresses on his conquest of numerous other women by lying to them. He acquires a wife called Patience and twin sons, but the marriage breaks up. Above all Exley recounts male friendships with often bizarre characters like Mr Blue, an ageing salesman and keep-fit fanatic obsessed with cunnilingus, or his grotesque brother-in-law, Bumpy Plumpton.

Exley's attitudes to women and homosexuals have already dated badly but he is no racist, which only underlines the gratuitous horror of the self-destructive fight he provokes at the climax of the book. Overhearing two men having a literary conversation, one white and one black, he begins hurling racist abuse at them until they fight him almost to the death ('I felt so many things go in his face,' he reports after landing a blow, 'that I almost became sick'). 'I fought because I understood, and could not bear to understand,' he says afterwards, 'that it was my destiny – unlike that of my father, whose fate it was to hear the roar of the crowd – to sit in the stands with most men and acclaim others. It was my fate . . . to be a fan.'

Most critics praised the book for its humour, honesty and style, although a few found it self-obsessed and overwritten. Exley often refers to literature (most frequently to Hawthorne, F. Scott Fitzgerald – with whom he has sometimes been compared – and Nabokov, while the critic Edmund Wilson is almost as much of a hero as Gifford), and the self-consciously literary nature of his style and sensibility is thematically integral. The reviewer of the *Times Literary Supplement* found that the book's eloquence gave a distance and perspective to the 'painfully funny' events narrated. It remains an extraordinary document of white-collar despair, male identity and mid-century American ugliness.

Barry HINES　　　　　　　**1939–**

A Kestrel for a Knave

Billy Caspar lives on a council estate in north-east England. His father has long since left home: his mother provides him with a series of 'uncles'. His older brother, Jud, works down the mines and soon, it is assumed, Billy will follow. No brighter future seems to beckon him, but the drabness of the present is relieved by his attachment to his kestrel, Kes.

A Kestrel for a Knave describes a critical day in Billy's life. (How he caught the hawk and raised her from an eyas is told in retrospect.) He gets up before seven for his paper round. En route he steals some orange juice, chocolate and eggs. Awkward and small, he is bullied at school by pupils and teachers alike. He is caned for falling asleep in assembly; later, he is provoked into a fight, which is stopped by Mr Farthing, the only teacher who tries to understand and encourage his charges (he comes to watch Kes being fed). After games, sadistic, infantile Mr Sugden forces Billy to stand beneath a cold shower.

Jud has given Billy five shillings to lay a double on the horses for him, but during lunch Billy decides to keep the money. From a classroom in the afternoon, Billy sees Jud angrily patrolling: Jud's horses have won and he has discovered that his bet was never placed. Billy spends the next few hours

evading his brother. He attends an interview with the Youth Employment Officer, which is a waste of time for both participants. He then rushes to the shed where he keeps Kes. The door is broken and the bird has gone. Guessing what has happened, Billy confronts Jud, who admits that he has killed Kes on account of the unplaced bet. The brothers argue violently, their mother joining in, and Billy flees with Kes's remains to a deserted cinema. Alone and in the dark, he imagines a picture in which he and Kes hunt and terrify Jud. He recalls the dreadful night when his father walked out. Then, emptied of emotion, he leaves the cinema, buries Kes in a field and goes to bed.

Because Billy's affections have been denied any human object, they focus the more fiercely on his kestrel. Mutually trusting, the hawk and the boy resemble each other. Both are savage, isolated creatures and, just as Billy has trained Kes, so this process has educated him. He has become an expert falconer and has found within himself the qualities of diligence and patience. Thus, the story suggests, Farthing's faith is justified: even this unpromising adolescent has a potential for development and fulfilment. There is, however, a negative side to the coin: Billy's dishonesty is unreformed. His theft brings on the novel's disaster. The tale's proper villain, however, is Billy's environment. In this respect, the novel harks back to the 'gritty realism' of the late 1950s and early 1960s. Although the terraces have been swept away by the decade's corporate planning, to be replaced by estates, the assumption that a boy will end up down the pit still prevails. In spite of passing years and social 'improvements' of various kinds, Billy's predicament is little different to that of the protagonists of **Billy Liar** (1959) or **A Kind of Loving** (1960). Like these novels, *A Kestrel for a Knave* was made into a highly regarded film. Directed by Ken Loach, who collaborated with Hines on the screenplay, it was released in 1969 as *Kes*, a title subsequently adopted for reissues of the novel.

P.H. NEWBY   1918–

Something to Answer For

Although he has written a number of novels, and had a reputation as one of the most promising novelists of his generation, Newby is a writer not as well known as he might be. Indeed, he is perhaps better known for his work at the BBC, where he was controller of the Third Programme and responsible for broadcasting stories by other writers. *Something to Answer For*, however, is remembered as the winner of the first ever Booker Prize, selected by the judges against considerable competition (Muriel Spark, Iris Murdoch, Nicholas Mosley). Set in Port Said during the Suez Crisis, it is more solemn in tone than some of his novels.

Townrow, an ex-army man, has returned to Nasser's Egypt at the request of 'Mrs K', the widow of his chance acquaintance Elie Khoury, a Lebanese arms dealer who has been found dead outside a convent. Mrs K is convinced she has somehow murdered her husband, but Townrow is more concerned with swindling her out of Elie's estate until his affair with Leah Strauss, daughter of the Khourys' lawyer, Abravanel, reveals to him a previously closed world of love and emotion.

On his way to Egypt Townrow is involved in an argument at Rome airport, where a stranger challenges British complacency and accuses Eden's government of failing to broadcast World Service warnings to European Jews about the concentration camps. Townrow assumes upon himself the collective guilt of the past, clinging to the fiction that, through his mother's family, his nationality is Irish. Confused about his own identity, he loses any sense of time and begins to inhabit a dream world of false memories and fake intuitions. Caught up in constant lies, he is imprisoned by the Egyptians and court-martialled as an Israeli agent, his old drinking companion Christou collaborating with the charge that, for all he knows, Townrow is Jewish.

Two visions haunt Townrow: the drunken departure of his father from the family car in the English countryside, and a requiem at sea for the body of Elie Khoury. Pursued by the Greek assassin Aristides, Townrow slowly switches his loyalty to the anti-British forces. Leah, a Jewish American, obsesses him and, with his passport confiscated, Townrow begins to wonder if he is in reality her crazy husband, committed to an expensive mental asylum in the USA. Then the deaths of two men, Amin, his gaoler on a repatriation train journey to Cairo, and Faint, an army officer he has duped into lending him money, shock Townrow into sudden action. With a party of drunken British troops shortly before the evacuation, Townrow goes to the cemetery and exhumes Elie's body, taking it out to sea amongst the battleships. Until the final moment he is not certain whether the coffin contains the corpse, a stash of arms or a booty of gold. Mrs K and Leah are persuaded on board the British fleet, imagining that Townrow will join them. But as Elie's coffin is opened up in the sunlight, Townrow intuits on the death's-head the message to break free, and sails off alone, too frightened and untrusting to engage in human relationships. 'When you were caught up in extraordinary events you did extraordinary things, even monstrous things like digging up a dead man and carrying him out to sea.' At best, Townrow's history seemed like a dream, or a false memory: 'What sort of reasonableness was that?'

Anthony POWELL   1905–

The Military Philosophers

see **A Dance to the Music of Time** (1975)

Simon RAVEN   1927–

The Judas Boy

see **Alms for Oblivion** (1976)

The Day of the Scorpion

see **The Raj Quartet** (1975)

The Sleep of Reason

see **Strangers and Brothers** (1970)

Couples

It is 1963 and in Tarbox, Massachusetts, eight upper-middle-class couples – mostly thirtyish – drink, dine, play games and holiday together. Thinking themselves brilliant and adventurous, they refer to their set as a 'charmed circle'. The circle, however, is crossed with adulterous lines. Piet Hanema, married to Angela, is having an affair with Georgene Thorne; Freddy, Georgene's husband, would like to reciprocate with Angela – but semi-frigid Angela will not have him. The Smiths and the Applebys form an irregular quadrilateral: the Saltzes and the Constantines likewise. Roger Guerin is rumoured to be homosexual, while Bea – his alcoholic wife – lusts after Piet.

Ken and Foxy Whitman – the perfect WASPish pairing, newly arrived – intend to stay aloof from all of this, but Foxy, who is expecting Ken's child, does not keep to her resolution. She hires Piet, a partner in a building firm, to refashion her house. Piet is allured by her pregnancy, and the two embark on a liaison. Georgene knows; Angela suspects; academic Ken, engrossed in biochemical researches, is ignorant. Only after Foxy's child is born do she and Piet agree to stop seeing each other. They make love one final, unsatisfactory time – and Foxy as a result becomes pregnant again. She and Piet decide on an abortion (illegal under state law), procuring it – with Freddy Thorne's assistance – from a dentist in Boston. As a reward for his co-operation, Freddy demands that Piet tell Angela to sleep with him. Piet does so. Angela consents, asking not to be told what Freddy's part in the arrangement was.

Angered by Piet's desertion of her for Foxy, Georgene apprises Ken of his betrayal. Ken summons Piet and Angela, announces that he will divorce Foxy and makes it plain that – according to his moral code – Piet should now divorce Angela and marry Foxy. Piet has transgressed the conventional amorality of Tarbox and must, therefore, be punished and excluded. Evicted by Angela from the family home, bought out by his partner and socially spurned, he dallies briefly with Bea and Carol before at last doing as Ken has proposed.

Quickly his wife was dead weight beside him. She claimed she never dreamed. Pityingly he put his hand beneath the cotton nightie transparent to his touch and massaged the massive blandness of her warm back, hoping to stir in the depths of her sleep an eddy, a fluid fable she could tell herself and in the morning remember. She would be a valley and he a sandstorm. He would be a gentle lion bathing in her river. He could not believe she never dreamed. How could one not dream?

COUPLES
BY JOHN UPDIKE

Around those central events, the rest of the couples squabble and swap and re-form. The novel suggests them to be hysterical boors. Harold translates his own phrases into French; Frank obsessively and showily quotes Shakespeare; Freddy keeps soft porn next to his bed. These symptoms apart, there is little cultural life. The protagonists' thoughts are as repetitive and limited as their patterns of relationship. (Their self-congratulatory introspection is particularly registered in the shallowness of their responses to Kennedy's death.) Thus, shifting from such dramatic paucity, Updike's attention rests on the physical world. He describes, in celebratory detail, the properties of things: the seaboard marsh; a church's weather vane; the slamming of a door; and, with a virtuoso's variation, the qualities of genitalia. Tarbox is full of vibrant metaphor; the folds of flesh of its inhabitants enclose comparison inside comparison. Shabby behaviour consorts with aesthetic glory and what Updike denies to his characters' intellect he grants to their senses. The cherishable experience of sex – and *a fortiori* of general existence – remains untainted by its means of achievement. If Updike disapproves of this doctrine, he simultaneously flirts with it – wedded to a sterner ethical code, yet drawn towards this fresh-faced hedonism.

Gore VIDAL   1925–

Myra Breckinridge

Consumed by her mission to 're-create the sexes and thus save the human race from extinction', the 'disturbingly beautiful' Myra Breckinridge arrives in Hollywood, 'source of all this century's legends'. Vidal's disaffected sex-comedy of Californian life (dedicated to Christopher Isherwood, and accurately described by Brigid Brophy as a masterpiece of 'the high baroque comedy of bad taste') takes the form of a journal kept by Myra for her psychoanalyst. Myra makes a fleeting reference to her husband, Myron, a film critic who has recently drowned while crossing to Staten Island on a ferry. She sees the completion of his book on films of the 1930s and 1940s as her life's work, and her obsessive references to old movies and their stars is a splendid parody of Hollywood 'culture'. In order to support herself, Myra has come to claim her widow's legacy of a fifty per cent stake in the Academy of Dancing and Modelling run by Myron's lecherous uncle, Buck Loner. The Academy is a thriving concern, and Myra temporarily accepts a position as a teacher of Empathy and Posture while her claim is legally validated. Buck is determined to thwart her, and entries from his own journal, dictated into a tape recorder, are interspersed with Myra's record of events.

Myra's caustic commentary upon her vacuous students is one of the novel's great strengths, and represents a satiric critique of West Coast 'counter-culture'. Vidal manipulates his readers' sympathies very cleverly: Myra's inflexible assertion of her own (eccentric) views mark her out as something of a monster, yet the narcissism and naïvety of her students provide her with a perfect foil. Two in particular arouse her tyrannical sexual interests, and she pursues parallel affairs with both. Rusty Godowski, a dumb but handsome hunk, is finally vanquished in the course of a medical examination which concludes in sexual role-reversal as Myra rapes him with a dildo ('borrowed yesterday on the pretext that I wanted it copied for a lamp base'), a scene which contributed substantially to the book's notoriety. Unmanned, Rusty is manipulated into a sadistic relationship with the masochistic Hollywood agent Letitia Van Allen, thus clearing the way for Myra to seduce Rusty's girlfriend, Mary-Ann Pringle.

When we arrived at the house, the door was opened by Clem, who wore nothing but glasses and a large door key on a chain about his neck. He is extremely hairy, which I don't like, and though he did not have an erection and so could not be fairly judged, his prick is small and rather dismal-looking as if too many people had chewed on it, and of course he is circumcised, which I find unattractive. Naturally, like so many physically underprivileged men, Clem regards himself as irresistible (no doubt some obscure psychological law of compensation is at work). He promptly took me in goatish arms, rammed his soft acorn against my pudendum, and bit my ear.

MYRA BRECKINRIDGE
BY GORE VIDAL

Meanwhile Buck has discovered that Myra was never married to Myron and threatens to withhold the inheritance on the strength of this technicality. Myra is forced to reveal her true identity: she *is* Myron, or was until she underwent a sex-change operation. She is subsequently run down by a car and when she wakes up in hospital she has resumed her male identity. Myra/Myron marries Mary-Ann and they convert to Christian Science and settle to a peaceful life farming in the San Fernando Valley.

The plot of *Myra Breckinridge* is clearly a preposterous fabrication, built around Vidal's determination to reverse, invert and recombine the genders and sexual roles of his characters. Male and female, active and passive, possessor and possessed all become less and less reliable co-ordinates as the novel progresses. Vidal negotiates between the weightiness of his theme and the book's gamey humour with considerable skill. Faced with the choice between pointing a moral or delivering a punch-line, he invariably opts for the latter. The result is a savagely funny, high-camp, Voltairean extravaganza, and a key text of the 1960s.

By a nice irony, the novel was made into a celebratedly dreadful film (directed by Mike Sarne from his own screenplay, 1970), featuring Raquel Welch, bizarrely cast in the title role, and a spectacularly elderly Mae West.

1969

Richard M. Nixon becomes US President ● Resignation of Charles de Gaulle ● Disturbances in Northern Ireland ● First men on moon ● Human egg cell fertilised in vitro ● Kenneth Clark, *Civilisation* ● Isaiah Berlin, *Four Essays on Liberty* ● Herbert Marcuse, *An Essay on Liberation* ● First Booker Prize for fiction awarded in Britain ● Rupert Murdoch buys the *Sun*, which is relaunched as a tabloid ● Films: *Oh! What a Lovely War, Z, Satyricon* ● Television: *Monty Python's Flying Circus, Sesame Street* ● Woodstock Music and Arts Fair in the US ● Karl Jaspers, Dwight D. Eisenhower, Walter Gropius, Ludwig Mies van der Rohe, Rocky Marciano, Ho Chi Minh die ● Ezra Pound, *Cantos* (1925–69)

Margaret ATWOOD   1939–

The Edible Woman

In North America Atwood is considered a major poet, but it was her novels which first brought her to the attention of the British public, in particular women who found much with which to identify in her feminist fables.

The protagonist of her first novel, *The Edible Woman*, is Marian MacAlpin, a perfectly 'ordinary' young woman, fresh out of university and happy with her allotted role in life. She has a job in market research with few prospects, but this hardly matters since her principal goal is marriage. Her conventional, well-mannered boyfriend, Peter, accepts Marian's self-appointed role as unquestioningly as she does herself, regarding her balance and pragmatism as an ideal. Marian is one of three female prototypes presented in the novel, contrasted with her flatmate, Ainsley, a boisterously independent, man-hunting young woman, and Clara, a friend from college, angelic and frail, at once part of the material world and yet somehow above it. Both women, however, see motherhood as their biological destiny: Ainsley's hunt is for an 'ideal' partner to father her children, while Clara is perpetually pregnant.

From the moment she agrees to marry, Marian's determination to cling to the down-to-earth image of herself everyone knows and accepts is challenged by intimations of self-fragmentation. She appears to oscillate between the need for masculine certainties and a feeling of feminine disembodiment, as at the women's Christmas party at the office: 'She felt suffocated by this deep sargasso-sea of femininity. She drew a deep breath, clenching her body and mind back into her self like some tactile sea-creature withdrawing its tentacles; she wanted something solid, clear: a man; she wanted Peter in the room.' At the same time as she yearns for a force which will bring her existence into sharp focus, something in her rebels against the logic of coherence. Appropriately, in a book so consistently peppered with images of eating, chewing, swallowing and cooking, her rebellion manifests itself as symbolic anorexia nervosa, a gradual rejection of all food. Her perception that she is 'being eaten' by the external world and its expectations (hence the book's title) leads to her attempting to reduce herself to the point where there will be nothing left for the world to feast upon.

The remedy is for Marian to magnify her role as an edible product and she bakes a woman-shaped cake, an icon of emotional cannibalism. The response of each character to this emblem gives a measure of his or her commitment to the real world. Peter refuses to touch it, but the half-starved Marian rediscovers the pleasure of eating. It is the 'mad' self-engrossed Duncan, the young man who stands in opposition to the stolid Peter, who can swallow with no qualms at all, one eye after another, one chocolate curl after another, and declare unashamedly: 'It was delicious.'

John FOWLES　　　　**1926–**

The French Lieutenant's Woman

Fowles's third and most widely acclaimed novel combines a pastiche of realistic Victorian fiction with the elaborate tricksiness of the modern French *nouveau-roman*. It begins as a fully fledged Victorian romance, though hedged about with an intrusive and recognisably modern commentary on the events portrayed. Later, the narrative is undermined from within as well, both by alternative versions of events and by the appearance of the author as a character in his own fiction.

At the centre of the Victorian romance is Charles Smithson, a progressive aristocrat, Darwinist and archaeologist, who becomes engaged to Ernestina Freeman, a wealthy businessman's daughter. This initial, modest romance of title intermarrying with trade is quickly overshadowed, however, by the appearance of another woman.

After all, he was a Victorian. We could not expect him to see what we are only just beginning – and with so much more knowledge and the lessons of existentialist philosophy at our disposal – to realize ourselves: that the desire to hold and the desire to enjoy are mutually destructive. His statement to himself should have been, 'I possess this now, therefore I am happy', instead of what it so Victorianly was: 'I cannot possess this for ever, and therefore am sad.'

THE FRENCH LIEUTENANT'S WOMAN
BY JOHN FOWLES

Sarah Woodruff is a governess, a caste even further removed from Charles's own, and is additionally compromised by a previous involvement with a French naval lieutenant. Charles meets her while visiting Ernestina at her aunt's house in Lyme Regis, and is beguiled by Sarah's combination of frankness, melancholy and mysterious independence. However, when Sarah is dismissed from her position as companion to the puritanical Mrs Poultney, she suddenly disappears.

Charles tracks her down to Exeter, but at this point the first alternative ending to the novel cuts short Charles's second romance: Charles repents, and returns to Ernestina and respectable marriage. But immediately the novel backtracks again, to the more exciting intrigue in Exeter. Charles and Sarah now consummate their affair in graphically un-Victorian fashion. Charles then returns to Lyme and breaks his engagement to Ernestina, only to find that Sarah has disappeared yet again, and this time without trace.

Years later, he gets news of her in London and, on calling round, discovers that the fallen woman has blossomed as a Victorian feminist, or 'New Woman', employed as an assistant to the painter, Dante Gabriel Rossetti. Far from requiring a rescue-operation by Charles, Sarah is apparently happy in her new-found independence.

Two more endings are now provided. In the first, Sarah reveals that she has borne and raised Charles's illegitimate child, and the reconciled couple begin a new relationship based on sexual equality. In the second, this modern romance is in turn displaced by postmodern, existential grimness: Charles leaves the house unreconciled, to lead a solitary life which will be relieved only by a new sense of personal freedom and individuality, the sole legacy of the woman he cannot forget.

Towards the end of the book, Fowles twice writes himself in as one of his own characters, with the result that responsibility for choosing the most desirable of the alternative narratives is shifted to the book's readers. Hence the novel's plea for freedom and choice is conducted by means of its ambivalent, teasing form, as well as through the more direct unfolding of the individual characters' experiences.

A film version of the book was directed by Karel Reisz in 1981, based on a slightly sanitised screenplay by Harold Pinter.

Graham GREENE 1904–1991

Travels with My Aunt

Greene's 'attempt ... to treat humorously a serious theme, old age and death' takes the form of an amoral caper in which a retired bank manager finds worldly salvation in the company of an elderly relative.

The novel is narrated in the first person by Henry Pulling, a middle-aged bachelor who has been retired early and is settling into the suburbs with little to enliven him except an enthusiasm for dahlias. At his mother's funeral he is greeted by an extraordinary septuagenarian who claims to be his Aunt Augusta. She confesses to a fondness for travel and invites Henry to accompany her upon a trip to Brighton, where she reveals something of her colourful past, in particular her involvement with a man called Curran,

who left the circus to found a church for dogs. They visit an old associate of Aunt Augusta, a tea-leaf reader called Hatty, who discerns 'a lot of travelling ... a lot of confusion too and running about' in the dregs of Henry's Lapsang.

'I have always disliked the unexpected,' Henry confesses when he bumps into Aunt Augusta's black factotum, Wordsworth, at the Gare du Lyon, 'whether an event or an encounter, but I was growing accustomed to it in my aunt's company.' Wordsworth has already disrupted Henry's sober life by mixing cannabis in with the late Mrs Pulling's ashes. On board the Orient Express (en route for Istanbul in order to do 'a little business' with a Gen. Abdul), Aunt Augusta relates more of her life story, and the saga of Mr Visconti, a minor war criminal, a major swindler, and the love of her life. Although naïve, Henry begins to realise that his aunt is not above a little smuggling herself, a fact which arouses the interest of Col. Hakim. Deported from Turkey, still in possession of her gold ingot, Aunt Augusta returns to England, but is persuaded by Henry to accompany him to Boulogne to visit his father's grave. There she makes short work of a Miss Paterson in whose arms Henry's father had died. Henry returns to England to face questions from the police about his aunt, who has travelled to Paris and then vanished.

Just as he is regretting his return to suburbia, Henry is summoned to Paraguay by his aunt. She has teamed up with Mr Visconti in a smuggling racket and the two old people manage to hoodwink Interpol and the CIA. The lovesick Wordsworth is sent packing, but returns to plead with Aunt Augusta, and is murdered by an ancient guard. Henry decides to remain in South America with Aunt Augusta (who finally admits to being his natural mother) and joins the flourishing Visconti 'import–export' business.

'Never presume yours is a better morality,' Aunt Augusta warns Henry. Her own taste for 'the unexpected character and the unforeseen event' is what keeps her young and vital. The attraction of Paraguay is that it is an unstable place where a young person is quite as likely to die as an old one and where 'every day you live will seem to you a kind of victory.' Henry's decision to join her is in itself a kind of victory.

Doris LESSING   **1919–**

Children of Violence

Martha Quest (1952)
A Proper Marriage (1954)
A Ripple from the Storm (1958)
Landlocked (1965)
The Four-Gated City (1969)

The *Children of Violence* sequence follows the progress of Martha Quest and, to varying degrees, registers world events from the 1930s to the year 2000. The unity of the sequence derives from Lessing's continuing use of the same central character, whose life she has acknowledged to be based largely upon her own. The changes which the separate novels of the sequence undergo mark changes in the world at large, in Lessing's own perception of it and in her self-perception as an artist. Martha Quest develops as the perceived subject and perceiving narrator of the novels. Lessing has described the sequence as 'a study of the individual conscience in its relations with the collective,' but as the novels progress, these terms undergo significant alteration. The traditional notion of the individual becomes more diffuse, its relations more complex and involved, the idea of the collective more fragile and truly comprehensible at a mystic rather than rational level. Martha's receding, intellectual goal is an adequate apprehension of the world which becomes steadily more resistant to understanding.

The events of the sequence are spread rather unevenly over the five novels. *Martha Quest* opens in 'Zambesia', Lessing's fictional version of colonial Rhodesia. Martha is fifteen years old and lives with her parents, who have scraped a living from the bush on their small, isolated farm. Martha cannot contemplate following their example and leaves home for the nearby town. Towards the end of the novel, she marries a govern-ment official and bears his child. The bulk of *A Proper Marriage* is taken up with her experience of marriage, her first affair and her introduction to politics.

Lessing charts Martha's development as a series of escapes into illusory freedoms which quickly relapse into new, constrictive environments. Both novels are permeated with Martha's feelings of dissatisfaction and longing. These are focused retrospectively on her vision of 'a noble city, set foursquare and colonnaded along its falling flower-bordered terraces. There were splashing fountains, and the sound of flutes; and its citizens moved, grave and beautiful, black and white and brown together ... running and playing among the flowers and terraces, through the white pillars and tall trees of this fabulous and ancient city.' This vision is a metaphor for a kind of authentic, unfettered existence that Martha (and perhaps Lessing too at the time of writing) has yet to find. The last novel of the sequence will take its title from this vision.

The political avenues to this Utopian scene are exhaustively explored in the third novel, *A Ripple from the Storm*. Newly liber-ated from her marriage, Martha joins a Marxist group dedicated to the cause of the 'working man' – in effect, the natives of Zambesia. All the group's members are white except one, and he is a government spy. Martha marries the group's leader, but neither group nor husband assuage her sense of incompleteness. The long political dis-cussions of the group are presented ironi-cally by Lessing, but the tedium of this material emerges all too clearly. The group is a dead-end both for Lessing and her heroine, who cannot articulate what she wants or needs through its well-meaning offices. When it is disbanded Martha remains locked in a futile marriage, while Lessing seems to have developed her character as far as the conventions of the realistic novel will allow.

In the seven years between the third and fourth volumes of the sequence, Lessing published her most ambitious and best-known work, *The Golden Notebook* (1962). The success of its technical experi-ments furnished Lessing with a new range of fictional devices. In *Landlocked* she de-veloped a new, more enveloping style with which to resume her heroine's story and probe more deeply into her psyche. When the novel begins, Martha is still unhappily married and her dreams are filled with images of isolation and aridity. A brief, passionate affair with Thomas Stern, a Polish Jew working as a gardener, brings her closer to realising what she wants. The two communicate telepathically (a fact which Lessing presents as a natural consequence of their intimacy) and share their most secret thoughts. The affair does not last and Thomas eventually dies insane, but Lessing presents his madness equivocally. Sanity, she implies, is only one psychological state amongst many and no better than any other. Allied to this, Thomas and Martha's non-intellectual, mystical communications are intentionally described in normal terms. The novel ends with Martha deciphering the scattered pages of Thomas's book and departing with it to England.

The last novel of the sequence is the longest and most ambitious. *The Four-Gated City* marks the culmination of the major themes of the four preceding volumes, but also acts retrospectively, accentuating and developing details that earlier seemed unimportant. The novel is organised concentrically rather than along a linear narrative and its structure militates against a single conclusion. At the centre of the novel is Martha herself as she becomes increasingly aware of her developing psychic potential. About her are the members of the Coldridge family with whom she lives in London, after the end of the war. About the Coldridges are all the upheavals and crises of world events from the 1940s to the projected year 2000. Lessing's increasing emphasis on the importance of paranormal psychic powers is here introduced through Lynda Coldridge, the 'mad' wife of Mark, Martha's lover. Martha grows increasingly close to Lynda, and Lessing's prose follows Martha's mind very closely in turn as its potential powers are unlocked through this relationship. The various members and generations of the Coldridge family comprise a wide social barometer of wider events in the world but fall away during the course of the book as Martha's telepathic and intuitive abilities usurp their function. The visions and dreams of the earlier books now come to seem the norm rather than aberrations and provide Lessing with the means to link Martha's subjective experience with the objective world in the widest sense. The sense of completeness she sought and did not find in her home, her marriages and involvements in politics is finally found within her own psychic resources.

The final section of the novel takes the form of letters exchanged between survivors of a nuclear holocaust. A network of people with powers of extra-sensory perception have foreseen and escaped this catastrophe. Their letters (one of them from Martha) tell of the struggle for survival and the emergence of a new race of children who are born with their psychic powers fully developed. They are well-equipped to live in and perhaps regenerate this changed world. The whole sequence ends optimistically with a belief in the possibility that they might prove equal to this challenge.

The central problem of *The Four-Gated City* is the article of faith it demands from its readers. Belief in the mixture of R.D. Laing, Sufism and paranormal psychology with which Lessing invests Martha is a basic premise of the novel. Once accepted, the transcendental climax of the sequence is something of a foregone conclusion. Without this acceptance, such a conclusion seems at best merely rhetorical. Emphasis on the ending is perhaps misplaced though, for the structure of *Children of Violence* is cyclical rather than linear. Martha Quest's situations are typically entered, outgrown, abandoned and re-entered in another guise. Her development is charted through her regenerations rather than a series of instructive events. Each encounter with the world at large brings her closer to understanding the vision of the four-gated city seen in the very first novel. Paradoxically, the wider her experience, the more her inmost psyche comes to seem the key to its comprehension. This constant referral of the world to the self sets up a cyclical pattern within the novels which is finally confirmed as a kind of unity between the two – a transcendental resolution of the social, political and sexual relations which *Children of Violence* charts through two continents, seventy years and all the upheavals of the twentieth century.

Joyce Carol OATES   1938–

them

them is a study of poverty and degradation, and of the strength needed to rise above it. The tone is set by the author's preface, and by the epigraph, taken from Webster's *The White Devil*: '... because we are poor / Shall we be vicious?'. According to Oates, she had known one of the protagonists of the novel, the girl Maureen. Her life story, told in many letters to Oates, obsessed the writer, who found herself unable to live her own life or to think of it. This obsession pervades the novel, which is claustrophobic, haunting, and unrelieved by brightness or humour. It won the National Book Award.

The story of the family's degradation begins many years before Maureen's birth. Her grandfather loses his job and finds it impossible to lead a human existence; his recourse to drink destroys the home. The adult life of Maureen's mother, Loretta, begins with sexual intercourse with a local boy who is killed in her bed by her brother. She is virtually raped by the local patrolman, but is happy enough to marry him to acquire the status of a married woman. The family 'curse' decrees that her husband should be as unsuccessful in life as her father had been, and that her son, Jules, should hate him, as her brother had hated their father. When Jules's pent-up violence, fed by the love he feels for his sister, Maureen, explodes, it reaches far greater proportions than the murder of just one boy. Jules becomes involved in riots and is an apostle of the healing violence, the cleansing fire. He is victim and perpetrator: his mistress tries to kill him and he kills a policeman.

Loretta's youngest child, Betty, becomes a street-wise delinquent, while Maureen goes through several metamorphoses. A quiet girl, she becomes a child-whore, is beaten up by her stepfather and retreats into a wordless trance of compulsive eating. From this state she is rescued by Jules. Eventually she goes to night school where she deliberately destroys her tutor's marriage and makes him marry her. Our last glimpse of her is one of almost paralysing normality after the obsessions and degradations of her earlier life. Married to the teacher she becomes a determined and contented housewife, anxious to avoid her family – her mother with her eccentric friends; her mother's brother, who had lost what little zest for life he had ever had; even her much loved brother Jules, who openly preaches violence and has become part of a race study group that seems to be doing more harm than good.

In her introduction, Oates calls Jules 'a strange young man' and says: 'One day he will probably be writing his own version of this novel, to which he will not give the rather disdainful and timorous title *them*.' Jules is the only character who tries to think beyond and outside the poverty and the seemingly endless unhappiness. He understands his family, loves them in his somewhat distant way, and pities them.

He and Maureen are successful in overcoming their family – she by disowning and forgetting, he by accepting his heritage of brutality and exaggerating it to include the whole world. Yet Jules, 'the murderer', as his mother calls him, is a saintly figure, whereas Maureen, 'the whore', is not.

Philip ROTH   1933–

Portnoy's Complaint

Dr Spielvogel was so delighted with Portnoy's complaint that he published it as a classic case study. Oedipal, anal, onanistic, impotent – Portnoy's subconscious yields a bumper crop of Freudian neuroses. Roth describes his novel as a protest against his Jewish detractors: 'They wouldn't let up, no matter what I wrote ... And out came Portnoy, apertures spurting.'

Portnoy's 'complaint' is a physical symptom of repression, interpreted through psychoanalytic paradigms with relentless precision. 'It was my mother who could accomplish anything,' he says, 'she could bake a cake that tasted like a banana.' Waging war on germs and bodily secretions, mother clamours to search his faeces for evidence of non-kosher hamburgers. 'I am the smartest and neatest little boy in the school,' he protests, 'I carry a comb and a clean hanky.' Father queues in an agony of constipation at the bathroom door while Portnoy discovers auto-erotic delights. Jerking off into his sisters' underwear or defiling the family's lunchtime liver, Portnoy protests that 'my wang was all I had that I could call my own'. Through a trail of creamed pyjamas and crumpled Kleenex, Portnoy progresses to adolescence and the ecstatic discovery that all women have the same thing between their legs.

And you, the implication is, when are *you* going to get married already? In Newark and the surrounding suburbs this apparently is the question on everybody's lips: WHEN IS ALEXANDER PORTNOY GOING TO STOP BEING SELFISH AND GIVE HIS PARENTS, WHO ARE SUCH WONDERFUL PEOPLE, GRANDCHILDREN? 'Well,' says my father, the tears brimming up in his eyes, 'well,' he asks, *every single time I see him*, 'is there a serious girl in the picture, Big Shot? Excuse me for asking, I'm only your father, but since I'm not going to be alive forever, and you in case you forgot carry the family name, I wonder if maybe you could let me in on the secret.'

PORTNOY'S COMPLAINT
BY PHILIP ROTH

As he becomes 'cunt crazy', however, Portnoy finds that even the erotic promise of tarty Bubbles Giradi cannot overcome the fear of the knife which lurks in his subconscious. The taste of bananas takes on new significance with The Monkey, but attempts at sex with her reveal him as a petrified college prig. Finally resorting to a symbolic refuge in the motherland, Portnoy makes a pilgrimage to Israel only to find that he thrives on persecution. With his tail between his legs, he retreats to the analyst's couch to absolve his inferiority complex. Repression is released in a primal scream, and Spielvogel believes he has made a breakthrough.

The sexually explicit nature of the novel caused considerable controversy, and ensured Roth a wide readership. His Jewish detractors may not have been pacified, but

Portnoy's Complaint established Roth's reputation as a major novelist who was held in high critical regard, and whose books became bestsellers.

Bernice RUBENS   1928–

The Elected Member

With this novel (published in the USA as *The Chosen People*), Rubens won the 1970 Booker Prize. Norman Zweck, son of Rabbi Abraham Zweck and his wife Sarah, is forty-one. Behind him lies a brilliant career as a child prodigy (fluent in nine languages before his teens) and as an accomplished barrister. After a breakdown in court, he has been hallucinating in his bedroom over the small grocery shop run by his father and sister Bella in the East End of London. The room appears to be swarming with silverfish which nobody but Norman can see. Norman is in fact a drug addict, sly and determined in his addiction.

Despairing of Norman's sanity, Rabbi Zweck and Bella commit him to a mental institution where another inmate, the 'Minister', who fancies himself a member of the Cabinet, keeps him surreptitiously supplied with drugs. These sustain Norman's paranoia despite the best efforts of the authorities and the tearful anxiety of his family to wean him back to normality. Normality for Norman is as problematic as madness: in his life, he has been accomplice to the heartbreak of his mother by leaving home; he has seduced his sister Bella, who remains (perhaps as a result) unmarried, a childlike figure in white ankle socks; and he has contributed to the marriage outside the Jewish faith of his sister Esther and to the suicide of his best friend David. Rabbi Zweck, in his distress at his son's addiction and committal, endlessly reflects on his own guilt and failure. Norman, though reluctantly conscious of his own errors, is cantankerously absorbed by the idea that he has been betrayed by his family, and plays upon their appalled emotions to rouse their pity and care. He is convinced that he has taken the entire suffering of the Jews upon himself.

Rubens has an unsparing, mordant approach to the human heart. Her sympathies are uniformly spread for those 'so loosely tacked onto life'. Norman is the scapegoat, the 'elected member' of the family, a notion Rubens derived from the work of R.D. Laing. He is left to his desolation after the death of his father and final realisation that he, and his companions in suffering, are 'the cold and chosen ones'.

Edward UPWARD   1903–

The Rotten Elements

see **The Spiral Ascent** (1977)

Kurt VONNEGUT Jr   1922–

Slaughterhouse-Five

The two world wars have achieved almost as much prominence in the literature of the twentieth century as they have in its history. A key experience in the lives of many writers, warfare has always demanded a literary response, but presented numerous obstacles in recreating the trauma of battle for the non-combatant reader. As a prisoner of the Germans during the Second World War, Vonnegut witnessed one of its most appalling episodes, the bombing of Dresden. He returned from the war determined to write about this, but found it very difficult to find a proper form in which to present it. His solution was to make an oblique approach to the subject by setting it within a science-fiction framework. The resulting novel was a bestseller, bringing Vonnegut instant fame and enduring critical and popular esteem.

Billy Pilgrim is a Second World War veteran who lives in Ilium, New York State. He was captured by the Germans in the Ardennes in late 1944, and released after the armistice in 1945. During his time in captivity, Billy was caught up in the fire-bombing of Dresden. As the book begins, in 1968, he is an optometrist, married with two children, Barbara and Robert. Billy's life suddenly changes when he is involved in an air-crash, of which he is the sole survivor. While he is recuperating, his wife dies of carbon monoxide poisoning in a freak accident. It is then that Billy's stories about being kidnapped by a flying saucer begin.

The visitor from outer space made a serious study of Christianity, to learn, if he could, why Christians found it so easy to be cruel. He concluded that at least part of the trouble was slipshod storytelling in the New Testament. He supposed that the intent of the Gospels was to teach people, among other things, to be merciful, even to the lowest of the low.

But the Gospels actually taught this: *Before you kill somebody, make absolutely sure he isn't well connected.* So it goes.

SLAUGHTERHOUSE-FIVE
BY KURT VONNEGUT Jr

Billy appears on a late-night radio chat show and explains that he was kidnapped by aliens from the planet Tralfamadore, who took him home and exhibited him in their zoo, mating him with Montana Wildhack, a Hollywood starlet. Billy's children are distressed by this, and ask why they didn't miss him, but Billy tells them that the Tralfamadorians took him through a time warp, so that he could be in two places at the same moment. The Tralfamadorians have explained to Billy that time is not a string of events, but that each moment has always existed and *will* always exist. This is borne out by the book's decidedly non-linear construction. In the time it takes to blink, Billy can find himself on Tralfamadore, in Ilium in 1968 and then back in the Battle of the Bulge, the final German offensive in 1944. Tagging along with two infantry scouts and a bloodthirsty artilleryman, Roland Weary, Billy is captured. The Germans, having killed the scouts, transport Billy and Weary to a prison camp in a train jam-packed with prisoners-of-war, many of whom die. The British officers who have been in the camp since 1940 cannot believe the degradation of the American soldiers, who were only captured days or weeks ago.

From the camp, Billy and the other American prisoners are taken to Dresden, and housed in a slaughterhouse – Slaughterhouse-Five. It is then that Billy observes the fire-bombing of the city on the night of 13 February 1945. Because of his time-travelling, Billy already knows this will take place. Finally, he is released and sent back to the USA to recuperate. It is then that Billy meets Eliot Rosewater, who has appeared before in Vonnegut's novels, and is introduced to the work of Rosewater's favourite novelist, Kilgore Trout, a purveyor of third-rate science fiction, and Vonnegut's mocking *alter ego*. The plots of several of Trout's books are related in the novel.

Explaining to his family about Tralfamadore, Dresden and his present state is not easy for Billy, and the only thing he can finally believe is that the universe is mechanically preordained, and that the only answer is to try to live only the happy moments. It is no accident that the leitmotif of this book is the phrase 'so it goes'.

1970

US offensive in Cambodia ● Palestinian terrorism ● Four anti-war demonstrators shot dead at Kent State University, Ohio ● Salvador Allende wins Chilean Presidential election ● First desegregated classes in southern US ● The Boeing 747 'jumbo jet' airliner begins scheduled flights ● IBM develops the 'floppy disc' for storing computer data ● Kate Millett, *Sexual Politics* ● T.S. Kuhn, *The Structure of Scientific Revolution* ● David Hockney, *Mr and Mrs Ossie Clark and Percy* ● World Trade Center, New York ● Yukio Mishima commits ritual suicide ● Dario Fo, *Accidental Death of an Anarchist* ● Films: *Kes*, *Five Easy Pieces*, *Love Story* ● The Beatles officially split up ● Bertrand Russell, Marc Rothko, E.M. Forster, Alexander Kerensky, François Mauriac, Jimi Hendrix, John Dos Passos, Abdel Nasser, Charles de Gaulle die ● Ted Hughes, *Crow* ● Peter Porter, *The Last of England*

John HAWKES   1925–

The Blood Oranges

Cyril and Fiona, Catherine and Hugh, 'a quartet of tall and large-boned lovers aged in the wood', are the main protagonists of Hawkes's sixth novel. Cyril and Fiona are 'sex aestheticians', Catherine and Hugh their initiates. The novel's epigraph is taken from Ford's ***The Good Soldier*** (1915) and asks: 'Is there then any terrestrial paradise where, amidst the whispering of the olive-leaves, people can be with whom they like and have what they like and take their ease in shadows and in coolness?' The answer to this question, as it was for Ford, is 'no'.

The facts of the novel are simple. Cyril and Fiona see Catherine and Hugh emerging from a motor bus which has crashed into a canal. Cyril embarks on an affair with Catherine, Fiona with Hugh. However, as this is not the terrestrial paradise invoked by Ford, the perfection of this symmetrical arrangement is denied them by Hugh's suicide towards the end of the novel. The novel is set in 'Illyria', although no specific geographical indications are given: the landscape is generalised, pastoral, and idyllic; the buildings 'moorish-looking' rather than specifically moorish. All of Hawkes's novels are intimately concerned with their settings. His work can be seen as a quest for the type of landscape envisioned by Ford, constructed from dreams to serve real needs.

Construction is crucial to Hawkes's vision and in *The Blood Oranges* is necessarily intricate. By far the most fully realised character is Cyril, who serves as the book's narrator. The dream-like story is constructed retrospectively from his musings upon and memories of the events. Hawkes suggests that Cyril's memory smoothes over the events which prompted them. Cyril's narrative is actually violently disjuncted (although seeming the opposite) and consists of perhaps a dozen incidents interspersed with more self-obsessed musings on his situation after the events of the novel have taken place. Isolated encounters are the basic building-blocks of the book: Hugh and Cyril photographing an almost sub-human peasant girl in a barn; Fiona, then Catherine, baring their breasts on the beach; the two men kissing a girl on a hillside. Dialogue is reported directly but used to convey mood rather than action.

Hawkes's unique prose style further homogenises the action and is seen at its best in the minute psychological notation which serves as a commentary on the events. Cyril's interior monologues pay detailed attention to the shifts and nuances which make up the relationship between the quartet. At the same time the occasionally disproportionate scrutiny of trivial incidents provides a subtly comic counterpoint to Cyril's ego. The question 'Can these people really matter this much?' lurks in the background, a brave artistic decision on Hawkes's part, for the worth of the novel depends upon the answer being 'yes'.

Sensuality pervades the world of these four lovers. Sex is never described directly, but is omnipresent; *The Blood Oranges* is profoundly *about* sex. Cyril and Catherine have found contentment, but Fiona and Hugh's liaison is riddled with tensions and suppressed violence which will prompt both Hugh's suicide and the disagreeable insight that 'Darkness can come to Illyria'. At the last, it is not the characters themselves but the landscape their thoughts and actions have drawn which remains. The players move on, the darkened stage remains and the question of how to construct the 'terrestrial paradise' envisioned by Ford and the characters of *The Blood Oranges* waits for another answer, another attempt.

Dan JACOBSON 1929–

The Rape of Tamar

The genesis of Jacobson's retelling of the biblical story recounted in II Samuel 13 lay some twenty years before he wrote his novel. In a 'youthfully self-conscious effort at intellectual self-improvement', Jacobson decided to read the Bible from start to finish. He failed in this endeavour, but the story of Tamar, Amnon and Absalom, children of King David, intrigued him: 'The compression of the tale, its startling reversals of direction, the truths about human nature hidden and revealed in the protagonists' terse words and violent actions – it was these that seemed worth exploring and enlarging on, for their own sake, as it were.' When he came to write the novel, however, he realised that the story could not be perceived 'in isolation either from the history that had preceded it and followed it, or from my own relationship to history. Indeed, the disparities and implausibilities of that relationship, as I saw them, had to be incorporated in the very structure of the novel itself. Hence the anachronistic utterances of ... Yonadab'. Jacobson retains and expands the Bible story, but abandons its impersonal perspective. He has as his narrator Yonadab, a minor figure in the biblical text, here a subtle, discontented courtier, malignly and opaquely motivated.

A nephew of King David, Yonadab describes how Amnon, David's oldest son, lusts after Tamar, his own sister. A misfit at court, sullen and unambitious, Amnon confesses his incestuous passion to Yonadab. Yonadab persuades him to indulge it and together the two of them plot the crime of the title: Amnon will feign illness and tell the king that a meal cooked by Tamar will cure him; David will send his daughter to Amnon's house, where Amnon, having dismissed his servants for the night, will violate her. The plan succeeds. (From a place of concealment, Yonadab watches and gloats.) The next morning Amnon throws Tamar out into the street. She, to shame him, parades her disgrace through Jerusalem; then she goes to Absalom, David's third son, for protection. Absalom swears to kill his brother. David, however, is grieved that Tamar has not come to him for help. What is more, he fears that her going to Absalom will augment the prince's prestige. (Already Absalom is gathering power and influence. His ultimate rebellion seems inevitable.) The king, therefore, decides to show mercy to Amnon, thus frustrating Absalom's revenge.

Two years pass. Amnon, declining into drunkenness and *ennui*, publicly seeks Absalom's forgiveness. Absalom withholds it. Yonadab shuttles between the households, both an intermediary and a spy. Via him – in an apparent gesture of reconciliation – Absalom invites his estranged brother to a spring festival beyond David's protection. Amnon accepts. All David's nineteen sons attend, while Yonadab – suspecting that his death as well as Amnon's is planned – stays in the city. Soon he learns how, at the festival's climactic banquet, Amnon was indeed murdered by Absalom's thugs. Yonadab views the corpse when it is brought from the country and finally accuses Tamar, 'All because of you! All of it! Yet who cares what will become of you now?'

More than once, Yonadab calls attention to the modernity of his tale. (He is a ghostly, resurrected figure, permitted to refer to Kant and Freud, alive just so long as the reader reads.) Stressing the similarities between ancient and contemporary behaviour, he means to demonstrate the validity of eternal cynicism. For him, self-interest governs human action and desire is always doomed to disappoint. Tamar's distress, even, is manipulative – an unconscious exploitation of her plight, not a genuine response to it. He fancies that he strips away pretence, though he does not apply his scrutiny to his own deeds. It escapes him that his habitual pleasure is voyeuristic and, consequently, that his cynicism is voyeurism at the ethical level. He must reveal the moral – if not the literal – nudity of the people around him. The truth is an instrument of humiliation and, uncovering it, he exposes himself.

'The King and I', an essay Jacobson wrote about the background to the novel, is collected in his *Adult Pleasures* (1988). *The Rape of Tamar* was later used by Peter Shaffer as a source for his play *Yonadab* (1985).

Iris MURDOCH   1919–

A Fairly Honourable Defeat

In the enigmatic Julius King, a Prospero-like figure who cynically and without any real motive exploits and manipulates the weakness and vanity of those about him, Murdoch created one of her most sinister puppet-masters. A Jewish-American biologist, who has been carrying out research into biological weapons, Julius represents the sort of force that really guides the world. In contrast, Rupert Foster, a complacent civil servant who is writing a philosophical book on the nature of goodness, is destroyed along with his work. The plot is complicated, involving all manner of deceptions and a considerable, but willing, suspension of disbelief on the part of the reader. (Melodramatic twists involve the disclosure that one character was in Belsen, and that the fourteen-year-old twin-sister of another was raped and murdered.)

Julius's former lover, the highly strung Morgan Browne, makes a wager with him that he can undermine the most stable partnership, little suspecting that he will use her to win the bet. With apparent relish, and by a combination of carefully engineered events and cleverly exploited situations, Julius is able to exaggerate existing weaknesses, destroy trust and demolish relationships. Employing skilfully forged letters, he tricks Morgan and her brother-in-law, Rupert, into thinking each is declaring love for the other. As a result, they really do fall in love, thus wrecking Rupert's twenty-year marriage to Morgan's sister, Hilda. Julius further arranges it so that Hilda discovers a cache of love-letters apparently written by Morgan to Rupert, although they were in fact written to Julius himself.

Two pawns in Julius's game are the Fosters' son, Peter, and Rupert's homosexual brother, Simon. Peter has dropped out of Cambridge, taken up shoplifting, and lives in the squalid flat of Morgan's saint-like former husband, Tallis. He is in love with Morgan, but after a passionate declaration agrees to her plea for 'innocent love ... a *happy* love'. When told by Julius of Rupert's affair with Morgan, Peter takes his revenge by destroying the manuscript of his father's book. Rupert is subsequently found drowned in a swimming pool with large amounts of alcohol and sleeping pills in his body; the coroner's verdict is 'death by misadventure'. Julius has also been preying upon the insecurity of Simon, involving him in his machinations, and making him appear deceitful to his lover, Axel Nilsson.

Julius had told Simon that his plot 'was a sort of midsummer enchantment and that he would unravel it all later on quite painlessly and no-one would be really hurt'. This is hardly true, but out of the chaos a new order is created: the constitutionally ineffectual Tallis is urged into action, and the exposure of an underlying lack of trust in Simon and Axel's relationship leads to a happier stability. During a philosophical debate with Rupert, Julius insists that goodness is dull and superficial, whereas evil 'reaches far far away into the depths of the human spirit and is connected with the deepest springs of human vitality'. The novel ends with him strolling happily through Paris, unscathed by the havoc he has wreaked, reflecting, in the novel's final words, that 'Life was good' (a neat authorial irony). However, goodness (personified in the weak Tallis) is seen to survive, even if it does not entirely prevail against evil, and this is the fairly honourable defeat of the title.

Shiva NAIPAUL　　　　　**1945–1985**

Fireflies

Naipaul's first novel instantly established him as an author in his own right, who was not going to live in the shadow of his older, established brother, V.S. Naipaul. The novel won the John Llewelyn Rhys and Winifred Holtby prizes as well as the Jock Campbell *New Statesman* Award. Set in Naipaul's native Trinidad, it tells the story of an indomitable optimist, Mrs 'Baby' Lutchman.

Baby is only an 'indirect descendant' of the Khojas, the self-appointed leaders of the island's Hindus, and therefore of little importance. She is married off to the unre-liable Ram Lutchman, a driver with the Central Trinidad Bus Company, and patronised by Govind Khoja, who is effectively head of the family now that his widowed mother is old and blind. Only her optimism carries Mrs Lutchman through the early years of her marriage; she supports Ram's successive enthusiasms for gardening, photography and Christmas. Though dependent on Mr Khoja in times of crisis, she stubbornly preserves her independence by blotting out anything she does not wish to see. Thus she is able to ignore her husband's youthful excesses and his protracted affair with the bogus anthropologist Doreen James. When he crashes the bus, however, it is Mr Khoja who finds him a job with the Ministry of Education, and Baby remains loyal to the family, attending innumerable functions and staunchly defending the posthumous reputation of Govind's mother. 'The woman dead now and you have no right to call she an old witch,' she tells her irreverent husband. 'She never do any harm to you. You should have a little respect for her memory.'

After Ram's death, all Mrs Lutchman's hopes rest upon her elder son, Bhaskar, who is training to be a doctor. (Her other son, Romesh, is a delinquent.) She takes in lodgers in order to support Bhaskar's studies, even though everyone else can see that he will never make the grade. Her faith is maintained by a neighbour, the Glasgow-born Mrs MacIntosh, who reads tea-leaves, and always has a ready explanation of Bhaskar's repeated failures: 'Somebody cheated him … Somebody with an "e" in their name.' Even when she has lost her home and has been deserted by Bhaskar, Mrs Lutchman maintains her faith and dignity.

A concurrent theme concerns the pompous Govind Khoja, who inherits the bulk of his mother's estate, and is consequently plotted against by his sisters. His political ambitions are ridiculed and his obsessive preoccupation with Rousseau's theories of education eventually lead to the failure of his school. He is supported throughout by his foolish wife, Sumintra. *Fireflies* is a masterly satire, which is funny, touching and biting by turns.

1970

Simon RAVEN   1927–

Places Where They Sing

see **Alms for Oblivion** (1976)

C.P. SNOW   1905–1980

Strangers and Brothers

George Passant (1940)
The Light and the Dark (1947)
Time of Hope (1949)
The Masters (1951)
The New Men (1954)
Homecomings (1956)
The Conscience of the Rich (1958)
The Affair (1960)
Corridors of Power (1964)
The Sleep of Reason (1968)
Last Things (1970)

When Snow was elevated to the peerage he chose for his crest the motto *Aut Inveniam Aut Faciam*: 'I will either find a way or make one.' The principle is as applicable to his literary work as to his controversial public life. Written over a period of thirty years, Snow's *roman fleuve*, *Strangers and Brothers*, spans half a century of life in England, forming a sequence of eleven volumes and some one-and-a-half million words. Although each of the novels is comprehensible as an independent work, *Strangers and Brothers* was conceived as a series, and achieves its effect through the gradual accumulation of detail. Like Anthony Powell's *A Dance to the Music of Time* (1951–75), the sequence portrays time passing both for individuals and for English society as a whole. In 1970 Snow revised *Strangers and Brothers* for an omnibus edition, rearranging the original order of publication so that the chronology of the fiction is less interrupted; this is, he states in the preface, 'the text which I should like to be read'. The following account of Snow's novels is accordingly based on the revised sequence, although it should be noted that some of the effect of 'compartmentalisation' identified by modern critics is a result of the varying periods of composition of the individual novels.

In a letter in 1961, Snow explained 'the phrase Strangers and Brothers is supposed to represent the fact that in a part of our lives each person is alone ... and in part of our lives ... we can and should feel for each other like brothers'. The balance between individual tragedy and social optimism is the recurrent theme of Snow's fiction, and the essential psychological dilemma for the narrator, Lewis Eliot, who is (in a phrase which Snow used about himself) a 'pious agnostic'. The character of Lewis Eliot, a blend of semi-autobiographical narrator and fictional device, has been the focus for the uneven critical reception of Snow's work. The literary style of *Strangers and Brothers* is a moral eschewal of the myth and symbol of the modern novel, a return to the 'realist' tradition of the nineteenth century with what Snow saw as its more valid concentration on social context rather than isolated psychology. The critical problem is that the reader is required to experience events solely through Eliot's eyes, but is also given sufficient indications of the narrator's psychological limitations to question his reliability. For the most part *Strangers and Brothers* portrays an exclusively masculine world in the upper echelons of society, in academia, scientific research establishments and the corridors of political power. Yet the all-consuming obsession with success rebounds into the emotional lives of the characters with disastrous results. There are few happy marriages in Snow's novels, and emotional fulfilment can be achieved only by relinquishing ambition in other spheres.

Chronologically opening the sequence at the start of the First World War, *Time of Hope* is the most concentrated exploration of the character of Lewis Eliot, dealing with his life until 1933. Bertie, Eliot's father, fails in business and plunges the family into poverty and its concommitant social humiliation. The boy softens the blows of the outside world for his dying mother, Lena, whose damaged pride dominates the house as she encourages her sons to fulfil her own thwarted ambitions. The effect on Eliot is to establish a pattern of behaviour dependent on a constant sense of self-repression, withholding himself as a detached spectator.

He had been manipulating the college for a generation. He was cunning, he knew all ropes, he did not invent dilemmas of conscience for himself. He wanted the mastership, and he would do anything within the rules to get it. But it had to be within the rules: and that was why men trusted him. Those rules were set, not by conscience, but by a code of behaviour – a code of behaviour tempered by robustness and sense, but also surprisingly rigid, surprising, that is, to those who did not know men who were at the same time unidealistic, political, and upright.

THE AFFAIR
BY C.P. SNOW

Throughout young adulthood, Eliot finds this psychological pattern damaging his relationships with others; his talent for intellectual analysis, whilst essential to professional success, proves a bar to emotional empathy. 'The next few years . . . are going to be a wonderful time to be alive,' rejoices the impervious optimist George Passant, the solicitor who becomes Eliot's patron; but as the youthful hopes of the 1920s slide into the disillusionment of the 1930s, Eliot finds his career at the bar threatened by illness and a tormented marriage. His love for the neurotic, frigid and possibly lesbian Sheila Knight takes a Proustian form, a parallel reinforced by the many references to *A la recherche du temps perdu*. Eliot is obsessed with Sheila but is unable to comprehend her. Their relationship takes on the one-sidedness of rape, and Eliot determines to end it; but he finds that only through such intensity can he 'lose vanity and caution', the metaphoric bondage answering aspects of his sado-masochistic psychology. Eliot's passion for Sheila reveals the man behind the public mask: she tells him that he's not as nice as people think.

Covering the same period of Eliot's adult life as *Time of Hope* (1925–33), *George Passant* presents the social and political context for Eliot's early years. Becoming a member of Passant's 'Group', Eliot witnesses the attempts of the young to escape the provincial morality of the town's 'bellwethers' (Passant's term for the Establishment); but the sexual liberation which the Group advocates proves increasingly to be another form of slavery, 'optimism gone mad'. Passant defends a member of the Group, Jack Cotery, against charges of immorality, as his scholarship at the technical college is threatened by hints of homosexuality. Passant celebrates the outcome as a great triumph, yet it is a hollow victory achieving only limited concessions.

More serious charges soon threaten Passant himself as he is indicted for conspiracy and fraud, and Eliot, as a young London lawyer, is called in as Chief Attorney for the Defence. The accused are acquitted, but Passant's reputation is tarnished. It is a mark of Eliot's growth into maturity that the man he once saw as 'one of the world's rebels' now appears to him as deliberately confining himself to a narrow sphere as an excuse for never achieving his potential.

Passant's early help in preparing Eliot for the bar examinations is succeeded in *The Conscience of the Rich* (which covers the period 1927–37) by the practical assistance of the March family, wealthy Anglo-Jews who give him access to powerful contacts. Ironically it is this very network of connections which drives Eliot's contemporary, Charles March, to reject the expectations of his father's world. Plagued by 'the conscience of the rich', he becomes a doctor in the East End of London, purportedly to be of more use to society; yet the fundamental choices in life can never be so unambiguously explained.

Charles's sister, Katherine, had eventually won the consent and support of their father over her marriage to a Gentile, the scientist Francis Getliffe. Mr March, however, finds it much more difficult to agree to his son's marriage, even though Ann Simon is Jewish; a tangled web of sexuality emerges as the father is also attracted to the girl. A further twist to the intrigue takes the form of Ann's involvement with an anti-Establishment magazine, since she intends to expose the shady dealings of Charles's uncle with the intention of damaging the Tory government. When dangerously ill as a consequence of the emotional pressure, Ann agrees to the family's requests to destroy the evidence, but passes on responsibility for the final action to her husband. Charles refuses to halt its publication, preferring to live with the contempt of his family rather than go against his principles.

The first of the 'Cambridge' novels, *The Light and the Dark* explores the manic-depressive character of Roy Calvert, whose tormented personality parallels Eliot's own experiences and the suffering of his wife, Sheila. Roy has received international acclaim for his study of Manichaean beliefs, and the heretical doctrine of this sect embodies one of the central themes of *Strangers and Brothers*; the dualism between soul (light) and body (dark), 'the most subtle and complex representation of sexual guilt'. Despite his worldly success, Roy is unable to obtain or sustain happiness: 'he was inescapably under the threat of this special melancholy, this clear-sighted despair in which ... he saw the sadness of man's condition'.

Roy's divided personality is further emphasised by the alternating settings of Cambridge and Berlin, where Eliot finds him living a Bohemian existence. An atheist who craves an ideal, Roy becomes sympathetic to the rising Nazi cause, and Eliot finds himself in the characteristic plight of loving someone whose deeply held views he finds morally repugnant. As war threatens, Roy returns to Cambridge, and Snow uses the high-table debates to analyse the complex responses of English intellectuals to the Munich crisis and the nature of power. Although Roy marries and has a daughter whom he adores, he is still torn by his melancholia, and deliberately chooses death by becoming a bomber pilot. Eliot guiltily remembers telling Roy that this was the most dangerous wartime job; to have known Roy was, he says, 'one of the two greatest gifts' of his life.

The setting of *The Masters* is Cambridge in 1937. In a college the old Master lies dying, the curtained windows of the lodge a silent reminder to his colleagues of the choices they face. The central question is the election of the new Master: Paul Jago, a scholar of literature, or Crawford, an honoured biologist. The cast of academic characters begins to group into two camps as the campaign gathers momentum. Rational argument plays little part in motivating allegiance, as support is determined by personal animosities or potential opportunities. In this confined sphere, Snow delineates a paradigm of the political process in action.

'I want a man who knows something about himself. And is appalled. And has to forgive himself to get along.' Roy Calvert's requirements of a candidate for the Mastership recall the wider debate of *The Light and the Dark*, and Jago manifests these qualities. Yet Jago's opponents manipulate his acknowledged faults through scurrilous rumour, exploiting the weakness of his candidacy, which is also his strength as a man – his devotion to an ambitious and unstable wife. By contrast Crawford remains an unknown quantity, avowedly above active self-promotion and indifferent to the outcome. The apparent civilities of high table and common room are a thin veneer glossing the power struggle, where every bottle of port presented uncorks further schemes and rivalries. Beneath the pictures of their predecessors the dons decide the future of their privileged world, and their final choice is the blankly portrayed Crawford; the modern age has no place for the faulty, humanist Jago.

The political processes of the Cambridge courts revealed in *The Masters* take on a sinister aspect in *The New Men* as the Second World War begins. The death of Calvert and defeat of Jago foreshadow the shifting of responsibility to the scientists, as the research establishments engineer the potential destruction of the human race with the development of the atomic bomb.

Lewis Eliot witnesses the increasing involvement of his brother Martin as the international struggle for supremacy perverts the discoveries of the physicists into a deadly game. Rapid career advancement propels Martin into authority, but when he prosecutes his treacherous colleague, Sawbridge, the unpleasant side of personal ambition takes on a moral weight which transcends the squabbling of the research staff. Although investigation reveals that Sawbridge's spying activities have not substantially damaged the national interest, Martin is determined on a relentless pursuit of his own power. Eliot sees in his brother 'the same closed mind, the two world-sides' which make him the counterpart of the traitor.

As the fission experiments proceed, however, Martin urges caution and thereby alienates himself from the Allies' strategy. Eventually refusing further involvement, he sacrifices career gains to an undistinguished life in academia. Yet although the history of the nuclear struggle develops through personality clashes and piecemeal discoveries, the individual who abdicates is not absolved from moral responsibility. Having gained unprecedented power, man is faced with the tragic ambiguities of its use; the Hiroshima bomb potentially radiates its cloud of destruction far beyond its target.

Encompassing the world events of *The New Men*, *Homecomings* focuses on the personal life of the narrator, not treated in depth since *Time of Hope*. It is now five years later in terms of fictional history, but nothing has changed in Eliot's marriage. Sheila's search for love has found inadequate outlets in a series of 'lame dogs', involvement with whom gives her 'a flash of hope'. Now sponsoring the flimsy business of an unscrupulous publisher, she is flattered into believing that she has literary talent. As the publisher spreads the rumour that Sheila is unbalanced and lesbian, however, the pace of her psychological disintegration quickens.

In a final attempt to make sense of her life, Sheila arranges to take a job, but as the starting date approaches she is terrified of going out into the world. Eliot forces the issue into a stern psychological test which she must pass to liberate them both; but pressure on Sheila results not in breakthrough but in suicide. Even as his wife dies, Eliot is distanced from grief and remorse by a feeling of relief, mingled with anxiety about the damage her death may have done to his career prospects.

Repressing his emotions by concentrating on success in government office, Eliot achieves rapid promotion. When he meets Margaret and they begin to live together, he again desires marriage but, learning of Sheila's suicide, Margaret concludes that 'in the end you'll break the heart of anyone who loves you'. Instead, Margaret marries George Hollis, a reliable young doctor, and has a child. After a while, however, the affair is renewed, and Eliot insists on its public recognition. Divorcing her husband, Margaret marries Eliot and has a second son. When the boy becomes dangerously ill as an infant, Eliot is finally forced into the 'true relation' which he had evaded for so long – he can no longer live 'above the battle of human involvement'.

The last of the Cambridge novels, *The Affair* covers events during the period 1953–4 and returns to the themes of *The Masters*, as Crawford's tenure draws to its close. The plot centres on a scientific publication in which the evidence has been distorted, and Snow comments that he took as a starting-point the Dreyfus affair, with its questions of moral justice. Donald Howard, the college fellow accused of the fraud, is widely disliked, and political and personal prejudice threatens to determine the outcome of the hearing before the Court of Seniors, as the academic elite teeters on the brink of tyranny.

Eliot argues that impartial justice must be the cornerstone of society, but the ideal is characteristically undercut as the affair ends in compromise. The evidence is insufficient to condemn Howard, and the apportionment of blame is confused by his loyalty to his now dead supervisor. Blatant injustice is avoided because Howard is not dismissed; but his fellowship will not be renewed at its expiry – a parallel verdict to the hollow victory of George Passant in *Time of Hope*.

Snow had used the phrase 'corridors of power' in *Homecomings*, and it had slipped into the national idiom to the extent that by using it as the title of the novel covering the period of the late 1950s he felt he was reappropriating his own cliché. The subject of *The New Men*, the future of nuclear weapons, has become an urgent political issue. Eliot is now at the apex of his government career; broadly sympathetic to the Labour movement, he is nevertheless chosen by the ambitious Tory minister Roger Quaife as a political associate. Finding himself privy to yet another privileged group, Eliot maintains an uneasy scepticism as he is accepted by the 'Basset set' – a fictionalised version of Lady Astor's Cliveden.

As in the earlier novels, Snow analyses the accidents which determine national policy; the rise and fall of powerful men depends on the same human weaknesses whether the political process is that of Cambridge or Whitehall. Quaife believes that the arms race must be stopped, but the international perspective of an American scientist, David Rubin, suggests that this is naïve idealism. Despite strong opposition Quaife pursues his conviction to the point of political suicide; but in his private life Quaife's conduct is less laudable, and Eliot becomes an embarrassed confidant for his duplicitous love-affair.

The title of *The Sleep of Reason* is taken from a Goya etching, 'The sleep of reason brings forth monsters'. Returning to his provincial roots in 1963, Eliot finds little cause for optimism about the human condition as he reassesses the past in the light of the present. Now a successful writer, Eliot has largely abdicated the official duties which he felt to be increasingly onerous; but political involvement takes on a different form as he is drawn into the in-fighting of the new university.

The vice-chancellor, Arnold Shaw, sternly upholds the educational ideals of a university, but his authoritarian élitism has made him many enemies. The immediate cause of contention is the dismissal of four students caught having sex in a hall of residence; Eliot argues that 'probably most people in the whole society ... didn't really regard fornication as a serious offence'. The incident takes on wider relevance, however, as Snow explores the conflicts between the generations – both within Eliot's own family and in the relationships of parents and children amongst his friends.

The question of the sexual revolution develops into sensational horror as Eliot is also drawn into a murder case as a legal advisor. Based on the Moors Murders trial of 1966, which Snow and his wife, Pamela Hansford-Johnson, attended, this strand of the plot explores the potential for tragedy as psychosexual drives spill over into insanity when acted out in real life. George Passant's niece and her lesbian room-mate are sentenced to life-imprisonment for the kidnapping, systematic torturing and murder of a schoolboy. The trial confronts Passant with the logical conclusion of his creed of libertinism, and he leaves England unable to face his past or continue to influence his protégés.

The dark tone of the novel persists in the sub-plots covering Eliot's personal life. Threatened with blindness – an autobiographical motif – he lies in the 'claustrophobic dark' of post-operation bandages. Margaret's father is thwarted in the attempt at suicide which would release him from suffering senility, and the novel closes with the funeral of Eliot's father, who has died with only the impersonal attendance of his lodger.

Concluding *Strangers and Brothers*, *Last Things* is really an afterword on the central themes and characters of the series rather than a novel in its own right. The plot is loosely organised and heavily allusive, and does not extend the philosophical unification achieved in *The Sleep of Reason*. Eliot's relationship with his son continues to re-enact the struggle between the generations; his failing sight again clouds his experience, and he decides not to accept a prestigious post with the new Labour government; and during a second operation he is pronounced technically dead, a brush with the afterlife conceived by an atheist. The nearness of death makes Eliot's constant sifting of his past a final acceptance of his own limitations, a resignation to the sad comedy of the human condition.

As his teenage son becomes embroiled in an affair, Eliot must acquiesce to a situation which would have been morally unacceptable in his own youth. The boy leaves for the Middle East with hopes of becoming a journalist, and Eliot declares that he will not be able to sleep easily until he returns; yet, as the links he had learned to forge throughout his life are broken, he finds instead that 'all was peace'. Through a lifetime's experience Eliot has gained wisdom, and looks forward to the optimistic future in the hands of the next generation. Personal tragedy is thus balanced by a final social optimism. For the reader, however, the perpetuation of the centres of power which Snow has occupied looks less certain, the deficiencies of the *status quo* counterbalancing the narrowly controlling hand of the narrator.

1971

War in Pakistan ● United Nations admits China ● Indira Gandhi wins in Indian general election ● US combat deaths in Vietnam exceed 45,000 ● First space station placed in orbit ● 'Greenpeace' founded ● Demise of agreed exchange rates ● First 'chip' introduced ● Germaine Greer, *The Female Eunuch* ● John Rawls, *A Theory of Justice* ● Edward Bond, *Lear* ● Films: *A Clockwork Orange, Death in Venice, The French Connection* ● Marvin Gaye, *What's Goin' On*, T. Rex, *Electric Warrior* – glam rock starts in Britain ● Coco Chanel, Stevie Smith, Igor Stravinsky, Nikita Khrushchev die ● Geoffrey Hill, *Mercian Hymns* ● George MacBeth, *Collected Poems 1958–1970*

Maureen DUFFY   1933–

Love Child

The narrator of this novel is Kit, a child of early teenage years, whose mother embarks, one hot Italian summer, on an affair with her husband's new secretary, Ajax. (The name was bestowed by Kit, a godlike act, to replace the never disclosed one which Ajax apparently dislikes.)

The household is, to say the least, cosmopolitan: a more or less English mother, a Scandiwegian husband, and a precocious offspring who between them can muster seven dead languages and five living, 'apart from ubiquitous English and smatterings of Cantonese, Urdu and Yoruba picked up on my father's travels. Foreign doesn't exist for us as a concept', and neither does Home. There are dwellings in London, Massachusetts, the Swiss Alps and Italy, between which Kit's diplomat-economist father and unspecifiedly influential mother shuttle on their way to and from international committees.

Just as the protagonists evince a general reluctance to be pinned down by geography and national identity, so too Kit's narrative withholds two pieces of information usually felt to be indispensable in the telling of a story of sexual love and jealousy. We never learn the gender of Ajax, nor, although discussions of this novel constantly assume that Kit is female, do we learn that of the child-narrator, either. Any description, simile or image which seems to indicate one sex is quickly followed by another which confounds the assumption. The technical challenge posed by the decision is considerable – how do you write without pronouns? – and it is one which Duffy has embraced in other, later novels, notably *Londoners* (1983). It is equally challenging to the reader, forced to consider the part gender plays in shaping both the story (any story) and the reader's response to it. Is Kit a boy or girl, jealous of a man or woman? At one level the uncertainty focuses attention purely on the emotions themselves; at another it keeps the reader in a state of perpetual disorientation, constantly confronting the supreme importance and irrelevance of gender.

As befits a citizen of the world – or eternally displaced person – Kit claims: 'If I don't feel myself greatly my parents' child I do, maybe more so because of that, feel myself the inheritor of civilization.' And it's true that the child's head is crammed with legends, myths, visual images and words drawn from several millennia of the world's cultures, many of which weave themselves into the fabric of the narrative. But Kit underestimates the destructive power of filial jealously unleashed when an adored mother takes a lover. The child, well versed to the point of ennui in sexual passion as it is depicted in the world's great paintings, poetry and music, is, in reality, utterly ignorant of its force and meaning. In part the ignorance stems from youthful inexperience, but only in part. Kit is dauntingly clever, formidably articulate, intellectually sophisticated, shrewdly observant, skilfully manipulative, and well informed about the mechanics of sexual activity – all qualities abundantly demonstrated by the glittering surface of a supremely accomplished narrative which ricochets between cynicism and despair. But intellectual development has, to a monstrous extent, outstripped emotional growth, and Kit's conduct as the affair progresses most closely resembles that of a malign Cupid who hovers, leers, spies, tricks, cheats, and eventually destroys the lovers as diligently as any of the maverick *putti* who people the Renaissance canvases alluded to throughout the novel.

You will be wondering, putative reader, why I have reported all this. The answer is quite simple: it interests me and you, forgive me, don't. I am not trying to tell you anything; I am at my childlike, priestlike task of creation. I am building sand castles for the tide to wash away or making mudpies that will never be eaten. You are privileged, if you ever exist, to look over my shoulder and study my re-creation but you mustn't interfere with your chatter about what you like.

LOVE CHILD
BY MAUREEN DUFFY

And, like a stricken Cupid who eventually stands tearful and repentant before Venus, his mother, Kit ends this narrative, all sophistication stripped away, with the anguished cry of a destructive child facing for the first time the true meaning of the devastation it has wrought: 'But I didn't know, I didn't know.'

E.M. FORSTER　　　　1897–1970

Maurice

'Publishable – but worth it?' Forster wrote on the typescript of this novel after he had prepared it for posthumous publication. The first draft had been written between 1913 and 1914 and 'Dedicated to a Happier Year'. It circulated amongst Forster's friends and underwent considerable revision throughout his life; in particular the ending was changed several times.

Maurice Hall is an unremarkable young man from the suburbs, whose education does little to break him of his habit of intellectual indolence. After prep school, where he is mystified by a sex lecture delivered on a beach, and Sunnington, where he is 'a mediocre member of a mediocre school', he enters Trinity College, Cambridge, and meets Clive Durham, an intelligent classics undergraduate. An affectionate friendship develops, the nature of which is unclear to Maurice until Clive unexpectedly declares his love. The priggish Maurice is appalled, and pompously rejects Clive's advances. Almost immediately he relents and has to persuade Clive of his love. In the exhilaration of the platonic romance which ensues, Maurice cuts lectures and insults the Dean, an episode which leads to his being sent down.

The friendship continues for three years, while Clive studies for the Bar and Maurice enters his late father's firm of stockbrokers. Then, during a solitary holiday in Greece, Clive undergoes a dramatic transformation: 'Against my will I have become normal.' He marries, and the shattered Maurice seeks a 'cure' from the uncomprehending Dr Barry, a family friend, and Mr Lasker Jones, a hypnotist who specialises in sexual difficulties. Salvation comes during a wet, dull week spent at Clive's estate at Penge in the arms of Alec Scudder, a gamekeeper. However, the divisions of class and Maurice's inherent snobbery besmirch this idyll, and Alec's letters suggesting that they should meet again strike Maurice as the threats of a blackmailer. The two men meet at the British Museum and Alec, stung by Maurice's behaviour, does indeed threaten to blackmail him. Once each has come clean about his feelings, they go to a hotel for a night. Alec, who is due to emigrate to Argentina, fails to catch the boat and Maurice discovers him in the boathouse at Penge. Maurice informs Clive of his decision to give up his job and live with Alec, thus fulfilling his dream: 'Two men can defy the world.' The novel ends with Clive planning to conceal the truth about his friend from his wife.

The novel has been accused of melodrama, sentimentality, proselytism and wish-fulfilment, but if *Maurice* is a failure, then it is an honourable one. Complementing the central, sexual theme are several familiar from Forster's other novels, in particular his concern with class, hypocrisy and bourgeois complacency. Like other characters, and like Forster himself, Maurice has to 'connect up and use all the fragments [he] was born with'. Whilst understandably wary of people's reactions to the novel, Forster felt that no assessment of his work would be complete without it.

In 1987, *Maurice* was made into a long, langourous, highly decorative and surprisingly successful film, directed by James Ivory from a screenplay by Kit Hesketh-Harvey.

Susan HILL   **1942–**

Strange Meeting

The First World War continues to exercise the imagination of British novelists, and Hill is one of several who have used the trenches as a setting for books about the relationships between men, their friends and families. The schism between the Home and Western Fronts provides a paradigm of the division between old and young, combatants and civilians, men and women. Hill's novel opens in the autumn of 1916 with a young officer, John Hilliard, on sick leave in an England he no longer understands or likes; he simply wants to return to the war zone. His parents are well-meaning but distant, while his sister Beth, the person to whom he has always been closest, has drifted away from him and announces that she is to marry a dull solicitor.

Back in France, Hilliard is billeted with a young subaltern, David Barton, fresh out of

training, cheerful and optimistic. At first mistrustful, Hilliard soon succumbs to Barton's openness: 'He thought, we need him, he has something none of us have, we need him to stay here, just as he is, to sit here night after night, telling us his stories, or nodding in that way he nods when someone else talks, sympathetic, happy to yield the floor – *liking us*.' Unlike Hilliard, Barton comes from a large, close family, with whom he keeps in constant touch by letter, and who automatically send their love to their son's friend, even though they have never met

---

The sweat was cooling on his back. The worst thing in his nightmares was always the smell, the sweet, rotten trench smell, of soil and chlorine and blood, and the mustard gas like garlic. His bedroom window was open and the room was full of the scent of roses, coming up from the warm garden. A sweet smell, and curiously like some cream or powder in a jar on his mother's dressing table. A sweet smell.

He pushed the bedclothes away, retching, leaned over the washbasin and felt the walls of his stomach clench uselessly.

STRANGE MEETING
BY SUSAN HILL

---

him. Barton becomes someone Hilliard can trust, confide in and, gradually, love: 'He was amazed at himself. That it was so easy'. Barton improves the morale of the entire battalion, even of Garrett, the CO who drinks so as to blot out the horror of having to pass on orders for useless and wasteful raids and attacks. Gradually, in spite of Hilliard's attempts to protect him, Barton loses his innocence in the face of death and destruction, but refuses to stop feeling and become cynical. In one extraordinary scene, he crawls out into no man's land in order to rescue a hedgehog, and then worries that he would not have done the same for one of the men. Much of the action is relayed through Barton's letters home; the rest is filtered through Hilliard's anxious, protective character. Inevitably, a futile attack is ordered – so futile that the CO refuses to carry it out and has to be replaced. Barton is killed. Hilliard survives, but loses a leg. The novel ends on an optimistic note, however, as Hilliard is driven to Barton's house, where the family wait to receive him as one of their number.

The title, of course, comes from Owen, and the entire novel is suffused with the poetry of the war, notably that of Sassoon, whose 'Lamentations' appears to have furnished Hill with a crucial scene in which a soldier breaks down. Essentially a love story, the novel is a moving, occasionally lyrical, evocation of the trenches written with great clarity and remarkable empathy.

Gladys MITCHELL   1901–1983

Lament for Leto

One of Mitchell's more than sixty distinguished, idiosyncratic detective novels, *Lament for Leto* is characteristic in combining a complex, coincidental plot, eccentric characters and a commonsensical but witty style. It features, as usual, Mitchell's celebrated detective, Dame Beatrice Adela Lestrange Bradley, Home Office psychiatric consultant, eldritch *grande dame* and sometime murderess.

Sheltering from the rain in the British Museum, Dame Beatrice meets an old acquaintance, the archaeologist Ronald Dick, who is getting up an expedition to Greece to visit various sites connected with the god Apollo, son of Leto by Zeus. The formal, slightly stylised tone of their encounter establishes the characteristic diction of the novel. Mr Dick persuades Dame Beatrice to join the group, which is the traditional ill-assorted, antagonistic holiday party of detective fiction: another middle-aged man, Henry, as devoted to botany as Ronald Dick is to archaeology; his two semi-criminal sons with their tutor; a pompous, difficult lady novelist, Chloe Cowie; her sullen niece, Mary; and two young half-Greeks, Hero and Simonides, who have been adopted by Ronald Dick.

The cruise sets off from Southampton for Piraeus, and soon the narrative is winding through alarming incidents such as theft of jewels, reports of murder plots, and ventriloquism at the Acropolis – events which drive a succession of angry, outraged characters to demand help from Dame Beatrice, only to be met by her 'crocodile leer' and pert replies ('What is the difference,' someone demanded, 'between a man and a botanist?' 'A botanist can be of either sex'). Gradually, the dislike of all becomes focused on the lady novelist, whose fall from the cliff called Sappho's Leap at Leukas is originally thought to be an accident, but turns out to be murder.

More *frissons* are added to the plot by the possibility that the dead woman may be not Chloe but her aunt, Megan Metoulides (*née* Hopkinson), conveniently living at Leukas and wanted by the Greek government for political agitation, and by the revelation that, unknown to them or herself, Chloe was the mother of the two young Greeks, although much disliked by them. Undeterred by these red herrings, Dame Beatrice, back in England, reveals the murderess to be the niece, Mary. Characteristically, although Dame Beatrice interviews the criminal in contemptuous style, there is no question of her informing the police: 'Your aunt was not an admirable woman.' As always in Mitchell, amiable amorality goes with a sense of nemesis and punishment fitting the crime.

V.S. NAIPAUL   1932–

In a Free State

Naipaul's short, Booker Prize-winning novel is set in a fictional, recently independent sub-Saharan African country in the throes of low-level tribally-based civil war. Naipaul explores realignments of both personal and political relationships caused by the retreat of colonialism in Africa and the overturning of the British ruling position.

Bobby, a colonial administrator, and Linda, the wife of one of his colleagues, share a two-day car journey through the civil war zone from the capital, controlled by the country's new president, to the 'Southern Collectorate', the tribal region of the now deposed king. They are an ill-matched pair. Bobby's promiscuous homosexuality and his supposed sympathy for the African cause alienate him from the English expatriate community. He values Africa for the freedom it allows him to escape English conventionalism. Although Linda shares his sexual indiscretions – she has a reputation as a maneater and uses the car journey as an opportunity to further a liaison with an American called Carter – her values and prejudices are those of the European compound. Her colonial mentality finds little affinity with emerging independent Africa, and her goal is to leave for South Africa.

Their initially tense relationship softens as they become aware of their common isolation amid an atmosphere of increasing menace outside the insulation of the car. A series of incidents contributes to their gloom and insecurity over their undefined position in the new Africa. A petrol attendant carelessly scratches the car and ignores Bobby's admonishment; a ride they are asked to give to an African turns sour when he insists they stop to pick up his friend and then go in a different direction; they are forced off the road by a military convoy. They stop for the night at a dilapidated lakeside hotel run by an aged and bitter colonel, who maintains his authority only by fear. In its deserted and decaying state, the once elegant resort is a pitiful anachronism. On an evening stroll outside the hotel grounds, they are intimidated by ragged soldiers from a nearby army camp and a pack of dogs, originally kept to guard white villas. Later that night Bobby's advances are rejected by the African barboy, and their car is almost stolen.

The next day, as they approach the Southern Collectorate zone, military activity increases. They pass what seems to be a car accident, but which afterwards they realise was the outcome of an ambush laid for the escaping king. The political life of the country is now moving at a level which does not involve them. With anxiety mounting, Linda becomes more colonialist and Bobby more assertive in his defence of African ways. They openly argue. At a checkpoint on the Southern Collectorate border, Bobby voluntarily stops at an army post to find out whether a curfew is operating. His attempted deference to the soldiers who, he soon realises, are out of control, rebounds on him, when he answers to the term 'boy'. He is

viciously beaten, but manages to drive the few miles back through the razed countryside into the European compound. As the final irony, rather than offering sympathy Bobby's houseboy Luke, who comes from the king's tribe, merely laughs at his misfortune.

Anthony POWELL — 1905–

Books Do Furnish a Room

see **A Dance to the Music of Time** (1975)

Simon RAVEN — 1927–

Sound the Retreat

see **Alms for Oblivion** (1976)

Paul SCOTT — 1920–1978

The Towers of Silence

see **The Raj Quartet** (1975)

Tom SHARPE — 1928–

Riotous Assembly

It all began when Miss Hazelstone of Jacaranda House, granddaughter of Sir Theophilus (one-time Viceroy of Matabeleland) and daughter of the late Judge Hazelstone, shot her Zulu cook, Fivepence, with an elephant gun. She freely admits not only to murder – the remains of Fivepence are spattered over the topmost branches of a tree – but, more seriously, to contravening South Africa's 'immorality' laws: 'Fivepence's death was a *crime passionelle*,' she tells the startled Kommandant Heerden. 'I was in love with him.' This is hard on the unfortunate Kommandant, quaintly an admirer of Britain's imperial past and of Miss Hazelstone as its last representative; covering up this 'lapse' is going to be the devil's own job. His first mistake is to send the homicidal Konstable

Els to the main gateway of Jacaranda Park with orders to shoot anyone trying to get in, and then to call out the riot squad. As a result, twenty-one policemen are killed. Luckily the Kommandant finds Miss Hazelstone's brother, the bishop, drunk on one of the beds upstairs. It is immediately obvious to him that the incapable bishop was responsible for all these tragedies, including that of Fivepence, and that his sister is covering up for him.

Mayhem follows in true Sharpe style. At one point the Kommandant finds himself tied to a bed, dressed in female rubber clothing, being raped (unsuccessfully) by Miss Hazelstone, dressed in male rubber clothing. This, and other incidents, overwhelm him. He diagnoses himself as suffering from a serious heart condition and, when the unhappy bishop is convicted of mass murder and sentenced to be hanged, persuades the condemned man to donate his heart for a transplant and reluctant doctors to carry it out. It is fortunate that the late Judge Hazelstone had decreed that any member of his family is to have the privilege of being hanged on the local gallows – unused for years and in a shaky condition – and that Konstable Els, promoted to Executioner Els, botches his first and last job. The manner in which the innocent bishop cheats the rope, the Kommandant keeps his perfectly good heart and Miss Hazelstone escapes from the mental home where she has started a riot, defies the art of synopsis.

Riotous Assembly is pure black farce and as such full of macabre scenes; there is, for example, the meeting in the prison chapel – with its stained-glass windows depicting martyrdoms – between the Kommandant and the bishop, conducted there in chains, to discuss the heart transplant question; and the 'hanging' itself, which ends in mass 'execution' with Els on top of his own funeral pyre. Sharpe lived in South Africa in the 1950s, and the comic fury of both *Riotous Assembly* and its sequel, *Indecent Exposure* (1973), was fuelled by his hatred of apartheid. Subsequent novels have satirised British life and, although their brand of bawdy farce has proved hugely popular, their targets have not inspired quite the same quality of savage humour.

Francis STUART   1902–

Black List, Section H

The theme of this novel is how H, a writer, reaches artistic maturity by experiencing pain and guilt. Deliberately placing himself beyond the social norm of acceptability, H eventually understands himself. It is partly autobiographical, but although 'real' names, such as W.B. Yeats and Liam O'Flaherty, are included, the details are not necessarily factual. *Black List, Section H* is written with a wry humour, and self-parody and comedy prevent the novel from becoming egocentric. Passages about Yeats, for example, might have been reverential, but are memorable because they do not conform to received opinion on the poet.

In 1918 H is sixteen and living comfortably with his Protestant landed uncles in Antrim. He is compelled to rebel and, because his family name will attract attention, publishes a letter supporting Home Rule in a Dublin newspaper.

This is his first conscious dissenting act, and it shocks his family. H, although he does not understand this at the time, wants to empathise with his dead father, Henry, of whom he knows nothing. Never spoken of, Henry is an outcast, rumoured to have died of alcoholism in Australia. To H, he is a hero.

Imaginatively, H gains support from Tolstoy's *Resurrection* and becomes obsessed by Masha, a prostitute and thief; another outcast, she is to him a heroine. H welcomed the Russian Revolution, and this and other factors convince him that he will be a dissident poet. H has only really ever loved his nurse, for he is remote from Lily his mother. Sent to Dublin to be tutored for Trinity College, H falls in love with Iseult Gonne, whom he marries. This, like much else H does, happens through a misunderstanding, and the marriage is disastrous. While imprisoned for Republican activities in the civil war, H meets Lane, a working-class prisoner who confirms H's philosophy. Lane insists that the only truth is to be 'countercurrent' to conformity. Life itself must be experienced fully, then re-created imaginatively through poetry or fiction.

This H tries to do and, to extend his

consciousness, embarks on a reckless course of drinking, gambling and womanising. He also goes through a deeply religious and reclusive phase, trying to resolve the contradictions of the sacred and the profane. Poetry and novels earn H praise from Yeats and others, but H is not content. He tries to repair his marriage, resenting the interference of Iseult's mother, Maud Gonne MacBride. She, as a revolutionary figure, is always certain of her ideals, whereas H is not. H and Iseult constantly quarrel, and the birth and death of a daughter force them further apart. Iseult's laziness and untidiness are ridiculed, but she and other characters are portrayed at their worst, to stress H's delinquent behaviour and apparent self-righteousness. His quest for spiritual satisfaction is paralleled by the appearance of various women to whom he is attracted, and marks significant sections of the novel.

With the setting in of the long nights, the war loomed directly overhead out of the moonless sky and street after street crumbled into hillocks of white rubble that soon weathered into a gritty mud that sprouted weeds, between which paths were trodden and new landmarks appeared to replace the old: the floor of a room balanced between two solitary upright walls; a cellar that they passed on their way to the only restaurant in the neighbourhood still serving fish off the ration, gaping from under a pile of bricks from which came the stench of death and disinfectant.

BLACK LIST, SECTION H
BY FRANCIS STUART

H's first love remains a fantasy, like his feelings for Masha. He believes he will fulfil his fantasy through Iseult, because of the poetic world she lives in, and when this fails, H turns to Polenskya, a dancer (and in reality Karsavina). Their love is platonic, but partially fulfilled, because she responds to poems H has written for her.

H spends some time in London in the 1930s, occasionally with an actress, Julia-,hoping to gain new insight about himself.

When he returns alone to Iseult, the Second World War is declared and, shortly afterwards, he accepts a post as a lecturer in Berlin University. H knows this is beyond the moral pale, although as an Irish neutral he can live undisturbed in Berlin. In all the wayward decisions of his life, H feels that he has not suffered enough as an outsider, but senses Germany will extend his imagination to embrace the self-awareness he desperately seeks.

A relationship with Susan, a neurotic would-be actress, fails because H has not yet fully understood that he creates fantasies about women rather than seeing them as they are. H willingly agrees to broadcast to Ireland, knowing this will finish his reputation as a writer, and begin his self-debasement. As the war ends, H becomes a refugee, but for the first time gives and receives intense love. He has met a Polish girl, Halka, and grows spiritually in her company. Their love is intensified by the physical danger they endure, including imprisonment. During this degradation, H at last achieves self-understanding, a state that for him is possible only through suffering.

Elizabeth TAYLOR   1912–1975

Mrs Palfrey at the Claremont

In 1967 Taylor read in *The Times* diary that the author of an acclaimed new novel, *At the Jerusalem*, was employed by Harrods. Curious that a young novelist should set his first novel in an old people's home, Taylor went to the store to take a look at the author, Paul Bailey. Some time later she met Bailey at a party, and she borrowed some of his circumstances for the character of Ludovic Myers, who is an actor turned writer, and has a connection with Harrods.

Mrs Palfrey at the Claremont is a painful, funny account of the depredations of old age, and the ruses adopted by the elderly to evade the grip of boredom and clutch at their few remaining scraps of dignity. The Claremont Hotel, a dowdy establishment in the Cromwell Rd, is grudging host to a group of decrepit residential guests, amongst whom are: the recently-widowed Laura Palfrey, a colonial who has the appearance of 'some famous general in drag'; the spiteful, inquisitive and arthritic Mrs Arbuthnot; fluttery Mrs Post, who is constantly knitting; Mrs Burton, a mauve-rinsed dipsomaniac; and Mr Osmond, the sole male, a peppery widower who spends his time writing letters to the *Daily Telegraph* and sharing risqué stories with the hotel staff. The residents perform a genteel *danse macabre* of social one-upmanship and points-scoring, bound together by their circumstances, rather than by any feelings of kinship or affection. Special occasions – such as the horrendous drinks party given by Mrs de Salis (a 'bird of passage' who had stayed at the hotel while convalescing) and the farcical Masonic Ladies' Night – are keenly anticipated but tend to disappoint.

When Mrs Palfrey becomes involved with Ludo Myers, who had come to her aid after she had fallen down in the street, she begins to break the three rules which have governed her life: 'Be independent; never give way to melancholy; never touch capital.' Because in the eyes of her fellow residents 'relations make all the difference', Mrs Palfrey passes off Ludo as her grandson. (Her real grandson is the dreary, pompous and rude Desmond, who works at the British Museum and is writing a book about Cycladic embroidery.) Ludo plays along with this, largely because Mrs Palfrey is good copy for a book he is writing, which he intends to call, after a remark made by Mrs Palfrey, *They Weren't Allowed to Die There*. (Guests who start to go to pieces, like Mrs Arbuthnot who becomes incontinent, are obliged to leave the hotel.) Mrs Palfrey forms a sentimental attachment to Ludo and lends him money, so that he can help his feckless mother, 'Mimsie', who has been deserted by her latest lover, 'the Major'.

Mr Osmond, who sees Mrs Palfrey as the still centre of almost masculine good sense amid the female swirl of the Claremont, proposes to her during the Masonic dinner. Appalled, she refuses him, and in a later attempt to evade his attention, falls and breaks her hip. When Ludo visits her in hospital in order to repay the loan, she confuses him with her real grandson as she slips into death. Her daughter, deciding that 'there was no-one left who would be interested', does not bother to announce the death in the *Telegraph*.

John UPDIKE   1932–

Rabbit Redux

see **The Rabbit Tetralogy** (1990)

1972

Richard Nixon visits China ● Direct Rule in Northern Ireland ● East–West Detente ● Tutankhamūn exhibition at the British Museum ● Tate buys André's bricks ● CAT (computerised axial tomography) scanning introduced ● Samuel Beckett, *Not I* ● Tom Stoppard, *Jumpers* ● Films: *The Discreet Charm of the Bourgeoisie*, *The Godfather* ● Television: *M.A.S.H.*, Ingmar Bergman, *Six Scenes from a Marriage* ● David Bowie, *The Rise and Fall of Ziggy Stardust and the Spiders from Mars* ● Reggae spreads from Jamaica ● Bobby Fischer beats Boris Spassky in World Chess Championship ● Maurice Chevalier, John Berryman, Kwame Nkrumah, J. Edgar Hoover, C. Day Lewis, Duke of Windsor, Louis Leakey, Ezra Pound, Harry S. Truman die ● Seamus Heaney, *Wintering Out*

John BERGER   1926–

G

Appearing after two earlier, fatally flawed novels, *G* caught contemporary reviewers by surprise. Its bold experimentation seemed calculated to provoke reaction and this it did. It was 'self-indulgent' to Bernard Bergonzi, 'a trendy mockery of a novel' to Graham Lord and 'imbecilic' to Auberon Waugh. *G* is set out in 'stanzas' ranging in length from a few pages to a single line, which announce at a glance its intention to get away from straightforward, linear narrative. Nevertheless, there is a plot which Berger clearly intends his readers to follow.

The novel opens by sketching in the immediate family history of its eponymous hero, G. The son of Umberto, a wealthy Italian merchant, and Laura, his mistress, G is born in 1888. Abandoned by his mother in England, he grows up on a farm with two adult cousins, one of whom will later sexually initiate him. The narrative then advances to 1910. G is wealthy and he drifts around Europe in search of sexual gratification. In Domdossola he witnesses the first flight across the Alps, but his attentions are more devoted to the seduction of a hotel maid. An affair with the wife of a motor magnate leads to his being shot in the shoulder by the outraged husband. He moves to England, then to Trieste just before Italy's entry into the First World War. Here G becomes embroiled in the political upheavals of the time and devotes much energy to the seductions of a banker's wife and a Slovenian working girl. Unsuccessful in both these enterprises, he is swept up in a riot and murdered on the orders of one of Trieste's many political players.

Berger intersperses this narrative with numerous digressions and asides to the reader, who is never allowed to forget that the book is first and foremost a fabrication. The story is told from several different angles, Berger's sophisticated collage-techniques being frankly rooted in those of the Cubist painters and the more radical elements within the French *Nouvelle Vague*. There are also literary antecedents for *G*. The 'New Novel' developed by Sarraute, Robbe-Grillet and others shares with *G* an acute awareness of its own fictiveness, a rejection of linear narrative and a tolerance of unresolved textual difficulties. The production of the story is at least as important as the story itself.

Many of these strategies, derived from contemporary Marxist theories, aroused the hostility of reviewers, but the novel was a popular success in Britain where readers saw it in the less rigorous context of gentler experimental works such as John Fowles's ***The French Lieutenant's Woman*** (1969). This popularity was endorsed when *G* won the Booker Prize. Perhaps recognising the co-option of his novel by mainstream (and, as he would see it, bourgeois) critical opinion, Berger attacked Booker McConnell for their colonial past in his acceptance speech and announced that he would donate half the prize money to the Black Panthers.

For Umberto madness is native to Livorno: he sees madness in the massive monolithic warehouses, eyeless and mute like deserted forts, in the four Moors chained cursing to the monument of Ferdinand I of Florence, in the conglomeration of stuffs with which the capacity of the city is overfilled, in the rectangular spaces of sky cut out by the massive regular buildings above the dark canals, in its shifting population, in the blankness of its walls, in the indeterminacy of its spaces, in its smell of poverty and superfluity, in its furtive opening to the sea.

G

BY JOHN BERGER

George Mackay BROWN   1921–

Greenvoe

Brown has always drawn upon his native Orkneys for his work, and in his first novel he creates an imaginary island called Hellya, which is required as a base for Operation Black Star, a military-technological project. The unbroken pattern of nature and the sea, impervious to human events and illustrated by poetic imagery, is a constant backcloth to the action. The book is divided into six chapters, the first five of which are set in 1968 and trace the last days of the village of Greenvoe. Although the trajectory of the novel seems pessimistic – the life of the island, and the lives of several islanders, are destroyed – each chapter ends with the rituals performed at dusk by a secret society, the Ancient Mystery of the Horsemen, which prepare the reader for an optimistic resolution in which the havoc caused by technology seems insignificant when set against nature's capacity for renewal. The rituals are conducted by the Lord of the Harvest, Mansie Anderson, who is admitting his son Hector to the society through the six initiation rites of the corn ceremony.

In spite of the novel's grim events, Brown's depiction of the villagers is often highly comic. They are parochial in outlook, dependent upon farming and fishing for their livelihood. Even before the arrival of Operation Black Star, the island is dying and its population is in decline; only the publican and the shopkeeper prosper. Joseph and Olive Evie, who own the village store, are a greedy, scheming couple, who connive at the community's destruction (finally retiring in considerable comfort) and provide their own commentary on local life. In the bar of the Greenvoe Hotel, and in the boats, the lives of three fishermen gradually evolve. Bert Kerston is constantly drunk, and consequently nagged by his wife. She becomes pregnant and Bert is beaten up by his son for neglecting her. The obsessively religious Samuel Whaness is hard-working but parsimonious and hypocritical, unwilling to thank his rival, Kerston, for saving his life. He shares Bible readings with his wife to compensate for their childlessness, while fathering a daughter upon another woman. 'The Skarf' is a Marxist and atheist who has abandoned fishing to write a local history and is bought drinks as he relates Hellya's past to anyone who will listen. The island has often been invaded – notably by the Vikings, who harvested the land and fed the inhabitants – and the Skarf's tales of feudalism and capitalism suggest that the Black Star project will be part of an established pattern.

The islanders have so far remained unaffected by progress, however, and this 'innocence' is represented by two harmless characters who are very much part of Hellya and are content to survive on their own terms. Timmy Folster is a meths drinker and beachcomber who lives in a ruined shack, while Alice Voar's pleasure in life is reflected in the large number of illegitimate children she bears (including Samuel Whaness's). To a 'civilised' society, however, their circumstances suggest decay. This decay also affects those who do not belong to the island, such as the minister, Simon McKee, a secret alcoholic who has been brought to Hellya from Edinburgh in a vain attempt to dry out, and his elderly mother, Elizabeth, who is racked by Calvinist guilt. Elizabeth believes she is responsible for her son's condition (she had dosed him with tonic wine as a child) and her conscience is assailed by imaginary 'Tormentors', who examine her past life and confront her with sins. She yearns to return to Edinburgh, and as she drifts into senile confusion, her wish is fulfilled.

Bella Budge presided over a diminishing republic of hens. Fewer wings beat at her skirt with every morning that passed. It became obvious to her that they were being stolen by the workmen at the site. 'Poor men,' said Bella, 'they'll be half starved up there, no doubt, and all that noise and gutter too.' She took Kitty inside and abandoned the rest to their fate. One by one, after nightfall, the hens were strangled or riddled with shotgun pellets.

GREENVOE
BY GEORGE MACKAY BROWN

The entire village is bought out by government representatives and every building is bulldozed. Anderson's cornfield, traditionally first planted by the Vikings, is dug up and the islanders are dispersed. Timmy is put into a home and although the Skarf is briefly employed as a clerk by Operation Black Star, he is later sacked because of his political sympathies, and subsequently drowns himself. After fifteen months, however, the Operation mysteriously fails, and Hellya is sealed off from the outside world. People and history have been needlessly destroyed, but not the hope that life will be renewed with the planting of seed. Ten years later, in the final chapter, Mansie Anderson returns to conduct the sixth rite of the Ancient Mystery of the Horsemen, at which the spirit of resurrection, which endures through time against the depredations of man, is reinvoked.

Susan HILL   1942–

The Bird of Night

In old age, Harvey Lawson remembers three of the twenty years he spent with Francis Croft, citing frequently Croft's diaries. The two meet, shortly after 1918, at a house-party. Francis already has a reputation as a poet: he has published some juvenilia and war poetry. He is also severely disturbed. He drags Harvey from his bed at two in the morning to take him on a walk through nearby woods. A tawny owl flies past. It is an omen of both wisdom and terror.

Harvey agrees to lunch with him in London. Francis does not appear. Alarmed. Harvey travels to Oxford, where Francis rents a dismal set of rooms. Francis is having a breakdown. His curtains are drawn; in his present state, he dreads the light. He weeps, raves, discordantly bangs the piano and begs Harvey to read Turgenev's *Fathers and Sons* to him. He declares that he is terrified of doctors – and of his parents, who wish to have him placed under medical care. His insanity did not begin in the trenches, he says: 'It began years ago, it began at school, it has always been there.' He tried to murder his brother when they were boys.

Sensing that Francis must 'have freedom and quiet to get on with his work', Harvey gives up his own career as an Egyptologist at the British Museum in order to become Francis's protector and companion. Together they travel to Kerneham, where Harvey has a cottage. There Francis writes *Janus*, a meditative, philosophical piece. Upon its publication, he is hailed as 'a major artist'. Whether in Kerneham, Venice or London, however, Francis is susceptible to depressive fits. With great forbearance, Harvey listens to his rantings and tolerates his whims, but following a tragi-comic episode in Hyde Park, Francis is interned in an asylum. His father, a farmer from Perth, visits him once, offering a common-sense perspective, as inappropriate as it is sympathetic.

Discharged, Francis teaches for a term at a Californian university. On his return to England, he attempts to kill first Harvey then himself. A further major poem is published:

'there was no doubt in anyone's mind now that Francis was a poet of genius'. The subsequent seventeen years – hurried by in Harvey's recollection – are spent in a castle on the Rhine and back at Kerneham. Finally, Francis commits suicide with a pair of garden secateurs. Harvey is pestered in his retirement by his friend's would-be biographers. He burns what papers remain, as he promised he would.

The fictional portrayal of a literary figure generates a peculiar difficulty: how is his work to be presented? Hill's strategy is to employ the conceit that Francis's work is in the public domain. If this approaches the limits of audacity, it also allows Hill to attend to her proper subject – namely, the relationship of genius to madness. It is, she suggests, a matter of oppositions: extreme neurosis and acute perception, prolonged tranquillity and profound distress. Like his Janus, Francis is always facing in two directions: towards the chaos of despair and the discipline of the aesthetic. The incompatibility of his vision with his art renders life in the end unbearable for him. He cannot assuage his agony in words, and so he resolves it in the silence of death.

Michael MOORCOCK   1939–

An Alien Heat

see **The Dancers at the End of Time** (1976)

Chaim POTOK   1929–

My Name is Asher Lev

In common with other of his works, Potok's first Asher Lev novel charts the rebellion of an orthodox Jewish son against the strictures of his religion into the forefront of western intellectual and artistic life. Potok draws on personal experience: he too was born into a doctrinaire Jewish family and reacted by becoming both a writer and a liberal rabbi.

Brought up in Brooklyn in the 1950s, Asher Lev is a genuinely gifted artist from a fundamentalist sect, the Ladover Hassidim.

Asher's childhood revolves around the traditional triple pillars of family, *yeshiva* (religious school) and synagogue. The Levs are one of the sect's foremost families with an exemplary history of service to their community: Asher's father works as emissary, sometimes undercover, for the hereditary leader of the sect, the Rebbe, to safeguard the interests of their followers in Europe, particularly Russia, then at the height of Stalinist anti-Semitism.

Asher Lev feels a heavy responsibility towards his Jewish heritage and his parents, but is interested only in his drawing. 'Playing, drawing, wasting time,' thunders a mythic ancestor who regularly haunts his sleep. Fearing his only child's talent is from the 'Other Side', as 'Jews don't draw', his father is openly hostile and uncomprehending. Asher is closer to his mother and while she both loves and supports him, she is deeply troubled by his gift. As his drawing develops, Asher is brought into conflict with his family. When his father is posted to Vienna, Asher refuses to leave New York, sending shockwaves throughout his family and the community. The Rebbe recognises Asher's talent and its importance to him and intervenes to allow him to stay with his uncle in Brooklyn, while his father, later followed by his mother, travels to Europe. The Rebbe also arranges for him to be tutored by Jacob Kahn, a famous painter and non-practising Jew. He becomes Kahn's protégé and Kahn educates him in the tradition of western and Christian art. Picasso becomes Asher's artistic model and the attainment of truth and integrity in his work becomes his highest ambition. His first exhibition is met with critical acclaim but boycotted by his family because of its inclusion of nudes.

Rejecting his culture only in so far as the rigid customs of the Ladover Hassidim impinge on his artistic vision, he continues to observe the customs of the group. His artistic egotism hangs awkwardly with his general desire to conform, and the drama of the book arises from the tension between these two conflicting loyalties. His second exhibition contains two artistic triumphs entitled 'Brooklyn Crucifixion I and II' which depict his mother on the cross. The blasphemy of the imagery ensures that his position in his community becomes impossible, and the Rebbe exiles him to Paris.

Asher Lev's story is taken up again in *The Gift of Asher Lev* (1990), in which he returns to Brooklyn and the Rebbe following his lengthy exile.

Simon RAVEN   1927–

Come Like Shadows

see **Alms for Oblivion** (1976)

Ishmael REED   1938–

Mumbo Jumbo

A detective novel, conspiracy thriller, black manifesto or alternative theological tract: these are all valid descriptions of Reed's third and arguably most demanding novel. Set in the Harlem Renaissance of 1920s' New York, *Mumbo Jumbo* tells the story of Papa LaBas, a detective in search of 'Jes Grew' and its texts. Jes Grew itself is part movement and part doctrine, almost an organisation and nearly a faith. Its nearest counterpart in contemporary fiction is the Trystero of Thomas Pynchon's **The Crying of Lot 49** (1966), which similarly manifests itself in 'outbreaks' rather than hard and fast proclamation. In his search, LaBas's main opponents are the followers of the Atonist Path and their military wing, the Wallflower Order. Their conflict is primarily a mythic one, for against the Atonist Path, Reed pits his own mythology of Neo-Hoodoo, while behind both lies the eternal struggle between Set and Osiris, darkness and light. The Wallflower Order, it gradually turns out, are descendants of Set through the Knights Templar of the twelfth century, and it is they who have appropriated the text of Jes Grew and are circulating it in fourteen sections as an approximation of Set's dismemberment of Osiris. Eventually, the text is assembled and translated, at which point it turns out to be a version of the myth of Thoth. In this manifestation, Jes Grew articulates itself through New Orleans ragtime before moving to New York from where LaBas will reassemble its mutilated history and tell the story that constitutes the novel.

Mumbo Jumbo is a determinedly experimental novel, performing its own dismemberment of the body of white novelistic tradition. The most obvious form this takes is Reed's parody of the detective genre, in particular an absurdly erudite explication scene. However, questions as to how history should or could be passed down, and how knowledge accumulates within stories (how it is then used to serve specific agendas), are also never far from the trail of LaBas's quest. The footnotes and 'Partial Bibliography' which adorn the novel's text are over-obvious postmodern adjuncts but serve to highlight Reed's reflexive sense of his own part in this process. He avoids the vacuous ironies such devices often generate by grasping an essential tension. That all texts are untrustworthy is a necessary condition of postmodern fiction; that all texts present their material to some degree as facts deserving attention is the contradiction which redeems this scepticism from tedious nihilism. Reed's experimentation is based upon exactly this insight. The mythology he part-manufactures, part-explicates, the elaborate and supple contortions of the plot and the arcane erudition he marshals behind his characters are all facts which demand attention and compel belief.

Reed was criticised for allowing the mythic elements of the novel to take precedence over the satire which had been a much-praised feature of his earlier works. However, his subsequent development of this feature in *The Last Days of Louisiana Red* (1974), in which LaBas continues his quest, has done much to vindicate the practice of the earlier work. With hindsight, it can be seen that *Mumbo Jumbo* marks a new phase in the work of one of America's foremost experimental writers.

Eudora WELTY   1909–

The Optimist's Daughter

Welty's last novel to date was originally written for the *New Yorker* in 1969 and then revised and extended for volume publication before winning the Pulitzer Prize in 1973. It is the most autobiographical of her novels and, like all her work, suffused with a sense of place.

Laurel McKelva Hand is living and working in Chicago when she learns that her elderly father, Judge McKelva, has gone into hospital in New Orleans for an operation. He has scratched his eye on a rose known to everyone as 'Becky's Climber' after his now-dead first wife. The symbolism is hard to ignore. In what seems an excessive gesture – expressive perhaps of a sense of guilt – Laurel flies to New Orleans to be at his side during the operation and necessarily long period of recuperation. Since the death of her mother, and her father's surprising remarriage to the vulgar, lower-class Fay, Laurel has seen little of him. A self-declared optimist throughout his life, the Judge goes into an unaccountable slump after the operation and dies. Fay accuses the doctor of being a murderer and spits at Laurel: 'Never speak to me again! That nurse dragged me and pushed me, and you're the one let her do it.'

Back in the Judge's home town of Mount Salus, old friends of the family gather for the funeral, recollecting the wonderful Becky and gossiping about the awful Fay. Fay's grotesque Texan family turn up for the funeral, further compounding the horror of the respectable southerners, and Fay indulges in a histrionic display of weeping over the open coffin. After Fay has returned home with her family for a few days, Laurel is left in the old house. She is appalled by the changes Fay has wrought and haunted by the ghosts of the past. The tragedy of her own life – she lost her young husband in the war – and of her mother's last years, when she became insane, are gradually revealed. Above all else, we are left with a typically Southern sense of the past as something irrecoverably lost in the face of a debased present.

1973

US withdrawal from Vietnam ● Yom Kippur War ● Energy crisis ● Britain, Ireland and Denmark become members of the EC ● Watergate hearings ● Military junta seizes power in Chile ● President Salvador Allende reportedly commits suicide ● Three-day week in Britain to save energy ● E.F. Schumacher, *Small is Beautiful, a Study of Economics as if People Mattered* ● Calf produced from a frozen embryo ● First NMR (nuclear magnetic resonance) image ● Edward Bond, *Bingo* ● Peter Shaffer, *Equus* ● Films: *O Lucky Man, Mean Streets* ● Television: *Kojak* ● L.B. Johnson, Elizabeth Bowen, Noël Coward, Pablo Picasso, Jacques Maritain, Otto Klemperer, John Ford, J.R.R. Tolkien, Pablo Neruda, W.H. Auden, Pablo Casals, David Ben-Gurion die ● Derek Walcott, *Another Life*

J.G. FARRELL   **1935–1979**

The Siege of Krishnapur

'It seemed to me that the really interesting thing that's happened during my lifetime has been the decline of the British Empire,' Farrell once commented. His three major novels, sometimes known as the 'Empire' trilogy, reflect this interest: *Troubles* (1970) is set in the crumbling Majestic Hotel in Ireland after the First World War; *The Siege of Krishnapur* depicts an imaginary mutiny in British India in the 1850s; and *The Singapore Grip* (1978) describes the fall of Singapore to the Japanese army in 1942, a surrender described by Winston Churchill as 'the worst capitulation in British history'. The novels are suitably ironic: woeful comedies into which, generally as a result of incompetence or miscalculation, tragedy erupts. They are scrupulously researched and highly readable, and Mary McCarthy spoke for many when she wrote that *The Siege of Krishnapur* (which won the Booker Prize) 'has everything you could expect to find in a big, old-fashioned novel or several of them – characters, suspense, military action, romantic attachments, satire, wit, philosophy'.

Mr Hopkins, the Collector (head of administration for the district), is a strong believer in the efficacy of 'civilisation'. While in England in 1851, he visited the Great Exhibition, and has since 'devoted a substantial part of his fortune to bringing to India examples of European art and science in the belief that he was doing as once the Romans had done in Britain'. However, he is under no illusions about the ungrateful Hindu sepoys, who are on the brink of mutiny, and far from ready for the march of progress. His pleasure-loving acquaintances in Calcutta watch with amusement as he stomps distractedly from house to house warning of the wrath to come.

When the sepoys attack, the British retire to the residency, where the Collector takes command, largely because there are not many under his charge to whom he could delegate. Some, indeed, prove to be positively disruptive. The atheistic magistrate, Tom Willoughby, his face 'raked, harrowed, even ploughed up by free-thinking and cynicism', despises everyone. The padre, exhausted by his efforts to root out sin and heresy from the garrison and point out the evils of materialism, proves singularly useless, haranguing the fighting men from the ramparts and delivering a largely incoherent sermon at the height of the siege. The opinionated Dr Dunstaple's rivalry with the 'upstart' Dr McNab is a running sore throughout the siege, and culminates in the great debate upon the treatment of cholera, at the conclusion of which Dunstaple dies after insisting that he takes his own cure.

Others prove more helpful, some unexpectedly so. George Fleury, recently arrived from England with his trunk full of books and his head full of poetry, seems to be interested only in Dr Dunstaple's popular daughter, Louise. Under the expert tuition of Louise's brother Harry, however, he becomes an effective if eccentric soldier. Harry himself, unsuitably in love with the 'fallen' Lucy Hughes, performs tremendous feats with cannon, saving many a desperate situation, while the railway engineer, Mr Ford, keeps a cool head and proves an excellent tactician (although he is killed before seeing his strategy's success). Even Lucy discovers a talent for making bullets. Meanwhile, Fleury's widowed sister, Miriam Lang, helps in the hospital, keeps up the morale of the other women, and becomes the Collector's chief solace.

By the time relief arrives, all the Collector's cultural treasures have gone, either destroyed by enemy action or transformed into ammunition. 'Civilisation' has proved useful in a way he could not have envisaged, while Victorian values have disintegrated. The general who rescues 'the heroes of Krishnapur' is appalled: 'He had never seen Englishmen get themselves into such a state before; they looked more like untouchables.' A final chapter reveals the subsequent fates of the characters. Harry marries Lucy, becomes a general, and remains in India, along with Dr McNab, who marries Miriam. Fleury returns to England, marries Louise and becomes stout and opinionated. The Collector also goes home, but resigns from all his artistic, antiquarian and philanthropic societies and shocks Fleury by declaring: 'Culture is a sham. It's a cosmetic painted on life by rich people to conceal its ugliness.' Farrell concludes that by the end of his life, the Collector had perhaps 'come to believe that a people, a nation, does not create itself according to its own best ideas, but is shaped by other forces, of which it has little knowledge'.

Graham GREENE   **1904–1991**

The Honorary Consul

Greene has described *The Honorary Consul* as his 'personal favourite' amongst his novels. 'It combines my religious interest with my political interest, and I think I have succeeded here as nowhere else in creating two characters – the Doctor and the Consul – who change and develop in the course of the novel.'

Dr Eduardo Plarr lives in a town in northern Argentina. His father – a liberal-thinking Englishman, whom Plarr remembers only from childhood – long ago disappeared into the maw of Paraguay's repression; his mother's sole surviving passion is cakes. Plarr conceives of himself as unloved and unloving. Nevertheless, he treats the peasants of the *barrio* for nothing. Among his paying patients is sixty-one-year-old Charley Fortnum, the honorary consul of the title. A harmless if bewildered alcoholic, Fortnum has recently married a teenage former prostitute, Clara, who has alleviated his loneliness and provided an object for his bumbling affection. She, though, is having an affair with Plarr; indeed, she is carrying Plarr's child. Naturally, Fortnum assumes that the child is his.

Plarr is approached by Paraguayan revolutionaries, one of whom, León Rivas, used to be a priest; now Rivas also has a wife. The revolutionaries ask Plarr to help them plan the kidnap of a US ambassador. The kidnappers mean to demand the release of twenty political prisoners – including Plarr's father – in exchange for the ambassador's life. Plarr gives his help. The kidnap, however, is bungled and Fortnum is taken and held in a *barrio* hut. He, of course, has little value as a hostage. At the same time, it would be dangerous to release him and callous to kill him – and these revolutionaries are not callous. Rather, they are lapsed and agonised Catholics. They have deserted the church because (they believe) the church has deserted the poor.

Through his literary and diplomatic contacts, Plarr tries to being pressure to bear on Fortnum's behalf. He gets nowhere. Summoned by the kidnappers to treat Fortnum, who has been wounded trying to escape, he next finds that he too is a prisoner: the kidnappers do not trust him not to betray them. As the police and military close in – finally surrounding the hut – Greene sharpens his characters' moral, ethical and religious dilemmas. The revolutionaries must determine what level of violence is acceptable in the struggle against a habitually brutal regime. For 'Father' Rivas, the difficulty of the question is accentuated by his strong and merely dormant faith. Plarr confronts the vacuity of his own lovelessness, and Fortnum, having learned of Clara's infidelity (and therefore that the child is not his), has to decide if he can forgive her and Plarr.

In the denouement, Plarr and Rivas are killed. Plarr's death is almost sacrificial – a replacement for Fortnum's; Rivas's seems an acknowledgement of the incontestable claims of the church. Fortnum is moved by both and survives to discover a renewed and deeper affection for Clara, transcending the jealous disgust he initially felt. Thus, the novel tells a tale of personal redemption even as it signals a measure of social distress. For the poor in spirit, such as Charley Fortnum, there is spiritual hope. For the materially poor, the half-starved masses of the *barrios*, Greene cannot create a material saviour: the way of revolution is self-defeating and the church remains complicit with a merciless state.

Erica JONG   **1942–**

Fear of Flying

Reviewing Jong's novel in the *New Yorker*, John Updike wrote: '*Fear of Flying* feels like a winner. It has class and sass, brightness and bite.' The novel went on to sell 6,000,000 copies and establish Jong as something of a notorious presence in the American literary scene. It begins with the heroine, Isadora Zelda White Stoller Wing, in flight en route to a convention of psychoanalysts in Vienna. She is accompanying her husband, Bennett, but leaves him shortly after their arrival to take up with Adrian Goodlove, an Englishman with whom she conducts a strenuous affair in various locations throughout Europe. Her erotic adventures with Adrian form the bulk of the picaresque narrative.

Fear of Flying is a book which means different things to different people. It can be read as the *Bildungsroman* of a woman poet, a rather visceral comedy of modern manners, or even as a gesture of Jewish rebellion against the Nazi legacy left behind in Europe. However, its notoriety derives from the Grail of Isadora's quest – the 'zipless fuck'. Jong's descriptions of sex from a specifically feminine point of view are graphic, free from any inhibiting sense of decorum, and often uncomfortably recognisable. The zipless fuck for which Isadora searches represents sex without constraint, without involvement and (therefore) without regret. On its publication in America, the novel was hotly debated by the feminist movement, some sections of which saw in Isadora's relentless search for fulfilment a barely veiled admission of womanly incompleteness and in her ready abandonment of sexual morals the reduction of herself to sex-object.

Although the charge that it reinforced sexual stereotypes is probably unjust (in one scene a male lover avoids Isadora's advances by feigning a headache), the book can be seen as the sexual glamorisation of what is still a serious debate – that is, the role of heterosexual feminists in their relations with (non-feminist) men. The correctives to this view have largely gone unremarked. Isadora's refusal to be passive is an important resistance, while Jong's oblique identification of the experience of women and Jews – both are seen as suffering, dispersed populations – refutes the charge of mere prurience. It is also worth noting that many of the more sexually explicit scenes play on the habitual scenarios of modern pornography in a very knowing way; in a plane, in a train, on a ship, in a church, etc.

Ultimately, the catalogue of locations where the zipless fuck might have taken place (but did not) becomes exhausting. Isadora realises belatedly that the sexual freedom she has found comes only at the price of confinement within her relationship with Adrian. The novel ends with Isadora returning to her husband, speculating on how their relationship will turn out. This formed the starting-point of her second novel, *How to Save Your Own Life* (1977), in which Jong confonted many of the issues left so successfully unresolved in *Fear of Flying*.

Anthony POWELL   1905–

Temporary Kings

see **A Dance to the Music of Time** (1975)

Thomas PYNCHON   1937–

Gravity's Rainbow

Pynchon's third novel is set in Europe during and after the closing stages of the Second World War. The action centres on Tyrone Slothrop, an American stationed in London with ACHTUNG (Allied Clearing House, Technical Units, Northern Germany), who is sent on a quest through 'the Zone' (post-war Europe) by the powers that be for the semi-legendary V2 rocket, serial number 00000. A synopsis can do justice to neither the scale of the plotting nor the intense plausibility invested in even the least likely connections within it. Slothrop, it is gradually revealed, has been subjected in infancy to a kind of Pavlovian conditioning by one Dr Laszlo Jamf, the remnants of which are reactivated during the V2 bombardment of London. The pattern of his sexual encounters predicts the pattern of V2 strikes and this conditioned correspondence is used by the officers at PISCES (Psychological Intelligence Schemes for Expediting Surrender) to track the progess of the rocket when Slothrop is turned loose in Europe. This basic quest-plot is gradually superseded in the course of the novel by a more complex confluence of many different plots. Slothrop becomes, in effect, the point at which all interests (more or less) converge.

The interests are many and various. Aside from his shadowy political masters in London, there is the statistician Roger Mexico, for whom Slothrop is a variable in a Poisson distribution; there is Tchitcherine, a Russian operative shadowing Slothrop through the Zone; Säure uses him as a dope-courier; for a German village he is a ritual pig; while for many in the Zone he is 'Rocketman', a mythic figure on a self-destructive trajectory. Add to these the major industrial interests (Krupp, IG Farben, General Electric, ICI and Siemens) of Europe and America who also have stakes in

the Rocket/Slothrop Project, and it is likely that everything will be connected, nothing will happen by accident and everyone will follow their own hidden agenda.

The vastness of the plot is not gratuitous. One of Pynchon's main interests in *Gravity's Rainbow* is the control and manipulation of information, and at one level the novel is a demonstration of these practices. Paranoia is the dominant psychological mode, the most paranoid characters being the least deceived. Pynchon describes the structures of control as a pair of systems: 'You're a novice paranoid, Roger ... Of course a well-developed They-system is necessary – but it's only half the story. For every They there ought to be a We. In our case there is. Creative paranoia means developing at least as thorough a We-system as a They-system.' The rocket itself is at once the centre of all these systems and their refutation, for in the indeterminacy of its flight the element of chance is introduced and chance is the closest cipher of freedom to which Pynchon will lend credence. It is also the Manichaean symbol *par excellence* for many key oppositions (We/They, Preterite/Elect, Masochist/Sadist, etc.) in the novel, 'a good rocket to take us to the stars, an evil rocket for the World's suicide, the two perpetually in struggle'.

Such large-scale themes demand a powerful prose style to connect them with the action. Pynchon's prose is allusive, highly worked and expansive. As one of few writers to include science within his frame of reference there is of necessity a huge amount of information continually being distributed. Correspondingly huge stylistic risks are taken on almost every page and almost all of them are made to work. Idioms range from the broadest slapstick to genuine (but rare) pathos, from technical explanation to the most lurid description.

In the context of Pynchon's own development, *Gravity's Rainbow* is the successor of *V* (1963) rather than **The Crying of Lot 49** (1966). Pig Bodine and Kurt Mondaugen reappear, and many of the earlier book's themes are recapitulated and extended. *Gravity's Rainbow* has been advanced as the Great American Novel. This it is not. If Pynchon belongs in the American literary tradition at all, it is as the greatest and most surreal heir of Henry James: an explorer of the American mind in Europe, witness to the European sensibility and its divisions. However, he is best seen as a truly international writer of truly international works. Its range of reference, compendiousness, erudition and fluid transitions between styles and subjects mean that *Gravity's Rainbow* simply covers more ground than any other post-war novel in either Europe or America.

Patrick WHITE   **1912–1990**

The Eye of the Storm

Published nearly twenty years after *The Tree of Man* (1955), in the year that White won the Nobel Prize for literature, *The Eye of the Storm* perfectly illustrates both his growing disgust with life and his technical mastery.

Elizabeth Hunter is very old, very rich and still a tyrant. She lives in her own world, as much in the past as the present, confined to her bedroom at the gloomy house in Moreton Drive, Sydney, attended round the clock by three nurses on whose susceptibilities she plays. The night-nurse is Sister Mary de Santis who is regarded as a saint. At one point Elizabeth reflects, 'Only yourself and de Santis are real. Only de Santis realizes that the splinters of a mind make a whole piece', a theme central to Patrick White from his first mature novel, **The Aunt's Story** (1948). In the morning Elizabeth endures the bossy, insensitive Sister Badgery, who announces herself at breakfast time 'with the hateful egg'. In the afternoon and evening the nubile Sister Flora Manhood massages her patient expertly and feeds her vanity by elaborately making up her ancient face, but has serious problems of self-disgust. There is also Mrs Lippmann, 'the small unhappy Jewess', guilty at having survived the Holocaust, who cooks deliciously and will dance for her employer when commanded – she was once a music-hall artiste. Attending devotedly to Mrs Hunter's affairs, as he has for decades, is old Arnold Wyburd, the solicitor, whose watchword is discretion. They await with trepidation the arrival of her children: from England her son, Sir Basil Hunter, the 'great'

actor with fading good looks and a tremendous ego inherited from his mother, who has never watched him act; from France her plain and embittered daughter, Princesse Dorothy de Lasabanes, who married unsuccessfully into the aristocracy there. Elizabeth has not seen her children for years – they are here now only to witness her death and get their hands on her money.

She has led a sinful life, indulging in numerous affairs and treating her mild, too forgiving husband abominably, though she did come home to nurse him through his last illness: '*I wanted very badly to love my husband . . . even after I knew I didn't – or couldn't enough.*' Even now, crawling obscenely towards death, she still wickedly dominates her household, who variously adore and hate her, by a mixture of bullying and flattery, dispensing gifts when expedient. Her ageing children, who are incapable of loving anyone – it is significant that they both hide behind their titles – know better than to stay in the house, and see as little of her as possible. They are, in truth, afraid of the old woman. They also hate each other; it is a mixture of lust and self-disgust that drives them to commit incest at 'Kudjeri', their childhood home in the country. On one thing they do agree: their mother, should she survive until a vacancy occurs, is to be sent to Thorogood Village, an old people's home, thus saving the considerable expense of maintaining the present establishment.

Much of the plot turns on this. The nurses and Mrs Lippmann are horrified when they discover the plan but lack the will to abort it. Flora goes to bed with Sir Basil, but though this exorcises her fantasies about the 'great' actor, she makes no real attempt to dissuade him from 'putting his mother away'. When Sister de Santis has lunch with him she discovers in herself a sensuality she hardly knew and Thorogood Village is not mentioned. It is Elizabeth herself who solves the problem by dying while sitting on her commode, having first been ritually made up by Flora Manhood and danced to by Mrs Lippmann. Whether or not the threat of Thorogood Village did kill Elizabeth Hunter, her children certainly encompassed her death in their minds. They attend the reading of the will but not the funeral, and fly home with enough money in the bank to enable Sir Basil to back a most improbable play, which will surely be his professional suicide, and Dorothy to go on being miserable in comfort. The household they leave behind disperses, each woman fulfilling her destiny, including the tragic Mrs Lippmann who cuts her wrists in the bath.

A strange and disturbing book, *The Eye of the Storm* is essentially a novel of character,

the separate lives of its people touching briefly at the bedside of the dying old woman. As one reviewer put it: 'On this basically simple structure, Patrick White has built a soaring baroque edifice of which no corner is left unvisited as he explores the lives of his characters, exposing their pitiful, comic humanity.'

1974

Resignation of President Richard Nixon ● Partition of Cyprus ● Discovery of the first Chinese emperor's 'terracotta army' ● Warning that chlorofluorocarbons may be damaging the atmosphere's ozone layer ● The first Grand Unified Theory about the origins of the Universe ● Joseph Beuys, *I Like America and America Likes Me* ● Ntozake Shange, *For Coloured Girls who have Considered Suicide when the Rainbow is enuf* ● Films: *Last Tango in Paris, Chinatown* ● Richard Crossman, Duke Ellington and Juan Peron die ● Philip Larkin, *High Windows* ● Anne Stevenson, *Travelling Behind Glass: Selected Poems 1963–1973*

Kingsley AMIS   1922–

Ending Up

The earlier, much shorter and more savagely pessimistic of Amis's two novels dealing exclusively with old age, *Ending Up*, which was shortlisted for the Booker Prize, is hardly inferior to **The Old Devils** (1986), which won it. The potent mixture of horror and comedy it generates makes it perhaps Amis's funniest, and certainly his cruellest, novel. Written firmly against the trend of social piety towards the aged, the book depicts senescence in grim colours.

Five people in their early seventies share Tuppeny-hapenny Cottage, an inconvenient house isolated in the countryside near Newmarket. Their inability to live happily with each other or to gain satisfaction from life now that they are old is the subject-matter of the novel.

Bernard Bastable is a former army officer who was forced to resign, and suffers from a gammy leg; his only pleasure is to find ways of tormenting and distressing the other people in the house. Adela is his sister, very

ugly and suffering from having lived an unloved life. Derrick Shortell, known as Shorty, is a former private soldier with whom Bernard once had an affair: he is part member of the household and part servant. Prof. George Zeyer, Bernard's brother-in-law, is a Czech academic who has had a stroke; he is confined to bed upstairs and suffers from nominal aphasia, so that he cannot remember the names of common objects. Marigold Pyke, a close friend of Adela, is a pretentious woman continually harking back to her romantic youth in the 1920s.

The novel develops almost entirely out of the painful interrelations of these five characters and, until the end, has little plot except various incidents illustrating their hostilities, *rapprochements*, moments of hope, and setbacks. On a visit to the cottage by Marigold's children, George is carried downstairs, which gives him renewed hope, and at Christmas he recovers his ability to name objects, his joy being a cause of sourness in Bernard. The latter, who is told by his doctor he has three months to live, occupies himself with plots to get George's dog, Mr Pastry, put down, to persuade Shorty that he is incontinent, and Marigold that she smells. Marigold uses the threat that she might leave the cottage as a means of getting her own way with Adela. A climax of rancour is reached at Christmas.

The novel ends, perhaps not entirely satisfactorily, with the deaths of all five characters on the same day early in the New Year. Bernard is on a ladder, cutting a telephone wire to cause further annoyance, and falls to his death, while the others die in a series of interrelated accidents and heart-attacks. Amis's unflinching depiction of the depredations of age suggests links with the work of his close friends John Betjeman and Philip Larkin, both of whom wrote some of their most powerful poetry on the topic of 'ending up'.

Beryl BAINBRIDGE 1934–

The Bottle Factory Outing

Brenda and Freda share a single-bedroom flat in north London. Brenda, thirty-two, is a refugee from a brutal marriage, excruciatingly timid and – according to Freda – 'a born victim'. Six years her junior, Freda is boisterous, domineering and wholly uninhibited by her obesity. The two work at Mr Paganotti's bottling plant where – apart from themselves and Patrick, the van driver – all the employees are from Bologna. Rossi, the manager, constantly harasses Brenda, pawing at her in the factory basement. Freda, for her part, is infatuated with Vittorio, Mr Paganotti's elegant nephew.

Because of a linguistic misconstruction, Vittorio believes that Freda has recently been orphaned. He visits the flat to offer condolences. Freda attempts to seduce him – and succeeds, unsatisfactorily. Brenda, meanwhile, is in the lavatory with Patrick (he having volunteered to mend the cistern). Her mad mother-in-law, Mrs Haddon, calls round and tries to shoot her (an incident taken from Bainbridge's own life). Mrs Haddon is led away, but Patrick's violent passions have been roused: he declares that he would 'swing' for Brenda.

The outing of the title is to Windsor. It has been planned by Freda and is from start to finish a disaster: the weather is cold and wet; the transport booked for the party does not arrive; the day is fraught with squabbles and picnicking misery. Freda thinks she has received a pledge of love from Vittorio (in fact, he is engaged to a beautiful and suitable Italian); Brenda fights off Rossi's obnoxious attentions; and Patrick sulkily prowls, at odds with Freda, who, in turn, has quarrelled with Brenda and Rossi.

When Freda goes for a walk in the woods nearby, Rossi looks for her and finds her murdered. It is plain that he, Vittorio or Patrick is the killer. The police are not summoned – the Bolognans are frightened of Patrick and too mutually loyal to incriminate one of their own – and Freda's body is ferried – via Windsor Safari Park – back to the factory, where it is hidden until Vittorio hits on an ingenious method for disposing of it. Later, one of the suspects confesses to Brenda. She is sceptical of the confession and sees a more likely solution. Then, dismally, she returns to her parents.

The tale pivots on the irony that it is the 'born victim' who survives. Whether survival is a good thing, however, is a question not decisively resolved. It is intimated that, with her foolish and insatiable appetites, Freda was bound only for frustration. Her attachment to the bottling plant is the attachment of the inadequate to the undemanding. That her life is spiced with fantasy renders it not less but more pathetic. In contrast, Brenda's meek pessimism – her capacity to absorb bullying – fits her for an undisappointed existence. Bleak though these alternatives may be, they are proposed with gusto and élan; and, if Bainbridge's portrayals are unsentimentally precise, she stays more kind than cruel to her cast. This widely praised black comedy was the second of Bainbridge's novels to be shortlisted for the Booker Prize and was awarded the *Guardian* Fiction Prize.

Wendell BERRY   1934–

The Memory of Old Jack

Like Berry's first two novels, *Nathan Coulter* (1960) and *A Place on Earth* (1967), *The Memory of Old Jack* is set in the farmland of Port Royal in North Kentucky. This area, where his family have farmed for five generations and where Berry himself now farms, is a central and constant fact in his fiction. Like Faulkner's Yoknapatawpha County, Port Royal is the terrain that all Berry's characters acknowledge, to which they all react and which acts upon them in turn.

The Memory of Old Jack is set during a single day in September 1952 but, as the title intimates, ranges much further back in time. The Jack of the title is an elder of Port Royal, a small-farmer much respected in the community there. Jack was born in 1860, and his memory provides Berry with the opportunity to reach back into the history of Port Royal in general and that of the small-farmers there in particular. Unsurprisingly, this is presented as a history of decline and Jack's passionate individualism stands in contrast to the family which now surrounds him. Having tried and failed to play the gentleman farmer for the benefit of his wife, Jack must now watch his only child and her husband (a banker) commit the same error. Their sentimental view of farming as a kind of quaint idyll represents for Jack (and Berry) a fatal error of perception leading to a divorce between man and land. The novel ends with Jack deciding to hand over the farm to his tenant, Andy Caplett, whose long stewardship suggests he shares Jack's own vision of the land.

The novel can be read as a long search for the right relationship between man and land, and the continuity which the novel urges is primarily the continuity of this relation. Jack must cut through the blood-ties of his own family to achieve this aim, and the order of Berry's priorities here places him in opposition to the agrarian movement of John Crowe Ransome and Robert Penn Warren. Although like these writers in finding common intellectual roots in Jefferson and Thoreau, Berry has consciously distanced himself from them and the regionalism which he has described as 'corrupt and crippling'. Berry emphasises the culture in agriculture as a device by which men might relearn their relationship with the land. In this he is a classicist, seeing a necessary link between rural and social decay and hubris as the human failing most responsible for both. Like all Berry's work (which includes several volumes of poetry and a number of influential essays) *The Memory of Old Jack* is ultimately didactic in intent. Berry's concern with environmental issues has alternately raised and depressed his public profile as these issues have risen or fallen in the public eye. His view that art should instruct as well as please is a traditional one and is largely borne out by this novel.

Anthony BURGESS   1917–1993

The Enderby Novels

Inside Mr Enderby (1963)
Enderby Outside (1968)
The Clockwork Testament (1974)

The Enderby novels form a series which illustrates and sums up Burgess's virtues as a writer. Written over a decade, they also illustrate Burgess's theories about the process of writing. Though the young Burgess had begun to establish a reputation for himself as a writer with his novels about the colonies and about provincial England, the threat of a suspected brain tumour caused a spurt of creativity. *Inside Mr Enderby* (published under the pseudonym Joseph Kell) was one of the novels which resulted from this. The second novel in the series, *Enderby Outside*, finished five years later, is more substantial but not essentially different. However, the third volume, *The Clockwork Testament*, consciously echoes the title of one of Burgess's best-known works, ***A Clockwork Orange*** (1962), and demonstrates that the character of Enderby, and that of his author, have changed and grown as a result of their experiences.

The central figure in all three volumes is the self-indulgent poet, F.X. Enderby, who has numerous relatives in Burgess's other novels: Denham in *The Right to an Answer* (1960), Christopher Howarth in *The Worm and the Ring* (1961), Richard Ennis in *A Vision of Battlements* (1965) and Edwin Spindrift in *The Doctor is Sick* (1960). All are inadequately sexual, and their asexuality becomes the butt of Burgess's imaginative humour. For example, Enderby rushes off from a female predator in mid-orgasm, to write a poem in his lavatory, his bathtub full of bits of paper. Of all the characters, Enderby is the one nearest to his author's heart, and we follow his amorous, digestive and literary triumphs and misfortunes from England to Rome, Tangier and New York. More at home with words than with life, his lack of involvement – in a world which he regards as being full of sharks, fools and exploiters – is what condemns him to remain a poet of merely minor stature.

In *Inside Mr Enderby*, the poet meets Vesta Bainbridge, a glamorous widow from the pop world, who persuades him into marriage. Alarmed by the level of sexuality and intimacy she expects, Enderby eventually flees. When poetic inspiration deserts him, Enderby attempts suicide, though he is discovered, restored to physical health and subjected to the attentions of Wapenshaw, a psychologist, who deprives him of his poetic gift and then certifies him as 'normal'. At the end of the novel, Enderby scrapes a living as Hogg, a bartender.

The other principal character in the novel is Rawcliffe, a poet like Enderby, but unlike him a decadent and deceitful one, whose worldly success is the result of expediency. He steals one of Enderby's ideas and achieves both money and fame by turning it into a film.

Enderby Outside begins with Enderby–Hogg just beginning to write poetry again. However, when he tells Wapenshaw the news, the good doctor is furious: Hogg had taken pride of place in a new book of case histories, and the reversal, of course, destroys the story. Meanwhile, Vesta has remarried; the victim this time is from her own world – a singer, Yod Crewsey. He appears to be receiving much idolatry as a poet, though the poems are Enderby's and were probably purloined by Vesta. When Crewsey is murdered, Enderby finds himself falsely accused, and escapes the country, arriving in Tangier where he meets a group of weird psychedelic poets.

In the final volume, *The Clockwork Testament* (1974), Enderby goes to Hollywood where, commissioned by a film producer to adapt Gerard Manley Hopkins's poem dedicated to five Franciscan nuns who died at sea, 'The Wreck of the Deutschland' (1876), into a screenplay, he is bullied by the producer into cranking up the non-existent sex and violence of the original. After the film's release causes the inevitable storm of outrage from the literary establishment and religious groups of Middle America, Enderby is forced to endure the obligatory round of US TV chat shows in which, parodying not too distantly Burgess's own ordeal at the hands of the media following Kubrick's 1971 film version of *The Clockwork Orange*, he is charged with immorality and obscenity.

All these novels show several different types of people trying to serve art in one way or another, while actually doing it a disservice so far as Burgess is concerned. Various women try, through sex, to help Enderby overcome his artistic failings. Enderby is funny, even pathetic, but he remains superior to all the other characters because he retains his integrity. His traditional workman-like poetry is not written out of any attempt to demonstrate anything; it is written because Enderby finds himself driven to write. Enderby is Burgess's equivalent in these novels, and gives him great occasion to exhibit his gift for pastiche, satiric social comment and verbal invention.

Nadine GORDIMER   1923–

The Conservationist

Although she has held posts abroad as visiting professor to a number of universities, Gordimer has always lived in South Africa. 'After my first trip out,' she has said, 'I realised that "home" was certainly and exclusively Africa. It could never be anywhere else.' Equally certain was that her own country would be her principal subject as a novelist, and that her books would be, in the broadest sense, political. Gordimer rarely writes pure polemic, but 'the influence of politics on people' is a constant and, given the setting, inevitable theme in her work. She writes as, and about being, a white, middle-class liberal, questioning the validity of that stance and often suggesting its ultimate impotence. Like several of her books, *The Conservationist*, which was joint winner of the Booker Prize, fell foul of the South African government's censorship board and was embargoed for some ten weeks before a decision was reached to release it.

Mehring, a rich businessman, fits ill into the society of white South Africa. He is contemptuous of narrow Afrikanerdom and of the white liberal intelligentsia alike but, as his sometime mistress, Antonia, observes, he is 'no ordinary pig-iron dealer' either, set apart from his own class by an unusual intelligence and emotional complexity. A romantic and a sensualist, he buys himself a dilapidated farm as a weekend refuge from the round of poolside barbecues: a place to bring a woman, he thinks to himself, but also where he hopes to find peace in communion with nature.

His love of the Transvaal is not a false emotion *per se*, but it is falsified all the same by the political context to which it is in part a response. For Mehring's political attitude is that of sceptics down the ages. Though he has no ideological commitment to the unequal distribution of power and wealth as he finds it, he dislikes dissent. Confident that the trickle-down effect of 'development' will make all come right in the end, he withdraws to his farm, acquiescing in the apartheid system but hoping there to shield his eyes from the worst of its effects.

This, however, turns out to be impossible. At the beginning of the novel, the body of a murdered black man is found on Mehring's farm. The police, refusing to investigate, bury the body, but it remains on the threshold of Mehring's consciousness as a constant witness. As the body is washed up once more by a flood, the brutality of apartheid forces itself on Mehring irresistibly: lured to a disused mine by a prostitute, he is robbed and killed by the same thugs – whether professional criminals or police it is not clear

– that killed the black man. Mehring dreamed of handing the farm on to his (as it happens uninterested) son. But only the black man is buried with due ceremony on the farm, apparently pointing to the impossibility of a permanent home for whites in South Africa.

Though Mehring is shown in the end to be the victim of his false consciousness, he remains an attractive character: Antonia is drawn to him, until forced into exile, away from her dry but politically pure academic husband, to her mixed amusement and unease. For if he is unable or unwilling to see the significance of his actions in any larger political picture, Antonia suffers from the opposite fault. In refusing to admit that anything has significance except insofar as it occupies a place on a larger political map, she is often no less out of touch with emotional realities than he is with political ones, thus exemplifying the emotional strain of avoiding collusion with the apartheid system.

Jennifer JOHNSTON   1930–

How Many Miles to Babylon?

As with all Johnston's spare, delicately written novels, *How Many Miles to Babylon?* is a straightforward story, told with unobtrusive skill, which sends out ripples from its still centre. The first-person narrative of a young Anglo-Irish officer in the First World War gradually expands to encompass the troubled history of Ireland, divisions of class and religion, the nature of male friendship and the dehumanising character of warfare.

Alexander Moore is the only child of a bleak marriage, growing up in a large house outside Dublin. His only friend is Jerry Crowe, an Irish boy, with whom he shares a passion for swimming and riding. Their friendship is concealed from Alexander's autocratic mother, who regards the Irish as treacherous and unspeakable peasants. In contrast, his father, ineffectual, unloved and dominated by his wife, is a conscientious landlord who sees himself as a mere caretaker of the land. When war breaks out it seems remote from Ireland until Mrs Moore decides that Alexander should enlist. Supported by his father, he resists her attempts at emotional blackmail, but when he discovers that Jerry has joined the Irish Rifles, he volunteers as well.

In France they come under the command of Major Glendinning, an English officer appalled to be given charge of Irishmen, whom he views much as Mrs Moore does. The relationship between the two men is viewed with suspicion in an army and a society where 'spontaneity and warmth were unknown, almost anarchic, qualities'. Only Alexander's English fellow officer, Bennett, accepts them. When Jerry is refused compassionate leave to search for his missing father, he goes AWOL and, although he voluntarily returns, he is court-martialled and sentenced to death. Like Mrs Moore, Glendinning epitomises lack of imagination and lack of feelings: 'How you damn Irish expect to be able to run your own country when you can't control your own wasteful emotions, I can't imagine,' he says when ordering Alexander to command the firing squad. Rather than leave his friend to the mercies of the squad, Alexander visits Jerry in his cell, shoots him dead with his revolver, and is thus condemned himself. This final act ironically echoes Jerry's remark about the 'warm heart' of an Englishman, which 'only means he'll cry as he shoots you'.

This vivid and subtle novel is suffused by the poetry of Yeats and by the history of Ireland. The parallel between Mrs Moore and Glendinning implies a connection between the First World War and the struggle in Ireland at that period, without making this an explicitly political book. Indeed, it is the claims of humanity, and the danger inherent in ignoring those claims, which guide the novel.

Alison LURIE   **1926–**

The War Between the Tates

It is 1969, and the cult of youth has reached even the small university town of Corinth, USA. At the Krishna Bookshop the guru Zed offers 'a way out of the system' to blue-jeaned students who seek the meaning of life through peace and love, astrology and marijuana. In a house near the campus Erica Tate, a faculty wife, sits at her kitchen table and weeps. Her two children Muffy and Jeffo have become teenagers, insolent and lazy strangers whose presence in the house means that she and her husband, Brian, 'dare not laugh or cry' at night. It is Erica's most carefully guarded secret that she hates her children. Drying her eyes, she tells herself that she is still luckier than the divorced Danielle, now an embittered feminist. Brian is still Erica's best friend, and she respects his conservative solidity. She alone is responsible for the failures of their children, Brian having assigned to her what he described as the domestic 'area of operations'.

Brian is a political scientist, who has spent his life trying to fulfil his destiny as a great man. Although a successful academic, his political ambitions have been thwarted, he feels, by his small size. Life is beginning to slip away from him, and he is worried about The Book which will earn him his reward for a lifetime of conformity. In the meantime, 'it is not beyond the bounds of possibility that one day Brian Tate will be the smallest person in his family'. Finding he is worshipped by a dim postgraduate student, Wendy, Brian embarks on an affair under the delusion that 'the way to cure a passion was by satiating it'. When Erica discovers this her image of her husband is shattered. The war between the Tates has begun.

Pregnant and planning an illegal abortion, Wendy visits Erica to apologise for the affair with her husband. Erica takes refuge in a self-martyrdom justified as 'doing the right thing'; Brian must divorce Erica and marry Wendy. To Brian, however, the affair is of no great importance, and he determines to persuade Wendy to have the abortion. Collecting his belongings from the family home, he feels nostalgia for the girl he married, but 'most men don't want to leave their wives half as much as they do their adolescent children'. An enemy within his own home, he temporarily adopts Erica's plan whilst initiating a counter-attack against the 'monstrous regiment of women' lining up against him under the banner of New Feminism.

Torn between an enjoyment of his freedom and the awareness that he looks rather foolish with his pregnant hippie girlfriend, Brian finds himself briefly allied with the youth cause against the administration. As theory becomes practice, however, Brian stages the dramatic deliverance of a male colleague from the hands of the women, and ironically achieves fame in the national press as a savage anti-feminist. Wendy's infatuation is indeed cured and she leaves with a new boyfriend for a Californian commune. Meanwhile Erica has learned a degree of independence. In an interrupted but erotic affair with Zed, she finds that Brian is literally a rather small man. Brian plans a comfortable reconciliation, but for Erica there can be no return to cosy domesticity.

The novel, which concludes with a march against American involvement in Vietnam, is amongst other things a wickedly funny compendium of late-1960s' fads and attitudes: peace protests, social sciences, radical feminism, Zen, recreational drugs, yoga, consciousness-raising and assorted hippie pursuits. In this anatomy of campus life, Lurie captures a whole era, just as Malcolm Bradbury was to do a year later with *The History Man* (1975).

Stanley MIDDLETON   1919–

Holiday

Middleton was born in Nottinghamshire and, apart from his war service, has remained there, becoming one of Britain's leading provincial novelists – certainly its most prolific. His first novel, *A Short Answer*, was published in 1958, and since then he has produced on average one novel a year, most of them set in the English Midlands. This combination of quiet industry and a staunch loyalty to the provinces as a setting and 'ordinary' lives as a subject has meant that Middleton's books have not, perhaps, received the sort of attention they deserve, although *Holiday* was joint winner of the Booker Prize in the days before the award generated so much media interest.

Middleton's novels often deal with people in crisis, and *Holiday* follows Edwin Fisher, a university teacher, through his first solitary week after splitting up with his wife, Meg. He books himself into a dismal boarding house in the east coast resort of Bealthorpe – 'Who came to these places, now that the package deals to Ibiza and Tangiers were so cheap?' – to endure a peculiarly British holiday of bed-and-breakfast, bangers and beer.

Between desultory chats in deckchairs and standing at the bar in pubs, Fisher relives the sentimental journey which brought him to his present situation. His father, a loud and vulgar shopkeeper called Arthur, had brought him to this town for his childhood holidays. Fisher remains embarrassed by this background and by his own upward mobility into the professional classes. Now he meets his father-in-law, a smarmy solicitor called David Vernon, who is determined to sort out the marital crisis.

Lena and Sandra, two working-class women, attempt to seduce Fisher under the noses of their husbands. Tactfully turning them down after a grope behind the chalets, Fisher is constantly summoned to talk to his father-in-law at the upmarket hotel of the Frankland Towers. Vernon stages a reunion between Fisher and Meg, and the holiday is over. But in the time of their separation Meg has become more independent, and she now announces that she is going on holiday to India. It makes Fisher seem more ineffective in his own eyes, and he begins to rue the time he wasted on his own unadventurous holiday.

Michael MOORCOCK   1939–

The Hollow Lands

see **The Dancers at the End of Time** (1976)

Iris MURDOCH   1919–

The Sacred and Profane Love Machine

One of the later and longer of Murdoch's novels, *The Sacred and Profane Love Machine* explores many themes common to the whole *œuvre*: the obsessive and unpredictable nature of human love; the barrier that self-deception forms to true awareness and morality; the drama of moral choice and the search for the good.

He was not all that odd, he early concluded. Most people were pretty odd. It was interesting. He had early discovered, partly by introspection, partly by intuition, partly by questioning others, and partly by a study of literature, that human minds, including the minds of geniuses and saints, are given to the creation of weird and often repulsive fantasies. These fantasies, he concluded, are in almost every case quite harmless. They live in the mind, like the flora and fauna which live in the bloodstream, and like these may be in some ways beneficial. They are of course symptoms of mental structure, but are not themselves usually causes, except perhaps in art.

THE SACRED AND PROFANE LOVE MACHINE
BY IRIS MURDOCH

Blaise Gavender is a middle-aged psychiatrist without formal qualifications for the job, who has been married happily for many years to Harriet, an emotionally powerful but not greatly intelligent middle-class woman: they have a son, David, who is on the point of reading Greats at Oxford. Blaise has also for several years been conducting an illicit affair with Emily McHugh, a young woman of lesser social status who lives in Putney and teaches French inadequately at an expensive girls' school. They have a retarded but compelling eight-year-old son, Luke, whose name has become Italianised to Luca. Blaise hopes to continue to carry on love-lives with both women, and he is portrayed as both selfish, hypocritical and untrustworthy but also as a representation of a sort of moral Everyman.

Luca begins making secret visits to Hood House in Buckinghamshire, where Blaise and Harriet live, and this prompts Blaise finally to confess his love-affair to Harriet. She attempts to cope with the situation, insists on meeting Emily and inviting her to Hood House, and develops an intense attachment to the neglected Luca. During one of her visits, Emily runs out of the house; Blaise follows her, and decides to commit himself to her, thus abandoning Harriet.

Harriet becomes more and more distraught, and is rejected in offering her love to her difficult neighbour, detective-novelist Montague ('Monty') Small. Eventually Harriet flees with Luca to join her soldier brother in Germany, but is shot dead in an accidental encounter with terrorists at Hanover airport as she shields Luca with her body. Blaise marries Emily, and a powerful scene shows them burning Harriet's possessions in an attempt to exorcise her from Hood House.

The novel contains a large number of minor characters and sub-plots, some of them only marginally connected with the main action. Minor characters include Edgar Demarnay, the ineffectual Master of an Oxford college, who yearns emotionally for most of the other characters; Constance Pinn, Emily's mysterious char and lodger; and the oversexed, half-French schoolgirl Kiki St Loy, who makes Emily's French lessons a misery ('She gravely corrected Emily's now even more frequent mistakes, pretending that she was the teacher and Emily the pupil').

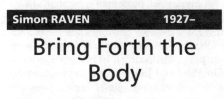

Simon RAVEN 1927–

Bring Forth the Body

see **Alms for Oblivion** (1976)

1975

Communist victory in Vietnam ● 'Emergency' in India ● First British North Sea Oil ● Civil war starts in Lebanon ● New York City narrowly averts bankruptcy ● The Spanish monarchy is restored and Juan Carlos becomes King ● Personal computers go on sale in the US ● Foundation of Ecology Party in Britain ● Sex Discrimination Act in Britain ● Adopted children over 18 granted right to information about their natural parents ● The first monoclonal antibodies produced ● Discovery of 'Lucy', the remains of a hominid about 3 million years old in Ethiopia ● 20,000 clay tablets with cuneiform texts found at Tell Mardikh (ancient Ebla) in Syria ● Paul Feyerabend, *Against Method* ● Trevor Griffiths, *Comedians* ● Films: *One Flew Over the Cuckoo's Nest, Jaws* ● Television: *Starsky and Hutch, The World at War* ● Bob Dylan, *Blood on the Tracks*, Queen, *Bohemian Rhapsody* – the first major rock video ● Arthur Ashe of the US becomes the first black Men's Singles champion at Wimbledon ● Julian Huxley, P.G. Wodehouse, Aristotle Onassis, Chiang Kai-shek, Barbara Hepworth, Dmitry Shostakovich, Haile Selassie, Eamon de Valera, Arnold Toynbee and General Francisco Franco die ● Charles Causley, *Collected Poems 1951–75*

Donald BARTHELME   **1931–1989**

The Dead Father

The Dead Father is 3,200 cubits (approximately one mile) long. He is being hauled across an unnamed country by a team of nineteen men under the supervision of Thomas and Julie. Their destination is a golden fleece. The Father believes this will rejuvenate him, for he is not actually dead; he is merely decaying. Querulous, libidinous and violent, he finds his commands degenerating to entreaties and his memories of omnipotence doubted.

A mysterious horseman is sighted, always following at a distance. A precocious child, met along the way, illustrates nineteen senses of the verb *to come*. Provocative Emma joins the party, distracting it from a pornographic film. She and Julie hold disjointed conversations; but, unlike Julie, Emma does not expose herself on demand. The men complain to Thomas about their treatment, then wish him a happy birthday. Next the procession passes through the land of the Wends, a father-hating people who insist that the Father's left leg be sacrificed. Since this is mechanical – containing *inter alia*

confessionals – it is not (except literally) a huge sacrifice. Julie and Thomas are given *A Manual for Sons* – an anti-paternalist pamphlet offering embittered and lunatic observations on the class: 'Two leaping fathers together in a room can cause accidents.' The Father is persuaded to draw up a will. He leaves everything to Edmund, a drunken and delinquent son of his. The mysterious horseman approaches. He, in fact, is she: the Mother. Thomas despatches her with a grocery list.

Shortly, the procession reaches its goal. There is no golden fleece, of course – except for Julie's pubic hair, which the Father pleads to be allowed to touch. Permission is refused him and instead he is buried alive in a gigantic grave. His final utterance is: 'One moment more!'

While Barthelme's novel has an appearance of collage, it is also meticulously structured. On the one hand, it charts the Father's decline; on the other – with its sequential mimicries and allusions – it sketches the development of literature. The Father's first soliloquy is a pastiche of Greek myth; his last a parody of **Finnegans Wake** (1939). As he relinquishes his powers – physical, sexual and legal – narrative mutates. From the clear fables of creation it lapses into the opacity of a senile mind: the centre cannot hold, so language falls apart. This 'modernist text' is thus a commentary on the genesis of works such as itself and a canny satirisation of them. Indeed, even though 'writing about writing', Barthelme is a fantasist in the tradition of Carroll. Full of learned lampoons, linguistic paradoxes and startling non-sequiturs, he contrives a story that can be read equally for its wicked humour and for its serious intent. If the Dead Father is a symbol of all authorities amalgamated, his tales dig the grave of his pretensions.

Saul BELLOW　　　　　　1915–

Humboldt's Gift

Charlie Citrine, Bellow's hero and first-person narrator, is a writer in late middle age. Behind him is a big Broadway success, *Von Trenck*. Nevertheless, he is more a man of ideas – part-journalist, part-philosopher, part-historian – than a popular playwright. His literary career dates from 1938, when he initiated a friendship with Von Humboldt Fleisher. Humboldt was then in his first flush of fame as a poet, but the rest of his life was a tale of failure. He never wrote the major work of which he, at least, believed himself capable; he failed to obtain an academic sinecure, declined into paranoid mistrusts and – after the ignominy of internment in an asylum – died broke and alone in a seedy hotel: at the end, Charlie withheld assistance. Humboldt now lies undistinguishedly buried 'far out in Deathsville, New Jersey'.

In 1970s' Chicago, Charlie's own life is less catastrophically threatened. His writing has stalled. Renata, his delightful mistress, is being courted by an undertaker. A minor hoodlum has vandalised Charlie's Mercedes in pursuance of a gambling debt. The debt is small: what really frightens Charlie is the judicial extortion of his divorce. His wife, both teams of lawyers and the judge all seem intent on pauperising him. The Internal Revenue Service is hounding him too, and Pierre Thaxter, a kind of modern-day Skimpole on a grand scale, touches him for loans

that will not be repaid.

Much of the novel looks back to its narrator's fraught relations with Humboldt, but, rude and persistent, Chicago intrudes. The hectic, philistine city is a hostile territory for the writer and (Bellow invites the reader to suspect) it may be because of its distractions that Charlie has chosen it for his home. The chaos he endures he also enjoys: it excuses him from confronting his creative block. Finally, however, he flies to New York. There he discovers that Humboldt has bequeathed him an original screenplay: a comedy of infidelity. Thinking it has no commercial potential, Charlie flies to Madrid to lend his prestige to a (bogus) project of Thaxter's. Renata, meanwhile, marries the undertaker. Distraught, Charlie spends weeks isolated in his hotel, only later to learn that a second screenplay – one jokily co-authored by Humboldt and him long ago – has become an international success. Charlie sues for royalties and next sells the film treatment willed to him. Thus, his financial plight is relieved. He helps arrange Humboldt's reburial in an exclusive German-Jewish cemetery and returns to New York to be at the service.

The country is proud of its dead poets. It takes terrific satisfaction in the poets' testimony that the USA is too tough, too big, too much, too rugged, that American reality is overpowering.

HUMBOLDT'S GIFT
BY SAUL BELLOW

Humboldt's material gift is the first film treatment. A twist in Bellow's plot renders it useful. Of greater significance, though, is Humboldt's spiritual legacy; for Humboldt posthumously teaches Charlie that reconciliation with the past is a necessity for peace in the present. Humboldt is the ghost to Charlie's Hamlet, and, if the resolution here is not tragic, there is yet some poison in the chalice. Neither writer is able to produce a work of magisterial vision: each, in the last resort, is rescued by unserious popular art. Diverse and boisterous, the USA offers a superfluity of experience. It cannot be organised in any one mind.

Humboldt's Gift was awarded the Pulitzer Prize in the year of its publication. The following year, Bellow won the Nobel Prize for Literature.

Malcolm BRADBURY   1932–

The History Man

'Things seem to be happening; back from Corfu and Sete, Positano and Leningrad, the people are parking their cars and campers in their drives, and opening their diaries, and calling up other people on the telephone.' It has become a tradition with that well-known couple the Kirks – if they were the sort of people who approved of traditions, which of course they are not – to have a party at this season, 'the turning-point when the new academic year starts, new styles are in, new faces about, new ideas busy'. Howard Kirk, a lecturer at the University of Watermouth, drives his mini-van to the supermarket, negotiating the one-way system and developing a strategy for making trouble, to buy the wine. His wife Barbara assumes her habitual role – though naturally the Kirks are a couple who reject sex stereotyping – and fills her shopping trolley with the French bread and cheeses, the olives and oranges, which form the props for the radically blended, carefully constructed social inter-changes which have made the Kirks' parties the high point of the college calendar.

Bradbury's satire on university life, which was made into a highly acclaimed television series (1981) scripted by Christopher Hampton, traces the affairs of an autumn term. This year, 1972, is much like last year, which was like the year before that. Howard is planning to shake up his colleagues and start student agitation by spreading the rumour that a neo-Nazi has been asked to speak, while Barbara is still committed to her liberal causes (Women for Peace, the Children's Crusade for Abortion, No More Sex for Repression), and is hoping to take a nightclass in commercial French so that she can read Simone de Beauvoir in the original. In the lives of the Kirks, only the partners change. This year Howard is being idolised

by a dim and scruffy student, Barbara makes her sporadic weekend trips to London to meet her lover, and the work in progress is Howard's book *The Defeat of Privacy*, which is much like his last book, though the modish jargon has been updated. 'The Kirks are very attractive, very buoyant, very aggressive people, and, even if you dislike or distrust them, or are disturbed by them (and they mean to be disturbing), very good company.' They did not, however, get this way without a lot of effort; indeed one could say that the Kirks have made themselves into new people.

Down below, in the Piazza, the students criss and cross, this way and that, in elaborate, asymmetrical patterns, ants with serious yet unguessable purposes.

How are they this year? Well, no longer do they look like an intellectual élite; indeed, what they resemble this autumn is rather the winter retreat of Napoleon's army from Moscow. For in the new parade of styles, which undergoes subtle shifts year by year, like the campus itself, bits of military uniform, bedraggled scraps of garments, fur hats and forage caps and kepis, tank tops and denims and coats which have lost their buttons have become the norm; the crowds troop along raggedly, avoiding the paths which have been laid out for them, hairy human bundles fresh from some sinister experience.

THE HISTORY MAN
BY MALCOLM BRADBURY

Among the *dramatis personae* on this autumn evening are Henry and Myra Beamish, old friends of the Kirks; Henry is a man on whom any passing football which just happens to be sailing over the wall of a schoolyard will land. The Beamishes have a nice kitchen, with pine fitted cupboards, and Myra is thinking of walking out, hoping to resume her affair with Howard. Henry acci-dentally – though nothing is accidental – slashes his wrist at the party by falling through a bedroom window. Also present are: Moira Millikin, 'unorthodox economist and unmarried mother'; Anita Dollfuss, a student, and her terrier dog Mao; Felicity, whose obsession with Howard follows her experimental lesbian phase, and whose sexual hangups are about to be treated by her tutor; Miss Callendar, the new Spenserian, an enigmatic Scotswoman with a bicycle; and Flora Beniform, Howard's mistress and therapeutic benefactor.

Minutely and exactly detailed, the term lumbers on through its meetings and seminars, tutorials and sexual encounters, though the much-vaunted visit of the controversial Dr Mangel turns into an anti-climax as he dies the night before his lecture. 'And now it is the winter again; the people, having come back, are going away again', and as they tend to do at this time of the year, the Kirks are planning another party. Upstairs in the bedroom a window smashes again: 'The cause is Barbara', though in fact no one hears, for 'as always at the Kirks' parties, which are famous for their happenings, for being like a happening, there is a lot that is, indeed, happening, and all the people are fully occupied'.

An academic himself, Bradbury has been one of the foremost fictional chroniclers of university life, from his first novel, **Eating People is Wrong** (1959), onwards. Indeed, *The History Man* might be described as the campus novel to end all campus novels. In a waggish 'Author's Note' Bradbury describes the book as 'a total invention with delusory approximations to historical reality, just as is history itself'. In fact, much of the novel's strength lies in the accuracy and fine detail of Bradbury's observations, which provide an almost anthropological report on life in a provincial university in 1972. As time has passed, *The History Man* has come to seem an historical novel, not so much a fiction that has dated, but a genuine period piece (which is how the television adaptation was presented).

E.L. DOCTOROW   1931–

Ragtime

Ragtime is set around the turn of the century and tells the story of a family, named only as Father, Mother, Grandfather, Younger Brother and Son, as they come into more or less close contact with the people and events which shaped those years. Henry Ford, J.P. Morgan, Houdini and Archduke Franz Ferdinand all play parts in the novel, as indeed they did in history. Doctorow the novelist has the edge on Doctorow the historian though. The private characteristics and obsessions of these mythologised figures are reconstructed by a novelistic imagination to show the intimate links between private idiosyncrasies and larger historical events. J.P. Morgan's belief in reincarnation emerges publicly as the dynastic ambitions of his financial empire; Houdini's incredible escapes are at once potent symbols of immigrants' attempts to escape the strait-jacket of their circumstances, and private acts of communion with his dead mother, unable to escape the strictures of mortality and her coffin. Doctorow perceives these oblique links between individual private activities and larger historical currents and obviously values the moments of contact by which both are illuminated.

In registering many historical currents in their effects on one family, Doctorow has to admit a high degree of coincidence and chance, which threatens his view of history as a causal relation between individual action and public, political events. The contradiction is avoided by an acute psychological perception of his characters which ferrets out their innermost drives and translates them as motives for their often mismatched contacts with real-life people and events. Thus Younger Brother's relationship with Emma Goldman, the anarchist leader, is primarily (and hilariously) sexual rather than political, and his relationship with a gang of black militants is an expression of his alienation from the family. Likewise, a black ragtime pianist's pride in his brand new car (manufactured by Henry Ford) ultimately becomes the motive for a series of anti-racist bombings, culminating in the siege of a private library (owned by J.P. Morgan). Ironies such as these accrue about most of the characters and spill over into the period itself, whose *Zeitgeist* Doctorow sets down in deadpan style: 'That was the style, that was the way people lived. Women were stout. They visited the fleet carrying white parasols.

Everyone wore white in summer. Tennis racquets were hefty and the racquet faces elliptical. There was a lot of sexual fainting. There were no Negroes. There were no immigrants.'

Morgan's intention in Egypt was to journey down the Nile and choose a site for his pyramid. He stowed in the safe in his stateroom the plans for this structure secretly designed for him by the firm of McKim and White. He expected that with modern construction techniques, the use of precut stones, steam shovels, cranes, and so forth, a serviceable pyramid could be put up in less than three years.

RAGTIME
BY E.L. DOCTOROW

Of course there were Negroes, there were immigrants. The family is framed in the novel by a black and an immigrant family on both of whom events bear down harshly. The immigrant father rising from cutting and selling silhouettes in a slum to producing movies is eventually absorbed into society. The Negro father sees his bride-to-be killed by the police and his car smashed by bigots before waging terrorist warfare on his oppressors and dying at the hands of police marksmen. In face of these parables, doubts arise as to whether Doctorow's angry irony is an adequate response to such events. These are political concerns rather than artistic ones through. In *Ragtime* Doctorow confronts and dissects the issues involved with an unpatronising use of historical hindsight and a full appreciation of their complexity. Much of this complexity was lost in the film adaptation (1981), directed by Milos Forman from a screenplay by Michael Weller.

William GADDIS   1922–

JR

Separated from the **The Recognitions** (1955) by twenty years, *JR* is similarly ambitious. In *The Recognitions* Gaddis attempted to write a novel from the unfeeling heart of post-war consumer culture. *JR* is the chaotic epic of the 1970s, a decade deracinated by information technology and depleted by corporate greed. The backdrop to the novel is the post-industrial city with its blandly uniform façades and unseen communications network. In this morally vacant environment institutions – hospitals, schools, banks, 'home' – merely define space; they enshrine no values. A superintendent defines the function of his Long Island school: 'It's here to keep those kids off the street until the girls are big enough to get pregnant and the boys are old enough to go out and hold up a gas station, it's strictly custodial and the rest is plumbing.' The school itself is run like a business and it is from its payphone that JR, an eleven-year-old with a pocketful of change and an appetite for 'dealing', begins to build up a multinational company with interests in shipping, pharmaceuticals, timber, funeral urns, pianola rolls, reject plastic flowers, condoms – in fact anything that can be traded over the phone.

The novel fields a huge cast of characters who are sucked into JR's illusory empire. A composer trying to compose his way out of his father's influence, a writer writing to escape his wartime experiences, and a painter trying to paint his way free of a society hostess's clutches are more serious and hopeless versions of the dilettantes in Gaddis's first novel; the redemptive possibilities of art seem to have receded in the intervening decades. The plot proliferates at the same exponential rate as the JR Corporation, and its extreme complication derives from the confusion of JR's finances.

The huge bulk of the novel is reported as direct speech, and the real action of *JR* happens 'off-stage'. The only real unifying threads are the telephone wires, telex cables and other informational conduits which disembody the characters' voices. Serious doubts arise as to what, if anything, lies behind these flows of information: reality, in Gaddis's vision, is held in thrall by the gadgets and apparatus of communication which constitute one of several deadening and prohibitory structures in the book. These are the ultimate targets of Gaddis's satire.

Towards the end of the novel, with JR Corp. crumbling, many of its interests coincide in a macabre, improbable incident. The failed painter has erected a large, mobile metal sculpture in which a boy from JR's school has contrived to ensnare himself. His parents file an insurance claim against the owner, the 'Modern Allies of Mandible Art' take out an injunction against the sculpture being damaged, firemen and protesters stand by. Hot-dog vendors and novelty-sellers vie for franchises as the accident becomes an 'event' and newspapers report it on a daily basis. The theme of humankind done to death by paperwork while caught in the arms of art bears some analogy with Gaddis's own activity. In a world where the peddlers of paperwork seek to replace human character with PR handouts and real industry with the rhetoric of leasebacks, writeoffs and losscarryforwards, the activity of novel-writing becomes problematic, if not compromised. At the end of the novel the boy is still hanging there, and the weather is getting worse ...

– And he found you something? this bass player?

– No well yes sort of indirectly, he said he wanted to help me out and sent me to a place over on the West Side where they said they wanted some nothing music, three minutes of nothing music it's for television or something, they said they had three minutes of talk on a track or a tape they needed music behind it but it couldn't have any real form, anything distinctive about it any sound anything that would distract from this voice this, this message they called it, they...

– But of all things how absurd, paying a composer to...

– Yes well they didn't, I couldn't do it I mean, they were in a hurry they would have paid me three hundred dollars and I tried and all I could, everything I did they said was too...

JR
BY WILLIAM GADDIS

Ruth Prawer JHABVALA   1927–

Heat and Dust

Jhabvala's economical, subtle and witty novel was published in the same year that Paul Scott finished his *Raj Quartet* (1975), and it is perhaps because of these two writers that the reading public became interested in British involvement in India, previously thought to be a hopelessly uncommercial subject for fiction. Both Scott's tetralogy and Jhabvala's novel use the device of a contemporary figure investigating (fictional) *causes célèbres* from the history of the Raj (both involving a relationship between a British woman and an Indian man); but whereas Scott's narrative tracks back and forth in time, looking at the events from every angle, Jhabvala's more concentrated book focuses upon two young women, separated in time but increasingly similar, despite the enormous changes that had taken place in India between 1923 and 1975.

An unnamed young woman comes to India to find out about the first wife of her grandfather, an assistant collector called Douglas Rivers. Olivia Rivers loved her husband, but soon became bored with the society of her fellow memsahibs. A glimpse of a more invigorating life is provided at a dinner-party given by the spoiled, handsome Nawab, where Olivia meets her host's house-guest, an amusing, insinuating, homosexual Englishman called Harry. The Nawab is not popular with the Anglo-Indian community, whose suspicions about political corruption are eventually proved correct. Olivia, however, finds convention-defying picnics in the glamorously *louche* company of the prince and Harry (who, it transpires, is as much a prisoner as a guest) more fun than staid evenings with the Saunders, Crawfords and Minnies. She becomes increasingly involved with the Nawab and, finding herself pregnant, is unsure whether he or Douglas is responsible. Both men are delighted by the news, each assuming (one from arrogance, the other from innocence) that he is the father. Olivia decides to have an abortion, a crude affair which is arranged for her by the Nawab's powerful mother, the Begum. Afterwards, she leaves Douglas for the

Nawab, who buys a house for her in the mountains.

Olivia's story is intercut with the diary of Douglas's granddaughter, who, looking for the India of Olivia, discovers the real country – its squalor, its disorder, the heat and dust of the title. Another false image of India is embodied by Chidananda ('Chid'), a young Midlander-turned-*sadhu*, who comes in search of spiritual fulfilment, but finds only disease, and has to be flown home. The young woman sleeps with both Chid and her landlord, Inder Lal, and becomes pregnant, but rather than submitting to the ministrations of a local midwife, she decides to have the child. The novel ends with her going to the hill-town where Olivia spent the rest of her life: 'There is no record of what [Olivia] became later ... More and more I want to find out; but I suppose the only way I can is to do the same she did – that is, stay on.'

Heat and Dust won the Booker Prize and was subsequently transformed into a successful but simplified film (1983) by the author with her frequent collaborators, Ismail Merchant and James Ivory.

| David LODGE | 1935– |

Changing Places

Lodge's fifth novel, and the first to feature Philip Swallow and the University of Rummidge, is essentially a comedy based on coincidence and gentle caricature. It also succeeds in interesting the reader in its characters' emotional development, and offers some temperate reflections on the virtues and limitations of 1960s' 'liberation'.

Set in 1969, it follows the parallel adventures of two Eng. Lit. academics who leave their wives behind to take part in a six-month exchange programme: Swallow, who flies out to the State University of Euphoria at Esseph (on the USA's west coast and not unlike San Francisco), and Morris Zapp of Euphoric State, who replaces Swallow at Rummidge (in the Midlands and not unlike Birmingham). True to national form, Zapp is ambitious and professional, with a long list of publications; Swallow is a gentlemanly dilettante who never publishes and does not have a 'field', or even a PhD. Each suffers severe culture-shock on arrival. Swallow is also dogged by Charles Boon, a failure as a student at Rummidge but lionised in Esseph, where he runs a no-holds-barred late-night phone-in. Campus protests and sit-ins, challenges to the old order, provide the setting, like a latter-day Forest of Arden, for the protagonists' emotional upheavals.

For Philip, these start by chance when Melanie Byrd – who turns out to be Zapp's daughter by his first marriage – initiates him into the mysteries of free dancing and foot massage. His libido, dormant for decades, suddenly revives, and he is unfaithful to his wife Hilary for the first time. But remorse sets in, and Melanie abandons him in favour of Boon, so it is left to Zapp's almost-estranged wife, Désirée, to complete his transformation. The remorse vanishes, he discovers a new freedom, and even becomes something of a local hero when he is arrested for his (accidental) involvement in the construction of a 'People's Garden' near the Euphoric State campus. Meanwhile in Rummidge, a chain of coincidences has brought Zapp and Hilary together and, at Zapp's hands, Hilary discovers there is more to life than washing and ironing. Though hesitant at first, after having things out with Philip – unwittingly over the air waves of Boon's talk show – she yields willingly to Zapp's advances.

When the six months are up, a complete role swap seems inevitable. Philip has no desire to return home; Zapp is offered the chair at Rummidge. Then at Désirée's suggestion, the two new couples converge on New York to sort themselves out. Though husbands and wives find their way back into each others' arms, the question of whether these pairings, or any, will last is left in the air. The two women wonder whether any real freedom has been won as Morris and Philip hog the conversation, talking the novel to a close on the theme of the aesthetics of endings.

The novel remains Lodge's best-known and was awarded both the Hawthornden Prize and the *Yorkshire Post* Fiction Prize.

Anthony POWELL   1905–

A Dance to the Music of Time

A Question of Upbringing (1951)
A Buyer's Market (1952)
The Acceptance World (1955)
At Lady Molly's (1957)
Casanova's Chinese Restaurant
    (1960)
The Kindly Ones (1962)
The Valley of Bones (1964)
The Soldier's Art (1966)
The Military Philosophers (1968)
Books Do Furnish a Room (1971)
Temporary Kings (1973)
Hearing Secret Harmonies (1975)

Powell's twelve-volume sequence of novels covers the period 1921 to 1968, with glimpses of the narrator's childhood before the First World War. The narrator, Nicholas Jenkins, is himself a novelist, and his admiration for Proust, to whose *A la recherche du temps perdu* this sequence bears a superficial resemblance, is alluded to in the ninth volume, *The Military Philosophers*, when he stays the night in a seaside resort he later discovers to be Proust's 'Balbec'.

Proust's and Powell's narrators have in common an interest in the workings of time and a field of observation mainly confined to the upper sections of their societies. Their approach, however, could hardly be more different: Proust's romantic and subjective, Powell's classical and objective. Powell is essentially – as the five separate novels he wrote before the sequence emphasise – a comic writer. Perhaps the most remarkable thing about *A Dance to the Music of Time* – what makes it unique in literature – is that its light comedy mode is not only able to consort with serious analysis and passages of great poignancy, but that it can sustain without any sense of strain or incompatibility the huge sweep and multifarious detail of half a century of history.

The key to this achievement is Powell's treatment of character. Unlike his contemporary and friend Evelyn Waugh, whose humour is broader and crueller, or the

gentler, less realistic light comedian of a previous generation, P.G. Wodehouse, to whom they are both a little indebted, Powell is never content with caricature. Every one of his characters, however minor, is minutely and tellingly observed. Their speech and actions may be laughable or even ridiculous, but never exaggerated. They retain the inconsequence, the opacity and above all the unpredictability of living people. 'All human beings,' remarks the narrator, 'driven as they are at different speeds by the same Furies, are at close range equally extraordinary.'

The multiple layers and threads of character and relationship in *The Music of Time* are shadowed and modified not only by discreet reminders of the cataclysmic events of the period (which naturally affect the characters' individual destinies), but also coloured and enriched by the narrator's subtle introduction of themes and references: literary, artistic, mythological and even astrological.

The major theme of the sequence, exemplified most vigorously by Widmerpool, but in varying degrees by most of the characters, is 'the will to power', whether exercised positively through sexual, political or business relationships or negatively through the unforeseen action of time. The grotesque, power-hungry Widmerpool has a hand in the wartime deaths of both the narrator's schoolfriends, Charles Stringham and Peter Templer, but Widmerpool's double sexual and political catastrophe in the penultimate volume, *Temporary Kings*, is caused by his marriage to Stringham's niece and his undercover connections with the East European country where Templer met his end. Behind the baroque classical structure and the comic tone of Powell's masterpiece there is a powerfully romantic story of time's revenge, of moral accounts settled.

A Question of Upbringing begins with a scene of workmen round a hole in the road. Flakes of snow falling on their brazier make the narrator, Nicholas Jenkins, think of the ancient world; and the men's physical attitudes recall Poussin's painting 'A Dance to the Music of Time'. Jenkins's thoughts plunge back thirty years to his schooldays.

His first memory is of an older boy called

Widmerpool, a solitary, awkward and even grotesque figure in that close-knit, conventional and rule-bound boarding-school society, encountered on his regular afternoon run in a misty drizzle. But the narrator is less concerned with the school itself than with his own intimations of the grown-up society beyond, sensed rather than understood through his friendship with two boys slightly senior to himself, Charles Stringham and Peter Templer. Stringham's mocking, subversive attitude to the world culminates in a practical joke played on the boys' housemaster, Le Bas. Templer boasts of picking up a prostitute in London and is eventually expelled from the school for smoking a pipe. A visit to the school by Jenkins's ne'er-do-well Uncle Giles, always on the move and without a fixed address, obsessed with increasing his small income from a family trust, hints at another dimension of grown-up life: the struggle for economic survival.

'How one envies the rich quality of a reviewer's life. All the things to which those Fleet Street Jesuses feel superior. Their universal knowledge, exquisite taste, idyllic loves, happy married life, optimism, scholarship, knowledge of the true meaning of life, freedom from sexual temptation, simplicity of heart, sympathy with the masses, compassion for the unfortunate, generosity – particularly the last, in welcoming with open arms every phoney who appears on the horizon. It's not surprising that in the eyes of most reviewers a mere writer's experiences seem so often trivial, sordid, lacking in meaning.'

BOOKS DO FURNISH A ROOM
BY ANTHONY POWELL

Before leaving school, Jenkins is briefly introduced to Stringham's princely home life with his rich, elegant mother and naval stepfather in a smart house in Mayfair. After leaving school, Jenkins visits Templer's home by the sea and makes contact with another section of society, almost as affluent as Stringham's, but connected with business

and the City instead of high society. Jenkins also begins to awaken to his own sexual feelings, stimulated by Templer's younger sister, Jean.

These feelings grow more insistent and are temporarily redirected towards a French girl during a visit *en pension* to Touraine. Here Jenkins meets Widmerpool again, is impressed by his success in reconciling two quarrelling Scandinavians, but notes (from his vantage-point in later years) that he still did not recognise in this ludicrous, humourless person 'the quest for power'. The novel ends with Jenkins, now at university, moving in a new circle of ambitious young men around the middle-aged Prof. Sillery, who uses them much as they use him for contacts with sources of power and influence in the world beyond the university. But Jenkins also re-encounters his old schoolfriends Stringham and Templer, and charts the widening gaps of experience, interest and temperament between all three.

During the course of *A Buyer's Market*, set in the late 1920s, Nicholas Jenkins, now living in London and working for a firm publishing art books, emerges fully into adult life. The main thread of his memories is provided by Edgar Deacon, an old friend of his parents, at one time a painter of large, late-Victorian male-figure paintings with classical themes, but now the owner of an antique shop. Jenkins meets Mr Deacon's lodger, Ralph Barnby, also a painter but of a more modern kind. Unlike Mr Deacon, Barnby is voraciously heterosexual and his knowledge and analysis of the female character form a subsidiary thread as Jenkins's own sexual experience increases. At the end of the novel, on the day of Mr Deacon's funeral, Jenkins sleeps with the grubby, waif-like Gypsy Jones, whom he had first met with Mr Deacon on the night of two great summer parties near Hyde Park Corner.

For Jenkins the chief element of both these parties is sexual. The first is a society ball given by Lord and Lady Huntercombe. Jenkins attends it in a party led by Sir Gavin and Lady Walpole-Wilson, old diplomatic friends of his parents, whose niece, Barbara Goring, attracts him without offering much encouragement. Widmerpool, still for Jenkins 'a kind of embodiment of thankless labour and unsatisfied ambition', is also one of the party and he too, it turns out, is attracted by Barbara Goring. But when he complains of her missing a dance with him, she empties a large sugar-caster over his head. The incident puts an end to Jenkins's as well as Widmerpool's infatuation with her.

The second party is less respectable. It is given by Milly Andriadis, former mistress of a 'Royal Personage' and now apparently (to Jenkins's astonishment) mistress of his old schoolfriend Charles Stringham. It also turns out that the house where the party is given has been leased by Mrs Andriadis from a friend of Peter Templer, and that this friend, Bob Duport, has recently married Templer's sister Jean, the girl Jenkins was first attracted to just after leaving school.

He meets her again in the country, on a lunch visit to the richly appointed Stourwater Castle, home of the industrial magnate Sir Magnus Donners. Already at the two parties Jenkins had been vaguely aware that behind the sexual jockeyings for position there was another world of intricate negotiations for political and financial power, centred on Sir Magnus Donners and the visiting prince of a Balkan country, Theoderic. Stringham, now employed as one of Donners's secretaries, is involved in this world; so, indirectly, is Prof. Sillery from Jenkins's university. Widmerpool, also visiting Stourwater, makes a fool of himself again by backing his car into a stone urn, but it is becoming clear that in this (to Jenkins) more shadowy world of power his influence and standing are on the increase, while Stringham's are on the wane. Widmerpool himself, who has also been briefly caught up with Gypsy Jones and has paid for her abortion (some other man's child) confides to Jenkins: 'No woman who takes my mind off my work is ever to play a part in my life in the future.'

In the third volume of the sequence, Widmerpool, still loosely connected with the powerful firm of Donners-Brebner, has entered the world of City broking, a shadowy arena known as 'The Acceptance World'. But, as Jenkins reflects, 'the whole world is the acceptance world, as one approaches thirty'. After the frenetic 1920s' atmosphere of *A Buyer's Market*, this novel – set in the Depression of the early 1930s – has a calmer

tempo as Jenkins and his fellow characters settle into more distinct patterns in their 'dance to the music of time'.

Nevertheless, the faintly ominous tone that pervades the book – a sense of stagnation combined with future threat – is heard from the beginning. Jenkins, visiting his Uncle Giles in a gloomy ship-like hotel called the Ufford ('riding at anchor on the sluggish Bayswater tides'), has his fortune told by the mysterious Mrs Erdleigh. Meeting her again at the Maidenhead home of his newly married friend Peter Templer, Jenkins and the other guests take part in a session of planchette. The garbled messages that come through seem to emanate from the spirit of Karl Marx and to be intended for one of the guests, a sullen, left-wing writer called J.G. Quiggin. Quiggin and his rival, the modernist but essentially right-wing and aesthetic poet Mark Members, have been encountered briefly in both the earlier novels of the sequence. In *The Acceptance World* their professional and political differences are highlighted by their struggle over the elderly, once bestselling novelist St John Clarke. Quiggin displaces Members as Clarke's secretary. Jenkins and Members witness a left-wing political demonstration in Hyde Park in which the procession includes Prof. Sillery and a group consisting of St John Clarke in a wheelchair, pushed by Quiggin and Templer's wife, Mona. Quiggin's triple triumph – sexual, professional and political – symbolises the apparent collapse of the old capitalist society represented by Clarke, Templer and in some degree Members and Jenkins also. But it is only a temporary triumph, for by the end of the book Quiggin has been himself replaced as Clarke's secretary by the German Trotskyist, Werner Guggenbühl, whose earnest speech on Brechtian theatre ('Drama as highest of arts we Germans know. No mere entertainment, please') is a comic *tour de force*.

It is matched towards the end of the book by speeches from Jenkins's two schoolfriends Widmerpool and Stringham. Widmerpool's, forced on the guests at a dinner in the Ritz for the old boys of Le Bas's house at their school, is a pompous and incomprehensible rigmarole on the economic situation of the time, and is interrupted when the housemaster suffers a mild stroke. Stringham's speech, or rather series of drunken monologues, includes a deadpan account of his recently broken marriage, followed by an outburst against Widmerpool who, with Jenkins's help, has taken Stringham to his flat and put him to bed. The shifting sexual and political patterns of the novel, including the consummation of Jenkins's own long-delayed affair with Templer's sister, Jean, at almost the same moment as the notorious womaniser Templer loses his wife to the unlikely Quiggin, are crowned by this last astonishing reversal of status. Widmerpool, the grotesque and accident-prone, lower-middle-class outsider, physically holds down on his bed the romantic, socially privileged Stringham, who has always mocked his pretentious awkwardness with wicked accuracy and even at this dire moment manages a joke: 'So these are the famous Widmerpool good manners, are they? This is the celebrated Widmerpool courtesy, of which we have always heard so much.'

Nicholas Jenkins, now (in 1934) a published novelist and employed as a writer of film-scripts, is introduced by his colleague Chips Lovell to the social no man's land *At Lady Molly's* in South Kensington. Daughter of an earl and widow of a marquess, Lady Molly is now the wife of plain Ted Jeavons (without job or money and widely considered a prize bore), whom she met when she was a nurse and he a wounded soldier in the First World War. Lady Molly's aristocratic relations and connections are legion, but her parties have no cachet at all, since she welcomes in every kind of human as well as animal stray.

Here Jenkins re-encounters the eccentric Gen. Conyers and his wife, old friends of his family, as well as Mrs Conyers's handsome and rackety sister Mildred Haycock, who, astonishingly, is now engaged to the increasingly successful Widmerpool. Here too Jenkins becomes involved with the enormous Tolland family and by the end of the book is himself engaged to Isobel Tolland, one of the younger sisters of the left-wing activist Earl of Warminster (known as 'Erridge' to his family).

Jenkins meets Erridge himself, a bearded, unkempt and ill-dressed figure ('his clothes gave off a heavy, earthen smell as if he had lived out in them in all weathers for a long time') at his country estate, Thrubworth Park, where Erridge's current political mentor, J.G. Quiggin, and his mistress, Mona, are installed in a cottage in the grounds. Jenkins, Quiggin and Mona are invited to the big house for a memorably dismal evening meal, served by the surly butler, Smith, who, having himself already drunk most of the contents of the cellar, is obstinately unwilling to serve what remains to Erridge's guests. In the ensuing battle of wills between master and servant, Jenkins is surprised to discover that for all Erridge's apparently almost paralysing diffidence, his will prevails.

Behind the comic and trivial merry-go-round of gossip and family relationships that fills the foreground of *At Lady Molly's*, Jenkins is becoming increasingly fascinated by the varied forms of egotism in those he observes. Hitler, by now Chancellor of Germany, is mentioned several times in the book – notably by Widmerpool who admires his social reforms and especially his belief in planning – and it can hardly be accidental that just at this juncture the theme of the exercise of the will begins to loom large for the narrator of *A Dance to the Music of Time*.

He even sees marriage as 'an assertion of the will', though in Widmerpool's case it is tinged with absurdity and eventually ignominy. Lunching Jenkins at his club, Widmerpool awkwardly seeks advice on whether he should sleep with Mildred Haycock before their marriage: 'I should not wish to appear backward in display of affection . . . in fact my fiancée . . . might even expect such a suggestion? . . . Might even regard it as *usage du monde*?' The outcome, taking place during a weekend in another country house, Dogdene, where Lady Molly had once been châtelaine and where she had originally met her present husband, is a disaster and the engagement is broken off. Jenkins, attending his own engagement party at Lady Molly's, hears the details from Mildred's brother-in-law, old Gen. Conyers (courtier, cellist and amateur psychoanalyst): 'It seems to me that he [Widmerpool] is a typical intuitive extrovert – classical case, almost.' Meanwhile, after another clash of wills, Erridge has gone off to China, taking with him Quiggin's ex-mistress, and Templer's ex-wife, Mona.

Most of the action of *Casanova's Chinese Restaurant* takes place in 1936–7, against a distant background of the Spanish Civil War and the British constitutional crisis over the abdication of King Edward VIII. Nicholas Jenkins's new brother-in-law, Erridge, has returned from China and now goes to Spain to help the Republican government (though not as a combatant). Widmerpool, lunching with Jenkins at the latter's club, drops heavy hints about 'moving in rather exalted circles ... not exactly royal yet ... You understand me?' but is later reported to be much 'put out by the Abdication. It might have been Widmerpool himself who'd had to abdicate.'

The novel's first chapter, however, is set back in time by three years in order to introduce another circle of Jenkins's acquaintances, a group of musical and theatrical people centred on his particular friend, the composer Hugh Moreland. Their friendship is consolidated in Casanova's Chinese Restaurant, whose 'recklessly hybrid name ... offered one of those unequivocal blendings of disparate elements of the imagination which suggest a whole new state of mind or way of life'. Not long after Jenkins's own marriage to Isobel Tolland, Moreland marries the actress Matilda Wilson, previously mistress to the tycoon Sir Magnus Donners.

The kaleidoscopic and indefinable nature of marriage is a major theme of the book. (King Edward's determination to marry the American divorcee Wallis Simpson was, of course, the reason for his abdication.) The narrator declines to investigate his own marriage, but he gives a closely observed account of the splenetic music critic Maclintick's dreadful relationship with his wife Audrey: 'When she opened the door to us, her formidable discontent with life swept across the threshold in scorching, blasting waves.' Nevertheless, when she puts an end to their mutual torment by disappearing with their lodger, the out-of-work violinist Carolo, Maclintick gasses himself. Moreland's marriage to Matilda is also strained when he falls in love with Priscilla, the youngest member of the Tolland family, and the two contrasting misalliances – the Maclinticks' coarse, overt, violent and irreparable; the

Morelands' publicly restrained, poignant and temporary – are observed during a party celebrating the first performance of Moreland's symphony.

This party, held at the grand Mayfair house of Stringham's mother, Mrs Foxe, which Jenkins last visited as a schoolboy, is also notable for the reappearance of Stringham, now a partly cured drunk under the control of 'Tuffy' Weedon, his sister's former governess. They live in the top flat of Lady Molly's house in Kensington, but Stringham escapes from time to time for drinking sessions with Lady Molly's husband, Ted Jeavons. The many sexual cross-currents sweeping through this party – including an apparently successful three-way relationship between Mrs Foxe, the homosexual actor Norman Chandler on whom she dotes and her husband, Buster – are dominated by Stringham's comic flirtation with the ghastly Mrs Maclintick (dressed in pink frills): ' "Hello, Little Bo-Peep," he said. "What have you done with your shepherdess's crook?" ' But although Mrs Maclintick 'was tamed, almost docile, under his treatment', Stringham is coolly removed from the party by his keeper, Miss Weedon, hastily summoned for the purpose by his stepfather, Buster Foxe.

At another gathering – a lunch-party mostly consisting of Jenkins's new Tolland relations-in-law – he meets for the first and last time the elderly, effete novelist St John Clarke, whose conversion to Marxism has brought him into a new friendship with Erridge ('we talked until the wee small hours'). Indeed, when Clarke dies at the end of the book, it transpires that he has left his money to Erridge, recently returned from the Spanish Civil War with dysentery but without having met Hemingway.

The Kindly Ones is a translation of the Ancient Greeks' euphemism for the Furies, who, as the narrator's governess taught him in his childhood: 'inflicted the vengeance of the gods by bringing in their train war, pestilence, dissension on earth; torturing, too, by the strings of conscience'. The novel's first chapter returns to the year 1914, when Nicholas Jenkins was a child of about nine living with his parents in the Surrey countryside near the military base of Aldershot.

He observes the characters and strained relationships of the servants with the same fervent curiosity as he displays in later life and, when his parents' old friends Gen. and Mrs Conyers come to lunch, is complimented by the General for the way he answers questions about the local children: '"An exceedingly well-informed report," said the General. "You have given yourself the trouble to go into matters thoroughly, I see. That is one of the secrets of success in life."' The General displays his own commanding personality when the distraught parlourmaid walks into the drawing-room completely naked: he wraps her in a Kashmir shawl taken off the piano and helps her to her room. That crisis resolved, however, a greater one immediately looms up. As the Conyers are leaving in their early motor car, Jenkins's Uncle Giles appears with the news that 'some royalty in a motor-car have been involved in a nasty affair today'. It is the assassination at Sarajevo, the spark that ignites the First World War.

The rest of the novel, set in 1938–9, is similarly shot through with hysteria and perturbation, glimpses of the re-aroused Furies. Against a background of threatening events in Europe (the Munich Treaty and the Russo-German Pact), Jenkins and his wife Isobel, with their friends Hugh and Matilda Moreland, go to dinner with Sir Magnus Donners at Stourwater Castle. The party also includes Donners's current mistress, Lady Anne Stepney, and Jenkins's old friend Peter Templer, now married to the beautiful but mentally unstable Betty. After dinner in a room surrounded with tapestries depicting the Seven Deadly Sins, Donners photographs the guests each portraying one of the Sins. When Templer, with Anne Stepney's help, gives his performance of 'Lust', Betty begins to weep uncontrollably; the scene takes on a still more sinister aspect when Widmerpool suddenly appears in the doorway wearing his uniform as a captain in the Territorial Army.

His Uncle Giles having died at a sleazy seaside hotel called the Bellevue, Jenkins goes there to organise the cremation. In the course of an eventful night he meets Bob Duport, ex-husband of his own former mistress, Jean, and learns of her affair with the fat and unprepossessing Jimmy Brent, which seems to have been almost simultaneous with the narrator's own. Later, Jenkins and Duport help to release another resident of the Bellevue, the once formidable mystic Dr Trelawney (first encountered during Jenkins's childhood, running over the Aldershot heather with a crowd of young disciples in Grecian tunics), who has locked himself in the bathroom. Dr Trelawney is finally given a calming injection by the clairvoyant Mrs Erdleigh, an old friend also of Uncle Giles and, as it turns out, his sole legatee.

The final chapter of the novel brings the outbreak of the Second World War ('I rather like the little man they've got in Germany now', had been Uncle Giles's frequent comment on Hitler). Now trying to get a job as an army officer, Jenkins unsuccessfully seeks help from the insufferably complacent Capt. Widmerpool: 'The General Staff of the Wehrmacht would be only too happy to possess even a tithe of the information I locked away before we quitted the Orderly Room.' Even the widowed Gen. Conyers, about to be married to 'Tuffy' Weedon, is too out of touch with the army to help. Visiting Lady Molly's blacked-out house in South Kensington, Jenkins discovers that the disconsolate Moreland – his wife Matilda has left him to marry Sir Magnus Donners – is Lady Molly's latest rescued stray. But it seems that Ted Jeavons's brother can help Jenkins to his army commission.

The Valley of Bones opens in 1940, and the Second World War is still in its early stages. Nicholas Jenkins, now a second lieutenant, joins his Welsh infantry battalion in a small Welsh town and begins to settle into army life, observing his fellow soldiers with his customary attention. But although he can trace his distant ancestors back to this part of the country, he has little else in common with all these new acquaintances, most of whom come from the same small community, where in civilian life the soldiers and NCOs tended to be miners, the officers bank employees. Preaching at a church service, the chaplain takes Ezekiel's text about the valley full of dry bones and asks: 'Must we not come together, my brethren, everyone of us, as did the bones of that ancient valley . . .?'

A sense of dislocation runs through the novel: the bones are being put together in new patterns, but the old patterns are still discernible. When Jenkins's battalion is transferred to Northern Ireland for training, he observes especially the contrast between the two officers closest to him, Capt. Rowland Gwatkin, his company commander, and Lt. Idwal Kedward, who commands another platoon. Both have previously worked in banks, both are ambitious to succeed as officers, but whereas Kedward is an entirely practical, hard-headed man, Gwatkin is a romantic with a secret admiration for the centurion in Kipling's *Puck of Pook's Hill*. After several disastrous errors during training, Gwatkin is relieved of his command and replaced by Kedward. Gwatkin's passionate love-affair with a local barmaid also ends disastrously when he and Jenkins come upon her embracing the irrepressibly cheerful Cpl Gwylt.

Back in England on a course at Aldershot (scene of his childhood), Jenkins encounters Jimmy Brent, who speaks frankly about his love-affair with Jenkins's former mistress, Jean. A new army friend called Odo Stevens drives Jenkins across country to spend a weekend's leave with his wife Isobel, who is expecting their first child. She is staying with her sister Frederica and the old patterns of pre-war life briefly resurface at what turns out to be a gathering of several Tolland relations and other characters from the past. The strait-laced Frederica is engaged to the elderly reprobate Dicky Umfraville (already four times married); Priscilla Tolland (now married to Chips Lovell) is obviously attracted to Jenkins's new friend Odo Stevens; Charles Stringham's stepfather, Com. Buster Foxe, looks in and reveals that Mrs Foxe, now living in much reduced circumstances with Norman Chandler, wants to divorce him.

By the end of the novel the so-called 'phoney war' is over: Germany has invaded Norway, the Netherlands and France; Italy has joined in on Germany's side. Jenkins, now a father, is posted to a new job at Divisional Headquarters, where he finds that he is to be dogsbody to the Deputy-Assistant-Adjutant-General, none other than Widmerpool, promoted to major: 'I saw that I was now in Widmerpool's power. This, for some reason, gave me a disagreeable, sinking feeling within.'

The second phase of Nicholas Jenkins's temporary career as a soldier is covered in *The Soldier's Art* and takes place in the darkest period of the Second World War, when Europe has been overrun by the Germans and Britain's major cities are subject to constant bombing. Jenkins himself, still in Northern Ireland but now attached to Divisional Headquarters, is subservient to the workaholic Major Widmerpool, continually active in the furtherance of his own interests. Widmerpool's Byzantine intrigues against other members of the headquarters' staff, especially Major 'Sunny' Farebrother, another City man whom Jenkins had once met long before at the Templers' house, are echoed by the rivalries and scorings-off between Cols Hogbourne-Johnson and Pedlar in the General's Mess and between Jenkins's unattractive table-companions in 'F' Mess. It is here that Pte Charles Stringham suddenly appears as a waiter (to Jenkins's embarrassment) and becomes a bone of contention between Soper, the Catering Officer who gave him the job, and the uncouth Biggs, in charge of Physical Training: ' "Don't like the look of this chap," said Biggs. "Gets me down, that awful pasty face." '

Meanwhile the mildly eccentric divisional commander, Major-Gen. Liddament (encountered briefly in the previous volume during a hilarious inspection of Jenkins's platoon), offers Jenkins the chance of a more congenial post in London. He fails the interview but makes use of his leave in London to see some old friends. At the Café Royal he hears Chips Lovell's side of the affair between Lovell's wife, Priscilla, and Odo Stevens. Lovell goes on to a party at the Café Madrid, leaving Jenkins to dine with Hugh Moreland and the woman who now shares his flat – the music critic Maclintick's widow, Audrey. They are joined later by Priscilla Lovell and Odo Stevens who happen to be dining at another table. 'A faint suggestion of distant gunfire' interrupts their conversation and as they discuss the German blitz (' "I have an impression of acute embarrassment when bombed," said Moreland,

"It's like an appalling display of bad manners one has been forced to witness" '), Priscilla suddenly becomes very distressed and leaves. When it turns out that the Café Madrid has been destroyed by a bomb, killing Chips Lovell as well as his hostess Bijou Ardglass, Jenkins rings Lady Molly, with whom Priscilla is staying, to give her the news. There is no reply. Jenkins goes round on foot and finds that Lady Molly's house too has been hit. The façade still looks untouched, but the back has been demolished, killing Lady Molly and Priscilla.

Returning to Northern Ireland, Jenkins learns that Widmerpool has had Stringham transferred from the Mess to the Mobile Laundry Unit, knowing that it is soon to be sent abroad, perhaps to the Far East. Jenkins tries to persuade Stringham to avoid being sent abroad 'on grounds of age and medical category', but Stringham takes melancholy pleasure in the prospect and quotes Browning:

I asked one draught of earlier, happier sights
Ere fitly I could hope to play my part
Think first, fight afterwards – the soldier's art;
One taste of the old time sets all to rights.

Widmerpool's plans for his own promotion are maturing, but he makes it clear that he is indifferent to Jenkins's future.

The grim events of this novel are capped grotesquely by the discovery that Capt. Biggs, the Physical Training Officer, has hanged himself ('in the cricket pav. of all places, and him so fond of the game', as Capt. Soper puts it).

In *The Military Philosophers* Nicholas Jenkins is now a captain (promoted to major in the course of this book) working in the War Office in London. Briefly attending a committee meeting of the Cabinet Offices 'in the bowels of the earth', Jenkins notes that 'the power principle could almost be felt here, humming and vibrating like the drummings of the teleprinter'. The rival Lt.-Cols Farebrother and Widmerpool are both present, so is Jenkins's other old schoolfriend, Peter Templer, working under Sir Magnus Donners in the Ministry of Economic Warfare. Templer is curt, preoccupied with, and

profoundly dejected by, the failure of his marriages (his second wife, Betty, 'went off her rocker' and is 'in the bin'), and his age. Jenkins, on the other hand, is distinctly more cheerful in his new job (liaison with Polish and later Belgian and Czech contingents of the Allied Forces) and, admiring both his immediate superiors – the strikingly ugly Lt.-Col. Finn, VC, and the debonair Major David Pennistone – relishes the intricacies and incipient hysteria of the huge administrative machine he is caught up in.

Beginning just after the fall of Singapore, this novel is set against a background of the German defeats at Stalingrad and in North Africa, the 1944 Normandy landings and the final victory for the Allies in 1945. As the war moves into its later phases, Widmerpool becomes a full colonel and Farebrother suffers a setback in his military career after organising the escape from detention of a delinquent Pole called Szymanski whom he wants for Secret Service work. Odo Stevens, too, is involved in this episode and he and Szymanski along with Peter Templer are sent to the Balkans to help the Resistance there. Stevens and Szymanski escape, but Templer is killed.

At the centre of *The Military Philosophers* is a major new character from a new generation. Pamela Flitton is the niece of Charles Stringham, whose death at the hands of the Japanese after the fall of Singapore is confirmed by Widmerpool. Pamela, noticed briefly in the second volume of the sequence as a bridesmaid at Stringham's wedding ('That child's a fiend'), is now about twenty, with black hair and a dead white complexion, and has a devastating effect on most of the men she meets. Jenkins re-encounters her first as an ATS driver for his section of the War Office, then at an opera, where she seems very familiar with the Balkan prince, Theoderic, and catches the attention of Widmerpool. A year and a half later she is among those taking refuge in the hallway of Jenkins's block of flats in Chelsea during a night raid by flying bombs. Her companion now is Odo Stevens, who confides to Jenkins that 'She's cross all the time. Bloody cross ... Makes her wonderful in bed. That is, if you like temper.' The clairvoyant Mrs Erdleigh, also sheltering from the raid, reads Pamela's palm and suggests she will be married in 'about a year'. Sure enough, by the end of the novel, Pamela is engaged to Widmerpool, whom she accuses in public of being responsible for Templer's death in the Balkans. Widmerpool remarks with satisfaction that his fiancée has inherited everything from her dead uncle, Stringham (for whose posting to Singapore he was responsible): 'With the right attention, Stringham's estate might in due course be nursed into something quite respectable.'

After an interlude accompanying the allied military attachés on a visit behind the front line in France and Belgium – during which he finds himself staying in Proust's 'Balbec', re-encounters his old Welsh company on active service (Kedward, the commander, does not remember him) and speaks to Field-Marshal Montgomery – Jenkins attends the Victory Service in St Paul's Cathedral. Afterwards he meets and at first fails to recognise his former mistress, Jean, now married to a smooth and genial Latin-American diplomat, Col. Flores.

The war over, Nicholas Jenkins picks up the threads of his old life in *Books Do Furnish a Room*. Revisiting his university, he pays a call on the retired Prof. Sillery, now preparing his diaries for publication with the help of an ambitious and pretty young woman called Ada Leintwardine. Revisiting his old school to arrange a place for his son, Jenkins encounters his former housemaster, Le Bas, also retired, but looking after the library.

Jenkins's brother-in-law, Erridge, the Earl of Warminster, has died suddenly and the funeral is held in the village church next to his country estate, Thrubworth Park. Most of the mourners are members of the large Tolland family, but there is also a contingent representing Erridge's left-wing, literary/political interests: J.G. Quiggin, Sir Howard and Lady Craggs (once Gypsy Jones) and Widmerpool with his new wife, the ferocious Pamela Flitton. Pamela causes a disturbance by leaving the church service in the middle and later, in the big house itself, is sick into a Chinese vase collected by one of Erridge's ancestors. In the churchyard after the burial Jenkins encounters Mona, once Templer's wife, former mistress of both Quiggin and Erridge, now happily married to an air vice-marshal.

Most of the novel, as its title implies, is concerned with Jenkins's re-entry into the London literary world. He is writing a book about the seventeenth-century author Robert Burton, whose *Anatomy of Melancholy* chimes well with Jenkins's own state of mind at this period, and he accepts the job of literary editor with the new magazine *Fission*, whose editor is the seedy and vaguely left-wing journalist 'Books-do-furnish-a-room' Bagshaw (his sobriquet dates from some obscure sexual episode in the past). The magazine occupies the same Bloomsbury premises as its parent publishing house, Quiggin and Craggs, and many old and new acquaintances are connected with it, including Odo Stevens (who has written a book about his wartime experiences called *Sad Majors*), Ada Leintwardine and Widmerpool, now an MP and parliamentary private secretary in the Labour government, who is one of the magazine's backers and contributes turgid articles on economic and political topics. Behind these more prominent figures appears a gallery of other literary and political characters, such as the rival critics Shernmaker and Sheldon, the novelists Alaric Kyd and Evadne Clapham, the Tory MP Roddie Cutts and the civil servant Leonard Short; but the book is dominated by another novelist, the indigent, egotistical but romantically sympathetic X. Trapnel, who habitually carries a swordstick with a death's-head knob.

Trapnel (who is drawn partly from the novelist Julian Maclaren-Ross) becomes the latest victim of Widmerpool's wife, Pamela. She leaves Widmerpool to live with Trapnel in a squalid flat near the Grand Union Canal in Maida Vale and Jenkins happens to be present when Widmerpool enters the flat, not so much to reclaim his wife as to upbraid Trapnel for his bad behaviour, which includes borrowing one pound from him and parodying his articles in *Fission* on top of taking away his wife. Trapnel draws his swordstick and drives Widmerpool out. Jenkins is also present some months later, helping Bagshaw to escort Trapnel home after a quarrel with Pamela, when they come upon the manuscript of Trapnel's unfinished novel half-submerged in the canal. Trapnel flings his precious swordstick after it: 'The scattered pages of *Profiles in String*, with the death's head swordstick, floated eternally downstream into the night. It was the beginning of Trapnel's drift too, irretrievable as they.' Pamela returns to Widmerpool.

More than ten years pass between the events of the previous novel and *Temporary Kings*; Nicholas Jenkins is now in his fifties. Attending an international cultural conference in Venice, he recalls the strange death of the novelist X. Trapnel, who collapsed and died after a magnificent spending spree at his favourite pub, the Hero of Acre: 'never had the Hero known a night like that for free drinks'. Trapnel's American biographer, Prof. Russell Gwinnett, is also attending the conference and is anxious to meet Trapnel's *femme fatale*, Pamela Widmerpool. It transpires that she, surprisingly still married to Widmerpool, is staying in the rich American-Venetian Jacky Bragadin's palazzo, famous for its Tiepolo ceiling. It is under this ceiling, depicting the story of Candaules and Gyges – the Lydian King Candaules exhibiting his naked wife to his concealed friend Gyges, who later kills him at the Queen's behest – that Gwinnett is introduced to Pamela by her current lover, the playboy and film tycoon, Louis Glober, also American.

Widmerpool himself, recently made a life peer after losing his seat at an election, arrives soon afterwards, much flustered by a government upheaval in the Balkan Communist state with which he has close connections for some time. The twin scandals – sexual and political – involving the Widmerpools intensify as the novel proceeds. Widmerpool comes near to being tried for spying and tax evasion, but the case against him is suddenly dropped. Pamela, meanwhile, pursues the mysterious Gwinnett (obsessed with death and characterised by his friend and fellow academic, Dr Emily Brightman, as 'a small fragment detached from the comparatively extensive and cavernous grottoes of gothic America'). Pamela makes a pass at Gwinnett inside St Mark's basilica in Venice and is seen one night naked in the hallway of a house in London where Gwinnett has taken lodgings. The house belongs to the journalist 'Books-do-furnish-a-room' Bagshaw (who has

become a television personality) and is otherwise occupied in shambolic circumstances by Bagshaw's large family.

The novel's climax comes after a charity performance of Mozart's opera *The Seraglio* in the Regent's Park mansion of Odo Stevens and his rich wife, Rosie Manasch. Waiting for a car on the pavement outside with a group including Widmerpool, Louis Glober, Stevens and others, Pamela is warned by the formidable and now very ancient clairvoyant Mrs Erdleigh that 'You are near the abyss. You stand at its utmost edge.' She responds by revealing that the French Marxist intellectual Ferrand-Sénéschal, who has recently been found dead in a Kensington hotel, was trying unsuccessfully to make love to her at the time he died and that her husband Widmerpool – as often on such occasions – was watching. She further reveals that Widmerpool escaped prosecution for spying by giving the authorities information about Ferrand-Sénéschal, who was a fellow agent. A scuffle ensues in which Widmerpool is punched in the face by Glober and has his glasses broken.

In the final chapter Jenkins visits the dying Hugh Moreland in hospital and discusses the Widmerpools in relation to the story of Candaules and Gyges (upon which Moreland would like to have based an opera). Glober has been killed in a car-crash, Gwinnett is finishing his biography of Trapnel in Spain, but Pamela is dead. She seems to have killed herself in Gwinnett's company, perhaps to satisfy his necrophilia. Widmerpool, encountered by the statue of Boadicea near Westminster Bridge, on his way to the House of Lords, refers to the subject enigmatically: 'The squalor – the squalor of that hotel ... The sheer ingratitude.'

The leitmotif of magic and mysticism, intermittently surfacing throughout Nicholas Jenkins's narrative, emerges as the major theme of *Hearing Secret Harmonies*, the final novel in the sequence. In Ariosto's epic *Orlando Furioso*, the hero's madness (for love of a woman who left him) is cured by his friend Astolpho making a journey to the 'Valley of Lost Things' on the moon and there recovering Orlando's lost wits. Reading a translation of this poem in 1968 (the year before the American astronauts landed on the moon), Jenkins reflects on the difference between Poussin's seated figure of Time in the painting 'A Dance to the Music of Time' and Ariosto's, the latter 'a writer's Time ... far less relaxed, indeed appallingly restless'. The word 'mage' in the translation leads Jenkins's thoughts on to the long-dead Dr Trelawney ('a mage if ever there was one') who, robed in white and followed by his young disciples, used to run over the heather near Jenkins's childhood home. Trelawney always seemed a mainly comic figure, but his mystical ideas, especially his ideal of 'Harmony', are now being revived among young people and a group of these, led by the sinister Scorpio Murtlock, who believes himself a reincarnation of Dr Trelawney, camp briefly in a field near Jenkins's house in the country.

Lord Widmerpool, after a period in America in academic circles, has returned to England as the chancellor of a new university and, condemning established institutions, is putting himself forward as a champion of youth. During his installation ceremony, shown on television, two students, the twin daughters of J.G. Quiggin and Ada Leintwardine, throw a pot of red paint over Widmerpool. He is perversely delighted ('Let me congratulate those two girls on being such excellent shots with the paint pot') and, attending a dinner for the Donners-Brebner Memorial Prize some months later, brings the Quiggin twins as his guests. The prize is being awarded to the American professor Russell Gwinnett for his biography of X. Trapnel, and Widmerpool insists on making an unscheduled speech 'in the name of contemporary counterculture', in which he not only castigates his old chief, Sir Magnus Donners, but reveals what was concealed in the biography, that he was the wronged husband of the woman who destroyed Trapnel, his manuscript and herself. The dinner breaks up in confusion when the Quiggin twins let off a stink bomb.

Jenkins next meets Widmerpool (whom he now recognises as a version of Ariosto's mad hero, Orlando) at the annual banquet of the Royal Academy, when he asks Jenkins's neighbour at table, the fashionable clergyman Canon Fenneau, to put him in touch with Scorpio Murtlock. Widmerpool is duly present (so is Gwinnett, as an observer) at the midsummer-night rites of Murtlock's sect around the ancient standing stones called 'The Devil's Fingers', near Jenkins's house. In the struggle for power between Murtlock and Widmerpool that follows a naked dance and an attempt at multiple sexual intercourse, Widmerpool is gashed with a knife. By the time he reappears nearly a year later, at Stourwater Castle (once home of Sir Magnus Donners, now a girls' school), wearing a blue robe and running through the playing-fields at the head of a dozen or more members of the sect, he looks 'ill, desperate, worn out' and has clearly become almost entirely subservient to Murtlock. Jenkins is at Stourwater to attend a family wedding and when Widmerpool finds out that the bride's grandfather, Sir Bertram Akworth, was the boy he reported at school for sending a homosexual letter, he rushes inside to kneel at Akworth's feet and ask forgiveness. When Murtlock arrives to reclaim his followers he refuses Widmerpool's pleas to be allowed to leave the sect.

However, another younger member of the sect, Barnabas Henderson, is allowed to leave it and some months later Jenkins visits the picture-gallery he has opened in London. The gallery is exhibiting paintings by Jenkins's old acquaintance Edgar Deacon ('a stupendous rescue job from the Valley of Lost Things'), together with Victorian seascapes collected by Bob Duport, first husband of Jenkins's old love, Jean. Duport himself (now in a wheelchair) is also visiting the show with Jean (whose Latin-American husband has been assassinated) and their daughter Polly, a well-known actress. Jenkins and Jean part with a polite handshake, but before Jenkins leaves the gallery he hears that Widmerpool is dead, having collapsed during an early-morning run through the woods, after a further power-struggle with Scorpio Murtlock. As Jenkins goes out into the street, he passes some road-menders round their brazier, a scene first described at the beginning of the whole sequence of novels.

James PURDY   **1924–**

In a Shallow Grave

The redemption of suffering through love, a frequent theme of Purdy's fiction, is developed in this novel through the experiences of a twenty-six-year-old Vietnam veteran, Garnet Montrose. The shallow grave of the title is a pile of corpses, under which Montrose had lain for days. He is rescued, and returns to his home on a Virginian plantation, where he lives alone. Although hideously disfigured, he has recovered from his physical injuries; his pain is caused by persistent memory. He cannot 'learn to forget', because he has never known any worldly happiness. He needs someone to look after him and to deliver letters to the Widow Rance, a neighbour he has loved since childhood and who has lost two husbands in Vietnam. She refuses to see him, but keeps his letters. Although he expects nothing in return, writing the letters gives Montrose some sense of purpose in his life; at the same time this falsely recreates the past, when he was handsome and well. Trapped in this past by his memory and his appearance, Montrose sees himself as dead, a conception perpetuated by his constant reading of archaic books about myth and philosophy.

Those applicants who reply to Montrose's advertisement do not take the job, because they are unable to bear the sight of his 'mulberry red' body, but the unexpected arrival of a willing young black called Quintus heralds a change in Montrose's life. Quintus takes care of Montrose's health and reads passages from books in order to inspire his employer's love-letters. His arrival is followed by that of Daventry, another 'non-applicant', who agrees to act as a messenger. Montrose, who has abandoned hope in expecting help from any earthly source, half-believes that Daventry has been sent from God, since he looks like a 'will o' the wisp'. Further mythical allusions gradually confirm that Daventry has some undefined purpose. When he confesses that he has killed two men in self-defence and lives in fear of divine retribution, Montrose hugs him and a process of mutual healing begins.

Barriers are also removed between Montrose and Quintus when the latter's mother dies and he shares his grief with his employer. Montrose starts to live again, required to give affection for the first time since his rescue from the grave. By rebelling against the protective roles Montrose originally intended them to play, both employees help towards his resurrection from the shallow psychological grave of the past. Montrose falls in love with Daventry and is astonished when Quintus tells him that this love is reciprocated. The Widow Rance is also in love with Daventry, however, and they marry. The men have accepted that their real being is in their spiritual union, unaffected by physical separation.

Shortly after the marriage, Daventry is killed in a hurricane. The Widow is revealed as a harbinger of death, from whom Daventry has protected Montrose. Hoping that Daventry will somehow appear and 'explain everything', Montrose visits an old dancehall he used to frequent in his past. Instead, he meets Georgina Rance (as she is now styled). He tells her that although he has achieved love, 'the longing was better, longing pains more, but it's more what you want'. This love of the unobtainable is also a longing for God, and by recognising this, Montrose will become completely healed. Georgina has also benefited from Daventry's role as spiritual messenger: previously unable to acknowledge her suffering as a widow, she is now able to begin her own process of renewal. In a reversal of roles, Georgina now admits to Montrose, in a series of letters, that grief had blocked her love for him.

Quintus, who had left his employer at the time of the wedding, now returns, having heard Daventry's discarnate voice instructing him to do so. As spring buds into new life, he tells Montrose that Daventry has no need to manifest himself, because his presence is always felt. Since Daventry always seemed otherwordly, it seems natural that he should continue to be a messenger after his death.

Paul SCOTT   1920–1978

The Raj Quartet

The Jewel in the Crown (1966)
The Day of the Scorpion (1968)
The Towers of Silence (1971)
A Division of the Spoils (1975)

The four independent novels which make up Scott's *Raj Quartet* cover roughly five years, 1942–7, the period leading up to Indian independence, the departure of the British and the horrifying slaughter which followed the partition of the country. The period extends twenty years or so on either side: back to the Jallianwallah Barg massacre of 1919, which sowed a ripening crop of fear and mistrust between the races, and forward to the early 1960s when the narrator – 'the stranger' who serves as a stand-in for Scott himself – returns to India to attempt some sort of reckoning with the past.

The time-span is important. Scott always insisted that he was not a historical novelist. Historians have learned much from him, but the past, in his view, was not divisible from the present or the future: the events he wrote about remained 'unfinished business'. Hence the peculiar cyclical pattern of the *Quartet*. It starts with a rape during the anti-British riots that swept India in 1942: the rape by Indian peasants of the English girl, Daphne Manners, who has already shocked her compatriots by falling in love with an Indian boy, Hari Kumar.

Kumar himself contains or re-enacts in his own person the split between the British and Indian peoples. His Hindu parentage and English public-school education have made him a black Englishman, belonging to neither race, rejected in the end by both. The story wheels and turns about this central dilemma, the rape and its repercussions. They occupy the foreground of *The Jewel in the Crown*, and are glimpsed again and again, receding in the course of the next three novels, through different eyes, recording at various angles the events which led up to and away from the assault on Miss Manners in the Bibighar Gardens. 'It is as if . . . time were telescoped and space dovetailed,' as some-

one says to the narrator. 'As if Bibighar had not happened and yet had happened, so that past, present and future are contained in your cupped hand.'

This quadripartite structure, in some ways closer to the rhythms of Indian music than any traditional western progression, is characteristic of a novelist who constantly confounds expectation. To write about India at all in the late 1960s and early 1970s, when the *Raj Quartet* came out, was to invite suspicion. India was tacitly agreed to be taboo, at any rate for anything more ambitious than adventure stories and historical romance. The political and social undercurrents which interested Scott were still too recent, too raw and painful to bear the kind of close scrutiny he had in mind. Although Scott posthumously achieved enormous popularity and a whole new readership in the wake of a hugely successful television adaptation of the novels (as *The Jewel in the Crown*, 1983), he was consistently underrated in his lifetime, and unfavourably compared with E.M. Forster.

Admittedly, he invited the comparison by starting *The Jewel in the Crown* where *A Passage to India* (1924), left off, with the actual or supposed rape of a white woman. But, as Scott said himself, the two were very different kinds of novelist. Scott's characters, for all their miraculous warmth and solidity, are also nearly always structural in concept. It is not that they do not exist in their own right, rather that each must bear a functional weight in a work which, through and beyond the fate of individuals, confronts what the historian Max Beloff called 'the full tragic significance' of British dealings with India.

Take, for instance, the retired missionary Barbie Bachelor, who supplies the emotional centre of Scott's third volume, *The Towers of Silence*. Drab, lonely, insignificant, dismissed by her compatriots as tiresome and too talkative by half, she provides the kind of intrinsically unpromising material Scott always treated with particular respect. Here she becomes the novel's pivot: it is through the eyes of the despised Barbie ('Blessed are the insulted and the shat-upon,' as she says herself) that the reader comes closest to comprehending the defensive self-absorption, racial fears and insular solidarity which underpinned so much of the public politics of British India.

Daphne Manners plays a comparable role in *The Jewel in the Crown*, as does Sarah Layton in later volumes. And so, in the overall pattern of the sequence, do Kumar and his opposite number, Ronald Merrick, the superintendent of police who arrests, incarcerates and torments him. Merrick – illiberal, authoritarian, sado-masochistic, a repressed homosexual, fatally convinced of his overriding racial superiority – emerges as perhaps the most powerful single character of the whole *Quartet*. Without ever minimising the dark, destructive forces he embodies, Scott somehow makes us see the world from Merrick's point of view. This is a compelling, even sympathetic, portrait of a vicious and manipulative character acting for the best, according to his lights, in the name of reason, patriotism and civic conscience.

Scott maintained that Forster was not primarily a comedian, not even a socio-realist, rather a prophetic writer. He was thinking of the dull, ominous, thudding echo in the Marabar Caves which seems to say: 'Everything exists. Nothing has value.' This was a nightmare which, Scott felt, had come increasingly to haunt the consumer societies of the west since Forster wrote *A Passage to India*; and Scott himself might also be described as, in this sense, a prophetic novelist. He saw things other people would rather not see, and he looked too close for comfort. His was a bleak, stern, insistent vision; and, like Forster's, it has come to seem steadily more accurate with time.

The Jewel in the Crown starts with an English girl running in the dark from her attackers, and an almost simultaneous assault by an Indian mob on the elderly British supervisor of mission schools in Mayapore, Edwina Crane. Both incidents are part of a wave of unrest sweeping India, to the consternation of the British civilian population, after the British Government's arrest of Congress leaders in August 1942.

The story is told by an Englishman returning to the subcontinent nearly a quarter of a century later. He pieces together Miss Crane's history (she is the owner of 'The Jewel in Her Crown', a coloured print of Queen Victoria with her Indian subjects), the background to her victimisation and her subsequent suicide. He reconstructs the course of the love-affair between Daphne Manners, a large, clumsy, but generous and warm-hearted VAD, and the Indian Hari Kumar. He explores Kumar's sad past: a happy English childhood and education at a great public school abruptly terminated when his father's death plunges Kumar back into the squalor and indignity of an India he never knew, and in which he feels an alien.

The only English people in Mayapore who so much as notice his existence are Daphne and the young police chief Ronald Merrick, who singles him out as a potential victim even before the night on which Daphne is torn from Kumar's arms by a gang of peasant rapists in the Bibighar Gardens. Merrick promptly arrests Kumar. British shock and outrage are exacerbated by Daphne's refusal to incriminate her lover, or the Indian boys with Nationalist sympathies wrongfully arrested with him. Indignation over Daphne's obstinacy deepens local sympathy with Merrick. The book ends with Kumar behind bars, and Daphne being packed off to die nine months later, having given birth to a daughter, Parrati, whom she firmly believes to be his.

Evidence of what happened that summer is collected long afterwards by 'the stranger' from various surviving witnesses: Daphne's loyal supporter, the sophisticated, cosmopolitan Lili Chatterji; Sister Ludmila, a white woman of indeterminate European origins who befriended the lovers at her sanctuary for the sick and dying in Mayapore; Robin White, the Deputy Commissioner, who puts the civilian authorities' case for patience and understanding; the army's Brig. Reid, arguing hotly in support of his own and Merrick's preference for forcible repression. Against a background of war in Europe, British defeat in Burma by the Japanese, and civil disturbance throughout India, the narrative has the urgency and tension of a detective story.

It also has a more ominous side, for events in Mayapore in 1942 foreshadow on a small scale the injustice, the random and indiscriminate violence, the irreconcilable clash of wills with which the British Empire in

India ended five years later. Or, as the narrator puts it: 'in such a way scenes and characters are set for exploration, like toys set out by kneeling children intent on pursuing their grim but necessary games'.

Asked by an Indian friend to explain the title of *The Day of the Scorpion*, Scott said that it referred to the old belief that scorpions would sting themselves to death if surrounded by a ring of fire: 'But actually they are shrivelled by the heat, and when they dart their tails they're not committing suicide but trying to attack. Well, that's what so much of the British in India was all about. They were driven out of their places in the end by a number of pressures – and were scorched by fires they had really set light to themselves.'

This second volume of the *Quartet* moves in accelerating patterns of death and dislocation. The time is still August 1942, but the scene has widened to include the national Quit India campaign, the growing split between Hindu and Muslim, and the failure of the Cripps mission which led to the internment of Congress leaders. The first of the book's two sub-sections, 'The Prisoner in the Fort', concerns the predicament of the ex-Chief Minister of Ranpur, Mohammed Ali Kasim (MAK), a Muslim leader highly respected by both his Hindu and his British colleagues. He is unable to endorse the policies of either religious party, preferring imprisonment to collabortion with the British, and his pessimistic reflections on India's divided future are reinforced by garbled reports of the Bibighar affair.

Meanwhile, Daphne's death, or rather the birth of her illegitimate, half-caste child, poses a problem for her compatriots. The Laytons of Ranpur and Pankot, a family with impeccable army connections, shelve it by snubbing the child's guardian, Daphne's even more impeccably connected aunt, the Governor of Ranpur's widow, Lady Manners. Susan, the younger and more attractive of the two Layton girls, marries the dull, conventional Capt. Teddie Bingham in the state of Mirat, whose Muslim ruler is a kinsman of MAK, and the employer of his handsome, debonair and apolitical younger son, Ahmed Kasim. Various embarrassing incidents involving the best man at the wedding, Ronald Merrick (now seconded to

the army), suggest a continuing campaign of harassment in the wake of Bibighar, and are noted by the *wazir*, a homosexual White Russian, Count Bronowski.

The book's second section, 'Orders of Release', begins in 1944 with the unofficial reopening of Kumar's case at Lady Manners' instigation. Kumar, still imprisoned without trial, describes under secret cross-examination the background and circumstances of his arrest, his relationship with Daphne and his interrogation by Merrick under torture.

Teddie Bingham is reported killed in an action from which Merrick emerges severely burned with an amputated arm. Grief, loss and the effects of pregnancy threaten Susan's already shaky mental balance. Responsibility for family affairs devolves upon her sister Sarah who, with Ahmed Kasim, comes increasingly to represent the cool voice of sanity and restraint in a perilously impassioned world. Ahmed's older brother Sayed, captured fighting with the Indian army, defects to the Japanese-backed force of would-be liberators, the Indian National Army (INA): MAK, bitterly distressed by his son's treachery, is released into the protective custody of the Nawab of Mirat. In a brief postscript, Susan, succumbing to pressures she can no longer withstand, ignites a ring of flame round her baby, which survives unharmed, while she is confined in a psychiatric ward.

The Towers of Silence is a work of consolidation and reflection, perhaps the least dramatic of the four volumes. Scott himself called it his 'slow movement'. It plays a set of variations on familiar themes, only this time we see the turbulence of 1942 mirrored in the reactions of the comparatively insulated British army community in the hill station of Pankot. Here the retired missionary Barbie Bachelor finds refuge at the beginning of the war as a paying guest at Rose Cottage with the elderly, liberally inclined military widow Mabel Layton. Mabel's stepdaughter-in-law, Mildred Layton, and her daughters, Sarah and Susan, and her stout, disapproving but kind-hearted sister, Fenny, form the centre of a group of army wives whose intolerance, prejudice and tendency to close ranks against outsiders are all intensified in time of trouble.

Barbie, having lost her job on the first page, finds herself after Mabel's sudden death spiralling slowly downwards in the course of the book through successive circles of redundancy, loneliness and humiliation to despair, madness and death. Her increasing isolation, her yearning to find useful work and the rebuffs it brings, her tendency to hallucinate, her gradual disintegration and gallant attempts to regain lost ground: all might be said in some sense to prefigure the forcible retirement and eviction facing the British collectively in the near future.

Meanwhile, life in Pankot becomes more and more unsettled. News arrives of the disturbing attacks on English women in Mayapore. Husbands are reported missing, captured, killed. The general atmosphere of uncertainty is reflected in Susan Layton's hasty engagement, disrupted wedding, premature widowhood and the birth of her baby, precipitated too soon by shock at Mabel's death. The Pankot community itself begins to break up in a round of wartime funerals, farewells and departures.

Barbie becomes a convenient scapegoat for the hard-drinking, adulterous Mildred Layton, who has her thrown out of Rose Cottage, to Pankot's dismay: 'A comic but horrifying thought took hold: of old Miss Bachelor, homeless, seated on a trunk in the middle of the bazaar, surrounded by her detritus ... to the amusement of Hindu and Muslim shopkeepers who would interpret such a sight as proof that the entire *raj* would presently and similarly be on its uppers.' But the British community's sympathy for Barbie is rapidly swallowed up by more pressing concerns, dread of the future and a vast indifference to anybody's fate except its own. Disposing of her possessions, Barbie briefly befriends Ronald Merrick and makes him a present of her reproduction of Miss Crane's 'The Jewel in Her Crown', showing Queen Victoria surrounded by her Empire's princes, potentates, soldiers and statesmen. The only person missing from this picture is the unknown Indian to whom, as Barbie says, she has devoted her life.

The book ends with Barbie's departure in an apocalyptic rainstorm on an overloaded tonga which capsizes, spilling her luggage on the road and finally destroying her mental equilibrium. She dies in hospital, having lost the faculties of speech, memory, recognition of anyone or anything save the vultures circling endlessly beyond her window above the Towers of Silence where the Ranpur parsees offer up their dead.

The *Quartet*'s majestic concluding volume, *A Division of the Spoils*, is almost as long as *War and Peace*, to which it might be said to be in some ways a twentieth-century equivalent. Both deal with the fate of nations as much as individuals caught in a convulsive upheaval they can neither control nor comprehend. In both a sweeping panoramic vision alternates with sharp, bright, close focus. Scott is as conscious as Tolstoy of the claims of history in a book which sees any number of ancient scores and long-outstanding bills presented for settlement.

A Division of the Spoils is divided into two books, each the size of a full-length novel. The first covers the crucial month of August 1945, after the German surrender in Europe, when the dropping of the atom bomb on Hiroshima finally reduces the Japanese to capitulation, and the election of a Labour government at home forces the British to acknowledge the imminent prospect of their leaving India.

Virtually every character is at some point obliged to confront the future head-on. Sharp distinctions are drawn. On the one hand are men of the past, like Merrick, whose rigid certainties, based on racial arrogance and a Kiplingesque sense of duty, produce an invincible self-confidence as dangerous as it is attractive in turbulent times. On the other stand those who, like Mildred Layton, combine an uncompromising allegiance to the present with a vigorous sense of history, 'vigorous because it pruned ruthlessly that other weakening sense so often found with the first, the sense of nostalgia, the desire to *live* in the past'.

Official British socialist policy lays down that India must be handed back without delay to her own people or, in the words of the professional economist, Capt. Purvis, written off as a wasted asset. The pragmatic and profoundly depressed Purvis tries to drown himself in his bath in the opening stages of this section, but is resuscitated by the alert young Sgt Perron. Guy Perron, himself a Cambridge history graduate, is appalled by the cynical expediency of the government line. He is still more depressed by the intransigence of the old guard represented by Major Merrick, who inspires in him an instinctive mistrust rapidly fanned to unconditional loathing.

Perron devotes much time and trouble to building up a dossier on Merrick and the unscrupulous methods – which range from physical and mental sadism to homosexual blackmail – used to impose his will on recalcitrants from Kumar onwards. Merrick himself is primarily concerned with the interrogation of returning INA suspects with a view to their prosecution and punishment. Chief among these is Sayed Kasim, MAK's son, facing one of the public show trials mounted by the British. In spite of Sayed's best efforts to persuade his father to espouse the patriotic Muslim cause by endorsing plans for an independent Pakistan, MAK remains aloof, filled with foreboding for his country's future.

One of the remarkable things about this long first section is the way in which the political momentum – India moving towards liberation, and the destruction it entails – matches a similar movement on a domestic scale within the Layton family. Col. Layton's return from a German prisoner-of-war camp produces an apparently united front, which is perpetually threatened by internal strains and tensions: Sarah's need for independence, her private hurts and fears, her father's vulnerability and inhibition, the hostility of her dominating mother, the basic instability behind Susan's precarious return to health. When Susan becomes engaged to Merrick (who has gained illicit access to her psychiatric files by terrorising a homosexual private), reactions within the family range from enthusiasm through ambivalence to frank alarm.

The second section of *A Division* covers the still more critical month of August 1947, when India gains her freedom. Perron, returning as a civilian to watch partition take effect, pays a visit to Mirat, an independent Muslim state now stranded in a Hindu heartland. Here he finds his old friend Nigel Rowan of the Political Department, supervising the reluctant Nawab's signing over of power. He also finds Sarah Layton (with whom he is falling in love), come to support her sister Susan, newly widowed for the second time by the apparently accidental death of Lt.-Col. Merrick, who had been seconded to sort out Mirat's police force.

Again, public events are inextricably entwined with private. Perron learns the full story of the persecution leading up to the dreadful circumstances of Merrick's murder, on the eve of the transfer of power, against a background of rioting and burning by Mirat's disaffected Muslims. The Laytons leave Mirat next day with Perron and Ahmed Kasim, who is murdered on the train by vengeful Hindus in the tidal wave of carnage sweeping the subcontinent in the wake of independence.

Perron leaves India after a last abortive effort to locate Hari Kumar, now discharged from prison and living quietly in Ranpur.

1976

Death of Chairman Mao • Jimmy Carter elected US President • IMF crisis in Britain • German terrorist leader Ulrike Meinhof commits suicide in prison • Amnesty for political prisoners in Spain • Race Relations Act passed in Britain • Concorde begins regular passenger service • Legionnaire's disease named • J.M. Roberts, *History of the World* • Louis Althusser, *Essays in Self-Criticism* • Noam Chomsky, *Reflections on Language* • Basil Hume appointed Roman Catholic Archbishop of Westminster • Denys Lasdun, National Theatre, South Bank Centre, London • David Edgar, *Destiny* • Barry Humphries, *Housewife – Superstar!* • Films: *Sebastiane*, *Marathon Man* • Television: *I, Claudius*, *The Muppet Show* • Abba become Sweden's biggest export earner after Volvo • The Eagles, *Hotel California* • Chou En-lai, Agathie Christie, Paul Robeson, L.S. Lowry, Luchino Visconti, Bernard V. Montgomery, Max Ernst, Howard Hughes, Alvar Aalto, Martin Heidegger, J.L. Monod, J. Paul Getty, Fritz Lang, Gilbert Ryle, Man Ray, Andre Malraux, Benjamin Britten die • Elizabeth Bishop, *Geography III*

Richard BRAUTIGAN　　　　1935–1984

Sombrero Fallout

Subtitled 'A Japanese Novel', Brautigan's book tells the story of an American humorist grieving over the loss of his Japanese girlfriend, Yukiko. He begins to write a story about a sombrero falling out of the sky to land in a small town, then tears it up in frustration. Involuntarily, he imagines his Japanese girlfriend sleeping with a new lover. Voluntarily, the torn-up story begins to piece its fragments together: 'They seemed to have a life of their own. It was a big decision but they decided to go on without him.' The story of the sombrero goes on: it turns the mayor on his cousin, the town turns on itself, its police force, the state police and finally an army of federal troops. Thus, a disagreement turns into a fight, a fight into a riot and a riot into a war. The crisis is so great that only Norman Mailer can report it. Meanwhile, back in the 'real' world, the writer still grieves over his Japanese girlfriend. He finds a strand of her hair, loses it, then finds it again. He remembers their first meeting, almost calls her, then restrains himself. Sixteen blocks away his Japanese girlfriend sleeps (alone) dreaming of her childhood.

These two stories are told in parallel in short sections ranging from a few lines to a few pages, each with a title ('Sombrero', 'Mailer', 'Flying Saucers', etc.) more or less relevant to the episode's action. Often, these sections, particularly those concerning Yukiko, have a lyricism independent of the action, while the two stories maintain a free-floating relationship, each touching lightly and intermittently on the concerns of the other. The fallout of the title refers not only to the sombrero but also to the falling-out of Yukiko and the writer and perhaps even to that larger US/Japan falling-out, the fallout over Hiroshima and Nagasaki.

There is no portentous political allegory though. Brautigan's broad interest is the question of what men and women do when the events of their lives exceed their capacity to understand them. To this end he supplies a fantasia of civic revolt and the ruminations of a writer over the loss of his girlfriend. Within these arbitrary choices he deploys an epigrammatic, ironising style which releases the humour of the situations in closely controlled fits and starts. A librarian with both ears shot off, the chance collision of helicopters carrying the State Governor and Chief of Police, and a humorist with no sense of humour are all occasions for comedy ranging from the apocalyptic to the elegiac. At the novel's close the reader is no wiser as to why a sombrero should fall from the sky or whether the writer will effect a reconciliation. Brautigan does not submit to the formal principle of comedy – that is, explanation and reconciliations at the end. Rather, he allows his material the autonomy, the seriousness, it demands; sombreros and girlfriends have existences outside the controllable domain of the novel. Likewise public strife and private grief.

Michael MOORCOCK 1939–

The Dancers at the End of Time

An Alien Heat (1972)
The Hollow Lands (1974)
The End of All Songs (1976)

Moorcock – writer, singer, editor and publisher – does not like to be pigeon-holed. He has described his novels as fantasy, rather than science fiction, and he uses extensively what he has called 'genre borrowings'. Echoing this stylistic flexibility, Moorcock's people and worlds are unstable. Identities shift, natural laws do not apply. The horizons of invention are unbounded and formal narrative structure is defied.

The Dancers at the End of Time portrays a world millennia from now. Its inhabitants can change appearance, create fresh landscapes and resurrect each other almost at will. Being immortal, they pursue their pleasures at a leisurely pace. Neither religion nor morality inhibits them. Behaviour is judged by aesthetic criteria and affectations have replaced emotions. Jherek Carnelian is fascinated by 'The Dawn Age' (our nineteenth century) and when Mrs Amelia Underwood is transported from Bromley, 1896, he decides to affect to fall in love with her. A visiting alien prophesies the end of the universe. His message is considered passé and he is claimed for a menagerie, as is Amelia. Jherek – assisted by Lord Jagged of Canaria – obtains Amelia by trickery for himself. She begins to teach him about duty, guilt and virtue but his education is interrupted when she is transported back to Bromley. Jherek follows her. He fails to find her, is implicated in a murder and hanged – a fate

which returns him to his time.

Undeterred, Jherek goes again to nineteenth-century London. H.G. Wells guides him to Amelia's house where Jherek encounters Harold, Amelia's husband. Jherek flees with Amelia, and Harold sets in train a police chase which finishes in the Café Royal. The fabric of time starts to decay. Some of Jherek's contemporaries appear – including the Lat, a band of vulgar inter-planetary brigands. In the ensuing chaos, the characters are dispersed across the centuries.

After a brief stay in the Palaeozoic Era, Jherek and Amelia make their way back to Jherek's age. They have not, though, escaped Mr Underwood and the British police; and, as the alien prophesied, the end of the universe is at hand. Jherek's love for Amelia, no longer an affectation, is consummated in the shadow of catastrophe. Lord Jagged, however, has been plotting to avert disaster and fix suitable destinies for the lovers, husband Harold and the police. The trilogy concludes with the outcome of Jagged's schemes and a second dispersal of Moorcock's creations.

The story is comic, and much of its comedy derives from the relationship of the distant future to known history. Jherek is hugely naïve about Victorian England, attempting a frank discussion of sex with Mr Underwood. H.G. Wells, by contrast, is sceptical of Jherek's time-travelling tales. A dynasty of tyrant film producers, the Pecking Pas, is evoked. Utopia is condemned by Li Pao, a twenty-seventh-century Maoist. Organised religion and sham mysticism are satirised. Moorcock's technique permits him a wide range of targets, but his mockery is gentle and its most frequent target is intolerance. The 'multiverse' allows a place for everything and the ultimate sin – perhaps the only one – is to exclude new possibilities and perspectives from it.

In this hospitable spirit, Moorcock's work is dense with allusions – to his friends, to music and literature and to his own fiction. (Characters from one book often cross to another. Jherek Carnelian, for instance, is a variation of Moorcock's Protean hero Jerry Cornelius.) Dedicatees of *The Dancers at the End of Time* include Hawkwind, a band of which Moorcock was a member.

R.K. NARAYAN   1907–

The Painter of Signs

The painter of signs is Raman – he paints the sorts of sign one sees in front of houses and offices, shops and factories. He falls in love with a woman who is unsuited to him in every way. Daisy is a rather modern young woman working at the Family Planning Centre, who lays down various conditions before she will marry Raman. The story goes through numerous farcical ups and downs before it leaves Raman not only unmarried but also without benefit of the care of his only close relative, his aged aunt, who has gone on a distant pilgrimage (from which she may never be able to return) in disapproval of the proposed marriage.

Raman represents contemporary India, that Daisy represents the forces of modernity, that the aunt represents those forces of religion and tradition which are disappearing under the impact of modernisation, and that the lack of a marriage represents the schizophrenia of the country today.

As usual, Narayan suggests parallels between the story that is being related and the myths and legends of India – in the *Mahabharata*, the goddess Ganga had also laid down conditions before she married King Santanu. Unlike most of his other stories, however, this one ends without any glimmer of a resolution in favour of tradition; modernisation has won, tradition has fled, and Narayan is bereft of anything more to say.

Yes, said his truthful conscience. You are absolutely right. You are now in a different category. You are an honest man. Examine your thoughts, assay the contents of your deepest thoughts, see what there is. You are preoccupied with her physical form inch by inch all the time you are discussing the measurements of your sign-board. The clothes on her simply do not exist for you, you are preoccupied with what you can accidentally glimpse at, hoping for a chance to see her clothes blown off; while she sits away at her desk, you fancy her on your lap; while she is conversing, you are sealing her lips with your kiss. That is the tragedy of womanhood – utility articles whether in bed or out. You never view them normally until they are past sixty and look shrunken-skinned.

THE PAINTER OF SIGNS
BY R.K. NARAYAN

The novels which are usually considered Narayan's best – *The Financial Expert* (1952), *The Guide* (1958) and *The Man-Eater of Malgudi* (1962) – variously assert the triumph of tradition over modernity. Indeed, the very form of these novels, in Narayan's hands, looks back towards ancient Indian moral fables. It is possible to argue that here

Simon RAVEN   1927–

Alms for Oblivion

The Rich Pay Late (1964)
Friends in Low Places (1965)
The Sabre Squadron (1966)
Fielding Gray (1967)
The Judas Boy (1968)
Places Where They Sing (1970)
Sound the Retreat (1971)
Come Like Shadows (1972)
Bring Forth the Body (1974)
The Survivors (1976)

Raven's ten-volume series of inter-connected novels is less a stately dance to the music of time than a raffish tango of post-war (1945–73) high and low life. 'If there is one theme which dominates the series,' Raven has written, 'it is that human effort and goodwill are persistently vulnerable to the malice of time, chance, and the rest of the human race.' The title is taken from a speech by Ulysses in Shakespeare's *Troilus and Cressida* in reply to Achilles' question, 'What, are my deeds forgot?':

Time hath, my lord, a wallet at his back,
Wherein he puts alms for oblivion ...

Which is to say that good deeds 'are devour'd / As fast as they are made, forgot as soon / As done.' An interesting background, both philosophical and biographical, to the series can be found in Raven's analysis of *The English Gentleman* (1961), in particular in the suggestion that, since the demise of 'defunct codes of gentility', 'tolerance, learning and scepticism . . . may inculcate an ethic which, flexible yet not dishonourable, will equip a man to deal with the present realities in a practical but seemly fashion'.

Unlike Powell's sequence, Raven's novels were not published in chronological order, and although they have subsequently been arranged like this, aficionados insist that the sequence should be read as written. For convenience, the following synopsis is arranged chronologically, starting with *Fielding Gray*.

In this volume we are introduced to several of the sequence's ten major characters, three of whom are public schoolboys. The time is 1945 and the narrator is Fielding Gray, intellectual and sporting star of the school, whose fall from grace comes as the result of his passion for another pupil, Christopher Roland. Their idyllic romance sours after consummation in a hayloft, and Christopher is transformed from an ideal into an object of gratification. Lonely and unsure of Fielding's affection, he is arrested for soliciting outside an army camp during the school holidays, after which he shoots himself.

Meanwhile, Fielding's obnoxious philistine father has become involved with a swindler called Tuck, who offers to take on Fielding as a tea-planter in return for heavy investment. Tuck's voluptuous young wife, Angela, is used as bait, but unfortunately Mr Gray dies while in bed with her. The previously timid Mrs Gray now controls the family money and plots with Angela to send Fielding to India, rather than to Cambridge,. using a signed photograph of Christopher in order to blackmail him. Fielding's school contemporary, the intolerably smug and power-hungry Somerset Lloyd-James, is implicated in these machinations. Fielding's classics master recognises that his pupil is doomed, like Antigone: 'Your star had turned hostile,' he observes, 'and its new malignity was reflected in your eyes.' Throughout the sequence Fielding will be dogged by remarkable ill-fortune, which will be compounded by his own personal failings. *Fielding Gray* ends with the eponymous

hero alone and miserable in the army, deserted by his erstwhile best friend, Peter Morrison.

It is Morrison who is the principal character of *Sound the Retreat* (set in 1945–6), posted to the Officers' Training School at Bangalore. He becomes friends with two other cadets, Alister Mortleman, a conservative Wykehamist, and Barry Strange, a gilded youth. As part of the preparations for the eventual handing back of India to the Indians, a Muslim officer, Capt. Gilzai Khan, becomes their instructor. He is staunchly pro-British and follows the army's code of love and loyalty, rapidly gaining the respect of all his cadets, except for Mortleman, who resents 'being ordered about by a bloody wog'. During a wake for Lord Canteloupe's son, Muscateer, a fellow cadet who has succumbed to jaundice, Mortleman accepts Gil'Khan's challenge of a public duel by sexual endurance. He wins and the two men become friends.

Meanwhile Morrison has entered a liaison with a versatile half-caste whore called Margaret Rose Engineer. Fearing that she will be massacred in the Muslim–Hindu violence which will follow the British withdrawal, she attempts to blackmail Morrison into marrying her. His commanding officer suggests that he will have to leave the course, but this is rendered unnecessary by the wily intervention of Gil'Khan, who plants papers upon the girl and has her arrested. Gil'Khan leaves the course under a cloud, after his affair with an unnamed cadet (in fact, Strange) is betrayed by 'Wanker' Murphy, an unwholesome and voyeuristic cadet.

At the end of the course, Strange persuades the CO of the Wessex Fusiliers, a regiment with family connections, to take on Mortleman and Morrison. To the disgust of the Colonel, their main duty is to await, and put down, riots. Morrison becomes an intelligence officer and receives a message from Delhi (via Murphy, now 'through some preposterous series of chances' promoted to 'Viceroy's Galloper') that Gil'Khan has become a Muslim agitator and must be killed. Morrison attempts to square his conscience ('Tell no lies and do as you would be done to') and arranges for Strange to kill the man in a fit of regimental and family pride during the monsoon-drenched siege of a railway. Major Giles Glastonbury and Capt. Detterling, both of whom are attached to Wavell, manage to pass the incident off as an accident.

The Sabre Squadron is set in 1952 in occupied Germany, where Daniel Mond, a brilliant young mathematician, has come to look at the papers of a German academic and attempt to discover their secret. Half Jewish, Mond is uneasy in Germany, and becomes more so as it becomes clear that Dortmund's papers were leading towards knowledge which has unlimited military potential, and that various parties are interested in his findings. Lonely, he is befriended by an American academic called Earle Restarick, but is later told that this man is an agent representing the neo-Nazis.

He then becomes involved with the 10th Sabre Squadron, an elite corps (commanded by Giles Glastonbury) attached to the dowdy Wessex Fusiliers, stationed in Göttingen, amongst whom is Fielding Gray. Leonard Percival, a British agent posing as a fusilier, intimates that he represents the Allied interest in Mond's work. Mond refuses to reveal his discoveries and is threatened from all sides. He confides in Gray and it is arranged that he should pose as a dragoon and be smuggled over the border in the course of a military exercise. He is captured by Percival and Restarick and informed that they are on the same side. They are not allowed to torture him, so threaten to have members of the Sabre Squadron cashiered and imprisoned because by allowing Mond to take part in the exercise they have made him privy to official secrets. Amidst multiple treacheries, the Squadron have been the only men who have remained loyal to Mond, following their regimental motto: *Res Unius, Res Omnium*. The novel ends with Mond cutting his throat in order not to betray their friendship and honour.

The Rich Pay Late is largely concerned with an attempt by an advertising printing company to take over *Strix*, 'A Journal of Industry and Commerce' edited by Somerset Lloyd-James. The bid is part of the personal ambition of the ruthless Jude Holbrook and is at first opposed by his ineffectual partner, Donald Salinger. Holbrook has

to get a majority vote from members of the Board of *Strix*, which include the eminently principled Prof. Constable, Vice-Chancellor of the University of Salop, who is opposed, Peter Morrison MP, who is wavering, and Henry Dilkes of the Institute of Political and Economic Studies, who is in favour. In order to pursue his ends, Holbrook threatens to discredit Morrison by dragging up an ancient, and, as it turns out, untrue scandal concerning under-age sex and illegitimacy.

Lloyd-James, guided by prayer and Jesuitical practice, is also pursuing his own ends. Apart from Morrison, who wishes to resign because of Suez but does not wish to damage the morale of the army by his actions, and Tom Llewyllyn, the socialist author of *The Bear's Embrace*, a topical book about the spread of Soviet power, everyone comes badly unstuck. In particular, Holbrook loses his job, reputation, wife, mistress (Angela Tuck) and his son, the only person he has ever loved, who dies of meningitis. The book contains a memorably disastrous party given by Salinger and his newly acquired, terminally promiscuous wife, Vanessa, at which Lloyd-James persuades the drunken Llewyllyn to tell Holbrook the scandal about Morrison.

Friends in Low Places is set in 1959 and concerns the attempts of Mark Lewson, a bisexual gigolo and con-man, to dispose of a letter which incriminates a number of senior members of the government with regard to the Suez Crisis. He acquires the letter (via Max de Freville, organiser of games of *chemin de fer* and obsessive collector of information) from Stratis Lykiadopolous, a sentimental Greek gambler with a taste for young men. Lewson's attempts to sell the letter are set against the backdrop of the selection of a new Conservative candidate for Bishop's Cross. The committee has to choose between Peter Morrison, attempting to return to politics after his resignation three years previously, and Somerset Lloyd-James. Sir Edwin Turbot, 'that egregious major-domo of the Conservative party', is implicated in the des Moulins letter, and is approached by Lewson and Lloyd-James, who threaten him with exposure unless he agrees not to promote Morrison. Turbot is concerned about his elder daughter's marriage to Tom Llewyllyn, a match of which he does not approve and which culminates in a disastrous reception at which the house is set on fire. Meanwhile, Lewson elopes with Isobel, Turbot's precocious and promiscuous younger daughter. His careless driving causes the death of a fireman, but no one has been able to identify him as the culprit. Turbot covers up and organises a search for the couple in which Llewyllyn, his journalist friend, Alfie Schroeder, his publisher, Gregory Stern, Fielding Gray (invalided out of the army after being maimed by a bomb in Cyprus, and now a novelist), Capt. Detterling and Peter Morrison all set off for the south of England.

The original of the letter has been stolen from Lloyd-James by Jude Holbrook, but is recovered by Gray. The novel ends at Westward Ho!, the first government-sponsored holiday camp, set up by Lord Canteloupe, and with the death of Lewson in a motor accident. Canteloupe destroys the letter.

The Judas Boy follows Fielding Gray on an assignment to Cyprus in 1962 in order to research American involvement in the civil war for a BBC documentary to be produced by Tom Llewyllyn. En route for Athens, Gray meets Leonard Percival, who warns him to be careful, after which an attempt is made on his life. It transpires that Earle Restarick, on behalf of the American Secret Service, stirred up trouble in Cyprus, providing the Greeks with arms, in order to discredit the English. To prove this, Gray goes to the grave of a Jewish Greek boy, murdered by Restarick, from whose decaying corpse he takes Restarick's handkerchief. This does not constitute proof, but it may be possible to gain confirmation from the Greek general, Grivas.

Meanwhile, Somerset Lloyd-James and Lord Canteloupe are attempting to stop Llewyllyn making his programme and try, unsuccessfully, to enlist the help of Maisie Malcolm, a discreet and kind-hearted whore whose clientele includes several of the sequence's characters. Restarick is also attempting to prevent the meeting between Gray and Grivas. Learning from Angela Tuck of Gray's obsession with Christopher Roland, Restarick finds a young Greek,

Nicos, who is similar to the dead schoolboy, and sends him to Gray. Obsessed by this apparent reincarnation, Gray fails to keep his appointment with Grivas. By this time, Llewyllyn has been sacked for failing to pay his National Insurance contributions, and his programme is scrapped. Restarick calls off Nicos, and Gray is left alone once more, consumed by alcohol and self-pity. Max de Freville sends Harriet Ongley, a rich American widow, to rescue him.

Places Where They Sing concerns student revolt in Cambridge in 1967. The Council of Lancaster College (which now includes both Daniel Mond, who survived his suicide bid, and Tom Llewyllyn) meets at the beginning of the summer term in order to decide how to spend the proceeds of a land sale, amounting to £250,000. Beyfus, a life peer and social scientist, urged on by his friend Mona Corrington, a radical anthropologist, suggests a brutal programme of reform, including vast buildings upon Scholar's Meadow and the sweeping away of college traditions and privileges. When this is firmly rejected, Tony Beck, a left-wing English don, stirs up revolt, using his prize student, Hugh Balliston, and bringing in an outside agitator called Mayerston. The Svengalian Mayerston grooms Balliston for martyrdom, persuading him to spend the night in the all-male college with his girlfriend, Hetta Frith. This protest is foiled by Robert Constable, the Provost, and attracts no publicity.

Llewyllyn's wife, Patricia, is bored and resentful of her husband's new academic life and his loss of interest in her sexually after a miscarriage. She seduces Balliston, after which she develops a taste for young flesh, embarking upon a career of promiscuity. (This marks the beginning of her decline. Later in the sequence she entertains incest fantasies, inspired by Bronzino's 'Venus, Cupid, Folly and Time'. She is placed in an asylum after biting off the ear of a seventeen-year-old boy.)

Meanwhile the Council is attempting to decide between two further proposals. The first, sponsored by Llewyllyn, is that a small, architecturally prestigious hall should be built to accommodate sixty new students. The second, sponsored by Balbo Blakeney, a biochemist aesthete, is that the splendid chapel should be preserved and that a large amount of money should be given in bursaries to the Choir School. This splits the Council between right and left factions.

Mayerston organises a spectacular disruption of the May Madrigals in which the students march into the chapel, bearing Balliston aloft, whilst Mayerston's henchmen attempt to destroy the founder's statue. Hetta, who has abandoned Balliston for Blakeney, is killed in the riot. The matter of the disposal of the money remains unresolved.

In *Come Like Shadows* it is 1970, and Fielding Gray is asked to work on the script of a film version of *The Odyssey*, being directed by Jules Jacobson for Pandarus/ Clytemnestra Films. The film is being co-financed by Max de Freville and the Oglander-Finkelstein Trust for 'cultural and creative projects', and the pay is lavish. Glad of an opportunity to get away from his keeper, Harriet Ongley, Gray flies to Corfu (where Max de Freville and Lykiadopolous are involved in 'tourist development'), and attempts to satisfy the demands of artistic integrity, Jacobson, Foxy Galahead (the libidinous producer), and the Og-Fink Trust which wants a left-wing reinterpretation of Homer. Distraction is provided by Sasha Grimes, a talented and committed young actress, cast as Nausicaa, whom Gray is determined to seduce. Angela Tuck suggests that he should appeal to Sasha by gaining her pity, a ruse which succeeds. Sasha has been appointed by the Trust as an overseer and Gray divides his time between interpolating into the script scenes depicting oppressed Greek peasantry and accommodating Sasha's Mary Magdalene fantasies.

While watching the antics of Gray and Sasha through a two-way mirror, the ageing and obese Angela Tuck suffers a fatal heart-attack. Max de Freville, who has contracted a sort of *marriage blanc* with Angela, blames Gray for this and so is willing to conspire with Galahead in thwarting his attempts to blackmail Pandarus Films into giving him more money. Earle Restarick is (mis-) informed that Gray is continuing his researches into Greek affairs, and becomes convinced that he has sold secrets to an anti-fascist organisation. He kidnaps Gray in

Athens and threatens to turn him into a heroin addict unless he provides information. Told that he has had a breakdown, Gray's colleagues leave him in Athens, but he is rescued by a British delegation, comprising Canteloupe, Detterling and Lloyd-James, who learn of the suspicious circumstances of his disappearance from a discontented starlet. Gray returns to England and, deserted by the possessive and wealthy Harriet, but with money in a Swiss account, begins work upon a study of Conrad.

In *Bring Forth the Body* Detterling joins forces with Leonard Percival in order to investigate the mysterious suicide of Somerset Lloyd-James in 1972. Their work is undercover, since Lloyd-James was Canteloupe's assistant (Parliamentary Under-Secretary at the Ministry of Commerce, involved in the underhand promotion of a British product over its European rivals) at the time of his death, and an official verdict of suicide owing to the pressure of work has been recorded. The two men interview a number of Lloyd-James's friends, following up a hunch of Detterling's that some small oversight, totally unexpected, led to this unexpected death. After pursuing various lines, including the revelation that at Cambridge Lloyd-James won the Lauderdale Essay Prize by foul means, they learn from Fielding Gray of a drunken end-of-term party, at which Lloyd-James was sick.

This means little until they learn from Peter Morrison, who has taken on the dead man's job, that Lloyd-James confessed to an old schoolfellow that in the course of the morning after the party, he had sex with one of the school maids and that a child was the result. The mother, Meriel Weekes, had kept the father's identity a secret, and married another man. However, after her husband's death she applied to Lloyd-James for help. He spent the last day of his life visiting the woman. Detterling and Percival discover her identity and seek an interview with her. She reveals that her son went to the bad, and was gaoled for burglary and for living off immoral earnings. In the course of a police chase, he had an accident which left him mutilated and brain-damaged. When Lloyd-James was confronted with his son he muttered: 'God is not mocked.' His school-boy maxim that 'Nothing which we ourselves have made can be beyond our control' has been exploded by this pitiful wreck, whose existence warns Lloyd-James that God has not made him special, or 'exempt from the system', and that he will have to pay for his behaviour, like everyone else.

Most of the surviving characters gather in Venice in *The Survivors*. Detterling and Fielding Gray are at the 1973 PEN Conference; Tom Llewyllyn is there with his daughter, Baby, doing research and waiting to be joined by Daniel Mond, who is coming there to die. Lykiadopolous has come to set up a casino in order to subsidise his Corfu interests. He is accompanied by his partner, Max de Freville, who is lending his time and money to the Venice in Peril appeal in order to appease the spirit of Angela Tuck, who loathed his desecration of Corfu.

A telegram arrives announcing the death of Lord Canteloupe, and the unforeseen passing of the title to Detterling. Before he leaves to take up his inheritance, Detterling persuades de Freville and Lykiadopolous to house Mond and Llewyllyn in a tower of the Palazzo Albani, which they have rented. He also becomes attached to Baby, and helps to sort out her educational problems. Fielding Gray has become intrigued by a figure in a family portrait at the Palazzo and persuades Piero, Lykiadopolous's young boyfriend, to search the attics in order to find a clue to the mystery. A manuscript is discovered which outlines the vicious career of the man in the picture, 'Humbert fitzAvon', and his infiltration and debauch of the Albani family – a tale of incest, paedophilia and murder. Llewyllyn and Gray follow the trail to a remote area where they find Jude Holbrook in residence. 'fitzAvon' was in fact a member of the Canteloupe family and the descendants of his enforced marriage to a peasant girl are the true heirs of the Marquisate.

Gray decides to use this story as a basis for a novel, but Piero, who has become devoted to Mond and spent time with him against the orders of Lykiadopolous, attempts to make amends by telling the Greek the whole story. During Mond's funeral, Lykiadopolous attempts to blackmail Detterling into giving him £500,000, because his casino is in danger of collapse due to extravagant Arab gamblers. Fortunately Detterling is in possession of evidence that Lykiadopolous has been using foul means to cheat his clients.

Piero has arranged for Mond to be buried in an island monastery, where Hugh Balliston has become a brother and where Piero himself will remain. Watching the funeral procession, Piero sees a dark stain spread all across the waters of the lagoon.

Perhaps inevitably, *Alms for Oblivion* is a little uneven – in particular, *Come Like Shadows* is for the most part fairly crude stuff about American philistinism – but at its best it is always stylish, civilised and intelligent, and often unexpectedly moving (the scene in *Sound the Retreat* in which Lord Canteloupe learns of his son's death, for example). The narrative is never less than gripping, springing numerous surprises upon the reader, and the comedy is splendidly gamey. The enormous cast is manipulated with great skill, and guessing which of Raven's Charterhouse contemporaries provided models for the characters supplies additional entertainment. (Somerset Lloyd-James shares certain characteristics with Sir William Rees-Mogg, former editor of *The Times* and much-derided moral watchdog of television, while Peter Morrison is allegedly to some degree inspired by the Conservative politician James Prior.) Fielding Gray is, of course, the author's own *alter ego*, as may be seen by comparing the novels with Raven's memoirs, one volume of which had to be withdrawn after a threatened libel action. Indeed, Gray becomes the author of the earlier books in Raven's sequence: his *Love's Jest Book* is clearly *Fielding Gray* and *Operation Armageddon* (winner of the Joseph Conrad Prize in 1964) is *The Sabre Squadron*. Max de Freville is another character who speaks for his creator. 'Corruption in high places is a hobby of mine,' he says. 'I like to be reminded that the world is run, even at the highest level, by petty-minded and venal men.' This is a view *Alms for Oblivion* exemplifies and endorses with great thoroughness.

The further (mis)fortunes of many of the characters, and of their progeny, are described in Raven's later sequence, *The*

First Born of Egypt, which starts with *Morning Star* (1984).

Saville

Storey's seventh novel, which won the 1976 Booker Prize, tells the story of the growing disaffection with his background of Colin Saville, a miner's son, and the slow disintegration of a small mining community in south Yorkshire.

The novel begins with the arrival of Colin's parents in the village some time between the wars. Ellen Saville is pregnant with Colin's elder brother, Andrew, who turns out a lively, talented boy who keeps running away from home. When Ellen is pregnant once more, this time with Colin, Andrew dies suddenly of pneumonia. One child is dead, another waiting to be born: 'nowt that happens could be worse than this,' cries his father. Colin is a placid child, physically very strong. He seems to accept everything and silently observe. At the seaside on their only holiday before the war, 'the boy, in watching, would grow quite still; it was like some cupboard door that had suddenly been opened, a curtain drawn aside to reveal things he'd never encountered or ever imagined could exist'. It is his first experience outside the village.

His friends inevitably belong to the small community. Colin remains detached from them, though he is accepted even by Batty and Stringer, the local tearaways who fail to involve him in crime. There are also the boys who live in the terrace with whom he goes to Sunday School: Michael Reagan, who plays the violin, and Ian Bletchley, who is fat and has inflamed knees. Michael's father works in the office in the colliery, is therefore superior, and despises his wife and son; Ian's father, a railway worker, is one of the few men in the terrace who does not work at the colliery. Colin endures placidly his father's inexpert tuition and wins an exhibition to the grammar school. 'Thy being at the school means everything to me. Whatever happens, I want you to know that, lad,' his father tells him. His sadistic form-master picks on him because his father is a mine worker; he is accused of foul play on the rugger field; he is ridiculed for writing poetry; and he is intermittently befriended by Stafford, scion of the local gentry, who later steals his first serious girlfriend.

On leaving school Colin becomes a teacher and begins to question the purpose of life. The defection of his girlfriend, Margaret, a doctor's daughter, is a serious blow; the legacy she leaves him is the full realisation of the squalor of his home. He has moved out of his own class without finding acceptance in the middle class. Out of sheer frustration he beats up his amiable young brother, Steven, who, content with his lot, rejects Colin's attempts to teach him; he quarrels with his headmaster and is sacked; he takes up with Elizabeth Walton, who is much older than himself and separated from her husband.

Meanwhile all around him is change and decay. His father grows shrunken and has lost interest in everything; Michael Reagan's father dies and his mother is taken off to a mental hospital; Michael himself finds that he can not earn a living as a musician and rapidly goes to seed; Ian Bletchley, who has gone to the 'varsity' and is now a graduate trainee in a cloth manufacturing company, has learnt to be ashamed of his background. In the end Colin breaks with Elizabeth, and decides to go to London: 'There was nothing to detain him. The shell had cracked.'

A long, leisurely novel, which relies heavily on its cumulative effect, *Saville* is principally a statement; at various moments Storey resists the temptation to develop dramatic possibilities in the interests of realism. The drama, he seems to be telling us, is implicit in his subject: the grinding effect of poverty and the impossibility of breaking free of the class structure within a narrow environment. Violence lurks beneath the surface, and sometimes it breaks out, but even passion is subordinate to the uniform greyness: 'Over half a century of soot appeared to draw the buildings, the people, the roads, the entire village into the ground.'

William TREVOR   1928–

The Children of Dynmouth

An ordinary Dorset seaside town, unruffled by anything apart from a small group of motorcyclists and the blustery wind on the cliff-path, is exposed by the furtive eaves-dropping and the corrupt imaginings of a solitary fifteen-year-old. The ubiquitous Timothy Gedge, pale, sharp-boned, with lemon hair and clothes, hangs around the town gathering impressions and information which he is to reveal with devastating conse-quences. He has been surprised to discover an aptitude for theatricals during a lesson at school in which he dressed up as Elizabeth I, and it becomes his ambition to win the Spot the Talent competition at the Easter fête, acting out the story of the Brides in the Bath. He indulges in fantasies of public recog-nition and appreciation, but to fulfil these he requires certain props; his method for acquiring them borders upon intimidation. He wrecks the marriage of the Abigails, a retired couple for whom he does odd jobs, by drunkenly questioning the Colonel's interest in the local scouts. He terrorises the Dasses by threatening to expose the fact that their pampered son left them because he felt they had ruined his life. He bribes the local publican because he has observed him using the public conveniences to seduce local women (amongst them Mrs Gedge). He destroys the idyll of two children, Kate and Stephen, by claiming that Stephen's father murdered his wife in order to marry Kate's mother. Each accusation has a grain of truth, enough to act as a constant source of anxiety for his victims, even if Timothy's interpret-ation is incorrect. As Lavinia Featherston, the vicar's wife, recognises: 'The truth was insidious, never blatant, never just facts.'

In desperation Kate approaches Mr Featherston insisting that Timothy must be possessed and should therefore be exorcised. The vicar's arguments sound hollow to him, particularly since he is guiltily aware that Timothy is different: 'He could feel no Christian love for him.' He manages to reassure Kate that Stephen's father was on a train when his wife died, but the doubts remain and the damage is done. This dark novel ends ironically at Easter, with Timothy embarking upon a new fantasy that he is the child of the local doctor and an infatuated spinster, and with the vicar reflecting that Timothy is 'a far better reminder of waste and destruction' than any number of Good Friday services. The truth is that Timothy *is* damned: damned by neglect and hopelessness.

Trevor's gifts for the comic and macabre are vividly combined in this novel, in which he pursues one of his favourite themes: that of the unexpected and bizarre which lurks just below the surface of everyday life. It won the Whitbread Award.

1977

Gary Gilmore is first person to be executed in the US since 1967 ● Baader-Meinhof terrorists assassinate the chief prosecutor of West Germany ● 'Gang of Four' expelled from the Chinese Communist Party ● Steve Biko is killed in police custody ● Anwar Sadat visits Israel and addresses the Knesset ● First AIDS victims in New York ● 800-mile trans-Alaska oil pipeline ● J.K. Galbraith, *The Age of Anxiety* ● Ronald Dworkin, *Taking Rights Seriously* ● 'Pompidou Centre', Paris ● In Britain, circulation of the *Sun* overtakes that of the *Daily Mirror* ● Films: *Annie Hall, That Obscure Object of Desire, Star Wars, Close Encounters of the Third Kind* ● The Sex Pistols' 'God Save the Queen' becomes Britain's top-selling single ● Anthony Eden, Werner von Braun, Vladimir Nabokov, E.F. Schumacher, Robert Lowell, Maria Callas, Bing Crosby, Charles Chaplin die ● Allen Tate, *Collected Poems* ● Howard Nemerov, *Collected Poems*

Robert COOVER   **1932–**

The Public Burning

The Public Burning started as a play which Coover began shortly after finishing *The Origin of the Brunists* in 1966. Finding that the form did not suit the subject, he abandoned the project only to take it up again, this time as a novella, for a proposed publication in 1972. The subsequent, huge expansion of the original idea quickly convinced Coover that the novella would have to become a long novel and work on it occupied him throughout the first half of the 1970s.

'It is the story of June 19, 1953', Coover wrote. 'On that day, the Rosenbergs are burned in Times Square and all the members of the tribe are drawn to the scene. All that has happened that day happens there, in a way; everything is condensed into one big circus event.' This is only half the story however, and the novel's long evolution through the Watergate years is directly relevant to the other half. The execution of the Rosenbergs is certainly the focal point of the book, but its subtitle might have been 'The Making of a President', for the main character and principal narrator is none other than the then vice-president (and president during much of its writing), Richard Milhous Nixon. Nixon's presence in the book also explains the year-long delay in its publication: almost every major publisher in the

US rejected *The Public Burning* after taking legal advice on the possibility of a libel suit. In the event, no charges were brought.

The novel attempts to tell the story of the whole of America on a single, crucial day in its moral history. The question of the Rosenberg's guilt is not addressed by Coover, who instead calls their accusers to account. As these accusers were, historically, almost the whole of the American people, the book is encyclopaedic in outlook and hugely ambitious in the techniques by which it attempts this inclusivity. The novel opens with 'Uncle Sam' (the word of populist American sentiment made flesh) initiating a 'Groun'-Hog Hunt' to find the culprits who gave 'The Phantom' (a personification of anti-American influence, or the Reds) the secret of the bomb. Uncle Sam's aggressive and banal outbursts are intercut with Richard Nixon's narration of the day the Rosenbergs died. These two voices form the bulk of the text but are interrupted three times by 'Intermezzo' sections: an Eisenhower speech presented as free verse; a dramatic dialogue between Eisenhower and Ethel Rosenberg; and, lastly, a 'Last-Act Sing Sing Opera' from Ethel and her husband, Julius.

Although all of these forms exhibit Coover's near-flawless literary technique, Nixon's narration is the most rewarding. The Vice-President emerges as a hapless, isolated buffoon, capable of inspiring sympathy despite his chicanery and hypocrisy. Nixon's slow odyssey through the twenty-four hours that lead to the executions can be seen as a prefiguring of his ascent to the presidency. He suffers agonies of humiliation and loneliness on this journey and, at the end of the novel, pays for his passage by being sodomised by Uncle Sam himself.

The Public Burning is a long and densely written novel. The sheer accumulation of detail by which Coover convenes the whole of America in Times Square on that night is often exhausting. On its appearance, reviewers (who had traditionally been deeply divided over the merits of Coover's work) united in condemning its obsessive detail and overt technical bravura. It is fair criticism, for Coover's prose takes enormous risks which do not always come off. The novel is certainly not an elegy for the Rosenbergs, but its sense of inhumanity derives from Coover's determination to fully and intimately expose the political machinery which offered up the Rosenbergs as a spectacle for the voters of America.

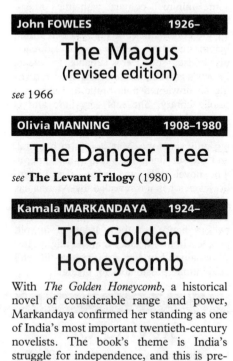

John FOWLES   **1926–**

The Magus
(revised edition)

see 1966

Olivia MANNING   **1908–1980**

The Danger Tree

see **The Levant Trilogy** (1980)

Kamala MARKANDAYA   **1924–**

The Golden Honeycomb

With *The Golden Honeycomb*, a historical novel of considerable range and power, Markandaya confirmed her standing as one of India's most important twentieth-century novelists. The book's theme is India's struggle for independence, and this is presented in an engrossing story focused upon three generations of an imaginary princely

family. 'The Golden Honeycomb' is the palace of Bawajiraj III, Maharajah of Devapur State. Bawajiraj, nicknamed 'Waji', inherits the kingdom from his father, who had been crowned by the British Raj as a convenient puppet ruler. To Waji, the vast palace and the fleets of limousines seem the natural privileges of his position; but to his illegitimate son Rabindranath ('Rabi'), such luxury is beginning to embody corruption and enslavement to the British Empire.

Growing up in the Golden Honeycomb, Rabi retains his plebeian sympathies, which bring him close to a servant-girl, Janaki, who gives him the only trinket she has, a cheap glass marble. Rabi is taken to the Great Durbar, a state ceremonial, and during his absence Janaki is accused of sexually corrupting him and is dismissed from the household. Rabi returns with a present for her, a purple sari, but the gift can never be given, and this abruptly broken friendship alienates Rabi further from the artificial life of the court.

Waji's wife, Shata Devi, has given birth to three daughters, but Rabi, the son and heir, is the child of Waji's mistress Mohini, 'the loyal concubine'. Waji tries to make Mohini marry him as his permitted second wife, but she values the independence which her irregular liaison provides, and, encouraged by Waji's mother, she refuses to let Rabi be educated and indoctrinated by the English. Rabi's resentment of the old regime increases, as an accidental involvement in a people's rebellion leads to his rescue by a prostitute, Jaya.

Back at the palace, Rabi's affections are divided between Sophie Copeland, daughter of the British Resident, and Usha, the anarchic daughter of Tirumal Rao, the Dewan. As the First World War breaks out, Waji is eager to play his part and take a regiment to fight for the British in Flanders. But Rabi refuses to contemplate entering the military academy and shares his mother's sense that Waji is a mere lackey to the British.

Rabi is faced with conflicting loyalties which had not troubled his more superficial father; to the British and the young Sophie Copeland, and to the Indian people and Usha Rao. The aftermath of the war sees Waji living it up at Buckingham Palace and the Ritz, while Rabi, encouraged by Usha's father, his old tutor, protests that the people must be better treated. The injustice of the Raj becomes focused on an unfair tax to be levied on a necessity, salt, and Rabi's stance is finally decided. He will take the side of his own nation and lead his people out of the Empire, preferring the rebellious spirit of Usha Rao to the sweet charms of Sophie Copeland: an individual choice which Markandaya portrays as representative of the long process of Indian liberation.

Barbara PYM — 1913–1980

Quartet in Autumn

'Have thought of an idea for a novel based on our office move –' Pym wrote in March 1972, 'all old, crabby characters, petty and obsessive, bad tempered – how easily one of them could have a false breast! But I'd better not write it till I have time to concentrate upon it.' When she retired from the International African Institute, she developed her idea and wrote what is perhaps her most highly regarded novel. After years of neglect, she suddenly found herself shortlisted for the Booker Prize and attending a grand literary banquet – wearing a 'long pleated black skirt, black blouse, Indian with painted flowers (C&A £4.90) and green beads'.

A novel at once richly comic and sombre, *Quartet in Autumn* depicts the approach of old age through the interrelated lives of four people – two women and two men – on the verge of retirement from a London office. (The book's original title was 'Four Point Turn'; learning of its theme, Philip Larkin had suggested it might be called 'Last Exit to Brookwood'.) The women are more strongly portrayed and contrasted than the men. Letty is a good-hearted spinster living in a bedsitter, who has fallen into a lonely and empty life; Marcia, also unmarried, has lived alone since the death of her mother and her cat, Snowy, in a large, neglected house in Clapham, and become increasingly eccentric and isolated with age. Of the men, Edwin is a widower who derives all necessary emotional support from his interest in the Anglican church, while Norman is an angry,

sarcastic and ineffectual bachelor who feels bereft of normal family life because he never knew his mother.

The novel has a minimal plot, but through a series of small incidents, and a few more calamities, a picture develops of restricted but not entirely hopeless lives. Letty has for some time been planning to join her friend Marjorie in the country after retirement, but almost simultaneously comes news that Marjorie is to marry a vicar, and that the house in which Letty lives has been sold to Mr Olatunde, the Nigerian leader of an exuberantly noisy Christian sect. Unable to cope with these changes, Letty goes to live in a house belonging to the fussy, censorious Mrs Pope. Marcia, meanwhile, who has been hoarding tins and milk bottles against some unspecified emergency, is determinedly repelling the attentions of Janice Brabner, a young, well-meaning but patronising social worker.

Halfway through the novel, the two women retire: Letty attempts to fill her time by studying sociology, but Marcia, despite some minimal contact with the others, becomes ever more marginalised. Obsessed by the consultant surgeon who performed her mastectomy, she goes to stand outside his house; she seems unable to locate Snowy's grave; and she insists upon returning an unwanted milk bottle to Letty in a public library. She eats very little and is eventually discovered seriously ill by Janice, Edwin and his vicar, Fr. Gellibrand. Weighing only six stone, Marcia dies in hospital, and the surviving trio meet at her funeral. The novel ends on a note of optimism, however, when it is revealed that Marcia has left Norman her house, and when Letty (despite the fact that Marjorie has been jilted by her fiancé for the warden of an old people's home) decides that she might after all remain in London, realising that 'life still held infinite possibilities for change'.

Paul SCOTT — 1920–1978

Staying On

Intended as a postscript or coda to *The Raj Quartet* (1975), this short novel takes up the story of two minor characters from the main work: the waspish Lucy Smalley and her ineffectual husband, Tusker (Col. Smalley of the Mahwar Regiment), who stayed on after nearly all their compatriots had left, struggling to keep afloat on inadequate funds with dwindling contacts in an almost unrecognisably changed independent India. It plays humorous variations on a melancholy theme, following a prescription laid down by one of the characters in the *Quartet*, talking about the intrinsic absurdity of the Raj: 'I suppose that to laugh for people, to see the comic side of their lives when they can't see it for themselves, is a way of expressing affection for them; and even admiration – of a kind – for the lives they try so seriously to lead.'

Lucy and Tusker muddle along in the shadow of advancing age, ill health and death, bickering, grumbling, worrying about money and where the next meal is coming from. They live in the shabby annexe of Smith's Hotel, Pankot, under intermittent fire from the proprietress, Mrs Bhoolabhoy, a vast voluptuary of volcanic temper and voracious lust, both of which she directs at her small, shrewd, patient husband who runs the hotel for her. 'On her bad days he walked on tiptoe and had the entire staff doing the same so that even the guests (when there were any) felt themselves under a cloud and got out of the place as soon as possible after breakfast.'

The relationship between the Bhoolabhoys – 'ownership' and 'management' – reverses the Smalleys' situation. Lucy, who never stops talking, natters away at her husband, wheedling and pleading in alternately apprehensive and exasperated appeal, making little or no impact on the taciturn and massively obdurate Tusker. Their mutual loneliness and obstinacy, their lack of contact exacerbated by the total dependence of each on the other, make this one of the most touching as well as among the funniest marriages in fiction. It is a sad and clumsy business (Tusker dies unexpectedly, after suffering a heart-attack on his thunderbox in the primitive lavatory). But it is also an absurdly inarticulate love song scored for two plaintive, gallant, rusty, old voices.

This was Scott's last novel, and the only one which had an immediate success in his lifetime. It won the Booker Prize in a strong year (Paul Bailey, Barbara Pym, Penelope Lively, Jennifer Johnston and Caroline Blackwood were also shortlisted), although Scott, mortally ill, was in America and unable to accept the award in person. The novel was subsequently made into a distinguished television film (1980), scripted by Julian Mitchell and starring Celia Johnson and Trevor Howard (reunited thirty-five years after their roles as the lovers in Noël Coward's *Brief Encounter*). *Staying On* was also in part a muffled love-letter to Scott's wife, Penny, who had been driven from the family home by her husband's destructive behaviour. Scott said that it was the quickest book he ever wrote: it took him barely twelve months after the ten years' hard labour of the *Raj Quartet*. For his readers, as well as for himself, it makes a gentle excursion in the foothills, to be taken for pleasure and refreshment before or after the long haul up to the commanding and exhilarating heights of the *Quartet* itself.

Edward UPWARD　　　　　　**1903–**

The Spiral Ascent

In The Thirties (1962)
The Rotten Elements (1969)
No Home But the Struggle (1977)

In spite of earning an almost legendary reputation in the 1930s, particularly amongst such contemporaries as Auden, Isherwood and Spender, to whom he was something of a literary hero and political mentor, Upward has never achieved a more widespread recognition of his particular merits and the important place he occupies in the literature of this century. The reasons for his neglect are undoubtedly political, for he was the most committed writer of his generation, remaining in the Communist Party from 1934 until after the Second World War. Although he remains best known for 'The Railway Accident' (1949), a short story written in 1929 in the surreal, visionary style he had developed at Cambridge, where he and Isherwood invented a 'fantastic village' called Mortmere, his most considerable achievement is this trilogy of twentieth-century grass-roots political life. Writing the book was an extremely painful process, since it forced Upward to confront his loss of political faith; attempting to publish it in the Cold War climate of the 1950s proved almost as traumatic. At one point, after *No Home But the Struggle* had been rejected by the publisher of the first two volumes, Upward seriously considered producing his own limited samidzat edition. Fortunately, after the intervention of Spender and Isherwood and a grant from the Arts Council, his publishers relented and the final volume was eventually published.

The principal theme of *The Spiral Ascent* is the struggle of its protagonist, a Marxist poet and teacher called Alan Sebrill, to reconcile his literary ambition with his political beliefs. The movement of the trilogy is from the youthful enthusiasm of the first volume, through the middle-aged disillusionment of the second, to the resolution in old age of the third. Alan (who is Upward's *alter ego*) spends the first volume attempting to rid himself of his inherited bourgeois romanticism by allying himself to the Communist Party. During an idyllic holiday on the Isle of Wight with his friend Richard Marple (based on Isherwood), he becomes involved with the sexually compliant Peg. The holiday ends in disaster, however, with the defection of both Peg and Richard, leaving Alan unable to write and contemplating suicide. He decides to return to the mainland and to teaching, and steels himself to begin working for a local cell of the Party. There he meets the worthy but plain Elsie Hutchinson, who encourages him to join the Party, and whom he eventually marries, after much prevarication, almost as a penance. He accepts a job at Condell's Secondary School, an establishment he describes as 'a cheap petty-bourgeois snob-imitation of a public school'. Unable to control his classes, he is almost sacked, but gradually overcomes the hostility of both pupils and staff and even wins one of his colleagues over to the Party. The book contains vivid descriptions of common-room life, a fascist rally and the workers' nature ramble with which the volume closes, and during which Alan realises that 'the poetic life' is not a romantic sham: 'There was only one way towards it, and that was a way of constant political effort, of Communist struggle for a struggleless world in which poetic living would at last become actual.'

In Communist parlance 'The Rotten Elements' were 'members who deviated seriously from the correct party line'. (The second novel was to have been called 'The Deviators' until Upward realised that this 'might raise the wrong kind of expectations'.) Subtitled 'A Novel of Fact', it depicts the isolation Upward himself experienced when he left the Communist Party after the war because he felt that it had abandoned its Marxist principles and become 'reformist and non-revolutionary'. The Sebrills have two children now and live in some comfort and contentment, but their life is disrupted by what they see as the Party's straying from its Leninist roots, in particular in its attempt to work within the framework of a post-war Labour government, which they believe to be as guilty of upholding capitalism as its Tory predecessors. Summoned to a meeting, the Sebrills are asked to explain their hostility to a Communist book, *Britain's Way Forward* (based on Harry Pollitt's *Looking Ahead*), and find themselves at odds with their comrades. When they state their case in print, they win some support, but they also attract enemies. The Party has been Alan's 'family' and he now begins to feel paranoid, particularly when he suspects a former ally of being an *agent provocateur*, and when he receives a mysterious letter. A meeting with the Party leadership confirms the Sebrills' worst fears and Elsie tears up her membership card. Alan, unable to write or even decide how to approach writing, becomes ill (as Upward himself did). However, the novel ends with his rededicating himself to poetry after an argument about the merits of *Soviet Literature* with a Party friend. The new poetic life will not be bourgeois and nostalgic, but forward-looking, 'the best contribution he could make towards expediting ... change' within the Party.

The third volume is narrated in the first person and is reflective in tone. Alan has retired from teaching to the Isle of Wight, where, no longer a Party member, he devotes his political energies to the CND and protest against the Vietnam War. As he attempts to write a long poem, he goes back over his life, recalling his 'imaginative activity during the years before [he] put politics first', a time when he had 'absolute belief ... in the validity of poety'. In considerable and comic detail, he recalls his childhood, and his education at a grim prep school, 'Rugtonstead' Public School and 'poshocracy'-ridden Cambridge. He attempts to find a poetic voice, veering between moonstruck romanticism and the savage and grotesque surrealism of the fictional world he creates with Richard Marple. He becomes a private tutor and a schoolmaster, and his poetic and political progress from decadence and neutrality to stern commitment is charted, as are several idealised and abortive love-affairs and the loss of his virginity to a Soho prostitute.

The trilogy ends where it began with a recapitulation of his holiday with Richard.

As this volume's title suggests, Alan comes to realise that his literary sterility was the result of his putting 'the Party not the political struggle first ... Only by making the struggle my first concern always, and by being prepared to lose everything for it if necessary, shall I become capable of finding more often in my surroundings here something of the same marvellousness that my imagination helped to give them years ago.' The poem he has been working on for six years, and the trilogy, are at last complete.

Upward's struggle to reconcile literature and political commitment (strikingly similar to Gerard Manley Hopkins's concern to make his poetry serve God and the Jesuits) is palpable in the trilogy, in which scenes of great lyricism are interpolated with somewhat rebarbative passages of Marxist dialectic. This provides an uneven but thoroughly rewarding read, one which gives a unique and historically invaluable account of the involvement of the British middle-class intelligentsia with communist politics in the 1930s.

1978

Year of three Popes ● Camp David Agreement ● Murder of former Italian Prime Minister ● In Guyana mass suicide in the Jim Jones cult centre ● Birth of the first 'test tube baby' ● First camera with automatic focus ● Footprints of a hominid, made 3.6 million years ago, found in Tanzania ● Christo, *Wrapped Walkways* ● David Hare, *Plenty* ● Tim Rice and Andrew Lloyd-Weber, *Evita* ● Films: *The Deer Hunter* ● Television: *Dallas, Pennies from Heaven* ● Disco music popular: the Bee Gees' *Saturday Night Fever* becomes the biggest-selling soundtrack album yet ● Hubert Humphrey, F.R. Leavis, Jomo Kenyatta, Margaret Mead, Giorgio de Chirico, Golda Meir die ● Hugh MacDiarmid, *Complete Poems 1920–1976*

Beryl BAINBRIDGE   1934–

Young Adolf

In 1912, Adolf Hitler's illegitimate half-brother came to Liverpool with his Irish wife and set up home in Upper Stanhope St. In 1913, possibly in an attempt to evade military service, Adolf Hitler left Vienna. Some nine months later he was traced to Munich

by the police, and ordered to present himself at a military board. From these facts Bainbridge constructs a bizarre fantasy whereby young Adolf flees to England to stay with his relatives in Liverpool. Much of the black comedy arises from Bainbridge's portrait of the future dictator as an inept, gawky young man, given to alternating spells of extreme lethargy and hyperactivity. Young Adolf is an unappetising hysteric, an utter failure who, in the book's grimly ironic last words, will 'never amount to anything'.

Alois Hitler had been expecting his sister, and is not pleased when Adolf arrives instead. Adolf's unwell appearance, shabby clothes, his appalling smell and erratic behaviour do little to endear him to his irascible half-brother and his Irish sister-in-law, Bridget, who is much preoccupied with her baby, Darling Pat. Adolf does not think much of Alois either, and seeks intellectual stimulus from Mr Meyer, the landlord. He also meets the hirsute housekeeper, Mary O'Leary, and Meyer's sinister associate, the festering Dr Kephaius. Adolf's sense of disorientation is increased when he sees a man with a bandaged head apparently leaping through the walls of the house, and he becomes aware that he is being followed and watched.

Alois procures him a job as a messenger boy at the Adelphi Hotel, where he proves of service to the mysterious M. Dupont. Adolf, who cannot afford to get into trouble with the authorities since his papers are not in order, becomes frightened when he discovers that Meyer is involved in subversive activities against the civil authorities, and that Dupont is a jewel thief. These experiences, together with the discovery that he is indeed being followed by Mary O'Leary's husband of one night, who is convinced that Adolf is his son, persuades the young visitor that he is amongst madmen, and confirms his desire to leave England as soon as possible.

The novel is strewn with intimations of Adolf's final destiny. He is extremely intolerant, rants against Jews, and unconsciously raises his hand when delivering an excitable speech at the dinner table. The origins of the Führer's trademarks are also to be found in his Liverpool sojourn: Bridget makes him a brown shirt from a curtain; he brushes his hair down across his forehead to conceal a bump received while involved with Meyer's subversives; and he decides to grow a moustache after being molested while disguised as a woman.

John BARTH   **1930–**

LETTERS

LETTERS brings the first phase of Barth's fiction-writing career to an appropriately reflexive close. Consisting of eighty-eight letters (divided into seven sections) exchanged between himself (as the 'Author'), five characters derived from his earlier works, and one new one, Lady Amherst, the novel begins with a determined attempt to exhaust the signifying potential of its own title. The letters which spell out the word 'LETTERS' on the title-page are made up of smaller letters which in turn spell out the book's sub-title, 'an old time epistolary novel by seven fictitious drolls and dreamers, each of which imagines himself actual'. In addition, each epistle is 'lettered' rather than numbered, and a complicated acrostic can be formed from these letters to give the sub-title again. Furthermore, each of the seven sections is prefaced with a calendar upon which the identifying letter of each epistle is superimposed over the date on which it was sent, and the pattern of identifying letters reveals the shape of the letter which heads that section.

Each of the characters (re-)invented for the novel are themselves 'Authors' and aware to differing degrees that they are participating in a novel. The most successful (and in the action most predominant) of these creations is Lady Amherst, who begins the novel by inviting 'John Barth' to accept an honorary doctorate from the Maryland university of which she is Acting Provost. The tale of this enticement quickly cedes to that of her own enticement by Ambrose Mensch, who was last seen as a Barth-like figure in the volume of short stories, *Lost in the Funhouse* (1968), and who now wishes to impregnate her. Lady Amherst is a straightforward personification of the modernist muse, comparing herself to, and even echoing, Molly Bloom. She has been tangentially involved with many of the major literary figures of the early twentieth century, and her impregnation by Mensch hints at a cross-fertilisation of the modernist and postmodernist movements, an enduring theme in Barth's work.

Barth's consideration of literary history is matched by a more general enquiry conducted largely through the character of Andrew Burlingame Cook IV (A.B.C.IV/D), a descendant of Henry Burlingame (the arch-manipulator of history in **The Sot-Weed Factor**, 1960), and Ebenezer Cooke (the would-be laureate of Maryland from the same book). Cook's letters to his unborn child (which will turn out to be twins) reveal the tamperings that that family has effected on the course of American history, and these tamperings are, often as not, presented as textual interventions, alterations to the story of history. Of the other correspondents, Jerome Bray, a descendant of Napoleon Bonaparte and Harold Bray (the false Grand Tutor of **Giles Goat-Boy**, 1966) partakes of both fictional and historical realms. As in *Giles Goat-Boy*, he represents the most serious threat to Barth's 'author-ity' through his project to write a novel called 'NUMBERS' – a counter-text to Barth's *LETTERS*. Jacob Horner (from *The End of the Road*, 1958) rights a losing oneiric battle with history in his attempt to 're-dream' it and Todd Andrews, who first appeared in *The Floating Opera* (1956), attempts a similar project on the smaller scale of his own life.

In the final letter of *LETTERS*, Barth addresses the reader directly, invoking a mixture of world events and mundane observations. The effect after the exhaustive games-playing and elaborate patterning of the novel is of a hard-won naturalism, legitimated by the rites of passage that the novel's difficulty represents. As it will turn out, the key to his next novel, *Sabbatical* (1982), and several themes from the one after, *Tidewater Tales* (1987), are contained in this envoi. In this, the last letter marks a Janus-like moment in Barth's career: a reluctant farewell to the programmatic excesses of the first phase of his writing, and a cautious prevision of the greater naturalism of the second.

David COOK   **1940–**

Walter

see **Winter Doves** (1979)

John IRVING   **1942–**

The World According to Garp

This is the novel which made Irving's name on both sides of the Atlantic, selling well over 3,000,000 copies. The novel attests to Irving's continuing fascination with such diverse subjects as marital infidelity, Vienna, wrestling, bears, violence and death, and to his belief in solid characterisation, clarity of prose and the importance of plot.

The story begins with T.S. Garp's mother, Jenny Fields, rebelling against received, male notions of womanhood: 'I wanted a job and I wanted to live alone. That made me a sexual suspect. Then I wanted a baby, but I didn't want to have to share my body or my life to have one. That made me a sexual suspect too.' Two results of this self-assertion are a bestselling book, *The Sexual Suspect*, which makes her a feminist figurehead, and the birth of the eponymous Garp, bizarrely conceived in the hospital where Jenny works, and raised by her at the Steering School in New England. It is here that Garp is educated, determines to be a writer, loses his virginity, takes up wrestling and falls in love with Helen Holm, the daughter of his coach. After a long, sexually adventurous trip to Vienna, 'a museum housing a dead city', Garp returns to America and marries Helen, who teaches English literature.

The bulk of the novel describes their life together and the events which affect it, such as Garp's occasional infidelities, his writing career, his mother's radical feminism, his children, about whom he is obsessively protective, and an abortive four-way affair with another couple. Irving also portrays mundane domesticity, but beneath this calm surface lurks the Undertoad ('undertow' as pronounced by his son, Walt), the current which sweeps your feet from under you. Irving punctuates the story of Garp's life with episodes of catastrophic violence, fateful conjunctions of circumstances which occur independent of human control, yet are specifically linked to human actions. For example, Garp's mother, who has set up a home for women from the proceeds of her book, is assassinated by a mad, wife-beating misogynist, and Garp's own assassination at the hands of a mad misandrist can be traced back to an incident in his past. The most striking example of such conjunctions occurs when Garp, returning home with his two sons, coasts down the drive with lights and engine off, a sensation the children love. His car collides with a station wagon in which Helen is fellating her lover. The younger child is thrown through the windshield to his death, the older gouges out an eye on the broken gear-stick (which Garp had been intending to fix for weeks), Garp breaks his jaw, and Helen bites off her lover's penis. The incident characteristically combines elements of tragedy, irony and black humour, and Irving manages to present it as both bizarre and credible.

'What do you mean, "This is Chapter One"?' Garp's editor, John Wolf, wrote him. 'How can there be any more of *this*? There is entirely too much as it stands! How can you possibly go on?'

'It goes on,' Garp wrote back, 'You'll see.'

'I don't want to see,' John Wolf told Garp on the phone.

THE WORLD ACCORDING TO GARP
BY JOHN IRVING

Such violent irruptions are explicitly linked to sexual desire, while underlying the story is a form of stoicism supported by sporadic quotations from Marcus Aurelius's *Meditations*. This is the weakest aspect of the novel, for Irving is no philosopher and the philosophy itself works against the determined energy of the characters and the darkly comic episodes of the story. Irving's real talents lie in the quite different discipline of storytelling. His characters are complex, fully imagined and convincing, however unlikely they may seem (Roberta Muldoon, standout tight end for the Philadelphia Eagles before her sex-change, or Ellen James, unwilling figurehead of a feminist movement whose members have their tongues cut out in imitation of their heroine, who was raped and mutilated as an eleven-year-old). The novel offers no abstract view of humanity and history, but deals with specific incidents as they happen to specific people, uncovering the truth by exaggeration rather than analysis.

Ian McEWAN   1948–

The Cement Garden

McEwan's first novel is more closely linked with his two earlier volumes of short stories than with later books. It shares their cool objectivity when describing macabre incidents and sexually unorthodox behaviour, refusing to pronounce any sort of judgement upon events that might commonly be regarded as immoral or repugnant. This is the novel's great strength, along with its graphic recreation of unlovely adolescence. It has been claimed that the book plagiarised an earlier novel, *Our Mother's House* (1963) by Julian Gloag, an accusation which caused considerable controversy. It is certainly true that, although the style of the two books is very different, the plots are remarkably similar. McEwan insisted that this was a coincidence and denied ever having read Gloag's book.

The novel is narrated by the adolescent Jack, who lives with his parents, his sisters, Julie and Sue, and little brother, Tom. The sudden death of their father, followed by the slow decline of their mother, dramatically hastens their introduction to adult responsibility. Aware that public knowledge of their mother's death will lead to their being taken into care and perhaps entrusted to separate institutions, the three older children decide to conceal her corpse in the cellar in a tin truck which they fill with cement.

During a long, stifling summer, the children affect nonchalance about their circumstances, pretending not to think about what lies in the cellar, not bothering with the washing-up or making any attempt to be domestic. Their grim secret is helped by the fact that their parents had no relatives or friends and that their house is in a derelict area, the rest of the street long since demolished to make way for a motorway that was never built. Jack spends much of his time masturbating, while Julie and Sue indulge in other fantasies, dressing up Tom as a girl. Tom gradually regresses to babyhood, demanding to be cradled by Julie, and sleeping in a cot beside her bed. Jack wants to recreate the sexual games he played with his sisters when they were younger. While they keep to themselves, their secret is safe, but Julie acquires a boyfriend, the snooker-playing Derek with his flashy sports car, who begins to ask questions. Meanwhile, the cement tomb has cracked open and a sweet smell begins to pervade the house. The children tell Derek that Jack had buried his dog in the trunk, a story he does not believe. One day he discovers Jack and Julie in bed together. Appalled – and piqued, since Julie had never allowed *him* such liberties – he smashes open the trunk and then drives off. The novel ends with the arrival of the police.

In 1993 the novel was successfully filmed, from his own screenplay, by Andrew Birkin.

Olivia MANNING   1908–1980

The Battle Lost and Won

see **The Levant Trilogy** (1980)

Armistead MAUPIN   1944–

Tales of the City

see 1990

Alice MUNRO   1931–

The Beggar Maid

Munro has emerged as one of Canada's most respected writers and one of the English-speaking world's leading exponents of the short story. Where other writers have deliberately rejected the comparative provincialism of Canada, Munro has made a virtue of parochialism, usually setting her fiction in small-town or rural communities, examining the lives of women less privileged, and more representative, than the customary heroines of more obviously feminist fiction. Originally published in Canada as *Who Do You Think You Are?* – a title Munro's American publishers thought their readers would fail to understand – *The Beggar Maid* is a series of interlinked stories which forms a discontinuous narrative of the life of Rose, its central character, from her childhood in the 1930s to maturity. Munro had originally been persuaded by her publishers to compile a book of stories featuring two different protagonists, Rose and Janet. In order to ensure that the chapters were divided equally between these two characters, she had to rewrite several of them. 'The stories were originally written with heroines of different names and appeared with different names in magazines,' she later explained. 'This isn't as important as it sounds ... I often write about the same heroine and give her a different name and a different occupation and a slightly different background because of something I want to do in the story. But her psychological make-up is not different.' When the book was set up in galley, however, she decided she was unhappy with it and completely revised it, adding two new stories and rewriting others, so that the entire book would be about Rose – her original intention. This cost her over $2,000, but the final book was highly praised and shortlisted for the Booker Prize.

Rose is brought up by her stepmother Flo at the poor end of Hanratty, a small town in rural Canada. Her father, who repairs furniture for a living, suppresses the sensitive side of his nature for Flo's benefit, herself a product of the big city, worldly and suspicious. This also means suppressing his fondness for Rose, who takes after him, and ironically it is Flo who becomes Rose's ally after the brutal beatings he periodically inflicts on her. But Flo reacts with disgust to Rose's love, in 'Privilege', for Cora, an older schoolgirl, since to her love means no more than dependence and self-deception. The subjection of emotional needs to the perceived demands of survival which Flo encourages sets a pattern for Rose's future.

After her father's death, Rose leaves home to go to college on a scholarship. Here she is courted by the prudish and spiritless Patrick Blatchford, the heir to a chain of department stores. While Rose's poverty is a source of acute discomfort to her, it adds to her charm in Patrick's eyes, offering him the opportunity to play the rescuing king to her 'Beggar Maid', the name he gives her after Burne-Jones's famous painting on this theme. Out of a mixture of pity and a growing capacity to distance herself from events, Rose agrees to marry him. A momentary flash of insight prompts her to break off the engagement, but she soon changes her mind, acting on an impulse she is unable to comprehend. Munro's treatment of this episode illustrates her typical refusal to impose an explanatory scheme on the indefiniteness of her main character's thought-processes.

Rose and Patrick have a child, Anna, but their marriage is emotionally sterile. Rose falls in love with Clifford, a violinist, but her hopes of fulfilment are disappointed as he decides to remain faithful to his wife ('Mischief'). Rose divorces Patrick and moves with the young Anna to a remote mountain town where she works for the local radio station. Here, a long-distance affair with another married man, Tom, comes to nothing ('Providence'). In 'Simon's Luck', we find Rose in a new place, working as an actress and drama teacher. Her relationship with Simon, an academic, seems to promise happiness. But once again she finds herself dwelling on the possibilities of dependence and betrayal by which, for her, love and hope have come to be inevitably overshadowed. When Simon does not appear for dinner one evening, she leaves town to recover the stability which solitude seems to offer.

In the last two stories, Rose, now a well-known television interviewer, returns to Hanratty to attend to Flo. Though she has come to see her life as a series of mistakes, a meeting with her schoolfriend Ralph Gillespie revives old feelings of closeness to him which enable her to look back on it with a measure of calm indifference.

Iris MURDOCH   1919–

The Sea, The Sea

Some 500 pages long and featuring a large cast of characters, *The Sea, The Sea* was awarded the Booker Prize and is considered by many to be Murdoch's finest novel. It concerns the experiences of Charles Arrowby, a famous and egotistical theatre director, who retires to an isolated house by the sea in order to write his memoirs. His intentions are soon disrupted, however, as people from his past converge upon the scene. Discovering that his first childhood love, Hartley – to whom several oblique references have been made – is living nearby in apparent misery, he becomes obsessed with the idea of rescuing her. He sets about it with little real concern for her feelings or the intricacies of her marriage to Ben Fitch, a man who is jealous of her past and who violently objects to Charles's presence in the neighbourhood.

Further interruption is caused by Rosina Vamburgh, a former girlfriend of Charles who takes her revenge by 'haunting' him (this explains some, though not all, of the strange occurrences in the house). Gilbert Opian, a homosexual actor whose attempts to set up house with another of Charles's former girlfriends, Lizzie Scherer, Charles has thwarted, arrives, offering to act as 'house-serf'. Seventeen-year-old Angela Godwin, step-daughter of Rosina's husband, unexpectedly writes to Charles and offers to have his baby. The Fitches' adopted son, Titus, comes to find out whether he really is, as Ben irrationally suspects, Charles's son, and drowns near the house.

Echoes of *The Tempest* resonate throughout the novel, and familiar Murdochian themes – obsessive love, the power of Fate, drowning and near-drowning – abound. These are juxtaposed with Charles's poetic observation of his surroundings (particularly the ever-changing sea) and meticulous accounts of his often bizarre diet (he is resolutely against *haute cuisine*). The reader's expectations are frequently undermined as Murdoch provides explanations for the seemingly irrational – even Charles's vision of a sea monster, it is suggested, is simply an after-effect of his taking LSD some years previously.

The inexplicable is provided, perhaps surprisingly, by James Arrowby, Charles's Buddhist cousin, whom he envied when they were children, and who saves Charles's life after Rosina's husband, Peregrine Arbelow, has tried to kill him. Only in retrospect does Charles begin to perceive the extraordinary powers James must have exerted to save him from drowning, and to interpret events in a more perspicacious manner. He acknowledges his responsibility – 'I had let loose my own demons, not least the sea serpent of jealousy' – and accepts the folly of his obsession with Hartley, 'a secret love which did not exist at all'.

Isaac Bashevis SINGER   1904–1991

Shosha

Growing up on Krochmalna St in an impoverished area of Warsaw, Singer's narrator, Aaron Greidinger, tells his first stories to Shosha – an innocent, introverted, unteachable girl. With her he shares his childhood fears and theories. But, in 1917, the Greidingers move away – to Austria and then Galicia – and two decades pass before Aaron returns. Anti-Semitism is rife in the city and German fascism menaces Poland's borders.

An obscure and virtually penniless writer, Aaron survives on work for magazines. His closest friend is a philosopher, Morris Feitelzohn, who propounds 'soul expeditions' – mystic indulgences of body and mind. Through Feitelzohn, Aaron meets Sam Dreiman and his actress mistress, Betty Slonim. Dreiman is a New York millionaire. In his old age, he wishes to produce a play in Yiddish: Aaron is commissioned to write it and Betty, of course, will take the leading role. Paid generously by Dreiman, Aaron rents a comfortable apartment. He seduces the maid, is seduced by Celia Chentshiner (whose mild husband, Haiml, does not object), continues an affair with Dora Stolnitz and commences a fresh one with Betty. Yet Shosha remains in his thoughts and he visits Krochmalna St to see if she still lives there. She does, alone with her mother. Although Shosha is mentally and physically stunted – the victim of disease and apparitions – Aaron's love for her awakes at once.

His play proves to be a disaster and is stopped in rehearsal. Dreiman and Betty plan to go back to New York and, since Hitler's ambitions are apparent, they offer to take Aaron with them. To circumvent the immigration laws, Aaron will marry Betty and Shosha will travel as the couple's maid. Spurning deliverance, Aaron proposes to Shosha. She accepts. Only Feitelzohn understands Aaron's decision. As Aaron says: '[Shosha] is the only woman I can trust.'

Some months later, Betty reappears and again tries to persuade Aaron to abandon Shosha and flee to the USA. Again, Aaron refuses. He and his friends draw closer together. The Holocaust by now is imminent.

Thirteen years on, Aaron is working for a Yiddish paper. During a trip to Israel, he meets Haiml Chentshiner and the two discuss the intervening time. Both have tales of suffering and escape. Shosha, Celia and Feitelzohn have been killed, directly or indirectly, by the Nazis. Reflecting on his people's tribulation, Haiml asks Aaron: 'What need was there for all this?' Aaron has no answer. In gathering darkness, Haiml waits for one.

Singer stated: 'This novel does not represent the Jews of Poland in the pre-Hitler years … It is a story of a few unique characters in unique circumstances.' Nevertheless, when in 1978 he received the Nobel Prize for Literature – an award sealed by the publication of *Shosha* – it was 'for his impassioned narrative art which … brings universal human conditions to life'. Thus, if *Shosha* lays no claim to general pertinence, its subject-matter is such that it can avoid neither metaphysics nor history. Through its memorialisation of a culture destroyed, it is continually seeking to know what it was that left the Jews defenceless. Some causes are easily named (and Singer names them): the Stalinist betrayal of Polish Communism, the failure of democratic opposition. Others are speculatively indicated – among these being the absence of God. And a further irresistible cause is suggested: the voluble passivity of the victims even as their torturers approach. Partly, Singer's protagonists are trapped by their naïvety – by their attachment to each other and place and by their capacity for self-deception until the final moment has arrived.

Shosha was initially published in 1974 in the *Jewish Daily Forward* as *Soul Expeditions*.

Written in Yiddish, it was translated into English by the author and his nephew, Joseph.

Fay WELDON   1931–

Praxis

Weldon made her reputation writing about inequality between the sexes, often in an oppressive domestic setting, and sometimes in the form of fables. *Praxis*, her sixth novel, is a characteristically powerful tale in which few escape blame: men who control women's lives; the mothers who punish their daughters for their own suffering; the women who allow emotion to conquer intellect; or the 'new women' who have become 'heartless, soulless, mindless' in the pursuit of their own desires. Also culpable are nature, which is 'on the man's side', and a society which believes in the 'natural law of male dominance' and teaches that women, being daughters of Eve, are 'responsible for leading men into sin, and for the loss of Paradise'. It is also a tale in which sex is rarely seen as a simple act of love, but as an exertion of power, a means of punishment or escape, a source of guilt or self-deception, an attempt to establish identity.

Middle-aged Praxis Duveen, self-confessed 'bastard, adulteress, whore, committer of incest, murderess', now unwilling darling of the Women's Movement, relates the extraordinary and depressing pattern of events which have led to two years' imprisonment for the murder of a mentally handicapped child. Against a background of sexual violence, instability and mounting madness in her family, Praxis is a surprisingly intelligent child who eventually wins a place at university, if only because her name is taken to be a man's when she applies. She soon meets the slovenly, mean and sexually inexhaustible Willy, however, and readily relinquishes her studies to become his slave. Considering sex to be her only talent, she is finally prompted by boredom to become a lunchtime whore, a role she is well able to cope with until she realises she has slept with her own father, who abandoned her as a child. Taking full responsibility for her unhappiness, she leaves Willy (who has

another girlfriend anyway), and after an unlikely first marriage and two children, she eventually marries Phillip, whom she has always loved.

I would rather be sorry a hundred times, thought Praxis than safe.
Well, I am, aren't I? Very sorry and not at all safe.

PRAXIS
BY FAY WELDON

There is no happy ending here, however, since her struggle to support husband, family and high-powered job is proved pointless when she discovers Phillip is having an affair and intends to leave her. A growing sympathy for feminism is kindled when she sees Mary Leonard, the illegitimate daughter of her now dead guardian, twice jeopardise a medical career – first, for a man who leaves her anyway, second to see through a pregnancy resulting from a casual meeting at a party. It is Mary who gives birth to a child with Down's syndrome, and it is this child whom Praxis, perhaps in an attempt to break a pattern of despair, smothers with a pillow when Mary leaves her hospital room for a moment. Such is the strength of Praxis's spirit, that she emerges from the experience transformed and sustained by the realisation that there is 'some kind of force which ... gives meaning and purpose' even to a life such as hers.

William WHARTON   1925–

Birdy

Wharton began writing as therapy after being discharged from the army in which he served during the Second World War until seriously injured in 1945. In 1977, a friend showed the manuscript of *Birdy* to a publisher, it was accepted for publication, won the American Book Award in 1979, and instantly established its author as a leading writer. Wharton is a secretive man, but it is known that he bred canaries and is assumed that he drew upon his own experience of combat trauma in depicting the character of his eponymous hero.

In the clinical and disorientating surroundings of a padded cell, Al, a tough, explosive Sicilian American, tries to make contact with his childhood friend Birdy, who has been incarcerated for his madness, his desire to become a bird, his wish to escape the world. The narrative flickers between Birdy's avian fantasies and Al's tales from their colourful, often desperate, past.

Weiss, the institution's 'doctor-major', attempts to pump Al for clues to Birdy's fallen state; for him Birdy is no more than a particularly fascinating case history, but Al is stubborn and canny as well as loyal to his friend, resenting Weiss's manic 'Santa Claus' stare. He colludes with Birdy in resisting the psychiatrist's probings.

During Al's visits the reader is drawn into Birdy's tangled mind and the reasons for his bird's-eye view of life. The boys grew up together, and raised pigeons: Birdy occasionally wore a home-made pigeon suit, something that 'definitely bugged his mother'. At high school they grow apart as Birdy's obsession with things avian (this time canaries) carries him further into a fantasy world. He begins to dream and in his elaborate dreams actually becomes the mate of a favourite canary, Perta. In real life his highly successful canary-breeding business reaps substantial financial rewards and keeps his father sweet. The dream bird world assumes greater and greater importance, so that Birdy can no longer draw a line between fact and fantasy. His dreams *are* his reality. As a bird he is cornered by a cat and it is this that has brought him to the asylum. Al has suffered in other ways, having seen his comrades die on the battlefields of Europe.

The novel opens with Al reminding Birdy of his fastidious mother who gathered stray baseballs that landed in the garden from an adjoining sportsfield. The mystery as to what she did with them remains unsolved. As if to outwit Birdy's madness Al suggests to Weiss that the baseballs are shipped to Birdy's cell to aid his recovery. Miraculously they arrive mouldy and decaying. In an orgy of retribution Birdy and Al, laughing inside themselves hysterically, pound Weiss with the baseballs. Although Al has said 'we're both impossibly screwed up', both men know the baseball 'madness' has brought about their own catharsis.

1979

Iranian Revolution • Margaret Thatcher elected British Prime Minister • Settlement of Rhodesia problem • USSR invasion of Afghanistan • Major accident at the Three Mile Island nuclear power station in Pennsylvania, US • The US Surgeon-general confirms that cigarette smoking causes cancer • Christopher Lasch, *The Culture of Narcissism* • Mother Teresa awarded the Nobel Peace Prize • Judy Chicago, *The Dinner Party* • Alban Berg, *Lulu*, first complete performance • Films: *Alien, Monty Python's Life of Brian* • Television: *Life on Earth, Tales of the Unexpected* • Popular music: The Clash, *London Calling*. The Police, *Message in a Bottle* and *Walking on the Moon* • Charlie Mingus, Nelson Rockefeller, Jean Monnet, Mary Pickford, John Wayne, Herbert Marcuse, Lord Louis Mountbatten, Gracie Fields die • James Berry, *Fractured Circles* • F.T. Prince, *Collected Poems*

André BRINK   1935–

A Dry White Season

'It was reported in a humdrum enough fashion – page four, third column of the evening paper. Johannesburg teacher killed in accident, knocked down by hit-and-run driver. Mr Ben du Toit (53).'

When Ben du Toit fears for his life, he asks an old university friend to keep his notes and journals, and to use them if necessary: 'they want to wipe out every sign of me, as if I'd never been here. And I won't let them.' Two weeks later Ben is dead. His friend – the narrator – is committed to telling Ben's story, the substance of *A Dry White Season*. Frankly, he is annoyed by this legacy, for he is a romantic novelist. 'Politics isn't my line,' he says, but Brink makes it clear that in South Africa just to exist is political. Ben realises towards the end of the novel that to be white is to be born into a state of privilege and – regardless of beliefs or actions – to be white is to be one of the oppressors in a system of apartheid.

Ben is an ordinary, good-natured man, remarkable only for seeming quite so unremarkable. He has a steady teaching job; a beautiful, ambitious, rather cold wife; two lovely daughters; and a doting son. He simply has no cause to question apartheid, until the black teacher at his school, Gordon Ngubene, dies in police custody. The inquest rules 'suicide', by hanging, but the circumstances are suspicious and the evidence contradictory. Ben embarks on a relentless search for justice, and finds his understanding of the world transformed. *A Dry White Season* shows an indifferent layman becoming a political activist – one man up against a corrupt, powerful and merciless regime.

Ben is regarded by his own people as a traitor; the Special Branch gradually but effectively eliminates his network of support, and Ben becomes completely isolated. At this point the novel asks: what for? Is the pain and loss of human life worth it? At least part of the answer comes in one of Ben's diary entries: 'If I act I cannot but lose. But if I do not act, it is a different kind of defeat, equally decisive and maybe worse. Because then I will not even have a conscience left.'

A Dry White Season is undoubtedly a novel of great power, but it is flawed. There is something smug about the writing – a complacent assumption that the system is wrong and that even if it triumphs, we, the intelligent liberals, are right. Old ground-rules go unchallenged. Black people are fundamentally helpless victims, while white men will inevitably be the vehicles of change that are going to count. There are stylistic weaknesses too. The diary extracts are unconvincing: it is difficult to reconcile Ben du Toit's intelligent and courageous public stance, as described by the narrator, with the simplistic (even soppy) voice of his private musings. A *Sunday Telegraph* journalist commented when the paperback of *A Dry White Season* was reissued: 'Political novels, by definition, are conceived within a specific context and it is only when times change that the writer's naked craft becomes apparent . . . in short, we needed the reporter in André Brink too badly to judge the writer in him.'

Brink has received many awards for his work, including the Martin Luther King Memorial Prize, and he has twice been nominated for the Booker Prize. *A Dry White Season* was made into a feature film in 1990. Brink is an important writer, and this is an important book – but its importance lies mainly in the time of its publication.

David COOK   1940–

Walter
1978

Winter Doves
1979

Cook's novels have often concerned themselves with the underprivileged and people on the margins of society: the elderly, the homeless, the handicapped, the sexually troubled and the dispossessed. 'I probably never shall, as a writer, get away from the walking wounded,' he has said; 'they press in so closely.' His writing is remarkable for the empathy he brings to characters whose lives are very different to his own, a skill he acquired as an actor. The character of Walter was suggested by two mentally handicapped people Cook had worked alongside in Woolworths and in the theatre. The first of the two novels was awarded the Hawthornden Prize.

Sarah's mother was penetrated once only by a man: he went to war and died at sea. She has narrow hips and had 'the most painful of all deliveries'. Now she listens at the dividing wall, sneering at Sarah and Eric in their attempts at love-making. Eric, whose father and mother were brother and sister, is as inadequate in bed as he is in most things. Backward, he is discovered to have packed comics for the honeymoon in Southport. Sarah works at Tyson's, the draper's shop, where she becomes manageress.

Seven years of marriage and practice in bed produce Walter. It is clear from the start that there is something the matter with him: he is 'one of Jesus's mistakes'. As a small child he becomes desperate for love, like a dog: 'Please! Please, show me I am loved and the world is a good place.' David Cook inhabits the unknowable interior worlds of his desperate characters: Sarah clings to a blind religious faith; Eric is traumatised by his childhood in the orphanage and the discovery of his bad blood; Walter is ever smiling and eager to please. At nineteen Walter has his first sexual experience when his ally, the thirteen-year-old Elaine, puts her hand on his crotch, but little develops in his life. He lives from moment to moment in an ever-shifting present. His only accomplishment is helping his father look after the pigeons, though he does manage to get a job of sorts, sweeping the floors at Woolworths.

When Walter's mother dies, the fragility of his existence in the larger world is exposed. 'He had not associated death with human beings', so he leaves her as he found her, in bed, and waits for her to wake up. He lets the pigeons into the bedroom so that he can feed them there without having to leave his mother's side. Only when the social services eventually arrive is the full horror of the scene fully described and the logic of his actions destroyed. This scene is one of several in the novel which are shocking but compassionate and, remarkably, comic. The chaos which follows when a pigeon is stolen at the Fancy Pigeon Show is the novel's one purely comic scene.

Once Walter has been taken into care, he becomes entirely disoriented, a state represented by goldfish swimming around inside his head. The novel ends bleakly with Walter institutionalised, living in a former workhouse, now a lunatic asylum, and performing degrading jobs. He is roughly seduced by one of the male inmates, and witnesses the murder in the washrooms of the 'shaking man' who 'stank of excreted brussel sprouts' and who 'had never spoken one word to anyone during the sixteen years in the hospital'. What little Walter had grasped of the reality of the world diminishes. The institution is now his reality. 'It had to be. He would never get out of it.'

Nevertheless, in the sequel, set in the late 1970s, Walter is released into the community. He meets the suicidal June and, on the floor of a church, he finally does 'what Mike at Woolworth's had done'. They meet Graham and move into a squat together. Walter sees Graham and June making love and overhears them:

'You're very good with him.'
'Good? Goodness doesn't come into it. I took him on when I needed someone, and he was the only one there.'

Graham and June abandon Walter and move to another squat. Graham meets another girl there and June returns to the first squat, where she kills herself.

Walter returns to the hospital where: 'He never speaks of pigeons. Only at night in his dreams the pigeons fly. Russian High-Fliers and Birmingham Rollers, Turbits, Long-Faced English Tumblers, White-Lace Fantails ... clapping their wings as they wheel above the solitary figure looking up at them from below.'

Cook later adapted both novels for the television. *Walter* was broadcast on the opening night of the new Channel 4 in 1982, and caused considerable controversy: in particular, the scene in which Walter (played by Ian McKellan) is homosexually assaulted was seen by the tabloid press as a portent of things to come on this 'alternative' channel. *Winter Doves* was broadcast two years later as *Walter and June*.

Penelope FITZGERALD   1916–

Offshore

'Are we to gather that *Dreadnought* is asking us all to do something dishonest?' asks the punctilious boat-owner Richard. '*Dreadnought* nodded, glad to have been understood so easily.' Fitzgerald's novel interweaves the lives of houseboat-owners on the Thames, where 'living on Battersea Reach, overlooked by some very good houses, and under the surveillance of the Port of London Authority, entailed, surely, a certain standard of conduct'.

Richard, the self-elected captain of this community amongst the flotsam, attempts to impose some order on their existence from the ex-battleship *Lord Jim*. His wife Laura, meanwhile, is angling for a house in the country, and becoming suspicious of Richard's feelings for the newly arrived Nenna James, who has taken refuge from a broken marriage on a leaky Thames barge with her two unruly daughters, truants from their convent school. Fitzgerald's narrative moves between their stories, as Laura deserts Richard and he becomes involved with Nenna, who is searching for her estranged husband. Their friendship is further complicated by the presence of Maurice, a rent-boy who is using his boat as the hiding-place for stolen goods.

On a night when disaster is about to befall them all, Nenna makes an epic journey across London to the dingy bedsit where her husband has holed up. Maurice's violent lover is caught with the stolen goods, landing Richard a knock on the head. Taking advantage of the situation, Laura disposes of *Lord Jim*, and drags Richard off to dry land, where she has bought a desirable residence which is the envy of her friends.

'The Thames barges, built of living wood that gave and sprang in the face of the wind, were as much at home as anything on the river. To their creaking and grumbling was added a new note, comparable to music. As the tide rose, the wind shredded the clouds above them and pushed a mighty swell across the water, so that they began to roll as they had once rolled at sea.' The community is disintegrating and Nenna's marriage is still on the rocks, a storm is brewing and the moorings are breaking loose: but Nenna's husband arrives to make his peace with her, and finds himself on the wrong boat getting drunk with Maurice. Leaning out perilously over the companionway looking for his wife, the mournful husband is once more adrift: 'the mooring-ropes, unable to take the whole weight of the barge, pulled free and parted from the shore. It was in this way that *Maurice*, with the two of them clinging on for dear life, put out on the tide.'

Fitzgerald has had other and better novels shortlisted for the Booker Prize, but it was *Offshore* which won the award in 1979. The judges' decision was controversial, since the novel was considered 'slight' by many commentators, particularly when set beside other shortlisted titles such as V.S. Naipaul's **A Bend in the River** and Fay Weldon's **Praxis**.

1979

Thomas FLANAGAN   1923–

The Year of the French

For the people of County Mayo, 1798 is the year of the French. That summer three warships anchor in Killala Bay and from them a small force under the command of Gen. Jean-Joseph Humbert disembark. Despite the example of cruel repressions in Antrim and Wexford earlier in the year it is the signal for the county to rise – for the 'Whiteboys of Killala' under the ferocious Malachi Duggan to burn the houses and maim the cattle of the landlords; for men, armed with pikes, to leave their cabins and answer the call of the United Irishmen, a patriotic society brought into being by Wolfe Tone and others of the legal fraternity of Dublin. It would have been over before it had started without Humbert's brilliant victory at Castlebar, but the second French fleet, bringing reinforcements, is caught by the British blockade and Humbert's campaign, using the Irish as cannon fodder, ends in a tragic rout. Grotesquely the scene of this massacre is a dismal hamlet in the Midlands called Ballinamuck, 'the place of the pig'; we are reminded that it is the best harvest for ten – some say twenty – years, and that the men who should be gathering it in are either lying dead on the battlefield or swinging from gibbets.

Such is the historical framework of this epic novel which, examining with equal compassion the layers of Irish society and its uncomprehending visitors through a fixture of historical and fictional characters, explores the eternal Irish predicament. At the centre is Owen MacCarthy, the drunken, licentious poet and hedge schoolmaster from County Kerry, caught up almost casually in events that he neither understands nor cares about; a man who is perfectly at home in a tavern, he is the link with the Irish peasantry. His death by hanging as a 'rebel' is the dramatic climax of the book. Other principal characters are George Moore, the reclusive aristocratic 'papist', friend of Burke, Fox and Sheridan, historian of the revolution in France, who watches impotently the disgrace and death in prison of his beloved younger brother, John; Dennis Browne, MP for County Mayo – his father defected to the Protestant persuasion – who punishes the 'rebels' with the full rigour of the law and is one of architects of Ireland's union with England; Arthur Broome, the English clergyman in Killala, whose 'impartial narrative' displays to the end a puzzled view of Ireland and its history; Malcolm Elliott, solicitor, landowner and member of the Society of United Irishmen, who goes to his death a disillusioned man; and Lord Cornwallis, his fame tarnished in America, who conducts the campaign with suave efficiency, displaying in the end considerable political skill without, like other Englishmen, understanding the issues involved.

All four of Thomas Flanagan's grandparents emigrated to America from County Fermanagh in the nineteenth century. Both his grandfathers were members of the Fenian Brotherhood, and his early knowledge of Ireland derives from his grandmother, Ellen Treacy Bonner, 'a witty, sardonic and large-hearted woman to whom I was much attached'. He brings to this complex novel, which was ten years in the making and is projected as the first of a trilogy covering the key moments in Irish history, a wealth of scholarship and humanity. It has many of the characteristics of the traditional epic, including the subtle use of repetition; his handling of a large cast of characters, including the various narrators of sections of the story, is masterly, as are his variations of style. It is a novel that deserves a wider readership outside the Irish on both sides of the Atlantic than it has so far received.

William GOLDING   1911–1993

Darkness Visible

In *Paradise Lost*, Milton described the gloom in which the fallen angels dwelt as 'darkness visible'. When Golding borrowed the paradox for the title of this novel, he provided a clue to the tone of his thinking – for he is concerned with the shadows and sparks of the divine as they manifest themselves in human nature.

The central character is Matty, a tiny child who staggers from the heart of a blitz fire. His survival is miraculous, his parents untraceable. One side of his face is disfigured and he is barely able to talk. He may be close to imbecilic; all the same, he memorises large sections of the Old Testament. He passes through a charity institution, where he fixes his loyalty on Sebastian Pedigree – a pederastic master. After the death of his favourite pupil, Pedigree is arrested and imprisoned. Believing himself to blame for this, Matty progresses through a number of jobs, sails to Australia, endures a kind of crucifixion and returns to England. Angelic figures start to appear to him. Convinced that he has a purpose, but unsure yet what it is, he finds employment as a gardener at an expensive public school.

Nearby, the Stanhope twins are growing up. Both are intelligent, wasteful and detached. Antonia joins a terrorist gang and Sophie drifts into sadism and crime. With her ex-army boyfriend, Sophie plans to kidnap a prince from Matty's school; Pedigree, released from gaol, tries to seduce small boys in the local park; and two elderly men, ill at ease in a world they no longer understand, seek spiritual enlightenment together. Matty practises a mute evangelism, consults the good spirits and fends off the bad. The kidnapping precipitates a crisis and the strands of narrative converge. Salvation, humiliation and damnation are distributed – perhaps inequitably – to Golding's extraordinary creations.

Light and darkness are the dominant images of the story. Matty dresses in black; he has his visions by night; his perceptions are dim. Against the dark, his flashes of illumination are the more clear. The twins, by contrast, are brilliant – and the effect of their brilliance is to dazzle, not to enlighten: they are Luciferian. Matty 'is despised and rejected of men, a man of sorrows, and acquainted with grief'. His obscure innocence, however, attains a qualified triumph while one half of the twins is defeated. The book ends enigmatically: one of its many achievements is that the reader never knows whether Matty is a modern-day Christ or merely a simpleton deluded.

The most overtly theological of Golding's novels, *Darkness Visible* none the less articulates its ideas within a contemporary framework. Satanic hubris assumes the shape of extremism; immigration makes the suburbs of London strange; the babble of mass communications is a counterpoint to Matty's almost speechless messages. Dialogue is often fragmented – as though to mark a dying attention span. But, if Golding records a depressing incoherence in his subjects – a sense that 'things fall apart' – the same incoherence is the source of much humour.

Elizabeth HARDWICK   1916–

Sleepless Nights

Hardwick's third novel, *Sleepless Nights*, is a fragmentary meditation on her own life, from her childhood in Lexington, Kentucky, to her later years in New York City. The tone is detached and aloof and she is more inclined to intellectual reference than to emotional revelation. Few of the people central to her life – even her husband of twenty years, the poet Robert Lowell, who is referred to throughout simply as 'He' – emerge as more than shadowy figures. 'Store clerks and waitresses are the heroines of my memories,' Hardwick once said with reference to the novel, a statement borne out by the emphasis upon seemingly peripheral characters.

From her childhood years she nostalgically remembers a man who drank himself to death, a prostitute of her acquaintance, and the Kentucky Derby: 'A *tristesse* falls upon the scene, down on the old memory. The horses are led away to their rest, their feelings about the race they have run unknown to us.' From her years as a postgraduate student at Columbia in the 1940s she recalls the sleazy Hotel Schluyer where she lived with a homosexual friend amidst scenes of domestic histrionics. She reflects on the effect of seeing Billie Holiday in a night-club, and then remarks: 'I have left out my abortion, left out running from the pale, frightened doctors and their sallow, furious wives in the grimy, curtained offices on West End Avenue.'

Her travels in Europe with Lowell in the early years of their marriage are described in a similarly evasive manner, with only fleeting glimpses of the marriage – 'How we fight after too much gin' – permitted to creep through. After their separation, 'alone here in New York, no longer a *we*', it is the lives of cleaning ladies and maids she dwells upon. She writes at the end of the novel: 'Sometimes I resent the glossary, the concordance of truth, many have about my real life, have like an extra pair of spectacles. I mean that such fact is to me a hindrance to memory.'

Penelope LIVELY   **1933–**

Treasures of Time

This, her second adult novel, won Lively the first National Book Award. Under its urbanity and acerbic social comedy lie her continuing preoccupations with the nature and veracity of memory and history, and the interaction of past and present. Tom Rider, a research student, is in love with Kate, daughter of Hugh Paxton, a flamboyant archaeologist now dead. Their relationship shows its first strains during a visit to her Wiltshire home, near the barrow whose excavation made Paxton's name. Kate, prickly and clever, is rendered even more uneasy by her beautiful mother Laura, shallowly self-absorbed and with 'the knack of putting everyone else at a disadvantage', but loves her astute, stroke-disabled aunt, Nellie. Tom, cheerful and opinionated, is regarded with disquiet by the snobbish Laura. But a young television director, Tony Greenway, has flattered her into co-operating in a programme about her husband, a project viewed with suspicion by Kate and Nellie. The sensibilities and memories of each main character provide the accumulating texture and background of the novel, assembling into a mosaic of love and betrayal: Nellie, herself an archaeologist, had loved Paxton before his defection to her pretty sister. Now Laura's patronising obstructiveness is hindering Nellie's preparation of Paxton's papers. Kate recalls her mother's coldness, and remembers witnessing her adultery as well as the desert of the marriage.

During preparations for the programme Tom gets friendly with the rootless, winning Tony. Kate, jealous and insecure, is persistently gruff about Tom's other friendships. When mishaps and misunderstandings separate them during an expedition Tom spends the night with an accommodating former girlfriend and, until the tremulous reconciliation, takes refuge in Tony's flat.

The novel culminates in the filming. To Tony's well-concealed irritation Laura turns the whole thing into a party. Kate arrives late and Tom gets royally drunk with Laura. Next morning, appalled, he fears he may have gone to bed with her. There is an eventful session at the site, followed by a party at the house. As Tom and Kate agree to part, as an interview proceeds indoors and a croquet game starts, the buried past disrupts the present. Laura loses her brittle composure. Nellie dies. The novel ends with Tom relinquishing his academic ambitions and taking a television job.

Lively's sense of landscape and inheritance root the novel in a carefully unsentimental Englishness, sparked with scenes such as Tom's inadvertent bear-leading of a party of tourists, and his friend Martin's contentment as a skilled craftsman. The skill with which Lively elicits sympathy for the difficult Kate is a reminder that her high reputation as a children's writer stems from sympathy and solid understanding for the young. Tom observes that Kate 'is dedicated to the belief that you can put the past safely away in a glass case ... And all the time she wears her own like an albatross.'

Norman MAILER   **1923–**

The Executioner's Song

Gary Gilmore was paroled in 1976 after spending twenty-two of his thirty-five years in gaol. Shortly after his release he shot and killed two men, a garage attendant and a motel manager, and was arrested. He was tried, found guilty and demanded the death penalty for himself. There followed one of the largest media circuses of the 1970s in

America with magazine journalists, television interviewers and writers jostling frantically for details of his life and thoughts on his impending, self-chosen death. Gilmore distributed the rights to his life and death right up until the moment when, nine months after his re-arrest, he was executed by firing squad.

The Executioner's Song attempts to tell this story with as much objectivity and thoroughness as possible. The book is over 1,000 pages long, but even this represents only a fraction of the interviews which Mailer both collected and conducted and which he estimated at about 15,000 pages. Mailer was contracted to write the book and took the assignment as a break from the even longer novel which eventually appeared as *Ancient Evenings* in 1983. However, once embarked on the project Mailer's commitment was wholehearted and the book went on to win the Pulitzer Prize in 1980.

The Executioner's Song is made up of thousands of short paragraphs through which the hundreds of characters involved in the case are presented *in propria persona*. Although he had experimented with the technique in an earlier book on Marilyn Monroe, this was the first large-scale, artistically ambitious work in which Mailer had subdued his own personality to the point of invisibility. The voices of the characters (several dozen occur with some frequency) convince with varying degrees of success but the story emerges with vivid authenticity from their conflicting accounts and perspectives.

Owing to the impersonality of its format, it is difficult to make sweeping claims for the book: there is no symbolism, no controlling structure through which authorial judgements can be passed down. Although Gilmore is the centre of the book, Mailer records reactions from all sections of American life, a life which is called into question by Gilmore's simultaneous extremity in relation to it (the murders were pointless and barbaric) and his centrality (for it is he, albeit as victim, who orchestrates the judges, lawyers, reporters, his family, friends and enemies). Mailer sees in Gilmore's case something quintessentially American, but his abstinence from direct comment means that this metaphorical dimension in the novel is never explicitly developed.

It is ironic that in this book, of all his books the one most concerned with 'judgement', one of the most opinionated novelists in America is left without an opinion; but Mailer's allegiance is to the facts. He has called *The Executioner's Song* 'a true-life novel' and this, coupled with its theme of violent killing, has inevitably led to comparisons with Truman Capote's ***In Cold Blood*** (1966). Mailer's book is less polished, more honest, than Capote's and it has the major difference of not betraying its subject. Gilmore emerges from its pages as a fully rounded character, but ultimately an inexplicable one. Mailer does not understand him, neither do Mailer's readers, nor the many people who knew him at first- or second-hand. It was an uncomprehending society that produced Gilmore and ultimately executed him. Mailer's testament is, in the last analysis, to the failures of communication within American society and the violence that ensues from them.

Bernard MALAMUD   1914–1986

Dubin's Lives

William B. Dubin, a middle-aged biographer, author of *Short Lives, Abraham Lincoln, Mark Twain* and *H.D. Thoreau*, is an inward-looking man, happier with the dead than the living. He has tried to enter the minds of his subjects knowing that 'there is no life that can be recaptured wholly; as it was. Which is to say that all biography is ultimately fiction.' Now he has embarked on D.H. Lawrence – 'a complex type with tormented inner life, that's who you felt you had to get involved with', Dubin tells himself. Given his own unsettled inwardness it may not have been a wise choice.

His marriage is sterile. A manifestation of his wife Kitty's insecurity is that she is neurotically obsessed with sniffing the gas burners to make sure that none is escaping; she compares Dubin often unfavourably with her first husband Nathaniel. He took her on when she was widowed with a small son, Gerald, whom Dubin loved. Now a deserter from the army hiding out in Stockholm, Gerald is lost to them. So, very nearly, is their daughter Maud – away at college, she seldom calls or comes home. Alone in a house that seems 'almost empty', Dubin and Kitty live increasingly apart.

Dubin feels guilty about her and is disturbed by Fanny Brick, a student – at twenty-two only a couple of years older than Maud – whom Kitty takes on briefly to do the cleaning. 'Fanny wore a faded denim wraparound skirt and black shirt without bra. Her abundant body, though not voluptuous, clearly had a life of its own.' She tells him, 'I think we're entitled to have sexual pleasure anywhere we want.' To her he says: 'what stays with me most, is that life is forever fleeting, our fates juggled heartbreakingly by events we can't foresee or control and we are always pitifully vulnerable to what happens next'.

Fanny certainly makes Dubin feel vulnerable when they go together to Venice and she winds up in bed with a gondolier. 'He needed me more than you ever did ... All you wanted was cunt,' she tells him. Back home, obsessed by his failure, he passes a bleak winter, making little progress with Lawrence, and Kitty worries about him. But in the spring, hope returns and so does a maturer Fanny. This time they make love idyllically, but when it comes to committing himself he cannot do it. He prefers to stay with Kitty even though he becomes impotent with her. In short, he is not to be depended on by either woman. Nor can he help the children: Maud becomes pregnant by a married man a few years older than himself, while Gerald, recruited by the KGB and now their victim, will die in the Soviet Union. Dubin and Kitty are stuck in an empty house in an empty relationship. She too has had affairs though she is hardly committed to them. In her diary she equates her husband with Carlyle: 'a narcissistic, nervous *obsessed, impotent biographer*'. They talk of divorce but nothing comes of it. When Fanny, having come to the conclusion that a career is preferable to relying on men, suggests that Dubin ought to stay three days with her and four with his wife, he does not think Kitty would agree. Perhaps if he has the courage to put it to her she might see it as the only option left.

Dubin's Lives may not have quite the hypnotic intensity of *The Assistant* (1957), but it is a psychological novel of extraordinary depth and complexity. If it meanders a little and sometimes turns back on itself, Malamud has nevertheless succeeded triumphantly in entering the mind of Dubin the biographer, living the lives of others, afraid to live his own. He emerges as a pathetic, lovable, sometimes comic, but finally dangerous figure, tragically aware of shattering the lives of those who are close to him, and helpless to prevent it. He sums up his problem with the remark, 'It isn't easy to give if you're anchored in involved subjectivity.' Poor Kitty tells him: 'If you could fuck your books you'd have it made.'

V.S. NAIPAUL   1932–

A Bend in the River

This novel is named after a bend in a fictional African river which has been the site of many civilisations: it had been a 'forest, a meeting place, an Arab settlement, a European outpost, a European suburb, a ruin like the ruin of a dead civilization, the glittering Domain of new Africa, and now this'. The words are those of the narrator, Salim, and 'this' is the disintegration that seems to be coming upon the now independent country, in spite of (or perhaps because of) the presidency of the 'Big Man' – a thinly disguised portrait of President Mobutu of the Congo. Conrad's *Heart of Darkness* (1902) was, of course, written about this place. Naipaul revisits the territory, with not very different assumptions or conclusions, though with a different style and feel.

Salim chronicles the development and break-up of his affair with Yvette, political events at this time of turmoil, his relationship as an outsider to his post-colonial society, his own identity as an Indian from a Muslim trading family who finds himself in an alien racial and geographical location inhabited by (a very different variety of Muslim) Arabs. He finds that his family's history is more or less non-existent, and that history (in which Naipaul places great faith) does not explain the place of his people. Although Salim seems to feel the lack of family and racial history, he makes no attempt to record his own history; he records only his present at the bend in the river, during which he draws what seem to be his final conclusions about the meaning of life. Though Salim's conclusions are not fundamentally dissimilar from the Buddha's, Salim's great illumination comes at the violent end of his affair with Yvette. The earlier growth of his dependence on Yvette parallels the growth of the nation's dependence on the Big Man, and the growth of Salim's rage, despair and violence parallels the growing chaos and violence of society and politics.

Later, I woke to the solitude of my bedroom, in the unfriendly world. I felt all the child's heartache at being in a strange place. Through the white-painted window I saw the trees outside – not their shadows, but the suggestion of their forms. I was homesick, had been homesick for months. But home was hardly a place I could return to. Home was something in my head. It was something I had lost. And in that I was like the ragged Africans who were so abject in the town we serviced.

A BEND IN THE RIVER
BY V.S. NAIPAUL

Salim's exploration of the direction in which his post-colonial world seems to be heading, and its implications for the ordinary individual, ends with Salim giving up trying to make sense of life; he stops trying to find fulfilling relationships, abandons his quest for his past and for a home. He finds himself certain of nothing but his physical existence. That is the penultimate step in a journey which must end with even that conclusion being thrown into doubt, if Salim continues thinking on the basis of his present assumptions. Moreover, such a conclusion, temporary as it may be, leaves one feeling terrifyingly vulnerable, especially in such a society as the one in which Salim finds himself. No wonder Salim finds himself concentrating simply on survival.

As usual, Naipaul is expressing his own view through his principal protagonist. Naipaul foresees the collapse of civilised society. The novel closes with an image that suggests animal activity continuing in darkness – but that is merely an existential assertion of the continuance of life at an animal level.

Naipaul seems to have moved away from his earlier belief that indigenisation provides for personal and communal integrity, and arrived at the bleaker conclusion that accepting a fragmented personality and abandoning the idea of home may be a means of survival and perhaps even some creativity – though what value such survival and creativity have is not clear, especially in view of his earlier opinions on these subjects.

Naipaul had earlier hoped that history would provide comfort by ordering chaos and explaining the present through investigation of the past. It now seems that the past provides no comfort; in fact, it appears unlikely ever to be fully known. Not only does history lack a consistent or even discernible meaning, truth may be just another form of fantasy – history and fiction appear to be often indistinguishable. People construct their own truth: the events of the past can be interpreted in various ways, depending on the perspective of the person doing the interpreting. Lacking meaning or stability, individuals do not even have much control over what is to happen to them.

What is most notable about the novel is the contradiction between Naipaul's attitude of despair and the quality of the creative act through which that despair is ordered and presented. The apparent optimism that remains in the novel, in other words, is an optimism of manner, not an optimism of substance. However, the symbols, philosophy and structure of the work cohere to produce a marvellous literary work from the pen of a man who despairs of contemporary human civilisation. It was the second of Naipaul's novels to be shortlisted for the Booker Prize.

Philip ROTH   1933–

The Ghost Writer

The Ghost Writer sees the first appearance of Nathan Zuckerman, the partially autobiographical narrator who is to haunt a whole series of Roth's novels. After the publication of *Goodbye, Columbus* (1959) and ***Portnoy's Complaint*** (1969), Roth was denounced by large sections of the Jewish community: *The Ghost Writer* is a fictional answer to these accusations.

Nathan Zuckerman is a young Jewish writer rejected by his family for his unflattering portrayal of their values. Bearing the burden of guilt towards his father, he finds an alternative patriarch in the figure of E.I. Lonoff, an established writer whom Zuckerman idolises. Visiting the great man up in the Berkshires of mid-America, Zuckerman finds his home an enigmatic habitat.

Lonoff's wife, Hope, seems the perfect companion, but her quiet manner dissolves into hysterical plate-smashing. The cause of her outburst is Amy Bellette, a puzzling young woman working on Lonoff's archive material. Amy is a European refugee from the Holocaust, the symbol of Jewish suffering. But her relationship with Lonoff is ambiguous – daughter, granddaughter or mistress – and Lonoff's affectionate behaviour does nothing to clarify matters.

Inhibited by his surrogate-son deference, Zuckerman waits for the evidence to emerge. In the early hours of the morning he is unable to sleep for pornographic fantasies of Amy, and, overhearing her and Lonoff talking, he strains to hear the creak of bedsprings. What he discovers is more astonishing than a love-affair: Amy's true identity is Anne Frank.

Imagination buoys him through the night as he fantasises about returning to his family with Anne Frank as his bride, the incontrovertible proof of his Jewish allegiance. In the morning, however, Amy's identity seems again uncertain, as the vision of the night is unsubstantiated. Hope leaves the house for a life of her own, and Lonoff's love for her takes precedence over his fictions as he sets off to find her. Zuckerman takes his place at Lonoff's typewriter, his writing consciousness liberated by the surreal experience.

Bernice RUBENS   1928–

Spring Sonata

While conducting a post-mortem examination, Dr Brown retrieves a notebook from the womb of a vastly pregnant woman, from which he reconstructs the fantastical story of Buster, a violin-playing foetal phenomenon. From the moment of his conception, Buster is a sentient being who eavesdrops upon the world into which he is destined to be born. In a previous life he had been one of the world's greatest child prodigies: 'Then on his way to a concert, he'd loosened his mother's hand, or it might have been the other way round, depending on whose synchronicity one has in mind. Whatever, a taxi had run over him violin and all.' The consequences of parents and children metaphorically holding on to or letting go of each other's hands lies at the heart of *Spring Sonata*.

When Sheila Rosen, a talented pianist, discovers she is pregnant, she is delighted, for she and her ineffectual husband, Bernard, have been trying to have a child for some time. 'So now I hope you'll give up smoking,' is her mother's predictable response to the news. The Jewish Mother to end all Jewish Mothers, Mrs Joseph is an amalgam of fierce pride, unassuageable guilt, lacerating self-pity and 'undeclarable love'; she can 'open fire even before the battle-field was agreed on'. Almost immediately Sheila, Bernard, Mrs Joseph and her mother (the unwilling occupant of an old people's home) quarrel about the baby's name, and begin to plan its future, burdening it with their own ambitions. Buster listens appalled: 'he ... winced at their crippling presumptions, their offensive trust, hoping all the time that between their outrageous syllables he could sniff a semblance of love. But there was nothing but their unbearable expectation. If he could fulfil that, then and only then, would he be entitled to loving.' The only member of the family who appeals to him is Sheila's brother, Robert, who ran away from its tensions to join an ashram. Unfortunately, Robert is arrested and imprisoned for pushing drugs and so will not be around when Buster is due to be born. When the day arrives, Buster decides to stay put, dodging the surgeon's hands as they grope towards him during caesarian section. Before his mother is stitched up again, Buster leans out to steal the violin and bow that Sheila's musical partner, Clarissa, was to have used to play him into the world. He also grabs a notebook and pencil in order to write his autobiography.

In spite of her doctor's insistence that she has had a phantom pregnancy, Sheila continues to believe in the existence of her child, and indeed she continues to swell until she is so vast that she is confined to a wheelchair. She refuses to co-operate with psychiatrists, and family relations deteriorate still further, finally exposed before a gathering of 'students, social workers, anthropologists, and even the odd priest' at a family therapy clinic. Buster eventually decides to reward his mother's faith by accompanying her on his violin as she plays the piano. Bernard immediately sees this miracle as a commercial prospect: 'What is the point in having a genius on your hands, if you can't show him off?' 'And that, ladies and gentlemen,' Buster declares, 'is the nitty gritty.' Bernard tricks his reticent, unborn son into playing in front of an audience, but when Buster discovers this deception, he uses the violin bow to saw through the umbilical cord, killing both himself and his mother.

Buster was woken by a prodding on the uterus wall. He yawned and looked about him. He felt refreshed as if he'd slept for a long while. The tapping continued and he wriggled a little in token of his presence. He heard a voice. 'Just lie back,' it was saying, 'and relax.' Buster was delighted. He was about to be acknowledged.

SPRING SONATA
BY BERNICE RUBENS

Dr Brown describes Beethoven's Spring Sonata for violin and piano as 'a passionate dialogue between the two instruments, at times echoing and confirming, at others, raging in passionate argument, and every bar an overt declaration of love'. Rubens's funny and affecting novel takes the basic Jewish joke about families and strips away the comedy to reveal the genuine but unadmitted love that lies beneath, a source of anguish as well as humour. Buster's conception takes place on Yom Kippur, the Jewish Day of Atonement, 'a day of a self-induced orgy of suffering, a repenting of one's past sins, a beating of breasts', and his putative family is crippled with (largely unnecessary) feelings of guilt, stubbornness, pride, resentment and inadequacy, 'the sterile accumulation of suffering'. It is, the doctor reflects, a 'strange and desolate story, and let it be a lesson to us all'.

William STYRON · 1925–

Sophie's Choice

Styron's sixth novel is based on the story of the relationship between Stingo and Sophie. Stingo is an innocent young man who has moved from the South to New York where he takes up residence in Brooklyn and begins to write his first novel. Sophie is a Jewish survivor of the concentration camps who has come to America at the end of the war. She lives in the apartment above Stingo's with her lover, Nathan. The stage is thus set for a meeting between American innocence and the darkest kind of European experience. Stingo's rite of passage is negotiated through his growing involvement with Nathan and Sophie and his deepening understanding of their damaged personalities. Nathan is a charming and intelligent but mercurial character, who is first presented as an experimental scientist. Later, Stingo learns that he is schizophrenic, prone to bouts of psychotic, jealous rage and obsessed with Sophie's past as a victim of the Holocaust.

Sophie is attractive and quite unlike any other woman Stingo has met. He falls in love with her without any real understanding of her past. In a revealing incident, Stingo walks into her room to find her face suddenly grotesquely misshapen: she lost all her teeth in the camp. When she replaces her dentures, Stingo is relieved; he cannot face this side of her experience directly. Nathan and Sophie's relationship grows more troubled, Nathan's rages more violent. Stingo eventually persuades Sophie to come away with him, but after a few days she leaves him to return to Nathan.

This is the main plot, and from it it would seem that Sophie's choice is between the two men in her life (a red herring exploited more by Alan J. Pakula's 1982 film adaptation than by the novel). However, in a very long flashback, Sophie recalls her life in Poland, her marriage and her children. When she is sent to the camp, her two children are sent with her. Once there, a sadistic officer tells her that she may keep only one and that she must choose between them. Sophie pleads for both, but in the end she chooses. Both are taken and she never sees them again. This is the experience that haunts her; her complicity in the murder of her children, her part in the process of the Holocaust.

When she leaves Stingo to return to Nathan she gives up on Stingo's hopes of a 'normal' life. Only Nathan, it seems, being similarly damaged, can help her. When Stingo wakes and finds her gone, he returns to the New York apartment. Nathan and Sophie have found some sort of peace together. They have committed suicide. This is Sophie's final choice. By the end of the novel, Stingo has thus been initiated into Sophie's experience, albeit at a considerable remove, and it is suggested that, in contradiction to his two friends, he has chosen life.

Typical reviews of the novel were headlined 'A Novel of Evil', 'Riddle of a Violent Century' and 'The Holocaust According to Styron'. Although Styron's research into the Holocaust was meticulous, and although Stingo's first-person narrative overtly centres on his development as a person, many critics took issue with Styron's inclusion of the death camps. Their objections raise the question of how (if at all) this darkest of historical episodes can be fitted into American experience. It would be unfair to say that Sophie's experience is trivialised as a type of obstacle race for Stingo to negotiate on his way to manhood, but it *is* put at the service of the novel to a certain extent. The mismatching of Stingo's personal experience against the enormity of the Holocaust is undoubtedly a deep, structural flaw in the work. However, placed as an American writer, Styron probably could not avoid this difficulty. *Sophie's Choice* is one of only a very few American novels to attempt to deal with the horror of the Holocaust and its depiction of how that event impinged on an ordinary American in the 1950s is, for better or worse, essentially accurate. *Sophie's Choice* is ultimately about the attempt to understand, and if Styron's greater goals are not fully achieved, then this has its own pathos.

Emma TENNANT   1937–

Wild Nights

'When the night came, the fissure of evil opened.' Through the eyes of her young narrator, Tennant weaves a fantastical world of childhood in a lonely Scottish valley, exploring 'the arena of memory and magic'. The arrival of the aunts has the inevitability of the seasons, as the mischievous Aunt Zita vies with the mild but moral Aunt Thelma.

'When my Aunt Zita came, there were changes everywhere.' The child observes the pressure of Zita's visit on the family, as the mother becomes neuralgic and the father shifts uncomfortably with the memory of his sensuous affection for his sister. To the villagers, Zita becomes a folklore scapegoat for the evils of world and war, burned in effigy at an annual ritual to exorcise the powers of darkness. The spirits of the dead whisper as Zita's elder brother, his suicide a local scandal, returns to make love to her on the carpet of their nursery. Aunt Thelma, supported by her shrine of the Madonna, looks on, powerless, with her less primitive ambitions.

Zita's arrival awakens the elements of nature, as 'the days outside, which were long and white at that time of year, closed and turned like a shutter, a sharp blue night coming on sudden and unexpected as a finger caught in a hinge'. Decked out in the fineries of her mother's dressing-table, the child rides with her aunt on the back of the wild north wind, twisting the worldly scandals of incest and suicide into the private mythology of witches and ghosts. A summer in the South, in the sun-filled house of Uncle Rainbow, seems to offer a vision of a brighter future; but the father, turning away from the forces of the light, chooses to return to the unkissed life of his gloomy ancestral home. 'We drove to the station, and waited for the train that would take us to the dark again ... It was possible that one year, at the ending of the world, winter would conquer spring.'

On the days after our evening flights it was my mother who looked pale and haggard, as if she had poisoned herself with thoughts about us. In the middle of her cheeks were pin-points of scarlet and a dank breath came out of her. She was as viciously and unexpectedly coloured as the red and white mushrooms she gathered under the birches.

WILD NIGHTS
BY EMMA TENNANT

In spite of its atmosphere of Gothic fantasy, *Wild Nights* draws upon real characters and incidents from Tennant's childhood during the Second World War. It was inspired by memories of a visit of the author's uncle, Stephen Tennant, to Glen, the family's vast, remote, mid-Victorian castle in Peebleshire. Uncle Stephen, a notorious aesthete who is also portrayed by Nancy Mitford in *Love in a Cold Climate* (1949), arrived in a large car filled with bird cages, just as Uncle Wilhelmina does in the novel. The anecdote about Wilhelmina painting his face and going out soliciting is based upon an incident in Stephen Tennant's chequered career as a professional beauty. The character of Uncle Rainbow is also partly drawn from the author's flamboyant relative, while Aunt Zita is based upon her Aunt Clare.

1980

Independence of Zimbabwe • Start of Iran–Iraq War • Emergence of 'Solidarity' in Poland • Archbishop Oscar Romero shot dead while celebrating mass in San Salvador • US commando mission to rescue hostages in Iran fails • Launch of the Sony Walkman • The 'Brandt Report', *North–South: A Programme for Survival*, calls for radical change in relations between rich and poor countries • Britain becomes a net exporter of oil • A gene is transferred from one mouse to another • Howard Brenton, *The Romans in Britain* • David Edgar (adaptor), *Nicholas Nickleby* • Mark Medoff, *Children of a Lesser God* • Films: *My Brilliant Career, Raging Bull* • Nigel Short, age 14, becomes youngest International Master in the history of chess • Cecil Beaton, Graham Sutherland, Oskar Kokoschka, Erich Fromm, Roland Barthes, Jesse Owens, Jean-Paul Sartre, Alfred Hitchcock, Tito, Henry Miller, C.P. Snow, Peter Sellers, Jean Piaget, Mae West, Oswald Mosley and John Lennon die • Gavin Ewart, *The Collected Ewart 1933–1980*

Paul BAILEY   1937–

Old Soldiers

This, Bailey's fifth novel, is an elliptical, moving and bleakly funny novel about two old men adrift in a vividly realised London, each burdened with memories of fathers who were failures, and of their experiences in the First World War. Victor Harker, recently

widowed and fleeing the scenes of past happiness, returns to the city of his birth. In St Paul's Cathedral he meets a man who introduces himself as Capt. Hal Standish. Despite the fact that Victor, a retired bank manager, recognises Hal as the sort of man to whom he would have refused an overdraft, the two drift into an uneasy friendship.

In fact, 'Standish' is really Eric Talbot, who served as a private in the war, became unhinged after deserting and entered an asylum. He subsequently created three seperate personae, for each of which he possesses a change of clothes and dentures. 'Hal Standish' is a bawdy and bigoted military type, obsessed by bodily functions and the number of Arabs in London. Devoted to the ladies and his 'joystick', he acts as an unreliable guide for gullible tourists at St Paul's. 'Tommy' is 'a gentleman of the road', who spends his days walking the streets, and his nights at a Salvation Army hostel run by Sgt Marybeth Myslawchuk, an American drawn to England by her love of Shakespeare, Blake and Wordsworth. The third of Talbot's personae is 'Julian Borrow', a raffish (and unpublished) old poet who lives in a room in Islington papered with rejection slips. He dons a purple fedora in order to address the crowds at Speakers' Corner on the subject of poetry. Talbot is haunted by a fear of death and after a particularly energetic encounter with a prostitute (in the character of 'Standish'), he collapses in the street and awakens in hospital, unsure of which of his characters he is.

He is visited by Victor, to whom he confesses his cowardice in the war, his spell in the asylum and his triple identity: 'The three of them kept boredom and Eric Talbot at bay for many, many years.' In possession once more of his true identity, he escapes from hospital and throws himself into the Thames. As he plunges to his death, he thinks of the precariously balanced statue of the Virgin on top of the cathedral at Albert, which soldiers believed would finally topple to the ground when the war ended: 'He saw the Golden Virgin break loose and knew that the war was over.' Similarly at peace with his past, Victor returns to his house in Newcastle, recalling his beloved wife and scenes from the war in which his best friend was killed, and shortly dies.

Just over 100 pages long, this is a novel at once economical and rich in character and atmosphere. The persistence of memory experienced by many veterans of the First World War is skilfully conveyed by sherds of recalled experience constantly rising to the surface of the characters' minds, while Talbot's fragmented character neatly suggests the trauma, loss of individuality and sense of dissolution inflicted upon a generation.

Anthony BURGESS   **1917–1993**

Earthly Powers

Narrated in the first person by a successful octogenarian homosexual writer, Kenneth Toomey, *Earthly Powers* is Burgess's most ambitious novel. Rumbustious, humorous and international, it provides a morally serious panorama of the twentieth century. It was tipped to win the Booker Prize, but the award eventually went to William Golding's **Rites of Passage** (1980), a decision which Burgess publicly and unsportingly deplored.

The book is highly self-conscious from a literary point of view, for example raising questions about 'the capacity of literature to cope with human reality'. Toomey (who is partly drawn from Somerset Maugham) discusses such literary issues with his friends James Joyce, Wyndham Lewis, Ford Madox Ford and Rudyard Kipling. As Toomey himself points out to us, the novel has one of the most arresting open sentences in fiction: 'It was the afternoon of my eighty-first birthday, and I was in bed with my catamite when Ali announced that the archbishop had come to see me.' The itinerant, eventful existence forced on Toomey by his homosexuality leads to his acquaintance with the key people and events of the age, such as 'Hitler … Mussolini and the rest of the terrible people this terrible century's thrown up'. His recollections stretch back over six decades to the First World War, and bring together the 'fragments of an individual vision' into a particular view of the twentieth century.

The moral centre of the novel is Toomey's lifelong acquaintance with Cardinal Carlo Campanati, who eventually becomes Pope Gregory. A liberal churchman who seeks to 'transform Christianity' with a vision of 'divine good to oppose to the growth of evil in our time', Campanati is apparently one of the strongest opponents of 'the terrible people'. But Toomey mocks his belief that man is inherently good, corrupted only by the devil and his agents such as Hitler. When he visits the concentration camps, shortly after they are liberated in 1945, Toomey says: 'I wanted to have Carlo with me there to smell the ripe gorgonzola of innate human evil.' Toomey therefore feels that Campanati, far from being the saint he is often supposed, is actually an inadvertent agent of the devil. This view is partly supported by events in the story, though Toomey's view of man as the meaningless offspring of the 'primordial dungheap' is also shown to be false – not surprising, because Burgess saw the world as a 'duoverse' locked in a universal struggle between good and evil.

William GOLDING   1911–1993

Rites of Passage

see **To the Ends of the Earth** (1989)

Russell HOBAN   1925–

Riddley Walker

Set in Kent in a distant future long after nuclear conflagration, and written in a rich invented dialect, Riddley's narrative describes the weeks after his twelfth birthday, or 'naming day'. His father dies, crushed under a piece of ancient machinery which is being excavated at Widders Dump, and Riddley is set to take over his father's role as 'connexion man' for the tribe, a job which requires him to cast light on performances of the Eusa Story, a theological puppet show performed by the Wes Mincer, Erny Orfing, and the Pry Mincer, Abel Goodparley.

The Eusa Story recounts events in the distant, dimly understood past when Eusa learnt the secret of the '1 Big 1' from 'the Littl Shynin Man the Addam': 'Eusa put the 1 Big 1 in barms then him & Mr Clevver droppit so much barms thay kilt as menne uv thear oan as thay kilt enemes. Thay wun the Warr but the lan wuz poyzen frum it the ayr & water as wel.'

When Riddley meets the Ardship of Cambrey, one of the mutant Eusa Folk who are believed to know the secret numbers of the 1 Big 1 – although they are unable to use them – he learns of Goodparley's efforts to exhume these secrets. For Goodparley and others this knowledge bears the promise of unrivalled power, but it is also intimately linked with the philosophical endeavours to understand the central mysteries of life: 'That's the woal idear of this writing which I begun wylst thinking on what the idear of us myt be.'

Riddley discovers an ancient document entitled 'The Legend of St Eustace', which is in fact a twentieth-century catalogue describing a series of paintings at Canterbury. From the words 'on a cross of radiant light, the figure of the crucified Saviour' he is sure he has deduced the 'the number of the salt de vydit in 2 parts in the cruciboal and radiating lite coming acrost on it' – the formula of the 1 Big 1. He gets hold of a consignment of 'yellerboy stoan' which is believed to be the missing ingredient, and with Goodparley and an old man called Granser mixes up the chemicals. The predictable explosion of gunpowder kills both his partners.

Riddley Walker describes a grim and bleak world, one where lives are short and frequently brutish (especially if you have a run-in with a 'hevvy' such as Fister Crunchman), and where packs of wild dogs roam a landscape of dead towns and post-nuclear destruction. It is a compelling and pessimistic prophecy, but Hoban's vision is redeemed by a profound sense of human endurance, and by the humour and radiant charm of Riddley.

Bernard MacLAVERTY   1945–

Lamb

A mere 150 pages long, *Lamb* is a dense and haunting first novel about the nature of love and the sacrifices it can involve, and about rebellious individuals at odds with an authoritarian world, where motives are misunderstood and affection is viewed with suspicion. MacLaverty's desperately moving story of a troubled priest and a foulmouthed boy on the run from the church and the state moves inexorably to its terrible conclusion. Even the characters' names predestine them to their fate, and throughout the book there are references to flying and falling: 'I see Lucifer on useless wings plummeting into the sea of Hell,' the Novice Master warns; and the boy and the priest read together the story of Icarus tumbling into the waves. The novel was subsequently made into an award-winning film (1986), directed by Colin Gregg from MacLaverty's own screenplay.

Michael Lamb has serious doubts about his vocation as Brother Sebastian, in particular about his role as a teacher in a grim approved school on the Atlantic coast in Ireland. He disagrees with the 'kill or cure' methods of the brutal, cynical principal, Brother Benedict, believing that they are of no help to such boys as the wayward, twelve-year-old Owen Kane. Funded by a small legacy from his recently deceased father, Michael decides to take Owen away from the school to England, determined to show the boy the sort of love he himself experienced as a child. Remarkably resilient in some respects, Owen nevertheless possesses a touching fragility, and suffers from eneuresis and epilepsy. He trusts no one, always expecting 'some cheat in the end', but gradually Michael gains his confidence.

They experience a brief moment of euphoria, but their disappearance has been reported as a kidnapping, and they begin to arouse suspicion in the hotels they frequent. A visit to a football match, planned as a treat, turns to a nightmare when Owen has a fit. The money begins to run out, Michael's solicitor will not forward funds, and the squat they move into proves to be a disaster when Owen is introduced to cannabis by a crudely characterised homosexual (whose possible interest in the boy is used by the author to emphasise the purely paternal instincts of the priest).

Michael's final solution echoes the 'kill or cure' philosophy of Brother Benedict, but with the crucial difference that it is 'motivated by love'. Returning to Ireland undetected, they drive to a deserted beach. Michael has substituted aspirins for Owen's epilepsy tablets and when the boy suffers the inevitable attack, Michael carries him into the sea and drowns him. His attempts to drown himself repeatedly fail, and he returns to the beach. It now becomes clear that the novel is partly a parable about the tragedy of Ireland:

> He had no luck. No faith. And now, no love. He had started with a pure loving simple ideal but it had gone sour on him, turned inevitably into something evil. It had been like this all his life, with the Brothers, with the very country he came from. The beautiful fly with the hook embedded. It was engrained like oil into the whorls and loops of his fingertips.

Olivia MANNING   1908–1980

The Levant Trilogy

The Danger Tree (1977)
The Battle Lost and Won (1978)
The Sum of Things (1980)

The Balkan Trilogy (1965) ended with Guy and Harriet Pringle aboard a rusty ship sailing for Egypt in the wake of the Nazi advance through Greece. *The Levant Trilogy* is shorter and was written more quickly than its leisurely predecessor, largely because Manning was seriously ill; indeed, the final volume was published posthumously. Once again, the Pringles are the principal characters, but a substantial part of the narrative is seen through the eyes of Simon Boulderstone, a young officer taking part in the Desert Campaign. One of the author's greatest achievements in the second trilogy is the convincing picture she gives of active service as seen through the eyes of a twenty-year-old man, a remarkable feat of imaginative reconstruction written thirty years after the war, largely guided, she said, by the memories of Field-Marshal Montgomery. Some of the background to *The Levant Trilogy* may be found in Manning's article, 'Cario: Back from the Blue' (*Sunday Times*, 17 September 1967).

The Danger Tree is the mango, a specimen of which shades the room the Pringles acquire in a flat belonging to the British Embassy in Cario's Garden City, through the intervention of 'Dobbie' Dobson, a diplomat they had known in Bucharest. The other residents include the startlingly beautiful but ineffably silly Edwina Little, who works as an archivist at the Embassy. It is Edwina whom Simon Boulderstone comes to see when he arrives in Cario, for he imagines that she is the girlfriend of his beloved elder brother, Hugo. In fact, Hugo is one of Edwina's many and transient conquests. The narrative alternatives between Simon and Harriet, the keen and naïve young soldier and the weary and disillusioned civilian. Simon has his first taste of battle and goes in search of Hugo, who, he discovers, has been killed in action.

Meanwhile, the Pringles have been separated by circumstances. Their old enemy, Colin Gracey, is in charge of the Organisation and unwilling to give Guy a job, the more so when Guy's scabrous rhyme about him reaches his ears. Guy is forced to accept an appointment as a tutor of commercial English at an Alexandrian business school (a worthless job he characteristically takes very seriously), while Harriet remains in Cairo, working for the American Embassy. When Gracey is sacked, Guy is appointed Director of the Organisation and, although back in Cairo, has little time to spend with Harriet, who loses her job when the Americans enter the war. She finds a new friend in Angela Hooper, who comes to live at the flat after the breakdown of her marriage in the wake of the death of her young son. To everyone's surprise, Angela takes up with Bill Castlebar, a dissolute poet and lecturer who is one of Guy's cronies. Harriet also befriends Aidan Pratt, a melancholy actor now 'incongruous' in the uniform of a captain in the Pay Corps, who is besotted with Guy.

The Battle Lost and Won opens with Simon coming to Cairo on leave, bringing news of Hugo's death to Edwina, who he innocently imagines will be as devastated as he is. Edwina, however, is in pursuit of Peter Lisdoonvarna, an Irish peer serving at GHQ. Harriet is feeling increasingly unwell, but refuses Guy's tactless suggestion that she should return to England.

Mistaken for a minister with a similar name, Prof. Lord Pinkrose is assassinated (possibly by two of Guy's tutors) whilst giving his long-awaited lecture at the Opera House. Edwina's plan to become Lady Lisdoonvarna evaporates when Peter manages to get transferred to active service, and informs her that he already has a wife. Meanwhile, through Lisdoonvarna's influence, Simon has become a liaison officer. While checking fuel supplies he becomes the victim of a booby trap when his jeep runs over a mine. His driver is killed and he is paralysed from the waist downwards.

To Castlebar's horror, his repulsive wife, Mona, arrives in Egypt to reclaim him from Angela Hooper. Harriet agrees to accompany Angela to Luxor, but once they are there, Angela is frightened by evidence of a cholera epidemic and decides to return to Cairo to be near Castlebar. Harriet returns later and enters the American Hospital where she is diagnosed as having amoebic dysentery. Weakened by illness, and exasperated by Guy's indifference towards her, she decides to join an evacuation ship, the *Queen of Sparta*. As she is about to board ship she comes upon Mortimer, a friend of Angela serving with the Motorised Transport Corps, and decides not to go to England, but to travel to Damascus.

A brief 'coda' records the loss of the *Queen of Sparta*, sunk by torpedoes with scarcely any survivors.

In *The Sum of Things*, news of the sinking has reached Cairo, but Guy is refusing to accept Harriet's death until he receives official confirmation. None the less, he is obliged to reassess their relationship and is moved to recognise some of his failings. It is only when he visits Simon in hospital that he breaks down. Edwina, aware that Guy is now available, decides to 'comfort' him, but is rapidly disillusioned when she joins him for what she had hoped would be a dinner *à deux*, but which soon turns into a party. She ends up being courted by Tony Brody, a coarse major at GHQ whom most people regard as too old and not good enough for her. Simon, in particular, resents Brody, but grimly recognises that his rival has 'two good legs', whilst he, although recovering feeling, is a 'cripple'.

Meanwhile, unaware of her 'death', Harriet is in Damascus, aimlessly sightseeing until her meagre supply of money runs out. A lawyer, Halal, offers her his protection and becomes a wearying guide to the city. He also finds her employment with Dr Beltado, a fellow guest at her *pension*, who is writing a vast book on comparative cultures. Beltado absconds without paying her and, eager to escape Halal, who clearly aspires to her hand, Harriet decides to travel to Lebanon. At Baalbek she is rescued by Lister, a kind but seedy intelligence officer she had known in Cairo. He takes her to a hotel, where she finds Angela and Castlebar, as well as Beltado whom Angela shames into paying Harriet her due.

In Cairo, Simon continues to make a good recovery, while Edwina announces her engagement to Tony Brody. Aidan Pratt makes a final appeal to Guy for his friendship, but Guy is unable to respond to these possibly homosexual overtures. Aidan subsequently shoots himself in the corridor of a train. While preparations are being made by the indulgent Dobson for Edwina's marriage, Harriet travels to Palestine with Angela and Castlebar; in Jerusalem they once again bump into Lister, who manages to get tickets for the Ceremony of the Holy Fire. During this riotous occasion, Harriet notices a woman in the crowd whom she recognises as Mrs Rutter, who was to have sailed with her on the *Queen of Sparta*. Puzzled that she should be in Jerusalem, Harriet approaches her and discovers the fate of the ship: Mrs Rutter was the sole surviving passenger. Realising that Guy must think she is dead, Harriet returns to Cairo with Angela and Castlebar, arriving in the middle of Edwina's wedding reception. Overcome to discover that Harriet is still alive, Guy nevertheless deserts her almost immediately, claiming that he has to 'meet some young Egyptians and give them a talk about self-determination'. Harriet realises that she would not like a possessive husband and recognises that: 'In an imperfect world, marriage was a matter of making do with what one had chosen.'

Simon, wholly recovered, achieves maturity when he sees Edwina married and somewhat tarnished; he manages to escape the shadow of both her and Hugo and returns to active service. Castlebar falls seriously ill and is eventually diagnosed as suffering from typhoid. As he lies in the American Hospital, Mona reasserts her claim as his legal wife. Angela has to be smuggled into the hospital, but is not with him when he dies and is excluded from Mona's elaborate and vulgar funeral arrangements. The trilogy ends with a coda: 'Two more years were to pass before the war ended. Then, at last, peace, precarious peace, came down upon the world and the survivors could go home. Like the stray figures left upon the stage at the end of a great tragedy, they had to tidy up the ruins of war and in their hearts bury the noble dead.'

As with *The Balkan Trilogy*, there is an enormous cast of characters who provide much of the comedy – Major Lister, with his air of 'innocent, almost infantile amiability', drunkenly recalling his nursemaid's brutality; and the unholy trio of Lush, Dudebat and Pinkrose – but Manning's principal theme is human relationships: the touching *amour fou* of Angela and Castlebar; the brief but intense camaraderie between Simon and his men; the relaxed lesbian relationship of Mortimer and her fellow driver, Phillips; and poor Edwina, of whom it is remarked that sexual relationships are 'the only sort that interested her'. All these reflect upon the central relationship between the Pringles. When they are together, Guy continues to ignore Harriet's needs, preferring to concentrate upon a forces revue to talk to Egyptians about the yoke of British imperialism. However, the doomed courtships of Harriet by Halal and Lister mirrors similar overtures made to Guy by Aidan and Edwina, emphasising that whatever its failings, the Pringles' marriage has an enduring quality.

Manning clearly delights in Guy's monstrous complacency and his disgraceful treatment of Harriet, and the trilogy is in part her funny and unembittered revenge upon her own husband for his treatment of her. When the Pringles are apart, and when Harriet appears to be dead, other characters can speak plainly. 'To tell you the truth,' Angela remarks to Harriet of Guy, 'I thought he was the most selfish man I've ever known. I often wondered why you didn't box his ears'; while Edwina, in a fit of pique, has a rare insight: 'Poor Harriet!' she says to Guy. 'You weren't all that nice to her when she was alive.' At a key moment, Guy appropriates a brooch which Angela has given Harriet, announcing that he wants it for Edwina to wear during the revue. Harriet is outraged and protests, but Guy pockets the brooch: 'Darling, don't be silly. You know you don't want it.' Later, after Harriet's 'death', Guy sees Edwina wearing the brooch, and cannot at first recall where it came from: an index of his absorption in anything and anyone but his wife. The reader is invited to share the author's relish in Guy's 'punishment', as he grieves for a wife, who is, of course, still alive. There is, however, no sentimental suggestion that Guy has learned anything from the experience or that he will mend his ways: 'For him the excitement was over. Harriet was safely back and there was no reason why life should not resume its everyday order.' The two trilogies prove one of literature's funniest and most moving portraits of a marriage.

Armistead MAUPIN   1944–

More Tales of the City

see **Tales of the City** (1990)

Julia O'FAOLÁIN   1932–

No Country for Young Men

The title of O'Faoláin's novel is adapted from the first line of Yeats's poem 'Sailing to Byzantium': 'that is no country for old men.' Set in Dublin in 1979, the novel concerns the effect of the Irish Civil War of 1922 on present and previous generations of two interrelated families, the Clancys and the O'Malleys. It was shortlisted for the Booker Prize, and is perhaps the best recent novel dealing directly with 'the Troubles', both as history and as a continuing part of Irish life.

At seventy-five, Sister Judith Clancy is the oldest survivor of the family. She has spent fifty-five years in a convent, and when her Order is disbanded she is taken in by her great-niece Grainne, who is married to

Michael O'Malley, a cousin. Judith is now senile, but for years she has been haunted by a recurring dream of gushing blood, a half-recollection of some violent event she is unable to identify. She remembers certain scattered details of the Republican struggle which followed the First World War. A brother was shot in 1919, and her sister Kathleen was lovelessly married to Owen O'Malley, a Republican now regarded as a great Irish patriot. Judith had been unswervingly loyal to Owen and his political cause, but for some reason she has now forgotten he had threatened to have her gaoled or committed if she ever attempted to leave the convent. Judith's memories are related in the present tense alongside the contemporary events in 1979, a device which emphasises the similarity of inherited attitudes in response to a comparable political situation.

> Cars were few in this residential district whose men were at work. It was the adultery hour she supposed, and wondered was any going on behind the blind windows pearly with reflections of sky. Her feet, held by the suction of the mud, came away with a sound like the kiss men on building-sites always threw at her. Did her need stick out a mile then?
>
> *NO COUNTRY FOR YOUNG MEN*
> BY JULIA O'FAOLÁIN

James Duffy arrives from America in order to make a film about Sparky Driscoll, an American fund-raiser for the Republican movement who was hacked to death by Orangemen in 1922. The naïve Duffy does not fully realise the problems of his assignment, who is to interview survivors of the 1920s for their memories of Driscoll. He is introduced to Grainne and Michael in the hope that he may get some information from Judith, who knew Driscoll. Michael is intermittently separated from Grainne and his fourteen-year-old son Cormac: he is a victim of his family's enduring political fame, and alcohol has wrecked both his marriage and a promising singing career. Differing loyalties, which destroyed families in the past, seem to persist. In defiance of his parents' wishes, Cormac has been recruited for IRA military training by his father's uncle, Owen Roe, an Irish senator with known IRA sympathies, who is encouraged in his plans by a loyal but half-witted supporter, Patsy Flynn.

Judith remembers that Owen O'Malley came to dislike Driscoll because he adopted a pro-Treaty stance. Kathleen had fallen in love with Driscoll and wanted to break off her long engagement to Owen, who was probably homosexual, had no love of individuals, and tolerated ruthless killing for his cause. Because of her narrow religious education, Judith had accepted Owen's stance, and could not understand why Kathleen, tired of violence and death, wanted a new life. This basic situation is now repeated: the American Duffy has an affair with Grainne, whose married life has been ruined by the effect politics has upon her husband.

Roe, who wants to preserve Owen O'Malley's status as a hero, is terrified that Judith will eventually recall her dark secret and reveal it to Duffy. He attempts to have the American deported, but when the plane is grounded, Duffy returns to Dublin, hoping to run away with Grainne. He drives to a canal to await her. Roe interrogates Judith and is satisfied that she is mad and would not have told Duffy that it was she, not the Orangemen, who murdered Driscoll. She had done it in order to prevent Driscoll betraying Owen O'Malley, who had been misusing American funds. Judith's own life has been sacrificed to perpetuate the false image of a hero who was corrupt and, for the sake of a political myth, continues posthumously to destroy love. The pattern of violence and betrayal is repeated. Unaware that Duffy is no longer a threat, Patsy murders him by pushing his car into the canal. This is witnessed by Judith from her window, but her senile confusion effectively prevents her story from being believed and the truth from being known.

Barbara PYM   **1913–1980**

A Few Green Leaves

'On the Sunday after Easter – Low Sunday, Emma believed it was called – the villagers were permitted to walk in the park and woods surrounding the manor.' Thus begins Pym's last, posthumously published novel, a chronicle of everyday events in an Oxfordshire village during part of a year in the 1970s. At first the novel seems to move rather aimlessly through bring-and-buy sales and flower festivals, 'Hunger Lunches' and history society meetings. Only later does it reveal its true character as a series of moving and profound meditations on death and its meaning of people leading apparently unremarkable lives.

The central character is Emma Howick, an anthropologist in her thirties, neither particularly intelligent nor attractive, who has come to live in a village with vague intentions of using it for a field study, tentatively entitled 'Some Observations on the Social Patterns of a West Oxfordshire Village'. She observes the ineffectual rector, Tom Dagnall, lost in studies of the village as it was in the seventeenth century; his unmarried sister Daphne, who keeps house for him, but dreams of living in Greece; the two village doctors, old Dr Gellibrand and thrusting young Dr Shrubsole; and all the other denizens of the village, its elders and the recent arrivals.

When she sees an old flame, the social studies academic and Africa expert, Graham Pettifer, appearing on the television, Emma invites him down to the village. An unconsummated relationship develops between them when Graham takes a cottage in the woods near the village for the summer. Daphne, who has long been mildly dissatisfied with life at the rectory, decides to go to live with a friend, the bossy, cold Heather Blenkinsop, outside Birmingham near a wooded common where they can exercise a dog. Tom is left rather helpless, hoping that his parishioners will invite him to their houses for meals.

The novel was written while Pym was mortally ill, and contains references to the deaths of characters from her other novels: Fabian Driver, from *Jane and Prudence* (1953), and, from *Less than Angels* (1955), Esther Clovis, whose memorial service Emma attends. Further intimations of mortality are provided by the mausoleum of the former lords of the manor, which is the setting for several encounters in the book. Miss Vereker, the former governess to the last daughters of the de Tankerville family to live at the manor, takes a nostalgic day-trip to the village. While wandering in the woods she gets lost, and is discovered by Emma and Mrs Shrubsole in a state of collapse, resting on some stones which turn out to be the DMV (deserted medieval village) for which Tom has been searching. That evening, during a power-cut, the eccentric Miss Lickerish dies: 'Some time during those dark hours the cat left her and sought the warmth of his basket, Miss Lickerish's lap having become strangely chilled'.

The novel ends in the early autumn, with Graham returning to his temporarily estranged wife and Emma (now working on 'Funeral Customs in a Rural Community') beginning to half-realise that a better possibility of emotional fulfilment lies with Tom. As always in Pym's novels, nothing is overstated, nothing portentous, yet the characteristic emotional atmosphere of her work – the feeling that mundane life is both necessarily incomplete and immensely valuable – comes over more strongly here, perhaps, than anywhere else in her work.

| John Kennedy TOOLE | 1937–1969 |

A Confederacy of Dunces

A Confederacy of Dunces was written between 1962 and 1963 but did not appear in print until 1980. The story of its publication is almost as convoluted as the plot of the novel itself. Toole wrote the book while serving in the US Army in Puerto Rico, and first approached Richard Gottlieb (then chief editor at Simon and Schuster) in late 1963. After a long correspondence between these two and several revisions of the novel, Gottlieb finally rejected *A Confederacy of Dunces* in 1966. On leaving the army, Toole worked as a teacher for several years before quitting in December 1968 to wander through the Southern states of America. In March 1969, he was found dead in his car in Mississippi after poisoning himself with exhaust fumes.

Toole's mother, Thelma, decided that her son's novel should be published, and during the next ten years she sent the manuscript to many publishing houses, all of whom rejected it. Her persistence paid off, however, when the American writer Walker Percy was badgered into reading it. His reactions, initial scepticism followed by unqualified praise, are recorded in the novel's preface. Percy took up the fight and, although his own publisher rejected the novel, it was finally published by the Louisiana State University Press in 1980.

Ignatius pulled his flannel nightshirt up and looked at his bloated stomach. He often bloated while lying in bed in the morning contemplating the unfortunate turn that events had taken since the Reformation. Doris Day and Greyhound Scenicruisers, whenever they came to mind, created an even more rapid expansion of his central region.

A CONFEDERACY OF DUNCES
BY JOHN KENNEDY TOOLE

The novel's action takes place in New Orleans where Ignatius Reilly lives with his mother. Ignatius is the novel's hero; overbearing and overweight, he does nothing but eat, eructate and rail against the modern world in which he is an outsize misfit. Mrs Reilly wants her son to find work and Ignatius's reluctant employment at Levy Pants, and later as a hot-dog salesman for Paradise Vendors Inc., provide occasions for the comic set-pieces which are one of the book's great strengths. Numerous sub-plots involve Patrolman Mancuso, who almost arrests Ignatius in the opening scene: Claude Robicheaux, who *is* arrested and later courts Mrs Reilly; and the denizens of the Night of Joy bar. These include Jones, a floor-sweeper on probation who is being exploited and loses no opportunity to make this known; Darlene, a stripper whose act involves being undressed by a cockatoo; and the bar's 'Nazi' owner, Miss Lee. A scheme to distribute pornographic postcards, Patrolman Mancuso's efforts to make more (or any) arrests, and the efforts of Myrna Minkoff,

Ignatius's much-maligned girlfriend, to perform unspecified erotic therapy on the reluctant Ignatius by post, are all somehow intertwined by Toole. At the novel's close, Mrs Reilly decides to have her monstrous son committed to a sanatorium. Ignatius gets wind of this and Myrna Minkoff turns up in the nick of time to rescue him. We last see him being driven away in her car as the sanatorium ambulance arrives, too late, at his home.

A Confederacy of Dunces takes its title from Swift's 'Thoughts on Various Subjects'. Ignatius's gross physical appearance, his function as a dubious moral centre and his rampant misogyny are all authentically Swiftian. Although Toole's satire provides many opportunities to judge his characters, few of these judgements remain in place for long. This is a side-effect of the strength of characterisation in the book. Ignatius is a monumental creation, but many lesser figures are fleshed out more fully than they need be. Jones, the floor-sweeper at the Night of Joy bar, for instance, plays little part in the action but is still wholly convincing as an articulate, powerless Negro, sharply aware and outraged by his situation. When the novel appeared in 1980, reviewers were unanimous in their praise and 40,000 copies were sold in hardback alone. By the end of that year *A Confederacy of Dunces* had been awarded the Pulitzer Prize and Thelma Toole's confidence in the worth of her son's hilarious comedy of bad, modern manners was finally vindicated.

A.N. WILSON   1950–

The Healing Art

One of Britain's most prolific authors, Wilson first made his reputation with comedies of manners such as *The Sweets of Pimlico* (1977), which won the John Llewelyn Rhys Memorial Prize. *The Healing Art*, which in character and setting owes something to the works of Barbara Pym, has a more sombre theme than his earlier novels and is often considered his best book, not least by prize juries: it won a Somerset Maugham Award, an Arts Council National Book Award and a Southern Arts Literature Prize.

Two women are being treated for breast cancer in an unnamed university town that is evidently Oxford. The devout Anglo-Catholic Pamela Cowper is an English don, while Dorothy Higgs is the wife of a printer. Their hospital consultant, Mr Tulloch, 'a rat-faced Scot in a white coat' who runs a large private abortion practice, mixes up the X-rays of the two women, telling Pamela that without chemotherapy she has only a few months to live, and Dorothy that she is cured. Exactly the opposite is the case.

> 'Look here, can I come in,' he said, in that fruity Ampleforth voice. What was it, she had often asked Sourpuss, that made one want to punch RCs? But Sourpuss, who in his youth would have worn a bucket on his head if it had been so decreed by the Holy Office, did not allow her to elaborate on this malice, even though it was from *him* that she had caught the trick of referring to them as 'our separated brethren'.
>
> *THE HEALING ART*
> BY A.N. WILSON

From this simple but striking initial idea the whole novel develops, tackling large questions of moral attitude and choice through an extensive cast of complexly related characters. Of the two women, Dorothy is treated in less detail. She quickly becomes very ill, although Tulloch, refusing to believe he has made a mistake, does not keep her in hospital. Her family life with her husband, George, and her son, Barry, to whom she is devoted and who is starting university in Swansea, is outlined in solid and moving detail. But, before she can die of cancer, she is killed in a car-crash with a friend, having just met Pamela by chance at what is evidently Blenheim Palace.

Pamela's life during these months is treated largely through her relationship with two men, her parish priest, Hereward Stickley, and a fellow don, John Brocklehurst, with whom her relationship has never quite taken off. During a visit to the shrine of

Our Lady at Walsingham, when she believes a miracle has rid her of her cancer, Pamela feels her attraction to the reticent Hereward increasing.

Pamela also has a relationship with an American girl, Billy, who is expecting a child which she says is John Brocklehurst's. Pamela wants to help her to bring up the child, but John persuades Billy to abort it and pays for Tulloch to perform the operation. The moral relativism in John which this and other actions reveals to Pamela leads to a weakening of their relationship: 'It was the difference between one who considers everything in life to be fluid and changeable and shifting, and one who searched, however fleetingly, for a point inside or outside what we call life where the motion and the half-truth ceased and the truth could be seen in stillness and face to face.'

The novel ends with a fire at John's college in which Billy and the real father of her child, her own adopted father, Mel, are killed. John later marries Mel's former wife, Gale, although realising at least that he loves Pamela. She, however, marries Hereward, and the last sentence of the novel reveals that, 'within the womb, she feels the mysterious stirrings of a new life'.

1981

Foundation of Social Democratic Party in Britain ● Socialist government in France ● Martial law in Poland ● Inauguration of Ronald Reagan as 39th President of US ● Mrs Sandra Day O'Connor appointed the first woman Justice of US Supreme Court ● France introduces its high speed train ● IBM launches its personal computer, using MS-DOS ● US Center for Disease Control recognises AIDS, thought to be caused by HIV virus ● Martin J. Wiener, *English Culture and the Decline of the Industrial Spirit, 1850–1980* ● Tom Wolfe, *From Bauhaus to Our House* ● Rupert Murdoch buys *The Times* and other *Times* newspapers in Britain ● Nell Dunn, *Steaming* ● Harvey Fierstein, *Torch Song Trilogy* ● Films: *Diva, Man of Iron* ● Television: *Brideshead Revisited* ● Samuel Barber, Omar Bradley, Marcel Lajos Breuer, Anwar Sadat and Moshe Dayan die ● D.J. Enright, *Collected Poems*

Robertson DAVIES　　　　**1913–**

The Rebel Angels

see **The Cornish Trilogy** (1988)

Maureen DUFFY　　　　**1933–**

Gor Saga

Duffy's science fantasy novel is written in the tradition of Mary Shelley's *Frankenstein* (1818) as a warning against the human desire to play God, not merely in the future realms of genetic engineering, but in the ever-present issue of animal rights. Its dystopian vision of a future where 'agribusiness' has raped the countryside, the gap between the haves and the have-nots ('nons') has become unbridgeable and where the inner cities have become uninhabitable areas of urban decay has moved uncomfortably nearer since the novel was first published.

The eponymous Gor (Gordon) is a Caliban figure, the result of an experiment carried out at the Primate Institute of the Ministry of Defence. The secret 'test tube product of a gorilla ovum and human sperm', he is the prototype in an unholy scheme dreamed up by his creator, Norman Forrester, who wants to produce a slave race of humanoid drudges. Gor has inherited more from his human father (Forrester himself) than from his ape mother, and he is farmed out to the 'non' Bardfields, foster parents whom Forrester regards as too dull to ask any awkward questions, but who in fact have an 'ingrained insubordination'. An operation upon Gor's larynx gives him the ability to speak and Forrester callously removes him from the Bardfield family and sends him to a prep school. His dark, simian appearance earns him the nickname 'Monkey', but his prowess in the gym and talent in the art school win him admiration and friends. Forrester persuades his wife, Ann, to look after Gor during the school holidays and she grows fond of him. When she tells Forrester that she has discovered Gor kissing their daughter, however, Forrester is horrified and in anger tells her the truth about Gor's origins. Unwilling to believe this, but sure that Gor is in danger, she gives him money and tells him to run away. After a grim time in Bristol, its centre decayed, its outskirts mere shanty-towns, he joins up with the 'ugs', a self-sufficient community blackened by the government as 'urban guerrillas'. Amongst the ugs are the Bardfields, who discover that Gor has no records. Gor and his foster-brother William, an ug leader, break into Forrester's office and steal Gor's file. Horrified by what he finds in it, Gor goes on the run, steals and becomes a wanted criminal. He is given sanctuary by a confused priest, then makes his way to his old home to confront his creator. He is saved from murdering Forrester by the arrival of William and Ann, who have fallen in love. They return to the ugs' Utopian settlement, where Gor is entirely accepted and crowned king of the day after helping peacefully to repel an unprovoked attack by government troops.

Beneath the gripping narrative, presented in saga form, is a parable about both human and animal rights and about freedom and captivity. Forrester regards the 'nons' (from which he came) as scarcely human, but the novel also makes a case for those animals which are *not* human. Forrester claims that in Gor he made 'a thing, an artefact', and Duffy suggests through Gor that animal experimentation does indeed reduce sentient creatures to *things*, 'vehicle[s] for [humans'] own pride and vanity'. Gor is proof that that the dividing line between human and non-human animals is thinner than many people care to believe. It is, of course, not Gor, but the chilling Forrester who really lacks 'humanity', a point ignored in a cavalier television adaptation, *First Born* (1988), which jettisoned the novel's moral framework and futurist setting, and transformed Forrester into a glamorous, philanthropic, Roman Catholic hero.

Molly KEANE   1905–

Good Behaviour

Under the pen-name M.J. Farrell, Keane had been a popular novelist and playwright (see *The Rising Tide* (1937)), but the sudden death of her husband in 1946 devastated her, and the failure in 1961 of her play *Dazzling Prospect* stopped her from writing altogether. During the late 1970s, however, she 'secretly' began writing a darkly comic novel with an Anglo-Irish setting about 'a fool who doesn't see what's happening', which was turned down by her publisher on the grounds that its humour was too black. Some years later Keane gave the manuscript to the actress Peggy Ashcroft, who had fallen ill while visiting her and had asked for something to read. Ashcroft's enthusiastic reaction prompted Keane to submit the book to André Deutsch, who subsequently published it to enormous acclaim. Shortlisted for the Booker Prize and serialised on television, *Good Behaviour* triumphantly relaunched the seventy-seven-year-old author's career.

The novel opens with fifty-seven-year-old Aroon St Charles having just whizzed up some rabbit quenelles for her bedridden mother. Lifting a forkful of this delicacy to her mouth, 'Mummie' cries out, vomits 'dreadfully', falls 'back into the nest of pretty pillows', dead. The servant, Rose, is hysterical, but Aroon reacts to this calamity with perfect composure: 'I do know how to behave – believe me, because I know. I always have known. All my life so far I have done everything for the best reasons and the most unselfish motives. I have lived for the people dearest to me, and I am at a loss to know why their lives have been at times so perplexingly unhappy.'

From this tremendous opening scene, the novel goes back in time as Aroon relates the events of her life. Vast in size and appetite, bosoms 'swinging like jelly bags', Aroon has lumbered through life, sustained by her unshakeable belief in the family's chilly 'code of manners', which prevents anything as vulgar as emotion. ('Her howling and screaming made all the glasses ring out,' Aroon's hateful mother exclaims with amused distaste, reporting how a maid responded to news of the death of her fiancé in the trenches.) Aroon is much influenced by her governess, Mrs Brock, who details the facts of life in revolting detail before drowning herself for love of Aroon's rakish, horse-mad father.

Aroon herself experiences an unfortunate infatuation with Richard Massingham, the bosom pal of her brother, Hubert. She is delighted when Richard noisily enters her bed, but the reader recognises that Richard merely wants to allay suspicions about his own sexuality. When Hubert is killed in a car-crash, Aroon hopes that this will bring her and Richard closer, but after weeks of silence, the letter which finally comes is rather remote. She later opens the *Tatler* to find a photograph of him with his new fiancée. This is a short-lived liaison; he later runs off to Kenya with Baby Kintoull. 'Married?' Aroon asks Richard's father. 'I don't think you quite have the riding of it,' she is told. 'They were in the same house at Eton.' Her romantic aspirations are overshadowed, however, when her beloved father suffers a stroke. The tyrannical Rose takes charge, and appears to massage rather more than Papa's foot beneath the bedclothes (a circumstance Aroon fails to register). When Papa dies, he leaves his entire estate to his daughter, and the novel ends with Aroon assuming control. Her voice 'humid with kindness', she promises her disbelieving mother: 'I'll always look after you.'

Salman RUSHDIE   1947–

Midnight's Children

'. . . no people whose word for "yesterday" is the same as their word for "tomorrow" can be said to have a firm grip on the time'. *Midnight's Children* is an ambitious attempt to encompass the whole of India in the life of one man, in particular to trace the country's post-independence history. While British critics (for the most part) felt that he had succeeded triumphantly, some Indian commentators were more sceptical. According to Rushdie, the book was approached differently in each country: 'To simplify: in England people read *Midnight's Children* as a fantasy, in India people read it as a history book' – so much so that the author was obliged to apologise to the ruling Gandhi family, members of which are depicted unflatteringly in the text. (Rajiv Gandhi later moved swiftly to ban Rushdie's **The Satanic Verses**, 1988, in India.)

The novel hinges on the interweaving of memories, which range from the mythical prehistory of India, through the country's arrival at independence on 15 August 1947 to the present moment, as the narrator, Saleem Sinai, struggles to make his recollections clear, urged on by Padma, his companion and muse. The central event of the novel is the 'midnight baby-swap'. On the stroke of midnight of the day on which India attains her independence, two babies are born in the same household: one is Shiva, the legitimate son of Amina and Ahmed Sinai, the other is Saleem, ostensibly the offspring of Vanita and the jester Wee Willie Winkie. In fact, Saleem's real father is the Englishman Methwold. Driven insane by her lover's desertion of her, the nurse, Mary Pereira, substitutes Saleem for Shiva, so that while Saleem becomes the exclusive object of his supposed family's attention, Shiva, neglected and despised as Willie's boy, has 'to fight for survival from his earliest days'. Moreover, while 'to Shiva, the hour had given the gifts of war', it had given Saleem 'the greatest talent of all – the ability to look into the hearts and minds of men'. His energy is thus channelled into using his mind as a 'forum' for the transmissions of the mental 'network' shared by the 581 children born on that famous midnight.

All the Midnight Children are characterised by powers of 'transmutation, flight, prophecy and wizardry'. One of them is Parvati-the-witch, whose introduction gives Rushdie an opportunity to mix realism with magic. Her surrealistic dominion is the 'ghetto of the magicians . . . to which the greatest fakirs and prestidigitators and illusionists in the land continually flocked'. Parvati has a son by Shiva, but it is Saleem who marries her, thus repeating a notorious paradox in their family history: 'Once again a child was to be born to a father who was not his father, although by a terrible irony the child would be the true grandchild of his father's parents'. The ghetto is invaded by Shiva's forces, Parvati crushed by the bulldozers, and Saleem captured and forced to reveal the names of the Midnight Children. They are then 'vas- and tubectomized' to prevent them reproducing.

My special blends: I've been saving them up. Symbolic value of the pickling process: all the six hundred million eggs which gave birth to the population of India could fit inside a single, standard-sized pickle-jar; six hundred million spermatozoa could be lifted on a single spoon.

MIDNIGHT'S CHILDREN
BY SALMAN RUSHDIE

Most of these key events are not located chonologically in the narrative. Rather, through hints, analogies and half-understood statements they are prepared for or recapitulated, temporarily abandoned or left floating in the air, finally picked up again and completed. Digressions, non-sequiturs, metafictional asides to the reader, govern the writing. Time itself is reshaped, as experiences are illustrated not in their entirety, as the result of some neat recollection, but in the very process of being remembered. Ultimately, it is only an illusion to believe that 'since the past exists only in one's memories and the words which strive vainly to encapsulate them, it is possible to create past events simply by saying that they occurred'. *Midnight's Children* was a phenomenal success and earned a great deal of publicity when it won the Booker Prize. Rushdie's previous novel, *Grimus* (1975), had attracted little attention, and his publishers had expected to sell some 2,500 copies of *Midnight's Children*; 32,000 hardback copies of the novel were eventually sold in the UK alone. The paperback edition proved equally marketable, selling some 45,000 copies in Britain. Rushdie had arrived, almost out of nowhere, as a leading contemporary novelist. In 1993, the novel was awarded 'The Booker of Bookers' – judged the best of the prizewinners since the prize was inaugurated in 1968.

Muriel SPARK — 1918–

Loitering with Intent

Wilde's assertion that 'Life imitates art far more than Art imitates life' might stand as an epigraph to this characteristically slim and elegant novel. Within Spark's witty narrative is a serious meditation upon this paradox, and upon the nature of fiction in general and the author's fiction in particular.

Set in 1949–50, the novel is narrated in the first person by Fleur Talbot, a waspish young woman with her first novel 'in larva' who lives 'on the grubby edge of the literary world'. The loiterer of the title, she finds employment with the Autobiographical Association, the members of which are committing their memoirs to paper under the guidance of the pathologically snobbish Sir Quentin Oliver. The finished products are to be kept locked away until anyone mentioned in them is dead, a circumstances which allows the 'complete frankness' by which Sir Quentin sets so much store. Fleur suspects that Sir Quentin himself lacks frankness and that there is some sinister purpose behind the Association. Her job is to edit and embellish the illiterate and dreary memoirs, and such is her contribution that she begins to 'consider them inventions of [her] own'. Her work is hindered by Sir Oliver's repellant and rapacious housekeeper, Beryl Tims, but enlivened by his ancient, outspoken and wilfully incontinent mother, Lady Edwina, whom she befriends.

When Sir Quentin and his entourage take on the personalities and pursue similar fates to the characters in Fleur's novel, *Warrender Chase*, she becomes suspicious. She discovers that Sir Quentin has acquired a copy of the book and is plagiarising parts of it to incorporate into his members' memoirs. He also halts the novel's publication on the grounds of libel and arranges for the proofs to be destroyed and the manuscript to disappear. Assisted by Lady Edwina, Fleur steals the memoirs from Sir Quentin's office to hold in ransom for the return of her manuscript. She now recognises that Sir Quentin's aim is not blackmail, but messianic power. He models himself not only upon the eponymous hero of her novel but also upon a twisted idea of J.H. Newman derived from the *Apologia*. Under the strain of his regime, which includes a course of appetite depressants, one of his disciples, Lady 'Bucks' Gilbert, commits suicide. Sir Quentin, like Warrender Chase, dies in a car-crash and the Association dissolves. Fleur finds a new publisher and becomes a successful novelist.

The book's distinctive flavour comes from Fleur's astringent narrative voice, and in this it resembles Spark's *A Far Cry from Kensington* (1988), which is set in 1950s literary London. Fleur has a healthy belief in her own gifts and an engagingly blunt way of summing up others: one character is described as 'a cripple and a bore', while another is 'small and wispy, about twenty, with arms and legs not quite uncoordinated enough to qualify him for any sort of medical treatment, and yet definitely ... not put together right'. Her original publisher, Revisson Doe, remarks that *Warrender Chase* is 'quite evil, especially in its moments of levity', and this gives some idea of the particular qualities of Spark's own sharply entertaining novels, of which *Loitering with Intent* is an outstanding example.

Paul THEROUX 1941–

The Mosquito Coast

Thirteen-year-old Charlie, the oldest of Allie Fox's children, narrates *The Mosquito Coast*. He describes the faith the family has in 'Father' and the limits to which Father's obsessions proceed. Father is an inventor. Disgusted by the USA's junk culture and convinced that God has botched creation, he builds gadgets marred neither by the market nor by nature. His employer, a Massachusetts farmer, fails to understand the worth of

Fat Boy – a non-electric ice-producing stove. This confirms Father's belief that North America is doomed, and he sails with his wife and children to Honduras.

There he buys Jeronimo – a piece of jungle on the Aguan river – and transforms it into a self-sufficient settlement. His energy is enormous and he has a gift for gaining the obedience of strangers: the natives willingly co-operate in the construction of a gigantic Fat Boy. Although the ice it makes is of scant value, Father treats it as a symbol of his triumph: he conceives more ambitious schemes. Fat Boy falls into disuse, but, when three enigmatic strangers seem to threaten, Father traps them in Fat Boy's innards and burns them alive. In the release of chemicals that follows, the Aguan is severely polluted.

Commandeering a boat, Father takes the Foxes deeper into Honduras. He declares that Jeronimo was a mistake. For a time, the family survives in primitive isolation beside a lagoon. Charlie starts to question Father's wisdom: Jerry, the younger son, is in open revolt. Floods come. Increasingly dictatorial, Father again insists on travelling inland. He claims that the apocalypse he predicted has by now destroyed the USA. Then, arriving at a mission, he sets it ablaze. A missionary shoots him through the neck, assuming him to be a Communist. Father is gravely wounded – but not killed. He dies at last on a deserted beach and Mother and the children struggle homeward.

As the book begins, Allie Fox is an eccentric protestor against the waste and exploitation of capitalism: before it ends, he is a ruthless and destructive paranoiac. What propels him from the one condition to the other is his fanatical search for an environment untouched by humans and forsaken by God – a place where he can better the designs of both without the hindrance of the former or the assistance of the latter. Yet, discarding the material relations of society, he discards its moral relations also. Only fear binds his family to him, while his firing of the mission is a symptom of his enraged contempt for property, religion – even life.

Civilisation, Theroux suggests, is indivisible: its ethical controls vary with the sophistication of its economic systems and, in rejecting these, Fox – like Nebuchadnezzar – descends to the level of a beast. His crazed idealism is a form of hubris and meets, accordingly, with retribution. Eaten by vultures, he has no chance to repent – so it is left to Charlie to disavow his father's blasphemous dreams. In the novel's final and conservative endorsement, Charlie judges: 'The world was all right ... It was glorious even here, in this old taxi-cab, with the radio playing.'

D.M. THOMAS — 1935–

The White Hotel

Like Salman Rushdie, D.M. Thomas had enjoyed little literary success before being shortlisted for the Booker Prize. Indeed, *The White Hotel*'s success in the UK was slow in coming, and it was only when the American rights were sold for a staggering $199,500 that the British began to take another look at a novel many critics had initially dismissed. Sales picked up further when the novel was shortlisted for the Booker Prize, and although it lost to **Midnight's Children** (1981), it sold even more copies: 20,000 in hardback, followed by a 100,000 paperback print run. In America the novel sold 100,000 in hardback, followed by a paperback print run of 1,000,000; foreign rights were sold to fourteen countries. Arguments about the novel's actual merits continue, some claiming that the novel is a truly European masterpiece, others dismissing it as at best hokum, at worst a misogynistic male fantasy which merely exploits the Holocaust. In particular, the fate of the female protagonist, who emerges from the death-pit only to be raped by bayonet – a culmination of the novel's preoccupation with sex and death – has been much criticised by feminist commentators.

The novel's genesis was a poem, 'The Woman to Sigmund Freud', narrated by Anna G, an imaginary patient of the famous psychoanalyst. In graphic detail, she recounts her erotic, masochistic fantasies about an affair with Freud's son. Thomas subsequently read a documentary account of the Nazi massacre, *Babi Yar*, and 'suddenly realised that the poems were, in fact, beginning a novel which would end in *Babi Yar*'. Anna G, it is revealed, is an opera singer called Lisa Erdman, who comes to a fictionalised Freud as an unhappy twenty-nine-year-old, almost crippled by pain and breathlessness. Returning from a health resort, she presents Freud with a poem which he describes as expressing 'an extreme of libidinous phantasy combined with an extreme of morbidity'. The poem, later extended into a more detailed narrative, frequently juxtaposes erotic images with images of death ('Eros in combat with Thanatos') and seems both to echo events from her past (her adulterous mother died in a hotel fire with her lover, Lisa's uncle), and to anticipate those in the future. She is a difficult and evasive patient, who often lies and conceals key facts about herself. The conclusions Freud reaches about her are somewhat tempered by her further revelations ten years later in reply to his request for permission to publish her case history.

> I now had the ludicrous sensation that I knew absolutely all there was to know about Frau Anna, except the cause of her hysteria. And a second paradox arose: the more convinced I grew that the 'Gastein journal' was a remarkably courageous document, the more ashamed Anna became of having written so disgusting a work. She could not imagine where she had heard the indelicate expressions, or why she had seen fit to use them. She begged me to destroy her writings, for they were only devilish fragments thrown off by the 'storm in her head'
>
> *THE WHITE HOTEL*
> BY D.M. THOMAS

Making a later and successful second marriage in 1934 (her first failed because she had to conceal the fact that she was half Jewish from her anti-Semitic husband), Lisa enjoys a short period of happiness in Kiev, even visiting her childhood home and achieving a sense of peace with the past. Believing herself 'cursed' with the gift of second sight, even she cannot foretell her husband's arrest and her own death and that of her stepson in the massacre of the Jews. The inexplicable pains she has suffered in the past, however, and her images of death, suddenly become the shocking premonitions of her horrific fate.

The extraordinary final section of the novel appears to depict a kind of after-life in which further images fall into place and further explanations arise when Lisa confronts her own mother, perhaps confirming the author's assertion that 'the soul of man is a far country, which cannot be approached or explored ... If a Sigmund Freud had been listening and taking notes from the time of Adam he could still not fully have explored even a single group, even a single person.'

John UPDIKE   1932–

Rabbit is Rich

see **The Rabbit Tetralogy** (1990)

A.N. WILSON   1950–

Who Was Oswald Fish?

A comic novel of modern manners, although not without serious and even disturbing undertones. *Who Was Oswald Fish?* is set in London and Birmingham during the 'winter of discontent' of early 1979, but has an important sub-plot taking place in the Victorian age. Fanny Williams is a former 1960s' pop star, now in operatic and highly sexed middle age, and running her highly successful 'Fanny shops', supermarkets for Victoriana. She runs a chaotic ménage in Chelsea which includes Marmaduke and Pandora, her outrageously sophisticated young children ('Good morning, Mr Willie,' Fanny greets Marmaduke when she sees an erection poking through his pyjamas), Charles Bullowewo, a black, homosexual Old Etonian barrister, and Fanny's layabout sister, Tracy, who takes up with a plumber called Sid.

Fanny has bought a disused church in a derelict area of Birmingham, St Aidan's,

Purgstall, which she plans to use as a warehouse for the 'Fanny shops'. This is the beautiful work of a late Victorian architect, Oswald Fish, who by coincidence turns out to be related to most of the main characters. His daughters, Jessica ('Nana') Owen, now an old women and Fanny's grandmother, discovers Oswald's diaries while setting out to visit Fanny in London, and extracts from the diaries placed throughout the novel provide a constant counterpoint between past and present.

Fred Jobling is a planner with Birmingham city council, locked into an unhappy marriage with Jen in Edgbaston, and devoted to the idea of knocking down St Aidan's to create a socially useful leisure centre. (His family, like that of Charles Bullowewo, turns out to be related to Oswald Fish.) He is initially hostile to Fanny, but, on a visit to the church, he meets her and her corgis down from London, and immediately falls in love with her. She is also attracted to him, they begin an affair in London, and eventually plan to marry.

As the novel goes on, it reveals more of the dark forces threatening the happiness and even the lives of the characters. David Matheson, an art historian, married but a repressed homosexual, has a brief affair with Charles Bullowewo. He is so disturbed by the fact that his wife has been given false information about his sexual tendencies by Marmaduke, that he throws himself under a tube train. The novel reaches a climax on the day Mrs Thatcher is elected to office in May 1979, a day of more death and destruction for the characters. On that day, Pandora has written to Jen telling her about the relationship between Fred and Fanny. Jen throws a poker at her husband's head and accidentally kills him. Fanny is in Birmingham that same day to remove a screen from St Aidan's, but during this process the church collapses and the two corgis are crushed to death. The novel ends with Fanny, not knowing Fred is dead, looking forward to married life with him as she and her family watch the election results on television.

1982

Anglo-Argentine War over the Falkland Islands • Israel invades Lebanon • Yuri Andropov becomes Soviet leader • In Britain, 20,000 women in protest against siting of US Cruise missiles at Greenham Common • CD (compact disc) players go on sale • Caryl Churchill, *Top Girls* • Michael Frayn, *Noises Off* • Films: *Gandhi, Bladerunner, ET* • Television: *Cheers, Boys from the Blackstuff* • Popular music: Michael Jackson, *Thriller*, Prince, *1999* • Ben Nicholson, Thelonious Monk, Rainer Fassbinder, Dame Marie Rambert, Grace Kelly, Jacques Tati, Leonid Brezhnev and Artur Rubinstein die • Derek Mahon, *The Hunt by Night*

Pat BARKER   **1943–**

Union Street

Barker wrote *Union Street* after taking a writing course during which Angela Carter advised her to write out of her own working-class experience. Her first published book, it won the Fawcett Society Book Prize (named after the suffragist Dame Millicent Fawcett and awarded annually to a book 'which has made a substantial contribution to the understanding of women's position in society today'), was a runner-up for the *Guardian* Fiction Prize, and led to the author being named as one of the 'Best of Young British Novelists' in an influential Book Marketing Council promotion. Set in an unnamed town in the north-east of England, it is made up of seven interlinked stories, each named after an individual woman. These go to make up a graphic account, in their own idiom, of women whose lives are circumscribed by poverty and violence.

The first chapter is the longest, describing a childhood typically shared by the older characters, and thus explaining their behaviour and attitudes as adults. Kelly Brown, an intelligent, streetwise eleven-year-old, is neglected by her mother, who cannot adequately feed or clothe her family since her husband left home. Playing truant, Kelly roams the streets until late at night and on one such occasion is raped. She conceals this from her mother for three weeks, and although haunted by the memory of the attack, she is even more strongly compelled to wander alone. Understood by no one, and filled with self-loathing, she expresses her anger by vandalising a middle-class villa and her school. Her own suffering gives Kelly the insight to sympathise with an old woman she meets in a park, who has abandoned her house to avoid being put in a local authority home. The woman will die in the cold, and this understanding of what life can be in the extreme, encourages Kelly to return to her mother.

There was only one memory she was sure of. Firelight. The smells of roast beef and gravy and the *News of the World*, and her father with nothing on but his vest and pants, throwing her up into the air again and again. If she closed her eyes she could see his warm and slightly oily brown skin and the snake on his arm that wriggled when he clenched the muscle underneath.

UNION STREET
BY PAT BARKER

'Joanne Wilson' takes place in a cake factory, where some of the female characters are employed, and where they unleash the frustration of their job through bitchy gossip, racism and assault. Interspersed with this are eighteen-year-old Joanne's thoughts about her unwanted pregnancy, and the prospect of a forced marriage. Her future husband is middle class and, although Joanne will be financially secure, she would rather remain in Union Street amongst the friends she knows.

The theme of childbirth is continued

through the story of Lisa Goddard. Beaten by her drunken husband, Lisa feels unable to cope with her family responsibilities, but the trauma of a difficult birth is assuaged by her surprised pleasure in holding her new daughter. For Lisa, nothing will improve, but she at least realises she is a survivor.

The effect of fatal illness on a family is represented by John Scaife, out of work and dying of a lung disease. Unlike many of the other characters, he is devotedly married to Muriel, a school cleaner. Because he is illiterate, he distrusts his clever son, Richard, and despite the knowledge of his own approaching death, is unable to communicate with him. Richard knows this, but is also too shy to talk. After Scaife's death, Richard understands his mother's devastation and loneliness, and is posthumously reconciled to his father, whom he learns was secretly proud of him.

Providing a thread through each chapter is Iris King, a buxom woman with husband, children and grandchildren. Spotlessly clean, kind-hearted and good-humoured, she helps her less-capable neighbours in their troubles. Through a tough life, she has learnt to judge situations, and does not shirk hard decisions. When her daughter Brenda becomes pregnant, and is too young to marry, she procures a backstreet abortion for her. Iris knows the moral issue and the risk, but is afraid that Brenda may become mentally ill, like Laura, her eldest daughter.

Beginning with the problems of childhood, *Union Street* ends with those of old age and solitary despair. Alice Bell, starved, cold and senile, has a horror of the workhouse, which she believes still exists, and has saved against the indignity of a pauper's funeral. Following a severe stroke, she at first refuses to go to a home for the elderly, then pretends to relent. Denied her hopes of ending her days in her own house, she decides to die in the open and struggles to the local park. She is the old woman that Kelly Brown met at the beginning of the novel. Alice's reflections on her past emphasise the cyclical pattern of life and death.

The novel was rendered unrecognisable when 'adapted' for the film *Iris and Stanley* (1989), in which the action was transferred to America and the title roles were played by Jane Fonda and Robert de Niro.

Saul BELLOW   1915–

The Dean's December

Albert Corde is an ex-journalist and dean of a Chicago college. He is in Bucharest with his wife, Minna, a Romanian-born defector whose mother, Valeria, is dying in the Party hospital. Valeria was the Minister of Health thirty years previously. Expelled from the Party, she was subsequently rehabilitated, although she refused to rejoin. Corde and Minna run into the veiled hostility of the Communist bureaucracy which refuses them more than one visit to Valeria because she is in intensive care. Corde attempts to pull strings at the American Embassy while Minna does likewise with old associates of her parents.

Back in Chicago, the absent Corde is the cause of controversy on two counts. First, he has recently published two articles in *Harper's* about the decaying inner-city ghettoes of Chicago, articles in which 'he gave up his cover, ran out, swung wild at everyone'. Secondly, he is involved as Dean in the murder trial of black Lucas Ebry following the suspicious death of a student. Corde is accused of racism by the campus radicals, one of whom is his dropout nephew, Mason Zachner. There is further family involvement in the form of Corde's cousin, Max Detillion, the aggressively ambitious lawyer who is defending Ebry. Both circumstances are a source of potential political embarrassment to the college authorities.

Corde is a liberal thinker, attempting to achieve clear moral focus in the 'moronic inferno', 'to recover the world that is buried under the debris of false experience or nonexperience'. Dewey Spangler, an old schoolfriend of Corde, now a celebrity journalist, arrives in Bucharest, and his conversations with Corde dramatise this world of false description. The press is characterised as a 'distortion furnace', and Spangler described as operating in 'a kind of event-glamour'. He accuses Corde of allowing himself 'the luxury of crying out about doom'. Meanwhile, Corde contemplates an invitation he has received to collaborate with a scientist on an article which claims that lead poisoning is the chief cause of urban problems and that 'chronic lead insult now affects all mankind'. Corde is dubious, but, characteristically, is intrigued by the potential metaphors – lead as 'the Stalin of the elements'.

Valeria dies after Corde and Minna have managed one more visit, and Ebry is convicted without much ado. Beyond this, there is little plot development in *The Dean's December*, the story serving chiefly as a vehicle for a sustained meditation on various moral and philosophical questions. There is throughout an implicit comparison between two political cultures – the ossified, censorship-controlled Communist state, and the free world as it is epitomised by Bellow's Chicago, which is the product of corruption and cynicism, false values and moral confusion.

William BOYD   1952–

An Ice-Cream War

Boyd made his name with his first novel, *A Good Man in Africa* (1981), which drew comparisons with the work of Evelyn Waugh and Kingsley Amis. *An Ice-Cream War*, which won the John Llewelyn Rhys Prize and was shortlisted for the Booker Prize, is once again set in Africa, where Boyd spent his early childhood, but goes back in time to an almost forgotten period in the continent's history, the First World War. The novel is meticulously researched, richly textured, fields an enormous cast of characters, and displays Boyd's narrative skills to the full. It is in some ways a rather old-fashioned adventure story, which none the less manages to convey something about the nature of war and the unpredictability of human life. Of his soldier protagonist Boyd writes: 'Gabriel thought maps should be banned. They gave the world an order and reasonableness which it didn't possess.' This is a theme which is to preoccupy Boyd in later, more ambitious, but perhaps less satisfying books, such as *The New Confessions* (1987) and *Brazzaville Beach* (1990).

Gabriel Cobb's fraught honeymoon is interrupted by the outbreak of the First World War. He rejoins his regiment and sails to East Africa, where, after a chaotic landing, he is severely wounded. Back in England, his adoring younger brother, Felix, an Oxford undergraduate in revolt against his military family, mixes in Bohemian society. He is obsessed by Amory Holland, the sister of his best friend, Philip Holland, but she brutally rejects him when he makes a drunken pass at her. He gradually drifts into an affair with Gabriel's new wife, Charis, who, racked with guilt and aware that the relationship cannot continue, drowns herself in a goldfish pond. She has written to Gabriel to confess her adultery, and Felix, sure that he has been named as her lover, determines to intercept the letter. He contrives to enlist and travel to Africa with an obscure regiment.

Meanwhile, Gabriel has been reinfecting his wound so as to remain at the German hospital where he is acting as an orderly. His colleague, the voluptuous nurse Liesl von Bishop, teaches him German, and he begins to compile a dossier from overheard remarks about troop movements. One evening he is discovered while waiting to spy upon Liesl as she showers, and is assumed to be eavesdropping on a group of German officers. He is imprisoned, but Liesl helps him to escape. Her genial Anglo-German husband, Erich, is sent to recapture him, but his native troops misunderstand Erich's orders and cut off Gabriel's head. Felix discovers the body and is determined to take revenge.

Erich is also being pursued by Temple Smith, an American sisal farmer and former neighbour. When war broke out, Erich was sent to requisition Temple's land and buildings, but promised not to touch Temple's prize possession, a pulping machine, or decorticator. Temple sets off in order to extract compensation for his losses and eventually persuades Reginald Wheech-Browning, a punctilious, pompous and infuriating ADC, and Mr Goolam Hoosam Essanjee, the plump, dapper and obliging manager of the African Guarantee and Indemnity Co., to accompany him to his farm, which is now in enemy territory, in order to make an assessment. The mission is a failure and Mr Essanjee is killed. Temple joins up, hoping to recapture his farm, but when he eventually returns there, he finds that the decorticator has vanished and the house has been used as a latrine. He spends the remainder of the war in pursuit of Erich.

Both Temple and Felix suffer from the incompetence of the army in general and Wheech-Browning in particular. Felix is severely injured when the accident-prone ADC accidentally fires a mortar, blowing the syphilitic Portuguese Capitao Pinto to irrecoverable pieces. While recuperating, Felix hears that the war has ended and that Erich is among the surrendering officers. He seeks him out, intending to discover why he killed Gabriel and then to shoot him. Erich has already died from influenza, however, and Felix and Temple reflect that the mystery surrounding Gabriel's death and the disappearance of the decorticator will never be solved.

Anita BROOKNER   1928–

Providence

The heroine of Brookner's second novel describes herself as a 'declassée woman'. Kitty Thérèse Maule is of mixed English, French and Russian descent. The duality of her personality is shown by the two names – Thérèse, used when she is with her French grandparents who had brought her up, and Kitty, reserved for her 'English' existence. She longs to understand English life, to identify with it, but her French nature shows itself in her practicality, her elegance and her cooking. At the same time she is a disappointment to her French family because at thirty she is still unmarried and because she has acquired a certain impractical sentimentality.

Kitty is a typical Brookner heroine – intelligent, intellectual, analytical. She is, in fact, a reflection of the author. Brookner has written about Romantic art, Kitty Maule studies Romantic literature. But Kitty dislikes her own intellectuality; she longs to be 'beautiful, demanding and irrational' – the sort of woman who, in her opinion, achieves what women really want. What this is remains slightly vague: probably marriage, children, a sense of belonging. Kitty, on the other hand, is described by her colleagues, and shown by the author, to be a talented scholar and an inspired teacher. Her analysis of Benjamin Constant's novel *Adolphe* and of Romantic literature as a whole is the second theme of the novel, a commentary on her own personality. Demoralised by her feeling of inadequacy, complimented by her colleagues, and exquisitely dressed by her couturier grandmother, Kitty achieves professional standing by giving a public lecture. Almost immediately she loses Maurice Bishop, the colleague she loves deeply, who, with his charm, complete assurance and a vague, sentimental piety, called by him 'belief in Providence', epitomises English life for her.

We last see Kitty (typically for Brookner's heroines) not at the moment of her triumph after her lecture, but at Maurice Bishop's dinner party. Having thoughtlessly led Kitty to believe that the party would symbolise their union, he equally thoughtlessly makes it clear that his real choice is Kitty's slovenly, intellectually lazy, stunningly beautiful student, Miss Fairchild. Ever the rational scholar, Kitty tells herself 'I lacked the information' and prepares to embark on another hopeless attempt to explain her complicated antecedents to the other English guests.

William COOPER   1910–

Scenes from Metropolitan Life

see **Scenes from Provincial Life** (1950)

Alice Thomas ELLIS   1932–

The 27th Kingdom

Aunt Irene, the forceful protagonist of *The 27th Kingdom*, has much in common with the strong female presences in Ellis's other novels. She is less malicious certainly than Rose in *The Sin Eater* (1977) and more homely than Lydia in *Unexplained Laughter* (1985) but like them she is a vessel in which the forces of good and evil vie. *The 27th Kingdom* is an intricate and subtle comedy of manners of a seeming simplicity, achieved only through careful and minute construction. The novel tantalises but eludes certain resolution. It was shortlisted for the Booker Prize.

Aunt Irene (pronounced Irina to rhyme with 'serener') lives in Chelsea, but senses about her still the spirit of the old country – Russia – from which her Boyar ancestors first fled as Roman Catholic converts; fleeing thence to Lithuania, Austria and on, until settling finally in the 27th kingdom of their travels – Britain.

Some minor official had been persecuting her for months with trivial enquiries about her means; but since she couldn't bear forms and had a profound conviction that her need of her own money was greater than the government's, she had ignored him. She felt the noble irritation of a fine spirit called from viewing the sunset to inspect the blockage in the kitchen sink. Her ancestors, she thought, would have had him boiled. In oil. Or, to be more culinarily precise, deep-fried.

THE 27TH KINGDOM
BY ALICE THOMAS ELLIS

Irene is basically good – 'she was tempted to give all her goods to the poor and see what happened, but decided against it' – but slothful and more interested in cooking and housekeeping than people. She is a 'partial, or flawed hedonist' whereas her handsome, malicious nephew, Kyril, with whom she lives at Dancing Master House, is a total hedonist, as is her cat, Focus.

Irene's sister is Revd Mother of a convent in Wales where Valentine, black, tall, slim and beautiful, a promising postulant from overseas, has become, owing to her absolute perfection, a conduit for the miraculous. There is 'absolutely nothing as tiresome, exhausting and troublesome as a fully functioning thaumaturge in a small community' and so Revd Mother decides upon placing Valentine with Irene until a sign comes to show that Valentine might return.

Nothing annoys devils more than the sight

of perfection. They are driven to taunt and surround its personification, usually nuns in convents. But Valentine, still perfect, is in Chelsea.

Sinful but goodhearted Mrs O'Connor, the cockney fence, witnesses the first miracle: Valentine levitating at the presbytery. There follows a whole series of them. Alcoholic and mad Major Mason, who virtually 'lives in a pub on the Kings Road' and has never been seen sober, sees the pigeons around St Luke's turned into devils and vows to give up drink. Mrs Mason, the reluctant and bitter daily help at Dancing Master House, 'not designed by nature or nurture to be a char', is affronted by the presence of Valentine and makes a racist remark so outrageous even she is shocked. Later, a miracle makes her understand, if only for a moment, how Valentine might have felt: 'loathed on sight for something you couldn't help'. Kyril sets his sights on Valentine only to be humiliated for the first time in his life. Poor Mr Sirocco, whose 'nerves did not permit spontaneity', has a morbid fear of pigeons and is in love with Kyril. The day after a party, at which he asks to dance with Valentine, he is found dancing at his window, hanging by the neck from his old school tie. On the floor is a scattering of pigeon feathers. Irene tells the story of the dancing master of Dancing Master House. Somewhere a woman is being hanged for a crime of passion.

Irene's disorganised life does not allow her time to open any official mail or declare her true income. Someone is phoning her and visiting the house. He seems to be the man from the Inland Revenue. She even thinks she saw him dancing at her party. Later, Irene has a premonition that the taxman is really phoning for Valentine. He is Stanley, whose wife, mother and sister were negligently drowned by Valentine's sister, Joan. Stanley came close to drowning then and does so again now. He is chased into the Thames by Mrs O'Connor's son and a gang of friends, out to put the frighteners on who they imagine to be Irene's persecutor. Valentine's last miracle is to rescue him, floating across the water.

Irene's sister receives the sign she has been waiting for.

1982

Alasdair GRAY   **1934–**

Lanark

Lanark is divided into four books and these are given in the order three, one, two and four. Book three, accordingly, begins the tale. Its central character, Lanark, is newly arrived in Unthank, an infernal Glasgow, where the sun is almost never seen. Unable to recall anything of his past, Lanark meets Sludden and falls clumsily in love with Rima, one of Sludden's many hangers-on. Lanark and Rima sleep together once. They find that they are both suffering from dragonhide: their skin is changing to insensate scales. Rima ends the affair. Miserable and frightened by his disease, Lanark wanders to a cemetery, where an enormous disembodied mouth invites him to hurl himself down its throat. Lanark does so. After a painful rebirth, he awakens, cured of dragonhide, in a subterranean institute. The scope of its activities is vague, but Lanark is appointed a physician. His task is to stop his patients turning into pure energy. He succeeds in the case of Rima, who has fallen into the hospital too. The pair renew their liaison and decide to return to the world above ground. Before they set out, an oracle describes Lanark's former existence – a biography which comprises books one and two.

Previously, Lanark was Duncan Thaw. Brought up in (the real) working-class Glasgow, his earliest memories are of the Second World War. To escape the bombing he is sent north to the Grampians. Lifted in spirit by their grim splendour, he is also depressed and troubled by asthma and sex. He develops into an artist and attends art school. Still, his life is bedevilled by illness and incompetence with women. Commissioned to decorate a church, he strives to convey in his murals his horror at history and his wonder at creation. His goal is beyond him: trying to attain it prompts nervous collapse and schizophrenic delusion. He kills (or believes he kills) a girl, then, a little later, drowns himself.

Book four tells of how Lanark and Rima leave the institute. They are forced to go back to Unthank and, on their nightmare journey, Rima discovers she is pregnant.

When she and Lanark reach their destination, they learn that it is under threat from apocalyptic pollution; furthermore, the council – an omnipotent global organisation closely connected to the institute – wishes to liquidate the city. Time's flow accelerates. Rima has a baby, Alexander, who grows to be a small boy in a matter of hours. Sludden, by now Unthank's political chief, makes Lanark Lord Provost of the region and tempts Rima away. She and Alexander settle with Sludden, while Lanark is despatched to beg the council to reprieve Unthank. He fails, baffled by the council's procedures. An old man, claiming to be the author/creator, explains that readers will not accept a happy ending. As Unthank subsides into the earth, Lanark is told of his approaching death. He says, 'I ought to have more love before I die. I've not had enough.'

'Lanark has been around for a long, long time,' said Wilkins, 'I think he deserves a three-syllable name, don't you?'
'Oh, he certainly deserves it,' said the other man. 'There's nothing wrong with a two-syllable name, I'm called Uxbridge, but Lanark has earned something more melodious. Like Blairdardie.'
'Rutherglen, Garscaden,' said Wilkins.
'Gargunnock, Carmunnock, Auchenshuggle,' said the other man.
'Auchenshuggle has four syllables,' said Wilkins.

LANARK
BY ALASDAIR GRAY

The novel includes an index of its own influences (described as 'an index of diffuse and imbedded Plagiarisms' in the table of contents). Listed amongst these are Marx, Freud and God: and, indeed, *Lanark* is a vast dystopian fiction. Often alluding to Old Testament narratives (notably, those of Sodom and Gomorrah, Job and Jonah and the Whale), it portrays a civilisation in the last stages of entropy. Material exploitation and moral cynicism finally bring destruction to fabulous Glasgow, whereas actual Glasgow – with its drabness, Philistines and fickle women – brings destruction to Thaw. Thus, books three and four are at once a prophecy to a modern Nineveh and a sado-masochistic dream. Inadequate, like Thaw, Lanark is punished – and Unthank is punished for punishing him.

Chapter 12 of *Lanark* was a runner-up in the 1958 *Observer* short-story competition. Each of the four books has an allegorical frontispiece designed by Gray.

Thomas KENEALLY   1935–

Schindler's Ark

While buying a briefcase in Los Angeles, Keneally got into conversation with the shop's proprietor, who told the author that one story he ought to write was that of Oskar Schindler, a German industrialist who saved the lives of hundreds of Jews during the Second World War by setting up his own 'benevolent' labour-camp. The shopkeeper, to whom the book is dedicated, had himself been saved by Schindler, and introduced Keneally to other survivors. Keneally interviewed fifty Jews from Israel, Australia, America and Europe, and *Schindler's Ark* is based upon their testimony and the author's own historical researches. It was the fourth of Keneally's novels to be shortlisted for the Booker Prize for Fiction and proved a controversial winner. Keneally insisted that he had 'attempted to avoid all fiction', but the judges decided that the book 'had a greater creative stature than the other books on the shortlist'.

The novel relates how Schindler, a self-made industrial tycoon, moves to Poland in 1939 in order to profit from the Nazi expro-·priation of Jewish businesses. A bonviveur, incorrigible womaniser, black-marketeer and arch-capitalist, he ingratiates himself with Nazi officials and makes a fortune from exploiting cheap Jewish labour in his enamel-ware factory in Cracow. He later spends that fortune (and risks his life) building his own labour-camp – the 'Ark' – in the backyard of his factory, and bribing SS officers, even gambling with them, in order to prevent his workforce being taken away to the death-camps.

The novel is a provocative study of good and evil, of the inexplicable quirk of fate that produced in Schindler a mission to save life, while producing in others a mission to exterminate. Schindler is not presented as a saint, and his recognition of a common humanity with the oppressed Jews is a re-action to a situation in which conventional codes of conduct play no part. The SS, chasing their dream of racial purity, slaughter the Jews indiscriminately, permitting Jewish witnesses because they believe all witnesses will eventually perish themselves. The strength of the Nazi obsession fuels Schindler's own, and between these two poles lies the fate of the Jews, whose obsession is to survive, armed only with an instinct for self-preservation and with the hope that Schindler inspires in them. The book documents acts of extreme courage and extreme baseness performed by ordinary people in extraordinary circumstances.

Keneally realised that 'by looking at Schindler, you looked at the Holocaust in a way that hadn't been – at least at the popular level – conceived before: the economic side, the bureaucratic side … It had never struck me before that it was so organised, that it had economic motivations, that Krupp and IG Farben had to pay the SS for their slave labourers.' He dismisses the word 'faction' as barbaric, and explains that he chose to write the book as a novel because he felt Schindler was 'the creation of some great cosmic novelist', and that his character was 'so suitable for what would appear to be the conventions of fiction'.

The book was published in the USA as *Schindler's List*, which is also the title of Steven Spielberg's dignified, monochrome, award-winning film adaptation (1994).

Armistead MAUPIN   1944–

Further Tales of the City

see **Tales of the City** (1990)

1982

Timothy MO   1950–

Sour Sweet

Mo's second novel, which was shortlisted for the Booker Prize and subsequently adapted by Ian McEwan for a film (1988) directed by Mike Newell, provides an authentically observed and richly comic portrait of the immigrant Chinese community in an alien 1960s' London. Mo, himself the product of an Anglo–Chinese marriage and education, explores the mutability of the traditional values as they are inevitably transformed by contact with the host community.

The novel centres on the fortunes and everyday life of a family of recent arrivals from Hong Kong: Chen, the husband, a waiter in a Chinese restaurant in Soho; his wife, Lily; their small son, Man Kee; and Lily's sister, Mui. In parallel, and tragically intertwined with their fate, the novel traces the ruthless infighting of the Hung 'family', or triad, a Chinese Mafia-style organisation, also originating from Hong Kong, in the process of developing the heroin trade in London.

When Chen naïvely becomes embroiled with the nascent triad gang after incurring gambling debts, the family moves to the virgin territory of South London, to evade the triad's blackmailing attentions. Urged on by Lily, who nurses business ambitions and who knows nothing of Chen's predicament, they open a Chinese takeaway. The business develops haphazardly under the matriarchal guidance of Lily supported by her sister, but the stability of the close family unit becomes strained by English influences. Mui becomes pregnant and Man Kee develops a taste for jam tart and custard at school. Lily finds herself distanced from the others as a result of her greater loyalty to their Chinese heritage. Their encounters with 'foreign devilry', as they regard various aspects of English life (tax inspectors, outings to the seaside), are packed with piquant observations of cultural misunderstanding and petty chauvinism.

Meanwhile, the Hung triad, under the leadership of the psychopathic and aptly named, Red Cudgel, launches a bloody, but only half-successful, surprise attack on the rival 14–K triad society. The Chens' dom-estic life is abruptly shattered when Chen is murdered on the orders of Red Cudgel, who mistakenly believes him to have been selling heroin on his own account. In fact Chen is entirely innocent, the culprit being a former colleague of Chen, the petty crook Roman Fok, who originally introduced Chen to the triad.

A group of younger and more astute triad members, named after their ranks of White Paper Fan, Night-brother and Grass Sandal, become dissatisfied with the crude out-moded methods employed by Red Cudgel, exemplified by Chen's unnecessary murder. They violently oust him, and make a deal with the 14–K society to share the heroin trade between them, a recognition that old ways, even in crime, must be adapted. Lily never finds out that Chen has been mur-dered, choosing to believe instead that he has temporarily deserted her to find better work, an illusion encouraged by the anonymous cheque she receives every month, a 'welfare' payment from the triad.

Anne TYLER   1941–

Dinner at the Homesick Restaurant

The novel considered by many people to be Tyler's most characteristic opens with eighty-one-year-old Pearl Tull on her death-bed, and tells her family's story through each member in turn. The young Pearl is a serious girl, who at thirty appears to have been left on the shelf. Then she meets Beck Tull, a good-looking twenty-four-year-old sales-man who courts her with chocolates and flowers and then – more seriously – with pamphlets describing the products of his firm. They marry and travel where Beck's job takes him. When their son Cody falls ill, Pearl wonders what she would do if he were to die, and decides she must have extra children, as 'spares'. So after Cody comes Ezra, and then Jenny. One night, when the children are aged fourteen, eleven and nine, Beck decides that he no longer wants to be married. Tyler vividly describes the details of how Pearl copes with the day-to-day struggle to bring up her family alone – meals, household chores, the childrens' pranks, having too little money – and uses them to show how the family members develop and interact.

At eighty-one Pearl recognises that the family is everything to her, but each child is closed off from her in some inexplicable way. Although they are all attractive, likeable people, each has a flaw in his or her character: Cody is prone to unreasonable rages; Ezra has not really lived up to his potential; and Jenny is too flippant. At family gatherings they tend to remember only poverty and loneliness, and Pearl wonders if they blame her for something. She is unable to fathom her children's mystery, and her last act is to be the focus of a final family gathering – her own funeral.

We learn of the children's lives largely from their own perspectives. Cody is the one who always ends up in trouble, both at school and at home. After college he decides to get rich, goes into business and becomes a great success as an efficiency expert. However, he is eaten up by a jealous rage against his brother. Everyone's favourite, Ezra proceeds through life absentmindedly, meditatively. Despite Pearl's conviction that he is destined for great things, he goes to work at Scarlatti's, the only decent restaurant in town. Mrs Scarlatti looks upon him as a substitute son, and for a nominal $1 she makes him a full partner. Attempting to wreak revenge, Cody steals Ezra's fiancée, only to find himself locked in an unhappy marriage, cut off by his own anger from the rest of the family. Ezra resigns himself to loneliness and continues to live with his mother. By now sole owner of Scarlatti's, he renames it 'The Homesick Restaurant' and serves dishes such as gizzard soup to a clientele of lonely people.

Jenny is skinny and severe-looking as a teenager. While all her classmates are wearing bouffant skirts and perky blouses with turned-up collars, she is in her mother's hand-me-downs. By the time she reaches medical school she has lost a little of her primness and developed a breathless, fly-away air. She marries and sets up a pediatric practice, but subsequently goes through two divorces and a nervous breakdown.

Although she has what seems to be a happy third marriage, surrounded by children, she can cope only by refusing to take anything seriously.

The novel ends with the family dinner that follows Pearl's funeral. As with all the other meals that have served as markers in the lives of the family members, this one begins to disintegrate. Cody turns on his long-absent father: 'You think we're some jolly situation comedy family when we're torn apart, torn all over the place.' Beck's first appearance at this point is somewhat jarring for the reader, and the ending of the novel, with its reconciliation between father and son, feels contrived, but Tyler's narrative voice is seductive. Her strength in depicting Pearl's determined attempt to survive is undeniable, and her (perhaps unfashionable) belief that families, despite and because of their problems, are the most important things we have is wholly convincing.

Alice WALKER   1944–

The Color Purple

Celie grows up in a black depressed area in the Southern states, with no hopes of a better future. Her biggest confusion is in trying to understand why. This is her first question to God in a series of simple but revealing letters which, together with those she receives from her sister, form the book's entire structure.

Her early letters describe how her brutish father, finding his wife unable to meet his sexual needs, turns to Celie and rapes her. She bears him two children and sees them taken away from her almost immediately. Her father forces her to marry 'Mister', a violent bully more interested in her younger sister. He uses her as housemaid and sexual outlet, and she also looks after his obnoxious children. Harpo, the eldest boy, eventually brings home a wife, Sofia; robust and self-confident, she forms a strong contrast to Celie's cowed humility.

Celie's life changes dramatically, however, when she meets the object of Mister's undying devotion. Singer and celebrity Shug Avery is everything that Celie is not – pretty, talented and able to speak her mind. She teaches Celie to be proud of herself, her character and her appearance, and with her Celie also finally discovers the pleasures of sexual love (this is detailed in one particularly frank communication with God). Through her love for Shug she also comes to a wider understanding of the God of love. As Shug explains, everything has to be loved, even the fields filled with the colour purple.

The only other person Celie has ever loved is her sister, Nettie, lost to her years ago when she left home in search of a better life. Shug discovers that Mister has been hiding Nettie's letters and the two of them go through the pile together, reading Nettie's detailed account of her life as a missionary in Africa.

This discovery leads to Celie's breaking out. She goes with Shug to the city, where she starts her own business designing trousers. The end of the novel sees her reunited with all those important to her. She has her own life and home, and a family for her love.

The Color Purple is about women, love, individuality and understanding. Walker packs in this multiplicity of themes by using the very simple but effective technique of the two letter writers. She avoids sentimentality or blatant feminism: both her writers are clearly honest and unpretentious. They serve as her commentators on two very different but intrinsically connected worlds. The novel won the Pulitzer Prize and was made into a much-heralded, but not altogether satisfactory film (1985), directed by Steven Spielberg from a screenplay by Menno Meyjes.

1983

Cruise Missiles deployed in Britain ● USSR shoots down Korean jumbo jet ● US invades Grenada ● Ronald Reagan proposes 'Star Wars' defence system for US ● Nobel Peace Prize awarded to Lech Walesa ● First US woman astronaut in space ● Wearing of seat belts by front-seat car passengers made compulsory in Britain ● IBM produces the first personal computer with a built-in hard disc ● Ayatollah Khomeini declares that Islam is a 'religion of the sword' and sends armed pilgrims to Mecca ● David Mamet, *Glengarry Glen Ross* ● Films: *The Big Chill* ● Television: Launch of 'TV AM' breakfast-time station ● Popular music: Frankie Goes to Hollywood, 'Relax' ● Tennessee Williams, Arthur Koestler, William Walton, Rebecca West, George Balanchine, Kenneth Clark, Buckminster Fuller, Luis Buñuel, Niklaus Pevsner and Joan Miró die ● Robert Creeley, *Collected Poems 1945–1975* ● James Berry (ed.), *Dance to a Different Drum* (anthology)

J.M. COETZEE   **1940–**

Life and Times of Michael K

When Michael K was born his 'lip curled like a snail's foot, the left nostril gaped'. Because of his disfigurement, and 'because his mind was not quick', he was taken out of school and spent the rest of his childhood at 'Huis Norenius' with other 'afflicted and unfortunate children'. From the start Michael K is marked out for solitude. At fifteen he joins the Parks and Gardens division of the municipal services of the City of Cape Town as 'Gardener, grade 3(b)'. Three years later, after a short spell of unemployment, he returns as 'Gardener, grade 1', the highest position he ever obtains. This suits him well as he is working on his own and growing things; but his mother, who used to 'polish other people's floors', becomes ill and expresses a desire to return to the farm where she was born, so Michael throws up his job to take her there. By now there is a war on.

They wait for their travel permits until they can wait no longer. The journey is hard, Michael's mother becomes weaker and they can not in the end escape the army road blocks. She dies in Stellenbosch hospital, where they have taken her. Michael takes her ashes in a box to what he hopes is the right farm – it is now deserted – and buries them. 'He hoped that his mother was more at peace now that she was nearer to her natal earth.'

The first fifty pages of the novel are concerned with mother and son's hopeless bid to escape. The remainder are also concerned with escape as Michael K, harassed by the world, progresses along the road to pure spirituality. His journey falls into five distinct phases. Michael's first spell alone in the veld sees him active, hunting the wild goats, and living on the birds he kills with his catapult. He is driven out by the arrival of an army deserter who claims to be the grandson of the man who owned the farm. The

---

He stood leaning against the frame of the pump, feeling the tremor that passed through it each time the piston reached the bottom of its stroke, hearing the great wheel above his head cut through the dark on its greased bearings. How fortunate that I have no children, he thought: how fortunate that I have no desire to father. I would not know what to do with a child out here in the heart of the country, who would need milk and clothes and friends and schooling. I would fail in my duties, I would be the worst of fathers. Whereas it is not hard to live a life that consists merely of passing time. I am one of the fortunate ones who escape being called.

LIFE AND TIMES OF MICHAEL K
BY J.M. COETZEE

'Visagie grandson ... had tried to turn him into a body-servant', but he cannot survive in the mountains. His time at Jackhalsdrif, one of the resettlement camps for refugees from the war where the policeman tells him, 'you work for your food like everyone else', is miserable. 'It was better in the mountains, K thought. It was better on the farm, it was

better on the road. It was better in Cape Town.' So he escapes. In his second period on the farm he is passive, fearful of discovery. 'A man who wants to live cannot live in a house with lights in the window. He must live in a hole and hide by day.' He catches lizards. His achievement is to grow pumpkins and melons. He is close to starvation when the army find him.

The next section is told by the doctor of the 'rehabilitation camp'. Michaels, as he is known there, fascinates and infuriates him. He cannot eat the food the camp provides, he refuses the saline drip, and he will die of starvation in great distress. There is nothing the doctor can do about it. Watching him, the doctor is driven to doubt everything that he has previously taken for granted. (It is only now that we learn what this futile war is about – 'so that minorities will have a say in their destinies'.) He burns to know the story that 'Michaels' will not tell him: 'I am the only one who sees you for the original soul you are.' But Michael escapes again, and still there is no peace for him. On the seafront at Cape Town, destitute and naked except for his overalls, he meets a strange man with his two 'sisters', who wants, unbidden, to help him; and one of the 'sisters' forces on him his first experience of sex. We leave him hiding from them in the room where his mother used to live.

Life and Times of Michael K, which won the Booker Prize and consolidated Coetzee's international reputation as a leading novelist, is a work of the purest intensity, and says more than any recent novel about the nature of freedom and spiritual exploration.

John FULLER   **1937–**

Flying to Nowhere

Fuller is best known as a poet, but he is also the author of tales for children in which everyday objects and events conceal fantastic happenings. Both these aspects of his work are apparent in this compact, richly imagined 'tale' for adults, which was short-listed for the Booker Prize. It takes its epigraph from the Emperor Hadrian's celebrated death-bed poem, which translates: 'Little soul, wandering and pleasant guest and companion of the body, into what places will you now depart, pale, stiff and naked; and you will sport no longer as you did!'

Some time in the Middle Ages, when miracles were possible and credible, St Llueddad drew water from rock on a Welsh island in order to quench the thirst of a dying bird, which was thus enabled to fly away. The miraculous well subsequently became a place to which pilgrims came to seek cures, and an abbey was founded on the island. When a bishop on the mainland receives complaints that pilgrims have failed to return from the island, he sends his emissary, Vane, to investigate. *Flying to Nowhere* opens with Vane's hazardous crossing to the remote island, during which, as a result of his carelessness, stubbornness and pride, his stallion dies horribly. The Abbot proves evasive when questioned about the fate of the missing pilgrims, directing Vane to a cemetery. Only three of the twenty-six pilgrims have graves there, but at the well Vane discovers a corpse dangling in the water, possibly poisoned by it. Vane decides to excavate the well's overflow system and gradually uncovers a channel which leads to the Abbot's house. Within this labyrinthine building, the Abbot has a dissecting room where, up to his elbows in putrescence, he conducts gruesome experiments in search of the seat of the soul.

Meanwhile, Vane's resentful young servant, Geoffrey, falls in love with Tetty, one of the girls who work on the island's farm. At the farmhouse, old Mrs Ffedderbompau lies dying, having fallen out of an apple tree: 'Why am I inside myself and not somewhere else?' she asks the Abbot. In preparation for the ceremonial ordeal which the monastery's noviates must undergo, the Abbot delivers a sermon about mushrooms, chastity, spirit and flesh: the monk must resist 'the temptation to forget that he is dust. It is the temptation to fly.' Wondering why the monastery has no older brothers, and only a handful of novices, Vane loses his footing in the underground chambers and drowns in the Abbot's corpse pit. As the waters flood the dissecting room, it seems as if they really do have revivifying powers, nourishing fungi on the walls and books. The Abbot imagines the woodwork becoming trees again and the

leather bindings becoming cattle, as a new life is born of corruption: 'on the edge of his hearing, louder than the stampede of his library, though reduced and distant, rose for an endless moment the purposeful clamour of tiny wings'.

Escape is an underlying and unifying theme of the novel, in particular the attempt to escape the restraints of the body, its infirmities and its mortality. The pilgrims attempt this by their faith in miraculous cures; watching a candle, a novice thinks: 'The spirit struggles to be free, vain as the wave struggling to attain the shape of a horse, or a horse struggling to fly'; Geoffrey and Tetty plan to leave the island to marry and have children, but feel they will never get away; and one of the farm girls, dreaming of escaping from a chrysalis, cries: I'm flying to nowhere. I'm just becoming myself.'

Howard JACOBSON   1942–

Coming from Behind

The comic hero of this, Jacobson's first novel, is Sefton Goldberg, a Manchester Jewish boy who reads English at Cambridge. From there his career goes steadily downhill, and we find him lecturing at Wrottesley Polytechnic, in an unlovely Midlands town, trying to hide from the greater successes of his contemporaries while he works half-heartedly on a book about failure. Nor is his own failure merely professional. Jacqueline, his occasional bedfellow, represents a missed opportunity to exchange his old, stale, world for her bright, fresh one. Jacobson himself, after reading English at Cambridge under F.R. Leavis, spent some time teaching at Wolverhampton Polytechnic, and *Coming from Behind* takes the campus novel – traditionally set at some remove from privileged Oxbridge in redbrick universities – even further downmarket academically.

Other comic themes are interwoven with that of Sefton's failure. His sophistication contrasts with the attitude of his parents, who live in a world where all Jewish boys become lawyers or dentists. At the polytechnic, structuralists battle it out with 'Common

Readers'; and in their personal lives literary academics pursue, with comic ineptitude, the romantic ideals they are used to lecturing on. At Cambridge, Sefton discovers two alien and characteristically Gentile traits which resurface throughout the novel: the obsession with Nature, and the curious high-mindedness which views the study of literature as a preparation for Ordinary Life. Behind both lurks the tutelary spirit of D.H. Lawrence, whom Sefton regards with extreme suspicion.

The graduation parade itself was as irresistible challenge to Sam. Not for as long as he lived would Sefton ever forget the sight of his father bounding along in his bowler in front of the procession on King's Parade, tearing from one side of the road to the other disappearing down alleys and materializing again in shop doorways, shouldering aside Proctors and Praelectors, in order to get the best possible snaps of that purple-faced ingrate, Sefton, in his gown and hood. Nor was he ever likely to erase from his memory the sound of strictly forbidden flash-bulb in the Senate House, the sharp blue light and the POP, just as he knelt to grasp the finger of the Vice-Chancellor, held out to him with waning elasticity like a well-milked teat. There was another boy holding another finger at the same time, but the Vice-Chancellor did not need to ask for whom the bulb flashed – it flashed for Goldberg, not for Goy.

COMING FROM BEHIND
BY HOWARD JACOBSON

The plot in the course of which these themes are played out is slight and farcical. Academic rivalries come to a head when the polytechnic's chief administrator, in league with the captain of Wrottesley Ramblers, arrange for the Department of Twentieth-Century Studies to transfer its premises to the local football stadium. Meanwhile Sefton finds he has been shortlisted for the Disraeli Fellowship at Holy Christ Hall, Cambridge. Anxieties about the Anglican exclusivity of that institution, dating from his under-

graduate days, are amply justified by his experiences at high table, and though he manages to give a good account of himself, the Fellowship goes to his rival, Ernest Weekley.

At the eleventh hour Sefton enjoys two unexpected triumphs. Chosen (by default) to give a speech at the official twinning ceremony of the poly with Wrottesley Ramblers, he has the pleasure of seeing his arch-enemy, Walter Sickert Fledwhite, dragged off by the police. And as the successful candidate for the Disraeli Fellowship moves on to higher things, it falls into Sefton's lap after all. But Sefton remains a lovable failure to the last: we learn in a postscript that while Sefton's book remains in manuscript, it is his friend Peter Potter who is to publish a novel.

William KENNEDY   1928–

Ironweed

Kennedy's fourth novel grew out of his previous work in terms of both style and content. His first novel, *The Ink Truck* (1969), was a black comedy set against the background of a newspaper strike which borrowed elements from both Joseph Heller and Thomas Pynchon. *Legs* (1975) mixed events from recent history with fictional elements to tell the story of the gangster 'Legs' Diamond. Both novels were set in Albany NY, although the role played by the setting is far more important to the latter book and the style is more determinedly realistic. *Billy Phelan's Greatest Game* (1978) confirmed Kennedy as 'an Albany writer' with its meticulous reconstruction of life there in the 1930s.

Ironweed takes in the fantastic elements from *The Ink Truck*, the gritty realism from *Legs* and its story from an undeveloped plot-line in *Billy Phelan's Greatest Game*. Its rootedness in Albany is held in common with all. It tells the story of Francis Phelan, Billy's wino father, as he moves towards atonement for the deaths he has caused. The first of these is that of his baby son, whom he dropped on his head by accident and whose death led to Francis's subsequent slide into alcoholism. There are others: a strikebreaker killed by a stone thrown by Francis, and a fellow wino who attacked him with a meat cleaver. These victims appear as ghosts to Francis in his wanderings about Albany and, although supernatural, they are presented naturalistically by Kennedy. Their appearances are supported in part by the novel's Catholicism, which also supplies the underlying basis of the plot. Guilt, expiation and grace are the stages through which Francis must pass. His guilt is given shape in the persons of the ghosts, his expiation takes the form of his deliberately elected wanderings as a bum, while grace is marked by his return to his family at the end of the book.

He found his way to a freight yard, found there an empty boxcar with open door, and so entered into yet another departure from completion: the true and total story of his life thus far. It was South Bend before he got to a hospital, where the intern asked him: Where's the finger? And Francis said: In the weeds. And how about the nose? Where's that piece of the nose? If you'd only brought me that piece of the nose, we might be able to put it back together and you wouldn't even know it was gone.

IRONWEED
BY WILLIAM KENNEDY

The greater part of the novel is set in the world of Albany's down-and-outs, which Kennedy depicts with great skill. The moral brutalisation of life on the street is matched by the more obvious physical hardships where freezing to death at night is a real possibility. The two winos closest to Francis, his friend Rudy and girlfriend Helen, are moving towards death when the novel opens and he cannot save them. Nevertheless, in the exchanges between the down-and-outs, Kennedy establishes Francis as more morally capable than the others, more perceptive of the needs of others and willing to help them when possible. At the same time, he remains an authentic wino, and the early scenes showing the edgy, mistrustful relations between Francis and the other derelicts are among the best in the book.

The publication of *Ironweed* was a troubled affair. After reading the first 100 pages and seeing that Kennedy's desire to write about Albany's low-life was as strong as ever, his editor at Viking recommended that he be dropped. Kennedy took the manuscript to thirteen other publishers, all of whom rejected it, until Saul Bellow, Kennedy's former teacher, heard of it. A strongly worded letter to Viking from the normally taciturn Bellow eventually led to the novel being published along with his three earlier works, repackaged as *The Albany Cycle*. The reviews of *Ironweed* were favourable almost without exception, and the book was awarded the Pulitzer Prize, the National Book Circle Critics' Award and a MacArthur Foundation grant.

Francis KING   1923–

Act of Darkness

Widely admired by critics, King's work has not had quite the popular success it deserves. Perhaps his most impressive achievements have been his short stories and novellas, but he has also written a great many novels which display his strong sense of narrative and his ability to go to the dark psychological heart of human experience. A characteristically compelling and disturbing book, *Act of Darkness* is based upon the famous case of Constance Kent, which has long fascinated writers from Dickens onwards. King transposes the action from mid-Victorian England to India of the 1930s, where he grew up, and draws upon his own family history in reimagining the story. The characters have different names from those of the real people involved, but their relationships are the same, except in one startling instance, which provides the author with his solution to the crime.

The novel is divided into five sections, the first of which, 'Omens', sets the scene before the murder takes place. The individual chapters reflect the lives and thoughts of the principal characters of the drama: Peter Thompson, the precocious six-year-old child who will be murdered; his father, Toby, a priapic government official whose first wife died young; Toby's newly pregnant second wife, Isabel, who had previously been the family governess; Helen, Toby's nineteen-year-old daughter from his first marriage; and Clare, the Eurasian nursemaid.

In 'Act', Peter vanishes, but is eventually discovered in the servants' privy, savagely murdered. The Assistant Inspector of Police, an English-educated Indian called Singh, is called in to investigate. The autopsy reveals that the child had been smothered with Clare's bra, then had his throat cut; Singh is convinced that a member of the household is the killer. It is possible that the murder had been an accident: that the child, who slept in the same room as Clare, had seen something, been about to cry out, and was smothered in an attempt to quieten him. Suspicion falls upon Clare (who claims to have slept through the child's abduction) and Toby, since it seems likely that they have been having an affair. Sing, however, believes that Peter's half-sister, the curiously unmoved Helen, is the murderer, although her motive remains obscure. Evidence provided by the *ayah* is insufficient to gain a conviction, and the crime remains unsolved.

In 'Darkness', Helen has returned to England where she trains as a doctor and lives with her Aunt Sophie, an eccentric but saintly philanthropist (based upon one of King's own aunts). During the war Helen meets a fellow doctor called Ilse, who is a Jewish refugee. Sophie dies as the result of doing a good turn for an ungrateful neighbour, and Ilse moves in with Helen. Toby, Isabel and their new child, Angela, return to England. Toby is a broken man, his life ruined by the gossip surrounding his involvement with Clare and the death of his son. Helen suffers nervous collapses, refusing to eat or communicate, but eventually confesses to Ilse that it was she who killed Peter. She gives herself up and is gaoled.

An 'Interlude' takes the form of a memoir written by the governor of the prison who, like many of the inmates, finds Helen curiously charismatic. A self-regarding bishop (who bears some resemblance to Lord Longford) attempts to secure Helen's release, but Helen is determined to atone for her crime.

'Illuminations' is set in Australia some years later, where Helen is attending a con-ference. After her release, she and Ilse (by that time her lover) had gone as doctors to Africa, where Ilse died. Helen is introduced to a commune in the home of a painter, now dead. Amongst the women there is Clare, now scarcely recognisable. Intermittent flashbacks reveal that in India Clare and Helen had been having an affair. Peter had woken up to see the two women making love and Clare had accidentally suffocated him, aided by Helen, who failed to call out a warning in time. Clare had been 'the implement of [Helen's] will'. Helen had disposed of the body, but the two women had been discovered by Toby, who agreed to collude with them to conceal the crime.

Brian MOORE   1921–

Cold Heaven

Moore is a novelist whose books attract the notice of both critic and general reader, one who, amid more self-consciously literary and experimental contenders for the Booker Prize (he has been nominated three times), remains eminently 'readable'. *Cold Heaven* is a characteristic example of his work, a gripping thriller written in clear, understated prose, which has an added dimension – in this case religious faith – which makes it more complex and profound than it at first appears.

The novel opens with an appalling accident in France, when Alex Davenport is mown down by a speed-boat while bathing in the sea. He is pronounced dead at the hospital, but when his wife, Marie, returns the next morning, to sign papers and make arrangements, Alex's body has disappeared. Marie, who has been brought up a Catholic, but since rejected all religion, begins to wonder whether she is being punished. Exactly a year earlier, she had experienced a vision of the Virgin Mary on the coast at Carmel, California, which told her a shrine must be built there. She had been staying there with her lover, Daniel, and decided to tell no one about her experience. At the time of the accident she was about to leave Alex for Daniel. She is convinced that some mistake has been made and that Alex is alive. After an eerie and prolonged chase, she eventually finds him at Carmel where Daniel has arranged to meet her. Confused and afraid, Alex is himself a phenomenon since he appears to experience clinical death a number of times and to recover miraculously.

To Marie, nothing that now happens is coincidence and when she relates at least part of her experience to the local priest she believes she is compelled to do so. Mgr Cassidy is inclined to write her off as hysteric, but Fr Niles, who has an interest in visions, cannot resist pursuing the case, in spite of the Monsignor's disapproval. He eventually establishes that an elderly nun from a nearby convent has been troubled by dreams which resemble the apparition. Accompanying Mother St Jude and her companion, Sister Anna, to the scene of the vision, Marie fiercely denies it when it occurs again, and does not contradict Sister Anna's claim that the shape of the cross has appeared in the rock, at the same moment, even though Marie witnesses its appearance after an earth tremor the previous day. Only Mother St Jude knows that Marie is not being entirely truthful. When a second vision appears to Sister Anna in the presence of the priests, Marie sees nothing. It seems to her that she, Alex and Daniel have all been 'pieces on a Chess board' but now, given that she has a right not to believe, the vision has been passed to someone else and she is allowed to return to ordinary life and the business of leaving her husband.

Graham SWIFT   1949–

Waterland

Swift's third novel confirmed his promise as a leading writer of the British post-war generation. More ambitious than the relatively small-scale *The Sweet-Shop Owner* (1980) and *Shuttlecock* (1981), *Waterland* was shortlisted for the Booker Prize, and won both the Winifred Holtby Prize and the *Guardian* Fiction Award. The Holtby Prize is awarded to a regional novel, and amongst other things *Waterland* is a minutely detailed evocation of the Cambridge Fens.

The narrative is directed by Tom Crick, a middle-aged history teacher struggling to engage the attention of pupils more interested in the contemporary nuclear threat than in the shaping of the past. He is also fighting for his job and for the status of his subject within the school curriculum ('We're cutting back History,' the headmaster blithely announces). Swift was himself a teacher, and much of the book is addressed directly to the reader, who is placed in the position of Crick's pupils. Abandoning the syllabus, Crick talks about the history of the Fens, a flat land steeped in tradition and superstition, where he and his wife Mary grew up.

But in those days the Ouse took a different course from that which it takes today. It is a feature of this footloose and obstinate river that it has several times during its brush with human history changed direction, taken short-cuts, long loops, usurped the course of other rivers, been coaxed into new channels and rearranged its meeting-place with the sea. All of which might be construed as a victory for history (for it is human ingenuity which in so many cases has effected these changes), yet which is more aptly to be interpreted as the continued contempt of the river for the efforts of men. Since without the old Ouse's perpetual if unhurried unruliness, without its ungovernable desire to flow at its own pace and in its own way, none of those cuts and channels and re-alignments, which are still being dug, and which ensnare the tortuous, reptilian Ouse in a net of minor waterways, would ever have been necessary.

WATERLAND
BY GRAHAM SWIFT

In particular, he reveals his own family's history. In doing so, he demonstrates that past and present are not separable entities, but are inextricably bound. Diverse elements – including the history of land reclamation, the creation of a Coronation Ale, and (most memorably) the life-cycle of the eel – are interwoven to create the background to three deaths which occur during Crick's adolescence. In order to protect him from his 'potato head' brother Dick, whose sexual initiation she has also undertaken, Mary blames her pregnancy upon a third party, whom Dick subsequently murders. Mary then has an abortion (an experience which leads to her present emotional problems), and Dick drowns himself after discovering, with Tom's help, the circumstances of his own birth: he is a 'bungle', the result of anincestuous relationship between his mother and her own father.

An almost documentary social history is combined with high melodrama in a story of Victorian richness and complexity spanning some 300 years. The book's scope is suggested by its two epigraphs. One it taken from the opening chapter of Dickens's *Great Expectations* ('Ours was the marsh country ...'), announcing a strong *genius loci* where events are shaped by place; the other comes from a Latin dictionary: '*Historia*, ae, f. **1.** inquiry, investigation, learning. **2.** a) a narrative of past events, history. b) any kind of narrative: account, tale, story.' *Waterland* is at once a story about storytelling and an investigation of the nature and making of history.

William TREVOR   1928–

Fools of Fortune

In his later fiction Trevor has become increasingly preoccupied with 'the Troubles' of his native Ireland. Early stories and novels such as *Mrs Eckdorf in O'Neill's Hotel* (1969) took an essentially comic view of Irish characters and life, but the divisions within the island have come to the fore in his work, just as they have in the public consciousness of mainland Britain. When he made a selection of his Irish stories in 1979, he gave the volume the ironic title *The Distant Past*, and the legacy of Ireland's history – and the sense of history repeating itself – is at the centre of *Fools of Fortune*, which won the Whitbread Award for the best novel of 1983.

The novel tells the story of the ill-fated love of Willie Quinton for his English cousin, Marianne. They first meet and fall in love following the murder of Willie's father, two sisters and most of the family servants in a vicious revenge attack by the Black and Tans. Their love is consummated a few years later, after the death of Willie's mother who, unable to support the memory of what has happened, has committed suicide. Discovering she is pregnant, Marianne resolves to leave her family and join Willie, but her arrival in Ireland is greeted by his resolutely unexplained absence. She eventually realises that he has murdered the leader of the Black and Tans and is now on the run. Determined to wait for him, Marianne lives with Willie's aunts at the family home, and raises her daughter, Imelda, in the belief that Willie is a hero and will one day return. Imelda becomes obsessed with every detail of her father's revenge, and goes insane.

Many years later, Willie returns to Ireland to visit a dying servant, and though he stays for the funeral, we learn that he neither acknowledges Marianne, nor accepts advice that his crime will by now be forgotten. It is another ten years before he comes home to spend his old age in peace with Marianne and their mute daughter who, perhaps a necessary sacrifice for past events, is regarded locally as something of a saint.

1984

Miners' strike in Britain ● IRA bombs British government ● Indira Gandhi assassinated ● President Ronald Reagan wins landslide victory ● Dr Alec Jeffreys invents 'genetic fingerprinting' ● Woman gives birth to a child created by in vitro fertilization of her husband's sperm with another woman's egg ● Woman gives birth to a child developed from a previously frozen fertilized embryo ● Wilson and Higuchi clone genes from an extinct animal, the zebra-like quagga ● Philip Johnson, John Burgee, AT&T Building, New York ● James Stirling, Nene Staatsgalerie, Stuttgart, Germany ● The Prince of Wales attacks Modern Architecture ● Philip Glass, *Akhnaten* (opera) ● Robert Maxwell buys the Mirror group of newspapers in Britain ● Films: *A Passage to India*; *Paris, Texas*; *Terminator* ● Television: *The Jewel in the Crown*, *Miami Vice* ● Popular music: Madonna, *Like a Virgin*, Bruce Springsteen, *Born in the USA* ● Count Basie, John Betjeman, Joseph Losey, Michel Foucault, Lillian Hellman, Richard Burton and François Truffaut die ● Craig Raine, *Rich* ● Samuel Beckett, *Collected Poems 1930–78*

Martin AMIS   1949–

Money

Amis subtitled his fifth, longest and most ambitious novel: 'A Suicide Note' and tell us that the note is addressed not to any of the book's characters, but to 'you out there, the dear, the gentle'. It is a nightmare evocation of a world in which the money of the title bestows power and replaces all other considerations as the motivating force of life. If all money corrupts, then absolute money corrupts absolutely.

The suicide note is a long, discursive narrative by John Self, a slobbish Anglo-American film director whose reputation has been made in 'nihilistic commercials' promoting the underbelly of consumerism, such as junk food, alcohol, tobacco and pornographic magazines. Self jets between London and New York, fuelled by fast food, drink and dollars, in pursuit of a film-deal, the movie in question being a partly autobiographical, semi-pornographic melodrama entitled *Good* (or *Bad*) *Money*. The gross Self is 'addicted to the twentieth century' and recognises the pornographic element in all his 'hobbies': 'The element of lone gratification is bluntly stressed. Fast food, sex shows, space games, slot machines, video nasties, nude mags, drinks, pubs, fighting, television, handjobs.' In New York, encouraged by the producer Fielding Goodney to spend as much money as he can and, nursing a near-permanent hangover, Self wrangles with a lesbian scriptwriter and a grotesque collection of narcissistic stars, each of whom has his or her own angle on the story: the ancient stud Lorne Guyland; the 'mature' Italian Caduta Massi; the neanderthal but clean-living Spunk Davis; and the not-so-dumb blonde Butch Beausoleil. His spare time is spent in bars and massage parlours or on the phone to London attempting to speak to his girlfriend, Selina Street, whom he suspects of infidelity. He is threatened by anonymous phone calls and given lessons in life and literature by an old friend, Martina Twain. The original scriptwriter is sacked and Martin Amis is hired to rewrite. Selina goes off with Martina's husband, Ossie, and Self is taken up by Martina herself. His anonymous persecutor turns out to be Goodney, and he realises that in his drink-fuddled state he has signed papers making him liable for all expenses. Dollarless, he escapes back to London where he plays chess with his creator, Amis, who discourses upon 'motivation' in the novel. The novel ends with Self's diary entry, outlining his new impoverished life with a fat nurse called Georgina.

Blackly, brutally, emetically funny, the novel batters language into submission in an exploration of the gross conspicuousness of late twentieth-century consumption and the 'close concordat' of pornography and money. The artistic concerns of 'Martin Amis' for the 'motivation' of the characters and the distance of author from narrator are set against the greed, vanity and lubriciousness of the film industry and the alcoholic delusions and myopia of the unfortunate Self.

J.G. BALLARD   1930–

Empire of the Sun

In Shanghai in 1941, the British community enjoy their privileged lives, the war a distant conflict in the battlefields of Europe. But to young Jim, a choirboy at the cathedral school, the flickering images of the newsreels are a fiction glossing over the horrors he has witnessed, 'the old battlegrounds at Hungjao and Lunghua where the bones of the unburied dead rose to the surface of the paddy fields each spring'.

Although best known as a writer of science fiction, Ballard had an enormous popular and critical success with this more conventional novel. It was shortlisted for the Booker Prize, won the *Guardian* Fiction Prize, and was made into an excellent film (1987), directed by Steven Spielberg from a screenplay by Tom Stoppard. Drawing on his own experiences during the Second World War, Ballard portrays the struggle for existence of a boy whose perceptions are filtered through his early knowledge of death. Cut off from his family, Jim suffers a confusion of loyalties; the Japanese pilots offer a glamorous vision of manhood, while rescue by the American allies belongs to the realm of delirium and fantasy. As the food stocks in the deserted houses begin to run low, Jim joins up with Basie, an untrustworthy American steward, and together they run a blackmarket in rations, stealing food from the dying in a makeshift refugee camp.

A transfer to a more permanent site provides Jim with what seems to be the only home he will ever know; a room shared with a hostile English family who are willing him to die. Forays into the surrounding countryside always end in his return to the camp, as he realises that it is the only place of safety in this dangerous land. 'In a real war no one knew which side he was on, and there were no flags or commentators or winners. In a real war there were no enemies.' As peace returns to the world, Jim is reunited with his family; but the nightmare of the war has made them strangers to each other, and Jim's childhood is over.

The ladies in silk dresses and their husbands in grey suits strolled through the debris of a war arranged for them by a passing demolition squad. To Jim the battlefield seemed more like a dangerous rubbish tip – ammunition boxes and stick grenades were scattered at the roadside, there were discarded rifles stacked like matchwood and artillery pieces still hitched to the carcasses of horses. The belt ammunition of machine-guns lying in the grass resembled the skins of venomous snakes. All around them were the bodies of dead Chinese soldiers. They lined the verges of the roads and floated in the canals, jammed together around the pillars of the bridges. In the trenches between the burial mounds hundreds of dead soldiers sat side by side with their heads against the torn earth, as if they had fallen asleep together in a deep dream of war.

EMPIRE OF THE SUN
BY J.G. BALLARD

Iain BANKS   1954–

The Wasp Factory

Banks's first novel, published on his thirtieth birthday, caused an outcry because of its perverse sexuality and its violence. Dismissed by some as repulsive and exploitative, the book was hailed by others as the debut of a writer of great imagination. This controversy ensured that *The Wasp Factory* remains Banks's best-known book, even though he has since published a number of other novels very different in tone and theme. Banks admits to a fascination for obsessive characters such as the book's teenage protagonist, Frank: 'They make sense to people, because people are more weird than they let on a lot of the time.'

Frank lives with his father, Angus, on a small Scottish island, and is a very unusual sixteen-year-old, already having three murders to his credit. In revenge for his cousin, Blythe, setting fire to his rabbit hutch, Frank placed a snake in Blythe's artificial leg, which resulted in a fatal bite. His second victim was his own younger brother, Paul. While walking with Paul on the beach, Frank discovered an unexploded Second World War bomb, which (once he had reached a safe vantage-point) he told Paul to hit with a piece of driftwood. His third victim was another young cousin, Esmerelda, whom Frank took kiting along the beach. When he suddenly attached the kite-strings to her wrists, she was carried out to sea, and oblivion.

On the island Frank has established various sacrificial zones as part of his deathly mythology, such as the Sacrifice Poles on the beach, which are totem poles bearing animals' skulls. The wasp factory is another creation, composed of an old clock face to which he has added various tunnels and compartments, each with its own significance. Into this he periodically introduces a live wasp, using the route it takes and the fate it endures as a source of advice and a prediction of the future. If a wasp enters the fire compartments, for example, this is a prelude to a fire; if it enters the water compartment, this may anticipate a drowning, and so on.

Frank's extraordinary psyche is partly explained by Angus's account of how his son was castrated as a tiny boy by old Saul, the family bulldog. Frank's stepbrother, Eric, shows similar traits, and in the course of the novel escapes from a mental hospital and makes his way home. Eric finally arrives with a flock of burning sheep, which he has ignited, and tries, unsuccessfully, to detonate a store of explosives which are kept in the cellar.

Disturbed by Eric's antics Angus falls into a drunken stupor, enabling Frank to take the keys to the study. To Frank's horror, he finds a box of tampons and a supply of male hormones. After jumping to conclusions and accusing Angus of being a transsexual, Frank learns that he is really Frances, a girl, and that Angus invented the story of his canine castration and used the male hormones to facilitate an experiment with his sexuality.

Julian BARNES   1946–

Flaubert's Parrot

Flaubert's Parrot is Barnes's third novel; it was shortlisted for the Booker Prize in 1984, and subsequently won the Geoffrey Faber Memorial Prize in 1985. The novel combines biography and literary criticism of Gustave Flaubert, and is narrated by Geoffrey Braithwaite, a retired English doctor, who is trying to come to terms with his wife's suicide. In researching Flaubert's life, and visiting the original site of Flaubert's house in Rouen, he wonders why a novel cannot in itself be enough for the reader, who usually tries to look beyond it into the private persona and possessions of the author. Why, he asks, are readers 'randy for relics'?

Geoffrey also questions the role of the literary critics, highlighting the critics' fallibilities in, for example, misunderstanding the context of a given passage or picking up on minutiae, such as Mme Bovary's eyes being described as three different colours in the course of Flaubert's novel. Geoffrey argues that details and inconsistencies such as these should not be important in the overall evaluation of a novel.

At a book fair Geoffrey meets Ed Winterton, an American writer who is researching a biography of Edmund Gosse, and in the process has discovered some interesting information on Juliet Harper, who was briefly governess to Flaubert's niece, before having a personal relationship with Flaubert. Furthermore, Ed has letters between Flaubert and Juliet Harper, which he says are deeply revealing. While Geoffrey is already planning their publication, he is shocked to learn that Ed has burnt them, supposedly in accordance with Flaubert's wishes. In revenge, Geoffrey tells him that he will not help to finance Ed's book on Gosse.

Continuing his research, Geoffrey makes regular ferry crossings from England to Dieppe and Rouen, and draws comparisons between his experiences when travelling there with his wife and incidents in Flaubert's own life, and those dramatised in *Madame Bovary* (1857). When he travels by train in Normandy, he relates this to the importance of railway travel in Flaubert's life. Flaubert despised trains, but they facilitated his affair with Louise Colet, who travelled from Paris, while he travelled from Rouen, so that they could meet in the middle. The relationship between Flaubert and Louise Colet is also described from her point of view, and she narrates the intense and fluctuating relationship she had with Flaubert, who loved her, mistreated her, and was unfaithful. Similarly, Geoffrey narrates the life he had with his wife, who was also unfaithful to him. And as he contemplates her suicide, he considers Flaubert's own death, and the implications of suicide.

Does the world progress? Or does it merely shuttle back and forth like a ferry? An hour from the English coast and the clear sky disappears. Cloud and rain escort you back to where you belong. As the weather changes, the boat begins to roll a little, and the tables in the bar resume their metallic conversation.
Rattarattarattaratta, fattafattafattafatta.
Call and response, call and response.

FLAUBERT'S PARROT
BY JULIAN BARNES

The unusual title of the novel comes from the stuffed parrot that Flaubert borrowed from the Museum of Rouen, and placed on his work-table while writing 'Un Coeur Simple' (1877). In that short story, a parrot assumes increasing importance in the heroine's life, and eventually, after it dies and has been stuffed, she prays to it as though it were a holy relic.

Anita BROOKNER 1928–

Hotel du Lac

Since 1981 Brookner has produced one novel a year and, like Jean Rhys before her, has created a recognisable character-type. The Brookner woman tends to be of a certain age, highly intelligent, vulnerable and doomed to disappointment. *Hotel du Lac* marks the culmination of the first phase of Brookner's fiction, since when the novels have become progressively gloomier, and the sharp humour which once enlivened them has gradually faded. An unexpected winner of the Booker Prize in a strong year, *Hotel du Lac* consolidated her growing reputation and brought her to a wider, and devoted, readership. It is undoubtedly one of her best novels, in which plangent melancholy and high comedy are held in perfect balance, and it was subsequently made into a distinguished film for television (1986), from a script by Christopher Hampton.

In a monochrome Swiss lakeside resort Edith Hope, a romantic novelist with the mien of Virginia Woolf, fills the out-of-season days in genteel fashion at the Hotel du Lac: changing for dinner, taking tea and *torte*, exchanging pleasantries with spinsters. Fellow guests, eager to invigorate their dreary sojourn, become enforced companions: the wealthy Mrs Pusey and her fawning, spoilt daughter, Jennifer; Mme de Bonneuil, eccentric dowager; Monica (Lady X); and the inscrutable Mr Neville. It rapidly becomes apparent that Edith is not on holiday. She pens passionate and lengthy missives to David Simmonds, her lover in London, and reveals through the mist of her sad thoughts the true circumstances that have brought her to such a backwater.

To escape the company of her own sex she perfunctorily agreed to marry the colourless but eligible Geoffrey Long. But intuition led her to abandon him at the last for David who, though needing her emotionally, cannot relinquish his marriage. To minimise scandal and mollify feelings, Penelope Milne, confidante and organiser of parties, 'suggests' Edith take a break abroad.

Ironically, the safe, grey Hotel du Lac brings further emotional complications when Mr Neville, a seeker after social respectability and collector of *famille rose* plates, offers marriage in name only. After a studied acceptance on a silent lake trip, Edith is jolted into reality by the sliding of a door in the hotel corridor. Mr Neville is seen to slip out of Jennifer's bedroom, wrapt in pleasure after his satisfaction, leaving Edith to return to London, to a clandestine love-affair and to the truth.

'Yes,' she went on, as Edith and Mr Neville made polite but sustained efforts to ignore this interruption. 'I'm afraid I'm a romantic.' With this pronouncement she smiled at them, as, reluctantly, they surrendered the *Observer*, the *Sunday Times*, the *Sunday Telegraph*. 'You see, I was brought up to believe in the right values.' Here we go, thought Edith, swallowing a tiny yawn.

HOTEL DU LAC
BY ANITA BROOKNER

Angela CARTER   1940–1992

Nights at the Circus

This novel unites two sharply contrasting modes: realism and surrealism. Although the events it depicts have a definite social and historical location – Victorian London and pre-revolutionary Russia – the reportage quality usually perceptible in the straight-forwardly historical novel is here subdued to a sense of magic which ultimately appears to mock history proper. The subversive, anarchic potential of fantasy transgresses the limits of rationality and transforms history into a vertiginous hall of distorting mirrors.

Fevvers, the 'Cockney Venus', is found in a basket of broken egg-shells and straw at the door of Ma Nelson's 'house', where she is reared as 'the common daughter of half-a-dozen mothers'. By the age of fourteen, the 'infernal itching' in her shoulders that had been tormenting her for some time turns out to be 'the herald to the breaking out of her wings'. She thus becomes ambiguously poised between being a freak and a symbol of female liberty, with a character at once idealistic and vulgar. Her grotesque odyssey, from her role of *tableau vivant* at Ma Nelson's, through the sadistic Mme Shreck's 'museum of woman monsters', to Mr Rosencreutz's devilish rituals, takes a new turn with her engagement at the Cirque d'Hiver as an *aerialiste*. As she travels around the world, applauded by the most powerful state leaders, Fevvers creates a new myth. Even time stands still as she and her faithful chaperon, the tiny, wizened Lizzie, perform their confidence tricks.

The reader is acquainted with this part of the Cockney Venus's story by the *aerialiste* herself in the course of an interview with the American journalist Jack Walser, eager to ascertain whether Fevvers, like Helen of Troy, was really 'hatched'. Her wings, the symbol of freedom, volition, the never-weary, make her an emblem of the New Woman: the novel appropriately ends on the first hour of the twentieth century. However revolutionary she may be as a symbol, Fevvers is also involved in a rather more conventional 'romance' with Walser, whose 'picaresque career' has made him suitable for all times and places. He thus joins Colonel Kearney's circus as a clown at St Petersburg, where Fevvers is performing; in this capacity he dances with a tigress and poses as the Human Chicken. But the next stage of the Circus's grand tour is never reached: a train accident forces the circus to disband and disperses its survivors in the Siberian wilderness. Walser is rescued by Olga Alexandrovna, an escapee from an ingenious house of correction designed to keep women apart from each other. Olga was the insti-gator of a revolutionary process whereby the prisoners were able to communicate with each other through surreptitious glances, touches and letters, which triumphs in the women's escape. Unable now to visualise a world that includes men, she reluctantly abandons him. Deprived of both his memory and his reason, he is taken up and 'bamboozled completely' by a Shaman, and his hypnotic drum.

Meanwhile Fevvers – who, with a broken wing, witnesses like a 'fallen angel' the fading away of her myth – also meets an escaped convict, a young, idealistic radical arrested for attempting to blow up a police station.

She learns from him that Walser is still alive, the escapee having met Olga and her companion, Vera, to whom he thoughtfully donated some sperm to freeze for the founding of their Utopia. Fevvers and Walser are reunited, and, as their love is at last being consummated, Fevvers' ecstatic laughter spreads around the globe. Walser himself joins in, although he may well be the butt of the joke, given that Fevvers is not (as she guaranteed) 'the only fully-feathered intacta in the history of the world'.

Anita DESAI 1937–

In Custody

Deven, the novel's central character, is a painfully timid teacher at Lala Ram Lal College, a struggling and desperately impoverished institution in Mirpore, a small town some miles from Delhi. 'More a poet than a professor' when he took the job and married Sarla, he has been forced to abandon his first, true love – the language the literature of Urdu – and earns his inadequate livelihood by teaching Hindi, the rival tongue.

His humdrum days are shattered by the irruption of an old school friend, Murad, a brash and domineering braggart. Restless son of a wealthy father, Murad owns and edits a ramshackle literary magazine (papa's gift) devoted to Urdu writing. He plans an issue devoted to Urdu poetry, and the jewel in its crown will be Deven's interview with Nur Shahjehanabadi, Urdu's greatest living poet. Nur's poetic voice has been silent for decades: he is ill now, and poor, but his remains 'a name that opened doors, changed expressions, caused dust and cobwebs to disappear, visions to appear, bathed in radiance'.

Beset as ever by crippling self-doubt, Deven tries – and fails – to resist Murad's urging. He undertakes the task, and in doing so embarks upon a series of tragi-farcical adventures, all of which lead him to a closer understanding of the relations between poets and poetry, lives and works.

Nur proves elusive and when Deven eventually runs him to earth he discovers the poet surrounded by a constantly changing entourage of raucous, drunken hangers-on, endlessly demanding 'loans', vowing allegiance and offering insult, filling the house with discord and clamour. Poetry could never be written here: no wonder Nur is silent now. Yet Nur's past achievement remains great, and for that Deven still reveres him.

But the house is a Babel and Nur's own powers of concentration have been eroded by years of drink and indiscipline. Moreover his willingness to co-operate with Deven is constantly undermined by his strong spirit of mischief and a fundamental distrust of those who 'teach' but do not practise poetry: 'I know your kind – jackals . . . trained to feed upon our carcases. Now you have grown impatient, you can't even wait till we die – you come to tear at our living flesh.' Desperately Deven falls deeper into debt as he travels constantly between Delhi and Mirpore, where Sarla, waiting mutinous and sullen, assumes he has found a mistress in the city.

Then there was silence. A long while later it was broken by the sharp, shrill whistle of the Janata Express from Assam clattering down the railway line. He bit down on a cigarette, cursing it: why was there always a train whistle in the dark, calling over vast spaces to all who longed to travel and move on? It promised nothing, it merely reminded prisoners of their bars, mocked them in their cells.

IN CUSTODY
BY ANITA DESAI

Deliverance seems at hand when Nur's first wife arranges for Deven to have free use of a room in a friend's house where he may take Nur to record conversations with him. But the tape-recorder Murad bullies him into buying is faulty, and the young shop-assistant foisted on him as an electronics expert is hopelessly incompetent. Nur's rabble accompanies him, and the result is a terrifying expensive mass of useless tape. And although Deven's own students salvage a few minutes of pitiably substandard material, they expect to be rewarded with magnificent examination results, and rapidly turn murderously unpleasant when baulked.

Yet embedded in the hours of recorded dross are moments of magic: Nur's recalling of previously unpublished poems, his flashes of brilliant insight and wisdom. For, as Nur insists and Devon must learn, 'sifting and selecting from the debris of our lives' is something that 'can't be done, my friend, it can't be done. I learnt that long ago.' Life and works are not extricable, and human greatness and abasement have the same roots. By the end of the book Devon's efforts to take a poet and his genius 'into custody' may well be about to cost him his life. The price, it seems, is one that he is willing to pay.

In Custody was shortlisted for the Booker Prize in 1984.

Joan DIDION   1934–

Democracy

In 1982 Didion spent two weeks in El Salvador. Her experiences were recorded in *Salvador,* which was published serially in the *New York Review of Books* later that year; 'experiences' rather than 'story', for the form of the book connoted the breakdown of narrative. Anecdotes, ad hoc analyses, communiqués and press briefings were juxtaposed in a restless mosaic of official versions, observable and incomprehensible facts. The discrepancy between political reality and rhetoric being so vast, ran Didion's implicit argument, storytelling would only add to an already vast falsification. There were causes, and there were effects, but any necessary relations between the two had been replaced by political violence. Even death became questionable – body-counts varied.

Didion's reportage in *Salvador* borrowed the techniques of modern fiction; *Democracy* repays the debt. The story is explicitly reported by the author ('Let the reader be introduced to Joan Didion ...'), who observes her characters though newspaper reports, television coverage and rumour, who interviews her heroine and so plays a part in the action. The plot centres on Inez Victor Christian, the wife of Sen. Harry Christian, and her lover, Jack Lovett, a shady international businessman. The year is 1975 and the long conflict in Vietnam is drawing

to a chaotic close. The Christians' marriage is in a similar state and events around the central trio occur in the way of sporadic violence. Inez's sister is killed by her father in a Hawaii apartment; her daughter sees heroin addiction as a consumer decision and gaining illegal entry into Vietnam as a career move. Inez leaves Harry for Jack, who finds himself under investigation for illegal arms trading. He dies in a swimming pool. Inez works as a nurse in a Cambodian refugee camp.

The plot of *Democracy* is quite consciously melodramatic. Didion's point is that individual lives are 'pumped up' by the escalation of public events until they became 'scandalous'. After the murder of Inez's sister, the networks ran the same clip of Inez dancing at a party over and over again – a surrogate for the 'real thing'. The press rarely intrudes into the action of the novel, but their presence is ubiquitous; questions are fielded rather than answered and Didion insistently points up the rhetorical nature of the meeting between public and private knowledge. Official euphemism is also subject to her critique and several passages in *Democracy* are clearly derived from experiences reported in *Salvador*: 'Inez remembered Harry giving a press conference and telling the wire reporters who turned up that the rioting in Surbaya reflected the normal turbulence of a nascent democracy.'

Democracy both suffers and gains from the diversity of Didion's talents. Her work as a screenwriter shows up in the discrete, immediate scenes which are the novel's basic unit. As a reporter, she presents these scenes 'on the run', as if only just having time to note them down (the novel was in fact carefully rewritten more than once). As critic, her *propria persona* intrusions are knowingly positioned, and place the novel in turn within a definite genre. The reflexive device is a gesture of solidarity with other sympathetically aligned books. The completion between these different idioms to present the facts reinforces a central theme of the novel: that different kinds of discourse will, from the same starting-point, produce very different versions of 'the truth'. But this competition stretches Didion's resources as a novelist. The narrative is sustained, but only just. In allowing her themes to dictate the form (or forms) of the novel, Didion is perhaps too true to her subject and *Democracy* emerges as a book which values its integrity more than the conventions of the well-made novel.

Christopher HOPE 1944–

Kruger's Alp

'Kruger saw it coming. His *Memoirs* make it clear that the discovery of gold was a catastrophe. It would "soak the country in blood."' Hope, a South African, explores the legacy of white hatred for the blacks in his savagely satirical 'dreambook', *Kruger's Alp*, which won the Whitbread Prize for fiction. The narrator witnesses events as a Bunyanesque pilgrim who falls asleep in the shade of a tree. 'People said it couldn't be true,' he says in his parable, 'until they remembered that anything you could think about could very easily be true.' He tells of the terrorist resistance of a group of youths who have been tutored in moral philosophy by Fr Ignatius Lynch, a liberal humanist. Their stories are truncated by the 'double-think' of police violence administered by a corrupt regime. Blanchaille, a priest, finally decides to leave the country, and is smuggled to England and then to Switzerland by his sympathisers, a move which fills him with guilt as this betrayal of his homeland. In London he meets Van Vuuren, a policeman but also a protestor, and together they begin their search for a mythical 'Promised Land'.

Their peers group includes 'Micky and Poet', 'known for four quite hopeless lines', which nevertheless lead to his arrest. In court he applies to call 'the entire prosecution team as witnesses for the defence and to cross-examine them carefully on all aspects of his case'. There is an adjournment while the judge considers this request, and Micky is discovered in his cell strangled with a piece of towel attached to the bedstead. This is explained away as suicide: 'What a miracle of athletic agility that had been,' comments the narrator, 'what a wonder of tenacity!' In London Tony Ferreira comes to an equally awful fate, tortured and murdered by South African security forces. The western econ-

omies are portrayed as being infiltrated by the 'Manus Virginis', a right-wing reactionary organisation with close links with the Mafia, which covertly control the South African establishment. Blanchaille, fleeing from his knowledge of this, seeks a fabled haven in Switzerland, a country house which is supposed to have been bought by Kruger to provide a refuge for righteous exiles. Magdalena, with whom Blanchaille has had a brief affair in London, is revealed as a double-agent who poses as a revolutionary while being in the pay of the establishment, and her betrayal of friends whom she has pretended to help is responsible for the deaths of several of them.

In 'Bad Kruger', the country mansion, Blanchaille is persuaded to believe that he has at last found a place of escape, but the mysterious, semi-religious rituals which govern life there mask the place's symbolic significance, which gradually emerges. It is a house of the dead, and the legend of Kruger's 'missing millions', rumoured to have been spirited away by him, turn out to be not gold but lost souls: 'You see, we are the Kruger millions.'

David LODGE 1935–

Small World

The subtitle to Lodge's *Small World* is 'An Academic Romance', and the adventures involving Persse McGarrigle, the young lecturer from Limerick, the beautiful and mysterious Angelica, the dullish English professor Philip Swallow, and his more significantly American counterpart, Morris Zapp (who, with several other characters, reappear from Lodge's *Changing Places* (1975)), share a number of important features with the medieval romance tradition. As we proceed through the book's ingeniously labyrinthine plot convolutions, and the texture of coincidences and correspondences grows denser, we are led into a world that bears more than a superficial resemblance to that of King Arthur and his knights. Lodge's contemporary romantic 'heroes' and 'heroines' are mostly university lecturers and literary critics, hunting for their Grail either within the cosy and self-important universe of the campus, or in their 'wanderings' across the globe to attend international conferences.

The whole academic world seems to be on the move. Half the passengers on transatlantic flights these days are university teachers. Their luggage is heavier than average, weighed down with books and papers – and bulkier, because their wardrobes must embrace both formal wear and leisurewear, clothes for attending lectures in, and clothes for going to the beach in, or to the Museum, or the Schloss, or the Duomo, or the Folk Village. For that's the attraction of the conference circuit: it's a way of converting work into play, combining professionalism with tourism, and all at someone else's expense. Write a paper and see the world! I'm Jane Austen – fly me!

SMALL WORLD
BY DAVID LODGE

Like Perceval in the romantic tradition, Persse comes from a primitive and unrefined background, is supremely chaste, innocent and idealistic, and never questions the rightness of his quest to find the lovely Angelica, whom he first spied at Rummidge University in the spring (the time of the year devoted by the romance to crucial encounters). But Persse also turns out to be the intellectual equal, if not superior, of academics with greater reputations, as is revealed by his ability to stir the stagnant atmosphere of the world conference with which the novel culminates by means of one simple question: 'I would like to ask each of the speakers ... what follows if everybody agrees with you?' That sudden intervention not only gains Persse unprecedented respect, but also transforms him into something of a saviour knight, able to heal and regenerate a barren and desolate land: 'some conferees patted his back and shoulders as he passed – gentle, almost timid pats, more like touching for luck, or for a cure'. In the romance legend of the Grail, there is a sexually maimed king (the Fisher King) and a waste land. It is the

saviour knight's task to restore the fertility of both the king and his land. The questor who arrives in the waste land must ask the meaning of what is shown to him: not until the question is uttered will the king's wound be healed, or the waters begin to flow again. Sure enough, Persse's question coincides with both an 'astonishing change in the Manhattan weather', whereby the icy wind turns into a warm breeze, and an equally pleasant and radical change in the sexual life of the world-famous (yet old and impotent) critic, Arthur Kingfisher.

At the same time as the romance conventions are explicitly invoked and rigorously followed, they are also occasionally mocked and subverted. For instance, the ideal of chastity which Persse upholds as staunchly as the perfect Arthurian knight, is parodied by his union not with the 'untainted' Angelica, but with her twin sister, Lily, a Soho strip-tease artiste. Similarly, the episode of the 'kidnapping', which in the romance world tends to pivot on the dire consequences of a faithful wife's abduction by an evil fairy, and the husband's attempt at rescue, is satirised by Lodge through the determination of Zapp's wife *not* to save her husband.

Like the romance form, *Small World* is eminently episodic, rambling and circular, so that the hero ends up where he started from, with the prospect of another quest ahead of him. This time his objective is the British Airways employee Cheryl Summerbee, whom he met while pursuing Angelica. By the time Persse returns to the Information Desk at Heathrow, however, Cheryl has left her job: 'The day's destinations filled four columns ... On the surface of the board ... he projected his memory of Cheryl's face ... and he wondered where in all the small, narrow world he should begin to look for her.'

Alison LURIE 1926–

Foreign Affairs

In the ironic tale of Anglo-American attitudes, which was awarded a Pulitzer Prize, Lurie presents shifting images of England as a pastiche of its literature, using the novel itself as a medium for a comedy of cultural exchanges. Through the antiphonal narratives the English experiences of two American academics are interwoven, as Vinnie Miner and Fred Turner arrive on sabbatical leave in 'the imagined and desired country'.

Humiliation in America dogs both of them. Vinnie devises horrible deaths for L.D. Zimmern, the critic who savaged her life's work on children's literature in her favourite magazine. Fred is fleeing in horror from the brash sexuality of his wife, Roo; in her recent exhibition, he had been confronted by a photo-study featuring a close-up of his own genitalia juxtaposed with a crop of dew-dappled woodland mushrooms. Vinnie seeks comfort in the cosy land of Beatrix Potter and Anthony Powell, whilst Fred paces the streets in search of the Grub St underworld of his literary quarry, John Gay.

In the playgrounds and drawing-rooms of Camden and Kensington lurk unsuspected disillusionments. Vinnie's quaint children's folklore has degenerated into the obscenities of a purple-haired punk. Instead, romance lies in the unlikely figure of Chuck Mumpson, a raw-hide American tourist earlier dismissed as 'a person without inner resources who splits infinitives'. Assisting his attempts to trace a mysterious ancestor, 'Old Mumpson', Vinnie hesitantly abandons her dowdy campus spinsterhood. Instead of an aristocratic pedigree, however, Chuck's researches reveal a wild man of the woods, a tamed hermit hired by whimsical landowners as a grotto ornament.

Exchanging American bluntness for English refinement, Fred becomes obsessed with the elusive star of a country-house soap opera, Rosemary Radley, whom he encounters at Vinnie's party. His passion cools as she proves to be less than the real thing; neurotic and alcoholic, she alternates between the roles of aristocrat and cleaning lady. Fred's wife is also regretting their separation, and in a desperate attempt to contact him she reveals to Vinnie that she is the daughter of the hated L.D. Zimmern.

Tempted to thwart their reconciliation, Vinnie is checked by the disappointed surprise with which Chuck would react to such

conduct. She uneasily embarks upon a dark journey to Hampstead Heath, to contact Fred in the midst of the magical solstice celebration. Emerging druid-like on Parliament Hill, she saves his marriage.

Alone again in London, Vinnie awaits Chuck's call, jealously imagining his reception in the Wiltshire countryside. When the phone rings, however, it is Chuck's daughter, announcing the death of her father in the village of his ancestor. Vinnie's final days in England are spent on a sentimental journey to the place where Chuck's ashes have been clumsily scattered by his 'barbarian' daughter. Returning to her American exile, Vinnie enumerates the sum of her life and finds it wanting.

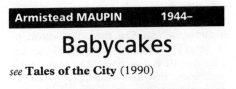

Armistead MAUPIN   1944–

Babycakes

see **Tales of the City** (1990)

Gore VIDAL   1925–

Lincoln

Vidal is well placed to write novels of political history. He grew up in the Washington home of his grandfather, Thomas Pryor Gore, Oklahoma's first US senator, and attended his first presidential convention at the age of fourteen. He has run for Congress, and in 1961 contributed a political column to *Esquire* magazine. An irreverent panorama of American political life is found in his 'American Saga' sequence of historical novels, of which *Lincoln* is chronologically the second. (The others, in order of publication, are *Washington D.C.*, 1967; *Burr*, 1973; *1876*, 1976; *Empire*, 1987; and *Hollywood*, 1989.)

The novel depicts Abraham Lincoln's administration, from inauguration to assassination focusing on three interlocking areas: Lincoln's private life (especially in regard to his extravagant, partially insane wife, Mary, who consults spiritualist mediums and hears voices); the political manœuvrings of his allies and rivals in Washington; and the conduct of the Civil

War. It is largely based on historical record, with few invented characters or incidents. The chief piece of invention concerns the life of David Herold, a real drugstore assistant in Washington who emerges into history only through his involvement with Lincoln's assassin, John Wilkes Booth. Vidal concocts for him a colourful low-life which gives the reader a view of the Confederate side of the war. Using his privileged position opposite the White House, he spies, runs messages over the Potomac into rebel Virginia, and attempts at one point to put poison in the President's constipation remedy.

Lincoln himself is not an abolitionist (though Mary is, and her Kentucky family are rebels), and his commitment to the war is founded on his belief that the secessionist Southern states are in breach of the Constitution. He carefully forms a cabinet which includes his chief rival, the radical abolitionist Sen. Chase, as Secretary of the Treasury. Chase's wife is dead and, spurred on by his ambitious daughter Kate, he schemes against Lincoln at every turn. Another former rival, Gov. Seward, is gradually drawn towards the improbable figure of 'Honest Abe' as he witnesses the subtle political acumen with which Lincoln outmanœuvres Chase. Engrossed in the business of government, Lincoln is unaware that Mary is spending wildly beyond their means on the refurbishment of the White House; she borrows money and risks a scandal when it is alleged that she has sold the contents of Lincoln's Address to the Congress to the newspaper proprietor James Gordon Bennett.

The long and bloody conduct of the Civil War sees a succession of generals rising and falling from grace. Gen. McDowell is replaced as Union commander by Gen. McClennan ('the Young Napoleon') following the unsuccessful invasion of Virginia. McClennan brilliantly organises a new army of 200,000 troops ('the United States was now the largest military power on earth'), but his indecisive strategy has some suspecting him of being a covert Confederate; he is replaced by Gen. Ulysses S. Grant, whose successful completion of the fight against Gen. Lee has many tipping him as a future president.

Mary WESLEY   1912–

The Camomile Lawn

Wesley's much-loved second novel was published when the pseudonymous author was seventy-two. It opens in Cornwall in August 1939, and it is the threat of war, followed by war itself, which gives the novel its vivid, febrile quality. Five young cousins have an annual ritual, the hair-raising Terror Run along steep cliffs at night, and this, along with a quickly abandoned lottery whose unknown 'winners' undertake to kill someone during the war years, haunts the narrative. Another darker note is struck as the local community is joined by the violinist Max Erstweiler and his wife Monika, Jewish refugees whose son is missing. Of the young people, handsome aspiring poet Oliver joins up first; he is to love his ravishing cousin Calypso for most of his life, but she rapidly marries the older, rich Scottish MP, Hector. Their cousin Walter is to die at sea, while his apparently sagacious sister Polly, who like naughty Calypso is emancipated by wartime life, is to astonish everyone by settling down to a lifelong *ménage à trois* with the identical Floyer twins, Cornish neighbours. Sophy, nine when the novel opens, seems haunted by the Terror Run, and we guess that she had taken the lottery in deadly earnest. Alone among the cousins she is disliked by forty-ish Helena, whose husband Richard is uncle to them all. Crusty and conventional, he is to remain in Cornwall throughout the war while Monika runs the household with extreme middle-European efficiency, despatching food parcels for Max and, later, Helena, in London. As the novel proceeds we flash forward to the 1980s to see many of the characters, including Helena, aged and acerbic, converge on Cornwall for a funeral – that of Max it transpires. Their memories and the third-person narrative evoke the terrors, tensions and griefs, as well as the hectically changing erotic relationships, of wartime London. Oliver, adored by Sophy, achieves one – rather unsatisfactory – night with Calypso. Although she has many lovers, her son Hamish is undoubtedly Hector's, and we realise at last that, in spite of her relentlessly frivolous facade, it is Hector she was always loved. Helena achieves astonishing sexual happiness with Max, now attaining fame and reputation, and he, at various times, makes love with all the female cousins, tenderly relieving the adolescent Sophy of her virginity. Meanwhile Richard and Monika evolve a comfortable alliance, briefly threatened by her attempted suicide when accused of spying. The quartet will last until death. When it becomes clear at Max's funeral that his unpleasant son Pauli –

miraculously and disconcertingly restored after the war – will sell the Cornish house and its nostalgically scented camomile lawn, twice-divorced Oliver is shattered. But the elegiac ending is lightened by a tentative *rapprochement* between him and Sophy.

Wesley's evocation of period is masterly, and the classy lightness of her tone sustains the narrative with surprising strength, deftly evading snobbery and superficiality and encompassing even the darker passages with assurance.

1985

Sinking of *Rainbow Warrior* ● TWA and *Achille Lauro* hijacks ● Anglo-Irish Agreement signed ● South Africa's first mixed marriage celebrated ● Riots in Tottenham, London, during which P.C. Keith Blakelock is murdered ● The British Antarctic Survey detects a hole in the ozone layer over Antarctica ● US film star Rock Hudson is the first celebrity to die of AIDS ● Jurgen Habermas, *The Philosophical Discourse of Modernity* ● P.F. Strawson, *Skepticism and Naturalism* ● Saatchi Collection opens in London ● Norman Foster, Hong Kong and Shanghai Bank Headquarters, Hong Kong ● Christopher Hampton, *Les Liaisons Dangereuses*, *Les Miserables* ● Films: *Ran*, *Kiss of the Spider Woman* ● Television: *EastEnders*, *The Golden Girls*, *Edge of Darkness* ● Popular music: The Pogues, *Rum, Sodomy and the Lash*, The Smiths, *Meat is Murder* ● Henry Cabot Lodge, Marc Chagall, Laura Ashley, Philip Larkin and Robert Graves die ● Stephen Spender, *Collected Poems* ● Allen Ginsberg, *Collected Poems 1947–1980*

Peter ACKROYD　　　　　1949–

Hawksmoor

Ackroyd's first two novels were widely praised, but it was *Hawksmoor* – bestseller and winner of both the Whitbread Novel of the Year and the *Guardian* Fiction Award – that placed him in the front rank of contemporary novelists, enabling him to command enormous publishers' advances for his books. As in *The Last Testament of Oscar Wilde* (1983), Ackroyd uses *Hawksmoor* to display his skills as a pasticheur, and subsequent novels would continue to explore the notion of the coexistence of different layers of time, first (and most successfully)

exploited here.

Hawksmoor is a mystery story which develops simultaneously on two distinct levels – the early eighteenth century and the twentieth century – by alternating chapters which deal with eighteenth-century events and describe them in the style of that time, and chapters whose temporal location and literary style are contemporary.

Nicholas Dyer, an architect in charge of the rebuilding of seven London churches (and clearly based upon Nicholas Hawksmoor), is the protagonist of the earlier section. Though surrounded by a general faith in scientific progress, typical of the Age of Reason, he does not share in its optimism, represented by a splendidly characterised

Christopher Wren. His inquisitive spirit, if no less intense than that of the Royal Society, is directed towards far less 'rational' ideas. Obsessively aware of the presence of 'darkness', of 'shadows', and of the 'dust' from which we all come and to which we will one day return, he has secret connections with a sect whose unorthodox faith is centred on the obscurest aspects of life and in the practice of rituals which do not exclude

---

The skie was getting wonderful Dark with a strong Winde which swirled around the Edifice: Do you see, *I said*, how the Architraves are so strangely set upon the heads of the Upright stones that they seem to hang in the Air? But the winde took my words away from him as he crouched with his Rule and Crayon. Geometry, *he called out*, is the Key to this Majesty.

HAWKSMOOR
BY PETER ACKROYD

---

human sacrifice. Indeed, the rebuilding of each of the seven churches coincides with a murder, which the architect regards as a propitiatory ceremony: 'When my Name is no more than Dust ... when even this Age itself is for succeeding Generations nothing but a Dreem, my churches will live on, darker and more solid than the approaching Night.'

The churches do survive. And so do their sanguinary associations. In the twentieth century, his seven churches become the theatres of seven murders whose nature and motives a detective called Nicholas Hawksmoor vainly struggles to unveil. Although he fails by *rational* standards, the complex web of interconnections between the two ages and places spun throughout the novel seems to suggest, both to Hawksmoor and to the reader, that responsibility for the eighteenth-century and the twentieth-century crimes alike stems from the same force – if not the same person. The murderer is clearly more than an individual in flesh and bone, firmly tied to one age or set of circumstances.

When reading *Hawksmoor* it is worth bearing in mind that Ackroyd's most recent book before this was his biography of T.S. Eliot. There seems to be evidence for the poet/critic's influence on this novel in several aspects: first, the intertextual amalgamation of disparate temporal levels; secondly, the employment of 'objective correlatives' (such as the recurring images of dust, street-singers, tramps dancing around the fire and dissected corpses, which feature in both narrative strands, or even whole sentences, which echo and rebound from one age to the other). But the closest connection probably lies in Ackroyd's powerful picture of evil. Dyer's hell-bent mind is strongly reminiscent of Eliot's words:

it is better, in a paradoxical way, to do evil than to do nothing: at least we exist. It is true to say that the glory of man is his capacity for salvation; it is also true to say that his glory is his capacity for damnation. The worst that can be said for most of our malefactors ... is that they are not men enough to be damned.

Margaret ATWOOD   1939–

The Handmaid's Tale

Atwood's novel is set towards the end of the twentieth century. The USA has become theocratic Gilead, its laws derived from the Pentateuch, distorted in the lens of tele-vangelism and applied to a rigidly structured system of castes. At the top are the Commanders – nebulous administrators, all of them men. They alone are permitted Wives and Handmaids. While the Wives preside discontentedly over their households, the Handmaids have a single function: to provide children for a nation which, through years of industrial pollution, has poisoned itself to the verge of sterility.

Offred, the narrator, is a Handmaid. Before the theocrats took power she had a job, a lover and a daughter – an equivocal independence, now taken away. Bullied into submissiveness, she wears the red yet modest habit of her class and endures regular couplings with her Commander. She does the household's shopping, pausing on her trips to consider the wall where executed dissidents are displayed. (Behaviour is regulated by a highly visible paramilitary and an eerily almost-invisible secret police.) She learns, via clandestine conversations, of a resistance; she sees, however, little sign of it. Permitted neither books nor entertainment, she mostly spends her days in solitude, wondering what has happened to her loved ones and fondly recollecting her previous life.

I lie in bed, still trembling. You can wet the rim of a glass and run your finger around the rim and it will make a sound. This is what I feel like: this sound of glass. I feel like the word *shatter*. I want to be with someone.

THE HANDMAID'S TALE
BY MARGARET ATWOOD

Her circumstances change. The Commander – illegally – invites her to his study at night. He asks her to play Scrabble, lets her read magazines that should have been burned (ancient copies of *Ms*, *Esquire* and *Vogue*) and takes her to Jezebel's, the illicit brothel-cum-night-club of the elite. He does not require sex. It seems, also, that he is infertile. This harms his status and endangers her. (If she can not produce a child, she will be condemned to clearing up nuclear waste.) Tacitly it is agreed that she will sleep with Nick, the Commander's chauffeur. Such an arrangement, of course, is forbidden by law. Nevertheless, Offred participates – and finds again the passion she shared with her lover. She becomes pregnant and is soon, apparently, betrayed. The police arrive. At her arrest, Nick indicates that she is being rescued: the police are members of the resistance come to smuggle her from Gilead. Uncertain as to the truth of Nick's claim, Offred is helped into a van. She ends, 'And so I step up, into the darkness within; or else the light.'

Atwood appends an epilogue: an aca-

demic paper delivered in 2195. Offred's story is under discussion. The speaker speculates on the fate of its protagonists and regrets that its emphasis is less on public fact than private emotion. The central narrative is not clarified, but Atwood implies that the era of evangelical reaction is brief. Indeed, when *The Handmaid's Tale* was first published, US televangelists were at the peak of their influence. One, Pat Robertson, was shortly to contest the Republican nomination for the presidency. He was humiliated at the polls and several others of his kind were ruined in a series of scandals. A Gileadine dystopia proved a transient threat. The fable, though, retains a pertinence. By no means a feminist tract, it pleads against the morbid effects of authoritarian and utilitarian culture: what deadens both Offred and the Commander is the lack of diversity in their experiences. If the latter's outlets are mundane, Offred's indeterminate escape hints at a hell or heaven still unmapped.

Atood has remained tactfully silent about Harold Pinter's adaptation of her novel for a film (1990), directed by Volker Schlöndorff. The critics did not share her reserve.

Paul AUSTER 1947–

City of Glass

see **The New York Trilogy** (1986)

Peter CAREY 1943–

Illywhacker

Narrator Herbert Badgery claims to be 139 years old, but then he is also a self-confessed liar, conman and spieler – an illywhacker. Book one opens in 1919 with Herbert, reluctant snake-handler and salesman of Model T Fords, landing his Morris Farman plane on Jack McGrath's land outside Geelong, Victoria. Soon he has inveigled his way into the lives of the McGrath family: he and Jack plan a factory which will produce Australian-designed planes; Molly dotes on him as the son she never had; and beautiful daughter Phoebe secretly seduces him on the roof of the house. After Herbert has outraged potential investors in the aircraft factory with his passionate advocacy of Australian nationalism and denunciation of their craven devotion to all things American, a depressed Jack kills himself by allowing one of Herbert's snakes to bite him. Free now to marry Phoebe, Herbert becomes head of the household, which moves to Melbourne with an epileptic 'poet', Horace Dunlop, in tow. Phoebe is obsessed with flying and feels tied down by the two children she has reluctantly borne. Taking the Morris Farman and her lover, Annette, she leaves her family, intent upon a career as aviationist and poet.

Book two commences with the 1922 Depression and finds Herbert trying to support his children by prospecting for gold. He teams up with snake-dancer Leah Goldstein to form Badgery & Goldstein (Theatricals), performing the disappearing act he purports to have learned as an orphaned child from his Chinese guardian, Goon Tse Ying. Like many of Herbert's assertions, this one must be regarded with some scepticism. This part of the novel also explores the background of Leah's parents, show-people and communists involved in the Unemployed Workers' Union and subjected to the brutality of the state.

In book three, Herbert's son, Charles, is an adult; following in his father's entrepreneurial footsteps (though somewhat more successfully), he builds up in Sydney what is to become the world's largest pet emporium. Touching upon history as the novel so often does, it is Charles's shop which provides Gen. MacArthur's cockatoo mascot with its immortal phrase 'Hello, Digger.' The shop is 'pure Australiana', until – again embodying the fate of the country – it is sold to Americans, and finally to the Mitsubishi Corporation. Herbert, Leah and others live in the shop, 'the rusted wrecks of lives'.

Illywhacker attempts to survey the history of Australia across the century, and does so via the devices of allegory, symbolism and fantasy. Its chief theme is lying – Herbert's lies both embodying and defying the greater lies which are at the heart of Australian history. Towards the close of the novel Herbert discovers the work of the historian M.V. Anderson and quotes approvingly:

Our forefathers were all great liars. They lied about the lands they selected and the cattle they owned. They lied about their backgrounds and the parentage of their wives. However it is their first lie that is the most impressive for being so monumental, i.e., that the continent, at the time of first settlement, was said to be occupied but not cultivated and by that simple device they were able to give the legal owners short shrift and, when they objected, to use the musket or poison flour, and to do so with a clear conscience. It is in the context of this great foundation stone that we must begin our study of Australian history.

Robertson DAVIES 1913–

What's Bred in the Bone

see **The Cornish Trilogy** (1988)

Don DeLILLO 1936–

White Noise

Jack Gladney is chairman of the Hitler Studies Department at Blacksmith College. His wife, Babette, takes care of the children accumulated from their previous marriages. Jack's ex-wife runs an ashram in Montana. Despite Jack's pre-eminence, he cannot speak German, and lives in fear of being exposed and professionally disgraced. One of his children is playing chess by post with a convicted murderer, while his best friend is training to beat the world record for staying unprotected in a cage with deadly snakes.

Two events break in upon this eccentric family. The first is an 'airborne toxic event':

It appeared in the sky ahead of us and to the left, prompting us to lower ourselves in our seats, bend our heads for a clearer view, exclaim to each other in half-finished phrases. It was the black billowing cloud, the airborne toxic event, lighted by the clear beams of seven army helicopters ... The enormous dark mass moved like some death ship in a Norse legend, escorted across the night by armoured creatures with spiral wings. We weren't sure how to react.

The airborne toxic event gradually recedes as inscrutably as it advanced. No one is sure where it came from or even where it went.

He picked up our bottle of extra-strength pain reliever and sniffed along the rim of the child-proof cap. He smelled our honeydew melons, our bottles of club soda and ginger ale. Babette went down the frozen food aisle, an area my doctor had advised me to stay out of.

'Your wife's hair is a living wonder,' Murray said, looking closely into my face as if to communicate a deepening respect for me based on this new information.

'Yes, it is,' I said.

'She has important hair.'

'I think I know what you mean.'

'I hope you appreciate that woman.'

'Absolutely.'

'Because a woman like that doesn't just happen.'

WHITE NOISE
BY DON DELILLO

The second is 'Dylar', an experimental drug designed to cure people's fear of death. Babette is enlisted by a maverick pharmacist to test this produce, and gradually becomes dependent on the drug. Jack shoots the pharmacist, but fails to kill him. Confused by this, he drives the injured man to a hospital, where a nun tells him that her order only pretends to believe in biblical miracles as it reassures people to think that someone keeps faith. She does not see her dedication as pretense, but her 'pretense as dedication'.

This admission is not presented as hypocrisy. It is a sentiment which lies close to the heart of DeLillo's novel for, in this world of malls, freeways, supermarkets and motels, existence is predicated very firmly upon being seen. When Jack is involved in a near-catastrophe aboard an aeroplane, the general panic is 'all for nothing' because the

media do not turn up to report it. A nearby tourist attraction is billed as 'The Most Photographed Barn in America'. DeLillo clearly finds the situation of *esse est percipi* a liberating one. His prose is at once mannered and interestingly awkward as it negotiates between 'objective' description and the vagaries of his characters' interpretations. Jack observes the parents of his returning students: 'The women crisp and alert, in diet trim, knowing people's names. Their husbands content to measure out the time, distant but ungrudging, accomplished in parenthood, something about them suggesting massive insurance coverage.' The drift from observation to (comic) assumption is made quickly and easily: a typical movement in *White Noise*.

The major failing of the novel is related to this 'ease'. In his choice of a large-scale industrial accident and the unlicensed experiments of a drug corporation as the two focal events of the book, DeLillo betrays a strong awareness that there is a price to pay for the quirky, small-town idyll he describes. The neglected industrial base returns to haunt the consumer society which has ignored it. The deregulated world described in *White Noise* cannot understand these bewildering and threatening intruders, and the techniques that DeLillo brings to bear are similarly helpless. At root, it is DeLillo's refusal to create a stable moral viewpoint which turns a potentially ferocious piece of satire into an enjoyable black comedy of postmodern life and manners.

Jane GARDAM   1928–

Crusoe's Daughter

Gardam made her reputation with books for teenage readers, although novels such as *A Long Way from Verona* (1971) are certainly sophisticated enough for most adult tastes. Of her novels for adults, the most widely acclaimed is *Crusoe's Daughter*, which is also the most ambitious, attempting to present the whole of the twentieth century through the life of one woman.

Polly Flint lives her life through fiction, in particular that 'great curiosity, the paradigm, *Robinson Crusoe* ... the novel elect'. Polly's desert island is the yellow house at Oversands where, as a small girl in 1904, she is sent from Wales by her improvident father, Captain Flint, to live with her spinster aunts, Mary and Frances Younghusband, their spiteful and mysterious companion, Mrs Woods, and the maid, Charlotte. Polly is soon orphaned when her father dies at sea. (Her mother – a flighty woman who seemed to bear no resemblance to her sisters, bleak Mary and gentle Frances – has died some time ago.)

Few relationships are what they seem in this comic novel about hidden lives. When Charlotte's 'nephew', Stanley – in reality her bastard son – dies from a cold that Polly may have given him, the distraught Charlotte disappears, awakening in Frances acute feelings of guilt. Frances undergoes a transformation when Fr Pocock (a sort of Mr Casaubon without the intellect) announces that he is to join a mission in India. Shocked that he has not taken into consideration his relationship with her – an attachment previously unsuspected by the reader – Frances forces him to marry her and take her with him. Fr Pocock dies at sea, and Frances's complete transformation is confirmed by a postcard that Polly receives showing her aunt dressed as a Pierrot next to a man with a monocle. Sent just after Pocock's death, it bears the message: 'High Jinks on Deck'.

When, shortly afterwards, Frances dies of amoebic dysentery, Mary goes into a retreat, and Polly travels to Yorkshire to stay at Thwaite Hall. Here she discovers that her mother was the daughter of Arthur Thwaite, a suitor of Mary, whom he would have married had he not impregnated her mother, Gertrude. Polly's mother was the issue. Further evidence of Gertrude's secret life is revealed by the discovery of her inscribed copy of *Fanny Hill*, from which Polly learns about an aspect of life not apparent in *Robinson Crusoe*.

Polly becomes the chatelaine of the yellow house when the mentally unstable Mary dies of a brain tumour. Mrs Woods, who has never recovered from the loss of Frances, with whom she was in love, has also died. 'Married' to the landscape (like Crusoe), Polly does not leave the unhappy house, but stays on with the new maid, Alice, both of

them declining into middle age. Increasingly eccentric, Polly starts to drink, and Alice takes charge.

Polly might have married Theodore Zeit, whom she had originally met, briefly but memorably, on the beach as a child, and had re-encountered at Thwaite Hall. Her hopes, however, are dashed when Theodore makes an unsuitable marriage to a childhood friend and returns to Germany. *Robinson Crusoe* has failed to provide her with the key to the lives of others; the people who seem to be superficially 'straightforward, strong and sexless', like Crusoe, are revealed to have deep and complex emotional lives. Defoe's book does, however, resonate with her own sense of isolation. Her insight into this novel in particular, and the novel in general, at last releases her into happiness and fulfilment. She becomes a teacher, writes a sort of 'spiritual biography', and becomes 'wife' and 'mother' in all but name to the refugee Theodore and his children, as the yellow house sails on into the chemical and nuclear age.

Keri HULME 1947–

The Bone People

The Bone People provided yet another example of the way in which the Booker Prize can confound expectations. The first novel of a Maori author virtually unknown outside New Zealand (where the book had already won two prizes), published by a feminist collective, it was the complete outsider in a year when novels by Doris Lessing, Iris Murdoch, Peter Carey and J.L. Carr were also shortlisted. Although it had its champions, this demanding, experimental and fragmented novel was pronounced unreadable by some critics. Further surprises came at the award ceremony, which Hulme was unable to attend, when the Prize was accepted on her behalf by a group of chanting women in feather cloaks and (male) evening dress.

The novel evolved from a short story over twelve years. Its female protagonist is Kerewin Holmes, a New Zealander of mixed Maori and Hebridean ancestry. A successful and wealthy artist, Kerewin is estranged from her family, and has built a recluse's tower as a home by the sea. One day, returning home from a walk on the beach, she finds a child, who has broken into her house. He cannot speak, but a label around his neck identifies him as the mute, Simon Peter Gillayley. Radioing the telephone operator to put a call through to his father, Kerewin learns that Simon Peter is regarded as a local eccentric, a truant with a reputation for hooliganism.

She is standing now at the far seaward end of the reef, on a black tongue of rock. A strange person in blue denims, sometimes obscured by mist from the waves that explode like geysers in the blowhole. She looks tense and desperately unhappy. Like she's at war with herself. Like a sword wearing itself out on its sheath. She doesn't look like a woman at all. Hard and taut, someone of the past or future, an androgyne. She hasn't moved from the rocks there for ten minutes. Still as a rock herself.

THE BONE PEOPLE
BY KERI HULME

When Simon's father, Joe, comes to collect the boy, Kerewin is drawn into a friendship which is to be the most important of her life. Joe's wife, Hana, and child, Timote, have died of influenza, and Simon ('Himi' in Maori) is a semi-mystical changeling-child, washed up on the beach after a shipwreck. Kerewin starts to trace his origins after Simon gives her a rosary with an aristocratic phoenix-emblem on a signet ring. Wanting to give Joe a break and provide them all with time to get to know each other, Kerewin drives the three of them up to her family's holiday huts on a distant coast. Simon catches his first fish but gets a fishing-hook caught in his thumb, and the minor amateur surgery that this necessitates reveals aspects of all their characters: Kerewin cannot bear to let her 'hard-boiled image' slip; Joe is deeply tender yet cruel to his surrogate son; and Simon is terrified of doctors and needles.

Kerewin is gripped by a searing pain in her stomach, and suspects that she has cancer, a diagnosis which is later confirmed. Rejecting the unholistic approach of modern medicine, she refuses treatment, and stoically sets about preparing to die. But a second crisis is sprung upon her, when she learns that Joe is regularly beating Simon. Kerewin becomes a complicit witness to the violence, realising that any interference will destroy all of them. When, after a particularly brutal attack, Joe is sent to prison, Simon is made a ward of court, but the boy refuses to emerge from the silence and withdrawal of autism now that he is separated from the substitute parents he loves.

On his release from prison Joe wanders off into the bush, and meets the keeper, who guards an ancient Maori god. The period of separation provides a catharsis for all involved, as they slowly return to life: Kerewin's cancer magically disperses; Simon is determined to be reunited with the adults; and Joe realises that, whatever its frustrations, his relationship with Kerewin and Simon is all he has. The shipwreck that orphaned Simon was caused by his heroin-addict father, a young man who sent to the keeper's home to die. In the final pages of the novel, the bizarre family are brought back together, as Kerewin rebuilds her tower in the shape of a seashell spiral, its rooms encompassing endless possibilities.

John IRVING   1942–

The Cider House Rules

At St Clouds Orphanage Dr Wilbur Larch performs the Lord's work and the Devil's. The Lord's work is delivering mothers of their unwanted children; the Devil's is abortion. Within this seemingly clear-cut distinction of good and evil, Irving inserts the qualifying, human contexts, the extenuating circumstances and implicit contradictions which reveal the simplicity of this distinction as merely simplistic. Dr Larch performs abortions on the victims of rape, incest and prostitution, women unable, unwilling, or both, to raise a child. The orphanage is run on a similarly pragmatic and human basis, its ultimate object being to place each orphan with a suitable family. These two aspects of the orphanage converge in Homer Wells, an orphan, assistant to Wilbur Larch and the book's hero.

Homer is unsuccessfully adopted three times before, at the age of eighteen, a young couple, Candy Kendall and Wally Worthington, arrive at St Clouds seeking an abortion. Homer returns with them to Ocean View Orchard to look after Candy. A bizarre *ménage-à-trois* develops. All three are clearly 'right' for each other, yet through the necessary evasions and secrets a clear sense of wrong emerges as love turns to adultery in Candy's affair with Homer. Parallel with this complex story of love and guilt, Irving relates local goings-on in general and the orchard business in particular. Each year a team of black apple-pickers is hired under the leadership of Mr Rose, who enforces the (unwritten) cider house rules of the title. As in the cases of adultery and abortion, these rules remain shadowy and unspoken until broken, when their full force becomes apparent. Mr Rose himself is a law-breaker who commits incest with his daughter, inciting her to swift and violent revenge. The incident brings matters to a head. Homer performs his first abortion on the girl, his long affair with Candy is admitted to and ended, and he returns to St Clouds to take the place of Wilbur Larch, who has died. The book closes with these multiple declarations of the truth, and a positive upholding of 'the rules'.

In several respects, the novel is a departure from Irving's previous work. Catastrophes are suppressed or underplayed, and the story proceeds by development of character rather than inventiveness of plot. The confrontation of serious issues (incest, abortion and infidelity) is not new in Irving's work, but here those issues are backed by substantial factual accounts which lend weight to the arguments. The obvious (and acknowledged) literary ancestor is Dickens, a writer Irving admires intensely and whose presence is felt in the expansive characterisation and broad social concern. Irving's readers will miss the fireworks of his earlier novels, for those episodes of tragi-comedy do not really fit into this new mode, but *The Cider House Rules* is a profound and serious novel, marking the beginning of a new phase in Irving's work.

Larry McMURTRY   1936–

Lonesome Dove

Set in the late 1870s, its pace leisurely, its structure complex and its cast huge. *Lonesome Dove* accommodates as much good humour as calamity. It won the Pulitzer Prize in the year of its publication.

Woodrow Call and Augustus McCrae – once captains in the Texas Rangers – are running a small livestock business in Lonesome Dove, a town just above the Mexican border. Terse and unemotional, Call works obsessively; McCrae drinks whisky on the porch, argues amiably with his employees or visits Lorena Wood, the local whore. Neither man is wholly satisfied with this placid existence and, when Jake Spoon – fleeing Arkansas because of a shooting – appears with tales of the rich pasture to be had in Montana, Call and McCrae decide to go north. They rustle several thousand head of cattle, gather a team of cowboys and begin a hazardous trek.

Two lovers, Lorena and Spoon, accompany them; Spoon, however, swiftly abandons Lorena and slides into a life of gambling and crime. For a while, July Johnson, an inexperienced sheriff, pursues him, but when he learns that his wife, Elmira, has left home, Johnson defers that search to look for her. Elmira proves elusive, travelling initially on a whisky boat and later with a strange pair of buffalo hunters.

Hence, four interwoven narratives comprise the greater part of the story: the cattle drive, Spoon's wanderings, Elmira's vagrancy and Johnson's quest. The first supplies the novel's central plot. Overcoming the savagery and indifference of nature – and, heroically, their personal fears – Call's and McCrae's recruits are strengthened by their epic journey. Distressing deaths are met with resignation: 'yesterday's gone on down the river and you can't get it back'. Lorena is kidnapped by a renegade Indian, sold, terrorised, beaten and repeatedly raped. Yet, rescued and cared for, she survives her traumatising violation. Newt Dobbs, the youngest of the party, passes from anxious adolescence to assured maturity. For Call and McCrae, the expedition engages temperament and ability: both men require a vast, capricious landscape – and the demands it makes – in order to sustain a sense of self. Even after the drive is concluded, and a cattle ranch established in Montana, Call feels compelled to fulfil a poignant mission that returns him to his Texas starting place.

If Spoon's fate is harsh, he has invited it by supposing that he is unlucky rather than weak; if Elmira's is savage, it is brought about by her deranged and callous folly; if Johnson's is unsensationally domestic, he has earned it on account of his persistence and integrity. Nevertheless, McMurtry does not present an easy correlation between behaviour and destiny: in his panoramic vision, the potency of evil, simple misfortune and the poor judgement of others can – and do – destroy the innocent. Thus, Call's and McCrae's triumph is partial at best and rendered possible only through adherence to a set of humane values: loyalty, honesty, disinterested courage. Indeed, the clarity of the protagonists' moralities and motivations endows the book with the aura of myth.

Harry MATHEWS   **1930–**

The Sinking of the Odradek Stadium

First published in *Paris Review* in 1971 and 1972, *The Sinking of the Odradek Stadium* is an epistolary novel consisting of letters exchanged between Zachary McCaltex, a Miami librarian, and his wife, Twang Panattapam. As the novel opens, Zachary is searching through the Miami Library's map collection in pursuit of clues as to the whereabouts of an Italian treasure trove. Twang has travelled to Rome to hunt for the treasure from its supposed origin. They are newly married, and their letters are filled with tender expressions of love, along with details of their respective finds. Twang is a native of 'Pan-Nam', Mathews's baffling approximation of Vietnam, and her letters are almost incomprehensible to begin with. Luckily, one of the many stories of the novel is her

gradual mastery of American English, although her spelling and word order remain eccentric throughout.

Zachary's and Twang's separation (and eventual convergence) are dependent upon the treasure hunt which involves them both in bizarre situations and alliances. Zachary joins a shadowy organisation, 'the Knights of the Spindle', whose members are also looking for treasure. Their rituals give Mathews the opportunity to present several preposterous *tableaux vivants*, involving porpoises, phantom baseballs and, nestling in a bed of black swans' feathers, Mallarmé's left ulna. Twang, meanwhile, immersed in a labyrinthine quest through the history of the Medicis, is assailed by unwanted suitors, degenerate aristocrats and bogus structural linguists. Realising that her letters are being intercepted (probably by the Knights of the Spindle) she sends word to Zachary that her communications will henceforth misrepresent both her true feelings and her progress in the search. But the letter is lost and Zachary agonises over his wife's sudden indifference to him.

Complications multiply as Twang's search uncovers more and more historical plot. A chest with three fishes (or balls) carved in it may have been stolen by a certain Amortenelli in the sixteenth century. It might contain the treasure, or coin-clippings, or silk, or silk woven with gold. It might lie sunk in a wreck off the Florida Keys, or in a cellar in Florence, or be lost entirely. A bastard son of the Medicis is involved, as is his mother, a slave brought from the Far East.

Mathews's resolution of the book's extraordinary plot involves the retrieval of the lost letter and the discovery that the slave-girl was Twang's ancestor: the treasure is rightfully Twang's. The last letter is from her to Zachary. In it she clears up all misunderstandings and delivers the happy news that she has found the treasure. It is already in transit, she writes, bound for the gold markets of Rangoon aboard a Panamanian cargo ship, the *Odradek Stadion* (sic).

The Sinking of the Odradek Stadium is Mathews's third novel. He is a prominent member of the OuLiPo group, and his interest in elaborate, structural games and patterns is everywhere apparent in the book. Having taken the most basic of all plots – the quest – Mathews complicates it to the point of unintelligibility before returning it to relative simplicity. The novel's central 'fact', the treasure itself, is questionable right up until the moment when it is simultaneously confirmed and consigned by the title to a watery grave. What remains is the love story of Twang and Zachary, which has competed with the treasure-quest all through the novel and only now wins out.

Published complete with an idiosyncratic index, *The Sinking of the Odradek Stadium* draws obvious inspiration from Raymond Roussel's novels in its principles of composition and bizarre inventions. Mathews's historical fakery is always convincing and his control over the seeming chaos of the plot is complete. But it is the characters of Zachary and Twang, separated victims of novelistic circumstance, which inject a human presence into the cold brilliance of Mathews's creation and thus give it a value beyond its undoubted technical sophistication.

John MORTIMER   1923–

Paradise Postponed

As well as having a full-time career as a barrister, Mortimer has been a prolific and versatile writer, with successes in print and on television and the stage. In later life he has combined the roles of novelist and screenwriter, often developing books and television series together. The genesis of *Paradise Postponed* occurred over lunch with a television producer who suggested that Mortimer should write a series of plays which would survey the whole course of history in post-war Britain. As this ambitious project developed, Mortimer wrote the novel in tandem with the television series; they were both critical and popular successes.

The novel has a large cast of characters who epitomise various aspects of post-war society, but the central plot concerns two brothers, Fred and Henry Simcox, and their rivalry with the Rt Hon. Leslie Titmuss, MP, whom they first encounter at a dismal boarding-school. Titmuss's lower-middle-class parents have scrimped and saved in order to buy him an education which comes to the Simcoxes as of right as part of their comfortable middle-class background. Titmuss's schooling is intended to provide the first rung on the social ladder. When the Simcoxes' father, a kindly rector who campaigns for nuclear disarmament, leaves a legacy to the unspeakably philistine and upstart Titmuss, Henry is determined to challenge the will. Fred would prefer to leave things as they stand and thus avoid the public embarrassment of litigation.

I grew up to understand the value of money because it took my father five years to save up for our first second-hand Ford Prefect. Every night he finishes his tea and says to my mother, 'Very tasty, dear. That was very tasty.' He always says the same thing. He falls asleep in front of the fire at exactly half past nine and at ten-thirty he wakes up with a start and says, 'I'll lock up, dear. Time for Bedfordshire!' Always the same. Every night. Just as he got to work at exactly the same time every morning for forty years. He's loyal to his job and my mother was loyal to the Stroves. You know what my parents are? They're true Conservatives!

PARADISE POSTPONED
BY JOHN MORTIMER

Why Simeon Simcox should leave his money to an already wealthy Member of Parliament is the central mystery of *Paradise Postponed*. Fred begins to suspect that Titmuss is Simeon's illegitimate son, but the truth proves to be more complex. Fred, who is now a GP with a country practice, becomes an unwilling detective: 'The truth would have to be shared and he supposed it would be better done in the sitting-room in Sunday Street than in a law court.' Henry tries to overturn the will by claiming that his father was insane, but, risking the scandal he has been so keen to avoid, Fred resists this notion. It transpires that it is not Titmuss who is Simeon's illegitimate child, but Titmuss's recently deceased wife, Charlotte. Feeling guilty about her sad life and the possibility that she committed suicide, Simeon left the money to Titmuss in order to educate their child.

Mortimer employs the contrast between the backgrounds of Titmuss and the Simcoxes to create a deft satire not only upon the whole of post-war Britain, but more particularly upon the Thatcher years. Titmuss is very much the thrusting, radical New Tory, while the Simcoxes represent decent, old-fashioned, liberal values. The further fortunes of the principal characters are pursued in a sequel having the equally Miltonic title *Titmuss Regained* (1990), in which Titmuss has risen to become secretary of state at the Ministry of Housing, Ecological Affairs and Planning and has acquired a smart new wife to go with his manor house.

Jeanette WINTERSON   1959–

Oranges Are Not the Only Fruit

'Like most people I lived for a long time with my mother and father', begins Winterson's semi-autobiographical comic first novel. Unlike most people, the central character – also called Jeanette – was being raised by her adoptive Lancashire evangelist mother to be a missionary who would save the world. 'My father liked to watch the wrestling, my mother liked to wrestle; it didn't matter what. She was in the white corner and that was that.'

The book begins when Jeanette is seven. Her time is spent mainly at church or prayer meetings or on religious roadshows, but there are early signs that she may not follow the path her mother has plotted. Jeanette becomes curious about 'Unnatural Passions' (which are, she is warned, what the women at the paper shop deal in) and she is told by a gypsy who looks at her palm, 'You'll never marry, not you, and you'll never be still.'

Jeanette is educated at home (taught to read from the Book of Deuteronomy), and only when a court order compels her, does her mother send her to school, a place she calls 'the Breeding Ground'. Try as she might, Jeanette cannot fit in. While other girls stitch 'TO MOTHER WITH LOVE' on their samplers, hers reads 'THE SUMMER IS ENDED AND WE ARE NOT YET SAVED'.

Events at school and home are interspersed with the thought-processes which lead Jeanette to wonder if oranges (the leitmotif representing her mother's view of the world) *are* after all the only fruit. She determines to reassess the options, secular and religious, that are available to her. She learns (from talking to her aunt and hiding in dustbins to overhear Next Door's conversations) that men are beasts. She therefore concludes that it is a good job she is destined to become a missionary because she certainly doesn't fancy marriage. But she has to question her vocation when she falls in love with Melanie. It feels good when she is with Melanie, so she thinks it cannot be 'Unnatural Passion'. She is surprised then when she is denounced by the church tribunal as one possessed of devils. Melanie is sent away, and Jeanette – after a bout of glandular fever and a brief return to the Lord – embraces her devils. She is discovered in bed with the newly converted Katie and is expelled from the church and her home. She works for a while driving an ice-cream van and painting the faces of corpses at a funeral parlour, and the book ends as Jeanette is about to journey out into the world.

It was in this way that I began my education: she taught me to read from the Book of Deuteronomy, and she told me all about the lives of the saints, how they were really wicked, and given to nameless desires. Not fit for worship; this was yet another heresy of the Catholic Church and I was not to be misled by the smooth tongues of priests.

'But I never see any priests.'

'A girl's motto is BE PREPARED.'

ORANGES ARE NOT THE ONLY FRUIT
BY JEANETTE WINTERSON

Her voyage has been foreshadowed by the fragmentary folk tales which are scattered through the text. These fanciful diversions tell of various knights and princes and their quests for perfection, and they give an indication of the direction Winterson's writing was to take. After *Oranges* and the unmitigated disaster *Boating for Beginners* (1986), she abandoned the comic novel to concentrate in both *The Passion* (1987) and *Sexing the Cherry* (1989) on blurring the lines between historical fact and fantastical storytelling.

Oranges Are Not the Only Fruit was lauded by the critics for its unconventional subject-matter, its often riotous humour and the sheer energy of the narrative voice. Winterson's style is shaped by the terseness of biblical prose (the novel is divided into chapters named after books of the Old Testament: Genesis, Exodus, Leviticus, Numbers, Deuteronomy, Joshua, Judges, Ruth), and her preacher training – to engage and persuade – shines through.

The book caused a rift between Winterson and her mother, but it won the Whitbread Award for a first novel in 1985, and in 1990 was adapted by the author for a controversial but extremely successful three-part television series.

1986

US bombs Tripoli ● Chernobyl nuclear power accident ● State of Emergency in South Africa ● Green Paper proposes introduction of poll tax ● US companies sell off or close subsidiaries in South Africa ● NSPCC reports doubling in number of reported cases of child sex abuse over year in Britain ● The first 'lap size' computer introduced in the US ● Explosion of US space shuttle *Challenger* leads to suspension of shuttle flights ● Return of Halley's Comet ● Lucian Freud, *Painter and Model* ● Arata Isozaki, Museum of Contemporary Art, Los Angeles ● Andreas Whittam-Smith and associates launch *The Independent* ● Films: *Jean de Florette, My Beautiful Laundrette, Platoon* ● Television: *Inspector Morse, The Singing Detective* ● Popular music: Paul Simon, *Graceland* ● Lord David Cecil, Christopher Isherwood, Joseph Beuys, Edmund Rubbra, Simone de Beauvoir, Jorge Luis Borges, Henry Moore and Harold Macmillan die ● Tom Paulin (ed.), *The Faber Book of Political Verse*

Kingsley AMIS   1922–

The Old Devils

'All of a sudden the evening starts starting after breakfast. All these hours with nothing to stay sober for.' Amis's sardonic study of retirement suggests that (in Wales at least) drink is the principal recreation of the elderly. The novel concerns a group of old friends who rarely draw sober breath: the men have their lunchtime sessions of beer and spirits at the Bible and Crown, while the women spend their days in each other's houses having coffee-mornings at which enormous quantities of white wine are consumed. The reappearance in their circle of Alun Weaver CBE, a professional literary Welshman who has been living in London with his wife Rhiannon, disturbs their precarious equilibrium.

Alun's return to his roots passes frustratingly unnoted by the press, but causes a stir of resentment amongst his cronies: 'We're all Welshmen here,' they point out. 'More's the pity.' Alun has been peddling the Celtic myth through coffee-table books and television chat shows, pandering to the cult of the Welsh poet Brydan, a revered figure whose life has inspired a tourist industry, and whose mantle Alun is widely believed to have inherited. While many of his friends recognise Alun as a fraud, they admire his panache, and some of the wives are pleased

to welcome him back into their lives. Addicted to the 'security-conferring streak of gratitude' to be found in elderly women, Alun rekindles the flames of old love with Gwen, the sharp wife of the ineffectual Malcolm Cellan-Davies, and Sophie Norris, whose husband Charlie has a part-share in a restaurant run by his homosexual brother, Victor. Malcolm is more pleased about the return of Rhiannon, of whom he has romantic memories, which she pretends to share. When this pretence is exposed, she weeps, something for which Malcolm is deeply grateful. (It inspires a poem at the novel's close.) Peter Thomas, vast in bulk in spite of the low-calorie tonics with which he dilutes large measures of gin, finds Rhiannon's reappearance rather more troubling. Years ago, while a student, he had had an affair with her, got her pregnant, procured an abortion and abandoned her. He now lives in bitter marital disharmony with Muriel, who comes from Yorkshire, and upon whom he is financially dependent. Muriel despises him: 'You emanate hopelessness and resentment and boredom and death.' It later transpires that Peter abandoned Rhiannon not for Muriel, but for another member of the circle, Angharad, who is now married to the parsimonious Garth Pumphrey and prematurely aged through ill-health.

Alun is planning a new series, *In Search of Wales*, and takes up his abandoned novel. Work on the first project gets little further than some pub-crawling. When asked his honest opinion of the novel, Charlie gives it, rather than the 'free-from-bullshit certificate' for which Alun had hoped. Alun's revenge leaves Charlie in a state of collapse. After a ferocious dispute between Alun and Tarc Jones, landlord of the Bible and Crown, the old devils are banished from the pub. They retire to Garth Punphrey's house, where Alun suffers a fatal seizure when Garth attempts to make his friends pay for their optic-measured drinks. (Amis's score-settling *Memoirs* [1990] make it clear that not standing your round is the one unforgivable sin.) The novel ends with Muriel Thomas announcing her intention to return to Yorkshire after her son, William, marries the Weaver's daughter, Rosemary. Peter confesses his continuing love for Rhiannon ('I'm sorry it sounds ridiculous because I'm so fat and horrible, and not at all nice or even any fun, but I mean it'), and moves in with her.

Most of those whose marriages have turned out less than well, say, might have been considered to have their ideas of how or why but not to know much about when. According to himself Peter was an exception. If challenged he could have named at least the month and year in which he and Muriel had been making love one night and roughly halfway through in his estimation, what would have been halfway through, rather, she had asked him how much longer he was going to be.

THE OLD DEVILS
BY KINGSLEY AMIS

The Old Devils returns to the territory of **Ending Up** (1974), but allows a glimmer of hope amidst the bleakness. 'I feel that any place where two people manage to fall in love can't be as bad as all that', Malcolm remarks of the unlovely Welsh town in which the novel is set. The novel was awarded the Booker Prize, a circumstance which Amis genially admitted altered his hitherto sceptical attitude towards literary prizes, and was made into a fine television series.

Paul AUSTER   1947–

The New York Trilogy

City of Glass (1985)
Ghosts (1986)
The Locked Room (1986)

The New York Trilogy comprises three short novels connected by theme and locale. In *City of Glass* a writer of mystery novels, Daniel Quinn, is drawn into the real world of investigation when he is mistaken for the private eye Paul Auster as a result of a wrong number. Assuming the role of Auster, he takes the job of tailing Stillman, a former Professor of Theology just released from prison. Stillman had been imprisoned for locking his son Peter in a darkened room for nine years; Peter's wife, Virginia, now fears Stillman will attempt to harm his son again. Quinn tries to divine significance from Stillman's walks around New York (do his journeys spell out the words 'THE TOWER OF BABEL'?), and when he loses Stillman he seeks the help of Auster, who turns out not to be detective at all but – surprise, surprise – a writer. Quinn installs himself in an alley opposite Peter's apartment to await Stillman, and remains there for many months sleeping in a dustbin to maintain his obsessive vigil. After learning of Stillman's suicide, he retreats into Peter's deserted apartment, but has disappeared when Auster and the narrator arrive there and find the red notebook in which he has recorded his investigation.

Ghosts concerns a similarly obsessive investigation. A man called White hires private eye, Blue, to tail a man called Black who lives on Orange Street. Black hardly ever leaves the room in which he sits writing. Blue compulsively monitors his activities until he discovers that his fiancée has deserted him and that he has been trapped 'into doing nothing, into being so inactive as to reduce his life to almost no life at all'. Adopting a variety of disguises he talks to Black on the street and eventually discovers that Black and White are the same person. White has hired Blue to watch him to prove his own existence to himself. Blue wrestles the gun from his hand and beats him to death.

The unnamed narrator of *The Locked Room* (who claims to be the writer of the previous two stories, and a friend of Paul Auster) learns of the mysterious disappearance of his childhood friend Fanshawe from Fanshawe's wife, Sophie. Fanshawe is presumed to be dead and has left a pile of manuscripts with the instruction that the narrator should assess whether or not they are worth publishing. He believes they are and arranges for the successful publication of *Neverland* and other works. He falls in love with Sophie and moves in with her, then receives a letter from Fanshawe: 'At the risk of causing heart failure, I wanted to send you one last word – to thank you for what you have done.' He obeys Fanshawe's instruction not to tell Sophie that her husband is still alive. Later, researching a biography of Fanshawe, he develops an obsessive hatred of him, and resolves to track him down, but the quest leads only to his own mental breakdown. Some years later, a final note leads him to Fanshawe, now an embittered recluse, who hands him a red notebook which explains his disappearance. The contents of the notebook are never divulged, and the narrator destroys it.

City of Glass was turned down by seventeen publishers, but the eventual success of trilogy established Auster's reputation on both sides of the Atlantic. The conventions of the detective story are wittily subverted – with suspense maintained, but the resolution withheld – and this is combined with postmodernist games-playing to produce what the *Times Literary Supplement* described as 'seductive metaphysical thrillers'.

Paul BAILEY   **1937–**

Gabriel's Lament

In Bailey's fragmentary autobiography, *An Immaculate Mistake* (1990), he recounts that, although he was eleven at the time, he was the only member of the family not to cry when his father died. Years later, in Fargo, North Dakota, he awoke to find his father at the foot of his bed, a 'visitation' which finally allowed him to mourn: 'I had accepted the gift of grief, and my acceptance, my complete acceptance, of it had released something in me of which I had been completely unaware – and might have stayed so always.' A similar acceptance forms the climax of *Gabriel's Lament*, which opens with the punning, resonant, declaration: 'I came to grief late in life – when I was forty, in Minnesota.' A tragi-comedy, the novel confronts the difficult subject of grief through the character of Gabriel Harvey, a clever boy from a working-class home whose mother disappears shortly before his thirteenth birthday.

The first, and longest, section of the book recounts, in the first person, Gabriel's relations with his parents. His childhood is happy until his elderly father, Oswald, inherits a large sum from a former employer and removes his family from modest rented accommodation to the grandiose 'Blenheim', where appearances must be kept up. Here, the marvellous stories Oswald told his son are replaced by self-important lectures on a wide variety of subjects, ranging from the wearing of suede shoes to the delights of 'rogering'. When Gabriel's mother, Amy, who is many years her husband's junior, fails to return from a mysterious and unexpected holiday, the suppressed grief her son feels at her continued absence manifests itself physically, earning him the ignominious nickname 'Piss-a-Bed' from his contemptuous father. As Gabriel recounts the course of his subsequent life, and his encounters with an extraordinary gallery of misfits and eccentrics, the memory of his mother is a constant undertow, evoked by snatches of remembered conversation, catchphrases and songs. In particular, Gabriel finds consolation in wearing one of her dresses and listening to a Handel aria she loved (a quirk which leads to his being thrown out of his lodgings). He attempts to keep his boastful, racist, foul-mouthed and mendacious father at a distance, but obeys the dying wish of his Swedenborgian aunt not to abandon the old man altogether. He takes on a number of menial jobs, including a spell as 'skivvy' at a home for old women, which (along with several of its inmates) is familiar from Bailey's first novel, *At the Jerusalem* (1967). It is here that Gabriel comes across an old ledger containing the death-bed thoughts of 'the Liverpool

Messiah' Roger Kemp. This inspires Gabriel to write *Lords of Light*, a book about evangelist-preachers, which becomes an unexpected success and is turned into a ludicrous film. Gabriel is invited to Sorg, Minnesota, to give a lecture, and it is here that he opens his legacy from Oswald, some letters which explain what really happened to Amy.

Although *Gabriel's Lament* is a painful book, it is also a baroquely comic one, peopled with splendid grotesques: Diana Sparey, the Communist landlady; Matthew, a former barrister turned tramp; Minnesotan Dale Armsted, 'the widest man in the world'; Oswald's crony, the odious Van Pelt, and his repellent elderly fiancée, Marge; and, above all, Oswald himself. At the novel's close Gabriel is forced to confess: 'I have not succeeded, Father, in what I set out to do. When I began this lament for my numb life three years ago, I meant to polish you off for ever, and I think I have failed. You are still here, curse you, in these pages.' Gabriel's failure is Bailey's triumph. Partly derived from the author's own father (whose name he shares), Oswald also appears to owe something to the father of the writer J.R. Ackerley, who was the subject of an 'Omnibus' television programme scripted by Bailey in 1980. (Ackerley *père* had a roguish manner, had risen in the world thanks to an aristocratic benefactor, had fathered illegitimate offspring, had a complicated sexual history, and lived at Blenheim House, Richmond.) Dense, allusive, and teeming with character and incident, *Gabriel's Lament* has been compared with the novels of Dickens, and was the second of Bailey's novels to be shortlisted for the Booker Prize.

<hr>

John BANVILLE   1945–

Mefisto

Banville, who has quietly emerged as one of Ireland's leading novelists, frequently used scientific themes in his novels some time before such subject-matter acquired literary chic. The novels often centre on a particular historical figure – *Dr Copernicus* (1976), *Kepler* (1981), *The Newton Letter* (1982) – and in *Mefisto* Banville introduces the imaginary Gabriel Swan, a man searching for the meaning of life, which he believes he will find through numbers. The attempt to impose order and find meaning in the face of chaos is a recurring theme of Banville's work, and a clear metaphor for the act of writing itself. *Mefisto* is bleaker than some of Banville's other novels, a fact he ascribes to the deaths of his parents.

Gabriel was born in the last year of the Second World War in a town on the coast of the Republic of Ireland, and with him a dead twin-brother: 'It seems out of all this somehow that my gift for numbers grew. From the beginning, I suppose, I was obsessed with the mystery of the unit.' He is obsessed, too, by nature's freaks – 'all those queer, inseparable things among which I and my phantom brother might have been'. The view of the world he presents is unsettling and vivid, filled with grim detail.

Because he is a prodigy, throughout the novel Gabriel is used by others for his talents, distracted from his own search for an answer, most notably by a rakish fixer named Felix. Lured away from home and school, Gabriel spends more and more time at Ashburn House. Felix and his odd companions – the mathematically minded Dr Kasperl and deaf-mute Sophie – have moved into the old country house with plans to re-activate the anthracite mine, Coolmine. Gabriel is needed to help Dr Kasperl; Sophie is the doctor's lover, and spends much of her time making marionettes. The novel's first part is titled 'Marionettes', and one of its themes is manipulation: Gabriel is a Pinocchio, a puppet only pretending to be 'a real boy'.

The need to make order out of chaos becomes imperative for Gabriel when his world falls apart. His grandfather dies; pockets of gas in Coolmine cause a series of explosions and fires; Ashburn House goes to seed; Gabriel's father and uncle are maimed and his mother killed in a car-crash. 'All this, it was not like numbers, yet it too must have rules, order, some sort of pattern. Always I had thought of numbers falling on the chaos of things like frost falling on water, the seething particles tamed and sorted, the crystals locking, the frozen lattice spreading outwards in all directions.' But Gabriel's hold on reality slips and he has a breakdown. In a scene that is hazy (because Banville does not explain when his narrator cannot), Ashburn House, Dr Kasperl, Sophie and Gabriel collapse into Coolmine and are consumed by fire.

The second part of the novel 'Angels', begins in a city hospital. Miraculously, Gabriel has survived; but he is horribly disfigured. Now fully initiated into the brotherhood of the maimed, he has gone through hellfire and been reborn, ready to find salvation.

The city I had thought I knew became transfigured now. Fear altered everything. I scanned the streets with a sort of passion, under the glare of it things grew flustered somehow, seemed to shrink away from me, as if stricken with shyness. They had never been noticed before, or at least not like this, with this fierce, concupiscent scrutiny. I saw pursuers everywhere, no, not pursuers, that's not it, that's too strong. But nothing was innocent any more.

MEFISTO
BY JOHN BANVILLE

It is no surprise to Gabriel or the reader that on his release he meets Felix. this time Felix needs Gabriel to help Prof. Kosok, who is conducting studies to prove that nothing can be proved. Kosok has a daughter, Adele, with whom Gabriel forms a relationship. Then, for a second time, Gabriel's world is shattered. The minister funding the research demands results, which the Professor cannot (and will not) give; Felix has to go into hiding; Adele overdoses and dies in hospital. 'A part of me, too, had died. I woke up one morning and found I could no longer add together two and two ... One drop of water will not make two drops but one.' With this realisation, Gabriel reaches a state of independence. He is able to abandon the search for security through order: 'In future, I will leave things, I will try to leave things, to chance.'

Christine BROOKE-ROSE   1926–

Xorandor

The daughter of a British father and a Swiss-American mother, born in Switzerland, brought up in Belgium, educated in Britain and latterly living and working in France, Brooke-Rose has developed as a truly European novelist. Indeed, in her more recent books the early satirical novels she wrote in England are no longer acknowledged amongst 'other works'. She has been influenced by the French *nouveau roman*, in particular by Alain Robbe-Grillet, whose work she has translated. Her novels have been described as 'concrete', 'experimental' and 'difficult', and are indisputably challenging, although often very funny. Her principal concern is with narrative and language, and in *Xorandor* she enters the world of computers (acknowledging her indebtedness to *Programming Languages – Design and Implementation* [1984] by Terrence W. Pratt). Xorandor himself is named from the logic of the program language he uses: exclusive OR (XOR) AND non-exclusive OR.

Computer-mad twins Jip (a boy) and Zab (a girl) are playing with their Poccom 2 computer by an old Cornish tin mine when a message appears on the screen. It seems that the stone they are sitting on is not a stone at all but an intelligent being they dub 'Xorandor', which is capable of communicating, even finding a tinny voice. They dictate their story into a computer, and the narrative becomes as much a debate on the necessary (and unnecessary) elements of storytelling as it is a relation of actual events. The interests Xorandor excites in the 'olders' he asks to meet has wide political implications. Given that he feeds on nuclear waste – the tin mine has become a secret site for nuclear dumping – they seem to have found the ideal solution, particularly since he has produced offspring which are subsequently sent to other waste-producing countries. At some point however, Xorandor has ingested circuit-damaging caesium-137, and when one of his offspring does the same, the results are catastrophic. Threats to blow himself up inside a reactor are appeased at the eleventh hour, but now all the world knows about the 'alphaguys' and intense debate ensues. Then the American 'alpha-nice-guy' neutralises a bomb on a nuclear test site and the disarmament versus deterrent issue reaches new heights.

After a period of self-imposed silence, Xorandor reveals to Jip and Zab that he is not from Mars at all, as is widely believed, but has merely confirmed the suppositions of the olders who questioned him. He is in fact a native of Earth, has lived for over 4,000 years and is one of many. Since 'MEN PREFER ULTIMATE DETERRENT TO NO DETERRENT ON EITHER SIDE', Xorandor realises that he and his known progeny will be sent 'back' to Mars, and pre-empts the decision by making the request himself, although it will mean certain destruction.

Only Jip and Zab know about this and the two post-caesium offspring secretly produced and perhaps programmed to neutralise warheads. A number of questions now remain. Why did Xorandor speak to them in the first place? Was it the result of his caesium binge, or was he trying to warn the world about the dangers of nuclear weapons? Why has he chosen to go to Mars? Is a computer-brain capable of 'moralising'? Since they alone are aware of his secret, should they destroy all recorded evidence so that no one else finds out? And do they?

Jim CRACE   1946–

Continent

Continent, Crace's first novel, won three prizes when it appeared in 1986: the *Guardian* Fiction Prize, the Whitbread First Novel Award and the David Higham Prize. It consists of seven short stories linked by the theme of the 'seventh continent', a Third World setting in which ancient rituals and taboos are in conflict with modern commercialism. In the first, 'Talking Skull', the wealthy native Lowdo is sent to college on the proceeds of his father's exploitation of peasant superstition, selling the milk of 'freemartins' (mutant calves) as a cure for impotence and infertility. The story attracts the attention of a film crew from Sweden, making Lowdo a temporary celebrity, and he plans to run the business from his cosmopolitan home in the city, only to find that his lethargic father is getting stronger and far from the death he had anticipated.

'The World With One Eye Shut' tells the story of a young prisoner whose gaoler, Beyat, is the lover of his mad sister, Freti. The anonymous prisoner tries to manipulate events by making Beyat 'the postman' for his messages to the outside world protesting against his wrongful arrest. The prison's warden, the Captain, makes Beyat push the discovered letter down the prisoner's throat, almost killing him until, in an unsuspected strength born of rage, the prisoner smashes the window of his cell and spews out the blood-spattered letter into the quadrangle.

'Cross-Country' is the tale of a young Canadian student teacher who becomes the unwilling rival of a local boy, Isra-kone. The teacher, Eddy Rivette, trains as a runner every evening, and Isra challenges him to a race in which Eddy's physical strength will be pitted against that of Isra's white mare. Eddy takes a long and arduous route along a dry stream-bed, while Isra rides over the mountains. In this Great Race Isra is thrown from his horse, which lands on top of him, crushing his leg, and Eddy is the ambiguous and guilty winner.

'On Heat', the fourth narrative, is a story of sexual temptation and inclination, as a scientific professor studies the breeding cycles of crabs and discovers a tribe who apparently all come into season at once. He takes a tribeswoman home with him, and the narrator, an elderly woman, discovers late in her life that the tribeswoman was really her mother. 'Sins and Virtues' tells of a Siddilic calligrapher, who becomes the authentic faker of his own work, in a tale of western exploitation of eastern folk-art. 'Electricity', the penultimate chapter, sees western technology running out of control in the introduction of electric energy to a small village; Warden Awni constructs the ultimate in air-conditioning fans, but it brings down the roof of his hotel. In the final story, 'The Prospect From Silver Hill', a company agent goes mad, separated from his own people and fantasising about the wife and children he never had, as, isolated in the schemes of mining for precious minerals, he begins to see the universe in a grain of sand.

Richard FORD   1944–

The Sportswriter

Narrator Frank Bascombe is a sportswriter who lives in the placid suburbia of Haddam, New Jersey – America at its most quintessentially ordinary. The novel recounts three days in his life over an Easter weekend in the early 1980s, and surveys the history of his previous thirty-eight years. During this period he married his wife – known simply as X in the novel – and published a book of short stories which was optioned by a movie company. His long-gestated first novel, *Tangier* ('where I had never been, but assumed was like Mexico'), somehow never got finished, and he was offered a job as a sportswriter. The couple had three children, and after the death of the eldest, Ralph, Frank felt a strange 'dreaminess' and a compulsion to philander. One night, after a burglary, X found a pile of letters from a woman, burned her hope chest and told Frank she wanted a divorce. She continues to live nearby and Frank sees her when he takes the children out. He thinks he still loves her, but is not sure.

'Joyce's epiphanies are a good example of falsehood', writes Frank, and Ford appears to share this view, as his novel examines Frank's life through a series of undramatic, largely unexceptional events. The novel opens early on the morning of Good Friday, Ralph's birthday, when Frank meets X in the cemetery where Ralph is buried. That evening he has a drink with Walter, a fellow member of the Divorced Men's Club. On Saturday he flies to Detroit with his girlfriend, Vicki, to interview a crippled football player who turns out to be mildly demented. On Sunday they go for lunch with her father and stepmother, and she decides to break off the affair. A phone call from X informs him of Walter's suicide, and he has to return to identify the body. For no particular reason he wants to see Walter's house and drives out there with X. This long, strangely poignant scene is suffused with a tone of quietly muted

mystery and desolation: 'I drive, an invisible man, through the slumberous, hilled, post-Easter street of Haddam. And as I have already sensed, it is not a good place for death. Death's a preposterous intruder. A breach.' Late that night he takes the train into New York and goes into his office where he meets a young girl. The epilogue, written in Florida in September, makes it clear: life continues as ever, without great epiphanies, without startling revelations.

Shena MACKAY   1944–

Redhill Rococo

Mackay began writing when she was very young: her first book contained two novellas written at the age of seventeen. She was immediately hailed as an exciting new talent. Early novels, such as *Music Upstairs* (1965) were Bohemian in setting, but later ones, written after she had reared a family, tend to be more domestic and suburban, though no less blackly comic, and no less consciously stylish. The unexpected juxtaposition of locale and style in the title of this novel neatly suggests Mackay's qualities as a writer – one who has been described as both 'rococo' and 'the supreme lyricist of daily grot'.

'Redhill was in essence a car park, or a series of car parks strung together with links of smouldering rubble and ragwort, buddleia and willowherb.' In this tarmac waste land, sexy Pearl Slattery struggles to keep up her family's spirits on a pittance from Snashfold's Sweet Factory and the gleanings of jumble-sale clothes, praying that her common-law husband Jack will not be released from prison to pester her too soon. The arrival as lodger of Luke Ribbons, an adolescent gaol acquaintance of Jack, is therefore less than welcome – especially when he falls in love with Pearl and solicits the advice of a witch-doctor.

Mackay's wry comedy catalogues the accumulation of everyday dross in the frustrated atmosphere of suburban England in the 1980s. Luke becomes immersed in Slattery life, learning the lore of Trail the Snail and a mug called 'Susan from Biology'. His efforts to seduce his landlady are frustrated by her own indifference and the interruptions of her children. Luke's family background in the vicarage of St Elmo's in Purley further embroils the Slatterys in cycles of misunderstanding, as his father's curate, Rick Ruggles, the trendy, guitar-playing leader of the Christian Youth Movement, begins to pursue Pearl's eldest daughter, Cherry, who is studying for her 'A' levels. Pearl's son, post-punk Sean, meanwhile is intent on corrupting Isobel Headley-Jones with vegetarianism and green politics, while her younger daughter, Tiffany, demands attendance at school open days and an expensive uniform for the Drum Majorettes.

As Pearl tries to muddle through, she finds herself in growing sympathy with 'the Weasel', Isobel's *Guardian*-reading, SDP-voting mother, a woman whose life is as antithetical to the Slatterys' squalor as it is possible to be. When Jack returns home unexpectedly to find his wife innocently in Luke's arms, Pearl takes a breather from her family's squabbles for an unlikely morning jog with the Weasel. 'Does the road wind up-hill all the way?' she puffs finally, pausing for a cigarette: 'Yes, to the very end.'

Redhill Rococo was an inspired choice as recipient of the 1987 Fawcett Prize, awarded to a book 'which has made a substantial contribution to the understanding of women's position in society today'.

Vikram SETH   1952–

The Golden Gate

Seth's affectionate satire of life amongst a group of friends and lovers in California is a rare but entirely successful example of the verse novel. Inspired by Charles Johnston's 1977 verse translation of Pushkin's *Eugene Onegin* (1831), Seth has chosen the difficult 'Onegin' stanza: fourteen tetrameter lines arranged as three quatrains and a couplet. Whereas Pushkin used a varying rhyme scheme, Seth has chosen a constant: *abab ccdd efef*, finishing with the couplet *gg*. The novel is divided into thirteen chapters, each containing around forty or fifty numbered stanzas. The entire book – including sections such as the acknowledgements, dedication,

contents list and a note about the author – follows the stanzaic pattern. It is an undoubted *tour de force*, which has been widely praised both in America and Britain. It was hailed by Gore Vidal as 'The Great Californian Novel', and was awarded the Commonwealth Poetry Prize.

The novel, set 'circa 1980', centres upon the love-affairs of John and Phil, former college room-mates who have remained friends ever since. They both live in California's 'Silicon Valley', where they have been employed in Defence. 'John kneels bareheaded and unshod / Before the Chip, a jealous God', but Phil, victim of a failed marriage, has left his job and joined the anti-nuclear peace campaign. John is a self-designated 'yuppie', at once hedonistic and workaholic, worried that there is no love in his life:

> I'm a young, employed, healthy, ambitious,
> Sound, solvent, self-made, self-possessed.
> But all my symptoms are pernicious.
> The Dow-Jones of my heart's depressed.

A former lover, the sculptor Janet Hayakawa, suggests John places an advert in lonely-hearts column, and when he refuses, she places one for him. As a result he eventually meets Liz Dorati, a lawyer.

Meanwhile, Phil broods upon his broken marriage and attempts to bring up his six-year-old son, Paul. At John and Liz's house-warming party, Phil meets the latter's brother, Ed:

> the image of
> El Greco's *Felix Paravicino*:
> The same pale, slender, passionate face,
> Strength and intensity and grace.

Phil gets very drunk and he and Ed end up in bed together. To his surprise, Phil has fallen in love, but Ed is a devout and troubled Catholic, a circumstance which eventually leads to the breakdown of their affair.

Meanwhile, Liz's cat, Charlemagne, wildly jealous of his mistress's affair with John, successfully contrives to break up the partnership. John is exposed as repressed and conventional when he reacts to the revelation of Phil and Ed's affair with disgust and dismay. Liz and Charlemagne join a peace protest at which Phil is arrested. Liz defies her firm and acts as Phil's attorney, and shortly afterwards they marry. John is greatly embittered and spends much of his time in singles bars. Eventually he and Janet get back together, but she is killed in a road accident. Liz and Phil have a son whom, in a gesture of reconciliation, they name John.

Reduced to mere plot, the novel sounds like a soap opera, but no prose outline can give any real impression of the immense wit, verve and skill of Seth's verse. He masters a tricky form with Byronic *brio*, delighting in unexpected and outrageous rhymes in an invigoratingly self-conscious performance. Quite apart from its sheer fun, the novel also contains lyrical descriptions of Californian scenery, meditations upon the pleasures and pains of love and family, an anti-nuclear sermon, and a complete recipe for preserving olives.

1987

Agreement to eliminate intermediate nuclear forces • Ronald Reagan accepts full responsibility for Iran–Contra scandal • 'Great Storm' sweeps across south-east England, reckoned to be worst storm in Britain for 300 years; • USSR leader Mikhail Gorbachev criticises Stalin for political errors • Allan Bloom, *Closing of the American Mind* • Superior Court Judge in 'Baby M.' case in US denies parental rights to surrogate mother • World population 5,000,000,000, double level of 1950 • Digging of Channel tunnel starts • First glass-fibre optic cable laid across the Atlantic Ocean • Simon Schama, *The Embarrassment of Riches* • Jacques Derrida, *Of Spirit: Heidegger and the Question* • Andres Serrano, *Piss Christ* • John Adams, *Nixon in China* (opera) • Judith Weir, *A Night at the Chinese Opera* (opera) • Caryl Churchill, *Serious Money* • Films: *The Last Emperor, Wings of Desire, Fatal Attraction, Wall Street* • Liberace, Danny Kaye, Primo Levi, Andrés Segovia, Fred Astaire, Jean Anouilh, Jacqueline du Pré and James Baldwin die • George Barker, *Collected Poems*

Chinua ACHEBE   1930–

Anthills of the Savannah

Christopher Oriko, Ikem Osodi and 'His Excellency' were educated together – first in Kangan, their native land, and then in the UK. Now His Excellency is President for Life of the vast West African republic of Kangan, Chris is his Commissioner for Information and Ikem the editor of the *National Gazette*. Chris is having an affair with Beatrice Okoh, a graduate of Queen Mary College, London; Ikem is affianced to the daughter of a local market trader, Elewa, who is expecting his child.

In the middle of a cabinet meeting, a delegation arrives from Abazon, the President's most troublesome province. Abazon has been systematically underdeveloped by central government as punishment for its disloyalty. Hence, the province's elders have travelled to Bassa, the capital city, to beg the President's forgiveness. He declines to see them, sending instead a mere minister. In his capacity as a journalist, Ikem also speaks to the visiting elders. Already (and always) fearful for his position, and conscious that Ikem and Chris are Abazonian, His Excellency treats the conference with suspicion. He has Ikem removed from his editorship. Ikem makes an eloquent but injudicious speech at the University of Bassa. Soon after, he is murdered by the Security Services.

I could read in the silence of their minds, as we sat stiffly around the mahogany table, words like: *Well, this is going to be another of those days.* Meaning a bad day. Days are good or bad for us now according to how His Excellency gets out of bed in the morning. On a bad day, such as this one had suddenly become after many propitious auguries, there is nothing for it but to lie close to your hole, ready to scramble in. And particularly to keep your mouth shut, for nothing is safe, not even the flattery we have become such experts in disguising as debate.

ANTHILLS OF THE SAVANNAH
BY CHINUA ACHEBE

Chris goes into hiding, knowing that he is likely to be the next victim. Along with Emmanuel Obete, a leading student activist and supporter of Ikem, and Braimoh, a friendly taxi driver, he flees north by bus to Abazon. The three are close to their destination when they come to a crowd of drinkers by the road. Chris learns that the drinkers are marking the President's 'disappearance': the Chief of Army Staff has taken control. During the roadside party, a policeman tries to rape a schoolgirl. Chris intervenes and the policeman kills him.

Emmanuel and Braimoh return to Bassa, where they break the news of Chris's death to Beatrice. Some months later, Elewa gives birth to a daughter, Amaechina, whose name means 'May the path never close.' The novel ends with the child's naming ceremony, where grief has been overwhelmed by indefinite hope.

Achebe narrates his novel from a number of different perspectives – first person and third – while his dialogue registers the shifts between the formal English of the politicians and the pidgin of the proletariat. Thus, he offers a story whose fragmentation is as significant as its coherence: on the one hand, the lives of his central actors are intimately damaged by the paranoia at the apex of the state; on the other, the changes in the composition of the elite are of no import to the common people. For the peasants and the urban masses, conditions will hardly change whoever the master. Furthermore, Achebe can propose only a limited terrain on which reformist intellectuals can fuse their aspirations with the oppressed. The extended family of relatives and friends, Christians and Muslims, men and women, privileged and deprived that gathers to celebrate Amaechina's birth guardedly offers an optimistic alternative to Kangan's perpetual cycle of bloodbath and purge.

Anthills of the Savannah was shortlisted for the Booker Prize.

Margaret DRABBLE 1939–

The Radiant Way
1987

A Natural Curiosity
1989

The Gates of Ivory
1991

The two central characters of *The Radiant Way*, Liz Headleand and Alix Bowen, have been friends since their undergraduate days and, by the time we first encounter them, have reached middle age. (Esther Breuer, the third of a close Cambridge trio, is a more marginal figure.) Set against the background of the social and political landscape of Britain, the novel recounts the changes they undergo in the first five years of the 1980s.

Liz grows up in the Yorkshire town of Northam but, unlike her sister Shirley, who settles there with Cliff Harper, a local businessman, she turns her back on her origins and on her ageing mother. After a medical degree and a disastrous first marriage, she marries Charles Headleand, a campaigning television producer. ('The Radiant Way' is an acclaimed documentary of his about education.) She acquires three stepchildren, followed by two of her own, her practice as a psychotherapist flourishes, and the Headleands' Harley St home becomes in the course of the 1960s and 1970s a focus for the fashionable, progressive London intelligentsia.

Alix, also from the North, marries the son of a wealthy artist and has a son by him, but he soon dies, and she brings up the child alone. Some years later she meets Brian Bowen another native of Northam and, like Alix, a committed socialist, who becomes her second husband. A former steelworker, he lectures at an adult education institute in London while Alix teaches English literature to women prisoners.

Their upward mobility and confidence in an ideal of social progress makes the experience of both women typical of a section of post-war British society. But 1980 heralds a new phase of political and social change. A new government turns its back on the post-war consensus, a new violence characterises public life (the novel provides regular bulletins on the progress of the miners' strike), a new awareness of Aids and widespread child abuse is just over the horizon. In this climate, the certainties of Liz and Alix's world come under strain. On the political level, Charles moves to the right, Brian further to the left, while his friend Otto Werner joins the new SDP. Alix, to whom none of these options is acceptable, is paralysed, typifying the recent political agony of the middle class. On the personal level, Charles leaves Liz for Lady Henrietta Latchett (though the marriage does not last); Alix, increasingly distant from Brian, hovers on the brink of an affair with Otto. But despite the decade's many novelties, Drabble locates the menace of the 1980s partly in a resurfacing of archaic, unsocialised aspects of human nature, which confront the optimistic belief in man's essential goodness with the disquieting thought that enlightenment cannot make evil go away. The suggestion is dramatised in the story of Paul Whitmore, a murderer who decapitates his female victims, and who was for years Esther's reclusive neighbour in Ladbroke Grove.

For Liz, however, the menace lies in her own past. Her sexual feelings had always been for her a source of unfathomable guilt. She had never linked this to her mother's pathological mysteriousness or to the disappearance of her father in her early youth.

But when her mother's death brings her back to Northam, she discovers the connection: her father was a paedophile who committed suicide after acquittal for a charge of indecent exposure. Repressed memories of how she was herself abused by her father, and of how she enjoyed it, crowd in upon Liz. They are overwhelming, but also liberating: her guilty secret, once out, ceases to threaten her. As Alix, Liz and Esther meets at the end of the novel to celebrate Esther's fiftieth birthday, the battered creed that knowledge brings freedom, and that there is no original sin, is tentatively reaffirmed.

The themes of the return to origins and of the reality of evil are pursued in *A Natural Curiosity*, which begins in 1987. Esther Breuer moves to Bologna and disappears from the narrative almost completely, and the scene shifts to Northam, where Alix and Brian now live. A new clutch of characters, burghers of Northam, are introduced. Outwardly they follow a provincial routine, but the archaeological remains of the Brigantes, an ancient British people given to human sacrifice who once inhabited the area, provide a metaphor for their repressed passions. However, these passions too are seen to be malign only to the extent that they are denied.

A party given by the eccentric Fanny Kettle gives Liz – now in Northam with Alix – the final clue to the riddle of her mother's mysteriousness. She discovers that one of the guests, Marcia Campbell, is in fact her half-sister, the illegitimate child of Liz's mother by a local aristocrat, whose servant she had once been. The atmosphere of guilt and concealment surrounding Liz's origins is finally dispelled, and she feels reborn into a 'guiltless world'.

The backbone of the narrative is supplied by Alix's dealings with Paul Whitmore, now serving a life sentence in Northam prison. Alix finds herself drawn to visit him first of all by a fascination with his violence, the dark side of human nature she had previously attempted to deny. She then becomes concerned to account for it and, at Whitmore's request, tracks down his warped and sadistic mother. In reconstructing his miserable upbringing, Alix is able partially to exonerate him: again, it seems that evil can be explained away as pathology. The political themes of *The Radiant Way* are recalled as, reunited once more with Esther and Liz, she manages, for all its faults, to forgive England too.

Each of the three novels, may be read independently of the others, but together they form a trilogy. This concluding volume begins with Liz receiving a mysterious package posted from Cambodia which contains fragments of writing by her friend Stephen Cox, a Booker Prize-winning novelist who has disappeared having travelled to the Far East, evidently to research a play about the Khmer Rouge leader Pol Pot. There are three interwoven threads to the narrative: an account of Stephen's experiences; the story of how Liz tries to make sense of the contents of the package and eventually resolves to look for Stephen; and a comical first person narrative by the gossip-mongering former actress Hattie Osborne, who is now Stephen's agent and who becomes pregnant by Liz's stepson Aaron Headleand.

Stephen arrives in Bangkok where he has an affair with the Thai beauty queen Miss Porntip, who has risen from poverty to the control of a diverse business empire, and lectures Stephen on the virtuous triumph of capitalism. His thoughts about Thailand extend the rather simplistic political message of the earlier novels: 'This was the lesson of the eighties. Avarice and greed have no natural limits.' He meets a photographer, Konstantin Vassilou, with whom he travels to Vietnam and Cambodia in search of the Khmer Rouge. He is taken prisoner in a hill village and contracts malaria; as he dies, one of his last fevered recollections is of his selections for *Desert Island Discs*. This narrative is given context by Liz's research as she tries to understand the Khmer Rouge atrocities which Stephen's fragmentary writings describe. Her conversations with Alix (who is largely absent from the novel) attempt to find a link between the private evil of Paul Whitmore and the public evil of Pol Pot's regime. In Bangkok she learns of Stephen's death from an American film-maker and discovers that the package was sent to her because she was named in Stephen's passport as his next of kin. The novel ends with Stephen's memorial service in London where Konstantin is reunited with his mother. There is little development of the characters from the earlier novels, except for Esther who is married to Robert Oxenholme, the Conservative Minister for Cultural Sponsorship, a concept for which Drabble shows a clear distaste.

Although no-one has doubted the ambition of Drabble's trilogy, critical opinion has been divided over whether or not it succeeds. Ostensibly a 'state of the nation' work, it started out as a portrait of late twentieth-century life amongst London intellectuals beleaguered by Thatcherism, and ended up considering genocide in the Far East. 'Hampstead set in the killing fields' was the headline of the *Guardian's* review of the final volume, and not everyone believed that Drabble had managed to integrate the numerous strands of her narrative into a cohesive whole. For others, Drabble's boldness has paid off, and the trilogy has been acclaimed as her most impressive work to date.

William GOLDING   1911–1993

Close Quarters

see **To the Ends of the Earth** (1989)

Penelope LIVELY   1933–

Moon Tiger

Claudia Hampton is old and dying in hospital of cancer of the gut. All her life she has been obsessed with words, becoming a writer of popular history: 'overblown, flashy stuff', according to her lover, Jasper. Now she is suffering from aphasia, and time is no longer flowing smoothly as she writes in her mind 'a history of the world ... The works, this time. The whole triumphant murderous unstoppable chute – from the mud to the stars, universal and particular, your story and mine.' The novel, like Claudia's history of the world, is kaleidoscopic: 'shake the tube and see what comes out. Chronology irritates me.' This universality and particularity is partly reflected in the structure, an alternating pattern of first and third-person narration. The voices of the noval often speak for themselves out of a third-person narrative. Claudia's own story is told in both first and third persons, or the third person rising out of a first-person narrative.

This reflexive quality is further strengthened by the device of repeated description of the same event from different perspectives. A totality is aimed at: 'Unless I am a part of everything I am nothing.' That the whole is unachievable is not in doubt. Claudia's world is breaking up and nothing is what she becomes.

It is left open whether or not the third-person Claudia is a creation of the first-person Claudia or of the novelist. If the former, then Claudia is 'guessing' at the 'secret' internal lives of others. If the latter, then it is the reader alone who sees the truly global picture of her life.

Secrets there certainly are. Lisa, Claudia's illegitimate and only child, has lived a life hidden from her mother. 'You are not omniscient,' Lisa says, and Claudia knows nothing of her lover, Paul. Conversely, Lisa knows nothing of one of the three passionate forces in Claudia's life: her affair in Cairo during the war with an officer, Tom. (Lively was born in Cairo in 1933 and lived there until the end of the war.) Tom is killed in action, but not before Claudia has become pregnant. She miscarries and learns that the foetus is of an indeterminate sex. The memory of Tom is embodied for Claudia in the image of the Moon Tiger which burned by their bed in the Winter Palace in Luxor: 'a green coil that slowly burns all night, repelling mosquitoes, dropping away into lengths of grey ash, its glowing red eye a companion of the hot insect-rasping darkness'. Tom has been 'picked off by history', and his story is forever without a conclusion. For dispassionate, anti-sentimental Claudia this is perhaps the hardest thing to bear.

The continuous force of her life is her brother, the economist Gordon. Their relationship as children was incestuous and has remained exclusive: 'We were an aristocracy of two.' Most excluded is Gordon's unsuitable and silly wife, Sylvia.

Lisa's father, Jasper, is a brilliant half-Russian wastrel, who flits from the Foreign Office to 'television mogul ... the power behind the lavish series recently screened which presented a dramatised history of the last war'. War and the aftermath of war permeate the book; from the death of Claudia's father in the trenches of the First World War, to her own war in Cairo, her war reports, her histories: 'Napoleon, Tito, the Battle of Edgehill, Hernando Corte ...', and the displaced of war. Laszlo comes to the forty-six-year-old Claudia as a refugee of the invasion of Hungary in 1956. The conflict of nations is echoed in the conflict of individuals – Lisa and Laszlo, Lisa and Claudia, Claudia and Jasper, Claudia and Sylvia. Lives are shown to be wasted and made empty by conflict.

Ian McEWAN   1948–

The Child in Time

The Child in Time was awarded the Whitbread Prize; it was also the subject of a provocative and persuasive 'CounterBlast' essay by Adam Mars-Jones entitled 'Venus Envy' (1990). Mars-Jones writes of McEwan:

> In his early stories the only rule about sex seemed to be that it should *not* take place with marital commitment and reproductive intent, within a fertile cleft – and the further removed it was from that situation the more it seemed to interest him. It would be hard to extrapolate this state of affairs backwards from *The Child in Time*, now that desire has been so completely mortgaged to the creation of new life.

An empathetic feminist note had been sounded in *The Comfort of Strangers* (1981); the concern of *The Child in Time* is with paternity, and (as the title of his essay suggests) Mars-Jones sees the novel as an attempt to claim for men the sort of qualities usually ascribed to women.

Set in the future, the novel describes the effects on the lives of Stephen and Julie Lewis when their three-year-old daughter, Kate, is mysteriously abducted in a supermarket. Julie, devastated by the loss, and resentful of Stephen, who had been in charge of Kate at the time, sinks into a state of speechless grief, while Stephen embarks upon a futile campaign to discover his daughter's whereabouts. The child appears to have vanished without trace and, unable to talk about the tragedy or meet each other's eyes, Stephen and Julie drift apart. Julie moves to the country and retreats into a private world of mourning, while Stephen continues to look for Kate, mistaking other children for her in the school playground.

The street in which his parents lived ran straight and shopless for a mile and a half, part of a single nineteen-thirties development, once despised by those who preferred Victorian terraces, and made desirable now by migrations from the inner city. They were squat, grubbily rendered houses dreaming under their hot roofs of open seas; there was a porthole by each front door, and the upper windows, cased in metal, attempted to suggest the bridge of an ocean liner. He walked slowly through the hazy silence towards number seven hundred and sixty-three. A lozenge of dog turd crumbled underfoot. He wondered, as he did each time he came, how there could be so little activity in a street where there were so many houses close together – no kids kicking a ball around or playing hopscotch on the pavement, no one stripping down a gear box, no one even leaving or entering a house.

THE CHILD IN TIME
BY IAN McEWAN

The novel is narrated two years after the disappearance: 'Kate's growing up had become the essence of time itself. Her phantom growth, the product of an obsessive sorrow, was not only inevitable – nothing could stop the sinewy clock – but necessary. Without the fantasy of her continued existence [Stephen] was lost, time would stop.' Set in counterpoint to this is the story of Stephen's friend and publisher, Charles Darke, who has become a politician, and has regressed to a fantasy boyhood. He lives in the country with his wife, Thelma, who sees no option but to humour him.

During a brief reunion, drawn together by their loneliness and grief, Stephen and Julie make love. Stephen then returns to London, numbed by his tangled feelings. When Thelma telephones to report Charles's disappearance, Stephen sets off to comfort her. Charles is found dead in a wood beneath a tree in which he had built a tree-house. This event marks a turning-point in Stephen's life, as his loss of Kate merges with the wider uncertainties of humanity. Anxious to see Julie again, he makes a desperate journey across country by mail train, and finds that she is heavily pregnant with the child conceived during their reunion. While awaiting the arrival of the midwife, Stephen is obliged to deliver the child, saving its life by untangling it from the umbilical cord. The novel, which is replete with images of real and imagined childhood, ends with the resolution of love through the birth of a new life, a symbol of hope and continuity following the strange death of Charles Darke, who had given himself up to childishness and despair.

Toni MORRISON   1931–

Beloved

'If the whitepeople of Cincinnati had allowed Negroes into their lunatic asylum they could have found candidates in 124.' Set in the mid-nineteenth century, *Beloved* is a ghost story which explores the psychological effects of murder, grief and loneliness. It begins in number 124 Bluestone Rd, the haunted house, with the arrival of Paul D, 'the last of the sweet home men', in search of Sethe, the woman he loved more than eighteen years ago. Sethe's daughter, Denver, resents this intrusion; her brothers have left home, and her only companion has been the ghost of Beloved, a maverick spirit which plays the tricks of a poltergeist. Paul D seems to have exorcised the ghost, and for a brief day he, Sethe and Denver live a life approaching normality, in a joyous day out at a carnival. On their return, however, a stranger is sitting at their gate: Beloved, reborn in the flesh as an eighteen-year-old girl, has returned.

In her disruption of the family's happiness, Beloved begins to claim all of Sethe's attention, and destroys her relationship with Paul D, who walks out. Denver, initially charmed by this strange girl, begins to sense the potential danger, as Beloved displays signs of her past familiarity with domestic details and her true identity starts to emerge. Sethe had escaped from a slave camp after being raped and savagely beaten by the pupils of her school teacher employer, and she now distrusts all whites. On a day which was supposed to have marked a celebration of her precarious freedom, she sees white men arrive in the yard and thinks they have come to reclaim her; determined to save her children from slavery, she embarks on the savage ritual of slaughtering them, and kills Beloved before she can be stopped.

Paul D had the feeling a large, silver fish had slipped from his hands the minute he grabbed hold of its tail. That it was streaming back off into dark water now, gone but for the glistening marking its route. But if her shining was not for him, who then? He had never known a woman who lit up for nobody in particular, who just did it as a general announcement.

BELOVED
BY TONI MORRISON

The narrative shifts between past and present as these details are revealed, taking us into the mind of Sethe and her terrible burden of guilt: 'She shook her head from side to side, resigned to her rebellious brain. Why was there nothing it refused? No misery, no regret, no hateful picture too rotten to accept?' Beloved now seems to have murderous intentions of revenge, and the black women of the suburb draw on folk magic for a cleansing ritual of exorcism, which returns Beloved to the dream-world of the dead. Paul D, filled with sorrow at the way he had deserted his mistress and given in to Beloved's sexual fascination for him, now comes home to 124 to reward Sethe's love by helping her to shake off the hold of the past over her life.

Beloved has its basis in a true story about a woman in the nineteenth century who murdered her baby in order to save it from the slavery she had herself escaped. 'I thought at first it couldn't be written,' Morrison commented, 'but I was annoyed and worried that such a story was inaccessible to art.' The novel's triumph is in the way it confronts an era of black history as tragic, important, and as problematic to treat fictionally, as the Holocaust has been in Jewish history. Rather than deal with the story in a realistic manner, Morrison brings to it the magical, supernatural elements she employed in her previous novels. The book became a bestseller, and has been described as an American masterpiece.

Edna O'BRIEN   1936–

The Country Girls Trilogy

see 1964

Tom WOLFE   1931–

The Bonfire of the Vanities

Prior to the publication of this, his first novel, Wolfe made his name as an acerbic social commentator and pioneer of the 'New Journalism', with books such as *Radical Chic and Mau-Mauing the Flak-Catchers* (1970). Before its much-hyped publication in book form, *The Bonfire of the Vanities* was published in instalments in *Rolling Stone* magazine: it was subsequently made into a film (1990), directed by Brian de Palma, which endured a storm of critical abuse.

Wall Street bond dealer Sherman McCoy secretly thinks of himself as a 'Master of the Universe' (he got the idea from 'a set of lurid, rapacious plastic dolls that his otherwise perfect daughter liked to play with'). Last year he earned $980,000, which still was not enough to stem the tide of 'hermorrhaging money'. Now he is putting together a massive, audacious deal which will earn him a

As the Mercedes ascended the bridge's great arc, he could see the island of Manhattan off to the left. The towers were jammed together so tightly, he could feel the mass and stupendous weight. Just think of the millions, from all over the globe, who yearned to be on that island, in those towers, in those narrow streets! There it was, the Rome, the Paris, the London of the twentieth century, the city of ambition, the dense magnetic rock, the irresistible destination of all those who insist on being *where things are happening* – and he was among the victors!

THE BONFIRE OF THE VANITIES
BY TOM WOLFE

personal commission of $1.75 million and pay off the loan on his lavish Park Avenue apartment. But one night, with his mistress, Maria, in the passenger seat of his Mercedes, he takes a wrong turning into the Bronx and finds the road blocked by a tyre. Thinking he is about to be mugged, he fights off two black youths, one of whom is fatally injured as Maria accelerates away. She dissuades him from reporting it to the police.

The incident feeds the ambitions of the other central characters: black leader and ghetto entrepreneur the Revd Reggie Bacon has a pretext for orchestrating demon-strations for the television cameras (Wall Street murderer! capitalist killer!); drunken English hack Peter Fallow finally has a big story to justify his big expense account; beleaguered Assistant District Attorney Larry Kramer has a high-profile case he can make his name from and so escape his tiny apartment in a West Side 'ant colony'; District Attorney Abe Weiss, running for re-election, welcomes the opportunity to prove that justice is colour-blind.

Sherman's ruin is heralded by the headline above Fallow's story in the tabloid *City Light*: 'HONOR STUDENT'S MOM: COPS SIT ON HIT 'N' RUN'. The police track him down from the first two initials of his registration plate, and Sherman is trawled ignominiously through the legal system in the full glare of publicity. He is systematically divested of job, family, friends, apartment and his $2,000 English suits. By the end he describes himself as a 'professional defend-ant', and Fallow has won a Pulitzer Prize for his coverage of the story.

Wolfe's New York is governed by racial antagonism, snobbery, greed, fear and van-ity. The novel is a social panorama, remi-niscent of Thackeray and Trollope, with swipes at many of the author's familiar targets: the art world, modern architecture, the Manhattan social set, and a warped legal system wherein the only people to thrive are the likes of Thomas Killian of the firm Dershkin, Bellavita, Fishbein & Schlossel.

1988

USSR withdrawal from Afghanistan • Election of George Bush as US President
• Margaret Thatcher becomes longest serving British prime minister in 20th century
• Ethnic violence in Azerbaijan • Nobel Peace Prize awarded to UN Peacekeeping forces
• Mikhail Gorbachev is elected President of USSR by Supreme Soviet • Benazir Bhutto is
sworn in as Prime Minister of Pakistan • First woman bishop in US Episcopal Church
• US 'Stealth bomber' goes on public display • Stephen Hawking, *A Brief History of
Time* • A French company markets the abortion-inducing drug RV486 • US scientists
announce a project to compile a complete 'map' of human genes • The Holy Shroud of
Turin is shown to date from the 14th century • David Henry Hwang, *M. Butterfly*
• David Mamet, *Speed-the-Plow* • Films: *Dead Ringers, The Last Temptation of Christ, A
Fish Called Wanda, Rain Man* • Popular music: k d lang, *Shadowland* • Ove Arup,
Richard Feynman, Alan Paton and Pietro Annigoni die • Philip Larkin, *Collected Poems*

Nicholson BAKER   **1957–**

The Mezzanine

Baker's first novel, portions of which originally appeared in the *New Yorker*, lacks any kind of plot in the conventional sense. Narrator Howie is an office worker in an unnamed American city who describes one ordinary lunch-hour, during which he urinates, descends the escalator from the mezzanine where he works, shops for a new shoe-lace, buys a hot-dog, a cookie and some milk, sits on a bench to read a few words of Aurelius's *Meditations*, and returns to the mezzanine. The novel comprises a series of lovingly obsessive descriptions of everyday objects: drinking straws, door knobs, office stationery, pharmaceutical products, milk cartons, Penguin paperbacks, the perforations in toilet paper, and so forth. There is, for example, a hymn to the corporate washroom, in which the 'four wall-mounted porcelain gargoyles' inspire him to think of Gerard Manley Hopkins. In another passage he remembers feeling 'like Balboa or Copernicus' on the day when he discovered how to apply underarm deodorant without having to remove his shirt. Towards the end of the novel Howie attempts to define his personality by drawing up a chart which lists the frequency with which he thinks about different things. His five most thought-about subjects are: his girlfriend – known only as 'L.' – 'Family', 'Brushing tongue', 'Earplugs', and 'Bill-paying'.

Baker's even shorter second novel, *Room Temperature* (1990), applied the same technique to a domestic setting (twenty minutes spent feeding his baby daughter with a bottle), while his third, the controversial *Vox* (1992) got down to the nitty-gritty detail of telephone sex. There is a similarly obsessive quality in *U and I* (1991), which describes with the same minute attention to trivial detail the author's bizarre preoccupation with John Updike. Baker is a droll cicerone of the quotidian, and part of the pleasure of his books is derived from the fact that he picks out and discusses what the reader has often noticed in life but never taken time out to consider. In his work, the familiar is made bizarre, yet remains instantly and reassuringly recognisable. While most commentators agreed that Baker's quirky pursuit of minutiae made unexpectedly compulsive reading, some have begun to wonder how many variations can be played on this particular theme.

Peter CAREY 1943–

Oscar and Lucinda

Carey's standing as one of the leading novelists of his generation was confirmed by this novel set in the mid-nineteenth century, which is Victorian in both scope and length. Meticulously detailed, teeming with character and incident, it has been described as 'Dickensian', although the consciousness of the narrator and the clarity of the writing is distinctly modern. It was awarded the Booker Prize.

Shuttling back and forth between England and Australia, the novel follows the youth and fortunes of its protagonists. Oscar Hopkins is a pale, skinny, red-haired child who is brought up in Devon by his strict widower father, a Plymouth Brethren naturalist (based upon Philip Gosse). Oscar rejects his father's tortured and joyless faith and goes to live with the Anglican vicar and his wife, Oxford luminaries now fallen upon hard times. The Revd Stratton arranges for Oscar to enter Oriel College to study theology, and whilst there the young man is taken to Epsom Races by a fellow student, Wardley-Fish. Oscar becomes an 'obsessive' gambler, keeping minute records of the form of horses and jockeys. His remote childhood has kept him a true innocent and he sees nothing wrong with betting, since he uses his winnings to pay for his education and pursue his career spreading God's word. Furthermore, faith itself is something of a gamble: 'We must gamble every *instant* of our allotted span. We must stake *everything* on the unprovable fact of [God's] existence.' After ordination he works in Notting Hill before deciding, upon the flip of a coin, to travel to Australia to become a missionary.

On board the *Leviathan*, whilst attempting to allay his pathological fear of water, he meets Lucinda Leplastrier and discovers that she shares his passion for gambling. Brought up in New South Wales, the daughter of a progressive friend of George Eliot and a feckless farmer, she inherits a great deal of money when she is orphaned and her father's land is sold off. She defies convention, wears bloomers, purchases a glass factory and becomes a 'compulsive' gambler, partly to combat her loneliness. She is returning to Australia, having failed to get a husband, and finds that her factory has fallen into disuse during her absence. She gets the furnaces going again and then once more bumps into Oscar, who is now a vicar.

The two are discovered playing cards and a scandal ensues. Oscar is obliged to work as a clerk and lodge with Lucinda. They fall in love but are unable to communicate this to each other. After visiting the factory, Oscar offers to build and transport a glass church through the Bush to a remote town where Lucinda's former friend, Dennis Hasset, is vicar. Oscar believes Lucinda to be in love with Hassett, and the journey will involve dangers and hardship, but Oscar has a wager with Lucinda that he can do it.

The expedition is led by the brutal and ambitious Mr Jeffris, who murders aboriginals and ensures that Oscar is regularly dosed with laudanum, supposedly to combat his phobia about water. Enraged and drugged, Oscar kills Jeffris, then sails up the river inside the glass church, which has been assembled upon a raft. When he arrives at Boat Harbour he is seduced by a disappointed spinster, Miriam Chadwick, feels obliged to marry her, but is drowned inside the church when it sinks into the river. Lucinda, who bet all her property against Oscar reaching Boat Harbour, is reduced to poverty when Miriam inherits it.

The unifying image of glass reflects in its brilliance, its ability to create wonder, its combination of strength and brittleness, the human aspirations of the characters. Oscar comes to learn that glass is 'the gross material most nearly like the soul, or spirit (or how he would wish the soul or spirit to be), that it was free of imperfection, of dust, rust, that it was an avenue for glory'.

Bruce CHATWIN 1940–1989

Utz

Chatwin's work, which combines history, travel, philosophy, anthropology and fiction, has always been difficult to categorise. He is pre-eminently a writer of ideas, but one who has an unerring eye for the quirky fact and the compelling story. He deliberately blurs the margins by which fiction is traditionally separated from non-fiction. As he put it in an introductory note to the posthumously published collection of pieces *What Am I Doing Here* (1989): 'The word "story" is intended to alert the reader to the fact that, however closely the narrative may fit the facts, the fictional process has been at work.' *Utz* is very short, written when Chatwin was mortally ill, but it is very far from being skimped, and was shortlisted for the Booker Prize. It draws upon the author's own knowledge of antiques (he was once a director of Sotheby's), and is a characteristic cabinet of Chatwin curiosities, crammed with beguiling objects and information.

His face was immediately forgettable. It was a round face, waxy in texture, without a hint of the passions beneath its surface, set with narrow eyes behind steel-framed spectacles: a face so featureless it gave the impression of not being there. Did he have a moustache? I forget. Add a moustache, subtract a moustache: nothing would alter his utterly nondescript appearance.

UTZ
BY BRUCE CHATWIN

The book opens with jackdaws wheeling in the dawn sky above Prague. Rather than the precious objects they are reputed to steal, they carry twigs with which to build their nests, and this access of domestic virtue reflects upon Kaspar Utz, who is receiving his cheerless obsequies below. Utz is one of their kind, an unprepossessing collector whose horde of Meissen figurines has been crammed into his minute two-roomed flat on the top floor of a squalid block on Široká Street. The narrator (Chatwin himself) had met Utz seven years before in 1967, while conducting research for a study of the psychopathology of the compulsive collector, with particular reference to the Emperor Rudolf II. He spends a day with Utz, and in this shabby aristocrat Chatwin finds an equally extraordinary example of '*Porzelankrankheit*'. Utz describes his mania, discusses the origins of porcelain and alchemy, and relates the story of Rabbi Loew's Golem.

He talks about his yearly trips to Vichy, where he finds that freedom has no savour, and describes the Swiss vault which houses a second collection of porcelain. Utz has been allowed to keep his collection on the understanding that it will pass to the state on his death, but when he dies, the shelves where once figurines stood six deep are found to be bare.

'When reconstructing any story,' Chatwin writes, 'the wilder the chase the more likely it is to yield results.' This is the excuse for, and the point of, Chatwin's inconclusive narrative. We discover that Utz's faithful maid, Marta, is in fact his wife, and that Utz had enjoyed curious erotic liaisons with numerous opera singers. Nobody knows exactly what happened to the collection. The importunate Prof. Orlík, an expert on both the mammoth and the house-fly, insists that Utz and Marta smashed every single piece. The motive remains obscure, but Chatwin suspects that Utz grew to dislike the porcelain because it reminded him of 'the grovelling and compromise' he had endured in his attempt 'to preserve in microcosm the elegance of Europe'. What at first seems to be an intriguing mystery story turns out to have been a provocative meditation upon the nature of freedom, and the tyrannies exerted by politics, human relations and possessions.

Robertson DAVIES 1913–

The Cornish Trilogy

The Rebel Angels (1981)
What's Bred in the Bone (1985)
The Lyre of Orpheus (1988)

Davies's non-sequential 'series of novels which explore the life and influence of Francis Cornish' bring a refreshing, not to say chilling, breath of the Gothic to the campus novel. Running in all to some 1,250 pages, crammed with eccentric characters and arcane information about alchemy, astrology, Canadian provincial history, gypsy lore and picture restoration, these books made the author's international reputation. Their preoccupation is with disguise and deception, art and artifice, instinct and creativity.

Francis Cornish, aesthetic member of a wealthy banking family, is the guiding spirit of the trilogy, although in the first volume, he is off-stage, having recently died. His executors are his nephew, Arthur, and three academics from his alma mater, the College of St John and the Holy Ghost (affectionately known as 'Spook') in Toronto. Cornish has been a noted collector and benefactor, and his executors are obliged to sort through his massive accumulation of rare books, manuscripts, paintings and sculptures. Clement Hollier, Professor of Paleo-Psychology, is keen to find a Rabelais manuscript amongst the treasures, since this will advance the career of his beautiful young research assistant Maria Magdalena Theotoky. Unfortunately, one of the other executors, a Professor of Renaissance Studies called Urquhart McVarish – a descendant of Thomas Urquhart, the translator of Rabelais – is supposed to have 'borrowed' this manuscript, a suggestion he denies. Hollier's life is further complicated by the reappearance of John Parlabane, an old friend whose brilliant academic promise was never fulfilled and who has been living in a monastery. In spite of his monkish habit, appetite has not been subdued in Parlabane, who is generally considered to be a bad lot. He substantiates his reputation by murdering McVarish with a pair of knitting-needles in the course of sexual games-playing, after which he commits suicide. His motive is that *Be Not Another*, the vast autobiographical *roman philosophique* he has been writing, will be given free publicity and thus published and recognised as the work of genius he believes it to be. *The Rebel Angels* is narrated alternately by Maria and by Cornish's other executor, Prof. the Revd Simon Darcourt, who is supposedly compiling a gossipy but serious account of university life entitled *The New Aubrey*. Almost everyone is in love with Maria, but she eventually marries Arthur Cornish.

What's Bred in the Bone takes its title from a medieval Latin proverb, 'What's bred in the bone will not out of the flesh', a concept that haunts Francis Cornish. The story of his life is narrated by the Lesser Zadkiel, Angel of Biography, and commented upon by the

Daimon Maimas, Cornish's Gnostic guiding spirit. Torn between the Protestantism of his absent parents and the Catholicism of the relatives who look after him, Francis endures a lonely and confused childhood in Ontario. Coming upon a manual of drawing by the caricaturist Harry Furniss, he develops his artistic skills in secret, assisted by the family groom and local undertaker, Zadok Hoyle, who allows Francis to sketch corpses in the embalming room. He also stumbles upon an older brother, a pinhead idiot confined in the attic, who is falsely reported to be dead. Francis goes to Oxford and falls in love with his feckless English cousin, Ismay, who, pregnant by a fellow undergraduate, allows him to marry her, then deserts him. Her family sponge on Francis, who inherits a large income from his grandfather, a timber merchant turned politician. His father, meanwhile, has ensured that Francis will follow him into the Secret Service, where he works in a minor capacity. He is sent to the castle of a Bavarian countess where he is apprenticed to Tancred Saraceni, a picture restorer. Saraceni is busy 'improving' minor paintings, which are smuggled out of the country then used to barter for important Italian canvases, which the Nazis are willing to exchange for German art. Francis exposes another faker and then paints an allegorical canvas of 'The Marriage at Cana' in the early sixteenth-century style, depicting people from his own life and overseen by the figure of his idiot brother, now dead. He also has an affair with Ruth Nibsmith, an English governess and astrologer, later killed in the blitz. He returns to England and works with the Allied Commission on Art, where he meets a handsome young Canadian called Aylwin Ross, the last person in his life whom he loves. Ross makes his reputation by researching 'The Marriage at Cana', which comes before the Commission. Francis allows the painting to be accepted as a genuine sixteenth-century canvas, attributed to an anonymous 'Alchemical Master'. Back in Canada he uses his money to collect paintings, but refuses to give money to Ross, who, as newly appointed Director of the National Gallery there, wants to buy 'The Marriage at Cana' which has come on the market. Ross is sacked and commits suicide.

In the final volume, the trustees of the Cornish Foundation are sponsoring the creation of an Arthurian opera by Hulda Schnakenburg, a recalcitrant but brilliant 'Spook' student, who is resurrecting the opera from a few notes left by E.T.A. Hoffman (who comments upon the action from Limbo). Under the guidance of the splendidly Sapphic Dr Gunilla Dahl-Soot, 'Schnak' composes *Arthur of Britain, or, The Magnanimous Cuckold*, with a libretto adapted from Sir Walter Scott by Darcourt. It is produced by the flamboyant, ambitious Geraint Powell in nineteenth-century style. The actions of the opera are mirrored in the life of the trustees as Powell, disguised as Arthur, impregnates Maria. Schnak, who has fallen unrequitedly in love with Powell, attempts suicide during the doctoral performance of her opera, but survives to attend the premiere and become a Doctor of Music. Assorted sub-plots include the appearance of a man claiming to be Parlabane's son who demands the manuscript of *Be Not Another* (which Maria has destroyed), and the machinations of Maria's gypsy mother and uncle, who are involved in the restoration of musical instruments.

As well as fulfilling the Arthurian concerns of the trilogy, the opera elaborates Davies's themes of deception, disguise and artifice. The illusions of nineteenth-century stagecraft and the transformations wrought upon the company reflect back upon Francis's careers as a painter and a spy. Simon Darcourt, friend and advisor to the troubled trustees of the Cornish Foundation, is revealed as the trilogy's real hero. He embraces his role as the Tarot pack's Fool (allotted to him by Maria's mother), 'marching merrily on his way, trusting to his intelligent nose and the little dog of intuition nipping at his rump to show him the path' towards completing his biography of Francis Cornish and solving the mystery of 'The Marriage at Cana'. If the trilogy is occasionally baggy, the reader forgives this in an author characterised by his generosity. Compared with the academic caution and meanness of spirit found amongst many of his contemporaries, Davies commends and exemplifies *sprezzatura*, which he defines as 'a noble negligence, a sudden leap in art towards the farther shore that could not be reached by the ferry-boats of custom'.

Elaine FEINSTEIN　　　　**1930–**

Mother's Daughter

Because of the poetic conciseness with which it is written, this ninth novel by Feinstein is dense and divisive: readers either take to it or take against it. In spite of the title, the novel tells the story of a girl, Halina, whose primary relationship has been with her glamorous and witty father, Leo. She had always regarded her mother, Tilde, as unbearably submissive; however, Tilde turns out to have been a war heroine and an opera singer, with a family background better than her husband's. By contrast, Leo turns out to have had only surface glitter. A flamboyant gambler and (possibly bigamous) gigolo, he reappears at intervals throughout the story. While his wife valiantly hid other Jews and resisted the Germans, he was smuggled ingloriously out of Budapest in a laundry basket. Halina always finds herself pulled towards him, even when he falls ill while living in England and subjects her to bitter diatribes as she is nursing him.

Not surprisingly, it was Halina's practical mother who had arranged to send her away to safety in England, and she never sees her again. Halina grew up in England, won a scholarship to Cambridge, and fell in love with a fellow exile, Janos, a freedom fighter turned philosopher, and a friend of Leo who shared many wartime secrets with him. If her father lit up her life unworthily, Janos robbed Halina of her self-esteem and ignored their son – a state of affairs that Halina colluded with, until she discovered that Janos was homosexual.

The whole story is told by Halina to her younger (American) stepsister, Lucy, who comes to London for Leo's funeral so that she can learn something of her father, though she hates him for his indifference to her and her mother. Lucy seems set to continue being a victim of her father but, at Leo's funeral, one of his old mistresses (whom Halina briefly but desperately hopes will turn out to be her mother) gives Halina a sort of absolution: by recognising her mother's worth, by longing for her, Halina has finally become her mother's daughter; she has finally achieved wholeness.

Penelope FITZGERALD   1916–

The Beginning of Spring

The third of four novels by Fitzgerald to be shortlisted for the Booker Prize, *The Beginning of Spring* is set in Moscow in 1913, but it is the passive and pastoral revolutionary figure of Tolstoy that overshadows the novel's characters and events, rather than the violent social and political revolution which is soon to befall Russia. Nellie Reid (*née* Cooper) has just left her husband, Frank. She had planned to take her children (Dolly aged ten, Ben nine, and Annie two and three-quarters), but mysteriously abandons them at Moscow railway station, and makes the long journey to Charing Cross on her own. We eventually learn that Nellie is heading for a Tolstoyan settlement in England called Bright Meadows, but we never entirely discover why she has left; whether or not she is to return is a question unanswered until the novel's very last sentence. Even the circumstances of her leaving are a surprise, one of many in Fitzgerald's characteristically short and deceptively simple story.

Born in Russia, where his father had founded the company of Reid's, a business devoted to the importing and assembling of printing machinery (and affectionately known as Reidka's), Frank had been sent to England to learn the family trade. In Norbury he met the twenty-six-year-old Nellie, who was a realist, forward and pragmatic: 'I'm not a dreamer,' she says. 'I have to look at things squarely, as they really are. That's one of the things you like about me.' Although Nellie is the novel's principal character, she never actually appears; she is presented entirely through the consciousness of other characters.

The novel's secret protagonist is Frank's closest colleague at Reid's, Selwyn Crane. The author of *Birch Tree Thoughts*, which Reid's has just printed, Crane is a bad poet and an idealist, a disciple of Tolstoy, with whom he has conversed and who fills all his conversation. He is made fun of, though not, significantly, by Nellie. When Nellie left Russia, Selwyn was supposed to keep a rendezvous with her at the station, but failed to turn up.

Selwyn is also responsible for introducing Lisa Ivanova into the Reid household, to look after the children. Mrs Graham, the chaplain's wife (who smokes rank shag cigarettes, which she rolls herself), describes Lisa as: 'Quiet, blank, slow-witted, nubile, docile.' Yet Lisa mesmerises everyone, particularly Frank, who eventually beds her. When Nellie's brother, Charlie, comes all the way from Norbury (where he lives with his wife, Grace, a woman who talks about damp) to take the children to England, they refuse to leave Lisa. The children themselves might be out of a novel by I. Compton-Burnett; indeed, Nellie often wished they were not so unlike other people's children:

'Are you going to live in our house?' asked Ben.
'I think so.'
'It would be better if you made up your mind.'

One night Dolly follows Lisa to a secret, mysterious meeting: 'Dolly saw that by every birch tree, close against the trunk, stood a man or a woman. They stood separately pressing themselves each to their own tree.'
Lisa disappears. The house at 22 Lipka Street is unsealed for the spring. Nellie returns.

Candia McWILLIAM   1955–

A Case of Knives

McWilliam's first novel takes its title from George Herbert's poem 'Affliction' ('My thoughts are all a case of knives / Wounding my heart / With scattered smart') and it tells the story of middle-aged heart surgeon Sir Lucas Salik. Lucas is deeply in love with an indifferent blond estate agent, Hal Darbo, who has become his protégé but does not sleep with him. When Hal wishes to get married, Lucas attempts to find him a suitably vacuous and unthreatening girl, and at a party given by his friend, Lady Anne Cowdenbeath, he thinks he has found one. Cora falls in love at first sight, not with Hal but with Lucas, and she has reasons of her own for getting married: she is already pregnant by another man. Lucas soon has Hal and Cora ('heart' is *cor* in Latin) on their way to the altar, but meanwhile he falls foul of some particularly unpleasant and anti-semitic animal rights terrorists, Angelica Coney and Dolores Steel. A cow's tongue on the seat of his car is an omen of worse to come, and before long he has been attacked and left for dead in a public lavatory, apparently by rough trade.

If it is possible to admire a person intellectually for their storage system, I did Anne. Her clothes cupboards were tall lozenges of space, obscure and cool. So all cupboards might be described, but hers were a cathedral of sartorial labour, its craftsmen Siamese mulberry tenders, French button piercers, Amazonian crocodile skinners, small men pulling shards off beetles in jungles below warring helicopters, men in palazzi weaving fields of cloth from dreams of past power.

A CASE OF KNIVES
BY CANDIA McWILLIAM

Structure is centrally important to this novel, which is narrated *Rashomon*-style in four sections by Lucas, Cora, Anne and Hal. These four points of view reveal that nothing is quite as it seems. Not only is Cora far more intelligent than she appears, and in love with Lucas and not Hal, but Lucas also fails to understand that Lady Anne is in love with him. Lucas's attacker is actually Hal, who has been put up to it by Angelica Coney. Lucas's love for Hal is based on the fact that despite his knighthood Lucas still feels deeply insecure about his own Polish-Jewish origins, and he believes Hal to be the flower of English public-school youth. In fact Hal is of Jewish origin himself, and – rather more seriously – he is a vicious and cynical social

climber who has poor, foreign Lucas completely fooled. Estate agents were a particularly conspicuous and disliked profession during the 1980s boom in which *A Case of Knives* was written, and the thoroughly ersatz Hal is a recognisable 1980s 'type': the upwardly mobile wide-boy. The last pieces in the story do not fall fully into place until we read Hal's narrative, the shortest of the four, which is told from prison.

Critics responded well to *A Case of Knives*, which won a Betty Trask Award and was shortlisted for the Whitbread First Novel Prize. Negative comments tended to centre on McWilliam's almost excessively articulate vocabulary (jokes were made about thesauruses), and what some saw as her meretricious attention to the trappings of the good life (this was connected by at least one critic with McWilliam having worked on *Vogue*). Others praised the structure, the wry social comment, and by the elegance of the writing and its 'hyper-aesthetic' detail. Above all, *A Case of Knives* is impressive for its insight into the unknowability of the human heart. McWilliam has commented: 'Really, it's four different truths. It's not done as a gimmick, but to show how arcane we are, even when we think we know each other.'

Armistead MAUPIN   1944–

Significant Others

see **Tales of the City** (1990)

Salman RUSHDIE   1947–

The Satanic Verses

Rushdie's fourth novel has the dubious distinction of having sparked an international crisis. It was regarded by Islamic fundamentalists as deeply and intentionally blasphemous, and initial protests rapidly escalated. Published on 26 September 1988, the novel was almost immediately banned in India and South Africa. In Britain it was shortlisted for the Booker Prize and won the Whitbread Novel of the Year Award, but copies of the novel were burned in the streets of Bradford, Yorkshire, which has a large Muslim community. In January 1989 there was a big demonstration in London but, despite petitions and further rallies, the publisher, Viking/Penguin, refused to withdraw the book and the Home Secretary declined to extend the UK's already discredited blasphemy laws to cover religions other than Christianity. Riots in Pakistan and India caused several deaths and widespread injuries, and on 14 February Ayatollah Khomeini, spiritual leader of Iran's Shia Muslims and founder of the Republic, pronounced a *fatwa*, or edict:

> I would like to inform all the intrepid Muslims in the world that the author of the book entitled *The Satanic Verses*, which has been compiled, printed and published in opposition to Islam, the Prophet and the Koran, as well as those publishers who were aware of its contents, have been sentenced to death.
>
> I call on all zealous Muslims to execute them quickly, wherever they find them, so that no one will dare to insult the Islamic sanctions. Whoever is killed on this path will be regarded as a martyr, God willing.

A price of £1,500,000 was placed on Rushdie's head. All Viking/Penguin books were banned in Iran, which called a day of national mourning and saw violent demonstrations outside the British embassy in Tehran. In Britain, Rushdie and his wife, the American writer Marianne Wiggins, went into hiding, under police protection; his apology, which was for causing offence rather than for writing the novel, was rejected, and the death-sentence confirmed. On 7 March Iran broke off diplomatic relationships with Britain.

The affair generated an enormous amount of news coverage and debate, giving literature a rare prominence in British public life. The *fatwa* was undoubtedly the most chilling example of a refusal to respect the imaginative freedom of the writer that had been seen in the twentieth century, and Rushdie's fellow authors (even those amongst whom he had been less than popular) organised petitions and protests in his support, both in Britain and abroad. The death of Ayatollah Khomeini in June 1989 eased pressure a little, diplomatic relations were re-established between Iran and Britain, and some sort of solution to what had seemed a wholly intractable problem was reached when Rushdie announced that he had re-embraced the Muslim faith (a step some of his supporters decried). The *fatwa*, however, remains in force and Rushdie, if no longer entirely a prisoner, is obliged to lead the life of a recluse. The cost in both human and financial terms had been extraordinary: the already strained relations between Islam and the West were severely damaged; the increased cost of security incurred by Viking/Penguin proved almost ruinous; Rushdie lost his freedom, and his marriage collapsed. It has been asked whether the novel was worth all the trouble it caused, but whatever its literary merits, the issues it raised were, and continue to be, of great importance.

The Satanic Verses plays with fictionality and unbelief, with the allusions to the Koran contained within dream sequences about the difficulties of faith against doubt. Although to Islamic fundamentalists it has seemed deeply blasphemous, the novel's theme is presented within the traditions of western humanism, in which every idea is open to question. Two friends from India make an unusual impact on English society. Salahuddin, now a passionate Anglophile with his name shortened to Saladin, is a passenger on board a hi-jacked jumbo jet. In the seat beside him is Gibreel Farishta, star of popular movies about the legends of the gods. The plane explodes, and in a rain of human limbs and airline trolleys the two men plummet towards England. Saladin hurtles down resignedly, his jacket tightly buttoned, 'taking for granted the improbability of the bowler-hat on his head', while Gibreel cavorts in the fall, 'swimming in air, butterfly-stroke, breast-stroke, bunching himself into a ball, spreadeagling himself against the almost-infinity of almost-dawn, adopting heraldic postures, rampant, couchant, pitting levity against gravity'. As London spreads below them like a wonderland through the clouds, Gibreel, clasped in Saladin's arms, begins to flap his hands and sing, wafting them gently towards the Channel.

Their miraculous survival is a symptom of a vast cosmic aberration. 'To be born again,' Gibreel crows, 'first you have to die.' But instead of being killed, the two men are transformed. Gibreel, his name a corruption of the Archangel Gabriel, develops a halo, which strikes awe into those who see him and embarrasses his friends; Saladin sprouts hairy shanks and horns spring from his forehead. Taking refuge in a London underworld of seedy cafes and night-clubs, the mutated humans are hailed as the heroes of rival youth cults, and the narrative builds up to a climatic concert in Wembley Stadium, which is both highly comic and mythologically significant.

Within the framework of the London narrative a second story is being told. Ayesha, a mystic, appears in Gibreel's dreams, taking the reader into the lives of villagers in India. The tale of Ayesha, clothed in a magical raiment made of living butterflies and leading a band of pilgrims towards Mecca, explores the superstitious aspect of religious belief and the power of faith to triumph over experience. To complete this trilogy of perspectives on faith and doubt, Rushdie takes the period of *Jahilia*, or ignorance, and turns it into a mystical landscape. The Prophet Mahound battles with the witch queen Hind in a further sequence of Gibreel's dreams, as Gibreel and the Prophet begin to penetrate one another's identities. In a contemporary parallel, a religious bigot is seen in exile in London, suggesting identification with Ayatollah Khomeini's exile in Paris, and inflaming the fundamentalists still further.

Throughout *The Satanic Verses* Rushdie emphasises the ambiguities inherent in his narrative. 'It is, it is not,' he reminds us, in the formula of the Arabian storyteller. The result of this technique is an endlessly fascinating novel, an intricate unfolding of themes on many levels, blending sharp social satire of a materialistic modern world with the conundrums of Rushdie's own inheritance of the traditions of East and West.

Anne TYLER   **1941–**

Breathing Lessons

The ramshackle, quirky world of *Breathing Lessons*, Tyler's eleventh novel, is very nearly the world of soap opera, but not quite. Instead, careful manipulation of language and plot ensures that the pre-packaged lectures on relationships always fall flat, and characters interact at subliminal levels which nobody can quite articulate. The narrative spans a day in the life of a humdrum marriage, intercut with parallel flashbacks, over the courtship of the couple and the marriage of their son.

Maggie and Ira Moran have ambled through twenty-nine years of wedlock, neither with tremendous happiness nor sufficient misery to necessitate the upheaval of divorce. Both lead disillusioned lives: Ira had wanted to be a doctor, but illness in his family and then marriage have ensured that he never leaves the picture-framing shop inherited from his father. Maggie was bright enough to leave her native Baltimore to go to college, but instead stayed there to marry Ira. At the time of the marriage, she had romanticised his quiet manner and part-Indian features, but she now sees him as a man of limited imagination and sympathies. Ira, conversely, had considered Maggie to be glamorously upper class, but now sees her as flighty, impractical and inefficient as a mother and housewife.

Their son's marriage has been even less satisfactory. Jesse married Fiona after she became pregnant, but his real ambition is to be a rock-star, and fatherhood proves too great a restraint. The young couple divorce after only eight months, but Maggie refuses to accept the situation. Maggie is an altruist without a cause, whose craving for happy endings is simultaneously aggravating and infectiously cheering. Investing her son's marriage with the romance absent from her own, she starts to make 'surprise' visits to Fiona's new home and tries to coax her back to Baltimore, much to Ira's disapproval.

The present-day story consists of a series of Maggie's impulsive decisions, all of which prove to be building only half-consciously towards yet another surprise visit to Fiona.

First stop is a memorial service for the husband of an old schoolfriend of Fiona's called Serena. At the service, Maggie finds herself singing a pulpy duet with Durwood Clegg, the heart-throb of her high-school graduation year. Later, she and Ira are expelled from the reception when they are found engaged in heavy, nostalgic petting in the widow's bedroom.

On the way home, Maggie gives an unnecessary lift to an old man who has run away from his wife, but this new adventure lands them serendipitously close to Fiona's house in Cartwheel. Ira cannot refuse a quick stop-off, and Maggie ingeniously manages to persuade Fiona to drive back with them to Baltimore to visit Jesse. The reunion is a disaster; but strangely, the realisation of the depressing truth about their son's marriage brings Maggie and Ira to a new mutual understanding.

Massacre in T'ien-an Men Square • Collapse of communism in Eastern Europe • Fatwa issued against British author Salman Rushdie, calling for his death for blasphemy in his book *The Satanic Verses* • Ayatollah Khomeini, spiritual and political leader of Iran, dies • Nelson Mandela makes first public statement since detention 25 years before • George Bush and Mikhail Gorbachev declare end of Cold War • National Curriculum is introduced into schools in England and Wales • Researchers identify a gene responsible for cystic fibrosis • Remains of the Rose and Globe Theatres uncovered in London, where Shakespeare's plays were originally performed • Mikhail Gorbachev and John Paul II agree to reestablish diplomatic relations • I.M. Pei, Pyramid, Louvre, Paris • John Tavener, *The Protecting Veil* • Films: *Do the Right Thing*, *sex, lies and videotape*, *Batman* • Television: *Twin Peaks*, launch of satellite station Sky TV in Britain • Popular music: Acid-house rave parties attract tens of thousands in England • Hirohito, A.J. Ayer, Laurence Olivier, Irving Berlin, Samuel Beckett, Nicolae Ceausescu and Lennox Berkeley die • Tony Harrison, *The Blasphemers' Banquet* • James Fenton, *Manila Envelope*

Martin AMIS　　　　1949–

London Fields

'Now I know the British Empire isn't in the shape it was. But you wonder: what will the *babies*' babies look like?' Sam Young, the writer/narrator of Amis's *London Fields*, is dying of radiation sickness, his speeded-up decay mirroring his sense of an apocalyptic world. Leaving his native New York background, Young arrives in London to live in the flat of Mark Asprey, an apparently successful writer. At Heathrow, Young is conned by Keith Talent, a cheat, whose life of semi-violent crime, darts and pornographic sexual tastes forms the staple of conversation in the London pub the Black Cross.

Lingering on the fringes of this low-life world, Young realises that he has stumbled upon the material for a story; all he needs to do is write it down. Nicola Six, a natural 'murderee', goes into the Black Cross to find her murderer (a narrative strand which echoes the plot of Muriel Spark's *The Driver's Seat*, 1970). Keith Talent, belligerent, belching, his brain able to run only on the clichés of the tabloids, seems to be the obvious candidate. 'Keith Talent was a bad guy. Keith Talent was a very bad guy. You might even say that he was the worst guy.' Guy Glinch, a wealthy financier, is the good guy, and Nicola sets out to draw the two men into her fate, determined to die before she suffers the ruination of middle age.

Sam Young sees Nicola dispose of her diaries in a park litterbin, and retrieves them as raw material for his novel. He begins to infiltrate the lives of his characters, becoming Nicola's 'confessional', and finding in the abused innocence of Keith's daughter Kim an optimistic testament to the renewed possibilities of love. But Nicola, sexually powerful, the orchestrator of events, is determined on her course of self-destruction. To Keith she plays the whore, to Guy, the virgin, as both men see in her only what they want to see, and turn her into the material of their fantasies.

When Nicola walked the streets she was lit by her personal cinematographer, nothing too arty either, a single spotlight trained from the gods. She had a blue nimbus, the blue of sex or sadness. Any eyes that were available on the dead-end street would find their way to her: builders in the gutted houses, a frazzled rep in a cheap car, a man alone at home pressing his face against the window pane with a snarl.

LONDON FIELDS
BY MARTIN AMIS

Nicola's *alter ego*, the imaginary Enola Gay, is the mother of 'Little Boy'. She pretends to Guy that they are missing Cambodian refugees, and the philanthropic cause of finding them appeals to his desire to do something worthwhile. With his wife, Hope, and their monstrous son, Marmaduke, Guy's life seems unreal, sealed off from that of the streets; but Keith Talent's dreams of fame as a TV darts champion are equally at odds with the reality of his squalid council flat, his prematurely ageing wife, Kath, and their bruised daughter, Kim.

Nicola lures the men into her story, and Young seems to be the passive recorder; but his jealousy of the successful Mark Asprey increases as he learns that he is the 'M.A.' of her diaries, the most gymnastic of her lovers.

Voyeuristic, misogynistic, Young's sexual life is over, as he sadly fights off the advances of Hope's sister, Lizzyboo, and begins to accept that his affair with his agent, Missy Harter, is over. Bonfire Night, the date of the dying sun's eclipse, and of Nicola's birthday, has been chosen as the night of her death. Choreographing the action, she flatters Keith's ambitions to be a TV star, financing his 'career' by money she has screwed out of Guy. The plot proceeds through its twists, as Guy and Keith both seem about to murder her, but the violence is left to Sam Young, who kills her just before his own death. The climactic aberrations resettle into normality as, coinciding with this catharsis, the end of the world recedes into the future, and Young's papers are left to the better writer, 'M.A.', for his own use. 'Enola Gay' and 'Little Boy', we learn in the final pages, are the names of the plane which flew to devastate Hiroshima, and of its deadly cargo, the atom bomb.

The sense of a disintegrating world, and the threat to the future in this novel develops a theme expounded in Amis's volume of short fictions, *Einstein's Monsters* (1987). These stories were prompted by the author's concern about nuclear issues, itself prompted by impending fatherhood. However, *London Fields* is also enlivened by Amis's black humour, his flashy style, and by his unerring skill (one he shares with his father, the book's dedicatee) of capturing and exaggerating the way people speak. In Keith Talent, who expresses himself in the debased argot of the London streets, Amis created a truly contemporary character. Widely admired, and chosen by many critics as the Book of the Year, *London Fields* failed to get shortlisted for the Booker Prize, a decision which proved controversial even by the standards of the usual wrangles over this prestigious award. It was claimed that Amis had been blackballed by the woman on the judging panel, who objected to his depiction of the death-wishing sex object, Nicola Six.

Beryl BAINBRIDGE   **1934–**

An Awfully Big Adventure

For many years, Bainbridge has been obsessed by the story of J.M. Barrie's *Peter Pan* (see *Peter and Wendy*, 1911). *An Awfully Big Adventure* takes its title from Peter's brave salute to death at the end of Act III, and a production of the play by a down-at-heel Liverpudlian repertory company is at the narrative and emotional centre of the novel. The result is a characteristically economical, horribly funny and beautifully plotted tragi-comedy, which was shortlisted for the Booker Prize in 1990. Bainbridge uses *Peter Pan* as a constant source of reference, studding her laconic narrative with unobtrusive quotations and allusions, and the novel is as resonant and as intimately concerned with the death of innocence as Barrie's own dark masterpiece. She is alert throughout to the disturbing intimations of sexuality which lurk beneath the surface of the play, and the 'corruption' of minors is an important strand of her story.

Like many of Bainbridge's protagonists, seventeen-year-old Stella Bradshaw stands at an odd angle to the world. She is not really in control of her feelings, at one moment overreacting wildly to some minor upset, at another remaining curiously impassive during experiences most people would regard as traumatic. Like Barrie's eponymous, other-worldly hero, she inhabits a disconcerting limbo, neither wholly a child, nor yet an adult. Another similarity is that she has apparently been abandoned by, and is looking for, her mother. She has been brought up by her Uncle Vernon, who is so alarmed by her character that he encourages her to become an actress. 'He says I'll do least harm if I'm allowed to go on the stage,' Stella informs the actor-manager, Meredith Potter, but she unwittingly causes havoc when she joins his theatre company as assistant stage manager and 'character juvenile'. 'It wasn't my fault,' she says of the tragedy she precipitates in her quest for experience. 'I'll know how to behave next time. I'm learning.' The novel plots her unsentimental education in the complexities of adult relationships.

The richly characterised company provides numerous instances of these complexities: tearful Babs Osborne refusing to believe that her Polish lover is giving her the brush-off; Dawn Allenby, peppermints and cologne not quite masking the reek of spirits, thrusting herself at the male lead; Meredith awaiting a letter from Hilary, who turns out not to be the young woman Stella imagines, but a man picked up at the BBC. Stella falls hopelessly in love with Meredith, blind to the fact that he is far more interested in Geoffrey, the other ASM. She allows herself to be seduced by the philandering P.L. O'Hara, who is playing Captain Hook and Mr Darling. O'Hara is puzzled by her lack of response, but becomes obsessed by her. 'I've a snapshot at home of myself aged ten wearing an Eton collar. It could be her,' he confides in the pianist. This resemblance is no coincidence: at the end of the novel, O'Hara discovers that he is in fact Stella's father, and subsequently drowns himself.

Bainbridge perfectly captures the atmosphere of theatrical life and provides luminous descriptions of an impoverished, post-war Liverpool. She has acknowledged that Stella is a partial self-portrait, and she draws upon her own experiences as a fifteen-year-old at the Liverpool Playhouse. Many of the incidents are taken directly from life, as may be seen by comparing the novel with Bainbridge's *English Journey* (1984).

Julian BARNES   **1946–**

A History of the World in 10½ Chapters

Barnes's fifth book hovers between being a novel and a collection of short stories. Its 10½ chapters include a worm's-eye view of the Flood, an essay on Géricault's 'The Raft of the Medusa', letters from a film actor working up the Amazon, and a disquisition on love. It begins with Noah's Ark and ends in heaven, stopping on the way to include the sinking of the *Titanic*, a terrorist massacre, and two very different expeditions to Mount

Ararat. The unity of these edited highlights from post-diluvian history is thematic rather than narrative. Woodworm are stowaways aboard the Ark, are put on trial in sixteenth-century France, and are suspected of infiltrating the frame of Géricault's painting. Voyages are another common emblem: Noah's Ark, the *Titanic*, Jonah's whale, the *Medusa*, an Apollo mission and a sailboat escaping nuclear holocaust, amongst others. Several chapters describe (in muted tones) incidents of political violence. A pleasure-cruiser, the *Santa Euphemia*, is boarded by Arab terrorists, who divide the passengers along political lines in readiness for slaughter, just as Noah divided the animals: clean and unclean, edible and inedible. Numerous points of correspondence suggest comparisons between otherwise diverse incidents and stories.

The problem of the book's unity is highlighted in the 'half-chapter' which appears two-thirds of the way through. Written *in propria persona*, Barnes's acute observations on the nature of love are clearly intended as a kind of emotional key to the surrounding stories, but the central thesis (that love, the act of love, affects the currents of history) is not supported by all the other chapters. Barnes's pursuit of the arcane and exotic often leads to an objectivity which resists 'love' only too successfully. Put simply, the trial of the woodworm and the vision of heaven as an endless shopping mall for hedonists may amuse and intrigue, but they do not admit any real emotional involvement. The antithesis of this view of omnipotent love – that altruism always gives way to self-interest – is likewise sustained by some, but not all, of the chapters. There is a community of related ideas (broadly derived from liberalism) but no sustained argument and, in the absence of a plot, the novel remains a collection of brilliant parts.

This flaw illuminates Barnes's unique position in contemporary British literature. His intellectual rigour and erudition, his faith in research and analysis, are the hallmarks of a continental writer: the essay on Géricault recalls Michel Foucault's similarity exhaustive treatment of Velasquez's *Las Meninas*, for example. On the other hand, his efforts to root arguments and debates within charac-

ters, and to establish emotional contexts for his fables, mark Barnes out as a British writer, part of an empirical rather than analytic tradition. Because these two traditions are genuinely separate it is unsurprising that the book falls somewhere between them, just as it falls somewhere between being a novel and a collection of stories. Love, in the British novel, has often been used as an all-purpose agent of resolution, but in *A History of the World in 10½ Chapters* Barnes expands the horizons to esoteric regions where British fiction rarely ventures, and where an old-fashioned faith in love is not quite enough.

Lindsay CLARK   1939–

The Chymical Wedding

Alex Darken, a minor poet and university lecturer in his late twenties, has borrowed a country cottage in rural Munding to try to overcome a writing block. His creative impulses have been deadened by marital breakup; his wife is having an affair with his best friend. One afternoon Alex is out for a rustic walk when he witnesses an almost mythological romp between an old man and a young nymph. In reality these two are old poet Edward Nesbit and his young American friend, Laura. Edward is a poet of 1940s' Fitzrovian vintage whose once brilliant career has come to an end, and whose work happens to be what first turned Alex on to poetry.

Laura studied parapsychology at an American university where Nesbit was a visiting professor, and they are keen dabblers in occult and hermetic areas such as tarot. Their occult interests are currently focused on Louisa Agnew and her life at Munding in the year 1848, when she was a student of alchemy. Louisa tried to help her embittered father with his interminable book on hermetic mystery, before mysteriously burning her own treatise on the subject, *An Open Invitation to the Chymical Wedding*. The Agnews are the local gentry, and the current Agnew has lent Nesbit the lodge on his estate. The two stories in *The Chymical*

Wedding, of Alex in the present and Louisa in the past, now interweave, as do the cast of the book in tormented triangular relationships: Louisa and young married rector Edwin Frere become lovers, and so do Alex and Laura, to the angry distress of Edward. Gypsy May, a crude female fertility carving at the local church, figures in both the Victorian and modern narratives.

It is suggested that alchemy has a great deal to do with a hermetic vision of sexual intercourse. Alex comes to believe that 'the real secret behind the burning of Laura's book was a sexual secret' and that perhaps 'the secret of the Art of Gold consisted in the male and female'. The book draws to a close with Edward's esoteric haranguing of Alex and Laura, after which he has a heart attack. He survives, and he and Edward become reconciled. The book is not strong on conventional plot resolution, but ends with notions such as 'the secret is there is no secret' and 'once you begin to admit the truth there is no ending'.

Critics were divided over *The Chymical Wedding*. Most enjoyed the richness of the writing, the conscious literariness, and the ambience of esotericism. Others disliked it for being too literary, and too 'safe' in its cosily arcane preoccupations. This critical reception was shaped in part by the fact that when *The Chymical Wedding* won the 1989 Whitbread prize, it did so at the expense of Alexander Stuart's controversial novel *The War Zone*. Writing about the choice between Clarke's 'contrived symmetries' and Stuart's 'hard-edged reality', a *New Statesman* commentator suggested that Stuart's book was the more urgent of the two, while Clarke's was 'the kind that wins literary prizes because it plays the game so well'.

| William GOLDING | 1911–1993 |

To the Ends of the Earth

Rites of Passage (1980)
Close Quarters (1987)
Fire Down Below (1989)

The central concern of Golding's first novel, *Lord of the Flies* (1954), is the struggle to establish and preserve codes of civilised behaviour when civilisation is at its most remote; this concern recurs in much of Golding's work. A secular Protestantism, Golding's *Weltanschauung* succeeds in being at once pessimistic and optimistic. Human beings are continually threatened by the hostility of the elements and the barbarity of their own nature; redemption is to be found in labour, invention and the construction of social relations.

The narrative of Golding's 'Sea Trilogy' takes place at the close of the Napoleonic Wars and, except for the final chapters of *Fire Down Below*, is set wholly at sea. Edmund Talbot, a self-important young man of aristocratic family, embarks for Australia where – by virtue of his godfather's patronage – he is to take up an administrative post. (As though to avoid burdening it with allegorical significance, Golding declines ever to name Talbot's ship.) The Revd Colley, a low-church parson whose demeanour and beliefs Talbot despises, is a fellow passenger – and it is Colley's story which provides the dramatic fulcrum for the first volume. Venturing down to the lower decks, the parson obscurely shames himself. At first it seems that Colley has been assaulted by the crew, but in the novel's closing pages it is revealed that he had performed fellatio upon a sailor when drunk. Unable to live with the memory of his actions, he wills himself to death, leaving a tortured confessional letter to his sister. Reading this, Talbot begins to understand the complexities of feeling that may exist in the most apparently squalid human beings.

In *Close Quarters* the emphasis of Golding's attention shifts from the passengers to the officers and crew. The story resumes with the ship becalmed. A second ship, the *Alcyone*, is encountered in mid-ocean and, to celebrate Napoleon's defeat, a ball is improvised. The *Alcyone* makes course for India, bearing with it Miss Chumley, the object of Talbot's infatuation. Although comical and jejune, Talbot's passion has its dark counterpoint in the violence of the sea. The elements threatened to overwhelm the vessel – now revealed to be as physically rotten as its crew is corrupt. Progress is denied by a growth of seaweed below the waterline; a tangle of rivalries on deck is a no less hazardous encumbrance. There is an enigmatic suicide and the book ends with the ship still in danger, Talbot separated from his love and little resolved.

The strength and resourcefulness of the protagonists are tested again in the final volume, *Fire Down Below*. Irresistible currents drag the ship to the edge of an ice field. The fire of the title is both the literal one smouldering in the timbers of the ship and the metaphorical one which Prettiman and Miss Granham, Utopian radicals amongst the human cargo, hope to ignite when they reach their destination. Talbot is invited to join this adventure. He chooses instead the personal ideal of marriage and, ultimately, a seat in Parliament for a rotten borough. If he regrets his decision, it is only in a dream he records at the end of his tale.

All three parts of the trilogy are told in the first person. *Rites of Passage* takes the form of a journal kept by Talbot for his 'honoured godfather'. Talbot is unsure for whom he writes the two remaining volumes. He fantasises that maybe his memoirs are worthy of publication; occasionally he congratulates himself on the excellence of his style. He grapples with what he calls 'tarpaulin' – the unfamiliar argot of the sea. Golding's pastiche of early nineteenth-century English is authentic-sounding without ever being pedantic or hampering the flow of events; and, while the author has a little fun at the expense of his narrative conventions, this is not 'a fiction about fictions' in the modernist way.

On its simplest level, the story describes Talbot's emotional education. It shows how his naïvety and pomposity are modified by the impact of experience. His vision is broadened, his sympathies deepened and his prejudices subverted. Nevertheless, larger issues swell and subside around the central character. Most importantly, Golding is addressing the problem: how do human beings control the twin enormities of their world and their nature? The advance of science and the strictures of social living are not, of themselves, unproblematic solutions. More clearly than in any of his previous novels, Golding proposes that the line divid-

ing invention from destruction, the divine from the demonic impulse, is almost as imperceptible as the Equator the ship crosses. Humanity, in applying its best efforts to save itself, thereby promotes the risk of its own destruction. Salvation and damnation are the inextricable north and south of Golding's final moral geography.

Rites of Passage won the Booker Prize in 1980.

Kazuo ISHIGURO   1954–

The Remains of the Day

Ishiguro was born in 1954 and left Japan at the age of five to come to England, where, since Japanese was spoken at home, he learned English by 'mimicry, copying'. This talent is exploited in *The Remains of the Day*, which, unlike his first two novels (both set in Japan), is quintessentially English in setting and character. It won the Booker Prize, and Ishiguro was particularly praised for the way in which he had recreated a period and a culture not his own. He insisted, however, that the novel is 'in many ways a consolidation of the sort of ideas' already explored in the earlier books. The novel was made into an award-winning film in 1993, directed by James Ivory from a screenplay by Ruth Prawer Jhabvala.

The events of the novel are viewed entirely from the perspective of Stevens, the butler of Darlington Hall, and the gulf between his self-image and the truth which his narrative unwittingly reveals is the novel's constant theme.

In the summer of 1956, Stevens sets off to visit the Hall's former housekeeper, whom he has not seen for twenty years. Miss Kenton (as he still thinks of her) has written to tell him she has left her husband. That she might have a more than merely professional motive for doing so does not occur to Stevens, whose declared purpose is to invite her back into service at the house. But it is clear to the reader that his real motive is an obscure desire to reckon with his past.

For Stevens, it is the mark of a great butler never to allow personal feelings to interfere with the performance of duty. But his uncomprising and self-destructive pursuit of professional 'dignity', as he calls it, serves deep unconscious purposes: to justify a terror of emotion, and to deal with the painful spectacle of his revered father's undignified decline. In the course of his journey, Stevens slowly and painfully sheds the illusions he has lived by. As he looks back on his pre-war years at Darlington Hall, it becomes obvious to the reader that Miss Kenton was in love with him. The very fierceness of Stevens's prolonged resistance to her approaches indicates that, at some subterranean level, he reciprocated; but, despairing that he would ever admit it, Miss Kenton eventually married another man.

It is sometimes said that butlers only truly exist in England. Other countries, whatever title is actually used, have only manservants. I tend to believe this is true. Continentals are unable to be butlers because they are as a breed incapable of the emotional restraint which only the English race is capable of. Continentals – and by and large the Celts, as you will no doubt agree – are as a rule unable to control themselves in moments of strong emotion.

THE REMAINS OF THE DAY
BY KAZUO ISHIGURO

This wasted personal opportunity has a political parallel, illustrated by Stevens's meeting with the villagers of Moscombe. Stevens's exaggerated deference contrasts sharply with the proud egalitarianism of Harry Smith, and Stevens escapes a bruising argument only by pretending, despite himself, to be a member of the aristocracy. The experience forces him to admit that he has no more come into his own as a citizen than he has as a human being. What gave meaning to Stevens's sacrifices was his belief that they were made in the service of a great man. But here too Stevens was deceived. For it

emerges that his master, Lord Darlington, was a fascist sympathiser and, far from being the leading statesman for which Stevens took him, was the unwitting tool of Nazi diplomats in their attempt to canvass British government support for appeasement.

When Stevens finally meets Miss Kenton, it is clear that she still loves him. But a lifetime's timidity prevails, and Stevens departs alone. In the course of his journey, he has learnt not only what real happiness is, but that the time for it has passed. No longer sustained by a false sense of superiority, he can look forward only to the limited pleasures which his declining years – the 'remains of the day' of the title – may afford.

James KELMAN   1946–

A Disaffection

Kelman has spent his life in Glasgow and, rejecting the London-based perspective of British literary life, has always written about his own culture. 'I don't earn much money,' he has said, 'so I'm involved in the culture I write about', and he gives a voice to the Scottish working class, writing in the vernacular, his prose generously salted with expletives. In spite of standing outside the literary establishment, he has won numerous awards in both Scotland and England, and *A Disaffection* was shortlisted for the Booker Prize.

The novel portrays a week in the life of Pat Doyle, a schoolteacher who has climbed out of his own social class through the benefits of university education. After a desultory evening at the local arts centre in Glasgow, Doyle 'found himself round the back of the premises for a pish, and discovered a pair of old pipes'. Taking these plumber's rejects home Doyle converts them into a handpainted set of musical instruments, their purity and absurdity symbolising for him his own thwarted dreams of taking off for another life on the Continent.

At twenty-nine Doyle teaches working-class Glaswegian kids whose dads are on the dole. He feels guilty about indoctrinating them with middle-class values and cannot condone the privileges of the state system he has enjoyed. Observing Glasgow life through an ironic patter of self-deprecating allusions to Tolstoy and Joyce, Proust and Marx, Pat's stream of consciousness flicks over dead-ends and skips from thought to thought, punctuated by empty obscenities. He despises himself as a thought-policeman, feeling more empathy with the lives of his downbeat family and their emotional tangles than with his own lonely existence.

A wee cafe and another chip shop down by the bingo hall, plus Chinese-style food carry-outs. No pubs. Patrick could buy another load of chips. He had finished with the last lot, and the rain was become a mere trickle. So he could buy another lot and maybe it would cease falling altogether, because here you had a case where there seemed a necessary connection, a contingency, between the purchase of potatoes chipped and fried in the fats of dead animals and the rainfall of a nation albeit a nation who knuckles under to another, and ships them all its freshest fish. But he just wasn't hungry. Not even for the sake of comparing notes for a very large oil painting he was thinking of doing on the whole damn racket, a sort of survey, entitled Chip Shops of Auld Glesgi Toon.

A DISAFFECTION
BY JAMES KELMAN

In an endless round of meaningless nights alone in his flat, or at the pub and the chip shop, or driving nowhere in his clapped-out car, Pat begins to believe he is in love with his school colleague Alison, who is married to a travelling salesman. Pat persuades her to meet him, but their sex-life goes nowhere, as is halted by tongue-tied embarrassment. Dreaming of driving away with her to East Anglia, France and Tuscany, Pat knows that his future is only a fantasy, and his hopes are deflated when Alison finally manages to tell him that she does not want to have an affair.

Getting drunk in his brother's flat with a band of unemployed workmen, Pat puzzles over his job transfer, of which he has just been told. Old resentments are relit, and he abandons the pool of carry-out booze he has

paid for, too drunk to drive his car home. The police begin to give chase as Pat lurches on foot through the rain.

They must have come running after him, to be shouting. What are they shouting. They're just shouting they hate him they hate ye we fucking hate ye, that's what they're shouting. It was dark and it was wet but not cold; if it had not been so dark you would have seen the sky. Ah fuck off, fuck off.

Hilary MANTEL   1952–

Fludd

Although witty and sharply observed, Mantel's ambitious fourth novel was a considerable departure from the high black comedy of her previous work. This boldness paid off; *Fludd* was very well received and awarded both the Winifred Holtby and the Cheltenham Festival of Literature Prizes.

Drawing upon Mantel's memories of her childhood in Derbyshire, the novel is set in 1956 in Fetherhoughton, a grim northern mill village on the edge of the moors. The landscape is as inhospitable as the people who lurk in its shadows. It seems the perfect setting for building Blake's Jerusalem, but in the experience of the local priest, Fr Angwin, the Countenance Divine does not shine forth upon the clouded hills. He has lost his faith and taken to the bottle. He is also at odds with his bishop, a plump, progressive figure who looks forward to the 1960s as 'the decade of unity': 'The ecumenical spirit. Don't you feel it in the breeze?' Angwin does not, but then, as he admits of his congregation: 'These people aren't Christians. These people are heathens and Catholics.' The bishop insists that Angwin clears his church of statuary, and the priest reluctantly buries the plaster saints in the graveyard.

When Fludd arrives, he is assumed to be the new curate and suspected of being an episcopal spy, but he also has an unforeseen effect upon the community. Angwin's housekeeper experiences a curious *frisson*: 'Miss Dempsey felt her mortality; but in the same instant, she felt her immorality too.' Others are similarly affected, notably Sister

Philomena, a strapping young Irish nun who was banished to the brutal moorland convent after her 'stigmata' turned out to be dermatitis. Peculiar noises are heard in the presbytery, whisky tumblers acquire the properties of the widow's cruse, warts appear and vanish, and a nun bursts into flames.

In a prefatory note, Mantel refers to 'the real Fludd', an Elizabethan physician, scholar and alchemist. Of her own Fludd, she writes:

> In these days, he no longer worked in metal, but practised on human nature; an art less predictable, more gratifying, more dangerous. The scientist burns up his experimental matter in the *anthenor*, or furnace, but no scientist, however accomplished, can light that furnace himself. The spark must be set by a shaft of celestial light.

Alchemy, Fludd explains, is 'releasing spirit from matter', a notion echoed in an eerie scene in which he oversees the exhumation of the church's statutes. A similar alchemy is wrought upon Sister Philomena, who reclaims her original identity as Roisin O'Halloran and absconds to London, where she is deflowered by Fludd in the Royal and Northwest Hotel. He subsequently vanishes, but leaves her with a bag of 'gold'.

'The time will come', Miss Dempsey muses at the end of the novel, 'when we will look back on what has occurred, and account it an Age of Miracles.' In this oblique, beautifully written novel, Mantel creates the sense of another world, ancient and mysterious, running parallel with the quotidian life of Fetherhoughton. Reminiscent of paintings in which medieval people go about their daily business beneath skies thronged with angels and demons, *Fludd* is a daring, exhilarating and wholly convincing evocation of the numinous.

Robert NYE   1939–

The Memoirs of Lord Byron

When Byron died in 1824, he left behind two copies of his memoirs: one with his executor, John Cam Hobhouse, and the other with the poet Thomas Moore. Four weeks later, Hobhouse, Moore, John Murray (Byron's publisher) and solicitors representing Lady Byron (his wife, Annabella) and Augusta Leigh (his half-sister) met at Murray's offices. There it was agreed that the memoirs be burned, and there both copies were torn up and thrown on a fire. Their contents were never disclosed by the few who had read them, but Hobhouse later characterised them thus: '[They] were fit only for a brothel, and would [have damned] Lord B. to everlasting infamy if published.' That they were pornographic is nearly certain; that they recorded the poet's homosexual encounters is probable: that they implicated Augusta in incest and Annabella in 'unnatural practices' is a strong likelihood, and accounts for the legal presences at their destruction.

Margarita just stalked in half-undressed to wish me goodnight. This she did with a Venetian benediction: '*Benedetto te, e la terra che ti fara!*' – 'May you be blessed, and the *earth* which you will *make.*' Is that not pretty? She can still occasionally please me with such expressions. A month ago I would have taken ten minutes out from my labours with my pen to belabour her a little with my penis. Now I merely record her wistful passage through my chamber, and pass back quickly to the story of my days.

THE MEMOIRS OF LORD BYRON
BY ROBERT NYE

Nye's novel is an imaginative reconstruction of the memoirs. In this account, Byron starts keeping his record in 1818. He has fled England after the rumours ignited on the collapse of his marriage. His death is six years away. From various rented villas in Italy, he tells of his Aberdeen childhood; of how he inherited (in 1798) his uncle's barony; of his inconsistent education, first by private tutor and then at Harrow and Cambridge; of his quarrels and friendships and, of course, of his affairs. The literary triumphs are mentioned too, but with an air of casual arrogance: for this Byron anyway, the poetry is 'dashed off', his gigantic reputation bores him, his public face is merely a pose. What occupies his attention is the chaos wrought by women in his life. To most of them his treatment has been callous. (He admits this with disarming flippancy.) His foolish, pretentious mother; the vengeful and volatile Lady Caroline Lamb; Claire Clairmont (Shelley's importuning sister-in-law); dull Annabella; even the latest Venetian mistress: each has proved (usually in the short run) insupportable, and each has exacted in turn an emotional price. Byron's deepest griefs, though, are caused by his exile from Augusta and by his separation from Allegra, his daughter by 'La Clairmont'.

The memoirs end in 1822. Allegra dies of a fever; Shelley, whom Byron loved and respected, drowns in the Bay of Spezia. Byron superintends a macabre cremation and foresees his own death. '[The] thought,' he observes, 'did not displease me.'

Kenneth Tynan once remarked that Nye, who has also written novels about Falstaff, Merlin, Faust and Gilles de Rais, 'writes like Rabelais reborn', which perhaps makes him the perfect ghost for Byron. Like Byron's letters, the novel is fluent, vivid, witty and idiomatic. The Byron it presents is wholly convincing, a figure of enormous energy (a lot of it self-destructive), impermanent passions and radical intellect. Further, it serves as a reminder that his relationship to Romanticism was at least ambiguous: he despised Keats and 'Turdsworth', reserving his highest admiration for Pope. He rejected the Shelleys' atheism and found their proto-feminism risible. He was a populist aristocrat and political maverick; his assistance to Greek nationalism was audacious, unequivocal and concrete. Adored and loathed in roughly equal measure, he was that rare thing: an egotist without pomposity. Nye's *Memoirs* do much to elucidate his contradictions and to evoke the charisma that briefly held Europe in thrall.

Alexander STUART 1955–

The War Zone

The War Zone was briefly celebrated as 'the book which won and lost the Whitbread Prize'. It is an emotionally raw and violent study of adolescence and incest, set during a long, hot summer holiday against a backdrop of crumbling 1980s Britain. Tom, the already embittered and rebellious teenage narrator, is uprooted from the anarchy of London when his middle-class family move to Devon; but life there is no quieter for him after he discovers his architect father and his older sister sharing an erotic bath together.

Stuart's treatment of incest was particularly strong because it was the sister herself (Jessie, or 'the Bitch' as Tom calls her) who instigated the relationship, seemingly just 'for kicks'; she is an emotionally blank and nihilistic character. Stuart said that he thought it would be 'interesting' to make the daughter the instigator of the father-daughter incest. Tom can hardly believe what he sees, and attempts to get evidence of the relationship on video. He also attempts to vandalise his father's phallic architecture, and his anger and his violent fantasies dominate the book, until at the very end he comes to see that his rage was an expression of his own jealous desire.

Much of the book was wilfully and symptomatically ugly, such as the scene in which father and the girl have anal sex in a rubbish strewn air-raid shelter, while Tom attempts to film them. Many critics praised it for the raw immediacy and power of the writing, but others were concerned that the sexual content was treated in a voyeuristic and exploitative manner, amounting to 'near-pornography'. Others voiced more discerning complaints about Stuart's attempts to make the book into a modish 'state of the nation' statement. Stuart's post-punk interest in depicting things like glue-sniffing, tanks in the streets of London, and Union Jack-clad skinheads was felt by some to over-freight the novel and to detract from what might otherwise have been the almost mythic power of the central scenario.

The War Zone was Stuart's second novel (not, as was widely believed, his first) and it represents the high point of his career to date. The limited and mannered aspects of the book have become more apparent in retrospect after he published *Tribes*, a frenzied novel about soccer violence and male sexuality which was almost universally panned by the critics.

Stuart was initially told that *The War Zone* had won the prize in the Best Novel category of the 1989 Whitbread, only to find ten days later that one of the judges, the novelist Jane Gardam, had successfully managed to stop the award because she objected to his treatment of incest in the book. The prize went instead to Lindsay Clarke's far cosier and less threatening novel *The Chymical Wedding*.

Rose TREMAIN   1943–

Restoration

Tremain is a novelist whose accomplishment is never in doubt, but whose work is difficult to categorise. She has said that what interests her most is communicating ideas through different forms of writing, and *Restoration* – a long, picaresque, historical novel which was shortlisted for the Booker Prize and won the *Daily Express* Book Prize – is certainly different from the novels which preceded it. Meticulously but unobtrusively researched, it is set during the seventeenth century, and narrated by Robert Merivel, son of Charles II's haberdasher. The restoration of the title refers not only to the king, but to Merivel himself, who, dazzled by meretricious splendour of his monarch, abandons a decent, useful career in order to enter the life of the court. All self-respect is abandoned in his pursuit of worldly glory, and he subsequently has to make a journey down through several circles of hell until he can regain a proper sense of values.

A student of anatomy, Merivel is taken by his father to the lavish court of Charles II, but his extreme nervousness makes him tongue-tied in the royal presence and he does not gain the hoped-for preferment. After the death of his parents in a fire, however, Merivel receives an unexpected command to leave Caius College Cambridge and become physician to the royal spaniels. The king captures Merivel's affection, which is rewarded by his popularity as an unofficial court jester. He tells Charles about a medical curiosity he witnessed at Cambridge, a man whose heart was exposed as the result of an ulcer, and as he touched the naked heart Merivel discovered that the organ had no feeling. The king interprets this as Merivel's philosophy and arranges for him to marry one of his mistresses, Celia, whose rival, Lady Castlemaine, is demanding that an appropriate husband be found. Merivel agrees to serve as paper bridegroom and not touch his wife, making way for the king on the night of the nuptials, and Charles elevates him to Sir Robert, granting him an estate, Bidnold Manor in Norfolk.

Merivel begins to fall in love with Celia, who has unwisely expressed jealousy over the other royal mistresses and has been temporarily exiled from the court. Merivel conceals from her the fact that the king still loves her, and leads her to believe that he has been appointed as her 'overseer' with authority to inform the king of her conduct. His amorous needs are partially satisfied by a barmaid, Meg Storey, and by Violet Bathurst, the wife of a neighbour, but his passion for Celia becomes so pressing that he finally declares it, giving great offence, and putting him out of royal favour when the scene is reported to Charles.

The truth is that, when the King returned, it was as if self-discipline and drudgery had exploded in a clap of laughter. I became much too excited by and greedy for life to spend much of it at work. Women were cheaper than claret, so I drank women.

RESTORATION
BY ROSE TREMAIN

The king dispossesses Merivel of his estate, and he wanders friendless until deciding to enter a Bedlam hospital run by his Cambridge friend John Pearce. Amongst the Quakers, Merivel begins to gain a sense of usefulness, but his lust leads him to lie with a mad woman, Katherine, who becomes pregnant. This breaking of the trust which has been placed in him results in Merivel's second exile, and he and Katherine return to London to live with her mother, Frances Elizabeth. The birth is painful and disastrous, as Merivel performs a Caesarean section, saving the life of their daughter, Margaret, but causing Katherine's death.

Still longing to be reunited with his beloved sovereign, Merivel resumes his medical practice during the Plague, and is invited to dine with the king on the night the Great Fire spreads across London. The memory of his parents spurs Merivel to a great act of bravery as he rescues a haber-

dasher's wife from the flames, and the deed reaches the ears of the king, unbeknown to Merivel. Months later, as memories of his past haunt Merivel's dreams, he decides to return to Bidnold for one last night in his old home, believing it now to be the property of a French aristocrat. King Charles has privately appropriated Bidnold as his retreat, a place to escape the pressures of state and his growing unpopularity, and he rewards the now morally worthy Merivel with a room in the manor, promising him a permanent restoration to the royal presence.

1990

Re-unification of Germany ● Iraq invades Kuwait ● Resignation of Margaret Thatcher ● David Dinkins becomes first black mayor of New York ● Nobel Peace Prize is awarded to Gorbachev ● Britain introduces separate taxation for married women ● Surgeons perform the first surgery on a baby in its mother's womb ● First human gene experiment ● Karl Popper, *A World of Propensities* ● Jenny Holzer is the first woman to represent the US at the Venice Biennale ● Damian Hirst, *My Way*, Jeff Koons, *Jeff and Ilona (Made in Heaven)* ● Films: *Jesus of Montreal, Cinema Paradiso* ● Popular music: Ice Cube, *Amerikkka's Most Wanted*, Concert held at Wembley, London, to celebrate the release of Nelson Mandela ● Lewis Mumford, Greta Garbo, A.J.P. Taylor, Alberto Moravia, Patrick White, Leonard Bernstein and Aaron Copland die ● Derek Walcott, *Omeros* ● Norman MacCaig, *Collected Poems*

A.S. BYATT — 1936–

Possession

As a novelist, Byatt has been overshadowed by her more prolific and commercially successful sister, Margaret Drabble, none of whose novels, however, has won the Booker Prize, as *Possession* did. It was also awarded the *Irish Times*/Aer Lingus Prize and was shortlisted in the novel category for the Whitbread Prize. Although the book has its detractors, who complain that, like Byatt's other novels, it is the work of an academic imagination (giving a professor the same name as a popular, comic television character was seen by some as characteristically unworldly), it proved a popular choice for the Booker, the proceeds of which would be

spent, Byatt announced, upon installing a swimming pool at her second home in France.

The novel's two central figures are the (fictitious) Victorian poets, Randolph Henry Ash, spurned by Leavis but now back in fashion, and Christabel LaMotte, less well known but who enjoys a small but devoted feminist following. A group of present-day literary scholars and critics reconstruct the truth about their relations with one another from a legacy of poems, letters and journals. (These learned pastiches constitute a good part of the text.) The scholars' researches provide the setting within which the Victorian drama unfolds.

The offices of Prof. James Blackadder of London University, editor of Ash's complete works, are known as 'the Ash Factory' or less affectionately, 'the Crematorium', and Ash studies in Britain are a morbid, dispiriting affair. The only person in the field able to thrive is Blackadder's American counterpart, the ghoulish Prof. Mortimer Cropper, child pornography enthusiast and master of inexhaustible funds with which grimly to reap Ash memorabilia for his university library in Harmony City, New Mexico.

Later, Maud stood in there, turning her long body under the hot hiss of the shower. Her mind was full of an image of a huge, unmade, stained and rumpled bed, its sheets pulled into standing peaks here and there, like the surface of whipped egg-white. Whenever she thought of Fergus Wolff, this empty battlefield was what she saw.

POSSESSION
BY A.S. BYATT

The action begins when Roland Michell, Blackadder's research assistant, turns up two drafts of an admiring letter to an unknown woman in Ash's hitherto untouched copy of Vico. Clever detective work reveals the woman to be Christabel LaMotte. Roland resolves to keep his discovery to himself until he has found what lies behind it, but he is soon obliged to enlist the help of the LaMotte scholar Maud Bailey. Clues from the journal of Blanche Glover, LaMotte's intimate friend, lead them eventually to an extensive cache of letters between her and Ash. The correspondence is breathlessly hyperbolic from the start. When love is finally mentioned, Christable's carefully nurtured self-sufficiency comes under threat as she is drawn away from Blanche to Ash (who is already married). Her feelings are echoed by Maud, hesitating between Roland and her American colleague Leonora Stern. For both Maud and Roland, the scholarly quest for the final truth about Ash and Christable has become a quest to know themselves.

A complex of jealousies, professional and sexual, ensure that Maud and Roland soon find others on the same trail. Cropper looks set to acquire the letters and spirit them away to America; meanwhile Blackadder and Leonora pursue Maud and Roland to Brittany, where it turns out Christabel fled in 1859 to bear Ash's child, conceived on a secret trip to Yorkshire. The journal of Ash's wife, Ellen, discloses that the final piece in the jigsaw – a much later letter from Christabel to Ash, which remained unopened at Ash's death – was buried with him. So Cropper joins forces with Ash's heir to exhume it; Maud, Roland, Blackadder and others converge on a country churchyard to intercept him. The document reveals that Christabel's child not only survived, but was Maud's great-great-grandmother. As Maud is the direct descendant of both Christabel and Ash, she is the rightful owner of the correspondence, and Cropper is foiled. Maud prepares to give up her solitude for Roland's sake, while Roland's moribund relationship with his girlfriend, Val, finally dissolves. Once its secrets are revealed, the literary past is transformed from a mere fetish into a source of new life.

J.M. COETZEE   1940–

Age of Iron

Coetzee's characteristically bleak sixth novel opens with Elizabeth Curren arriving back at her Cape Town home after a doctor's diagnosis that she is dying of cancer, to find a black tramp, Vercueil, living rough in her garden. While acting as a symbol of decay throughout the novel, Vercueil also becomes Elizabeth's companion and confidant, as she tries to make peace with her own world, and with all the strife of a country in turmoil. As Elizabeth tries to continue her daily life and faces the inevitability of her fate, she thinks: 'Death is the only truth left.' Her own vivid imagination and the effect of various pills cause her to experience contrasting images of the afterlife, visualising it as a lobby with armchairs and music, or a bus with standing room only on its way from nowhere to nowhere.

When Florence, Elizabeth's black domestic, returns with her son, Bheki, they are soon joined by the latter's friend, John. Both boys are victimised by policemen, who eventually cause them to have a bad car accident. John is hospitalised but, to Elizabeth's surprise, Florence and Bheki do not want to be involved in John's predicament, saying they are afraid of the police, and claiming that there is a conspiracy between police and ambulancemen. So Elizabeth and Vercueil go without them to check that John is being treated properly. A telephone call late one night tells Florence that Bheki is in trouble, so Elizabeth drives Florence out to a shanty town to try and find him. The search proves to be a harrowing experience, as Elizabeth is caught up in terrible riots, only to discover that Bheki has been killed by the army. This experience intensifies her anger at the injustices all around her, and increases her conviction that her life is worthless.

Discharging himself from hospital, John returns to Elizabeth's house to recover a hidden weapon, but is killed during a police raid there. Elizabeth is subsequently ordered to leave her house for the night by the police, and wanders bewildered in her nightgown, sleeping rough on the street, before being found by Vercueil.

She returns to spend her final days at home, and reaches a new height of intimacy with Vercueil, who is constantly by her side, as she struggles for atonement and mourns the turmoil and the loss of life during South Africa's cruel and relentless 'age of iron'. She poignantly sums up her own position against the apartheid regime: ' "Your days are numbered" I used to say to those who will now outlast me.'

The novel takes the form of a letter written by Elizabeth to her daughter in America, to be opened after her death, and it is suffused with the guilt of white South Africa. Elizabeth's illness is symbolic not only of her own life and her response to the country in which she lives, but also of the whole country's malaise: 'I have cancer from the accumulation of shame I have endured in my life.' Elizabeth has no future, and South Africa's is grim, with the age of iron being followed by an age of bronze: 'How long, how long before the softer ages return in their cycle, the age of clay, the age of earth?'

Nadine GORDIMER   1923–

My Son's Story

My Son's Story is set in the Johannesburg of the 1970s and 1980s. It is a powerful political novel in which a 'coloured' family is destroyed by an extramarital relationship and by the fight against apartheid.

It begins with the discovery that father and son are deceiving each other: young Will, the narrator, is going into a cinema when he should be studying; his father, Sonny, is just leaving, with Hannah, his blonde blanc-mange-pink lover. Will is angry and bitter about this betrayal (his commentary is at times hysterical), but he must for the sake of his mother and sister become his father's accomplice. The books shows that in South Africa learning to live a lie is a normal part of growing up.

Sonny is a pillar of the community – a man who had worked hard to become a schoolteacher and who cherishes his wife and children – until the firmness of his beliefs leads to his involvement in a banned revolutionary organisation. This new life of secrecy makes it easy for him to have an affair with the human-rights activist who attends his political trial and visits him in prison. Their passion is fuelled by the aphrodisiacal atmosphere of conspiracy and idealism (which Gordimer conjures up vividly), and Sonny's family begins to break up. His daughter, Baby, goes over the border – without Sonny's consent or approval – to join a militant group. His beautiful, quiet, respectable wife, Aila, leaves him and, although she returns, she is changed. Her long, oiled hair is shorn and, in a twist to the tale that surprises, she is arrested for having terrorist arms and explosives hidden in the yard. As part of the process of change he himself fought for, Sonny's status as patriarchal hero is overturned, and he is left alone (Hannah is also gone, her ambition fulfilled with a high-flying UN job).

All the characters have been politicised, even Will, who has up to now been considered marginal, as only an observer of action. His role is to be a writer – to record 'what it was really like to live a life determined by the struggle to be free'. He realises too that in writing his father's story, Will, the child, becomes the father of the man.

He looked slowly round my room; I suppose it must be a year, more, since he's been in it. There was the beginning of a crinkling round his eyes, affirmation rather than recognition, at what I've kept, and he went over and stood a moment, head back, as if he were in an art gallery, not his son's bedroom, before a poster of a desert. That's new. Just space. I don't know what desert, where – I hoped he wasn't going to expect me to say.

MY SON'S STORY
BY NADINE GORDIMER

At the end of the novel Will's home, the last remaining physical evidence that the family had existed, is petrol-bombed, an act which serves as a reminder that there is still plenty of passion for destruction left in Gordimer's country, and that there is no simple solution. *My Son's Story* does not polemicise, but raises questions, presenting a situation that is morally and politically extremely complex.

Hanif KUREISHI   1954–

The Buddha of Suburbia

Expanded from 'With Your Tongue Down My Throat', a story originally published in 1987 in *Granta*, and subsequently adapted (by Kureishi and Roger Michell) for a controversial but acclaimed four-part television series (1993), *The Buddha of Suburbia* is narrated by Karim Amir, a young man on the make. Son of an Indian father and an English mother, he is 'an Englishman born and bred, almost'. Karim's parents separate and his civil servant father Haroon – the Buddha of the title – takes up with another Englishwoman, Eva, and becomes something of a guru at her parties. Haroon preaches an 'internal return' to India, even

though – as Karim points out – it is all he can do to find his way to Beckenham without getting lost.

Karim's adolescence in the South London surburbs is equally devoted to rock music and sex. He is bisexual and takes a keen interest in his good-looking friend Charlie, Eva's son, as well as in his sexy cousin Jamila, who is influenced by the black militant Angela Davis. Unfortunately Jamila's father – Karim's grocery-owning uncle Anwar – is a Muslim who pushes her into an arranged marriage with the fat, balding and lecherous Changez. Changez, from Bombay, is a brilliant and surprisingly likeable comic creation who sees England as a 'gold mine' of sexual possibilities. Far from gaining access to a gold mine, the unfortunate Changez finds that Jamila refuses to have anything to do with him. Instead she contents herself with the delectable Karim, whom she nicknames 'creamy'. Relations between Changez and Anwar deteriorate into farcical hostility, leading to Anwar's accidental death.

There is a major change of gear in the second half, when Karim's parents move from the suburbs into West Kensington. This more metropolitan location coincides with Karim becoming an actor, which allows Kureishi to explore west London Bohemian life and its sexual opportunities. Acting also allows him to present the irony of Karim playing Mowgli in the *Jungle Book*. Karim falls painfully in love with a beautiful upper-class girl, Eleanor, and also becomes a protégé of theatre director Matthew Pyke, who sets up a foursome with his wife, Karim and Eleanor. Karim's theatre career takes off in a big way (in a role modelled on Changez) and he takes the capital by storm.

The scene shifts again when Karim's stage success takes him to New York. He meets up with his old wanking chum Charlie, now a hedonistic and cynical rock star who despises grey old Britain. Karim, in contrast, is nostalgic for it, and he returns to London, where Kureishi ties up a few loose ends. Karim's mother has found herself a new boyfriend; Eva and 'Daddio' the Buddha are thinking of getting married; and Jamila is now a single parent who lives in a commune and has come out as a lesbian. It is just left to Karim to ponder over what a chaotic mess everything has been, and yet express his passionate faith in the future.

Those who liked *The Buddha of Suburbia* responded chiefly to its exuberance, farcical humour, and its evocative picture of a time and place: 1970s London with its rock music, Paki-bashing, dustmen's strike and 'turdy parks'. Its larger significance rested on a timely exploration of racial and sexual identities in contemporary Britain, and Kureishi's championing of a multi-racial, polysexual future. Those who disliked it criticised it as chaotic, overblown, politically facile and superficial, as well as negative in its Asian stereotypes. Kureishi predictably angered commentators on the political right, especially when he described London's poll-tax riots as 'terrific'. Kureishi's best work is generally felt to be his film *My Beautiful Laundrette* (1985), and his most cogent and penetrating discussion of UK-Asian identity, and the emptiness of conventional Britishness, is to be found in the essay which prefaces the published text of the film (1986).

John McGAHERN 1934–

Amongst Women

Moran is a veteran guerilla. During the war for Irish independence he led his men to legendary triumphs. But now, a widower of many years, his sole remaining command is over his children – and they have started to elude his grip.

Luke, the eldest, has already quit the Republic for London. The rest – Maggie, Mona, Sheila and Michael – live at Great Meadow, Moran's small farm. Maggie manages the household until Moran, after a coolly executed courtship, remarries. His bride is Rose Brady. Considerably younger than her husband, she has rejected all her previous suitors. The reception is inexpensive, at Moran's insistence. Luke does not attend and, through his absence, reflects his father's bitter temperament.

Time passes, marked by the taking of meals, the recitation of the rosary and the occasional playing of cards after prayers. Moran and Rose quarrel infrequently and never irreconcilably. Maggie joins Luke in London to train as a nurse. Sheila and Mona are successful at school; both enter the Civil Service and settle in Dublin. One winter, Michael has an affair with a waitress; Moran learns of it and tries to beat his boy. Michael flees to his sisters, who effect a truce between father and son. Soon, however, Michael too has gone to London. He works for a while on the building sites then, with Luke's assistance, begins to study for accountancy. Maggie marries an alcoholic labourer; Sheila a fellow civil servant. Luke flies from England for the latter's wedding and treats Moran with distant courtesy. Mona stays single. Apart from Luke, the children return regularly to Great Meadow, where they help with the farm and where they witness their father's decline into death. His last words are, '*Shut up!*', directed at Maggie. The IRA is represented at his funeral: again, Luke does not attend. After the service, there is a sense of poignant relief. Maggie remarks. 'He'll never leave us now.'

The novel's subject is Moran's control of the people around him. He uses neither violence nor – except rarely – overt bullying. Rather, his power derives from the peculiar sympathy he evokes. (Often, he is referred to as 'poor Daddy'.) His unpredictable sensitivities, his sulks and his silences prompt his nearest to accuse themselves: the patriarch's severe dignity in some clumsy way has been offended and, therefore, appeasement must be made. Moran thus becomes an emotional minefield. Still, he dare not explode and risk destroying his familial bonds: he needs them for the exercise of power. For a similar reason, he chooses to live isolated from his community and former friends: to participate in society might be to measure a diminished status. Indeed, the narrative is notable for the extent to which – as if in deference to its protagonist – it avoids discussion of topics upsetting to him: the world beyond Great Meadow, the cause of the rift with Luke, the previous marriage. By omitting so much, it shows the shape of Moran's consciousness – the limits determined by its protective repressions.

McGahern's fourth novel, in a career spanning thirty years, *Amongst Women* was shortlisted for the Booker Prize. Though it failed to win, it attracted vigorous support from *inter alios* John Banville and earned an increased public for its author.

Three Times Table

Maitland's themes of woman-magic, myth and an idiosyncratically heterodox Christianity are here successfully synthesised with sustained and lively narrative and an assured mingling of naturalism and magic. Three women, Rachel, her daughter Phoebe and granddaughter Maggie, live in the same London house. In the course of the novel and in a deftly plausible synchronicity each woman meets and masters an acute crisis. The novel's time of a few days unfurls, through their memories, to show the full texture of each woman's life in a skilful accumulation of flashback and incident. Rachel has won distinction as a palaeontologist with a triumphantly vindicated theory of the survival of carnivorous dinosaurs into the mammalian era. Now that the catastrophe theory has superseded her passionately espoused Darwinian gradualism, scholarly pride enforces her recantation. As she steels herself to announce this to her colleagues, she recalls her husband, a scientist whose grudging attitude towards her work soured the marriage, and reflects how her love for her rich, encouraging father led her to injustice to her mother.

Phoebe, now gardening for a London borough, abandoned Oxford and mathematics for a protracted hippie fugue after her father's sudden death, returning to bear Maggie in a comfortless commune. Now, doggedly, and usually contentedly, she gardens, rejecting the ideological gloss her friends ascribe to her work. But, feeling some profound failure of love, she cherishes a breast tumour as her secret dark lover, postponing the diagnosis which could bereave her.

Maggie at fifteen is captivating, clever and skinny (indeed amenorrhoeic and virtually anorexic), whose secret is the glorious dragon Fenna, who comes to her at night and has taught her the secret of flight. She has instinctively recognised that in order to achieve maturity she must renounce her ecstatic flights and the adored, manipulative dragon. The night before her mother and grandmother are to face their life-challenging crises, Maggie resists Fenna's fiercest seductions and, as the skylight of her attic buckles and implodes, dismisses him, asking him to go to her grandmother who, Maggie obscurely realises, needs comfort. And indeed, while Phoebe accepts the possibility of cure and its attendant struggles, Rachel, who has achieved her own renunciation, glimpses a friendly dragon near the Natural History Museum.

Rachel soon discovered that while she liked the early stories of her profession, Martin did not. It irritated him somehow, for instance, that Mantell's first iguanadont tooth, which had led to the realisation of the whole nature of vanished species, species which had to be saurian whatever the Parisian experts were trying to say, had been found not on some well-structured dig, but casually, accidentally, by the wife of a country doctor.

THREE TIMES TABLE
BY SARA MAITLAND

The Fenna strand, though handled with beguiling assurance as a naturalistic feature, should be read by the sceptic as allegory or metaphor; but in any case it sorts comfortably with the 'daily' narrative, richly rooted in contemporary London life and encompassing a host of plausible and varied friendships. The themes – dragons and dinosaurs, love and work, friendship and loss, cosmology, faith and learning – are handled confidently in a novel which can be enjoyed at many levels.

Armistead MAUPIN   1944–

Tales of the City

Tales of the City (1978)
More Tales of the City (1980)
Further Tales of the City (1982)
Babycakes (1984)
Significant Others (1988)
Sure of You (1990)

Maupin's sequence of novels celebrating San Francisco first appeared as a daily serial in that city's *Chronicle* newspaper in 1976. This format resulted in stories of considerable pace and allowed Maupin to incorporate contemporary events, fads and fashions as they happened. The great success of the series resulted in its being republished, 'after significant revisions', in book form, where the immediacy it had on first publication was transformed into detailed social history. Apart from being a witty and touching record of life in the city over fourteen years, the novels provide a large cast of characters whose fortunes wholly engage the reader. Maupin has also managed to write a mainstream book about homosexual experience: he recalls that when the principal homosexual character fell ill, he received a letter threatening: 'I'm nothing but a middle-aged housewife from Moraga, with two little machos of my own, but if you kill Michael Mouse, I'll never subscribe to the *Chronicle* again.' An enormous success with both the critics and the reading public, Maupin's sequence has been compared with the work of Dickens, Wodehouse and Waugh, although it remains quintessentially American.

The novels are centred upon 28 Barbary Lane, a ramshackle rooming-house on San Francisco's Russian Hill, owned by Mrs Anna Madrigal, a woman of a certain age who affects kimonos and welcomes new tenants by taping to their door a joint made from her own home-grown cannabis. Mary Ann Singleton, a naïve young woman from Cleveland, Ohio, arrives in the city and rents a room at Barbary Lane. The other tenants are Mona Ramsey, an ageing hippie; the eternally promiscuous Brian Hawkins, formerly a liberal lawyer, but now working as a waiter; and the hopelessly romantic Michael 'Mouse' Tolliver, forever seeking Mr Right. Mary Ann joins the advertising agency Halcyon Communications as secretary to the chairman, Edgar Halcyon, who describes himself as 'the biggest son-of-a-bitch in town'. This accolade rightly belongs to Edgar's son-in-law, the dazzlingly handsome, bisexually adulterous Beauchamp Day. We are also introduced to Beauchamp's unhappy wife, DeDe; her mother, Frannie, a 'sweet, but rummy, society dowager'; Mona's former lover, D'orothea Wilson, a model as exotic as her name; tacky Connie Bradshaw, an old schoolfriend of Mary Ann, now an air stewardess; the desirable society gynaecologist Jon Fielding; and the distinctly creepy Norman Neal Williams, a new tenant at Barbary Lane, who claims to be a salesman.

Michael Tolliver had spent rush hour in the Castro, the time of day when the young men who worked in banks came home to the young men who worked in bars. He watched from a window seat at the Twin Peaks as they spilled from the mouth of the Muni Metro, stopping only long enough to raise the barrels of their collapsible umbrellas and fire at the advancing rain. Their faces had the haggard, disoriented cast of prisoners who had somehow tunneled to freedom.

BABYCAKES
BY ARMISTEAD MAUPIN

Any attempt to give a plot synopsis is doomed to failure, since the novels are packed with incident and surprise. In *Tales*, Mary Ann shakes off Beauchamp Day and becomes involved with Norman Neal Williams, who turns out to be even more sinister than his appearance. Michael has a brief affair with Jon; Mrs Madrigal has an affair with Edgar Halcyon; DeDe becomes pregnant by the Chinese grocery boy; and Mona uncovers D'or's dark secret.

More Tales is principally concerned with Mary Ann's attempts to help her new boyfriend, Burke Andrew, discover the root of his amnesia, and with Mona's discovery of Mrs Madrigal's true identity. Michael takes up with Jon again, but becomes seriously ill; and Brian develops a sexual relationship with a woman he never meets.

In *Further Tales* it is revealed that DeDe and her half-Chinese twins have not after all perished in the Jonestown massacre. They are, however, being pursued by a man who may or not be Jim Jones. DeDe wants to retain editorial control over the sensational story of her escape and enlists the help of Mary Ann, who is frustrated in her job in television introducing the afternoon movie. She and Brian (who is still waiting at Perry's) are now living together as lovers. Michael has once again parted from Jon and is employed as manager of God's Green Earth nursery. He is pursuing an unsatisfactory affair with a cop, and has a rather more stimulating fling with '———', a famous movie star. Mona has greeted the new decade by moving to Seattle, and Prue Giroux, an ambitious society columnist, embarks upon an affair with a hermit she encounters in Golden Gate Park.

In spite of the odd eruption of psychopathology, queer-bashing, murder and cannibalism, the first three volumes are light-hearted and optimistic, largely fulfilling the formulation provided by Miss Prism, describing her own three-volume novel in *The Importance of Being Earnest* (1895): 'The good ended happily, and the bad unhappily. That is what Fiction means.' The last three novels, although equally funny, are much darker in tone, once again accurately reflecting the times in which they were written. In March 1983 the *San Francisco Chronicle* was running more stories on Aids than any other newspaper, which was appropriate given the high incidence of San Franciscans who had contracted the virus. In *Babycakes* we discover that one of the characters has already died; Michael is HIV-positive. The novel opens with Queen Elizabeth II and the Duke of Edinburgh visiting San Francisco. A crew member of the *Britannia* jumps ship and swaps accommodation with Michael, who spends a month in England tracking down Mona and becoming involved with a half-Aboriginal teenager called Wilfred. Brian and Mary Ann are now married, and

the latter is torn between her hugely successful career in television and her husband's desire to have a child, hence the novel's title. ('Babycakes' is also Michael's principal term of endearment.) Maupin's normally infallible ear is not wholly attuned to the patterns of British speech, with the result that the chapters set in England are rather unconvincing. The plot, however, is as exhilarating as ever, and the ending is a happy one, with both Brian and Mary Ann's wishes fortuitously granted.

Significant Others introduces three new characters: Wren Douglas, 'the world's most beautiful fat woman'; Polly Berendt, an attractive young lesbian employed by Michael at Plant Parenthood (as the nursery has been renamed); and Thack Sweeney, a renovator of antebellum houses in Charleston, whom Michael meets during a tourist trip to Alcatraz. The Hawkinses have moved to 'The Summit', as befits Mary Ann's status, and 28 Barbary Lane is crumbling. Much of the novel is set at two camps: the Bohemian Grove, 'a secret haven where captains of industry and pillars of government could let down their guard and indulge in the luxury of first-name-only camaraderie', and Wimminwood, where women and young children go to escape from men and attend consciousness-raising groups ('Body and Facial Hair: In slides, stories and song. Bonnie Moran, facilitator'). DeDe is an unwilling participant at Wimminwood, dragged there by her lover, the born-again feminist D'or; at the Grove, but escaping for sex with Wren Douglas, is Booter Manigault, DeDe's stepfather. Nearby, at a friend's cabin, Michael is spending time with Brian, who fears he has Aids, and Thack, who seems the ideal candidate to become Michael's significant other.

In *Sure of You* (a title which becomes increasingly ironic), Brian has become Michael's partner at the nursery. Burke Andrews reappears to offer Mary Ann a job in New York, and she has to chose between her husband and her career. Michael has moved out to live with Thack, who gives him some sharp lessons in gay politics (clearly articulating some of the author's own anger), and – Polly apart – 28 Barbary Lane is peopled by strangers. Mrs Madrigal is persuaded to take a holiday and joins Mona on Lesbos. While the sequence ends on an upbeat note, the novel itself is painful, particularly in its depiction of the crumbling of old alliances and loyalties.

Further Tales bears an epigraph from Sir Thomas Browne's *Religio Medici* which might stand for the entire sequence: 'Surely there are in everyone's life certain connections, twists and turns which pass awhile under the category of Chance, but at the last, well examined, prove to be the very Hand of God.' As controlling deity, Maupin is generous and affectionate, and his plotting is both dazzling and audacious. Only in the last volume does this sympathy falter. It is as if Maupin changes course about halfway through the sequence. It begins as a *Bildungsroman*, with Mary Ann apparently set upon a similar trajectory to that of the hero of *Wilheim Meister's Apprenticeship* (1795–6, and the prototype of the genre), blundering innocently towards maturity with the help of her friends. In the process, however, she loses more than her innocence, and by the fifth volume, even the fiercely loyal Michael is conceding: 'She just hasn't ... responded well to being famous.' Because Maupin makes the reader care about his characters, however, the use to which Mary Ann is put in the final volume (an embodiment of yuppie ambition and the failure of heterosexual imagination in the Aids crisis) seems like a betrayal, and, in consequence, is emotionally and artistically unsatisfactory. Brian, on the other hand, starts out as a narcissistic womaniser and ends up as the ideal house-husband, embodying most of the liberal virtues of 1990s New Man. This shifting of emphasis and sympathy, a reflection of the times in which the stories were written, is one of the most fascinating aspects of the largely triumphant and endearing sequence.

Although the novels would seem obvious source material for cinema or television, neither Hollywood nor any of the American networks would touch a project in which homosexuality played so prominent and approved a role. The end result was that this quintessentially American story was eventually made into a television series (1993 onwards) by Britain's Channel 4, which translated the novels to the screen with considerable success and without bowdlerisation. The series was subsequently sold back to American television, where it was watched by the largest audience ever for a drama series on public service broadcasting.

Nicholas MOSLEY   1923–

Hopeful Monsters

The critical reaction to Mosley's novels has always been mixed and, although *Hopeful Monsters* had been highly praised by some reviewers, this demanding, uncompromising book was an unexpected choice for the Whitbread Novel of the Year Award, which most people thought would go to A.S. Byatt's ***Possession*** (1990). It then went on to win the Whitbread Book of the Year Award, which the modest author (who claims to write books 'for a minority') gratefully acknowledged as 'recognition on a larger scale than I had grown used to'. The book comprises two halves, 'We know thepredicament' and 'So what do we do'. Both sections are divided into chapters, and these (with the exception of the final one) are narrated alternately by Eleanor Anders and Max Ackerman, who meet initially in 1928 at a performance of *Faust* in the Black Forest. Their stories converge, diverge and – ultimately – interlock. Coincidence figures large, as do thematic assonances. Indeed, although the novel is not epistolary, its narrators frequently address each other.

Eleanor's narrative begins in Berlin in the winter of 1918–19. It is a time of putsch, purge, inflation and near-starvation. Her father is a philosopher of science and an admirer and interpreter of Einstein's theories. Her mother, a Jew, is a member of Rosa Luxemburg's Spartacist Party. Hence Eleanor has a choice of perspectives: speculative detachment or dogmatic commitment; and, if she is baffled by the internecine duplicities of German Communism, the rise of Nazism simultaneously galvanises her into political activity.

Max, meanwhile, is being educated in

Cambridge. His father is a biologist and his mother a psychoanalyst. The one is preoccupied with the ecosystem, the other with the ego system – but Max is heir to the conundrum common to both their disciplines: What is it that determines the development of an individual or a species? More particularly: how does a mutation survive in a world that seems to lag a little behind it? Max's provisional solution is dalliance – academic and sexual. His choice is neither as stark nor as dangerous as Eleanor's. Nevertheless, he shares her basic 'predicament': he cannot adapt to the intellectual and social formation in which he finds himself.

Eleanor escapes from Germany in 1934. Her mother is in a concentration camp and soon to die. Max is in the Soviet Union (where Lysenko's lunacy is rampant). He and Eleanor have met only twice, but they meet on a third and decisive occasion in Spain during the Civil War. Max is a prisoner of the fascists and Eleanor (ironically) a fascist nurse. They marry in Pamplona and return to England. Max researches in the field of nuclear physics (eventually working with Oppenheimer). Revisiting Germany on the eve of the Second World War, Eleanor discovers that her father is employed by the Nazis in similar research; but he, she guesses, is operating as a saboteur. In a messy and unconvincing epilogue, Mosley tries to summarise the rest of the couple's lives, to integrate *Hopeful Monsters* with the balance of the *Catastrophe Practice* quintet (1979–90), of which it is the final volume, and to answer the titular question, 'So what do we do?'

A certain messiness is unsurprising, for Mosley's subject is nothing less than intellectual and political revolution in Europe. Marx, Darwin, Freud and Einstein provide the novel's compass points: Stalin and Hitler are its twin dark stars. Its rarefied (and mainly cerebral) drama derives from the impact of these turmoils on personal consciousness. Eleanor and Max must construct modes of thought and behaviour capable of negotiating the terrors and contradictions of their era. Thus, they are themselves the 'hopeful monsters': 'things born perhaps slightly before their time; when it's not known if the environment is quite ready for them'.

Mordecai RICHLER   1931–

Solomon Gursky Was Here

Along with the equally rumbustious work of Robertson Davies, Richler's fiction defies Canada's reputation as a land of quiet understatement. Richler's Canada is seen from a Jewish perspective, and he has frequently drawn upon his own family background for his novels. The long, teeming *Solomon Gursky Was Here*, with its enormous cast of characters and fragmented chronology, is an almost parodic family saga, which attempts to encompass Canada in some 400 pages. There was disagreement as to whether or not it succeeded. His first novel in ten years, it was shortlisted for the Booker Prize, but disappointed some critics, who found it ill-disciplined and unwieldy. The novel is a complex tapestry spanning five generations of the Gursky family, from the wild nineteenth-century forebear Ephraim – sometimes Durham miner, pickpocket, Newgate prisoner, cabin-boy on Franklin's ill-fated expedition to discover the Northwest Passage, conman, fur-trader and bogus religious leader – to the present generation of heirs ('the new Jewish royalty in America') squabbling over the family fortune. The family wealth was created by the three brothers, Bernard, Solomon and Morrie, out of bootlegging in the years of prohibition. Solomon's gambling and philandering led to a bitter rivalry between him and the avaricious Bernard. His disappearance and apparent death in a Gypsy Moth plane in 1934 during the brothers' trial is the central mystery of the novel, one now obsessively investigated by the drunken writer Moses Berger (whose poet father worked for Bernard).

Moses has dedicated the best part of his life to researching the history of the Gursky family, in which everything is radically uncertain and liable to misinterpretation. Solomon's journals come into his possession, and from them he learns of the occasion when Ephraim 'kidnapped' the boy Solomon from outside his school and took him on a sledge up to the edge of the polar sea. Rumours of how Ephraim managed to survive Franklin's expedition and joined up with the Eskimos as a faith-healer filter through the text. Likewise are there questions about Solomon's disappearance: was the plan blown up by Bernard, or did Solomon stage his own disappearance? What is the significance of the mysterious figure of a raven which recurs throughout? Was the dead raven on Bernard's grave put there by the surviving Solomon? None of these questions is ever conclusively answered.

Much of the novel's comedy derives from the ludicrous extremes of the rags-to-riches saga: at Bernard's seventy-fifth birthday junket Zero Mostel (one of many famous real people to appear in the text) sings 'If I Were a Rich Man', and a film montage shows the sod hut on the bleak prairie where Bernard was born, with the hated Solomon neatly air-brushed out of the picture.

John UPDIKE   1932–

The Rabbit Tetralogy

Rabbit, Run (1960)
Rabbit Redux (1971)
Rabbit is Rich (1981)
Rabbit at Rest (1990)

Updike's second novel was to turn into the first book of an eventual tetralogy. However, there is no evidence that this was his original intention, and *Rabbit, Run* can be read as a self-contained work. It was published to considerable acclaim, and several reviewers saw in it a knowing attempt to capture the mass market of middle-brow American readers. Of all Updike's previous fiction (which included a large number of short stories) *Rabbit, Run* was the most sexually explicit and seemed, in the context of its time, designed to shock.

The novel is divided into sections and tells the story of Harry 'Rabbit' Angstrom, a twenty-six-year-old who believes his finest hour (as a star of the college basketball team) is behind him. He is married to Janice, who drinks, watches TV and who, at the start of the book, is expecting their first baby. On impulse, Harry takes flight from this life into the arms of Ruth Leonard, a prostitute. The two set up house together despite the efforts of Eccles, the Angstroms' minister, to reconcile Harry with Janice. The second section takes up the story two months later, and offers insights into Harry's irresponsibility from Eccles's viewpoint when he visits Harry's unhappily married parents. Harry and the reader learn more about Ruth's sexual background, but Harry is humiliated by the knowledge, and humiliates Ruth in turn. Learning that Janice is in labour, he rushes to the hospital; this prompts a brief reunion, which ends when Janice rejects his sexual advances a few days later. Harry storms out, Janice turns to the bottle and inadvertently drowns their baby. The third section is shorter than the previous two and tells of Harry's reaction to this news. The novel's climactic scene at the cemetery finds Harry running again. He tries to take refuge in the arms of Ruth as before but, when he finds out that she is pregnant, he cannot accept the responsibility. The novel ends with Harry obeying the title's command, running mindlessly into the night.

Rabbit, Run is written in the present tense – appropriately, for Harry rarely registers past or future. His lack of responsibility is in part a lack of objectives and traditions. Neither good nor evil, Harry is simply unanchored and runs without any idea of direction. He is only successful sexually in spur-of-the-moment encounters, and cannot usefully link his own sexuality to any form of love or commitment.

Rabbit Redux finds Harry settled and reconciled with Janice. This time it is she who embarks on an affair, leaving Harry for Charlie Stavros, a car salesman. Harry takes care of their son, Nelson, begins an affair with Jill, a young runaway from a wealthy background, and shelters Skeeter, a black Vietnam veteran who introduces him to pot and politics. Harry gradually adapts to this anti-establishment view of America, but Skeeter becomes sexually involved with Jill and the *ménage-à-trois* is broken up by outraged neighbours who set fire to the house. Jill dies in the fire and at this time

Janice decides to leave Charlie for health reasons – their passion is a dangerous strain on his heart-condition. Janice and Harry are reconciled again, and the last scene of the book finds them sleeping together (without making love) in a motel.

Updike sets his characters amidst the 'big stories' of the time: Vietnam, race relations, inner-city decay and the increase of the drugs trade. Various specific events are alluded to: the race riots in York and Reading, the death of Mary Jo Kopechne at Chappaquiddick and the first manned space flight. All these link Harry's story to much larger events and movements and also provide lines of imagery and metaphors which contribute to a real sense of continuity in the book. *Rabbit Redux* is a more optimistic book than *Rabbit, Run*, for Harry seems more capable of learning from the events which befall him, but behind both is the same sense of America's slow social disintegration.

Rabbit is Rich opens in 1979. Harry is forty-six, reconciled with Janice once more, and moderately well off. The bad luck that dogged him in the first two books seems to have transferred itself to his son, Nelson, who is plagued by accidents, from car-wrecks to the pregnancy of his girlfriend. *Rabbit is Rich* is in part a book about one generation handing over the reins to the next. Harry has become more aware of the past and is afflicted by ghostly remembrances of his drowned baby daughter and Jill. When he meets Ruth Leonard again, she is described as a ghost, and the graffito slogan 'Skeeter Lives' haunts him from the walls of his town. The plot of *Rabbit is Rich* is the slightest in the tetralogy, chiefly concerned with Nelson's misadventures and the Angstroms' holiday in the Caribbean. It ends with Nelson's flight from the responsibilities of fatherhood (the suggestion of a grim cycle) and the birth of his daughter. As in the preceding novel, 'background' events are kept well to the fore: Carter's administration lurches with good intentions from crisis to crisis and Skylab falls towards its burn-up in the upper atmosphere.

Rabbit at Rest finds Harry and Janice in semi-retirement in Florida. It is 1988 and Bush is in the White House. From their condominium, the Angstroms contemplate a landscape of golf-courses, bingo-halls and malls. Nelson visits with his wife, Pru, two children and a cocaine habit fed by swindling the Toyota concession entrusted to him by Harry. Over Lockerbie, Pan Am flight 103 is blown up by a terrorist's bomb and Harry begins to brood on this event. A sailing accident precipitates a heart-attack and he and Janice head north once more to salvage their son and what remains of the car lot. Their efforts seem in vain. Nelson enters a drug rehabilitation programme, but the concession is lost and Harry flees back to Florida after cuckolding his son with Pru. The heart he carries with him is a bomb, however. Trying to recapture a moment of his youth on a basketball court, the bomb goes off, Harry succumbs fatally to his second heart-attack and Rabbit at last stops running.

The final book in the tetralogy takes congestion as its theme and death as its conclusion. Harry's arteries are as clogged with fat as his son's nostrils are with cocaine. The airwaves are clogged with junk TV and America itself with cheap imports. The political landscape is befogged by the legacy of Reagan's amiable haze; the economic bomb left unattended in the nation's hold has gone off, and the weight of the National Debt descends over the country.

Fittingly, Rabbit's last run is from the scene of his last coupling. His infidelity with Pru is flushed out and Rabbit along with it. The incident forces the truth of their situation on Nelson, Pru and even Janice, but it is no more than a last flurry for Harry. Of all the Rabbit tetralogy, *Rabbit at Rest* is most concerned with the body: Harry betrays him doubly by springing unbidden into life with Pru, then into death on the basketball court.

Sex is a troubled central focus in all four books and supplies fertile ground for Updike's moralising. The disintegrating landscape of America's urban and suburban sprawl, with its motels, fast-food joints and bars, is the result of a failure to cohere between its citizens. Sex, Updike implies, could be the elemental union leading to cohesion on a larger, social scale, but it has become detached from its function: pregnancies and births are always difficult, always getting in the way of the characters aspirations. Sex for pleasure alone is not 'evil', but it is not enough either. This traditional treatment of sex as a possible restorative of cohesion and tradition itself finds its counterpart in Updike's technique. Updike is an exceptionally gifted writer who chooses to write traditional novels about a world that rejects tradition. As such, form and content tend to work against one another. The discontinuities and disruptions that this mismatching produces are invariably smoothed over through Updike's skills as a novelist. This makes his novels resolutely contemporary (there is no nostalgia for 'the good old days'), easily readable (there is no experimentation) and, perhaps most importantly, immensely popular.

1991

Disintegration of Yugoslavia ● Coup against Gorbachev ● Demise of the USSR ● Republics of Croatia and Slovenia declare independence from Yugoslavia ● Robert Maxwell, publishing tycoon, dies after falling overboard from his yacht off the Canary Islands ● British government announces appointment of Mrs Stella Rimington as head of M.I.5 ● Repeal of South Africa's apartheid legislation ● British Prime Minister, John Major, launches the 'Citizen's Charter' ● British House of Lords rules that a husband can be guilty of marital rape ● Astronomers at Mt. Palomar, California, announce the discovery of the most distant object yet seen, a quasar ● Discovery in the Italian Alps of the preserved body of a man from c.3,300 BC, with clothes, bow, arrows, axe and other implements ● Michael Dummett, *The Logical Basis of Metapysics* ● 'Pop Art' exhibition at the Royal Academy, London; Sir Norman Foster, Sackler Galleries, Royal Academy, London; Sir Norman Foster, Stansted Airport, Essex; Robert Venturi and Denise Scott Brown, Sainsbury Wing, National Gallery, London ● Harrison Birtwistle, *Sir Gawain and the Green Knight* (opera) ● Alan Bennett, *The Madness of George III* ● Ariel Dorfman, *Death and the Maiden* ● Films: *Thelma and Louise, Silence of the Lambs* ● Television: Sonic the Hedgehog leads Sega's computer game war against Nintendo ● Popular music: Nirvana, *Nevermind* – emergence of grunge music from Seattle, US ● Margot Fonteyn, Graham Greene, Olivier Messiaen, Angus Wilson and John Cage die ● Robert Creeley, *Selected Poems*

Pat BARKER　　　　　**1943–**

Regeneration

see 1993

Harold BRODKEY　　　　**1930–**

The Runaway Soul

Few books have been as long awaited as *The Runaway Soul*. Brodkey's first collection of short stories came out in 1958, and was only followed by a second in 1989, but meanwhile reports circulated of a fabulous work-in-progress: a gigantic Proustian novel, which was rumoured to open with the line 'Imagine a mind shaped like a human body' (which is not in fact the first line of the published work, although a related phrase occurs in the second paragraph). This anticipation fuelled Brodkey's arcane but nonetheless towering reputation, and when the book was prematurely thought to be finished in 1977, the*New York Times* front page declared: 'Brodkey Delivers'.

The Runaway Soul is a book of massive

and virtually unprecedented introspection, a work of almost 850 pages that excavates the younger life of narrator Wiley Silenowicz from his uterine memories onwards. It is largely autobiographical, and both Wiley and another character, Ora, have already figured in Brodkey's short stories. Wiley's mother dies and he is adopted by foster-parents S.L. and Lila, both of whose voices are commemorated in the book at length. They already have a daughter, Nonie, who is to be a key figure in Wiley's life. His adopted father dies slowly of cancer when Wiley is fourteen and Lila takes up with a new man, Abe, who also subsequently dies.

It is a game like hopscotch to catch bits of sensation in memory and to advance in knowledge of the sensations while backing off from them. It is like some weird sport that isn't a famous sport.

THE RUNAWAY SOUL
BY HAROLD BRODKEY

The book proceeds plotlessly, in a series of examinations of moments and relationships. These include birth; masturbation; inter-course with the beautiful and almost super-human Ora ('to see her at the dinner table was to see Marxism die'); Wiley's sexualised relationship with his dying father; his quasi-homosexual relationships with his friend Remsen and with Abe's son Daniel; and his adopted mother's long deathbed mono-logue. Throughout the book Wiley is acutely aware of Nonie, the difficult and unlovable sister whom he has displaced in the family's affections, and who at times seems almost the incarnation of interpersonal evil. The book ends with a reprise of his feelings towards her after her death in a fire, for which the adult Wiley still feels almost grateful. No summary of *The Runaway Soul* could do justice to its texture, which can expand an act of intercourse or masturbation to the length of a novella with microscopic, moment-by-moment abstract analysis. Brodkey's writing is monumentally over-articulate, and yet at the same time spon-taneously colloquial and unpolished: 'I can see, as I curve through a mental arc inwardly (in an actual moment), the shape of how I

sort of kind of *hate* people who make me say no.'

The book's eventual appearance was something of an anti-climax. Many critics saw it as a white elephant and ridiculed Brodkey's style and egotism, although a few stood by it as a work of authentic greatness. Brodkey had been HIV positive for almost twenty years (by his own account, although this is an abnormally long incubation) and he attributed his development of full-blown Aids to the book's reception.

Margaret DRABBLE   1939–

The Gates of Ivory

see 1987

Mary GAITSKILL   1954–

Two Girls, Fat and Thin

Nothing prepares the reader for the impact of Gaitskill's first novel, a disquieting journey into the darkest corners of the female psyche. Its only precursor was a collection of short stories called *Bad Behaviour*, published in 1989 to great acclaim on both sides of the Atlantic.

I rode to school with five other girls, whom I remember mainly as knees tinted beige by pantyhose and arms clasped around books. Four of us were ugly and unpopular. When the car pulled up at the school, the one pretty, popular girl would leap out and walk ahead with frantic briskness so that no one would suspect that she had any connection with us.

TWO GIRLS, FAT AND THIN
BY MARY GAITSKILL

Two Girls is as far away from a story of cosy female friendship as it is possible to get. Told in alternating first and third person narratives, it opens with thirty-four-year-old Dorothy Never, a lonely, obese Wall Street proofreader, finding a message on the bulletin board at her local laundromat, placed by a journalist who is seeking followers of the cult novelist and thinker Anna Granite, for whom Dorothy once worked. The journalist is Justine Shade, a skinny, pretty twenty-eight-year-old who is writing a feature on Granite and her right-wing movement, 'Definitism', for one of New York's alternative magazines.

Their first meeting is awkward, with Dorothy as friendly and helpful as Justine is cool and defensive, although they awaken something in each other. The middle section of the book backtracks over their unhappy suburban childhoods, oppressed by the boorish complicity of their weak, posturing parents, and the cruel power games of the playground. Both have been sexually and emotionally abused – Dorothy by her father, Justine by a friend of her father. Dorothy suffered all the classic fat girl rejections, which drove her to the solitary consumption of huge quantities of junk food in front of the TV, the only place she felt safe. Justine became a streetwise manipulator who, at the age of ten was happily masturbating over books about SS torture, and who once sexually assaulted a classmate. Both women are masochists, but whereas Dorothy punishes herself by over-eating, Justine indulges in sado-masochistic sex. There is a constant aura of danger around her.

The last section of the book details Dorothy's discovery, at the age of seventeen of Anna Granite's greatest novel *The Bulwark*, and her conversion to the Granite faith. She joins the organisation as secretary and falls in love with one of its stalwarts, a kind man who loves her, but loves his fiancée more. Yet Granite's fundamental belief in the freedom of the individual becomes the rock on which Dorothy has built her life.

Justine works part time as an 'assistant secretary for a doctor of internal medicine' while nourishing her journalistic ambitions with sporadic feature writing. When the Anna Granite piece appears, Dorothy is horrified by Justine's betrayal of all she holds sacred. It is only when she spontaneously decides to confront Justine that she and we discover the depths of sexual degradation to which Justine has subjected herself with a sadistic boyfriend. Instead of berating her, Dorothy is forced to mount a rescue opera-

tion, and in doing so realises how spiritually parallel they are, locked inside prisons of their own making. Dorothy is without friends or lovers, Justine unable to form any relationships apart from unhealthy sexual encounters with strangers. Brought together by fate, a glimmer of hope suggests that they might be able to release each other from their obsessions.

Reviewers seemed a little awestruck by the force of the book's subject matter and language. The words 'disturbing' and 'perversely erotic' occurred over and over again, referring to the hard-core descriptions of Justine's sex life, but most critics were united in their praise for the book's scope, and for its muscular, vivid prose style. Some assumed (rightly) that Anna Granite was based on the cult novelist Ayn Rand (1905–82), and *The Bulwark* on her novel *The Fountainhead* (1943). Rand was a staunch advocate of individualism, and the intense, cranky atmosphere that surrounds charismatic cult heroes, is brilliantly evoked.

Two Girls is a disturbing book, desolate and affecting, but not without humour, and Gaitskill's skill at describing the isolation of damaged lives is as sharp as her eye for the vicissitudes of both New York life and Middle America. Every pretence is stripped away from her characters to reveal them in all their ugliness, beauty, pain and longing. Each relationship in the book is exploitative and suffocating, yet we know that the women will be victims only for as long as they continue to act out the role, and their stories are enmeshed with such authority that the element of doubt remains until the very last page.

Norman MAILER   1923–

Harlot's Ghost

Harlot's Ghost is a long and ambitious novel, the product of seven years' work, in which Mailer seeks to describe the history and inner workings of the CIA. He portrays an institution which is self-serving, incestuous and insanely bureaucratic, and ends the novel with the words of Lenin: 'Whom? Whom does all this benefit?' The novel incorporates many real people and draws on a number of historical incidents (with varying degrees of imaginative licence). Narrator Harry Hubbard is a CIA agent whose father, Cal, and godfather, Hugh Tremont Montague – code-named Harlot – were founding members of the agency. In later years, Harlot (who is loosely based on James Jesus Angleton) is the *éminence grise* of the agency, and Harry continues to serve him, often against the interests of his own bosses. Harry's wife Kittredge was married to Harlot until shortly after the death of their son in the early 1970s.

The first part of the novel describes one night in 1983 when Harry returns from his mistress to his home in Maine and learns of Harlot's apparent death and Kittredge's affair with another agent, Dix Butler. With fellow agent Arnie Rosen he concludes that Harlot has probably defected to Moscow. After Rosen is found dead – possibly killed by Butler – Harry goes into hiding. A year later he flies to Moscow in search of Harlot to discover if his mentor had been a Soviet infiltrator throughout his career. Like many other questions in the novel, this is one which remains unanswered.

The rest of the novel comprises Harry's memoirs of the years between 1956 and 1965. Much of the painstakingly detailed account of CIA operations comes in illicit letters to Kittredge. In 1956 he witnesses the exposure of the Berlin tunnel (possibly the work of Harlot) whilst working for the paranoid Bill Harvey and pretending to seek the identity of an agent he knows to be himself. Later he works for Howard Hunt during the disastrous Bay of Pigs invasion of Cuba, and at Harlot's instigation becomes involved with an air stewardess called Modene Murphy, who is the lover of J.F. Kennedy, Frank Sinatra and the Mafia boss Sam Giancana. (This episode draws explicitly on the life of Judith Campbell Exner.)

None of these operations ever seems to resolve itself, and Harry reflects: 'I found myself reading spy novels on evenings I spent alone, and they were satisfying in a manner that was never true of work with all its partially glimpsed projects, ops, capers, researches, stunts and scenarios, but then, the spy novels were never true to life.' *Harlot's Ghost* is itself perhaps too true to life to be wholly successful as a novel.

Ben OKRI   **1959–**

The Famished Road

Okri's narrator, Azaro, is a spirit child. He explains that beings such as he are able to will their own deaths and, before they enter the world of the living, always swear not to be too long absent from the paradise where they normally dwell. Those who break this promise are 'assailed by hallucinations and haunted by their companions.'

The world was still, as if it had momentarily become a picture, as if God were The Great Photographer. The clearing turned into a new world. Out of the flash came the sharp outlines of spirits rising into the air with weary heads. And then they fell down and bounced and floated over the stillness of the world. The spirits passed me, passed through me, their eyes like diamonds. And when the next explosion came, followed by another blinding flash, the spirits were obliterated. The heaviness of the air settled, the clouds opened, and the first torrential drenching of the land began.

THE FAMISHED ROAD
BY BEN OKRI

Azaro is born into an impoverished family in an unspecified West African city. He has no brothers and no sisters; his mother is a street trader, his father is a load carrier, humiliated and worn down by his menial work; home is a rat-infested room. All the same, Azaro chooses to stay with humankind. This decision angers his fellow spirits, and supernatural emissaries are sent to lure him back to 'the place of fountains'. Azaro's quotidian vision often dissolves; in its stead appear monsters or strangely deformed men and women who take the boy through disturbing, mutative lands. Conveyed with the rambling vividness of fever, these episodes provide a frequent interruption of the novel's naturalistic portrayal of an oppressed and violent existence where sickness is frequent, hunger habitual and drunkenness common.

Azaro's country is soon to be independent

of its white rulers and two parties – the Party of the Rich and the Party of the Poor – are vying for support in the imminent elections. Dad lets it be known that he will vote for neither organisation: each is as corrupt as the other. After this announcement. Dad finds it hard to gain employment, Mum's stall is attacked and the family's landlord raises the rent. A photographer is forced to flee the neighbourhood after recording the thuggery of the electioneers. In contrast, Madam Koto – proprietress of the local bar – prospers by cultivating a clientele drawn from the new political élite. She is the first person in the area to own a car and to have electricity supplied to her premises. Once, believing that he brought good luck, she encouraged Azaro to visit her and, once, Dad used to drink in her den. However, hostile to her powerful friends, Dad instructs his son to treat her as an enemy.

Isolated and desirous of prestige, Dad trains to be a boxer. He calls himself Black Tyger and fights three fierce challengers, one of whom is probably a phantom. His final encounter occurs at a party thrown by Madame Koto. A badly battered victor, Dad scarcely survives; but the book concludes with his recovery and with Azaro's declaration that: 'A dream can be the highest point of life.'

Thus, Azaro remains in the world. What detains him supremely is his affection for his parents, and it is this aspect of *The Famished Road*, rather than its excursions into the realm of the spirits, which supplies the story's most convincing theme. Indeed, Azaro's community – caught at the moment of colonialism's ending, when modernity collides with ancient custom – is sufficiently vibrant to require no fantastic support. Okri's triumph is to show the entrancing oddity of the mundane: against that measure, his ghostly diversions fail to persuade or entertain.

The Famished Road won the Booker Prize in 1991.

Iain SINCLAIR   1943–

Downriver

Downriver is a novel only in the expanded sense that a book like William Burroughs's **The Naked Lunch** (1959) is a novel, being a collection of twelve loosely-linked stories exploring the seamier and more arcane aspects of East London. It is held together by the questing presence of Sinclair himself (modestly described as looking like Lord Longford on amphetamines). A central incident in the book, which is strung on the Thames, is the sinking of the Victorian pleasure steamer *Princess Alice* with over six hundred deaths. One of those affected was an Elizabeth Stride, who lost her husband and children, became a prostitute, and fell victim to Jack the Ripper; the Ripper has long been one of Sinclair's imaginative obsessions. Also prominent in *Downriver* is a performance artist named Edith Cadiz, who works by day as a nurse and whose night-time striptease act includes a dog and some street maps (a neat image, perhaps, of Sinclair's feeling for topographical sleaze). Edith's disappearance is one mystery among many in the book; another is the vanishing of a synagogue caretaker named David Rodinsky, apparently leaving his abandoned Whitechapel room as a densely packed time-capsule from twenty-five years ago.

Sinclair also records his real-life quest after the memory of the poet Nicholas Moore. Other adventures include an involvement with repulsive media men (like the vapidly enthusiastic television producer Sonny Jaques, in a tale called 'Living In Restaurants'), and a nightmare journey on the Thames with a drug-ravaged boatman sporting the kind of US Confederate style favoured by a certain section of the British working class: 'The war had been lost. But they fought on: as electrical contractors, respray jockeys, pine strippers. The surviving remnant of Robert E. Lee's greybelly cavalry is hiding out in the swamps of East London.'

Sinclair has written that the original intention of *Downriver* was to 'redeem' certain aspects of his earlier novel *White Chappell Scarlet Tracings* (1987), a book which explored the 'psychotic geography' of the Ripper murders and which Sinclair later came to feel was 'dangerously phallic, predatory, unsated'. *Downriver* was originally to be called 'Vessels of Wrath', taking its punning title from an illustration of demons in Francis Barrett's 1820 magical primer *The Magus*: Sinclair took these very loosely to represent malign forces threatening the city of London. There are no shortage of malign forces in *Downriver*, from a train bearing nuclear waste through Hackney to the presiding presence of the Thatcheresque figure 'The Widow'.

The context of *Downriver* lies in the late Thatcher era, the yuppified late 1980s. Under a boom economy, excitement about the City and 'docklands' caused the pivotal point of London to shift Eastwards. This, combined with property gentrification, created enormous media and public excitement about East London. Sinclair has dealt with this situation in the book in chapters such as 'Isle of Doges [sic]: VAT City PLC' and 'Riverside Opportunities', and he has also written: 'These were evil days: so grim was the atmosphere that even the most determined hermit noticed something stinking in the cellar. When the tabloids preach banner headline celebration, look out.'

Critics and other writers, such as Michael Moorcock and Angela Carter, for the most part received *Downriver* with admiration for its satire, its reportage, its visionary qualities and its writing. As with the rest of Sinclair's work, cavilling was directed at its arcane qualities, its almost prurient preoccupation with things low and sinister, and its apparent faith in the visionary and supernatural as something more than metaphors. All these were equally regarded as strengths in other quarters, and there was similar disagreement about Sinclair's prose, which some found over-written.

1992

Bill Clinton wins US Presidential election ● British Conservatives elected for fourth consecutive term ● Serb and federal army forces begin bombardment of Sarajevo ● Four white policemen in Los Angeles are acquitted of beating a black motorist, 58 people die in riots and looting ● Five nights of serious rioting marks resurgence of anti-foreigner violence in eastern Germany ● US troops arrive in Mogadishu, Somalia, in operation 'Restore Hope' ● First woman speaker of British House of Commons ● First woman Director of Public Prosecutions ● Lloyd's of London announces record loss of £2 billion on one year's trading ● Polytechnics in Britain are given university status ● The UN holds a Conference on Environment and Development in Rio de Janeiro ● The first transplant of a baboon liver into a human ● The Church of England General Synod votes to allow women to be ordained to the priesthood ● Roman Catholic Church introduces new catechism ● Closure of *Punch* in Britain ● John Guare, *Six Degrees of Separation* ● Films: *The Player, Howard's End, Malcolm X, Orlando, Strictly Ballroom* ● Television: Cable TV reaches 60 per cent of US households ● Popular music: AIDS awareness benefit concert held at Wembley, in memory of Queen's Freddie Mercury ● Menachem Begin, Friedrich von Hayek, Isaac Asimov, Satyajit Ray, Francis Bacon, Marlene Dietrich, John Piper and Willy Brandt die ● Thom Gunn, *The Man with Night-Sweats*

Martin AMIS   1949–

Time's Arrow

Amis's seventh novel borrows its central conceit from a passage in Kurt Vonnegut's **Slaughterhouse-Five** (1969) in which the bombing of Dresden is imagined as a film running backwards. The narrator of *Time's Arrow* is a 'passenger or parasite' inside the mind of Tod Friendly ('I have no access to his thoughts – but I am awash with his emotions') who experiences Tod's life in strict reverse order. In this inverted world people walk backwards, absorb excrement, and regurgitate food which they assemble and take to the supermarket. Love affairs and financial transactions are particularly strange and comic seen from this point-of-view.

The early chapters of the novel chronicle Tod's years as a doctor in an American hospital where, from the narrator's perspective, healthy people are injured and made sick. In New York Tod changes his identity and becomes John Young, and in 1948 he

sets sail for Portugal where he becomes Hamilton de Souza. From intimations in Tod's dreams the narrator senses the imminent disclosure of a guilty secret: 'I *will* know how bad the secret is. I will know the nature of the offence.' In Rome Tod becomes Odilo Unverdorben and travels to Auschwitz 'shortly after the Bolsheviks had entrained their ignoble withdrawal'. In the chapter describing Unverdorben's work as a Nazi doctor the time-reversal works its greatest ironic effect in making the Holocaust appear as an act of benign creation: 'To make people from the weather. From thunder and from lightning. With gas, with electricity, with shit, with fire.' The barbarous experiments of the doctor known as Uncle Pepi are reversed and lauded by the narrator: 'In this new lab of his he can knock together a human being from the unlikeliest odds and ends.'

The novel's 'alternative title' is 'The Nature of the Offence' and much of the critical debate the book aroused was concerned with whether the book was of itself offensive. There were arguments amongst

Amis's readers about whether or not the Holocaust was a suitable subject for the essentially comic device of ironic reversal, and the author was accused by some of exploiting his subject. Indeed, some felt that the gentile Amis had merely 'appropriated' the Holocaust, and his printed acknowledgements to Jewish writers and friends proved particularly contentious, seen by some as tellingly defensive. Further offence was caused in some quarters by Amis's suggestion that the unexplained suicide of Primo Levi, who had survived the concentration camps to become one of their most distinguished chroniclers, was an act of 'ironic heroism'. While the novel had its share of articulate champions, perhaps more memorable were the comments and actions of its detractors. The *Spectator* flagged its damning review: 'Designer Gas Ovens', while the poet Tom Paulin savaged both book and author on television. After a ferocious argument which ended with the resignation of one of the judges, the novelist Nicholas Moseley, *Time's Arrow* was shortlisted for the Booker Prize.

Janet HOBHOUSE   **1948–1991**

The Furies

The Furies was the fourth and last of Janet Hobhouse's critically acclaimed novels, and arguably her best. The book was published posthumously, just a year after her death from cancer at the age of forty-two. She was still editing and revising the text when she died, so that although ostensibly the novel is complete, there are sections towards the end that might seem a little hurried or insubstantial.

Hobhouse was the daughter of a beautiful, unstable American mother, and a conventional English father. They separated when she was a baby, and she was brought up by her mother in New York until she came to England at the age of sixteen anxious to meet her father. Subsequently, she went to Oxford, married an Englishman, and became a writer, dividing her life between two countries and cultures. *The Furies* is almost entirely autobiographical, and the

book she had undoubtedly been working towards through the previous three: *Nellie Without Hugo* (1982), *Dancing in the Dark* (1983) and *November* (1987) – all of them sharply focused examinations of the complexities and contradictions of love, marriage and friendship in the late twentieth century.

The Furies moves us in even closer. Written in the first person by Helen and divided into four sections, it is essentially the story of a daughter–mother love of heartbreaking intensity. Helen and her wayward mother Bett, who was divorced from her English husband after only two years of marriage and forever after at the mercy of her search for love and happiness, are the survivors of a Jewish immigrant dynasty dominated by matriarchs. The history of the family's fall from 'plutocracy in Frankfurt to poverty in 50's New York' across seventy-five years is beautifully observed in a lengthy prologue. Helen's childhood years with Bett in New York – where both characters are made vulnerable through financial hardship, philandering men and a desparate need for respectability – are sustained by their mutual love and dependence, but darkened by Bett's unreliability and depressions. At sixteen, Helen escapes to England in search of her father, and although she lives with his family for a year, he remains cold and unapproachable. On her own initiative, she carves a brilliant career for herself out of Oxford, the 'right' contracts, and love and finally marriage (after a stormy on/off relationship) to Ned, an old Etonian. For a while, they are the archetypal charmed couple, cushioned by material wealth; but increasingly they grow out of step with each other, as passion and disenchantment collide, and the marriage disintegrates in the wake of Bett's suicide – a tragedy from which Helen never really recovers.

The last section of the book, 'Alone', is a stampede of tragedy. Helen has an affair with a famous actor, and loses her home, her car and her possessions, which have been in storage, through a swift succession of eerie accidents. She retreats to Cape Cod to finish a book, and when she feels healed enough to return to the world, it is to find that she has cancer. Although Hobhouse neatly reverses Helen's diagnosis in a coda, we already know what wishful thinking this must have been on her part.

The Furies was almost universally praised. The only reservations were chirrups of unease about the searing intensity of the emotional content (some English critics are never comfortable with this) and the unevenness of tone resulting from hasty completion – scant criticism for a writer who excelled in describing the vicissitudes of human behaviour, and wore her heart proudly on her sleeve. The prologue, like some of her previous work, has been compared to the writing of Edith Wharton and Henry James, because of its 'irony, distance and control'. Self-pity is markedly absent, although self-loathing is evident alongside an indefatigible love of life. Her famous detachment only flies apart after her mother's death, and the last third of the book is an angry, painful, disquieting post mortem on Helen and Bett's fractured life together, an elegy to what was and what might have been. *The Furies* has been likened to a love letter from the author to her mother, and an attempt to come to terms with the guilt she felt about the harshness of their final exchange, which happened in life the way it happens on the page. Eloquent, haunting, and deeply moving, this is a unique and important book.

Cormac McCARTHY 1933–

All The Pretty Horses

The year is 1949, and John Grady Cole, aged sixteen, sees no reason to stay in San Angelo, his Texas home town. He and his friend, Lacey Rawlins, saddle their horses and ride south. They meet an enigmatic younger boy whose name may or may not be Jimmy Blevins and whose horse may very possibly be stolen. Rawlins feels uneasy about Blevins from the first, but the three cross together into Mexico and continue south, hoping to find work beyond the Sierra del Carmen. Blevins's behaviour, however, becomes increasingly erratic and, after a sequence of bizarre events, he is separated from the other two.

Cole and Rawlins ride on alone. Soon they are employed on a ranch, breaking wild horses: this is the kind of life that they had hoped to discover. Then, in spite of being warned against it, Cole has an affair with Alejandra, the ranch owner's daughter. Brief and intense, the liaison is interrupted when Alejandra is sent to Mexico City and Cole and Rawlins are arrested for no immediately clear reason. In a village gaol, they are reunited with Blevins, who has killed a man. The sadistic police captain in charge of the case interrogates Cole and Rawlins, insinuating that they are guilty of murders and theft. With Blevins, they are to be transferred to Castelar prison but, during the transfer, Blevins is shot. Cole and Rawlins barely survive Castelar's brutalities, and their release is only secured when Alejandra's great aunt, Alfonsa, pays a suitable bribe.

Rawlins returns to Texas; Cole, however, returns to the ranch. Alfonsa tries to explain to him something of the tragedy and allure of her family. She implies that Alejandra's emancipated rebellion is doomed amid her country's atavisitic pride. Undaunted, Cole arranges a rendezvous with the girl and asks her to marry him. She rejects his proposal. He returns to the village where he was initially incarcerated, and regains possession of his, Rawlins's and Blevins's mounts, which the police captain had confiscated. He takes them back to Texas, where he tries unsuccessfully to find the owner of Blevins' horse. His future uncertain, he finally rides into the West Texas desert and 'the world to come'.

Detached from his family and from his North American surroundings, Cole is a paradigm of rootlessness: what saves him from complete deracination is his loyalty to Rawlins and his unsentimental affinity with horses. If the former provides a moral bond, the latter is a practical attachment; yet, at a more profound level, the horse connects the novel's leading characters to the history of the lands they traverse. Blevins's fatal error is to act within the traditions of a gunslinging past, whereas Cole and Rawlins conform to the heroism of determination and endurance. To that extent, the three protagonists are anachronistic in their deeds and values. Nevertheless, in its fractured modernity McCarthy's prose is capable of registering the tension between a simple ethical system and the complexities with which it must cope.

All The Pretty Horses is the first volume of *The Border Trilogy*.

Toni MORRISON   1931–

Jazz

Set mainly in Harlem in 1926, *Jazz* describes the causes and effects of a fatality. Most of the story is told by an 'I', 'silent and unobservable', neither a participant in the action nor a reliable judge of it. Thus, whereas this 'I' declares herself a 'know-it-all', she is plainly not equivalent to Morrison, who on occasions allows the characters to utter their own opinions, observations and deeds.

Joe Trace is fifty-two and he has just shot Dorcas, his eighteen-year-old lover. He escapes prosecution because 'nobody actually saw him do it' and because Alice Manfred, the dead girl's aunt, believes that Joe's sorrow is 'as bad as jail'. Violet, Joe's wife, in a state of temporary derangement, attends Dorcas's funeral and attempts to slash the corpse's face. Ejected from the service, she returns to her apartment and sets free her collection of pet birds. Later, she tries to punish Joe by taking a boyfriend. When that strategy fails, she tries to regain Joe's affection by pampering him. That, too, is unsuccessful – so Violet decides to find out what she can about Joe's victim.

She visits the beauty parlour that the girl used; she talks to the girl's former teachers; she even learns the steps that the girl danced. She overcomes Alice's initial hostility and discovers from her that 'this neice had been hardheaded as well as sly'. Nevertheless, it is only when Felice, Dorcas's contemporary and closest acquaintance, visits Violet and Joe that the truth of the shooting seems to be settled. The judgement passed on Dorcas is harsh; but Felice's presence and perspective bring relief to Joe, who all this while has stayed fixed in his grief.

The novel, however, does not restrict itself to a linear version of events, for, in a sequence of retrospections, it discloses aspects of its protagonists' pasts. Dorcas's parents were killed in a riot; oppressed by poverty, Violet's mother went mad; Joe never knew his father and never persuaded his (probable) mother to admit that he was her son. Each of the three has directly experienced racism and each has left the rural south for the comparative prosperity and safety of New York. Dorcas has received an education. Joe and Violet have entered the lower middle class – he as a waiter and a cosmetics salesman, she as a hairdresser. What befalls the trio, though, is in part a consequence of the demands that the city makes and the freedoms that it offers: unlike the labour of the fields, urban life is both improvised and organised, both individual and collective. Dorcas, Violet and Joe mishandle these contradictions, yet the manner in which they do so is determined by the sufferings they have earlier endured.

A secondary narrative, its connection more thematic than mechanical, is arranged within the primary one. It is tantalisingly inconclusive and digressive: an element of its function is to link Morrison's structure to her title. If it also rephrases and reinterprets several of the novel's sharpest questions, then it does so at the risk of dissipating in tension what it displays in virtuosity.

Michael ONDAATJE   1943–

The English Patient

The English Patient takes place in a half-ruined Tuscan villa north of Florence, in which four isolated characters are stranded during the last days of the Second World War. Hana is a young Canadian nurse; the cynical and superficially dashing David Caravaggio is a shady thief and spy; and Kirpal Singh is a Sikh bomb-disposal expert. 'The English patient', meanwhile – who may not even be English – is a silent and bedridden man who has been unrecognisably burned in a plane crash over the Libyan desert.

Before the war, the English patient's life was the stuff that Boys' Own tales of Empire are made of. He was an intrepid desert explorer, and his story is narrated in flashbacks under the influence of morphine. Ondaatje's poetic descriptions of desert exploration – and his own psychological exploration of the rather warped personality that finds deserts attractive – are among the most impressive parts of the novel. It is put in a very different perspective by the implicit contrast with the far more useful heroism of Singh (known as 'Kip', with its echo of Kipling), the colonial underdog who builds bridges, clears mines and defuses bombs. The difference between the two provides a study in the nature of heroism.

All four characters are uncertain in their identities and identifications. Caravaggio has been psychologically damaged by German torture, while Hana is looking for a father figure and projects this role on to the unknowable English patient, whom she nurses like a dutiful daughter. She falls in love with Kip after seeing him defuse a bomb, and it is Kip's own crossed identifications which, along with his contrast to the maimed explorer, are at the heart of the book. Kip has turned his back on Indian culture and embraced the civilisation of his Imperial masters, like his old mentor Lord Suffolk, who taught him bomb disposal. His brother, meanwhile, is an Indian nationalist who hates the English.

In the arboured bedroom the burned patient views great distances. The way that dead knight in Ravenna, whose marble body seems alive, almost liquid, has his head raised upon a stone pillow, so it can gaze beyond his feet into vista. Farther than the desired rain of Africa. Towards all their lives in Cairo. Their works and days.

THE ENGLISH PATIENT
BY MICHAEL ONDAATJE

Kip's consciousness is violently transformed when radio news comes through that the atomic bomb has been dropped on Hiroshima. He is so shocked by this – and there is an elegant irony about his being a bomb-disposal expert, but powerless in the face of this one – that he turns violently against the 'White man' and his civilisation. He realises that his brother has been right all along, almost attacks the English patient, and leaves the villa.

Ondaatje's novel is a very powerful evocation of a ruined post-war landscape, with mines and booby traps everywhere and Italian churches reduced to shelled wreckage. Even more than that, though, it is an evocation of the ruins of Empire. As a Sri Lankan-born Canadian, Ondaatje is particularly sensitive to questions of identity in a post-Imperial world. This post-colonial theme made the novel particularly suited to the Booker Prize, which is often perceived to value non-European and Commonwealth concerns. *The English Patient* shared the 1992 Booker Prize with Barry Unsworth's **Sacred Hunger** (a novel about Empire and slavery), and the splitting of the prize between the more meditative Ondaatje novel and the relatively swashbuckling one by Unsworth only heightened the already extensive cynicism and weariness around the Prize.

The English Patient was much praised for the poetic intensity of its writing and imagery, for the haunting beauty of its desert scenes, and for the incisive and elegantly constructed post-Imperial thematics. Its detractors, on the other hand, believed it to be over-written (Ondaatje's likening of a penis to a seahorse achieved a certain notoriety) and a little simplistic in its censure of European civilisation. To agree fully with the book's indictment one would have to accept that the English were culpable for the (American) bombing of Hiroshima, and that such a bomb would never have been dropped on a white enemy; although in fact the ultra-intensive British fire-bombing of civilian Dresden and the ensuing 'fire storm' is probably more outrageous and less defensible. Critics in the middle ground detected elegiac mixed feelings in the book itself, and an ambivalent respect for the strange and perhaps ultimately futile beauty of English values.

Donna TARTT   1964–

The Secret History

Few first novels receive the sort of pre-publication hype that surrounded *The Secret History*. Here was a 524-page novel that had taken a twenty-eight-year-old unknown eight years to write, and then sold for a staggering $450,000, plus lucrative foreign and film rights. A seven-page profile in *Vanity Fair* ensured that Tartt was a star before her book even went on sale. It became the novel everyone wanted to hate, but many readers were surprised by how much they liked it.

The Secret History could be described as a campus murder mystery, but, more accurately, it is a literary thriller that is elegant and erudite and totally beguiling. From the first page we know that a murder has been committed and by whom, but to discover why we are admitted into a sealed, privileged world of intellectual and material snobbery which, in the course of the book, slowly comes apart at the seams. The story is narrated by Richard Papen, an impoverished Californian classics student who aspires to an academic life that he knows only an old established East Coast college can provide. He wins a scholarship to Hampden, an exclusive liberal arts college in Vermont (loosely based on Bennington, Tartt's own *alma mater*), and becomes infatatuated by an elite coterie of five sophisticates studying classical Greek under the tutelage of the charismatic and eccentric scholar, Julian Morrow. Papen manages to ingratiate himself into the group by demonstrating his versatility with Greek semantics, but he is an impressionable, flawed innocent among these exotics.

Rich, spoilt and arrogant, the group's natural leader is Henry Winter, a manipulative charmer who wears 'dark English suits'. The other members are Francis, a wealthy homosexual with a country estate; the beautiful, incestuous twins, Charles and Camilla; and the preppy odd man out, Bunny. Richard is allowed only so far into the group, sharing idyllic country-house weekends and parties. Unknown to him, Henry, Francis, Charles and Camilla have been testing out Julian's theories on the pleasures of losing control, by indulging in a series of bacchanals to evoke the spirit of Dionysus. Fuelled by alcohol, drugs and euphoria, they finally reach a state of frenzy which results in the murder and mutilation of a local farmer. As the outsiders, Richard and Bunny are excluded from this bloodfest, but

when Bunny discovers what has happened, he threatens to tell, and Richard is drawn into a collective decision to silence him.

The rest of the book relates the aftermath of the murder: the search for and discovery of the body, questioning by the FBI, Bunny's funeral (a splendid set piece), and the slow disintegration of brilliant young minds cast adrift from their intellectual moorings into the real world of guilt, remorse and moral recrimination for which none of them is equipped, and from which there is no return.

In the hands of a less skilful writer, the narrative – with its bacchanalian origies, 1980s aesthetes out of Evelyn Waugh, and a proselytising mentor claiming to have been on first name terms with Ezra Pound, the Sitwells *and* Marilyn Monroe – might have deteriorated into whimsy. But Tartt juggles plot, setting and characters (including some marvellous cameos who occasionally upstage the main cast) with the authority of a veteran.

For the most part the critics ran out of superlatives, but the dissenters were united in their grievances: the book was too long, over-written and pretentious, with a surfeit of literary allusions and Greek and Latin phrases. Such critics overlooked the fact that as an ambitious, large-scale concept, the book demands an overview, and it was interesting that more women than men seem to have been impressed. If it reads like Barbara Vine's '*A Fatal Inversion* rewritten by Scott Fitzgerald', as one critic said, does it really matter? Tartt's cunning balance of erudition and suspense is seductive. She wants us to be flies on the wall. She invites us to sample the best and worst of American WASP society in the bud, and if she interrupts the story with a little Nietzsche here, and a touch of Plato there, first novels are allowed a few indulgences. Furthermore, she displays a wonderful gift for black humour when she focuses attention on the campus hoi polloi surrounding her gilded group.

The Secret History is an astonishing debut that defied all the laws of publishing gravity. Whether Tartt stays the course as a novelist hardly matters. Years from now, her book will still have an originality, bravura and brilliance all too rare in literary novels that dare to be popular.

Jeff TORRINGTON   1935–

Swing Hammer Swing!

A former factory worker, Torrington was fifty-seven when this, his first novel, was published. *Swing Hammer Swing!*, which won the Whitbread Book of the Year Award, is set in Glasgow at the end of the 1960s. The twenty-eight-year-old narrator Tam Clay is a would-be writer currently on a self-allocated sabbatical from paid employment and living in the 'sub-world of means testees, tribunalees, and appeals panelees'. He is one of the last residents of the soon-to-be-demolished Scobie Street in the Gorbals, and angrily laments the destruction of the old community which 'they'd bundled off into the asylum of history with all the furtive shame of a family of hypocrites dumping Granny in Crackpot Castle'.

Snow rasped under my tread as I made my way down Scobie Street. I passed Greasy Tam's knackered fish'n chip shop. The place hissed with fractured water pipes and although it was too dark to see I knew that the big salmon that'd swum above the frying pans lay now in a shatter of blue scales on the floor. Tam'd kicked the bucket shortly after he'd clinched the padlock on his premises. Rumour has it that he died from a 'sudden bath' though this isn't very likely. He'll be down in Hell right now peeling totties for Heaven's endless round of purveys. It was odds-on too that he still hadn't changed his apron.

SWING HAMMER SWING!
BY JEFF TORRINGTON

In a richly inventive prose, which mixes the Glasgow demotic with allusions to Plato, Nietzsche, Pascal and Sartre, Tam describes five shambolic days in the run-up to Christmas. His wife Rhona is in the maternity ward awaiting the arrival of their first child, and her respectable sister Phyllis denounces Tam as 'a tramp mooching from one day to the next, no pride in his appearance, not an ounce of ambition in his head'. Tam is already being chased by debt-collectors, and various mysterious events increase his paranoid sense that some malign force is pursuing him. He scours the Glasgow pubs for a man who has disappeared with his £50 winnings from a horse, and narrowly escapes a severe beating when tricked into the house of four card-playing heavies who mistake him for a Catholic. On a museum trip Phyllis's young son temporarily disappears with an apparent Tam *doppelgänger*. Pretending to be his friend Matt Lucas, Tam picks up Becky McQuade, and after a raucous night of passion learns that her enormous lorry-driver husband is pursuing Lucas. In the event, Lucas is accidentally run down by a bubble car while dressed as a mummy as a promotional stunt for the Planet Cinema. The episode embodies Tam's sense of Glasgow as a surreal vaudeville act:

as wee Matt lay groaning on the slushy pavement with authentic slow-welling bloodstains creeping out amongst the spurious red-ink ones, a pair of fat wifies in the Gossip and Grumbles mould had stopped to do a bit of rubbernecking. 'My, would you look at that, Senga!' says one to the other. 'They've got'm wrapped in bandages already. By jings, that was quick, was it no?'

Barry UNSWORTH   1930–

Sacred Hunger

Like many of Unsworth's novels, *Sacred Hunger* (his tenth) has an exotic historical setting – in this case, the eighteenth-century slave trade. Unsworth began researching the book in 1985, when he was a writer-in-residence at the University of Liverpool. His reaction to working in a depressed inner city in Thatcher's Britain – he had spent much of his adult life abroad – was to suffer writer's block. He overcame this and used it by writing *Sugar and Rum* (1988), a novel in a contemporary setting about a man who finds himself unable to write his novel about the Liverpool slave trade. *Sugar and Rum* cleared the way for the massive *Sacred*

Hunger, which was joint-winner of the Booker Prize.

Matthew Paris is a doctor who has recently been bankrupted and imprisoned for having published his seditious views on evolution. After the death of his wife he seeks escape by becoming ship's surgeon on his uncle William Kemp's newly built slaver the *Liverpool Merchant*. William's son Erasmus nurses a grudge against his cousin dating from a childhood incident when the older Paris lifted him from a dam he was trying to build on the beach. As the boat sails, Erasmus reluctantly becomes involved in a country house production of *The Enchanted Isle* (a version of *The Tempest*) in order to court Sarah Wolpert, and is unaware of his father's financial difficulties.

On the Guinea Coast, Paris tries to alleviate the suffering of the slaves but does not question the morality of the trade. He rebels against Captain Thurso's authority only when the ship is becalmed on its passage to the West Indies and Thurso begins to throw the slaves overboard so as to be able to claim the 30% insurance on their value (which he could not do were they to die of dysentery). When the ship fails to return it is presumed lost; Kemp is ruined and commits suicide, and Erasmus is forced to call off his engagement to Sarah.

Twelve years later, in 1765, Erasmus – now a prosperous sugar merchant who has married for money – learns that the *Liverpool Merchant* has been sighted beached in Florida. Sensing what may have happened, he travels to Florida, where he gains the assistance of the garrison to track down and take his revenge upon Paris, whom he blames for his father's ruin and death. Inspired by the free-thinking Frenchman Delblanc, who took passage with the ship, the surviving crew and slaves from the *Merchant* have established an egalitarian community; but Paris now sees the profit motive once again setting man against man. When he is taken prisoner by Erasmus, he explains his mutiny thus: 'It was impossible not to see that we had taken everything from them and only for the sake of profit – that sacred hunger, as Delblanc once called it, which justifies everything, sanctifies all purposes.' Paris dies from the wounds he received during his capture, and Erasmus, denied the hanging he desired, realises that for all these years 'the thought that Paris might still be in the world had given his life meaning and purpose'.

The success of the novel led to Peter Hall's commissioning Unsworth to adapt it for a projected eight-part television series. Unsworth also plans to write a sequel set against the background of the Industrial Revolution.

1993

President George Bush and President Boris Yeltsin sign START II Nuclear Arms Pact • Inauguration of President William Jefferson Clinton • Russian troops put down revolt against President Boris Yeltsin • NAFTA (North American free trade agreement) signed • Bosnian peace talks stalled • North Korea pulls out of nuclear non-proliferation treaty and refuses UN on-site inspection • Israel and Vatican formally establish ties • IBM announces loss of $4.97 billion, largest single-year loss in US corporate history • Branch Davidian religious cult compound in Texas burned after 51-day siege with loss of 72 lives • Key cancer gene identified • Films: Steven Spielberg, *Jurassic Park*, Martin Scorsese, *The Age of Innocence*, Jane Campion, *The Piano* • Michael Jackson accused of child sexual abuse • Arthur Ashe, Willy Brandt, Sammy Cahn, Alexander Dubcek, Dame Elizabeth Frink, William Golding, Audrey Hepburn, Kenneth MacMillan, Bobby Moore, Rudolf Nureyev, Anthony Perkins, Freya Stark, Mary Wells die • Adrienne Rich, *Collected Early Poems, 1950–1970*

Pat BARKER   1943–

Regeneration
1991

The Eye in the Door
1993

'The Somme is like the Holocaust,' Barker has said. 'It revealed things about mankind that we *cannot* come to terms with and cannot forget. It never becomes the past.' The earliest novels about the First World War were largely written by men who were attempting to come to terms with their own experiences of the trenches. Later in the century, however, women writers of a younger generation have been drawn to the war, notably Susan Hill and Jennifer Johnston, both of whom Barker admired when she began her writing career. Although her own two novels about the war (volumes of a projected trilogy) are largely concerned with officers, they are clearly connected – principally through the experiences of one character, Billy Prior – to her earlier novels of contemporary working-class life, such as **Union Street** (1982). In dealing with political and sexual heterodoxy (specifically pacifism and homosexuality), they are also part of what might be called an undercurrent of fiction written by women about the war, which includes Rose Macaulay's **Non-Combatants and Others** (1916) and, particularly, A.T. Fitzroy's *Despised and Rejected* (1918), a morally brave but artistically negligible novel, which was successfully prosecuted and suppressed under the Defence of the Realm Act.

Regeneration takes its title from a medical experiment carried out before the war by two doctors, William Rivers and Henry Head, two of the real people Barker includes amongst her characters. 'The nerve supplying Head's forearm had been severed and sutured, and then over a period of five years they had traced the process of regeneration.' This process, with its two phases – the 'protopathic' and the 'epicritic' – is linked by Barker to the work Rivers did during the war, when it was his job to assist the regeneration of soldiers suffering from 'neuraesthenia', a blanket term which covered many types of breakdown resulting from battle trauma. Rivers worked at Craiglockhart War Hospital outside Edinburgh, where his most famous patient was Siegfried Sassoon, who had been sent there after making his 'Soldier's Declaration', in which he publicly announced his refusal to participate further in the war. It had been thought expedient to declare Sassoon neuraesthenic in order to defuse his declaration and avoid the scandal of having to court martial a notably brave officer.

The action of the novel is set at Craiglockhart, where patients are encouraged to dredge up and confront the traumas which brought them to the hospital. They may have escaped the trenches physically, but, like Christopher Marlowe's Mephostophilis, they carry hell with them: there is the army doctor who can no longer stand the sight of blood; an officer who has convinced himself that he is paralysed; Capt. Burns, who is unable to eat; and Second-Lieutenant Billy Prior, who has become mute. One of the many ironies of the war is that, unlike a civilian doctor, Rivers is obliged to 'cure' these men so that they can be sent back to the front and almost certain death, and the novel ends with Rivers writing 'Discharged to duty' on Sassoon's file.

Apart from Sassoon (and Wilfred Owen and Robert Graves, who make cameo appearances), the patients are fictional, although Barker has drawn upon Rivers's accounts of real cases. Rivers's empathy with his patients (like many of them he has a stammer, almost certainly the result of psychological trauma), and his gentle coaxing are contrasted with the efficacious but barbarous treatment meted out in London by (the real) Dr Yealland, who relies upon electrical shock therapy. The principal fictional character is Prior, an officer from a northern working-class background, who soon recovers his voice, but takes much longer to come to terms with the horrors of what he has witnessed in the trenches, specifically the occasion he was obliged to pick up the eyeball of one of his men who had been blown to pieces by a shell. His relationship with Rivers, which is frequently antagonistic, grows more equivocal as the story progresses: which is the doctor, which the patient?

In *Regeneration*, Prior falls in love with a munitions worker, Sarah Lumb. *The Eye in the Door*, however, opens in London with his engaging in a graphic bout of sex with Charles Manning, an officer whose leg has been severely damaged. Bisexual, a 'temporary gentleman', now working in intelligence supposedly against his pacifist friends, Prior inhabits a sexual, social and moral no-man's land. The novel's epigraph is taken from *The Strange Case of Dr Jekyll and Mr Hyde* (1886), in which R.L. Stevenson's protagonist is forced to admit that he is present in both incarnations. During the novel, Prior suffers from extensive blackouts during which the sadistic impulses he usually manages to suppress (except in dreams of battle from which he awakes ejaculating) come to the surface. Rivers, who is now based in London, likens Prior's 'split personality' to that of Sassoon (the recklessly brave soldier and bitter pacifist) and doctors such as himself and Head, who have to maintain a balance between the genuine sympathy they feel towards their patients and the detached scientific interest they take in them. Similarly, combatants have to be trained to suppress their humane and moral instincts in order to kill the enemy. A further split, which develops the related theme of masculinity from *Regeneration*, is that officers are expected to display all the aggressive qualities traditionally associated with manliness, but also to develop a maternal attitude towards the men in their charge. Rivers notes that many of the mental disorders these men develop are strikingly similar to those suffered by women in peacetime.

This apparently straightforward, but in fact highly complex and sophisticated novel (which was awarded the *Guardian* Fiction Prize) takes its title from a glass eye placed in the doors of prison cells to remind the inmate that he or she is under permanent scrutiny. Falsely convicted of plotting the assassination of Lloyd George, Beattie Roper languishes in a cell, where she is visited by Prior, whom she was partly responsible for bringing up. Prior wants to expose Spragge, the *agent provocateur* upon whose evidence Roper was arrested, but, as a former combatant, his attitude to pacifism is equivocal. It later transpires that during one of his blackouts Prior has betrayed another figure from his past, the pacifist agitator, Mac.

Others, too, are being watched. Prior is apparently being trailed by Spragge, while someone is sending Manning newspaper clippings concerning the famous 'Cult of the Clitoris' case, in which the maverick MP,

Noel Pemberton Billing, used his own newspaper to publicise the preposterous allegations of a man called Harold Spencer, who was later declared insane. Spencer stated that the Germans had a notebook containing the names of 47,000 prominent British men and women, whose sexual 'perversions' laid them open to blackmail. Billing suggested that many of those attending the actress Maud Allen's performance in Wilde's *Salomé* would be in the book. Allen's subsequent libel action (which she lost) resulted in one of the most scandalous and farcical trials in the annals of British justice.

Barker's deft weaving together of the real and the invented leaves the reader unsure of what is fact and what is fiction in the novels, although she appends notes giving sources for her principal borrowings. 'Though I know I'm walking a moral tight-rope,' she has said, 'I'm very careful to be accurate. I think I'm actually more restrained than a lot of biographers. I don't change facts.' Indeed, what she has done is to reimagine history, and in this skilful balancing act she scarcely falters.

The third volume of the trilogy will follow the story of Prior and his return to France, and will deal with Rivers's quest, while treating a patient who has regressed to early childhood, to discover what it was in his own childhood that has robbed him of his powers of visualisation. Barker intends to contrast the 'primitive' warfare Rivers had studied as an anthropologist in Melanesia with the 'civilised' warfare in Europe, and to trace the effect of the war upon the Europeans' faith in the moral authority of their culture.

Joan BRADY   **1939–**

Theory of War

Published quietly at the beginning of 1993, and reviewed enthusiastically, though not as widely as it deserved, *Theory of War* made such an impact on everyone who read it that it quickly earned a considerable reputation: it was eventually shortlisted for the *Guardian* Fiction Prize and won the Whitbread Book of the Year award.

Brady is an American who has been domiciled in Britain for over twenty years;

Theory of War is her second novel. Based on fact, it is an extraordinary fictional recreation of the life of her grandfather who, although a white child, was sold into slavery at the end of the American Civil War – a 'boughten boy' who suffered such cruelty at the hands of his guardians that, two generations later, his family are still unable to shake off his legacy. The selling of white children to poor farmers looking for cheap labour was a sordid and now conveniently forgotten fact of American history, occurring at a time when black slaves had just been liberated but impoverished widows and ex-soldiers who were unable to provide for their children were forced to sell them off.

Jonathan Carrick is four years old when he is sold for $15 to a brutish Midwest tobacco farmer called Alvah Stoke. He is a bright, warm, friendly child who is systematically beaten, tethered, taunted and starved by Alvah and his family, particularly Alvah's wife and eldest son George. It is George who senses everything good and clever in Jonathan and sets out to humiliate and destroy him. But Jonathan's natural intelligence and will to survive eventually outwit his tormentors, and at the age of sixteen, following a particularly sadistic episode when he has had all his teeth forcibly removed, he escapes westwards and finds work on the burgeoning railroads as a brakeman. After the death of his only friend, he travels east to Maine, where the friend's family of teachers take him in and educate him. He converts to Christianity, and for a time, becomes a low-church preacher. At thirty-six, he marries and finds a semblance of peace – but only just, and not for long. George Stoke is a senator by this time – corrupt politics have served him well – and Jonathan is able to follow his career and his whereabouts through newspapers. Jonathan realises that his entire life has been fuelled by hatred, and the need to revenge himself upon George, but revenge, when it becomes possible, is swift and final, and does not bring the hoped-for release. The scars are more ingrained than even Jonathan knew.

Jonathan's story is told by his middle-aged, wheelchair-bound English granddaughter, who has flown from England to Washington State to visit her alcoholic Uncle

Atlas, Jonathan's last surviving son. She needs to piece together the facts through his memories and Jonathan's own account in his coded diaries. Few of his relatives have remained unscathed: four of his seven children have committed suicide. His granddaughter needs to reconcile herself with some irrefutable truth, in spite of her uncle's insistence that impermanence is the order of things. Brady's narrative bristles with passion and anger, breathing life into the landscape, characters and language of pioneer America. Jonathan's flight west and his career as a brakeman (one of the most dangerous jobs of all, requiring nerves of steel and split-second timing) are particularly vivid, contrasting the barren prairies with the squalor of the boom towns, where drinking and whoring are the main pastimes.

Theory of War is a disturbing book. The catalogue of cruelty that comprises the first half is a careful preparation for Jonathan's planned escape and the cat-and-mouse stalking of George that becomes part of his life, and which Brady likens to the strategic planning of a military campaign: 'A war between two people is not all that different from a war between two countries.' Cynics might argue that a background like Brady's is something of a gift to a novelist, but few writers could manage the juggling act between intensity and objectivity that she sustains with such assurance and dexterity. There was much praise for her precise, original prose, keen ear for dialogue and her marvellous evocations of time and place.

Adverse criticism was reserved for her metaphorical use of the narrator's infirmity, and particularly for the von Clausewitz philosophy of war, whose rules she rigorously applies to Jonathan's campaign of revenge against George. The narrator and her uncle (and it is as much their story as Jonathan's) can seem intrusive at times, but they are necessary lynchpins – not just as examples of the Carrick legacy, but because without them, the reverberations of Jonathan's tragedy would not be felt. Brady carefully sidesteps any temptation to sentimentalise or play to the gallery, and there are few opportunities for tears, although plenty for shock. No tidy, happy endings either, and few philosophical musings on the nature of life and death, apart from Atlas's own impatience with anything even bordering on certainty. 'Life is a whirlpool, a vortex,' he says. '... you were born and then you die. If that's tragedy, life is tragic. But what's important is the rotating centre. Where's your sense of wonder? The pattern is alive. Infinitely complex.' Explanations are, finally, absent. Only the grossness of human brutality and the dignity of the human spirit remain.

The Stokes family are a cast of grotesques. There is no emotional centre to Alvah Stoke. He is a blank page, an automaton without fear, knowledge or feeling. The apple that hasn't fallen very far from the tree is George – Alvah with an education, yet infinitely more dangerous and deadly. George learned early on that if you 'lie, scheme, cheat – break every rule you've ever been taught, cross every line that's ever been drawn – you win'. Brady adds: 'This is the secret of life. What greater secret can anybody possess?' But George's secret has been intuited by Jonathan many years before. George is intelligent, crude, streetwise, but what he lacks is Jonathan's independence of spirit which can't be taught or bought at any price. George wants to possess and destroy that quality in his 'brother'. There is even an element of sexuality in his obsession, from which Jonathan recoils. We begin to understand why Jonathan thinks he will never be free until George is dead, little dreaming that it will cost him his soul.

Roddy DOYLE　　　　1958–

Paddy Clarke Ha Ha Ha

Doyle's remarkable evocation of an Irish childhood was a popular winner of the 1993 Booker Prize. Two years previously. *The Van* had been shortlisted for the Prize, and the two other novels in his 'Barrytown Trilogy', *The Commitments* (1987) and *The Snapper* (1990), had been made into highly successful films. Even before it was on the shortlist, *Paddy Clarke* was selling well and swiftly emerged as favourite to win against such strong competition as Carol Shields'

The Stone Diaries and David Malouf's *Remembering Babylon*.

Like much of Doyle's other work, *Paddy Clarke Ha Ha Ha* is set in the fictional Barrytown, drawn from Kilbarrack, a northern suburb of Dublin, where the author grew up and where he worked as a teacher for fourteen years. Accepting the Booker Prize, Doyle said that Kilbarrack 'offered me a huge, deep pool of inspiration: I dived in and very happily drowned.' The novel is set in 1968, and seen entirely through the eyes of ten-year-old Patrick Joseph Clarke. Although Paddy is the same age as his creator, Doyle insists that the book is not autobiographical. Prompted by 'a rush of memories' which came to him after the birth of his own first child, Doyle started writing the book by making notes about his central character, imagining himself into the world of a ten-year-old by compiling lists of his likes and dislikes, his vocabulary and expressions, and the larger part of the novel takes the form of a mosaic made up of brief, vivid scenes from this boy's life. We see him in school under the genial tutelage of Mr Hennessey ('Henno'); messing about on building sites with his gang of friends – Kevin Conroy, Ian McEvoy, James O'Keefe, and Liam and Aidan O'Connell, who have 'a dead ma'; lighting fires; playing football; shoplifting; bullying his little brother Francis ('Sinbad'); snubbing the boys from the Corporation houses; administering 'dead legs'; and fantasising about becoming a missionary in a leper colony.

Gradually, it becomes apparent that the marriage of Paddy's parents is breaking down. He cannot understand why this should be: 'She was lovely. He was nice.' He tries to defuse the tension within the house by making his parents laugh, but realises that he is powerless to provide anything other than temporary distractions. The previously cosy atmosphere of the house begins to seethe with a suppressed violence far more frightening than anything indulged in by the gang. The novel ends with Paddy's father, who has moved out, visiting the house to deliver Christmas gifts. Father and son greet each other with heartbreaking formality, and one senses that, to some extent, Paddy Clarke has grown up. When his heartless

schoolfellows, who have ostracised him after his bloody but inconclusive fight with Kevin, set up the chant from which the novel gets its title – 'Paddy Clarke – / Has no da. / Ha ha ha!' – Paddy comments: 'I didn't listen to them. They were only kids.'

It must have been great being mental. You could do anything you wanted and you never really got into proper trouble for it. You couldn't pretend you were mental though; you had to be that way all the time. No homework either and you could slobber your dinner as much as you wanted.

PADDY CLARKE HA HA HA
BY RODDY DOYLE

The novel's great strength lies in its unnerving recreation of what it is like to be ten years old. Readers who grew up in the late 1960s will recognise much of the carefully integrated period detail – *The Man from U.N.C.L.E.*, George Best, plastic sandals, the boy with polio – as very accurately observed. The language is extremely vivid, reflecting Doyle's observation that 'the way to really get characters alive is to get them to open their mouths'. Doyle also conjures up some startling images entirely apt for a child: a jelly fish, for example, is 'a runny umberella'. His laconic narrative perfectly captures the destructiveness and the casual brutality of children running wild in an anarchic world of the imagination, as when Aidan is gradually sinking into the mud at the bottom of a building-site trench:

> I checked Aidan's leg. His knee was covered now. He was sinking. He was trying to hold onto the side, trying not to fall, trying not to cry. He'd been crying earlier; you could tell from his face. I thought about throwing stones at him, but there was no need.
>
> We sat on our school bags.
>
> – Can you drown in mud? Ian McEvoy asked.

The novel is a remarkable work of ventriloquism, extremely funny and occasionally very moving, as in the shifting relationship between Paddy and 'Sinbad' (who proves unexpectedly and gallingly skilled at football). While some readers felt that the book had more charm than substance, Lord Gowrie, Chairman of the Booker judges, predicted that it would come to be seen as one of the great classic novels of childhood.

David MALOUF 1934–

Remembering Babylon

Set in Queensland, Australia – and, for the most part, in the middle of the nineteenth century – *Remembering Babylon* describes the startling arrival of Gemmy Fairley in a community of colonising farmers. His first contact is with three children: Lachlan Beattie and his cousins, Janet and Meg McIvor. Stammeringly, Gemmy declares himself to be British. It is plain that he is illiterate, while his capacity for speech is so limited that it is only with difficulty that the settlement can assemble an account of his life. He is around thirty years old. Until he was about ten, he was a rat-catcher's assistant in London; he spent a period at sea until he was 'put overboard' on the North Australian coast; he has spent the last sixteen years wandering with a tribe. Thus, he is referred to as 'the black white-feller'. His ambiguous status troubles the farmers, who fear that he may be a spy for malevolent natives. Furthermore, in the midst of a civilising project, he is a reminder of mankind's essential barbarity and an intrusion of that very wilderness which it is the colonisers' task to conquer.

Gemmy's tale is written down by the local schoolmaster and Gemmy is granted accommodation by Janet's and Meg's parents. Nevertheless, his presence in the community provokes suspicion, division and – after he is visited by myalls – violence. He is sent to stay with the enigmatic Mrs Hutchence but, in spite of her kindness, he feels himself isolated and his situation oppressive. Having obtained from the schoolmaster what he thinks are the pages of his biography, he returns to the wilderness. His eventual fate remains uncertain. Nearly a decade later, though, Lachlan Beattie comes to believe that Gemmy was killed by cattlemen in 'too slight an affair to be called a massacre'.

The novel's closing chapter gives a glimpse of Lachlan and Janet McIvor in old age. The First World War is being fought in Europe and Lachlan, a Minister of the Crown, has embarrassed the Government through his peripheral involvement in an absurd espionage scandal which has also touched Janet. Lachlan means shortly to resign his post.

Around the central narrative, Malouf constructs an allusive portrait of nascent Australia. He delineates the insecurities and ignorance of the settlers, the inappropriate ambitions of the administrators and the detached passivity of the aboriginals. Yet the book's principal concern is neither historical nor sociological; rather, it is to suggest – from a number of perspectives – the ways in which being, consciousness and environment may infuse each other. If Gemmy has been brutalised by his London upbringing, he possesses a transcendent ability to recognise the spiritual in nature. Simultaneously, he is 'a ragged fragment' of the landscape, and he arouses in Lachlan and Janet a love which encompasses much more than any one person or any one place. Complex, dense and poetic, *Remembering Babylon* was shortlisted for both the Booker Prize and the *Irish Times* International Fiction Prize.

E. Annie PROULX 1935–

The Shipping News

Proulx entered the public consciousness in 1992 with her first novel *Postcards*, of which the *New York Times Book Review* declared: 'She has come close to writing the "Great American Novel"'. Similar accolades flowed from critics on both sides of the Atlantic, and in 1993 she became the first woman ever to win the coveted PEN/Faulkner Award. In the same year, her second novel, *The Shipping News*, created equal excitement and won the 1993 *Irish Times* International Fiction Prize, the National Book Award, and the Pulitzer Prize, catapulting its fifty-eight-year-old author from backwoods obscurity into a literary celebrity.

The Shipping News is a novel that defies

categorisation, as Proulx's unique voice unravels the tale of Quoyle, a lonely, gauche, red-haired giant of a man, reared for failure by a boorish father, yet still clinging to expectations while lacking the impetus to change anything. Living in a small town in upstate New York as a hopeless newspaper reporter when the book opens, Quoyle stumbles into a disastrous marriage with a nymphomaniac. Six years later he finds himself first orphaned and then widowed when his absconding wife and her lover are killed in a car crash after she has sold their two young daughters to a paedophile photographer.

Into the breach, after the children's rescue, steps his father's sister, Agnis Hamm, a tough, independent, long-lost aunt who proves to be the family's salvation. Nursing private griefs of her own, she decides they should all decamp to Newfoundland, their ancestral home, and begin life afresh. Quoyle, at first reluctant, realises he has nothing to lose, and travels north hopefully to a storm-lashed, fog-wreathed, inhospitable new world.

'Now, what I want you to do. I want you to cover local car wrecks, write the story, take pictures. We run a front-page photo of a car wreck every week, whether we have a wreck or not. That's our golden rule. No exceptions. Tert has a big file of wreck pictures. If we don't have a fresh one, we have to dip into his file. But we usually have a couple of good ones.

THE SHIPPING NEWS
BY E. ANNIE PROULX

Quoyle's metamorphosis from wimp to hero provides Proulx with the core of her book, which is as much about Newfoundland and its wonderfully eccentric, gamey populace as it is about Quoyle and his getting of wisdom. As the odd quartet of Quoyle, Aunt Agnis, and daughters Bunny and Sunshine settle into the harbour town of Killick-Claw, Agnis sets up her own yacht upholstery business and Quoyle finds work reporting the shipping news for the local newspaper, the *Gammy Bird*, which specialises in tabloid sleaze and the gory details of automobile smashes. Thus Quoyle is perfectly placed for a crash course in community relations.

In Killick-Claw, Proulx introduces us to an inbred fishing colony, desperately trying to come to terms with the late twentieth century and a government who have ensured that fishing rights have been distributed to the world, killing off the locals' livelihood, and leaving unemployment and poverty in its wake. Most of them are acclimatised to a life locked in skirmish with the elements, but none of them is prepared for the barren wastes of redundancy. Proulx creates a rich gallery of old and young salts with names as emblematic as their lives: Tert Card, Nutbeem, Billy Pretty, Wavey Prowse, Dennis and Beety Buggit, Alvin Yark, Diddy Shovel, Bayonet and Silver Melville. Each of them has a history as faceted and complex as Newfoundland's coastline (similarly studded with aptly-named locations like Capsize Cove, Misky Bay, Cape Despond and Port Anguish), yet each in their turn assists Quoyle's jerky passage towards upward mobility and self-fulfillment.

These are lives forever teetering on the brink of disaster, both natural and man-made. In due course, Quoyle's old family home is literally wrenched from its moorings and blown away; he discovers a dismembered body; almost drowns at sea; and along with the rest of the population, assumes that his boss at the newspaper *has* drowned, until against all odds, the old man revives at his own funeral. Proulx has such a direct, unsentimental way of dispatching these details that they become as integral to the story as the town's high incidence of incest, and its addiction to a dreary diet of squidburgers and fried bologna. In the midst of such turbulence, Quoyle's growing attraction to the widow Wavey Prowse, with whom he will finally find peace, and the enjoyment he begins to find in his children, offer a touching and welcome diversion. Everyone in this book is battling with demons, and we are allowed to eavesdrop on them all. 'We face up to awful things because we can't go around them, or forget them,' Aunt Agnis tells Quoyle, and he does, overcoming every shock and setback, until he reaches a truce with his soul.

Proulx's rich and innovative prose fuels her narrative. Her short sentences, often shorn of conjunctions and verbs, stab straight into the heartland of her characters. They are equally successful in her evocations of a Newfoundland so vividly alive that you can feel its weather, smell the salt and snow and the swirling fogs, hear the crack of ice floes. It was interesting to note, however, that academically trained reviewers had trouble accepting Proulx's idiosyncratic prose style. Some called it 'challenging', as if reading it might be the literary equivalent of scaling the Matterhorn, while those who were able to set grammatical rules aside and accept the book on its own terms found a work of constantly unfolding delights, crackling with invention, humour and skill.

Proulx prefaces each chapter with an illustration and quote from *The Ashley Book of Knots*, which served as her metaphorical inspiration for the events of Quoyle's life and beyond. Far from homespun, the knot philosophy is a constant pleasure and surprise – like signal flares guiding us away from Quoyle's despair into the warmer waters of possibility.

The Shipping News is an exhilarating book brimming with ideas, stories, places and people – the narrative can barely contain them all. Proulx is that rare thing, a writer suffused with the joy of life *in spite of* its tragedies. As she says: 'I came to writing late, and I'm racing against the clock to get everthing down. My head is jammed with stories.' Not only that, but she is a true original in love with the potency of language.

Vikram SETH · 1952–

A Suitable Boy

Beginning in 1950, and spanning a period of about eighteen months, Seth's vast novel describes the fortunes of four families: the Mehras, the Kapoors, the Chatterjis and the Khans. In Brahmpur, capital of Purva Pradesh, the wedding of Savita Mehra to Pran Kapoor is celebrated. There, Mrs Rupa Mehra, Savita's widowed mother, is already concerned that her younger daughter, Lata, should find – or have found for her – an appropriate spouse: one who shares the Mehras' middle-class standing and Hindu religion. Lata, who is studying English at Brahmpur University, is not quite as pliant as her mother would wish. Her first infatuation is with a fellow student, Kabir Durrani, who proves to be a Muslim. Appalled at the prospect of such an alliance, Mrs Mehra takes her daughter first to Calcutta then to Delhi. In the latter city, Lata is introduced to Haresh Khanna, a supervisor in a shoe factory. Haresh, Mrs Mehra decides, is the husband-to-be. Lata, though, has not forgotten Kabir; also, while staying in Calcutta, she has met Amit Chatterji. The oldest of the five Chatterji children, Amit is a published poet and, after prompting from his glamorous sisters, he becomes the third of Lata's suitors.

Each of the three has much to commend him: mentally and temperamentally, Amit is the most compatible; Haresh is determined, candid – if naïve – and energetic; but Kabir has stirred the deepest passion. Thus, Lata must choose not merely between individuals but between different types of marriage: for, whereas Amit offers intellectual companionship, and Haresh security and kindness, Kabir implicity proposes the mysteries of unchartable romance.

Eventually, Lata makes her choice, and the story concludes, as it began, with a wedding. Around this simple and central plot, however, Seth arranges a number of other narratives, the cumulative effect of which is to present a portrait of a volatile moment in India's past. Independence is just a few years behind; Pakistan is bellicose; Nehru's Congress Party is divided over policy towards the minority Islamic population. Some, like Pran's father, Mahesh Kapoor (Minister of Revenue for Purva Pradesh), wish to promote tolerance: others, like Mahesh's rival, Agarwal, favour repression. At the same time, Mahesh sees the need for reforms which will adversely affect many Muslim zamindari. He guides the relevant bill through Purva Pradesh's Legislative Assembly, yet manages to retain the friendship of the Nawab Sahib of Baitar, owner of vast estates amd patriarch of the Khans. When a rift occurs, causing Mahesh to lose his political power, its origin is a bizarre and bloody misunderstanding involving Mahesh's and the Nawab's elder sons and a courtesan, Saeeda Bai.

That the story should contain a thread of melodrama is hardly surprising since, in scope and execution, it is reminiscent of Victorian realism: it locates its protagonists' private concerns within a public panorama; it is alert to the nuances of etiquette, manner and taste; its author's voice is assured; its egregious length (1,349 pages) assumes the reader's leisure. Wilfully anachronistic, it runs the risk of drawing attention more to its virtuosity than to its content, less to its substance than to its facsimile charm. Paradoxically, its great achievement is to revivify a dying form.

Although praised by one reviewer as 'the most fecund as well as the most prodigious work of the latter half of this century', *A Suitable Boy* was not nominated for the Whitbread or Booker Prizes: Lord Gowrie, chairman of the Booker judges, commented: 'It's like a good movie that's been uncut ... All the wrong bits are in and the right bits are out.' This opinion notwithstanding, the novel won the W.H. Smith award in 1994.

Carol SHIELDS   **1935–**

The Stone Diaries

Although well known in North America as the author of four novels and two volumes of short stories, Shields remained unpublished in the UK until 1990, when her fifth novel, *Mary Swann*, was launched to considerable critical acclaim. For someone who had not published her first novel until the age of forty, her elevation into the front rank of Canadian writers, alongside such contemporaries (and admirers) as Margaret Atwood and Alice Munro, was swift. *Happenstance* (two earlier novels bound together in one volume, 1991) and *The Republic of Love* (1992) were rapturously received. *The Stone Diaries*, the seventh and most ambitious of her novels, was shortlisted for the Booker Prize and won the Governor General's Award.

At first glance it looks like a conventional biography of one woman's life, but it is a novel that imparts its secrets like a series of Chinese boxes. Daisy Goodwill, who thinks herself an unexceptional Canadian 'Mrs Average', feels the necessity to authenticate a life in which she feels nothing much has happened. Yet, as occasional narrator and full-time participant, she reveals a life rich in quirky incident and history.

Born in rural Manitoba in 1905 to a mother who dies in childbirth, Daisy selects passages from ninety years of her life and times, while musing on the limitations of autobiography. She eschews the more obvious events such as education, two world wars and the birth of her children, in favour of a childhood fostered out to a neighbour called Clarentine Flett, who later leaves her husband to live with her botanist son, Barker, who will become Daisy's second husband. Her first, the eligible Harold Hoad, leaves the marriage abruptly unconsummated on their honeymoon when he is killed falling out of a window. Marriage to the much older Barker Flett, is long and happy. There are three children, an unexpected late flourish as a gardening columnist, and old age in a retirement condominium in Florida, before illness and death. The exigencies of change flow through the book like a *leitmotif*. Lives follow no patterns or rules, and everyone behaves, more or less, out of character.

This is no morbid tale of domestic unease, however. Daisy's autobiography is enlivened by album photographs, a family tree, letters, lists, recipes and asides and opinions from her family and friends. 'In a sense,' Shields says, 'I see her as one of life's fortunates, a woman born with a voice that lacks a tragic register. Someone who's learned to dig a hole in her own life story.'

By the end of Daisy's life, her three children and eight grandchildren have their own messy lives and marriages, whose eccentricities are recorded with the detailed flourishes at which Shields excels. She relishes the minutiae of everyday life. These are the poignant clues which provide the key to her characters. Thus we learn that Daisy's mother, Mercy Stone, blissfully unaware that she is even pregnant, is making a Malvern pudding when she goes into labour. We know the fashion and fabric details of Daisy's bridesmaids' dresses, her own exhaustive list of wedding lingerie. In later life, she outlines her routine as the gardening columnist Mrs Green Thumb. Even when she is dying, she writes out recipes and makes lists – of addresses where she has lived, favourite books, things she 'Must Do'. There is not an ounce of wastage in Daisy's life until her children grow up and leave home, her husband dies, and she loses her job on the paper. Only then, as her sense of purpose fades, does she give in to a breakdown, which is an accumulation of depression stifled by the years she set aside for other people. It is a grief she is unable to share, which returns her, for the first time, to the sadness of her lonely birth.

Infirm at the end of her life and haunted by feelings of insubstantiality, Daisy feels betrayed by the cranking down of her body into a decline for which her mind is not prepared. Trapped in a hospital bed, desperate to discuss life's remaining imponderables to which she can find no answers, she is at the mercy of staff and visitors, while her children send useless gifts such as inflatable giraffes, and bedjackets ('a bedjacket speaks of desperation, and what it says is: toodle-oo'). Yet that rich inner life is dismissed by her familiars. 'She just let life happen to her,' one of her daughters says after her death, but only we can decide if Daisy's life has been a triumphant act of resistance or a surrendering to circumstances.

Reviewers were unanimously consistent in their praise for the book. All were impressed by Shields' autobiographical gymnastics and by the resonance of a voice which provided distance and intimacy in equally baffling measure. With percipience and wit, *The Stone Diaries* captures perfectly the mood and language of its times in an irresistibly seductive prose style, which is both hospitable and confiding. It is as much a loss to the reader when Daisy writes herself out of the story, as it is to her uncomprehending family. Shields' territory has always been the human heart and its awkward, unpredictable geography. Where she differs from her contemporaries is in her gift of elevating the commonplace into something vivid and extraordinary. In Daisy Goodwill, landlocked by happenstance, she has produced an unforgettable character and a novel of subtle warmth and charm.

Author Biographies

Achebe, [Albert] Chinua[lumoga] (1930–)

Achebe studied medicine and literature and was director of the Nigerian Broadcasting Company's external services. Committed to the Biafran cause during the Nigerian Civil War, he and his family narrowly escaped assassination when their flat was bombed. He has edited *Okike*, Nigeria's leading journal of new writing, since 1971, and has taught English and African studies at various universities. In 1990 a car accident left him paralysed from the waist down. His novels include *The African Trilogy*, which established his reputation (*Things Fall Apart*, 1958; *No Longer at Ease*, 1960; *Arrow of God*, 1962) and *Anthills of the Savannah* (1987), whose themes reflect his own experience of being torn between European and African culture. He has also written poetry, essays and children's books.

Ackerley, J[oe] R[andolph] (1896–1967)

As long-serving literary editor of *The Listener* (1935–1959), Ackerley attracted both controversy and the very best contributors, amongst whom were Leonard and Virginia Woolf, E.M. Forster, Wyndham Lewis and Christopher Isherwood. His own work, most of which is strongly autobiographical, includes drama, poetry, travel writing and fiction: *Hindoo Holiday* (1932) describes his experiences as private secretary to an eccentric maharajah; *My Dog Tulip* (1956) and *We Think the World of You* (1960) describe his relationship with his Alsatian dog; and *My Father and Myself* (1968) recounts the discovery of his father's extraordinary double life.

Ackroyd, Peter (1949–)

At the age of twenty-three, Ackroyd became literary editor of (and later film critic for) the *Spectator*; since 1986 he has been chief book reviewer for *The Times*, and has been called one of the last English men-of-letters. He is a master of pastiche and has developed a form of historical fiction combining past and present and creating intriguing 'dialogues with the dead'. His novels include *The Great Fire of London* (1982), *The Last Testament of Oscar Wilde* (1983) and *Hawksmoor* (1985), a popular and critical success. He has also written poetry and an acclaimed biography of T.S. Eliot (1984), although the reception for his biography of Dickens (1990), which included passages of imaginary conversation between Dickens, his characters, Ackroyd, Wilde and Blake, was somewhat mixed.

Agee, James [Rufus] (1909–1955)

Born of a poor family in Tennessee, Agee graduated from Harvard and worked principally as a journalist. One of his best-known books, originally intended as an article for *Fortune*, was *Let Us Now Praise Famous Men* (1941), a study of Alabama sharecroppers during the Depression. His screenplays, including the one for John Huston's *The African Queen* (1951) have been published with his film criticism as *Agee on Film* (1960, two volumes). His Pulitzer Prize-winning unfinished semi-autobiographical *A Death in the Family* (1957) was published two years after his death, which came after a series of heart attacks.

Aldington, Richard [i.e. Edward Godfrey Aldington] (1892–1962)

Imagist poet, novelist and contentious biographer, Aldington was a collaborator with Ezra Pound and was married to H.D. His experience of the trenches resulted in the rancorous novel *Death of a Hero* (1929), which remains perhaps his best book. His biographies of both D.H. and T.E. Lawrence proved controversial, while too many of his other books were written principally to make money, of which he was always short.

Amis, Kingsley [William] (1922–)

After Oxford and service in the army with the Royal Corps of Signals, Amis taught English at Swansea and Peterhouse, Cambridge, before becoming a full-time writer. He first became known as one of the Movement poets along with Larkin, Conquest and Jennings. The same anti-romantic spirit was adopted in his first novel, *Lucky Jim* (1954) which established Amis amongst the 'angry young men'. He has adopted some of Jim Dixon's virulent philistinism for his increasingly reactionary public persona, delighting in lampooning pretentious, 'arty' liberals and the cultural establishment. His other novels include *Take a Girl Like You* (1960), *The Green Man* (1969), *Ending Up* (1974), *Jake's Thing* (1978) and *The Old Devils* (1986). Amis has also written short stories, science fiction, and books on food and drink. He is the father of the writer Martin Amis.

Amis, Martin [Louis] (1949–)

The son of Kingsley Amis, he immediately established his own literary credentials with his first novel, *The Rachel Papers* (1973), an hilarious account of adolescent sexual obsession. He shares his father's preoccupation with the way people express themselves and has created numerous grotesque characters, notably in *Dead Babies* (1975) and *Money* (1984). Later novels concern themselves with large moral issues: nuclear weapons in *London Fields* (1989) and the Holocaust in *Time's Arrow* (1992), and these have proved even more controversial than his earlier books. He has also been a literary editor and journalist, collecting stories and articles in such volumes as *The Moronic Inferno* (1986).

Anand, Mulk Raj (1905–)

Art historian, critic and one of India's most important novelists, Anand made history in 1935 with *Untouchable*, the first Indian novel to have a member of the lowest caste as a hero. It was influenced by Gandhi, and subsequent novels have further explored the lives of those on the lower rungs of Indian society. His other books include a trilogy drawing upon his father's experiences as a soldier in the First World War: *The Village* (1939), *Across the Black Waters* (1940) and *the Sword and the Sickle* (1942).

Anderson, Sherwood (1876–1941)

Best known for his volume of interconnected short stories, *Winesburg, Ohio* (1919), which went to make up a portrait of a Midwestern town, Anderson was an important influence on American writers of a later generation. He wrote several novels, such as *Poor White* (1920), but his stories, collected in such volumes as *The Triumph of the Egg* (1921) and *Horses and Men* (1923), are considered his finest achievements.

Arlen, Michael [i.e. Dikran Kouyoumdjian] (1895–1956)

Born in Bulgaria of Armenian parents, Arlen was educated in Britain and went on to write several novels of Mayfair life, the only one of which to have had a lasting reputation is *The Green Hat* (1924), which was considered vary daring in its time. He also published numerous short stories set in a similar milieu and collected in such volumes as *These Charming People* (1923). Although a naturalised British subject, he spent much of his life in France and eventually settled in New York.

Arnim, Elizabeth [Mary Annette] von (1866–1941)

Born in Sydney, Australia, a cousin of Katherine Mansfield, von Arnim made her name in 1898 with the anonymous and hugely successful *Elizabeth and her German Garden*, based upon her life as châtelaine of her Prussian husband's *Schloss* in Pomerania. She also wrote a number of novels, including the epistolary *Fräulein Schmidt and Mr Anstruther* (1907) and *The Enchanted April* (1922).

Ashford, Daisy [i.e. Margaret Mary Julia Ashford] (1881–1972)

Ashford wrote her comic masterpiece, *The Young Visiters* (1919), as a nine-year-old child in Lewes. Much of the humour may have been unconscious and although she also wrote *Where Love Lies Deepest* (1920) and *Love and Marriage* (1920), her talent did not mature into adult creativity.

Atwood, Margaret [Eleanor] (1939–)

Atwood made her reputation as one of Canada's leading poets before writing the feminist novels, such as *The Edible Woman* (1969), *Surfacing* (1972) and *The Handmaid's Tale* (1985), for which she is internationally known. She is also a critic, short-story writer and an editor of both poetry and prose, this varied work forming a programme of analysis and polemic in which she explores the key issues of feminism and nationalism.

Auster, Paul (1947–)

The New Jersey-born author of a series of highly original, post-modernist 'metaphysical thrillers', such as *The New York Trilogy* (1985–1986), Auster is concerned in all his writing with the nature of identity and obsession. His other novels include *The Invention of Solitude* (1982), *Moon Palace* (1989) and *The Music of Chance* (1990).

Bagnold, Enid [Algerine] (1889–1981)

Best known for her popular novel of horse racing, *National Velvet* (1935), Bagnold established her reputation with an unflinching account of her experiences as a VAD nurse during the First World War, *Diary without Dates* (1917). She wrote several other novels, including *The Loved and the Envied* (1951), the tragic story of a society beauty, but had more success as a playwright, notably with *The Chalk Garden* (1955).

Bailey, Paul [i.e. Peter Harry Bailey] (1937–)

The unexpected child of elderly parents, Bailey was brought up in Battersea, as he records in his memoir *An Immaculate Mistake* (1990). He started out as an actor, then wrote his first novel, *At the Jerusalem* (1967), which is set in an old people's home and won a Somerset Maugham Award and an Arts Council Prize for the best first novel published between 1963 and 1967. His comic and compassionate novels, which have won many awards, are usually set in a richly evoked London, often among the dispossessed. He is also a critic and broadcaster and has written *An English Madam: The Life and Work of Cynthia Payne* (1982).

Bainbridge, Beryl [Margaret] (1934–)

Born in Liverpool and originally an actress, Bainbridge writes with dramatic economy and pace. Acutely observed, peppered with ironic black humour and often dealing with the tragedy and comedy of human self-delusion, her novels include *The Dressmaker* (1973) and *The Bottle Factory Outing* (1974), which depict with wry compassion the pains and complexities of domestic life, and *Watson's Apology* (1984) and *The Birthday Boys* (1991), which draw upon historical figures. The winner of several literary awards, she has also written for television and been a newspaper columnist.

Baker, Dorothy (1907–1968)

Born in Montana, Baker lectured and taught at various American universities. The classical simplicity she admired in the writing of others is a hallmark of her own in novels, such as *Young Man with a Horn* (1938), *Trio* (1943), (dramatised in 1944 with her husband, the novelist and poet Howard Baker), and *Cassandra at the Wedding* (1962), which, with its punchy, immediate prose, is acknowledged as her best.

Baker, Nicholson (1957–)

Baker's early novels, such as *The Mezzanine* (1988), were plotless but beautifully vivid poems on the minute superficialities of modern life. Later works, such as *Vox* (1992), a witty novel about telephone sex, and *The Fermata* (1994), have been criticised for their obsessive sexual voyeurism. Baker is a great admirer of John Updike, about whom he wrote an idiosyncratic book, *U and I* (1991).

Balchin, Nigel [Marlin] (1908–1970)

A versatile British writer with an enormous range of interests. Balchin's best books are the popular thriller about a psychoanalyst and his patient, *Mine Own Executioner* (1945), and the laconic novels he based on his wartime experience as a psychologist and scientific advisor in Whitehall: *Darkness Falls from the Air* (1942) and *The Small Back Room* (1943).

Baldwin, James [Arthur] (1924–1987)

Brought up in poverty in Harlem, Baldwin found his first audience as a boy preacher, and much of his work is polemical. Hailed in 1953 for his strongly autobiographical first novel, *Go Tell It On the Mountain*, this black, homosexual writer became an increasingly marginalised figure during the 1960s, his work having alienated white liberals by its bleak pessimism and disappointed younger black radicals with its perceived defeatism. As well as novels and short stories, Baldwin also wrote plays and essays.

Ballard, J[ames] G[raham] (1930–)

Although he became familiar to mainstream readers only in 1984 with *Empire of the Sun*, a novel based upon his childhood experiences of Japanese-occupied Shanghai, Ballard already had a distinguished record as a leading British writer of science fiction and fantasy, notably with such prophetic novels as *The Drowned World* (1962) and *High Rise* (1975).

Banks, Iain [M.] (1954–)

The author of a growing number of science-fiction novels, published under the name Iain M. Banks, this Scottish writer achieved instant notoriety in 1984 with his first book, a violent fantasy entitled *The Wasp Factory*. In both his sci-fi and his mainstream novels, Banks remains consistent in his concern to explore the possibilities for perversity and develop his dystopian vision.

Banks, Lynne Reid (1929–)

Banks started out as a playwright in the 1950s. She made her name, however, with her first novel, *The L-Shaped Room* (1962), which captured life in London's bed-sitters in the 1960s, was a great popular success and remains her best-known book. Two sequels were among the other novels which followed. She has also written non-fiction books on Israel and a large number of books for children.

Banville, John (1945–)

Born in Wexford, Ireland, Banville has pursued a career in journalism, eventually becoming literary editor of *The Irish Times*. The scientific interest of earlier novels such as *Doctor Copernicus* (1976) and *Kepler* (1981) has been maintained in later novels like *The Newton Letter* (1989), though a growing bleakness has been noted in the picture of the universe Banville presents.

Barker, Pat (1943–)

After a number of unsuccessful attempts at the middle-class comedy of manners, Barker found her true voice writing tough, humorous fiction rooted in the working-class life she had herself experienced during her Teesside childhood, notably in *Union Street* (1982). She subsequently embarked upon a trilogy set during the First World War.

Barnes, Djuna [Chappell] (1892–1982)

Barnes studied art in New York, aquiring a taste for the Bohemian life which was not fully satisfied until she moved to Paris in the 1920s, when the expatriate American 'lost generation' was just beginning to find itself. Influenced by friends such as Gertrude Stein and F. Scott Fitzgerald, she began to write her own, highly poetic fiction, notably the novel *Nightwood* (1936). She also wrote poetry and plays, publishing a collection of prose, drama and verse as *A Book* in 1929. Her other most highly-rated work is the verse drama *The Antiphon* (1958).

Barnes, Julian (1946–)

Barnes's novels demonstrate the quintessence of English poise and ironic humour, together with a preoccupation with obsessive emotions. He has been hailed as an English writer whose work embodies the experimental attitudes and multi-cultural resonance which are the requirement for truly international success. His novels include *Metroland* (1980), *Flaubert's Parrot* (1984, Prix Médicis), *Staring at the Sun* (1986), *A History of the World in 10½ Chapters* (1989) and *Talking It Over* (1991). He has also written detective novels under the pseudonym of Dan Kavanagh.

Barrie, J[ames] M[atthew] (1860–1937)

Scottish playwright and novelist, whose popular dramas for adults include *Quality Street* (1901), *The Admirable Crichton* (1902) and *What Every Woman Knows* (1906). He had his greatest success, however, with his play for children *Peter Pan* (1904), which was influenced by his friendship with a family of small boys named Llewellyn Davies, whom he eventually adopted after their parents' premature deaths. The play was later recast in the form of a novel, *Peter and Wendy* (1911), and Barrie's preoccupation with what he saw as the tragedy of growing up may be seen in such plays as *Mary Rose* (1920).

Barstow, Stan[ley] (1928–)

The son of a Yorkshire miner, Barstow came to fame with his first novel, *A Kind of Loving* (1960). Other vivid portrayals of Yorkshire life, such as *Ask Me Tomorrow* (1962) and *Joby* (1964), followed, contributing to the development of the regional novel associated with John Braine, Alan Sillitoe and Keith Waterhouse. Other novels include *Watchers on the Shore* (1966), *Raging Calm* (1968), *Brother's Tale* (1980) and *Just You Wait and See* (1986).

Barth, John [Simmons] (1930–)

For many years a professor at Johns Hopkins University, Baltimore, Barth has been accused of insular academicism, a charge belied by the success his enormous, elaborate novels of playfulness and pastiche, such as *The Sot-Weed Factor* (1960), have had in promoting avant-garde views of fiction to a mass audience.

Barthelme, Donald (1931–1989)

American novelist and short-story writer lionized in the 1960s for an outrageous, collage-based style which seemed, in such books as *Snow White* (1967), to offer a literary equivalent for the then-fashionable pop art. The quiet rigour and craftsmanship behind it all ensured that Barthelme's reputation would outlive his vogue.

Bates, H[erbert] E[rnest] (1905–1974)

British novelist and short-story writer, best known today for the irreverent country comedies of the *Darling Buds of May* sequence, made into a resoundingly popular British television series in the early 1990s. Bates is arguably more interesting, however, for the government-commissioned war fiction in which he transcended his propagandist brief to produce real literature, chiefly about the RAF.

Beckett, Samuel [Barclay] (1906–1989)

Irish writer, who settled in France and wrote in both English and French. He is best known for his plays, including the absurdist *Waiting for Godot* (1953), *Krapp's Last Tape* (1958) and *Happy Days* (1961), but considered himself first and foremost a novelist: *Murphy* (1938) and the trilogy of novels beginning with *Molloy* (1955). In both his drama and fiction, his vision is bleak, but is enlivened by black humour. For a long time misleadingly associated with existentialism, Beckett is now regarded as being, with his mentor, James Joyce, one of the most important early postmodernists.

Bedford, Sybille (1911–)

Although she was born into the German aristrocracy, it was as an English writer that Bedford would make a career spanning over forty years, and embracing both fiction and non-fiction. Her early success as a writer of autobiographical novels was renewed by that of *Jigsaw*, shortlisted for the 1989 Booker Prize.

Beerbohm, [Henry] Max[imilian] (1872–1956)

One of the most elegant and colourful figures of the London of the 1890s, 'the incomparable Max' was also one of its most versatile. He was gifted and successful as a caricaturist, critic, essayist, novelist and, latterly, a broadcaster.

Bellow, Saul (1915–)

The Canadian-born US novelist, short-story writer and lecturer has typically set his naturalistic narratives in Chicago, inhabiting the consciousness of a central character, frequently a Jewish-American intellectual, to portray an individual's frustration with the ongoing events of an indifferent society. His finely styled works and skilled characterizations won him the Nobel Prize for Literature (1976). His novels include *Dangling Man* (1944), *The Adventures of Augie March* (1953), *Herzog* (1964), *Humboldt's Gift* (1975) and *The Dean's December* (1982).

Bennett, [Enoch] Arnold (1867–1931)

Prolific novelist, playwright, diarist and critic, born in the Staffordshire Potteries, the setting of his best known fiction: *Anna of the Five Towns* (1902), *The Old Wives' Tale* (1908) and the 'Clayhanger' novels (1910–1916). The realist aesthetic of this most regional of writers was formed by the literary novelists of France, where Bennett lived for several years – in particular by Zola. Thoroughly professional, Bennett kept a meticulous record of the number of words he managed to write when working on a novel. He was also a highly influential book critic for the *Evening Standard* during the last years of his life.

Benson, E[dward] F[rederick] (1867–1940)

One of the eccentric brood of F.W. Benson, Archbishop of Canterbury, Benson made his name in 1893 with his novel *Dodo*. Enormously prolific, he was to produce almost 100 books, including biographies, memoirs, ghost stories and novels, the best known of which are the sentimental public-school romance *David Blaize* (1916), and the books about two fearful women, Mapp and Lucia, rival queens of Tilling, Benson's version of Rye, where he lived and of which he was mayor.

Benson, Stella (1892–1933)

Benson was educated privately and abroad, which gave her a lifelong taste for travel. Her first novel, *I Pose* (1915) was inspired by her experiences in the West Indies in 1912. She became involved in women's suffrage and social reform and sailed to California in 1918, where she worked in a wide range of jobs. She married in China during the revolution and settled in Yunnan until her death from pneumonia at the age of forty. Her most famous novel, *Tobit Transplanted* (1931) (published in the US as *The Faraway Bride*), is about a colony of White Russian exiles in China, which also provided the setting for several of her short stories. She also wrote poetry, travel books and essays.

Bentley, E[dmund] C[lerihew] (1875–1956)

The classic detective novel *Trent's Last Case* (1912) apart, Bentley's fiction has been overshadowed by the verse form he invented and which he used in several volumes of comic

'biography', the clerihew. He also wrote short stories and was a journalist, working for many years as a leader writer for the *Daily Telegraph*.

Berger, John (1926–   )

Educated at the Chelsea College of Art and Central School of Art, Berger established his reputation as an outspoken art critic, broadcaster and journalist holding Marxist views. *Ways of Seeing* (1972) is a much admired book on art. He has also published several novels including the experimental *G* (1972), and the trilogy based on his experiences among French peasant farmers, starting with *Pig Earth* (1979).

Berridge, Elizabeth (1921–   )

After a brief spell in the Bank of England, Berridge became a journalist and was encouraged to write fiction by her husband, the writer and editor Reginald Moore. She made her reputation with the three darkly comic novels she wrote in the 1960s, starting with *Across the Common* (1964), set in Wandsworth, where she grew up.

Berry, Wendell (1934–   )

Although Berry spent some time in California and New York, he returned to his birthplace in Kentucky to farm, as his family had done for generations, and novels such as *A Place on the Earth* (1960) and *The Memory of Old Jack* (1974) are set amongst the farming communities of Port Royal. He is also a poet and essayist, concentrating in his work upon the relationship between man and the land.

Bowen, Elizabeth [Dorothea Coles] (1899–1973)

Dublin-born Anglo-Irish novelist and short-story writer, whose style and subject-matter owe something to Henry James. Her early novels, such as *The Hotel* (1927), are fairly straightforward comedies of manners. Later books become more complex, both in their prose and in their emotional nuances, though they remain sharply witty. The setting for her stories is usually the country houses of the Anglo-Irish or the elegant London homes of the English upper-middle class. Among her best novels are *The House in Paris* (1935), *The Death of the Heart* (1938), *The Heat of the Day* (1949), *A World of Love* (1955) and *The Little Girls* (1964). Her equally regarded short stories were published in several volumes and collected in 1980.

Bowles, Jane [Stajer Auer] (1917–1973)

An unconventional life at the heart of New York's Bohemia of the 1930s and 1940s, followed by bisexual entanglements in Morocco, make Bowles's story as intriguing as any novel. Plagued by ill health, she wrote very little, her only novel being the short, defiant tale of *Two Serious Ladies* (1943).

Bowles, Paul [Frederick] (1910–)

American novelist, short-story writer, translator and poet, who has spent much of his life in Morocco. He started as a composer, then embarked upon a literary career in 1945 with short stories, collected in *The Delicate Prey* (1950), and his first and best known novel, *The Sheltering Sky* (1949). He achieved some notoriety because of the cool depiction of brutality which was an occasional feature of his fiction. After moving to Tangier with his wife, the writer Jane Bowles, he translated the works of Moroccan writers and began writing his own Arab-influenced fables, collected in such volumes as *Midnight Mass* (1985). He has also written travel books and memoirs.

Boyd, William [Andrew Murray] (1952–)

Boyd was born in Ghana and educated at Gordonstoun, two formative influences upon his work as a novelist, short-story writer and television dramatist. His first novel, the rumbustious *A Good Man in Africa* (1981), drew comparisons with the work of Evelyn Waugh and Kingsley Amis. *An Ice-Cream War* (1982) was a gripping, subtle and intelligent adventure story set in Africa during the First World War, a conflict which surfaced again in *The New Confessions* (1987). Later novels, such as *Brazzaville Beach* (1990), have been more ambitious, but perhaps less successful. His screenplays, *Good and Bad at Games* and *Dutch Girls* (both 1985) portray public-school life in comic and convincing detail.

Boyle, Kay (1903–1992)

Born in Minnesota, Boyle lived for a long time in France, but returned to the US in 1941. Her lyrical first novel, *Plagued by the Nightingale* (1930) tells the Jamesian story of a young American woman encountering the jaded culture of the Old World. She published many other impressionistic stories, novels, poems and essays, including *Gentleman, I Address You Privately* (1933), an early novel with a homosexual theme, *Collected Poems* (1962) and *Fifty Stories* (1980).

Bradbury, Malcolm [Stanley] (1932–)

A leading exponent of the British campus novel, Bradbury has been almost as important for founding a campus for novelists at the University of East Anglia, where the creative writing programme he founded has helped a number of leading young writers on their way. His novels, in which satire is heightened by acute observation, include *Eating People Is Wrong* (1859), *The History Man* (1975) and *Rates of Exchange* (1983). He has also written extensively on twentieth-century

fiction, and adapted the work of other novelists for television.

Bradbury, Ray[mond Douglas] (1920–)

Bradbury's midwestern down-to-earthness and his suspicion of technology make him an unlikely candidate for the position of America's top science-fiction writer, yet he has occupied that position for several decades, latterly crossing over into mainstream literary fiction and drama. He has published several volumes of short stories, including *The Illustrated Man* (1951) and *I Sing the Body Electric!* (1969), as well as novels, the best known of which are *The Martian Chronicles* (1950) and *Fahrenheit 451* (1953).

Brady, Joan (1939–)

Born in the USA, Brady trained in ballet and at the age of twenty was a dancer with the prestigious New York Ballet Company. At twenty-two, she quit dancing – this process, and the events leading to it, are dealt with in her autobiographical book, *The Unmaking of a Dancer* (1992). She took a philosophy degree and moved to the UK. *Theory of War* (1993), her second novel, won the Whitbread Book of The Year award.

Braine, John [Gerard] (1922–1986)

The son of a Bradford works superintendent, Braine was a librarian in Bingley, after wartime service in the Navy. With the success of *Room at the Top* (1957), combining Northern aggression, political radicalism, rebellion and comparatively explicit sex for its time, Braine became a full-time writer. Other novels such as its sequel *Life at the Top* (1962) and *The Jealous God* (1964) consolidated his popularity, but with his conversion to a reactionary political stance and the falling quality of his work, his reputation suffered. The events of his last years are chronicled in the novels *One and Last Love* (1981) and *These Golden Days* (1985).

Brautigan, Richard [Gary] (1935–1984)

With best-selling books such as *Trout Fishing in America* (1967) and *In Watermelon Moon* (1968), Brautigan was lionized as the laureate of American flower-power counter-culture. Although later novels such as *Sombrero Fallout* (1976) are equally impressive, he fell out of fashion, and this may have been a contributing factor to his suicide. His comic-surreal books tend to incorporate elements of fact and fiction, and he also published several volumes of poetry.

Brink, André [Philippus] (1935–)

Born of Afrikaner stock in the Orange Free State, Brink became the leading spokesman for the generation of writers known as the 'Sestigers' ('men of the sixties'). He has combined the career of a distinguished academic with that of a major, and much awarded, novelist. His work was vital in the development of the modern Afrikaner novel, radical in its attitudes to sex, religion and politics. His novels, which he wrote in both Afrikaans and English, include *A Dry White Season* (1979) and *A Chain of Voices* (1982). He is also a playwright and has written numerous translations and volumes of criticism and travel writing.

Brodkey, Harold [Aaron Roy Weintraub] (1930–)

Brodkey made a name for himself with the short stories he published in the *New Yorker* and which were collected as *First Love and Other Sorrows* (1957). There then followed a long silence as Brodkey, whose reputation as a literary great had mushroomed, worked upon his promised novel, which eventually appeared in 1991 as *The Runaway Soul*. A massive tome, principally about a Brodkey-like character, it divided the critics. It had been preceded by two volumes of short stories, *Women and Angels* (1985) and *Stories in an Almost Classical Mode* (1988) and was followed by a second novel, *Profane Friendship* (1994) which charted the relationship between two men in Venice.

Brooke, [Bernard] Jocelyn (1908–1966)

Best known for the autobiographical novels which made up *The Orchid Trilogy* (1948–1950), the sensitive, dreamy, country-loving Brooke was an unlikely recruit to the Royal Army Medical Corps. His ambivalence about his military career surfaced in the Kafkaesque novel, *The Image of a Drawn Sword* (1950). His native county of Kent haunts much of his work, which is peopled by lively eccentrics (notably the four portraits of *Private View*, 1954). His books combine fiction and non-fiction and, a keen botanist, he is also the author of a standard work on the orchid.

Brooke-Rose, Christine (1926–)

British writer, academic and critic, who has spent much of her life in France. She began her career as a novelist improbably enough with a satirical account of philologists, *The Languages of Love* (1957), but soon became dissatisfied with conventional fictional forms. Influenced by the French *nouveau roman*, some of which she has translated, she developed new forms, combining this with an interest in the discoveries of modern science in such novels as *Xorandor* (1986).

Brookner, Anita (1938–)

Brookner's work as an art critic is reflected in her skills as a novelist, which show her subtle sense of visual detail. Her novels, which are often centred around lonely single women undergoing mid-life crises, whose complex hopes and fears are gently

unravelled, include *A Start in Life* (1981), *Providence* (1982), *Look at Me* (1983), *Hotel du Lac* (1984), *Family and Friends* (1985) and *A Misalliance* (1986).

Brophy, Brigid [Antonia Susan] (1929–)

Novelist, critic, biographer and campaigner for the rights of animals and authors, Brophy is the daughter of the writer John Brophy. Celebrated as a cultural commentator in the 1960s, she is also a distinctive and underrated novelist, whose fiction is wide-ranging and closely related to her non-fiction work. Her novels include *The Finishing Touch* (1963), a pastiche of Firbank, about whom she has written a pioneering and celebratory study (*Prancing Novelist*, 1973), and *In Transit* (1969), an experimental work set in an airport lounge concerning gender and identity. She has also written studies of Beardsley (1968 and 1976), *Mozart the Dramatist* (1964) and Freudianism (*Black Ship to Hell*, 1962).

Brown, George Mackay (1921–)

Born in the Orkney Islands, Brown read English at Edinburgh University before starting to write in the 1950s. He has produced many plays and novels, as well as fine short-story collections such as *The Masked Fisherman. Beside the Ocean of Time*, a time-travel fantasy novel set in The Orkneys was published in 1994. *A Portrait of Orkney* is a documentary view of the place which is the setting for most of his work. Brown's *Selected Poems 1954–1983* (1991), drew praise from the leading Irish poet Seamus Heaney. A number of his poems have been set to music by Sir Peter Maxwell Davis.

Buchan, John (1875–1940)

A Scottish minister's son, Buchan worked variously as journalist, barrister, MP and public official. Serving both in Foreign Office and Intelligence, he eventually rose to a peerage and to the position of Governor-General of Canada. Altogether, it was a background which equipped Buchan ideally for writing the 'shockers' of international espionage for which he is now chiefly remembered, featuring Richard Hannay: *The Thirty-Nine Steps* (1915), *Greenmantle* (1916) and others.

Burgess, Anthony [i.e. John Anthony Burgess Wilson] (1917–1993)

Marked for life by a Catholic upbringing whose theological tenets he rejected but whose cultural influence he carried with him, Burgess wrote throughout the 1950s, but his productivity leapt in 1959 after doctors told him he had twelve months to live. Thereafter he kept up a steady stream of new work: not only novels but nonfiction, journalism, memoirs and even symphonic compositions. His best known books include *A*

Clockwork Orange (1962), *The Wanting Seed* (1962), *The Enderby Novels* (1963–1974) and *Earthly Powers* (1980).

Burnett, Frances Hodgson (1849–1924)

English-born, Burnett emigrated with her family to the USA in 1865. Her novels for children include the sentimental *Little Lord Fauntleroy* (1886) and the less cloying *The Secret Garden* (1911), which has its values anchored in nature mysticism.

Burns, John Horne (1916–1953)

Burns is best known for his acclaimed novel *The Gallery* (1947), an episodic story set in Naples and North Africa and based on his service in the US army during the Second World War. He wrote two other novels while an expatriate in Italy, where he died: *Lucifer with a Book* (1949) and *A Cry of Children* (1952).

Burroughs, William S[eward] (1914–   )

Notorious for their graphic depictions of the often sordid, sometimes violent, world of drug addiction, Burroughs's frank, innovative novels, such as *Junkie* (1953), *The Naked Lunch* (1959), *Nova Express* (1964) and *Cities of the Red Night* (1981), may be viewed either as dreary exercises in narcissism or strikingly original satires on the spiritual emptiness of contemporary life. His work is noted for its experimental methods, black humour, explicit homoeroticism and apocalyptic vision.

Butler, Samuel (1835–1902)

The English author, painter, poet, anti-Darwinist and travel writer made his name in 1872 with a satiric attack on contemporary utopianism, *Erewhon* ('nowhere' reversed), but is chiefly remembered for his revealing semi-autobiographical novel *The Way of All Flesh* (1872–1885), published posthumously in 1903.

Byatt, A[ntonia] S[usan] (1936–   )

The sister of Margaret Drabble, Byatt is a novelist, short-story writer and critic. Her novels, which are dense and allegorical, rich in artistic and cultural speculation, include *Shadow of a Sun* (1964) and *The Game* (1967), both of which explore the influence of art on life; *The Virgin in the Garden* (1978) and its sequel *Still Life* (1985), which chronicle the life of a northern family in the 1950s with allusions to both the first and second Elizabethan ages; and *Possession* (1990), which won the Booker Prize.

Cain, James M[allahan] (1892–1977)

A journalist and author of several ambitious historical and psychological novels, Cain remains famous as co-founder with Dashiell Hammett and Raymond Chandler of the 'hard-boiled' school of American crime-writing, with such novels as *The Postman Always Rings Twice* (1934), *Double Indemnity* (1936) and *Mildred Pierce* (1941). Several of these were adapted for the cinema and Cain himself was also a screenwriter.

Caldwell, Erskine [Preston] (1903–1987)

A nomadic childhood as the son of an itinerant preacher gave Caldwell that deep knowledge of rural Southern life so evident in such novels as *Tobacco Road* (1932) and *God's Little Acre* (1933), besides fostering a wanderlust he would be able to indulge as foreign correspondent and husband of news photographer Margaret Bourke White. He also wrote travel books and several volumes of short stories.

Calisher, Hortense (1911–   )

Born in New York City, Calisher has taught English at various universities and is known for stories that subtly interpret character, in such collections as *In the Absence of Angels* (1951) and *Extreme Magic* (1964). Her novels include *False Entry* (1961) and *Textures of Life* (1963), a study in relationships between parents and adult children and their respective marriages. Like Henry James, Calisher evokes emotional responses from her reader where the material, dispassionately considered, treats only the mundane.

Callaghan, Morley (1903–1990)

Canadian realist novelist and short-story writer, deeply influenced both by the Depression and the teaching of the Catholic modernist Jacques Maritain. Callaghan characteristically writes about social misfits and outcasts and the possibilities of personal salvation through the power of love. His novels include *More Joy in Heaven* (1937) and *A Fine and Private Place* (1975).

Capote, Truman [pseud. of Truman Steckfus Persons] (1924–1984)

Born in New Orleans, Capote made his reputation with *Other Voices, Other Rooms* (1948), an oblique exercise in Southern Gothic. He worked for the *New Yorker* and contributed stories and articles to numerous magazines, publishing a metropolitan comedy, *Breakfast at Tiffany's* in 1958. He pioneered the 'non-fiction novel' with his account of a multiple murder in the controversial *In Cold Blood* (1966). His creativity undermined by celebrity, drink and drugs, he died without completing a projected Proustian novel of New York society.

Carey, Peter (1943–   )

After a science education, Carey turned to writing, living in Sydney and combining a copywriting job with his own creative writing. After two books of short stories, *The Fat Man in History* (1980) and *Exotic Pleasures* (1981), he began to establish himself as the most prominent young Australian writing today with such novels as *Illywhacker* (1985), *Oscar and Lucinda* (1988) and *The Tax Inspector* (1991). He has been described as a fabulist who mixes many traditions in his fiction, weaving together elements of realism, fantasy, surrealism and macabre comedy.

Carter, Angela (1940–1992)

This English magic-realist writer's works are marked by elements of Gothic fantasy, a fascination with the erotic and the violent, tempered by a complex lyricism and a comic touch. She co-wrote the script for the film *The Company of Wolves* (1984), based on one of her stories. Her novels include *The Magic Toyshop* (1967), *Nights at the Circus* (1984) and *Wise Children* (1991). Her non-fiction includes *The Sadeian Woman* (1979), while *Black Venus* (1985) is an outstanding volume of short stories.

Cary, [Arthur] Joyce [Lunel] (1888–1957)

Born in Londonderry, Cary used his experiences gained in the Colonial Service in Nigeria as a backdrop to early novels such as *Mister Johnson* (1939). His major work is the trilogy centred upon the painter Gulley Jimson, exploring the theme of individual freedom and choice, opening with *Herself Surprised* (1940). *Charlie Is My Darling* (1940) is a fine evocation of the life of an evacuee from London during the Second World War. Cary also wrote poetry and short stories.

Cather, Will[ell]a [Sibert] (1873–1947)

Brought up in the Shenandoah Valley and later Nebraska (the background to half of her fiction) Cather worked as journalist and teacher, moving to New York to edit the influential *McLure's Magazine* (1906–1911), before devoting herself to writing poetry, short stories, novels, journalism and comment, living with her lifelong companion Edith Lewis. Her twelve novels fall into three groups: the early Jamesian tales, such as *Alexander's Bridge* (1912); those dealing with immigrant life in the West, such as *O Pioneers!* (1913); and a late group of historical novels showing her growing interest in Roman Catholicism, such as *Death Comes for the Archbishop* (1927).

Chandler, Raymond [Thornton] (1888–1959)

Chicago-born, but educated at Dulwich College, England, Chandler turned the pulp detective-mystery form into a successful literary genre and created the quintessential private eye in the tough but chivalric loner, Philip Marlowe. His major works are *The Big Sleep* (1939), *Farewell, My Lovely* (1940), *The Lady in the Lake* (1943) and *The Long Goodbye* (1954).

Chatwin, [Charles] Bruce (1940–1989)

In a short but crowded life, Chatwin worked as a valuer at Sotheby's, studied archaeology, worked as a journalist and above all travelled – from the mountains of Afghanistan to the deserts of Patagonia, from the Australian outback to the teeming cities of India. Books such as *The Viceroy of Ouidah* (1980) and *The Songlines* (1987) occupy a middle ground between fiction and non-fiction, while *On the Black Hill* (1982) and *Utz* (1988) are more conventional, though none the less idiosyncratic, narratives.

Cheever, John (1912–1982)

Born in Massachusetts, Cheever focuses in his stories and novels on the ironies of upper-middle-class life in suburban America. His short stories were frequently published in the *New Yorker* magazine and *The Stories of John Cheever* (1978) won the Pulitzer Prize. His novels include *The Wapshot Chronicle* (1957), *The Wapshot Scandal* (1964) and *Falconer* (1977), set in the eponymous 'Correctional Facility'.

Chesterton, G[ilbert] K[eith] (1874–1936)

Enormously prolific English author of novels, short stories, poems and essays, who also pursued a busy career as a journalist. This large amount of work allowed him to indulge to the full his astonishing range of literary, theological and political interests and his consuming passion for paradox. Novels such as *The Napoleon of Notting Hill* (1904) and *The Man Who Was Thursday* (1908) are bizarre fantasies and perhaps less well known than his popular series of books featuring the detective–priest, Father Brown, beginning in 1911 and collected in 1929 as *The Father Brown Stories*.

Childers, [Robert] Erskine (1870–1922)

A Clerk in the House of Commons. Childers served in the Boer War before involving himself in Irish affairs. In 1921 he was appointed director of publicity for the Irish Republicans and in 1922 was court-martialled and shot by firing squad. He is known as a writer for his thriller, *The Riddle of the Sands* (1903).

Christie, Agatha [Mary Clarissa] (1890–1976)

Born in Torquay and educated at home and in Paris, Christie began writing detective stories during her unhappy first marriage. Her first novel, *The Mysterious Affair at Styles*, was published in 1920 and inaugurated a prolific career as the 'Queen of Crime'. She created two much-loved detectives, the Belgian Hercule Poirot and the English spinster, Miss Marple, and her many, hugely popular novels are valued more for the ingenuity of their plots than the quality of their prose. She also wrote a number of plays, of which *The Mousetrap* (1952) has the distinction of being the longest running play in London's West End.

Clark, Walter van Tilberg (1909–1971)

Having spent his early childhood on the East Coast of America, Clark moved to Nevada at the age of eight and there found the landscape in which he would set his novels. His first and best, *The Ox-Bow Incident* (1940), is a sophisticated and moving Western; subsequent novels were less successful, although Clark won the O. Henry Prize for one of his short stories.

Clarke, [Victor] Lindsay (1939–)

Clarke came late to writing after a career as a teacher in England and Africa. A dream prompted him to start writing fiction and he published his first novel, *Sunday Whiteman*, in 1987. He won the Whitbread Prize in 1989 for *The Chymical Wedding*, a novel which demonstrates his preoccupation with arcana, green issues and 'New Age' ideas.

Coetzee, J[ohn] M[ichael] (1940–)

One of South Africa's leading novelists, who gained an international reputation in 1983 with *Life and Times of Michael K*, Coetzee worked for some years as a computer programmer before turning to literature, both as a teacher and, increasingly, as a writer. His work, inevitably, deals with those marginalized by apartheid, and he has won several major prizes. His other novels include *Waiting for the Barbarians* (1980) and *Age of Iron* (1990).

Collier, John [Henry Noyes] (1901–1980)

Poetry editor of *Time and Tide* in the 1920s, Collier was a writer of fantastic, macabre and satirical stories, the best known of which is the novel *His Monkey Wife* (1930), describing the marriage between a man and his pet chimpanzee. He became a Hollywood screenwriter in the 1930s. A selection of his work was published in 1972 as *The John Collier Reader*.

Compton-Burnett, I[vy] (1884–1969)

The daughter of a homoeopathic doctor, Compton-Burnett grew up in the same atmosphere of domestic tyranny that she describes so alarmingly in her novels. After the false start of *Dolores* (1911), she wrote books which combine highly-wrought dialogue, dark, ironic humour and unashamedly melodramatic plots. Most of her novels are set in the late-Victorian period, and her principal subject is the family and its ills (she remained unmarried herself). Her output was regular and entirely *sui generis: Daughters and Sons* (1937), *Manservant and Maidservant* (1947) and other, similarly-titled volumes are some of the most distinctive and witty novels of the century.

Comyns, Barbara (1909–)

The British author of several novels in which comedy and horror are held in fine balance, Comyns started writing at the age of ten, but had a number of jobs and exhibited as a painter before publishing her first, autobiographical, novel *Sisters by a River* (1947). She is particularly good at describing the world through children's eyes, as in *The Skin Chairs* (1962) – a characteristic title, referring to some extremely macabre furniture. Amongst her other novels are *Our Spoons Came from Woolworths* (1950), *The Vet's Daughter* (1959) and *The Juniper Tree* (1985). She lived for many years in Spain, described in her non-fiction book, *Out of the Red into the Blue* (1960).

Connolly, Cyril [Vernon] (1903–1974)

Undoubtedly one of the great literary journalists – and literary personalities – of the century, Connolly liked to give the impression that had he been so minded he might have been one of the great novelists too. He only completed one novel, however: *The Rock Pool* (1936). In 1938 he provided a classic dissection of his classical education and literary influences in *The Enemies of Promise*, and during the Second World War he founded and edited the magazine *Horizon*. His other major work was his introspective 'Word Cycle', *The Unquiet Grave*, published in 1944 under the pseudonym of 'Palinurus'.

Conrad, Joseph [i.e. Teodor Josef Konrad Korzeniowski] (1857–1926)

Born in Russian Poland (now Ukraine), Conrad was exiled to Siberia with his father, a dissident writer. In 1874 he joined a French ship at Marseilles, the start of two decades at sea. It was not until his retirement that he began writing seriously, drawing upon his seafaring experiences to produce novels and stories written in the language of his adopted country, England. Amongst the best known of these are *Lord Jim* (1900), *Typhoon* (1902) and *Heart of Darkness* (1902). His best novels set on dry land are *The Secret Agent* (1907), which displays Conrad's rich ironic humour, and *Under Western Eyes* (1911).

Cook, David (1940–)

A 'fairly successful' actor since the early 1960s, Cook embarked upon his second career, as a novelist, during a period of unemployment, and he has always brought an actor's feel for character to his written work. His first novel, *Albert's Memorial* (1972), concerns the relationship between a young homosexual and a bag-lady, and he has always written about those marginalized by society. Amongst his other novels are *Walter* (1978) and its sequel *Winter Doves* (1979) and the historical *Sunrising* (1984).

Cooper, Lettice [Ulpha] (1897–1994)

Born in Lancashire, brought up in Leeds, and indelibly marked by her experience of the Depression, Cooper frequently concerned herself in her novels with the special identity and particular problems of the north of England, as in *National Provincial* (1938). She was also strongly influenced by Freud, hoping to use traditional fictional methods to investigate the importance of the subconscious. She spent much of her time in Italy, the setting of *Fenny* (1953), worked as a journalist, and wrote several books for children.

Cooper, William [pseud. of Harry Summerfield Hoff] (1910–)

Trained as a scientist, and working for much of his life as a civil servant, Cooper wrote his first four novels in the 1930s under the name H.S. Hoff. He adopted his pseudonym for the ground-breaking and defiantly titled *Scenes from Provincial Life* (1950), which influenced the work of younger writers, who later in the decade were dubbed 'angry young men'. This novel turned out to be the first of a disrupted trilogy about the engagingly caddish Joe Lunn, and Cooper has also written several other comedies of modern life, notably *The Struggles of Albert Woods* (1952) and *The Ever-Interesting Topic* (1953), as well as a memoir, *From Early Life* (1990).

Coover, Robert [Lowell] (1932–)

Brought up in a typical small midwestern town, Coover went on to become an academic, teaching philosophy at a number of American universities. As a novelist, he is a leading exponent of postmodernist fiction, using allegory and satire to depict trends in contemporary American life, and introducing real people into books such as *The Public Burning* (1977). His short stories are collected in the volume *Pricksongs and Descants* (1969).

Crace, Jim (1946–)

Crace spent much of his working-class London childhood drawing maps of imaginary places, and this instinct for fantasy endures in his fiction. He spent two years travelling and working in Africa and the Middle East, then returned to England to work in journalism and write plays for radio. The great success of his first novel, *Continent* (1986), allowed him to devote himself entirely to writing. Subsequent books include *The Gift of Stones* (1988) and *Arcadia* (1992).

Crompton, Richmal [i.e. Richmal Crompton Lamburn] (1890–1969)

From *The Innermost Room* (1923) to *The Ridleys* (1970), Crompton published some fifty books of adult fiction: none is remembered. Yet her stories for children, of an eleven-year-old schoolboy and his gang of four, are one of the great comic achievements of the century. Between 1922 and 1970 thirty-eight volumes of *William* stories appeared: *Just – William* to *William the Lawless*. Although the books were rooted in the 1920s, some details were updated by Crompton's niece to make them more accessible to today's children. Their enduring popularity is not, however, due to this tinkering but to Crompton's storytelling skills. At the level of both adult and child, the stories are wildly funny, with a satirical bite reminiscent of P.G. Wodehouse.

cummings, e[dward] e[stlin] (1894–1962)

Famous for his abandonment of the upper case – even (until the 1930s) when writing his own name – cummings is principally known as a poet, although he first made his reputation with his only novel, *The Enormous Room* (1922), which drew upon his experiences in the First World War. The following year, he published the first of twelve volumes of poetry, *Tulips and Chimneys*. Influenced by jazz, incorporating modern slang, and characterized by typographical inventiveness, his poetry is wide-ranging in style and subject. It was collected in 1972 as *The Complete Poems 1913–1962*. He also wrote plays, an impressionistic account of a visit to Russia (*Eimi*, 1933), and was an accomplished painter.

H.D. [i.e. Hilda Doolittle] (1886–1961)

An American writer who spent most of her life in Europe, Doolittle came to prominence as a leader of the imagist movement in poetry. (She was at one time engaged to Ezra Pound, and – in spite of being predominently lesbian – married Richard Aldington.) Her verse was published in *Poetry* and collected in several volumes, including *Sea Garden* (1916) and *Hymen* (1921). Her boldly self-exploratory fiction, often set partly or wholly in the ancient world, includes such novels as *Palimpsest* (1926) and *Hedylus* (1928), and the transparently autobiographical *Bid Me to Leave* (1960), and has recently been rediscovered by feminist critics.

Dark, Eleanor (1901–   )

Leading Australian novelist, whose fictional preoccupations reflect her family background as the daughter of a Labour politician in Sydney, and her political activism during the 1930s. *The Little Company* (1945) is her response to the Second World War, while her other books include *Slow Dawning* (1932), *Prelude to Christopher* (1934), the trilogy beginning with *The Timeless Land* (1941), and *Lantana Lane* (1959).

Davies, [William] Robertson (1913–   )

Canadian novelist, journalist, playwright and academic, who spent some time in England as a student and an actor, before returning to his native country to become an editor and columnist for several journals. His journalism and essays have been collected in such volumes as *The Table Talk of Samuel Marchbanks* (1949) and *The Enthusiasms of Robertson Davies* (1990). He then became an academic and a novelist. His expansive, energetic novels – which often take the form of trilogies and are packed with magic, arcana and Jungian theories – challenge the prejudice that his native country is dull and safe, and the success of *What's Bred in the Bone* (1985), brought him to a new international audience.

Delafield, E.M. [pseud. of Edmée Elizabeth Monica de la Pasture] (1890–1943)

The daughter of the popular novelist Mrs Henry de la Pasture, Delafield published her first novel, *Zella Sees Herself*, in 1917. Married and settled in Devon, she wrote many novels, of which the gently satirical *Diary of a Provincial Lady* (1930) seems most likely to survive. The novel began life as a serial in *Time and Tide*, and Delafield also wrote a great deal of journalism, as well as three plays.

de la Mare, Walter [John] (1873–1956)

Though the renown which gained his ashes their resting place in St Paul's Cathedral was won largely by his haunting, fine-crafted verses, de la Mare's fiction was every bit as beautiful, and perhaps even more entirely original, encompassing ghost stories, more conventional tales and novels such as *Memoirs of a Midget* (1921). He was also one of the century's best editors of poetry anthologies.

DeLillo, Don (1936–   )

A reclusive novelist, DeLillo was born a Roman Catholic in New York and studied history, philosophy and theology at a Jesuit university. His first novel, *Americana* (1971), was, like subsequent ones, influenced by film. His principal subjects, wryly observed, are various aspects of contemporary American life: football in *End Zone* (1972), rock culture in *Great Jones Street* (1973), space research in *Ratner's Star* (1976), and so on. Several of his novels, including *White Noise* (1985) and *Libra* (1988), have won major awards, and he has also published plays and short stories.

De Morgan, William [Frend] (1839–1917)

De Morgan started out as a ceramicist, working alongside William Morris and producing some of the finest pottery of the Arts and Crafts movement. Obliged by ill-health to retire, he embarked upon a second career as a fiction writer, publishing his first novel *Joseph Vance* (1906) at the age of sixty-seven. Several further novels followed, the most interesting of which is the curious 'dichronism', *Alice-for-Short* (1907).

Desai, Anita [Mazumdar] (1937–   )

Brought up in Delhi, the daughter of a Bengali businessman and his German wife, Desai spoke English at home and writes in that language. Most of her novels and short stories are set in India and deal particularly with the lives of women. Both *Clear Light of Day* (1980) and *In Custody* (1984) were shortlisted for the Booker Prize. She has also written books for children.

Didion, Joan (1934–   )

Californian-born novelist, journalist and essayist, who made her reputation in the 1960s with pieces on contemporary life and culture for *Vogue*, *Mademoiselle* and other journals. These were collected in 1968 as *Slouching Towards Bethlehem*. Later journalistic assignments have included war reporting in El Salvador and led to a volume of reportage, *Salvador* (1983), and political novels such as *Democracy* (1984). She has also written screenplays, often with her husband, the novelist John Gregory Dunne.

Doctorow, E[dgar] L[awrence] (1931–   )

A writer of novels which elaborately weave together history and fiction, Doctorow achieved critical and commercial success with his third novel, *The Book of Daniel* (1971), the story of the Rosenbergs' trial and execution as told by their fictional son, which established him as an imaginative and experimental revisionist of American history. His other novels include the equally successful *Ragtime* (1975), and *Loon Lake* (1980).

Donleavy, J[ames] P[atrick] (1926–   )

Born in New York to Irish immigrants, Donleavy attended Trinity College, Dublin, after which he became a painter. His first novel, *The Ginger Man*, was rejected by numerous publishers because of its sexual frankness and was eventually published by the Olympia Press in 1955. A bawdy farce, it set the tone for Donleavy's subsequent books, including *The Beastly Beatitudes of Balthazar B* (1969) and *The Destinies of Darcy Dancer, Gentleman* (1977). Donleavy has also written successfully for the stage and written an entertaining autobiography, *The History of the Ginger Man* (1994).

Dos Passos, John [Roderigo] (1896–1970)

After a comfortable, East-Coast, middle-class upbringing, Dos Passos was radicalized by his experience as an ambulance driver on the Western Front, the setting of his novel *Three Soldiers* (1921). He was strongly identified with the political factionalism of the 1920s and 1930s, and it was only afterwards that the considerable artistic merits of his work, marked by stylistic innovations derived from the cinema, could be fully appreciated. His novels include *Manhattan Transfer* (1925) and the trilogy *USA*, beginning with *The 42nd Parallel* (1930).

Douglas, [George] Norman (1868–1952)

Scottish diplomat, essayist, novelist and travel writer, whose travel books include *Siren Land* (1911) and *Old Calabria* (1915), both dealing with Italy, where he had gone to evade prosecution for homosexual offences. His best-known novel, the satiric *South Wind* (1917), is set in his adopted island of Capri. He also published an informal autobiography, *Looking Back* (1933). He was the 'editor' (i.e. author) of the celebrated collection of aphrodisiacal recipes, *Venus in the Kitchen* (1952).

Doyle, A[rthur] Conan (1859–1930)

Conan Doyle wrote the first Sherlock Holmes stories as light squibs to keep busy and supplement his income while waiting for trade at his new Portsmouth medical practice to pick up. On their publication in the *Strand Magazine*, however, he found greatness thrust upon him, the public clamouring for new tales long after he himself had tired of his detective hero.

Doyle, Roddy (1958–)

Doyle was born in Dublin and has spent his life in the area as a teacher and writer. He had considerable success with *The Barrytown Trilogy*, a saga of working-class Dublin life which opened with *The Commitments* in 1987 and was followed by *The Snapper* (1990) and *The Van* (1991). The first two novels were made into popular films, and were followed by *Paddy Clark Ha Ha Ha* (1993), which was a best-seller even before winning the Booker Prize. He has also written plays.

Drabble, Margaret (1939–)

The sister of novelist A.S. Byatt, Drabble is a novelist and critic, who has published studies of Wordsworth and Arnold Bennett and is the editor of the revised *Oxford Companion to English Literature* (1985). Her novels, sometimes criticized for concentrating too narrowly on middle-class domestic crisis in north London, include *Summer Bird-Cage* (1963), *The Millstone* (1965), *Jerusalem the Golden* (1967), and the trilogy beginning with *The Radiant Way* (1987).

Dreiser, Theodore [Herman Albert] (1871–1945)

The founder of American naturalism, Dreiser was born into a large German immigrant family beset by poverty and upheaval, circumstances which clearly influenced the view of American society he put forward in his controversial novels. Books such as *Sister Carrie* (1990) offered an unwelcome critique of the American way of life, and were later augmented by volumes of socialist non-fiction, including *Tragic America* (1931). The title of this book echoed that of Dreiser's most highly praised novel, *An American Tragedy* (1925). He also wrote drama, short stories, poetry and essays.

Duffy, Maureen [Patricia] (1933–)

The British novelist, poet, playwright, biographer and campaigner was born illegitimate into an impoverished, working-class environment, vividly evoked in her autobiographical first novel, *That's How It Was* (1962). Her fiction is wide-ranging and characterized by stylistic innovation in such books as *The Microcosm* (1966) and *Love Child* (1971). Her non-fiction work includes biographies of Aphra Behn and Henry Purcell, an animal rights handbook (*Men and Beasts*, 1984) and a study of *The Erotic World of Faery* (1972). Her several volumes of poetry were gathered in 1985 as *Collected Poems 1949–1984*. A co-founder of the Writers' Action Group, she has for many years been actively involved in fighting for authors' rights.

Du Maurier, Daphne (1907–1989)

Du Maurier was the granddaughter of the artist and novelist George Du Maurier, and the daughter of the actor–manager Gerald Du Maurier, and moved in literary circles in childhood. Novels such as *Jamaica Inn* (1938) and *Rebecca* (1938) draw upon the Cornish landscape, where she lived and about which she wrote a book, *Vanishing Cornwall* (1967). Categorized as 'romantic fiction', her novels were always more artistically ambitious than this label would suggest. She also published several volumes of short stories, wrote a number of plays, and biographies of her father, Francis and Anthony Bacon and Branwell Brontë.

Durrell, Lawrence [George] (1912–1990)

Born in India, Durrell joined the foreign service and lived mainly in the eastern Mediterranean, the setting of his novels, including *The Alexandria Quartet* (1957–1960). The heady prose and bizarre characters of the quartet reflect his exotic sources of inspiration and established his reputation as one of the most original novelists of the post-war period. A poet, writer of topographical books, verse plays, and farcical short stories, he was the brother of Gerald Durrell, the naturalist.

Ellis, Alice Thomas [pseud. of Anna Haycraft] (1932–)

A former columnist for the *Spectator*, Ellis writes subtle, inventive social comedies, coloured by bizarre, mysterious events with portentous overtones. These include *The 27th Kingdom* (1982) which, in common with *The Sin Eater* (1977) and *Unexplained Laughter* (1985), possesses a strong female protagonist who is a vessel through which the forces of good and evil do battle. A Roman Catholic, she is married to Colin Haycraft of the publishers Duckworth, for whom she works as an editor.

Ellison, Ralph [Waldo] (1914–1994)

Over forty years after completing his first novel, *Invisible Man* (1952), Ellison died without producing his (reportedly enormous) second, fragments of which had been published as work-in-progress. His reputation as the grand old man of black American letters rests upon this one novel, a few uncollected short stories, and a volume of essays, *Shadow and Act* (1964). He was co-founder and co-editor of *Negro Quarterly* and taught at a number of American universities.

Exley, Frederick [Earl] (1929–)

The defining experience of Exley's life was his inability to live up to the memory of a sporting-hero father, who died tragically young. Exley has transformed that failure into success in a series of 'fictional memoirs', starting in 1968 with the award-winning *A Fan's Notes*.

Farrell, J[ames] G[ordon] (1935–1979)

Of Anglo-Irish stock and hence himself the descendant of colonists, Farrell ironically questions the certainties of Britain's imperial past in his historical novels, notably *Troubles* (1970), *The Siege of Krishnapur* (1974) and *The Singapore Grip* (1978). Large, comic books, packed with character and incident, they combine fine story-telling and acute social dissection. Accidentally drowned, he left behind an unfinished novel set in India, published in 1981 as *The Hill Station*.

Farrell, James T[homas] (1904–1979)

A child of Chicago's Irish community, Farrell was brought up by grandparents because his parents could not afford to keep him. This impoverished background ensured that he would never lose his driving sense of social mission, even after he had won success and financial security as a writer with his 'Studs Lonigan' trilogy (1932–1935).

Farrell, M.J. *see* Molly Keane

Faulkner, William [Cuthbert] (1897–1962)

Faulkner's works are noted for their experimental, stream-of-consciousness narrative style, complex structure and epic mapping of a quasi-imaginary Southern region, Yoknapatawpha County. His major novels are *The Sound and the Fury* (1929), *As I Lay Dying* (1930), *Sanctuary* (1931), *Light in August* (1932), *Absalom! Absalom!* (1936), *The Unvanquished* (1938), *Intruder in the Dust* (1948) and *Requiem for a Nun* (1951). He was recognized as one of America's greatest writers only after the Second World War, and was awarded the Nobel Prize for Literature in 1949. Two late books, *A Fable* (1954) and *The Reivers* (1962) both won the Pulitzer Prize.

Feinstein, Elaine (1930–)

Though Feinstein was born in Bootle and spent much of her youth in Leicester, her concerns are very much not those of a 'regional', or even an 'English', writer. Her highly sophisticated poetry and fiction are both fiercely internationalist in outlook – a consequence, perhaps, of her Russian–Jewish origins. Apart from such novels as *Mother's Daughter* (1988) and volumes of her own poetry, Feinstein has also published translations of several writers, in particular Marina Tsvetayeva, whose biography she wrote in 1987.

Firbank, [Arthur Annesley] Ronald (1886–1926)

Firbank was for many years dismissed as a minor purveyor of high-camp, post-1890s froth, although a younger generation of writers, including Evelyn Waugh, acknowledged their debt to him. His genuinely innovative tragi-comic novels, meticulously worked up from notebooks, tend to be impressionist in style, exotic in setting and character and boldly celebratory in the delineation of sexual heterodoxy. He accurately described his own style as 'aggressive, witty, & unrelenting', and a convincing case for his importance was made by Brigid Brophy in her critical biography, *Prancing Novelist* (1973). His novels, almost all of which were published at his own expense, include *Vainglory* (1915), *The Flower Beneath the Foot* (1923) and *Concerning the Eccentricities of Cardinal Pirelli* (1926), and he also wrote a play, *The Princess Zoubaroff* (1920).

Fitzgerald, F[rancis] Scott [Key] (1896–1940)

Chief chronicler of the American Jazz Age in such short stories as 'The Diamond as Big as the Ritz' (1922) and such novels as *The Great Gatsby* (1925), Fitzgerald pursued a high life similar to that of his characters. He spent much of the 1920s shuttling back and forth between America and Europe, while money troubles, battles with alcoholism, and the deteriorating mental health of his wife Zelda (reflected in *Tender is the Night*, 1934), led to his having to churn out work in order to keep afloat. He endured three unhappy spells as a screenwriter in Hollywood, which resulted in his final (unfinished) novel, *The Last Tycoon* (1941).

Fitzgerald, Penelope (1916–)

Fitzgerald's novels are notable for their exquisite prose style and purity of feeling, but are also often humorous. They include *The Bookshop* (1978), *Human Voices* (1980, drawing upon her experiences working for the BBC), *The Beginning of Spring* (1988) and *The Gate of Angels* (1990). She has also written books about Burne-Jones and Charlotte Mew and a study of her father and uncles, *The Knox Brothers*.

Flanagan, Thomas [James Bonner] (1923–)

Born in Connecticut, of Irish descent, Flanagan has spent his adult life living in both America, where he has taught at assorted universities, and in Ireland. Ireland is the subject of his historical novels, *The Year of the French* (1979) and *The Tenants of Time* (1988), and his most important work of criticism, a study of *The Irish Novelists: 1800–1850* (1959).

Flecker, James [Herman] Elroy (1884–1915)

Like many of his generation, Flecker sought beauty and found it in the East, studying Persian and Arabic and serving as a diplomat in Constantinople, Smyrna and Beirut. In 1911 he married the Greek poet Hellé Skiadaressi, but within four years, having contracted tuberculosis in Turkey, he was dead. He was the author of several volumes of poetry, the most celebrated of which was *The Golden Journey to Samarkand* (1913), the verse play, *Hassan* (1922), and the curious fantasy novel, *The King of Alsander* (1914).

Fleming, Ian [Lancaster] (1908–1964)

Although Fleming worked during the 1930s as a Reuters correspondent abroad, it was his wartime service as assistant to the Director of Naval Intelligence that made him as a writer, providing the experiences and insights which would – albeit exaggerated and embellished beyond recognition – form the basis for his novels about the secret agent, James Bond: *Casino Royale* (1953), *From Russia with Love* (1957), *Goldfinger* (1959), *On Her Majesty's Secret Service* (1963), and so on.

Ford, Madox Ford [formerly Ford Hermann Hueffer] (1873–1939)

British critic, editor and novelist of Anglo-German descent. The author of more than eighty books, he is best known for the short masterpiece, *The Good Soldier* (1915), and the multi-volume sequence based upon his experiences in the First World War, *Parade's End* (1924–1928). He wrote two books with Joseph Conrad and in 1909 founded and became editor of the influential *English Review*, to which Conrad, Hardy, D.H. Lawrence and other leading figures of the period contributed. He spent part of the 1920s in Paris, where in 1924 he founded the *Transatlantic Review*, a short-lived but important forum for Modernist writing.

Ford, Richard (1944–)

The son of a travelling salesman, Ford has spent more than twenty years travelling. Not surprisingly, nomadism – which he considers central to the American psyche – is a major theme of his writing. An exponent of so-called 'dirty realism', Ford has made his reputation with novels such as *The Sportswriter* (1986) and *Wildlife* (1990), and short story collections.

Forster, E[dward] M[organ] (1879–1970)

As a novelist, essayist and critic of social, political and moral attitudes, Forster has had an important influence on literary studies, though he ceased working as a novelist after the publication of *A Passage to India* (1924). His other novels are *Where Angels Fear to Tread* (1905), *The Longest Journey* (1907), *A Room with a View* (1908), and *Howards End* (1910). Much of his work is concerned with the interplay of personality, the creative process, and the conflict between convention and instinct.

Fowles, John [Robert] (1926–)

An English novelist who sometimes uses elements of magic realism, Fowles teases and provokes the reader into questioning the nature of elaborately conjured fictional realities in such novels as *The Magus* (1966, revised 1977), in which a British schoolmaster is subjected to a series of mysterious apparitions on a Greek island, and *The French Lieutenant's Woman* (1969), a 'historical' novel of a forsaken Victorian woman whose sensibility is viewed simultaneously from a period and modern viewpoint.

Franklin, [Stella Maria Sarah] Miles (1879–1954)

The daughter of New South Wales farmers, Franklin described the poverty of her bush childhood and her work as a governess in the guise of the narrator Sybilla in *My Brilliant Career* (1901). A sequel, *My Careeer Goes Bung* (1946), written in 1902 but withheld as being too 'audacious', exploded the romances concocted earlier by Sybilla. Franklin became involved in feminist circles in Sydney, Chicago and London. Promoting Australian literature, she published among others a six-volume cycle of pastoral novels.

Fuller, John [Leopold] (1937–)

The son of the poet Roy Fuller and best known himself as a poet, whose volumes of poetry include *The Illusionists* (1980), a satiric verse narrative of contemporary life, Fuller also writes children's stories. His novels include the compact, richly imagined novel, *Flying to Nowhere* (1983) and *The Burning Boys* (1989). He is also the author of *A Reader's Guide to W. H. Auden* (1970). His *Selected Poems* appeared in 1985.

Gaddis, William (1922–)

Born in New York, Gaddis is a distinctive and satirical stylist of non-psychological work, often written on a vast, perplexing scale. His three highly innovative novels to date, *The Recognitions* (1955), *JR* (1975, National Book Award) and *Carpenter's Gothic* (1986), written at long intervals, are gargantuan tours de force of verbal playfulness and black humour, in which he sees the contemporary world as essentially bleak and loveless.

Gaitskill, Mary (1954–)

Born in Lexington, Kentucky, Gaitskill moved to New York after graduating from the University of Michigan. She made her name in 1988 with her first book, a highly praised volume of short stories titled *Bad Behaviour*, which sold well in both America and England. The promise she showed was confirmed by her highly accomplished and disturbing first novel, *Two Girls, Fat and Thin* (1991).

Galsworthy, John (1867–1933)

An idler from a wealthy family, the young Galsworthy was urged to write by the mate of a clipper he was taking on a trip to Cape Town – 'a Pole called Conrad'. On his return he threw himself into writing with increasing seriousness, and with a passion for social justice which had come late but would never be lost. For all his commitment, the author of *The Forsyte Saga* (1922) would not be taken seriously by the critics, being early branded as a 'middlebrow'.

Gardam, Jane (1928–)

British novelist and short-story writer, who made her name with books for teenage readers, several of which were subsequently reissued as adult novels. Her books are marked by a keen yet compassionate observation of people – particularly of children, as in *A Long Way from Verona* (1971) and *God On the Rocks* (1978) – and subtle comedy. While her most ambitious book is the novel *Crusoe's Daughter* (1985), much of her best work is to be found in her volumes of short stories: *The Sidmouth Letters* (1980), *The Pangs of Love* (1983) and *Showing the Flag* (1989).

Garner, Alan (1934–)

Garner grew up in the village of Alderley Edge, speaking a local dialect. He has lived in Cheshire nearly all his life, and its history, archaeology and folklore are the mainsprings of his work. His first book *The Weirdstone of Brisingamen* (1960) was an outstanding debut and through the 1960s and early 1970s he became the most fashionable children's writer. His fiction is subtle and imaginative, complex and ambiguous enough to satisfy adult audiences. The dividing line between past and present, myth and reality often becomes blurred in his novels and stories which include *Holly for the Bongs* (1965), written for his local school but highly rated, *The Owl Service* (1967, Carnegie Medal and Guardian Fiction Prize), *Red Shift* (1973), perhaps his most ambitious, and *The Stone Book Quartet* (1976–1978), judged by many to be his best work.

Garnett, David (1892–1981)

The son of Edward and Constance Garnett, he trained as a botanist, became a prominent member of the Bloomsbury Group, was a conscientious objector and farm labourer in the First World War, and ran a bookshop in Soho. His novels include two fables, *Lady into Fox* (1922) and *A Man in the Zoo* (1924), *The Sailor's Return* (1925) and *Aspects of Love* (1955), which reached a new audience when used as the basis for Andrew Lloyd-Webber's musical. He was a translator, edited the letters of T.E. Lawrence, T.H. White and Dora Carrington, and wrote a biography of Pocohontas.

Gass, William H[oward] (1924–)

Born in North Dakota, Gass went to Cornell after three years in the Navy and wrote a dissertation on metaphor. Profoundly influenced by Wittgenstein, whose seminars he attended, he has taught philosophy since 1969. His first novel, *Omensetter's Luck* (1966) is highly stylized, treating words as 'musical notes'. Rejecting realism, books such as *Willie Masters' Lonesome Wife* (1968) blur the boundary between fiction and philosophical speculation. In 1975 he received the National Institute for Arts and Letters Prize for Literature.

Gellhorn, Martha [Ellis] (1908–)

Born into a wealthy St Louis family, Gellhorn was a journalist in the States and in Paris. Returning to the USA, she published a highly autobiographical first novel, *What Mad Pursuit* (1934), and a book of four didactic novellas, *The Trouble I've Seen* (1936), based on her work on the Federal Emergency Relief programme. In 1936 she met Hemingway in Florida and in 1937 joined him in Madrid (they married in 1940, divorced in 1946), where she worked as a war correspondent. Between 1937 and 1946 she covered five major wars and has continued to write, living in Mexico, Kenya, and latterly London and Wales. Other novels include *A Stricken Field* (1940), *Liana* (1944), *His Own Man* (1961), *The Lowest Trees Have Tops* (1967) and *The Weather in Africa* (1978). Among her collections of outstandingly vivid journalism and memoirs is *Travels with Myself and Another* (1978).

Gerhardie, William [Alexander] (1895–1977)

Brought up in St Petersburg in a household where English, French and German were spoken, Gerhardie made his reputation in 1925 with a novel drawing upon his background, appropriately titled *The Polyglots*. Educated in Russia and at Oxford, he was the author of the first ever study of *Anton Chekov* (1923), an influence upon his own fiction. He wrote several other novels, including *Futility* (1922) and *Of Mortal Love* (1936). *God's Fifth Column*, a history of the half century, 1890–1940, was incomplete at his death, but edited for publication in 1981.

Gibbon, Lewis Grassic [pseud. of James Leslie Mitchell] (1901–1935)

The son of a Scottish farmer, Gibbon ran away from school, becoming first a junior reporter then an army and RAF clerk. With the help of H.G. Wells, he published short stories in *Cornhill* magazine and the first volume of his trilogy, *A Scots Quair* (1932–1934), a lyrical, realistic account of the transition from rural to urban and industrial society of a Kincardineshire community. Published under his pseudonym, it attracted international praise. His reputation as a writer of genius, tragically curtailed by his death at the age of thirty-four, stands on this trilogy. He also wrote a life of the Scottish explorer Mungo Park, and published several novels under his own name.

Gibbons, Stella [Dorothea] (1902–1989)

The British author of numerous novels, Gibbons is chiefly remembered as the author of one of literature's funniest parodies, *Cold Comfort Farm* (1932), which satirized the rural works of Mary Webb and others. She worked for much of her life as a journalist and launched her literary career with a volume of poetry, *The Mountain Beast* (1930). Her *Collected Poems* appeared in 1950, and she also wrote several volumes of short stories, including *Christmas at Cold Comfort Farm* (1959).

Glasgow, Ellen [Anderson Gholson] (1873–1945)

Born in Richmond into one of Virginia's leading families, Glasgow wrote novels which offered the first sustained fictional critique of the modern South, its problems and its people. A prolific author, who wrote poetry and short stories as well as novels, she is best known for such books as *Barren Ground* (1925), *The Sheltered Life* (1932) and *In This Our Life*, which won a Pulitzer Prize in 1941.

Godden, [Margaret] Rumer (1907–)

Godden based *The River* (1946) on her upbringing on a remote river station in Bengal where her father ran a steamship company. Sent to convent school in England, which she hated, she returned to India to open a dancing school in Calcutta and began writing as an escape from her compatriots' snobbery and philistinism. After paying off her absconding husband's debts when he embezzled his firm's funds, she removed herself to a life of extreme frugality in the Himalayan foothills. She has written over twenty novels – including *Black Narcissus* (1939) and *The Greengage Summer* (1958) – autobiographical works, poetry and books for children.

Golding, William [Gerald] (1911–1993)

Though essentially religious, Golding's works are sombre allegories on man's capacity for self-deception and evil. His first book, *Lord of the Flies* (1954), concerns the degeneration into savagery of a group of English schoolboys marooned on a Pacific island (Golding worked for many years as a schoolmaster), and *Pincher Martin* (1956) is a study of greed and self-delusion. He was awarded the Nobel Prize for Literature in 1983. His other novels include *The Inheritors* (1955), *The Spire* (1964) *Darkness Visible* (1979) and the seafaring trilogy *To the Ends of the Earth* (1980–1989).

Gordimer, Nadine (1923–)

Gordimer has always been one of the more vocal and active white liberal opponents of the South African regime, yet she has remained in South Africa. Her work blends fiction with a factual depiction of her society to register the human consequences of apartheid. Her short stories are particularly vivid in describing the 'small horrors and tendernesses' of existence in her country. Her novels include *Burger's Daughter* (1979), *The Conservationist* (1974), *A Guest of Honour* (1970), *July's People* (1981) and *My Son's Story* (1990).

Some of her best short stories are contained in *Selected Stories* (1975).

Gosse, Edmund [William] (1849–1928)

The relentlessly pious atmosphere of a Devon home presided over by Gosse's rigid Plymouth Brethren father was memorably recreated in his remarkable *Father and Son* (1907). Gosse triumphed over a narrow, bookless upbringing to become a leading light in the London literary world, writing numerous biographies and critical studies.

Grahame, Kenneth (1859–1932)

Born in Edinburgh, Grahame was brought up mostly in England, the countryside of which he celebrated in his books. He spent his working life in the Bank of England, eventually becoming Secretary, but found the work uncongenial and took (or was forced into) early retirement. He was a contributor to *The Yellow Book* in the 1890s and works such as *The Golden Age* (1895) and *Dream Days* (1898) are concerned with childhood. The book for which he is remembered, however, is the children's classic, *The Wind in the Willows* (1908), which was derived from stories he had devised for his young son.

Graves, Robert [von Ranke] (1895–1985)

Graves first published his poetry while serving in the First World War, the horrors of which he recorded in his memoir *Goodbye to All That* (1929). Although his output as essayist, novelist, biographer, children's writer, classicist and translator was prodigious, he saw himself primarily as a poet. He wrote on mythology and interpreted myths (*The White Goddess*, 1948; *The Greek Myths*, 1955, etc.) and was an imaginative and witty historical novelist (*I, Claudius* and *Claudius the God*, 1934). He lived for much of his life in Majorca. His *Collected Poems* appeared in 1975.

Gray, Alasdair (1934–)

Gray is a profoundly innovative and uncompromising writer who breaks conventions in style, layout, location and plot. Deceptions, tricks and allegories abound in his books, which he also illustrates. Among his novels are *Lanark* (1982), which ranges from descriptions of Gray's Glasgow to fantastical adventures in the city of Unthank, *1982, Janine* (1984) and *The Fall of Kelvin Walker* (1985).

Green, F[rederick] L[awrence] (1902–1953)

After a spell in an accountant's office, Green renounced respectability for an itinerant life in a variety of casual jobs. He went to Ireland, initially for a holiday, in 1933, but it was there that he started to write, and he ended up staying for eighteen years, producing a classic novel about 'the troubles', *Odd Man Out* (1945).

Green, G[eorge] F[rederick] (1910–1977)

A Derbyshire childhood, penury in London and army service in Ceylon which ended with a prison sentence for homosexual offences, all contributed to Green's ideas about the relationships between men of different social and cultural backgrounds. This became a major concern of his work in his stories, collected in *Land Without Heroes* (1948) and *A Skilled Hand* (1980), and in the novel *The Power of Sergeant Streater* (1972). He also wrote a novel of childhood, *In the Making* (1952), and edited a volume of stories on this theme, *First View* (1950).

Green, Henry [pseud. of Henry Vincent Yorke] (1905–1973)

The most innovative and enigmatic novelist of his generation, Green pursued twin careers as an industrialist and a writer. His earlier books, such as *Living* (1929) and *Party Going* (1939) rely upon elaborately descriptive, impressionistic and elliptical prose; later books, such as the witty *Nothing* (1950) and *Doting* (1952), are written almost entirely in dialogue. *Caught* (1943) perfectly captures the atmosphere of London during the Blitz, when Green served in the Auxiliary Fire Service. He also wrote an autobiography of his youth, *Pack My Bag* (1940), but published nothing after 1952.

Greene, [Henry] Graham (1904–1991)

One of the most important English novelists of the century, Greene was preoccupied with personal and political moral dilemmas. His novels of guilt, despair and penitence are set in a milieu of urban seediness or political corruption in many parts of the world. He converted to Catholicism in 1926 and many of his works reflect his religious views. His lighter novels and screenplays, which he called 'entertainments', include thrillers, also concerned with the problem of good and evil. His novels include *Brighton Rock* (1938), *The Power and the Glory* (1940), *The Third Man* (1950), *The End of the Affair* (1951), *The Quiet American* (1955), *Travels With My Aunt* (1969) and *The Honorary Consul* (1973).

Greenwood, Walter (1903–1974)

A working-class childhood in Salford and an early adulthood in and out of work at a wide variety of menial jobs gave Greenwood abundant qualification to describe the proletarian condition in his fiction, notably the Depression classic, *Love on the Dole* (1933).

Hall, [Marguerite] Radclyffe[-] (1883–1943)

Although she published four volumes of verse and eight novels, Hall is remembered for *The Well of Loneliness* (1928) whose open treatment of lesbianism brought a notorious court case and a banning until it was republished in 1949. Her other novels include *The Unlit Lamp* (1924) and *Adam's Breed* (1926), which won the Prix Femina Vie Heureuse and the James Tait Black Memorial Prize.

Hamilton, [Anthony Walter] Patrick (1904–1962)

The son of an eccentric clergyman, Hamilton was a repertory actor turned playwright and novelist. His plays include *Rope* (1929, filmed 1948) and *Gaslight* (1939, filmed 1940). His novels, largely set in pubs and boarding-houses, include *Hangover Square* (1941), *The Slaves of Solitude* (1947) and the trilogy, *Twenty Thousand Streets Under the Sky* (1935). The pleasures and dangers of drink and Hamilton's compassionate brand of Marxism are characteristic themes.

Hammett, [Samuel] Dashiell (1894–1961)

After a string of dead-end jobs, Hammett's career took a more interesting turn in 1915, when he started at Pinkerton's Detective Agency. This seems to have been just another job to Hammett, and he would soon leave to do a business course, but when he began writing he would draw upon his experience as a detective in such 'hard-boiled' novels as *The Maltese Falcon* (1930) and *The Glass Key* (1931).

Hampson, John [i.e. John Hampson Simpson] (1901–1955)

Prevented by ill health from having a proper education, Hampson drifted in and out of numerous jobs, many of them in the catering trade, and spent some time in prison for stealing books. He eventually found a time-consuming but congenial job looking after a mentally handicapped boy and started to write in his spare time, becoming a member of the 'Birmingham Group' of writers. His first and best known novel, *Saturday Night at the Greyhound* (1931), was followed by six others, as well as some short stories. He also wrote wartime documentaries for the BBC.

Han Suyin (1917–)

Chinese historian and biographer whose epic, convincing and intelligent novels include *Destination Chungking* (1943), *A Many Splendoured Thing* (1952), *The Mountain is Young* (1958), *Winter Love* (1962), *The Four Faces* (1963), *Till Morning Comes* (1982) and *The Enchantress* (1985). Her non-fiction includes studies of Asia, Tibet, China, Mao Tse-tung and recent Chinese history, as well as several autobiographical works.

Hanley, James (1901–1985)

Born into an Irish seafaring family, Hanley was brought up in Liverpool, going to sea as a stoker at the age of thirteen. After serving in the First World War he did a series of menial jobs ashore, in time gaining sufficient education to become a journalist and finally a full-time writer, occasionally arousing controversy with such books as *Boy* (1931). His other books include a trilogy set amongst the Dublin poor, starting with *The Furys* (1935).

Hardwick, Elizabeth (1916–)

Founder and advisory editor of *The New York Review of Books*, Hardwick writes of the American South in the tradition of Carson McCullers but with her own distinctly individual voice, inclined more towards intellectual reference than emotional revelation. Her novels include *The Ghostly Lover* (1945) and *Sleepless Nights* (1979), a fragmentary, shattered meditation on her own life from a Kentucky childhood to her later New York years.

Harland, Henry (1861–1905)

Anglophile American author, often compared to Walter Pater and Henry James. Harland wrote several novels about Jewish immigrant life under the pseudonym Sidney Luska. In 1890 he moved to London where he edited *The Yellow Book* and published several romances, of which the most popular was *The Cardinal's Snuff-Box* (1900).

Harris, [Theodore] Wilson (1921–)

Though Harris has been based in London since 1959, his highly poetic, strongly 'un-English' prose style still testifies to his origins in British Guiana, where he was born and spent the first four decades of his life. He began his writing career as a poet, but made his major reputation with *The Guyana Quartet* of novels (1960–1963). His subsequent career as a novelist has been both varied and prolific.

Hartley, L[eslie] P[oles] (1895–1972)

Hartley's novels, notably the *Eustace and Hilda* trilogy (1944–1947), which traces the development of an intense, complex brother-and-sister relationship, and *The Go-Between* (1953), are characterized by a powerful evocation of childhood through which the adult world comes to be seen as sinister and corrupt. His other novels include *The Brickfield* (1964), *Poor Clare* (1968) and a science-fiction novel, *Facial Justice* (1960).

Hawkes, John [Clendennin Burke] (1925–   )

Born in Connecticut, Hawkes has led a comfortable, conventional life as an academic. However, two periods stand out as crucial in inspiring the bleak terror and violence of his fiction: an early adolescence in the wilds of Alaska and a tour of duty as an ambulance driver in war-ravaged Europe. Of his best-known novels, *The Cannibal* (1949) and *Adventures in the Alaskan Skin Trade* (1985) draw directly upon these experiences, while *The Blood Oranges* (1970) is a more pastoral book, albeit one tainted by death. Hawkes's books are much concerned with place and form, and he has also written short stories.

Heller, Joseph (1923–   )

A combat flyer, advertising copywriter and university lecturer, Heller drew on his experiences in the US air force in the Second World War to write his best-selling *Catch-22* (1961), satirizing war, the conspiracy of bureaucratic control, and the absurdism of history. His later novels, often as bleakly pessimistic, include *Something Happened* (1974), *Good as Gold* (1979) and *God Knows* (1984), for which he was awarded the French Prix Médici Etranger in 1985.

Hemingway, Ernest [Miller] (1899–1961)

The son of an Illinois doctor, Hemingway worked as a reporter, serving in 1918 as a volunteer with an ambulance unit on the Italian front where he was wounded. In Paris, among the expatriate literary group he describes in *A Moveable Feast* (1964), he published collections of stories and poems. *The Sun Also Rises* (1926, UK title: *Fiesta*) caught the post-war mood of the 'lost generation' and *A Farewell to Arms* (1929) confirmed his influential status. His dissatisfaction with contemporary culture and cultivation of the brutal and primitive was expressed in his later writing. He actively supported the Republicans during the Spanish Civil War, setting *For Whom the Bell Tolls* (1940) against its background, and was a war correspondent before moving to Cuba, where he wrote *The Old Man and the Sea* (1952). He was awarded the Nobel Prize in 1954 and, after a period of serious illness, shot himself.

Heppenstall, [John] Rayner (1911–1981)

The son of a draper and domestic servant, Heppenstall grew up in the industrial Midlands. A lifelong Francophile after a visit to France, he began as a poet but remains best known for his first novel, *The Blaze of Noon* (1939). His later fiction,

especially *The Woodshed* (1962) is reminiscent of the work of Alain Robbe-Grillet whom he championed. His other published work includes criminology and crime history, memoirs, literary criticism, poetry and translations.

Herbert, A[lan] P[atrick] (1890–1971)

A practising barrister, regular contributor to *Punch* and sometime MP for Oxford University, Herbert was celebrated for his humorous legal sketches, large quantities of light verse and for his novels, of which *The Secret Battle* (1919) questioned the conduct of courts martial in the First World War and *Holy Deadlock* (1934) formed part of the author's long campaign for the reform of the divorce laws. He also wrote a number of plays, while a ceaseless stream of letters to *The Times* amounted to a popular work in themselves.

Highsmith, Patricia (1921–   )

US writer of mixed German and English-Scots parentage. The best known of her stylish crime novels feature the amoral anti-hero Tom Ripley who, like his author, is a French resident. Highsmith excels in creating tension and psychological exploration of character. Her novels include *The Talented Mr Ripley* (1956) and *Deep Water* (1957).

Hill, Susan (1942–   )

Hill's highly polished novels and stories, rooted in the traditions of Dickens and Hardy, have a powerful sense of the symbolic potential of the English landscape; her prose is able to penetrate apparent conventionalities of character and narrative to reveal emotional undercurrents. They include *Strange Meeting* (1971) and *The Bird of Night* (1972, Whitbread Award). She also writes children's fiction and books on country life.

Hilton, James (1900–1954)

Born in Lancashire, Hilton settled in Hollywood as one of its most successful scriptwriters. His books include *Lost Horizon* (1933), envisaging Shangri-la, a remote district of Tibet where time stands still; *Goodbye, Mr Chips* (1934), the portrait of an old schoolmaster; and *Random Harvest* (1941).

Hines, Barry [Melvin] (1939–   )

Hines is an author with a deep engagement in contemporary British social issues. His work depicts with compassion the pathos and troubled existence of modern urban communities. His novels include *Blinder* (1966), *A Kestrel for a Knave* (1968), *The Gamekeeper* (1975), *The Price of Coal* (1979), *Looks and Smiles* (1981) and *Unfinished Business* (1983). He has often collaborated with the realist film director Ken Loach.

Hoban, Russell [Conwell] (1925–)

An illustrator before he was a writer, Hoban was a successful author of children's books (an occupation he has never entirely renounced) before he turned to adult fiction. His most celebrated book for children is *The Mouse and His Child* (1969), while of his novels for adults, *Ridley Walker* (1980), with its disintegrating language and philosophical and moral complexity, is considered a major work.

Hobhouse, Janet [Konradin] (1948–1991)

Her childhood marred by her parents' difficult marriage and acrimonious divorce, Hobhouse went on to a turbulent adulthood of her own, her tragedies informing her darkly emotional, though unsentimental and frequently comic, fiction. Her novels include *Nellie Without Hugo* (1982) and *The Furies* (1992). She also wrote an acclaimed biography of Gertrude Stein, *Everybody Who Was Anybody* (1985).

Holtby, Winifred (1898–1935)

Holtby's novels, which vividly describe the kind of Yorkshire village society she knew as a girl, have recently been rediscovered: *South Riding* (1936) remains the most widely read of her books. For a long time, she was overshadowed by Vera Brittain, who commemorated their friendship, and Holtby's work as a journalist and political speaker, in her *Testament of Friendship* (1940).

Hope, Christopher [David Tully] (1944–)

Born in Johannesburg, Hope started his writing career as a poet, producing several volumes of verse, the first of which, *Cape Drives*, was published in 1974 (the year before he moved to England) and won the Cholmondeley Award. Growing up in a white South Africa ostracized by the world, he has constantly striven – in both his poetry and his novels – to articulate the feeling of isolation and rejection this community has internalized. *The Hottentot Room* (1986), for example, is set in a club for exiles in Earls Court. He has won numerous prizes, including the Whitbread for *Kruger's Alp* (1984), and has published travel books, volumes of short stories and children's books.

Household, Geoffrey [Edward West] (1900–1988)

Born in England, Household travelled the world for many years – first tamely enough as a banker and banana salesman, then (more appropriately for an adventure novelist) as a commando and an intelligence officer – before settling down to a writer's life. His second novel, *Rogue Male* (1939), was a great success and was followed by numerous other thrillers in the tradition of John Buchan.

Hudson, W[illiam] H[enry] (1841–1922)

An ornithologist whose explorations in Patagonia had won him fame, the Argentine-born Hudson still found himself destitute when he came to London in 1874. He duly turned to writing, his tales – somewhere between fiction and travelogue – providing an escape to the wide open spaces for countless workbound city-dwellers. Apart from the fantasy novel *Green Mansions* (1904), Hudson wrote several volumes of fiction, as well as numerous books of ornithology and the classic account of rural Wiltshire, *A Shepherd's Life* (1910).

Hughes, Richard [Arthur Warren] (1900–1976)

Although Hughes was celebrated as a poet, playwright and political intriguer while still at Oxford, his subsequent career was less obviously brilliant. The novels he produced so slowly and painfully, such as the uncompleted trilogy *The Human Predicament* (starting in 1961 with *The Fox in the Attic*), nevertheless earned him a considerable reputation.

Hulme, Keri (1947–)

An artist as well as a writer, Hulme has always sought to capture in her fiction and poetry a wider, fuller sense of her native New Zealand, incorporating the rhythms, accents and textures of the Maori idiom. She made her international reputation in 1985 when *The Bone People* became the unexpected winner of the Booker Prize.

Hutchinson, R[ay] C[oreyton] (1907–1975)

Educated at Oxford, Hutchinson went into advertising, writing short stories for *The English Review*. A family man, who disliked literary society and worked for the church, he regarded his life as uneventful. His novels, researched with meticulous accuracy, include *Thou Hast a Devil* (1930), *Testament* (1938, Sunday Times Gold Medal), and *A Child Possessed* (1964, W.H. Smith Literary Annual Award). His final complete novel, the comic *Origins of Cathleen* (1976), was an instant bestseller. Though personally modest, he had many admirers, including Edwin Muir and I. Compton-Burnett.

Huxley, Aldous [Leonard] (1894–1963)

From the disillusionment and satirical eloquence of his early novels to the Utopianism, mysticism and search for spirituality of his later work, Huxley's novels of ideas helped to liberate a generation by shedding light 'in dark places'. They include *Crome Yellow* (1921), *Antic Hay* (1923), *Point Counter Point* (1928), *Brave New World* (1932), *Eyeless in Gaza* (1936) and *Island* (1962). A prodigious and versatile writer of essays, historical studies, travel books, verse and drama, he was the grandson of Thomas Henry Huxley and the brother of Julian Huxley.

Irving, John [Winslow] (1942–   )

Irving was a teacher of literature and a wrestling coach before the success of *The World According to Garp* in 1978 allowed him to devote himself to writing full time. His more recent work has taken him to the heart of public controversies over abortion and sexual politics in such novels as *The Cider House Rules* (1985).

Isherwood, Christopher [William Bradshaw-] (1904–1986)

Best known for his witty, observant novels set in late Weimar Berlin, *Mr Norris Changes Trains* (1935) and *Goodbye to Berlin* (1939), Isherwood had a long and varied career. In the 1930s he formed an influential triumvirate of writers with W.H. Auden and Stephen Spender, collaborating with Auden on three verse plays and a book about the Sino-Japanese War, then emigrating with him to America in 1939. Most of his books were directly or indirectly autobiographical and later novels, such as *Down There on a Visit* (1962), *A Single Man (1964)* and *A Meeting by the River* (1967) were influenced by Vedanta, about which he wrote several non-fiction books.

Ishiguro, Kazuo (1954–   )

His oceanographer father having come to England to assist with the development of the North Sea oilfields, Ishiguro grew up in Guildford, Surrey, speaking Japanese at home and English outside. His fictional work has explored mythical views of both cultures in such novels as *An Artist of the Floating World* (1986) and *The Remains of the Day* (1989).

Jacobson, Dan (1929–   )

Jacobson grew up in South Africa, and his early, naturalistic fiction (including the novellas *The Trap*, 1955, and *A Dance in the Sun*, 1956) is set there, although he had moved to England in the mid-1950s. Apparently exhausting that theme, there was a gap in his career before, in the late 1960s, he began to rediscover his Jewish identity, and with it a new way of writing, in such novels as *The Rape of Tamar* (1970) and *Her Story* (1987), both of which draw upon Biblical stories. He has

also been an academic, producing volumes of essays and a book of 'autobiographies', *Time and Time Again* (1985).

Jacobson, Howard [Eric] (1942–   )

Born in Manchester, Jacobson read English at Cambridge under F.R. Leavis and subsequently taught in universities and polytechnics, and came to notice with a campus novel, *Coming from Behind* (1983). He also spent some time in Australia, the setting of his equally funny novel *Redback* (1986) and his travel book, *In the Land of Oz* (1987). He has also written journalism, notably as a television critic, and written documentaries on Australia and Jewish life.

James, Henry (1843–1916)

Born in New York, James lived in Europe from 1875 and became a naturalized British subject 1915. His novels deal with social, moral, and aesthetic issues arising from the complex relationship of European to American culture. Initially a master of psychological realism, noted for the complex subtlety of his prose style, James became increasingly experimental, writing some of the essential works of early Modernism. His major novels include *Roderick Hudson* (1876), *The Europeans* (1878), *Daisy Miller* (1879), *Washington Square* (1881), *The Bostonians* (1886), *The Portrait of a Lady* (1881), *The Aspern Papers* (1888), *The Wings of the Dove* (1902), *The Ambassadors* (1903) and *The Golden Bowl* (1904).

Jameson, [Margaret] Storm (1891–1986)

Born in Whitby, the daughter of a sea captain, Jameson published novels, poetry, essays and biographies. Her major work is the trilogy *The Mirror in Darkness* (1934–1936). She also published several volumes of autobiography, including *Journey from the North* (1969), describing her time in the last war as president of English PEN.

Jarrell, Randall (1914–1965)

Harrowing poet, mordant critic and well-known editor (of among other magazines *The Partisan Review*), Jarrell wrote only one novel, *Pictures from an Institution* (1954), a satire which drew upon his experiences as an academic at various American universities. His *Complete Poems* were published in 1968.

Jenkins, [Margaret] Elizabeth [Heald] (1905–   )

A schoolmaster's daughter, Jenkins herself spent nearly two decades as a teacher. Since the war, she has devoted herself to writing. Her novels include *The Tortoise and the Hare* (1954) and *A Silent Joy* (1992), which meticulously trace the shifting balances of power in human relationships and, in particular, examine the possibilities of friendship between adults and children. She is also the author of distinguished biographies of (amongst others), Jane Austen and Elizabeth I.

Jesse, F[ryniwiyd – pseud. of Wynifried Margaret] Tennyson (1888–1958)

Despite losing her right hand in a flying accident, Jesse taught herself to write again and travelled the world as a journalist, even reporting from behind enemy lines during the First World War. Criminology became a consuming interest, leading her to edit volumes of the *Notable British Trials* series and write *A Pin to see the Peepshow* (1934) based on the Thompson-Bywaters case. Other novels include *The Lacquer Lady* (1929), and she also wrote poetry, plays (some in collaboration with her husband, H.M. Harwood), and a history of Burma.

Jhabvala, Ruth Prawer (1927–)

Jhabvala was born into a Polish family in Germany, brought to England in 1939, emigrated with her husband to his native India, and now lives in New York. Her central concern in such novels as *A Backward Place* (1965) and *Esmond in India* (1958) and stories collected in *How I Became a Holy Mother* (1976) has not surprisingly been the often comic conflicts and tensions which grow up across racial and cultural boundaries. In addition to her novels, Jhabvala has had an important career as screenwriter for a number of Merchant-Ivory films.

Johnson, B[ryan] S[tanley William] (1933–1973)

A London-born writer whose novels experimented with typography and style, Johnson was more interested in form than content. Novels such as *Albert Angelo* (1964), *The Unfortunates* (1969, which consisted of twenty-seven unbound sections in a box) and *Christie Malry's Own Double-Entry* (1973) show the influence of Concrete Poetry and his admiration for Joyce and Beckett. He committed suicide.

Johnson, Pamela Hansford (1912–1981)

The daughter of Henry Irving's manager, Johnson was brought up in Clapham, which is evoked in many of her novels. She married C.P. Snow in 1950 and is best known for her 'Dorothy Merlin' trilogy, satirizing the pretensions of literary life, which comprises *The Unspeakable Skipton* (1959), *Night and Silence, Who is Here* (1962) and *Cork Street, Next to the Hatter's* (1965).

Johnston, Jennifer [Prudence] (1930–)

Marked for life when her father, playwright Denis Johnston, left home, Johnston has always written about yearning, abandoned figures. An avowed Republican, she lives in Derry, and Ireland's longing for freedom is another recurring theme in her work. Amongst her novels are *The Captain and the Kings* (1972), *How Many Miles to Babylon?* (1974) and *The Old Jest* (1979).

Jones, James (1921–1977)

Jones was the son of a dentist whose family was hit hard by the Depression, and his life at home and in the army (which he joined in 1939, well before America entered the war) was as harsh and tough as his fiction. Despite his bourgeois origins, therefore, he would with some justice describe himself as 'the last of the proletarian novelists'. Remembered chiefly for his award-winning first novel, *From Here to Eternity* (1951), which is set amongst GIs in Hawaii, Jones also wrote *Some Came Running* (1957), about a writer returning from the war to his small home town, short stories, and a non-fiction *Viet Journal* (1974).

Jong, Erica (1942–)

Born in New York, Jong was educated at Columbia, lectured in English at various universities and in 1971 published her first book of verse. Her first novel *Fear of Flying* (1973) became an international bestseller and a milestone in its honest approach to female sexuality. Two sequels, *How to Save Your Own Life* (1977) and *Parachutes and Kisses* (1984), take her heroine Isadora Wing to Hollywood and back to her Russian-Jewish roots. Other novels include *Fanny, Being the True History of the Adventures of Fanny Hackabout Jones* (1980), a riposte to Cleland's *Fanny Hill*, and *Any Woman's Blues* (1990).

Joyce, James [Augustine Aloysius] (1882–1941)

The Dublin-born writer revolutionized the form of the English novel. His first published work was *Chamber Music* (1907), a book of verse, followed by *Dubliners* (1914), a collection of short stories. *A Portrait of the Artist as a Young Man* (1916), was a largely autobiographical novel. Of his two major works, *Ulysses* (1922, UK edition 1936) records the events of a single day in Dublin, experiments with language and combines direct narrative with the unspoken, unconscious reactions of the characters, and *Finnegans Wake* (1939) continued Joyce's experiments with language, attempting a synthesis of many forms of consciousness.

Keane, Molly [i.e. Mary Nesta Keane] (1905–)

Irish novelist whose early comic novels of Anglo-Irish life, under the pseudonym M.J. Farrell, were popular in the 1930s and 1940s. They include *Taking Chances* (1929), *Devoted Ladies* (1934), *Full House* (1935), *The Rising Tide* (1937) and *Two Days in Aragon* (1941), which portrays the endangered ease of Anglo-Irish life, treating political complexity with wit and understanding. After a damning review of her last play in the 1950s, she stopped writing for twenty years until she wrote her masterful black comedy, again of Anglo-Irish life, *Good Behaviour* (1981). Other novels under her own name include *Time After Time* (1983) and *Loving and Giving* (1988).

Kelman, James (1946–)

Having lived and worked all his life in the working-class Glasgow he describes in his fiction, Kelman is a committed and consistent enemy of everything he regards as metropolitan, English and imperialistic. He was first noticed for his collection of short stories, *Not Not While the Giro* (1983). His novels, which have won several awards, make striking use of Glaswegian demotic, notably in *A Disaffection* (1989). He has also written plays.

Keneally, Thomas [Michael] (1935–)

An Australian from an Irish background, Keneally established his reputation in the Commonwealth long before the international success – and controversy – of *Schindler's Ark* (1982). He has always been interested in historical themes, rooting his fiction strongly in real people and events, such as the signing of armistice of the First World War (*Gossip from the Forest*, 1975), or the American Civil War (*Confederates*, 1979). Other novels include *The Chant of Jimmy Blacksmith* (1972) about historical Australian racism, and *Toward Asmara* (1991), about the Eritrean war.

Kennedy, Margaret [Moore] (1896–1967)

British author, now remembered for one book, her second novel, *The Constant Nymph* (1924). Although nothing she wrote thereafter matched the popular and critical success of this novel, later in her career she won the James Tait Black Memorial Prize for *Troy Chimneys* (1952). She also wrote for the stage, notably an adaptation of *The Constant Nymph* in 1926, and published works of criticism.

Kennedy, William (1928–)

Born in Albany, the only child of working-class Irish-American parents, Kennedy was introduced through his father's involvement in the Democratic Party to the political clubs and gaming rooms which feature in his fiction. After working as a sports writer and serving in the Army, Kennedy became Puerto Rico correspondent for Time-Life and studied under Saul Bellow. Returning to Albany, he later became professor there and it is the setting for much of his fiction. His first novel, *The Ink Truck* (1969), is a sardonic, comic story of a newspaper strike. *Ironweed* (1983,

Pulitzer Prize) was published only after Bellow's intervention, but became a literary and popular success. Kennedy has also written essays and screenplays.

Kerouac, Jack [i.e. Jean Louis Kerouac] (1922–1969)

Kerouac named and epitomized the 'Beat Generation' of the 1950s. His work, which includes the novels *On the Road* (1957), *The Dharma Bums* (1958), *The Subterraneans* (1958), *Big Sur* (1962) and *Desolation Angels* (1965), was written with freewheeling, rhapsodic if sometimes undisciplined energy, inspired by his interest in jazz and Buddhism. A legendary symbol of youthful rebellion from the late 1950s, Kerouac was unable to cope with his fame while his innate political conservatism alienated him from the Bohemian lifestyles of his followers. He died an alcoholic semi-recluse.

Kesey, Ken [Elton] (1935–)

Declaring after the success of *One Flew Over the Cuckoo's Nest* (1962) and *Sometimes a Great Notion* (1964) that he intended to pursue life rather than literature, Kesey became a notorious countercultural figurehead until, equally abruptly, he decided in 1969 to settle down on the family farm in Oregon, where he has lived, a lot more quietly (though still unconventionally) ever since.

King, Francis [Henry] (1923–)

King wrote his first novel while still an undergraduate and has had a prolific career as a novelist and critic ever since. He set much of his early fiction in the various countries where he worked for the British Council, notably Japan, Italy and Greece, taking up writing full time when he settled in England in 1964. In both his stories, which are perhaps his finest achievements, and his novels, he explores the darker recesses of the psyche, combining an unblinking investigation of human (often homosexual) relationships with great narrative drive. Amongst his best books are *The Brighton Belle and Other Stories* (1968), *Hard Feelings and Other Stories* (1976), *The Firewalkers* (published under the pseudonym Frank Caudwell, 1956) and *Act of Darkness* (1983).

Kipling, Rudyard (1865–1936)

Kipling was born in Bombay and brought to England in 1871. His schooldays at Westward Ho! were later depicted in *Stalky & Co.* (1899). A journalist in India, he published many early stories and poems which later appeared in such collections as *Plain Tales from the Hills* (1888) and *Soldiers Three* (1890). In 1889 he achieved instant literary celebrity with the publication of his poems in the *Scots Observer*, later collected as *Barrack-Room Ballads* (1892). He travelled extensively, spending much time in South Africa. Widely regarded as unoffical poet laureate he was the first English writer to receive the Nobel Prize. With his vast ouput, Kipling was regarded as the poet of the Empire but later was accused of jingoism. His tales for children such as *The Jungle Book* (1894) and *Just So Stories* (1902) are perhaps his most popular and enduring legacy, and *Kim* (1901), his picaresque novel of India, is generally considered his masterpiece.

Koestler, Arthur (1905–1983)

Born in Budapest and educated in Vienna, Koestler was a foreign correspondent and Communist Party member in the 1930s and was imprisoned under Franco in the Spanish Civil War. From 1940 he lived in England writing fiction (*Darkness at Noon*, 1940) and books and essays on politics, science and literature. *The Roots of Coincidence* (1972) reflects his later interest in parapsychology. A firm believer in the right to euthanasia, he died by suicide, together with his wife.

Kosinski, Jerzy (1933–1991)

Polish-born US author, in the USA from 1957. His childhood experiences as a Jew in Poland during the Second World War are recounted in *The Painted Bird* (1965). The novel that established his cult status, the comic media satire *Being There* (1971, filmed 1979), was followed by increasingly violent works. He was found dead in his bath.

Kureishi, Hanif (1954–   )

Having established his reputation as a dramatist and screenwriter with *My Beautiful Laundrette* (1985) and *Sammy and Rosie Get Laid* (1987), dealing with flair and humour with Anglo-Asian culture in Britain, Kureishi in his first novel, *The Buddha of Suburbia* (1990), exuberantly evokes London in the 1970s, exploring racial and sexual identities in that setting.

Larkin, Philip [Arthur] (1922–1985)

Born in Coventry, Larkin was one of the leading English post-war poets. From 1955 he was librarian at Hull University, and his literary output, though very distinguished, remained small. His perfectionist, pessimistic verse includes *The North Ship* (1945), *the Whitsun Weddings* (1964) and *High Windows* (1974). He also wrote two novels, *Jill* (1946) and *A Girl in Winter* (1947), and published two volumes of collected reviews (one of them about jazz, a passion). He was offered, but turned down, the laureateship. On his instructions, his voluminous diaries were destroyed after his death, and a collection of his letters and a biography by fellow poet Andrew Motion both proved controversial in their revelations of the dark, intolerant and misanthropic aspects of his character.

Laverty, Maura [Kelly] (1907–1976)

Her father a gentleman-farmer grown poor through gambling, Laverty had to support herself, going to Spain to work as a governess and secretary, later becoming a journalist. After her marriage she returned to Ireland and published her first novel, *Never No More*, in 1942. A popular success, it was followed by three more, as well as plays and children's books. An interest in food, already evident in her fiction, also resulted in two cookery books.

Lawrence, D[avid] H[erbert] (1885–1930)

The son of a Nottinghamshire miner, Lawrence expresses in his work his belief in emotion and the sexual impulse as creative and true to human nature. His stormy personal life – a volatile marriage to Frieda von Richthofen, increasing ill health and constant anxiety about money – was reflected in the controversy surrounding his work, caused by his frank treatment of sexual themes and his sustained criticism of industrial society. His major novels include *Sons and Lovers* (1913), *The Rainbow* (1915), *Women in Love* (1920), *Aaron's Rod* (1922), *Kangaroo* (1923), *The Plumed Serpent* (1926) and *Lady Chatterley's Lover* (1928).

Lee, Harper [Nelle] (1926–   )

The youngest child of an Alabama lawyer, Lee abandoned her own legal studies to become a writer in New York. She has written only one book, *To Kill a Mockingbird* (1960), which drew upon her Southern background and her experiences of the law, and became an international best-seller. She assisted her childhood friend, Truman Capote, in gathering interviews for his *In Cold Blood* (1966). A total recluse, she now lives back in Monroeville, Alabama.

Lee, Laurie (1914–   )

Lee transmuted the Gloucestershire rural poverty in which he was born and brought up into a lyrical classic, *Cider With Rosie* (1959). His wanderlust took him first to London then to Spain on the brink of the Civil War and on to the eastern Mediterranean, experiences recounted in *As I Walked Out One Midsummer Morning* (1969) and *A Moment of War* (1991). He has also published several volumes of verse in which he celebrates the countryside.

Lehmann, Rosamond [Nina] (1901–1990)

Born into a rich literary family, Lehmann had an easy entrée to the London salons of the inter-war years, yet her fiction was more adventurous than this implies, being written with utter conviction, and genuinely audacious in its tackling of taboos like homosexuality and abortion in such novels as *Dusty Answer* (1927) and *The Weather in the Streets* (1936).

Lessing , Doris [May] (1919–)

Born in Iran and brought up in Rhodesia, Lessing moved to London in 1949, but returned to Rhodesia for a visit described in her book *Going Home* (1957), after which she was declared a 'prohibited immigrant' by the Rhodesian authorities. Her work documents the development of her concern with social, political and spiritual themes, particularly the place of women in society. Her other novels include *The Children of Violence* sequence (1952–1969) and *The Golden Notebook* (1962).

Leverson, Ada [Esther] (1862–1933)

A journalist and writer of sketches and parodies during the 1890s, Leverson was a loyal friend of Oscar Wilde, who dubbed her 'the Sphinx'. After the collapse of her marriage, she wrote six Edwardian comedies of manners, the best of which form a trilogy, *The Little Ottleys* (1908–1916). These novels display a wit almost equal to Wilde's, and dramatize the difficulties she experienced with her own philandering husband.

Lewis, C[live] S[taples] (1898–1963)

Lewis's books include the medieval study *The Allegory of Love* (1936) and the space fiction trilogy including *Out of the Silent Planet* (1938). A committed Christian, he wrote essays in popular theology such as *The Screwtape Letters* (1942), and *The Chronicles of Narnia* (1950–1956), a series of books for children. A Fellow of Magdalen College, Oxford (1925–1954), he was later professor of Medieval and Renaissance English at Cambridge (1954–1963).

Lewis, [Henry] Sinclair (1885–1951)

Brought up in Sauk Center, Minnesota, where his father was a doctor, Lewis learned all he needed to know about the mediocrity and hypocrisy of American small-town life, a central theme of his fiction, and one that caused much controversy at the time. *Main Street* made his name in 1920, and was followed by *Babbit* (1922) and *Arrowsmith* (1925), the satirical account of a doctor's career, for which Lewis was awarded, but turned down, a Pulitzer Prize. In 1930 he became the first American author to win the Nobel Prize, but the quality of his fiction declined thereafter.

Lewis, [Percy] Wyndham (1886–1957)

Born off Maine, on his father's yacht, Lewis pioneered the new spirit of art that Ezra Pound called Vorticism and which sought to reflect the age of industry. He had a hard, aggressive style both in his writing and painting. Although he has been assessed as a leading force of the early twentieth century (notably as founder-editor of the magazine *Blast*, 1914–1915), his support in the 1930s of fascist principles alienated critics. His books include the novels *Tarr* (1918), *Childermass* (1928) and *The Apes of God* (1930), and the memoir *Blasting and Bombadiering* (1937).

Linklater, Eric [Robert Russell] (1899–1974)

It was only towards the end of his life that Linklater revealed to a public he had not hitherto troubled to disillusion that he had been born in South Wales rather than in Orkney, where he had lived all his adult life and with which his fiction had so strongly associated him. He started out as a poet and playwright, and his first, autobiographical, novel was *White-Maa-'s Saga* (1929), which shared an Orkney setting with subsequent books such as *The Men of Ness* (1932) and *Magnus Merriman* (1934). Other novels were set further afield and include *Juan in America* (1931) and *Private Angelo* (1946). His prolific output also included short stories, histories, anthologies, children's books and biographies.

Lively, Penelope [Margaret] (1933–)

Having begun her literary career as a children's writer, Lively has pursued much the same themes – the importance of memory, individual and collective; the past-in-the-present – in her adult fiction, such as *According to Mark* (1984) and *Moon Tiger* (1987).

Lockridge, Ross (1914–1948)

The son of a prominent Indiana historian, Lockridge was uniquely qualified to write a classic of grassroots historical fiction. Successful as *Raintree County* (1948) was, the effort of researching and writing it proved exhausting: Lockridge had a breakdown, and committed suicide two months after its publication.

Lodge, David [John] (1935–)

The two central influences upon Lodge's life and work have been Roman Catholicism and the academic study of English literature. He was for many years a lecturer at Birmingham University, and both *Changing Places* (1975) and *Small World* (1984) have academic settings, while *How Far Can You Go?* (1980) confronts the moral dilemmas of English Catholics during a quarter of a century's social upheaval. Early novels were realistic accounts of lower-middle-class life, but later ones are comic, drawing upon critical theory and contemporary social politics. He has also written for the stage and the television, notably an adaptation of his own novel, *Nice Work* (1988).

London, Jack (1876–1916)

Few writers have led such fell, adventurous lives as London, who, starting out in the San Francisco slums, had been a sailor, hobo, political activist and Klondyke prospector by the time he was twenty. After a furious process of self-education he was ready to begin turning all this experience to literary account, publishing numerous novels and volumes of short stores, most famously *The Call of the Wild* (1903) and *South Sea Tales* (1911).

Loos, Anita (1888–1981)

The writer of a prodigious number of screenplays, particularly in the silent period, Loos spent most of her working life in Hollywood, where she was also a producer, and about which she wrote several evocative memoirs. She is best remembered for her satirical novels about gold-diggers, *Gentlemen Prefer Blondes* (1925) and its sequel *But Gentlemen Marry Brunettes* (1928), but she also had some success as a playwright, notably with *Gigi* (1951), adapted from the novel by Colette.

Lowry, Malcolm [Clarence] (1909–1957)

The son of a Methodist cotton broker, Lowry went to sea after leaving school, this experience furnishing him with material for his first novel, *Ultramarine* (1933). He returned to England to go to university, but spent much of his adult life abroad, battling with alcoholism and attempting to write. His best book, set and largely written in Mexico, is the visionary *Under the Volcano* (1947). The majority of his work was published after his mysterious death (almost certainly by suicide), and edited by his widow. *Hear us O Lord from Heaven Thy Dwelling Place* (1961) is a collection of stories, and *Lunar Caustic* (1968) is based upon his time as an inmate of New York's Bellevue Hospital. A volume of his *Collected Poetry* appeared in 1991.

Lurie, Alison (1926–   )

One of America's leading novelists, Lurie is also an academic and critic. She teaches studies in children's literature, the subject of her non-fiction book, *Don't Tell the Grown-Ups* (1990). Her novels are particularly good at catching times and places, which she observes with a fine satiric eye: *The Nowhere City* (1965) is a sharp portrait of 1960s' Los Angeles; *The War Between the Tates* (1974) is at once a campus novel and a portrait of a disintegrating marriage; the Pulitzer Prize-winning *Foreign Affairs* (1984) concerns an unlikely freindship between two American visitors to London. She has also written a non-fiction book *The Language of Clothes* (1982).

Macaulay, Rose [Emilie] (1881–1958)

Cambridge-born novelist, essayist and travel writer, the serious vein of whose early novels such as *Non-Combatants and Others* (1916) changed to light satire in *Potterism* (1920) and *Keeping up Appearances* (1928). Macaulay's later books, including *The Towers of Trebizond* (1956), followed her return to the Anglican faith from which she had long been estranged through her love for a married man who died in 1942.

McCarthy [Jr], Cormac [Charles] (1933–   )

McCarthy was born in Providence, Rhode Island, but brought up in the South, the backdrop for much of his fiction. His first novel, *The Orchard Keeper* (1965), won the William Faulkner Foundation Award, and although subsequent novels were highly praised, he did not reach a wide audience until *All the Pretty Horses* (1992) won both the National Book Award and the National Book Critics Circle Award, and thus became a best-seller. His often violent books, which deal with justice and retribution, are written in an idiosyncratic, sparsely punctuated but immediate prose.

McCarthy, Mary (1912–1989)

Orphaned at the age of six, McCarthy was brought up by strictly Catholic relatives, whose views she would never accept, and much of her fiction pivots upon the conflicts of faith. Since, however, she was an active crusader and factional fighter on the left, such conflicts would more often be ideological than theological. She was encouraged to write by her second husband, Edmund Wilson, and produced many volumes of fiction, criticism and travel writing. *Memories of a Catholic Girlhood* (1957) describes her upbringing, *The Group* (1963) draws upon her student days at Vassar, while *The Groves of Academe* (1952) was inspired by her teaching jobs in universities.

McCullers, Carson [Smith] (1917–1967)

One of the leading Southern writers, McCullers was born in Georgia. She regarded herself as being vaguely freakish, somehow akin to the grotesques who populate much of her fiction, most of which is set amongst misfits and outcasts in the Deep South. The quirky, melancholy tone of her work occasionally borders on the sentimental, but such novels as *The Member of the Wedding* (1946) show a tough and witty writer in total command of her material. Other novels include *The Heart is a Lonely Hunter* (1940) and *Reflections in a Golden Eye* (1941). She won numerous awards and also wrote for the stage.

McEwan, Ian (1948–)

English novelist, screenwriter and short-story writer, whose work often has sinister or macabre undertones and contains elements of violence and bizarre sexuality, as in the short stories in *First Love, Last Rites* (1975). His novels include *The Cement Garden* (1978), *The Comfort of Strangers* (1981) and *The Child in Time* (1987). *Black Dogs* (1992) was shortlisted for the Booker Prize.

McGahern, John (1934–)

Although McGahern denies that his work is autobiographical, his novels are clearly rooted in his West of Ireland childhood. He was obliged to go into exile after his second book, *The Dark*, was banned for 'obscenity' in 1965 and he lost his job as a schoolteacher. He returned to County Leitrim in the 1970s and consolidated his position as one of Ireland's leading novelists with *Amongst Women* (1990).

MacInnes, Colin (1914–1976)

Son of the novelist Angela Thirkell (1890–1916), MacInnes was brought up in Australia before returning to England. After serving in the army during the last war, he became a journalist and novelist. He is best known for his novels of teenage and black immigrant culture of the late 1950s, including *The London Trilogy* (1957–1960), in which he describes with vivid documentary clarity the Bohemian underworld of coffee bars and jazz clubs around Notting Hill.

Mackay, Shena (1944–)

Mackay's career as a prodigy, who wrote her first fiction at fifteen and published it at nineteen, came to an end with her marriage, after which she chose to sideline her writing while she reared three daughters. She continued to write, however, and success in a 1980 literary competition prompted her return to full-time writing. A superb stylist, she frequently explores the underbelly of contemporary urban life, transforming the mundane with both wit and lyricism in highly acclaimed novels and volumes of stories such as *Redhill Rococo* (1986) and *Babies in Rhinestones* (1983).

Mackenzie, [Edward Morgan] Compton (1883–1972)

Mackenzie was a prolific writer of books of travel, biography, essays, poems, journalism, and novels, including *Sinister Street* (1913–1914), the massive work charting the spiritual and material education of a naive young man at the turn of the century. His many other novels include the six volumes of *Four Winds of Love* (1937–1945) and *Whisky Galore* (1947). His enormous powers of recall are revealed in his ten-volume autobiography, *My Life and Times* (1963–1971).

Maclaren-Ross, J[ulian] (1912–1964)

Novelist, journalist, short-story writer, film writer and chronicler of Fitzrovia in his classic though unfinished *Memoirs of the Forties* (1965), Maclaren-Ross was also the model for X. Trapnel in Anthony Powell's *A Dance to the Music of Time*. His first novel, *Bitten by the Tarantula* (1945) is just over 100 pages and is set in the South of France in the 1930s; his second, *Of Love and Hunger* (1947) drew upon his experiences as a vacuum-cleaner salesman. His droll, laconic style is perhaps best seen in his stories, collected under such titles as *The Stuff to Give the Troops* (1944) and *Better than a Kick in the Pants* (1945).

MacLaverty, Bernard (1945–)

MacLaverty sets his novels and short stories in his native Belfast, writing sensitively of the difficulties of family relationships, viewed against the violence of the Ulster troubles. His novels include *Lamb* (1980) and *Cal* (1983), both of which have been movingly filmed. His stories have been collected in such volumes as *A Time to Dance* (1982) and *Walking the Dog* (1994), and several have been adapted by him for television.

McMurtry, Larry [Jeff] (1936–)

Born in Wichita Falls, Texas, McMurtry mostly writes about the Southwest and about his home state of Texas in particular. Among his many novels are *Horseman, Pass By* (1961, filmed as *Hud*, 1963), *The Last Picture Show* (1966, filmed 1971) and *Lonesome Dove* (1985), for which he won a Pulitzer Prize.

McWilliam, Candia (1955–)

McWilliam arrived on the literary scene in considerable style with her sophisticated and dazzlingly written first novel, *A Case of Knives* (1988). The book was shortlisted for a Whitbread Prize and won a Betty Trask Award, and was followed by the equally well wrought *A Little Stranger* (1989), an unsettling tale in which a mother describes what happens when she hires a nanny to look after her small son. *Debatable Land* (1994) is set aboard a yacht sailing the South Seas, though principally concerned with the influence of the author's native Scotland on the characters. McWilliam's novels are characterized by wit, startling imagery and elegant prose.

Mailer, Norman (1923–)

One of the most prominent figures of post-war American literature, Mailer gained wide attention with his first, bestselling book *The Naked and the Dead* (1948), a naturalistic war novel. His later works, such as *An American Dream* (1964) and *The*

Executioner's Song (1979), show his personal engagement with history, politics, and psychology. A pugnacious and controversial writer, his polemics on the theory and practice of violence-as-sex brought him into conflict with the feminist Kate Millet in a series of celebrated debates during the 1970s.

Maitland, Sara (1950–)

Living in London's East End and married to an Anglican priest, Maitland is a writer of novels and short stories that are intense and lyrical, suffused with myth and magic. Her books include *Daughter of Jerusalem* (1978), *A Book of Spells* (1987) and *Three Times Table* (1990).

Malamud, Bernard (1914–1986)

Malamud was the Brooklyn-born son of Russian-Jewish immigrants – a heritage which would be the primary concern of his fiction, although his first novel, *The Natural* (1952), deals with that most all-American of themes, baseball. Subsequent novels with Jewish themes include *The Assistant* (1957), *The Fixer* (1966 and winner of a Pulitzer Prize and a National Book Award) and *The Tenants* (1971). He also published several volumes of short stories, including *The Magic Barrel* (1958) and *Rembrandt's Hat* (1973).

Malouf, David [George] (1934–)

Born in Australia, Malouf worked for ten years as a teacher in Britain. He returned to teach English at Sydney University. An imagistic writer and poet, he is concerned with man's search for self-knowledge. *An Imaginary Life* (1978), an invention of the last years of Ovid, celebrates the poet's own theme of metamorphosis. His dense and poetic *Remembering Babylon* (1993, shortlisted for the Booker Prize and the *Irish Times* International Fiction Prize), is set in mid-nineteenth century Queensland. His *Selected Poems* appeared in 1981, and *Child's Play* (1982) is a volume of short stories. He now lives in Italy.

Manning, Frederic (1882–1935)

Born in Australia, Manning came to England in 1898 and, of independent means, set about becoming a man of letters. His collection of prose pieces, *Scenes and Portraits*, was published in 1909 the year that he met Ezra Pound, with whom he planned to edit a new journal. This came to nothing, but Pound remained an influence upon Manning's imagist poetry, the first volume of which, *Poems*, was published the following year. Manning is now remembered as the author of *The Middle Parts of Fortune* (expurgated as *Her Privates We*, 1929), in which he recalled with merciless immediacy and astonishing frankness his experiences as a private on the Somme. *Eidola* (1917) includes some distinctive war poetry.

Manning, Olivia [Mary] (1908–1980)

Best known for her sextet of novels based upon her experiences as a young newly-married woman in Romania, Greece and Egypt during the Second World War (*The Balkan Trilogy*, 1960–1965, and *The Levant Trilogy*, 1977–1980), Manning also wrote novels set in London (*The Doves of Venus*, 1955) and the Indian Ocean (*The Rain Forest*, 1974), as well as two volumes of short stories and a biography of the explorer, Sir Henry Stanley.

Mantel, Hilary [Mary] (1952–)

British novelist and short-story writer, who made her reputation with two very black comedies – *Every Day is Mother's Day* (1985) and its sequel *Vacant Possession* (1986) – which drew upon her experiences as a social worker. She is also a very fine stylist, a quality particularly notable in the mysterious *Fludd* (1989). She spent part of her early adulthood in the Middle East and Africa, and these places form the background of *Eight Months on Ghazzah Street* (1988) and *A Change of Climate* (1994). *A Place of Greater Safety* (1992) is a vast historical novel which imaginatively recreates the French Revolution.

Markandaya, Kamala [Purnaiya] (1924–)

Markandaya established her reputation as the first notable female novelist from India, where she was born and educated, with *Nectar in a Sieve* (1954). In novels like *A Silence of Desire* (1960) and *A Handful of Rice* (1966) she showed the impact of new economic and political ideas on traditional Indian society. *Two Virgins* (1973) marked a radical change, being more experimental and more sanguine about modernization. *The Golden Honeycomb* (1977), her most ambitious novel, charting how the coming of Independence affected a native princely state, confirmed her considerable reputation.

Masefield, John [Edward] (1878–1967)

Masefield was a poet and novelist, whose early years in the navy inspired *Salt Water Ballads* (1902) and several adventure novels. Fame came with the verse narrative of a drunkard's conversion, *The Everlasting Mercy* (1911), and his early poetry was a great influence upon the poets of the First World War. He also wrote children's books, such as *The Box of Delights* (1935), and plays. He was poet laureate from 1930.

Mason, A[lfred] E[dward] W[oodley] (1865–1948)

After a spell as an actor, Mason embarked upon a long and prolific career as a writer of historical novels, romances and detective stories. His best known novel of Empire is *The Four Feathers*

(1902), while the cases investigated by Inspector Hanaud of the Sûreté in such novels as *At the Villa Rose* (1916), *The House of the Arrow* (1924) and *The Prisoner in the Opal* (1929) show Mason's gift for bizarre detail and macabre invention.

Mathews, Harry (1930–)

Mathews studied music at Harvard and in Paris, where he met the poet John Ashbery, with whom he founded and edited the short-lived magazine *Locus Solus* in 1960. His fiction is ludic and postmodernist and he was the first American writer to be elected a member of the Paris-based OuLiPo (or 'Workshop for Poetential Literature.'), founded by Raymond Queneau. Mathews' books include the novel, *The Sinking of the Odradek Stadium* (1985) and a 'short fiction' about masturbation, *Plaisirs Particuliers* (1983).

Maugham, W[illiam] Somerset (1874–1965)

Born in Paris, Maugham studied medicine at St Thomas's, London. During the First World War he was a secret agent in Russia and his 'Ashenden' spy stories are based on this experience. He was a prolific novelist, and a master storyteller; his fiction, often set against exotic backgrounds, includes novels such as *Of Human Bondage* (1915), *The Moon and Sixpence* (1919) and *Cakes and Ale* (1930), and several short-story collections. He also had considerable success as a dramatist.

Maupin, Armistead (1944–)

Born in Washington, Maupin celebrates his adopted city, San Francisco, in all its diversity in his sequence of novels, *Tales from the City* (1978–1990). This hugely popular serial originally started as a serial in the *San Francisco Chronicle* and is remarkable both for its terrific pace and the polymorphous perversity of its characters. His first novel since ending the sequence, *Maybe the Moon* (1992), takes the form of a journal kept by 'the world's shortest woman', a diminutive Hollywood actress called Cadence Roth.

Mayor, F[lora] M[acdonald] (1872–1932)

Mayor's father was Professor of Classics and Moral Philosophy at King's College, London; her mother an outstanding linguist who translated a Zulu grammar from Danish into English. Between successive failed attempts at acting, Mayor transmuted the tedium of her upper-middle-class existence in novels such as *Mrs Hammond's Children* (published under the pseudonym Mary Stafford in 1901) and *The Third Miss Symons* (1913). After her fiancé died suddenly of malaria shortly before their wedding, she suffered a physical and nervous breakdown. *The Rector's Daughter* (1924), relating the self-sacrificing life of a spinster, is the best known of her other novels.

Middleton, Stanley (1919–)

Described as the 'Chekhov of suburbia', Middleton is a retired schoolmaster who lives near Nottingham. Since the 1960s he has written over thirty novels which obliquely and with intuitive sympathy chronicle provincial Middle England, bringing the mundane to vivid life. His novels include *Holiday* (1974), *Entry into Jerusalem* (1983), *After Dinner's Sleep* (1986) and *Catalysts* (1994).

Miller, Henry [Valentine] (1891–1980)

Born in New York, Miller spent the 1930s enduring semi-starvation under the roofs of Paris, doing poorly-paid hackwork to stay alive. He achieved notoriety in 1934 with the publication of his sexually explicit and largely autobiographical novel, *Tropic of Cancer*. He returned to America in 1939, and found his work slowly gaining acceptance there, although he later came in for much criticism from feminist critics. His other work includes travel books, art criticism, a study of Rimbaud and a trilogy of novels, *The Rosy Crucifixion* (1949–1960).

Milne, A[lan] A[lexander] (1882–1956)

His once successful West End comedies now long forgotten, Milne's reputation rests on the poems and stories he wrote to entertain his young son Christopher Robin: *When We Were Very Young* (1924), *Now We Are Six* (1927) and the books about Winnie-the-Pooh. He also wrote *Toad of Toad Hall* (1929), a hugely successful stage adaptation of Kenneth Grahame's *The Wind in the Willows*, and worked for *Punch*.

Mitchell, Gladys [Maude Winifred] (1901–1983)

Graduating from University College, London, in 1921, Mitchell embarked on a forty-year career as teacher, which she managed to combine with an output of at least one crime novel a year – as well as assorted short stories and children's books – until her death. Her detective stories feature the eccentric Dame Beatrice Adela Lestrange Bradley and include *Speedy Death* (1929), *Laurels are Poison* (1942) and *Here Lies Gloria Mundy* (1982).

Mitchison, Naomi [Mary Margaret] (1897–)

Born in Edinburgh into the Haldane family, Mitchison is a prolific writer of fiction, plays, biography, travelogue and children's books. A committed socialist, her work often has political themes. She is best known for her historical novels in which she brings alive the past by the use of modern idiom. She has experimented in several fictional genres and amongst her best novels are *The Corn King and the Spring Queen* (1931), *To the*

Chapel Perilous (1955) and *Memoirs of a Spacewoman* (1962). She recalled her work with Mass Observation in *Among You Taking Notes* (1985), and has written a lively memoir of the 1920s and 1930s, *You May Well Ask* (1979).

Mitford, Nancy [Freeman] (1904–1973)

Born into one of the aristocracy's more eccentric families and educated at home with a clutch of more or less outrageous siblings, Mitford used childhood experience, lightly fictionalized, in her comic novels: *The Pursuit of Love* (1945), *Love in a Cold Climate* (1949), *The Blessing* (1951) and others. She also wrote biographies, translated from the French (she spent much of her adult life in Paris) and edited a celebrated symposium on 'The Identifiable Characteristics of the English Aristocrat', *Noblesse Oblige* (1956).

Mo, Timothy (1950–)

Mo is the son of a wealthy Chinese father and an English mother, and his first language was Cantonese, picked up from the household servants. Coming to England when his mother remarried, he forgot most of this, and despite the Chinese themes and settings of his fiction, *The Monkey King* (1978), *Sour Sweet* (1982) and *An Insular Possession* (1986), he feels, he declares, thoroughly British. He also writes journalism about boxing.

Moorcock, Michael [John] (1939–)

A former rock musician, Moorcock is a leading British exponent of science fantasy. He was an influential editor of the science-fiction magazine *New Worlds*, and has himself written a large number of short stories and several 'sword and sorcery' epics. His most considerable achievement is the trilogy *Dancers at the End of Time*, which opened with *An Alien Heat* (1972), but a less mystical (albeit still highly eccentric) interest in history is evident in his increasingly respected 'mainstream' novels, such as *Mother London* (1988).

Moore, Brian (1921–)

An Irish-born novelist, Moore emigrated to Canada in 1948, and then to the USA in 1959. His earliest books were published under the pseudonym Michael Bryan. Catholicism, obsession, and the contrast between dreams and reality are recurrent and powerful themes, depicted with stylistic economy and realism in such novels as *The Lonely Passion of Judith Hearn* (1955) and *Cold Heaven* (1983).

Moore, George [Augustus] (1852–1933)

Born in Co. Mayo, Moore studied painting in Paris. Returning to England, he exploited Balzac and Zola's naturalistic, realist techniques to revitalize the Victorian novel. *The Modern Lover* (1883), banned by libraries, fuelled Moore's battle against censorship. Later novels included *A Mummer's Wife* (1885), set in the Potteries and influencing Bennett, *Esther Waters* (1894), his most successful, using the background of his father's racing stables, *Evelyn Innes* (1898) and its sequel *Sister Teresa* (1901); later novels such as *The Brook Kerith* (1916) and *Héloïse and Abélard* (1921) were epic in mode. His several autobiographical works include *Hail and Farewell* (1911–1914), an important source for the history of the Irish revival. Moore also collaborated in the planning of the Irish National Theatre.

Morrison, Toni [pseud. of Chloe Anthony Wofford] (1931–)

Born in an Ohio steel town, Morrison was a voracious reader in girlhood. A student and then professor at the highly regarded black college, Howard University, she spent some years in publishing before leaving in 1983 to write full time. Morrison won the 1993 Nobel Prize for Literature. Her novels include *Sula* (1973), *Tar Baby* (1981) and *Jazz* (1992).

Mortimer, John [Clifford] (1923–)

The son of an eccentric barrister (the subject of his play, *A Voyage Round My Father*, 1971), Mortimer for many years pursued twin careers, one at the bar, the other as a writer, these two sides coming together in the popular 'Rumpole' series of courtroom stories, starting with *Rumpole of the Bailey* (1978). In recent years he has concentrated on his writing, not only of novels and plays, but also of work for television, including a hugely successful adaptation of Evelyn Waugh's *Brideshead Revisited*. He has also published volumes of interviews and autobiography.

Mortimer, Penelope [Ruth] (1918–)

Her own experience of married life (including a couple of decades with the barrister-writer John Mortimer) and a two-year stint as an agony aunt on the *Daily Mail* helped prepare Mortimer for a career as a writer specializing in the tensions threatening middle-class marriages. Her novels include *The Pumpkin Eater* (1962), *My Friend Says It's Bulletproof* (1967) and *The Handyman* (1983). She has also published short stories, a controversial biography of the Queen Mother and two volumes of highly frank autobiography.

Mosley, Nicholas (1923–)

Many of Mosley's novels are experimental, intellectual and metaphysical. *Accident* (1965) is perhaps the best known, because of the 1967 Pinter-Losey film. Other novels include *Impossible*

Object (1968) and *Catastrophe Practice* (1979), a complex work based on the mathematical 'catastrophe theory' of the 1970s, and *Hopeful Monsters* (1990). His biographies include a two-volume study of his father, Oswald Mosley.

Mottram, R[alph] H[ale] (1883–1971)

Born in Norwich into a long line of bank clerks, Mottram seemed all set to continue the tradition. As for so many young men of his generation, however, the First World War changed everything, and his powerful fictional recollections of the conflict, collected in 1927 as *The Spanish Farm Trilogy, 1914–1918*, inaugurated his career as the author of some sixty, mostly forgotten, books, mainly fiction set in East Anglia, but also including a memoir of his literary mentor, John Galsworthy.

Munro, Alice (1931–)

One of the most acclaimed short-story writers of the century, Munro was born, lives and sets most of her fiction in Canada. A remarkable sensitivity to suppressed or unrecognized feeling in small-town life is a notable feature of her work, as is her humour and her concise, understated prose. Many of her stories first appeared in the *New Yorker*, and they have been collected in such volumes as *Dance of the Happy Shades* (1968), *The Progress of Love* (1987) and *Friend of My Youth* (1990). *Lives of Girls and Women* (1971) is a novel, and *the Beggar Maid* (1978) is a collection of interrelated short stories.

Murdoch, Iris [Jean] (1919–)

Dublin-born English novelist, critic and philospher. Murdoch's novels combine philosophical speculation with often outrageous situations and tangled human relationships. Though often comic and bizarre, their underlying concern is how love, freedom, and goodness can survive moral and intellectual blindness. These themes are analysed formally in her philosophical studies of Sartre and Plato. She has won numerous awards, including the Booker Prize for *The Sea, The Sea* (1978). Her other novels include *The Bell* (1958), *A Fairly Honourable Defeat* (1970) and *The Sacred and Profane Love Machine* (1974).

Nabokov, Vladimir[ovich] (1899–1977)

Russian novelist, poet and literary scholar, who moved to the US in 1940. From then onwards all his novels were written in English. After the controversial success of *Lolita* (1955), he devoted himself full-time to writing. One of the most original prose writers of the century, he wrote with great narrative skill and linguistic invention. His other novels include *The Real Life of Sebastian Knight* (1941), *Bend Sinister* (1947) and *Pale Fire* (1962). He also published poems, short stories and a translation of *Eugene Onegin*.

Naipaul, Shiva[dhar Srinivasa] (1945–1985)

Naipaul was the younger brother of novelist V.S. Naipaul, but his first novel, *Fireflies* (1970, John Llewelyn Rhys and Winifred Holtby prizes), a masterly, funny, touching and biting satire, established him as an author in his own right. A retrospective selection of his work since his departure from Trinidad in 1964, *Beyond the Dragon's Mouth* (1984), includes stories and an autobiographical sketch. He also wrote political and travel studies.

Naipaul, V[idiadhar] S[urajprasad] (1932–)

Born in Trinidad of a Brahmin family, Naipaul moved to England, working as a literary journalist. His first novels were comedies of manners, set in Trinidad, but from *The Mimic Men* (1967) onwards his work became more overtly political and pessimistic. Recurrent themes of political violence, endemic homelessness and alienation have allowed comparisons with Conrad. His other novels include *A House for Mr Biswas* (1961), *In a Free State* (1971, Booker Prize) and *A Bend in the River* (1979).

Narayan, R[asipuram] K[rishnaswami] (1906–)

Although he published his first novel in 1935, Narayan came of age as a writer after the death of his young wife in 1939, a tragedy he attempted to work out in *The English Teacher* (1945), which is generally considered to be the first mature work of India's greatest living writer. Much of his fiction is set in the imaginary town of Malgudi, whose inhabitants he observes with a gently ironic eye. Amongst his novels are *The Financial Expert* (1952), *The Man-Eater of Malgudi* (1962), *The Sweet-Vendor* (1967), *The Painter of Signs* (1976) and the fable, *A Tiger for Malgudi* (1983). He has also published several volumes of short stories, a translation of the *Mahābhārata*, and his memoirs, *My Days* (1975).

Nesbit, E[dith] (1858–1924)

Prolific author best known for children's classics such as *The Railway Children* (1906) and the series about the Bastable family, including *Five Children and It* (1902). Married to Hubert Bland, a philandering minor Fabian, with whom she collaborated on novels and stories for adults, she was obliged to write in order to support her family and much of her output was mere hackwork.

Newby, P[ercy] H[oward] (1918–)

Wartime service in Egypt and a postwar secondment to teach English literature at Cairo

University gave Newby a long-standing interest in the Middle East, the setting for several of his novels, one of which, *Something to Answer For*, was the winner of the first ever Booker Prize in 1968. After his return to England, Newby combined his career as the writer of some twenty novels with that of a producer on BBC radio. Amongst his other novels are *A Journey to the Interior* (1945) and *Coming In with the Tide* (1991). He has also written books for children, as well as several volumes of non-fiction.

Ngugi Wa Thiong'o [formerly James T. Ngugi] (1938–)

Kenyan novelist whose books include *Weep Not, Child* (1964), based on his own childhood and mission-school education, *The River Between* (1965) and *A Grain of Wheat* (1967), which combines the political quest of his first two novels with the compassionate realism of his short stories. In 1977 Ngugi was arrested by the Kenyan government after he began writing in his own language. He wrote his first Gikuyu novel in prison, later translated and published as *Devil on the Cross* (1982). *Detained: a Prison Writer's Diary* (1981) was published in English.

Norris, Frank[lin] [Benjamin] (1870–1902)

Born in Chicago and brought up in California, Norris embarked on an abortive career as a painter and poet before buckling down to more realistic vocations as a newspaperman and publisher's reader. A spell studying art in Paris had brought him under the influence of Zola, and he became one of the greatest exponents of American naturalism in such novels as *McTeague* (1899) and the projected triology, 'The Epic of the Wheat', only two volumes of which – *The Octopus* (1901) and *The Pit* (1903) – were completed at his early death. He wrote several other novels (including *Vandover and the Brute*, (1914), a naturalistic tale of lycanthropy), two volumes of short stories, and a non-fiction work, *The Responsibilities of a Novelist* (1903).

Nye, Robert (1939–)

British poet, novelist, playwright and short-story writer, best known for his novels which imaginatively and entertainingly recreate the lives of historical, literary and mythical figures: *Falstaff* (1975), which won the *Guardian* and Hawthornden Prizes, *Merlin* (1978), *The Memoirs of Lord Byron* (1989), and so on. His poetry has been published in several volumes and was gathered in 1989 as *A Collection of Poems 1955–1988*. He has also published volumes of short stories and books for children, written plays and

edited the works of, amongst others, Swinburne and William Barnes.

Oates, Joyce Carol (1938–)

For some years a professor of English at Detroit, Oates uses that city as the setting for much of her work. A prolific writer, she has employed a wide range of genres in her novels and numerous short stories, which are often aggressive, realistic descriptions of the forces of darkness and violence in modern culture. Her novels include *A Garden of Earthly Delights* (1969, National Book Award) and *them* (1969). She has also written poems, reviews, and critical essays.

O'Brien, Edna (1932–)

Born in the west of Ireland, O'Brien made her reputation with *The Country Girls Trilogy* (1960–1964). The themes of female sensuality, male treachery, and the pursuit of experience are characteristic of her work, and have brought comparisons with Colette. Later novels include *August is a Wicked Month* (1964), *A Pagan Place* (1971), *Night* (1972) and *Johnny I Hardly Knew You* (1977). She has also published several books of short stories.

O'Brien, Flann [pseud. of Brian O'Nolan] (1911–1966)

Irish civil servant, journalist, novelist and playwright, who wrote in both English and Gaelic. Of the many noms-de-plume adopted by O'Nolan, the best known are Flann O'Brien, under which he wrote surreal novels such as *At Swim-Two-Birds* (1939) and *The Third Policeman* (1967), and Myles na Gapaleen, under which he wrote a long-running column for the *Irish Times*, collecting his pieces in *Cruiskeen Lawn* (1943). He was hailed as 'Ireland's funniest genius' and his work, influenced by Joyce, is studded with multilingual wordplay.

O'Brien, Kate (1897–1974)

Born in Limerick, O'Brien discovered Spain after graduating from university, when she spent a year as a governess in Bilbao. Her novel *Mary Lavelle* (1936) draws on this experience. During the 1930s she lived in Spain for several years with a woman companion, and the country became a second homeland and a recurrent setting for her fiction. Novels such as *The Last of Summer* (1944) are set in Ireland, where O'Brien, like many writers, occasionally fell foul of the Censorship Board.

O'Connor, [Mary] Flannery (1925–1964)

O'Connor was born in Georgia into an old Catholic family in the chiefly Protestant Bible-Belt, a circumstance which informs all her work. A major figure in the Southern Gothic tradition, she is known chiefly for her first novel, *Wise Blood* (1952), a dark and grotesque tale of religious fanaticism. Similar in theme is *The Violent Bear It Away* (1960), and O'Connor was working on a volume of short stories at her early death from lupus, a debilitating and painful tubercular skin disease. This was published posthumously as *Everything That Rises Must Converge* (1965) and was followed by a volume of occasional prose, *Mystery and Manners* (1969), one of letters, and an edition of *Collected Stories* (1971).

O'Faoláin, Julia (1932–   )

O'Faolain is the author of *No Country for Young Men* (1980, shortlisted for the Booker Prize), perhaps the best recent novel dealing directly with 'the Troubles', both as history and as a continuing part of Irish life. Other novels include *Women in the Wall* (1973), a vivid portrait of sixth-century Gaul, and *The Irish Signorina*, a love story set in Ireland and Italy. She is the daughter of Sean O'Faoláin.

O'Faoláin, Sean (1900–1991)

Born John Whelan in Cork, the son of a police constable, O'Faoláin was a member of the IRA during the Troubles. As a novelist, short-story writer and biographer, he explored the frustrations of Irish society and the doomed aspirations of nationalists. He wrote three novels, *A Nest of Simple Folk* (1934), *Bird Alone* (1936) and *Come Back to Erin* (1940), but is best known for his short stories, which evoke the missed opportunities of characters limited by their environment and show his debt to Chekhov.

O'Flaherty, Liam (1896–1984)

Described by Sean O'Faoláin as an 'inverted romantic', O'Flaherty is best known for his short stories published in many periodicals and several volumes, and dealing unsentimentally with life, and often death, from an animal's point of view. An Irish nationalist, he published three volumes of flamboyant memoirs of his turbulent life. His novels include *The Neighbour's Wife* (1923), *The Informer* (1925), *The Puritan* (1931) and *Famine* (1937).

O'Hara, John [Henry] (1905–1970)

Destined for Yale had it not been for his father's early death, O'Hara found middle-class prosperity by another route: as a fiction writer chronicling the lives of America's bourgeoisie in his short stories and novels. His stories, many of which first appeared in the *New Yorker*, were collected in several volumes, and stage and film adaptations of *Pal Joey* (1940) and *Butterfield-8* (1935) brought his work to a larger audience. Of his other novels, the most renowned are *Appointment in Samarra* (1934) and *A Rage to Live* (1949).

Okri, Ben (1959–

Born in Nigeria, Okri went to primary school in London but returned to Lagos when he was seven. He completed his first novel, *Flowers and Shadows* (1980), at the age of nineteen. He returned to England but, unable financially to continue a degree in comparative literature at Essex University, he moved to London, sleeping rough. Two story collections and a further novel, *The Landscapes Within* (1981), followed and though he won literary prizes and awards he was unable to find an immediate audience. The turning-point came with his third novel, *The Famished Road*, which won the Booker Prize in 1991. He published his first volume of poetry, *An African Elegy*, in 1992.

'Olivia' [pseud. of Dorothy Bussy] (1865–1960)

The sister of Lytton Strachey, Bussy wrote only one novel, *Olivia* (1949), a partly autobiographical account of love and passion in a girls' school. Married to the French painter Simon Bussy, she spent much of her life in France, working as a translator. In particular, she translated the works of André Gide, a close friend, and their relationship is commemorated in a volume of *Selected Letters* (1983).

Ondaatje, Michael [Philip] (1943–   )

Born in Sri Lanka, of mixed origins, Ondaatje did not speak English until he came to school in England at the age of ten. He emigrated to Canada at nineteen and taught at the University of Western Ontario while establishing himself as a poet. His first volume of poetry, *The Dainty Monsters*, was published in 1967, but a wider reputation was made with *The Collected Works of Billy the Kid* (1970), which tells the outlaw's story in both poetry and prose. *Running in the Family* (1982) was equally difficult to classify: part memoir, part family history and part travel book. He continued to publish highly praised volumes of poetry, but his international reputation as a fiction writer was made in 1992, when his fourth novel, *The English Patient*, was joint-winner of the Booker Prize.

Orwell, George [pseud. of Eric Arthur Blair] (1903–1950)

Born in India and educated in England, Orwell served for five years in the Burmese police force. He worked as dishwasher, was a tramp, and fought for the Republican cause in the Spanish Civil War. His novels include *Burmese Days* (1934), *A Clergyman's Daughter* (1935), the satirical fable *Animal Farm* (1945) and the prophetic *Nineteen Eighty-Four* (1949). A deep social conscience and an antipathy towards political dictatorship characterize his work. He worked for the BBC during the Second World War and was a regular contributor to *Tribune* and the *Observer*. His essays and other journalism have been collected in several volumes.

Panter-Downes, Mollie (1906–   )

Panter-Downes published her first novel, *The Shoreless Sea* (1923), at the age of seventeen, and her second, *The Chase*, two years later. Although she has published five novels in all, *One Fine Day* (1947) is the only one she rates, and she sees herself principally as a journalist. She was for many years the London correspondent for the *New Yorker*, and published two collections of the letters she wrote for the magazine during the Second World War. She is also the author of *Ooty Preserved* (1967), about a Victorian hill station in India, and *At the Pines* (1971), a moving, funny, evocative and beautifully written account of Swinburne's declining years with Theodore Watts-Dunston in Putney.

Paton, Alan (1903–1988)

Born in Pietermaritzburg, Paton became a schoolmaster, an enlightened principal of a Johannesburg reformatory, and National President of the South African Liberal Party until it was declared illegal in 1968. His novel *Cry, the Beloved Country* (1948) focused movingly on racial inequality in South Africa. He also wrote political and social studies, short stories and an autobiography, *Towards the Mountain* (1980).

Peake, Mervyn (Lawrence) (1911–1968)

Born in China, son of a medical missionary, Peake studied art at the Royal Academy. Teaching art and later commissioned as a war artist, he was profoundly affected by a visit to Belsen. He illustrated most of his own work, including his extraordinary comic gothic fantasy, *The Gormenghast Trilogy* (1946–1959), and books of poetry and nonsense verse. He died tragically early of Parkinson's disease.

Plath, Sylvia (1932–1963)

Plath's powerful, highly personal poems, often expressing a sense of desolation, are distinguished by their intensity and sharp imagery. Her *Collected Poems* (1981) was awarded a Pulitzer Prize. Her autobiographical novel *The Bell Jar* (1963) deals with the events surrounding a young woman's emotional breakdown. Married to poet Ted Hughes, she committed suicide while living in London.

Plomer, William [Charles Franklyn] (1903–1973)

Born in Pietersberg, of British parents, Plomer set his most famous novel, *Turbott Wolfe* (1926), in racially segregated South Africa. Educated in England, he spent some time in Japan, the setting of his stories collected as *Paper Houses* (1929) and his discreetly homosexual novel, *Sado* (1931). He settled in England in 1929 and became a reader for Jonathan Cape, discovering and editing the diaries of Francis Kilvert. He was also a distinguished poet, specializing in the ballad form, which he often used to tell macabre stories, and he collaborated with Benjamin Britten on his 'Church Parables' (1964–1968) and the opera, *Gloriana* (1953).

Porter, Katherine Anne (1890–1980)

Born in Indian Creek, Texas, Porter began her writing career after high school, working on newspapers and as a freelance, and lived in Mexico and then in Europe. Her short stories are notable for a depth and complexity usually found only in novels. She also wrote a collection of essays, and the allegorical novel *Ship of Fools* (1962). Her *Collected Short Stories* (1965) won a Pulitzer Prize.

Potok, Chaim [Icyck] (1929–   )

Born in New York of Polish immigrant parents, Potok trained as a rabbi and subsequently served as an army chaplain in Korea. His first novel, *The Chosen* (1967) and its sequel, *The Promise* (1970) are set in the Hasidic community in Brooklyn and follow the fortunes of two boys, both sons of rabbis, one liberal, one orthodox. He has written several other novels, all on Jewish themes – including *My Name Is Asher Lev* (1972) and its late sequel, *The Gift of Asher Lev* (1990) – and volumes of non-fiction, including *Wanderings* (1978), a history of the Jews.

Potter, [Helen] Beatrix (1866–1943)

A lonely, introverted child, Potter came early to love the Lake District, where her London family went for summer holidays, and where she would settle herself as soon as she could. In 1893 she wrote *The Tale of Peter Rabbit* in a letter to a young acquaintance, and this became, in 1902, the first of her small but perfect books for children, all illustrated by her watercolours.

Powell, Anthony [Dymoke] (1905–   )

Powell's pre-war reputation as a writer of light comedy and satire has given way to that of the much more ambitious tragi-comic chronicler of the century through the sequence of twelve novels,

A Dance to the Music of Time (1951–1975) that begins shortly after the Second World War and charts a period of fifty years in the lives of Nicholas Jenkins and his upper-class friends. Powell's memoirs, *To Keep the Ball Rolling* (1976–1982), shed much light on the creation of his fictional characters.

Powys, John Cowper (1872–1963)

Powys was the son of a Dorset country parson and brother of writers Llewelyn and Theodore. His mystic and erotic books include *Wolf Solent* (1929), a remarkable *Autobiography* (1934) and the huge, pantheistic epic *A Glastonbury Romance* (1933). Prodigious also as an essayist, poet, and lecturer, Powys has been charged with self-indulgence and verbosity. Others regard him as a unique and major writer on the scale of Dickens and Dostoevsky.

Powys, T[heodore] F[rancis] (1875–1953)

Powys was born in Dorchester, the brother of the writers John Cowper Powys and Llewelyn Powys. He lived most of his life in Dorset, and the local countryside forms the backdrop to the majority of his fiction. His best book, *Mr Weston's Good Wine* (1927), is a genial if bizarre Christian allegory, and religion and a concern with good and evil inform most of his highly idiosyncratic novels and short stories.

Priestley, J[ohn] B[oynton] (1894–1984)

Born in Bradford, Priestley began work as a journalist and critic, before major successes with such novels as the rambling comedy *The Good Companions* (1929), and the realist *Angel Pavement* (1930). As a playwright he was often preoccupied with theories of time. Among his best known plays are *Time and the Conways* (1937), *When We are Married* (1938) and *An Inspector Calls* (1947). He was a popular and influential wartime broadcaster, essayist, and literary and social critic. A prolific writer, Priestley inherited H.G. Wells's mantle as English Man of Letters for the mid-century.

Pritchett, V[ictor] S[awdon] (1900–)

Born in Ipswich, the son of a Christian Scientist travelling salesman, Pritchett has made the lower-middle-class provincial condition and the varieties of religious enthusiasm central themes of his fiction. A master of the short story, Pritchett has concerned himself only secondarily with the novel – one reason for the comparative neglect with which his writing has been treated. His *Collected*

Stories appeared in 1990 and two volumes of autobiography in 1968 and 1971. He has also published several critical studies, notably of Balzac and Turgenev.

Proulx, E[dna] Annie (1935–)

Proulx came to writing late, but arrived with prize-winning panache. Her first book was *Heart Songs and Other Stories* (1988), but it was with her first novel, *Postcards* (1991) that she really made her mark, becoming the first woman to win the PEN/Faulkner Prize. Proulx's story of a Vermont hill farmer on the run from murder was told in a terse, punchy prose, and her extraordinarily vivid style also marked out *The Shipping News* (1993), which won three major literary prizes.

Purdy, James (1923–)

Born in Ohio, Purdy deals in his fiction, as in his poetry and plays, with extreme emotional states and dramatic transformations. His first novel, *Malcolm* (1959), concerns the quest of a teenage boy for his lost father and is set in a world of bizarre depravity. His work typically treats homosexual subjects with stylistic elegance and a flair for strange or grotesque images. Other novels include *The Nephew* (1960), *Cabot Wright Begins* (1964) and *In a Shallow Grave* (1975).

Pym, Barbara [Mary Crampton] (1913–1980)

Born in Shropshire, Pym was educated at Oxford, where she imagined her circle in later life, eventually using this notion as the basis for her first novel, *Some Tame Gazelle* (1950). She subsequently worked as an editor at London's International African Institute. She produced many novels, in which the quiet lives of 'excellent women' (devoted largely to the church, good works and spinsterly yearnings for curates), are observed with a sharp eye and wit. In spite of having a devoted following, she failed to find a publisher for her seventh novel. Years of neglect came to an end, however, when encomiums from David Cecil and Philip Larkin led to a revitalized career and the publication of several new novels.

Pynchon, Thomas (1937–)

Born in Glen Cove, New York, Pynchon has created, with great stylistic verve, a bizarre, labyrinthine world in his books, the first of which was *V* (1963). It was followed by the shorter comic quest novel, *The Crying of Lot 49* (1966), before his gargantuan tour-de-force *Gravity's Rainbow* (1973), which represents a major achievement in twentieth-century literature, with its fantastic imagery and esoteric language, drawn from mathematics and science.

Rao, Raja (1909–)

Rao was educated in India and at the Sorbonne. His first novel, *Kanthapura* (1938), draws upon his own involvement in the independence movement of the 1930s, and is set in the eponymous southern Indian village, creating an Indian idiom and tempo through his use of the English language. *The Serpent and the Rope* (1960) describes the spiritual quest of a Brahmin intellectual. Rao has also published two volumes of short stories.

Raven, Simon [Noël] (1927–)

Classicist, journalist and scurrilous memoirist, best known for his series of ten interconnected, richly comic novels, *Alms for Oblivion* (1964–1973), which chart the fortunes of a group of friends and enemies from the end of the Second World War to the 1970s. This unflattering portrait of an era has been followed by another sequence featuring some of the same characters, *The First Born of Egypt*, starting with *Morning Star* (1984), and Raven has also written several unconnected novels, all of which depict enduring human virtues assailed by the malignities of fate and the powers of corruption. He has also published volumes of plays and essays.

Reed, Ishmael [Scott] (1938–)

Reed is an editor and publisher of minority ethnic writers, and a teacher of creative writing at the University of California at Berkeley. His novels parody and satirize notions of historical fact, exploiting traditions taken from jazz and voodoo. They include *The Free-Lance Pallbearers* (1967), *Mumbo Jumbo* (1972), and *Reckless Eyeballing* (1986). His poetry includes the collection *Chattanooga* (1973).

Reid, Forrest (1875–1947)

Anglo-Irish by birth, Reid was born and spent most of his life in Belfast. His principal subject as a novelist was boyhood, about which he wrote many books, notably *The Garden God* (1905), *Peter Waring* (1937) and *The Tom Barber Triology*, which was published out of chronological order, starting with the final volume, *Uncle Stephen*, in 1931. He also wrote critical studies of W.B. Yeats and Walter de la Mare, and two volumes of autobiography, *Apostate* (1926) and *Private Road* (1940).

Renault, Mary [pseud. of Mary Challans] (1905–1983)

The daughter of a doctor, Renault worked as a nurse before emigrating to South Africa in 1948 with her lover, Julie Mullard. She drew upon her medical experiences in her early fiction, little of which is now read, as well as in her pioneering novel of (male) homosexuality, *The Charioteer* (1953). In 1956 she published *The Last of the Wine*, the first in a series of remarkable novels in which she brought the ancient world to vivid life. Amongst these later novels are *The King Must Die* (1958), *The Bull from the Sea* (1962), *The Praise Singer* (1978) and *Funeral Games* (1981).

Rhys, Jean [pseud. of Ella Gwendolen Rees Williams] (1894–1979)

Born in Dominica, where her Creole mother's family had been planters, Rhys came to England at the age of sixteen, but was left in poverty after her father's death. She had to leave drama school to work as a mannequin and chorus girl, the start of a precarious, drifting existence around Europe, which would see her become herself the lonely, vulnerable woman familiar from such novels as *After Leaving Mr Mackenzie* (1930), *Voyage in the Dark* (1934) and *Good Morning, Midnight* (1939). She subsequently vanished from the literary scene, but her career was revitalized in 1966 by *Wide Sargasso Sea*, which drew upon her childhood and *Jane Eyre*. She also published several volumes of highly praised short stories and a fragment of autobiography, *Smile Please* (1979).

Richardson, Dorothy [Miller] (1873–1957)

After working as governess, secretary, translator and journalist, Richardson became an intimate friend of H.G. Wells and was encouraged to write. She was the first English novelist to use the 'stream of consciousness' method in *Pointed Roofs* (1915). She was credited by Virginia Woolf with having invented 'the psychological sentence of the feminine gender'. Her novels were collected under the title *Pilgrimage* (1938).

Richardson, Henry Handel [pseud. of Ethel Florence Lindesay Richardson] (1870–1946)

The Melbourne-born novelist was the first Australian writer to win a reputation abroad. In 1888 she went to study piano in Leipzig, Germany, and her first novel, *Maurice Guest* (1908), is based on these years. *The Getting of Wisdom* (1910), based on her schooldays, was filmed in 1977.

Richler, Mordecai (1931–)

Brought up in Montreal's Jewish ghetto and strongly influenced by his grandfather, a Hasidic scholar, Richler went to university in Montreal but dropped out to live in Paris. Returning to Montreal he published his first novel *The Acrobats* (1954). After writing film and television scripts in England he found his fictional form with his second novel, *Son of a Smaller Hero* (1955), the first of a trio drawn from his youth, escape from the ghetto and sexual confrontation with the Gentile; the others are *The Apprenticehip of Duddy Kravitz* (1959) and *St Urbain's Horseman* (1971). He has also written collections of essays and books for children.

Richter, Conrad [Michael] (1890–1968)

Born in Pennsylvania, Richter was fourteen when his father was ordained a Lutheran minister. If this background, along with his work on the land after leaving college, influenced the themes of his fiction, his terse, economic style was formed by his subsequent experiences as a journalist in Pittsburgh. His novels are largely set in pioneer farming communities and include *The Awakening Land*, a trilogy beginning with *The Trees* (1950). Later works, such as *The Waters of Kronos* (1960), introduce elements of fantasy to his largely historical fictional territory.

Robertson, E[ileen] Ar[buth]nott (1903–1961)

The daughter of a Surrey country doctor, Robertson was educated at Sherborne, Paris and Switzerland. She published nine novels, the first of which, *Cullum* (1928), caused a minor sensation with its sexual frankness and the last of which, *The Strangers on My Roof* (1964), was published posthumously. Her first real success came with *Four Frightened People* (1931), a story of endurance in the Malayan jungle. Revived by Virago in the 1980s, her work has aroused interest for its atmosphere, outstanding descriptions of landscape, sailing and bird-life, and the bright-eyed wit and cruelty of her young middle-class heroines.

Rolfe, Fr[ederick William Serafino Austin Lewis Mary] (1860–1913)

An English Roman Catholic convert, Rolfe styled himself 'Baron Corvo'. Frustrated in his desire to enter the priesthood, he wrote the novel *Hadrian the Seventh* (1904), in which the character of Hadrian rose from being a poor writer to become pope. In *Desire and Pursuit of the Whole* (1934) he wrote about his homosexual fantasies and friends. Although he had admirers during his lifetime, he alienated most of them by his paranoia and requests for financial support, earning W.H. Auden's description of him as 'a master of vituperation'. His other work includes *Stories Toto Told Me* (1898), *Chronicles of the House of Borgia* (1901) and *Don Tarquinio* (1905).

Roth, Henry (1906–)

Until his late eighties, Roth had a large reputation which rested upon one book, *Call It Sleep* (1934), a novel of immigrant Jewish life in New York, clearly based upon his own background. He apparently wrote and destroyed a second novel,

and occasionally published short stories and fragment of memoir in such magazines as the *New Yorker*. In 1994, however, there appeared *A Star Shines Over Mt Morris Park*, the first volume of a projected sequence of six novels.

Roth, Philip (1933–)

Roth's witty, sharply satirical, and increasingly fantastic novels depict the moral and sexual anxieties of twentieth-century Jewish-American life, most notably in *Goodbye Columbus* (1959) and *Portnoy's Complaint* (1969). He has also written a series of semi-autobiographical novels about a writer, Nathan Zuckerman, a memoir about his father's death, *Patrimony* (1991) and several allegories, including *The Breast* (1972).

Rubens, Bernice (1928–)

Born in Cardiff, Rubens is noted for her bizarre, comic touch and her deft ability to suggest the complex undercurrents of human behaviour. Her novels, which often deal with Jewish life and themes, include *The Elected Member* (1969), *Five Year Sentence* (1977), *Spring Sonata* (1979), *Brothers* (1983) and *Birds of Passage* (1987).

Rushdie, Salman [Ahmed] (1947–)

Born and brought up in Bombay, Rushdie came to England as a student. He is a leading exponent of magic realism and *Midnight's Children* won the Booker Prize in 1981, putting Rushdie firmly on the Commonwealth literary map. It was *The Satanic Verses* (1988) that brought worldwide celebrity, however, largely because the novel offended Muslim susceptibilities. Accused of blasphemy, Rushdie saw his books being burned publicly, and then was obliged to go into hiding when Iran's Ayatollah Khomeini issued a *fatwah*, urging Muslims to murder the author.

Saki [pseud. of H[ector] H[ugh] Munro] (1870–1916)

Burmese-born British writer best known for his savagely witty, laconically narrated short stories. His earliest volumes, in the tradition of Wilde, featured a young man-about-town called Reginald; later ones, such as *Beasts and Super-Beasts* (1914), are often macabre, concerned with revenge, and involve Pan, in various guises, causing havoc in Edwardian drawing rooms. He also worked as a foreign correspondent for the *Morning Post*, wrote satirical sketches for the *Westminster Gazette* and published two novels. He was killed in action in the First World War.

Salinger, J[erome] D[avid] (1919–)

Born in New York City, Salinger studied writing at Columbia University. After *The Catcher in the Rye* (1951), his classic novel of adolescent angst, became a success, particularly among college students, he moved to New Hampshire, and became increasingly reclusive, developing his lyrical Zen themes in *Franny and Zooey* (1961) and *Raise High the Roof Beams, Carpenters and Seymour: An Introduction* (1963), short stories about a Jewish family named Glass, after which he stopped publishing.

Sansom, William (1926–1976)

The London-born short-story writer and novelist made his reputation in the Second World War with stories about the London Fire Service, collected in *Fireman Flower* (1944). Numerous volumes of stories followed. His novels are less well-known, apart from *The Body* (1949), his first and best. He also wrote books on travel, screenplays and television drama.

Sargeson, Frank (1903–1982)

Arguably New Zealand's greatest twentieth-century writer, Sargeson was also the only important literary figure to have spent his working life in that country, where he gave generous encouragement to younger writers, notably Janet Frame. In his fiction he often adopts a relaxed New Zealand idiom in first-person narratives, as in his first volume of stories, *Conversations with My Uncle* (1936). Homosexual himself, his characters are often those on the margins of society. His short stories and novellas are perhaps his greatest achievements, but he also wrote several novels, including the comic, picaresque *Memoirs of a Peon* (1965), plays and three volumes of autobiography.

Sassoon, Siegfried [Lorraine] (1886–1967)

Brought up as a horse-loving country gentleman, Sassoon was ideally placed to chart the loss of innocence experienced by a generation in the First World War. A minor lyric poet, he was transformed by the trenches, began writing savagely satirical verses (collected in such volumes as *Counter-Attack*, 1918), and eventually attempted to resign his commission in protest. Haunted by the war throughout his life, he meditated upon his past in two more or less autobiographical trilogies. His poetry gradually became more spiritual, particularly after his conversion to Roman Catholicism in 1957.

Sayers, Dorothy L[eigh] (1893–1957)

Notorious at Oxford as an unconventional bluestocking, and always highly eccentric, Sayers had a brief but influential advertising career before turning to writing. She remains best known as a principal writer of the 'Golden Age' of British detective fiction: her aristocratic amateur sleuth, Lord Peter Wimsey, solved numerous complex cases, including one involving *Strong Poison*

(1930) and that of *The Nine Tailors* (1934). Tiring of detective fiction, she devoted her later years to Christian apologia, radio plays and her (at times superb) translation of Dante's *Divine Comedy*.

Scott, Paul (Mark) (1920–1978)

Scott served in the Indian Army in India in 1943–1946 but otherwise lived in north London, working for many years as a literary agent. Yet in the complex interwoven narratives of *The Raj Quartet* (1966–1975) he created a vivid portrait of the political, personal, racial and religious conflicts in India in the period leading up to Independence. The quartet received wider popularity after his death when televised in 1984 as *The Jewel in the Crown*. Other novels include *Staying On* (1977), set in post-Independence India.

Selby Jr, Hubert (1928–)

Selby was born in Brooklyn and his acclaimed first novel, *Last Exit to Brooklyn* (1964), vividly depicted urban vice and violence and was the subject of obscenity trials in Britain 1966 and 1967. Similar portrayals followed in such novels as *The Room* (1971) and *Requiem for a Dream* (1978), and in the stories of *Song of the Silent Snow* (1986).

Seth, Vikram (1952–)

Seth studied in his native India and in England, California and Communist China. As a writer he has balanced these various perspectives in such books as *From Heaven Lake: Travels through Sinkiang and Tibet* (1984) and *The Golden Gate* (1986), the widely acclaimed verse novel. *A Suitable Boy* (1993) is his most ambitious, certainly longest, novel – an epic 1350-page 'love story' set in post-Independence India. He has also published several volumes of poetry, including *The Humble Administrator's Garden* (1985) and *All You Who Sleep Tonight* (1990).

Sharpe, Tom [i.e. Thomas Ridley Sharpe] (1928–)

Going to South Africa in 1951, Sharpe had careers as a social worker, teacher and photographer before being deported back to his native England in 1961 to become a history teacher. The popular success in 1971 of his first South African comedy, *Riotous Assembly*, the energy of which was fuelled by what he had seen of apartheid, enabled him to become a full-time writer. His speciality is rumbustious and bawdy farce and several of his novels – including *Porterhouse Blue* (1974), *Blott on the Landscape* (1975) and *Wilt* (1976) – have been adapted for cinema and television.

Shields, Carol (1935–)

American-born Canadian novelist, short-story writer, playwright and poet. Although she had a considerable reputation in Canada and the US, she did not arrive on the British literary scene until 1990, when her fifth novel, *Mary Swann*, became the first to be published in the UK. Her standing as one of Canada's leading novelists was rapidly consolidated by the republication of earlier novels and short stories, and confirmed by *The Republic of Love* (1992) and *The Stone Diaries* (1993). Shields is a writer whose sophisticated, beautifully written books make a virtue of the domestic and small-scale, demonstrating the extraordinary qualities to be found in everyday lives.

Sillitoe, Alan (1928–)

One of five children of an illiterate, often unemployed Nottingham labourer, Sillitoe served in the RAF in Malaya, contracting tuberculosis after being demobbed. While in hospital he began to read widely and write. The hero of *Saturday Night and Sunday Morning* (1958), unlike other provincial novels of the 1950s, was not middle-class but a working-class Nottingham man. Sillitoe is also the author of *The Loneliness of the Long Distance Runner* (1959) and many other novels, poems, plays, screenplays and children's books.

Sinclair, Iain [Macgregor] (1943–)

A noted poet and documentary film director, Sinclair has also been a book dealer, specialising in the Beat poets and *noir* thrillers. His work is intimately concerned with London and its history and his poem 'Lud Heat' inspired Peter Ackroyd's novel, *Hawksmoor* (1985). His first novel *White Chappell, Scarlet Tracings* (1987), draws upon the Jack the Ripper murders, which also surface in his second, *Downriver* (1991), a grim evocation of the capital in the 1980s.

Sinclair, May (1863–1946)

The youngest daughter of a bankrupt shipowner, Sinclair wrote reviews, translations and novels. A supporter of Women's Suffrage, she was interested in psychoanalysis, and was influenced by imagist poetry. Among her twenty-four novels were *The Three Sisters* (1914), a study in female frustration, and *Life and Death of Harriet Frean* (1922), a masterpiece of economy examining a Victorian story from a Modernist viewpoint. Highly regarded by her contemporaries, Sinclair's novels were rediscovered by feminists in the 1970s.

Sinclair, Upton [Beall] (1878–1968)

The son of an inebriate liquor salesman, an aristocratic Southern heritage no more than a humiliating memory, Sinclair had to support himself through college by writing voluminously for pulp magazines. His experience of penury helped form his social conscience, while the productivity levels he had set heralded a career in which he would write almost 100 books of fiction and non-fiction, as well as involve himself in social projects and stand for the governorship of California. His best novels are the pioneering *The Jungle* (1906), and *Boston* (1928), based on the Sacco-Vanzetti murder case.

Singer, Isaac Bashevis (1904–1991)

Brought up in Warsaw and in a rural *shtetl*, Singer worked as a journalist for the Yiddish press in Poland for some years before emigrating to the United States in 1935. In New York, he continued to work in Yiddish, both as a journalist and as a fiction writer, but co-operated so closely with the translators of his novels and short stories that he was practically writing in English. His novels, all of which concern the Jewish experience in Europe and (less frequently) America, include *The Magician of Lublin* (1960), *The Slave* (1962) and *Shosha* (1978). His numerous short stories have been collected in such volumes as *Gimpel the Fool* (1957) and *The Image* (1985). He also wrote several volumes of autobiography and a number of books for children. He was awarded the Nobel Prize for Literature in 1978.

Smart, Elizabeth (1913–1986)

The daughter of one of Ottawa's leading families, Smart became involved in the 1930s Bohemian world of Henry Miller and Anais Nin. Her most famous novel, *By Grand Central Station I Sat Down and Wept* (1945) is based upon her relationship with the poet George Barker. She settled in England in 1934 and became involved in the poetry scene, publishing two volumes of her own verse as well as a second, equally brief, novel, *The Assumption of the Rogues and Rascals* (1978).

Smith, Stevie [i.e. Florence Margaret Smith] (1902–1971)

Born in Hull, Smith was brought up in the north London suburbs, where she spent the rest of her life. She worked as a secretary and began writing furtively on the company's time and stationery, producing her *Novel on Yellow Paper* in 1936. Although she published two other novels, she considered herself primarily a poet, and had only turned to fiction when her early poems had failed to find a publisher. Her droll verses – which often tackle such subjects as death and religion, are distinctly idiosyncratic, and were usually accompanied by her drawings – soon became popular, and she published several volumes, starting with *A Good Time Was Had By All* (1937), and including *Not Waving But Drowning* (1957).

Snow, C[harles] P[ercy] (1905–1980)

A physicist who held government scientific posts in the Second World War and the 1960s. Snow's sequence of novels *Strangers and Brothers* (1940–1964) portray English life from 1920 onwards. His *The Two Cultures and the Scientific Revolution* (Cambridge Rede lecture 1959) discussed the absence of communication between literary and scientific intellectuals in the West, and added the phrase 'the two cultures' to the language.

Somerville, Edith [Œnone] (1858–1949) and **Ross**, Martin [pseud. of Violet Martin] (1862–1915)

Second cousins of Irish families, Somerville and Ross separately and together wrote some thirty books, mainly set in Ireland, as well as articles, letters and diaries. Their works include *The Real Charlotte* (1894) and the internationally successful *The Irish R.M.* (1928). After Ross's death, Somerville wrote many more books including the historical romance, *The Big House at Inver* (1925).

Soyinka, Wole (1934–   )

Soyinka, a Nigerian, was the first black African to win the Nobel Prize for Literature. Though he regards himself as a playwright, it is his fiction rather than his other work (plays, poems, essays, translations, autobiographical works, journalistic pieces) that stand out. His inventiveness of symbol and language have prevented his work from being widely popular. An essentially conservative writer, Soyinka is concerned in such novels as *The Interpreters* (1965) with personal survival rather than wider issues.

Spark, Muriel [Sarah] (1918–   )

Born in Edinburgh of Scottish-Jewish descent, Spark spent some years in Central Africa and worked for the Foreign Office during the war. She later edited the *Poetry Review* before turning to fiction. A Catholic convert, she is the author of *The Ballad of Peckham Rye* (1960), *The Prime of Miss Jean Brodie* (1961), *Girls of Slender Means* (1963), *Loitering with Intent* (1981) and *Symposium* (1990), novels that are witty and finely structured, blending fantasy and sharply observed realism to express essentially spiritual themes. She has also written biographies and poetry.

Stead, Christina [Ellen] (1902–1983)

Born in Australia, Stead lived in Europe and the USA for much of her life. An exploratory, psychological writer, imaginatively innovative in form and style, and sensitive to the expressive possibilities of fantasy, she disclosed elements of the irrational, even the grotesque, in the subconscious of her characters. Her novels

include *The Man Who Loved Children* (1940), *Dark Places of the Heart* (1966) and *The People with the Dogs* (1952).

Stein, Gertrude (1874–1946)

Born and educated in America, Stein forsook a medical career to join her brother in Paris, where in the 1920s she set up a literary salon with her lifelong companion, Alice B. Toklas. She befriended artists such as Picasso and Matisse and writers such as Hemingway, while making her own career as a novelist, poet and critic. Her idiosyncratic, incantatory style, frequently innocent of punctuation and conventional syntax, was considered unreadable by some, but she achieved a great success with *The Autobiography of Alice B. Toklas* (1937). Amongst her other books are the novel *Three Lives* (1909), the verse drama *Four Saints in Three Acts* (1929), and *Wars I Have Seen* (1945), an account of her life in occupied France during the Second World War.

Steinbeck, John [Ernst] (1902–1968)

Steinbeck was born in California and became the chief chronicler of the Depression's rural poor in his short stories and such novels as *Tortilla Flat* (1935) and *The Grapes of Wrath* (1939). After studying inconclusively at Stanford, he worked as a labourer and fruitpicker, thus gaining an insight into the lives of migrant workers, whose stories he told in his fiction, much of which is set in his native region of Monterey Bay. Other novels include *Of Mice and Men* (1937), *Cannery Row* (1945) and *East of Eden* (1952). He won the Nobel Prize in 1962.

Stephens, James (1882–1950)

Born in poverty, Stephens was an Irish poet and story-writer, whose best-known work includes the prose fantasy *The Crock of Gold* (1912) and his first novel *The Charwoman's Daughter* (1912). Later novels include *The Demi-Gods* (1914) and *Deirdre* (1923). He also wrote many volumes of stories, including *Irish Fairy Tales* (1920) and *Etched in Moonlight*(1928) and was a popular broadcaster of stories and verse.

Storey, David [Malcolm] (1933–   )

A Wakefield-born miner's son, Storey was an art student at the Slade, a professional footballer, teacher, farm worker and erector of show tents, before becoming a novelist and playwright. His first novel, *This Sporting Life* (1960), established him as a grittily realistic writer of working-class life. Later novels include *Radcliffe* (1963) and

Saville (1976). Both his plays and novels combine documentary naturalism with symbolism, and a preoccupation with the turmoil caused by social mobility.

Strachey, Julia [Frances] (1901–1979)

The niece of Lytton Strachey and Dorothy Bussy ('Olivia'), Strachey was born in India and brought up in England in chaotic Bloomsbury style by a succession of aunts. A perfectionist, she published only two slim novels, *Cheerful Weather for the Wedding* (1932) and *The Man on the Pier* (1951, later reissued as *An Integrated Man*), and a handful of uncollected sketches and short stories. After her death, a self-portrait was assembled from her papers by Frances Partridge and published as *Julia* (1983).

Stuart, Alexander (1955–)

Stuart achieved some notoriety when his controversial novel of incest, *The War Zone*, was prevented from winning the Whitbread Prize in 1989. He started out in film, working as a producer, and his first book, *Joe, Jo-Jo and the Monkey Masks* was written for his son, whose harrowing illness and early death were recorded in *Five and a Half Times Three* (1991), written with the child's mother, Ann Totterdell. His other books include *Tribes* (1992), a novel about football hooliganism.

Stuart, [Henry] Francis [Montgomery] (1902–)

Of Protestant Irish landed gentry, Stuart was born in Australia, brought up in Northern Ireland, and educated at Rugby. Influenced by the Russian revolutionary poets, he published verse and was interned for Republican activities in the 1922 Civil War. After his release and still only twenty-one he was awarded a poetry prize by the Royal Irish Academy. He moved to Berlin, where he was dubbed Ireland's 'Haw-Haw' by Cyril Connolly for broadcasting to Ireland; Freiburg, where he was imprisoned in 1946 by the occupying French; finally returning to Ireland in 1958. His early novels draw on his pursuits of horse racing, women and religion; his wartime experiences in Germany are reflected in *The Pillar of Cloud* (1948) and *Redemption* (1949). *Black List, Section H* (1971) re-established his reputation and a series of experimental novels has followed. With dry humour he celebrates his lifelong obsessions in *A Compendium of Lovers* (1990).

Sturgis, H[oward] O[vering] (1855–1920)

The son of an architect and banker, Sturgis was born into Boston high society but went to Eton where he was overshadowed by his handsome, athletic brother Julian, who also became a novelist. After Cambridge he settled in Eton with his lifelong companion William Haynes Smith, entertaining a wide circle of friends such as Henry James and Edith Wharton. Although considered a dilettante, he published three novels, *Tim* (1891), a sentimental story of friendship between Eton boys, *All that was Possible* (1895) and *Belchamber* (1904), a minor classic of twentieth-century literature.

Styron, William [Clark] (1925–)

Styron was born in Virginia and much of his work has a Southern setting. Novels like *The Confessions of Nat Turner* (1967, Pulitzer Prize), and *Sophie's Choice* (1979) won critical and popular acclaim. Styron also wrote the short memoir of a descent into depression and madness, *Darkness Visible* (1991).

Swift, Graham (1949–)

Swift made a name for himself with short stories, published in various magazines and collected in 1982 as *Learning to Swim*. His novels often have historical settings and are occasionally didactic in tone, notably, *Waterland*, which confirmed the promise of *The Sweet Shop Owner* (1980) and *Shuttlecock* (1981), when it was published to huge acclaim in 1983, winning several literary prizes. Later books include *Out of This World* (1988), a novel, and *The Magic Wheel* (1985), an anthology of fishing he co-edited with the novelist and poet David Profumo.

Tartt, Donna (1964–)

Tartt was born in Mississippi, a precocious child who wrote poems from the age of five. She began writing her first, phenomenally successful, novel, *The Secret History* (1992), while a student at Bennington College, Vermont, an establishment that has subsequently been identified with the book's setting. A stylish, erudite thriller, the book engendered enormous interest and was both a popular and critical success.

Taylor, Elizabeth (1912–1975)

A former governess and librarian, Taylor became one of the subtlest and wittiest writers of the century. Her territory (which goes some way to explaining her comparative neglect) is the post-war middle classes, observed with an often shockingly sharp eye, particularly for people's capacity for self-deception. Exquisitely written, mordantly funny, and very shrewd about the human heart and its frailties, novels such as *A View of the Harbour* (1947), *Angel* (1957) and *A Game of Hide-and-Seek* (1951) are matched by several outstanding volumes of short stories, including *Hester Lilly* (1954), the long title-story of which is a miniature masterpiece.

Tennant, Emma [Christina] (1937–)

Born into the aristocratic Glenconner family – eccentric members of which occasionally appear in such novels as *Wild Nights* (1979) – Tennant was brought up in London and Scotland. She has worked extensively as a journalist, founding and editing the magazine *Bananas*. She was general editor of the *Penguin Lives of Modern Women* series, and has written for children. Remarkable for its versatility, her fiction has embraced feminism, comedy, satire and science fantasy, and includes such novels as *The Bad Sister* (1978), *Alice Fell* (1980) and *The Adventures of Robina* (1986).

Theroux, Paul [Edward] (1941–)

American novelist and travel writer whose accounts of his journeys by train, notable for their sharp depiction of the socio-economic divides, include *The Great Railway Bazaar* (1975), *The Old Patagonian Express* (1979), *The Kingdom by the Sea* (1983), and *Riding the Iron Rooster* (1988). Among his novels are *Saint Jack* (1973), *The Mosquito Coast* (1981), *Doctor Slaughter* (1984) and *Chicago Loop* (1990).

Thomas, D[onald] M[ichael] (1935–)

Thomas, a poet, novelist and translator, was born in Cornwall and educated there, in Australia and at Oxford. While doing National Service he learned Russian, whose literature was later to influence his work. He lectured in English and translated volumes of Russian poetry including Pushkin. He has published several volumes of his own poetry and his first novel, *The Flute Players* (1979) is dedicated to the persecuted creative spirit of modern Russian poets. *Birthstone* (1968), set in Cornwall, blends fantasy, comedy, realism, eroticism and magic. He achieved international literary success with *The White Hotel* (1981). *Ararat* (1983) is also concerned with the relationship between sex and death and has a similar narrative complexity.

Tolkien, J[ohn] R[onald] R[euel] (1892–1973)

Merton Professor of Anglo Saxon (1925–45) and English language and literature (1945–59), Tolkien created the fictional world of Middle Earth in *The Hobbit* (1937) and the trilogy *The Lord of the Rings* (1954–1955), fantasy novels peopled with hobbits, dwarves and strange magical creatures. His work developed a cult following in the 1960s and had many imitators.

Toole, John Kennedy (1937–1969)

Born in New Orleans, living in obscurity and finally committing suicide in despair, Toole achieved literary success only posthumously. His doting mother pestered critics and publishers remorselessly until her son's (genuinely remarkable) talent was at last recognized with the publication of his only mature novel, *A Confederacy of Dunces* (1980).

Torrington, Jeff[rey] (1935–)

Torrington was born in the Gorbals in Glasgow, spent part of his childhood in a tuberculosis sanatorium, and received his real education at the local library. He worked in a number of jobs and started submitting short stories to newspapers and magazines in his twenties. His first story was accepted in 1960, but he achieved real success with his first novel, *Swing Hammer Swing!*, which took almost thirty years to write and was eventually published in 1992, when it won the Whitbread Book of the Year Award. *The Devil's Carousel* (1994) is a volume of interconnected short stories set in a car factory.

Tremain, Rose (1943–)

British novelist, short-story writer and playwright, who studied English literature under Angus Wilson and made her mark in 1977 with her first novel, *Sadler's Birthday*. As in subsequent books, the central character, a butler, is a solitary and dispossessed figure. Her reputation was consolidated by *Letter to Sister Benedicta* (1978) and *The Cupboard* (1981), and in 1983 she was chosen as one of the 'Best of Young British Novelists'. Her short-story collections include *The Colonel's Daughter* (1984) and she has written many plays for both the radio and television.

Tressell, Robert [pseud. of Robert Noonan] (1868–1911)

A house-painter of Irish extraction, Tressell is remembered for his posthumously published novel, *The Ragged Trousered Philanthropists* (1918, revised 1955), edited from a manuscript left in his daughter's care. Drawing on his experiences working for a builder in Hastings, it gives a detailed account of the poverty of working people's lives and has become a classic text of the Labour movement.

Trevor, William [i.e. William Trevor Cox] (1928–)

Born in Co. Cork, Trevor is an Anglo-Irish writer whose work includes the novels *The Old Boys* (1964), *The Children of Dynmouth* (1976) and *Fools of Fortune* (1983), and short-story collections such as *The Ballroom of Romance* (1972), the title story of which poignantly evokes rural Ireland. His more recent work has reflected the effects of Irish terrorism. He has adapted many of his stories for radio and television.

Trumbo, Dalton (1905–1976)

A successful writer of stories and novels, Trumbo was also earning $200,000 a year as an MGM scriptwriter when he fell foul of the House Un-American Activities Committee. After losing his job and spending a year in prison, he wrote under pseudonyms from Mexico until the 1960s saw him gradually rehabilitated. His best known novel is the savage anti-war fable, *Johnny Got His Gun* (1939).

Tyler, Anne (1941–)

Tyler's novels are both acidly funny and sad, exploring the disruptions and irrationality of Baltimore family life, and the consequent isolation of individuals. *Dinner at the Homesick Restaurant* (1982), perhaps her most characteristic, examines the loneliness and grief of a deserted wife who is now on her deathbed. Her other novels include *The Accidental Tourist* (1985), *Breathing Lessons* (1988) and *Saint Maybe* (1991).

Unsworth, Barry (1930–)

Many of Unsworth's books, such as *Stone Virgin* (1985), a multi-layered novel about the mystery of a statue of the Madonna, and *Sugar and Rum* (1988), about a man unable to write his novel about the Liverpool slave trade, have an exotic historical setting. Unsworth himself suffered writer's block while writer-in-residence in 1985 at Liverpool University but went on to write his tenth novel, *Sacred Hunger* (1992, joint Booker Prizewinner), the story of an eighteenth-century ship's surgeon who organises a mutiny on a slaver.

Updike, John [Hoyer] (1932–)

Updike established his reputation with stories for the *New Yorker*, and has published several collections, including *The Music School* (1966) and *Trust Me* (1987). Two characters recur in novels preoccupied with the erotic and sacred, expressed in the tensions and frustrations of middle-class life: former baseball player 'Rabbit' Angstrom who matures in the tetralogy beginning with *Rabbit Run* (1960), and novelist Henry Bech, who appears in *Bech: A Book* (1970) and *Bech is Back* (1982). Other major novels are *The Centaur* (1963), *Couples* (1968), *The Witches of Eastwick* (1984) and *Roger's Version* (1986). A writer of polished prose, he is also an essayist and poet.

Upward, Edward [Falaise] (1903–)

The son of a doctor, Upward joined the Communist Party in 1934 and was the most politically committed of the Auden Generation. The story of his struggle to reconcile his art and his political beliefs is told in the autobiographical trilogy, *The Spiral Ascent* (1962–1977). Apart from this, Upward is best known for the stories he wrote in the 1920s in collaboration with Christopher Isherwood. Scatalogical, surrealist and set in an imaginary village, they remained largely unpublished until the appearance of *The Mortmere Stories* (1994), although the most famous of them was the title story of *the Railway Accident and Other Stories* (1969). Upward's other books include the novella *Journey to the Border* (1937) and two further volumes of stories, *The Night Walk* (1987) and *An Unmentionable Man* (1994).

Van Vechten, Carl (1880–1964)

Born in Iowa, Van Vechten was a novelist, portrait photographer and music and drama critic. As a white American, he did much to widen public interest in black culture from the 1920s onwards – his novel *Nigger Heaven* (1926) was a controversial attempt in this direction – and he became an important figure in the Harlem Renaissance. Other books include stylish novels such as *Peter Whiffle* (1922) and *The Blind Bow-Boy* (1923), which combine satire and fantasy.

Vidal, Gore [Eugene Luther] (1925–)

Vidal, born at West Point, New York, followed the family custom by joining the army: his writing career began with *Williwaw* (1946) based on his army experience. He has made a reputation as an iconoclast; nevertheless, he received a reasonable poll when he ran (1960), unsuccessfully, for Congress. His many works include *Julian* (1964), *Myra Breckinridge* (1968), *Duluth* (1983), and the historical trilogy, *Burr* (1974), *1876* (1976) and *Lincoln* (1984). Vidal has also written plays, short stories, film scripts and essays.

Vonnegut Jr, Kurt (1922–)

A prisoner of war in Dresden during the notorious raid of 1945, Vonnegut was afterwards employed digging out corpses from the rubble, an experience that would mark him for life and form the background to his novel *Slaughterhouse-Five* (1969). Like much of his other work – which includes short stories and novels such as *Cat's Cradle* (1963) and *Slapstick* (1978) – this book mixes science-fiction, realism and satire. He has also written essays and a volume of memoirs.

Waddell, Helen [Jane] (1889–1965)

The daughter of a Presbyterian missionary, Waddell had already established a formidable reputation as a medieval scholar when she wrote her only novel, *Peter Abelard* (1933). Her most substantial book was her study of *The Wandering Scholars* (1927), but she also wrote plays and translations of medieval lyrics.

Wain, John [Barrington] (1925–1994)

In his youth, Wain was prone to attract labels: 'angry young man' and 'Movement' poet. He thereafter worked to find his own voice, not only in his poetry and novels, but also in criticism and scholarship. His novels include *Hurry On Down* (1953) and *Young Shoulders*, which won the Whitbread Prize in 1982, and his poems were collected in 1981. He also published volumes of short stories, memoirs and a life of Dr Johnson. From 1973 to 1978 he was Professor of Poetry at Oxford.

Walker, Alice [Malsenior] (1944–)

Walker's father defied death threats to cast the first black vote in Georgia, and this tradition of resistance has continued both in Walker's life and in her writing. One of black American feminism's most powerful literary voices, she has written poems, plays, short stories, essays and novels, including the popular and Pulitzer Prize-winning *The Color Purple* (1982). Almost all her work concerns the plight, but also celebrates the strength, of women oppressed not only by sexism but also by racism.

Walpole, Hugh [Seymour] (1884–1941)

Six months in a seaman's mission having convinced him that he did not want to follow his father into the priesthood, Walpole put in a few months as a schoolmaster before devoting himself entirely to writing. The experience of schools led to his best novel, *Mr Perrin and Mr Traill* (1911), as well as *Jeremy at Crale* (1927), the third volume of a sentimental trilogy about a boy. In spite of the New Zealand-born Walpole being a protégé of Henry James, most of his novels – including 'The Herries Chronicles', a four-volume family saga opening with *Rogue Herries* (1930) – were popular and distinctly middlebrow. Extremely prolific, he also wrote short stories, *belles lettres* and studies of Trollope and Conrad.

Warner, Rex [Ernest] (1905–1986)

A poet, novelist and translator, Warner was a close friend of Auden and Day Lewis. His later novels such as *The Young Caesar* (1958) and *Imperial Caesar* (1960), are based on classical themes, but he is better remembered today for his earlier works, such as *The Wild Goose Chase* (1937), *The Professor* (1939) and *The Aerodrome* (1941), which are disturbing parables based on the political situation of the 1930s.

Warner, Sylvia Townsend (1893–1978)

The daughter of a Harrow School housemaster, Warner was a poet, novelist and editor of *Tudor Church Music* (1922–1929), her love of early music being reflected in her later fiction. She had a great success with her first novel, the feminist fable, *Lolly Willowes* (1926), and with the short stories she published in the *New Yorker*. Other novels often have historical settings and include *Mr Fortune's Maggot* (1927), *Summer Will Show* (1936) and *The Corner That Held Them* (1948), a remarkable recreation of life in a fourteenth-century Norfolk convent.

Warren, Robert Penn (1905–1989)

Born in Kentucky, Warren is the only author to have received a Pulitzer Prize for both poetry and prose – for the two volumes, *Promises* (1957) and *Now and Then* (1978), and for the novel *All the King's Men* (1946). He was one of the founder-editors of *The Fugitive* magazine and *The Southern Review*, became a senior figure of the New Criticism, and was appointed America's first poet laureate in 1986. Other novels, mostly based in the South and dealing with its history, include *Night Rider* (1939), *World Enough and Time* (1950) and *Band of Angels* (1955).

Waterhouse, Keith [Spencer] (1929–)

Born and educated in Leeds, Waterhouse came to notice in 1957 with his fine first novel, *There is a Happy Land*. He was a leading member of the generation of Northern writers who came to prominence in the 1950s and 1960s, not only as a novelist, but also (usually in collaboration with Willis Hall) as a screenwriter in the new wave of British film-making. His best known novel, *Billy Liar* (1959, filmed 1960), was a significant work of this period. A humourist and social commentator, he worked for many years as a journalist, and has written numerous plays, including the very successful *Jeffrey Bernard is Unwell* (1989). Later novels include *Maggie Muggins* (1981) and *Bimbo* (1990).

Waugh, Alec [i.e. Alexander Raban Waugh] (1898–1961)

Expelled from Sherborne, Waugh brilliantly exposed the hypocrisies of the public-school system in his first novel, *The Loom of Youth*, which was published in 1917 on the day he set off for the trenches. After the war he worked for his father's publishing company and wrote a study of *Public School Life* (1922), but his chief career was as a frankly popular novelist, always overshadowed by his younger brother, Evelyn. He also wrote travel books and several volumes of memoirs.

Waugh, Evelyn [Arthur St John] (1903–1966)

Chief chronicler of the 'Bright Young Things' in early novels such as *Decline and Fall* (1928) and *Vile Bodies* (1930), Waugh was one of the century's most accomplished and funny novelists.

His comic vision always had its black side and this darkened after the collapse of his first marriage, achieving moral weight when he converted to Roman Catholicism in 1930. *Brideshead Revisited* (1945) and *The Sword of Honour Trilogy* (1952–1961) are concerned with the workings of faith. He also worked as a foreign correspondent, which led to travel books and such novels as *Scoop* (1938) and the exquisitely tasteless *Black Mischief* (1932), in which a principal theme of the clash between barbarism and civilisation emerges, recast in English and American locations in *A Handful of Dust* (1934) and *The Loved One* (1948).

Webb, Mary [Gladys] (1881–1927)

Born in Shropshire, Webb became one of the century's leading regional novelists, celebrating the county and its countryside in all her novels, though living and working in London. Her reputation was made in 1924 when *Precious Bane* was publicly acclaimed by the Prime Minister, Stanley Baldwin, although earlier novels, such as *The Golden Arrow* (1916) and *Gone to Earth* (1917), are also admired. Her melodramatic plots, use of dialect and passages of lyricism were effectively parodied by Stella Gibbons in *Cold Comfort Farm* (1934).

Welch, [Maurice] Denton (1915–1948)

Born in Shanghai, Welch was educated unhappily in England, until he went to art school. While a student, he was knocked off his bicycle by a motorist and sustained appalling injuries which both circumscribed and shortened his life. Although he continued to paint and draw, he began to develop as a writer of remarkable sensitivity, particularly in his descriptions of the physical world and the sensations of adolescence. His first book, *Maiden Voyage* (1943) describes his return to China, having run away from school. *In Youth is Pleasure* (1945) is an autobiographical novel, while *A Voice Through a Cloud* (1950 and unfinished at his death) describes his accident and its aftermath. He also wrote some sixty short stories, collected in several volumes. His remarkable *Journals*, running to some 200,000 words, did not appear in an unexpurgated edition until 1984.

Weldon, Fay (1931–)

Weldon, born in Worcester, deals with feminist themes, often employing irony, fantasy and tragi-comedy. Her gift for realistic dialogue is displayed in novels such as *The Fat Woman's Joke* (1967), *Praxis* (1978), *The Life and Loves of a She-Devil* (1984, filmed 1990), and *The Hearts and Lives of Men* (1987), and in her plays for the stage, radio, and television.

Wells, H[erbert] G[eorge] (1866–1946)

Wells is best-known for 'scientific romances' such as *The Time Machine* (1895); later novels such as *Love and Mr Lewisham* (1900), *Kipps* (1905), *Tono-Bungay* (1909), *The History of Mr Polly* (1910) and *The New Machiavelli* (1911) had an iconoclastic humour remarkable for their time. Wells was a writer, thinker, historian, lively advocate of socialism, feminism, evolutionism, nationalism and scientific advancement, and his literary output and influence were vast. His many other books include *The Shape of Things to Come* (1933), prophecies from which have since been fulfilled.

Welty, Eudora (1909–)

Leading Southern writer who was born, and has spent much of her life, in Mississippi, the setting of her best-known novel, *Delta Wedding* (1946). Her principal achievement has been her short stories, published in several volumes and collected in 1980, but she has also written acclaimed novels, including *The Robber Bridegroom* (1942), *The Ponder Heart* (1954) and the Pulitzer Prize-winning *The Optimist's Daughter* (1972), all of which deal with the South. She has taught at various universities and published a collection of essays, *The Eye of the Storm* (1978), and an autobiography, *One Writer's Beginnings* (1984).

Wesley, Mary [i.e. Mary Wellesley] (1912–)

Wesley wrote her first books for children, but although she started writing adult novels late in life, she has rapidly achieved popularity for her warm, humorous and vividly paced stories, such as her second novel, *The Camomile Lawn* (1984). Her other novels include *Jumping the Queue* (1983), *The Vacillations of Poppy Carew* (1986) and *A Sensible Life* (1990).

West, Nathanael [pseud. of Nathan Weinstein] (1904–1940)

Born in New York, West is noted as an idiosyncratic black-humour parodist. His Surrealist-influenced novels, such as *Miss Lonelyhearts* (1933), capture the absurdity and extremity of American life and the dark side of the American Dream. His most powerful novel, *The Day of the Locust* (1939), is a vivid exploration of the apocalyptic violence given release by the fantasies created by Hollywood, where West had been a screenwriter.

West, Rebecca [pseud. of Cecily Isabel Fairfield] (1892–1983)

Daughter of an Anglo-Irish family who grew up in straitened circumstances in Edinburgh, West trained for the stage, then became a feminist and journalist. Her outspoken review of H.G. Wells's *Marriage* (1912) led to a ten-year love affair and the birth of a son. Her novels include *The Return of the Soldier* (1918), *The Judge* (1922), *The Strange Necessity* (1928), *Harriet Hume* (1929), *The Thinking Reed* (1936) and, after a long gap, *The Fountain Overflows* (1956) and *The Birds Fall Down* (1966). She also wrote studies of Yugoslavia, the Nuremberg trials, the Vassall and Ward spy scandals, William Joyce ('Lord Haw-Haw') and others. She continued to write and review with vigour until her death, and has continued to be admired by 1980s feminists for her craftsmanship and strong, unconventional heroines.

Wharton, Edith [Newbold] (1862–1937)

Born of wealthy New York parents, Wharton is distinguished for the subtlety and form of her work, which was influenced by her friend Henry James, and was mostly set in New York society. It includes *The House of Mirth* (1905), which made her reputation; the grim, uncharacteristic New England novel *Ethan Frome* (1911); *The Custom of the Country* (1913) and *The Age of Innocence* (1920) which won the Pulitzer Prize, and brought her work to a new audience when it was filmed in 1993. She also published several volumes of short stories.

Wharton, William (1925–   )

The pseudonym of a writer and painter who lives in Paris, cultivating a well-guarded anonymity, though he gives interviews. Born to working-class Philadelphia parents, he was a captain in the US Army in Germany during the war and was seriously injured by shrapnel. After cosmetic surgery in Paris, he studied painting in California and received a doctorate in psychology, moving to Paris in 1961. After the success of *Birdy* (1979), in which a combat-traumatised war veteran believes he is a bird, he continued to draw on his own experience in such novels as *Dad* (1981), *A Midnight Clear* (1982), *Scumbler* (1984) and *Last Lovers* (1991).

White, Antonia (1899–1980)

The daughter of a classics teacher, White was convent-educated, a childhood described in the first of her four autobiographical novels, *Frost in May* (1933) begun at sixteen but not completed until after her father's death in 1929. It is upon these books that her reputation rests. She worked for much of her life as a journalist, and wrote a number of short stories. She also translated many novels from the French, notably works by Maupassant and Colette .

White, Patrick [Victor Martindale] (1912–1990)

White did more than any other writer to put Australian literature on the international map. His partly allegorical novels explore the lives of early settlers in Australia and often deal with misfits or inarticulate people. They include *The Aunt's Story* (1948), *The Tree of Man* (1955), *Voss* (1957), *A Fringe of Leaves* (1975) and *The Twyborn Affair* (1979). He also wrote short stories and a frank autobiography, *Flaws in the Glass* (1981). White won the Nobel Prize for Literature in 1973.

White, T[erence] H[anbury] (1906–1964)

Born in Bombay and educated at English public schools, White began to write with a collection of poetry, published 1926. He is known as a novelist, however, and especially for his imaginative and enchanting reworking of the legend of King Arthur, the tetralogy *The Once and Future King* (1958). *The Book of Merlyn* (1977) although thematically linked, was a departure stylistically, lacking the magic of the earlier work. His other books include the fantasy *Mistress Masham's Repose* (1947), historical works, a translation of a Latin bestiary, and three autobiographical works, the finest, *The Goshawk* (1951) detailing his experience of training and working with the eponymous bird.

Wilder, Thornton (1897–1975)

Born in Wisconsin, Wilder is best-known for his philosophical novel *The Bridge of San Luis Rey* (1927), in which the meaning of the lives lost when a bridge breaks is explored to divine God's purpose. Other novels include *The Ides of March* (1948) and *Heaven's My Destination* (1935). Of his plays, the comedy, *Merchant of Yonkers* (1938), was revised as *The Matchmaker* (1954) and as the musical *Hello, Dolly!* (1963).

Williams, Tennessee [i.e. Thomas Lanier Williams] (1911–1983)

The son of a travelling salesman, Williams was born and brought up in Mississippi and St Louis. His prolific career as a major playwright includes *A Streetcar Named Desire* (1947), *The Rose Tattoo* (1950), *Camino Real* (1953) and *Cat on a Hot Tin Roof* (1955). His masterpiece in prose fiction is the novella, *The Roman Spring of Mrs Stone* (1950). Williams's work, much of which is set in the Deep South, had a deep warmth and insight into the foibles of human passion, and is often explicit in the exploration of sexuality.

Williamson, Henry (1897–1977)

Profoundly scarred by his experiences as a private in the First World War, Williamson began writing nature notes in order to supplement his war pension. He remains most celebrated for his two books of animal life, *Tarka the Otter* (1927) and *Salar the Salmon* (1935). He recalled his experiences of the trenches in *A Patriot's Progress* (1930) and in several volumes of his *Chronicles of Ancient Sunlight* (1951–1969), a fifteen-volume, autobiographical *roman fleuve*. His reputation suffered because of his espousal of fascism, but the majority of his work is far removed from the world of politics, more concerned with the English countryside, where he spent most of his life.

Wilson, A[ndrew] N[orman] (1950–)

Prolific novelist, critic and biographer (of Belloc, Milton, Scott and Tolstoy), Wilson is a witty and stylish writer of ironic social comedies of contemporary English middle-class life. His novels include *The Sweets of Pimlico* (1977), *The Healing Art* (1980), *Who Was Oswald Fish?* (1981), *Scandal* (1983) and *Love Unknown* (1986).

Wilson, Angus [Frank Johnstone] (1913–1991)

Born in Durban of Scottish-South African parents, Wilson established himself as a writer with short stories while working as a deputy superintendent of the British Museum Reading Room. His stories, which perfectly capture the era in which they were written, were collected in such volumes as *The Wrong Set* (1949). Equally acute were his novels, among which are *Hemlock and After* (1952), *Anglo-Saxon Attitudes* (1956), *The Old Men at the Zoo* (1961) and *Late Call* (1964). In his detailed portrayal of English society he extracted high comedy from its social and moral grostesqueries, displaying brilliant satiric wit, but also an overriding humanity and compassion. He was also a lecturer and critic and wrote studies of Dickens and Kipling.

Wilson, Edmund (1895–1972)

The most eminent American man of letters of the century, Wilson was acclaimed not only for his fiction – two novels, *I Thought of Daisy* (1929) and *Galahad* (1957), and a volume of short stories, *Memoirs of Hecate County* (1946) – but also for his plays, his poetry, his journalism and above all his literary criticism: *Axel's Castle* (1931), *The Wound and the Bow* (1941), *Patriotic Gore* (1962) and numerous other volumes. His memoirs, each given the title of a decade and starting with *The Twenties* (1975), have been published posthumously.

Winterson, Jeanette (1959–)

Winterson broke out of a difficult and constrained childhood in Lancashire to read English at Oxford, and her autobiographical first novel,

Oranges Are Not the Only Fruit (1985), describes in humorous and horrifying detail the clash between her upbringing as an Evangelical Pentacostalist and her evolving homosexuality. Subsequent novels have divided critics and readers, some seeing her as one of the most talented writers of her generation, others dismissing hers as a minor talent vitiated by self-regard. *The Passion* (1987) is perhaps the best of her other novels.

Wodehouse, P[elham] G[reville] (1881–1975)

The son of a civil servant who became a Hong Kong judge, Wodehouse spent much of his childhood in England in the care of various aunts. He began writing short stories for boys' magazines and later published in *Strand Magazine*, *Punch* and others to establish himself as one of the most popular humourists of his time. His prolific output of novels and stories includes the Jeeves and Wooster series of stories: *My Man Jeeves* (1919), *The Inimitable Jeeves* (1923), *Carry On Jeeves* (1925) and so on. Other books feature Lord Emsworth and his prize sow, the Empress of Blandings (*Galahad at Blandings*, 1965), Psmith (*Psmith in the City*, 1910), Mr Mulliner, a number of aunts and patrons of the Drones Club. He also worked in musical comedy, the theatre and Hollywood. During the war he was captured by the Germans and perhaps foolishly agreed to broadcast to America. In spite of the innocuous nature of his broadcasts, the scandal caused in England led him to settle in America and he became a US citizen in 1955. His reputation as a master of comic writing continues to grow.

Wolfe, Thomas [Clayton] (1900–1938)

The journey from small-town Carolina via Harvard to literary New York is told in Wolfe's two autobiographical novels, *Look Homeward, Angel* (1929) and *Of Time and the River* (1935). He also wrote plays and short stories, and a considerable amount of work was published posthumously.

Wolfe, Tom [i.e. Thomas Kennerly Wolfe Jr] (1931–)

Long before he wrote his first novel, the long, satirical *The Bonfire of the Vanities* (1987), Wolfe had won renown for bringing the techniques of fiction to non-fiction writing as a pioneer of the 'New Journalism', with such books as *The Electric Kool-Aid Acid Test* (1968), about psychedelia, and *The Right Stuff* (1979), about the American space programme.

Woolf, Virginia [Adeline] (1882–1941)

As a key member of the Bloomsbury Group Woolf produced a succession of novels, short stories, and critical essays to become one of the major English writers of the twentieth century. In such novels as *Mrs Dalloway* (1925), *To the Lighthouse* (1927), *Orlando* (1928) and *The Waves* (1931) she developed an experimental fiction, using her 'stream of consciousness' technique to capture the subtle, ceaseless flow of everyday experience. In several influential essays she argued that women had to develop their own form of fiction. Plagued by bouts of depression, she committed suicide by drowning in 1941.

Wright, Richard (1908–1960)

Brought up in Memphis and largely self-educated, Wright was one of the first writers to depict the condition of black people in twentieth-century US society with his powerful and tragic novel *Native Son* (1940) and the autobiography *Black Boy* (1945). Between 1932 and 1944 he was active in the Communist Party. Shortly thereafter he became a permanent expatriate in Paris.

Wyndham, John [i.e. John Wyndham Parkes Lucas Beynon Harris] (1903–1969)

Although Wyndham was publishing science-fiction stories in American pulp magazines as early as 1931, he did not make his name as a leading English exponent of the genre until two decades afterwards, when his novel *The Day of the Triffids* (1951) was published. One of the most accessible of sci-fi writers, whose books generally introduce malign alien forces into everyday life, he reached a wide general audience with such books as *The Kraken Wakes* (1953) and *The Midwich Cuckoos* (1957), and produced several volumes of short stories.

Young, E[mily] H[ilda] (1880–1949)

A novelist whose work is mainly set in Bristol, Young offers exquisite and witty portraits of class, love and provincial life in the 1920s. Her work has been undergoing a revival of interest in recent years since Virago began to reprint her novels, which include *The Misses Mallett* (1922), *Miss Mole* (1930), *Jenny Wren* (1932), *The Curate's Wife* (1934) and *Chatterton Square* (1947).

Index of Authors and Titles

The date of publication for each novel follows the title in column three. Where that novel is part of a series, the date of publication is that of the final book in the series. Where a year appears in brackets directly following the title (and not in column three) this year indicates the publication date of an individual book in a series.

Index of Titles